Year	Name	Country	Contribution
1858	Rudolf Virchow	Germany	States that cells come only from preexisting cells.
1858	Charles Darwin	Britain	Presents evidence that natural selection guides the evolutionary process.
1858	Alfred R. Wallace	Britain	Independently comes to same conclusions as Darwin.
1865	Louis Pasteur	France	Disproves the theory of spontaneous generation for bacteria; shows that infections are caused by bacteria and develops vaccines against rabies and anthrax.
1866	Gregor Mendel	Austria	Proposes basic laws of genetics based on his experiments with garden peas.
1869	Friedrich Miescher	Switzerland	Discovers that the nucleus contains a chemical he called nuclein, now termed DNA.
1878	Joseph Lister	Britain	Devises a method of sterilizing the operating room to prevent infection in surgical patients.
1880	Walther Flemming	Germany	Studies the movement of chromosomes during mitosis.
1882	Robert Koch	Germany	Establishes the germ theory of disease and developes many techniques in bacteriology.
1888	Wilhelm Roux	Germany	Founds the science of embryology by performing experiments on embryos.
1892	August Weismann	Germany	Formulates the germ plasma theory which states that only germ plasma is passed from generation to generation.
1897	Eduard Büchner	Germany	Extracts enzymes from yeast and uses them to bring about fermentation.
1900	Hugo De Vries Erich von Tschermak Karl Correns	Holland Austria Germany	Independently rediscover Mendel's laws.
1900	Walter Reed	United States	Discovers that yellow fever virus is transmitted by a mosquito.
1901	Hugo De Vries	Holland	States that mutations account for the presence of variations among members of a species.
1901	Santiago Ramón y Cajal	Spain	Suggests that neurons are separated by synapses.
1902	W. S. Sutton	United States	Suggests that genes are on the chromosomes, after noting the similar behavior of genes and chromosomes.
1902	William M. Bayliss E. H. Starling	Britain	Found the study of endocrinology by demonstrating the action of the hormone secretin.

Continued on inside back cover

Fifth Edition

INQUIRY INTO LIFE

Sylvia S. Mader

wcb
Wm. C. Brown Publishers
Dubuque, Iowa

INQUIRY INTO LIFE

Cover Image

Zebras are closely related to the horse. There are three species of these black-and-white striped mammals which are native to southern and eastern parts of Africa. These are Burchell's zebras (*Equus guagga*) that prefer rich grasslands and will migrate long distances to find suitable feeding grounds. They need to drink only every 36 hours. Usually there is a small family group consisting of one stallion and several mares but zebras may temporarily mingle with antelopes who benefit from their general alertness to predators.

All three types of zebras are disappearing from the wild because of habitat loss due to human activities. They will breed in captivity, however, and it is hoped that substantial numbers can be maintained, especially by employing embryo transfer techniques. A zebra embryo is removed from it's mother's womb and placed in the womb of a horse, where it develops normally. This technique will allow zebras to reproduce at a faster rate than usual.
Cover photo © Carol Hughes/Bruce Coleman, Inc., New York

Book Team

Editor *Kevin Kane*
Designer *Mary K. Sailer*
Production Editor *Kay J. Brimeyer*
Photo Research Editor *Shirley Charley*
Permissions Editor *Mavis M. Oeth*
Visuals Processor *Renee Pins*
Marketing Manager *Matt Shaughnessy*

wcb group
Chairman of the Board *Wm. C. Brown*
President and Chief Executive Officer *Mark C. Falb*

wcb
Wm. C. Brown Publishers, College Division

President *G. Franklin Lewis*
Vice President, Editor-in-Chief *George Wm. Bergquist*
Vice President, Director of Production *Beverly Kolz*
National Sales Manager *Bob McLaughlin*
Director of Marketing *Thomas E. Doran*
Marketing Information Systems Manager *Craig S. Marty*
Executive Editor *Edward G. Jaffe*
Manager of Design *Marilyn A. Phelps*
Production Editorial Manager *Julie A. Kennedy*
Photo Research Manager *Faye M. Schilling*

The credits section for this book begins on page 784, and is considered an extension of the copyright page.

Library of Congress Catalog Card Number: 87–70678

ISBN 0–697–05115–3

Printed in the United States of America by Wm. C. Brown Publishers
2460 Kerper Boulevard, Dubuque, IA 52001

10 9 8 7 6 5 4 3 2

For my children

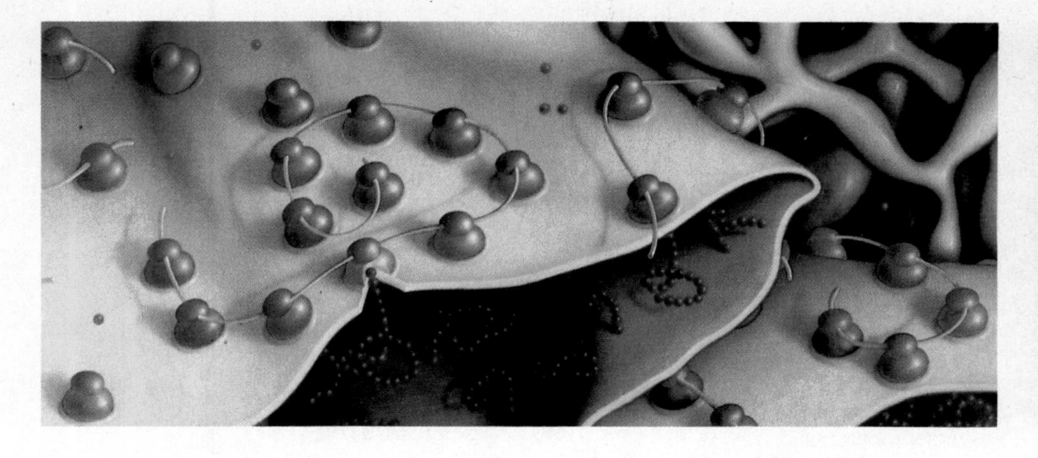

Brief Contents

Contents

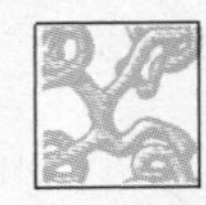

4 Cell Division 74

5 Cellular Metabolism 92

Part Two
Plant Biology 113

6 Photosynthesis 114

7 Plant Form and Function 131

Part Three
Human Anatomy and Physiology 161

8 Human Organization 162

9 Digestion 181

10 Circulation 207

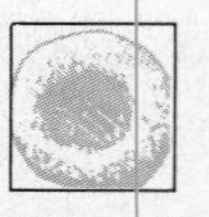

11 Blood 230

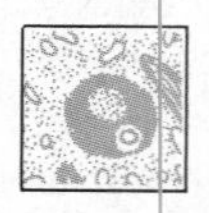

12 Immunity 250

13 Respiration 268

14 Excretion 289

15 Nervous System 308

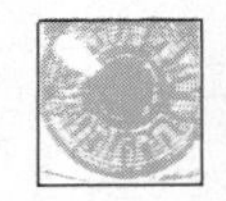

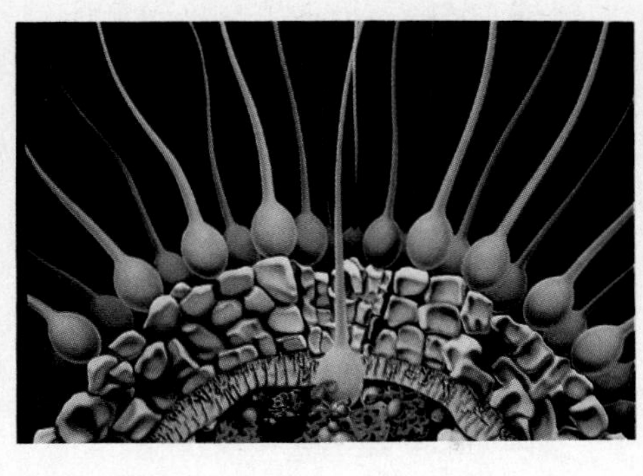

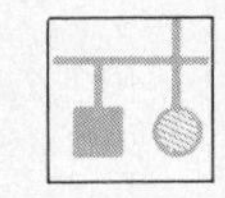

Part Six

Behavior and Ecology 643

29 Behavior within Species 644

30 Behavior between Species 668

31 Ecosystems 687

32 The Biosphere 714

33 Human Population Concerns 740

Tables

Readings

*These readings were written by the author to highlight and expand upon topics discussed in the text.

Preface

Inquiry into Life is written for the young adult who would like a working knowledge of biology. Educational theory tells us that adults are most interested in knowledge of immediate practical application. This book attempts to remain true to this idea, and its basic theme is knowledge about and understanding of human biology. Plants and other animals are included, however, because human beings cannot understand themselves unless they understand other living things. All organisms face the same problems, but through the evolutionary process, they solve these problems differently. Human beings can better understand themselves when they see the unity of life while at the same time seeing the diversity. Concerned citizens need to realize that humans are not the pivot point, nor even the culmination, of life but are a part of a great, overall, interrelated network.

While this text covers the whole field of basic biology, it emphasizes the application of this knowledge to aspects regarding humans themselves and the relation of humans to other organisms. Students will thereby be able to see that biology is truly relevant. Concurrent with this approach, the student is led to a discovery of biological concepts and principles. Detailed, high-level scientific data and terminology are not included because the author believes that true knowledge consists of working concepts rather than technical facility. The latter can always be added by more advanced study after acquisition of a core of knowledge upon which to build.

The fifth edition of *Inquiry into Life* has the same organization and style as the previous four editions. Each chapter presents the topic clearly, simply, and distinctly so that the student will feel capable of achieving an adult level of understanding. Although it is quite possible for succeeding editions of a text to become more lengthy and erudite, this tendency has been strenuously avoided, and adopters will find the fifth edition just as to the point and straightforward as the other editions.

In addition to chapter concepts, each chapter now begins with an outline. Outlining the chapters helped to assure that they all were logically organized. Frequent summary statements within each chapter now supplement the summary at the end of each chapter. It is hoped that these statements will highlight important concepts for students. The achievement of this goal is aided by the addition of objective questions and a glossary of key terms at the end of each chapter.

Organization of the Text

An introductory chapter precedes the thirty-three chapters of *Inquiry,* which are grouped into six parts.

Introduction

The introductory chapter reviews the characteristics of life. This discussion is designed to have students see that humans are but one kind of living thing and that all types of organisms share the same characteristics. The scientific method is outlined in order to show that the biological view of life is a product of this method. There follows an overview of the book that indicates the manner in which its various parts are related to one another. The overview provides students with an opportunity to see the whole before beginning their study of the individual parts.

In this edition, the scientific method has been given greater emphasis. Also, this presentation uses the discovery of the AIDS virus as the basis for discussion.

Part One, The Cell, a Unit of Life

The first part introduces basic biological principles and serves as a foundation for the parts that follow. The overall organization and content remain essentially as they were in the fourth edition but rewritten sections will be found throughout. The chemistry chapter has been retooled and refreshed with new illustrations. The cell chapter has been reorganized in that divided coverage of the cytoskeleton has been brought together. The metabolism chapter now begins with a revised discussion of enzyme function.

Part Two, Plant Biology

Coverage of photosynthesis precedes that of plant structure and function. The photosynthesis chapter has been reorganized and includes a summary illustration that will assist instructors in presenting an overview of this complex topic. The entire discussion of transport in plants has been rewritten and new illustrations are provided. Similarly the section devoted to photoperiodism has been revamped and expanded.

Part Three, Human Anatomy and Physiology

Just as the flowering plant is used as the basis for discussing plant physiology, so humans are used as the basis for discussing animal physiology. Reorganization of chapters and updating of material characterize the revision of this part. The latest information in regard to dieting, hypertension, the inflammatory reaction, AIDS, and urinary formation are included, for example.

Part Four, Reproduction, Development, and Inheritance

Human reproduction, development, and inheritance are the topics for this part. The reproduction chapter has been reorganized, the development chapter has been rewritten, and molecular genetics has been revised. There are additional biotechnology applications and new information about the eukaryotic chromosome which have been included at an appropriate level of presentation. Recent findings regarding oncogenes are also given.

Part Five, Evolution and Diversity

This part not only covers the principles of evolutionary theory including evidences for evolution, but it also contains a survey of the classification of organisms. A new and updated classification system is used. Portions were rewritten throughout but the most significant changes include an expanded and updated section on viruses. Also, the chapter on human evolution was rewritten to achieve a more modern approach to this topic.

Part Six, Behavior and Ecology

In this part there are two chapters devoted to behavior before ecosystems, biomes, and human population are discussed. Readers will find a stronger human emphasis including a more extensive coverage of sociobiology in chapter 29. The ecology principles chapter puts more stress on community ecology than previous editions. The need to preserve the tropics is also included. As before, the text ends with a chapter on Human Population Concerns but the last few pages now outline the characteristics of a steady state.

"History of Biology" End Sheets

For those instructors who like to include history of biology references in their lectures, the front and back cover now list major contributions to the field of biology in a concise, chronological manner. Students may be referred to these whenever it is felt to be appropriate.

Comparison of Texts

Biology: Evolution, Diversity, and the Environment

As many may know, I am also the author of a general biology text, *Biology: Evolution, Diversity, and the Environment.* This text approaches the study of life from the perspective of the three themes in the title—evolution, diversity, and environment. In contrast to *Inquiry, Biology* explains general principles primarily in reference to forms of life other than the human form and secondarily in reference to the human organism. For example, the animal physiology section emphasizes comparative animal physiology in an evolutionary context. Even so, the discussion includes all aspects of human physiology. Ecology is a constant theme that runs throughout the text. Several readings highlight various modern ecological problems and they, along with suggested solutions, are also discussed at length in the last few chapters.

Inquiry into Life

Inquiry into Life, as has been discussed, focuses on human biology. The human biology focus is an emphasis that immediately appeals to many students. General principles are explained primarily in reference to the human organism and secondarily in reference to other forms of life. A significant portion of *Inquiry into Life* is devoted to human physiology. The text also includes plant and animal diversity and ecological principles and their applications to modern ecological problems.

Both *Inquiry into Life* and *Biology: Evolution, Diversity, and the Environment* cover all aspects of general biology, giving nonmajors a review of the entire field of biology and majors a firm basis on which to build. The availability of both of these texts gives instructors a choice of emphasis, according to their own preference and knowledge of the students at their particular institution.

Human Biology

My new book, *Human Biology,* is a short text designed for students in one-semester human biology courses. It emphasizes basic biological concepts, human anatomy and physiology, the human's role in the biosphere, and ethical human issues as they relate to biology. *Human Biology* may also be used in two-semester general biology courses that take a human emphasis. However, *Human Biology* does not include coverage of plant physiology or taxonomy of plants and animals.

Aids to the Reader

Inquiry into Life includes a number of aids that have helped students study biology successfully and enjoyably.

Text Introduction

The introductory chapter discusses the characteristics of life, the scientific method, and presents an overview of the book. This chapter surveys the field of biology as a whole and prepares the student for an examination of its individual portions.

Part Introductions

An introduction for each part highlights the central ideas of that part and specifically tells the reader how the topics within each part contribute to biological knowledge.

Chapter Concepts

Each chapter begins with a list of concepts stressed in the chapter. This listing introduces the student to the chapter by organizing its content into a few meaningful sentences. The concepts provide a framework for the content of each chapter.

Chapter Outlines

In addition to the chapter concepts, each chapter now has a chapter outline. These will allow students to tell at a glance how the chapter is organized and what major topics have been included in the chapter. The chapter outlines include the first and second level heads for the chapter.

Readings

Two types of boxed readings are included in the text. Readings chosen from popular magazines illustrate the applications of concepts to modern concerns. These spark interest by illustrating that biology is an important part of everyday life. The second type of reading, usually written by the author, is designed to expand, in an interesting way, on the core information presented in each chapter. Topics such as plant growth regulators and U.S. population problems are addressed in these readings.

Tables and Illustrations

Numerous tables and illustrations appear in each chapter and are placed near their related textual discussion. The tables clarify complex ideas and summarize sections of the narrative. Once students have achieved an understanding of the subject matter by examining the chapter concepts and the text, these tables can be used as an important review tool. The photographs and drawings have been selectively chosen and designed to help students visualize structures and processes.

Boldfaced Words

New terms appear in boldface print as they are introduced within the text and are immediately defined in context. If any of these terms are reintroduced in later chapters, they are italicized. Key terms are defined in the end-of-chapter glossary and all boldface terms are in the text glossary, where a phonetic pronunciation is given with the appropriate page reference.

Internal Summary Statements

New to this edition, summary statements are placed at strategic locations throughout the chapter. These immediately reinforce the concept that has just been discussed. The summary statements will aid student retention of the chapter's main points.

Chapter Summaries

Chapter summaries offer a concise review of material in each chapter. Students may read them before beginning the chapter to preview the topics of importance, and they may also use them to refresh their memories after they have a firm grasp of the concepts presented in each chapter.

Chapter Questions

Objective questions and study questions are at the close of each chapter. The objective questions allow students to quiz themselves with short fill-in-the-blank objective questions. Answers to these questions appear on the same page. The study questions allow students to test their understanding of the information in the chapter.

Chapter Glossary

New to this edition is a chapter-ending glossary. Major boldfaced terms within the chapter are now defined in the back matter of each chapter for more convenient review. Selected key terms are listed with their phonetic pronunciation, carefully defined, and page referenced. All boldface terms are still listed alphabetically with their pronunciations, definitions, and page references in the text glossary at the end of the book.

Further Readings

For those students who would like more information about a particular topic or are seeking references for a research paper, each part ends with a listing of articles and books to help them get started. Usually the entries are *Scientific American* articles and speciality books that expand on the topics covered in the chapter.

Appendix and Glossary

The appendix contains optional information for student referral. It includes an expanded Periodic Table of the Elements, and a review of the metric system. An important part of the appendix is the Classification System of Organisms used in the text.

The text glossary defines the terms most necessary for making the study of biology successful. By using this tool, students can review the definitions of the most frequently used terms.

Index

The text also includes an index in the back matter of the book. By consulting the index it is possible to determine on what page or pages various topics are discussed.

Additional Aids

Instructor's Binder

An instructor's binder is now available with *Inquiry into Life*. The binder includes the text, Instructor's manual with test items, and Lecture Enrichment units collated by chapter. This will facilitate preparation of lectures and help instructors utilize the many additional aids available with *Inquiry into Life*.

Instructor's Manual

The *Instructor's Manual* is designed to assist instructors as they plan and prepare for classes using *Inquiry into Life*. Possible course organizations for semester and quarter systems are suggested, along with alternate suggestions for sequencing the chapters. An outline and a general discussion are provided for each chapter; together these give the overall rationale for the chapter. A large number of objective test questions and several essay questions are provided for each chapter. A list of suggested audiovisuals for the various topics and a list of suppliers are included at the end of the *Instructor's Manual*.

Student Study Guide

To ensure close coordination with the text, the author wrote the *Student Study Guide* that accompanies this text. Each text chapter has a corresponding study guide chapter that includes a listing of behavioral objectives, a pretest, study exercises, and a posttest. Answers to study guide questions appear at intervals throughout each study guide chapter, giving students immediate feedback.

Laboratory Manual

The author also wrote the *Laboratory Manual* that accompanies *Inquiry into Life*. With few exceptions, each chapter in the text has an accompanying laboratory exercise in the manual (some chapters have more than one accompanying exercise). In this way, instructors will be better able to emphasize particular portions of the curriculum if they wish. The thirty-two laboratory sessions in the manual were designed to further help students appreciate the scientific method and learn the fundamental concepts of biology and the specific content of each chapter. All exercises have been tested for student interest, preparation time, and feasibility.

Laboratory Resource Guide

More extensive information regarding preparation can be found in the *Laboratory Resource Guide*. This guide, developed by the author and Dr. Trudy McKee, is a new addition to the package and will assist the instructor in making the laboratory experience a meaningful one for the student. The guide includes suggested sources for materials and supplies, directions for making up solutions and otherwise setting up the laboratory, expected results for the exercises, and suggested answers to all questions in the laboratory manual.

Transparencies

This edition is accompanied by 203 transparencies in two and four color. The transparencies feature text illustrations with oversized labels facilitating their use in large lecture rooms.

Slides

Often instructors prefer to use slides rather than transparencies. These are available upon special request. It is possible to acquire slides of all transparency art pieces.

Slides of Photomicrographs and Electron Micrographs

This addition to the ancillary program features 46 slides of high interest photomicrographs and electron micrographs, most of which are SEMs.

Lecture Enrichment Kit

The Lecture Enrichment Kit is a series of optional lecture notes to accompany the text's transparencies. For each transparency there is a corresponding lecture unit. Each unit contains a summary of all text material pertinent to the process or phenomenon depicted in the transparency. The summary is followed by three to five "extensions," topics not discussed in the text. "Extensions" are drawn from popular periodicals of general interest, scientific periodicals, or more advanced texts. They vary in detail and degree of rigor.

Computerized Ancillaries

wcb QuizPak

Student computer software programs are now available with this edition of *Inquiry into Life*. **wcb QuizPak** provides students with true-false and multiple choice questions for each chapter in the text. Using these programs in the Learning Resource Center will help students to prepare for examinations. They are available on Apple® and IBM® PC diskettes.

wcb TestPak and wcb GradePak

wcb TestPak, a computerized testing service, provides instructors with either a mail-in/call-in testing program or the complete test item file on diskette for use with the Apple® and IBM® PC computers. **wcb** TestPak requires no programming experience. **wcb** GradePak is a computerized grade management system for instructors. This program tracks student performance on examinations and assignments. It will compute each student's percentage and corresponding letter grade, as well as the class average. **wcb** GradePak is user friendly: instructors can use it after only twenty minutes of direction.

Acknowledgments

The personnel at Wm. C. Brown Publishers have always lent their talents to the success of *Inquiry into Life,* and I especially want to thank Kevin Kane, biology editor, for guiding the book from its inception to its completion. Mary Sailer designed the book with skill and Shirley Charley picked just the right photographs. Kay J. Brimeyer was the production editor who coordinated the efforts of many.

There are many new illustrations in this edition of *Inquiry into Life.* Kathleen Hagelston, as always, provided most of these; the clarity and beauty of her work is well known to all those familiar with the text. In addition, Thomas Waldrop redid several of the anatomical illustrations in a way that is immediately appealing. Also, I was pleased to have the assistance of Francis Leroy who provided many new strikingly beautiful illustrations.

Finally I wish to express appreciation to my family for their constant support. My children Karen and Eric and my sister, Rhetta, were always ready to offer advice and encouragement. My nephew Terry, although new to the endeavor, was very patient and understanding.

Many instructors have contributed not only to this edition of *Inquiry* but to previous editions. The author is extremely thankful to each one, for we have all worked diligently to remain true to our calling and to provide a product that will be the most useful to our students.

In particular, it seems proper to acknowledge the help of the following individuals:

Ken Abbott
Yavapai College

Steve Adams
Lake City Community College

James Averett
Nassau Community College

Paul Biersuck
Nassau Community College

Clyde Bottrell
Tarrant County Junior College, South Campus

Kathy Burt-Utley
University of New Orleans

Roy B. Clarkson
West Virginia University

John Cowlishaw
Oakland University

Ron Daniel
Cal Poly, Pomona

Russell Davis
University of Arizona

John Dixon
University of Wisconsin at Eau Claire

Albert Gordon
Southwest Missouri State University

Martin Hahn
William Paterson College

Laszlo Hanzely
Northern Illinois University

Mark Levinthal
Purdue University

John F. Lyon
University of Lowell

Priscilla Mattson
University of Lowell

Joyce Maxwell
California State University at Northridge

Betty McAtee
Prince George's Community College

Heather McKean
Eastern Washington University

Steven N. Murray
California State University at Fullerton

Ethel Sloane
University of Wisconsin at Milwaukee

Warren Smith
Central State University

Tommy Wynn
North Carolina State University

Gordon Ashcroft
San Juan College

Karen Belcher
Lake City Community College

Michael Bell
Richland College

Barbara Berkley
St. Paul's College

John Biehl
Riverside City College

Dick Birkholz
Sheridan College

Hessel Bouma III
Calvin College

Sheila Brown
University of S. Mississippi

Donald Butler
Duquesne University

Joseph Cancannon
St. Johns University

William Carden
Grossmont College

Eldon Carins
Auburn University at Montgomery

William Carr
Alexander City State Junior College

John M. Chapin
St. Petersburgh Junior College

David Cherney
Azusa Pacific College

Al Chiscon
Purdue University

Barbara Clarke
American University

Jerry A. Clonts
Anderson College

Arthur A. Cohen
Massachusetts Bay Community College

Don Collier
John C. Calhoun State Community College

Charles Cottingham
South Carolina State College

Shirley Crawford
SUNY Agricultural & Technical College

Robert Dahm
William Wright College

Patrick Daley
Lewis & Clark Community College

George DeHullu
Northern Essex Community College
Glen Drews
Chaffey College
Carlos Estol
New York University
Kathy Fergen
Miami-Dade Community College
Fred First
Wytheville Community College
Marie Fitzgerald
Holyoke Community College
Gary Fungle
Santa Barbara City College
Carol Gerding
Cuyahoga Community College, West Campus
Lewis Grey
Truman City College-Chicago
Laird Hartman
University of South Dakota-Springfield
William R. Hawkins
Mt. St. Antonio College
Jerry Henderson
St. Louis Community College
John Hoagland
Danville Junior College
Sadako Houghten
Los Angeles Pierce College
Ivan Huber
Farleigh Dickenson University-Madison
Frank Johnson
Massasoit Community College
Neil Johnson
N. Dakota State School of Science
James Kane
Muskegon Community College
Vincent A. Kissel
CUNY-Bronx Community College
George Klee
Kent State University-Stark
Mary Lou Longo
American International College
Benjamin Lowenhaupt
Edinboro State College
Edward Lucier
Becker Junior College
Madhu Narayan Mahadeva
University of Wisconsin–Oshkosh
George Mason
East Central Junior College
Jack Dennis McCullough
Stephen F. Austin State University
Harry S. McDonald
Stephen F. Austin State University
Ellen McGloflin
Sanford University
James McIver
Idaho State College
Roger McPherson
Clarion University of Pennsylvania
Jim Milek
Casper College
Jean Monier
College of Notre Dame
Milton Nathanson
CUNY Queens College
Hossam A. Negm
Grambling State University
Robert Neher
University of LaVerne
William Neilson
Harrisburg Area Community College
Edwin Lane Netherland
Cameron University
Lynne Osborn
Middlesex Community College
Ann Robinson
Carnegie-Mellon University
J. Rosko
St. Thomas Aquinas College
Jim Royce
Iowa Central Community College
James Sandoval
LA City College
Robert Schodorf
Lake Michigan College
A. Floyd Scott
Austin Peay State University
John Sharp
John Tyler Community College
Ted Sherill
Eastfield College
Jean Shuemaker
Gallaudet College
Judith Slon
Hilbert College
James Smith
Lawson State Community College
David Sulter
College of the Desert
Maurice Sweatt
Canada College
Virginia Teagarden
Lees-McRae College
Wesley Thompson
Rio Hondo College
Doris Tingle
North Greenville College
Mario A. Vecchiarelli
Housatonic Regional Community College
Ben Van Wagner
Bethany Bible College
George Washington
Jackson State University
Stephen Wheeler
Alvin Junior College
William Whittaker
Greenville Technical College
Billy Williams
Dyersburg State Community College
Michael Willig
Texas Tech University
Gary Wrinkle
West Los Angeles College
W. Brooke Yeager
Luzerne Community College

I would also like to thank the following adopters of the fourth edition for their help in preparing the current one. Each contributed greatly to our understanding of the relative strengths and weaknesses of that edition by responding to a user's survey of the text.

Dr. Qamar A. Abbasi
Olive-Harvey College

William Barnes
Clarion University

Marie Casciano
Pace University

S. Chidambaram
Canisius College

Joseph W. Cliborn
Pearl River Junior College

James V. A. Conkey
Truckee Meadows Community College

Lesta J. Cooper-Freytag
University of Cincinnati-Raymond Walters College

Neil Crenshaw
Indian River Community College

James M. Davenport
Washtenaw Community College

Lester Davis
Sandhills Community College

H. Douglas Dean
Pepperdine University

Joyce Denham
University of Minnesota

Steven A. Fink
West Los Angeles College

James E. Forbes
Hampton University

Bernard L. Frye
University of Texas-Arlington

Donald C. Giersch
Triton College

Matthew R. Gilligan
Savannah State College

Sushil K. Gilotra
Delgado College

Jack L. Gottschang
University of Cincinnati

Malkiat S. Guram
Voorhees College

Harry R. Holloway, Jr.
University of North Dakota

Isidore A. Julien
Roxbury Community College

Dean H. Kruse
Rancho Santiago College

W. Donald McGavock
Tusculum College

Raymond A. Miller
University of Kentucky - Somerset Community College

Thomas H. Milton
Richard Bland College

John Munford
Wabash College

Malcolm Nason
North Shore Community College

Diane R. Nelson
East Tennessee State University

Jerry Nielsen
Western Oklahoma State College

Carl Oney
Ononoga Community College

David C. Robinson
Albany State College

Kenneth S. Saladin
Georgia College

Robert C. Simpson
Lord Fairfax Community College

Morton Sinclair
Erie Community College, City Campus

Jeffrey L. Smith
Dillard University

Dixie Stone
Washington Technical College

R. Bruce Sundrud
Harrisburg Area Community College

Alexander Varkey
Liberty University

Miryam Z. Wahrman
William Paterson College

Bernard L. Woodhouse
Savannah State College

INQUIRY INTO LIFE

Introduction

Introduction Concepts

1 Although life is difficult to define, it can be recognized by certain common characteristics.

2 The information in this book has been gathered by using the scientific method, a process based on certain identifiable procedures.

3 There are various approaches to the study of biology. (The approach chosen in this text stresses human biology.)

Introduction Outline

This book is about living things (fig. I.1) and therefore it is appropriate to first define life. Unfortunately this is not so easily done—life cannot be given a simple one-line definition. Since this is the case, it is customary to discuss life in terms of its characteristics; the following five characteristics are commonly attributed to living things:

1. Living things have a structure that is ultimately made up of cells.
2. Living things grow and maintain their structure by taking chemicals and energy from the environment.
3. Living things respond to the external environment.
4. Living things reproduce and pass on their organization to their offspring.
5. Living things evolve, or change, and adapt to the environment.

Figure I.1
Impalas are found in the less fertile woodlands of central and southern Africa where they graze on grasses, browse on leaves, and also eat flowers, fruits, and seeds. The basic social unit is a herd of females controlled by a single male during mating season. Only the males have horns.

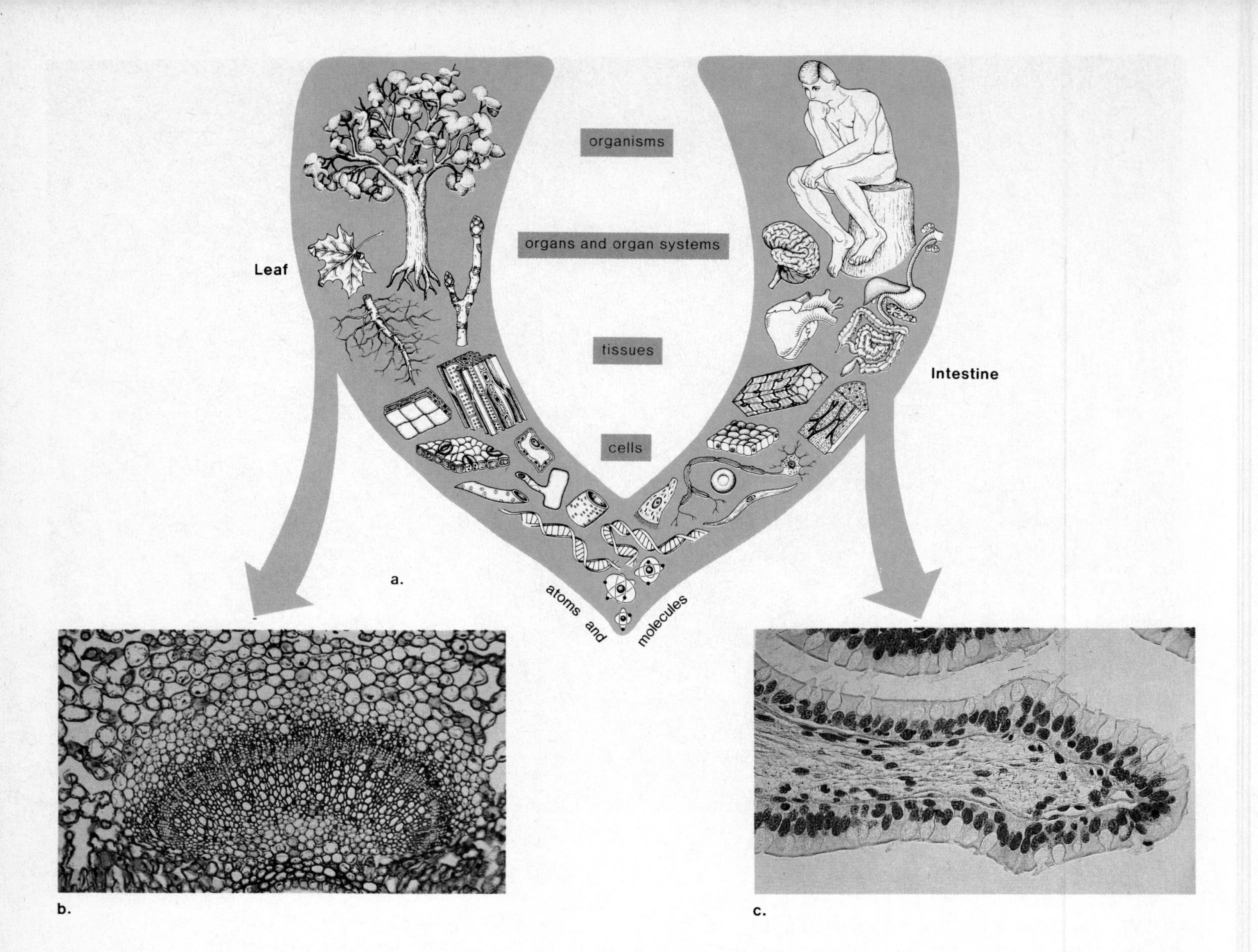

Figure I.2
Organization of plants and animals. a. *Cells are composed of chemicals; tissues are made up of cells; organs are composed of tissues; and organisms contain organs.* b. *Cross section of plant (privet) leaf showing individual cells.* c. *Longitudinal section of villi from small intestine showing individual cells. Some are goblet cells discharging mucus into the gut cavity.*
(b.) © Carolina Biological Supply Company.

Characteristics of Life

Living things have a structure that is ultimately made up of cells. Figure I.2 shows that both plants and animals are organisms, which are composed of organs, which in turn are composed of tissues, and tissues are composed of cells. Cells are made up of chemicals (molecules), nonliving substances that contain atoms, the smallest units of matter nondivisible by chemical means.

Living things grow and maintain their structure by taking chemicals and energy from the environment. Only plants and plantlike organisms are capable of utilizing inorganic chemicals and the energy of the sun in the process of making their food. Other organisms, such as animals, must take in preformed food as a source of chemicals and energy (fig. I.3). The chemicals and energy obtained from the environment are used in part to maintain the organization of living organisms. Each living organism has its own particular organization that is maintained only because energy can be acquired from the environment.

Figure I.3
Summer tanagers feeding their offspring. In birds, feeding behavior is often elicited by the gaping mouth of the fledglings whose immature state requires parental care.

Living things respond to the external environment. This characteristic is commonly recognized by people who acknowledge a living thing by its ability to move. A multicellular animal can move because it possesses a nervous system, but other living things, including plants, possess a variety of mechanisms by which they respond to the physical and biological environment. Responses to the environment constitute the behavior of organisms (fig. I.3).

Living things reproduce and pass on their organization to their offspring. A unicellular organism reproduces asexually simply by dividing into two new organisms. Because the new organisms have the same hereditary factors, or genes, as the original organism, they also possess the same structure and function. In contrast, multicellular organisms often reproduce by means of sexual reproduction. Male and female organisms each contribute one half the total number of genes to the new organism, which often resembles the parents even as it is maturing (fig. I.3).

Living things evolve, or change, and adapt to the environment. While the preceding characteristics of life pertain to individual living things, this fifth characteristic is concerned with the **species,** a group of similarly constructed organisms that share common genes. It is the species, rather than the individual, that evolves. Evolution begins when certain organisms happen to inherit a genetic change that causes them to be better suited to a particular environment. These organisms, which are said to be better adapted, tend to survive and have more offspring than those that are not as well suited. In this way, evolution produces successive generations of organisms that are better adapted to the environment. If the environment should change, they may no longer possess the genetic capability of adapting to the new environment, and extinction can follow.

Figure I.4
All these animals are closely related but their appendages are adapted in various ways. a. *In this fishing bat, the forelimbs are adapted for flight and the hindlimbs for catching fish.* b. *In the gibbon, both the hands and feet have opposable thumbs, making it easy for them to grasp tree limbs.* c. *This cat, called a serval, is stalking through tall grass—its forelimbs and hindlimbs are adapted for running and the paws have claws that pin down prey.*

The evolutionary process causes life to have a history. The belief today is that a chemical evolution produced the first cell or cells, and from these all other forms of life have evolved. There is a variety of life forms on this planet because they are adapted to various ways of living in specific environments (fig. I.4).

All living things display certain characteristics: they are made up of cells, maintain their structure by taking chemicals and energy from the environment, respond to external stimuli, and reproduce. As species, living things evolve and change. Evolution accounts for the diversity of life we see about us.

Scientific Method

Science helps human beings understand the material world and is concerned solely with information gained by observing and testing that world. Scientists try to be objective and not let personal feelings influence the outcome of their investigations. The general approach for gathering information by scientists, including biologists, is known as the **scientific method.** The scientific method is as varied as scientists themselves, but even so there are certain processes that are typically characteristic of science. First of all, scientists ask only *causality questions,* such as what caused this or how does this occur, rather than *teleological questions* that ask for what purpose something occurs. For example, a scientist might address the question, What causes this particular illness? but would not ask the more philosophical question, Why should humans get sick?

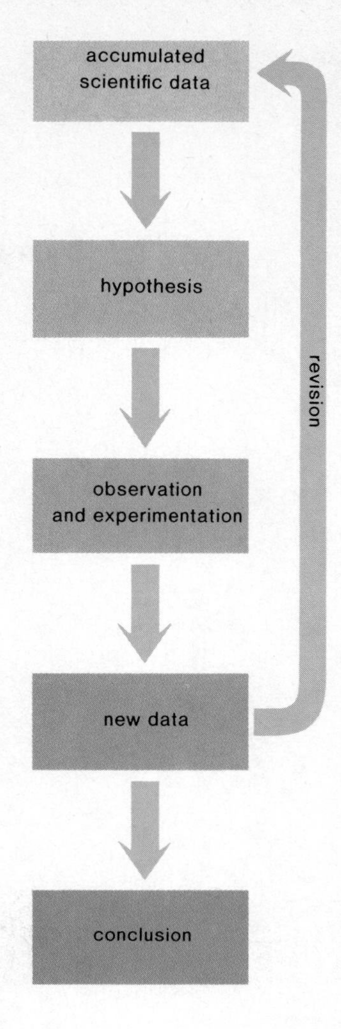

Figure I.5
Steps involved in the scientific method. Inductive reasoning is used to formulate the hypothesis from accummulated scientific data and deductive reasoning is used to decide what observations and experiments would be appropriate in order to test the hypothesis.

Formulation of the Hypothesis

Observations permit scientists to form **hypotheses,** tentative explanations of observed phenomena (fig. I.5). To arrive at a hypothesis, scientists use various methods of reasoning, especially *inductive reasoning.* Inductive reasoning allows scientists to arrive at a generalization after observing specific facts. For example, in 1981 a small number of homosexual men in New York and California became ill with a condition that weakened their immune system, causing them to die from a fairly uncommon form of pneumonia and skin cancer. As the number of victims steadily increased, scientists began to consider the possibility that this condition, now called AIDS, acquired immune deficiency syndrome, was caused by an infectious agent.

Robert Gallo of the National Cancer Institute had previously isolated a retrovirus called HTLV (human T-cell leukemia virus) I and II that caused T cells, a type of white blood cell, to proliferate. This work prompted Gallo and others to hypothesize that HTLV-I was the cause of this new syndrome.

Experimentation

After a hypothesis such as this has been stated a second type of reasoning, *deductive reasoning,* comes into play. Deductive reasoning begins with a general statement that infers a specific conclusion. It often takes the form of an if . . . then statement. In science, deductive reasoning allows a scientist to determine the type of experiment and/or observation necessary to support or refute a hypothesis. For example, it follows that if AIDS is caused by the HTLV-I virus then it should be possible to find evidence of this virus in the blood of AIDS victims. Scientists made this deduction, and they began to test the blood of AIDS victims for antibodies to HTLV-I. The results were not conclusive enough to support the hypothesis. Then, Gallo made the observation that whereas an HTLV-I infection causes T cells to proliferate, the T cells die off in persons with AIDS. Therefore, Gallo began to hypothesize that a different but related retrovirus was the cause of AIDS. Luckily, about this time an associate discovered a line of T cells (from a leukemia patient) that would not die off when infected with the virus. Now the virus could be grown in quantity in the laboratory. This breakthrough allowed Gallo's group to study and identify the virus they at first called HTLV-III and it also paved the way for the development of a blood test for the presence of the virus. Now it was time to test the blood of patients with AIDS (fig. I.6).

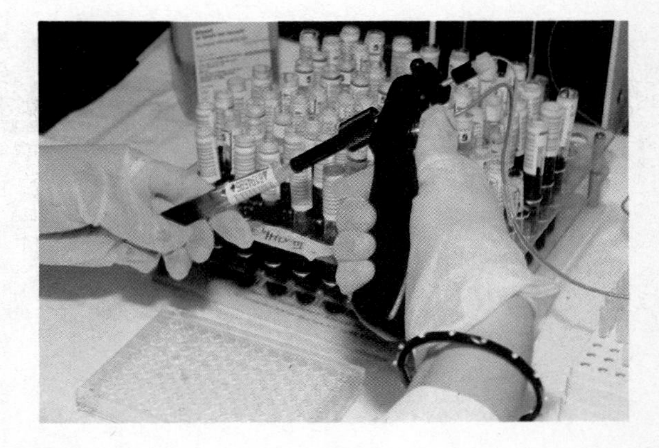

Figure I.6
A scientist often prefers to work in a laboratory. This scientist is testing blood samples for the presence of the AIDS virus.

Controlled Experiment

Very often when scientists perform an experiment, they prefer to include a **control** sample (a sample that undergoes all the steps in the experiment except the one being tested). The controlled experiment allows scientists to determine that they have indeed found a cause and effect relationship. For example, in the May 4, 1984, issue of *Science,* Gallo reported that antibodies were detected in almost 90 percent of AIDS patients and 80 percent of patients with a pre-AIDS condition. Only one out of 186 normal persons in the control sample tested positive. Scientists much prefer mathematical **data** (pertinent facts) such as these because mathematical data is highly objective and not subject to individual interpretation.

Our example illustrates that scientists most often report their findings in *journals* (magazines) that are studied by the scientific community. They do so because all experimental results must be repeatable; that is, other scientists using the same procedures are expected to get the same results. If they do not, the original data is not considered to have supported the hypothesis.

Conclusions

The data collected from experiments either supports or fails to support the hypothesis. Earlier, we mentioned that Gallo's original hypothesis was not supported by the data. In such instances, the hypothesis is considered invalid and is discarded. However, Gallo's second hypothesis, that the HTLV-III virus was the cause of AIDS, was supported by the data. Often, too, the hypothesis is supported by additional experiments. For example, researchers have been able to produce antibodies and even early symptoms of AIDS in chimpanzees by injecting them with the HTLV-III virus, now called the HIV (human immunodeficiency virus). Therefore, it is stated today that HIV is the cause of AIDS. Even so, we have to realize that hypotheses are never actually proven true. Instead, scientists are always aware that the present body of information represents the truest available at the moment and that further observations and experiments could lead to changes in prior conclusions.

Concepts

The ultimate goal of science is to understand the natural world in terms of *concepts,* interpretations that take into account the results of many experiments and many observations. For example, the theory of evolution is one such conceptual scheme. It allows scientists to understand the history of life, the variety of living things, the anatomy and physiology of organisms, embryological development, and so forth. When the designation *theory* is used in science it means that scientists have the utmost confidence in a concept whose broad scope gives it fundamental importance. This is contrary to the way that the word "theory" is used in everyday language.

Bioethics

It is not the role of science to answer bioethical decisions. While scientists can inform and even recommend, it is up to all individuals to decide to what degree and in what manner the conclusions of science are to be used. For this reason it is extremely important for you to possess a knowledge of human biology, a knowledge that will allow you to contribute intelligently to shaping the future destiny of human beings.

The scientific method permits scientists to objectively collect information concerning the material world. It is up to all individuals to decide how this information is to be used.

Inquiry into Life

There are many different ways to approach the study of life. This text, while covering all aspects of biology, focuses on human biology. For example, human anatomy and physiology is studied as representing vertebrate anatomy and physiology; and the chief environment of humans, the country and city, is studied as an example of an ecosystem.

A brief introduction to the many topics discussed in the six units of this text follows.

The Cell, a Unit of Life

Multicellular organisms, including humans, are composed of many cells (fig. I.2). In order to understand how multicellular organisms function, it is necessary to understand how the cell functions. Since cells are made up of chemicals, we begin our study by considering some basic chemistry essential to the cell. Our study of cells also includes how cells grow and reproduce, two vital functions in the life of the cell.

Plant Biology

There are two major types of higher organisms—plants and animals. Flowering plants are the most recently evolved of the plants, and they serve as our basis for the study of plant biology in this unit. Plant cells carry on photosynthesis, the process by which they make their own organic food after capturing energy from sunlight. Photosynthesis is extremely important because it ultimately provides food for all living things (fig. I.7).

Figure I.7
Sun streaming down on this cornfield will provide the energy for these plants to grow and produce corn, to be used as food for animals such as humans and their domesticated cattle.

Human Physiology

As in our study of plant biology, our study of animal biology centers on the most recently evolved animal—humans. The human body is composed of many systems, each designed to perform a particular life function (table I.1). The unit on human physiology discusses the anatomy and, in particular, the physiology of each of these systems while it also highlights related areas of interest, such as dieting, drugs, smoking, and important illnesses. It is hoped that the knowledge you gain about the human body will assist you in understanding the workings of your own body.

Table I.1 *Animal Organ Systems*

Name	Function
Digestive	Convert food particles to nutrient molecules
Circulatory	Transport of molecules to and from cells
Immune	Defense against disease
Respiratory	Exchange of gases with environment
Excretory	Elimination of metabolic wastes
Nervous and sensory	Regulation of systems and response to environment
Muscular and skeletal	Support and movement of organism
Hormonal	Regulation of internal environment

Reproduction, Development, and Inheritance

The unit on human reproduction and inheritance explores topics of extreme interest to young people, who are just beginning the reproductive years of their lives. The anatomy and physiology of the reproductive system is considered, as well as birth control, infertility, and sexually transmitted diseases. The stages of development are reviewed, giving you an opportunity to see how a fertilized egg becomes the newborn infant. Inheritance is an extremely important topic today for several reasons. Much more is known today about human genetic diseases than was known even as little as a decade ago. Science has made it possible to sometimes predict the chances of a child being born with a genetic disease. Also, we are entering an era of genetic engineering, that may soon make it possible to correct genetic abnormalities.

Evolution and Diversity

After an introductory chapter that presents the principles of evolution, the unit on evolution and diversity surveys living things from the origin of life to human evolution. Particularly today, it is important for you to become acquainted with the diversity of life not only because it illustrates our relationship with other living things but because it may also enhance your awareness of the need to preserve and protect all forms of life.

The Everglades

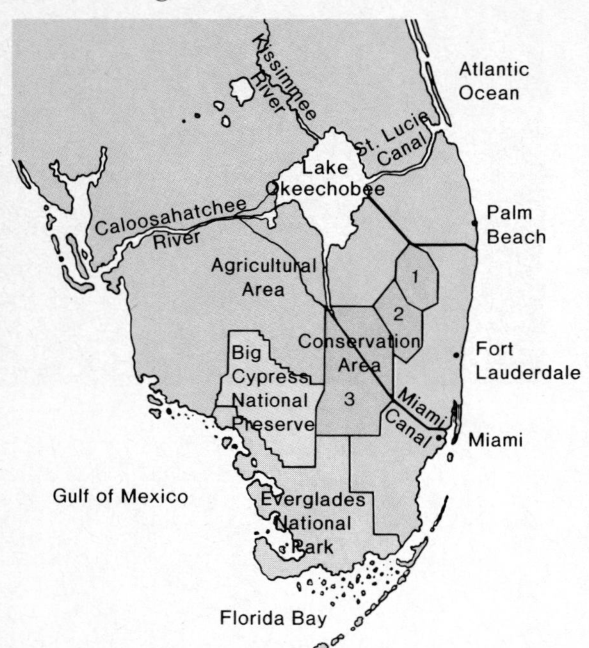

The Everglades once extended from Lake Okeechobee south to Florida Bay. Now the glades are found only in three conservation areas and Everglades National Park.

Originally, the Everglades encompassed the whole of southern Florida from Lake Okeechobee down to Florida Bay (see the accompanying map). Now largely in the Everglades National Park alone do we find the vast saw grass prairie interrupted occasionally by a cypress dome or hardwood tree island. Within these islands, both temperate and tropical evergreen trees grow amongst the dense and tangled vegetation. Mangrove, or salt tolerant trees, are found along sloughs (creeks) and at the shoreline. Only the roots of the red mangrove can tolerate the sea constantly. The prop roots of this tree protect over forty different types of juvenile fishes as they grow to maturity. During the wet season, from May to November, animals are dispersed throughout the region, but in the dry season, from December to April, they congregate wherever pools of water are found. Alligators are famous for making "gator holes" where water collects, and fish, shrimps, crabs, birds, and a host of living things survive until the rains come again. Almost everyone is captivated by the birds that find a ready supply of fish they need for daily existence at these holes. The large and beautiful herons, egrets, roseate spoonbill, or anhinga fill one with awe. These birds once numbered in the millions; now they number only in the thousands. Why is this?

At the turn of the century, settlers began to drain the land just south of Lake Okeechobee in order to grow crops on the soil that had been enriched by partially decomposed saw grass. The large dike that now rings the lake prevents the water from taking its usual course: over the banks of Lake Okeechobee and slowly southward. In times of flooding, water can be shunted through the St. Lucie Canal to the Atlantic Ocean or through the canalized Caloosahatchee River to the Gulf of Mexico. In times of drought, water is contained not only in the lake, but also in three so-called conservation areas established to the south of the lake. Water must be conserved for the irrigation of the farmland and to recharge the Biscayne aquifer (underground river) that supplies drinking water for the cities on the east coast of Florida. Containing and moving the water from place to place has required the construction of over 1,400 miles of canals, 125 water control stations, and 18 large pumping stations. Now the Everglades National Park receives water only when it is discharged artificially from a conservation area. This disruption of the natural flow of water has affected the reproduction pattern of the birds, which is attuned to the natural wet-dry season turnover.

It took considerable human effort and a huge financial investment to control nature and to establish the "Everglades Agricultural Area." Has this attempt to bend nature to human will been worthwhile? The area does, in fact, produce more sugar than Hawaii and a large proportion of the vegetables consumed in the United States each winter. But this has not been without a price. The rich soil, built up over thousands of years, is disappearing and most likely will be unable to sustain conventional agriculture after the year 2000. It has been suggested that *at that time* we might use the Everglades Agricultural Area for the growth of aquatic plants. Perhaps it should have been decided in the beginning to *work with nature* by growing aquatic plants instead of conventional plants. Then all the canals and pumping stations would have been unnecessary, the water would still flow from Lake Okeechobee to the glades as it had for eons, and the birds yet today would number in the millions.

The heart of the Everglades is a vast saw grass prairie interrupted occasionally by hardwood tree islands that contain a variety of tropical and temperate trees.

An alligator beside his "gator hole," which also supplies food for the great egret in the background.

Proproots of the red mangrove provide protective cover for fishes and other sea life during maturation.

Figure I.8
Classification of organisms. In this text, organisms are classified into the five kingdoms described here.

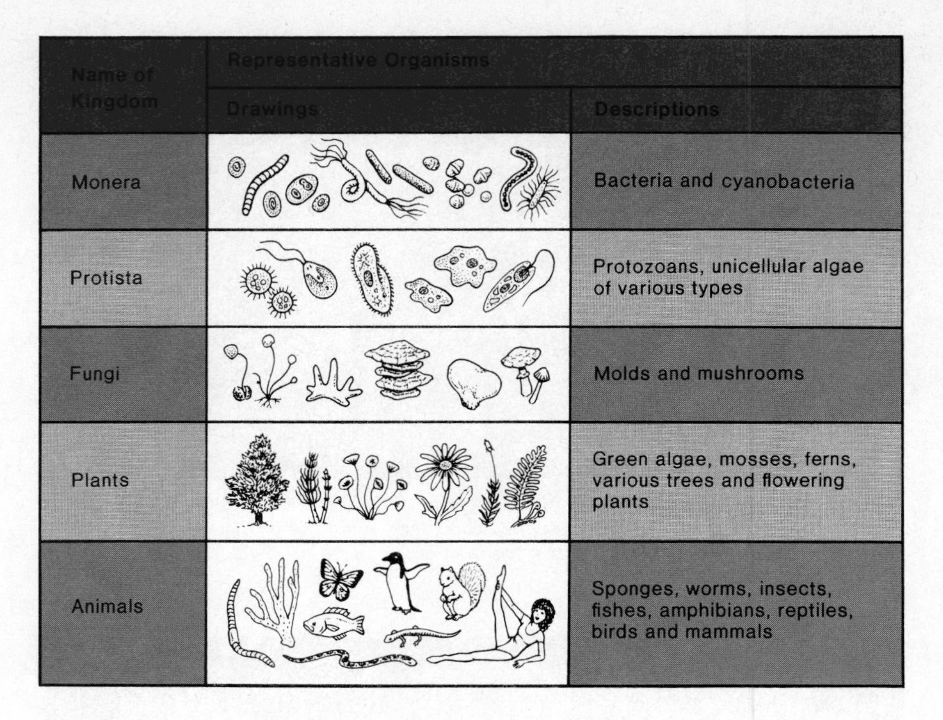

Name of Kingdom	Representative Organisms	
	Drawings	Descriptions
Monera		Bacteria and cyanobacteria
Protista		Protozoans, unicellular algae of various types
Fungi		Molds and mushrooms
Plants		Green algae, mosses, ferns, various trees and flowering plants
Animals		Sponges, worms, insects, fishes, amphibians, reptiles, birds and mammals

Biologists classify living things according to their evolutionary relationships. Since these relationships are not known exactly, various classification systems are in use. This text classifies organisms into five major groups, or kingdoms, as outlined in figure I.8.

Behavior and Ecology

The behavior of organisms can be studied just as their anatomy and physiology can be. In fact, much of an organism's behavior can be attributed to its genetic inheritance. Certain behavior patterns are particularly applicable to the study of ecology, which is defined as the interactions of organisms with each other and with the physical environment (fig. I.9).

Figure I.9
Painted lady butterfly probes for nectar from a black-eyed Susan flower. Coloration in butterflies is often for the purpose of protection—the wing colors we see here breaks them up into seemingly unrelated parts.

In many cases, as with the Florida Everglades, p. 10, humans have drastically altered the environment and are just now coming to realize their potential for destroying nature.

The future existence of human beings is dependent on preserving the natural world, and it is the goal of this unit to make you aware of this dependence and what should be done to protect the balance of nature. The last chapter in this unit on behavior and ecology considers the growth and size of the human population. Only humans have increased their number to the extent that they dominate the planet. Our use of fossil fuel energy (coal, natural gas, petroleum), especially as it is used to grow food, has made this inordinate increase possible. The last chapter in the book also considers energy resources and food production problems.

Summary

All living things display certain characteristics: they are made up of cells; maintain their structure by taking chemicals and energy from the environment; respond to external stimuli; and reproduce. As species, living things evolve and change. Evolution accounts for the diversity of life we see about us.

When studying the world of living things, biologists and other scientists use the scientific method. Inductive reasoning based on previous and current observations and data allow the formulation of a hypothesis. Now deductive reasoning is used to decide what new experiments and observations should be done to test the hypothesis. A hypothesis can be supported by collection of new data but it can never be proven true. All conclusions are subject to revision whenever new findings dictate. Still, there are some concepts, such as the cell theory and the theory of evolution, that have been supported for so long and by so many observations and experiments that they are generally accepted as true.

Science does not answer ethical questions; we must do this for ourselves. Knowledge provided by science, such as the contents of this text, can assist you in making decisions that will be beneficial to human beings and other living things.

Objective Questions

1. All living things are composed of ____________, the smallest unit of a tissue.
2. All living things need a source of ____________ and ____________ to maintain themselves.
3. When a plant turns toward the light it is ____________ to an external stimulus.
4. When living things ____________, the offspring resemble the parents.
5. Living things are suited, that is, ____________, to their environment.
6. A type of reasoning called ____________ reasoning helps scientists formulate hypotheses.
7. Very often the next step after formulation of the hypothesis is ____________, a type of testing that usually includes a control sample.
8. Scientists try to be objective; therefore they prefer ____________ data.
9. In science the word ____________ is often used to stand for a broad conceptual scheme that shows how various types of information are related.

Answers to Objective Questions
1. cells 2. chemicals, energy 3. responding
4. reproduce 5. adapted 6. inductive
7. experimentation 8. mathematical
9. theory

Study Questions

1. Name the five characteristics of life and discuss each one. (pp. 3–6)
2. Support the statement "All living things are organized." (p. 4)
3. Food provides what two necessities for living things? (p. 4)
4. Explain the process by which living things become increasingly more adapted to the environment. (p. 5)
5. List a series of steps to explain the scientific method. (p. 7)
6. Explain the difference between inductive and deductive reasoning. (p. 7)
7. What is the ultimate goal of science? Give an example that supports your answer. (p. 8)
8. Give two reasons why humans should study the biology of other living things. (p. 9)

Further Readings for Introduction

Asimov, I. 1960. *Wellsprings of life.* New York: Abelard-Schuman.

Baker, J., and G. Allen. 1971. *Hypothesis, prediction, and implication in biology.* Reading, Mass.: Addison-Wesley.

Grobstein, C. 1965. *The strategy of life.* San Francisco: W. H. Freeman.

Luria, S. E. 1973. *Life: The unfinished experiment.* New York: Charles Scribner's Sons.

Szent-Gyorgyi, A. 1972. *The living state.* New York: Academic Press.

Part One

The Cell, a Unit of Life

The cell is the smallest of living things, and all the characteristics of life are found here. An understanding of cell structure, physiology, and biochemistry serves as a foundation to an understanding of multicellular forms.

Part 1 studies each aspect of cellular biology in detail and thereby covers the fundamental concepts of biology.

Principles of inorganic and organic chemistry are discussed before a study of cell structure is undertaken. The cell is bounded by a membrane and contains organelles, many of which are also membranous. It is membrane that regulates the entrance and exit of molecules and determines how cellular organelles carry out their functions.

Cell reproduction is dependent upon mitotic cell division; whereas animal and plant reproduction are dependent upon meiotic cell division. Both types of cell division are introduced in part 1. Cells require energy for growth and reproduction, and the biochemical means by which this energy is provided is considered in chapter 5.

The intricate structure of the cell includes the endoplasmic reticulum, membranous canals that wander throughout the cytoplasm. This is an artist's representation of rough endoplasmic reticulum, so called because it is studded with ribosomes (shown in blue) which help bring about protein synthesis.

1

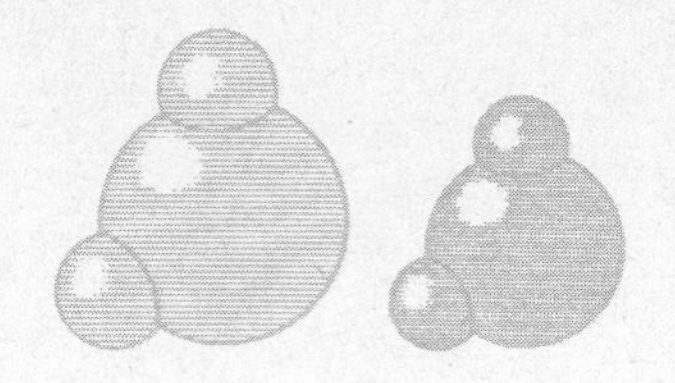

Chemistry and Life

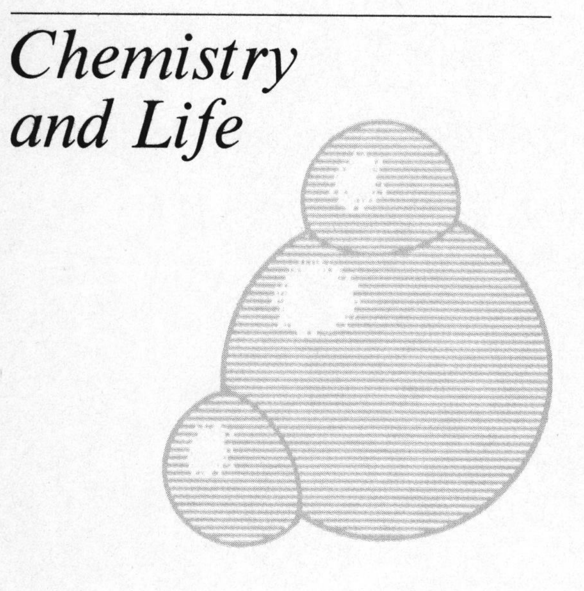

Chapter Concepts

1 Atoms, the smallest units of matter, react with one another to form compounds.

2 All living organisms are composed only of inorganic and organic chemicals.

3 Some important inorganic molecules in living organisms are water, acids, bases, and salts.

4 Some important organic molecules in living organisms are proteins, carbohydrates, fats, and nucleic acids, each of which is composed of smaller molecules joined together.

Chapter Outline

Figure 1.1
All living things are composed only of chemicals.

It is not always easy to understand that all living things are composed of chemicals, especially since it is not possible to see the chemicals that make up an organism's body (fig. 1.1). However, a few minutes' reflection regarding modern advances in biology and medicine usually convinces one that humans are indeed made of chemicals. Various genetic diseases, some of which cause an early death if untreated, can now be chemically controlled. Drugs are often administered to correct abnormal physical conditions: adrenalin is given for a heart attack; tranquilizers reduce nervous tension; and iron is used in the treatment of anemia. Good nutrition is based on our knowledge of the everyday chemical requirements necessary to keep the body in good running order. For example, the reading on page 35 suggests that the diet can help prevent heart disease.

Since living things, including humans, are composed only of chemicals, it is absolutely essential for a biology student to have a basic understanding of chemistry.

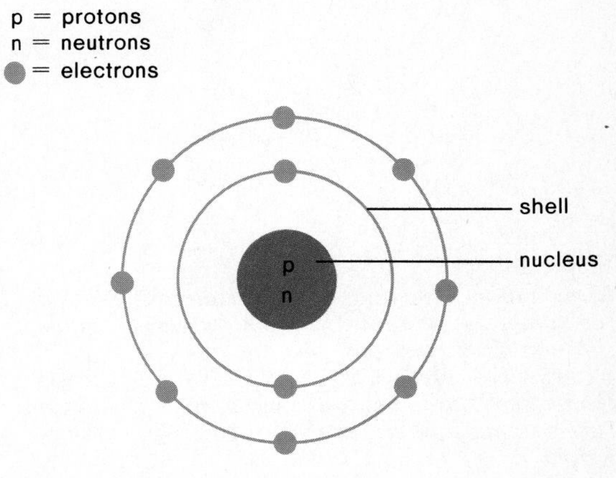

Figure 1.2
Representation of an atom. The nucleus contains protons and neutrons; the shells contain electrons. The first shell is complete with two electrons, and every shell thereafter may contain as many as eight electrons.

Atoms

An **atom** is the smallest unit of matter, nondivisible by chemical means. While it is possible to split an atom by physical means, an atom is the smallest unit to enter into chemical reactions. For our purposes, it is permissible to think of an atom as having a central **nucleus** where subatomic particles called **protons** and **neutrons** are located and *shells* where **electrons** orbit about the nucleus (fig. 1.2). Electrons have varying amounts of energy. Those with the greatest amount of energy are located in shells farthest from the nucleus. Another important feature of protons, neutrons, and electrons is their weight and/or charge, which is indicated in table 1.1.

Table 1.1 Subatomic Particles

Name	Charge	Weight
Electron	One negative unit	Almost no weight
Proton	One positive unit	One atomic unit
Neutron	No charge	One atomic unit

Figure 1.3
Periodic Table of the Elements (simplified). See the Appendix for a complete table. Each element has an atomic number, an atomic symbol, and an atomic weight. The elements in a darker color are the most common and those in a lighter color are also common in living things. The dark line separates metals (to the left) from the nonmetals (to the right).

I	II	III	IV	V	VI	VII	VIII
1 H hydrogen 1							2 He helium 4
3 Li lithium 7	4 Be beryllium 9	5 B boron 11	6 C carbon 12	7 N nitrogen 14	8 O oxygen 16	9 F fluorine 19	10 Ne neon 20
11 Na sodium 23	12 Mg magnesium 24	13 Al aluminum 27	14 Si silicon 28	15 P phosphorus 31	16 S sulfur 32	17 Cl chlorine 35	18 Ar argon 40
19 K potassium 39	20 Ca calcium 40						

Atomic Number
Atomic Symbol
Atomic Weight

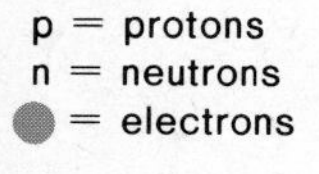

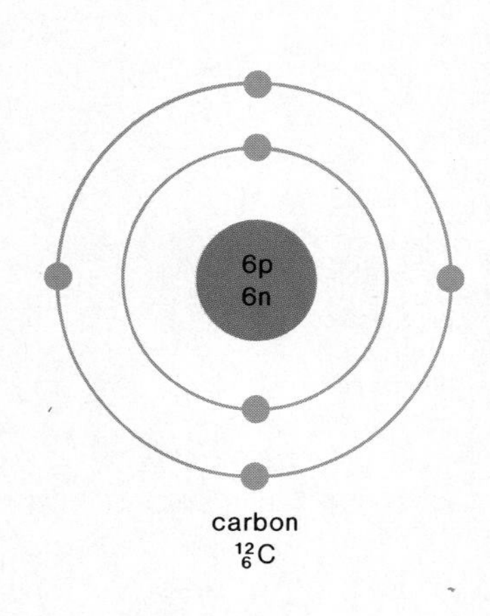

Figure 1.4
Carbon atom. The diagram of the atom shows that the number of protons (the atomic number) equals the number of electrons when the atom is electrically neutral. Carbon may also be written in the manner shown below the diagram. The subscript is the atomic number, and the superscript is the weight.

The Periodic Table of the Elements in Appendix A shows all the atoms that are presently known. An **element** is any substance that contains just one type atom. Figure 1.3 gives a simplified table that highlights the elements most common to living things. Notice that in the table each specific atom has a symbol; for example, C = carbon and N = nitrogen. Also, each type atom has an **atomic number;** for example, carbon is #6 and nitrogen is #7. *The atomic number equals the number of protons.* Also, each type atom has an **atomic weight** or mass. Carbon has an atomic weight or mass of 12[1] and nitrogen has an atomic weight of 14. *The atomic weight equals the number of protons plus the number of neutrons.*

Now, it is possible to diagram a specific, electrically neutral atom (fig. 1.4). In an *electrically neutral* atom the number of protons (+) is equal to the number of electrons (−). The first shell of an atom can contain up to two electrons; thereafter, each shell of those atoms in the simplified table (fig. 1.3) can contain up to eight electrons[2].

It is readily apparent that the elements in the Periodic Table are linearly arranged in order of increasing atomic number and weight, but they are also vertically arranged according to similar chemical properties. For example, the atoms in the first column all have one electron in the outermost shell, and we shall see that they give up this electron in chemical reactions.

Isotopes

The atomic weights listed in the Periodic Table are actually the average weight of each type atom. Individual atoms can vary as to weight. When they do, they are called **isotopes** of one another. The isotopes of carbon may be written in the following manner in which the subscript stands for the atomic number and the superscript stands for the weight:

$$^{12}_{6}C \qquad ^{13}_{6}C \qquad ^{14}_{6}C^*$$

[1]Atomic weights are relative weights. The most common isotope of carbon has been assigned a weight of 12, and the other atoms are either lighter or heavier than carbon.

[2]The maximum number of electrons for any shell except the outer shell is $2n^2$ where n is the shell number. However we will consider only atoms 1–20 in which all shells may have eight electrons except the first shell, which has two electrons.

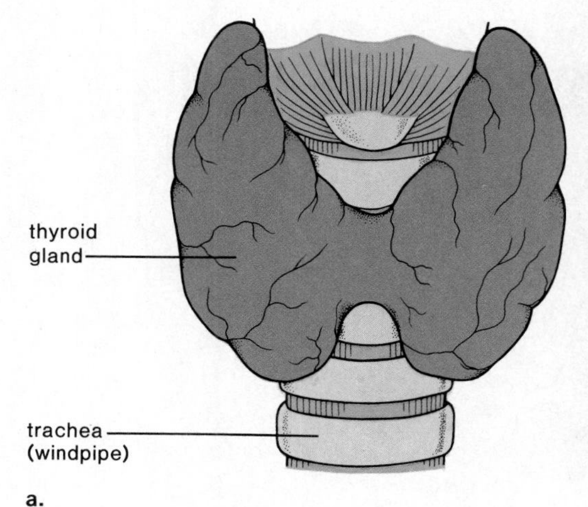

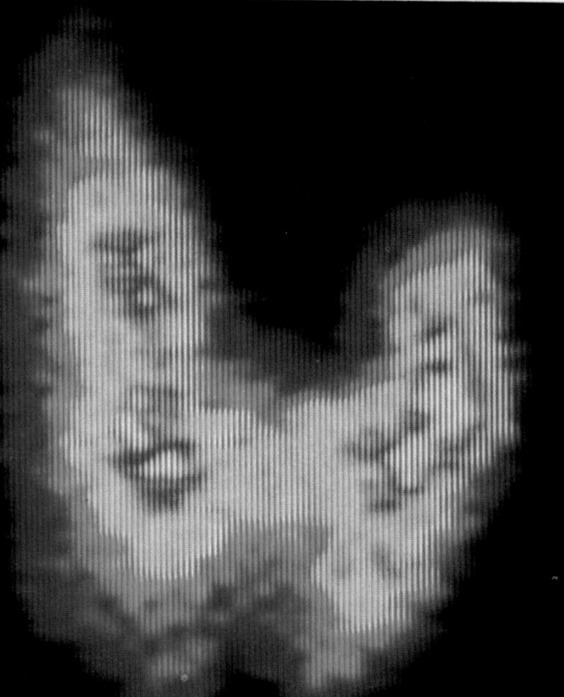

Figure 1.5
Use of radioactive iodine. a. *Drawing of the anatomical shape and location of the thyroid gland.* b. *A scan of the thyroid gland 24 hours after patient was administered radioactive iodine.*

The number of protons in these isotopes does not vary, but the weight does; this indicates that the number of neutrons must be responsible for the weight difference since electrons have almost no weight.

Certain isotopes, called **radioactive** isotopes, are unstable and emit radiation, which may be detected with a special counter or photographically. Among those isotopes of carbon listed, only Carbon-14 is radioactive, as the asterisk indicates. Radioactive isotopes are widely used in biological research and medical diagnostic procedures. For example, because the thyroid gland uses iodine, it is possible to administer radioactive iodine and then view a scan of the thyroid gland sometime later (fig. 1.5).

All matter is composed of atoms that are arranged in the Periodic Table of the Elements according to increasing weight. The weight of an atom is dependent on the number of protons and neutrons in the nucleus, while the chemical properties are dependent on the electrons in the shells.

Reactions between Atoms

Usually reactions between atoms involve the electrons in their outer shells. The *octet rule,* based on chemical findings, states that atoms react with one another in order to achieve eight electrons in their outer shells. An exception to this rule occurs when the outer shell is the first shell, which is complete with two electrons only.

Ionic Reactions

In one type of reaction, atoms give up or take on electrons in order to achieve a completed outer shell. Such atoms, which thereafter carry a charge, are called **ions** and the reaction is called an **ionic reaction.** In ionic reactions, atoms lose or gain electrons to produce a compound that contains ions in a fixed ratio to one another. The *formula* for the compound shows the proper ratio of atoms in the compound.

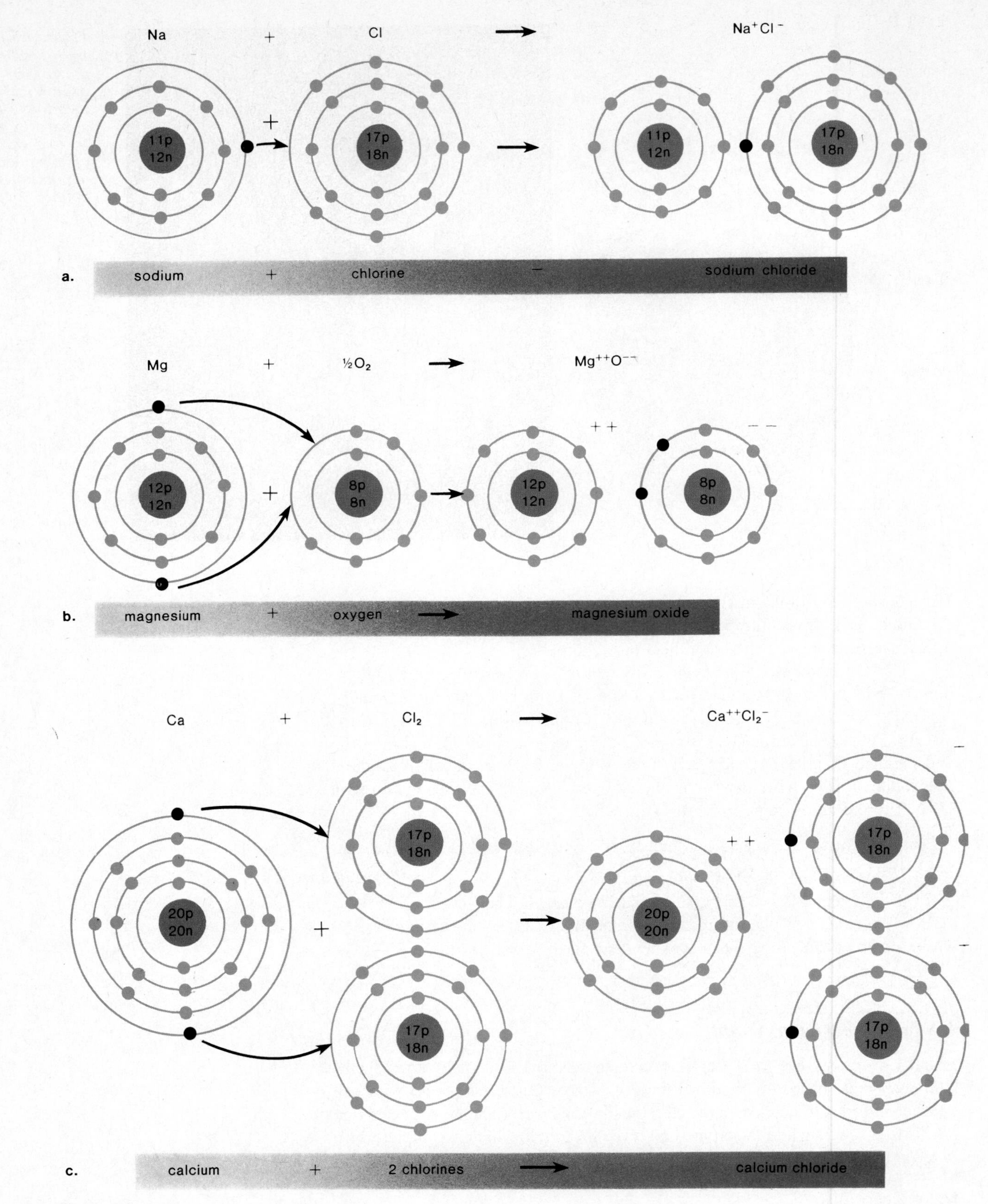

Figure 1.6
Ionic reactions. In ionic reactions nonmetals take electrons from metals and, in that way, each atom achieves eight electrons in the outer shell. a. *During this ionic reaction, an electron is transferred from the sodium atom to the chlorine atom. Each resulting ion carries a charge as shown.*
b. *Formation of magnesium oxide.* c. *Formation of calcium chloride.*

For example, figure 1.6a depicts a reaction between sodium (Na) and chlorine (Cl) in which chlorine takes an electron from sodium. The resulting ions in the compound sodium chloride, Na^+Cl^-, have eight electrons each in the outer shell. Notice that when sodium gives up an electron, the second shell with eight electrons becomes the completed outer shell. Chlorine, on the other hand, needs an electron to achieve a total of eight electrons in the outer shell.

It is easy to understand the charge of any ion by realizing that while the number of electrons can vary, the number of protons must stay constant. A minus charge indicates that the ion has more electrons (−) than protons (+) and a plus charge indicates that the ion has more protons (+) than electrons (−). Oppositely charged ions are attracted to one another and this attraction is called an ionic bond.

Ionic bond formation occurs when a **metal** reacts with a **nonmetal.** Metals are those atoms that appear to the left of the dark line in the simplified Periodic Table (fig. 1.3), and nonmetals appear to the right of the dark line. Metals lose electrons and become positively charged. In contrast, nonmetals gain electrons and become negatively charged. Two additional ionic reactions are shown in figures 1.6b and c.

Figure 1.7
Sodium chloride is a crystal with which we are very familiar, since it is used daily to enliven the taste of food. Unfortunately it has been linked to hypertension, possibly due to its action on the kidneys.

With the exception of hydrogen, atoms react with each other in order to achieve eight electrons in the outer shell. In an ionic reaction, positively and negatively charged ions are formed when (an) electron(s) is (are) transferred from a metal to (a) nonmetal(s). The attraction between ions is called an ionic bond.

Covalent Reactions

When nonmetals react with nonmetals, a covalent compound results (fig. 1.8). In a covalent compound, the atoms share electrons instead of losing or gaining them. The overlapping outer shells in figure 1.8 indicate that the atoms are sharing. Sharing is usually equal; each atom contributes one electron to each pair that is shared. These electrons spend part of their time in the outer shell of each atom; therefore they may be counted as belonging to both atoms. When this is done, each atom will have eight electrons in the outermost shell.

At this point it is convenient to introduce the term molecule because this term is often used when referring to covalently bonded atoms. A **molecule** is the smallest combination of atoms that retains the properties of a compound. No doubt you have often heard of a water molecule, for example.

Instead of drawing complete diagrams of molecules, electron dot diagrams are sometimes used. For example, in reference to figure 1.8a, each chlorine atom can be represented by its symbol, and the electrons in the outer shell can be designated as dots. The shared electrons are placed between the two sharing atoms, as shown here:

$$:\ddot{\underset{..}{Cl}}\cdot + \cdot\ddot{\underset{..}{Cl}}: \qquad :\ddot{\underset{..}{Cl}}:\ddot{\underset{..}{Cl}}:$$

Electron dot diagrams are a bit cumbersome, and covalent bonds are often indicated simply by straight line structural formulas.[3] At times, even the lines are omitted, and molecular formulas that indicate only the number of each type atom are given:

$$Cl-Cl \quad \text{or} \quad Cl_2$$

[3]Structural formulas show the orientation of the atoms to one another and try to reflect how the atoms are arranged in space.

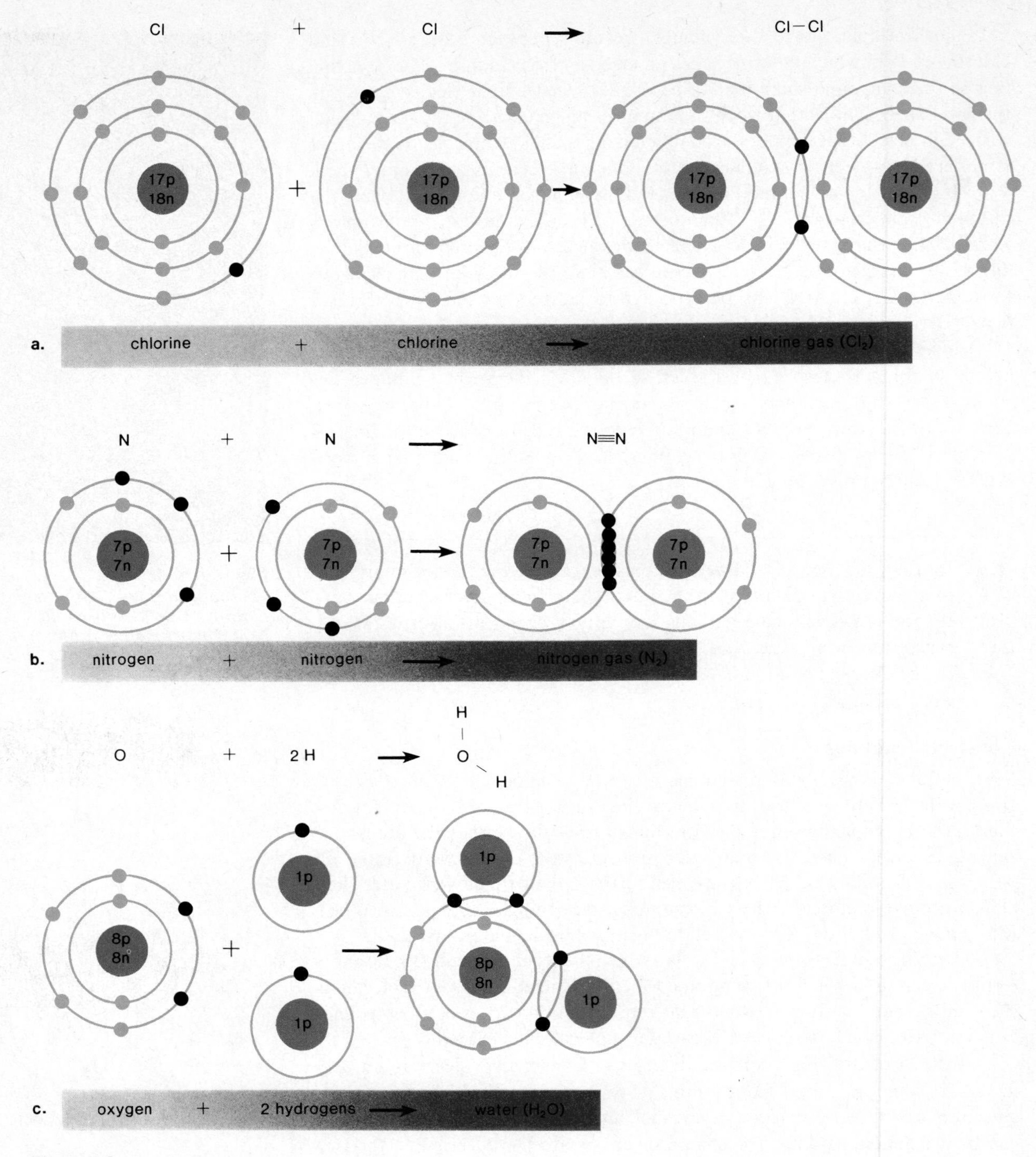

Figure 1.8
Covalent reactions. After a covalent reaction, the atoms share electrons so that each atom has eight electrons in its outer shell. To show this it is necessary to count the shared electrons as belonging to both bonded atoms. a. *Formation of chlorine gas.* b. *Formation of nitrogen gas.* c. *Formation of water.*

Even if the molecule is written in this way, it is easy to tell that the two chlorines are sharing electrons because (1) they are both nonmetals and (2) no charge is indicated. Additional examples of electron dot formulas versus structural formulas versus molecular formulas are shown in figure 1.9.

Double Bonds

Besides a single bond, like that between the two chlorine atoms, a double bond or triple bond may form in order for two atoms to complete their octets. In a triple bond, two atoms share three pairs of electrons between them. For example, in figure 1.8b the reaction between two nitrogen atoms results in a triple bond because each nitrogen needs three electrons in order to achieve a total of eight electrons in the outermost shell. Notice that in the diagrammatic representation six electrons are placed in the outer overlapping shells and that three straight lines are indicated in the structural formula for nitrogen gas.

Electron Dot Formula	Structural Formula	Molecular Formula
:Ö::C::Ö: carbon dioxide	O=C=O carbon dioxide	CO_2 carbon dioxide
H : N : H H (with dot pairs) ammonia	H–N–H with H below ammonia	NH_3 ammonia
H : O : H (with dot pairs) water	H–O–H water	H_2O water

Figure 1.9
Electron dot, structural, and molecular formulas. In the electron dot formula, only the atoms in the outermost shell are designated. In the structural formula, the lines represent a pair of electrons that are being shared between two atoms. The molecular formula indicates only the number of atoms found within a molecule.

In a covalent reaction, nonmetals share electrons so that all atoms involved can have completed outer shells. Each shared pair of electrons is a covalent bond; double bonds and even triple bonds are possible.

Oxidation-Reduction

When oxygen combines with a metal, oxygen receives electrons and becomes negatively charged; the metal loses electrons and becomes positively charged. For example, consider the reaction that is illustrated in figure 1.6b.

$$Mg + \frac{1}{2}O_2 \longrightarrow Mg^{++}O^{--}$$

In such cases, it is obviously appropriate to say that the metal has been oxidized and that because of oxidation, the metal has lost electrons. Then we need only admit that the oxygen has been reduced because it has gained electrons, or minus charges.

Today, the terms **oxidation** and **reduction** are applied to many ionic reactions whether or not oxygen is involved. Very simply, *oxidation refers to the loss of electrons, and reduction refers to the gain of electrons.* In our previous ionic reaction, $Na + Cl \longrightarrow Na^+ Cl^-$, the sodium has been oxidized (loss of electron) and the chlorine has been reduced (gain of electron).

The terms *oxidation* and *reduction* are also applied to certain covalent reactions. In this case, however, oxidation is the loss of hydrogen atoms, and reduction is the gain of hydrogen atoms. A hydrogen atom contains one proton and one electron; therefore when a molecule loses a hydrogen atom, it has lost an electron; when a molecule gains a hydrogen atom, it has gained an electron. We will have occasion to refer to this form of oxidation-reduction again in chapter 5.

When oxidation occurs, an atom becomes oxidized (loses electrons). When reduction occurs, an atom becomes reduced (gains electrons). These two processes occur concurrently in oxidation-reduction reactions.

Table 1.2 Inorganic versus Organic Chemistry	
Inorganic Compounds	**Organic Compounds**
Usually contain metals and nonmetals	Always contain carbon and hydrogen
Usually ionic bonding	Always covalent bonding
Always contain a small number of atoms	May be quite large with many atoms
Often associated with nonliving elements	Often associated with living organisms

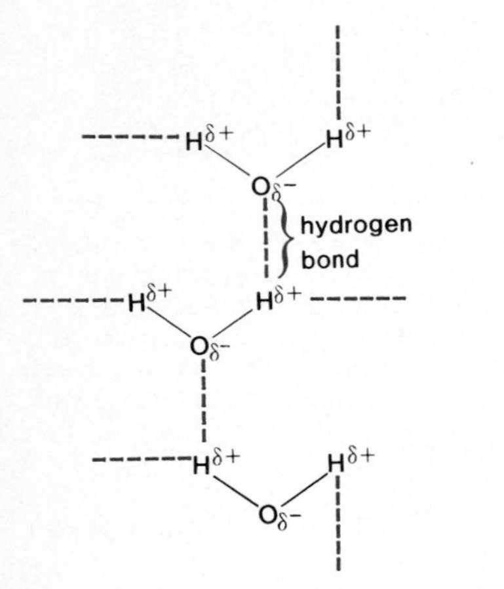

Figure 1.10
Water molecules are polar; each hydrogen carries a partial positive charge and each oxygen carries a partial negative charge. The polarity of the water molecules brings about hydrogen bonding between the molecules in the manner shown. The dotted lines represent hydrogen bonds. (δ = partial.)

Inorganic versus Organic Chemistry

There are two types of chemistry pertinent to our study, **inorganic** chemistry and **organic** chemistry. Ionic reactions are common to inorganic chemistry, while covalent reactions are common to organic chemistry. Table 1.2 lists the important differences between inorganic and organic compounds. As will be apparent in the following discussion, both types of compounds are necessary to the proper functioning of the living organism. The inorganic compounds discussed in the following paragraphs affect the well-being of all organisms.

Some Important Inorganic Molecules

Water

Water, or H_2O, is not an organic molecule because it does not contain carbon; but as figure 1.8c shows, water is covalently bonded. Often the atoms in a covalently bonded molecule share electrons evenly but in water the electrons spend more time circulating the larger oxygen than the smaller hydrogens. Therefore, there is a slight positive charge on the hydrogen atoms and a slight negative charge on the oxygen atom. For this reason water is called a **polar** molecule and hydrogen bonding occurs between water molecules (fig. 1.10). A **hydrogen bond** occurs whenever a partially positive hydrogen is attracted to a partially negative atom. The hydrogen bond is represented by a dotted line in figure 1.10 because it is a weak bond that is easily broken.

Characteristics of Water Hydrogen bonds are relatively weak, but they still cause water molecules to cling together. Without hydrogen bonding between molecules water would boil at −80° C and freeze at −100° C, making life impossible. Instead, water is a liquid at body temperature. It absorbs a great deal of heat before it becomes warm and evaporates and gives off this heat as it cools down and freezes. This property allows great bodies of water, such as the oceans, to maintain a relatively constant temperature. It also helps keep an animal's body temperature within normal limits and even accounts for the cooling effect of sweating.

The cohesiveness of water allows it to fill tubular vessels, and water is an excellent transport medium for distributing substances and heat throughout the body. Its cohesive property is obvious whenever we observe the surface tension of bodies or containers of water.

Because of hydrogen bonding, liquid water is more dense than ice. Therefore ice floats on liquid water, and bodies of water always freeze from the top down making ice-skating possible (fig. 1.11). Furthermore, the layer of ice protects the organisms below and helps them survive the winter.

Figure 1.11
The presence of hydrogen bonding causes ice to be less dense than liquid water. Therefore, lakes freeze from the top down and this is a protection to the organisms that survive in the water beneath the layer of ice. It also permits humans to enjoy ice skating.

Water, being a polar molecule, acts as a solvent and dissolves various chemical substances, particularly other polar molecules. This property of water greatly facilitates chemical reactions in cells.

---

*L*ife is dependent on the various characteristics of water.

Dissociation Polarity also causes water molecules to tend to **dissociate,** or split up, in this manner:

$$H—O—H \longrightarrow H^+ + OH^-$$

The hydrogen ion (H^+) has lost an electron; the hydroxide ion (OH^-) has gained the electron. Very few molecules actually dissociate; therefore few hydrogen ions and hydroxide ions result (fig. 1.12).[4]

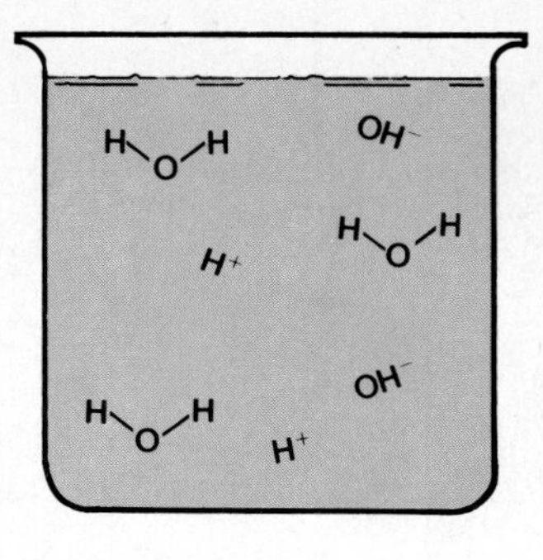

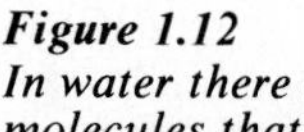

Figure 1.12
In water there are always a very few water molecules that have dissociated. Dissociation produces an equal number of hydrogen and hydroxide ions.

Acids and Bases

Acids are compounds that dissociate in water and release hydrogen ions (or protons)[5]. For example, an important inorganic acid is hydrochloric acid (HCl), which dissociates in this manner:

$$HCl \longrightarrow H^+ + Cl^-$$

Dissociation is almost complete, and this acid is called a strong acid. If HCl is added to a beaker of water (fig. 1.13), the number of hydrogen ions increases.

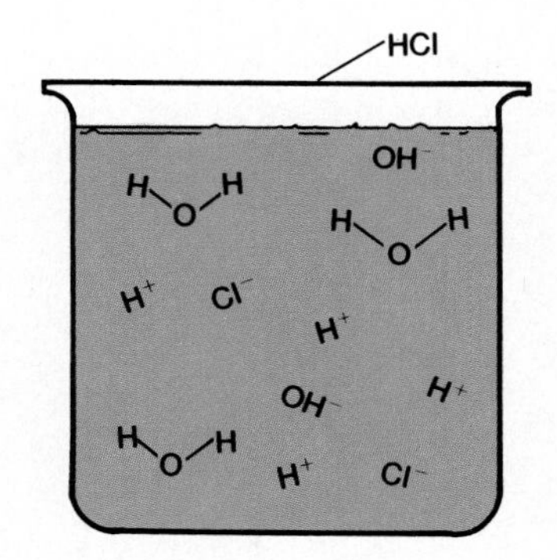

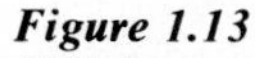

Figure 1.13
HCl is an acid that releases hydrogen ions as it dissociates in water. Notice that the addition of HCl to this beaker has caused it to have more hydrogen ions than hydroxide ions.

Bases are compounds that dissociate in water and release hydroxide ions (OH^-). For example, an important inorganic base is sodium hydroxide (NaOH), which dissociates in this manner:

$$NaOH \longrightarrow Na^+ + OH^-$$

Dissociation is complete, and sodium hydroxide is called a strong base. If NaOH is added to a beaker of water (fig. 1.14), the number of hydroxide ions increases.

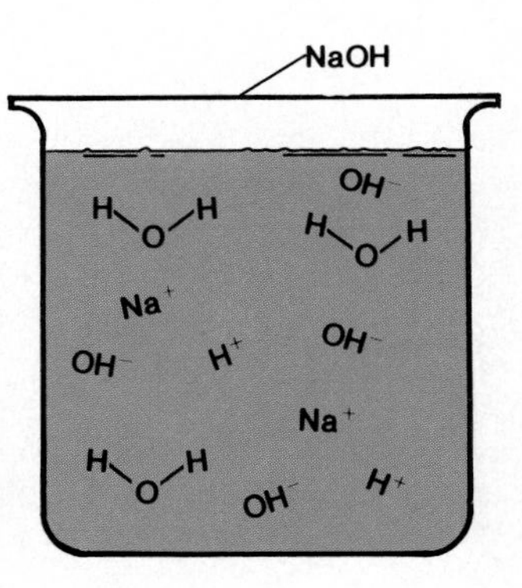

Figure 1.14
NaOH is a base that releases hydroxide ions as it dissociates. Notice that the addition of NaOH to this beaker has caused it to have more hydroxide ions than hydrogen ions.

pH

The **pH**[6] scale ranges from 0–14. Any pH value below 7 is acid, with ever-increasing acidity toward the lower numbers. Any pH value above 7 is basic (or alkaline), with ever-increasing basicity toward the higher numbers. A *pH of exactly 7 is neutral.* Water has an equal number of H^+ and OH^- ions, and therefore, one of each is formed when water dissociates. The fraction of water molecules that dissociate is 10^{-7} (or 0.0000001), which is the source of the pH value for neutral solutions. The pH scale was devised to simplify discussion of the hydrogen ion concentration $[H^+]$, without using cumbersome numbers. For example:

a. $1 \times 10^{-6}\ [H^+] = \text{pH } 6$
b. $1 \times 10^{-7}\ [H^+] = \text{pH } 7$
c. $1 \times 10^{-8}\ [H^+] = \text{pH } 8$

Each lower pH unit has ten times the amount of H^+ as the next higher unit

[4]The figure is for illustration purposes and is not mathematically accurate.

[5]A hydrogen atom contains one electron and one proton. A hydrogen ion is only one proton and is often called a proton.

[6]pH is defined as the negative logarithm of the hydrogen ion concentration.

Figure 1.15
The pH scale. The proportionate amount of hydrogen ions (H^+) to hydroxide ions (OH^-) is indicated by the diagonal line. Any pH above 7 is basic, while any pH below 7 is acidic.

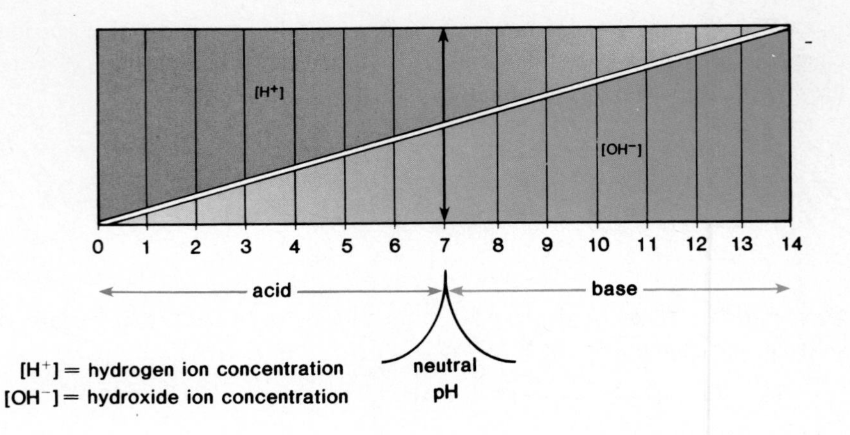

Which of the items on the preceding page (*a* or *c*) indicates a higher concentration of hydrogen ions and therefore refers to an acid? The numbers with the smaller negative exponents indicate a greater quantity than those with a larger negative exponent. Therefore, *a* refers to an acid. Bases add hydroxide ions to solutions and increase the OH^- ion concentration of water. Basic solutions, then, have fewer H^+ ions compared to OH^- ions. In the preceding list, *c* refers to a base because it indicates a lesser concentration of H^+ than OH^- ions compared to water. Figure 1.15 gives the complete pH scale with proper notations.

Buffers All living things need to maintain the hydrogen ion concentration, or pH, at a constant level. For example, the pH of the blood is held constant at about 7.4 or we become ill. The presence of buffers helps keep the pH constant. A **buffer** is a chemical or a combination of chemicals that can take up excess hydrogen ions or excess hydroxide ions. When an acid is added to a solution, a buffer takes up excess hydrogen ions, and when a base is added to a solution, a buffer takes up excess hydroxide ions. Therefore a buffer resists changes in pH when either H^+ or OH^- ions are added. Eventually, however, a buffer can be overwhelmed and the pH will change.

Salts

When an acid reacts with a base, a salt and water result. In the case of a strong acid and strong base, the reaction is complete:

$$\underset{}{HCl} + NaOH \longrightarrow \underset{\text{salt}}{Na^+Cl^-} + \underset{\text{water}}{HOH}$$

Further, if an equal quantity of both take part in this reaction, neutralization occurs; the resulting solution will be neither acid nor base. In the neutralization process, the H^+ from the acid and the OH^- from the base combine to form water. The salt consists of the positive ion of the base and the negative ion of the acid.

Acids have a pH that is less than seven and bases have a pH that is greater than seven. The presence of buffers helps keep the pH of internal body fluids constant at neutral or pH seven because a buffer can absorb both hydrogen and hydroxide ions.

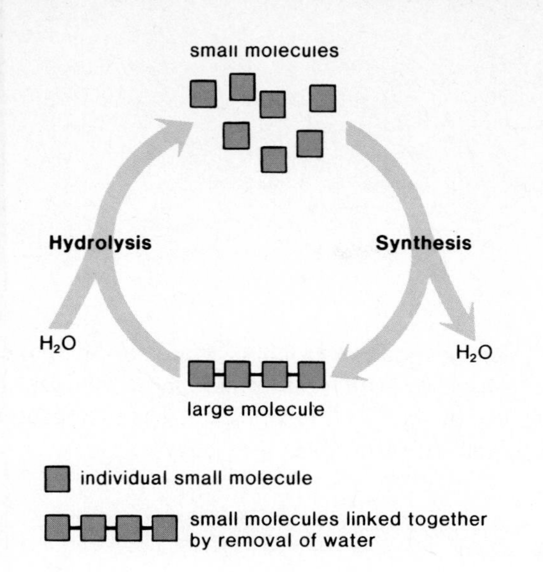

Figure 1.16
Synthesis and hydrolysis of an organic polymer. When the unit molecules join together to form the polymer (synthesis), water is released, and then when the polymer is broken down (hydrolysis), water is added.

Some Important Organic Molecules

Unit Molecules

The chemistry of carbon accounts for the formation of the very large number of organic compounds we associate with living organisms. Carbon is a nonmetal with four electrons in the outer shell. In order to achieve eight electrons in the outer shell, it must share with other nonmetals. Carbon shares electrons with as many as four other atoms. Many times, carbon atoms share with each other to form rings or chains of carbon atoms. These act as a skeleton for the unit molecules found in the life molecules—carbohydrates, fats, and nucleic acids. Therefore, the properties of carbon are essential to life as we know it.

Synthesis and Hydrolysis Figure 1.16 diagrammatically illustrates that the life molecules are synthesized or made when small unit molecules join together. A bond that joins two unit molecules together is created after the removal of H^+ from one molecule and OH^- from the next molecule. As water forms, **dehydration synthesis** occurs.

Life molecules are often **polymers,** or chains of unit molecules, joined together. They can be broken down in a manner opposite to synthesis: the addition of water leads to the disruption of the bonds linking the unit molecules together. During this process, called **hydrolysis,** one molecule takes on H^+ and the next takes on OH^-.

Proteins

Functions Proteins are large, complex macromolecules that sometimes have mainly a structural function. For example, in humans, *keratin* is a protein that makes up hair and nails, while *collagen* is a protein found in all types of connective tissue, including ligaments, cartilage, bone, and tendons. The muscles (fig. 1.17) contain proteins that account for their ability to contract.

Some proteins function as **enzymes,** necessary contributors to the chemical workings of the cell and therefore of the body. Enzymes are organic catalysts that speed up chemical reactions. They work so quickly that a reaction that might normally take several hours or days will take only a fraction of a second when an enzyme is present.

Figure 1.17
Most athletes have well-developed muscle cells containing many protein molecules. A major controversy today is the taking of steroids to promote the buildup of muscles.

Figure 1.18
An amino acid may be drawn in either of the two ways shown. The proteins of living organisms contain about 20 common amino acids that differ only in their R *groups. The* R *stands for the remainder of the molecule that can vary from a single hydrogen atom to a complicated ring compound.*

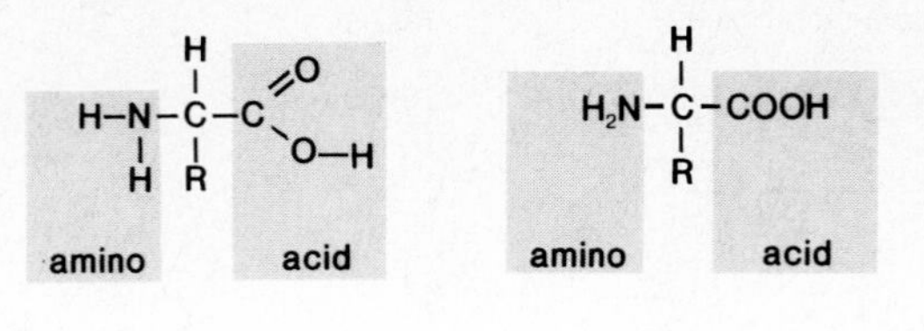

Name	Structural Formulas	
alanine	H H H-N-C-C=O OH H-C-H H	H_2N-CH-COOH CH_3
valine	H H H-N-C-C=O OH CH H-C-H H-C-H H H	H_2N-CH-COOH CH CH_3 CH_3
cysteine	H H H-N-C-C=O OH H-C-H SH	H_2N-CH-COOH CH_2 SH
phenylalanine	H H H-N-C-C=O OH H-C-H C H-C C-H H-C C-H C H	H_2N-CH-COOH CH_2
tyrosine	H H H-N-C-C=O H-C-H OH C H-C C-H H-C C-H C OH	H_2N-CH-COOH CH_2 OH
histidine	H H H-N-C-C=O H-C-H OH C=C-H H-N N C H	H_2N-CH-COOH CH_2 C=CH HN N C H

Figure 1.19
Representative amino acids. Notice that each amino acid is drawn twice and that the second is more simplified than the first.

Amino Acids The unit molecules found in proteins are called **amino acids.** Most amino acids have the structural formula shown in figure 1.18. The name amino acid refers to the fact that the molecule has two functional groups: an amino group and an acid (also called a carboxyl) group:

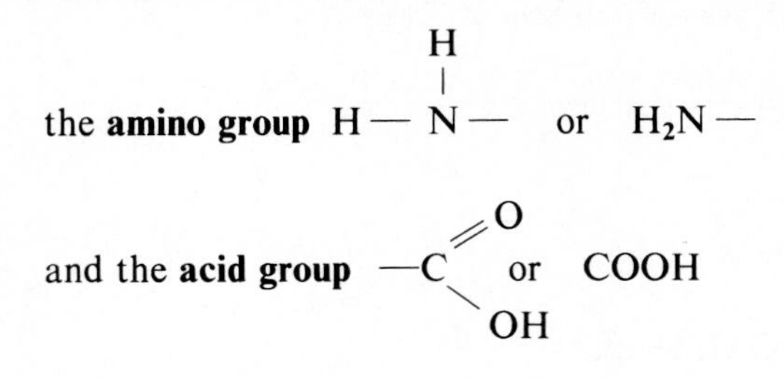

Amino acids differ from one another by their *R groups,* the Remainder of the molecule. In amino acids, the R group varies from being a single hydrogen atom to those that are complicated rings (fig. 1.19). There are about 20 different common amino acids found in the proteins of living things. Therefore there are about 20 different types of R groups.

Peptides The bond that joins two amino acids together is called a **peptide bond.** As you can see in figure 1.20, when synthesis occurs, the acid group of one amino acid reacts with the amino group of another amino acid and water is given off. A **dipeptide** contains only two amino acids, but when up to 10 or 20 amino acids have joined together, the resulting chain is called a **polypeptide.** A very long polypeptide of approximately 75 amino acids or more is called a **protein.**

The atoms associated with the resulting peptide bond, namely oxygen, carbon, nitrogen, and hydrogen, share the electrons in such a way that the oxygen carries a partial negative charge (fig. 1.21a). Therefore, the peptide bond is a polar bond and hydrogen bonding, represented by the dotted lines in figure 1.21b, is a frequent occurrence in polypeptides and proteins.

Levels of Structure Proteins commonly have three levels of structure (fig. 1.22), although some have a fourth level. The *primary structure* is the sequence of the amino acids joined together by peptide bonds. Any number of the 20 different amino acids may be joined together in any particular sequence just as if one were making a necklace from assorted beads.

The *secondary structure* of a protein comes about when the polypeptide chain takes a particular orientation in space. One common arrangement of the chain is the **alpha helix,** or right-handed coil, with 3.6 amino acids per turn. Hydrogen bonding between amino acids, in particular, stabilizes the helix.

The *tertiary structure* of a protein is its final three-dimensional shape. In a structural protein like collagen, the helical chains lie parallel to one another. But enzymes are globular proteins in which the helix bends and twists in different ways. The tertiary shape of a protein is maintained by various types of bonding between the R groups. Covalent, ionic, and hydrogen bonding are all seen.

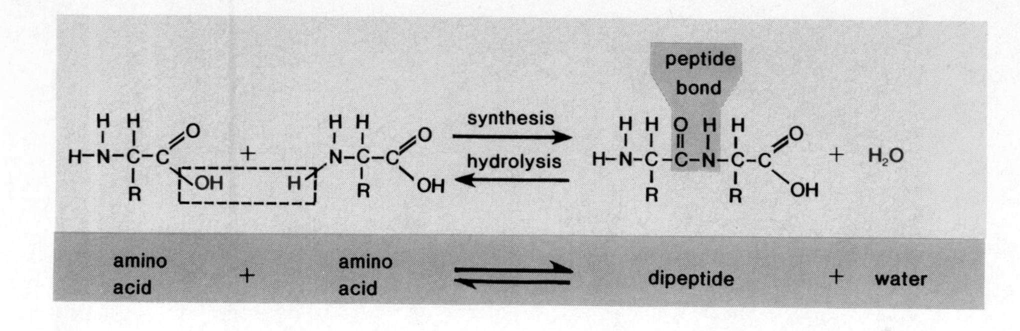

Figure 1.20
Notice that as the peptide bond forms, water is given off—the water molecule on the right-hand side of the equation is derived from components removed from the amino acids on the left-hand side. During hydrolysis, water is added as the peptide bond is broken.

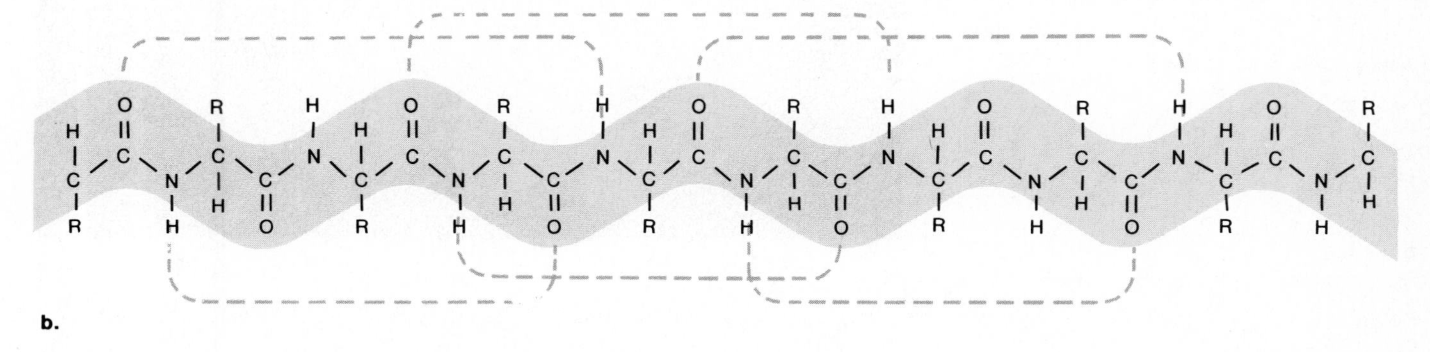

Figure 1.21
Hydrogen bonding in a peptide. a. *In a peptide bond, the oxygen is partially negative and the hydrogen indicated is partially positive. (δ = partial)* b. *Therefore hydrogen bonding occurs along the length of a polypeptide as indicated by the dotted lines. Notice that a polypeptide has a backbone of repeating N — C — C — N atoms and side chains of "R" groups.*

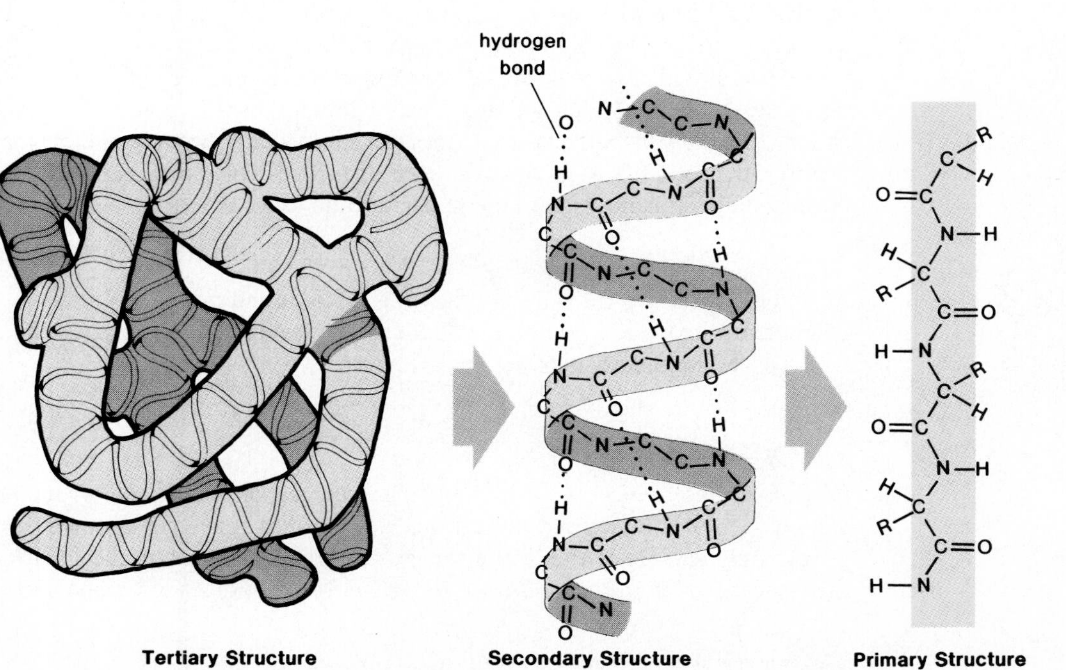

Figure 1.22
Proteins have at least three levels of structure. Primary structure is the order of the amino acids; secondary structure is often an alpha helix; and in globular proteins, the tertiary structure is the twisting and turning of the helix that takes place because of bonding between the R groups. Enzymes are globular proteins.

The four polypeptide chains of hemoglobin are arranged as shown. Each has a tightly bound nonprotein heme group (represented here as a plane) that contains an oxygen-carrying iron atom (represented as a sphere).

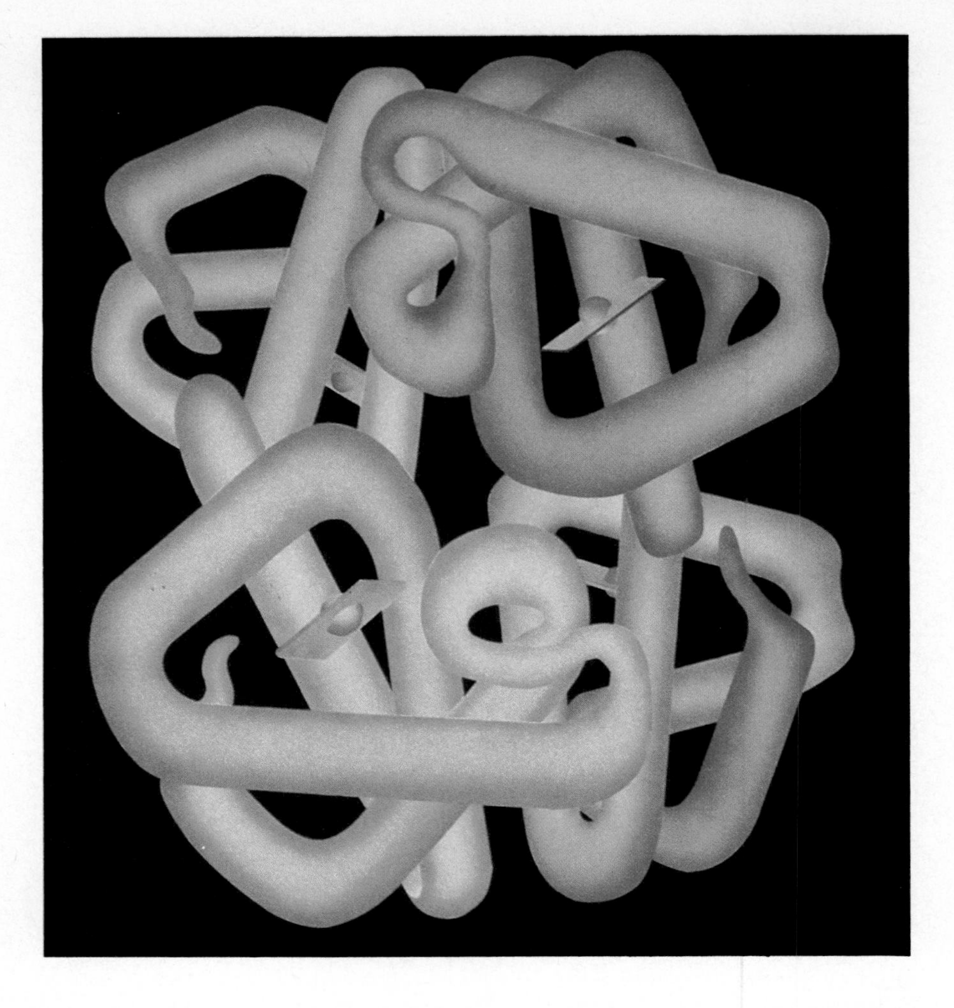

Some proteins have more than one type of polypeptide chain, each with its own primary, secondary, and tertiary structures. Within the protein, these separate chains are arranged to give a fourth level of structure termed the *quaternary structure*. Hemoglobin is a complex protein having a quaternary structure.

The final shape of a protein is very important to its function, as will be emphasized again when discussing enzyme activity. Both temperature and pH can bring about a change in protein shape. For example, we are all aware that the addition of acid to milk causes curdling; heating causes egg white, a protein called albumin, to coagulate. When a protein loses its normal configuration, it is said to be denatured. Denaturation occurs because the normal bonding between the R groups has been disturbed. Once a protein loses its normal shape, it is no longer able to perform its usual function.

Amino acids are the unit molecule for peptides and proteins. Proteins have both structural and metabolic[7] functions in cells. All enzymes which speed up metabolic reactions are proteins.

Carbohydrates

Carbohydrates are characterized by the presence of the atomic grouping H—C—OH in which the ratio of hydrogen atoms to oxygen atoms is approximately 2:1. Since this is the same as the ratio in water, it accounts for their name, which means *hydrates of carbon*. If the number of carbon atoms

[7]Metabolism is all the chemical reactions that occur in a cell.

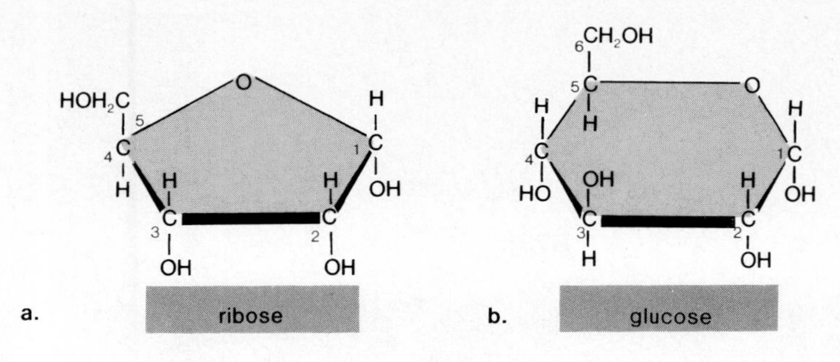

Figure 1.23
Common monosaccharides. a. *Ribose, a pentose—a five-carbon sugar. Deoxyribose has one less oxygen atom attached to the second carbon compared to ribose.* b. *Glucose, a hexose—a six-carbon sugar. The small numbers count the carbon atoms.*

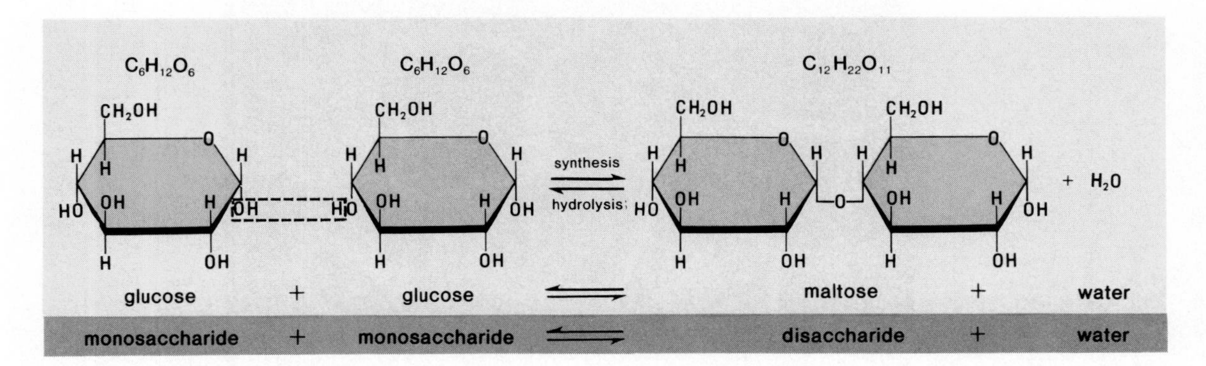

Figure 1.24
Synthesis and hydrolysis of maltose, a disaccharide containing two glucose units. During synthesis, a bond forms between the two glucose molecules as the components of water are removed. During hydrolysis, the components of water are added as the bond is broken.

in the compound is low (from about three to seven), then the carbohydrate is a simple sugar, or monosaccharide. Thereafter, larger carbohydrates are created by joining together monosaccharides in the manner described in figure 1.16 for the synthesis of organic compounds.

Monosaccharides As their name implies, **monosaccharides** are simple *sugars* having only one unit. These compounds are often designated by the number of carbons they contain; for example, **pentose** sugars have five carbons (e.g., ribose, fig. 1.23) and **hexose** sugars have six carbons (e.g., glucose, fig. 1.23). **Glucose** is the primary energy source of the body, and most carbohydrate polymers can be broken down into monosaccharides that either are or can be converted to glucose. All these monosaccharides have the molecular formula $C_6H_{12}O_6$. However, in this text we will use the molecular formula $C_6H_{12}O_6$ to mean glucose since glucose is the most common six-carbon monosaccharide found in cells.

Disaccharides The term **disaccharide** tells us that there are two monosaccharide units joined together in the compound. When two glucose molecules join together, **maltose** (fig. 1.24) is formed. The chemical equation for this reaction indicates that the forward direction is a dehydration synthesis and the backward reaction is a hydrolysis. You may also be interested in knowing that when *glucose* and another monosaccharide, *fructose,* are joined together, the disaccharide called sucrose is formed. Sucrose is derived from plants and is commonly used at the table to sweeten foods.

Polysaccharides A **polysaccharide** is a carbohydrate that contains a large number of monosaccharide molecules. There are three polysaccharides that are common in organisms: starch, glycogen, and cellulose. All of these are polymers, or chains, of glucose, just as a necklace might be made up of only one type bead. Even though all three polysaccharides contain only glucose, they are distinguishable from one another.

Figure 1.25
a. *Starch is a relatively straight chain of glucose molecules.* b. *Electron micrograph showing starch granules in plant cells. Starch is the storage form of glucose in plants.*

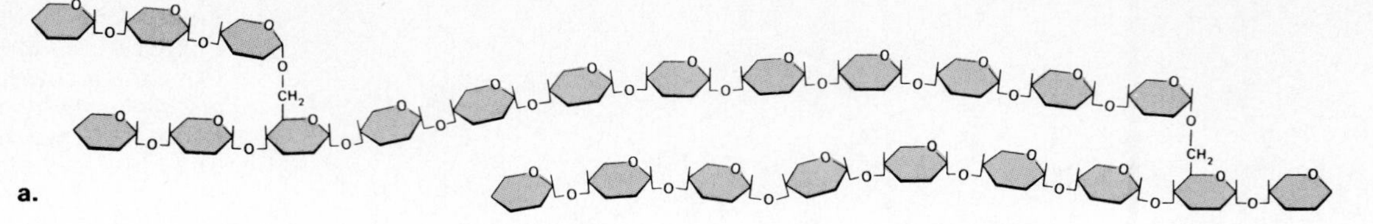

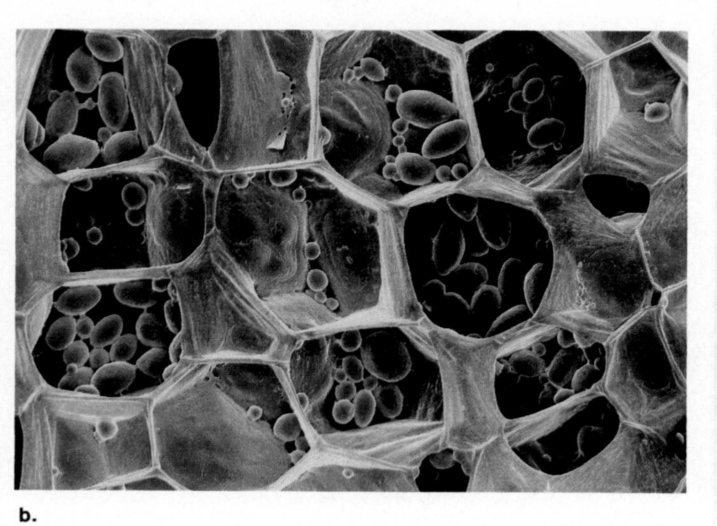

As figure 1.25a shows, **starch** has few side branches or chains of glucose that branch off from the main chain. Starch is the storage form of glucose in plants. Just as we store orange juice as a concentrate, plants store starch as a concentrate of glucose. This analogy is particularly apt because, like the synthetic reaction described in figure 1.16, water is removed when glucose molecules are joined together to form starch. The following equation also represents the synthesis of starch:

$$\text{n glucose} \underset{\text{hydrolysis}}{\overset{\text{synthesis}}{\rightleftharpoons}} \text{starch} + (\text{n} - 1)H_2O$$

n = some large number

Glycogen is characterized by the presence of side chains of glucose (fig. 1.26). Glycogen is the storage form of glucose in animals. After an animal eats, the liver stores glucose as glycogen; then in between eating, the liver releases glucose so that the blood concentration of glucose is always 0.1 percent.

The polymer **cellulose** is found in plant cell walls and accounts in part for the strong nature of these walls. In fact it may be said that cellulose is the primary structural component of plants. In cellulose, the glucose units are joined by a slightly different type of linkage compared to that of starch and glycogen (fig. 1.27a). While this might seem to be a technicality, actually it is important because we are unable to digest foods containing this type of linkage; therefore cellulose passes through our digestive tract as roughage. Recently it has been suggested that the presence of roughage in the diet is necessary to good health and may even help prevent colon cancer.

Carbohydrates are hydrates of carbon. The monosaccharide glucose is frequently used as an energy source in cells. The polysaccharides starch and glycogen are storage compounds in plant and animal cells respectively and the polysaccharide cellulose is found in plant cell walls.

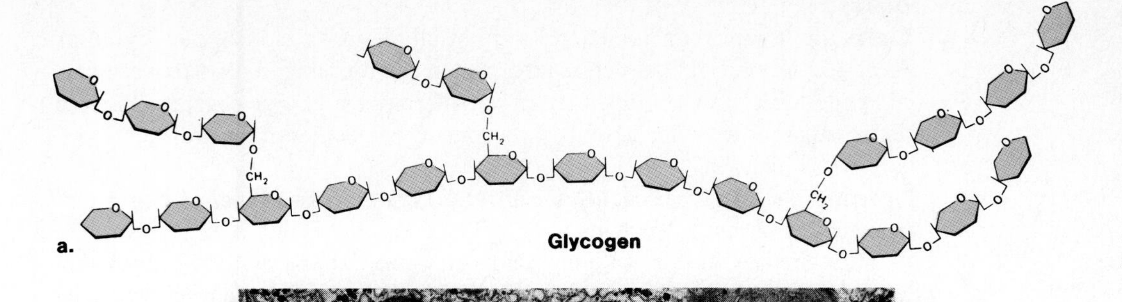

b.

Figure 1.26
a. *Glycogen is a highly branched polymer of glucose molecules. The branching allows breakdown to proceed at several points simultaneously.*
b. *Electron micrograph showing glycogen granules in liver cells. Glycogen is the storage form of glucose in animals.*

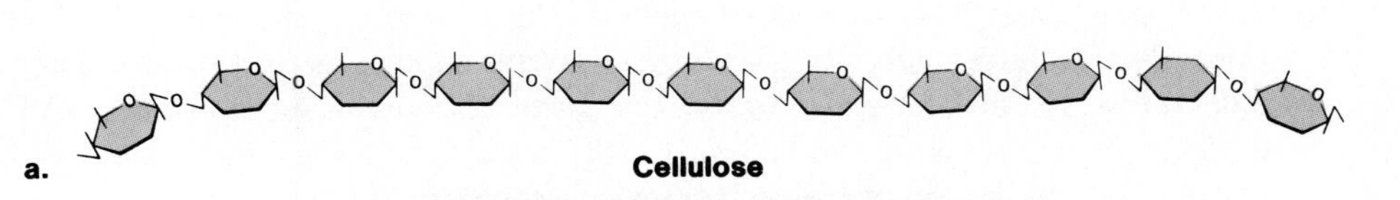

b.

Figure 1.27
a. *Cellulose contains a slightly different type of linkage between glucose molecules compared to starch and glycogen.*
b. *Plant cells have walls containing cellulose. The rigidity of the cell walls permits nonwoody plants to stand upright as long as they receive an adequate supply of water.*

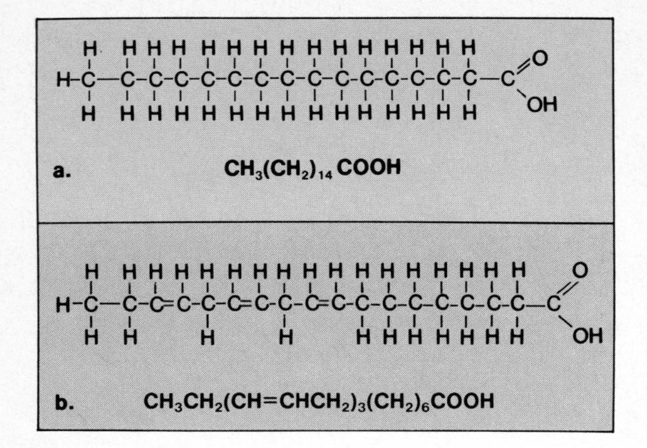

Figure 1.28
Fatty acids are either saturated (having no double bonds) or unsaturated (having double bonds). a. *In a saturated fatty acid the carbons carry all the hydrogen atoms possible.* b. *In this unsaturated fatty acid, there is a double bond at the third from last carbon (called omega-3) and at other carbons. Research indicates that omega-3 fatty acids prevalent in fish oil help prevent heart disease.*

Lipids

Lipids are organic compounds that are insoluble in water. The most familiar lipids are the neutral fats such as lard, butter, and oil, which are used in cooking or at the table. In the body, fats serve as long-term energy sources. Adipose tissue is composed of cells that contain many molecules of neutral fat.

Neutral Fats (Triglycerides) A **neutral fat** contains two types of unit molecules: glycerol and fatty acids.

Each **fatty acid** has a long chain of carbon atoms, with hydrogens attached, and ends in an acid group (fig. 1.28). Most of the fatty acids in cells contain 16 or 18 carbon atoms per molecule, although smaller ones are also found. Fatty acids are either saturated or unsaturated. *Saturated fatty acids* have no double bonds between the carbon atoms. The carbon chain is saturated, so to speak, with all the hydrogens that can be held. *Unsaturated fatty acids* have double bonds in the carbon chain wherever the number of hydrogens is less than two per carbon atom. Unsaturated fatty acids are most often found in vegetable oils and account for the liquid nature of these oils. Vegetable oils are hydrogenated to make margarine. Polyunsaturated margarine still contains a large number of unsaturated, or double, bonds.

For several years now it has been known that the intake of saturated fatty acids is associated with high blood pressure and heart disease. Recent research has suggested that high doses of polyunsaturates can promote cancer, among other diseases. But there is good news as the reading on page 35 tells us. Fish oil, having only a few unsaturated bonds, has been found to reduce the risk of heart disease and, as discussed in the reading, it is only necessary to eat fish once or twice a week to achieve a benefit. The beneficial effects of fish oil are believed to be due to the presence of unsaturated fatty acids that have a double bond on the third-to-last carbon in the chain. This type of fatty acid is called an "omega minus 3" or "omega-3."

Glycerol is a compound with three hydrates of carbon. Notice in figure 1.29 that glycerol has three —OH groups. When fat is formed, the acid portions of three fatty acids react with these groups so that fat and three molecules of water are formed. Again, the larger fat molecule is formed by dehydration synthesis in the forward direction. The backward direction represents how fat can be hydrolyzed to its components.

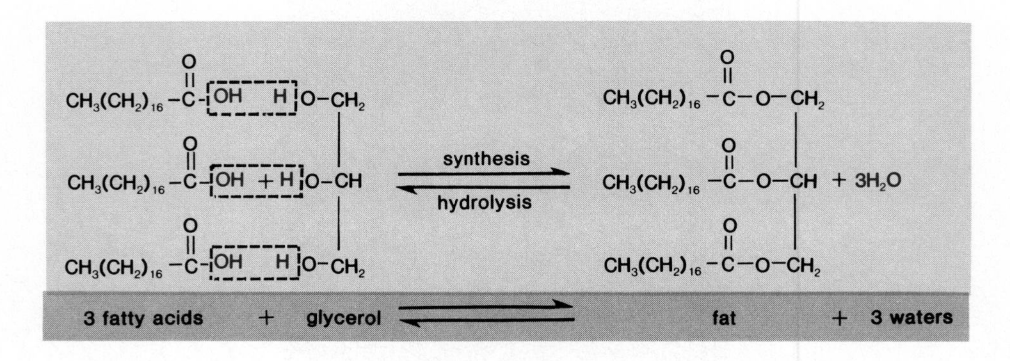

Figure 1.29
Synthesis and hydrolysis of a neutral fat. Three fatty acids plus glycerol react to produce a fat molecule and three water molecules. A fat molecule plus three water molecules react to produce three fatty acids and glycerol.

Heart Disease: Let Them Eat Fish

Evidence suggests that eating fish reduces the risk of heart disease. This effect has been attributed to the presence of omega-3 unsaturated fatty acids in fish oil.

If fish had the ability, they might be blushing from all the plaudits heaped on them . . .

In 1960, Dutch researchers from the University of Leiden asked 852 men and their wives what the men were eating, then kept track of the men for the next 20 years. Though the researchers found no relationship between fish consumption and such established heart disease risk factors as blood cholesterol level and blood pressure, they found the more fish a man ate, the less likely he was to die of heart disease.

The death rate from heart disease was more than 50 percent lower among men who ate at least 30 grams (1 ounce) of fish per day compared with men who ate no fish. Just one or two fish dishes a week, the researchers say, "may be of value in the prevention of coronary heart disease," and they suggest that dietary guidelines include this recommendation.

The probable key is the action of fatty acids in fish oil (also found in leafy vegetables and soy, walnut and rapeseed oils) called omega-3. These fatty acids alter metabolic pathways in the body, discouraging the formation of heart-attack-causing blood clots. What remains to be studied, notes John A. Glomset of the University of Washington in Seattle "is whether the consumption of fish also correlated, perhaps unfavorably, with mortality from cancer and other diseases."

A second report comes from the Oregon Health Sciences University in Portland. Twenty people with high triglyceride levels—a factor in heart disease—rotated among diets containing fish oil, polyunsaturated vegetable oil or low-fat foods. The fish oil, which is a type of polyunsaturated fat, significantly lowered both cholesterol and triglyceride levels, leading the researchers to conclude that fish oils and fish may be useful treatments for high triglyceride levels. Omega-3 fatty acids reduce the synthesis of a molecule that carries triglycerides and cholesterol through the blood, and increase cholesterol excretion, they say.

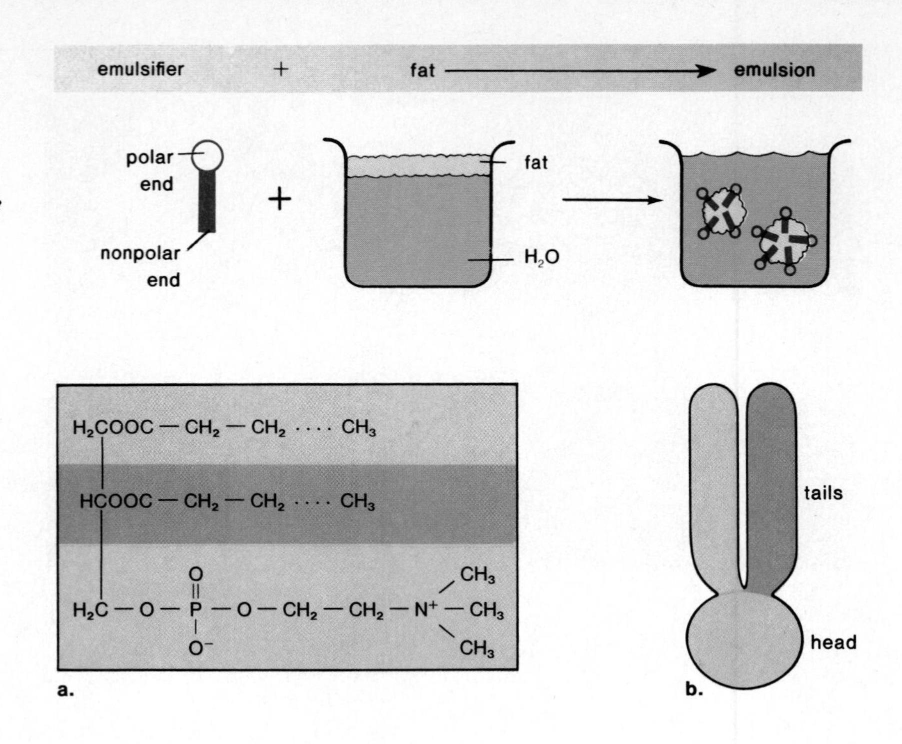

Figure 1.30
Fat molecules, being nonpolar, will not disperse in water. An emulsifier contains molecules that have a polar end and nonpolar end. When an emulsifier is added to a beaker containing a layer of nondispersed fat molecules, the nonpolar ends are attracted to the nonpolar fat, and the polar ends are attracted to the water. This causes droplets of fat molecules to become dispersed.

Figure 1.31
Phospholipids are constructed similarly to fats except that they contain a phosphate group.
a. *Lecithin, shown here, has a side chain that contains both a phosphate group and a nitrogen containing group.* b. *The charged portion of the molecule is soluble in water, whereas the two hydrocarbon chains are not soluble in water. This causes the molecule to arrange itself as shown.*

Soaps A soap is a salt formed by a fatty acid and an inorganic base. For example:

$$\underset{\text{sodium hydroxide}}{NaOH} + \underset{\text{fatty acid}}{RCOOH} \longrightarrow \underset{\text{soap}}{RCOO^-Na^+}$$

Fats do not mix with water because they are nonpolar; a soap, being polar, will mix with water. When soaps are added to oils, then oils too will mix with water. Figure 1.30 shows how a soap positions itself about an oil droplet so that the polar ends project outward. Now the droplet will be soluble in water. This process of causing an oil to disperse in water is called **emulsification,** and it is said that an emulsion has been formed. Emulsification occurs when dirty clothes are washed with soaps and detergents. Also, prior to the digestion of fatty foods, fats are emulsified by bile. Usually a person who has had their gallbladder removed has trouble digesting fatty foods because the gallbladder stores bile for use at the proper time during the digestive process.

Phospholipids Phospholipids contain a phosphate group and this accounts for their name:

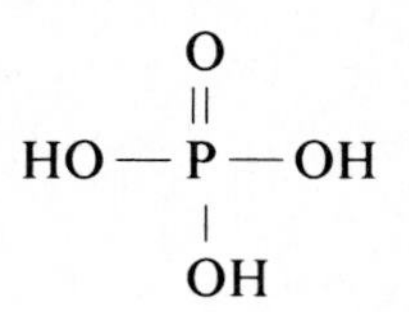

Essentially, phospholipids are constructed as neutral fats are, except that in place of the third fatty acid there is a phosphate group or a grouping that contains both phosphate and nitrogen (fig. 1.31). These molecules are not electrically neutral as are the fats because the phosphate group can ionize. Notice, then, that phospholipids have both a nonpolar (uncharged) region and a polar (charged) region. Therefore, phospholipids are soluble in water. This latter property makes them very useful compounds in the body, as we will see in the next chapter.

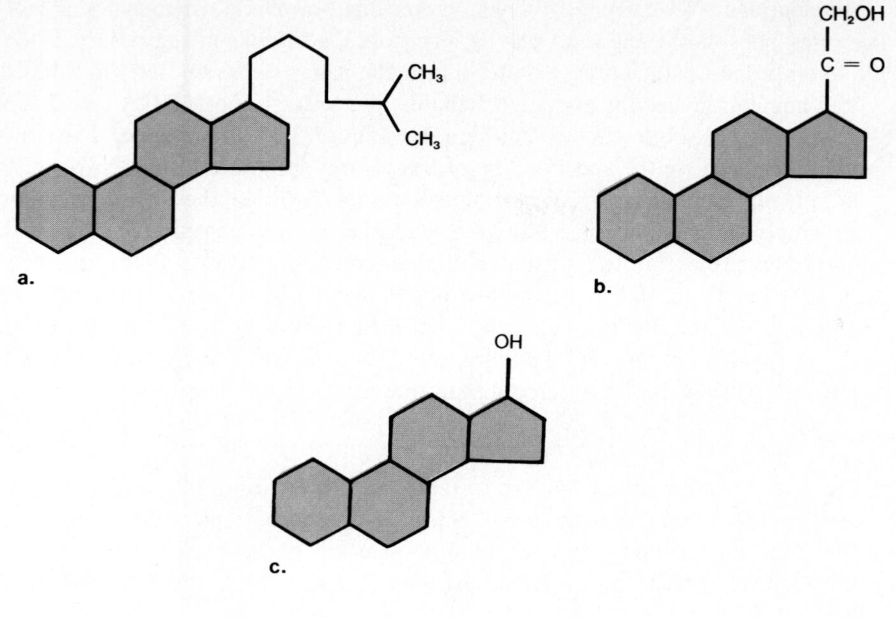

OH

c.

Figure 1.32
Like cholesterol in a, *steroid molecules have four adjacent rings, but their effects on the body largely depend on the type of chain attached and the location indicated. The chain in* b *is found in aldosterone, which is involved in the regulation of sodium and water metabolism, while the chain in* c *is found in testosterone, the male sex hormone.*

Steroids The steroids have a structure that is related to the structure of **cholesterol.** They are constructed of four fused rings of carbon atoms to which is usually attached a chain of varying length (fig. 1.32). For many years it has been suggested that a diet high in saturated fats and cholesterol can lead to circulatory disorders due to reduced blood flow caused by the deposit of fatty materials on the inner linings of blood vessels. In a 10-year study of 3,806 men conducted by the National Heart, Lung, and Blood Institute it was found that when, by the daily dose of a drug, the cholesterol level of the blood was reduced by 25 percent, the risk of heart disease was cut by 50 percent.

Despite adverse publicity, the steroids are very important compounds in the body; for example, the sex hormones are steroids.

Lipids include neutral fat, a long term energy storage molecule, that consists of glycerol and three fatty acids and the related phospholipids in which one of the sidechains is charged. The steroids have an entirely different structure similar to that of cholesterol.

Nucleic Acids

Nucleic acids are huge, macromolecular compounds with very specific functions in cells; for example, the genes are composed of a nucleic acid called **DNA** (deoxyribonucleic acid). Another important nucleic acid, **RNA** (ribonucleic acid), works in conjunction with DNA to bring about protein synthesis.

Both DNA and RNA are *polymers of nucleotides* and therefore are chains of nucleotides joined together. Just like the other synthetic reactions we have studied in this section, these units are joined together to form nucleic acids by the removal of water molecules:

$$\text{n nucleotides} \underset{\text{hydrolysis}}{\overset{\text{synthesis}}{\rightleftharpoons}} \text{nucleic acid} + (\text{n}-1)\ H_2O$$

n = some large number

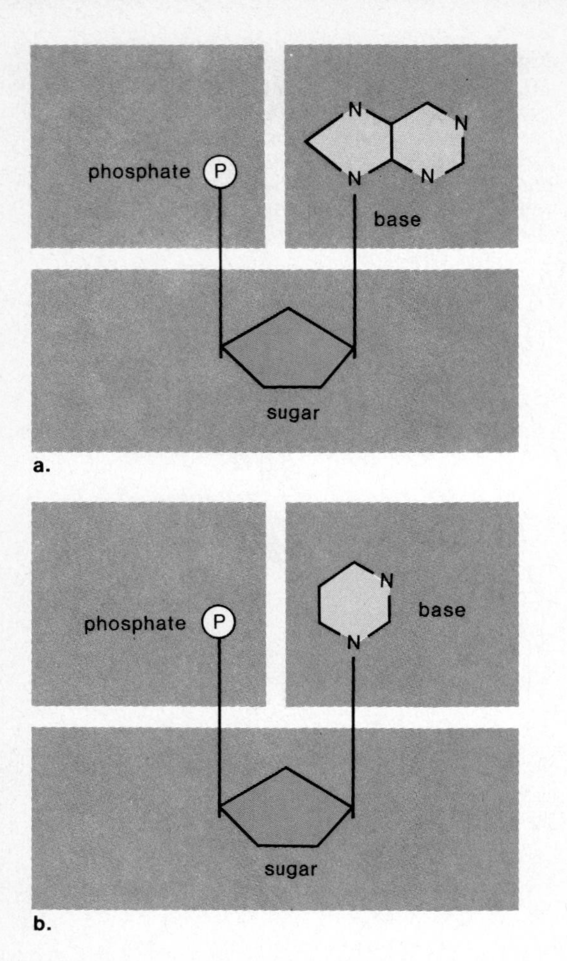

Figure 1.33
Generalized nucleotides. All nucleotides contain a phosphate molecule, a pentose sugar, and a base. The two types of nucleotides differ as to whether the base has a. *two rings or* b. *one ring.*

Nucleotides Every **nucleotide** is a molecular complex of three types of molecules: phosphoric acid (phosphate), a pentose sugar, and a nitrogen base. DNA is composed of nucleotides that contain the sugar deoxyribose, while RNA has nucleotides having the sugar ribose. The bases in both DNA and RNA have either a single ring or double ring. Figure 1.33 shows generalized nucleotides because the specific type of base is not designated; the phosphate is simply represented as Ⓟ. When nucleotides join together, they form a polymer in which the backbone is made up of phosphate-sugar-phosphate-sugar, with the bases projecting to one side of the backbone (fig. 1.34). Such a polymer is called a strand. RNA is single-stranded and DNA is double-stranded, the two strands being held together by hydrogen bonding between the bases (fig. 22.3). This is only a brief description of the structure and function of DNA and RNA; they are considered again in more detail in chapter 22.

Nucleic acids are of two types, DNA and RNA. Both of these are polymers of nucleotides; a DNA is composed of nucleotides having the sugar deoxyribose, while RNA nucleotides contain the sugar ribose. DNA makes up the genes and, along with RNA, controls the metabolism of the cell.

Adenosine triphosphate (**ATP**) This molecule is a nucleotide that is used as a carrier of energy in cells. The structure of ATP is similar to that shown in figure 1.33a. Adenine is the base; the sugar is ribose; and there are three phosphate groups instead of one. It is customary to draw the molecule as shown in figure 1.35 so that the three phosphate groups appear on the right. ATP is known as the energy molecule because the triphosphate unit contains two high-energy bonds, represented in figure 1.35 by wavy lines.

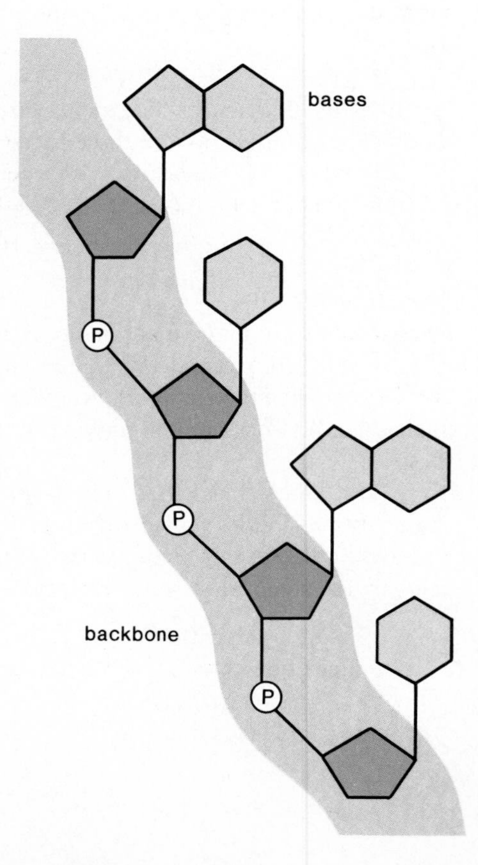

Figure 1.34
Generalized nucleic acid strand. Nucleic acid polymers contain a chain of nucleotides. Each strand has a backbone made of sugar and phosphate molecules. The bases project to the side.

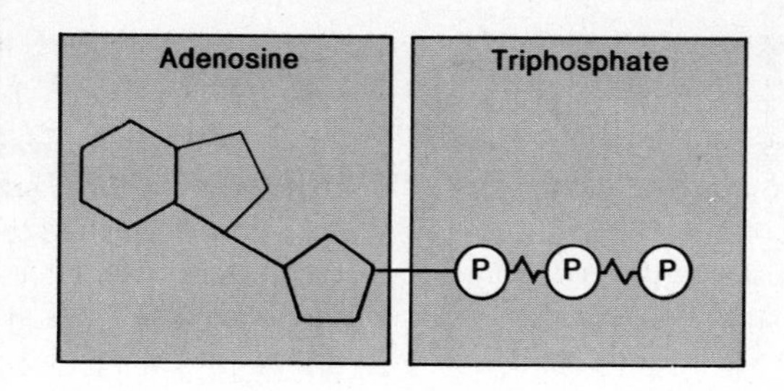

Figure 1.35
ATP is a nucleotide with three phosphate units; two of the phosphate bonds are high energy bonds, indicated by wavy lines.

Summary

All matter is made up of atoms each having a weight that is dependent on the number of protons and neutrons in the nucleus and chemical properties that are dependent on the number of electrons in the outermost shell. Atoms react with one another in order to acquire eight electrons in the outermost shell. In ionic reactions, metals give electrons to nonmetals and in covalent reactions nonmetals share electrons.

Water, acids, and bases are important inorganic compounds. Water has a neutral pH; acids decrease and bases increase the pH of water. The organic molecules of interest are proteins, carbohydrates, lipids and nucleic acids, each of which has (a) particular unit molecule(s) (table 1.3). Dehydration synthesis joins unit molecules together and hydrolytic degradation releases them. All enzymes are proteins; carbohydrates serve as immediate energy sources, and fats are a long-term energy source for the individual. Nucleic acids are of two types, DNA and RNA. Both of these function in protein synthesis which will be discussed in chapter 22.

***Table 1.3** Organic Compounds of Life*

Macromolecules	Unit Molecule	Usual Atoms
Protein	Amino acid	C, H, O, N, S
Carbohydrate, e.g., starch	Glucose	C, H, O
Lipid	Glycerol and fatty acids	C, H, O
Nucleic acid	Nucleotide	C, H, O, N, P

Objective Questions

1. The atomic number is equal to the number of ______________ in an atom.
2. An ion is negatively charged when it has more ______________ than ______________.
3. In a covalent bond, the atoms ______________ electrons.
4. ______________ take up either H^+ or OH^- and therefore act to stabilize the pH.
5. When an acid is added to water, the number of H^+ ______________ and the pH ______________.
6. ______________ are organic catalysts that speed up chemical reactions.
7. The bond that joins two amino acids together is called a ______________ bond.
8. The sequence or order of amino acids in a protein is termed its ______________ structure.
9. The simple sugar ______________ is the primary energy source of the body.
10. The polymer ______________ is found in plant cell walls.
11. Fatty acids having no double bonds are said to be ______________.
12. All neutral fats are composed of ______________ and fatty acids.
13. Nucleic acids are composed of ______________.
14. DNA is ______________ -stranded while RNA is single-stranded.

Answers to Objective Questions
1. protons 2. electrons, protons 3. share 4. Buffers 5. increases, decreases 6. Enzymes 7. peptide 8. primary 9. glucose 10. cellulose 11. saturated 12. glycerol 13. nucleotides 14. double

Study Questions

1. Name the subatomic particles of an atom; describe their charge and weight and their location in the atom. (p. 17)
2. Draw the atomic diagram for calcium. (p. 18)
3. State the octet rule and explain how it relates to chemical reactions. (p. 19)
4. Give an example of an ionic reaction and explain it. Mention in your explanation: compound, ion, formula, and ionic bond. (p. 19)
5. Give an example of a covalent reaction and explain it. (p. 21)
6. Explain oxidation-reduction in terms of loss or gain of electrons. (p. 23)
7. Name four general differences between inorganic and organic compounds. (p. 24)
8. On the pH scale, which numbers indicate a basic solution? An acidic solution? Why? (p. 26)
9. What are buffers and why are they important to life? (p. 26)
10. Explain dehydration synthesis of organic compounds and hydrolytic breakdown of organic compounds. (p. 27)
11. What are some functions of proteins? What is the unit molecule of protein? What is a peptide bond, a dipeptide, a polypeptide? (p. 27)
12. Discuss the primary, secondary, and tertiary structures of a protein. (p. 28)
13. Name some monosaccharides, disaccharides, and polysaccharides, and state appropriate functions. What is the most common unit molecule for these? (p. 31)
14. Name some important lipids and state their function. What is a saturated fatty acid? An unsaturated fatty acid? How is fat formed? (p. 34)
15. What are the two important nucleic acids? What is the unit molecule? (p. 37)

Key Terms

acid (as′id) a solution in which pH is less than 7; a substance that contributes or liberates hydrogen ions (protons) in a solution. *25*

amino acid (ah-me′no as′id) a unit of protein that takes its name from the fact that it contains an amino group (NH_2) and an acid group (COOH). *28*

atom (at′om) smallest unit of matter nondivisible by chemical means. *17*

ATP adenosine triphosphate; a compound containing adenine, ribose, and three phosphates, two of which are high-energy phosphates. It is the "common currency" of energy for most cellular processes. *38*

base (bās) a solution in which pH is more than 7; a substance that contributes or liberates hydroxide ions in a solution; alkaline; opposite of acidic. Also, a term commonly applied to one of the components of a nucleotide. *25*

buffer (buf′er) a substance or compound that prevents large changes in the pH of a solution. *26*

DNA deoxyribonucleic acid; a nucleic acid, the genetic material that directs protein synthesis in cells. *37*

electron (e-lek′tron) a subatomic particle that has almost no weight and carries a negative charge; travels in an orbital, called a shell, about the nucleus of an atom. *17*

emulsification (e-mul″sĭ-fĭ-ka′shun) the act of dispersing one liquid in another. *36*

enzyme (en′zīm) a protein catalyst that speeds up a specific reaction or a specific type of reaction. *27*

hydrogen bond (hi′dro-jen bond) a weak attraction between a hydrogen atom carrying a partial positive charge and another atom carrying a partial negative charge. *24*

ion (i′on) an atom or group of atoms carrying a positive or negative charge. *19*

isotopes (i′so-tōps) atoms with the same number of protons and electrons but differing in the number of neutrons and therefore in weight. *18*

lipid (lip′id) a group of organic compounds that are insoluble in water; notably fats, oils, and steroids. *34*

neutron (nu′tron) a subatomic particle that has a weight of one atomic mass unit, carries no charge, and is found in the nucleus of an atom. *17*

nucleic acid (nu-kle′ik as′id) a large organic molecule made up of nucleotides joined together; for example, DNA and RNA. *37*

pentose (pen′tōs) a 5-carbon sugar; deoxyribose is a pentose found in DNA; ribose is a pentose found in RNA. *31*

peptide bond (pep′tīd bond) the bond that joins two amino acids. *28*

pH a measure of the hydrogen ion concentration; any pH below 7 is acid and any pH above 7 is basic. *25*

polysaccharide (pol″e-sak′ah-rīd) a macromolecule composed of many units of sugar. *31*

protein (pro′te-in) a macromolecule composed of one or several long polypeptides. *28*

proton (pro′ton) a subatomic particle found in the nucleus of an atom that has a weight of one atomic mass unit and carries a positive charge; a hydrogen ion. *17*

RNA ribonucleic acid; a nucleic acid that assists DNA in the production of proteins within the cell. *37*

2

Cell Structure and Function

Chapter Concepts

1 A cell is highly organized and contains organelles that carry out specific functions.

2 The nucleus controls the metabolic functioning and structural characteristics of the cell.

3 Endoplasmic reticulum, Golgi apparatus, vacuoles, and lysosomes are all membranous structures concerned with the production, digestion, or transport of molecules.

4 Chloroplasts absorb the energy of the sun in order to produce glucose. Mitochondria are concerned with conversion of glucose energy into ATP energy.

5 Centrioles and related structures are concerned with the shape and/or movement of the cell.

Chapter Outline

Figure 2.1
Comparison of micrographs of red blood cells within a blood vessel. a. *Light micrograph.* b. *Transmission electron micrograph.* c. *Scanning electron micrograph.* d. *Technician operating a transmission electron microscope.*

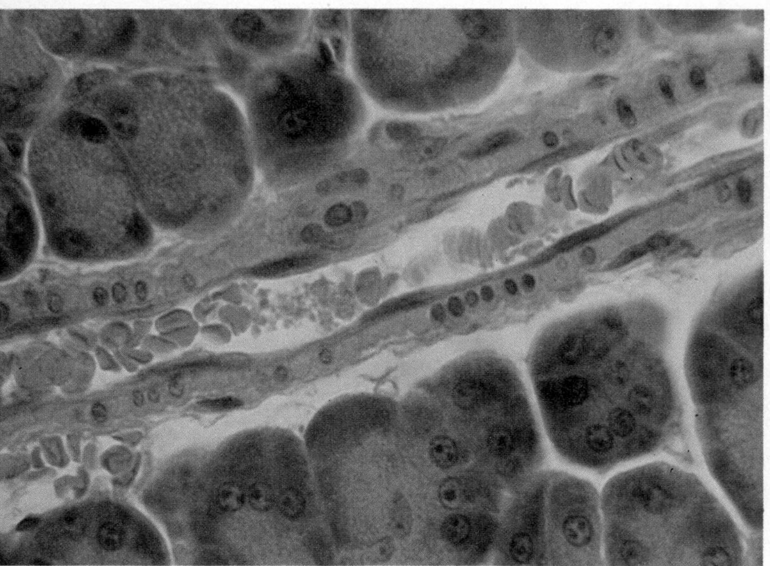

a.

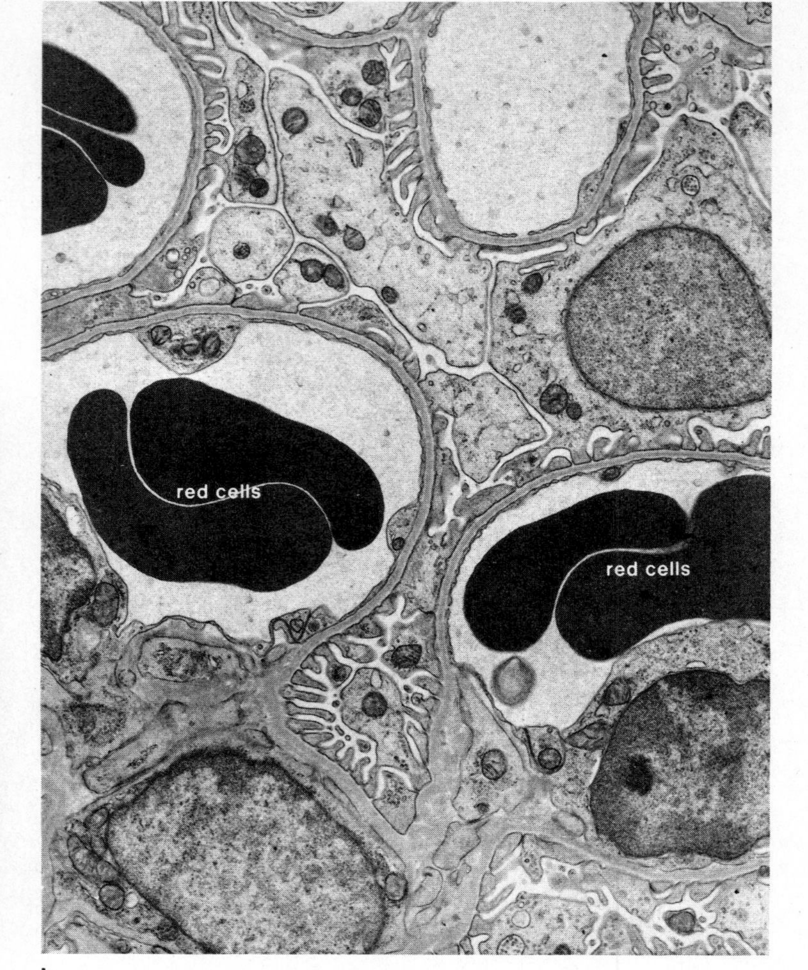

b.

c.

d.

Cell Theory

All living things are made up of **cells.** Cells come in many different shapes and sizes; but no matter what the shape or size, each one carries on the functions associated with life, interacting with the environment, growing, and reproducing.

The cell marks the boundary between the nonliving and the living. The molecules that serve as food for a cell and the organic molecules that make up a cell are not alive and yet the cell is alive. The answer to what life is will have to be found within the cell, because the smallest living organisms are single cells while larger organisms are **multicellular** and composed of many cells. The statement that all living things are composed of cells is called the **cell theory.**

Microscopy

With some exceptions, such as various types of eggs, cells are not readily visible to the eye; therefore a microscope is needed to view them. *Light microscopes* utilize light and electron microscopes utilize electrons to view the object. A *transmission electron microscope* gives us an image of the interior of the object while a *scanning electron microscope* provides a three-dimensional view of the surface of an object.

Transmission electron microscopes have a much greater **resolving power** than do light microscopes. Resolving power is the capacity to distinguish between two points. If two points are seen as separate, then the image appears more detailed than if the two points are seen as one point. Resolving power of a microscope is improved as the wavelength of the illumination becomes shorter, and an electron beam has a much shorter wavelength than a visible light ray. At the very best, a light microscope can distinguish two points separated by 200 nm (nanometer $= 1 \times 10^{-6}$ mm), but the transmission electron microscope can distinguish two points separated by only .5 nm.[1] Therefore, this microscope gives a much more detailed image. Although the scanning electron microscope has a resolving power of only 10 nm, the images produced still provide useful information.

Pictures obtained by using the light microscope are sometimes called photomicrographs, and pictures resulting from the use of the electron microscopes are called (transmission) electron micrographs and scanning electron micrographs respectively (fig. 2.1).

Generalized Cells

Cells are usually divided into two main groups called **prokaryotic** cells and **eukaryotic** cells. We will begin with an examination of eukaryotic cells and will then compare them to prokaryotic cells. Even though there is a great variety of eukaryotic cells, differing in regard to specific structure and function, they all have the same basic organization. This chapter will stress the generalized animal cell depicted in figure 2.2a and the generalized plant cell depicted in figure 2.2b. These drawings are a composite based on electron micrographs of animal and plant cells.

Eukaryotic cells are surrounded by an outer membrane, or *cell membrane*. **Cytoplasm,** the substance of the cell between the nucleus and the cell membrane, contains various **organelles,** small bodies with specific structures and functions. In addition to a cell membrane, plant cells have an exterior cell wall (fig. 2.2b).

Organelles

The various organelles are specialized for particular functions (table 2.1). For the sake of discussion, we will divide the organelles into five categories: (1) the nucleus, (2) membranous canals and vacuoles, (3) energy-related organelles, (4) the cytoskeleton, and (5) centrioles and related organelles.

[1]See appendix B on p. 763.

Figure 2.2
Animal and plant cells. These generalized representations are based on electron micrographs. a. *Animal cell.* b. *Plant cell (p. 45). In the average mature plant cell, the vacuole actually occupies a greater percentage of the cell's volume than it does in this drawing.*

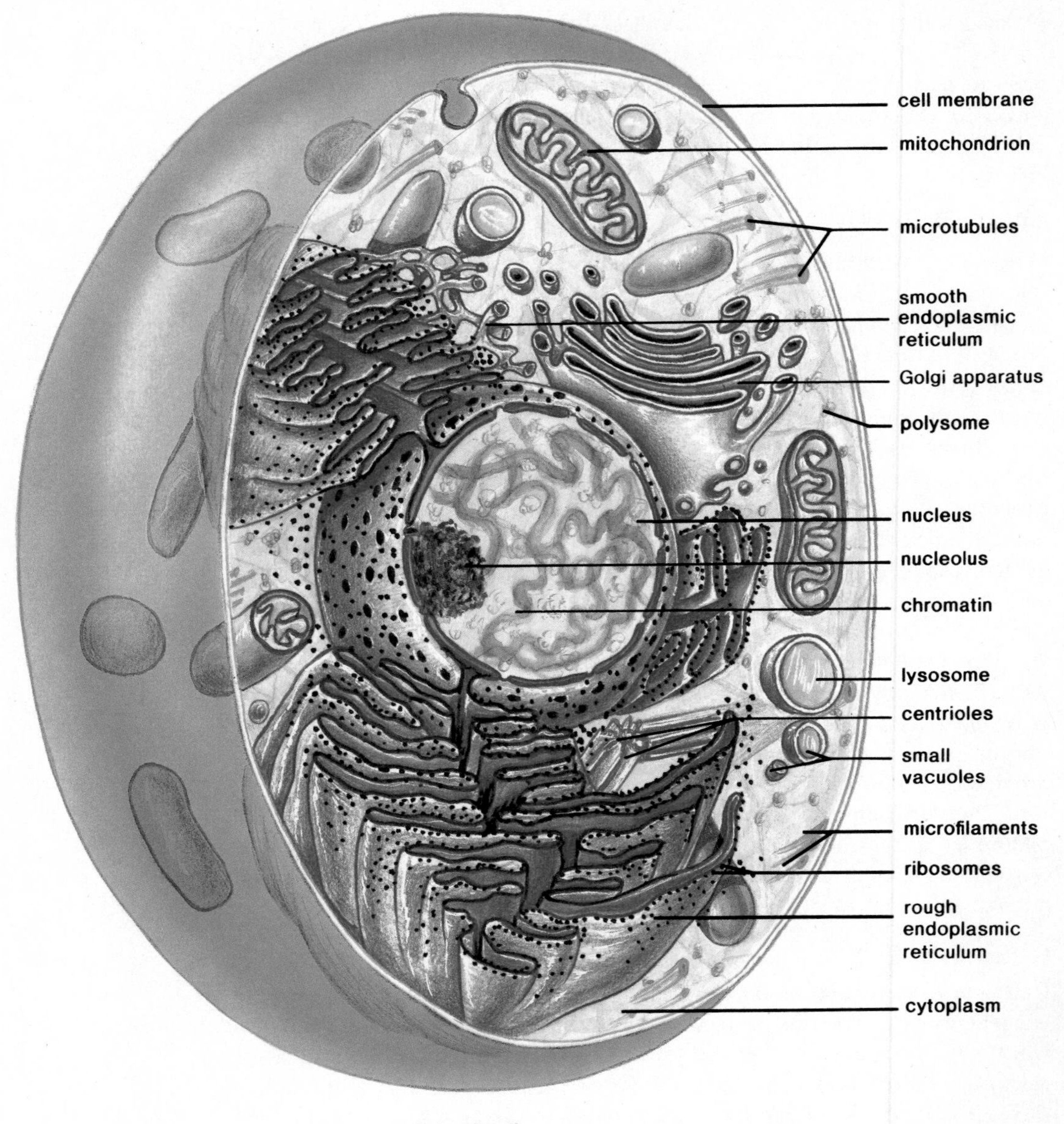

a. Animal cell

***Table 2.1** Eukaryotic Cell Structures and Functions (Simplified)*

Name	Structure	Function
Cell membrane	Bilayer of phospholipid and globular proteins	Passage of molecules into and out of cell
Cell wall	Cellulose fibrils	Provides rigidity to plant cells
Nucleus		
Nucleus	Nuclear envelope surrounding chromosomes (DNA) and nucleoli	Cellular reproduction and control of protein synthesis
Nucleolus	Concentrated area of RNA in the nucleus	Ribosome formation
Membranous Canals & Vacuoles		
Endoplasmic reticulum (ER)	Folds of membrane forming flattened channels and tubular canals	Transport by means of vesicles
Rough ER	Studded with ribosomes	Protein synthesis
Ribosome	Protein and RNA in two subunits	Protein synthesis
Smooth ER	Having no ribosomes	Lipid synthesis
Golgi apparatus	Stack of membranous saccules	Packaging and secretion

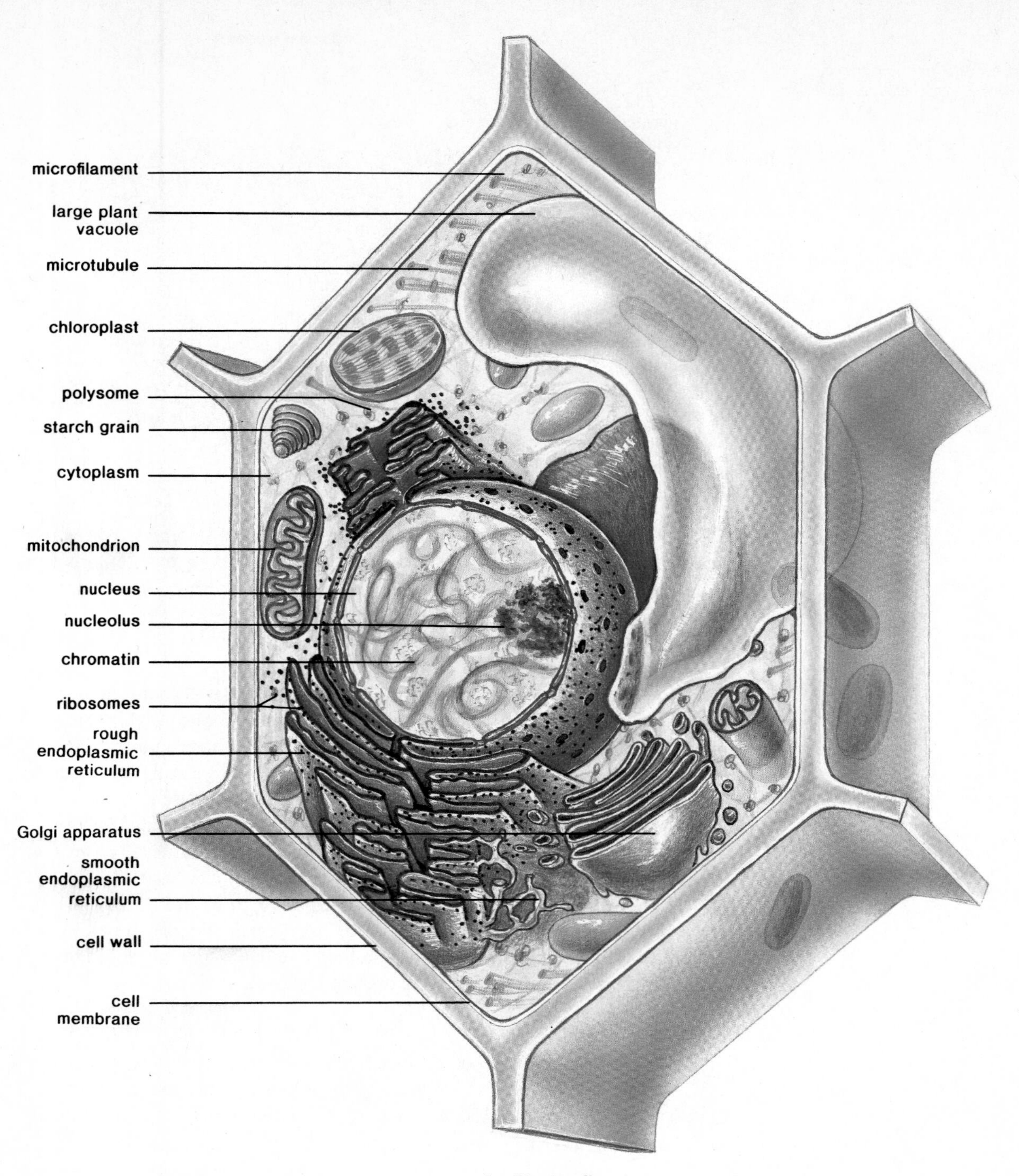

b. Plant cell

Name	Structure	Function
Membranous Canals & Vacuoles continued		
Vacuole and vesicle	Membranous sacs	Containers of material
Lysosome	Membranous container of hydrolytic enzymes	Intracellular digestion
Energy-related Organelles		
Mitochondrion	Inner membrane (cristae) within outer membrane	Cellular respiration
Chloroplast	Inner membrane (grana) within outer membrane	Photosynthesis
Cytoskeleton		
Microfilament	Actin or myosin proteins	Movement and shape of cell
Microtubule	Tubulin protein	Movement and shape of cell
Centrioles and Related Organelles		
Centriole	9 + 0 pattern of microtubules	Organization of cilium and flagella
Cilium and flagellum	9 + 2 pattern of microtubules	Movement of cell

Figure 2.3
Fluid-mosaic model of intracellular membrane. Proteins float wi.hin a phospholipid bilayer and are also located at the inner side.

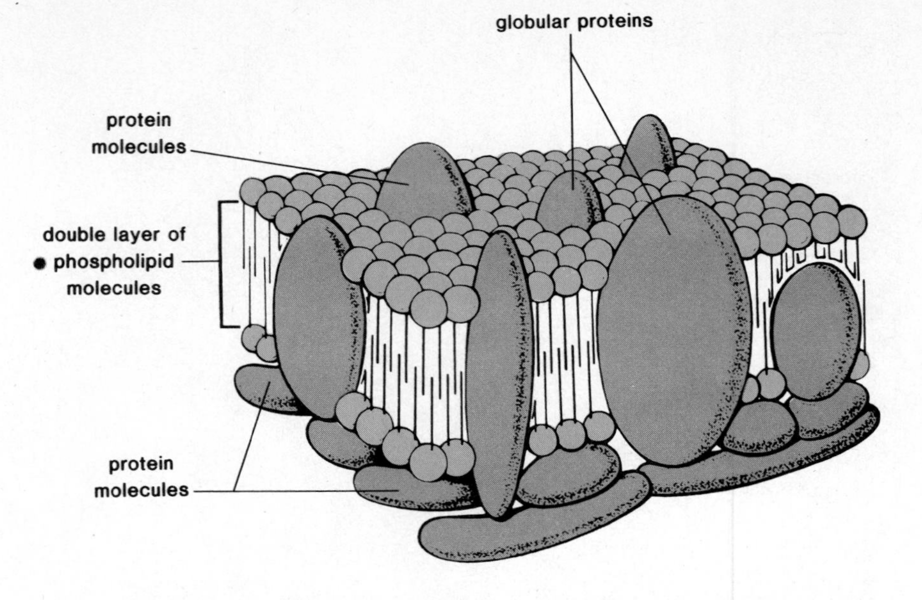

The organelles vary in size, but most are very small and only the electron microscope can make out their inner detail. While the nucleus, chloroplast, and mitochondrion are within the range of the light microscope, the electron microscope is required to distinguish the others.

Intracellular Membrane

Most investigators today support the **fluid-mosaic model** of membrane structure (fig. 2.3). The protein molecules form a pattern (mosaic) within the phospholipid bilayer, which is in a liquid (fluid) state the consistency of light oil. Notice the manner in which the phospholipid molecules arrange themselves. Their structure, discussed in chapter 1, causes each molecule to have a polar head and nonpolar tails. Within the phospholipid bilayer, the tails face inward and the heads face outward where they are likely to encounter a watery environment.

The protein portion of membrane is believed to be more variable than the phospholipid portion. Depending on the membrane, the protein molecules may reside above or below the phospholipids, extend from top to bottom of the membrane, or simply penetrate a short distance.

All organisms are composed of cells whose internal structure has been revealed by electron microscopy. All eukaryotic cells contain a nucleus and various other organelles, each having a detailed structure and function.

Eukaryotic Cell Organelles

Table 2.1 states the structure and function of the organelles we will be discussing.

Nucleus

The **nucleus** (N) (fig. 2.4), a large, often centrally located organelle, is enclosed by a double-layered membrane (fig. 2.2), called the **nuclear envelope** (NE). There are pores in this envelope through which certain molecules are believed to pass from the cytoplasm to the nucleoplasm or vice versa.

The nucleus is of primary importance in the cell because it is the control center that oversees the metabolic functioning of the cell and ultimately determines the cell's characteristics, as experimentation has shown. In one type

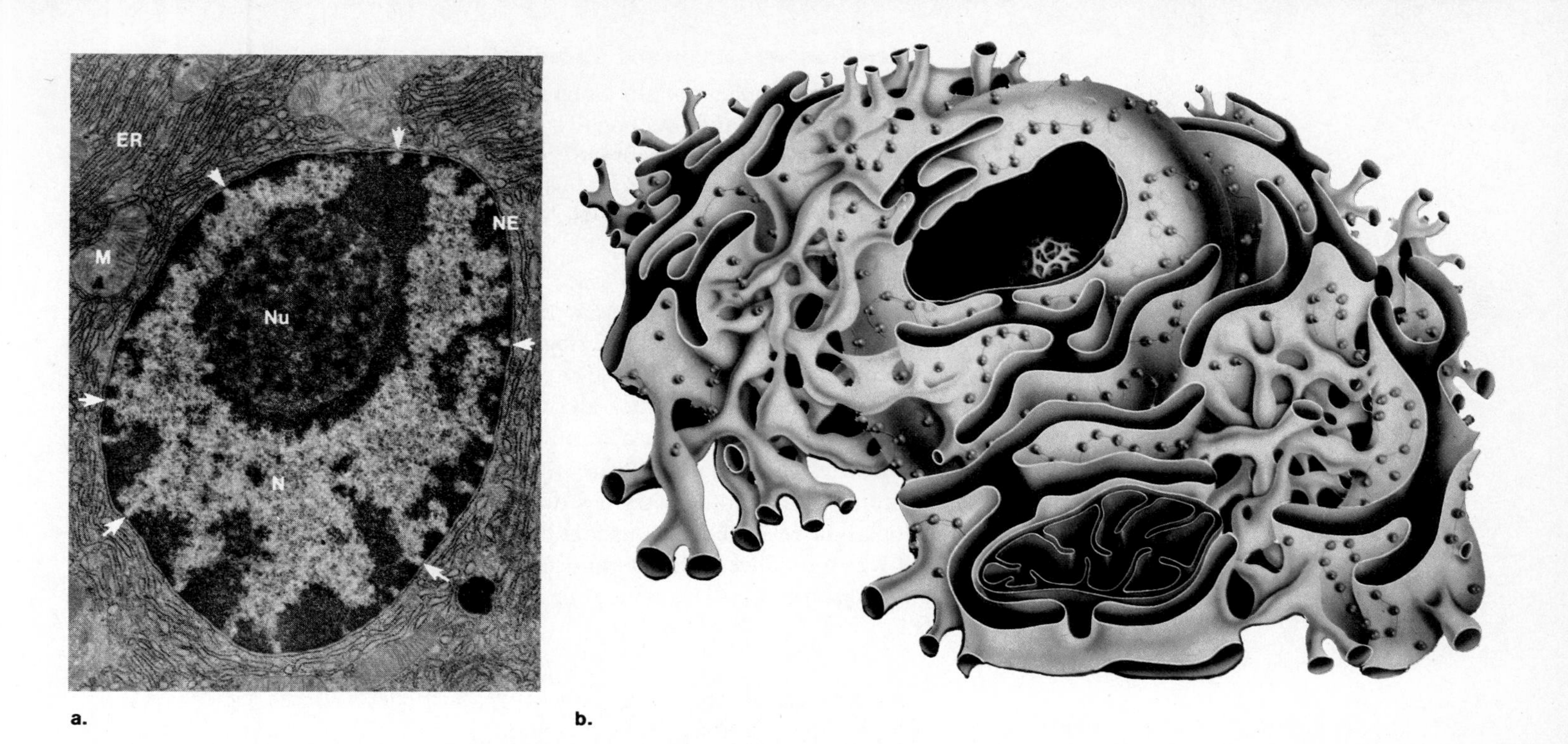

of green alga, *Acetabularia,* the organism consists of a base, stalk, and cap (fig. 2.5). If the stalk and cap are removed from the base as in figure 2.5a, the stalk and cap die but the nucleus-containing base develops into a new organism. The importance of the nucleus is further exemplified in figure 2.5b when the base is combined with the stalk of a different species, and the cap regenerated is appropriate to the species of the nucleus rather than to the cytoplasm of the stalk. This demonstrates that the nucleus controls both the function and stucture of the cell.

Within the nucleus there are masses of threads called **chromatin,** so called because they take up appropriate stains and become colored. Chromatin is indistinct in the nondividing cell, but it condenses to rodlike structures called **chromosomes** (fig. 4.2) at the time of cell division. Chemical analysis shows that chromatin, and therefore chromosomes, contain the chemical DNA (deoxyribonucleic acid) along with certain proteins and some RNA (ribonucleic acid). It is now known that DNA, with the help of RNA, *controls protein synthesis* within the cytoplasm and that it is this function that allows DNA to control the cell.

Figure 2.4
An electron micrograph of a nucleus (N) *with a clearly defined nucleolus* (Nu) *and irregular patches of chromatin scattered throughout the nucleoplasm. The nuclear envelope* (NE) *contains pores indicated by the arrows.* b. *A nucleus is surrounded by endoplasmic reticulum* (ER) *and its size may be compared to a mitochondrion* (M) *as is also shown in* a.

Nucleoli

One or more **nucleoli** are present in the nucleus. These dark-staining bodies are actually specialized parts of chromosomes in which a special type of RNA called ribosomal RNA (rRNA) is produced. Ribosomal RNA joins with proteins before migrating to the cytoplasm where it becomes part of the ribosomes, organelles to be discussed in the following sections.

The nucleus contains chromatin which condenses into chromosomes just prior to cell division. Chromosomes contain DNA that, with the help of RNA, directs protein synthesis in the cytoplasm. Another type of RNA, rRNA, is made within the nucleolus before migrating to the cytoplasm where it is incorporated into ribosomes.

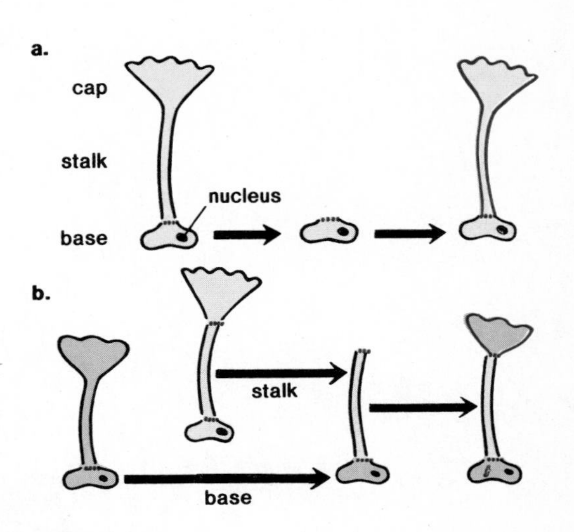

Figure 2.5
Experiments with the large single-celled alga, Acetabularia. a. *In this experiment, the stalk and cap without a nucleus die, but the base with a nucleus regenerates.* b. *In this experiment, the regenerated cap resembles that of the species of the nucleus and not that of the cytoplasm. These experiments show that the nucleus controls the rest of the cell.*

Membranous Canals and Vacuoles

Endoplasmic reticulum, the Golgi apparatus, vacuoles, and lysosomes (fig. 2.2) are structurally and functionally related membranous structures. Ribosomes are not composed of membrane but are included in this category because they are often intimately associated with the endoplasmic reticulum.

Endoplasmic Reticulum

The **endoplasmic reticulum** (ER) forms a membranous system of tubular canals that begins at the nuclear envelope and branches throughout the cytoplasm. Small granules, called ribosomes, are attached to some portions of the endoplasmic reticulum. If they are present, the reticulum is called **rough endoplasmic reticulum;** if they are not present, it is called **smooth endoplasmic reticulum.** Figure 2.6 illustrates rough endoplasmic reticulum and figure 2.7 illustrates the smooth. Apparently, smooth endoplasmic reticulum contains, within its membrane, enzymes that synthesize lipids. Therefore, smooth endoplasmic reticulum is abundant in cells of the testes and adrenal cortex, both of which produce steroid hormones. Also, it is known that the administration of drugs increases the amount of smooth endoplasmic reticulum in the liver.

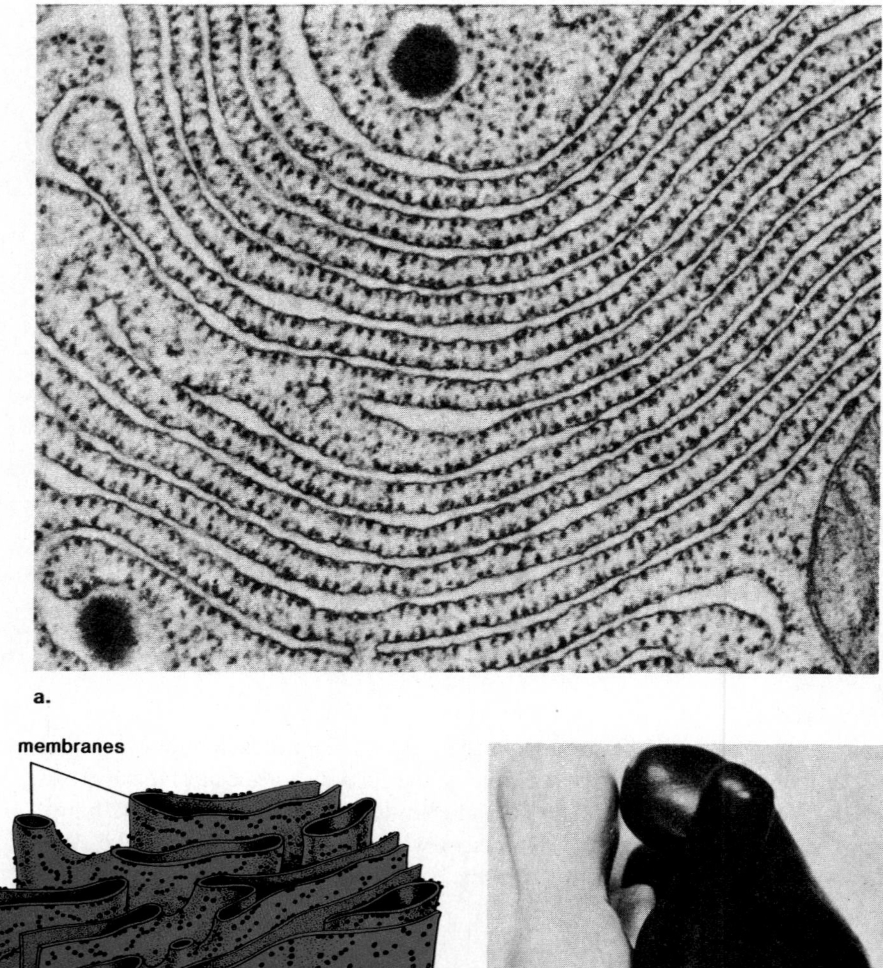

a.

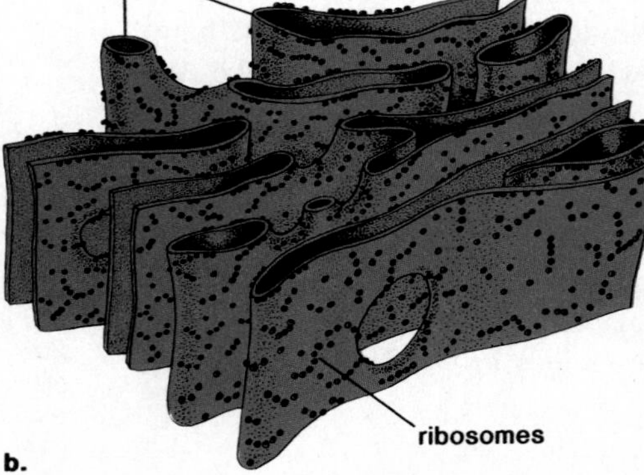

b.

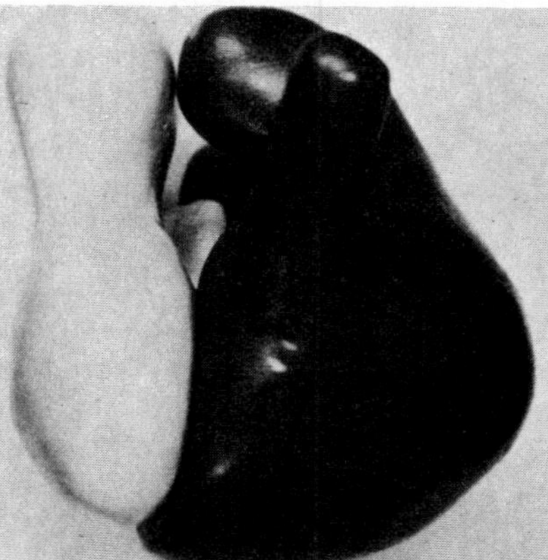

c.

Figure 2.6
Rough endoplasmic reticulum. a. *Electron micrograph showing that, in some cells, the cytoplasm is packed with this organelle.* b. *A three-dimensional drawing gives a better idea of the organelle's actual shape.* c. *This three-dimensional model of a ribosome is based on high-power electron microscopy studies that indicate that a ribosome is composed of a small and a large subunit.*

It would seem then that the reticulum has enzymes that detoxify drugs. It is quite possible that these are detoxified within structures called **peroxisomes,** membrane-bound vacuoles often attached to smooth ER that contain enzymes capable of carrying out oxidation of various substances including alcohol.

Ribosomes **Ribosomes** look like small, dense granules in low-power electron micrographs (fig. 2.6a), but a higher resolution shows that each one contains two subunits (fig. 2.6c). As their name implies, ribosomes contain RNA, but they also contain proteins. The larger of the two subunits contains at least thirty different proteins, and the smaller unit contains at least twenty different proteins.

Ribosomal RNA produced in the nucleolus joins with proteins before migrating to the cytoplasm. Once the ribosomes are fully assembled within the cytoplasm, they function in the process of protein synthesis. Synthesis, as discussed in chapter 1, refers to the joining together of small organic molecules to make larger ones. In this case, amino acids are joined together to make a protein.

Ribosomes are very often attached to endoplasmic reticulum (fig. 2.6) but are also found unattached within the cytoplasm. In these instances, several ribosomes, each of which is producing the same type protein, are arranged in a functional group called a **polysome.** Most likely these proteins are for use inside the cell. But when the ribosomes are attached, it most likely means that the protein is for export outside the cell. For example, certain pancreatic cells that produce digestive enzymes for use in the small intestine have extensive amounts of rough endoplasmic reticulum.

Protein that is destined for export outside the cell is prepared at the ribosomes and temporarily stored in the channels of the reticulum. Small portions of the endoplasmic reticulum then break away to form membrane-enclosed vesicles (small vacuoles) that migrate to the Golgi apparatus, where the product is received and repackaged for export.

Golgi Apparatus

The **Golgi apparatus** (fig. 2.8) is named for the person who first discovered its presence in cells. It is composed of a stack of about a half-dozen or more *saccules* that look like flattened vacuoles. In animal cells, one side of the stack faces the nucleus and the other side faces the cell membrane. Vesicles are seen especially at the rims of the saccules, but vesicles also occur along the length of the saccules at either face of the stack.

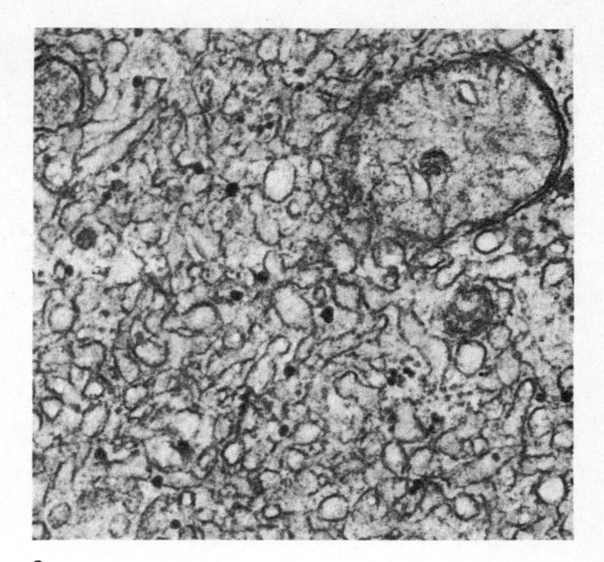

a.

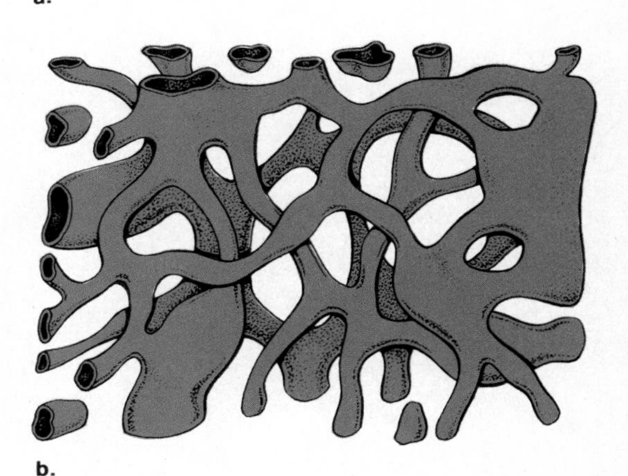

b.

Figure 2.7
Smooth endoplasmic reticulum. a. *Electron micrograph showing the abundance of this organelle in cells that secrete steroids.* b. *A three-dimensional drawing illustrates that this organelle lacks ribosomes.*

Figure 2.8
Golgi apparatus structure. A frozen cell was fractured (split) and then subjected to scanning electron microscopy. In this case, the treatment revealed the interior of the saccules and shows that vesicles appear along both sides of the Golgi apparatus and at the rims of the saccules.

Figure 2.9
Golgi apparatus function. The Golgi apparatus receives vesicles from the endoplasmic reticulum and thereafter forms at least two types of vesicles, lysosomes and secretory vesicles. Lysosomes contain hydrolytic enzymes that can break down large molecules. Sometimes lysosomes join with vesicles, bringing large molecules into the cell. Thereafter, any nondigested residue is voided at the cell membrane. The secretory vesicles formed at the Golgi apparatus also discharge their contents at the cell membrane.

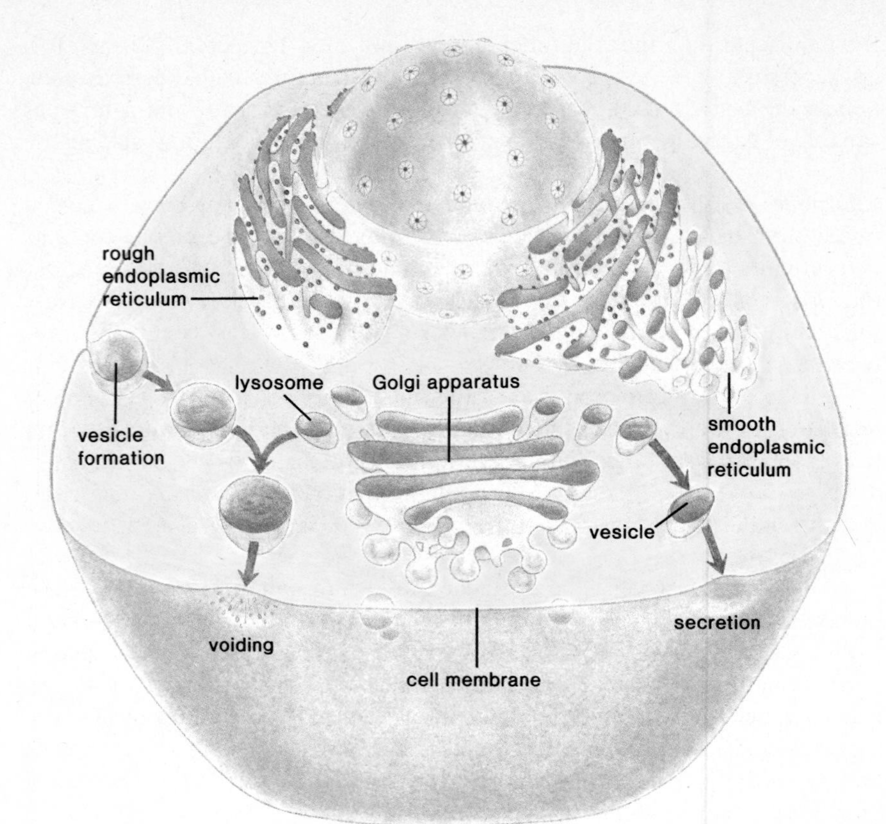

These observations along with biochemical evidence suggest that the Golgi apparatus receives protein-filled vesicles from the endoplasmic reticulum at its inner face. After the proteins are sorted out, they are packaged into vesicles at the rims of the saccules and at the outer face. Thereafter, the vesicles move to different locations in the cell (fig. 2.9).

Vacuoles

A cross section of a **vacuole,** which is called a *vesicle* when small in size, most often shows a clear area bounded by a membrane. Vacuoles are storage sites for various kinds of molecules in solution or suspension. Vesicles can be made by the Golgi apparatus, as previously described, or they can arise by an infolding of the cell membrane. The large, central plant vacuole (fig. 2.2b) is attached to the endoplasmic reticulum, and it seems that this particular vacuole might be a part of this system. The central vacuole helps support a plant cell when it is full of water. When a plant wilts, its cells' vacuoles are not filled with water.

Lysosomes

A **lysosome** is a special type of vesicle (fig. 2.9), most likely formed by the Golgi apparatus. All lysosomes are concerned with intracellular digestion and contain *hydrolytic enzymes* (p. 27). Following formation the lysosome may fuse with a vesicle that contains a substance to be digested. While the products of digestion enter the cytoplasm, the nondigested residue is expelled from the cell.

Occasionally, a person is unable to manufacture an enzyme normally found within the lysosome. In these cases, the lysosome fills to capacity as the substrate for that enzyme accumulates. The cells may become so filled with lysosomes of this type that it brings about the death of the individual.

Lysosomes also carry out autodigestion, or the disposal of worn-out or damaged cell components such as mitochondria, which have a short life span in the cell. This is an essential part of the normal process of cytoplasmic maintenance and turnover. By turnover it is meant that the cell is constantly breaking down and remaking its parts. The lysosomes no doubt play a role in this process, and electron micrographs sometimes show mitochondria enclosed within these structures (fig. 2.10).

When a cell dies, or for some other reason is destined for total destruction, the lysosome sometimes releases its contents. The term *suicide bag* has been used in connection with the lysosome because of its ability to bring about a complete breakdown of the cell itself. Normally, complete cell destruction occurs when a change of shape is required by the organism. For example, the disappearance of a tadpole's tail may be brought about by lysosome action.

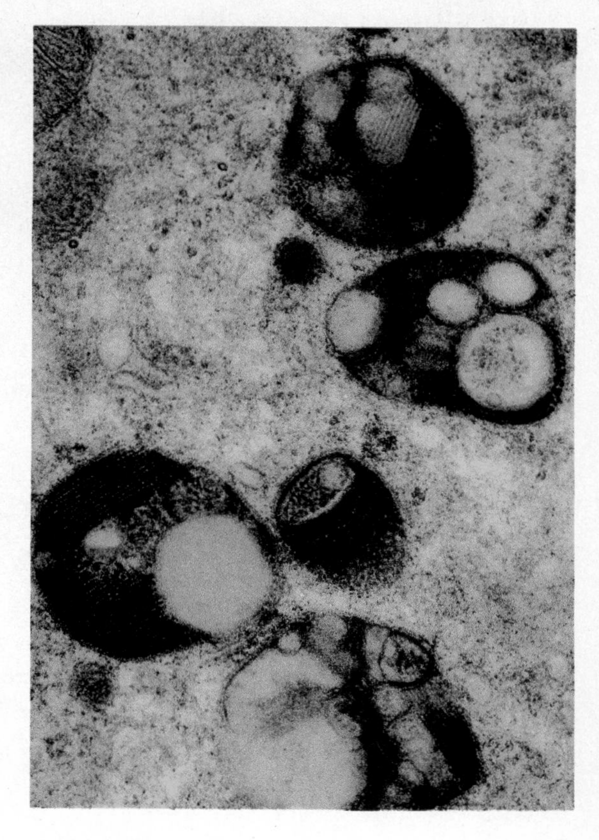

Figure 2.10
Lysosomes within a cell. Lysosomes contain hydrolytic enzymes that digest macromolecules and even parts of cells.

The endoplasmic reticulum is a membranous system of tubular canals that may be smooth or rough. Proteins synthesized at the rough ER are processed and packaged in vesicles by the Golgi apparatus. Some vesicles discharge their contents at the cell membrane and some are lysosomes that digest any material enclosed therein.

Energy-Related Organelles

The energy-related organelles, mitochondria and chloroplasts, are transformers, changing one form of energy into another. While chloroplasts are unique to plant cells, mitochondria are found in both plant and animal cells.

Mitochondria

A **mitochondrion** is a rather complex organelle that produces ATP energy. Every cell needs a certain amount of ATP energy to synthesize molecules, but many cells need ATP to carry out their specialized functions. For example, muscle cells need it for muscle contraction and nerve cells need it for the conduction of nerve impulses.

Mitochondria are extremely efficient at what they do. One investigator, after calculating the weight of mitochondria in a horse's legs, suggested that the "magnitude of the effect of the mitochondria per weight unit is the same as the one delivered by the engines in a jet plane in vertical ascent. . . . In other words, the mitochondria are admirably effective machines."[2]

Mitchondria are often referred to as the powerhouses of the cell because, just as a powerhouse burns fuel to produce electricity, the mitochondria burn glucose products to produce ATP molecules, the chemical energy needed by cells. In the process, mitochondria use up oxygen and give off carbon dioxide and water. The oxygen you breathe in enters cells and then the mitochondria; the carbon dioxide you breathe out is released by the mitochondria. Since gas exchange is involved, it is said that mitochondria carry on **cellular respiration.** A shorthand way to indicate the chemical transformation associated with cellular respiration is:

carbohydrate + oxygen ⟶ carbon dioxide + water + energy

[2]Bjorn Afzelius, *Anatomy of the Cell* (Chicago: University of Chicago Press, 1966), p. 11.

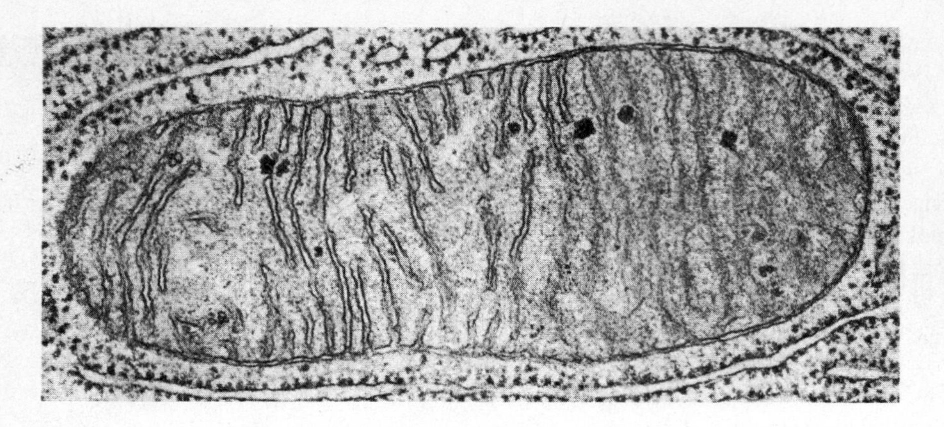

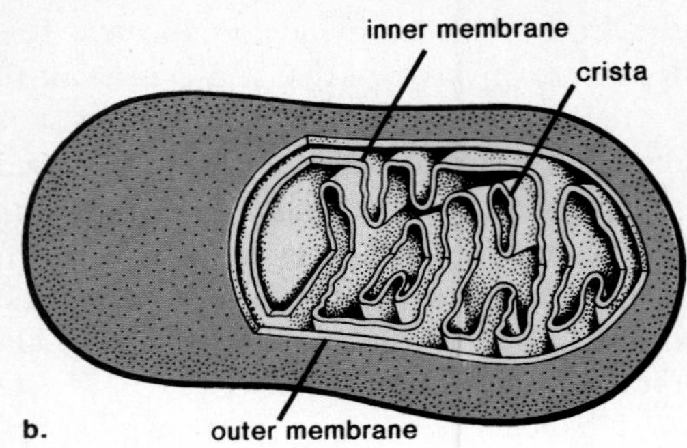

Figure 2.11
Mitochondrion structure. a. *An electron micrograph of a mitochondrion surrounded by rough endoplasmic reticulum. Note the shelflike cristae formed by the inner membrane.* b. *Diagrammatic drawing shows outer and inner structure more clearly.*

Each mitochondrion is composed of two membranes, an outer membrane and an inner membrane (fig. 2.11). The inner membrane is convoluted into shelflike projections called *cristae.* The molecules that aid in the production of energy are located in an assembly-line fashion on these membranous shelves. The membrane is divided into functional units, and a very small area of each crista contains one respiratory unit. The inner membrane lends itself to this type of arrangement and thus we see that structure aids function.

Chloroplasts

Plastids are membranous structures that often contain pigments and give plant cells their color. Some plastids, however, are colorless and act as storage bodies for starch, protein, or oils.

The most familiar and abundant plastid is the **chloroplast** (fig. 2.12). In higher plants, chloroplasts have an ovoid, or disklike, shape and may be even larger than mitochondria.

Within a chloroplast there are stacks of membranous sacs called *grana.* The green pigment, chlorophyll, is found within the grana, and thus chloroplasts are green. The inner portion of a chloroplast is called the *stroma.*

Chlorophyll is a chemical molecule that can absorb the energy of the sun. This energy allows photosynthesis to take place. *Photosynthesis* (synthesis by means of light energy) refers to the production of food molecules, glucose, for example. The glucose molecules may later be joined together to form starch. Chloroplasts take in carbon dioxide, water, and radiant energy from the sun in order to produce glucose. They give off oxygen, which leaves the plant as a gas. Again, we can use the shorthand method to describe what has been said:

$$\text{energy} + \text{carbon dioxide} + \text{water} \longrightarrow \text{carbohydrate} + \text{oxygen}$$

The equation for photosynthesis is the opposite of cellular respiration, as you can see by comparing the shorthand statements for each.

Chloroplasts also illustrate that structure facilitates the function of an organelle. Within the grana, light energy is used to form ATP molecules, the type of energy that can be used by the enzymes within the stroma to make food molecules. It is believed that the grana are highly organized, just as the cristae of the mitochondria are organized. Electron micrographs show particles that are believed to be photosynthetic units.

Mitochondria are the powerhouses of the cell because the inner cristae have enzymes that carry on cellular respiration, a process that converts carbohydrate energy to ATP energy. Chloroplasts contain stacks of grana where chlorophyll captures the energy of the sun starting the photosynthetic process that is complete when carbohydrates are produced in the stroma.

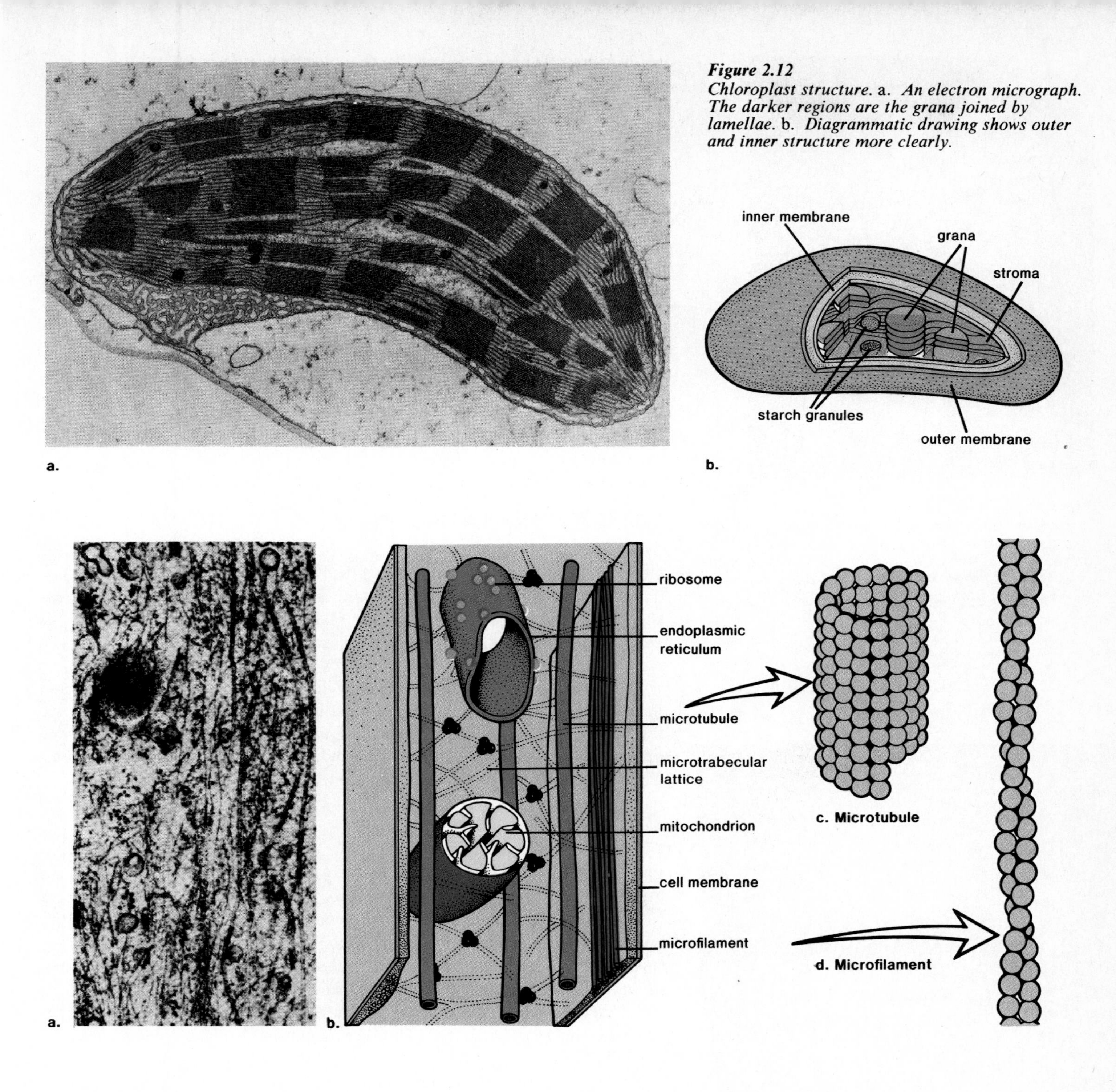

Figure 2.12
Chloroplast structure. a. *An electron micrograph. The darker regions are the grana joined by lamellae.* b. *Diagrammatic drawing shows outer and inner structure more clearly.*

Figure 2.13
Cytoskeleton of cell. Notice that the various organelles are suspended by a cytoplasm that includes microtubules, microfilaments, and a lattice composed of very fine protein strands. a. *Electron micrograph.* b. *Drawing of microtubules and microfilaments.* c. *Detailed structure of a microtubule.* d. *Detailed structure of a microfilament.*

Cytoskeleton

Several types of filamentous protein structures form a **cytoskeleton** (fig. 2.13) that helps maintain the cell's shape, anchors the organelles, or allows them to move as appropriate. The cytoskeleton includes microfilaments and microtubules and there is also some evidence for an irregular, three-dimensional network, or lattice (termed the microtrabecular lattice), in the cytoplasm. The lattice is believed to be dynamic and its exact shape is constantly changing. While the lattice itself is protein rich, the spaces between the strands are water rich.

Microfilaments are long, extremely thin fibers (approximately 4–7 nm in diameter) that usually occur in bundles or other groupings. Microfilaments have been isolated from a number of cells. When analyzed chemically, their composition is similar to that of actin or myosin, the two proteins responsible for muscle contraction.

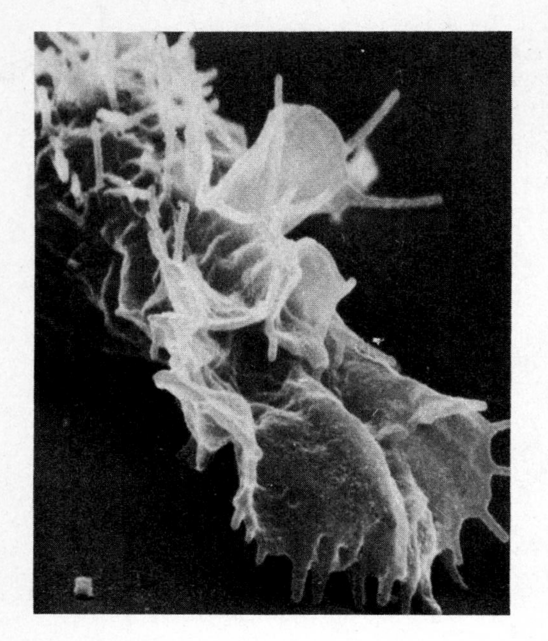

Figure 2.14
A scanning electron micrograph of an individual cell in a tissue culture. Notice the fingerlike projections on the "ruffle," which marks the leading edge of the cell. Microfilaments and microtubules are most likely present in this ruffle.

Microtubules are shaped like thin cylinders and are several times larger than microfilaments (about 25 nm in diameter). Each cylinder contains thirteen rows of tubulin, a globular protein, arranged in a helical fashion. Aside from existing independently in the cytoplasm, microtubules are also found in certain organelles, such as cilia, flagella, and centrioles.

Remarkably, both microfilaments and microtubules assemble and disassemble within the cell. When they are assembled, the protein molecules are bonded together and when they are disassembled, the protein molecules are not attached to one another. When microfilaments and microtubules are assembled, the cell has a particular shape and when they disassemble, the cell can change shape (fig. 2.14).

The cytoskeleton contains microfilaments and microtubules and possibly a lattice. Microfilaments, thin actin or myosin strands, and microtubules, 13 rows of tubulin protein molecules arranged to form a hollow cylinder, maintain the shape of the cell and also direct the movement of cell parts.

Centrioles and Related Organelles

Centrioles

Centrioles are short cylinders, with a 9 + 0 pattern of microtubule triplets. There is always one pair (fig. 2.2) lying at right angles to one another near the nucleus in animal, but not higher plant, cells. Before a cell divides, the centrioles replicate and the members of each pair are also at right angles to one another.

Centrioles give rise to basal bodies that direct the formation of cilia and flagella. Centrioles may also be involved in other cellular processes that use microtubules, such as the movement of material throughout the cell or the appearance and disappearance of the spindle apparatus (p. 81) in animal cells. Their exact role in these processes is uncertain however.

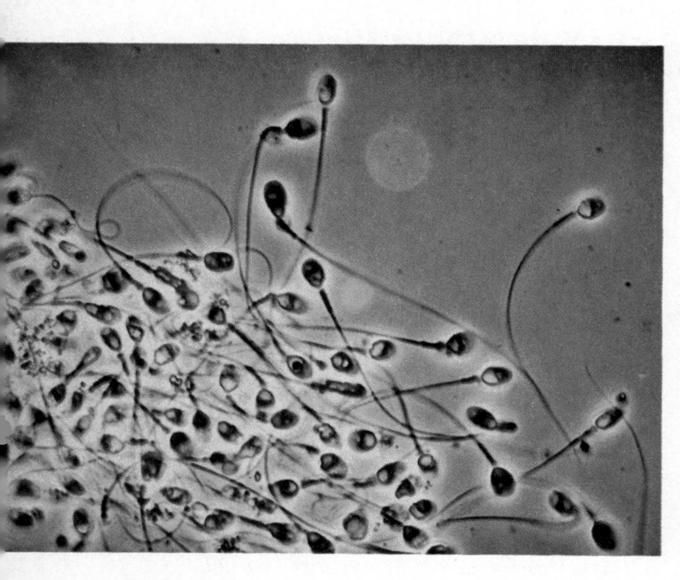

Figure 2.15
Sperm cells use long whiplike flagella to move about. Cilia and flagella have the structure depicted in figure 2.16.

Cilia and Flagella

Cilia and **flagella** are hairlike projections of cells that can move either in an undulating fashion, like a whip, or stiffly like an oar. Cells that have these organelles are capable of movement. For example, single-celled paramecia (p. 549) move by means of cilia; sperm cells, carrying genetic material to the egg, move by means of flagella (fig. 2.15). The cells that line our upper respiratory tract are ciliated. The cilia sweep debris trapped within mucus back up into the throat and this action helps keep the lungs clean.

Cilia are much shorter than flagella, but even so they both are constructed similarly (fig. 2.16a). They are membrane-bound cylinders enclosing a matrix area. In the matrix are nine microtubule doublets arranged in a circle around two central microtubules (fig. 2.16b). This is called the 9 + 2 pattern of microtubules. Each doublet also has pairs of arms projecting toward a neighboring doublet and spokes extending toward the central pair of microtubules. Recent evidence indicates that cilia and flagella move when the microtubule doublets slide along one another. The clawlike arms and spokes seem to be involved in causing this sliding action, which requires ATP energy.

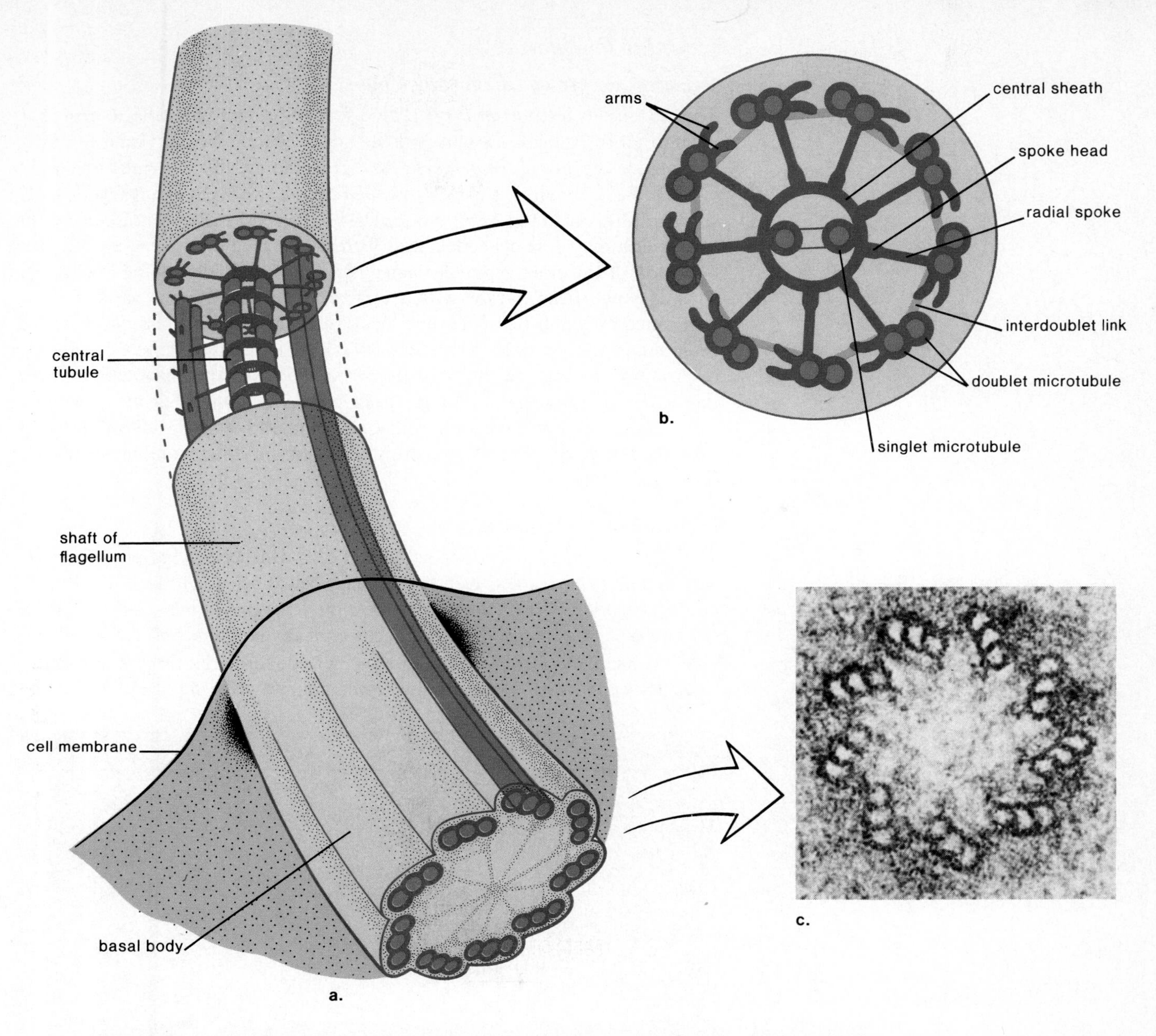

Figure 2.16
Anatomy of cilia and flagella. a. *Drawing showing that the 9 + 2 pattern of microtubules within a cilium or flagellum is derived (in some unknown way) from the 9 + 0 pattern in a basal body.* b. *Cross-section drawing of the 9 + 2 pattern shows the exact arrangement of microtubules. Notice the clawlike arms of the outer doublets and the spokes that connect them to the central pair.* c. *Electron micrograph of the cross section of a basal body.*

Each cilium and flagellum has a basal body (fig. 2.16c) lying in the cytoplasm at its base. **Basal bodies,** which are short cylinders with a circular arrangement of nine microtubule triplets called the 9 + 0 pattern, are believed to organize the structure of cilia and flagella.

Centrioles have a 9 + 0 pattern of microtubules and give rise to basal bodies that organize the 9 + 2 pattern of microtubules in cilia and flagella. Centrioles may be connected in some way to the origination of microtubules and the spindle fibers that are seen during cell division.

Cellular Comparisons

Prokaryotic versus Eukaryotic Cells

Thus far in this chapter we have been discussing eukaryotic cells, a term that means that the cell has a membrane-bound, or "true," nucleus (karyote). Prokaryotic cells evolved before eukaryotic cells and they lack a true nucleus.

Table 2.2 compares prokaryotic cells to the two types of eukaryotic cells studied in this chapter. You'll notice that prokaryotic cells, represented only by bacteria including cyanobacteria (formerly termed blue-green algae), also lack most of the other types of organelles we have been discussing (fig. 2.17). This does not mean, however, that their cells do not carry on the functions performed by organelles. The functions simply occur within the cytoplasm of these much smaller cells. Therefore, they have chromosomes but not nuclei; respiratory enzymes but not mitochondria; and when they have chlorophyll, there are no chloroplasts, although the chlorophyll is associated with lamellae.

Prokaryotes do have a cell wall and if motile, they possess flagella. However, the structure of these cell features differs from those found in eukaryotic cells.

Plant versus Animal Cells

Table 2.2 also compares plant and animal cells. It is wise to keep in mind that while all cells have a cell membrane, plant cells have a cell wall in addition; and that while all eukaryotic cells have mitochondria, plant cells also have chloroplasts. However, higher plant cells typically lack centrioles and the cilia and flagella of animal cells. Even so, plant cells do form mitotic spindles; structures associated with centrioles in dividing animal cells (fig. 4.9).

Table 2.2 *Comparison of Prokaryotic and Eukaryotic Cells*

	Prokaryotic	Eukaryotic	
		Animal	Plant
Cell membrane	yes	yes	yes
Cell wall	yes	no	yes
Nuclear envelope	no	yes	yes
Mitochondria	no	yes	yes
Chloroplasts	no	no	yes
Endoplasmic reticulum	no	yes	yes
Ribosomes	yes, small	yes, large	yes, large
Vacuoles	no	yes, small	yes
Lysosomes	no	yes, usually	no, usually
Cytoskeleton	no	yes	yes
Centrioles	no	yes	no

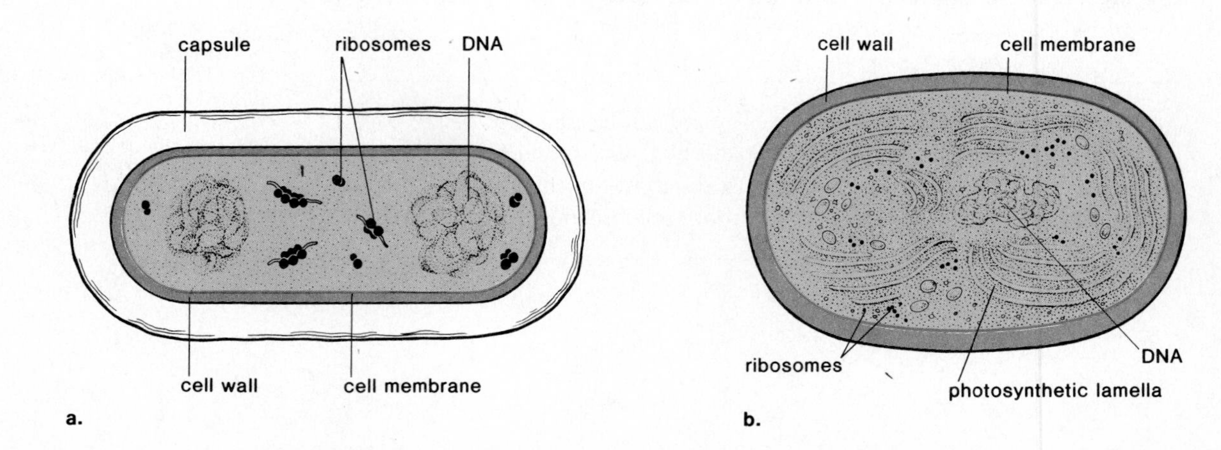

Figure 2.17
Prokaryotic cells. Notice the lack of discrete organelles. a. *Generalized nonphotosynthetic bacterium.* b. *Generalized cyanobacterium, a photosynthetic bacterium.*

Summary

Figure 2.18 summarizes the animal cell organelles we have studied in the chapter. The nucleus is a large organelle of primary importance because it controls the rest of the cell. Within the nucleus lies the chromatin which condenses to become chromosomes during cell division.

Proteins are made at the rough endoplasmic reticulum before being packaged at the Golgi apparatus. Golgi-derived lysosomes fuse with incoming vacuoles to digest any material enclosed within and lysosomes also carry out autodigestion of old parts of cells.

Mitochondria and chloroplasts are the energy-related organelles. During the process of cellular respiration mitochondria convert carbohydrate energy to ATP energy, and during photosynthesis chloroplasts form carbohydrates.

Microfilaments and microtubules are found within a cytoskeleton that maintains the shape and permits movement of cell parts. Centrioles are associated with the spindle apparatus during cell division and they also produce basal bodies that give rise to cilia and flagella.

Prokaryotic cells lack the organelles typically found in eukaryotic cells such as plant and animal cells. Nevertheless, they carry on all the same functions as eukaryotic cells.

Objective Questions

1. Electron microscopes have a greater ________ power than do light microscopes.
2. The fluid-mosaic model of membrane structures says that ________ molecules drift about within a ________ bilayer.
3. Chromosomes are located within the ________, a structure that controls the rest of the cell.
4. Rough endoplasmic reticulum has ________ but smooth endoplasmic reticulum does not.
5. Lysosomes contain ________ enzymes.
6. Vesicles derived from endoplasmic reticulum make their way to the ________, an organelle that functions in packaging and secretion.
7. Both plant and animal cells have ________, where glucose products provide energy for ATP formation.
8. Photosynthesis takes place within ________.
9. Microfilaments and microtubules are a part of the ________, the framework of the cell that provides its shape and regulates movement of organelles.
10. Basal bodies that organize the microtubules within cilia and flagella are derived from ________.

Answers to Objective Questions

1. resolving 2. protein, phospholipid 3. nucleus
4. ribosomes 5. hydrolytic 6. Golgi apparatus
7. mitochondria 8. chloroplasts 9. cytoskeleton
10. centrioles

Study Questions

1. Briefly define the cell theory. (p. 43)
2. Describe the structure and biochemical makeup of membrane. (p. 46)
3. Describe the nucleus and its contents, including the terms *DNA* and *RNA* in your description. (p. 46)
4. How does the experiment with *Acetabularia* illustrate the function of the nucleus in the cell? (p. 47)
5. Describe the structure and function of endoplasmic reticulum. Include the terms rough and smooth ER and ribosomes in your description. (p. 48)
6. Describe the structure and function of the Golgi apparatus. Mention vacuoles and lysosomes in your description. (p. 49)
7. Describe the structure and function of mitochrondria and chloroplasts. (pp. 51–52)
8. Describe the structure and function of microfilaments, microtubules, centrioles, cilia, and flagella. (p. 53)
9. What are the two main types of cells and how do they differ structurally? (p. 56)
10. What are the structural differences between animal and plant cells? (p. 56)

Figure 2.18
Anatomy of an animal cell. Various organelles are depicted in electron micrographs and appropriate drawings.

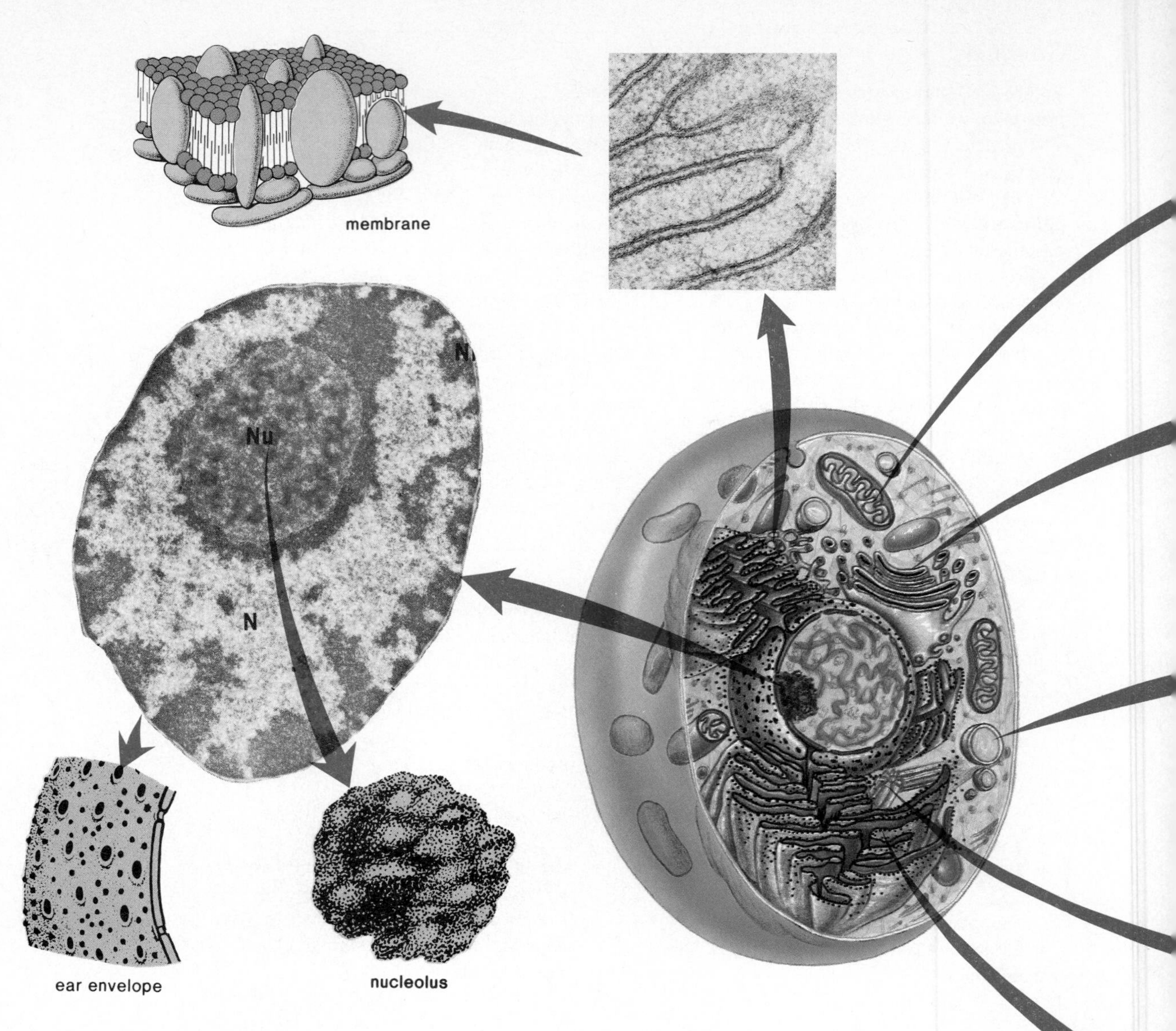

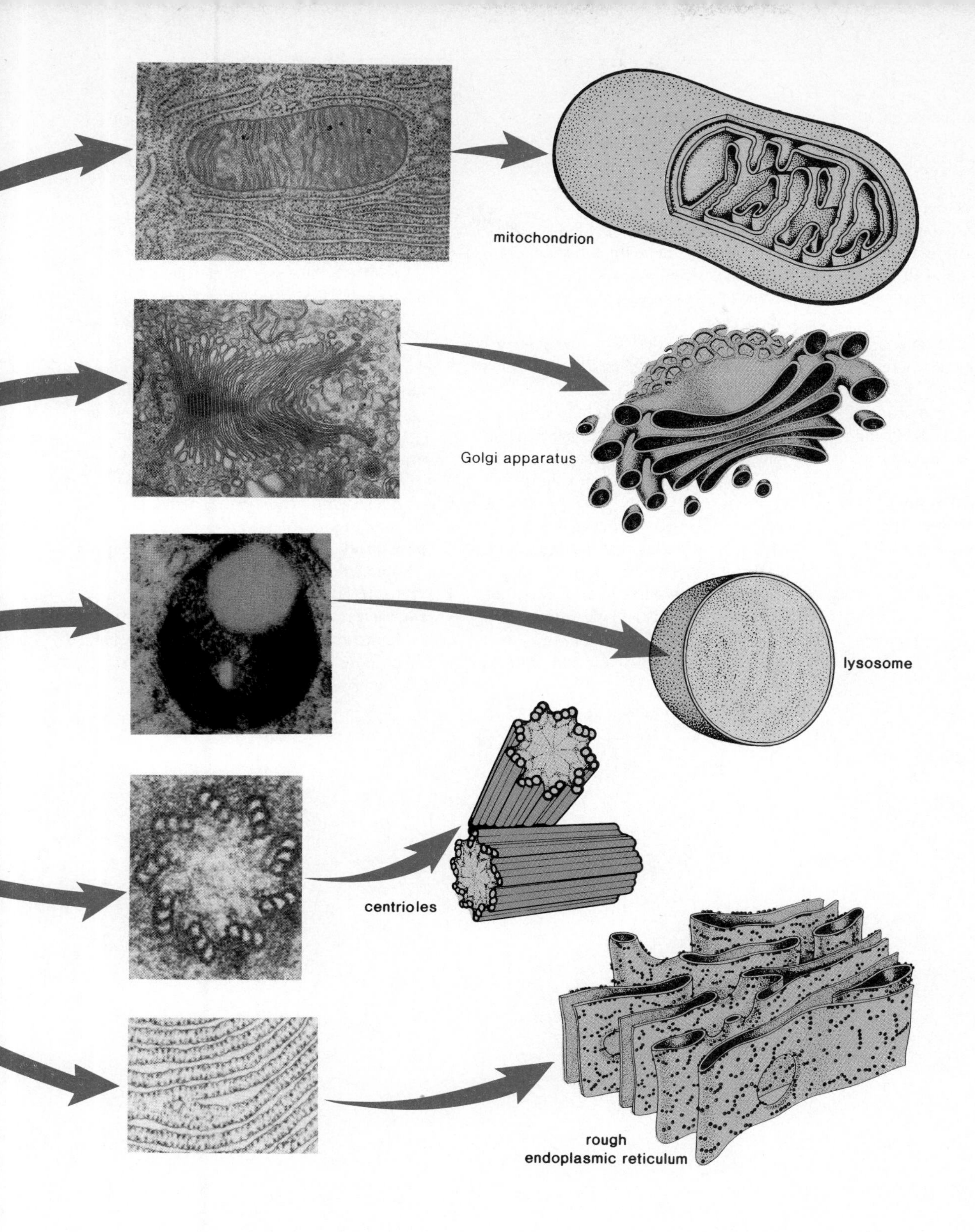
mitochondrion
Golgi apparatus
lysosome
centrioles
rough
endoplasmic reticulum

Key Terms

cell (sel) the structural and functional unit of an organism; the smallest structure capable of performing all the functions necessary for life. *43*

centriole (sen'tre-ōl) a short, cylindrical organelle in animal cells that contains microtubules in a 9 + 0 pattern and is associated with the formation of the spindle apparatus during cell division. *54*

chloroplast (klo'ro-plast) an organelle which contains chlorophyll and where photosynthesis takes place. *52*

chromosomes (kro'mo-sōmz) rod-shaped bodies in the nucleus, particularly during cell division, that contain the hereditary units or genes. *47*

cytoplasm (si'to-plazm'') the ground substance of cells located between the nucleus and the cell membrane. *43*

cytoskeleton (si''to-skel'ĕ-ton) filamentous protein structures found throughout the cytoplasm that help maintain the shape of the cell. *53*

endoplasmic reticulum (en-do-plaz'mic rĕ-tik'u-lum) a complex system of tubules, vesicles, and sacs in cells; sometimes having attached ribosomes. *48*

eukaryotic (u''kar-e-ot'ik) possessing the membranous organelles characteristic of complex cells. *43*

Golgi apparatus (gol'ge ap''ah-ra'tus) an organelle that consists of concentrically folded membranes and functions in the packaging and secretion of cellular products. *49*

lysosome (li'so-sōm) an organelle in which digestion takes place due to the action of hydrolytic enzymes. *50*

microfilament (mi''kro-fil'ah-ment) an extremely thin fiber found within the cytoplasm that is involved in the maintenance of cell shape and movement of cell contents. *53*

microtubule (mi''kro-tu'būl) an organelle composed of 13 rows of globular proteins; found in multiple units in several other organelles such as the centriole, cilia, and flagella. *54*

mitochondrion (mi''to-kon'dre-on) an organelle in which aerobic respiration produces the energy molecule, ATP. *51*

nucleolus (nu-kle'o-lus) an organelle found inside the nucleus; composed largely of RNA for ribosome formation. *47*

nucleus (nu'kle-us) a large organelle containing the chromosomes and acting as a control center for the cell. *46*

organelles (or''gan-elz') specialized structures within cells such as the nucleus, mitochondria, and endoplasmic reticulum. *43*

prokaryotic (pro''kar-e-ot'ik) lacking the organelles found in complex cells; bacteria including cyanobacteria. *43*

ribosomes (ri'bo-somz) minute particles, found attached to endoplasmic reticulum or loose in the cytoplasm, that are the site of protein synthesis. *49*

3

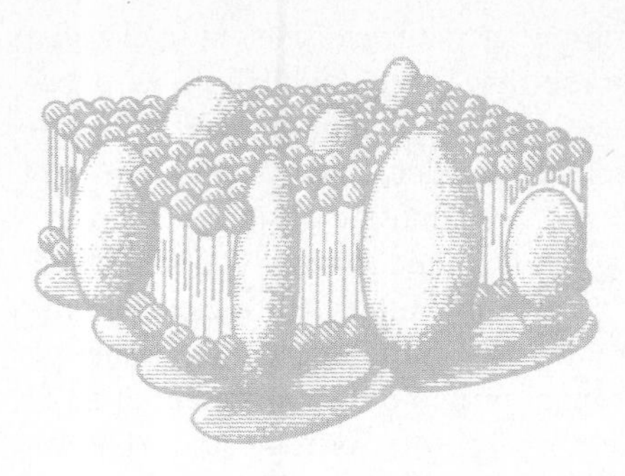

Cell Membrane and Cell Wall Function

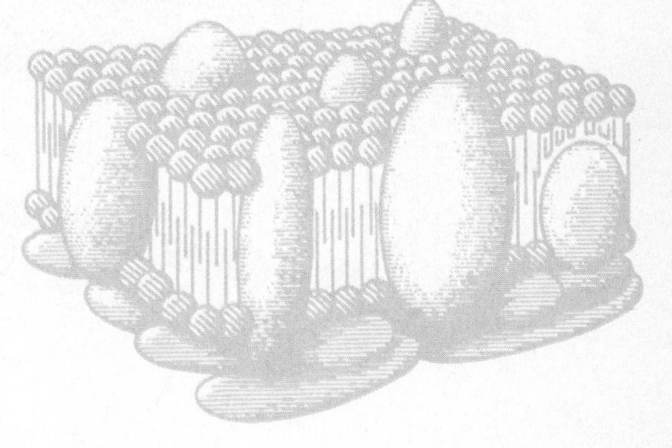

Chapter Concepts

1 The cell membrane is differentially permeable, allowing some substances to pass through freely, and restricting the passage of other substances.

2 Small molecules can pass through a cell membrane along a concentration gradient either by diffusion or facilitated transport, the latter requiring protein carriers.

3 Active transport, the passage of molecules against a concentration gradient, requires protein carriers and an expenditure of cellular energy.

4 Vesicle formation can take large molecules into the cell or discharge substances from the cell.

Chapter Outline

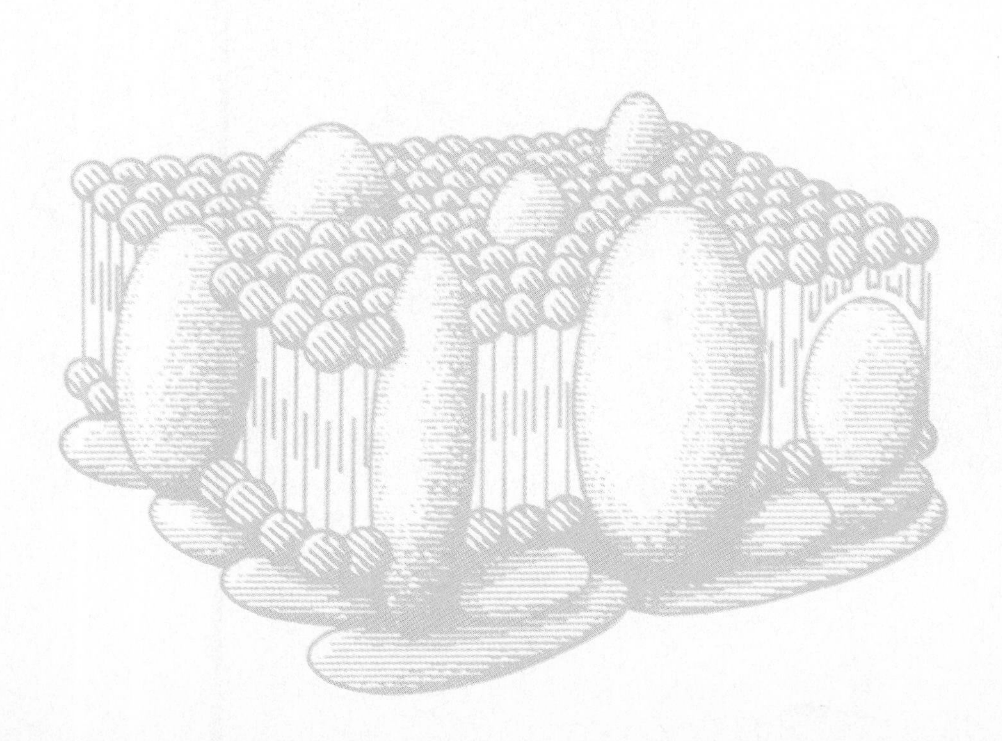

Outer Boundaries

Cell Membrane

The fluid mosaic model of membrane structure, discussed in chapter 2, also applies to the **cell membrane.** There is a double layer of phospholipid molecules, having the consistency of light oil, in which protein molecules are either partially or wholly embedded. Since the proteins are scattered throughout the membrane, they form a mosaic pattern. The presence of proteins in the membrane is revealed when the membranes of red blood cells are frozen and then fractured before they are subjected to electron microscopy (fig. 3.1). Different types of proteins are found in the membranes of different cells. Some of these are *carriers* that promote the movement of molecules into and out of the cell, and some are *receptors* for molecules that influence the metabolic activity of the cell.

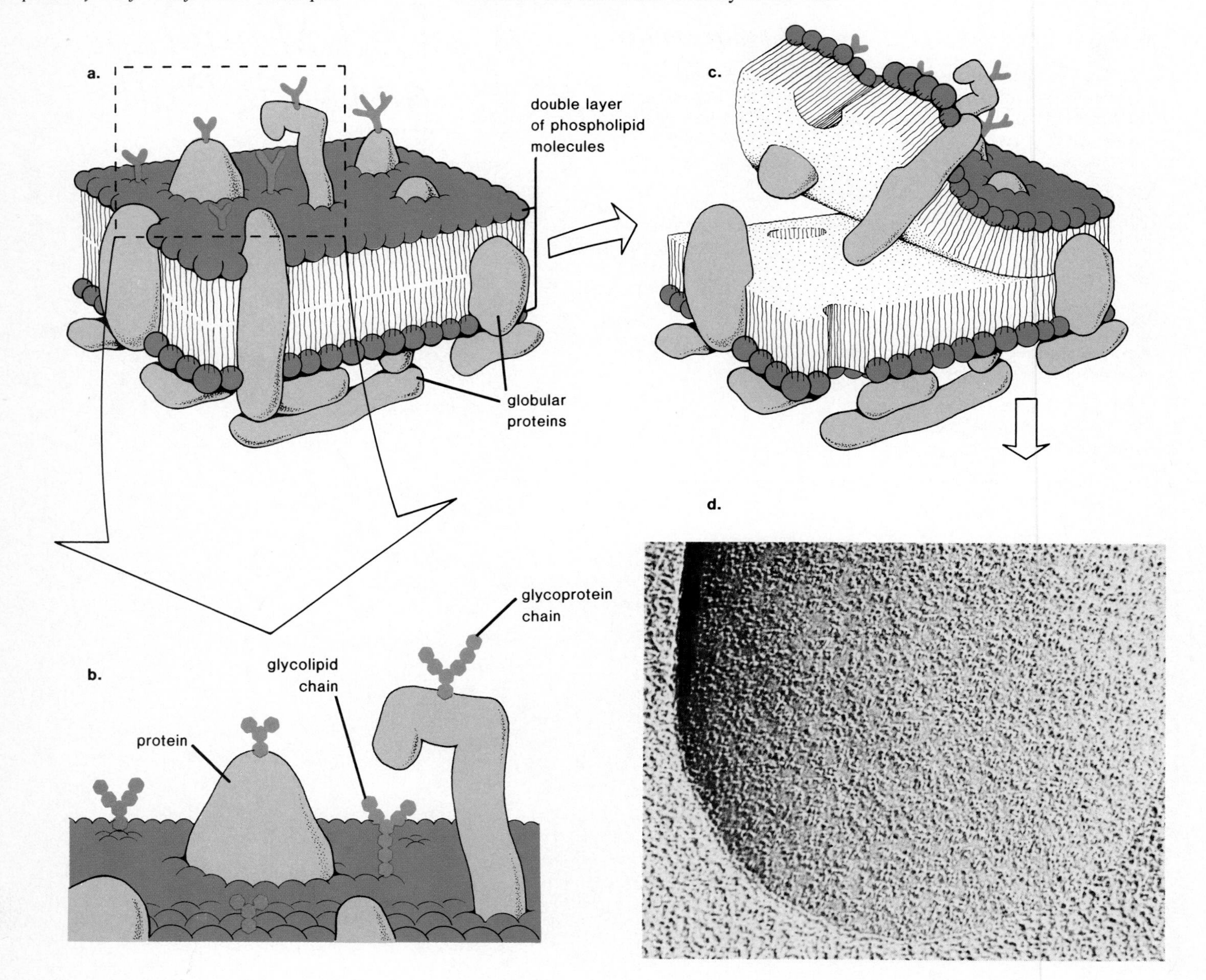

Figure 3.1
Fluid mosaic model of a cell membrane. a. *Protein molecules are embedded in and project to either side of a double layer of phospholipid molecules.* b. *Enlargement of a section of the membrane to show glycoproteins and glycolipids.* c. *A frozen cell that is fractured before being viewed by a scanning electron microscope reveals proteins that are embedded, along the fracture line, in the membrane.* d. *Electron micrograph of red blood cell membranes prepared by the freeze-fracture technique.*

The animal cell membrane, unlike the intracellular membrane discussed previously, also contains carbohydrates. Simple sugars are strung together in chains that are attached to proteins (glycoproteins) and lipids (glycolipids). The chains are always on the outside of the membrane and probably function as markers to identify the cell. Cells that are not recognized by the body are apt to be rejected, as sometimes happens when individuals receive organ transplants. On the other hand, cell recognition processes may be faulty when cancerous cells are permitted to grow and reproduce.

Plant Cell Wall

In addition to a cell membrane, plant cells are surrounded by a cell wall (fig. 3.2a) that varies in thickness depending on the function of the cell. All plant cells have a primary cell wall whose main constituent is cellulose molecules united into threadlike microfibrils. There is a sticky substance, called the middle lamella, that lies between adjacent plant cells and keeps them bound together. Some cells in woody plants have a secondary cell wall that forms inside the primary cell wall. The secondary wall has alternating layers of cellulose microfibrils and is reinforced by the addition of lignin, a substance that adds to their strength. Plant support cells have secondary walls; the cell dies and the strong wall remains as support material.

Humans make use of the cellulose from plant cell walls. Cotton, rayon, flax, hemp, paper, and even wood are derived from plant cell walls. Because lignin causes paper to turn yellow with age, it is usually removed during the manufacturing process. Today, the lignin is used in the production of various synthetic products such as rubber, plastics, pigments, and adhesives.

Animal cells are surrounded by a cell membrane whose structure is similar to intracellular membrane except that the surface has chains of sugar molecules attached extracellularly to lipid and protein molecules. Plant cells have a cell wall in addition to a cell membrane.

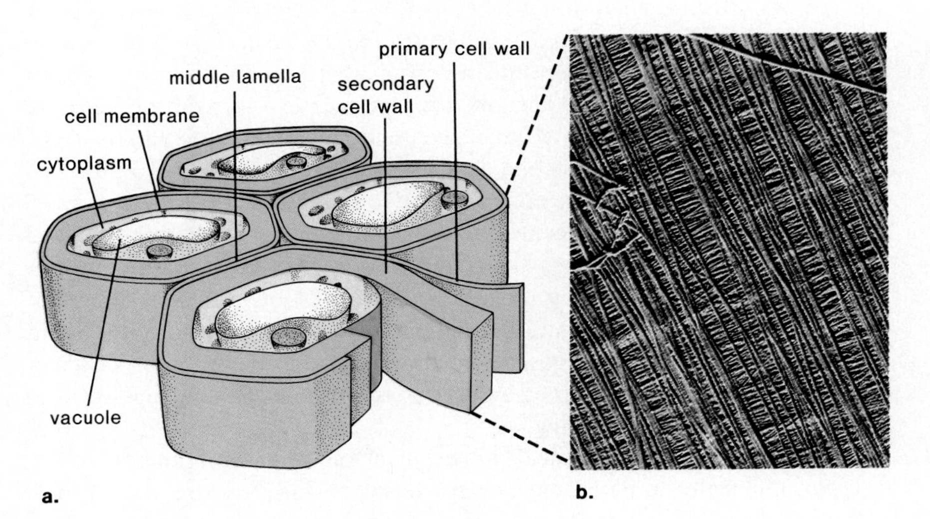

Figure 3.2
Plant cell wall. a. *All plant cells have a primary cell wall and some have a secondary cell wall.* b. *Electron micrograph of alternating layers of cellulose microfibrils that make up the cell wall.*

Table 3.1 Passage of Molecules into and out of Cells

Name	Direction	Requirements	Examples
Diffusion	Toward lesser concentration	———	Lipid-soluble molecules Water Gases
Transport			
Facilitated	Toward lesser concentration	Carrier	Sugars and amino acids
Active	Toward greater concentration	Carrier plus energy	Sugars, amino acids, and ions
Endocytosis and exocytosis			
Pinocytosis	Either	Vacuole formation plus energy	Macromolecules
Phagocytosis	Either	Vacuole formation plus energy	Cells or subcellular material

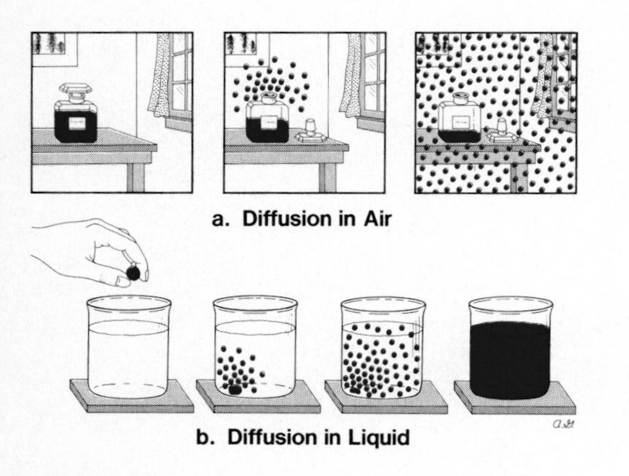

Figure 3.3
Diffusion occurs when a. *a perfume bottle is opened and the scent fills a room because the molecules have moved away from the bottle and* b. *a tablet of dye is placed in a beaker, and water becomes colored because the molecules have moved away from the original area of the tablet.*

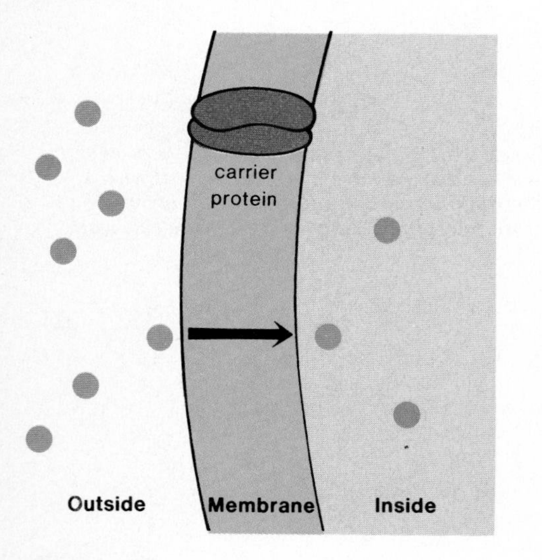

Figure 3.4
Certain molecules move freely across the membrane by means of diffusion. Notice that a carrier protein is not utilized for diffusion.

Permeability

The plant cell wall is freely permeable but the cell membrane in both animal and plant cells is not and instead it regulates the entrance and exit of molecules into and out of the cell.

Sometimes the cell membrane is said to be semipermeable. A permeable membrane would allow all molecules to pass through; an impermeable membrane would allow no molecules to pass through; and a semipermeable membrane allows some molecules to pass through. Certain small molecules can cross a cell membrane, while large molecules cannot cross a membrane. However, some small molecules pass through the cell membrane quickly, while others have difficulty in passing through or fail to pass through at all. Therefore, a cell membrane is often regarded as selectively or differentially permeable rather than semipermeable.

As listed in table 3.1, there are three general means by which substances can enter and exit cells from the surrounding medium: (1) diffusion, (2) transport by carriers, and (3) endocytosis and exocytosis.

Diffusion

Diffusion is a physical process that can be observed with any type of particle. Diffusion is such a universal phenomenon that there is a physical law called the law of diffusion, which states that *particles move from the area of greater concentration to the area of lesser concentration until equally distributed.* To illustrate diffusion, imagine opening a perfume bottle in the corner of a room (fig. 3.3a). The smell of the perfume soon permeates the room because the molecules that make up the perfume have randomly drifted to all parts of the room. Another example is putting a tablet of dye into water (fig. 3.3b). The water eventually takes on the color of the dye as the tablet dissolves.

The movement of molecules by diffusion alone is a slow process. The rate of diffusion is affected by the **concentration gradient** (the difference in concentration of the diffusing molecules between the two regions), the size and shape of the molecule, and the temperature. Also, diffusion in a liquid medium is slower than in a gaseous medium; however, distribution of molecules in cytoplasm is sped up by an ever-constant flow of the cytoplasm that is called cytoplasmic streaming.

The chemical and physical properties of the cell membrane allow few types of molecules to enter and exit by means of diffusion (fig. 3.4). Lipid-soluble molecules, such as alcohols, can diffuse through the membrane simply because phospholipids are the membrane's main structural components.

Gases can also diffuse through the membrane; this is the mechanism by which oxygen enters cells and carbon dioxide exits cells. As an example, consider the movement of oxygen from the air sacs (alveoli) of the lungs to the blood in lung capillaries (fig. 3.5). After inspiration (breathing in), the concentration of oxygen in the alveoli is greater than the concentration of oxygen in the blood; therefore oxygen diffuses into the blood.

Water passes into and out of cells with relative ease. Since water is not lipid soluble, it has been hypothesized that the membrane contains protein-lined pores large enough to allow the passage of water and perhaps some ions. Other molecules cannot utilize these pores because they are either too large or they carry a charge that prevents passage. The fact that water can penetrate a membrane has important biological consequences, as described in the discussion that follows.

The cell membrane is semipermeable, or preferably differentially permeable. A few types of small molecules can diffuse through the cell membrane from the area of greater concentration to the area of lesser concentration.

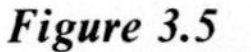

Figure 3.5
Oxygen (dots) *diffuses into the capillaries of the lungs because there is a greater concentration of oxygen in the air sacs than in the capillaries.*

Osmosis

The diffusion of water across a differentially permeable membrane has been given a special term: it is called **osmosis.** Osmosis is defined as the *net movement of water molecules from the area of greater concentration of water to the area of lesser concentration of water across a differentially permeable membrane.*

To illustrate osmosis, a thistle tube containing a protein solution and covered at one end by a membrane is placed in a beaker of distilled water. Inside the tube is a solution containing both a **solute** (particles) and a **solvent** (water), while outside the tube is pure water (fig. 3.6a). Obviously, the lesser solute and greater concentration of water is outside the tube, and therefore there will be a net movement of water from outside the tube to inside the tube across the membrane. The solute (protein) is unable to pass out of the tube because the membrane is impermeable to it; therefore the solution within the tube rises (fig. 3.6b).

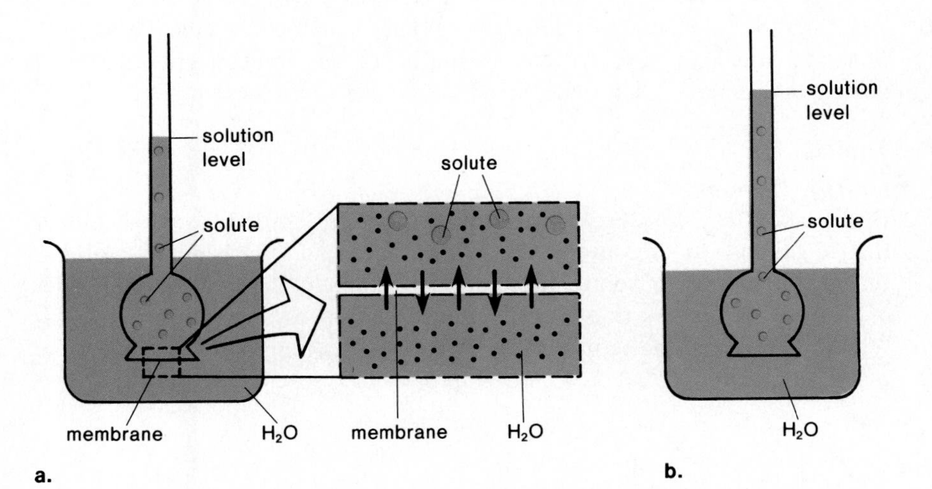

Figure 3.6
Osmosis demonstration. a. *A thistle tube, covered at the broad end by a membrane, contains a solute* (large circles) *in addition to a solvent* (small circles). *The beaker contains only solvent. The solute is unable to pass through the membrane, but the solvent passes through in both directions.*
b. *There is a net movement of solvent toward the inside of the thistle tube. This causes the solution to rise in the thistle tube until a back pressure develops that prevents any further net gain of solvent.*

Table 3.2 Effect of Osmosis on Cell

Tonicity of Solution	Concentrations Solute	Water	Net Movement of Water	Effect on Cell
Isotonic	Same as cell	Same as cell	———	———
Hypotonic	Less than cell	More than cell	Cell gains water	Swells, turgor pressure
Hypertonic	More than cell	Less than cell	Cell loses water	Shrinks, plasmolysis

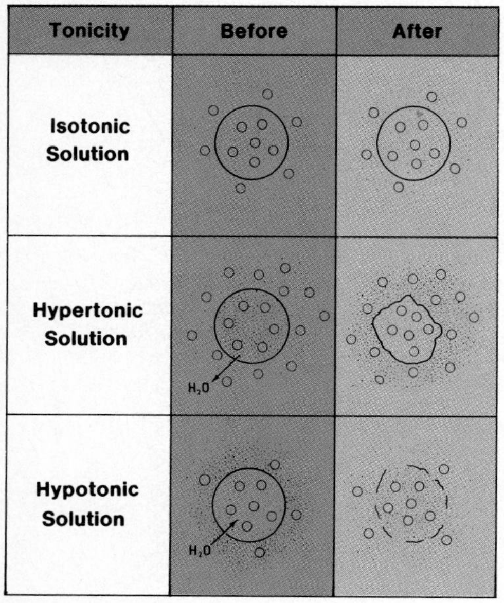

The pressure due to the flow of water from the area of greater concentration to the area of lesser concentration is called **osmotic pressure.** In this case, osmotic pressure will eventually be counterbalanced by a hydrostatic pressure (pressure exerted by liquids) caused by the height of the solution. When hydrostatic pressure equals osmotic pressure, equilibrium is reached, the net flow of water ceases, and the solution rises no higher in figure 3.6b because as many water molecules now exit the tube as enter the tube.

Notice that in this illustration of osmosis

1. a membrane separates a solution from pure water;
2. a difference in solute and water concentrations exists on the two sides of the membrane;
3. the membrane is impermeable to the solute particles, which therefore do not move through the membrane;
4. the membrane is permeable to water, which therefore moves from the area of greater concentration to the area of lesser concentration;
5. due to the process of osmosis or, if you prefer, due to osmotic pressure, the amount of liquid increases on the solution side of the membrane.

These considerations will be important as we discuss osmosis in relation to cells placed in different solutions. Recall that solutions are made up of two parts: the solute and the solvent. The solute is the solid substance dissolved or suspended in the solvent, which is usually water. The concentrations of solutions are usually described in terms of percentages of solute; for example, a 10 percent salt solution. This solution would contain 90 percent water.

Cells may be placed in solutions that contain the same number of solute molecules per volume, a greater number of solute molecules per volume, or a lesser number of solute molecules per volume than does the cell. These solutions are called isotonic, hypertonic, and hypotonic, respectively. Figure 3.7 depicts and table 3.2 describes the effects of these solutions on cells.

Tonicity	Before	After
Isotonic Solution		
Hypertonic Solution	H_2O	
Hypotonic Solution	H_2O	

Figure 3.7
Effect of tonicity on an animal cell. In an isotonic solution, there is no net movement of water and the appearance of a red blood cell remains the same. In a hypertonic solution, there is a net movement of water (arrow) *to the outside of the cell and the cell shrinks. In a hypotonic solution, there is a net movement of water* (arrow) *to inside the cell and the cell swells to bursting. (Circles = solute; stipples = solvent)*

Tonicity

Isotonic Solutions

In the laboratory, cells are normally placed in solutions that cause them neither to gain nor to lose water. Such solutions are said to be **isotonic solutions;** that is, the number of solute molecules per volume is the same on both sides of the membrane, and thus there is no net gain or loss. The term *iso* means "the same as" and the term *tonicity* refers to the strength of the solution.

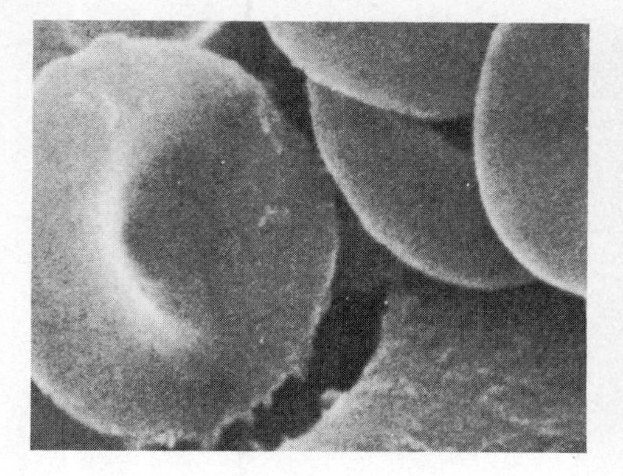

Figure 3.8
Scanning electron micrograph of normal-appearing red blood cells. If red blood cells are placed in an isotonic solution, they have this appearance.

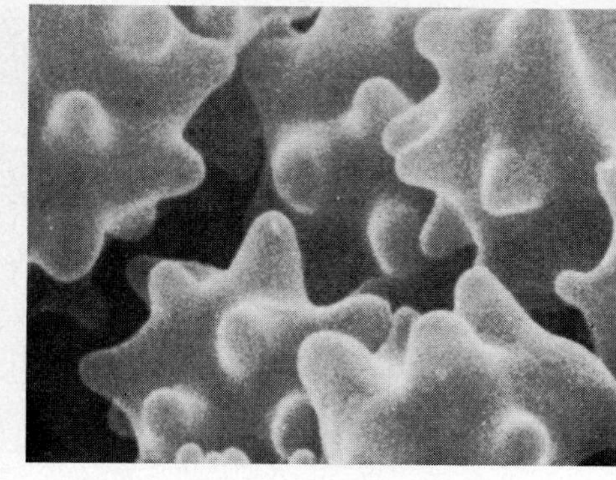

Figure 3.9
Scanning electron micrograph of red blood cells that were placed in a hypertonic solution. The cells are crenated due to the loss of water.

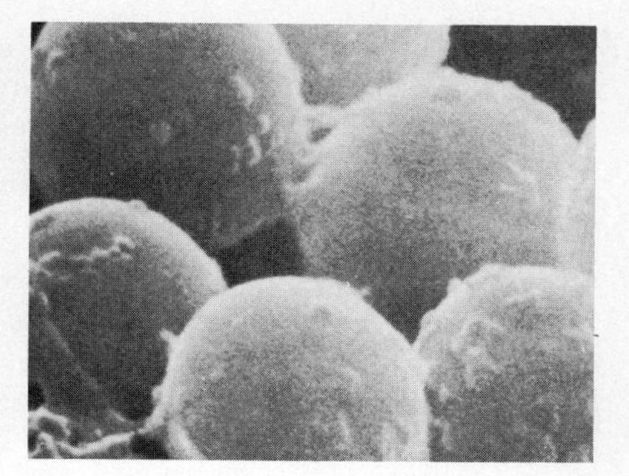

Figure 3.10
Scanning electron micrograph of red blood cells that were placed in a hypotonic solution. The cells are swollen due to the uptake of water.

It is possible to determine, for example, that a 0.9 percent solution of the salt sodium chloride (Na^+Cl^-) is isotonic to red blood cells because the cells neither swell nor shrink when placed in such a solution (fig. 3.8).

Hypertonic Solutions

Solutions that cause cells to shrink or shrivel due to a loss of water are said to be **hypertonic solutions.** The prefix *hyper* means "greater than" and refers to a solution with a *greater concentration of solute* (lesser concentration of water) than the cell. If a cell is placed in a hypertonic solution, water will leave the cell—the net movement of water is from the inside to the outside of the cell (fig. 3.7).

A 10 percent solution of sodium chloride is hypertonic to red blood cells. In fact, any solution with a concentration higher than 0.9 percent NaCl is hypertonic to red blood cells. If red blood cells are placed in this solution, they will shrink (fig. 3.9). The term *crenated* is used to refer to red cells in this condition.

Hypotonic Solutions

Solutions that cause cells to swell or even burst due to an intake of water are said to be **hypotonic solutions.** The prefix *hypo* means "less than" and refers to a solution with a *lesser concentration of solute* (greater concentration of water) than the cell. If a cell is placed in a hypotonic solution, water will enter the cell—the net movement of water is from the outside to inside of the cell (fig. 3.7).

Any salt solution less than 0.9 percent is a hypotonic solution to red blood cells. Red blood cells placed in such a solution will expand (fig. 3.10) and even burst due to the buildup of osmotic pressure. The term *lysis* is used to refer to disrupted cells.

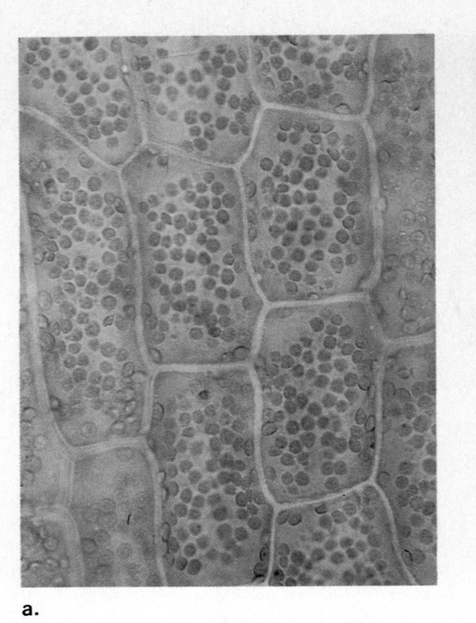

a.

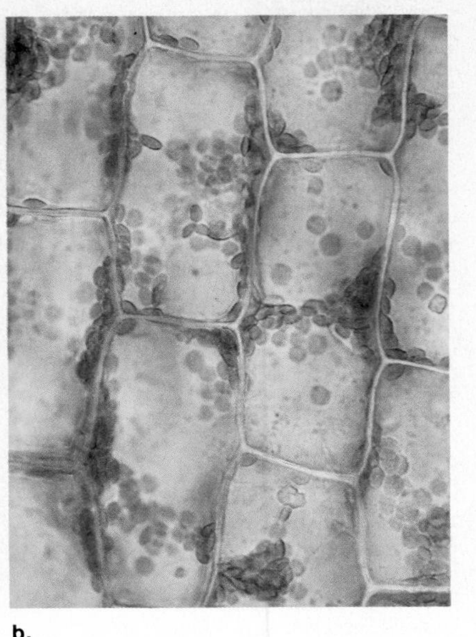

b.

Figure 3.11
Response of plant cells to varying tonicity. a. *In a hypertonic solution, plasmolysis is obvious because the cell contents (note chloroplasts) have pulled away from the cell wall.* b. *In a hypotonic solution, turgor pressure is obvious because the cell contents (note chloroplasts) are next to the cell wall.*

Tonicity	Before	After
Isotonic Solution	cell wall; water vacuole; cell membrane	
Hypertonic Solution	H_2O	cell membrane
Hypotonic Solution	H_2O	vacuole

Figure 3.12
Effect of tonicity on a plant cell. In an isotonic solution, the appearance of a plant cell remains the same because there is no net movement of water. In a hypertonic solution, the cell membrane withdraws from the cell wall as the plant vacuole shrinks because of a net movement of water to outside the cell (arrow). *In a hypotonic solution, the cell membrane is pressed up against the cell wall and the vacuole expands due to a net movement of water to inside the cell* (arrow).

Plant Cells

The effect of solutions of various tonicity is easily observed by viewing plant cells under the microscope. A hypertonic solution brings about **plasmolysis,** shrinking of cytoplasm due to osmosis. When a plant cell is placed in a hypertonic solution, we can observe the shrinking of the cytoplasm because the large central vacuole loses water, and the cell membrane pulls away from the wall (figs. 3.11a and 3.12).

A hypotonic solution brings about **turgor pressure,** due to swelling of the cell. When a plant cell is placed in a hypotonic solution, we can observe expansion of the cytoplasm because the large central vacuole gains water, and the cell membrane pushes against the rigid cell wall (figs. 3.11b and 3.12). The hydrostatic pressure increases until it is equal to the osmotic pressure; the plant cell does not burst because the cell wall does not give way. Turgor pressure in plant cells is extremely important to maintaining the erect position of the plant.

Importance

Osmosis occurs constantly in living organisms. For example, due to osmosis, water is absorbed from the human large intestine, retained by the kidneys, and taken up by the blood. Since living things contain a very high percentage of water, osmosis is an extremely important physical process for their continued good health.

Osmosis is the diffusion of water across a cell membrane. When a cell is placed in an isotonic solution, it neither gains nor loses water. When a cell is placed in a hypertonic solution (greater solute concentration than isotonic), the cell loses water and the cytoplasm shrinks. When a cell is placed in a hypotonic solution (less solute concentration than isotonic) the cell gains water.

Transport by Carriers

Facilitated Transport

The concept of **facilitated transport** explains the passage of molecules, such as glucose and amino acids, which are observed to cross the cell membrane even though they are lipid insoluble. There is evidence that the passage of these molecules is facilitated by a reversible combination with proteins that in some manner transport the smaller molecules through the cell membrane. These proteins, called **carriers,** are highly specific and combine only with one particular type of molecule. For example, various sugar molecules of identical size might be present inside or outside the cell, but certain ones cross the membrane hundreds of times faster than others. As stated earlier, this is the reason that the membrane may be called differentially permeable.

A model of a facilitated transport system (fig. 3.13) shows that after a carrier has assisted the movement of a molecule to the other side of the membrane, it is free to assist the passage of other similar molecules. Neither simple diffusion, explained previously, nor facilitated transport by means of protein carriers requires an expenditure of energy by the cell because the molecules are moving down a concentration gradient in the same direction they would tend to move anyway. Sometimes, therefore, facilitated transport is called facilitated diffusion, or passive transport. In passive transport, a substance moves only from the area of higher concentration to the area of lower concentration.

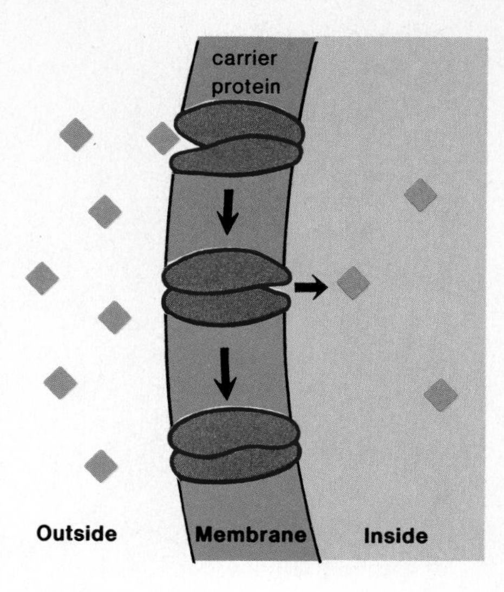

Figure 3.13
Facilitated transport is apparent when certain molecules easily cross a cell membrane toward the area of lesser concentration. A protein carrier is presumed to transport these molecules across the membrane.

Active Transport

Due to **active transport,** molecules or ions move through the cell membrane, accumulating either inside or outside the cell in the region of *higher* concentration. For example, iodine collects in the cells of the thyroid gland; sugar is completely absorbed from the gut by the cells lining the digestive tract; and sodium (Na^+) is sometimes almost completely withdrawn from urine by cells lining the kidney tubules.

Both protein carriers and an expenditure of energy (fig. 3.14) are needed to transport substances from an area of lesser concentration to an area of greater concentration. In this case, energy (ATP molecules) is required to cause the carrier to combine with the substance to be transported. Therefore, it is not surprising that cells primarily involved in active transport, such as kidney cells, have a large number of mitochondria near the membrane where active transport is occurring (fig. 14.13).

One type of active transport present in all cells, but especially important in nerve and muscle cells, is the active accumulation of sodium ions outside the cell and a corresponding accumulation of potassium ions within the cell. These two events are presumed to be linked, and a *sodium/potassium pump* has been hypothesized. There is evidence that the "pump" is a protein molecule capable of combining with both sodium and potassium ions. Internally the protein combines with sodium, assisting its passage to the outside of the cell; externally the protein combines with potassium, assisting its passage to the inside of the cell. Presumably the shape of the carrier changes so that it can alternately move sodium to the outside of the cell and move potassium to the inside of the cell (fig. 3.15). ATP energy released by an ATPase (an enzyme that breaks down ATP) is believed to be necessary to bring about the necessary change in shape.

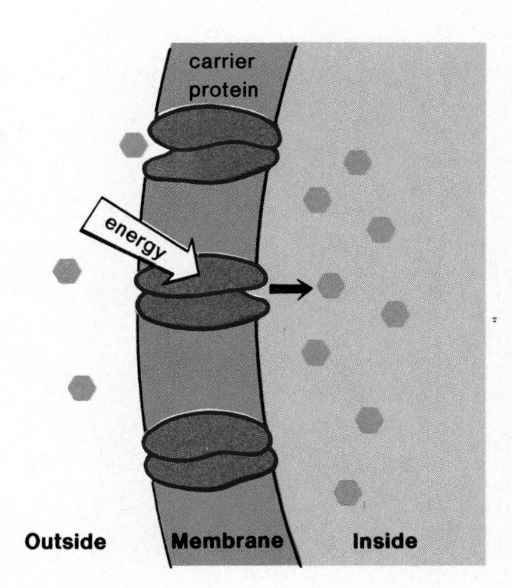

Figure 3.14
Active transport is apparent when a molecule crosses the cell membrane toward the area of greater concentration. An expenditure of ATP energy is required, presumably to allow a protein carrier to transport molecules across the cell membrane.

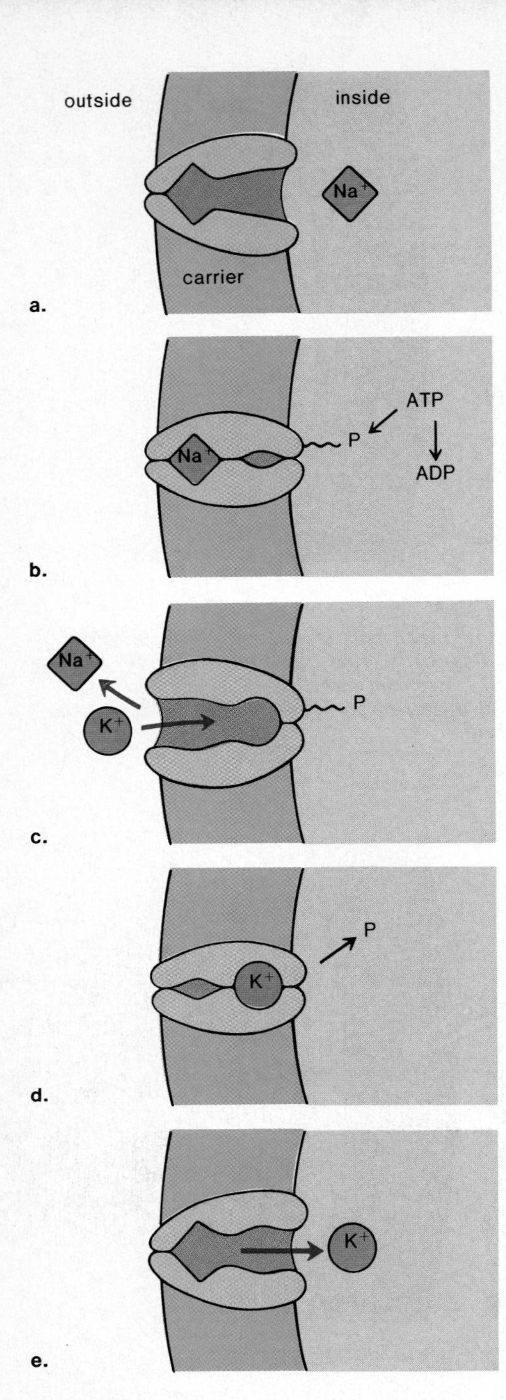

Figure 3.15
Sodium/potassium pump moves sodium (Na^+) to outside the cell and potassium (K^+) to inside the cell by means of the same type protein carrier.
a. *Presumably, after Na^+ is taken up by the carrier,* b. *a phosphate group (P) attaches to the carrier as ATP splits.* c. *This changes the shape of the carrier so that Na^+ is released and K^+ is taken up.* d. *After the phosphate group (P) is released* e. *so is the K^+.*

The passage of salt (Na^+Cl^-) across a cell membrane is of primary importance in cells. The chloride ion (Cl^-) does not cross the cell membrane unless it is attracted by positive charges. Once sodium (Na^+) has been pumped across a membrane, chloride follows passively through channels that allow its passage. As the reading on page 72 indicates, it is hypothesized that chloride channels in persons with cystic fibrosis malfunction, leading to the symptoms of this inherited (genetic) disease.

Many molecules are transported across the cell membrane by protein carriers. During facilitated transport (no energy required), small molecules follow their concentration gradients. During active transport (energy required), small molecules go against their concentration gradients.

Endocytosis and Exocytosis

Endocytosis

At times, large molecules or other matter become incorporated into cells by the process of **endocytosis** (fig. 3.16a), which requires the formation of a vesicle. Endocytosis, even when not moving substances toward a greater concentration, requires energy.

When the material taken in by the process is quite large, the process is called **phagocytosis** (cell eating). Phagocytosis is common to amoeboid-type cells, such as macrophages, a large phagocytic cell found in humans. These cells phagocytize bacteria and worn-out red cells, for example (fig. 3.17).

Vesicles also form around large-sized molecules such as proteins; this is called **pinocytosis** (cell drinking). Whereas phagocytosis can be seen with the light microscope, pinocytosis requires the use of the electron microscope.

Once formed, vesicles or vacuoles contain a substance enclosed by membrane. In order that these substances might be broken down and incorporated into the cytoplasm, digestion is required. Therefore, it is believed that lysosomes probably fuse with these bodies in order that digestive enzymes may begin to break down the molecules they contain.

Exocytosis

Exocytosis (fig. 3.16b) is the reverse of endocytosis and requires that a vesicle fuse with the membrane, thereby discharging its contents. As we saw in chapter 2, vesicles formed at the Golgi apparatus transport cell products out of the cell; this entire process is called *secretion.* Also, residues remaining after digestion by lysosome enzymes may be discharged from the cell by fusion of a vesicle with the cell membrane.

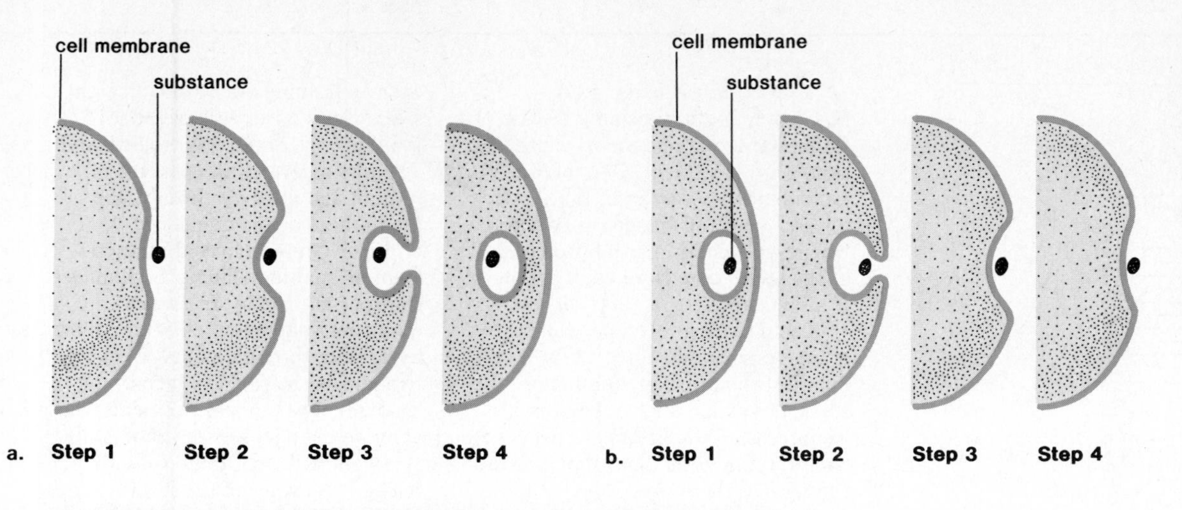

Figure 3.16
a. *Endocytosis. In both phagocytosis and pinocytosis, the cell membrane forms a vesicle around the substance to be taken in.* b. *Exocytosis. A substance within a vacuole is deposited outside the cell when the vesicle fuses with the cell membrane.*

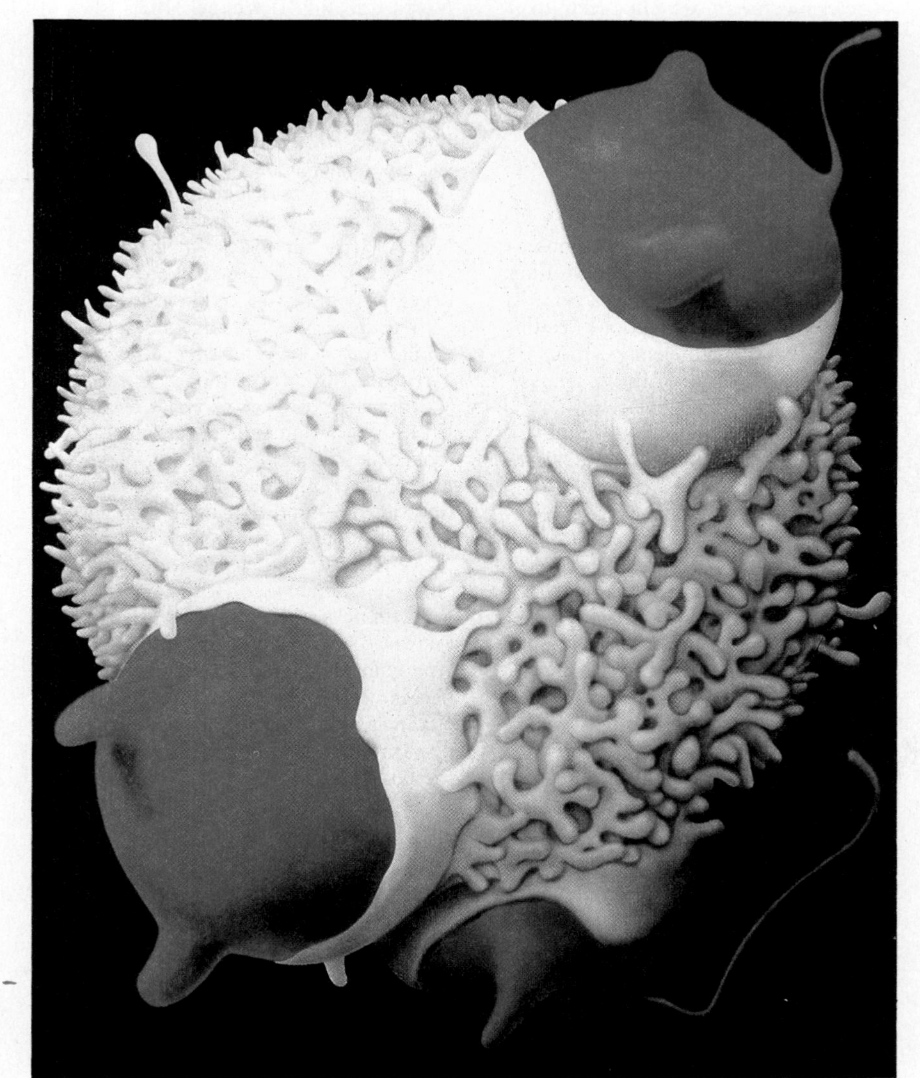

Figure 3.17
Macrophages are the body's scavengers. These "big eaters" take in (phagocytize) all sorts of debris, including worn out red cells, as shown here.

Cystic Fibrosis

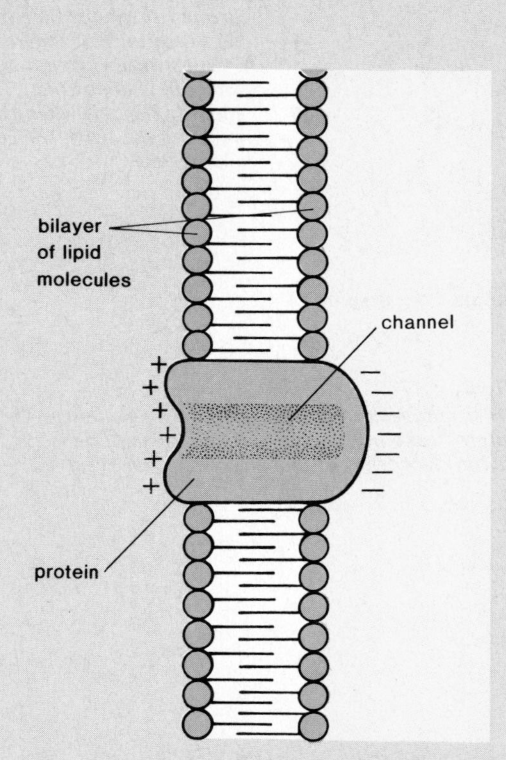

Normally chloride is repelled from moving through a cellular channel because it is negatively charged. When sodium is pumped to one side, chloride is attracted to the positive charges and does flow through the channel.

Cystic fibrosis is the most common fatal genetic disease of Caucasians, and one for which there is no good treatment. One in 20 individuals is a carrier of the cystic fibrosis gene and the disease occurs whenever a child inherits two copies of the gene—one from each parent. It afflicts about 1 in 1800 whites.

The disease affects many different organ systems, but three of its manifestations are predominant. First, the patients have lung disease. As infants, cystic fibrosis patients get bacterial lung infections that persist throughout their lives. "Once the bacteria come in, we can't seem to kill them even with high doses of antibiotics," says Pamela Davis of Case Western Reserve University. Along with the lung infections, the patients secrete thick mucus that clogs the airways of their lungs. These two conditions lead to the breakdown of the walls of their airways. Eventually, most patients die of respiratory failure. In addition, the ducts of their pancreases get clogged by viscous secretions, leading to pancreatic insufficiency. This results in too few digestive enzymes in the gut, leaving the patients vulnerable to malnutrition. Finally, the patients have high concentrations of salt in their sweat—a condition used to diagnose the disease.

Three groups of investigators have suggested that it [cystic fibrosis gene] might be a gene affecting chloride ion channels in cell membranes. The channels are the way that chloride enters and leaves cells—a process which seems to be completely passive, requiring no energy.

The most extensive evidence for a chloride channel defect has been reported by Paul Quinton of the University of California at Riverside. Quinton looked at sweat ducts, which he gets from small pieces of skin, each containing about four to eight sweat ducts. (People have about 2 to 5 million sweat ducts on their body.) Then he measures the passage of sodium and chloride through the membranes of the ducts.

First, Quinton looked at ducts from normal individuals. Sodium is transported first by an energy requiring system and chloride follows by passing through a channel. The two ions both cross the membrane easily. In cystic fibrosis patients, the chloride channel was not functioning.

In addition to Quinton's finding, Richard Boucher of the University of North Carolina at Chapel Hill reported that the chloride flux in the nasal epithelia of cystic fibrosis patients is half that of controls. He removed nasal polyps from cystic fibrosis patients and cultured them in the laboratory before looking at how they transport chloride ions. As controls, he used polyps cut from the noses of hypoallergic hay-fever sufferers, who also are prone to develop these benign tissue growths.

The third piece of evidence implicating chloride channels in cystic fibrosis is from Jonathan Widdicombe of the University of California in San Francisco. He looked at tracheal cells from a person who had recently died from cystic fibrosis and found that their transport of chloride ions was blocked.

What is striking about the chloride channel results is that they were found in three different tissues known to be affected by the disease. And chloride channels are so very basic to cell functioning that blocked channels may cause multiple and disparate effects in different parts of the body.

Summary

Plant cells are surrounded by a cell wall in addition to a cell membrane. Substances cross cell membranes by diffusion, transport by carriers, and vesicle formation. The diffusion of water across a membrane is called osmosis. When a cell is placed in an isotonic solution, there is no net gain or loss of water. In a hypertonic solution, cells shrink and in a hypotonic solution, cells swell. The presence of a cell wall makes it possible to observe plasmolysis and turgor pressure when plant cells are placed in a hypertonic and hypotonic solution, respectively.

Facilitated transport moves molecules toward a lesser concentration while active transport, an energy requiring process, moves molecules toward a greater concentration. Vesicle formation during endocytosis permits molecules (pinocytosis) and debris (phagocytosis) to be taken into cells. Exocytosis describes the process of secretion.

Objective Questions

1. Both plant and animal cells have a cell membrane, but in addition plant cells have a cell __________ .
2. Molecules diffuse from the area of __________ concentration to the area of __________ concentration.
3. In a hypertonic solution, cells __________ water and the cell contents __________ .
4. When plant cells are placed in a hypotonic solution, __________ is obvious because cell contents __________ against the cell wall.
5. Glucose and amino acids utilize __________ to cross the cell membrane during __________ transport.
6. During __________ transport, a carrier moves molecules from the area of __________ concentration to the area of __________ concentration.
7. Sodium and potassium ions move across the membrane in opposite directions due to the action of the __________ .
8. During endocytosis, __________ formation takes substances into the cell.

Answers to Objective Questions

1. wall 2. greater, lesser 3. lose, shrink 4. turgor pressure, press 5. carriers, facilitated 6. active, lesser, greater 7. sodium/potassium pump 8. vacuole

Study Questions

1. How does the cell membrane differ from intracellular membrane? (p. 62)
2. Why can a cell membrane be called semipermeable? differentially permeable? (p. 64)
3. What are the three mechanisms by which substances enter and exit cells? (p. 64)
4. Define diffusion and give an example. (p. 64)
5. Define osmosis. (p. 65) Define isotonic, hypertonic, and hypotonic solutions, and give examples of these concentrations for red blood cells. (p. 67)
6. Draw a simplified diagram of a red blood cell before and after being placed in these solutions. What terms are used to refer to the condition of the red blood cell in a hypertonic and hypotonic solution? (pp. 66–67)
7. Draw a simplified diagram of a plant cell before and after being placed in these solutions. Describe the cell contents under these conditions. (p. 68)
8. How does facilitated transport differ from simple diffusion across the cell membrane? (p. 69)
9. How does active transport differ from facilitated transport? Give an example. (p. 69)
10. Draw diagrams that show endocytosis and exocytosis. Give an example for each of these. (p. 70)

Key Terms

active transport (ak'tiv trans'port) transfer of a substance into or out of a cell from lesser to greater concentration by a process that requires a carrier and expenditure of energy. *69*

carrier (kar'e-er) a molecule that combines with a substance and transports it through the cell membrane. *69*

cell membrane (sel mem'brān) a membrane that surrounds the cytoplasm of cells and regulates the passage of molecules into and out of the cell. *62*

diffusion (dī-fu'zhun) the movement of molecules from an area of greater concentration to an area of lesser concentration. *64*

endocytosis (en''do-si-to'sis) a process in which a vesicle is formed at the cell membrane to bring a substance into the cell. *70*

exocytosis (eks''o-si-to'sis) a process in which an intracellular vesicle fuses with the cell membrane so that the vesicle's contents are released outside the cell. *70*

facilitated transport (fah-sil'ĭ-tāt-ed trans'port) transfer of a substance into or out of a cell along a concentration gradient by a process that requires a carrier. *69*

hypertonic solution (hi''per-ton'ik so-lu'shun) one that has a greater concentration of solute, a lesser concentration of water than the cell. *67*

hypotonic solution (hi''po-ton'ik so-lu'shun) one that has a lesser concentration of solute, a greater concentration of water than the cell. *67*

isotonic solution (i''so-ton'ik so-lu'shun) one that contains the same concentration of solute and water as does the cell. *66*

osmosis (oz-mo'sis) the movement of water from an area of greater concentration of water to an area of lesser concentration of water across a semipermeable membrane. *65*

osmotic pressure (oz-mot'ik presh'ur) pressure generated by and due to the osmotic flow of water. *66*

phagocytosis (fag''o-si-to'sis) the taking in of bacteria and/or debris by engulfing; cell eating. *70*

pinocytosis (pin''o-si-to'sis) the taking in of fluid along with dissolved solutes by engulfing; cell drinking. *70*

plasmolysis (plaz-mol'ĭ-sis) contraction of the cell contents due to the loss of water. *68*

solute (sol'ūt) a substance dissolved in a solvent to form a solution. *65*

solvent (sol'vent) a fluid such as water that dissolves solutes. *65*

turgor pressure (tur'gor presh'ur) osmotic pressure that adds to the strength of the cell. 68

4

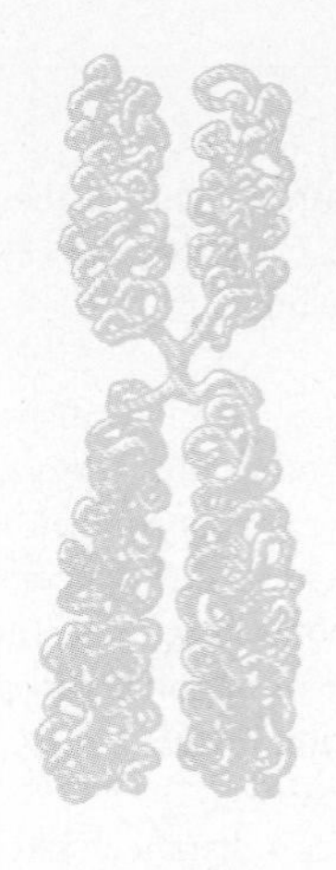

Cell Division

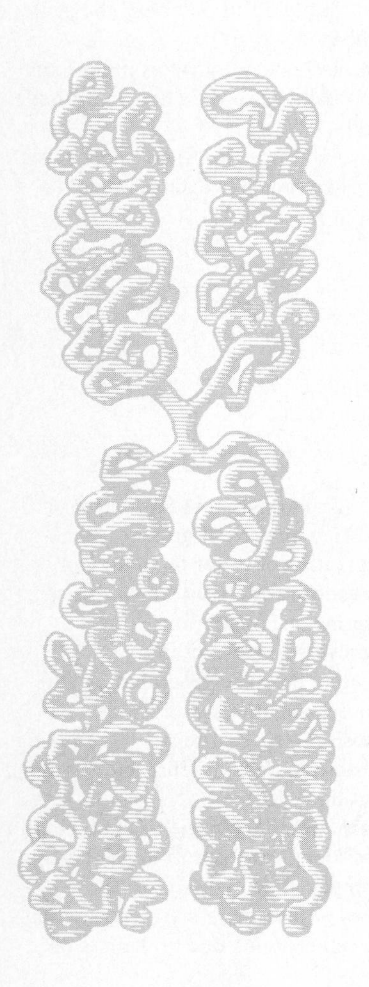

Chapter Concepts

1 Cell division includes the division of both the cytoplasm and the nucleus.

2 The nucleus contains the gene-bearing chromosomes. Ordinary cell division assures that each new cell will receive a full complement of chromosomes and genes.

3 In animals, reduction division is required to produce the sex cells, which contain one-half the full number of chromosomes.

4 In plants, reduction division produces spores, structures that precede the formation of the sex cells in the life cycle of these organisms.

Chapter Outline

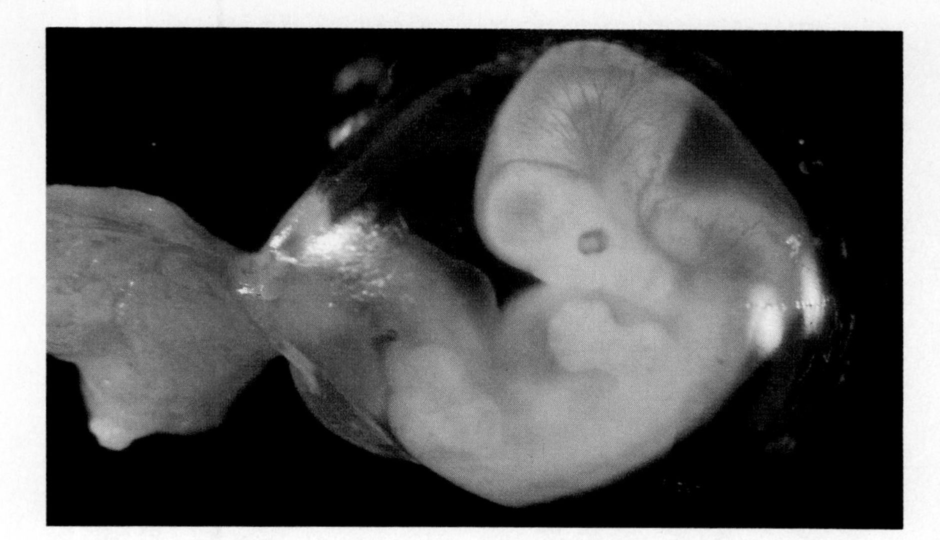

Figure 4.1
Already this human embryo is made up of millions of cells. Due to the process of division, each cell contains the same number and kinds of chromosomes, copies of the very ones that were inherited from its parents. It is possible to retrieve some embryonic cells and view these chromosomes by preparing a karyotype as described next.

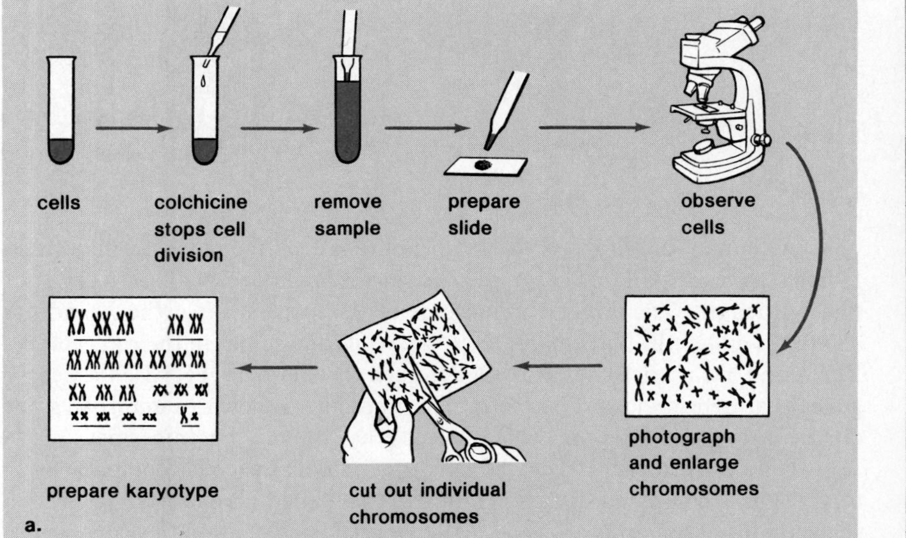

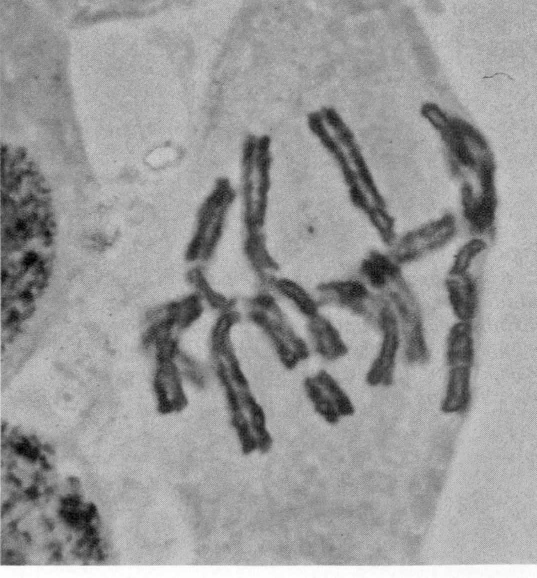

Figure 4.2
a. *Preparation of a karyotype. Cell division is halted by chemical means and then the cells are microscopically magnified and photographed.* b. *An enlargement of the photograph permits the chromosomes to be cut out and arranged by pairs.*

Cell division is necessary for growth (fig. 4.1) and repair of multicellular organisms, and for reproduction of all organisms. Cell division requires that not only the nucleus but also the cytoplasm be divided. An examination of the body cells of a multicellular organism shows that all the nuclei have the same number of chromosomes. This number is characteristic of the organism—corn plants have 20 chromosomes, houseflies have 12, and humans have 46. The particular number has nothing to do with the complexity of the organism. For example, hydras (fig. 27.11), which are very simply organized microscopic organisms, have 32 chromosomes, many more than do houseflies with 12.

In order to clearly view the chromosomes so that they can be counted, a cell is treated and photographed just prior to division, as described in figure 4.2. The chromosomes may then be cut out of the photograph and arranged by pairs. (Pairs of chromosomes have the same size and general appearance.) The resulting display of chromosome pairs is called a **karyotype.**

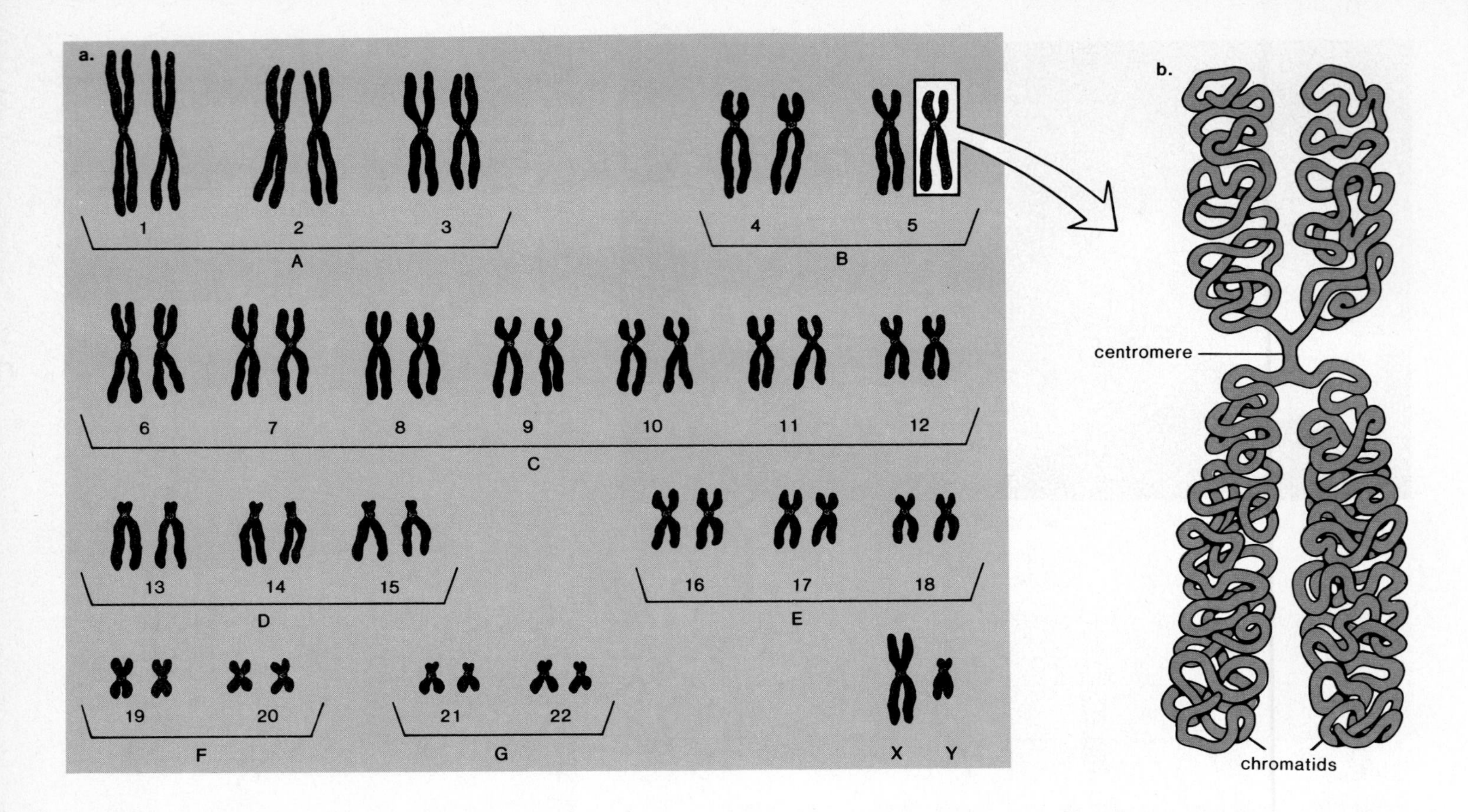

Figure 4.3
a. *Karyotype in a male. Note pairs of autosomes, numbered from 1 to 22, and one pair of sex chromosomes, X and Y. The chromosomes are also grouped (by letters) according to their length and the location of the centromere.* b. *Enlargement of one chromosome.*

A human karyotype is shown in figure 4.3. Although both males and females have 23 pairs of chromosomes, in the male one of these pairs is of unequal length. The larger chromosome of this pair is called the X and the smaller is called the Y. Females have two X chromosomes in their karyotype. The X and Y chromosomes are called the **sex chromosomes** because they carry genes that determine sex. The other chromosomes, known as **autosomes,** include all the pairs of chromosomes except the X and Y chromosomes.

Notice, as further illustrated in figure 4.3b, that each chromosome prior to division is composed of two identical parts, called **chromatids.** These two sister (twin) chromatids are genetically identical and contain the same *genes,* the units of heredity that control the cell. The chromatids are held together at a region called the **centromere.**

Each organism has a characteristic number of chromosomes; humans have 46. A human karyotype shows 22 pairs of autosomes and one pair of sex chromosomes. The sex pair is an X and a Y chromosome in males and two X chromosomes in females. Each chromosome in a karyotype is composed of two sister chromatids, held together at the centromere.

Life Cycle of Animals

Advanced multicellular animals, including humans, typically have a life cycle (fig. 4.4) that requires two types of cell division: **meiosis** and **mitosis.**

Meiosis occurs during the production of the *sperm* and *egg,* which are the sex cells, or **gametes.** A new individual comes into existence when the sperm of the male fertilizes the egg of the female. In humans, the resulting **zygote** contains 46 chromosomes, and as the zygote grows to become the adult, *mitosis* occurs so that each and every cell has 46 chromosomes. In this way each

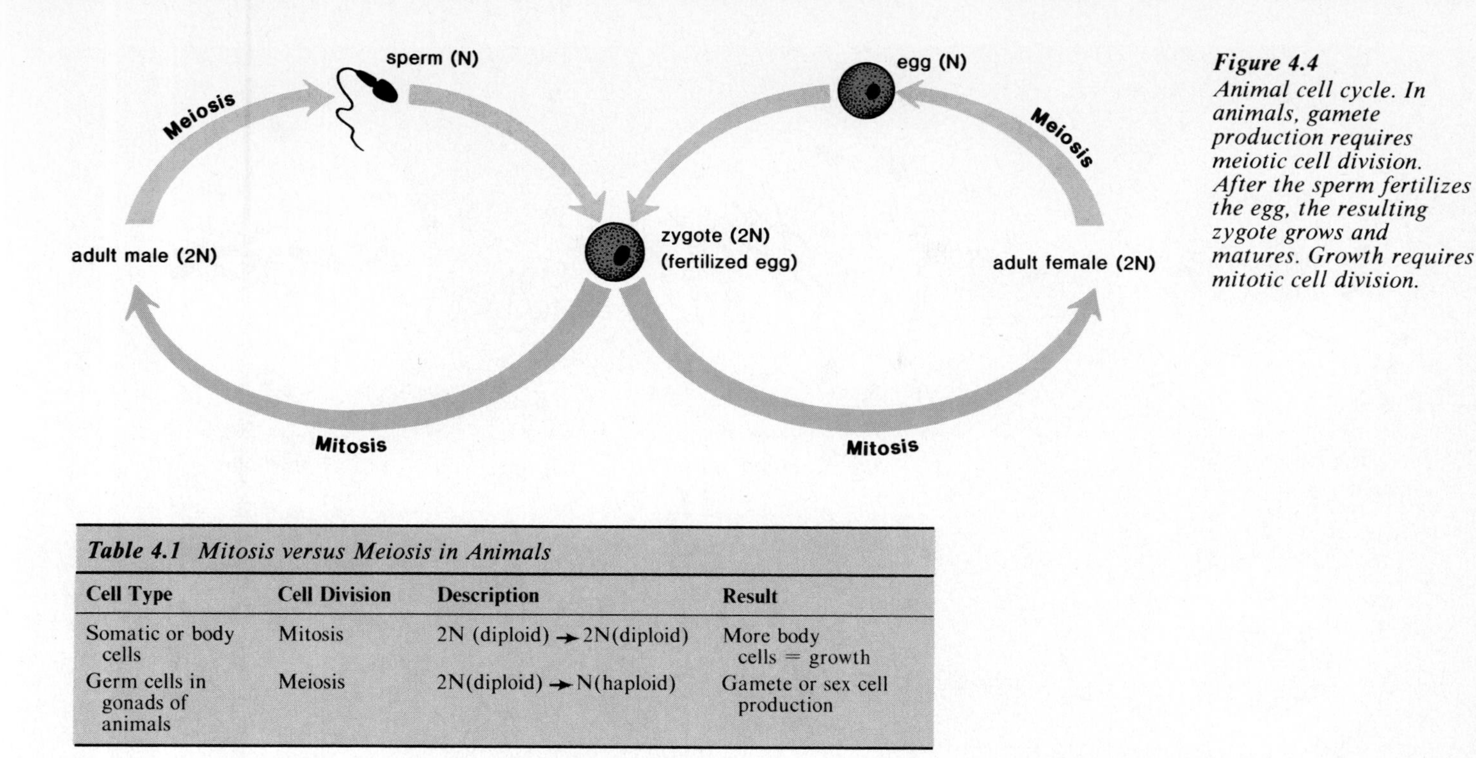

Figure 4.4
Animal cell cycle. In animals, gamete production requires meiotic cell division. After the sperm fertilizes the egg, the resulting zygote grows and matures. Growth requires mitotic cell division.

Table 4.1 *Mitosis versus Meiosis in Animals*

Cell Type	Cell Division	Description	Result
Somatic or body cells	Mitosis	2N (diploid) → 2N(diploid)	More body cells = growth
Germ cells in gonads of animals	Meiosis	2N(diploid) → N(haploid)	Gamete or sex cell production

body cell contains the full complement of chromosomes and genes. The full complement of chromosomes is called the **2N,** or **diploid,** number of chromosomes.

Meiosis occurs only in the sex organs, or **gonads**—the testes in males and the ovaries in females. Here diploid cells called germ cells develop into the gametes, which have half the total number, called the **N,** or **haploid,** number of chromosomes. *The haploid number always has one from each of the pairs of chromosomes. Therefore, the haploid number has one of each kind of chromosome.* For example, in humans the germ cells have 46 chromosomes, or 23 pairs, but gametes contain only 23 chromosomes—one of each of the pairs.

When a haploid sperm fertilizes a haploid egg, the new individual has the diploid number of chromosomes, half of which came from the father and half of which came from the mother. Thus, each parent contributes one of each of the pairs of chromosomes possessed by the new individual. Table 4.1 summarizes the major differences between mitosis and meiosis in multicellular animals.

The life cycle of humans requires two types of cell divisions: mitosis and meiosis. Mitosis is responsible for growth and repair, while meiosis is required for gamete production.

Mitosis

Overview of Animal Mitosis

Mitosis is *cell division in which the daughter cells retain the same number and kinds of chromosomes as the mother cell.*[1] Therefore the mother cell and the daughter cells are genetically identical. The **mother cell** is the cell that divides, and the **daughter cells** are the resulting cells. Figure 4.5 is an overview of mitosis; each cell in the diagram contains four chromosomes. (In determining the number of chromosomes it is necessary to count only the number

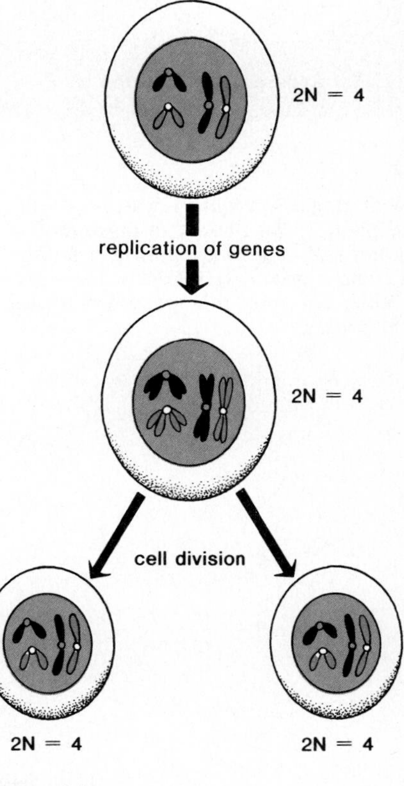

Figure 4.5
Mitosis overview. Following replication of the genes, each chromosome in the mother cell contains two sister chromatids. During mitotic division, the sister chromatids separate so that daughter cells have the same number and kinds of chromosomes as the mother cell.

[1]The term *mitosis* technically refers only to nuclear division but for convenience is used here to refer to division of the entire cell.

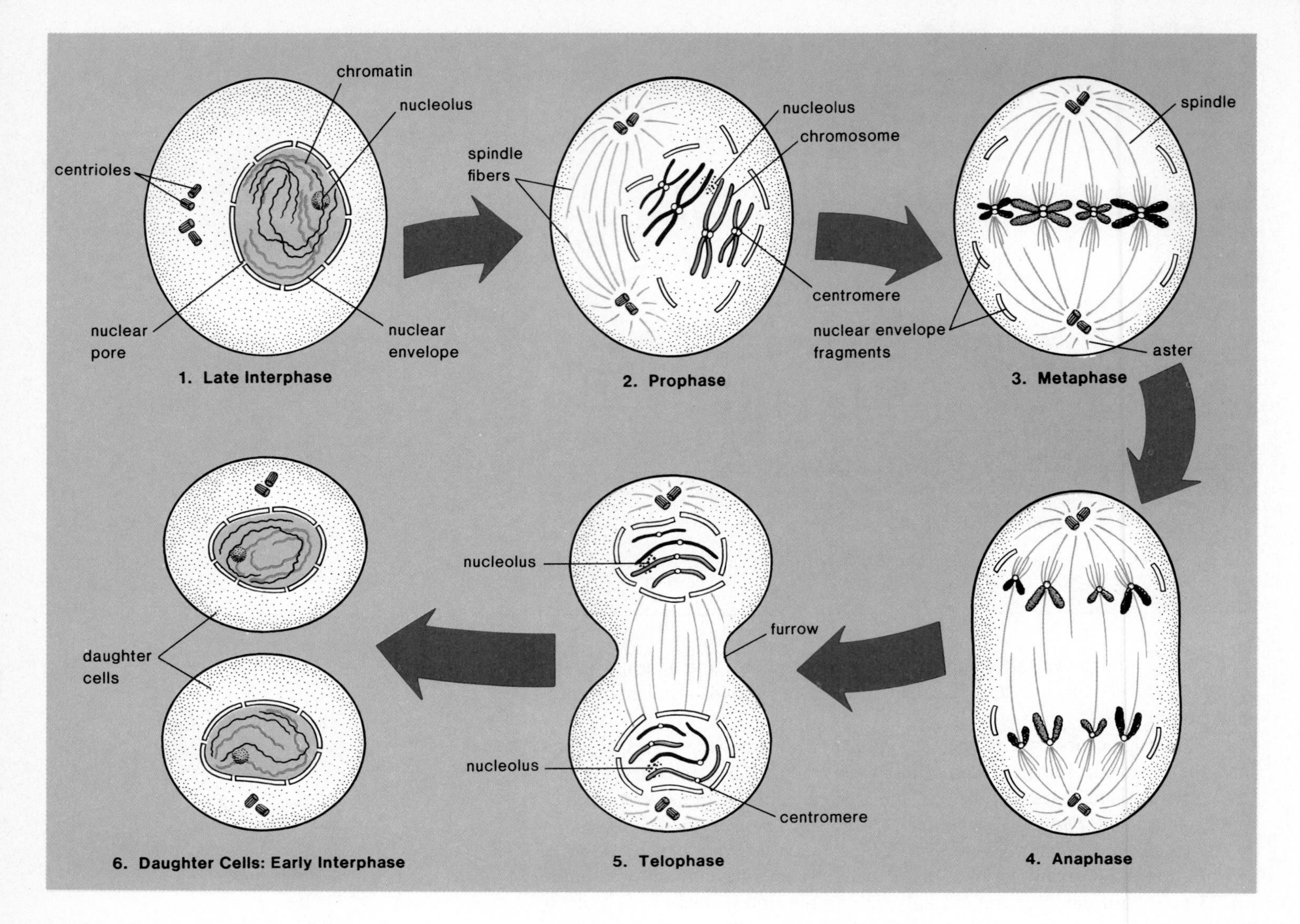

Figure 4.6
Mitosis has four stages, excluding interphase and daughter cells. Notice that these drawings are for mitosis in animal cells and that, because centrioles double in cells about to undergo mitosis, there are two pairs of centrioles in the late interphase cell at the start of the process.

of independent centromeres.) Figure 4.5 points out that a cell prepares for mitosis by replication of the genetic material contained within each chromosome. **Replication** is the process by which DNA makes a copy of itself as is described in detail in chapter 22. Because of replication, each chromosome in the mother cell contains two sister chromatids, sometimes called a chromatid pair, or **dyad.** During mitosis, sister chromatids separate and go to newly forming cells, ensuring that each cell receives a copy of each chromosome rather than two copies of one chromosome and none of another. Different genes are on different chromosomes, and it is necessary for each cell to receive a copy of each chromosome in order to have a full complement of genes.

Stages of Mitosis

As an aid in describing the events of mitosis, the process has been divided into four phases: prophase, metaphase, anaphase, and telophase (fig. 4.6). Although it is necessary to depict the stages of mitosis as if they could be separated, they are continuous and flow from one to the other with no noticeable interruption. Between cell divisions, the cell is said to be in interphase. At this point, the cell may mature and become specialized or it may continue in the **cell cycle** and prepare to divide again (fig. 4.7).

Interphase is the longest of the phases in the cell cycle. The length of time required for the entire cycle varies according to the organism and even the type of cell within the organism, but 18 to 24 hours is typical for animal cells. Mitosis is usually the shortest portion of the cycle, lasting from less than an hour to slightly more than two hours.

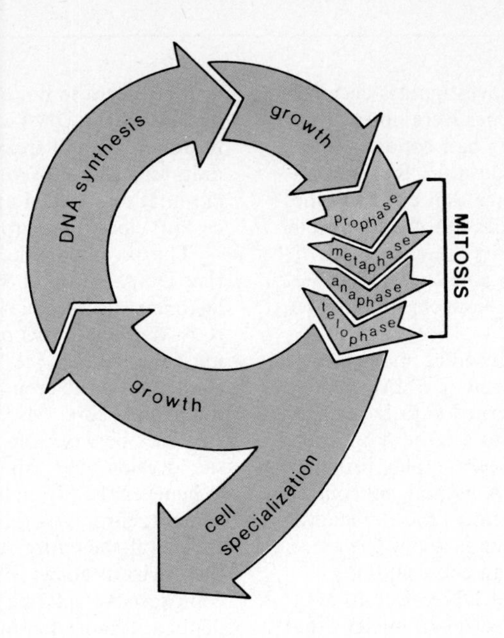

Figure 4.7
The cell cycle consists of mitosis and interphase. During interphase, there is growth before and after DNA synthesis. DNA synthesis is required for the process of replication by which DNA makes a copy of itself. Some daughter cells "break out" of the cell cycle and become specialized cells performing a specific function.

There is a limit to the number of times an animal cell will divide before degenerative changes lead to death. Most will divide about 50 times and only cancer cells retain the ability to divide repeatedly. Cell aging in the individual appears to be a normal process most likely controlled by the nucleus. Cancer cells have abnormal chromosomes and other irregularities of cell structure. In the laboratory, aging of cells can be delayed by environmental circumstances such as reduced temperature, but it cannot be postponed indefinitely.

Cell division is made up of four stages: prophase, metaphase, anaphase, and telophase. The cell cycle includes an additional stage termed interphase. During interphase, centriole duplication occurs and DNA replication causes each chromosome to have sister chromatids.

Interphase

During **interphase** an animal cell resembles figure 2.2a. The nuclear envelope and the nucleoli are visible. The chromosomes, however, are not visible because the chromosomal material is dispersed into fine threads called *chromatin.*

It used to be said that interphase was a resting stage, but we now know that this is not the case. During interphase the organelles are metabolically active and are carrying on their normal functions. When and if the cell is going to divide, two duplication events occur: the centrioles duplicate and DNA replicates. Whereas a nondividing cell has one pair of centrioles outside the nucleus, a cell that is about to divide has two pairs. When centriole duplication occurs, a new centriole forms at right angles to each member of the original pair. Also during this phase, DNA is replicating by a process to be described in detail (p. 480) so that as mitosis begins each chromosome consists of sister chromatids.

Prophase

It is apparent during **prophase** that cell division is about to occur. The chromatin material shortens and thickens so that the chromosomes are readily visible. The reading on page 80 describes the manner by which chromatin can condense to form the chromosomes.

What's in a Chromosome?

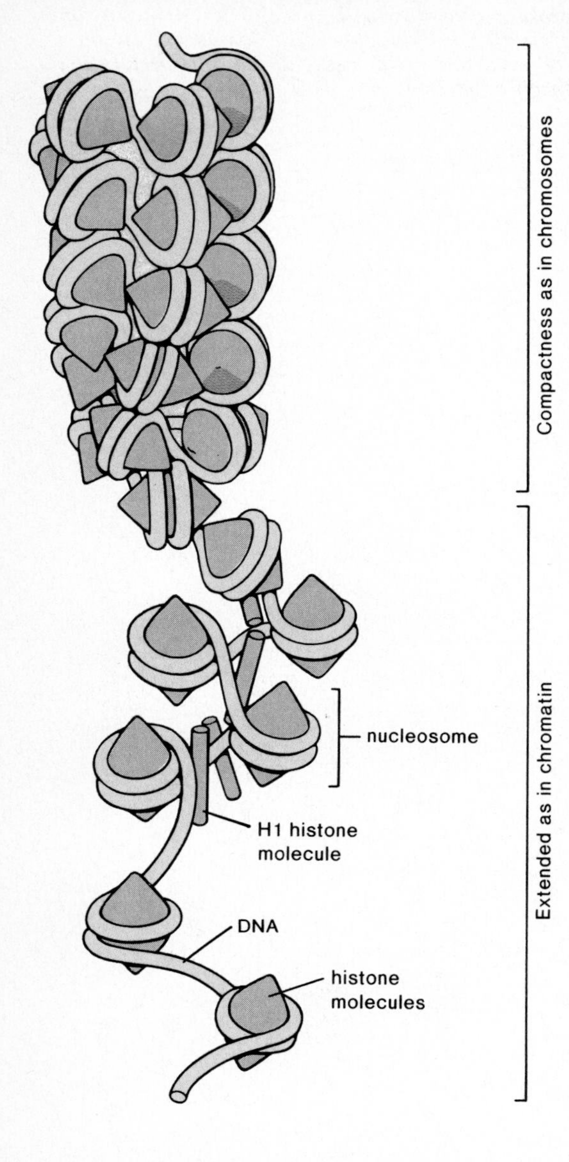

When early investigators decided that the genes were on the chromosomes, they had no idea of chromosome composition. By the mid-1900s, it was known that chromosomes are made up of both DNA and protein. Only in recent years, however, have investigators been able to produce models suggesting how chromosomes are organized.

A eukaryotic chromosome is more than 50 percent protein. Many of these proteins are concerned with DNA and RNA synthesis, but a large proportion, termed histones, seem to play primarily a structural role. A human cell contains forty-six chromosomes, and the length of the DNA in each is about five cm. Therefore, a human cell contains at least two meters of DNA. Yet all of this DNA is packed into a nucleus that is about five μm in diameter. The histones seem to be responsible for packaging the DNA so that it can fit into such a small space. The packing unit, termed a *nucleosome,* gives chromatin a beaded appearance in certain electron micrographs.

The accompanying drawing shows that DNA is wound around a core of histone molecules in a nucleosome. To fully appreciate this packing unit, you must realize that the DNA strand is twisted on itself even as it winds about the histone core. Also, notice how DNA stretches between the nucleosomes at the location of H_1 histone molecules. Whenever the H_1 molecules make contact, chromatin would shorten. Indeed, if the entire structure should then twist as shown, even more compactness could be achieved. No doubt still more folding processes occur as chromatin condenses to form the chromosomes.

Chromosome versus chromatin structure.

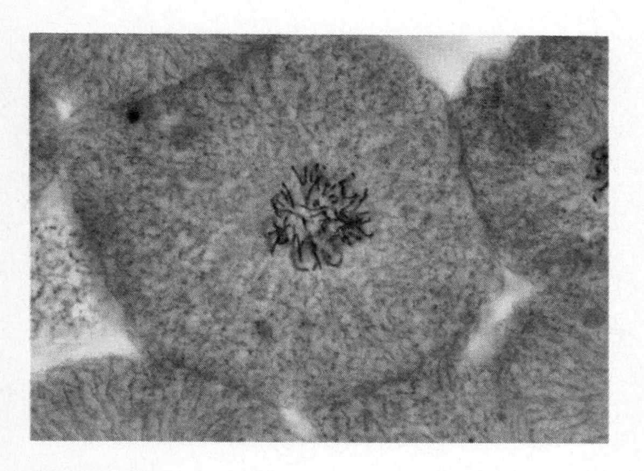

Figure 4.8
Micrograph of late prophase. The chromosomes are moving toward the equator of the spindle located midway between the asters found at the poles of the spindle.

As prophase continues, the pairs of centrioles begin separating and moving toward opposite ends of the nucleus. **Spindle fibers** appear between the separating pairs of centrioles. As the spindle appears, the nuclear envelope and nucleolus begin to disappear. Figure 4.8 is a micrograph of an animal cell at the time of late prophase. The chromosomes are randomly placed even though the spindle appears to be fully formed.

Function of Centrioles The entire spindle apparatus is shown in figure 4.9. It consists of asters, spindle fibers, and centrioles. Both the short **asters** radiating from the centrioles and the long spindle fibers are composed of microtubules. It is known that microtubules are capable of assembling and disassembling (fig. 4.10), which would account for the appearance and disappearance of asters and spindle fibers. It is possible that the centrioles are organizing centers for spindle formation but it could also be that their location at the poles simply ensures that each daughter cell will have a pair of centrioles.

Figure 4.9
Artist's representation of a spindle apparatus in an animal cell. The asters at each pole contain a pair of centrioles. The chromosomes in blue have not yet attached to the spindle fibers.

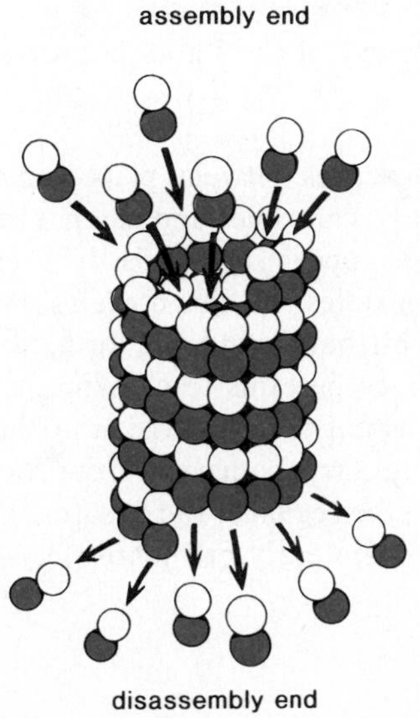

assembly end

disassembly end

Figure 4.10
Microtubules can assemble at one end and disassemble at the other. During assembly, protein dimers join together, and during disassembly the protein dimers separate from one another.

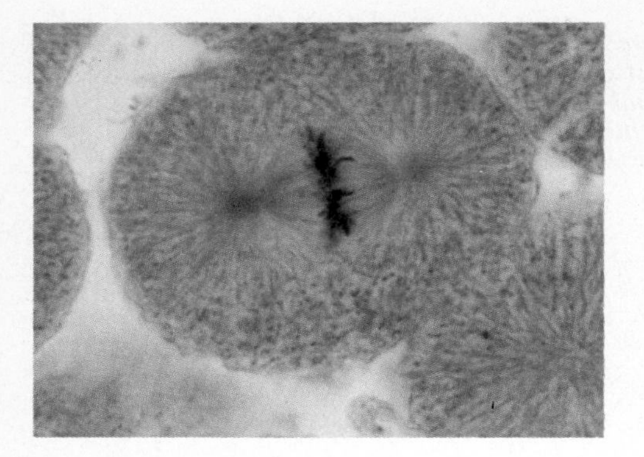

Figure 4.11
Micrograph of metaphase. The chromosomes are now lined up along the equator of the spindle.

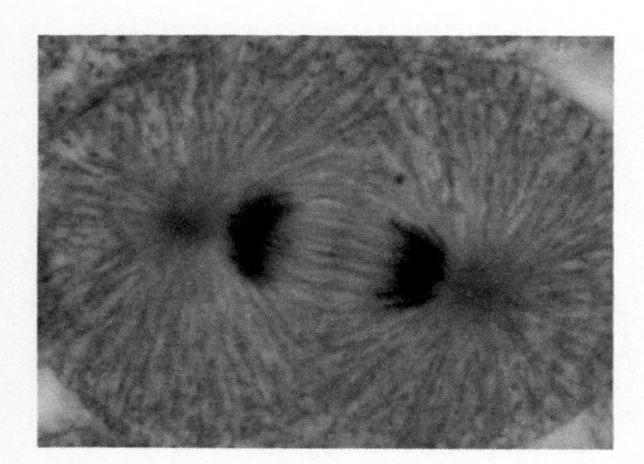

Figure 4.12
Micrograph of anaphase. Separation of sister chromatids results in chromosomes that are pulled by spindle fibers to opposite poles of the spindle.

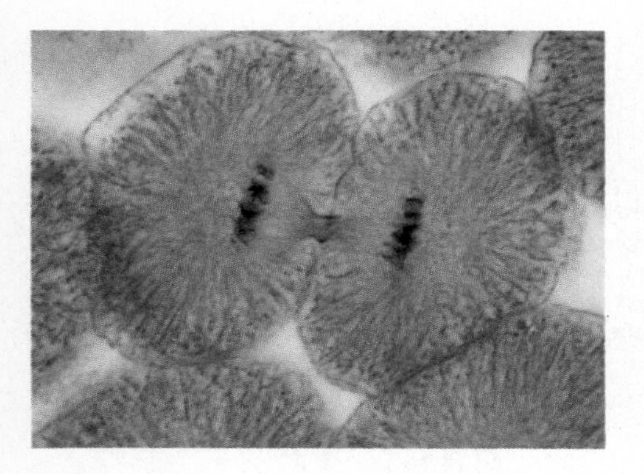

Figure 4.13
Micrograph of late telophase. Furrowing has resulted in two daughter cells separated by membrane. Remnants of the spindle are still seen, but these will disappear as the daughter nuclei reform.

Metaphase

As **metaphase** begins, the nuclear envelope has disappeared and the spindle now occupies the region formerly occupied by the nucleus. Each chromosome is attached to the spindle and moves to the equator (center) of the spindle. Metaphase is characterized by a fully formed spindle, with the chromosomes, each composed of two chromatids, arranged along the equator (fig. 4.11).

Anaphase

During **anaphase** each centromere divides. Now the sister chromatids separate and each moves toward an opposite pole of the spindle (fig. 4.12). *Once separated, the chromatids are called chromosomes.* Separation of the sister (twin) chromatids ensures that each cell will receive a copy of each type chromosome and thus have a full complement of genes.

Movement of the Chromosomes The mechanism of chromosomal movement during anaphase is the subject of much investigation and speculation. Two types of spindle fibers have been identified. The so-called *polar fibers* extend from the poles to the equator of the spindle where they overlap one another. These fibers elongate during anaphase, causing the spindle apparatus to increase in length. During elongation, the fibers are pulling away from each other and creating *a push* that moves the chromosomes toward the poles. Since ATP is required, it suggests that the microtubules making up the polar fibers are sliding past one another.

The chromosomes themselves are not attached to the polar fibers. Each is attached to the second type of spindle fibers, the *centromeric fibers,* that extend from the region of the centromere to each of the poles. These fibers get shorter and shorter as the chromosomes move toward the poles, and eventually they disappear. Undoubtedly disassembly (fig. 4.10) is the mechanism by which these fibers shorten, providing a *pull* that moves the chromosomes toward the poles.

Telophase

During **telophase,** (fig. 4.13) the spindle disappears, possibly due to disassembly of the microtubules making up the spindle fibers. As the nuclear envelopes form and nucleoli appear in each cell, the chromosomes become indistinct chromatin again. Following nuclear division, cytoplasmic division, sometimes called **cytokinesis,** usually occurs. In animal cells, **furrowing,** or an indentation of the membrane between the two newly forming cells, divides the cytoplasm (fig. 4.14). Furrowing is complete when each cell has a complete membrane enclosing it. Microfilaments are believed to take part in the furrowing process since they are always present in the vicinity.

Table 4.2 is a summary of the stages of mitosis.

Mitosis in Plants

There are two main differences between plant and animal cell mitosis. First of all, in higher plant cells, centrioles and asters are not seen during mitosis. However spindle fibers do appear (fig. 4.15). It is interesting to note that animal cells deprived of centrioles will also form a spindle. It may be, therefore, that centrioles do *not* contribute to spindle formation.

The second difference pertains to cytokinesis. The rigid cell wall that surrounds plant cells does not permit division of the cytoplasm by means of furrowing. Instead vesicles largely derived from the Golgi body travel down the polar spindle fibers to the region of the equator. These vesicles fuse to form a membranous structure, the early **cell plate** (fig. 4.15) that spreads to the sides and marks the boundary of the two daughter cells. A daughter cell is complete once it forms a new cell membrane and cell wall next to the cell plate. No furrowing is observed in plant cells.

Mitosis in Protists

Single-celled organisms, such as some protozoans and algae, (termed protists because they are in the phylum Protista), normally reproduce by means of mitosis. For example, when an amoeba (fig. 25.12) undergoes mitosis, it divides into two amoebas, whereas before there was only one. Since an adult amoeba has only a haploid number of chromosomes, it is obvious that mitosis, on occasion, can be represented by $N \longrightarrow N$, rather than $2N \longrightarrow 2N$.

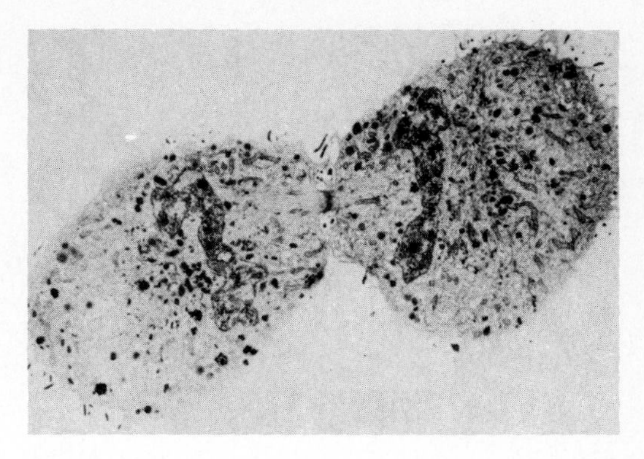

Figure 4.14
Electron micrograph of isolated human cell in advanced stage of cytokinesis. In animal cells, furrowing divides the cytoplasm.

Table 4.2 *Stages of Mitosis*

Stage	Events
Prophase	Replication has occurred and each chromosome is composed of a pair of sister chromatids
Metaphase	Chromatid pairs (dyads) are at the equator (center) of the cell
Anaphase	Chromatids separate and each one is now termed a chromosome
Telophase	At each pole there is a diploid number of chromosomes, the same number and kinds of chromosomes as the mother cell

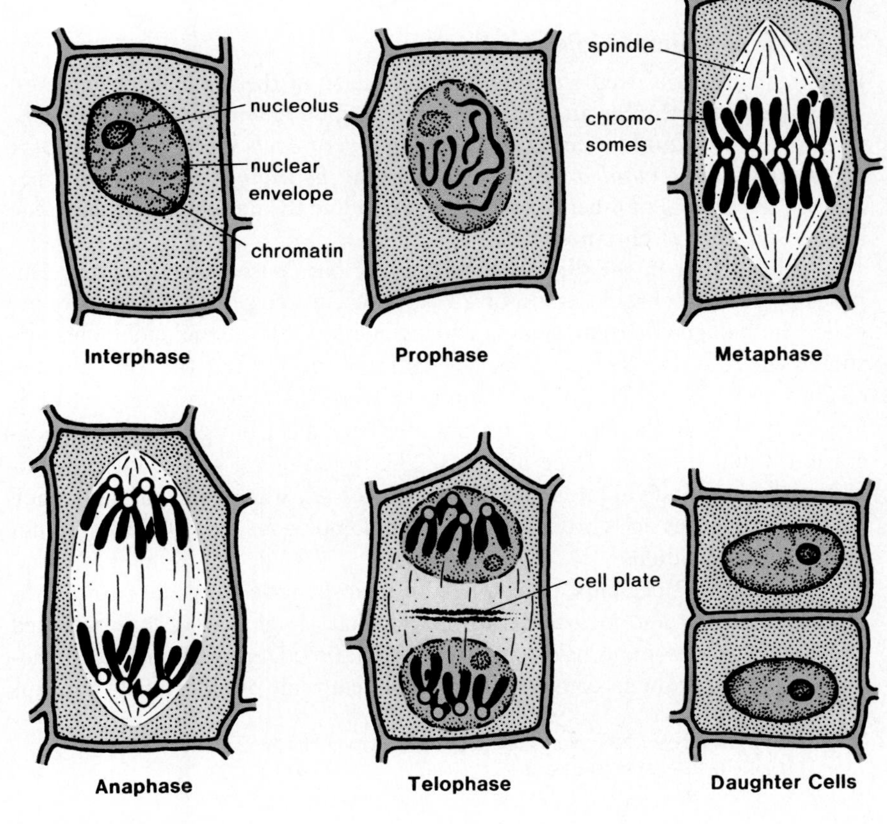

Figure 4.15
Plant cell mitosis. Notice the absence of asters and the square shape of the cells. In telophase, a cell plate develops in between the two daughter cells. The cell plate marks the boundary of the new daughter cells where new cell membrane and new cell wall will form for each cell.

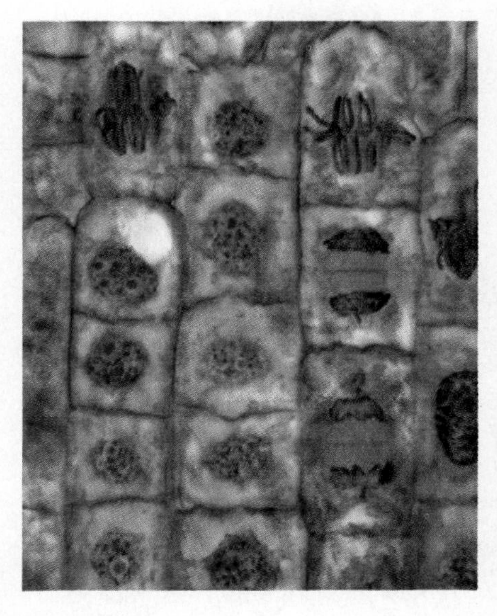

Figure 4.16
Longitudinal section of an onion root tip. Many of the cells are undergoing cell division because this portion of the onion contributes to root growth. Compare these plant cells to the animal cells in figures 4.11 through 4.13. Note that these plant cells have no asters.

Importance of Mitosis

Mitosis assures that each daughter cell receives the same number and kinds of chromosomes as the mother cell, and thus mitosis assures that each daughter cell is *genetically identical* to the mother cell.

Mitosis is important to the growth and repair of multicellular organisms. When a baby develops in its mother's womb, mitosis occurs as a component of growth. As a wound heals, mitosis occurs to repair the damage.

Mitosis also occurs during the process of asexual reproduction. In protists, one division results in two organisms whereas before there was only one. Some lower animals (fig. 27.11) and many plants (fig. 7.19) are also capable of reproducing asexually. In these instances, several divisions are required to produce a replica of the organism.

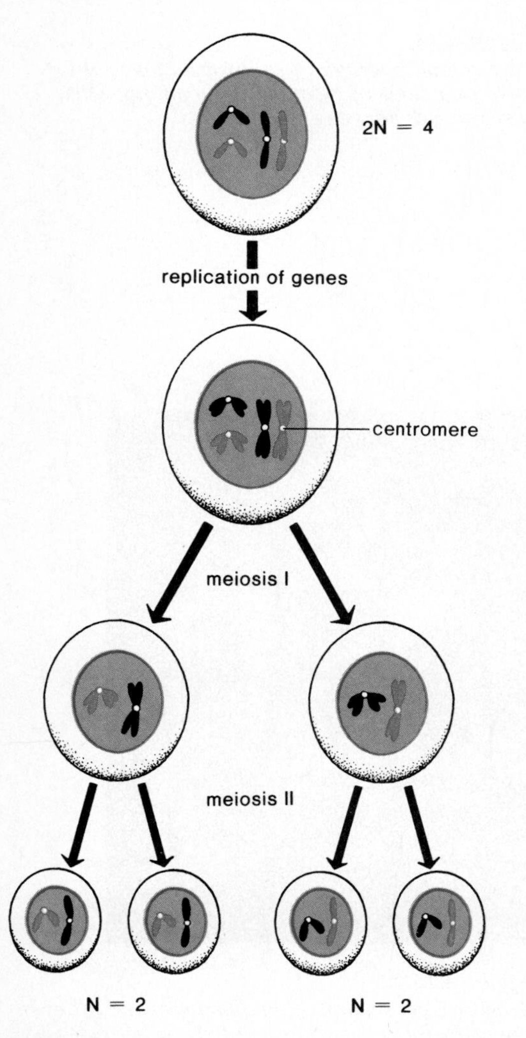

Figure 4.17
Overview of meiosis. Following replication of genes, the mother cell undergoes two divisions, meiosis I and meiosis II. During meiosis I, homologous chromosomes separate, and during meiosis II, chromatids separate. The final daughter cells are haploid.

Mitosis is found in animal, plant, and protist cells. It assures that each cell will have the same number of chromosomes. In a dividing cell each chromosome consists of two chromatids. When the sister chromatids separate during anaphase, each newly forming cell receives the same number and kinds of chromosomes as the original cell. In multicellular organisms mitosis typically occurs as a part of growth and repair; in unicellular organisms, it occurs during asexual reproduction.

Meiosis

Overview of Animal Meiosis

In animals, meiosis occurs during the production of the sex cells or gametes. The gametes are the egg and sperm. Meiosis, which requires two cell divisions, results in *four daughter cells, each having one of each kind of chromosome and thus half the number of chromosomes as the mother cell.*[2] The mother cell has the diploid number of chromosomes, while the daughter cells have the haploid number of chromosomes.

Recall that in the diploid condition the chromosomes are paired. Humans possess 46 chromosomes, or 23 pairs of chromosomes. These pairs are called **homologous chromosomes.** During meiosis, the homologous chromosomes separate. In this way, each new cell will receive half the total number of chromosomes but one of each kind. Halving the chromosome number is necessary to keep the chromosome number constant from generation to generation. Each sperm and egg has only 23 chromosomes, so that after fertilization the new individual has 46 chromosomes. By way of the gametes, each parent contributes one chromosome of each homologous pair of chromosomes in the new individual.

Figure 4.17 presents an overview of meiosis, indicating the two cell divisions, **meiosis I** and **meiosis II.** Prior to meiosis I, replication has occurred and each chromosome consists of sister chromatids. During meiosis I, the homologous chromosomes come together and line up side by side, due to a means

[2]The term *meiosis* technically refers only to nuclear division but for convenience is used here to refer to division of the entire cell.

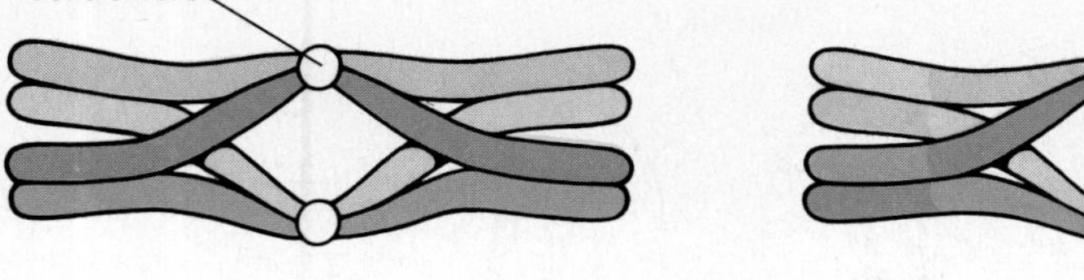

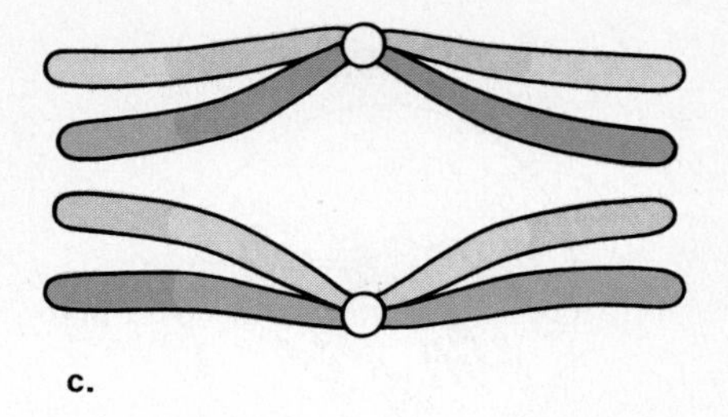

Figure 4.18
During crossing over, pieces of chromosomes are exchanged between chromatid pairs. a. *Chromatid pairs before crossing over has occurred.*
b. *Chromatid pairs during crossing over.*
c. *Chromatid pairs after crossing over. Notice the change in chromosome structure.*

of attraction still unknown. This so-called **synapsis** results in **tetrads,** an association of four chromatids that stay in close proximity during the first two phases of meiosis I. During synapsis, the chromatids exchange genetic material as illustrated in figure 4.18. The exchange of genetic material between chromatids is called **crossing over** and is an additional means by which new combinations of genes occur so that an offspring will have a different genetic makeup than either of its parents.

Following synapsis, the homologous chromosomes separate during meiosis I. This separation means that one chromosome of every homologous pair will reach each gamete.[3] There are no restrictions to the separation process; either chromosome of a homologous pair may occur in a gamete with either chromosome of any other pair.[4]

Notice that at the completion of meiosis I (fig. 4.17), the chromosomes still consist of sister chromatids. During meiosis II, the chromatids separate, resulting in four daughter cells, each of which has the haploid number of chromosomes. Although two cell divisions have taken place, replication has occurred only once. Remembering that one counts only the number of independent centromeres verifies that the mother cell has the diploid number of chromosomes, while each of the four daughter cells has the haploid number.

Stages of Meiosis

First Division

The stages of meiosis I are diagrammed in figure 4.19. During *prophase I,* the spindle appears while the nuclear envelope and the nucleolus disappear. Homologous chromosomes undergo synapsis, forming tetrads. At *metaphase I,* tetrads line up at the equator of the spindle. During *anaphase I,* homologous chromosomes separate and the chromosomes (still composed of two chromatids) move to the poles of the spindle. Each pole receives one-half the total number of chromosomes. In *telophase I,* the nuclear envelope and the nucleolus reappear as the spindle disappears. In certain species, the cell membrane furrows to give two cells and in others, the second division begins without benefit of complete furrowing. Regardless, each daughter nucleus now contains only one from each homologous pair of chromosomes.

[3]See Mendel's law of segregation, p. 448.

[4]See Mendel's law of independent assortment, p. 452.

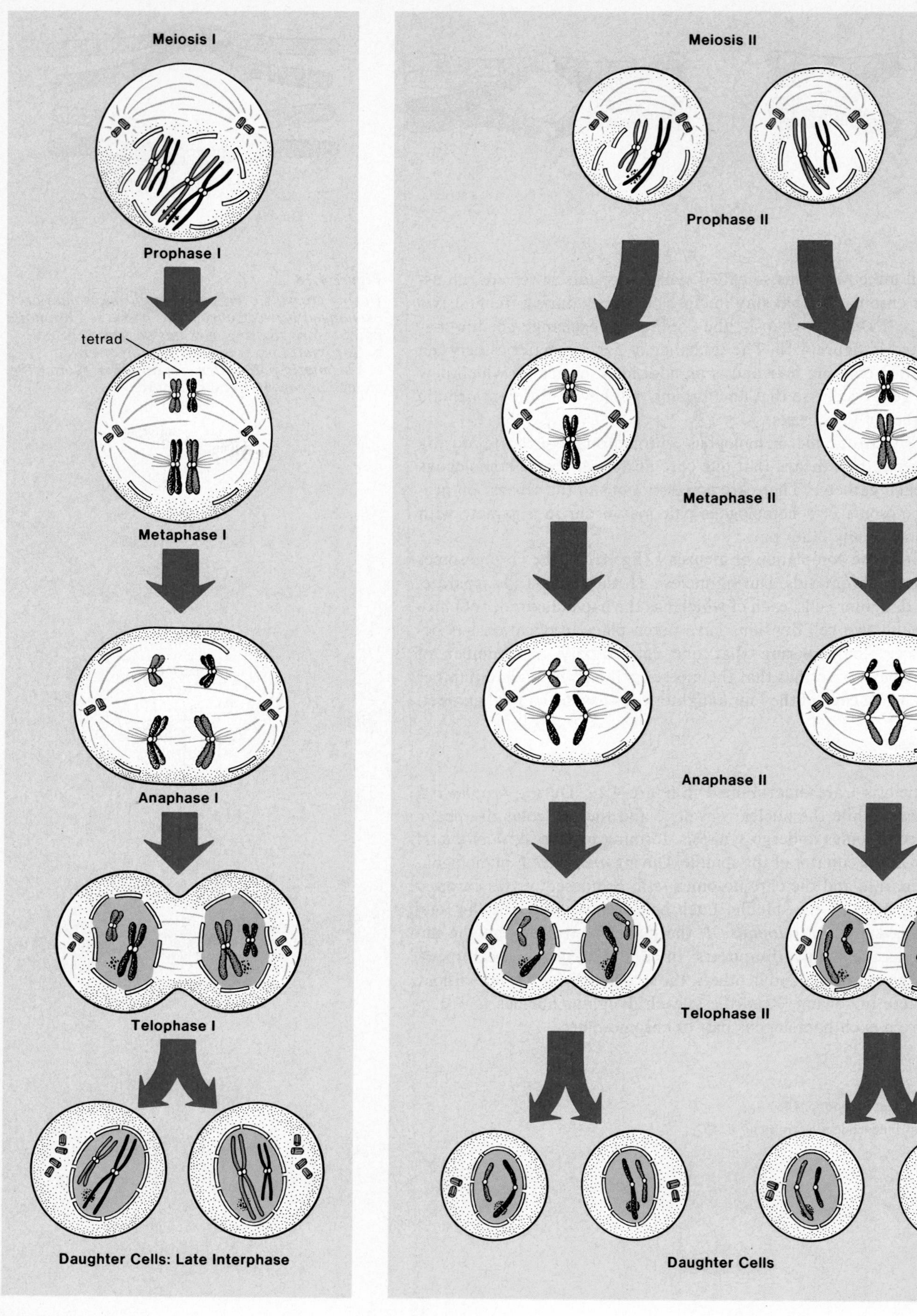

Figure 4.19

Figure 4.20

Table 4.3 *Stages of Meiosis I and Meiosis II*

	Meiosis I	Meiosis II
Prophase	Homologous chromosomes, each one composed of chromatids, synapse	Each chromosome is still composed of sister chromatids
Metaphase	Tetrads (chromosome pair = four chromatids) are at the equator	Chromatid pairs (dyads) are at the equator
Anaphase	Homologous chromosomes separate	Chromatids separate and each one is now termed a chromosome
Telophase	At each pole there is one from each pair of homologous chromosomes	At each pole there is the haploid number and one of each kind of chromosome

Second Division

The stages of meiosis II are diagrammed in figure 4.20. At the beginning of *prophase II,* a spindle appears while the nuclear envelope and nucleolus disappear. Each chromosome with its two chromatids attaches to the spindle independently. During *metaphase II,* the chromosomes are lined up at the equator. During *anaphase II,* the centromeres divide and the chromatids separate and move toward the poles. Each pole receives the same number of chromosomes. In *telophase II,* the spindle disappears as the nuclear envelopes form. The cell membrane furrows to give two complete cells, each of which has the haploid, or N, number of chromosomes. Since each cell from meiosis I undergoes meiosis II, there are four daughter cells altogether. Figure 4.20 shows only two of these.

Table 4.3 is a summary of the stages of meiosis I and meiosis II. This summary is appropriate to both plants and animals. Our preceding discussion refers, in general, to animal cell meiosis because centrioles are present in the figures, the cells are rounded without cell walls, and furrowing occurs to divide the cells. In animals, meiosis is specifically involved in either spermatogenesis or oogenesis.

Meiosis involves two cell divisions. During meiosis I, homologous chromosomes come to lie side by side during synapsis. The chromatids making up the resulting tetrad exchange chromosome pieces; this is called crossing over. When the homologous chromosomes separate during meiosis I, each daughter cell receives one from each pair of chromosomes. Separation of sister chromatids during meiosis II then produces a total of four daughter cells, each with the haploid number of chromosomes.

Figure 4.19
Stages of meiosis I. During meiosis I, chromosome pairs separate so that each daughter cell has only one chromosome from every original homologous pair of chromosomes. Note that each chromosome still contains two chromatids and that no replication of genetic material occurs during interphase.

Figure 4.20
Stages of meiosis II. During meiosis II, chromatids separate so that each daughter cell has the haploid number of chromosomes in single copy.

Spermatogenesis and Oogenesis

During **spermatogenesis** in males, sperm are produced, and during **oogenesis** in females, eggs are produced. The final product of meiosis, as well as the process itself, is different in the two sexes as you can see in figure 4.21. In males, meiosis results in four viable sperm from each original cell. In females, the first meiotic division produces two cells, but one is much larger than the other. The smaller nonfunctional cell is called a **polar body.** The second division also results in two cells of unequal size, but again one is much larger than the other which is another nonfunctional polar body. Therefore, meiosis in females produces only one functional egg that contains a rich supply of cytoplasmic components. The polar bodies are a way to discard unnecessary chromosomes while retaining much of the cytoplasm that will serve as a source of nutrients for the developing embryo. Figure 4.21 shows how the sperm and egg are adapted to their function. The sperm is a tiny flagellated cell that will contribute only chromosomes to the new individual while the large egg will contribute most of the cytoplasm.

There is another difference between spermatogenesis and oogenesis. Spermatogenesis, once started, goes to completion and mature sperm result. In contrast, oogenesis does not necessarily go to completion. Only if a sperm fertilizes the maturing egg does it undergo meiosis II, otherwise it simply disintegrates. Regardless of this complication, however, both the sperm and egg contribute the haploid number of chromosomes to the new individual. In humans each contributes 23 chromosomes.

Meiosis in Plants

While the events of meiosis are essentially the same in plants as animals, the resulting cells called spores are not gametes and have a different function, which will be brought out in the discussion of plant life histories in chapter 26.

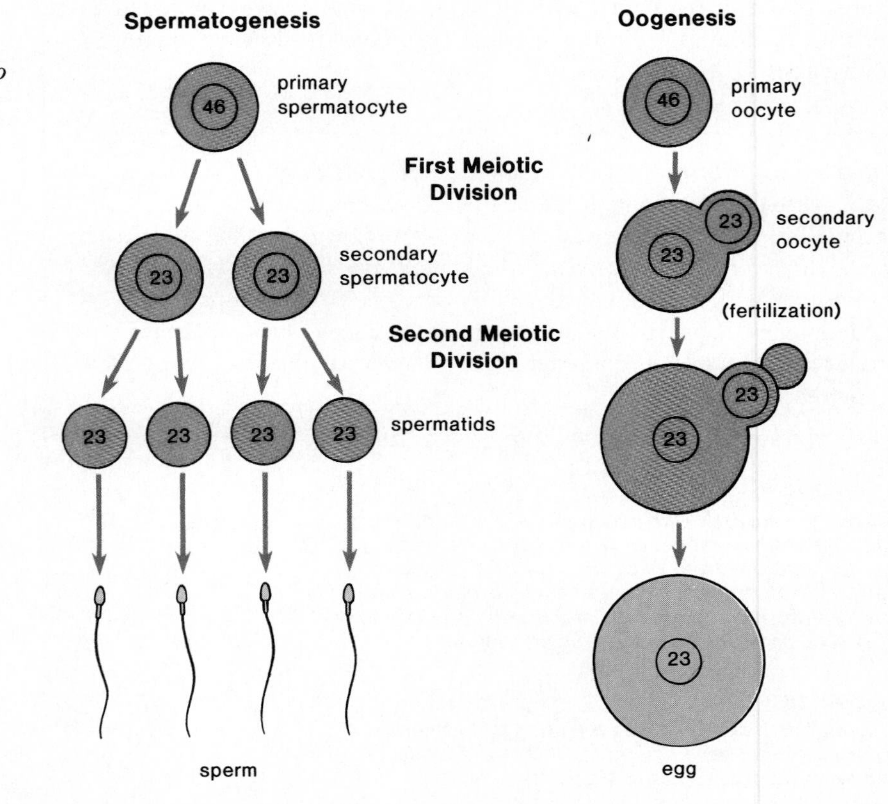

Figure 4.21
Spermatogenesis produces four viable sperm, whereas oogenesis produces one egg and at least two polar bodies. In humans, both sperm and egg have 23 chromosomes each; therefore, following fertilization the zygote has 46 chromosomes.

Importance of Meiosis

Meiosis is nature's way of keeping the chromosome number constant from generation to generation. In animals, it occurs prior to maturation of the egg and sperm. In higher plants, it occurs when spores are formed.

Meiosis assures that the next generation will have a different genetic makeup than that of the previous generation. As a result of crossing over, the chromosomes carry a new combination of genes. In animals, the egg carries one-half the genes from the female parent and the sperm carries one-half the genes from the male parent. When the sperm fertilizes the egg, the zygote has a different combination of genes than either parent. In this way, meiosis assures *genetic variation,* generation after generation.

Meiosis is an important part of sexual reproduction in both animals and plants. It assures that the gametes have the haploid number of chromosomes; the diploid number is restored when fertilization occurs. In humans, meiosis occurs in the sex organs; spermatogenesis produces four viable sperm, while oogenesis produces one egg and at least two polar bodies.

Summary

The life cycle of higher organisms requires two types of cell divisions, mitosis and meiosis. Mitosis assures that all cells in the body have the diploid number and same kinds of chromosomes. It is made up of four stages: prophase, metaphase, anaphase, and telophase. The cell cycle includes an additional stage termed interphase. During interphase, DNA replication causes each chromosome to have sister chromatids. When the chromatids separate, each newly forming cell receives the same number and kinds of chromosomes as the original cell. The cytoplasm is partitioned by furrowing in animals and cell plate formation in plants. Mitosis is required for growth and repair of body parts in both animals and plants.

Meiosis involves two cell divisions. During meiosis I, the homologous chromosomes (following crossing over between chromatids) separate and during meiosis II the sister chromatids separate. The result is four cells with the haploid number of chromosomes. Meiosis is a part of gamete formation in animals and spore formation in plants.

Mitosis is contrasted to meiosis in figure 4.22.

Objective Questions

1. During interphase the chromosomes are not visible because they are extended into fine threads called ____________ .
2. If an organism has twelve chromosomes, it would have __________ homologous pairs.
3. Every chromosome just prior to division is composed of two identical parts called ____________ .
4. If the mother cell has twenty-four chromosomes, the daughter cells following mitosis will have ____________ chromosomes.
5. As the organelles called ____________ separate and move to the poles, the spindle fibers appear.
6. ____________ is the stage of mitosis during which the chromatids separate and become chromosomes.
7. Cytokinesis in an animal cell occurs by a ____________ process, while in a plant cell it involves the formation of a ____________ .
8. Whereas mitosis results in two daughter cells, meiosis produces ____________ daughter cells.
9. During anaphase I of meiosis, the ____________ separate. This means that eventually the gametes will have the haploid number of chromosomes.
10. Meiosis ensures that the zygote will have a ____________ combination of genes than has either parent.

Answers to Objective Questions
1. chromatin 2. six 3. chromatids 4. twenty-four 5. centrioles 6. Anaphase 7. furrowing, cell plate 8. four 9. homologous chromosome pairs 10. different

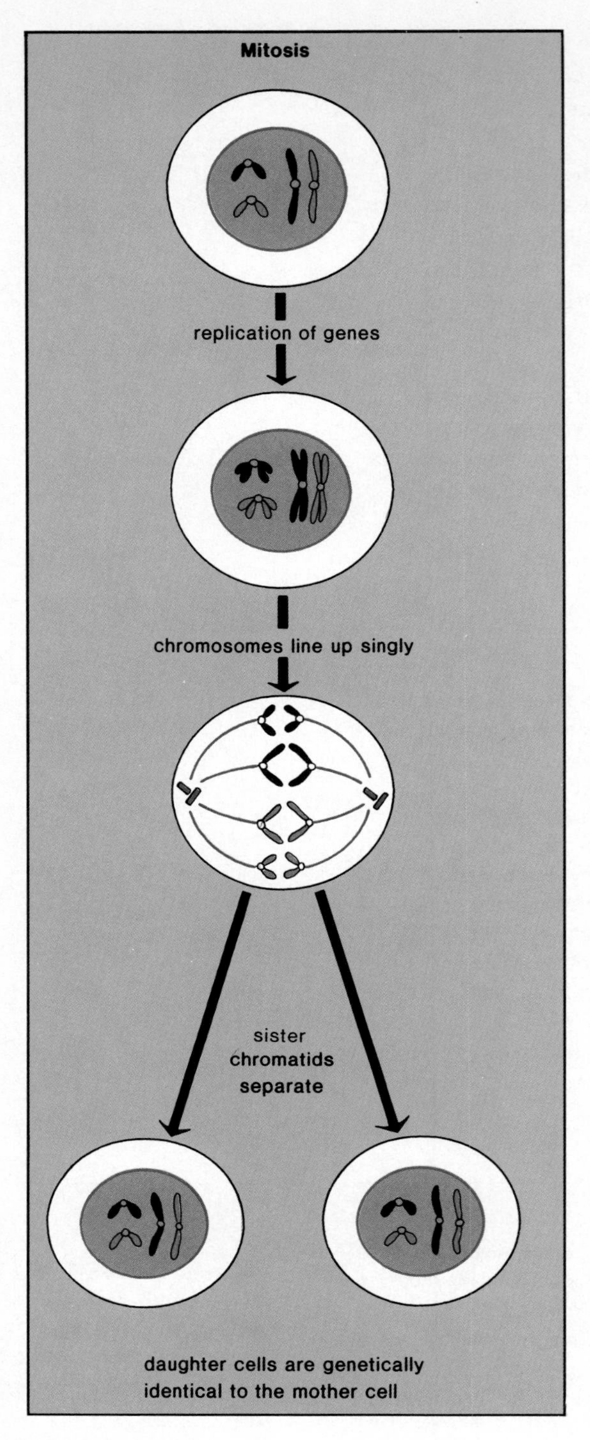

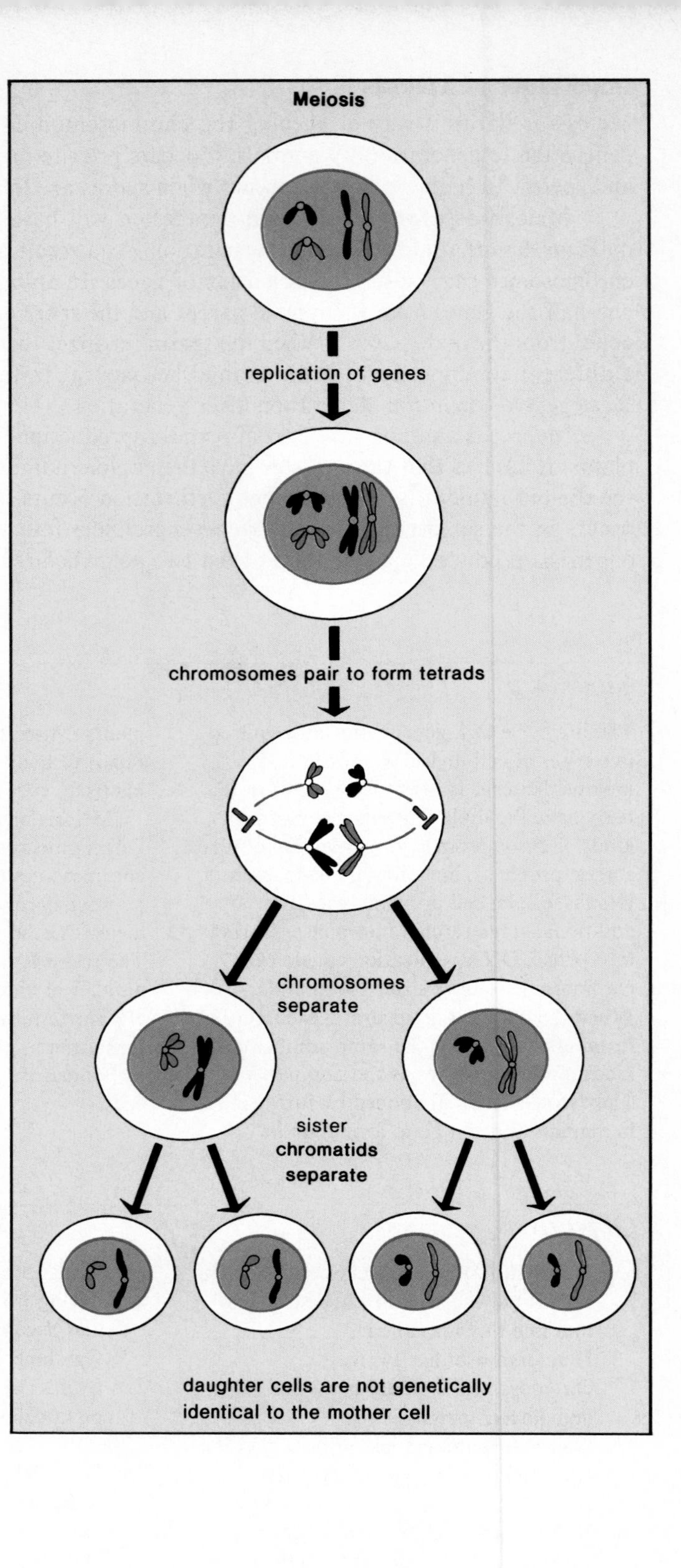

Figure 4.22
Mitosis compared to meiosis.

Study Questions

1. Describe the normal karyotype of a human being. What is the difference between a male and female karyotype? (p. 75)
2. Relate the terms *diploid (2N)* and *haploid (N)* to mitosis in somatic cells and meiosis in germ cells. (p. 77)
3. Explain the makeup of a chromosome prior to cell division. (p. 75)
4. Describe the stages of animal mitosis, including in your description the terms *centriole, nucleolus, spindle,* and *furrowing.* (p. 78)
5. Name two differences between plant cell mitosis and animal cell mitosis. (p. 82)
6. Give several instances when mitosis occurs in humans. (p. 84)
7. Describe the stages of meiosis I, including in your description the term *tetrad.* (p. 85)
8. Compare the second series of stages of meiosis to a mitotic division. (p. 87)
9. Explain the fact that oogenesis produces one mature egg, but spermatogenesis results in four sperm. (p. 88)
10. What is the importance of meiosis in the life cycle of any organism? (p. 89)
11. Give several differences between mitosis and meiosis. (p. 90)

Key Terms

asters (as'terz) short rays of microtubules that appear at the ends of the spindle apparatus in animal cells during cell division. *80*

autosomes (aw'to-sōmz) chromosomes other than sex chromosomes. *76*

centromere (sen'tro-mēr) a region of attachment for a chromosome to spindle fibers that is generally seen as a constricted area. *76*

chromatids (kro'mah-tidz) the two identical parts of a chromosome following replication of DNA. *76*

crossing over (kros'ing o'ver) the exchange of corresponding segments of genetic material between chromatids of homologous chromosomes during synapsis of meiosis I. *85*

cytokinesis (si''to-ki-ne'sis) division of the cytoplasm of a cell. *82*

diploid (dip'loid) the 2N number of chromosomes; twice the number of chromosomes found in gametes. *77*

gametes (gam'ets) reproductive cells that join in fertilization to form a zygote; most often an egg or sperm. *76*

gonads (go'nadz) organs that produce sex cells; the ovary, which produces eggs, and the testis, which produces sperm. *77*

haploid (hap'loid) the N number of chromosomes; half the diploid number; the number characteristic of gametes that contain only one set of chromosomes. *77*

homologous chromosomes (ho-mol'o-gus kro'mo-sōmz) similarly constructed; homologous chromosomes have the same shape and contain genes for the same traits. *84*

karyotype (kar'e-o-tīp) the arrangement of all the chromosomes within a cell by pairs in a fixed order. *75*

meiosis (mi-o'sis) type of cell division that occurs during the production of gametes or spores by means of which the daughter cells receive half the number of chromosomes as the mother cell. *76*

mitosis (mi-to'sis) type of cell division in which daughter cells receive the exact chromosome and genetic makeup of the mother cell; occurs during growth and repair. *76*

oogenesis (o''o-jen'ĕ-sis) production of an egg in females by the process of meiosis and maturation. *88*

sex chromosomes (seks kro'mo-sōmz) chromosomes responsible for the development of characteristics associated with maleness or femaleness; an *X* or *Y* chromosome. *76*

spermatogenesis (sper''mah-to-jen'ĕ-sis) production of sperm in males by the process of meiosis and maturation. *88*

spindle fibers (spin'd'l fi'berz) microtubule bundles involved in the movement of chromosomes during mitosis and meiosis. *80*

synapsis (sĭ-nap'sis) the attracting and pairing of homologous chromosomes during prophase I of meiosis. *85*

tetrads (tet'radz) a set of four chromatids resulting from the pairing of homologous chromosomes during prophase I of meiosis. *85*

zygote (zi'gōt) diploid cell formed by the union of two gametes, the product of fertilization. *76*

Cellular Metabolism

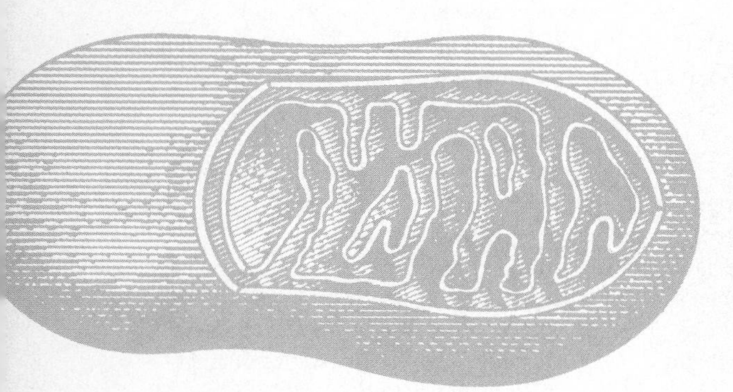

Chapter Concepts

1 A metabolic pathway is a series of reactions controlled by enzymes.

2 Enzymes are protein molecules that speed up chemical reactions and only function properly when they retain their normal shapes.

3 Cells require the energy molecule ATP to drive forward synthetic reactions and for various other functions.

4 Cellular respiration, a metabolic pathway involving mitochondria, provides the necessary energy to form ATP molecules.

5 During cellular respiration, glucose products are oxidized and concurrently ATP is formed from ADP + Ⓟ molecules.

Chapter Outline

Figure 5.1
Multicellular organisms contain millions of cells that are actively carrying out metabolic reactions. Among these, cellular respiration provides the energy needed to allow one to swim or perform any manner of physical activities.

Metabolism

Cells are not static; they are dynamic. Drawings of cells and even microscopic slides of cells give us the impression that cells are inactive; actually, cells are constantly active. Pinocytotic and phagocytotic vesicles are constantly being formed, organelles are moving about, and division may be taking place. A vital part of this activity is constantly occurring chemical reactions, which collectively are termed the **metabolism** of the cell.

Metabolic Pathways

Reactions do not occur haphazardly in cells; they are usually a part of a metabolic pathway. *Metabolic pathways* begin with (a) particular reactant(s) and terminate with (an) end product(s). While it is possible to write overall equations for metabolic pathways, the actual pathway itself proceeds by many minute steps. One reaction leads to the next reaction, which leads to the next reaction, and so forth in an organized, highly structured manner. This system makes it possible for one pathway to lead to several others, especially since various pathways have several substances in common. Also, metabolic energy is more easily captured and utilized if it is released in small increments rather than all at once.

Metabolic pathways can be represented by the following diagram, as long as we realize that side branches may occur at any juncture:

$$A \xrightarrow{1} B \xrightarrow{2} C \xrightarrow{3} D \xrightarrow{4} E \xrightarrow{5} F \xrightarrow{6} G$$

In the pathway represented, the letters are products of the previous reaction and the reactants for the next reaction. *A* is (are) the beginning substance(s) and *G* is (are) the end product(s). The numbers in the pathway refer to different enzymes. *Every reaction in a cell requires a specific enzyme.* Enzymes are protein molecules (p. 27) that speed up chemical reactions, and their concentration controls the rate at which reactions occur. In effect, no reaction occurs in a cell unless its enzyme is present. For example, if enzyme number 2 in the diagram is missing, the pathway cannot function; it will stop at *B*. Since enzymes are so necessary in cells, their mechanism of action has been studied extensively.

---

*M*etabolic pathways are series of reactions that proceed in an orderly step-by-step manner. Each reaction within a cell requires a specific enzyme.

Enzymes

Each and every enzyme typically speeds up only one particular reaction and therefore is said to be *specific* for that reaction. The specificity of an enzyme can be explained in reference to the following equation:

$$E + S \longrightarrow ES \longrightarrow E + P.$$

In this reaction, *E* = enzyme, *S* = substrate, *ES* = enzyme — substrate complex, and *P* = product.

***Table 5.1** Enzymes Named for Their Substrates*

Substrate	Enzyme
Lipid	Lipase
Urea	Urease
Maltose	Maltase
Ribonucleic acid	Ribonuclease
Lactose	Lactase

Substrate

The **substrate(s)** is (are) the reactant(s) in an enzyme's reaction. Table 5.1 indicates that the name of an enzyme is often formed by adding *ase* to the name of its substrate. Some enzymes are named for the action they perform, as, for example, a dehydrogenase is an enzyme that removes hydrogen atoms from its substrate.

Enzyme

Notice in figure 5.2 that the enzyme is unaltered by the reaction. Only a small amount of enzyme is actually needed in a cell because enzymes are used over and over again.

Enzyme-Substrate Complex

For a reaction to occur, the reactants must be brought into close proximity. Therefore, in figure 5.2, the substrates are seemingly attracted to the enzyme because their shapes fit together like a *key fits a lock*. However, it is now thought that the enzyme may very well undergo a slight change in shape

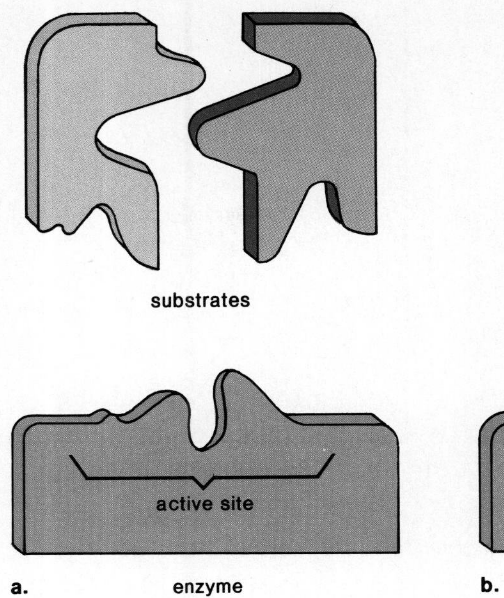

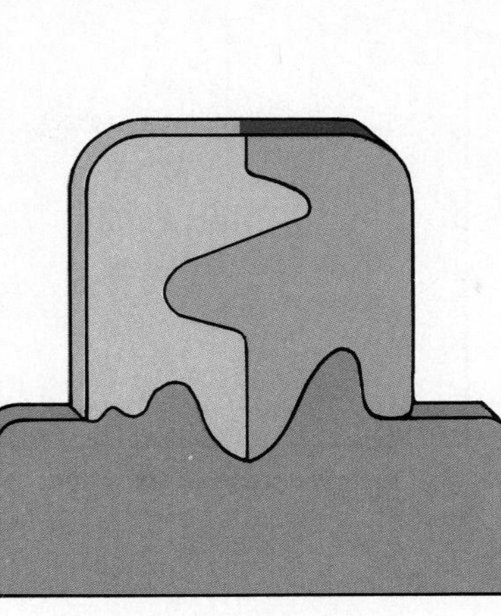

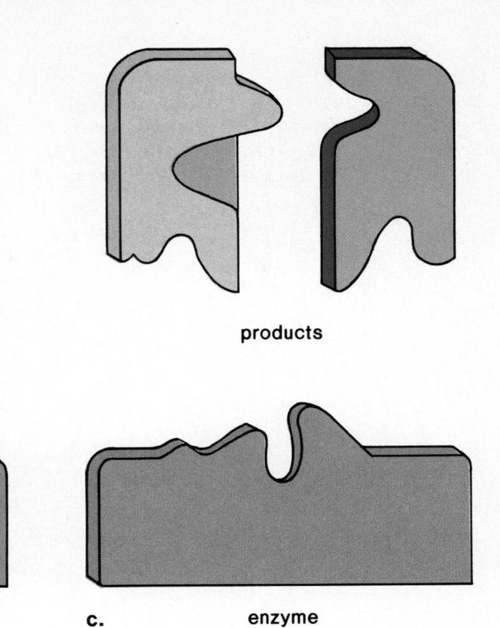

in order to more perfectly accommodate the substrates. This is called the *induced-fit hypothesis* because as binding occurs the enzyme is induced (undergoes a slight alteration) to achieve maximum fit. The region where the substrate(s) attach(es) is called the **active site,** and it is here that the reaction takes place. After the reaction has been completed, the product(s) is (are) released and the active site returns to its original state.

Figure 5.2
a. & b. *Enzymatic action. An enzyme has an active site where the substrates and enzyme fit together in such a way that the substrates are oriented to react.* c. *Following the reaction, the products are released and the enzyme assumes its prior shape.*

Product

Only (a) certain product(s) can be produced by any particular reactant(s), therefore, an enzyme cannot bring about impossible products. However the presence of an active enzyme can determine whether a reaction takes place or not. For example, if substance *A* can react to either *B* or *C,* whichever enzyme is active, 1 or 2, will determine which product is produced.

$$A \begin{cases} \xrightarrow{1} B \longrightarrow D \\ \xrightarrow[2]{} C \longrightarrow F \end{cases}$$

Since enzymes cannot bring about impossible reactions or products, it is usually emphasized that enzymes only speed up their particular reaction. They do this by lowering the energy of activation.

Energy of Activation

Organic molecules frequently will not react with one another unless they are activated in some way. For example, wood does not burn unless it is heated to a moderately high temperature. In the laboratory, too, activation is very often achieved by heating the reaction flask so that the number of effective collisions between molecules increases, allowing the molecules to react with one another.

Figure 5.3
a. *Energy of activation (E_a) that is required for a reaction to occur when an enzyme is not available.*
b. *Required energy of activation (E_a) is much lower when an enzyme is available. Notice that the energy level of the entire system is always lower following a reaction.*

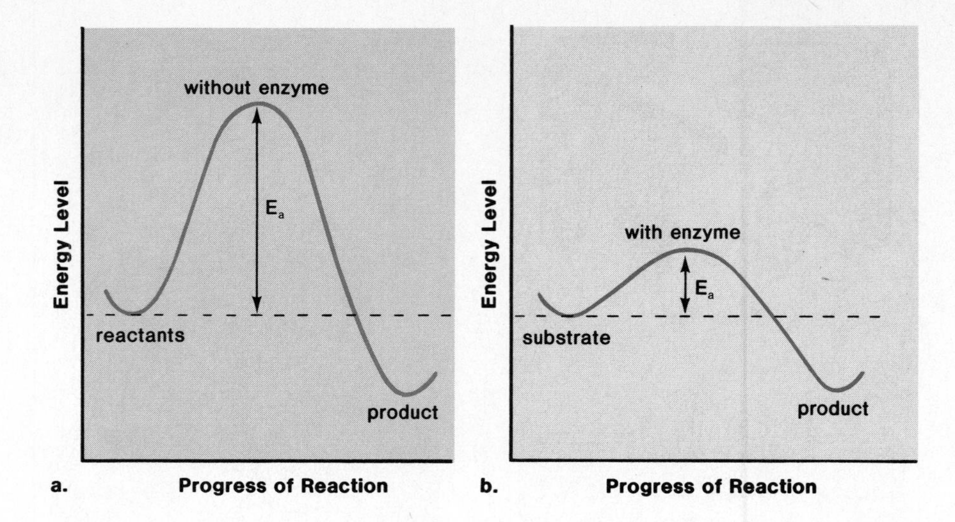

The energy that must be supplied to cause molecules to react with one another is called the *energy of activation.* Figure 5.3 compares the energy level of nonreactive molecules, reactive ones, and the energy level of the products after the reaction. The difference between the nonreactive energy level and the reactive energy level is the necessary energy of activation. Enzymes lower the necessary energy of activation. For example, the following data are available for the hydrolysis of casein, the protein found in milk.

	Energy of activation cal/mole[1]
Without an enzyme	> 20,600
With an enzyme	12,000

This means that when an enzyme is present, less energy, and, therefore, less heat, is needed to bring about a reaction. Enzymes lower the energy of activation by bringing the substrates together so that a reaction can occur.

Conditions Affecting Yield

Certain factors influence the yield of enzymatic reactions such as the following:

Concentrations If the concentration of the substrate or enzyme is increased, the amount of product increases; that is, the more *S* or *E* available, the more *P* there is within a certain amount of time. In many instances, the substrate is plentiful within the cell, but the enzyme is present only in small amounts. The amount of enzyme limits the overall rate of the reaction. In other words, if there is only a small amount of enzyme present, there will be less product in a given unit of time.

Inhibition It happens that on occasion another molecule is so close in shape to the enzyme's substrate that this molecule can *compete* with the true substrate for the active site of the enzyme. However, since this molecule is not a reactant in the enzyme's reaction, the product is never realized. Such a molecule is designated as *I* for inhibitor in these reactions:

$$\mathrm{I + E \longrightarrow EI \longrightarrow \text{no further reaction}}$$
$$\mathrm{I + E \longrightarrow EI \longrightarrow E + I}$$

[1]Cal/mole = Calories per mole. A calorie is a common method of measuring heat, and a mole is 6.02×10^{23} molecules of the substance being considered.

The first reaction is irreversible and the enzyme and inhibitor never become unlocked. This situation is very serious since the enzyme is thereafter incapacitated. The second reaction is reversible and the enzyme is only temporarily out of action. Death of bacteria due to the drug penicillin is an example of irreversible competition for the active site of an enzyme. When penicillin binds with this enzyme it cannot perform its normal function necessary to the formation of the cell wall and the bacteria die. On the other hand, the sulfonamide drugs only reversibly bind to a bacterial enzyme that is necessary for folic acid production. When sulfa drugs are taken, bacteria die but the human body is unaffected. In humans, hydrogen cyanide is an inhibitor for a very important enzyme (cytochrome oxidase) in all cells, and this accounts for its lethal effect on the human body.

Some substances can inhibit an enzyme by combining with it at a site other than the active site. This *noncompetitive* inhibition causes the enzyme to assume a shape that inactivates it. This is the manner by which lead (Pb) causes lead poisoning.

Not all cases of inhibition lead to death of the organism. Reversible inhibition is often used as a normal means to control the metabolic activity of the cell. Consider, for example, that it would be possible to slow down the metabolic pathway depicted on page 94 if the end product *G* were to inhibit enzyme 3. Such feedback inhibitions are quite common in cells.

Denaturation A denatured protein is one that has lost its normal configuration and therefore its ability to form an enzyme-substrate complex. These environmental factors can cause denaturation:

1. *Temperature.* Cold temperatures slow down chemical reactions, and warm temperatures speed up chemical reactions (including enzymatic reactions). High temperatures, however, affect hydrogen bonding and can cause the enzyme to restructure in a way that inactivates it.
2. *pH.* Each enzyme has a preferred pH at which the speed of the reaction is optimum. Presumably, any other pH affects the ionic bonding between the side chains of the molecule, leading to denaturation.
3. Some *other factors* that cause denaturation are inhibitors of enzymes such as those mentioned previously; for example, heavy metals (e.g., Pb^{++} and Hg^{++}) cause enzymes to be denatured. Exposure to ionizing radiation will also inactivate enzymes by denaturing them.

Coenzymes

Enzymes are very often composed of two portions: a protein called an **apoenzyme** and a nonprotein group called a **coenzyme.** The protein portion (apoenzyme) of the enzyme accounts for its specificity; that is, the ability of the enzyme to speed up only one particular reaction. The coenzyme portion of the enzyme may actually participate in the reaction by accepting or contributing atoms to the reaction:

Coenzyme	**Apoenzyme**
nonprotein	protein
helper	specificity

Coenzymes are generally large molecules that the body may be incapable of synthesizing. Many are vitamins that are required organic molecules in our diet. Vitamins are needed in only small amounts for efficient metabolism. Niacin (or nicotinic acid), thiamine (or vitamin B), riboflavin, folic acid, and biotin are all well-known vitamins that are parts of coenzymes.

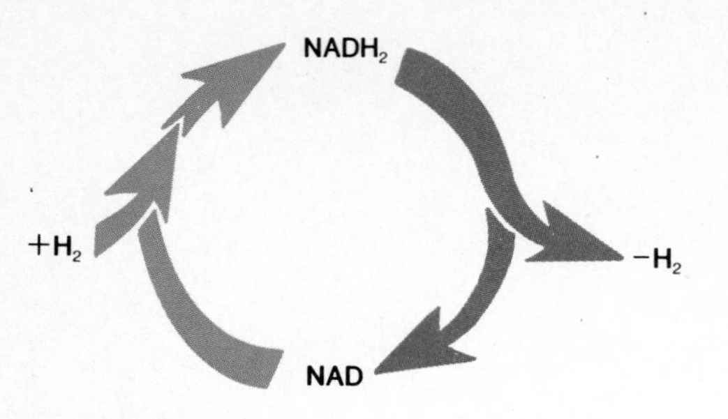

Figure 5.4
The NAD cycle. NAD is reduced and becomes $NADH_2$ when it accepts hydrogen atoms and becomes oxidized to NAD again when the hydrogens are passed to another acceptor.

NAD Cycle

Presently we wish to consider a coenzyme, known as **NAD**[2], that contains the vitamin niacin. **Dehydrogenase** enzymes often incorporate the coenzyme NAD which removes and passes H atoms from one substrate to another. Therefore dehydrogenases are agents of oxidation (removal of hydrogen atoms) and reduction (addition of hydrogen atoms). Only a small amount of a coenzyme, like NAD, is present in a cell because the same coenzyme molecule is used over and over again, just as an enzyme is used over and over again. Figure 5.4 illustrates this point. After NAD accepts hydrogen atoms and becomes (reduced) $NADH_2$, $NADH_2$ is apt to turn around and pass the hydrogen atoms to another acceptor, becoming (oxidized) NAD again.

Enzymes are protein molecules that have a shape appropriate to their substrate. Any environmental factor that affects the shape of a protein also affects the ability of an enzyme to speed up its reaction. Most enzymes require a coenzyme. NAD is a coenzyme of oxidation-reduction.

Energy

When cells require energy, a certain kind of energy must be available. Electricity is the type of energy we use to light our homes and run our electric appliances. Cells, through the process of evolution, have come to depend on the molecule *ATP* (adenosine triphosphate) whenever they require energy, just as our society has come to depend on electricity.

ATP Reaction

ATP (fig. 5.5) is a nucleotide composed of the base adenine and the sugar ribose (together called adenosine) and three phosphate groups. The wavy lines in the formula for ATP indicate high-energy phosphate bonds; when these bonds are broken, an unusually large amount of energy is released. Because of this property, ATP is the energy currency of cells; when cells "need" something, they "spend" ATP.

ATP is used in body cells for synthetic reactions, active transport, nervous conduction, and muscle contraction. When energy is required for these processes, the end phosphate group is removed from ATP, breaking down the molecule to ADP (adenosine diphosphate) and Ⓟ (phosphate) (fig. 5.5).

[2]Nicotinamide adenine dinucleotide.

Figure 5.5
ATP, the energy molecule in cells, has two high-energy phosphate bonds (indicated in the figure by wavy lines). When cells require energy, the last phosphate bond is broken and a phosphate molecule is released.

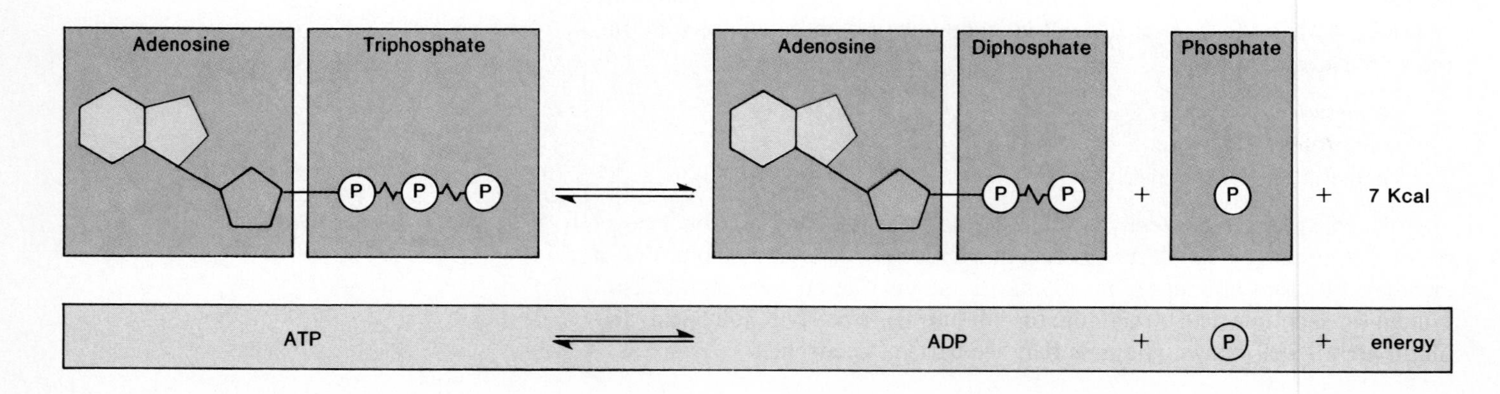

ATP Cycle

The reaction shown in figure 5.5 occurs in both directions; not only is ATP broken down, it is also built up when ADP joins with Ⓟ. Since ATP breakdown is constantly occurring, there is always a ready supply of ADP and Ⓟ to rebuild ATP again.

Figure 5.6 illustrates the ATP cycle in a diagrammatic way. Notice that when ATP is broken down, energy is released and when it is built up, energy is required. We shall see that aerobic cellular respiration, a metabolic pathway that takes place largely within mitochondria, produces energy needed for ATP buildup.

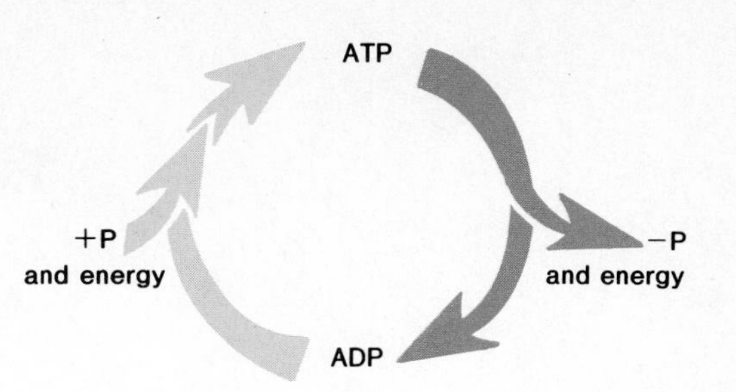

Figure 5.6
The ATP cycle. When ADP joins with a P group, energy is required; but when ATP breaks down to ADP and a P group, energy is given off.

ATP is the energy molecule in cells because it contains high-energy phosphate bonds. ATP breaks down to ADP + Ⓟ + energy, and can be built up from these same components.

Aerobic Cellular Respiration (Simplified)

Both plants and animals, whether they reside in the water or on land, carry on aerobic cellular respiration (fig. 5.7). During this process, organic molecules are oxidized by the removal of hydrogen atoms. Oxidation releases the energy needed to cause ATP buildup (fig. 5.8).

Figure 5.7
Almost all organisms, whether they reside in the water or on land, take in oxygen and carry on aerobic cellular respiration. During aerobic cellular respiration, metabolites are oxidized and the energy released is used to form ATP molecules.

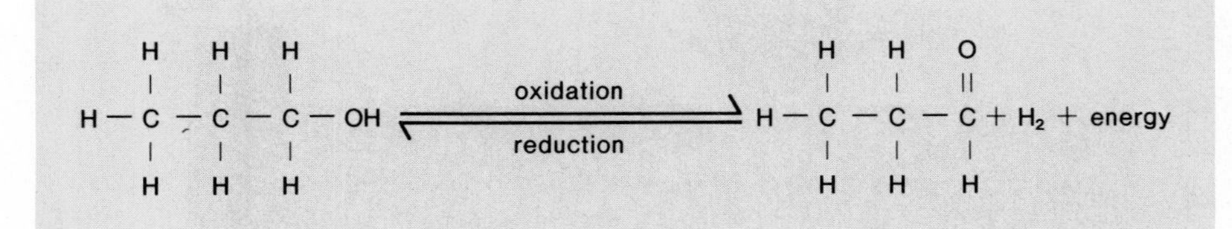

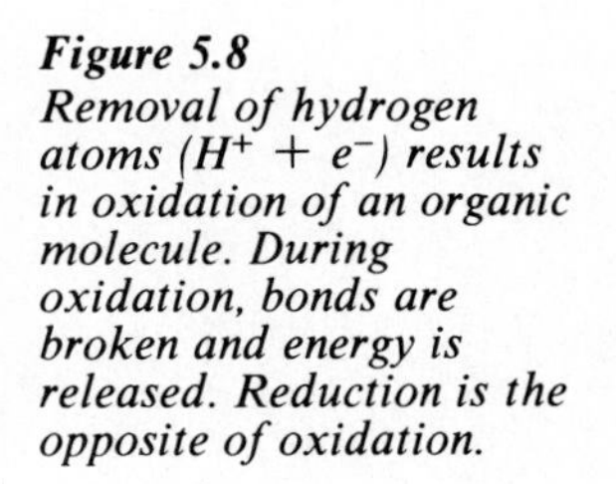

Figure 5.8
Removal of hydrogen atoms ($H^+ + e^-$) results in oxidation of an organic molecule. During oxidation, bonds are broken and energy is released. Reduction is the opposite of oxidation.

Overall Equation

An overall equation for aerobic cellular respiration is shown in figure 5.9. The term **aerobic** means that oxygen is required for the process. As the equation suggests, cellular respiration most often begins with glucose but, as we shall see later, other molecules can also be used. Most importantly, we want to realize that as glucose is broken down, ATP molecules are produced, and this is the reason the ATP reaction is drawn using a curved arrow above the glucose reaction arrow.

Subpathways

During cellular respiration, glucose is oxidized to carbon dioxide and water (fig. 5.9). The oxidation of glucose does not occur in one step; instead it occurs bit by bit. The entire process requires three subpathways (*glycolysis, the Krebs cycle, and the respiratory chain*). The transition reaction is an intermediate step connecting glycolysis with the Krebs cycle, as illustrated in figure 5.10.

$$C_6H_{12}O_6 + 6O_2 \xrightarrow{38\,ADP + 38P \;\to\; 38ATP} 6CO_2 + 6H_2O$$

glucose + oxygen ⟶ carbon dioxide + water

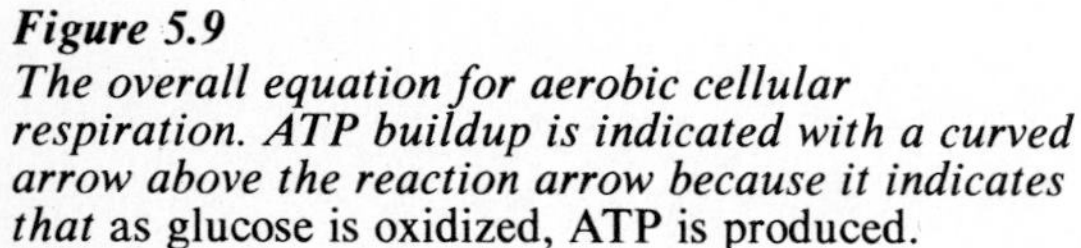

Figure 5.9
The overall equation for aerobic cellular respiration. ATP buildup is indicated with a curved arrow above the reaction arrow because it indicates that as glucose is oxidized, ATP is produced.

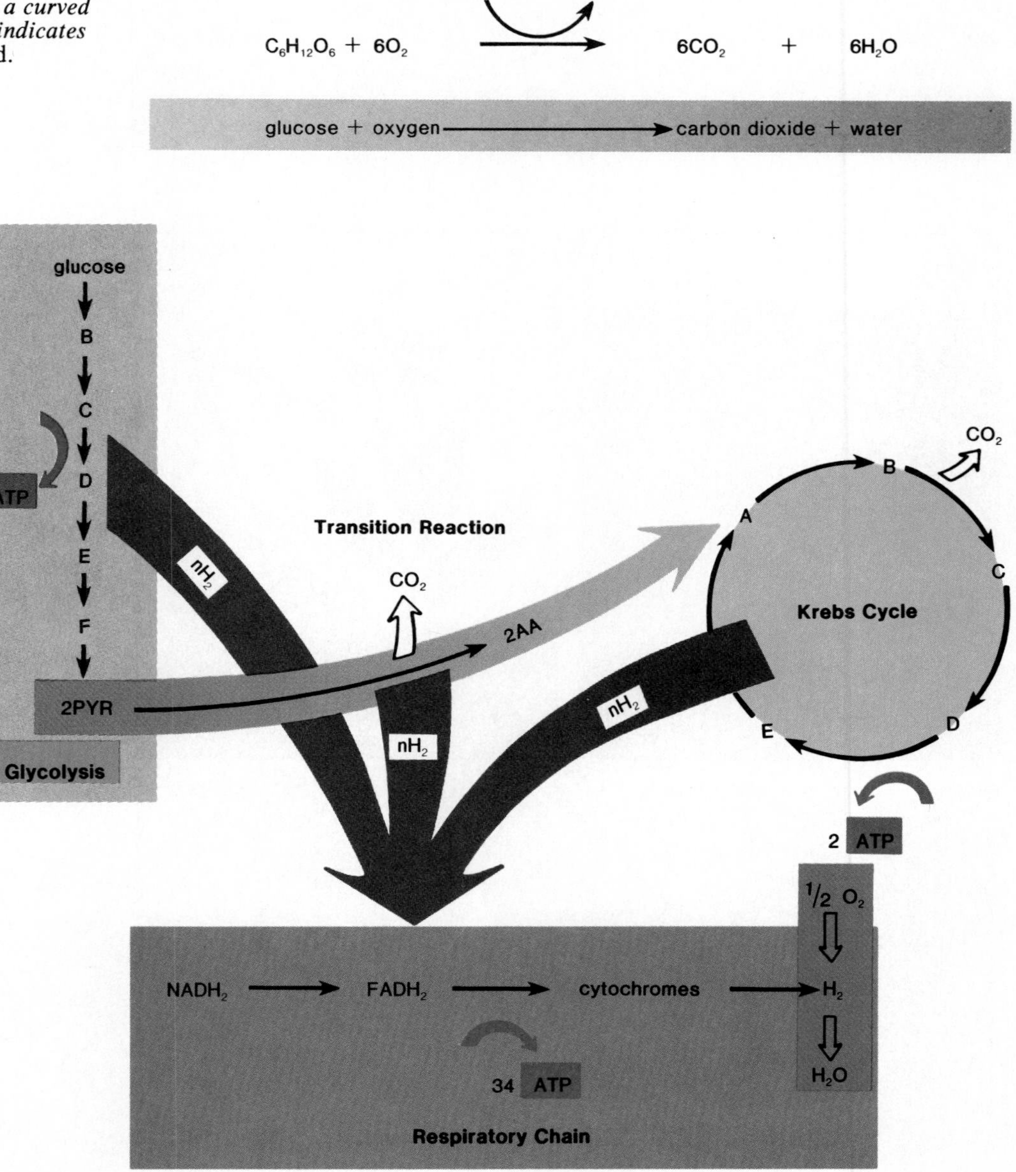

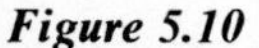

Figure 5.10
Aerobic cellular respiration contains three subpathways: glycolysis, Krebs cycle, and the respiratory chain which are discussed on page 101. As the reactions occur, a number of hydrogen atoms (nH_2) and carbon dioxide molecules are removed from the various substrates. Oxygen acts as the final acceptor for the hydrogen atoms.

In the figure, each arrow represents a different enzyme, and the letters represent the product of the previous reaction and the substrate for the next. Notice how each pathway resembles a conveyor belt in which a beginning substrate continuously enters at the start and, after a series of reactions, end products leave at the termination of the belt. It is important to realize, too, that all three pathways are going on at the same time. They can be compared to the inner workings of a watch in which all parts are synchronized.

It is possible to relate the reactants and products of the overall reaction (fig. 5.9) to the subpathways in figure 5.10:

1. Glucose, $C_6H_{12}O_6$, is to be associated with **glycolysis,** the breakdown of glucose to two molecules of pyruvic acid (PYR). Oxidation by removal of hydrogen atoms provides enough energy for the buildup of two ATP molecules.
2. Carbon dioxide, CO_2, is to be associated with the transition reaction and the Krebs cycle. During the **transition reaction,** PYR is oxidized to active acetate (AA). AA enters the **Krebs cycle,** a cyclical series of oxidation reactions that give off CO_2 and produce one ATP molecule. Notice that since glycolysis ends with two molecules of PYR, the transition reaction and the Krebs cycle occur twice per glucose molecule. Altogether, then, the Krebs cycle accounts for two ATP molecules per glucose molecule.
3. Oxygen, O_2, and water, H_2O, are to be associated with the respiratory chain. When the coenzyme $NADH_2$ enters the chain, it carries with it most of the hydrogen atoms removed during glycolysis, the transition reaction, and the Krebs cycle. The **respiratory chain** (often referred to as the electron transport system, ETS, or the cytochrome system) is a series of molecules that pass hydrogen and/or electrons from one to the other until they are finally received by oxygen, which is then reduced to water. As the electrons pass from one molecule to the next, oxidation occurs and this releases the energy needed for ATP buildup.
4. ATP is to be associated with glycolysis, the Krebs cycle, and the respiratory chain. Most ATP, however, is produced by the respiratory chain. The chain can produce as many as 34 ATP molecules per glucose molecule.

Table 5.2 summarizes our discussion of aerobic cellular respiration. This chart assumes that aerobic cellular respiration produces 38 ATP per glucose molecule.

***Table 5.2** Overview of Aerobic Cellular Respiration*

Name of Pathway	Result
Glycolysis	Removal of H_2 from substrates Produces 2 ATP molecules
Transition reaction	Removal of H_2 from substrates Releases CO_2
Krebs cycle	Removal of H_2 from substrates Releases CO_2 Produces 2 ATP after two turns
Respiratory chain	Accepts H_2 from other pathways and passes them on to O_2 producing H_2O Produces 34 ATP

Figure 5.11
Structure versus function in a mitochondrion. A mitochondrion has both an outer and inner membrane. The inner membrane is folded into cristae and surrounds the matrix. The enlargement shows that the enzymes responsible for the Krebs cycle are located in the matrix and the carriers of the respiratory chain are located along the cristae. These molecules pass hydrogen atoms from one to the other and deposit them in the membranous space between the inner and outer membrane. This use of the energy of oxidation has been associated with the eventual buildup of ATP molecules.

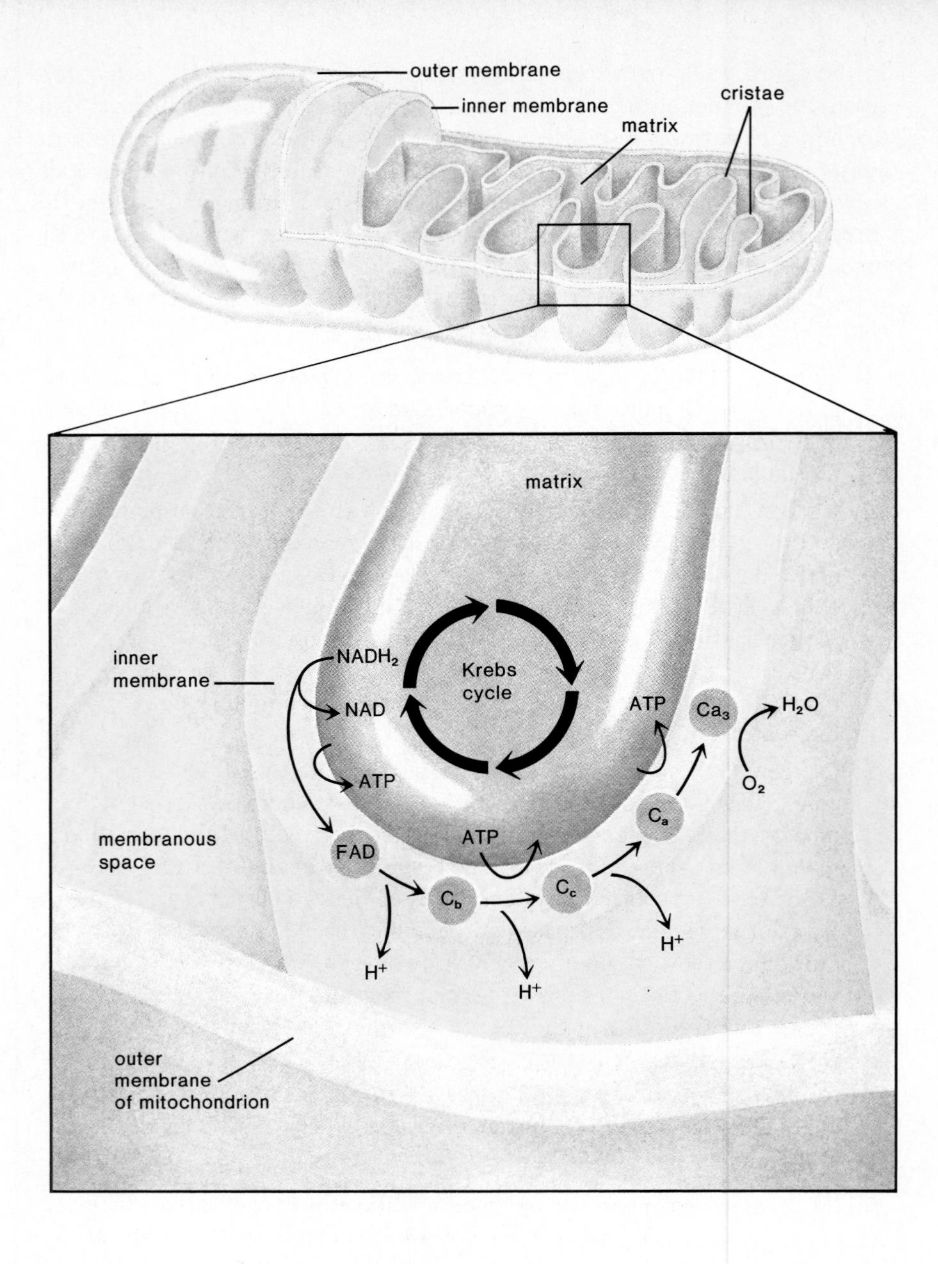

Mitochondria

A significant portion of aerobic cellular respiration takes place in mitochondria. Glycolysis occurs outside the mitochondria, but the transition reaction, Krebs cycle, and respiratory chain occur within the mitochondria. Evidence suggests that the enzymes of the Krebs cycle are located in the **matrix,** an innermost compartment filled with a gelatinlike substance, while the molecules of the respiratory chain are located along the **cristae,** shelflike projections of the inner membrane (fig. 5.11).

The breakdown of glucose to carbon dioxide and water, termed aerobic cellular respiration, is accompanied by the buildup of ATP molecules. Glycolysis and the Krebs cycle produce only a total of 4 ATP directly. Electrons carried by NAD to the respiratory chain, however, result in an additional maximum number of 34 ATP. The Krebs cycle and respiratory chain are located within mitochondria.

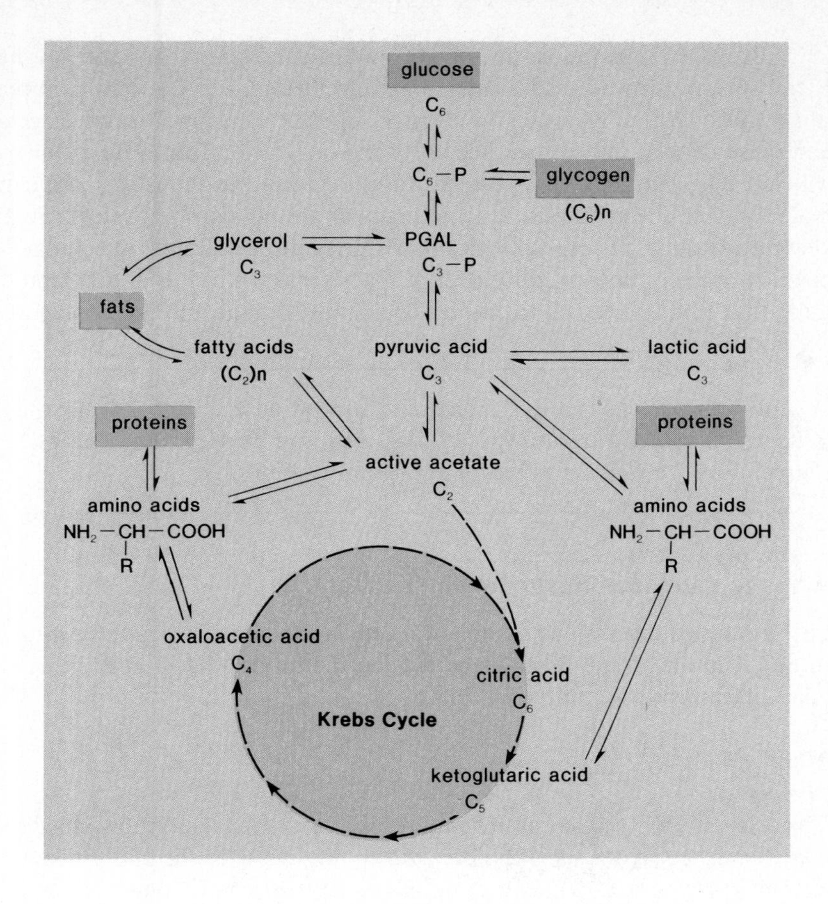

Figure 5.12
The major metabolic pathways within a cell are interrelated as shown. Glycolysis and the Krebs cycle can serve to both break down or build up the molecules noted.

It is interesting to think about how our bodies provide the reactants for aerobic cellular respiration and how they dispose of the products. The air we breathe contains *oxygen,* and the food we eat contains *glucose.* These enter the bloodstream that carries them to the body's cells, where they diffuse into each and every cell. In mitochondria, glucose products are broken down to carbon dioxide and water as ATP is produced. All three of these leave the mitochondria. The *ATP* is utilized inside the cell for energy-requiring processes. Carbon dioxide diffuses out of the mitochondria and out of the cell into the bloodstream. The bloodstream takes the *carbon dioxide* to the lungs, where it is exhaled. The *water* molecules produced, called metabolic water, only become important if by chance this is the organism's only supply of water. In these cases, it can help prevent dehydration of the organism.

Other Metabolites

Other molecules besides glucose can be oxidized by the cell to release energy. When fats are used as an energy source they break down to glycerol and three fatty acids. As figure 5.12 indicates, glycerol is easily converted to PGAL, a metabolite in the glycolytic pathway. The fatty acids are converted to active acetate, which enters the Krebs cycle. A fatty acid that contains 18 carbons will result in nine active acetates. Calculation shows that respiration of these can produce a total of 216 ATP. For this reason, fats are an efficient form of stored energy because there are three long fatty acid chains per fat molecule.

The carbon skeleton of certain amino acids can enter the Krebs cycle (fig. 5.12). Before the skeleton enters, the amino acid must undergo **deamination,** or the removal of the amino group. Later, this group is converted to ammonia (NH_3), an excretory product of cells. Just where the carbon skeleton enters the Krebs cycle is dependent on the length of the R group since this determines the number of carbons left following deamination.

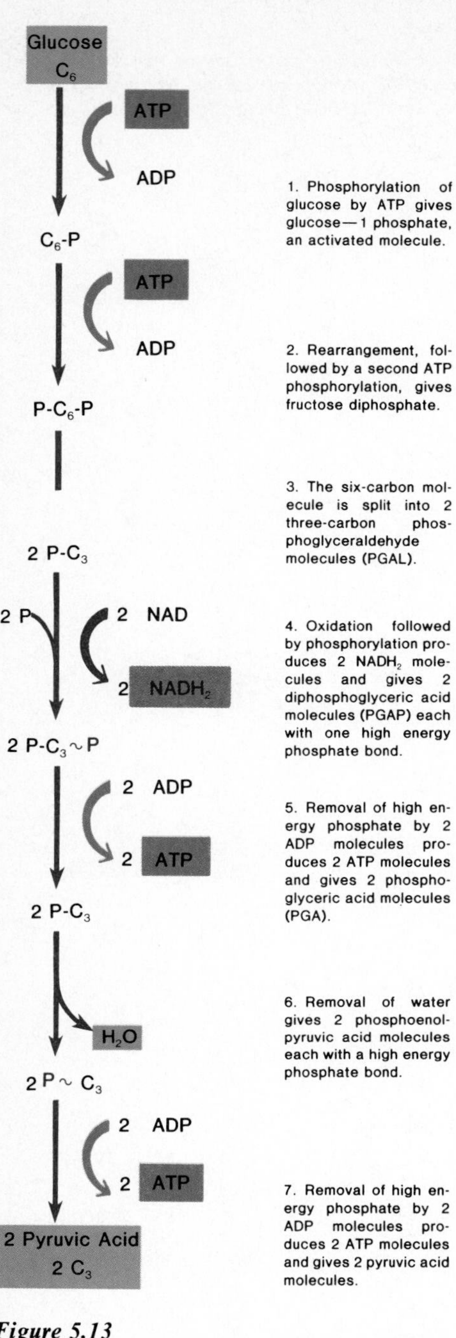

Figure 5.13
Glycolysis is a metabolic pathway that begins with glucose and ends with pyruvic acid. Net gain of two ATP molecules can be calculated by subtracting those expended from those produced. Print in the boxes explains each reaction.

Table 5.3 Glycolysis Summary

Reaction	$NADH_2$	ATP
1		−1 ATP
2		−1 ATP
3		
4	2 $NADH_2$	
5		+2 ATP
6		
7		+2 ATP
Net gain	2 $NADH_2$	+2 ATP

Because the reactions are reversible, the substrates that make up these pathways can also be used as starting molecules for various synthetic reactions. Consider that glycogen (or starch) can be broken down to active acetate and these may be joined together to form a fatty acid. This is the manner in which the ingestion of carbohydrates causes us to gain weight. Also, the starting molecules for the synthesis of some kinds of amino acids is oxaloacetic and ketoglutaric acids. These examples illustrate that the reactions of cellular respiration do more than produce energy. They form the center of a "metabolic mill" that enables the cell to convert one kind of organic molecule to another.

All the reactions involved in cellular respiration are part of a metabolic mill in which various types of molecules can be used as an energy source or as the starting point for synthetic reactions.

Aerobic Cellular Respiration (in Detail)

In the detailed discussion that follows, cellular respiration is divided into two stages. The first stage takes place outside the mitochondria and the second takes place inside the mitochondria.

Stage One

Glycolysis

Glycolysis is that portion of cellular respiration that takes place outside the mitochondria. Glycolysis is the breakdown of glucose to the end product **pyruvic acid** (pyruvate). During these reactions, hydrogen atoms are removed from the substrates of the pathway and are picked up by NAD. When the hydrogen atoms are removed, oxidation of the substrates has occurred. Oxidation releases energy and, in this case, the energy is first captured within the substrate molecules and then used to supply enough energy to form two ATP molecules. The net result of glycolysis is the formation of two $NADH_2$ molecules and two ATP molecules. This can be better appreciated by examining the series of reactions that comprise glycolysis (fig. 5.13). In this figure, each of the molecules involved is represented by the number of carbon atoms it contains (each molecule also contains hydrogen and oxygen atoms). Table 5.3 summarizes the reactions of glycolysis.

As we shall see on page 108, glycolysis is also a part of anaerobic respiration, a process that is often called fermentation.

Stage one of aerobic cellular respiration takes place outside the mitochondria. During glycolysis, glucose is broken down to pyruvic acid with a net gain of 2 ATP and 2 $NADH_2$.

Stage Two

This portion of aerobic cellular respiration occurs within the mitochondria (fig. 5.11) and consists of a transition reaction, the Krebs cycle, and the respiratory chain.

Transition Reaction

The transition reaction connects glycolysis to the Krebs cycle, as is apparent in figure 5.10. In this reaction, pyruvic acid (PYR) is converted to a molecule called **active acetate** (AA), or acetyl coenzyme A, and carbon dioxide is given off in the process.

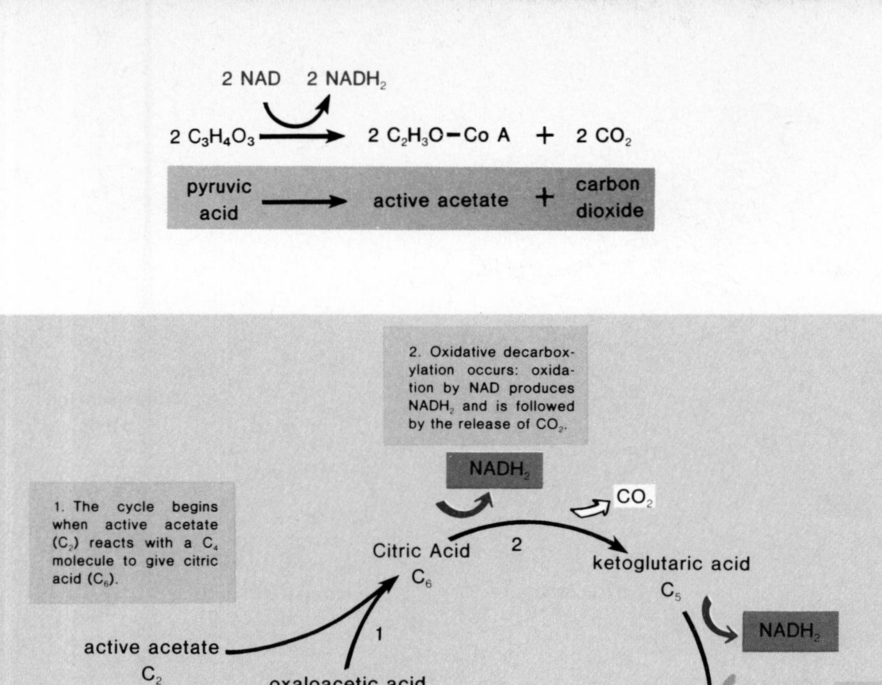

Figure 5.14
During the transition reaction, pyruvic acid is oxidized to a two-carbon acetyl group attached to coenzyme A.

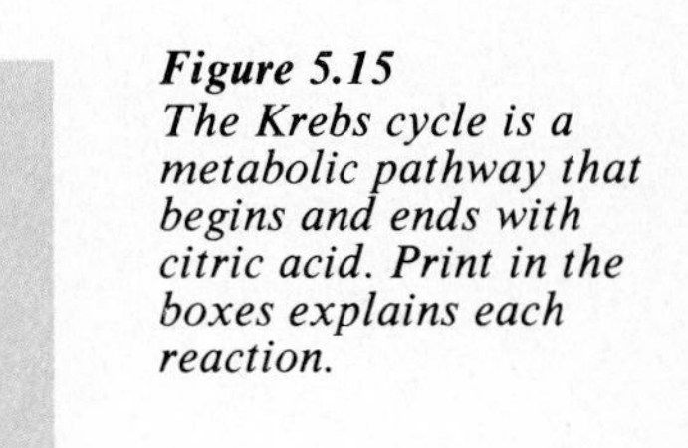

Figure 5.15
The Krebs cycle is a metabolic pathway that begins and ends with citric acid. Print in the boxes explains each reaction.

As figure 5.14 shows more clearly, this is an oxidation reaction in which hydrogen atoms are removed from pyruvic acid by NAD. Also, active acetate is actually an acetyl group attached to a coenzyme called **coenzyme A.** This coenzyme activates the acetyl group, hence the name *active acetate.* Notice that since glycolysis results in two pyruvic acid molecules, the transition reaction occurs twice per glucose molecule. Altogether, the reaction produces two molecules of carbon dioxide, two molecules of active acetate, and two $NADH_2$ per glucose molecule.

Krebs Cycle

The Krebs cycle, represented in detail in figure 5.15, is named for the person who discovered it. It is called a cycle because it is a series of reactions, or steps, that begin and end with **citric acid.** Sometimes the cycle is called the *citric acid cycle.* During this series of reactions which occur in the matrix of mitochondria (fig. 5.11), *oxidative decarboxylation* occurs. **Oxidative decarboxylation** means that hydrogen atoms and carbon dioxide are removed from the substrates at the same time. Oxidation results in energy that is partially captured by the molecules within the cycle and this energy is used to form 1 ATP per turn. The carbon dioxide given off is a metabolic waste and is excreted by cells. Most of the hydrogen atoms are picked up by NAD, but a few are taken by FAD. **FAD** is another coenzyme of oxidation-reduction, which is used infrequently compared to NAD. Altogether, the Krebs cycle turns twice per glucose molecule and produces 4 CO_2, 6 $NADH_2$, 2 $FADH_2$, and 2 ATP. Table 5.4 summarizes the reactions of the Krebs cycle.

***Table 5.4** Summary of Krebs Cycle**

Step	CO_2	$NADH_2$	$FADH_2$	ATP
1				
2	CO_2	$NADH_2$		
3	CO_2	$NADH_2$		ATP
4			$FADH_2$	
5		$NADH_2$		
	2 CO_2	3 $NADH_2$	$FADH_2$	ATP

*Per turn. Cycle turns twice per glucose molecule.

Figure 5.16
Respiratory chain. Hydrogen atoms enter the chain attached to either NAD or FAD. Thereafter, certain molecules of the chain accept only electrons and deposit hydrogen ions between the inner and outer membrane of the mitochondrion (see fig. 5.11). This creates an electrochemical gradient that supplies the energy necessary for ATP synthesis.

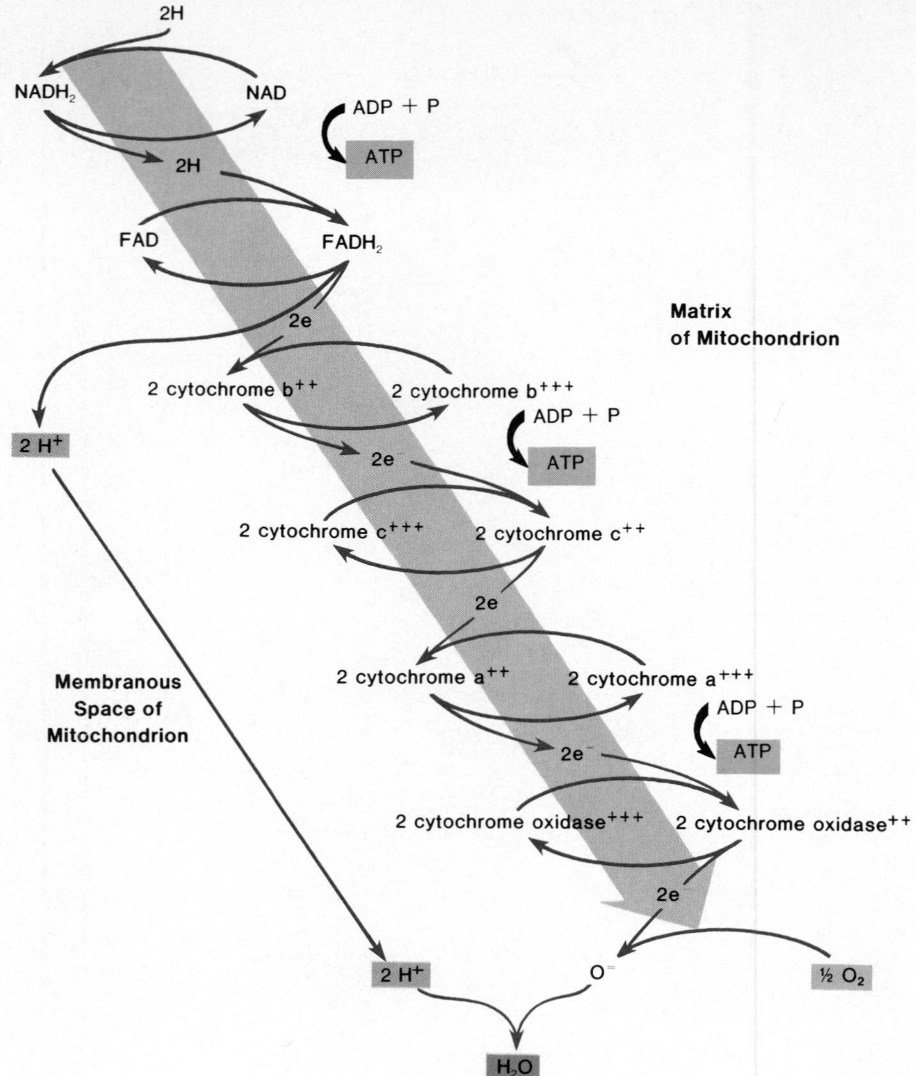

Respiratory Chain

The respiratory chain (figs. 5.10 and 5.16) is a series of molecules that pass electrons from one to the other. The electrons are at first a part of the hydrogen atoms ($H^+ + e^-$) attached to NAD or FAD. These are the same hydrogen atoms that were removed from the molecules of glycolysis and the Krebs cycle. When the electrons enter the chain located on the cristae of mitochondria (fig. 5.11), they are at a high-energy level, but as they are passed "downhill" from one molecule to another, they lose energy as oxidation occurs. For example, at the top of figure 5.16, as $NADH_2$ becomes NAD and is oxidized, the two hydrogens are passed to FAD which then becomes reduced as $FADH_2$. At this point the cytochromes accept only electrons and the hydrogens are pumped into the membranous space of the mitochondrion. Later this established electrochemical gradient is used to produce ATP. In this way the energy of oxidation is eventually converted to ATP energy. If the electrons enter attached to NAD, there is often enough energy to produce 3 ATP for every two hydrogen atoms. However, if they enter by way of $FADH_2$, only 2 ATP molecules are produced.

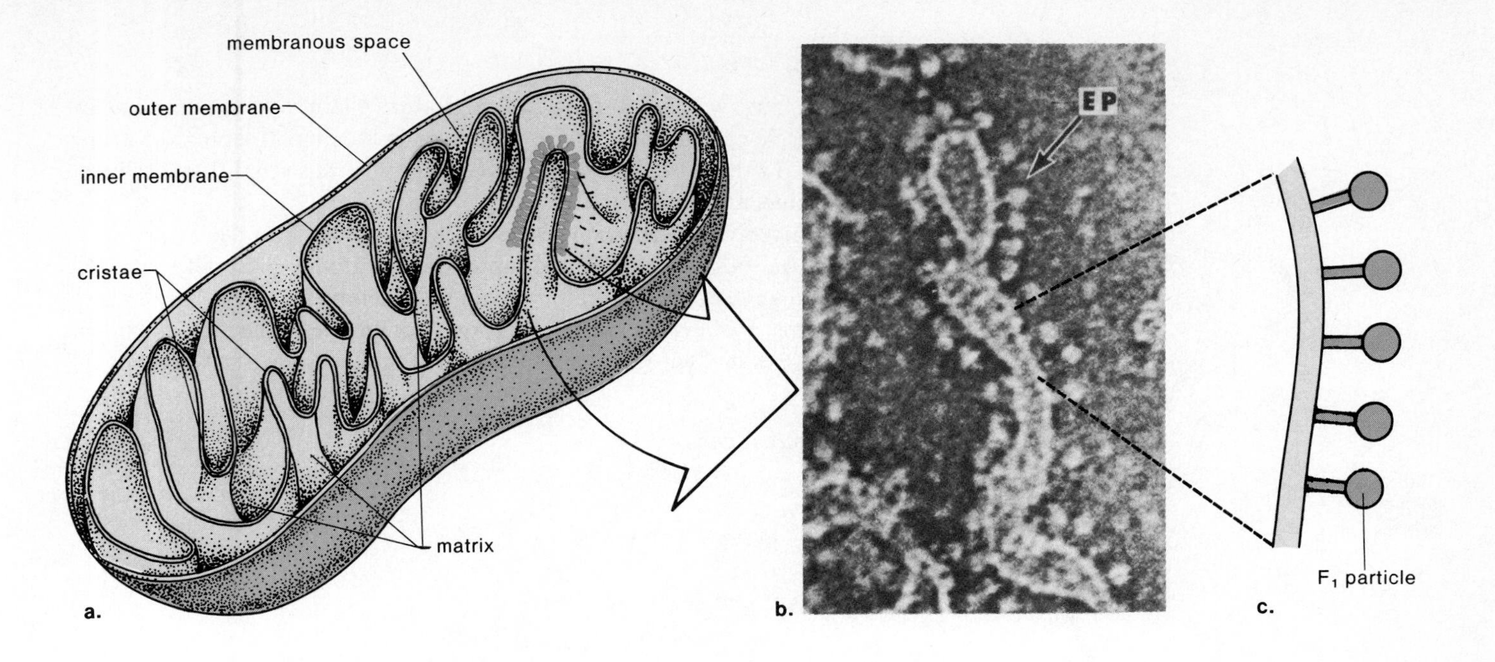

Figure 5.17
Inner membrane of a mitochondrion contains F_1 particles. ATP production occurs at these particles. a. *Longitudinal section of mitochondrion.* b. *Micrograph showing inner membrane.* c. *Drawing of F_1 particles.*

$FADH_2$ passes electrons to a series of cytochrome molecules. Because of this, another name for the respiratory chain is the *cytochrome system.* It happens that the cytochromes, and other respiratory chain molecules not mentioned, accept only electrons and deposit the hydrogen ions (H^+) outside the inner mitochondrial membrane (fig. 5.11). This uneven distribution of hydrogen ions is believed to be highly significant since it creates an electrochemical gradient that supplies the necessary energy for ATP production at F_1 particles (fig. 5.17).[3]

The passage of electrons down the respiratory chain gives it still another name, the **electron transport system.** Oxygen is the final acceptor for both electrons and hydrogen ions. Therefore, water is an end product of cellular respiration.

As table 5.5 indicates, the respiratory chain accounts for most of the ATP produced during aerobic cellular respiration and since the chain is present in mitochondria, this makes them the powerhouses of the cell. For the sake of discussion, it is usually calculated that the chain produces 34 of a possible total 38 ATP molecules per glucose molecule. These numbers are for discussion purposes because the chain produces a varying amount of ATP that is usually less than 34 ATP per glucose molecule.

Stage two of aerobic cellular respiration takes place within the mitochondria. Following the transition reaction, the two revolutions of the Krebs cycle in the matrix produce 6 NAD_2, 2 $FADH_2$, and 2 ATP. Once the hydrogen atoms have entered the electron transport system located on the cristae, a maximum of 34 ATP are produced in addition to the four already made.

Table 5.5 *Summary of ATP Produced by Cellular Respiration*

	Direct	By Way of Respiratory Chain
Glycolysis	2 ATP	2 $NADH_2$ = 6 ATP*
Transition reaction		2 $NADH_2$ = 6 ATP
Krebs cycle	2 ATP	6 $NADH_2$ = 18 ATP 2 $FADH_2$ = 4 ATP
Subtotal	4 ATP	34 ATP
Grand total	38 ATP*	

*The numbers in this column and the total number of ATP are usually less because the chain does not always produce the maximum possible number of ATP per $NADH_2$.

[3]This is known as the chemiosmotic theory.

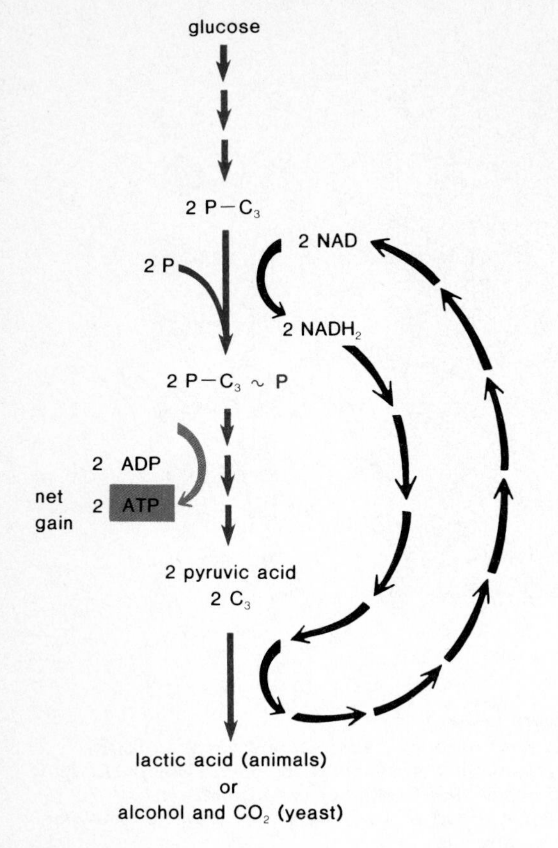

Figure 5.18
Fermentation consists of the glycolytic pathway plus one additional reaction in which the end product of the glycolytic pathway, pyruvic acid, accepts hydrogens and becomes reduced. This "frees" NAD so that it may return to pick up more hydrogen atoms.

Table 5.6 *Aerobic versus Anaerobic Respiration*		
	Aerobic	**Anaerobic**
Maximum number of ATP per glucose molecule	38 ATP	2 ATP
Final acceptor for hydrogen atoms	oxygen	Pyruvic acid
End product	CO_2 & H_2O	Lactic acid (animal) Ethyl alcohol & CO_2 (plant)

Anaerobic Cellular Respiration

If oxygen is not available to cells, the respiratory chain soon becomes inoperative because electrons plug up the system when the final acceptor, oxygen, is not present. In this case, most cells have a safety valve so that some ATP can still be produced.

The glycolytic pathway will run as long as it is supplied with "free" NAD; that is, NAD that can pick up hydrogen atoms. Normally, $NADH_2$ passes hydrogens to the respiratory chain and thereby becomes "free" of hydrogen atoms. However if the chain is not working due to the lack of oxygen, $NADH_2$ passes its hydrogen atoms to pyruvic acid, as shown in the following reactions:

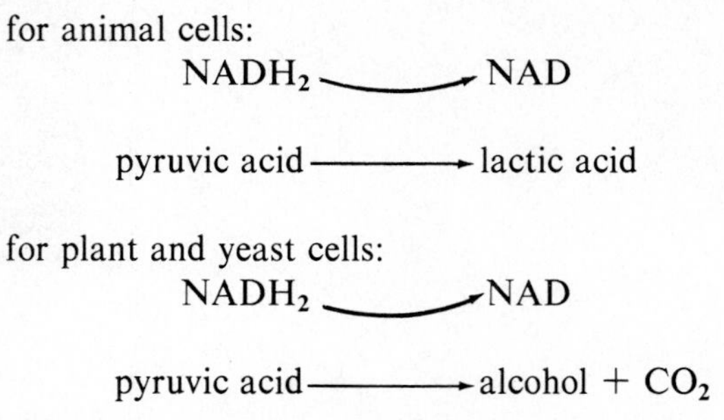

The whole purpose of this reaction, as shown in figure 5.18, is to keep the limited amount of NAD available in the cell free of hydrogen atoms so NAD can continue to oxidize PGAL (designated as 2 P—C_3 in the diagram). Without this reaction the glycolytic pathway would also stop occurring.

I f oxygen is not available in cells, the respiratory chain is inoperative and fermentation, anaerobic respiration, occurs. Pyruvic acid, from glycolysis, is reduced by $NADH_2$ to lactic acid (animal cells) or alcohol and CO_2 (yeast cells), harmful end products. This reaction frees NAD so that glycolysis can occur once more. Fermentation has an energy yield of only 2 ATP per glucose molecule compared to a possible 38 from aerobic respiration (table 5.6).

Notice that anaerobic respiration, also called **fermentation,** consists of the glycolytic pathway plus one additional reaction in which the end product of the glycolytic pathway, pyruvic acid, accepts hydrogens and becomes reduced. The Krebs cycle and respiratory chain do not function as part of fermentation, but when oxygen is available again, lactic acid can be converted back to pyruvic acid and metabolism can proceed as usual.

Fermentation in animals is an impractical process for two reasons. First, it produces only 2 ATP per glucose molecule compared to a much larger number for aerobic respiration. Second, it results in lactic acid buildup. Lactic acid is toxic to cells and causes muscles to cramp and fatigue (p. 346). If fermentation continues for any length of time, death will follow.

Yeast fermentation is useful to humans in at least two ways: the baking of bread and the production of alcohol. The reading on page 109 discusses the latter process.

Alcoholic Beverages

Wine, beer, and whiskey production all require yeast fermentation. To produce wine, grape juice is allowed to ferment. After the grapes are picked, they are crushed in order that the juice may be collected. In the old days, wine makers simply relied on spontaneous fermentation by yeasts that were on the grape skins, but now many add specially selected cultures of yeast. Also, it is common practice to maintain the temperature at about 20° C for white wines and 28° C for red wines. Fermentation ceases after most of the sugar has been converted to alcohol. Various methods are used to clarify the wine, that is, the removal of any suspended materials. Also many fine wines improve when they are allowed to "age" during barrel or bottle storage.

Brewing beer is more complicated than wine production. Usually grains of barley are first *malted;* that is, allowed to germinate for a short time so that amylase enzymes are produced that will break down the starch content of the grain. After the germinated grains have been crushed and mixed with water, the *malt wort* is separated from the spent grains and traditionally boiled with hops (an herb derived from the hop plant) to give flavor to the beer. Now the *hop wort* is seeded with a strain of yeast that converts the sugars in the wort to alcohol and carbon dioxide. At the end of fermentation, the yeast is separated from the beer, which is then allowed to mature for an appropriate period. After filtration and pasteurization, the beer is packaged.

The production of whiskey (from grains), brandy (from grapes), and rum (from molasses) differs from wine and beer production chiefly in that the alcohol is removed from the fermented substance by distillation. Most often, in the United States, corn or rye is used in the production of whiskey. These grains are ground up and mashed to release their starch content. Then amylase enzymes are added to convert the starch to fermentable sugars. Now yeast is added so that fermentation can occur. Following fermentation, the alcohol is concentrated by distillation. A warm temperature causes the alcohol to become gaseous and rise in a column where it condenses to a liquid before entering a collecting vessel. The alcohol content of the collecting vessel is much higher following this distillation process. The distillate is usually stored, quite often in an oak barrel, to improve the aroma and taste of the final product.

Fermentation of grapes (above right), barley (above left), and corn produces wine, beer, and whiskey respectively.

Summary

Metabolic pathways are a series of reactions that proceed in an orderly step-by-step manner. Each of these reactions requires a specific enzyme. Any environmental factor that affects the shape of a protein also affects the ability of an enzyme to do its job. Sometimes enzymes require coenzymes, a nonprotein portion that participates in the reaction. NAD is a coenzyme.

Aerobic cellular respiration (the breakdown of glucose to carbon dioxide and water) requires three subpathways: glycolysis, the Krebs cycle, and the respiratory chain. As a pair of electrons (usually contributed by $NADH_2$) pass down the chain, 34 ATP (maximum) are generated (in addition to 4 ATP produced directly by glycolysis and the Krebs cycle). Other types of molecules such as amino acids and neutral fats can also be respired in cells.

If oxygen is not available in cells, the respiratory chain is inoperative and fermentation (anaerobic cellular respiration) occurs. Pyruvic acid from glycolysis is reduced by $NADH_2$ to either lactic acid (animal cells) or alcohol and CO_2 (yeast), substances harmful to these cells. Now that NAD is available once again, glycolysis can reoccur. Fermentation only gives a net gain of 2 ATP and it will eventually cause the death of animal and yeast cells.

Objective Questions

1. Every reaction that occurs in a cell requires a(n) ____________ .
2. If more substrate or more enzyme is added, you would expect an enzymatic reaction to ____________ .
3. NAD is a coenzyme of ____________ .
4. Glycolysis is the breakdown of glucose to two molecules of ____________ .
5. The Krebs cycle begins and ends with ____________ .
6. The respiratory chain is located on the ____________ of the mitochondria.
7. The final acceptor for hydrogen molecules at the end of the respiratory chain is ____________ .
8. Carbon dioxide formation should be associated with the ____________ , one of the three major pathways involved in aerobic cellular respiration.
9. The immediate acceptor of hydrogen molecules during anaerobic cellular respiration is ____________ .
10. Anaerobic cellular respiration only results in a net gain of ____________ ATP.
11. Fatty acids are broken down to ____________ molecules which enter the Krebs cycle.

Answers to Objective Questions

1. enzyme 2. speed up 3. oxidation-reduction
4. pyruvic acid 5. citric acid 6. cristae
7. oxygen 8. Krebs cycle 9. pyruvic acid
10. two 11. active acetate

Study Questions

1. Discuss and draw a diagram to describe a metabolic pathway. (p. 94)
2. Discuss and give a reaction to describe the specificity of enzymatic actions. (p. 94)
3. Name several factors that affect the yield of enzymatic reactions and explain the effect of these factors. (pp. 94–97)
4. Define coenzyme and give several examples. (p. 97)
5. Which molecule is known as the energy molecule of cells? Why is this molecule appropriate to the task? (p. 98)
6. Give the overall equation for cellular respiration and discuss the equation in general. (p. 100)
7. Name and describe the events within the three subpathways and transition reaction that make up aerobic cellular respiration. (p. 101)
8. How do our body cells obtain glucose and oxygen? What happens to the carbon dioxide given off by these cells? (p. 103)
9. Calculate the number of ATP that are formed as a *result* of glycolysis and the Krebs cycle. (pp. 104, 105)
10. Explain the term *oxidative decarboxylation,* which occurs in the Krebs cycle. (p. 105)
11. The respiratory chain is composed of what type molecules? (p. 106)
12. Calculate the number of ATP that are actually produced as hydrogens pass down the respiratory chain. (p. 107)
13. Why is fermentation wasteful and potentially harmful to the human body? (p. 108)

Key Terms

active acetate (ak'tiv as'ĕ-tāt) an acetyl group attached to coenzyme A; a product of the transition reaction that links glycolysis to the Krebs cycle. *104*

active site (ak'tiv sīt) the region on the surface of an enzyme where the substrate binds and where the reaction occurs. *95*

aerobic (a''er-ōb'ik) growing or metabolizing only in the presence of oxygen as in aerobic respiration. *100*

apoenzyme (ap''o-en'zīm) protein portion of an enzyme. *97*

coenzyme (ko-en'zīm) a nonprotein molecule that aids the action of an enzyme, to which it is loosely bound. *97*

coenzyme A (ko-en'zīm) a coenzyme which participates in the transition reaction and carries the organic product to the Krebs cycle. *105*

deamination (de-am''ĭ-na'shun) removal of an amino group ($-NH_2$) from an amino acid or other organic compound. *103*

dehydrogenase (de-hi'dro-jen-ās) an enzyme that accepts hydrogen atoms, speeding up the process of dehydrogenation. *98*

FAD a coenzyme of oxidation; a dehydrogenase that participates in hydrogen (electron) transport within the mitochondria. *105*

fermentation (fer''men-ta'shun) anaerobic breakdown of carbohydrates that results in organic end products such as alcohol and lactic acid. *108*

glycolysis (gli-kol'ĭ-sis) the metabolic pathway that converts sugars to simpler compounds and ends with pyruvate. *101*

Krebs cycle (krebz si'kl) a series of reactions found within the matrix of mitochondria that give off carbon dioxide. Also called the citric acid cycle because the reactions begin and end with citric acid. *101*

metabolism (mĕ-tab'o-lizm) all of the chemical changes that occur within a cell. *93*

NAD a coenzyme of oxidation; a dehydrogenase that frequently participates in hydrogen transport. *98*

oxidative decarboxylation (ok''sĭ-da'tiv de''kar-bok''sĭ-la'shun) a reaction that involves the release of carbon dioxide as oxidation occurs. *105*

pyruvate (pi'roo-vāt) the end product of glycolysis; pyruvic acid. *104*

respiratory chain (re-spi'rah-to''re chān) a series of carriers within the inner mitochondrial membrane that pass electrons one to the other from a higher energy level to a lower energy level; the energy released is used to build ATP; also called the electron transport system; the cytochrome system. *101*

transition reaction (tran-zish'un re-ak'shun) a reaction within aerobic cellular respiration during which hydrogen and carbon dioxide are removed from pyruvic acid; connects glycolysis to Krebs cycle. *101*

Further Readings for Part One

Allen, R. D. 1987. The microtubule as an intracellular engine. *Scientific American.* 256(2):42.

Avers, C. J. 1981. *Cell biology.* 2d ed. New York: D. Van Nostrand.

Baker, J. J. W., and G. E. Allen. 1981. *Matter, energy, and life.* 4th ed. Reading, Mass.: Addison-Wesley.

Berns, M. W. 1983. *Cells.* 3d ed. New York: Holt, Rinehart & Winston.

Bretscher, M. S. 1985. The molecules of the cell membrane. *Scientific American* 253(4):100.

Darnell, J. E. 1985. RNA. *Scientific American* 253(4):68.

Dautry-Varsat, and H. F. Lodish. 1984. How receptors bring proteins and particles into cells. *Scientific American* 250(5):52.

Dickerson, E. 1980. Cytochrome and the evolution of energy metabolism. *Scientific American* 243(3):136.

Dickerson, R. E. 1983. The DNA helix and how it is read. *Scientific American* 249(6):94.

Doolittle, R. F. 1985. Proteins. *Scientific American* 253(4):88.

Dustin, P. 1980. Microtubules. *Scientific American* 243(2):66.

Felsenfield, G. 1985. DNA. *Scientific American* 253(4):58.

Grivell, L. A. 1983. Mitochondrial DNA. *Scientific American* 248(3):78.

Lake, J. A. 1981. The ribosome. *Scientific American* 245(2):84.

Osborn, M., and K. Weber. 1985. The molecules of the cell matrix. *Scientific American* 253(4):110.

Porter, K. R., and J. B. Tucker. 1981. The ground substance of the living cell. *Scientific American* 244(3):56 (offprint 1494).

Sharon, N. 1980. Carbohydrates. *Scientific American* 243(5):90.

Sheeler, P., and D. E. Bianchi. 1983. *Cell biology: structure, biochemistry, and function.* 2d ed. New York: John Wiley.

Sloboda, R. D. 1980. The role of microtubules in cell structure and cell division. *American Scientist* 68(3):290.

Stryer, L. 1981. *Biochemistry.* 2d ed. San Francisco: W. H. Freeman.

Swanson, C. P., and P. L Webster. 1985. *The cell.* 5th ed. Englewood Cliffs, N.J.: Prentice-Hall.

Unwin, N. 1984. The structure of proteins in biological membranes. *Scientific American* 250(2):78.

Weber, K., and M. Osborn. 1985. The molecules of the cell membrane. *Scientific American* 253(4):100.

Weinberg, R. A. 1985. The molecules of life. *Scientific American* 253(4):48.

Wolfe, S. L. 1981. *Biology of the cell.* 2d. ed. Belmont, Calif.: Wadsworth.

Part Two

Plant Biology

Plants capture radiant energy from the sun and use this energy to convert inorganic chemicals to organic chemicals. Photosynthesis is therefore a two-step process. The first step, which requires light energy, produces ATP and $NADPH_2$. The second step uses these molecules to reduce CO_2 molecules to carbohydrate molecules.

Plant form and function can be related to a plant's ability to carry out photosynthesis. Flowering plants, representative of plants in general, are composed of a root system and a shoot system; the latter includes the stems, leaves, and flowers. Photosynthesis occurs primarily in leaves containing the green pigment chlorophyll, which is capable of absorbing the energy of the sun.

Photosynthesis is essential to the continued existence of all living things. Its by-product, oxygen, makes cellular respiration possible, and its primary product, carbohydrates, is food for the entire biosphere.

Beams of sunlight provide this plant with the energy to photosynthesize, producing its own organic food. The pigment chlorophyll makes these leaves green and able to absorb solar energy.

6

Photosynthesis

Chapter Concepts

1 Photosynthetic organisms produce the organic molecules that are used as a source of food and chemical energy by all living things.

2 Photosynthesis takes place primarily within chloroplasts. Recent scanning electron micrographs suggest the location of photosynthetic subpathways.

3 Photosynthesis has two subpathways. The first drives the second by providing the energy and the hydrogen atoms needed by the second to reduce carbon dioxide to carbohydrate.

4 Photosynthesis can be compared to cellular respiration; both similarities and differences exist.

Chapter Outline

Only plants, algae, and a few bacteria carry on photosynthesis. As the name implies, **photosynthesis** refers to the ability of these organisms to make their own food in the presence of sunlight.

Radiant Energy

The food produced through photosynthesis eventually becomes the food for the rest of the living world (fig. 6.1). For example, humans and all other animals either eat plants directly or eat animals that have eaten plants, and so forth. Another way to express this idea is to say that **autotrophs,** possessing the ability to synthesize organic molecules from inorganic raw materials, feed heterotrophs, which must take in preformed organic molecules as food. Therefore, autotrophs produce organic food, while **heterotrophs** consume it.

After food has been eaten and digested, the resulting small molecules may be used either as building blocks for growth or as a source of energy (for the production of ATP). Therefore the food provided by autotrophs supplies energy for metabolism.

Plants also supply energy in another sense because their bodies became the fossil fuel coal that is still in abundant supply. For several reasons, then, it is correct to say that all life is ultimately dependent on the energy of the sun:

1. Sunlight supplies the energy needed for photosynthesis.
2. The food made by photosynthetic organisms becomes the food of the biosphere. This food may be used not only for growth but also for metabolic energy purposes.
3. The bodies of plants became the fossil fuel coal, upon which we are still dependent today.
4. Sunlight can be an energy source, called solar energy, for private and industrial use.

Figure 6.1
a. *Autotrophs represented by green plants, algae, and a few bacteria (the latter two represented in circles) produce food that feeds* b. *herbivores which feed directly on plants or plant products,* c. *carnivores, which feed on herbivores or other carnivores and* d. *omnivores that feed on all of these.*

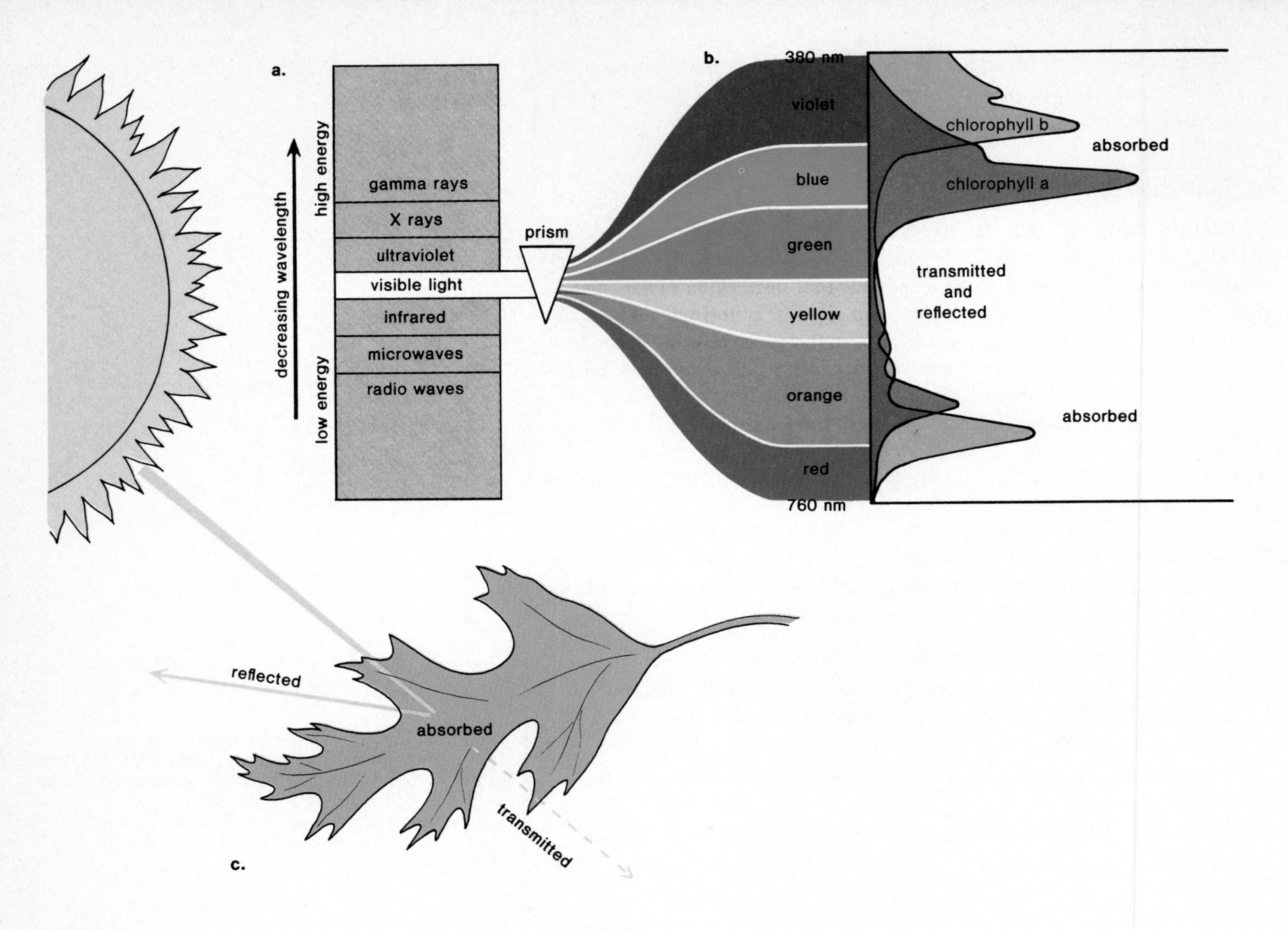

Figure 6.2
a. *Energy from the sun is categorized into the types listed in the gray portion. Gamma rays have the shortest wavelength and the most energy content. Radio waves have the longest wavelength and the least energy content. Visible (white light) actually contains many colors of light as can be detected when the light is passed through a prism.*
b. *Absorption spectrum of chlorophyll* a *and* b *shows that these two pigments absorb violet and red light the best.* c. *Leaves appear green to us because the color green is reflected or transmitted by chlorophyll.*

Sunlight

Radiant energy from the sun can be described in terms of its wavelength and its energy content. Figure 6.2 lists the different types of radiant energy from the shortest wavelength, gamma rays, to the longest, radio waves. The shorter wavelengths contain more energy than the longer ones. *White,* or *visible, light* is only a small portion of this spectrum. Visible light itself contains various wavelengths of light, as can be proven by passing it through a prism; then we see all the different colors that make up visible light. (Actually, of course, it is our eyes that interpret these wavelengths as colors.) The colors range from violet (the shortest wavelength) to blue, green, yellow, orange, and red (the longest wavelength). The energy content is highest for violet light and lowest for red light.

The pigments found within photosynthesizing cells are capable of absorbing various portions of visible light. The absorption spectrum for chlorophyll *a* and chlorophyll *b* is shown in figure 6.2b. Both chlorophyll *a* and chlorophyll *b* absorb violet, blue, and red better than the other colors. Because the color green is only minimally absorbed and is primarily reflected (fig. 6.2c), leaves appear green to us.

---

*P*hotosynthesis is absolutely essential for the continuance of life because it supplies the biosphere with food and energy.

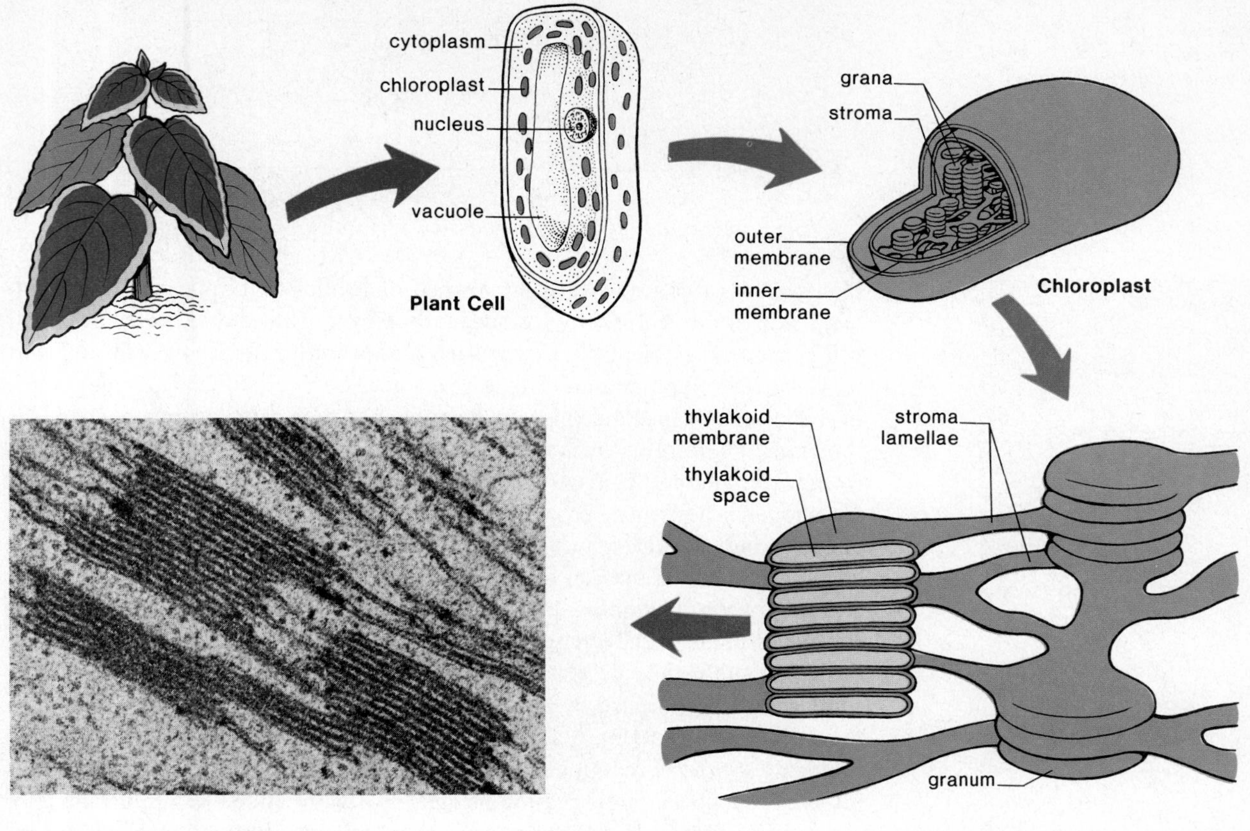

Figure 6.3
Leaves are the primary photosynthetic organs of a plant. Top. *The cells in a leaf contain many chloroplasts, the organelles that carry on photosynthesis.* Bottom. *The stroma within a chloroplast contains grana, each a stack of flattened membranous sacs called thylakoids. The electron micrograph shows stacking of thylakoids to form a granum.*

Chloroplasts

Chloroplasts are the organelles, found in a plant cell (fig. 6.3) in which photosynthesis takes place.

Anatomy

The chloroplast is membranous. A double membrane or envelope surrounds a large central space called the **stroma.** The stroma contains many different soluble enzymes that function to incorporate CO_2 into organic compounds. Here, too, membrane forms the **grana.** Each granum is a stack of flattened sacs or disks called **thylakoids.** Grana are connected with each other by way of **stroma lamellae. Chlorophyll** is found within the membrane of the grana, making them the energy-generating system of the chloroplast. Each thylakoid, being saclike, contains a space. Movement of ions from this space across the thylakoid membrane is believed to be important in the photosynthetic process.

Physiology

The overall equation for photosynthesis may be written in this manner to indicate that carbohydrates in general are the end product of photosynthesis:

$$CO_2 + H_2O + \text{light energy} \longrightarrow (CH_2O) + O_2$$

Multiplying this equation by 6 shows that glucose is often an end product of photosynthesis:

$$6\ CO_2 + 6\ H_2O + \text{light energy} \longrightarrow C_6H_{12}O_6 + 6\ O_2$$

Figure 6.4
Equation for photosynthesis showing the relationship between reactants and products. Oxygen given off by photosynthesis is derived from water.

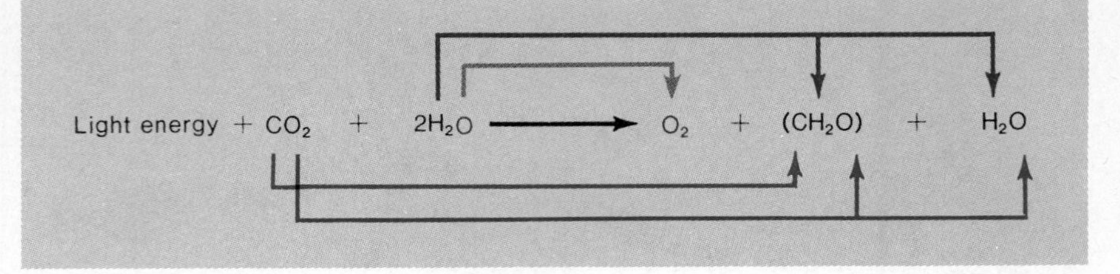

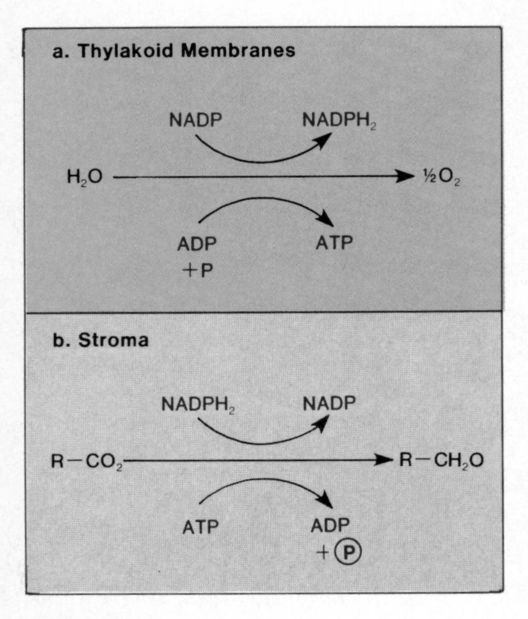

Figure 6.5
Photosynthesis involves two pathways. a. *The overall reaction for the pathway that occurs in the thylakoid membranes: ATP and $NADPH_2$ are formed as water is split, releasing oxygen.* b. *The overall reaction for the pathway that occurs in the stroma: ATP and $NADPH_2$ from the first reaction are used to reduce carbon dioxide.* (R = *remainder of the molecule).*

However, we now know that the oxygen molecules that appear on the right-hand side of the equation were originally a part of the water molecules. This was proven experimentally by exposing plants first to carbon dioxide and then to water that contained an isotope of oxygen, O^{18}, called heavy oxygen. Only in the latter instance did this isotope appear as molecular oxygen given off by the plant. Therefore, it was shown that oxygen released by chloroplasts comes from water and not from carbon dioxide. To indicate this, the overall equation for photosynthesis may be written as shown in figure 6.4. The arrows in the equation indicate the relationship between the molecules on the left and those on the right. If we analyze this equation, we see that photosynthesis provides a link between the nonliving and living world. The small inorganic molecules (on the left side of the equation) are converted to a much larger organic molecule (on the right side of the equation). The carbohydrate molecule represents the food produced by photosynthesis.

The equation (fig. 6.4) also indicates that photosynthesis is an energy requiring synthetic reaction. As with some synthetic reactions, photosynthesis involves reduction: hydrogen atoms ($H^+ + e^-$) are added to a molecule. Notice that hydrogen is added to the carbon-oxygen combination when a carbohydrate is formed. This addition of hydrogen requires the formation of new bonds and therefore requires energy. It is not surprising to learn that carbon dioxide and water are low-energy molecules and that carbohydrates are high-energy molecules.

Two Subreactions

It was discovered some years ago that the process of photosynthesis involves two subreactions. It was proposed that these subreactions be called the "light" reaction, which requires that light be present, and the "dark" reaction, which does not require light. These two subreactions were detected because when light is being maximally absorbed by a photosynthetic system, a rise in temperature still increases the rate of photosynthesis. This indicates that the first subreaction (or pathway) is primarily dependent on light, while the second subpathway is dependent primarily on temperature. Today the terms "light" and "dark" reaction are being phased out because they do not adequately point out the most important difference between the two subpathways of photosynthesis. The first pathway represented by the reaction given in figure 6.5a captures the energy of sunlight and uses this energy to produce $NADPH_2$ and ATP, and the second pathway represented by the reaction given in figure 6.5b uses these molecules to reduce carbon dioxide to a carbohydrate.

Chloroplasts carry on photosynthesis, an energy-requiring reaction that results in carbohydrate synthesis. First the energy of sunlight must be captured and then CO_2 must be reduced to CH_2O.

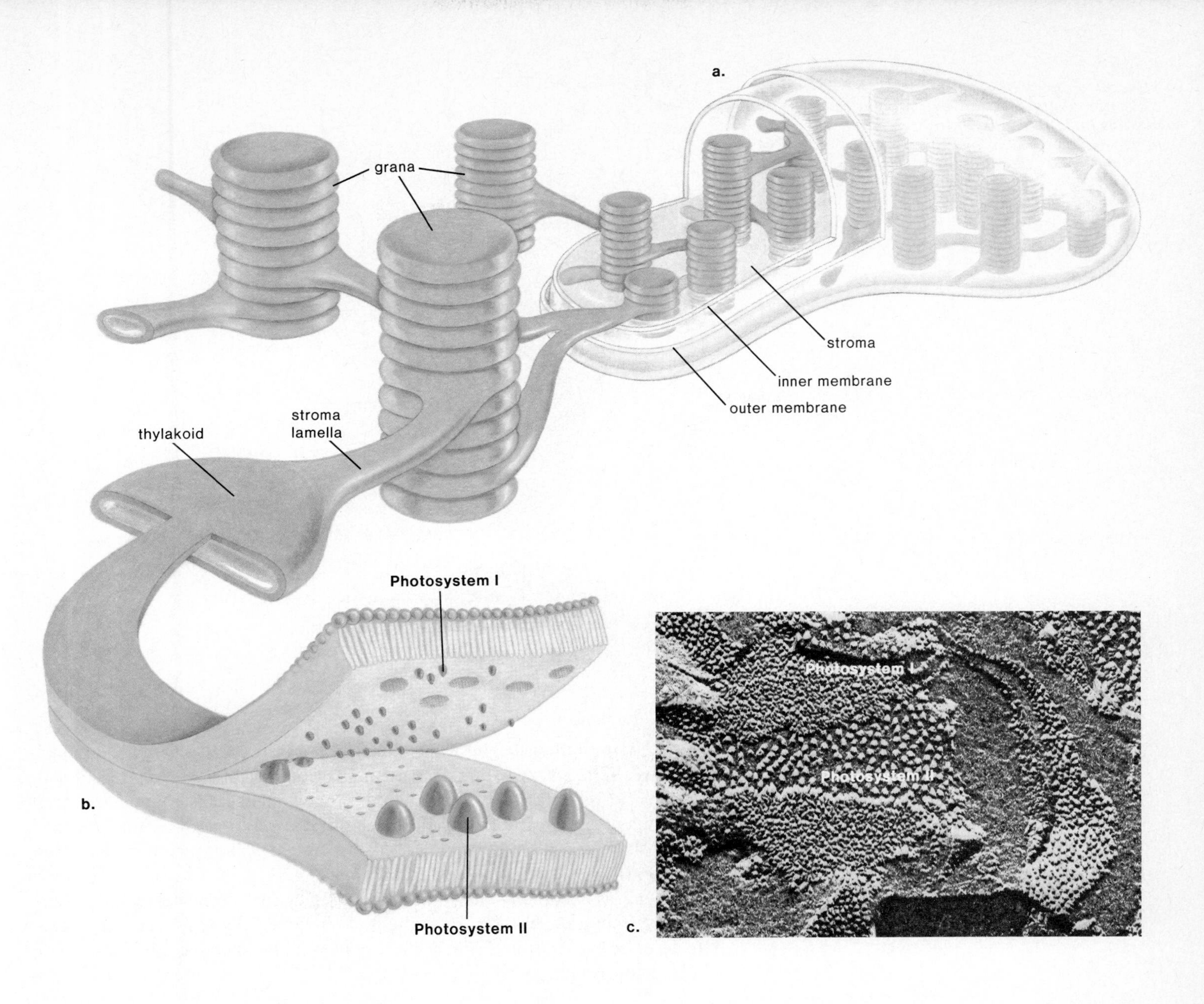

Capturing the Energy of Sunlight

This portion of photosynthesis requires the participation of two photosystems, called **Photosystem I** and **Photosystem II.** The freeze-fracture method of preparing thylakoid membrane shows two types of particles (fig. 6.6). Smaller particles that appear in the outer half of the membrane are believed to be locations for Photosystem I, while the larger particles that appear in the inner half of the membrane are believed to be locations for Photosystem II. Each photosystem consists of a pigment complex plus an electron acceptor and donor. The pigment complex contains several hundred molecules of chlorophyll and also carotenoid pigments. The pigment complexes have been termed *antennae* because just as TV antennae are aimed to pick up signals, so the leaves of a plant turn to allow the pigment complexes to pick up solar energy.

Figure 6.6
Grana anatomy. a. *Location of a grana within a chloroplast. Each granum is composed of thylakoids.* b. *Drawing of an enlarged thylakoid membrane to show location of particles in the inner half and outer half of the membrane.* c. *Drawing is based on electron micrographs, such as this one that was made following freeze-fracture of thylakoid membrane. The smaller particles are believed to be the location of Photosystem I and the larger particles are believed to be the location of Photosystem II. Light energy is captured by the photosystems.*

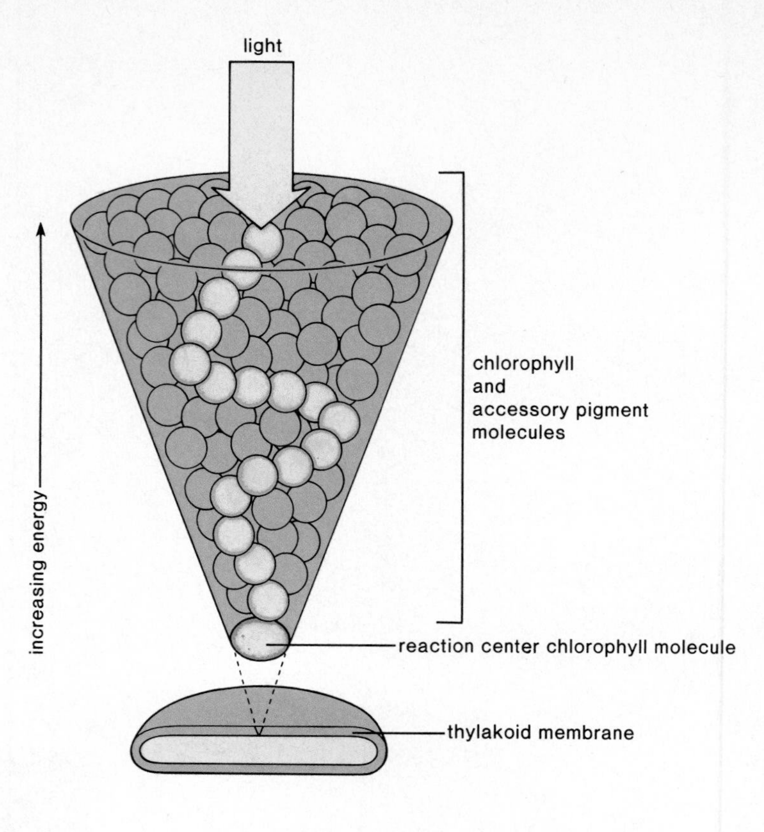

Figure 6.7
Pigment complex in each photosystem consists of a collection of pigment molecules. Light energy is absorbed and then passed from molecule to molecule (represented by the balls) *until it reaches a reaction center chlorophyll* a *molecule. Thereafter this chlorophyll molecule enters an excited state.*

Photosynthesis begins when the pigments absorb light energy and funnel it to *reaction centers* within the antennae of both Photosystem I and Photosystem II (fig. 6.7). The antenna within Photosystem I has a reaction-center chlorophyll *a* that absorbs maximum light at a wavelength around 700 nm and is therefore called P700 (P stands for pigment). The antenna within Photosystem II has a reaction-center chlorophyll *a* that absorbs slightly shorter wavelengths maximally and is called P680. In figure 6.8 you can see that each photosystem probably absorbs sunlight at the same time. However, it is easiest to describe events as if they occur in a sequential manner and as if they began with Photosystem II.

Events

1. As energy is received by P680, its electrons become so highly charged that a few actually leave the chlorophyll molecule. P680 would soon disintegrate if it did not receive replacement electrons. It receives these electrons from water, which splits in the following manner:

   $$H_2O \longrightarrow 2\ H^+ + 2e^- + \frac{1}{2}O_2$$

   This freed oxygen is the oxygen gas given off during photosynthesis.
2. The electrons that leave P680 are received by an acceptor molecule before being passed down an *electron transport system* consisting of a series of carriers, some of which are **cytochrome** molecules. For this reason, the system is sometimes referred to as a *cytochrome system.* As the electrons pass "downhill" from one cytochrome to another, energy is made available for ATP formation.

   $$ADP + (P) + \text{energy} \longrightarrow ATP$$

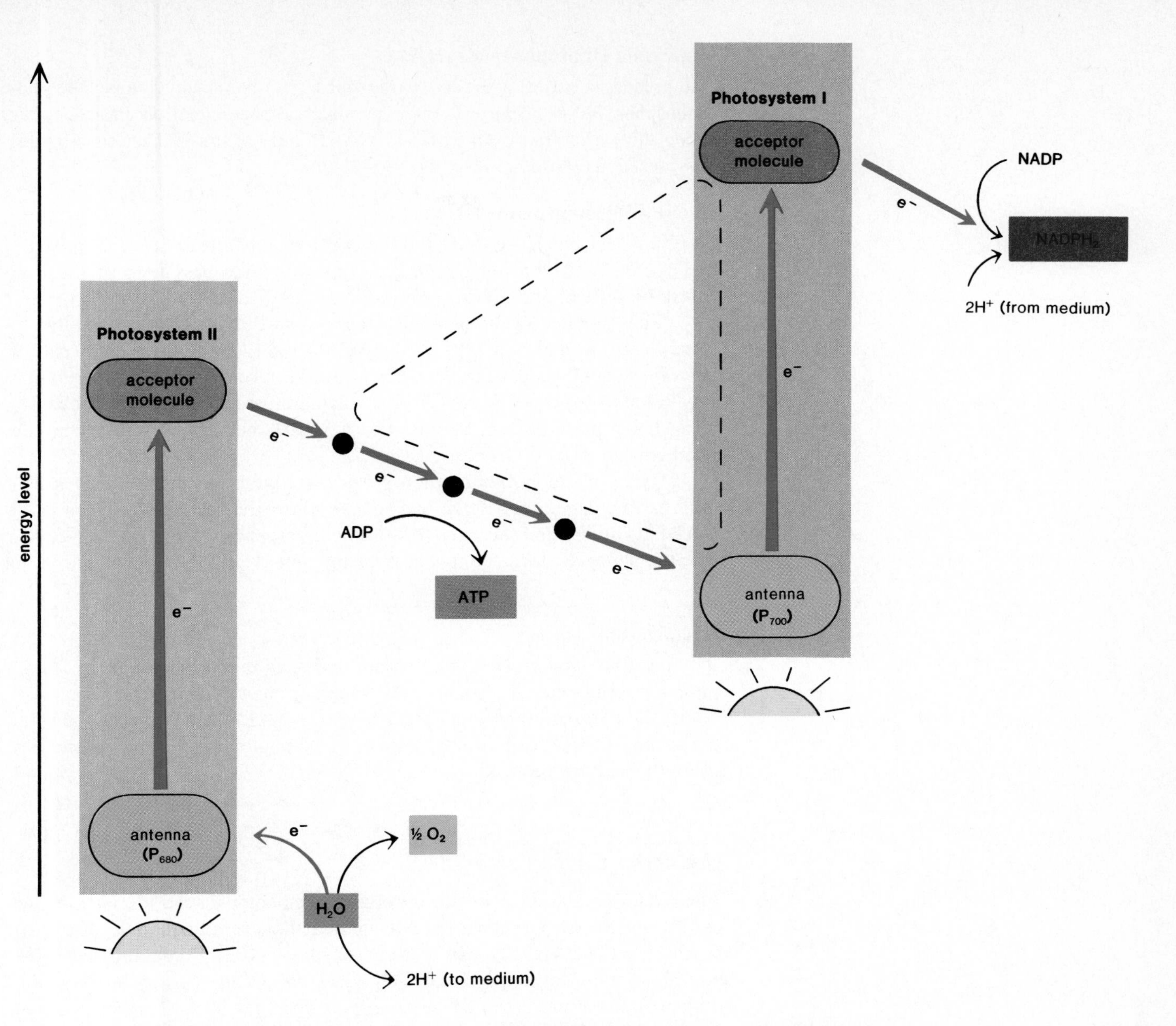

Figure 6.8
Photophosphorylation (ATP formation) occurs in thylakoid membranes. In noncyclic photophosphorylation, electrons move from water to P680, to an acceptor molecule that passes them down a transport system, to P700, which sends them to another acceptor molecule, before they are finally sent to NADP. In cyclic photophosphorylation (dotted lines), *electrons pass from P700 to an acceptor molecule that sends them down the electron transport system before they return to P700.*

Just how this occurs is still under investigation. It is believed that some of the members of the electron transport system deposit hydrogen ions within the thylakoid space. This creates an electrochemical gradient that serves as a source of energy for ATP formation when these ions flow from the thylakoid space to the stroma (fig. 6.3).[1]

3. After the electrons leave the cytochrome system, they are in a lower energy state and may be received by P700, the reaction-center chlorophyll *a,* in an antenna within Photosystem I. These electrons replace some that previously left P700. Prior to their arrival, the pigment complex of Photosystem I has absorbed light energy and funneled it to P700. A few of its electrons were so highly energized that they left the chlorophyll molecule to be eventually taken up by NADP, as shown in the following reaction. (**NADP** is similar in structure to NAD except that it carries an extra phosphate group, as indicated by the additional ***P.***)

$$NADP + 2e^- + 2H^+ \longrightarrow NADPH_2$$

[1]This is known as the chemiosmotic theory.

Noncyclic Photophosphorylation

The preceding series of events (paragraphs 1–3) is known as **noncyclic photophosphorylation** because (1) it is possible to trace electrons in a one-way (noncyclic) direction from water to NADP and (2) sunlight energy (photo) has caused ATP production (phosphorylation).

Cyclic Photophosphorylation

At times electrons leave P700 and then return to it instead of being taken up by NADP (dotted line in fig. 6.8). Before arriving, they pass down the electron transport system, and ATP is produced.

This is called **cyclic photophosphorylation** because (1) it is possible to trace electrons in a cycle from P700 to P700 and (2) sunlight energy (photo) has caused ATP production (phosphorylation). Cyclic photophosphorylation is believed to occur whenever CO_2 is in such limited supply that carbohydrate is not being produced. At these times, $NADPH_2$ buildup prevents noncyclic photophosphorylation from occurring.

Cyclic photophosphorylation provides an independent means by which ATP can be generated. But it does not result in the release of oxygen or in $NADPH_2$ production. Perhaps this form of photophosphorylation evolved before the noncyclic form, simply as a means to make ATP.

Photosynthesis begins when pigment complexes within Photosystem I and Photosystem II located within thylakoid membranes absorb radiant energy and pass it to a reaction-center chlorophyll *a* molecule. Noncyclic photophosphorylation produces the $NADPH_2$ and ATP used in the stroma. If CO_2 is in limited supply, cyclic photophosphorylation occurs and only ATP is produced.

Reducing Carbon Dioxide

The $NADPH_2$ and ATP produced by noncyclic photophosphorylation are now used to reduce carbon dioxide. Look again at the overall equation for photosynthesis on page 117. Notice that carbohydrate (CH_2O) contains hydrogen atoms, whereas carbon dioxide (CO_2) does not. When carbon binds to hydrogen atoms, it has accepted electrons plus hydrogen ions. Therefore, carbon has been reduced. The reduction process is a building-up, or a synthetic, process because it requires the formation of new bonds. Hydrogen atoms and energy are needed for reduction synthesis, and these are supplied by $NADPH_2$ and ATP.

Reduction of CO_2 occurs in the *stroma* of the chloroplast. It is a several-step process that is catalyzed by a series of soluble enzymes, with a specific enzyme for each reaction.

C_3 Photosynthesis

Figure 6.9 is a simplified diagram of the reactions that occur as carbon dioxide is taken up and reduced so that carbohydrate synthesis occurs. Figure 6.10 illustrates the same reactions in more detail. Altogether, the reactions make up a cycle called the **Calvin-Benson cycle,** named for the men who discovered it. Carbon dioxide is combined with the five-carbon sugar, **ribulose biphosphate (RuBP).** The resulting six-carbon molecule immediately breaks down to

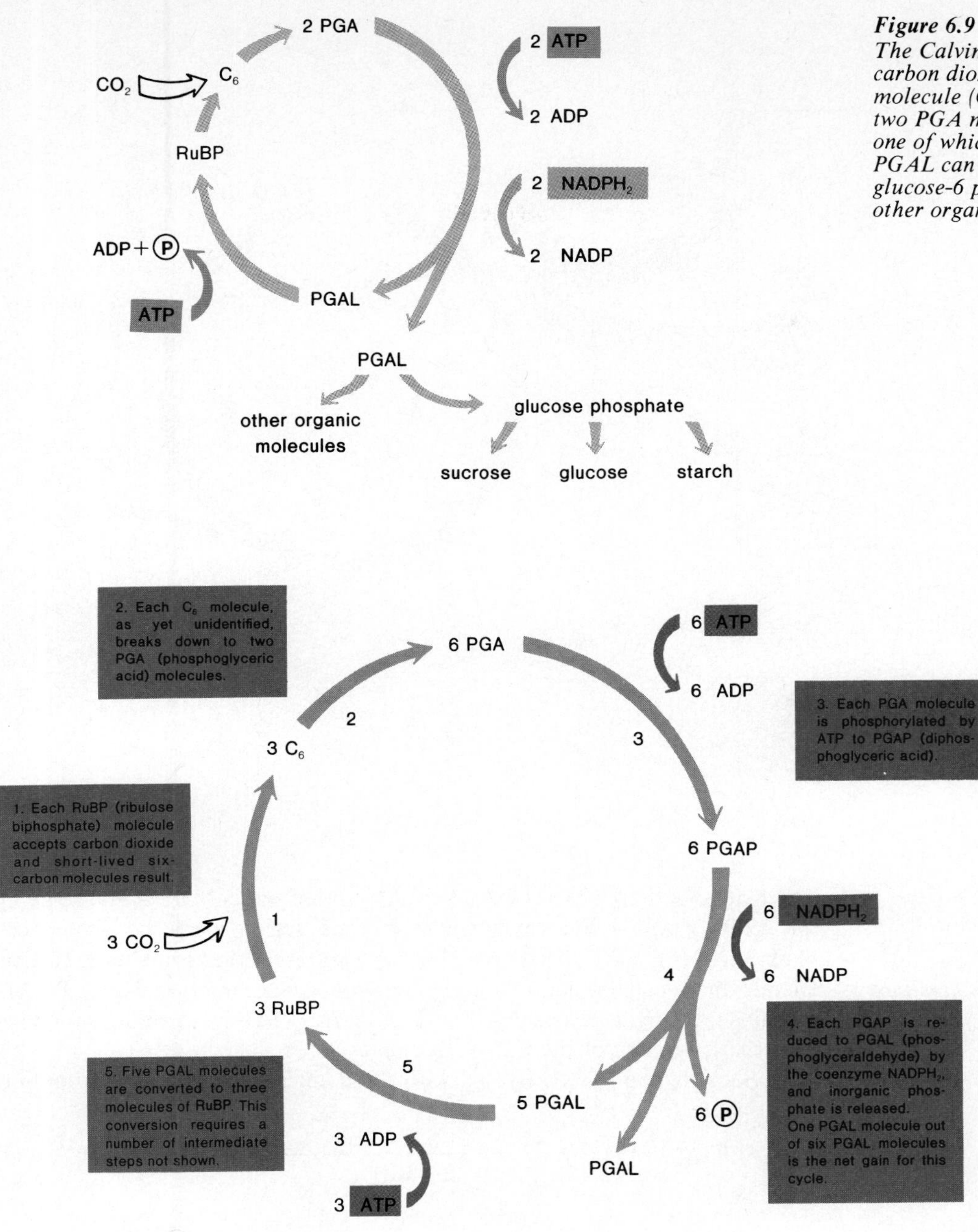

Figure 6.9
The Calvin-Benson cycle (simplified). RuBP accepts carbon dioxide (CO_2), forming a six-carbon molecule (C_6), which immediately breaks down to two PGA molecules that are reduced to 2 PGAL, one of which represents the net gain of the cycle. PGAL can be combined with another PGAL to give glucose-6 phosphate that can be metabolized to other organic molecules.

Figure 6.10
The Calvin-Benson cycle (in detail). The molecules in the cycle have been multiplied by three because after five molecules of PGAL (15 carbons) are converted to three molecules of RuBP (15 carbons), one molecule of PGAL will remain. Since it takes 2 PGAL to produce one six-carbon sugar, the numbers shown must be remultiplied by two in order for a monosaccharide to be the end product. Compare steps 3 and 4 to steps 5 and 4 in figure 5.13.

two molecules of PGA (phosphoglyceric acid). This is called **C_3 photosynthesis** because the first molecule that can be detected, namely PGA, is a three-carbon molecule. PGA immediately undergoes reduction to PGAL (phosphoglyceraldehyde):

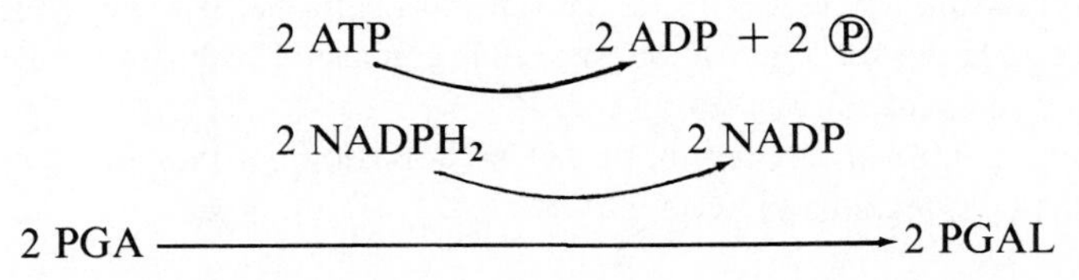

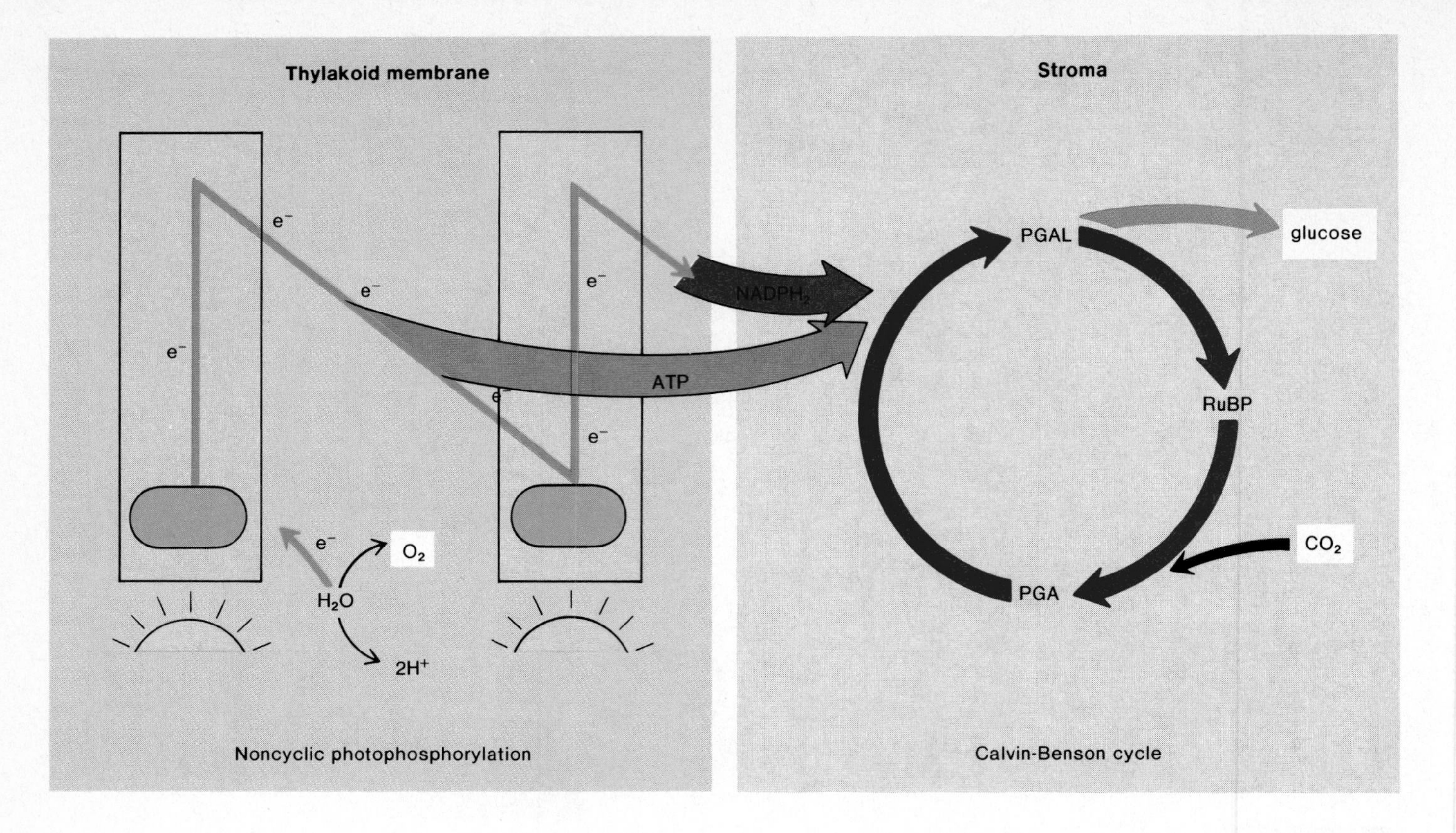

Figure 6.11
Photosynthesis contains two sets of reactions. Noncyclic photophosphorylation, which takes place in the thylakoid membrane, produces the $NADPH_2$ and ATP required for carbon dioxide reduction that takes place in the stroma.

The reactions that reduce PGA to PGAL use up the $NADPH_2$ and some of the ATP formed in the thylakoid membrane during noncyclic photophosphorylation (fig. 6.11). These reactions also represent the reduction of carbon dioxide and its conversion to a high-energy molecule. In other words, PGAL contains more hydrogen atoms than does PGA. PGAL is the immediate photosynthetic product of the Calvin-Benson cycle.

Some of the PGAL is used within the chloroplast to re-form ribulose biphosphate.

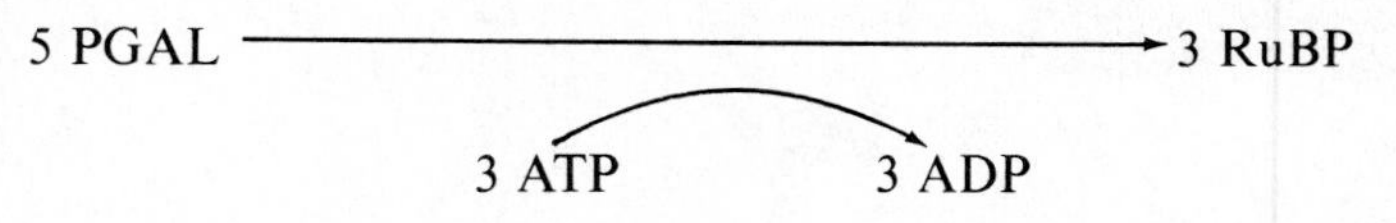

This reaction also utilizes some of the ATP produced by photophosphorylation. Altogether the ATP and $NADPH_2$ consumed in converting CO_2 to the level of a hexose (a six-carbon sugar) results in about a 30 percent energy-efficiency rate for photosynthesis.

PGAL, the product of the Calvin-Benson cycle, can combine with another PGAL to form glucose phosphate, which can be converted to glucose. The overall equation for photosynthesis often shows glucose as the end product for the process. However, glucose phosphate is converted to sucrose for transport in plants or to starch for storage.

Table 6.1 summarizes the function of each participant in photosynthesis, and figure 6.11 represents an overview.

Table 6.1 *Thylakoids versus Stroma*

	Participant	Role
Thylakoids	Sunlight	Provides energy
	Chlorophyll	Absorbs energy
	Water	Donates electrons and releases oxygen
	ADP + Ⓟ	Forms ATP
	NADP	Becomes $NADPH_2$
Stroma	RuBP	Takes up CO_2
	CO_2	Reduced to PGAL
	ATP	Provides energy for reduction
	$NADPH_2$	Provides electrons for reduction
	2 PGAL	Becomes glucose phosphate

C_4 Photosynthesis

The more the Calvin-Benson cycle functions, the greater the amount of food produced. Recently, it has been discovered that plants adapted to a hot, dry environment, having so called **C_4 photosynthesis,** produce food more efficiently than those adapted to a temperate climate having so-called *C_3 photosynthesis.* The reason for this is based on their relative abilities to fix (reduce) carbon dioxide despite its low concentration in a leaf. This is discussed in the reading on page 126.

During C_3 photosynthesis, CO_2 is reduced to PGAL during the operation of the Calvin-Benson cycle in the stroma. Two PGAL molecules join to form glucose phosphate which can be converted to all organic molecules needed by a plant. C_4 photosynthesis, a modification of C_3, is more efficient because of better capture of CO_2.

Carbohydrate Utilization

The carbohydrates (PGAL, glucose phosphate, sucrose, and so forth), synthesized by means of photosynthesis in the leaves, are transported to other parts of the plant, usually in the form of sucrose. Although it may be temporarily stored in the form of starch, much of this carbohydrate will eventually be broken down by means of cellular respiration (just as in animals) to produce the ATP needed for cell metabolism.

Plants not only have the enzymatic capability to produce sugars from carbon dioxide and water but they also have the ability, by way of the metabolic pathways diagrammed in figure 5.12 and by even more complex pathways, to produce all the various types of organic molecules they require:

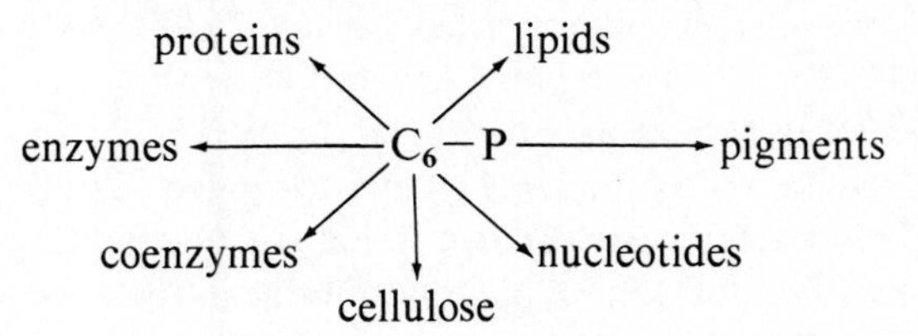

In comparison to the animal cell, algal and plant cells possess enormous biochemical capabilities. Biochemically speaking, plant cells are due our utmost respect because they are capable of making all necessary organic molecules, using only inorganic molecules as nutrients.

Increasing Crop Yields: C_3 versus C_4 Photosynthesis

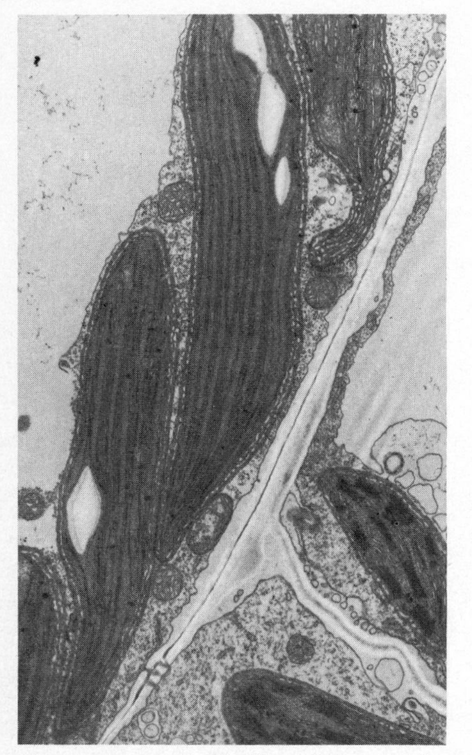

Two corn plant leaf cells at bottom right capture CO_2 and pass it to the bundle-sheath cells on the left. Notice that the chloroplasts in the bundle-sheath cells have stroma lamellae but lack grana. These cells are specialized to carry on the Calvin-Benson cycle.

The world's population continues to increase and, as it does so, encroaches on agricultural land. In the United States where there were once fields of grain, there are now towns, suburbs, and shopping malls. More food must be produced on less land, therefore agricultural yields must be constantly increased. Despite advanced technology and the development of mutant plants, the average yield per acre has not significantly increased in recent years. New ideas are needed and one may have been discovered.

Observation has shown that temperate zone plants, such as wheat, alfalfa, and potatoes, do not take up CO_2 as efficiently as do plants such as corn, which are adapted to climates of high light intensity and high temperature. The fault lies with the first enzyme of the Calvin-Benson cycle. This enzyme can catalyze one of two reactions:

1. $CO_2 + RuBP \rightarrow\rightarrow C_3$
2. $O_2 + RuBP \rightarrow\rightarrow CO_2 + H_2O$

The first reaction is a normal part of photosynthesis. The second reaction is called **photorespiration** because oxygen is taken up and CO_2 is given off to the environment. Photorespiration, which competes with the photosynthetic reaction, accounts for the lower yield of temperate zone plants because it obviously does not lead to carbohydrate synthesis.

How is photorespiration avoided in plants adapted to a hot dry climate? First, the Calvin-Benson cycle only operates in certain cells called the **bundle-sheath cells.** Second, other cells specialize in capturing CO_2 and passing it to the bundle-sheath cells. CO_2 is captured by a reaction that occurs despite the low concentration of CO_2 in the leaves:

3. $CO_2 + PEP \rightarrow C_4$

These C_4 molecules are transported into the bundle-sheath cells where CO_2 is released.

Notice that it is possible to make a distinction between the two groups of plants according to the first molecule detected following CO_2 uptake. In temperate zone plants, C_3 is always the first molecule detected following CO_2 uptake (see reaction #1), and in the tropical zone plants under discussion, C_4 is the first molecule detected (see reaction #3). Therefore, it is now customary to speak of *C_3 versus C_4 photosynthesis*—C_4 being much more efficient. Scientists are now exploring the possibility of transforming temperate zone plants into C_4 photosynthesizers. If this can be accomplished, agricultural yields will be greatly increased.

Autotrophic Bacteria

Bacterial Photosynthesis

Although the cyanobacteria carry on photosynthesis much like plants do, some other photosynthetic bacteria do not release oxygen when they photosynthesize because they do not use water as a hydrogen donor. The green sulfur and purple sulfur bacteria use hydrogen gas (H_2) and hydrogen sulfide (H_2S) as hydrogen donors. These bacteria usually live in anaerobic conditions like the muddy bottoms of marshes and cannot photosynthesize in the presence of oxygen.

Chemosynthesis

Some bacteria can oxidize inorganic compounds, such as ammonia, nitrates, and sulfides, and can trap the small amount of energy released to synthesize carbohydrates. While this capability is not believed to contribute greatly to the support of life on land, it has recently been found capable of supporting

Figure 6.12
Giant tube worms are a part of the community of organisms found near the deep ocean ridges and vents. These communities have been given such names as Dandelions, the Rose Garden, and the Garden of Eden.

an entire community a mile and a half below sea level. By means of deep-diving research submarines, scientists have been examining the mid-ocean ridge system where hot minerals spew out from the inner earth. Here, bacteria in the vent water oxidize hydrogen sulfide and carry on **chemosynthesis.** Living off the resulting organic molecules are giant tube worms (fig. 6.12), clams, and crabs. There is evidence that such bacteria may also reside within the bodies of the worms. Prior to finding these organisms it was thought that such communities could not exist on the ocean floor where light never penetrates.

Except for cyanobacteria, photosynthetic bacteria do not evolve O_2 as do plants because they utilize hydrogen from a source other than water. Chemosynthetic bacteria produce their own food by oxidizing inorganic compounds.

Comparison of Cellular Respiration and Photosynthesis

Differences

Only certain cells carry on photosynthesis but all cells, including photosynthesizing cells, carry on cellular respiration, either aerobic or anaerobic. The cellular organelle for cellular respiration is the mitochondrion, while the cellular organelle for photosynthesis is the chloroplast.

The overall equation for aerobic cellular respiration is the opposite of that for photosynthesis:

$$\text{energy} + \text{carbon dioxide} + \text{water} \rightleftharpoons \text{carbohydrate} + \text{oxygen}$$

The reaction in the forward direction represents photosynthesis, and the energy is the energy of the sun. The reaction in the opposite direction represents cellular respiration, and the energy then stands for ATP.

Obviously, photosynthesis is the building up of glucose, while cellular respiration is the breaking down of glucose. See table 6.2 for a summarized list of differences between these processes.

***Table 6.2** Cellular Respiration and Photosynthesis*

Cellular Respiration	Photosynthesis
Mitochondrion	Chloroplast
Oxidation	Reduction
Releases energy	Requires energy
Requires oxygen	Releases oxygen
Releases carbon dioxide	Requires carbon dioxide

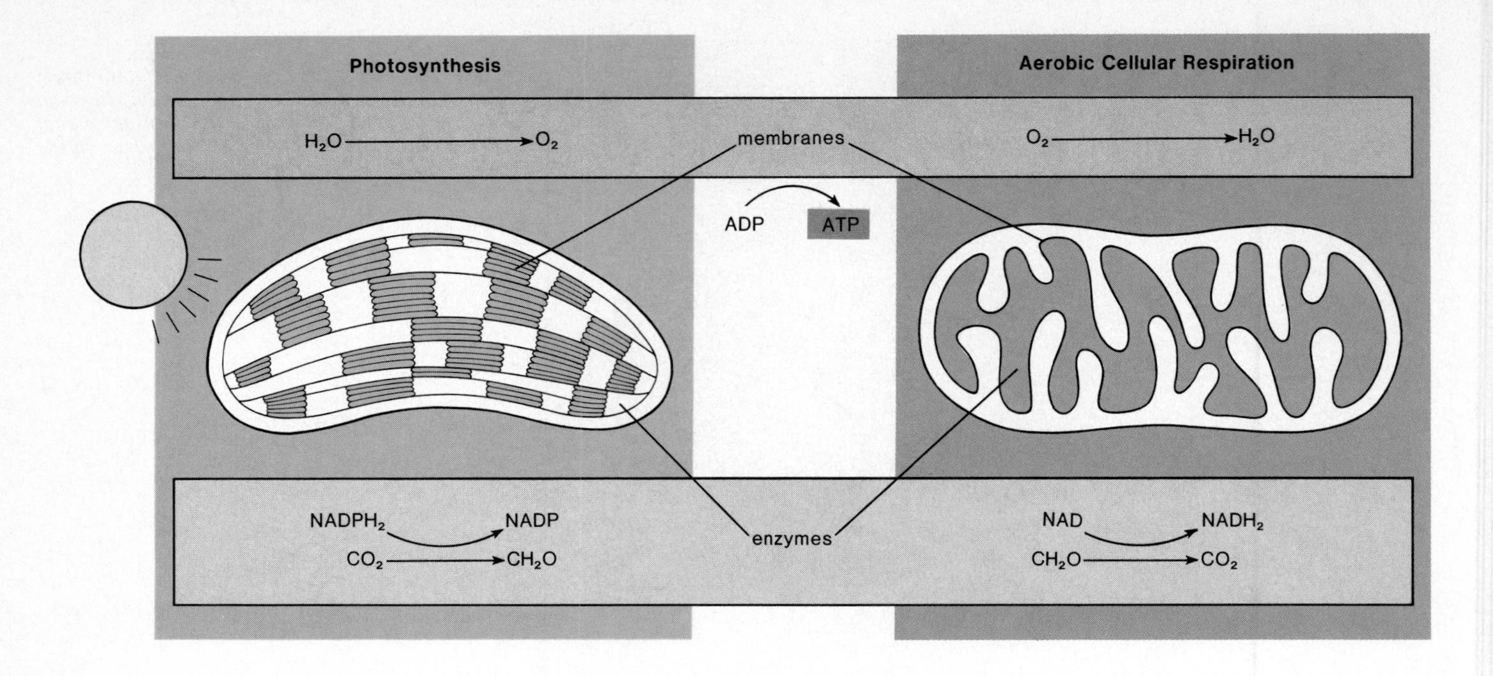

Figure 6.13
Diagram illustrating similarities and differences between photosynthesis that takes place in chloroplasts and cellular respiration that takes place in mitochondria. Both have a cytochrome system located within membranes where ATP is produced. Both have enzyme-catalyzed reactions in solution. The coenzyme NAD(P) operates in the membrane and the solution. Photosynthesis releases oxygen and reduces carbon dioxide into carbohydrates; cellular respiration reduces oxygen and releases carbon dioxide.

Similarities

Both photosynthesis and cellular respiration are metabolic pathways within cells and therefore consist of a series of reactions that the overall reaction does not indicate. Within the pathways, both make use of a cytochrome system located in membrane (fig. 6.13) to generate a supply of ATP, use both make use of a hydrogen carrier: cellular respiration uses NAD and photosynthesis uses NADP.

Both pathways utilize this overall reaction but in opposite directions. For photosynthesis, read from left to right in the following diagram, and for cellular respiration, read from right to left.

$$\begin{array}{c} \mathrm{NAD(P)H_2} \quad \curvearrowright \quad \mathrm{NAD(P)} \\ \mathrm{PGA} \longleftrightarrow \mathrm{PGAL} \\ \mathrm{ATP} \quad \curvearrowright \quad \mathrm{ADP} + \text{Ⓟ} \end{array}$$

Both photosynthesis and cellular respiration occur in plant cells. While both of these occur during the daylight hours, only cellular respiration occurs at night. During daylight hours, the rate of photosynthesis exceeds the rate of cellular respiration, resulting in a net increase and storage of glucose. The stored glucose is used to support cellular metabolism that continues during the night.

Summary

Photosynthesis, carried out largely by plants and algae, is absolutely essential for the continuance of life because it supplies the biosphere with food and energy. Photosynthesis takes place in chloroplasts and requires two subpathways: photophosphorlyation occurs within thylakoid membranes of the grana, and the Calvin-Benson cycle occurs within the stroma. Noncyclic photophosphorlyation produces the ATP and $NADPH_2$ needed to reduce carbon dioxide. During C_3 photosynthesis, carbon dioxide enters the Calvin-Benson cycle and PGAL molecules exit from it. Two PGAL molecules join to form glucose phosphate, a molecule that can be metabolized to all the substances a plant needs. Table 6.3 summarizes the roles played by the various participants in photosynthesis.

***Table 6.3** Participants in photosynthesis*

	Participant	Role
Thylakoid membranes	Sunlight	Provides energy
	Chlorophyll	Absorbs energy
	Water	Donates electrons and releases oxygen
	ADP + Ⓟ	Forms ATP
	NADP	Accepts electrons and H^+ and becomes $NADPH_2$
Stroma	RuBP	Takes up CO_2
	CO_2	Reduced to PGAL
	ATP	Provides energy for reduction
	$NADPH_2$	Provides electrons for reduction
	2 PGAL	Becomes glucose phosphate

Cyanobacteria carry out photosynthesis like plants but other types of bacteria do not release oxygen when they photosynthesize. Chemosynthetic bacteria remove hydrogen atoms from a number of substances in order to capture energy. A surprising find has been deep ocean communities supported by chemosynthetic bacteria.

Objective Questions

1. Life requires a continual supply of energy. Ultimately, this energy comes from ____________ .
2. An organism that makes its own food is called an ____________ .
3. The large central space within a chloroplast is called the ____________ .
4. The photosystems are located within the ____________ membranes of the grana.
5. Chlorophyll molecules are located within the ____________ of the photosystems.
6. Lying between Photosystem I and Photosystem II, the ____________ system converts ADP to ATP.
7. In addition to ATP, noncyclic photophosphorylation produces the ____________ needed by the Calvin-Benson cycle.
8. The Calvin-Benson cycle takes up ____________ and reduces it to a carbohydrate.
9. The Calvin-Benson cycle is found in the ____________ of the chloroplast.
10. The thylakoid membranes within chloroplasts can be compared to the ____________ within mitochondria because they both contain a cytochrome system.

Answers to Objective Questions

1. sunlight 2. autotroph 3. stroma 4. thylakoid 5. antennae 6. cytochrome 7. $NADPH_2$ 8. carbon dioxide 9. stroma 10. cristae

Study Questions

1. Why are all living things dependent upon the process of photosynthesis and the energy of the sun? (p. 115)
2. Which rays of light are most important for photosynthesis and why? (p. 116)
3. Give the overall equation for photosynthesis and the equations for the two subpathways involved in the process. (p. 118)
4. Describe the structure of a chloroplast, and indicate where photophosphorylation and the Calvin-Benson cycle occur. In what way are these two processes connected? (pp. 117, 119, 122)
5. Trace the path of electrons during noncyclic photophosphorylation. (pp. 120–122)
6. Give the primary steps of the Calvin-Benson cycle, indicating the reaction that represents the reduction of carbon dioxide. (p. 122)
7. Describe the role of each participant in photosynthesis. (p. 129)
8. Why would it be correct to say that a plant cell is more biochemically competent than an animal cell? (p. 125)
9. Discuss bacterial photosynthesis. (p. 126)
10. Contrast cellular respiration and photosynthesis in at least five ways. How are the two cellular processes similar? (pp. 127–128)

Key Terms

autotroph (aw'to-trōf) an organism that is capable of making its food (organic molecules) from inorganic molecules. *115*

C_3 photosynthesis (se thre fo''to-sin'thĕ-sis) photosynthesis that utilizes the Calvin-Benson cycle to take up and then fix carbon dioxide; so named because the first molecule detected after CO_2 uptake is a C_3 molecule. *123*

C_4 photosynthesis (se for fo''to-sin'thĕ-sis) photosynthesis in which the first detected molecule following CO_2 uptake is a C_4 molecule. Later, this same CO_2 is made available to the Calvin-Benson cycle. *125*

Calvin-Benson cycle (kal'vin ben'sun si'kl) a circular series of reactions by which CO_2 fixation occurs within chloroplasts. *122*

chemosynthesis (ke''mo-sin'the-sis) the process of making food by using energy derived from the oxidation of reduced molecules in the environment. *127*

chlorophyll (klo'ro-fil) the green pigment found in photosynthesizing organisms that is capable of absorbing energy from the sun's rays. *117*

cyclic photophosphorylation (sik'lik fo''to-fos''for-ĭ-la'shun) the synthesis of ATP by a cytochrome system within thylakoid membranes utilizing electrons received from and returning to P700 chlorophyll molecules within Photosystem I. *122*

cytochrome system (si'to-krōm sis'tem) a series of cytochrome molecules present in thylakoid membranes and inner mitochondrial membranes that pass electrons one to the other from a higher energy level to a lower energy level; the energy released is used to build ATP. *120*

grana (gra'nah) stacks of flattened membranous vesicles in a chloroplast where chlorophyll is located and photosynthesis begins. *117*

heterotroph (het'er-o-trof'') an organism that cannot synthesize organic compounds from inorganic substances and therefore must acquire food from external sources. *115*

NADP a coenzyme of reduction; a hydrogenase that frequently donates hydrogen atoms to metabolites. *121*

noncyclic photophosphorylation (non-sik'lik fo''to-fos''for-ĭ-la'shun) the passage of electrons within thylakoid membranes from water (by way of Photosystem II, a cytochrome system, Photosystem I) to NADP; results in the generation of ATP and $NADPH_2$. *122*

photosynthesis (fo''to-sin'thĕ-sis) in plants the process of making carbohydrate from carbon dioxide and water by using the energy of the sun. *115*

Photosystem I and II (fo''to-sis'tem) molecular units located within the membrane of a thylakoid that capture solar energy making photophosphorylation possible. *119*

ribulose biphosphate (ri'bu-lōs bi-fos'fāt) (RuBP) the molecule that acts as the acceptor for carbon dioxide within the Calvin-Benson cycle. *122*

stroma (stro'mah) the interior portion of a chloroplast. *117*

stroma lamellae (stro'mah lah-mel'e) membranous connections between adjacent thylakoids of the grana. *117*

thylakoid (thi'lah-koid) an individual flattened vesicle found within a granum (pl. grana). *117*

7

Plant Form and Function

Chapter Concepts

1. The plant body, including the roots, stems, leaves, and flowers, is made up of a few major types of specialized cells and tissues derived from an ever-present source of embryonic cells.

2. Theories have been formulated to explain the transport of water and minerals and the transport of organic nutrients.

3. The sex organs of a flowering plant are located in the flower. Flowering is controlled by the length of daylight (photoperiod) in some plants.

4. Seeds within fruits are the products of sexual reproduction in flowering plants. Germination of a seed results in another plant.

Chapter Outline

The body of a flowering plant is divided into two portions, the *root system* and the *shoot system* (fig. 7.1). The roots, which lie below ground level, anchor the plant and absorb water and minerals. Within the shoot system, the stem lifts the leaves to catch the rays of the sun. The leaves receive water and minerals that are sent from the roots up through the stem and take in carbon dioxide from the air. Therefore, the leaves are provided with the light energy, water, and carbon dioxide they need to carry on photosynthesis.

This chapter will begin with the basic anatomy of the roots, stems, and leaves of a flowering plant. These are vegetative organs, meaning they are not concerned with reproduction. Some of the tissues found within each organ are listed in table 7.1. In addition to the tissue types listed, plants also contain embryonic tissue called **meristem tissue,** which is unspecialized and continually capable of cell division. Cell division is followed by differentiation into the cell types depicted in figure 7.2. Parenchyma and sclerenchyma cells are found in most tissues. **Parenchyma** cells are relatively unspecialized and correspond best to the generalized cell of a plant (fig. 2.2b). **Sclerenchyma** cells are hollow, nonliving cells with extremely strong walls that give support to plant tissues and organs. Two tissue types, epidermis and endodermis, do not have parenchyma and sclerenchyma cells. **Epidermis,** composed of epidermal cells, covers the entire body of nonwoody and young woody plants. Epidermis functions to protect inner body parts and to prevent the plant from drying out.

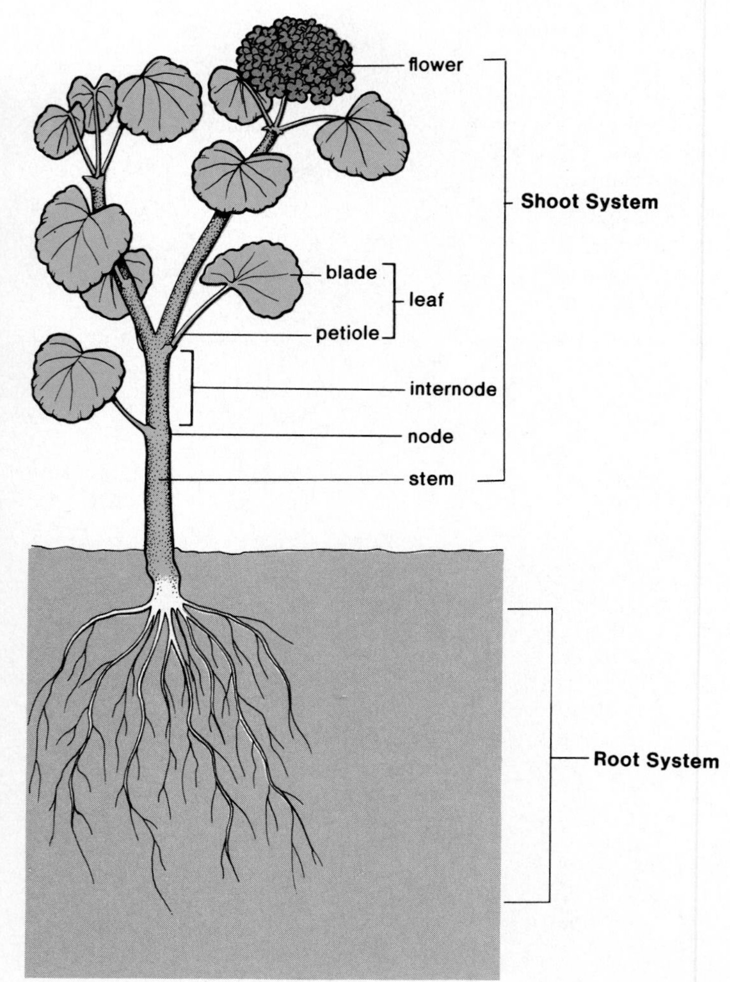

Figure 7.1
A land plant is divided into two main portions: the root system below ground and the shoot system containing the stems and leaves above ground. A node is a region of the stem where one or more leaves are attached.

In addition, the epidermis has specialized structures and functions that are discussed when each organ is considered. In contrast to epidermis, **endodermis,** which is nearly universally found in roots, contains only endodermal cells.

The other cell types in figure 7.2 are found in vascular (transport) tissue, which is considered in detail on page 145.

The body of a plant is composed of tissues differentiated to perform various functions and containing specialized cells. One tissue, the meristem, remains undifferentiated and forever capable of dividing and producing new cells.

***Table 7.1** Vegetative Organs and Major Tissues*

	Roots	Stems	Leaves
Function	Absorb water and minerals Anchor plant	Transport water and nutrients Support leaves and flowers	Carry on photosynthesis
Tissue			
Epidermis*	Root hairs absorb water and minerals	Protect inner tissues	Stomata carry on gas exchange
Cortex †	Store products of photosynthesis	Carry on photosynthesis, if green	———
Vascular ‡	Transport water and nutrients	Transport water and nutrients	Transport water and nutrients
Pith†	———	Store products of photosynthesis	———
Mesophyll †	———	———	
Spongy layer			Gas exchange
Palisade layer			Photosynthesis

Plant tissues belong to one of three tissue systems
*Dermal tissue system
† Ground tissue system
‡ Vascular tissue system

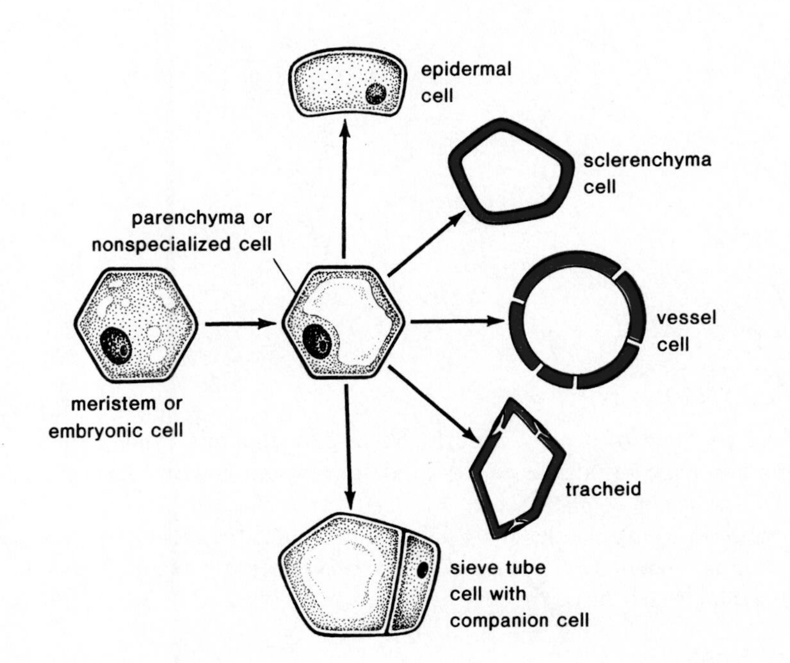

Figure 7.2
Plant cell types. A meristem cell is embryonic and, by the process of maturation, may become a relatively nonspecialized parenchyma cell or one of the highly specialized cells shown. Nonliving cells that lack cytoplasm are shown in color. Sclerenchyma cells are support cells. Vessel cells and tracheids are in the transport tissue called xylem (fig. 7.14). The sieve-tube cell and the companion cell are in the transport tissue called phloem (fig. 7.16). The epidermal cell is found in epidermis, the outermost tissue in all organs of a plant.

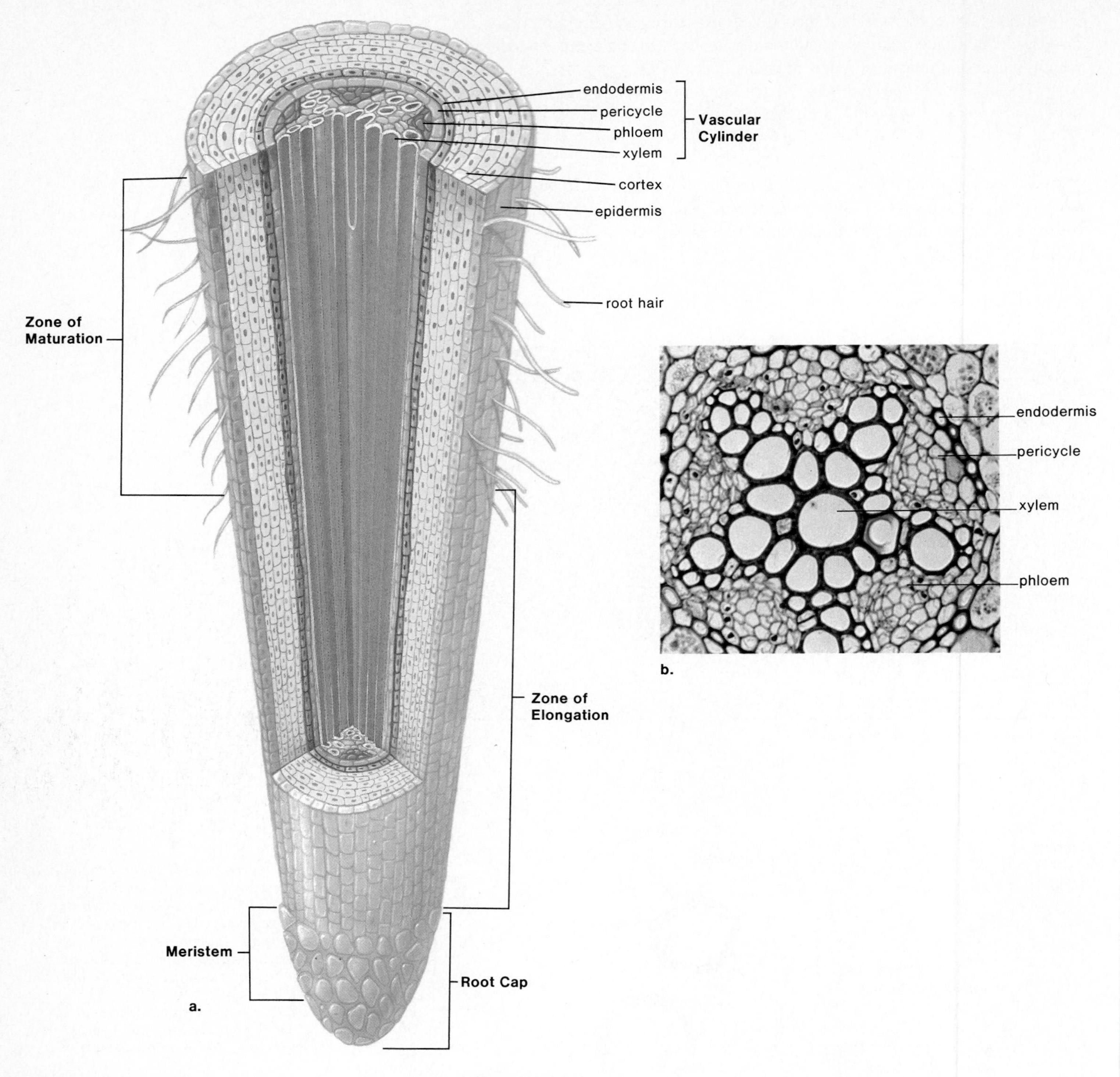

Figure 7.3
Dicot root tip. a. *Root tip is divided into four zones best seen in a longitudinal section such as this.*
b. *Vascular cylinder of a dicot root contains the vascular tissue; xylem is typically star-shaped, and the phloem lies between the points of the star.*

Root System

The root system of a plant absorbs water and minerals from the soil. The reading on page 135 discusses this vital activity and how it relates to the extensive formation of plant roots.

Flowering plants are divided into two groups, the dicots and monocots. This chapter points out various differences between dicots and monocots including the tissues of their roots.

Dicot Root

A longitudinal section of a dicot root (fig. 7.3) shows apical meristem located at the tip, protected by a **root cap** composed of a thimble-shaped mass of parenchyma cells. Outer root cap cells contain a slimy substance that assists

Plants as Food

Plants need only inorganic nutrients in order to produce all the organic molecules that make up their bodies. Aside from carbon, hydrogen, and oxygen (obtained from carbon dioxide and water), the mineral elements listed in table 7.A are required nutrients for plants. The mineral elements are classified as **macronutrients** when they are used by plants in greater amounts and **micronutrients** when they are needed by plants in very small amounts. Both types of minerals are found in the soil, but in low concentrations; not only must a plant be able to take them up, it must also be able to concentrate them. Fortunately, the root system of a plant is designed for just this purpose. As the root system grows, it branches and branches again so that the roots are exposed to a tremendous amount of soil. It has been estimated that a rye plant has roots totaling about 900 kilometers (more than 650 miles) in length. Further, because of the extensive number of root hairs, the total surface area is about 635 square meters, or more than 7,000 square feet! Water, and possibly minerals too, enters root hairs by diffusion, but eventually active transport is used to concentrate the minerals within the organs of a plant. A plant uses a great deal of ATP for active transport.

It is lucky for us that plants have the capability of concentrating minerals, for we are often dependent on them for our basic supply of such ions as sodium to maintain blood pressure, calcium to build bones and teeth, and iron to help carry oxygen to our cells. Once plants have taken the minerals up, they are incorporated into proteins, fats, and vitamins when these substances are formed from carbohydrates, the product of photosynthesis. When we eat plants, we are supplied with minerals and all types of organic molecules, some of which become building blocks for our own cells and some of which are used as an energy source.

***Table 7.A** Inorganic Nutrients Necessary for Plant Life*

Compound	Element supplied
Macronutrients	
KNO_3	K;N
$CaNO_3$	Ca;N
$NH_4H_2PO_4$	N;P
$MgSO_4$	Mg;S
Micronutrients	
KCl	Cl
H_3BO_3	B
$MnSO_4$	Mn
$ZnSO_4$	Zn
$CuSO_4$	Cu
H_2MoO_4	Mo
Fe-EDTA	Fe

From Epstein, E., *Mineral Nutrition of Plants: Principles and Perspectives.*

***Table 7.B** Food from Plants*

Plant Part	Foods
Roots	Sweet potato, beets, radish, carrot, turnip, parsnip
Stems	White potato, sugar cane, asparagus
Leaves	Cabbage, kale, spinach, lettuce, tea leaves
Petioles*	Celery, rhubarb
Seeds †	Peas, navy beans, lima beans, nuts, coffee beans
Fruits †	Wheat, rice, corn, oats, rye, string beans, apple, orange, peach, tomato, squash

*Part of a leaf
† Derived from flower parts

Table 7.B lists examples of foods that humans consume directly from plants. Each type of food is associated with a particular organ of a flowering plant: the root, stem, leaf, or flower.

movement through the soil. Above the meristem are nonspecialized cells that eventually elongate and develop into specialized tissues. Specifically, the root has four zones: (1) root cap, (2) meristem, or zone of cell division, (3) zone of elongation where the cells increase in length, and (4) zone of maturation where the cells are differentiated into specialized tissues.

The outer epidermal cells display **root hairs,** which tremendously increase the absorption surface of the root. The absorbed water and minerals pass through the **cortex,** a tissue composed largely of parenchyma cells. The

Figure 7.4
Movement of solutes from the cortex to the vascular cylinder by way of endodermal cells. a. *Diagram illustrating the relationship of root tissues, and the position of the Casparian strip.* b. *Endodermal cells in detail also showing location of the Casparian strip. Because of the Casparian strip, water and solutes* (large arrow) *must pass through the cytoplasm of endodermal cells. In this way the endodermal cells regulate the passage of materials into the vascular cylinder.*

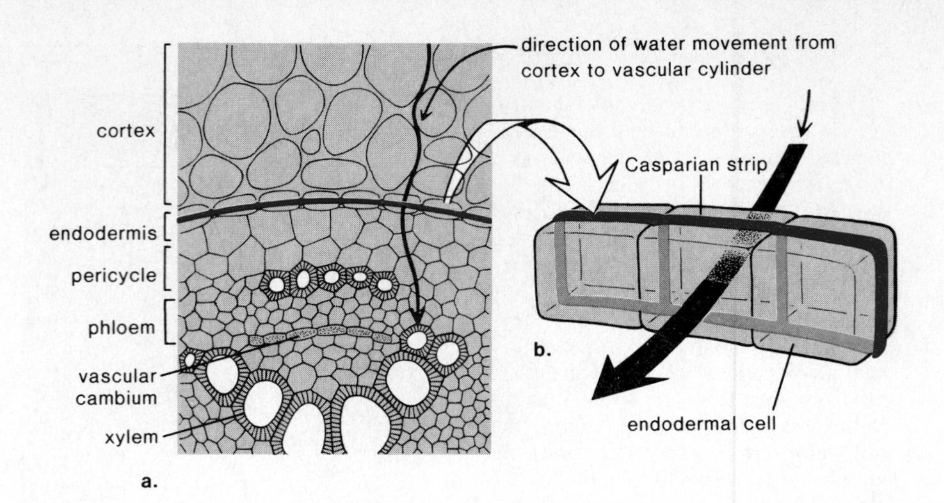

water and minerals must then cross through a layer of endodermis before entering the **vascular cylinder.** The endodermal cells fit snugly together and are bordered only on four sides by a strip of waxy material known as the **Casparian strip** (fig. 7.4). This strip will not permit water and solutes to pass *between* adjacent endodermal cells; therefore, the only access to the vascular cylinder is *through* the endodermal cells themselves as shown by the arrow in figure 7.4b.

Within the vascular cylinder (fig. 7.3), water and minerals are transported upward by way of the **xylem** and the products of photosynthesis are most often transported downward by way of the **phloem** for storage in the cortex. Lying between the endodermis and the vascular tissue is the **pericycle,** composed of parenchyma cells, that retains the ability to undergo cell division and on occasion produces branch roots. The pericycle also contributes to the formation of vascular cambium, meristematic tissue lying between xylem and phloem that is capable of producing new vascular tissue.

Monocot Root

Monocot roots often have **pith,** centrally located parenchymal tissue that is surrounded by xylem and phloem. They also have pericycle, endodermis, cortex, and epidermis the same as dicot roots.

Root Diversity

In some plants, notably dicots, the first or primary root grows straight down and remains the dominant root of the plant. This so-called **taproot** (fig. 7.5a) is often fleshy and stores food. Carrots, beets, turnips, and radishes have taproots that we consume as vegetables.

In other plants, notably monocots, a number of slender roots develop, and there is no single main root. These slender roots and their lateral branches make up a **fibrous root** system (fig. 7.5b). Even the branches of fibrous roots are sometimes fleshy and store food, in which case they are called tuberous roots. Sweet potatoes have tuberous roots, for example.

Sometimes a root system develops from an underground stem or from the base of an above-ground stem. These are **adventitious roots** whose main function may be to help anchor the plant. If so, they are called *prop roots* (fig. 7.5c). Mangrove plants have large prop roots (p. 11) that spread away from the plant to help anchor it in the marshy soil where mangroves are typically found.

a.

b.

c.

Figure 7.5
Three types of roots. a. *A taproot has secondary roots in addition to a main root.* b. *A fibrous root has many secondary roots with no one main root.* c. *Adventitious roots, such as prop roots, extend from the stem.*

The root system of a plant absorbs water and minerals that cross the epidermis and cortex before entering the endodermis, the tissue that regulates the entrance of molecules into the vascular cylinder.

Shoot System

The shoot system consists of stems and their numerous appendages, the leaves. Although leaves have a shape and arrangement on the stem that is typical of the particular species, leaves are always arranged so that each one is exposed to the rays of the sun. The area or region of a stem where a leaf or leaves are attached is called a **node,** and stem regions between nodes are called **internodes** (fig. 7.1).

The shoot has apical meristem but no cap. Instead, the apical meristem produces leaves that grow up and around it, forming the apical bud (fig. 7.6). At the junction of a leaf and the stem is a group of meristem cells that may become branches or may develop into next year's flowers. This group of cells is called an **axillary,** or **lateral bud.**

Beneath the apical and lateral bud meristem, newly formed cells gradually elongate and differentiate into the various kinds of cells characteristic of the mature tissues of the plant (table 7.1). Plant growth is regulated by hormones, chemicals produced by one part of a plant that affect another part. As the reading on page 139 indicates, it is hoped that appropriate use of natural and synthetic plant hormones will contribute much to crop yield.

The shoot system of a plant consists of stems and leaves.

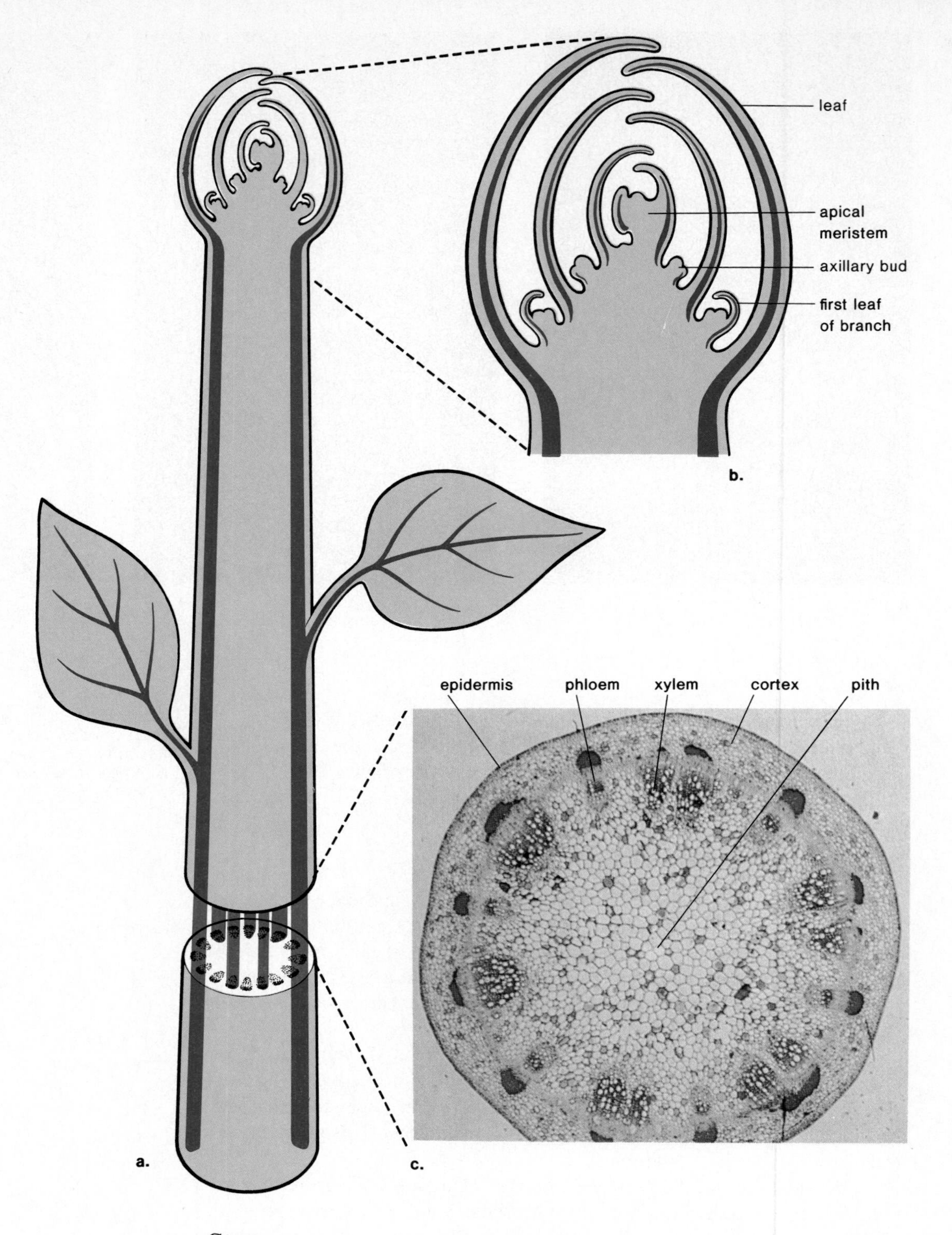

Figure 7.6
Dicot herbaceous shoot system. a. *Portion of shoot system in which the location of vascular tissue is shown diagrammatically.* b. *The apical bud contains the apical meristem protected by many immature leaves.* c. *A micrograph of a cross section of the stem shows that the vascular bundles occur in a ring.*

Stems

In nonwoody stems, the epidermis is covered by a waxlike substance, the **cuticle,** that prevents water loss. Beneath the epidermis a small circular region of *cortex* often surrounds a *pith.* Both cortex and pith are typically composed of parenchyma cells, but the former may carry on photosynthesis, while the latter stores the products of photosynthesis. Instead of a vascular cylinder, stems have **vascular bundles** that contain xylem and phloem. The arrangement of the vascular bundles varies for **monocot** and **dicot** plants. In dicots (fig. 7.6), the bundles are usually arranged in a circular pattern around the central pith.

Plant Growth Regulators

Plant growth involves production of cells by means of cell division, enlargement of these cells, and finally, differentiation as the cells take on specific functions. Three types of hormones are known to promote plant growth: the **cytokinins** stimulate cell division, the **auxins** bring about cell division and enlargement of plant cells; the **gibberellins** promote enlargement of cells and to a lesser extent cell division. A fourth class of plant hormones, termed **inhibitors,** retard or prevent growth in general. Plant growth regulators include natural hormones and related synthetic hormones. Today plant growth regulators are used to bring about an increase in crop yields just as fertilizers, irrigation, and pesticides have done in the past.

Plants bend toward light, and experiments with oat seedlings have shown that bending occurs because auxin is transported to the shady side of the shoot. This can be proven by removing the tip of a shoot and placing an auxin-containing agar block on one side of the stump. The cells on this side elongate causing bending to occur. Since the time these experiments were first performed, many commercial uses for auxins have been discovered. Auxins can cause the base of a shoot to form new roots so that new plants can be started from cuttings. When sprayed on trees, auxins can prevent fruit from dropping too soon. Auxins also inhibit the growth of lateral buds; potatoes sprayed with auxin will not sprout and thus can be stored longer. In high concentrations, auxins are used as herbicides that prevent the growth of broad-leaved plants. The synthetic auxins known as 2,4D and 2,4,5T were used as defoliants during the Vietnam war.

Gibberellins cause the entire plant, including all its parts, to grow larger. Before World War II, the Japanese studied a disease they called "foolish seedling disease" because the young plants grew rapidly, became spindly, and fell over. They found that this disease was caused by gibberellins secreted by a fungus that had infected the plants. Since this time it has been discovered that the application of gibberellins can cause seeds to germinate and plants, such as cabbages, to bolt (meaning rapid stem elongation) and flower. Gibberellins are used commercially to increase the size of plants. Treatment of sugarcane with as little as two ounces per acre increases the yield of cane by more than five tons.

Cytokinins were discovered when mature carrot and tobacco plant cells began to divide when grown in coconut milk. Testing revealed the presence of cytokinins. Later, scientists were able to grow entire plants from single cells in test tubes when various plant hormones were present in correct proportions.

Nurseries now culture all sorts of plants with assembly-line efficiency. Plant breeders are extremely interested in utilizing a modification of the tissue culture technique in which most often leaf cells are treated to produce *protoplasts,* cells that have been chemically stripped of their outer wall. (A single protoplast will give rise to a new plant, identical in nature to the original plant.) It is quicker and easier to test protoplasts instead of entire plants for desired characteristics such as resistance to bacteria and fungi, high temperatures, and drought. Also, protoplasts can be made to fuse together. In one experiment, a potato and a tomato protoplast were fused and a hybrid plant was eventually grown. Perhaps protoplast fusion will eventually allow botanists to alter the genetic makeup of a variety of plants.

The hormone ethylene, which is classified as an inhibitor, causes fruit to ripen. Fruits are commonly kept in cold storage to prevent the release of ethylene. Many synthetic inhibitors simply oppose the action of the natural stimulatory hormones (auxins, gibberellins, and cytokinins). The application of synthetic inhibitors can cause leaf and fruit drop. Removal of the leaves of cotton plants by chemical means aids harvesting; thinning the fruit of young fruit trees produces larger fruit from the trees as they mature; and retarding the growth of some plants increases their hardiness. For example, an inhibitor has been used to reduce stem length in wheat plants so that they do not fall over in heavy winds and rain. Other synthetic inhibitors mimic the action of ethylene and cause ripening of fruit and other crops. Fields and orchards are now sprayed with synthetic growth regulators just as they are sprayed with pesticides.

Crops are sometimes sprayed with plant growth regulators to increase the yield per acre.

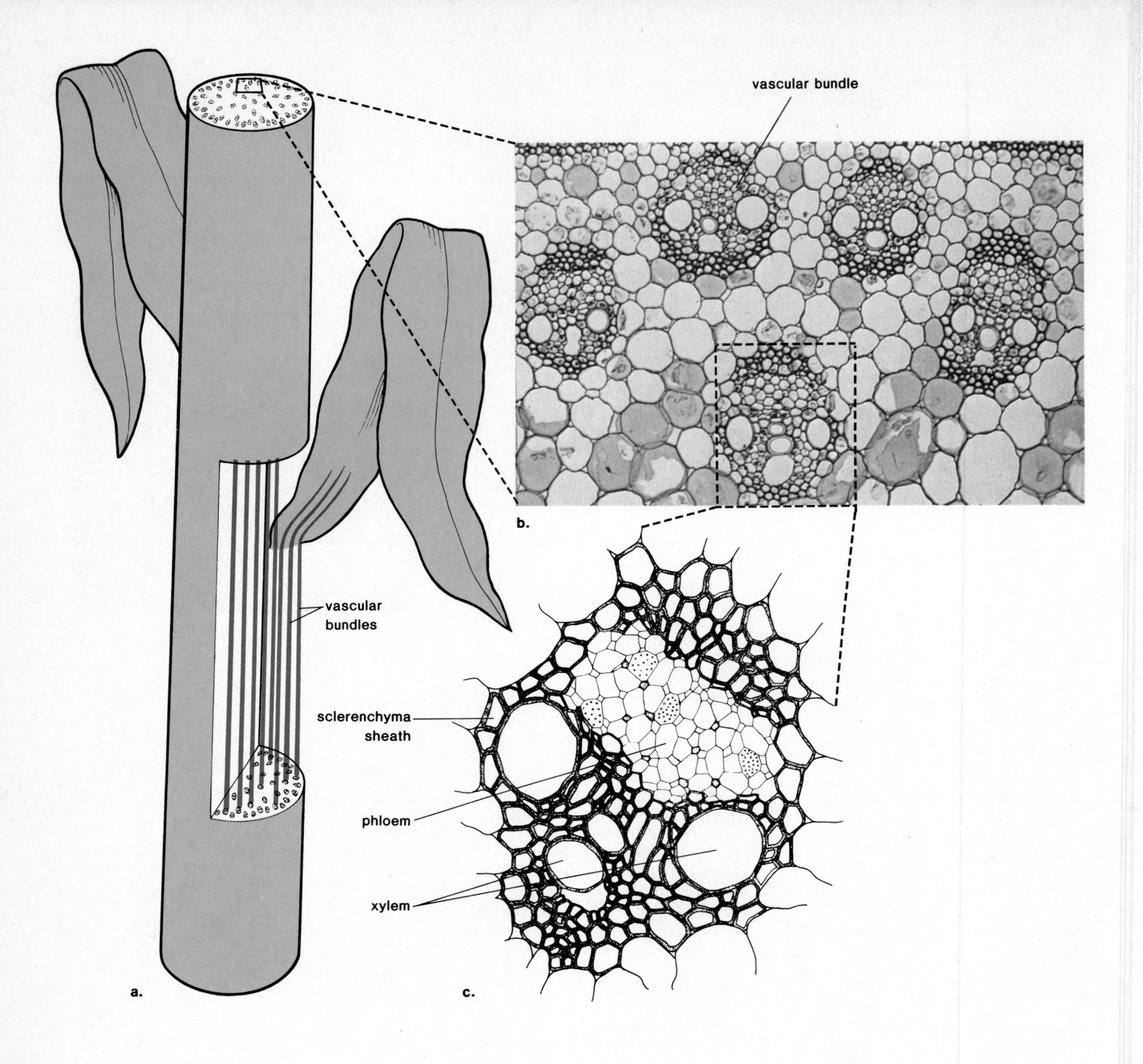

***Figure** 7.7*
Monocot shoot system. a. *Portion of stem and leaves in which the location of vascular tissue is shown diagrammatically.* b. *A micrograph of a cross section of several vascular bundles shows that they are scattered about in the stem.* c. *A drawing of a vascular bundle shows the location of the xylem and phloem.*

In monocots (fig. 7.7), the bundles are randomly scattered, without a definite pattern. This and other differences between the two major types of flowering plants are illustrated in figure 7.8.

So far we have been discussing nonwoody **herbaceous** stems, such as those of common garden plants; trees have **woody stems,** of course. Woody stems increase in diameter due to secondary growth. Between rings of primary xylem and phloem there is a layer of meristematic tissue called the **vascular cambium.** Cells produced by the vascular cambium become *secondary xylem* to the inside of the meristem and to the outside they become *secondary phloem.*

Monocot

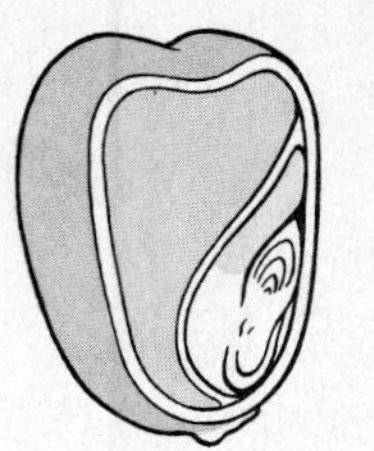

One cotyledon in seed

Flower parts in threes and multiples of three

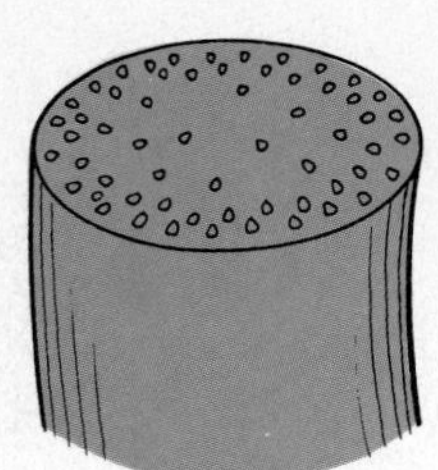

Vascular bundles scattered in stem

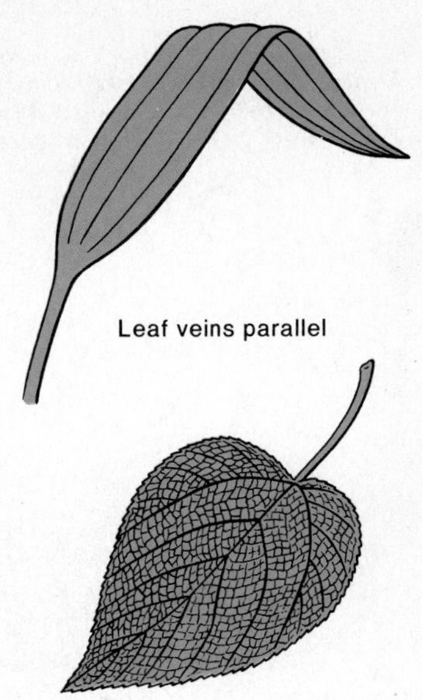

Leaf veins parallel

Dicot

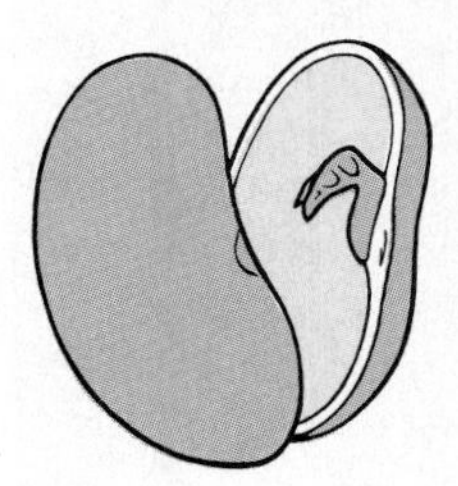

Two cotyledons in seed

Flower parts in fours or fives and multiples of four or five

Vascular bundles in a definite ring

Leaf veins form a net pattern

Figure 7.8
Comparison of monocots and dicots. These four features are used to distinguish monocots from dicots.

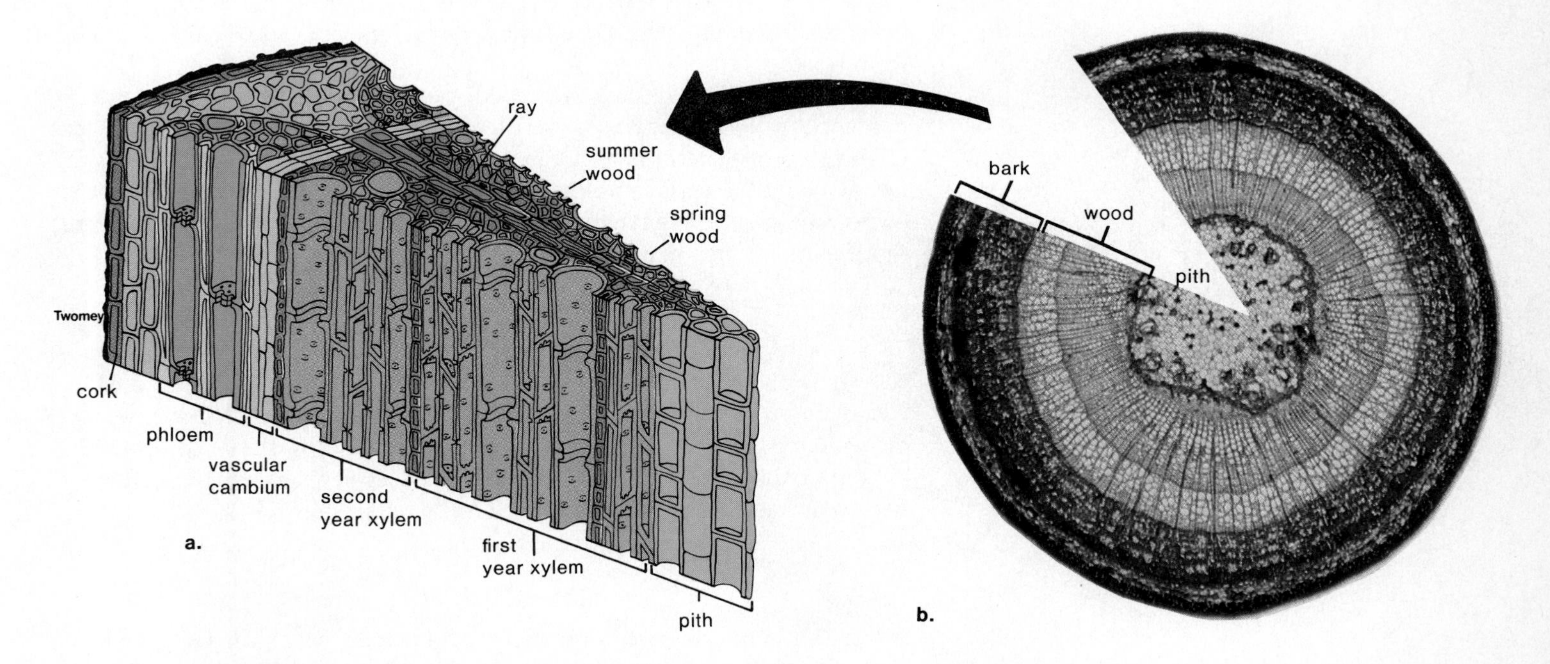

Figure 7.9
Woody stem. a. *A drawing indicating the location of cork, phloem, vascular cambium, and xylem that accumulates to give annual rings.* b. *A photomicrograph of a cross section shows the three parts of a woody stem: bark, wood, and pith.*

Secondary xylem builds up and forms the **annual rings** (fig. 7.9), which can be counted to tell the age of a tree. It is easy to tell where one ring begins and another ends; in the spring, when moisture is plentiful, the xylem cells are much larger than in the late summer, when moisture is scarcer.

The stem of a tree can be divided into three parts: bark, wood, and pith (fig. 7.9b). The **bark** contains cork, cork cambium, cortex, and phloem. Since the phloem is in the bark of a tree, even partial removal of the bark can seriously damage a tree. The **wood** contains the annual rings of xylem. In older

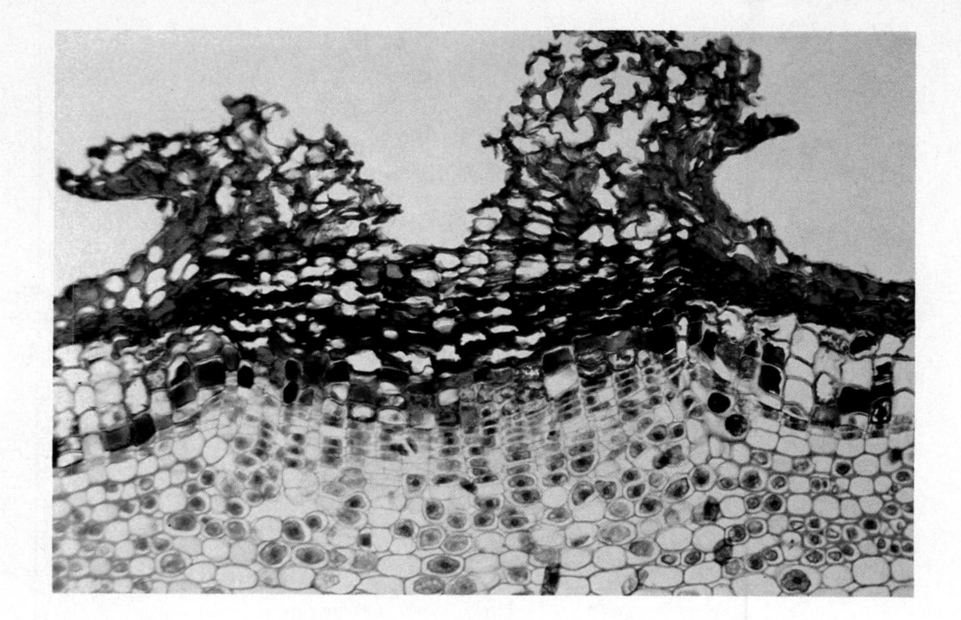

Figure 7.10
Cross section of a portion of the bark of a woody stem in which the cork is stained a reddish color. A lenticel is a portion of the cork where an indentation among the cells permits gas exchange to occur.

stems, most of the xylem is nonliving inner *heartwood* that no longer transports water but does function to support the tree. The most recent annual rings contain functioning cells that do transport xylem sap, the watery contents of vessel cells, and are therefore collectively called *sapwood.* Wood is used to make all sorts of products. For example, lumber, furniture, and paper are made from wood.

As the girth of a woody stem increases, the epidermis is replaced by cork produced by **cork cambium,** a meristem tissue derived from the cortex. **Cork** is made up of dead cells impregnated with suberin, a waterproof material. Occasionally the cork cambium forms pockets of loosely arranged cells that do not become impregnated with suberin. These are the **lenticels** (fig. 7.10) that function in gas exchange.

Dicot herbaceous stems have vascular bundles in a ring and monocots have them randomly arranged. In a cross section, a nonwoody dicot has an outer epidermis, narrow band of cortex, and an inner pith. A woody dicot stem has three parts: the bark containing phloem, the wood or xylem, and pith. A layer of vascular cambium produces new phloem and xylem each year and the secondary xylem forms annual rings.

Leaves

A leaf is usually composed of a **blade** and a **petiole,** which connects the blade to the stem. Between the upper and lower epidermis, the blade contains mesophyll tissue and veins. In the leaves of plants adapted to the temperate zone (fig. 7.11a) the **mesophyll** contains two layers of cells: the **palisade layer,** composed of compact but elongated cells, and the **spongy layer,** with irregular cells bounded by air spaces. The parenchyma cells of the palisade layer have many chloroplasts and carry on most of the photosynthesis for the plant. Water needed in the process of photosynthesis is transported to these cells by the **leaf veins,** the final extensions of vascular tissue. Leaf veins have a *net pattern* in dicot leaves and a parallel pattern in monocot leaves (fig. 7.8). The larger leaf veins

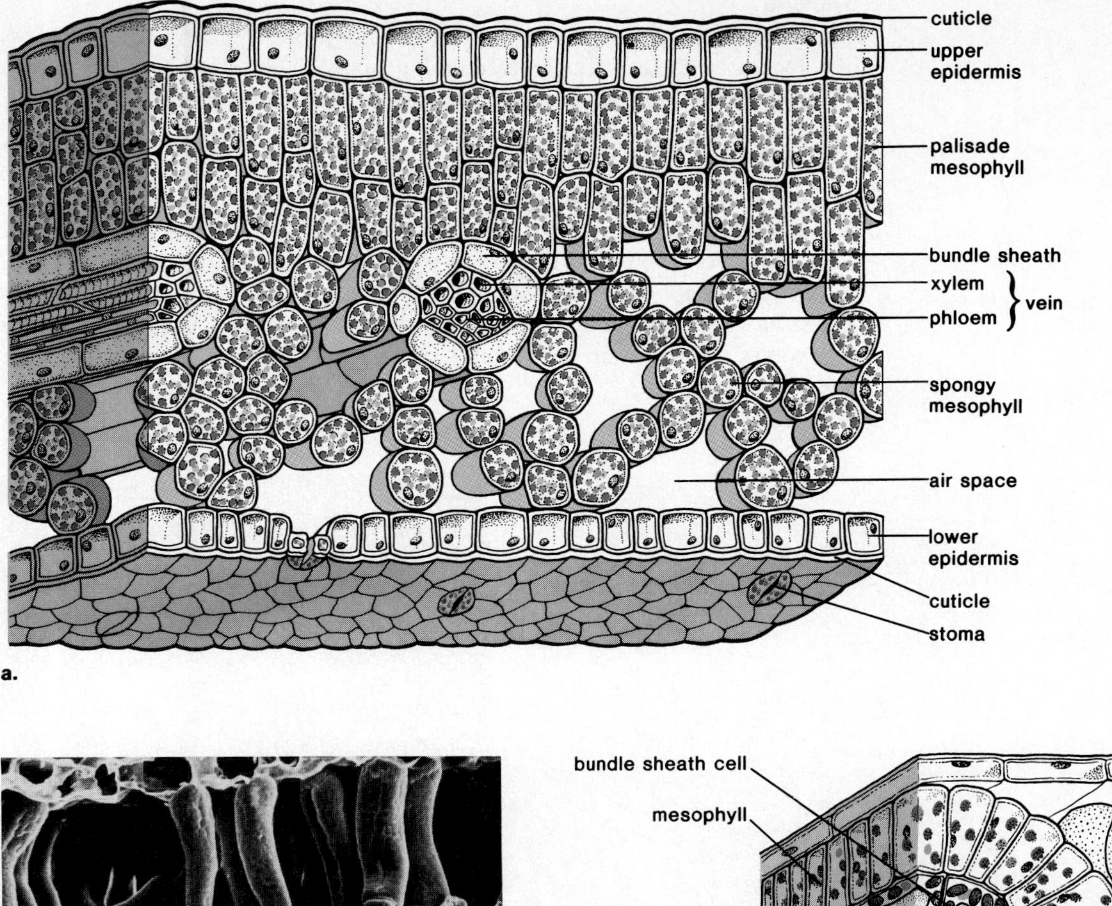

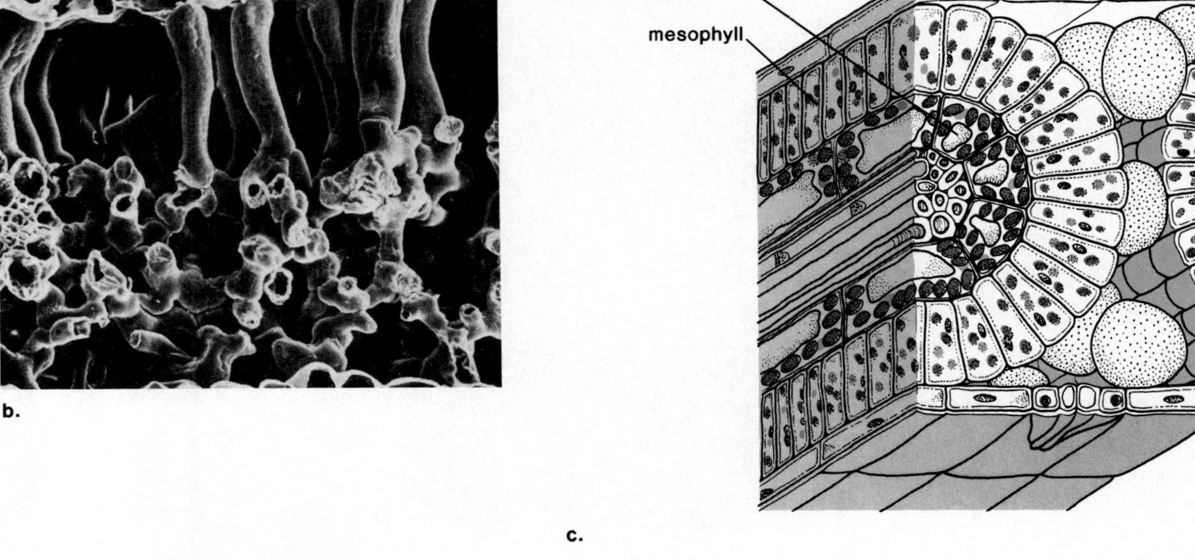

are enclosed by a circular layer of tightly packed parenchyma cells known as the bundle sheath. Notice in figure 7.11a that the bundle sheath cells in plants adapted to the temperate zone do not contain chloroplasts.

The anatomy of a leaf in a plant adapted to a hot, dry, sunny environment is slightly different from the one just described. The mesophyll cells that correspond to the palisade layer are arranged circularly about the bundle sheath cells (fig. 7.11c). As discussed in the reading on page 126, these cells are specialized to pass CO_2 to the bundle sheath cells. This keeps a high concentration of CO_2 in the bundle sheath cells and allows photosynthesis to continue even when the air within the leaf contains a low concentration of CO_2.

Figure 7.11
Leaf anatomy. a. *Drawing of a leaf from a C_3 (photosynthesis) plant that is adapted to temperate climate. In a C_3 leaf the mesophyll is divided into a palisade and spongy layer. The bundle sheath cells lack chloroplasts.* b. *A scanning electron micrograph of the mesophyll.* c. *Drawing of a leaf vein from a C_4 (photosynthesis) plant that is adapted to a hot, dry climate. In a C_4 leaf the mesophyll is arranged around the bundle sheath, and the bundle sheath cells have chloroplasts.*

Stomata

Carbon dioxide, which is necessary for photosynthesis, enters the air spaces of a spongy layer by diffusion through openings, called **stomata,** present particularly in the lower epidermis. When a leaf receives a plentiful amount of water and carbon dioxide and is carrying on photosynthesis, it gives off oxygen, which also exits by way of the stomata. Stomata are tiny pores, each surrounded by two specialized epidermal cells called **guard cells** (fig. 7.12). Due to thickened inner cell walls, guard cells expand outward only when they fill with water. This outward expansion thereafter causes the stomata to open. When guard cells lose water, the loss of turgor causes the stomata to close. Water has been shown to enter the guard cells by osmosis because of a high K^+ (potassium) concentration (fig. 7.13). Since the stomata are open when a

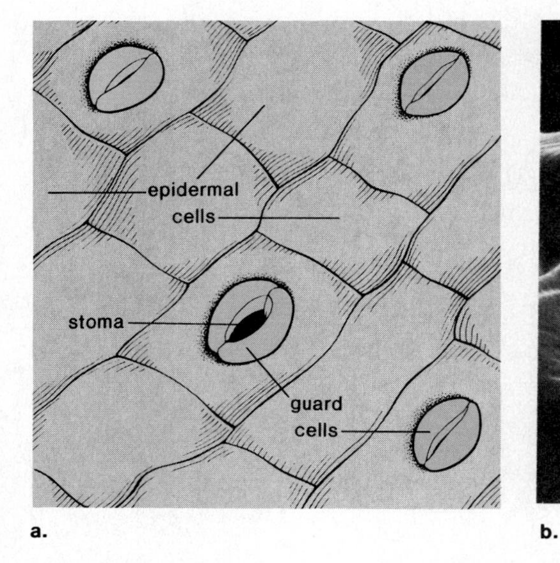

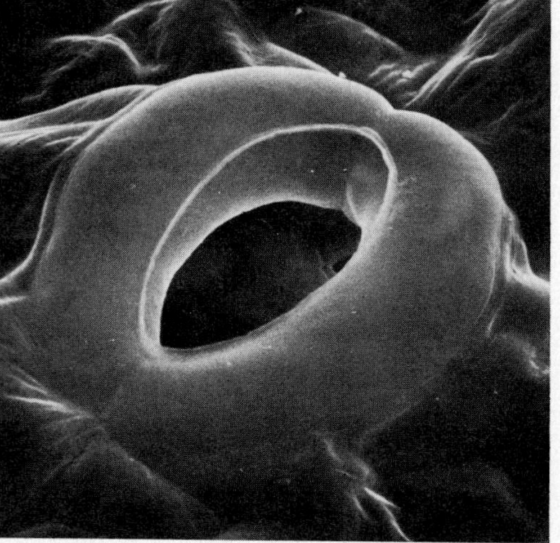

Figure 7.12
Stomata. a. *Drawing of the lower layer of the epidermis of a leaf showing the presence of several stomata. Each stoma is surrounded by two guard cells that regulate whether the stoma is open or closed.* b. *Scanning electron micrograph of one stoma.*

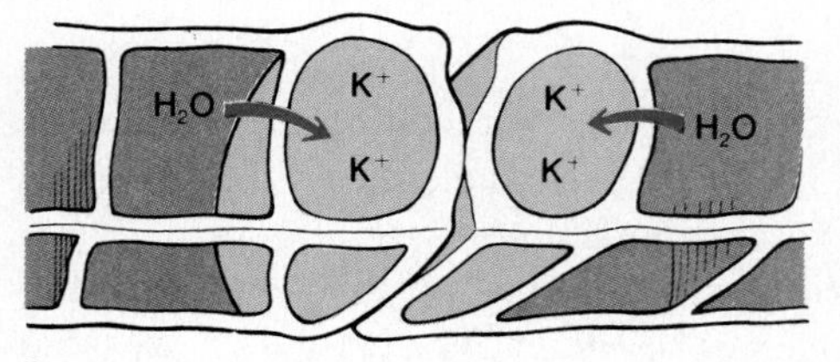

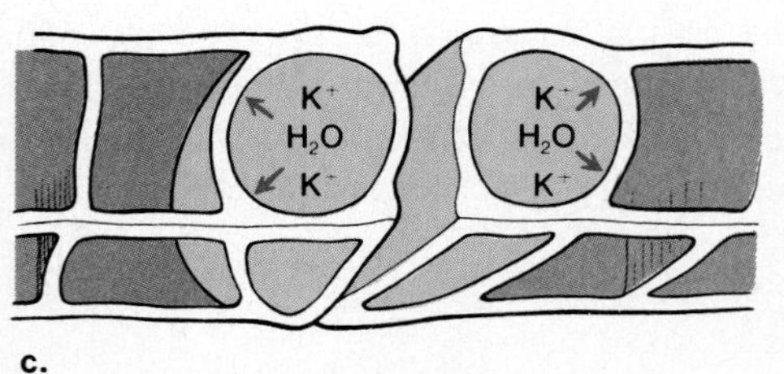

Figure 7.13
Opening of a stoma requires a sequence of events. a. *K^+ enters guard cells and this creates an osmotic pressure that causes* b. *water to enter the guard cells.* c. *Guard cells have thick inner walls but weak outer walls. Therefore the entrance of water creates a pressure that causes the cells to bulge outward, opening the stomata.*

plant is photosynthesizing, it is assumed that photosynthesis in some way triggers a potassium pump that actively transports K^+ into the guard cells, creating an osmotic pressure that eventually opens the stomata. Another possibility is that light directly initiates opening of the stomata but, as you might predict, stomata will not open unless there is a plentiful supply of water.

Much of the water that is transported from the roots to the leaves evaporates and escapes from the leaf by way of the stomata. The evaporation of water from a leaf is called **transpiration;** table 7.2 gives the transpiration rates for a number of plants. The amount of water lost in this manner is truly phenomenal, but it is believed that evaporation of water keeps the leaf cool enough to function in the bright sun and also aids the transport of water. Stomata close at night to prevent water loss.

Table 7.2 *Transpiration Rates*

Per day midsummer		
Ragweed		6–7 quarts
10 foot apple tree		10–20 quarts
12 foot cactus		0.02 quarts
Coconut palm		70–80 quarts
Date palm		400–500 quarts
Per growing season		
Tomato	100 days	30 gallons
Sunflower	90 days	125 gallons
Apple tree	188 days	1800 gallons
Coconut	365 days	4200 gallons
Date palm	365 days	35,000 gallons

In C_3 plants, photosynthesis occurs mostly in the palisade cells of the mesophyll. Water for photosynthesis is transported in leaf veins (having a net pattern in dicots and a parallel pattern in monocots) and carbon dioxide enters the mesophyll's spongy layer at the stomata. Stomata, found mostly in the lower layer of epidermis, open when K^+ ions are pumped into the guard cells which then take up water.

Transport

Vascular tissue, composed of xylem and phloem, is responsible for transport in plants. Xylem, present in all parts of a plant, transports water and minerals primarily from the roots to the leaves, and phloem, also present in all parts of a plant, transports the products of photosynthesis from the leaves to the roots during the growing season.

Xylem

Xylem contains two types of conducting cells: vessel cells and tracheids, each of which is a cell specialized for transport (fig. 7.14). The nonliving **vessel cells** are sometimes called vessel elements because piled one on top of the other they form a continuous pipeline that stretches from the root to the leaves.

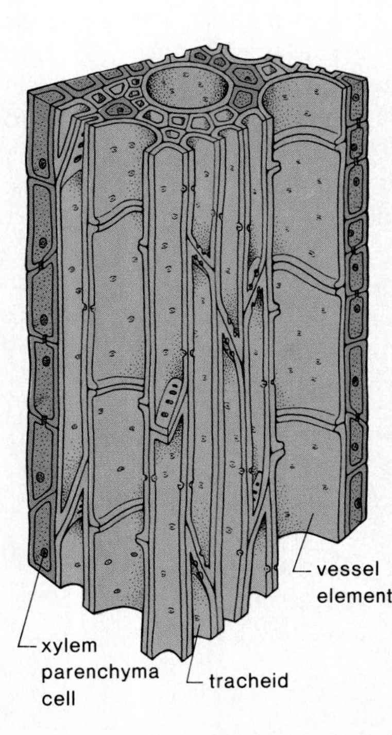

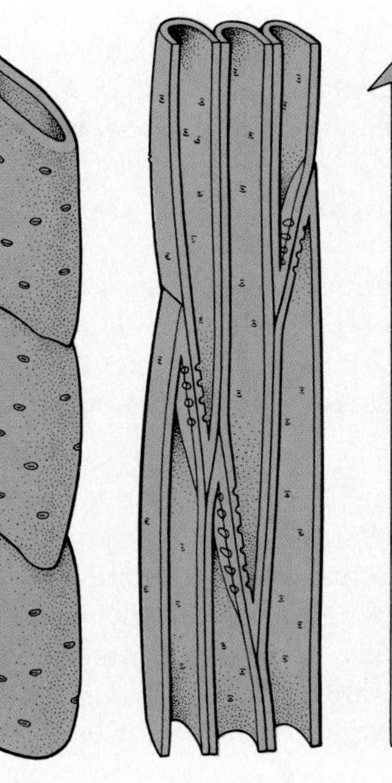

Figure 7.14
Xylem structure. General organization of xylem at the far left, followed by an external view of vessel elements stacked one on top the other and a longitudinal view of several tracheids. Tracheids and vessel elements usually conduct water in the direction shown.

Figure 7.15
Cohesion-tension theory of water transport. Water enters a plant at the root hairs and evaporates (transpires) at the leaves. Xylem forms a continuous pipeline from the roots to the leaves, and this pipeline is completely full of water. Therefore, transpiration exerts a pull on the water column that causes it to move upward.

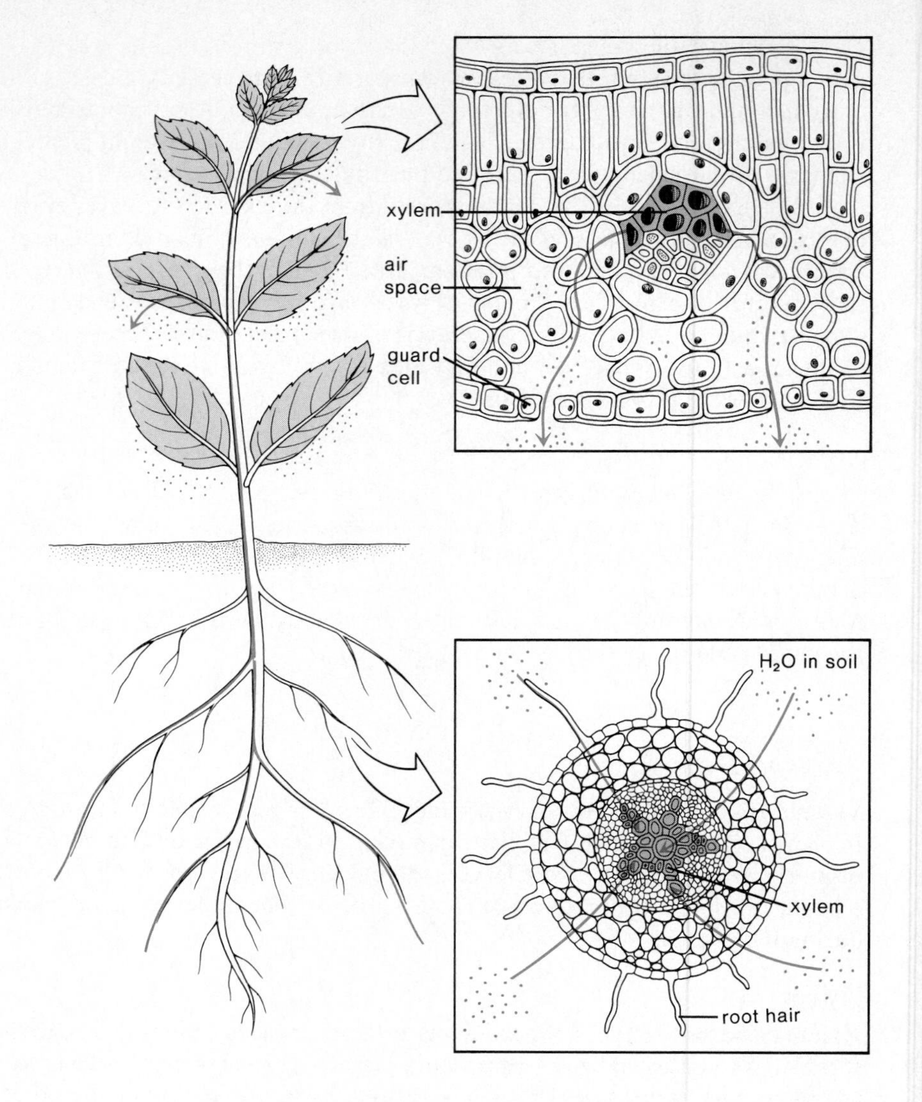

It is an open pipeline because the elements have no end walls separating one from the other. The elongated **tracheids** are also dead at maturity, but they have tapered end walls that are unperforated. Both types of cells have secondary walls that contain lignin, an organic substance that makes the walls tough and hard. Even so, water can move between the lateral walls of both types of cells and the end walls of tracheids because of the presence of pits, depressions where the secondary wall does not form.

Cohesion-Tension Theory of Water Transport

How is it possible for water absorbed by root hairs to be transported from the roots to the leaves within the xylem of even very tall trees? The two factors necessary to this process have already been mentioned:

1. The vessel elements (and tracheids) within xylem form a continuous pipeline from the roots to the leaves (fig. 7.15). Water fills this pipeline. Water enters a root primarily at the roots hairs and makes its way across the cortex until it enters xylem within the vascular cylinder. Within the vessel elements, polar water molecules adhere to the walls and, because of hydrogen bonding, they cling together. *Cohesion* of water molecules within the xylem pipeline is absolutely necessary for water transport in plants.

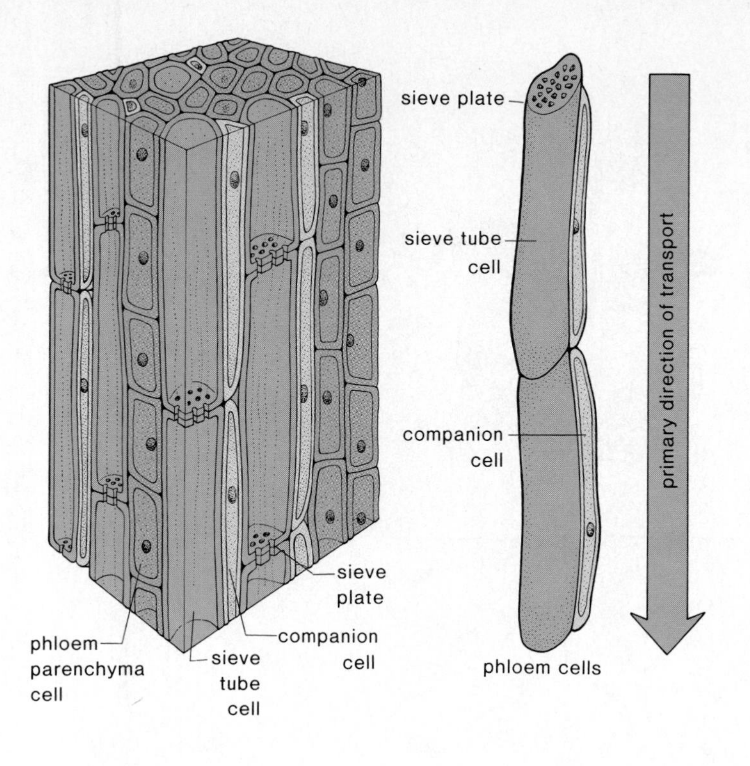

Figure 7.16
Phloem structure. General organization of phloem at the far left, followed by an external view of two sieve-tube cells and their companion cells. Sieve-tube cells usually transport organic nutrients in the direction shown.

2. Water is transpiring (evaporating) from the leaves as long as the stomata are open (table 7.2). More water molecules continuously take the place of water molecules that have just evaporated. In this way, transpiration exerts a pull—creates a *tension*—that draws the column of water up the vessel elements from the roots to the leaves.

The tension created by transpiration would not be effective if it were not for the cohesive property of water. Therefore the theory of water transport in xylem is called the **cohesion-tension theory.**

Xylem contains vessel elements and tracheids where water and minerals are transported from the roots to the leaves. Transpiration of water at the leaves pulls on the column of water molecules that exhibit cohesion.

Figure 7.17
Aphids on the stem of a plant. These tiny insects feed on phloem sap by injecting a stylet mouthpart into the sieve-tube cells and withdrawing some of the nutrient-laden liquid.

Phloem

The conducting cells in phloem are sieve-tube cells, each of which typically has a companion cell (fig. 7.16). **Sieve-tube cells** contain cytoplasm but no nucleus; as their name implies, these cells have pores in their end walls that make these walls resemble a sieve. Through these pores, strands of cytoplasm extend from one cell to the other. The smaller **companion cells** are more generalized cells and have nuclei. It is speculated that the nucleus may control and maintain the life of both cells.

Chemical analysis of phloem sap shows that it is composed chiefly of sucrose and that the concentration of nutrients is 10 to 13 percent by volume. Samples for chemical analysis are most often obtained by the use of aphids (fig. 7.17), small insects that are phloem feeders. The aphid drives its stylet, a short mouthpart that functions like a hypodermic needle, between the epidermal cells and withdraws sap from a sieve-tube cell. If the aphid is anesthetized by ether, the body may be carefully cut away, leaving the stylet, which exudes the phloem contents that can be collected and analyzed.

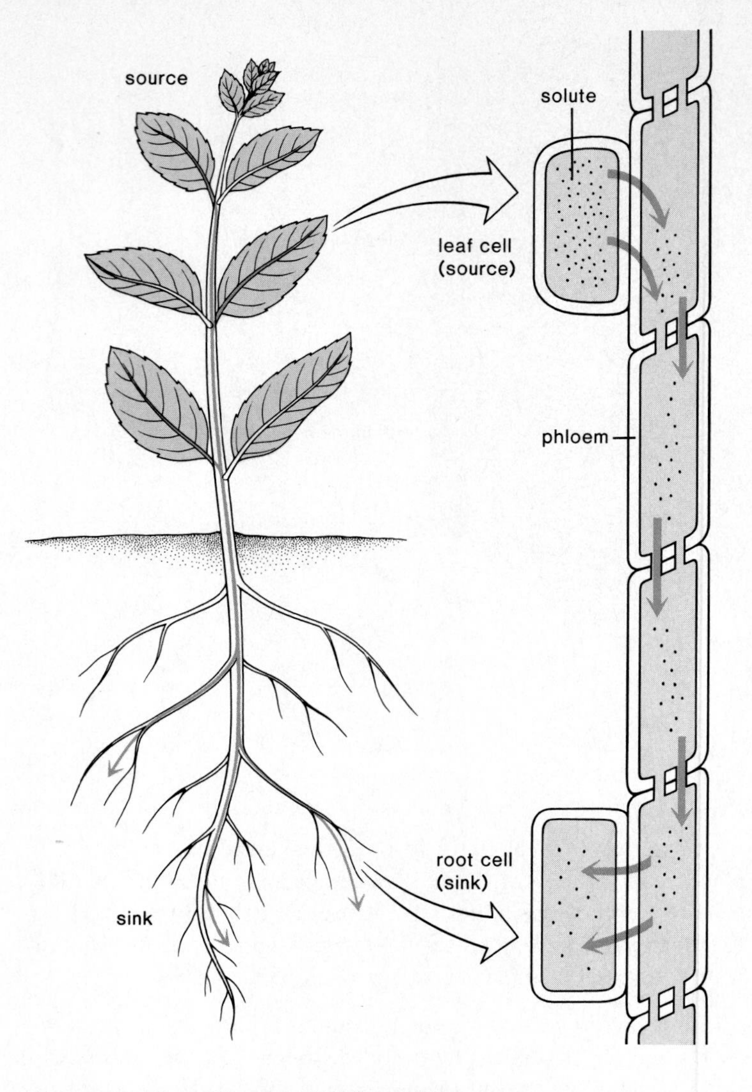

Figure 7.18
Pressure-flow theory of phloem transport. Sieve-tube cells form a conducting system for solutes from the leaves to the roots. Sucrose is actively transported into the sieve-tube cells at the leaves, and water follows passively by osmosis. This creates a pressure. At the roots, sucrose is actively transported out of the sieve-tube cells, and water follows passively by osmosis. Now the pressure created at the leaves causes water to flow toward the roots and as it flows it brings along the sucrose.

Pressure-Flow Theory of Phloem Transport

Phloem is present in a plant from the leaves to the roots of a plant (fig. 7.18). The sieve-tube cells are connected by strands of cytoplasm that extend through the adjoining sieve-tube plates. In order to understand movement of solute within phloem, it is important to envision this continuous stream of cytoplasm from the leaves to the roots. Also, it should be remembered that there is a companion cell for each sieve-tube cell.

1. During the growing season, the leaves are photosynthesizing and producing sugar. This sugar is transported into phloem, and water follows passively by osmosis. Transport and osmosis are possible because sieve-tube cells have a living cell membrane. Also, the energy needed for sucrose transport is provided by the companion cells. The buildup of water within the sieve-tube cells at the leaves creates a *pressure.*
2. At the roots (or other places in the plant) the sugar is transported out of the phloem and water follows passively by osmosis. This exit of sucrose and water at the roots means that the pressure created at the leaves will cause a *flow* of water from the leaves (where pressure exists) to the roots. As the water flows along, it brings the sucrose with it.

The **pressure-flow hypothesis** can account for the observed reversal of flow in the sieve-tube system; for example, in the spring before the leaves are out. When the plant is not photosynthesizing, the roots serve as a source of sucrose, and the other parts of the plant remove it from the sieve-tube cells. Therefore water will tend to enter the sieve-tube cells in the roots with a greater force and will flow toward the leaves, carrying sucrose with it.

Phloem contains sieve-tube cells where organic solutes are transported. In the leaves, a pressure is created when sucrose is transported into phloem and water follows by osmosis. The pressure causes a flow of sucrose from the leaves to the roots where sucrose is transported out of phloem.

Figure 7.19
Each dish shows several shoots of Douglas Fir growing from a single cotyledon. Such cultures are part of a research project for cloning genetically improved trees. Subsequently, the shoots will be cut, rooted, and planted in the forest. Tissue culture propagation is expected to play an important part in bringing forest yields toward their theoretical maximum.

Reproduction

Plants can reproduce both asexually, without the need for gametes, or sexually, which does require gametes.

Asexual Reproduction

Asexual reproduction, also known as **vegetative propagation,** is common in plants. In vegetative propagation, a portion of one plant gives rise to a completely new plant. Both plants now have identical genes. As an example, some plants have above-ground horizontal stems, called *runners,* and others have underground stems, called *rhizomes,* that produce new plants. To take a concrete example, strawberry plants grow from the nodes of runners and violets grow from the nodes of rhizomes. White potatoes can be propagated in a similar manner. White potatoes are actually portions of underground stems, and each eye is a node that will produce a new potato plant. Sweet potatoes are modified roots and may be propagated by planting sections of the root. You may have noticed that the roots of some fruit trees, such as cherry and apple trees, produce "suckers," small plants that can be used to grow new trees.

Asexual reproduction has a great deal of commercial importance. Once a plant variety with desired characteristics has been developed through vegetative propagation, new plants can be supplied to gardeners and farmers. Cuttings can be taken from the plant, and the cut end can be treated to encourage it to grow roots, or a cutting can be grafted to the stem of a plant that has a root. Budding is a form of grafting most often used commercially. In this procedure, just the axillary buds are grafted onto the stem of another plant. Today, entire plants can also be produced by tissue culture (fig. 7.19), a technique that may eventually replace the older methods thus far discussed. Usually an embryonic *tissue* is removed from a plant and placed in a special *culture* medium. After the tissue has grown for a while, it is subdivided so that many identical plants are produced from a very small amount of starting cells.

Plants reproduce both asexually and sexually. Asexual reproduction occurs when a portion of one plant gives rise to an entirely new plant. Grafting, particularly budding, and tissue culture propagation have commercial importance today.

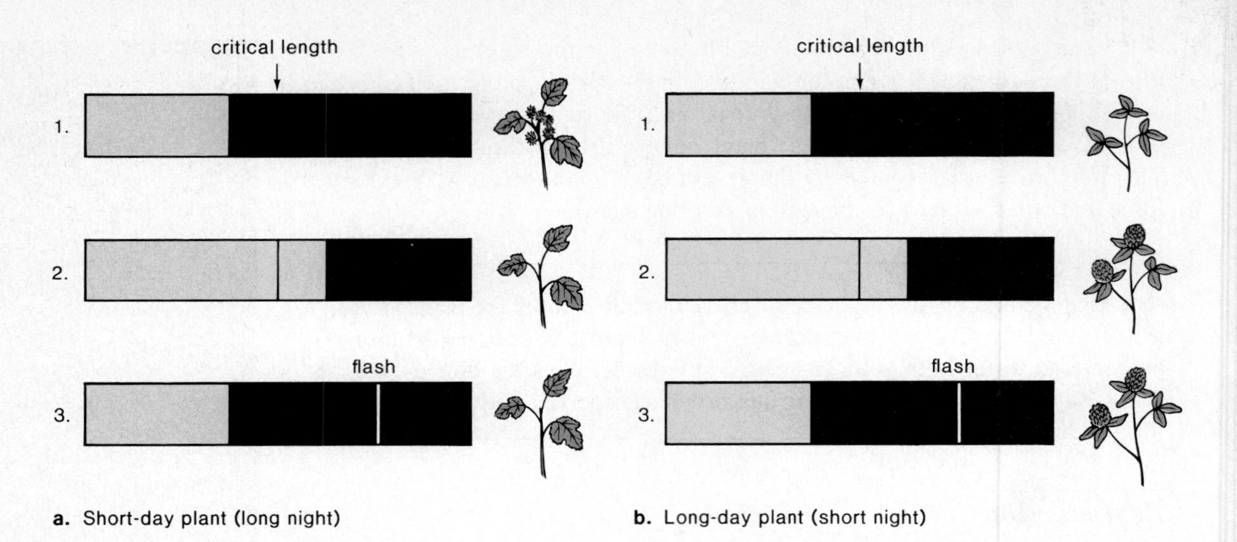

Figure 7.20
Day length (night length) effect on two types of plants. a. *Short-day (long-night) plant.* 1. *When the day is shorter (the night longer) than a critical length, this type of plant flowers;* 2. *it does not flower when the day is longer (night is shorter) than the critical length;* 3. *it also does not flower if the longer-than-critical-length night is interrupted by a flash of light.* b. *Long-day (short-night) plant.* 1. *When the day is shorter (the night longer) than a critical length, this type plant does not flower;* 2. *it does flower when the day is longer (the night shorter) than a critical length;* 3. *it does flower if the longer-than-critical-length night is interrupted by a flash of light.*

Sexual Reproduction

Plants also reproduce sexually. This may come as a surprise to those who never thought of plants as being male and female. Sexual reproduction is properly defined as reproduction requiring gametes, often an egg and a sperm. In flowering plants, the sex organs are located in the flower, and we will digress to discuss flowering, an event that occurs only at a particular time of year in certain plants.

Flowering

The effect of photoperiod (length of daylight compared to the length of darkness) on plants is particularly obvious in the temperate zone. In the spring, plants respond to increasing day length by initiating growth and in the fall they respond to decreasing day length by ceasing growth processes. Among its effects, day length also regulates flowering in some plants; for example, plants such as violets and tulips flower in the spring and others such as asters and goldenrods flower in the fall. Investigators who first studied the effect of **photoperiodism** on flowering came to the conclusion that plants could be divided into three groups (fig. 7.20):

Short-day plants—flower when the photoperiod is shorter than a critical length. (Good examples are cocklebur, poinsettia and chrysanthemum.)

Long-day plants—flower when the photoperiod is longer than a critical length. (Good examples are wheat, barley, clover, and spinach.)

Day-neutral plants—flowering is not dependent on a photoperiod. (Good examples are tomatoes and cucumbers.)

Later, experiments were done with artificial time periods of light and dark that did not necessarily correspond to a normal 24-hour day. It was discovered that the cocklebur, a short-day plant, would flower as long as the dark period was continuous for at least eight and one-half hours, regardless of the length of the light period. Further, if this critical length dark period was interrupted

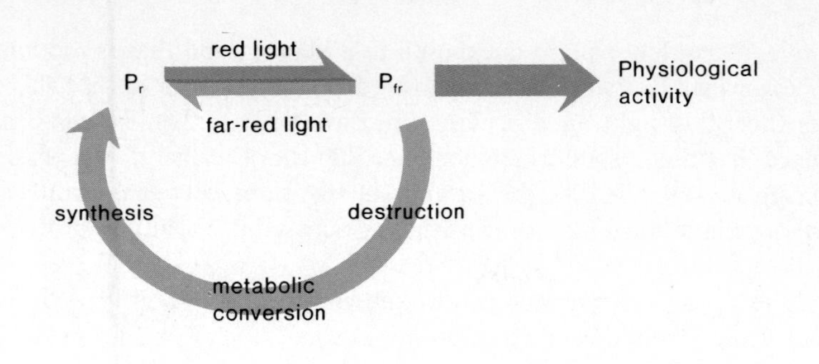

Figure 7.21
The $P_r \rightleftarrows P_{fr}$ conversion cycle. As indicated by the coloring, the inactive form P_r is prevalent during the night. At sunset or in the shade when there is more far-red light, P_{fr} is converted to P_r. Also during the night metabolic processes cause P_{fr} to be replaced by P_r. As indicated by the coloring, P_{fr}, the active form of phytochrome, is prevalent during the day because at that time there is more red light than far-red light.

by a brief flash of light, the cocklebur would not flower. (Interrupting the light period with darkness had no effect.) Similar results have also been found for long-day plants. They require a shorter dark period than a critical length regardless of the length of the light period. However, if a longer-than-critical-length night is interrupted by a brief flash of light, long-day plants will flower. We must conclude then that it is a critical (specific for each plant) length of the dark period that controls flowering and not the length of the light period. Of course, in nature, shorter days always go with longer nights and vice versa.

Phytochrome If flowering is dependent on day/night length, plants must have some way to detect these periods. Many years of research led to the discovery of a plant pigment called phytochrome. **Phytochrome** is a blue-green leaf pigment that alternately exists in two forms (fig. 7.21).

P_r (phytochrome red) absorbs red light (of 660 nm wavelength) and is converted to P_{fr}.

P_{fr} (phytochrome far-red) absorbs far-red light (of 730 nm wavelength) and is converted to P_r.

Sunlight contains more red light than far-red light; therefore, P_{fr} is apt to be present in plant leaves during the day. In the shade and at sunset there is more far-red light than red light; therefore P_{fr} is converted to P_r as night approaches. There is also a slow metabolic replacement of P_{fr} by P_r during the night. It was thought for some time that this slow reversion might provide a means for the plant to measure the length of the night. However, it is now assumed that the active form of phytochrome, P_{fr}, signals a biological clock, an internal system that controls the timing of certain physiological and behavioral responses; in this case, flowering. (Biological clocks are discussed in more detail in chapter 29.) Once the clock notes the presence of P_{fr} during a certain time span, it is presumed to initiate the hormonal (followed by enzymatic) changes that promote flowering.

What is the possible hormonal control of flowering? For a very long time, researchers have been searching for a flowering hormone called florigen. They have been able to extract substances from a plant that will cause a vegetative shoot to produce flowers but no one has yet been able to identify chemically a specific hormone present in these extracts. Experiments do indicate that the hormone (if it exists) is transported from the leaves to the shoot apex. For example, a cocklebur plant is exposed to a photoperiod that is too long to induce flowering. If a single leaf is covered by black paper so that this leaf receives short-day lighting, the plant will flower (fig. 7.22). This is considered evidence that there is a florigen hormone. On the other hand, it is possible that flowering is controlled by the balance in the plant between stimulatory and inhibitory plant hormones, such as those discussed in a reading for this chapter (p. 139).

The P_r to P_{fr} conversion is now known to control other growth functions in plants aside from flowering. It promotes some seed germination and inhibits stem elongation, for example. The presence of P_{fr} apparently indicates to photosensitive seeds that sunlight is present and that conditions are favorable for germination. Following germination, the presence of P_r indicates that stem elongation may be needed in order to reach sunlight. Seedlings that are grown in the dark etiolate, that is, the stem increases in length and the leaves stay small (fig. 7.23). Once the seedling is exposed to sunlight, and P_r is converted to P_{fr}, the seedling begins to grow normally—the leaves expand and the stem branches.

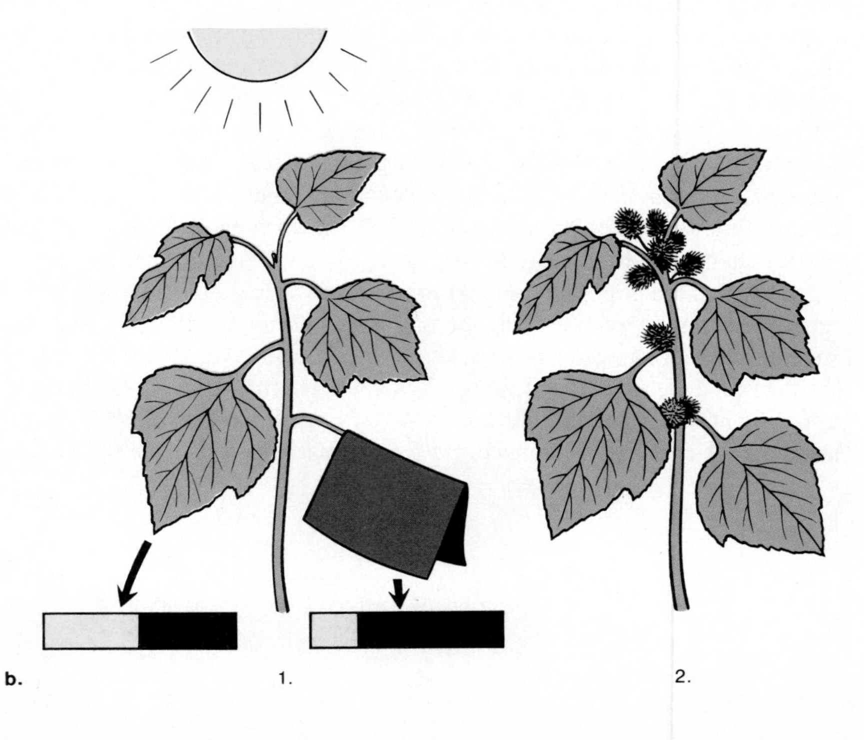

***Figure** 7.22*
a.*Cocklebur plant. This plant is often used in flowering experiments because it will respond to a single exposure of appropriate lighting. Other plants require appropriate lighting for several days and nights in succession.* b. *Flowering experiment.* 1. *A cocklebur plant is kept on an inappropriate long-day exposure except for a single leaf that is covered so it alone receives a short-day exposure.* 2. *The cocklebur plant flowers, indicating that a substance has passed from the single leaf to the flower buds.*

Some plants that exhibit photoperiodism flower only when the days are long (nights are short) and others flower only when the days are short (nights are long). Flowering is dependent on the pigment phytochrome, whose active form P730 communicates with a biological clock presumably causing it to initiate hormonal changes that bring about flowering.

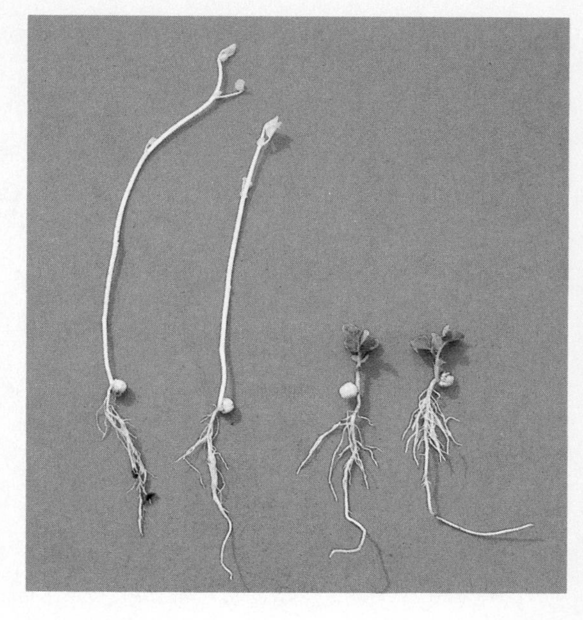

Figure 7.23
Phytochrome has other functions besides regulating flowering. For example, if far-red light is prevalent as it is in the shade, the stem of a seedling elongates and the leaves remain small (left-hand side of photo). However, if red light is prevalent as it is in bright sunlight, the stem does not elongate and the leaves expand. These effects are due to phytochrome.

Flower

Figure 7.24 shows the parts of a typical flower. The **sepals,** most often green, form a whorl about the **petals,** whose color accounts for the attractiveness of many flowers. In the center of the flower is a small vaselike structure, the **pistil,** which usually has three parts: the **stigma,** an enlarged sticky knob; the **style,** a slender stalk; and the **ovary,** an enlarged base. The ovary contains a number of ovules that play a significant role in reproduction. Grouped about the pistil are a number of **stamens,** each of which has two parts: the **anther,** a saclike container, and the **filament,** a slender stalk.

Flowering plants have a life cycle called **alternation of generations** (p. 561) because it contains two generations: the sporophyte generation and the gametophyte generation. The sporophyte generation is a diploid (2N) generation that produces haploid spores by meiosis. The spores give rise to a gametophyte (N) generation. The gametophyte generation produces gametes that join to give rise to the sporophyte generation once again.

A plant that flowers is actually a **sporophyte generation** that produces spores. Within the ovary, each **ovule** contains a megaspore (*mega* means large) mother cell (fig. 7.25) that undergoes meiosis to produce four haploid megaspores. Three of these disintegrate, leaving one functional megaspore that divides mitotically. The result is the *female gametophyte generation,* which typically consists of eight nuclei embedded in a mass of cytoplasm that is partly differentiated into cells. One of these cells is an egg.

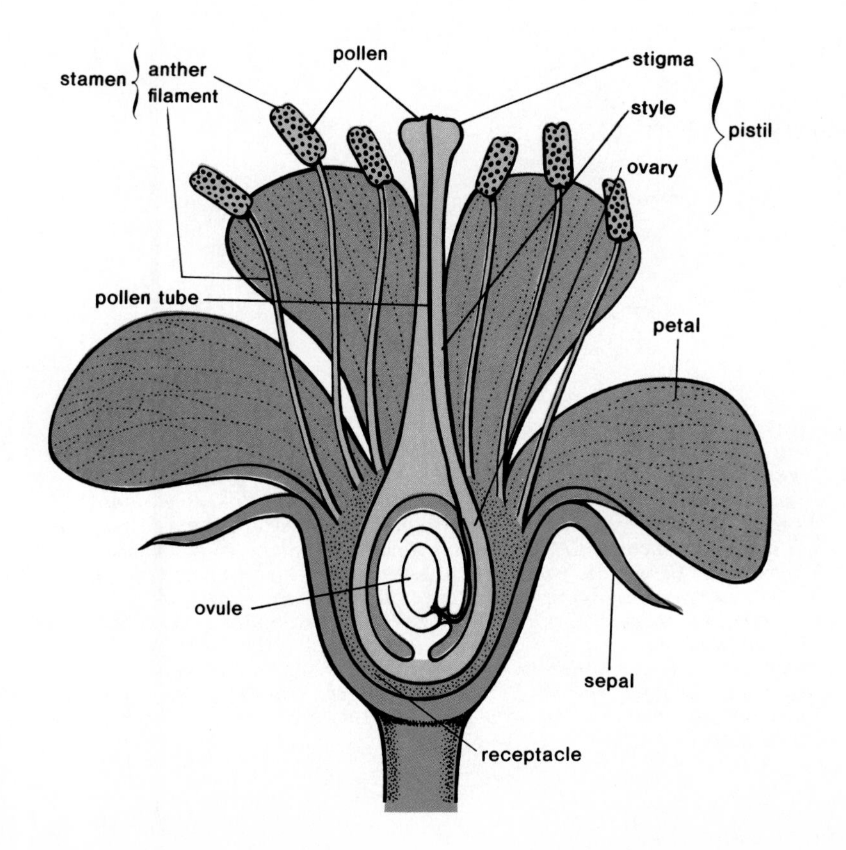

Figure 7.24
Diagram illustrating flower parts. The mature pollen grain contains two sperm that travel down a pollen tube to the ovule. If conditions are correct, one sperm fertilizes the egg cell within this structure.

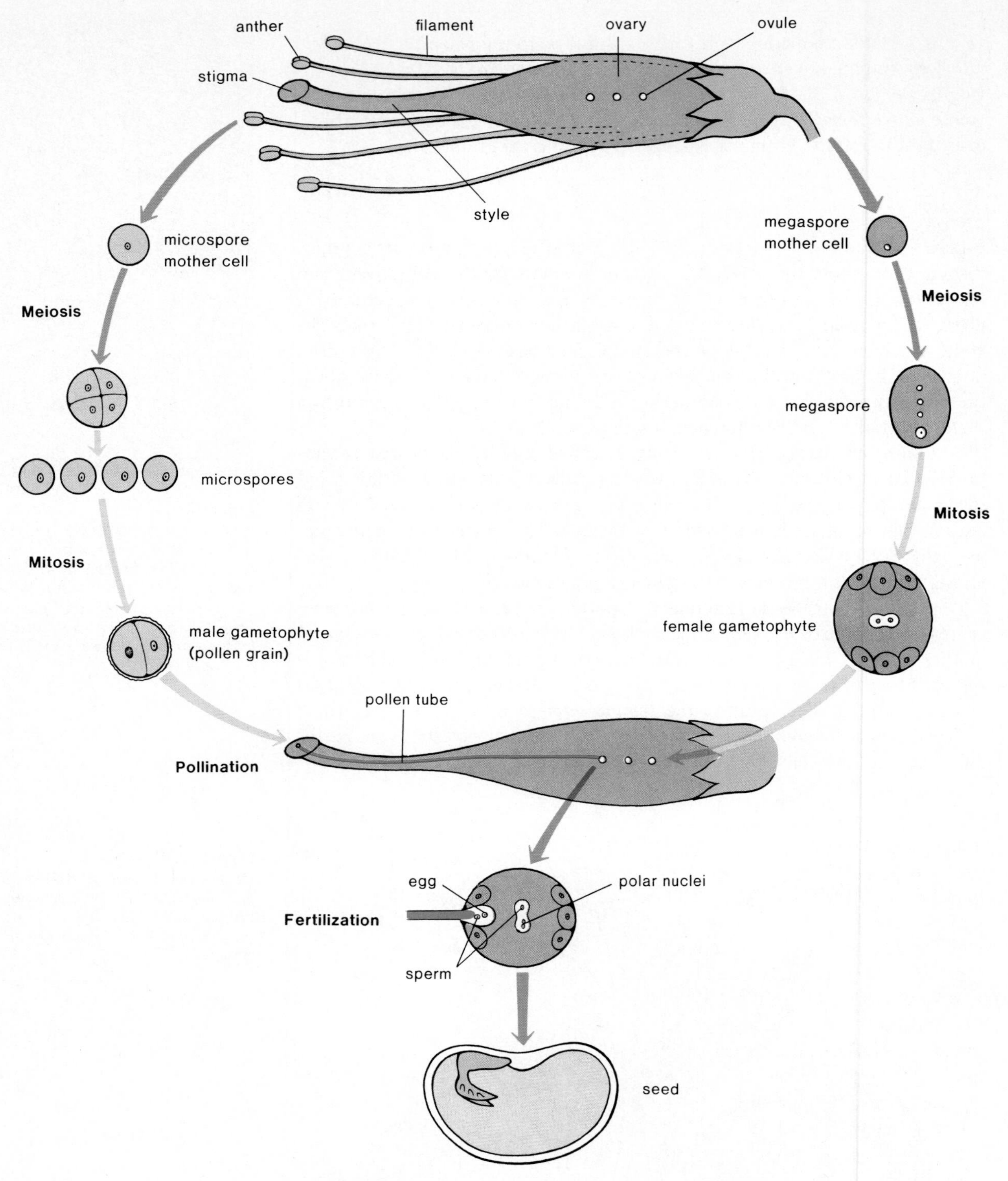

Figure 7.25
Events leading to fertilization and seed formation in a flowering plant. There is a megaspore mother cell in each ovule found in the ovary. The megaspore mother cell undergoes meiosis to give one functional megaspore which undergoes mitosis. The result is the female gametophyte generation, which contains an egg cell. The anther contains microspore mother cells that undergo meiosis to give four functional microspores. Following mitosis, each is a pollen grain containing two cells. At about the time of pollination, one of these cells divides to give two sperm which travel down a pollen tube to the ovule. During double fertilization, one sperm nucleus joins with the egg nucleus, and the other joins with two polar nuclei. The ovule now matures and becomes a seed that contains an embryonic plant and stored food.

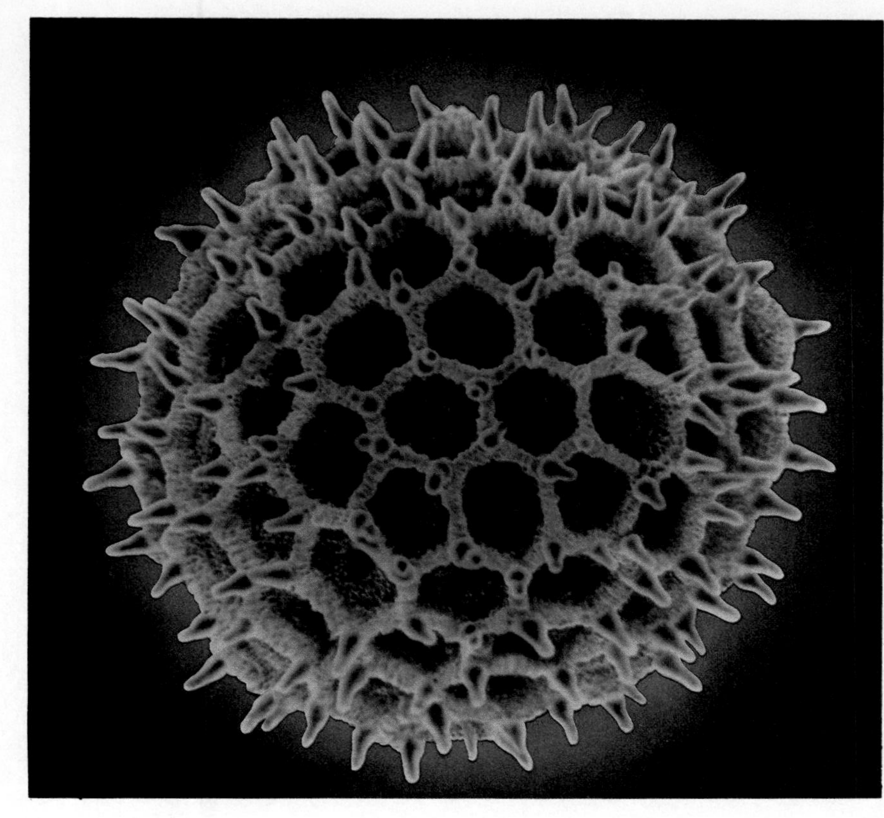

Figure 7.26
Scanning electron micrograph of a pollen grain. The appearance of pollen grains varies from plant to plant.

The anther contains numerous microspore (*micro* means small) mother cells, each of which undergoes meiosis to produce four haploid cells called microspores. The microspores usually separate and each one becomes a **pollen** grain (fig. 7.26), or male gametophyte generation. At this point, the young male gametophyte generation contains two nuclei, the *generative nucleus* and the *tube nucleus*.

Pollination (fig. 7.27) occurs when pollen is windblown or carried by insects, birds, or bats to the stigma of the same type plant. Only then does a pollen grain germinate and produce a long pollen tube that grows within the style until it reaches an ovule in the ovary. Before fertilization occurs, the generative nucleus divides, producing two sperm that have no flagellae. On reaching the ovule, the pollen tube discharges the sperm. One of the two sperm migrates to and fertilizes the egg, forming a zygote; the other sperm migrates to and unites with the polar nuclei, producing a 3N (triploid) endosperm nucleus. The endosperm nucleus divides to form endosperm, food for the developing plant. Note that flowering plants have a *double fertilization.* One fertilization produces the zygote, the other produces endosperm.

In a flower the ovary, a part of the pistil, contains eggs within ovules and the pollen produced by the anther contains sperm. Upon pollination, the pollen is transported to the pistil and germinates. The sperm pass down the pollen tube and one fertilizes the egg; the other fuses with the polar nuclei, which then divides to give endosperm. The ovule now matures to produce a seed and the ovary becomes a fruit.

a.

b.

Figure 7.27
Most flowering plants depend on insect pollinators to carry pollen from one flower to the other so that cross-fertilization can occur. In order to lure insects, plants often produce nectar, a sugary substance secreted by little glands in the center of the flower. The flower parts are arranged so that the insect must brush up against the stamens and pistil(s) in order to reach the nectar. Cross-fertilization occurs when pollen from the stamens of one flower sticks to the hairs of the insect and is rubbed off onto the pistil(s) of the next flower. A plant tends to attract only one type of pollinator. a. *Bees seem to like violet, blue, yellow or pink flowers, and* b. *butterflies prefer red, blue, purple, or yellow flowers.*

Figure 7.28
Monocot seed and germination. a. *Longitudinal section of mature seed. Notice the large amount of endosperm in addition to the cotyledon in the monocot seed.* b. *When the seed germinates, the plumule becomes the leaves and the radicle becomes the roots.*

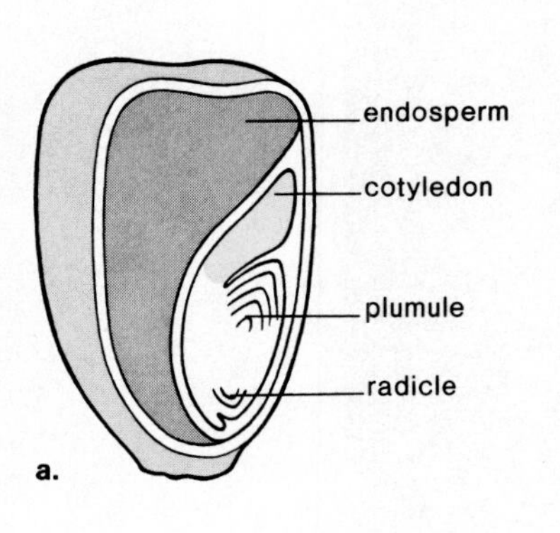

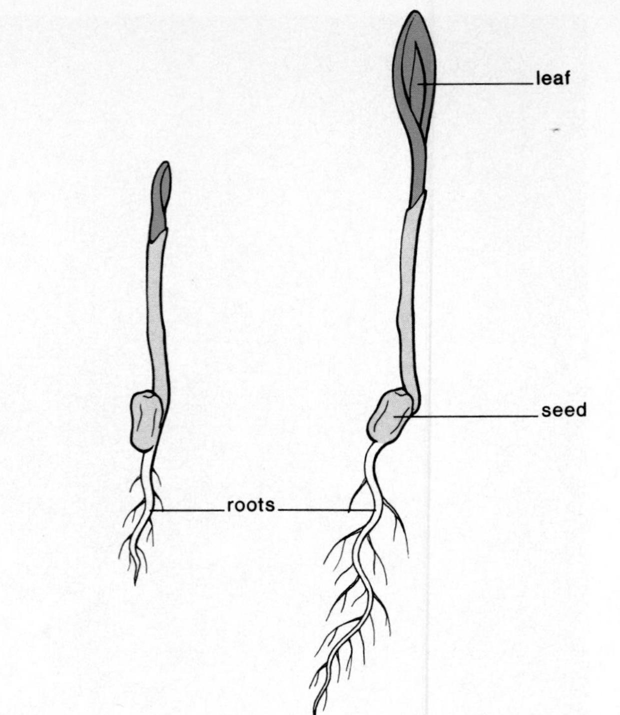

Figure 7.29
Dicot seed and germination. a. *Longitudinal section of a dicot seed shows two large cotyledons, one on either side of the embryo.* b. *Germination of a dicot seed makes it easier to detect that the epicotyl gives rise to the leaves; the hypocotyl becomes a portion of the stem and the radicle becomes the roots.*

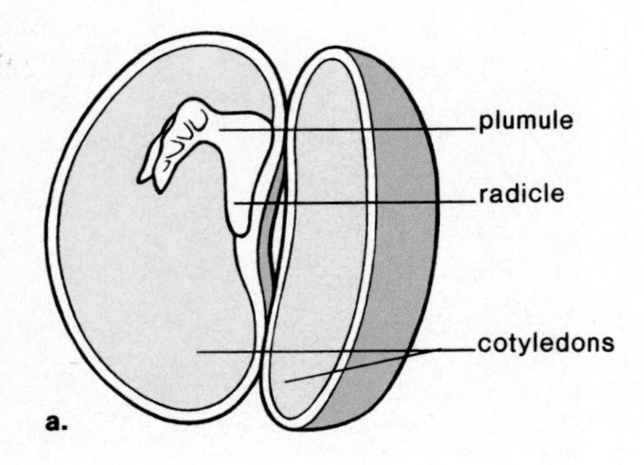

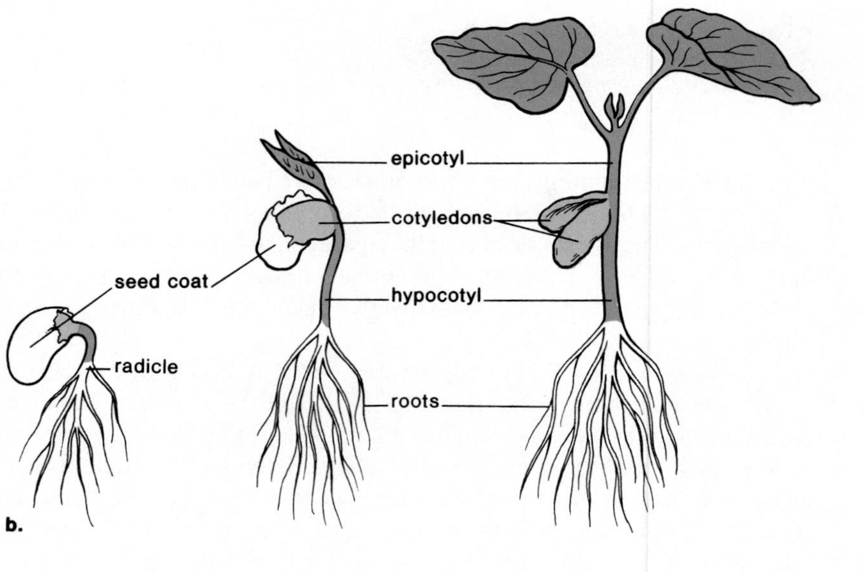

Seed

Following fertilization, each ovule becomes a mature seed containing an embryo that has at least three parts: the **cotyledon(s),** or seed leaf (leaves); the **epicotyl,** that portion of the embryo above the attachment of the cotyledon(s); and the **hypocotyl,** which lies below the attachment of the cotyledons, becomes a portion of the stem. The epicotyl contains the apical meristem of the shoot and sometimes bears young leaves, in which case it is called a **plumule.** A **radicle,** the embryonic root, may be at the lower end of the hypocotyl.

Monocots have seeds with one cotyledon and dicots have seeds with two cotyledons (fig. 7.28a and fig. 7.29a). Cotyledons typically provide nutrient molecules for the growing embryo. In monocot embryos, the cotyledon rarely stores food; rather, it absorbs food molecules from the endosperm and passes them on to the embryo. In many dicot embryos, the cotyledons replace the endosperm, which typically has already transferred its nutrients to the cotyledons.

Figure 7.30
Development of fruit from a flower. Once seed formation begins, the ovary and accessory parts begin to enlarge and grow larger until only remnants of the other flower parts remain and finally disappear entirely, leaving only the mature fruit.

In addition to the embryo and stored food either within cotyledon(s) or within endosperm, a seed is covered by a *seed coat.* In flowering plants, all seeds are also enclosed within a fruit that develops from the ovary (fig. 7.30) and, at times, from other accessory parts. Although peas, beans, tomatoes, and cucumbers are commonly called vegetables by laymen, botanists categorize them as fruits. Fruits protect seeds, sometimes provide extra nourishment, and also sometimes aid dispersal. For example, winged dry fruits like that of a maple tree are adapted to distribution by the wind while fleshy fruits like that of a cherry are eaten by birds and the seeds are deposited some distance away.

The mature seed typically contains an embryo consisting of the cotyledon(s), epicotyl and hypocotyl. Upon germination, a radicle (root) appears and the epicotyl gives rise to the leaves.

Seed Germination A seed will not normally germinate unless environmental factors are favorable. Water and oxygen are essential to the completion of germination in nearly all seeds, but light is not necessarily a requirement because some prefer darkness. Most seeds need a temperature that is above freezing but below 45° C. The first event normally observed in a germinating seed is the emergence of the radicle, the first indication of the roots (figs. 7.28b and 7.29b). This is followed shortly by the appearance and expansion of the seedling shoot formed from the elongating epicotyl. In dicots, the cotyledons degenerate after their nourishment is consumed by the developing plant. As the seedling emerges from the soil, the shoot may be hook-shaped, protecting the delicate leaves. But once the seed is above ground, the stem straightens out and the leaves expand as photosynthesis begins. The plant continues to grow as long as it lives because of the presence of meristem tissue.

Summary

Water and minerals enter root hairs at the maturation zone of a root; crossing the epidermis, cortex, and endodermis before entering the vascular cylinder. Thereafter water and minerals pass up the shoot system of the plant. Herbaceous stems have vascular bundles arranged in a ring and monocots have scattered vascular bundles. In woody stems xylem and phloem are separated by a vascular cambium that produces secondary xylem and phloem each year. Wood is composed of rings of secondary xylem.

Finally the water and minerals arrive at the leaf veins within leaves. Photosynthesis occurs within the palisade layer and gas exchange occurs within the spongy layer of the mesophyll. Gases enter and exit a leaf at the epidermis where stomata open when guard cells take up potassium and water. When the stomata are open, water evaporates from a leaf. This process of transpiration is an important part of the cohesion-tension theory of water transport. Once photosynthesis has occurred, sucrose is transported within the phloem, another vascular tissue. The pressure-flow theory explains the transport mechanism.

Flowers contain reproductive organs. Flowering in some plants is controlled by the photoperiod detected by the pigment phytochrome. Stamens are reproductive organs that produce pollen grains within anthers; the pistil produces eggs within ovules. Double fertilization occurs following pollination. One sperm from the pollen grain fertilizes the egg cell and another joins with polar nuclei to give endosperm. The ovule now becomes a seed enclosed within a fruit derived from the ovary. Upon germination, the epicotyl and hypocotyl of the seed contribute to stem and leaf formation and the root develops from the radicle. The cotyledons of a seed are involved in storage of food.

Objective Questions

1. The ____________ tissue is unspecialized and forever capable of cell division.
2. In the root the zone of ____________ follows the zone of ____________ .
3. Adventitious roots develop from the ____________ , a portion of the plant.
4. In a monocot stem, the vascular bundles are said to be ____________ .
5. In a woody stem, the secondary xylem builds up and forms the ____________ , which can be counted to tell the age of a tree.
6. The mesophyll contains two layers, the ____________ layer and the spongy layer.
7. The transport of water and minerals is dependent upon ____________ which occurs whenever the stomata are open.
8. The conducting cells in xylem are called ____________ and ____________ .
9. The ____________ hypothesis explains the transport of solutes in sieve-tube cells.
10. ____________ is the pigment that is believed to signal a biological clock in plants that exhibit photoperiodism.
11. Plants have a life cycle called ________ .
12. Monocots have seeds with one ________ while dicots have a seed with two.

Answers to Objective Questions

1. meristem 2. maturation; elongation 3. stem 4. scattered 5. annual rings 6. palisade 7. transpiration 8. vessel elements; tracheids 9. pressure-flow 10. Phytochrome 11. alternation of generations 12. cotyledon.

Study Questions

1. Name the two major divisions of a plant and the organs that each contains. (p. 132)
2. Discuss the anatomy of a root, stem (nonwoody and woody), and leaf. (pp. 134, 138, 142)
3. What are the three types of roots? Give examples. (p. 136)
4. Explain how stomata open and close. (p. 144)
5. Describe the structure of xylem, and discuss water and mineral transport. (pp. 145–146)
6. Describe the structure of phloem, and discuss organic nutrient transport. (pp. 147–148)
7. Give examples of asexual reproduction in plants. (p. 149)
8. Explain periodic flowering in some plants. (p. 150)
9. Trace the production of seeds in flowering plants, starting with a megaspore mother cell in the ovule and microspore mother cells in the anther. (pp. 153–154)
10. Describe the structure and germination of a monocot and dicot seed. (pp. 156–157)
11. Name four differences between dicots and monocots. (p. 141)

Key Terms

Casparian strip (kas-par'e-an strip) band of waxy material found on endodermal cells of plants; prevents passage of molecules outside of cells. *136*

cotyledon (kot''ĭ-le'don) the seed leaf of the embryo of a plant. *156*

dicot (di'kot) dicotyledon; a type of angiosperm distinguished particularly by the presence of two cotyledons in the seed. *138*

endodermis (en''do-der'mis) a plant tissue consisting of a single layer of cells that surrounds and regulates the entrance of materials into particularly the vascular cylinder of roots. *133*

epicotyl (ep''ĭ-kot'il) the plant embryo portion above the cotyledons that contributes to stem development. *156*

epidermis (ep''ĭ-der'mis) the outer layer of cells of organisms including plants. *132*

herbaceous stem (her-ba'shus stem) nonwoody stem. *140*

hypocotyl (hi''po-kot'il) plant embryo portion below the cotyledons that contributes to stem development. *156*

lenticel (len'tĭ-sel) pocket of loosely arranged cells in cork that permit gas exchange. *142*

meristem tissue (mer'ĭ-stem tish'u) a plant tissue that always remains undifferentiated and capable of dividing to produce new cells. *132*

mesophyll (mes'o-fil) the middle tissue of a leaf made up of parenchyma cells. *142*

monocot (mon'o-kot) monocotyledon; a type of angiosperm in which the seed has only one cotyledon, such as corn and lily. *138*

palisade layer (pal'ĭ-sād la'er) the upper layer of the mesophyll of a leaf that carries on photosynthesis. *142*

phloem (flo'em) the vascular tissue in plants that transports organic nutrients. *136*

photoperiodism (fo''to-pe're-od-izm) a response to light and dark; particularly in reference to flowering in plants. *150*

phytochrome (fi'to-krōm) a plant pigment that is involved in photoperiodism in plants. *151*

plumule (ploo'mūl) the shoot tip and first two leaves of a plant. *156*

pollination (pol''ĭ-na'shun) the delivery by wind or animals of pollen to the stigma of a pistil in flowering plants and leading to fertilization. *155*

radicle (rad'ĭk'l) the embryonic root of a plant. *156*

spongy layer (spun'je la'er) the lower layer of the mesophyll of a leaf that carries out gas exchange. *142*

stoma (sto'mah) opening in the leaves of plants through which gas exchange takes place (pl. stomata). *144*

vascular bundle (vas'ku-lar bun'd'l) tissues that include xylem and phloem enclosed by a sheath and typically found in herbaceous plant stems. *138*

vascular cambium (vas'ku-lar kam'be-um) a cylindrical sheath of meristematic tissue that produces secondary xylem and phloem. *140*

vascular cylinder (vas'ku-lar sil'in-der) a central region of roots that contains the vascular and other tissues. *136*

xylem (zi'lem) the vascular tissue in plants that transports water and minerals. *136*

Further Readings for Part Two

Alberts, Bruce et al. 1983. *Molecular biology of the cell.* New York: Garland Publishing, Inc.

Childress, J. J. et al. 1987. Symbiosis in the deep sea. *Scientific American* 256(5):114.

Cronquist, A. 1982. *Basic botany,* 2d ed. New York: Harper & Row Publishers, Inc.

Dickerson, Richard E. 1980. Cytochrome *c* and the evolution of energy metabolism. *Scientific American* 242(3):136.

Evans, M. L. 1986. How roots respond to gravity. *Scientific American* 255(6):112.

Galston, A. W., P. J. Davies, and R. L. Slater. 1980. *The life of the green plant,* 3d ed. Englewood Cliffs, N.J.: Prentice-Hall.

Miller, Kenneth R. 1979. The photosynthetic membrane. *Scientific American* 241(4):102.

Nassau, Kurt. 1980. The causes of color. *Scientific American* 243(4):124.

Raven, Peter, H. et al. 1986. *Biology of plants,* 4th ed. New York: Worth Publishers, Inc.

Rayle, D., and H. L. Wedberg. 1980. *Botany: A human concern.* Boston, Mass.: Houghton Mifflin.

Shepard, J. F. 1982. The regeneration of potato plants from leaf-cell protoplasts. *Scientific American* 246(5):154.

Youvan, D. C., and B. L. Marrs. 1987. Molecular mechanisms of photosynthesis. *Scientific American* 256(6):42.

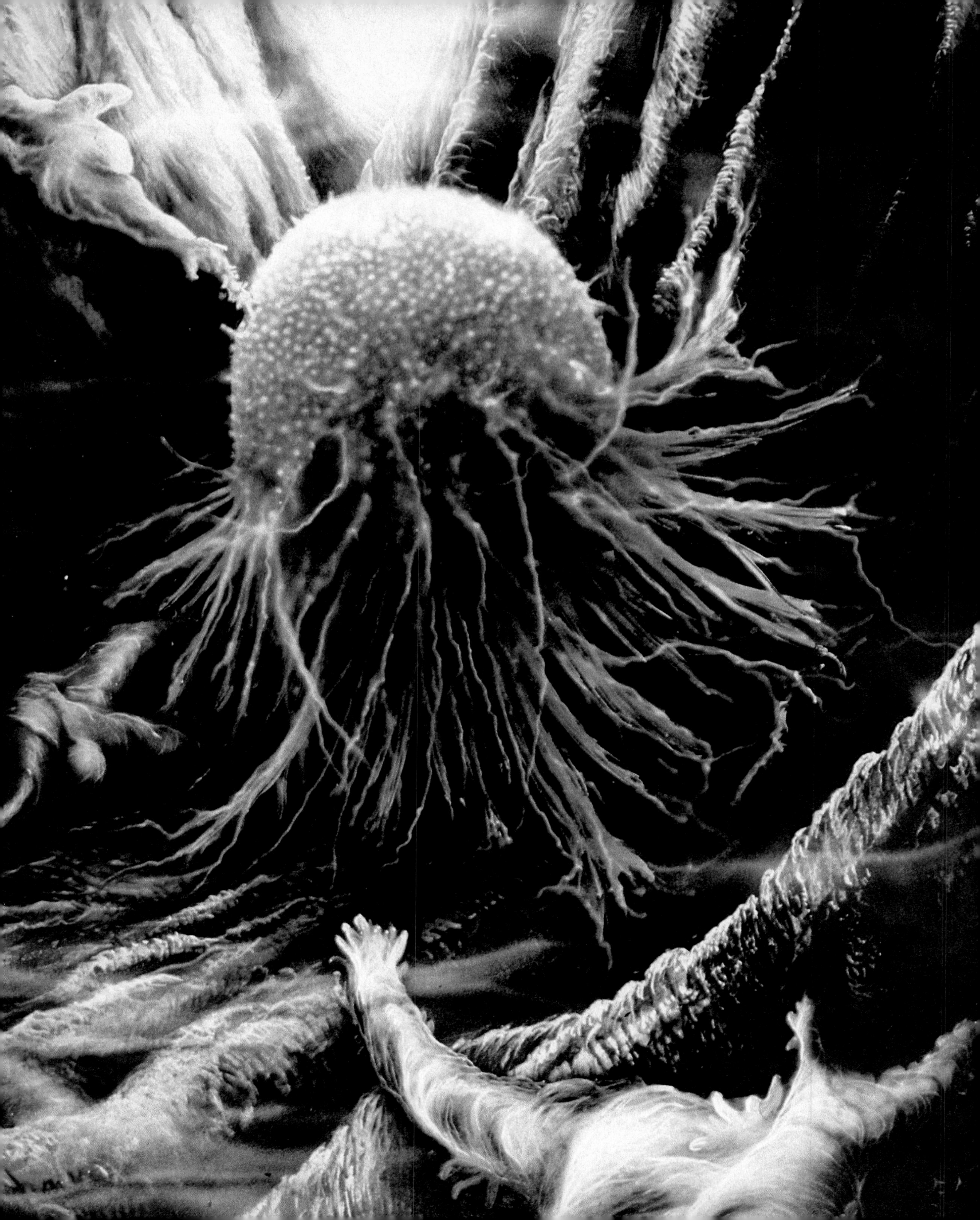

Part Three

Human Anatomy and Physiology

The study of human anatomy and physiology serves as a guide to an understanding of the vertebrate body. A limited number of tissues make up organs, which form systems to carry out the functions assumed by the cell in less complex animals. All body systems help maintain a relatively constant internal environment so that the proper physical conditions exist for each cell.

The digestive system provides nutrients, and the excretory system rids the body of metabolic wastes. The respiratory system supplies oxygen but also eliminates carbon dioxide. The circulatory system carries nutrients and oxygen to and wastes from the cells so that tissue fluid composition remains constant. The immune system helps protect the body from disease. The nervous and hormonal systems control body functions. The nervous system directs body movements, allowing the organism to manipulate the external environment, an important life-sustaining function.

A disruptive cancer cell coursing through a lymphatic vessel is a threat to the body's integrity. The lymphatic system aids in the body's defense but may, on occasion, help spread cancer in the manner depicted here by an artist.

Human Organization

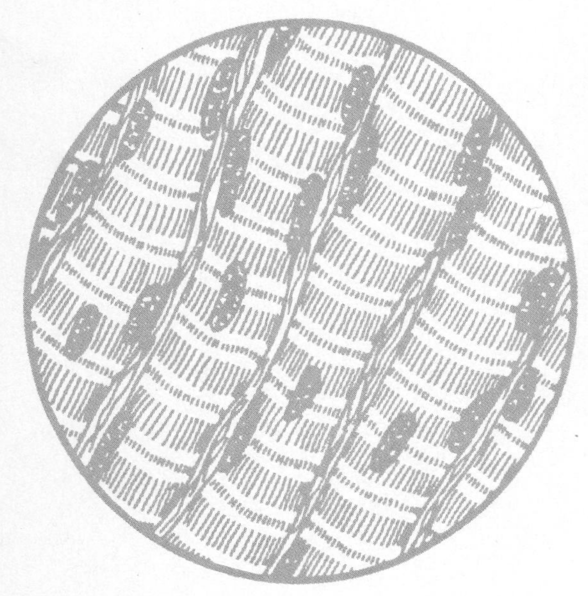

Chapter Concepts

1 Animal tissues can be categorized into four major types: epithelial, connective, muscle, and nervous tissues.

2 Organs usually contain several types of tissues. For example, although skin is primarily composed of epithelial and connective tissue, it also contains muscle and nerve fibers.

3 Organs are grouped into organ systems each of which has specialized functions.

4 Mammals exhibit a marked ability to maintain a relatively constant internal environment. All organ systems contribute to homeostasis.

Chapter Outline

In the chapters to follow, human physiology is studied as representative of vertebrate physiology. Our study will be more meaningful if we first review human organization. Figure I.2 shows that the human body, like that of other organisms, has levels of organization. Cells of the same type are joined together to form a tissue. Different tissues are found in an organ, and various types of organs are arranged into an organ system. Finally, the organ systems comprise the organism.

Tissues

The tissues of the human body can be categorized into four major types: epithelial tissue that covers body surfaces and lines body cavities; connective tissue that binds and supports body parts; muscular tissue that causes parts to move; and nervous tissue that responds to stimuli and transmits impulses from one body part to another (fig. 8.1).

Epithelial Tissues

Epithelial tissue forms a continuous layer, or sheet, over the entire body surface and most of the body's inner cavities. On the external surface, it forms a covering that, like the epidermis in plants, protects the animal from injury and drying out. On internal surfaces, this tissue may be specialized for other functions in addition to protection; for example, it secretes mucus along the digestive tract; it sweeps up impurities from the lungs by means of hairlike extensions called cilia; and it efficiently absorbs molecules from kidney tubules because of fine cellular extensions called microvilli.

There are three types of epithelial tissue. **Squamous epithelium** (fig. 8.2a) is composed of flat cells and is found lining the lungs and blood vessels. **Cuboidal epithelium** (fig. 8.2b) contains cube-shaped cells and is found lining the kidney tubules. In **columnar epithelium** (fig. 8.2c), the cells resemble pillars or columns, and nuclei are usually located near the bottom of each cell. This epithelium is found lining the digestive tract. Each type of epithelium may have microvilli or cilia as appropriate for its particular function. For example, the oviducts are lined by ciliated cells whose beat propels the egg and embryo toward the uterus or womb (fig. 8.3).

Each of these types of epithelium may be stratified. **Stratified** means to exist as layers piled one over the other. The nose, mouth, anal canal, and vagina are all lined by stratified squamous epithelium. As we shall see, the outer layer of skin is also stratified squamous epithelium, except here the cells have been reinforced by keratin, a protein that strengthens cells. Some epithelium cells are **pseudostratified** and appear to be layered; because each cell touches the base line, however, true layers do not exist. The lining of the windpipe, or trachea, is *pseudostratified ciliated columnar epithelium* (fig. 8.4).

Epithelial cells sometimes secrete a product, in which case they are described as glandular. A gland can be a single cell, as in the case of the mucous-secreting goblet cells found within the columnar epithelium lining the digestive tract (fig. 8.2c), or a gland can contain numerous cells. Glands that secrete their products into ducts are called **exocrine glands,** and those that secrete directly into the bloodstream are called **endocrine glands.**

Epithelial tissue is classified according to the shape of the cell. There can be one or many layers of cells and the layer lining a cavity can be ciliated and/or secretory (table 8.1).

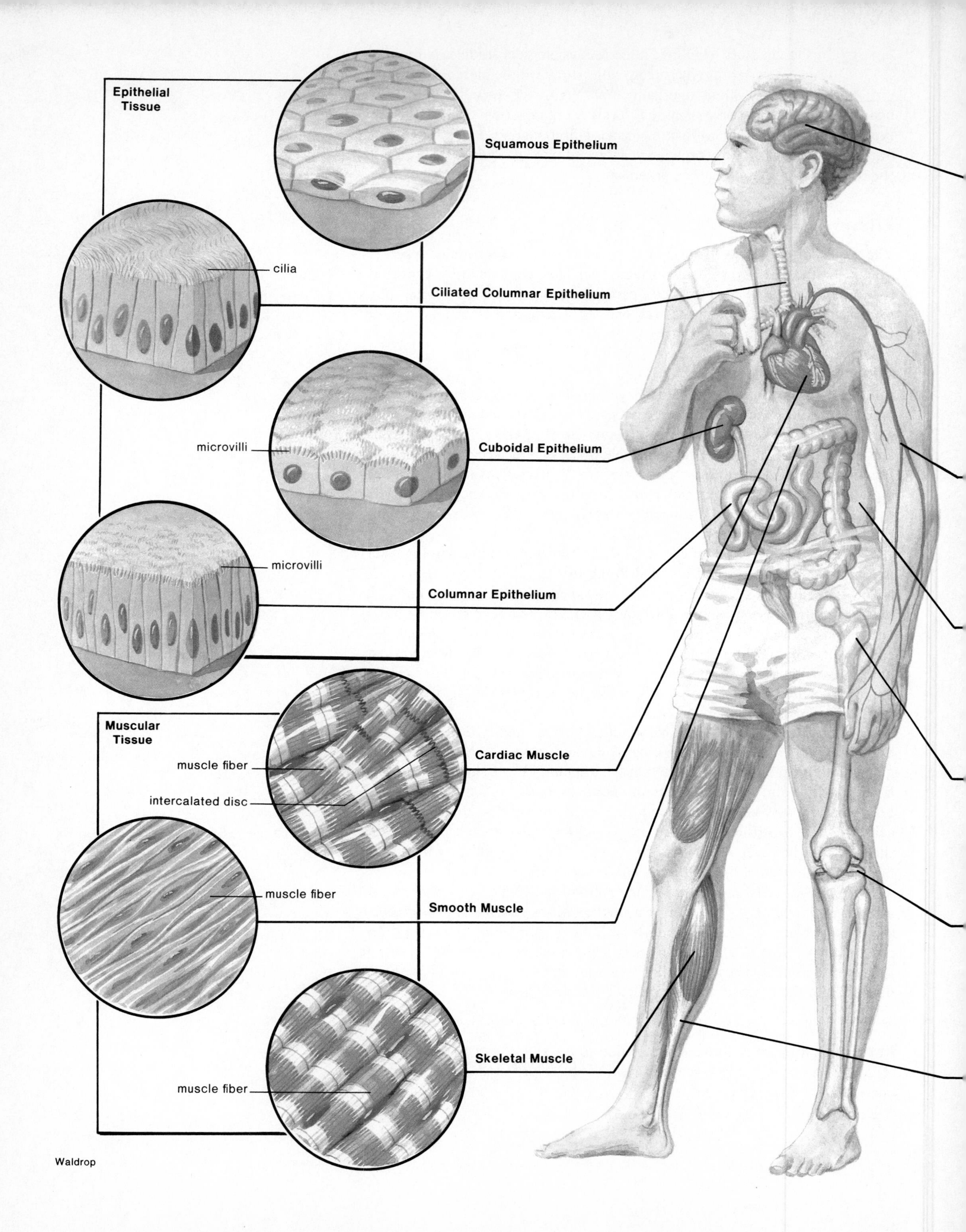
Epithelial Tissue
Squamous Epithelium
cilia
Ciliated Columnar Epithelium
microvilli
Cuboidal Epithelium
microvilli
Columnar Epithelium
Muscular Tissue
muscle fiber
intercalated disc
Cardiac Muscle
muscle fiber
Smooth Muscle
muscle fiber
Skeletal Muscle
Waldrop

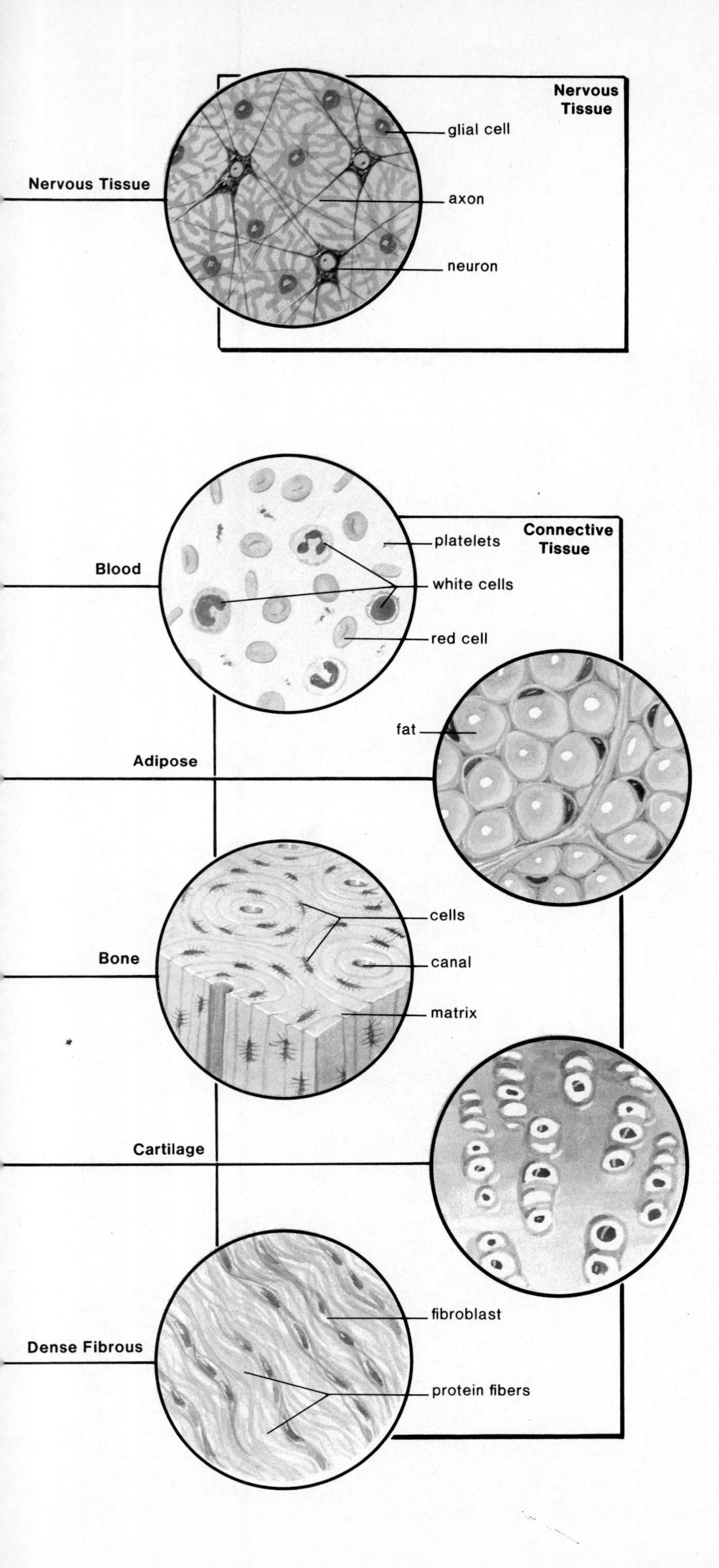

Figure 8.1
The major tissues in the human body. Reading clockwise, observe that the nervous tissue contains specialized cells called neurons. Connective tissue includes blood, adipose tissue, bone, cartilage, and dense fibrous tissue. Muscular tissue is of three types: skeletal, smooth, and cardiac. Epithelial tissue includes columnar, cuboidal, ciliated columnar, and squamous epithelium.

Figure 8.2
Simple epithelial tissue. a. *Simple squamous consists of a single layer of thin cells.* b. *Simple cuboidal is composed of cells that look like cubes.* c. *Simple columnar cells resemble columns because they are elongated. (The arrow points to a goblet cell.)*

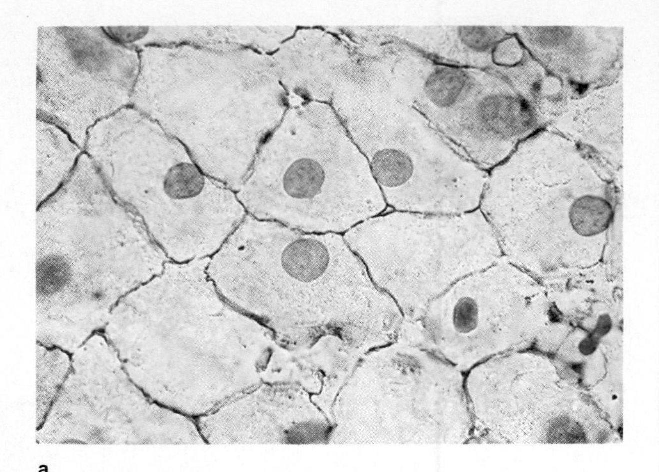

a.

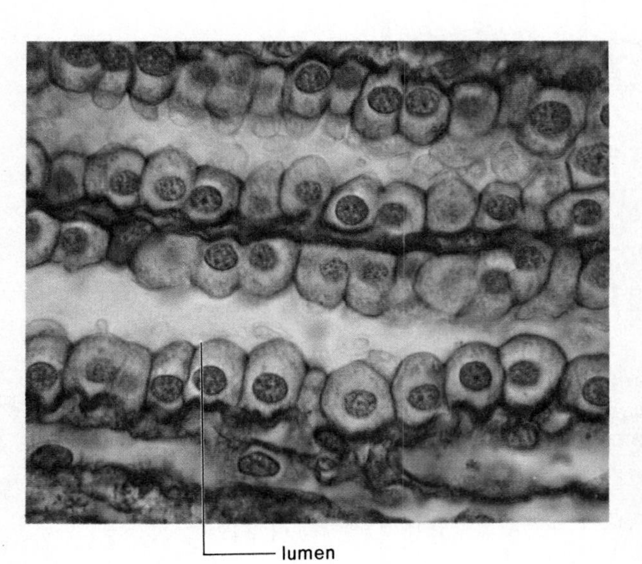

b.

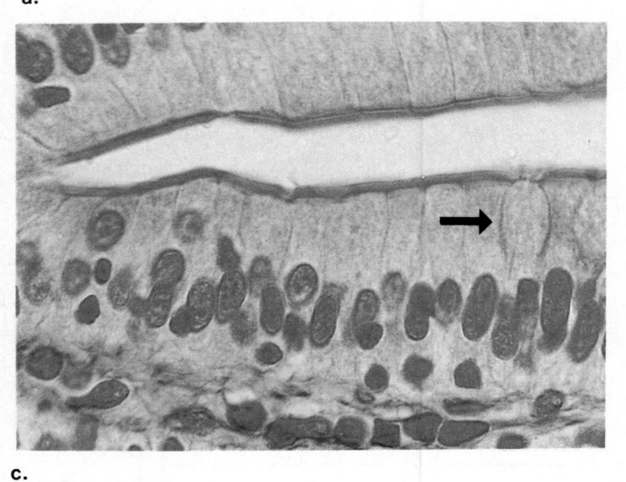

c.

Figure 8.3
Scanning electron micrograph of ciliated columnar epithelium that lines the oviducts. The beating of the cilia propels the egg toward the uterus. It has also been suggested that the return motion of the cilia sets up currents that help move the sperm toward the egg. (Ci-cilia; Mv-microvilli)

From Kessel, R. G. and R. H. Kardon. Tissues and Organs: A Text-Atlas of Scanning Electron Microscopy. © W. H. Freeman and Company, 1979.

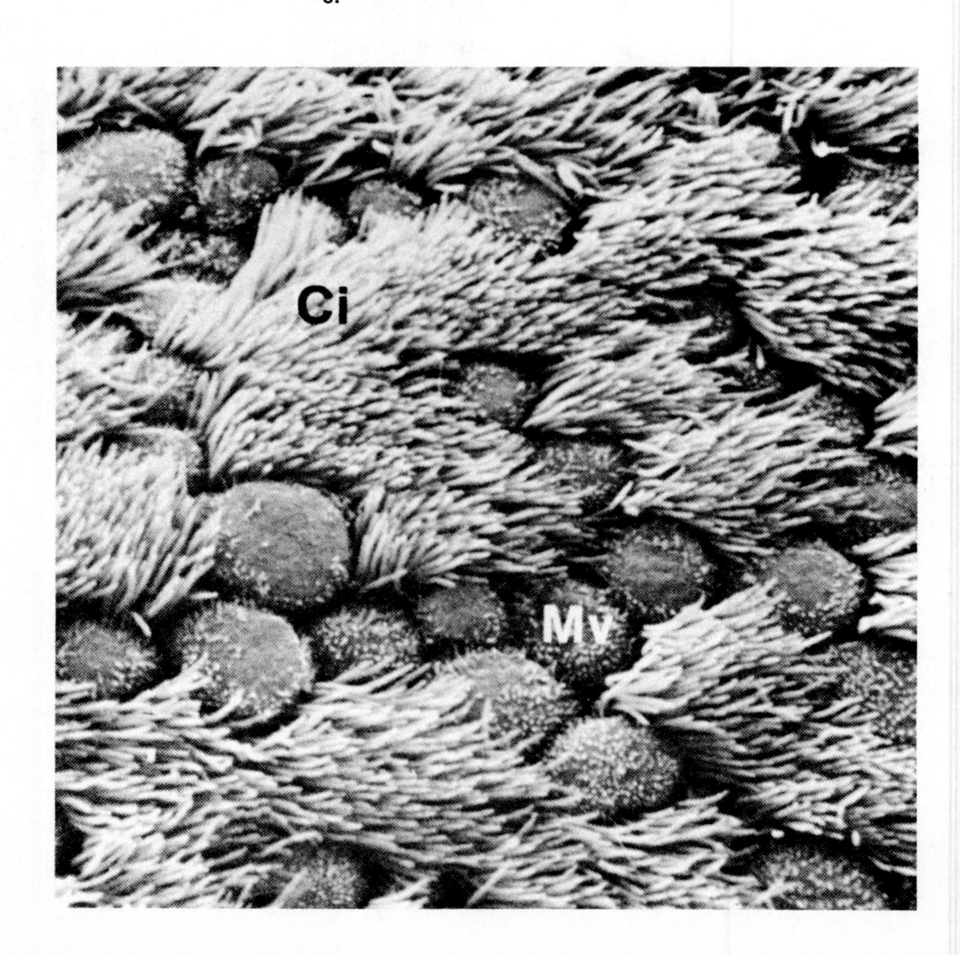

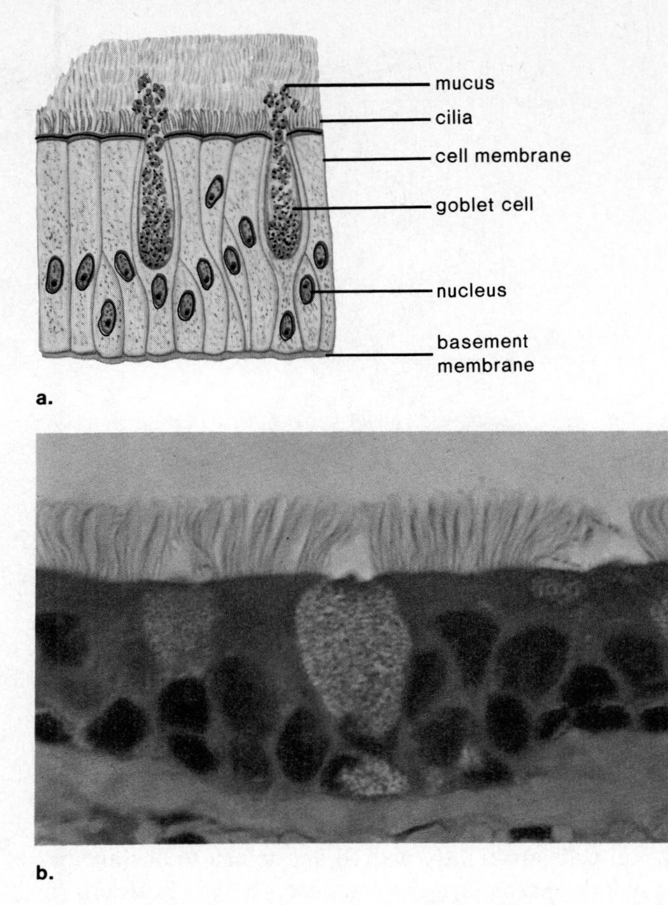

Figure 8.4
a. *Pseudostratified ciliated columnar epithelium from the lining of the windpipe. When you cough, material, trapped in the mucus secreted by goblet cells, is moved upward to the throat where it can be swallowed.* b. *Photomicrograph of pseudostratified ciliated columnar epithelium.*

***Table 8.1** Epithelial Tissues*

Type	Function	Location
Simple squamous	Filtration, diffusion, osmosis	Oral cavity, lining of blood vessels
Simple cuboidal	Secretion, absorption	Surface of ovaries; linings of kidney tubules
Simple columnar	Protection, secretion, absorption	Linings of uterus; tubes of the digestive tract
Pseudostratified columnar	Protection, secretion, movement of mucus and sex cells	Linings of respiratory passages; various tubes of the reproductive systems
Stratified squamous	Protection	Outer layers of skin, vagina, and anal canal

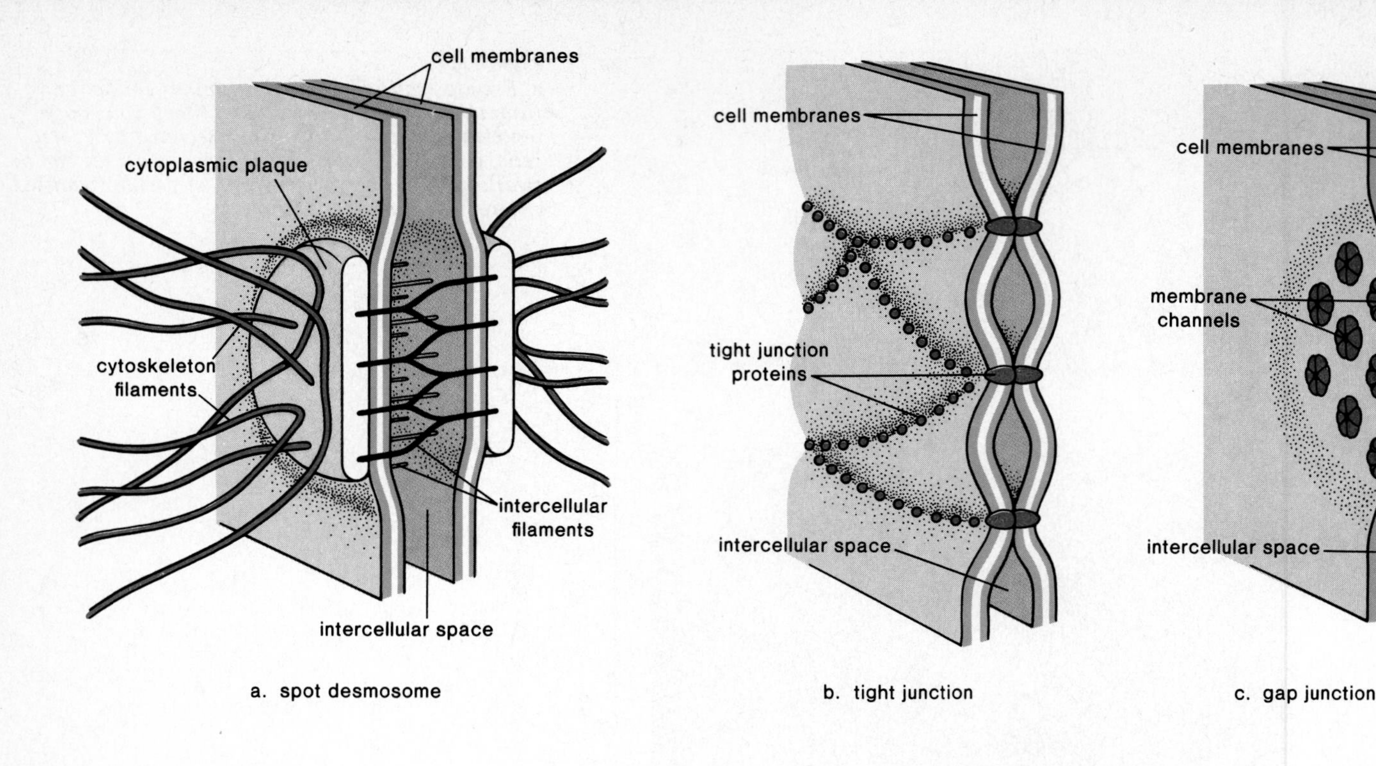

Figure 8.5
Epithelial cells are held tightly together by a. *desmosomes and* b. *tight junctions.* c. *Gap junctions allow materials to pass from cell to cell.*

Junctions between Cells

Epithelial cells are tightly packed and joined to one another in one of three ways: spot desmosomes, tight junctions, and gap junctions (fig. 8.5). In a **spot desmosome** internal cytoplasmic plaques, firmly attached to the cytoskeleton within each cell, are joined by intercellular filaments. The cells are more closely joined in a **tight junction** because adjacent cell membrane proteins actually attach to each other producing a zipperlike fastening. A **gap junction** is formed when two identical cell membrane channels join. This does lend strength but it also allows substances to pass between the two cells.

As mentioned earlier, some epithelial tissues have many layers. In any case, the bottom layer is often joined to underlying connective tissue by a so-called **basement membrane.** We now know that the basement membrane is glycoprotein reinforced by fibers supplied by the connective tissue.

Connective Tissues

Connective tissues (table 8.2) bind structures together, provide support and protection, fill spaces, and store fat. As a rule, connective tissue cells are widely separated by a noncellular **matrix.** The matrix may have fibers of two types. White fibers contain collagen, a substance that gives them flexibility and strength. Yellow fibers contain elastin, a substance that is not as strong as collagen but is more elastic.

Loose Connective Tissue

Loose connective tissue joins tissue layers and holds organs in place (fig. 8.6). The cells of this tissue, which are mainly **fibroblasts,** are located some distance apart from one another and are separated by a jellylike intercellular material that contains many white and yellow fibers. The white fibers occur in bundles

Table 8.2 Connective Tissues		
Type	**Function**	**Location**
Loose connective	Binds organs together	Beneath the skin, beneath most epithelial layers
Adipose tissue	Insulation; storage of fat	Beneath the skin, around the kidneys
Fibrous connective tissue	Binds organs together	Tendons; ligaments
Hyaline cartilage	Support; protection	Ends of bones, nose, walls of respiratory passages
Elastic cartilage	Support; protection	External ear and part of the larynx
Fibrocartilage	Support; protection	Between bony parts of backbone and knee
Bone	Support; protection	Bones of skeleton

Adapted from Hole, John W., Jr., *Human Anatomy and Physiology 4th ed.* © 1978, 1981, 1984, 1987 Wm. C. Brown Publishers, Dubuque, Iowa.

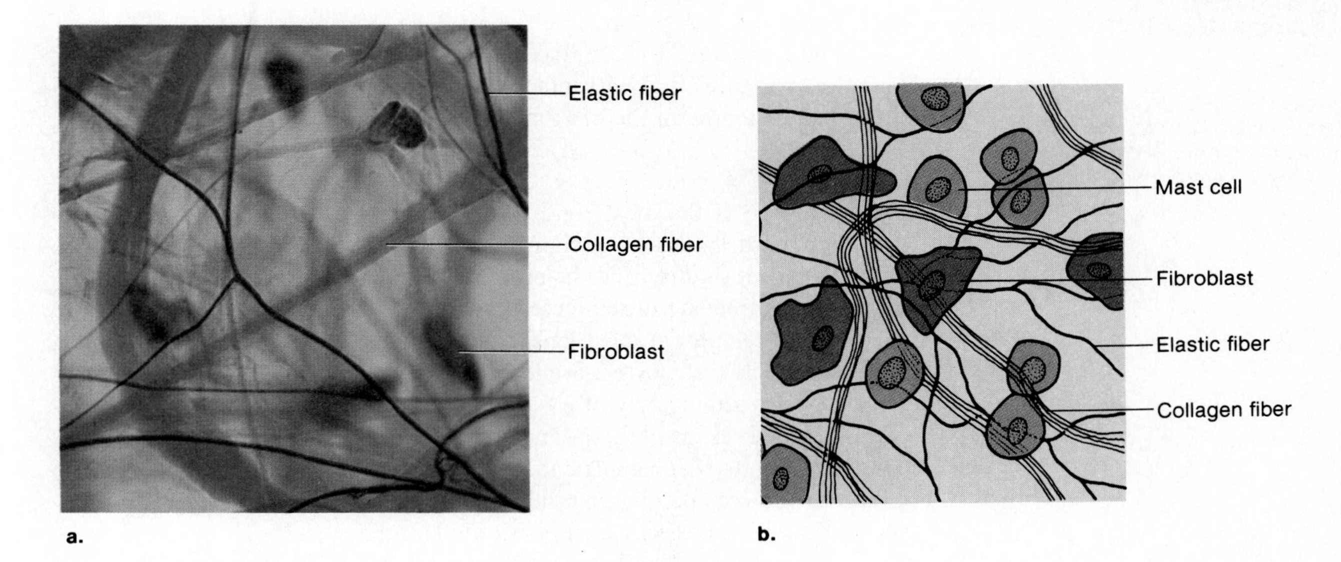

Figure 8.6
Loose connective tissue has plenty of space between components. This type tissue is found surrounding and between the organs. a. *Photomicrograph.* b. *Drawing.*

and are strong and flexible. The yellow fibers form networks that are highly elastic—when stretched they return to their original length. As discussed previously, loose connective tissue commonly lies beneath epithelium. In certain instances, epithelium and its underlying connective tissue form body membranes (p. 178). In addition, adipose (fig. 8.7) tissue is a type of loose connective tissue in which the fibroblasts enlarge and store fat and the intercellular matrix is reduced.

Fibrous Connective Tissue

Fibrous connective tissue contains large numbers of collagenous fibers that are closely packed together. This type of tissue has more specific functions than does loose connective. For example, fibrous connective tissue is found in **tendons,** which connect muscles to bones, and **ligaments,** which connect bones to other bones at joints. Tendons and ligaments take a long time to heal following an injury because their blood supply is relatively poor. (See chapter 16 for more information on this subject.)

Loose and fibrous connective tissues, which bind body parts together, differ according to the type and abundance of fibers in the matrix.

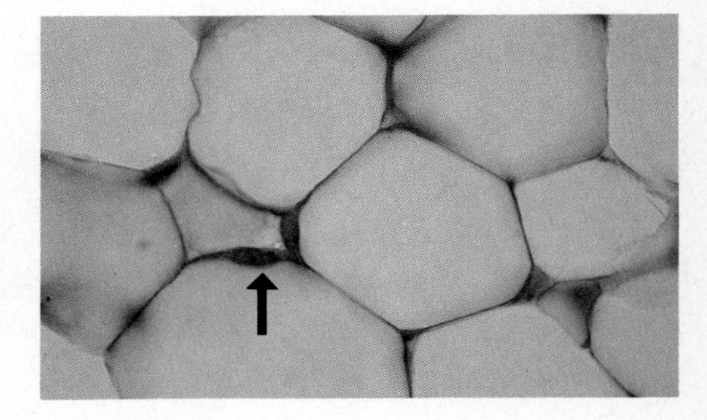

Figure 8.7
Adipose tissue cells look like white "ghosts" because they are filled with fat. The nucleus of one cell is indicated by the arrow.

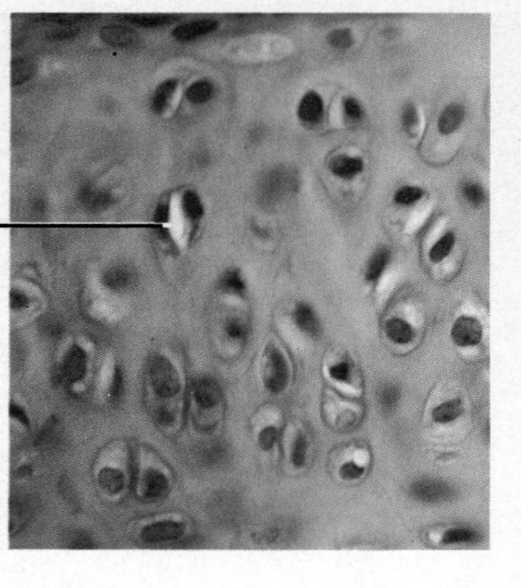

Hyaline cartilage cells, located in lacunae, are separated by a flexible matrix rich in protein and fibers. This type cartilage forms the embryonic skeleton, which is later replaced by bone.

Cartilage

In **cartilage** the cells lie in small chambers called **lacunae,** separated by a matrix that is solid yet flexible. Unfortunately because this tissue lacks a direct blood supply, it heals very slowly. There are three types of cartilage according to the type of fiber in the matrix.

Hyaline cartilage (fig. 8.8) the most common type, contains only very fine collagenous fibers. The matrix has a clear, milk-glass appearance. This type of cartilage is found in the nose, at the ends of the long bones and ribs, and in the supporting rings of the windpipe. The fetal skeleton is also made of this type of cartilage. Later, the cartilaginous skeleton is replaced by bone.

Elastic cartilage has more elastic fibers than hyaline cartilage. For this reason it is more flexible and is found, for example, in the framework of the outer ear.

Fibrocartilage has a matrix containing strong collagenous fibers. It is found in structures that withstand tension and pressure such as the pads between the vertebrae in the backbone and the wedges found in the knee joint.

Bone

Bone (fig. 8.9) is the most rigid of the connective tissues. It consists of an extremely hard matrix of calcium salts deposited around protein fibers. The minerals give it rigidity, and the protein fibers provide elasticity and strength, much as steel rods do in reinforced concrete.

The outer portion of a long bone contains compact bone. In **compact bone,** bone cells (osteocytes) are located in lacunae that are arranged in concentric circles around tiny tubes called Haversian canals. There are nerve fibers and blood vessels in these canals. The latter bring the nutrients that allow bone to renew itself. The nutrients can reach all of the cells, because there are minute canals (canaliculi) containing thin processes of the osteocytes that connect them with one another and with the Haversian canals.

The ends of a long bone contain spongy bone (fig. 16.6), which has an entirely different structure. Spongy bone contains numerous bony bars and plates separated by irregular spaces. Although lighter than compact bone, **spongy bone** is still designed for strength. Just as braces are used for support in buildings, the solid portions of spongy bone follow lines of stress.

Cartilage and bone are support tissues. Cartilage is more flexible than bone because the matrix is rich in protein and not calcium salts like that of bone.

Blood

Blood (fig. 8.10) is a connective tissue in which the cells are separated by a liquid called plasma, whose contents are listed in table 8.3. Blood cells are of two types: **erythrocytes (red)** which carry oxygen, and **leukocytes (white),** which aid in fighting infection. Also present are *platelets,* which are important to the initiation of blood clotting. Platelets are not complete cells; rather, they are fragments of giant cells found in the bone marrow.

Blood is a connective tissue in which the matrix is plasma.

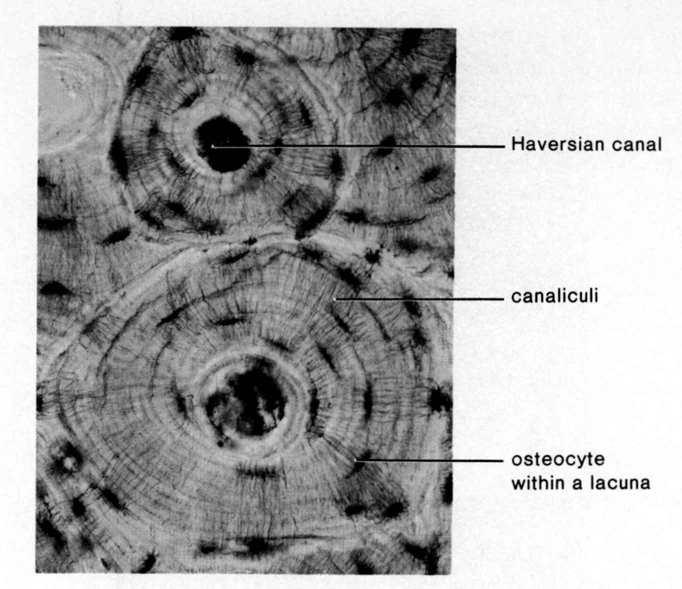

Figure 8.9
Compact bone is highly organized. The cells are arranged in circles about a central (Haversian) canal that contains a nutrient-bearing blood vessel.

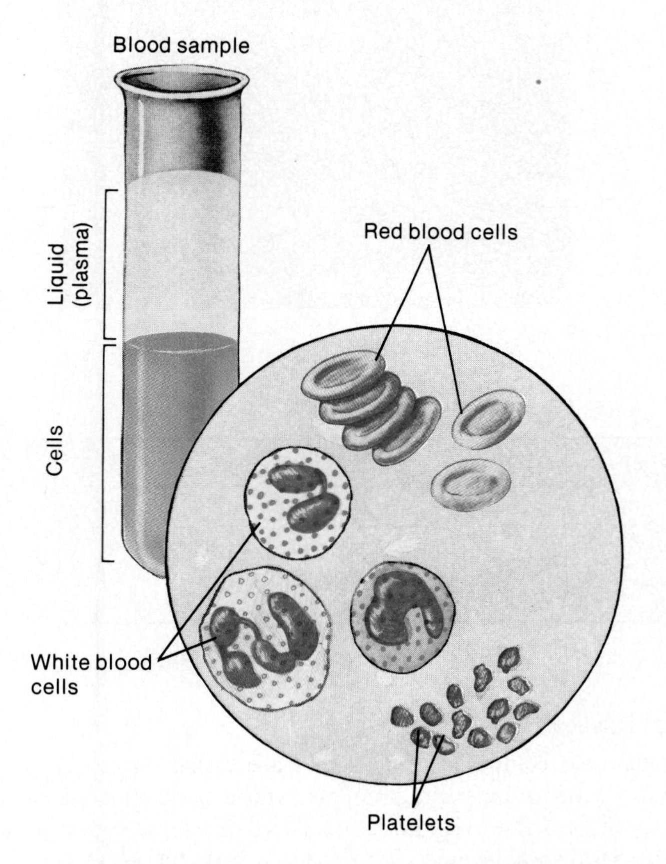

Figure 8.10
Blood is classified as connective tissue. Plasma, the liquid portion of blood, contains the formed elements (red cells, white cells, and platelets).

Table 8.3 Blood Plasma	
Water	92% of plasma
Inorganic ions (salts)	Na^+, Ca^{++}, K^+, Mg^{++}; Cl^-, HCO_3^-, HPO_4^-, SO_4^{--}
Gases	O_2 and CO_2
Plasma proteins	Albumin, globulin, fibrinogen
Organic nutrients	Glucose, fats, phospholipids, amino acids, etc.
Nitrogenous waste products	Urea, ammonia, uric acid
Regulatory substances	Hormones, enzymes

Figure 8.11
How do you distinguish a plant from an animal? One way is to detect motion—only animals have contractile fibers that permit movement. a. *Skeletal muscle is found within the muscles attached to the skeleton.* b. *Smooth muscle cells are found in the walls of internal organs.* c. *Cardiac muscle permits the pumping of the heart.*

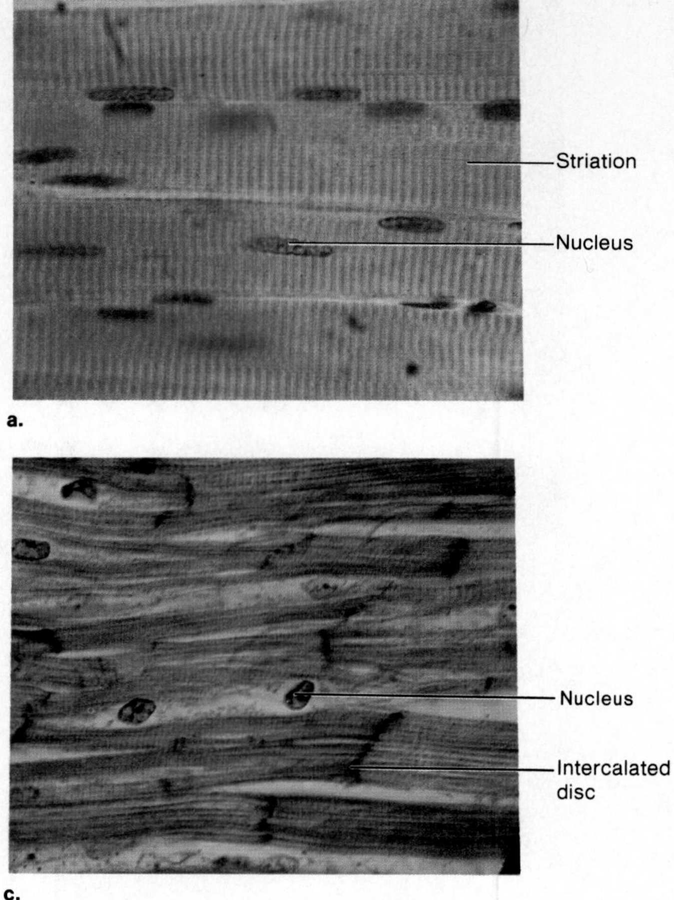

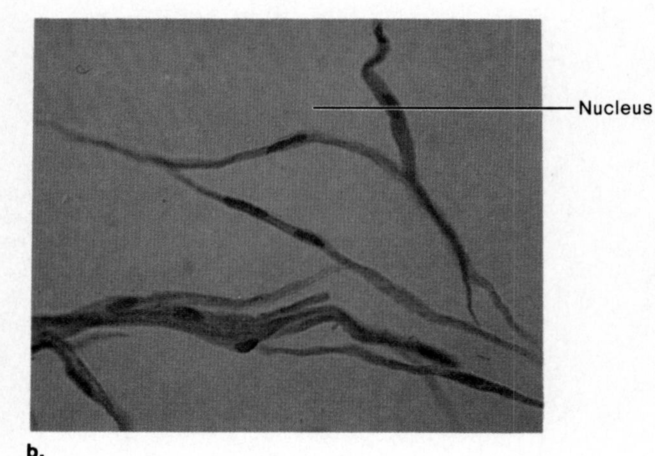

Table 8.4 Muscular Tissues			
Type	**Fiber Appearance**	**Locations**	**Control**
Skeletal	Striated	Attached to skeleton	Voluntary
Smooth	Spindle shaped	Internal organs	Involuntary
Cardiac	Striated and branched	Heart	Involuntary

Muscular Tissues

Muscular tissue is composed of cells that are called *muscle fibers.* Muscle fibers contain actin and myosin filaments, whose interaction accounts for the movements we associate with animals. There are three types of vertebrate muscles: *skeletal, smooth,* and *cardiac* (table 8.4) (fig. 8.11).

Skeletal muscle is attached to the bones of the skeleton and functions to move body parts. It is under our *voluntary control* and has the fastest contraction of all the muscle types. The cylindrical cells of this muscle have characteristic light and dark bands perpendicular to the length of the cell or fiber. These bands give the muscle a **striated** appearance.

Smooth muscle is so named because it lacks striations. The spindle-shaped cells that make up smooth muscle are not under voluntary control and are said to be *involuntary.* Smooth muscle, found in the viscera (intestine, stomach, and so on) and blood vessels, contracts more slowly than skeletal muscle but can remain contracted for a longer time. The cells tend to form layers in which the thick middle portion of one cell is opposite the thin ends of adjacent cells. Consequently, the nuclei form an irregular pattern in the tissue.

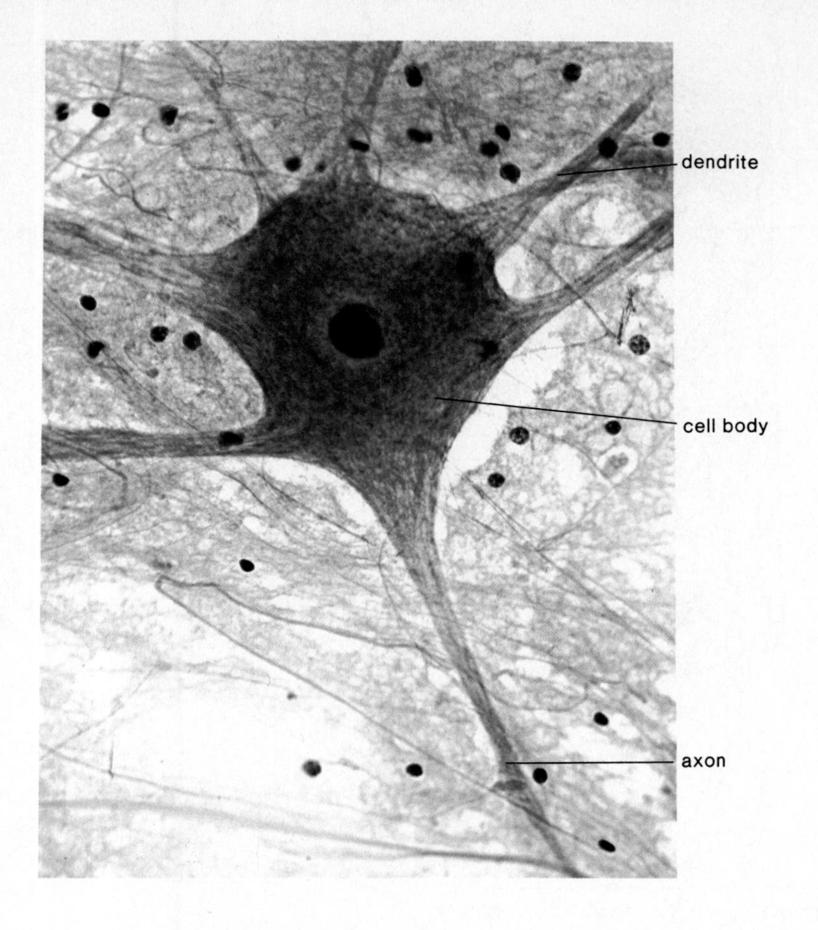

Figure 8.12
Conduction of the nerve impulse is dependent on neurons, each of which has the three parts indicated. A dendrite takes nerve impulses to the cell body, and an axon takes them away from the cell body.

Cardiac muscle seems to combine features of both smooth and skeletal muscle. It has *striations* like those of skeletal muscle, but the contraction of the heart is involuntary for the most part. Heart muscle cells also differ from skeletal muscle cells in that they are branched and seemingly fused, one with the other, so that the heart appears to be composed of one large, interconnecting mass of muscle cells. Actually, however, cardiac muscle cells are separate and individual.

All muscle tissue contains actin and myosin microfilaments; these form a striated pattern in skeletal and cardiac muscle but not in smooth muscle.

Nervous Tissue

The brain and nerve cord (also called the spinal cord) contain conducting cells termed neurons. A **neuron** (fig. 8.12) is a specialized cell that has three parts: (1) *dendrites* conduct impulses (send a message) to the cell body; (2) the *cell body* contains most of the cytoplasm and the nucleus of the neuron; (3) the *axon* conducts impulses away from the cell body.

When axons and dendrites are long, they are called *nerve fibers*. Outside the brain and spinal cord, nerve fibers are bound together by connective tissue to form **nerves.** Nerves conduct impulses from sense organs to the spinal cord and brain, where the phenomenon called sensation occurs. They also conduct nerve impulses away from the spinal cord and brain to the muscles, causing them to contract.

In addition to neurons nervous tissue contains **glial cells.** These cells maintain the tissue by giving support and protection to neurons. They also provide nutrients to neurons and help keep the tissue free of debris.

Figure 8.13
Skin anatomy. Skin is composed of three layers: the epidermis, the dermis, and the subcutaneous layer. Most cells in the epidermal layer are no longer living. Skin cancer brought on by UV radiation from the sun starts in the lower epidermal cells because they are mitotically active. The dermis contains the various structures depicted.

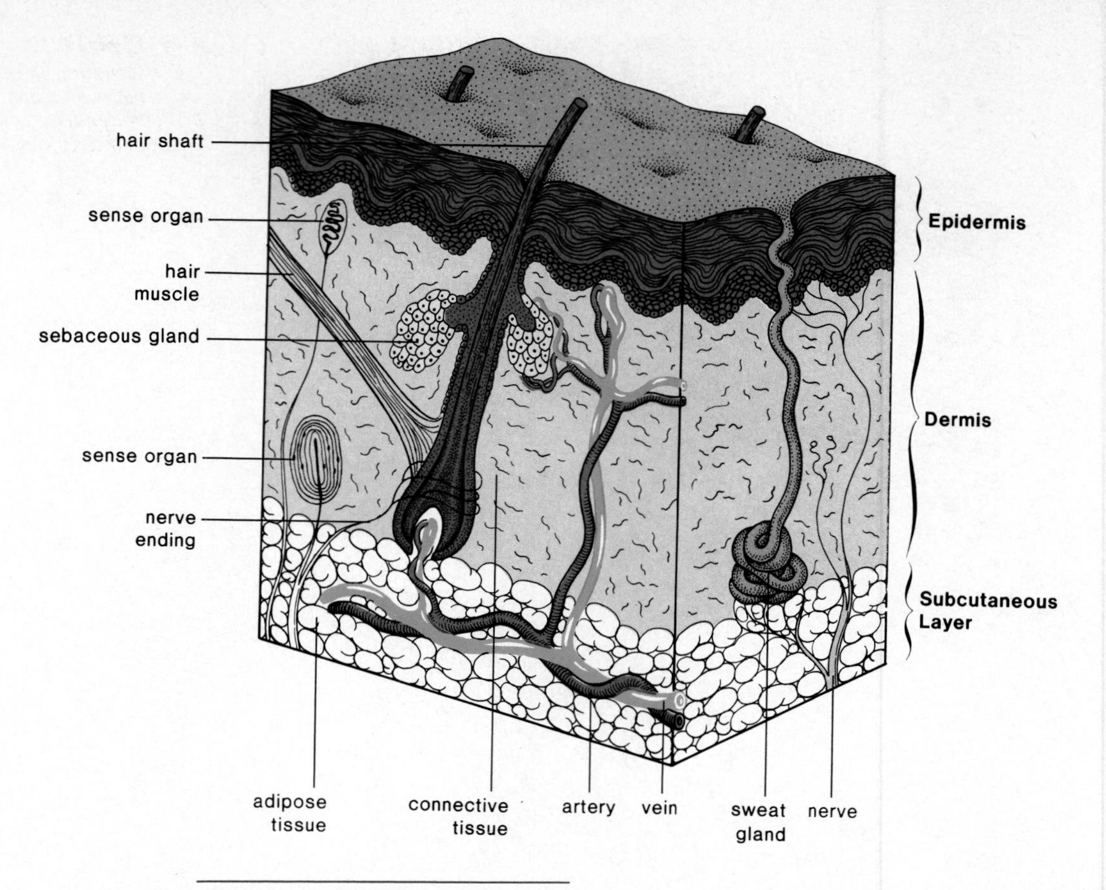

Organs and Organ Systems

Some organs in the body are composed largely of one type of cell. For example, muscles are made up of muscle cells; and glands are most often made up of epithelial cells. Many organs, however, are a composite of different types of tissues. We will consider skin as an example.

Skin, an Organ

Skin (fig. 8.13) covers the body, protecting it from loss of water and deterring invasion by microorganisms. It also helps regulate body temperature and contains sense organs for touch, pressure, temperature, and pain. Skin has an outer *epidermal* and the inner *dermal layer.* Beneath the dermis there is a *subcutaneous layer* that binds the skin to underlying organs. A burn can affect one or all of these layers. A first-degree burn, which affects only the epidermal layer, is characterized by redness, pain, and swelling. As in a sunburn, the skin usually peels in a few days. A second-degree burn, affecting both the epidermis and dermis, usually causes blistering. A third-degree burn is the most serious because it leaves underlying parts with no protection at all. When a person is burned over a large portion of the body, it is sometimes difficult to find enough skin to make autographing (graph from remaining skin) possible. Under these circumstances, physicians can now make use of artificial skin, consisting of two layers. The inner layer is a lattice made from shark cartilage and collagen fibers from cowhide. The outer layer is a rubberlike silicone plastic. After the artificial skin is sewn in place, the lattice is slowly digested away and replaced by the patient's own cells. Then the silicone layer can be safely removed.

Epidermis

The **epidermis** is made up of stratified squamous epithelial cells that are continually produced by a germinal layer lying next to the dermis. Here, constant cell division of *basal cells* produces many new cells that gradually rise to the surface when outer layers of cells are worn off. Specialized cells in this layer called melanocytes produce **melanin,** the pigment responsible for skin color. As the reading on page 176 describes, a suntan is caused by the appearance of melanin granules in the outer surface of the skin. Unfortunately, ultraviolet (UV) radiation from the sun can cause dividing skin cells to become cancerous. In recent years, there has been such a high increase of skin cancers due to sunbathing that physicians are strongly recommending that everyone stay out of the sun—or at least, use sunscreen lotions to protect the skin. The most dangerous type of skin cancer is malignant melanoma, recognized by dark brown or black patches that look like moles. The more common types of skin cancer are basal cell cancer and squamous cell cancer, often appearing as rough scaly patches for reasons described following.

Cells gradually become flattened and hardened as they are pushed toward the surface of the skin. Hardening is caused by keratinization, cellular production of a fibrous, waterproof protein called **keratin.** Over much of the body, keratinization is normally minimal and patches of overly keratinized cells, called spots of keratosis, can be early indications of basal cell or squamous cell cancer. In contrast, both the palm of the hand and the sole of the foot normally have a particularly thick outer layer of dead keratinized cells arranged in spiral and concentric patterns. These patterns, unique to each person, are known as fingerprints and footprints. Hair is present wherever the skin is less keratinized, although hair and nails are composed of tightly packed keratinized cells.

Dermis

The **dermis** is largely loose connective tissue with many elastic fibers. Sunlight contributes to the aging process because it causes this tissue to lose its elasticity, producing "wrinkles" in the skin.

The epidermis regularly dips down into the dermis, serving as a location for sweat glands and hair follicles, with their associated sebaceous glands. The sweat glands help regulate body temperature. They become active and release fluid onto the skin when the body temperature rises. The evaporation of this fluid cools the body. The sebaceous glands lubricate the adjoining hairs and skin. At the time of puberty, they are the site of acne.

Blood vessels and nerve endings are also present in the dermis. Some of the latter control a small muscle attached to each hair follicle. When this muscle contracts, the hair stands on end, causing what is commonly known as "goose bumps." Encapsulated (membrane-surrounded) nerve endings are found in the microscopic *sense organs* for touch, pressure, and temperature. Stimulation of free nerve endings produces pain. Regulation of the size of the arteries in skin helps maintain a constant internal body temperature. When the arteries increase in size, more blood is brought to the surface of the skin where it is cooled off. At the same time, sweat glands become active and the person perspires.

Subcutaneous Layer

The **subcutaneous** layer is a layer of loose connective tissue containing adipose cells. A well-developed subcutaneous layer gives a rounded appearance to the body. Excessive development accompanies obesity.

Skin, composed of the epidermis, dermis, and subcutaneous layers, is an organ that performs various functions including protection, sense reception, and temperature regulation.

Bring Back the Parasol

There is a higher incidence of skin cancer among sunbathers.

With summer just around the corner, the pale of face throughout the Northern Hemisphere will soon be hitting beaches in pursuit of a deep, dark and sexy tan. The Victorian ideal of delicate, camellia-white skin has long since been supplanted by the bronzed-god look. But the trend has taken a mortal toll. Sun-related skin cancer is rapidly on the rise in the U.S. and Europe, and afflicting younger and younger people. The incidence of the most lethal form, malignant melanoma, though less directly linked to sunshine, has jumped tenfold in the past 20 years. . . . [Dermatologists suggestion]: bring back the parasol.

Doctors have long known that ultraviolet (UV) radiation from the sun produces profound changes in human skin. "Even one day's exposure can cause damage," says Dermatologist Fred Urbach of Temple University in Philadelphia. The most insidious rays are the short wave length UVB, which prevail during the peak sun hours (between 11 a.m. and 3 p.m.). But new research has shown that even longer UVA waves, which are present all day, can promote skin cancer. The damage caused by these invisible rays ranges from ordinary sunburn, to the wrinkles and liver spots caused by years of sunbathing, to the precancerous dark patches known as actinic keratosis and, finally, cancer. Each of these is part of the same process, says Urbach. "First you look old, then if you've had a lot more sun, you get keratosis, and after that skin cancer. If we all lived long enough, we would all get skin cancer."

The process begins when solar UV damages basal cells near the surface of the skin, causing them to swell. The pain and redness, which appear a few hours after exposure, are caused by the dilation of blood vessels in the damaged area. The ensuing tan is the body's desperate effort to save its skin from further injury. Tiny granules of melanin, a brownish pigment made in specialized skin cells, rise to the surface in response to UV radiation and act as sunlight deflectors. Over the years, however, the beachgoer pays for this glamorous natural shield. The buildup of melanin, combined with UV damage to the elastic fibers in underlying layers, gives the skin the texture of an old baseball mitt.

Like X rays, UV radiation can alter cell DNA, producing the mutations associated with cancer. "Both UVA and UVB are carcinogenic," says Harvard Photobiologist Madhu Pathak. UV also appears to suppress the body's immune system. This may explain why certain viral infections, such as chicken pox and fever blisters, become more severe in the sun. And since the immune system is believed to play a role in preventing tumor growth, its suppression "may also be an aggravating factor in the development of skin cancer," says Dr. Margaret Kripke of the National Cancer Institute.

About 80% of the skin cancers caused by the sun are basal-cell carcinoma. Usually occurring on the head or neck, they are the most common and curable form of cancer in the U.S. . . . Skin cancers that appear elsewhere on the body are usually squamous-cell carcinoma, also easily cured by surgery.

Far more lethal are the darkly pigmented spots of malignant melanoma, which strikes more than 15,000 Americans a year, killing 45% of them. Though melanoma tends to occur on such sun-exposed areas as the chests of men and legs of women, its relationship to the sun remains unclear. A history of severe sunburns may play a role; pregnancy and birth control pills have also been implicated.

The evils of ultraviolet are easily escaped. The Food and Drug Administration has published guidelines recommending a sunscreen of certain strength for each type of skin. For fair-skinned people who never tan and always burn (Type I skin), sunscreens labeled with the number 15 are best. The number indicates that it will take at least 15 times longer to burn when the product is used than when the skin is unprotected. People who sometimes burn and never tan should use sunscreens in the 6-to-8 range; those who occasionally burn but tan well need only 4-to-8 protection; olive-complected Type IVs, who never get scorched, can manage with a factor of 2-to-4. In the future, however, a sunscreen pill being developed in Pathak's lab may make the lotions obsolete.

Organ Systems

In this text, we are going to study the organ systems listed in table 8.5. Each of these systems has a specific location within the body. The central nervous system is located dorsally (toward the back); the brain is protected by the skull, and the spinal cord, which gives off spinal nerves, is protected by the vertebrae (fig. 8.14). The repeating units of vertebrae and spinal nerves show that humans are segmented animals, meaning that body parts reoccur at regular intervals.

Within the musculoskeletal system, the skeleton provides the surface area for attachment of striated muscles that are well-developed and powerful. The musculoskeletal system makes up most of the body weight and is specialized for locomotion.

The other internal organs are found within a body cavity called the **coelom.** In humans and other mammals, the coelom is divided by a muscular diaphragm that assists breathing. The heart, a pump for the closed circulatory system, and the lungs are located in the upper (thoracic or chest) cavity. The major portion of the digestive system, the entire excretory system, and much of the reproductive system are located in the lower (abdominal) cavity. The major organs of the excretory system are the paired kidneys, and the accessory organs of the digestive system are the liver and pancreas. Each sex has characteristic sex organs.

The preceding attributes are vertebrate characteristics and, in this text, human physiology is studied as representative of vertebrates in general. Of all the vertebrates, mammals (animals who nourish their young by means of mammary glands) and birds are best able to maintain constancy of internal body conditions. No doubt this has contributed greatly to their success.

Table 8.5 Human Organ Systems

Name	Function
Digestive	Convert food particles to nutrient molecules
Circulatory	Transport of molecules to and from cells
Immune	Defense against disease
Respiratory	Exchange of gases with environment
Excretory	Elimination of metabolic wastes
Nervous and sensory	Regulation of systems and response to environment
Muscular and skeletal	Support and movement of organism
Hormonal	Regulation of internal environment
Reproductive	Production of offspring

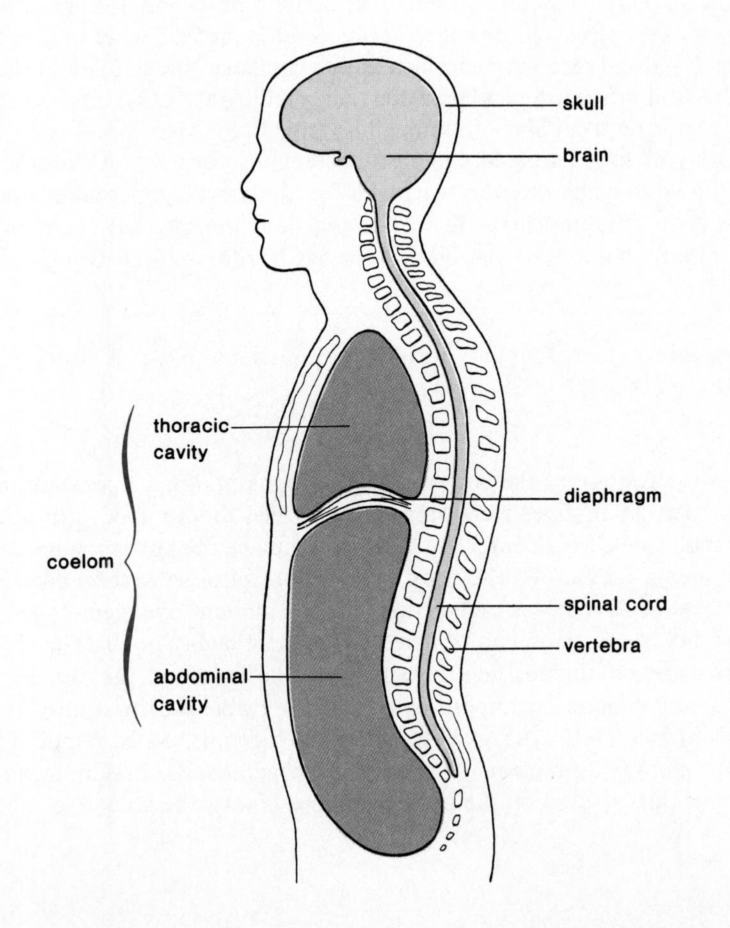

Figure 8.14
Organization of the human body. Like other mammals, humans have a dorsal nervous system and well-developed coelom that contains the internal organs. The coelom is divided by the diaphragm into the thoracic and abdominal cavity.

Body Membranes

The term membrane on the organ level generally refers to a thin lining or covering composed of a layer of epithelial tissue overlying a layer of loose connective tissue. For example, mucous membrane lines the organs of the respiratory and digestive systems. This type membrane, as its name implies, secretes mucus. Serous membrane lines enclosed cavities and covers the organs that lie within these cavities such as the heart, lungs, and kidneys. This type membrane secretes a watery lubricating fluid.

The body is divided into cavities within which the organs are found. These cavities are usually lined with membrane. Some even call the skin cutaneous membrane.

Homeostasis

Homeostasis means that the internal environment remains relatively constant regardless of the conditions in the external environment. In humans, for example:

1. Blood glucose concentration remains at about 0.1 percent.
2. The pH of the blood is always near 7.4.
3. Blood pressure in the brachial artery averages near 120/80.
4. Blood temperature averages around 37° C (98.6° F).

The ability of the body to keep the internal environment within a certain range allows humans to live in a variety of habitats, such as the arctic regions, deserts, or the tropics.

This internal environment includes a tissue fluid that bathes all the tissues of the body. The composition of tissue fluid must remain constant if cells are to remain alive and healthy. Tissue fluid is created when water (H_2O), oxygen (O_2), and nutrient molecules leave a capillary (the smallest of the blood vessels), and it is purified when water, carbon dioxide (CO_2), and other waste molecules enter a capillary from the fluid (fig. 8.15). Tissue fluid remains constant only as long as blood composition remains constant. Although we are accustomed to using the word *environment* to mean the external environment of the body, it is important to realize that it is the internal environment of tissues that is ultimately responsible for our health and well-being.

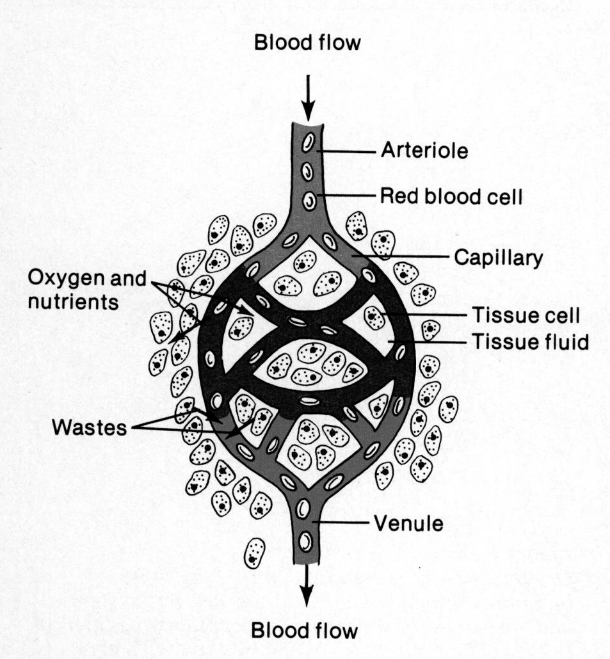

Figure 8.15
The internal environment of the body is the blood and tissue fluid. Tissue cells are surrounded by tissue fluid which is continually refreshed because nutrient molecules constantly exit from and waste molecules continually enter the bloodstream as shown.

The internal environment of the body consists of blood, and tissue fluid that bathes the cells.

Most systems of the body contribute to maintaining a constant internal environment. The digestive system takes in and digests food, providing nutrient molecules, which enter the blood and replace the nutrients that are constantly being used up by the body cells. The respiratory system adds oxygen to the blood and removes carbon dioxide. The amount of oxygen taken in and carbon dioxide given off can be increased to meet bodily needs. The chief regulators of blood composition, however, are the liver and the kidneys. They monitor the chemical composition of plasma (table 8.3) and alter it as required. Immediately after glucose enters the blood, it can be removed by the liver for storage as glycogen. Later the glycogen can be broken down to replace the glucose used by the body cells; in this way, the glucose composition of the blood remains constant. The hormone insulin, secreted by the pancreas, regulates glycogen storage. The liver also removes toxic chemicals, such as ingested alcohol and drugs and nitrogenous wastes given off by the cells. These are converted to molecules that can be excreted by the kidneys. The kidneys are also under hormonal control as they excrete wastes and salts, substances that can affect the pH level of the blood.

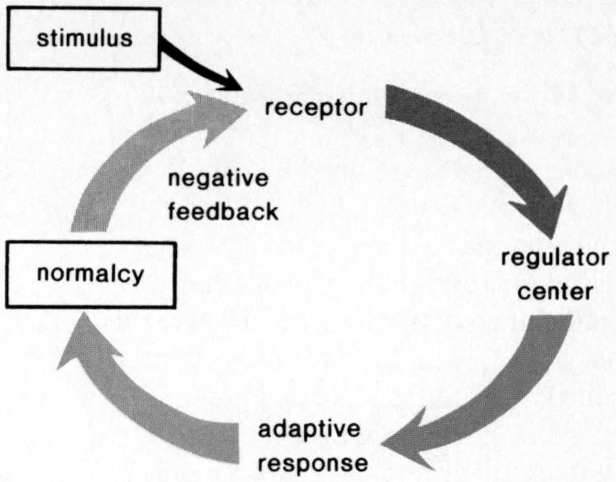

stimulus
receptor
negative feedback
regulator center
normalcy
adaptive response

Figure 8.16
Diagram illustrating the principle of feedback control. A receptor (sense organ) responds to a stimulus such as high or low temperature and notifies a regulator center that directs an adaptive response such as sweating. Once normalcy such as a normal temperature is achieved, the receptor is no longer stimulated.

All the systems of the body contribute to homeostasis, that is, maintaining the relative constancy of the internal environment.

Although homeostasis is, to a degree, controlled by hormones, it is ultimately controlled by the nervous system. The brain contains centers that regulate such factors as temperature and blood pressure. Maintaining proper temperature and blood pressure levels requires sensors that detect unacceptable levels and signals a control center. If a correction is required, the center then directs an adaptive response (fig. 8.16). Once normalcy is obtained, the receptor no longer stimulates the center. This is called control by **negative feedback** because the control center shuts down until it is stimulated to be active once again. This type of homeostatic regulation results in fluctuation between two levels, as illustrated for temperature control in figure 8.17. We may also note that feedback control is a self-regulatory mechanism.

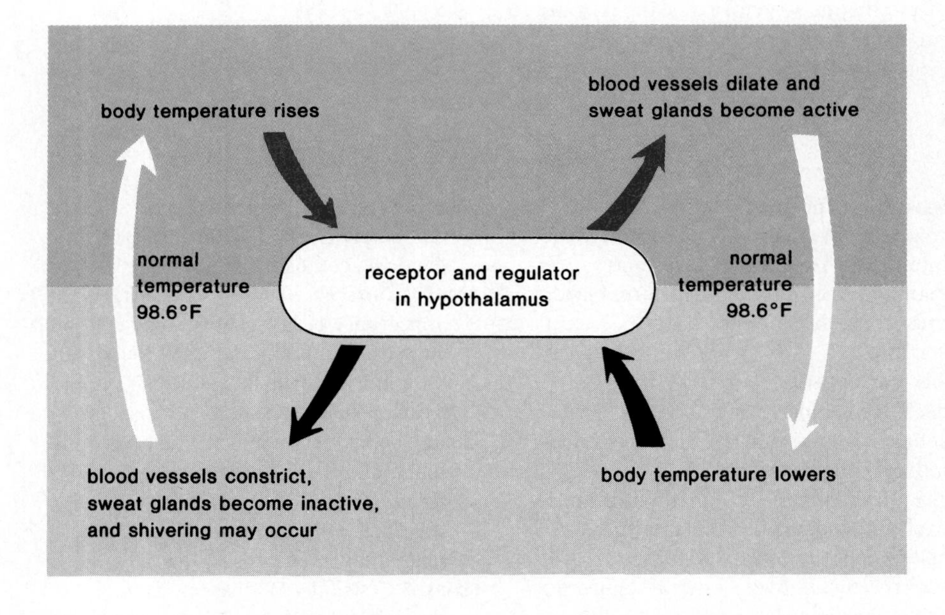

Figure 8.17
Temperature control. When the body temperature rises, the blood vessels dilate and the sweat glands become active. When the body temperature lowers, the blood vessels constrict and shivering may occur. In between these extremes the receptor is not stimulated and, therefore, body temperature fluctuates above and below normal.

Summary

Human tissues are categorized into four groups. Epithelial tissues cover the body and lines its cavities. Connective tissues often bind body parts together. Contraction of muscle tissues permit movement of the body and its parts. Nerve impulses conducted by neurons within nervous tissue help bring about coordination of body parts.

Tissues are joined together to form organs, each one having a specific function. Organs are grouped into organ systems. In vertebrates, the brain and spinal cord are dorsally located and the internal organs are located in the coelom composed of both the thoracic and abdominal cavities.

All organ systems contribute to the constancy of the internal environment. The nervous and hormonal systems regulate the other systems. Both of these are controlled by a feedback mechanism which results in fluctuation above and below the desired level.

Objective Questions

1. Most organs contain several different types of ________ .
2. Kidney tubules are lined by cube-shaped cells called ________ epithelium.
3. Pseudostratified ciliated columnar epithelium contains cells that appear to be ________ , have projections called ________ , and are ________ in shape.
4. Both cartilage and blood are classified as ________ tissue.
5. Cardiac muscle is ________ but involuntary.
6. Nerve cells are called ________ .
7. Skin has three layers: epidermis, ________ , and subcutaneous.
8. Outer skin cells are filled with ________ , a waterproof protein that strengthens them.
9. Mucous membrane contains ________ tissue overlying ________ tissue.
10. Homeostasis is maintenance of the relative ________ of the internal environment, that is the blood and ________ fluid.

Answers to Objective Questions

1. tissues 2. cuboidal 3. layered (stratified), cilia, columnar (elongated) 4. connective 5. striated 6. neurons 7. dermis 8. keratin 9. epithelial, loose connective 10. constancy, tissue

Study Questions

1. Name the four major groups of tissues. (p. 163)
2. What are the functions of epithelial tissues? Name the different kinds and give a location for each. (pp. 163–167)
3. What are the functions of connective tissue? Name the different kinds and give a location for each. (pp. 168–171)
4. What are the functions of muscular tissue? Name the different kinds and give a location for each. (p. 172)
5. Nervous tissue contains what type cell? Which organs in the body are made up of nervous tissue? (p. 173)
6. Describe the structure of skin and state at least two functions of this organ. (pp. 174–175)
7. In general terms, describe the location of the human organ systems. (p. 177)
8. List at least four vertebrate characteristics of humans. (p. 177)
9. What is homeostasis and how is it achieved in the human body? (pp. 178–179)

Key Terms

bone (bōn) connective tissue having a hard matrix of calcium salts deposited around protein fibers. *170*

cartilage (kar'tĭ-lij) a connective tissue, in which the cells lie within lacunae embedded in a flexible matrix. *170*

coelom (se'lom); a body cavity of higher animals that contains internal organs such as those of the digestive system. *177*

compact bone (kom'pakt bōn) hard bone consisting of Haversian systems cemented together. *170*

connective tissue (kō-nek'tiv tish'u) a type of tissue, characterized by cells separated by a matrix, that often contains fibers. *168*

dermis (der'mis) the thick skin layer that lies beneath the epidermis. *175*

epidermis (ep"ĭ-der'mis) the outer skin layer composed of stratified squamous epithelium. *174*

epithelial tissue (ep"ĭ-the'le-al tish'u) a type of tissue that lines cavities and covers the external surface of the body. *163*

homeostasis (ho"me-o-sta'sis) the constancy of conditions, particularly the internal environment of birds and mammals: constant temperature, blood pressure, pH, and other body conditions. *178*

hyaline cartilage (hi'ah-līn kar'tĭ-lij) cartilage composed of very fine collagenous fibers and a matrix of a clear milk-glass appearance. *170*

lacuna (lah-ku'nah) a small pit or hollow cavity, as in bone or cartilage where a cell or cells are located. *170*

ligament (lig'ah-ment) dense connective tissue that joins bone to bone. *169*

muscular tissue (mus'ky-lar tish'u) a type of tissue that contains cells capable of contracting; skeletal muscles are attached to skeleton; smooth muscle is found within walls of internal organs and cardiac muscle comprises the heart. *172*

negative feedback (neg'ah-tiv fēd'bak) a mechanism that is activated by a surplus imbalance and acts to correct it by stopping the process that brought about the surplus. *179*

neuron (nu'ron) nerve cell that characteristically has three parts: dendrite, cell body, axon. *173*

pseudostratified (su"-do strat'e-fīd) the appearance of layering in some epithelial cells when actually each cell touches a base line and true layers do not exist. *163*

spongy bone (spun'je bōn) porous bone found at the ends of long bones. *170*

stratified (strat'ĭ-fīd) layered, as in stratified epithelium, which contains several layers of cells. *163*

striated (stri'āt-ed) having bands; cardiac and skeletal muscle are striated with bands of light and dark. *172*

subcutaneous (sub"ku-ta'ne-us) a tissue layer found in vertebrate skin that lies just beneath the dermis and tends to contain adipose tissue. *175*

tendon (ten'don) dense connective tissue that joins muscle to bone. *169*

9

Digestion

Chapter Concepts

1 Small molecules, such as amino acids, glucose, and fatty acids, that can cross cell membranes are the products of digestion that nourish the body.

2 Regions of the digestive tract are specialized to carry on specific functions; for example, the mouth is specialized to receive and chew food, and the small intestine is specialized to absorb the products of digestion.

3 Digestive enzymes are specific hydrolytic enzymes and have a preferred temperature and pH.

4 Proper nutrition requires that the energy needs of the body be met and that the diet be balanced so that all vitamins and essential amino acids and fatty acids are included.

Chapter Outline

Digestion takes place within a tube, often called the gut, which begins with the mouth and ends with the anus (table 9.1 and fig. 9.1). Digestion of food in humans is an extracellular process. Digestive enzymes are secreted into the gut by glands that reside in the lining or lie nearby. Food is never found within these *accessory glands,* only within the gut itself.

Table 9.1 *Path of Food*

Organ	Special Features	Functions
Mouth	Teeth	Chewing of food; digestion of starch to maltose
Esophagus		Passageway
Stomach	Gastric glands	Digestion of protein to peptides
Small intestine	Intestinal glands Villi	Digestion of all foods and absorption of unit molecules
Large intestine		Absorption of water
Anus		Defecation

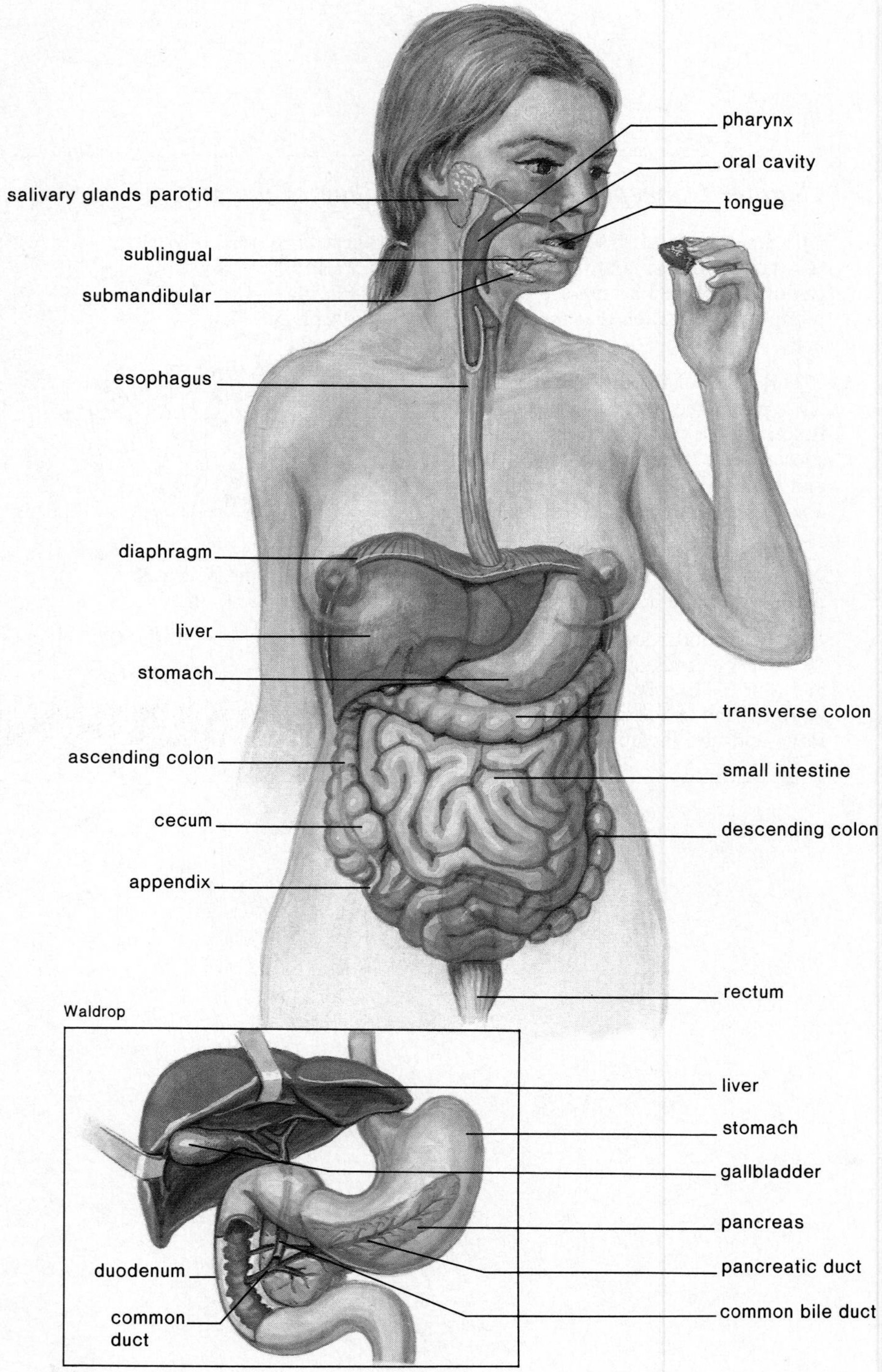

Figure 9.1
The human digestive system. In the insert, the liver has been moved back to show the gallbladder and to expose the stomach and duodenum.

While the term digestion, strictly speaking, means the breakdown of food by enzymatic action, we will expand the term to include both the physical and chemical processes that reduce food to small, soluble molecules. Only small molecules can cross cell membranes and be absorbed by the gut lining. Too often we are inclined to think that since we eat meat (protein), potatoes (carbohydrate), and butter (fat), these are the substances that nourish our bodies. Instead, it is the amino acids from the protein, the sugars from the carbohydrate, and the glycerol and fatty acids from the fat that actually enter the blood and are transported throughout the body to nourish our cells. Any component of food, such as cellulose, that is incapable of being digested to small molecules leaves the gut as waste material.

Digestion of food requires a cooperative effort between different parts of the body. We shall see that the production of hormones and the performance of the nervous system achieve the cooperation of body parts.

Digestive System

Mouth

The mouth receives the food in humans. Most people enjoy eating because of the combined sensations of smelling and tasting food. The olfactory receptors, located in the nose, are responsible for smelling; tasting is, of course, a function of the taste buds, located on the tongue. (See chapter 17 for a description of these sense organs.)

Normally, adults have 32 teeth (fig. 9.2) which chew the food into pieces convenient to swallow. One-half of each jaw has teeth of four different types: two chisel-shaped incisors for biting; one pointed canine for tearing; two fairly flat premolars for grinding; and three molars, well flattened, for crushing. The last molar, or wisdom tooth, may fail to erupt, or if it does, it is sometimes crooked and useless.

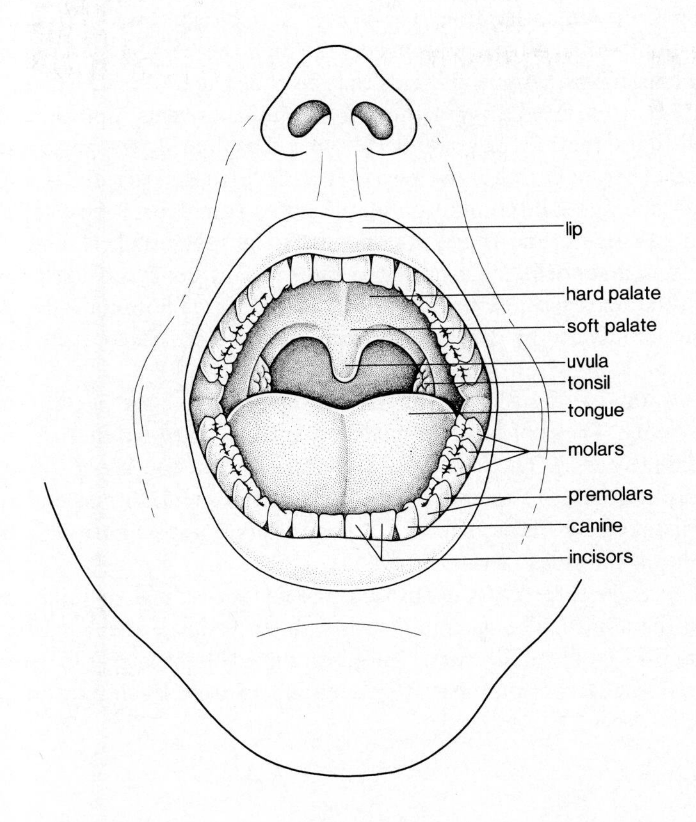

Figure 9.2
Diagram of the mouth showing the adult teeth. The sizes and shapes of teeth correlate with their functions.

Figure 9.3
Longitudinal section of a canine tooth. A tooth contains nerves and blood vessels within the pulp.

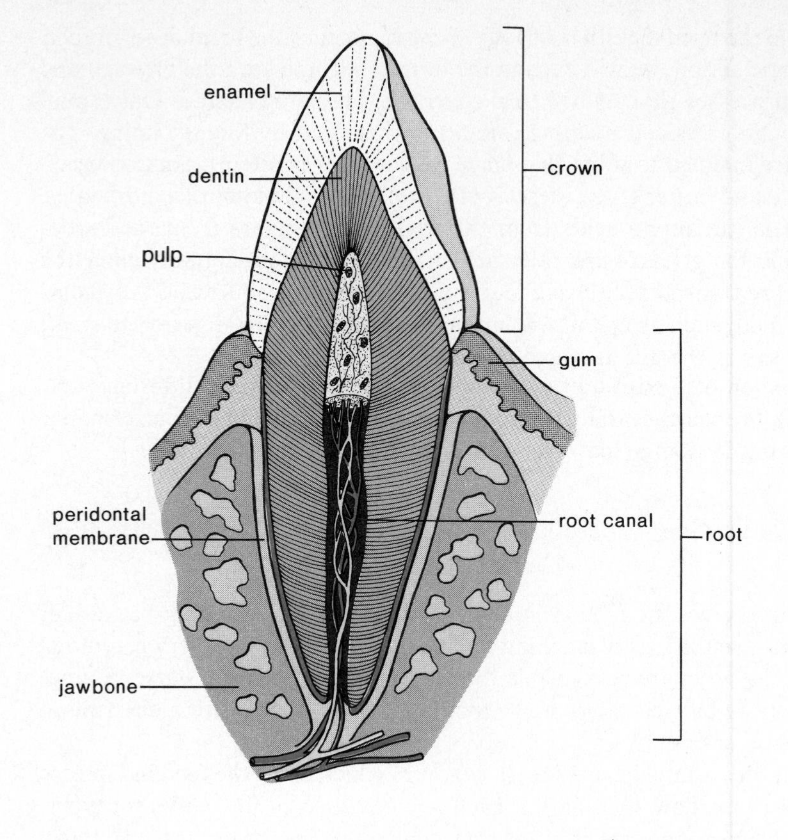

Each tooth (fig. 9.3) has a layer of enamel, an extremely hard outer covering of calcium compounds; dentin, a thick layer of bonelike material; and an inner pulp which contains the nerves and blood vessels. Tooth decay or *caries,* commonly called a cavity, occurs when the bacteria within the mouth metabolize sugar and give off acids that corrode the tooth. Two measures may prevent tooth decay: eating a limited amount of sweets, and daily brushing and flossing of teeth. It has also been found that fluoride treatments can make the enamel stronger and more resistant to decay. Gum disease is more apt to occur as one ages. Inflammation of the gums (gingivitis) may spread to the periodontal membrane (fig. 9.3) that lines the tooth socket. The individual then has **periodontitis,** characterized by a loss of bone and loosening of the teeth so that extensive dental work may be required. Stimulation of the gums in a manner advised by dentists has been found helpful in controlling this condition.

In humans, the roof of the mouth separates the air passages from the mouth cavity. The roof has two parts: an anterior **hard palate** and a posterior **soft palate** (fig. 9.2). The hard palate contains several bones, but the soft palate is merely muscular. The soft palate ends in the *uvula,* a suspended process often mistaken by the layperson for the tonsils, but as figure 9.2 shows, the tonsils lie to the sides of the throat.

There are three pairs of **salivary glands** that send their juices by way of ducts to the mouth. The parotid glands lie at the sides of the face immediately below and in front of the ears. They become swollen when a person has the mumps, a viral infection most often seen in children. Each parotid gland has a duct that opens on the inner surface of the cheek at the location of the second upper molar. The sublingual glands lie beneath the tongue, and the submandibular glands lie beneath the lower jaw. The ducts from these glands open into the mouth under the tongue. You can locate all these openings if you use your tongue to feel for small flaps on the inside of your cheek and under your tongue.

Saliva, secreted by the salivary glands, contains mostly water, mucus, and the digestive enzyme **salivary amylase.** This enzyme acts on starch. Like all of the digestive enzymes, salivary amylase is a **hydrolytic enzyme.** This means that its substrate is broken down on the addition of water:

$$\text{starch} + H_2O \xrightarrow{\text{salivary amylase}} \text{maltose}$$

In this equation, salivary amylase is written above the arrow to indicate that it is neither a reactant nor a product in the reaction. It merely speeds up the reaction in which its substrate, starch, is digested to many molecules of maltose. Maltose is not one of the small molecules that can be absorbed by the gut lining. Additional digestive action is required to convert maltose to glucose. This occurs farther along the digestive tract.

No other digestive process occurs in the mouth. The tongue takes the chewed food and forms it into a mass called a *bolus,* in preparation for swallowing.

In the mouth, food is chewed and acted upon by salivary amylase before it is swallowed.

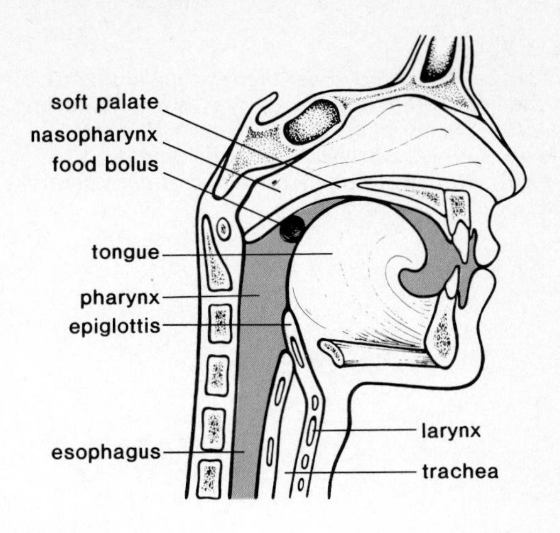

Figure 9.4
When food is swallowed, the soft palate covers the nasopharyngeal openings and the epiglottis covers the glottis so that the food bolus must pass down the esophagus. Therefore, one does not breathe during swallowing.

Pharynx

Swallowing (fig. 9.4) occurs in the **pharynx,** a region between the mouth and the esophagus, a long muscular tube that leads to the stomach. Swallowing is a *reflex action,* which means the action is usually performed automatically and does not require conscious thought. Normally, during swallowing, food enters the esophagus because the air passages are blocked. Unfortunately, we have all had the unpleasant experience of having food "go the wrong way." The wrong way may be either into the nose or into the windpipe (trachea). If it is the latter, coughing usually forces the food up out of the trachea into the pharynx again. Food usually goes into the esophagus because the openings to the nose, called the *nasopharyngeal openings,* are covered when the soft palate moves back. The opening to the **larynx** (voice box) at the top of the trachea, called the **glottis,** is covered when the trachea moves up under a flap of tissue, called the **epiglottis.** This is easy to observe in the up-and-down movement of the *Adam's apple,* a part of the larynx, when a person eats. Notice that breathing does not occur during swallowing because air passages are closed off.

The air passage and food passage cross in the pharynx. When one swallows, the air passage is usually blocked off and food must enter the esophagus.

Figure 9.5
Wall of the esophagus. Like the rest of the digestive tract, several different types of tissues are found in the wall of the esophagus. (Lu = *central lumen;* Mu= *mucous membrane;* Su = *submucosa;* Me=*muscular layer; and* Ad=*adventitia, or serous layer.*)

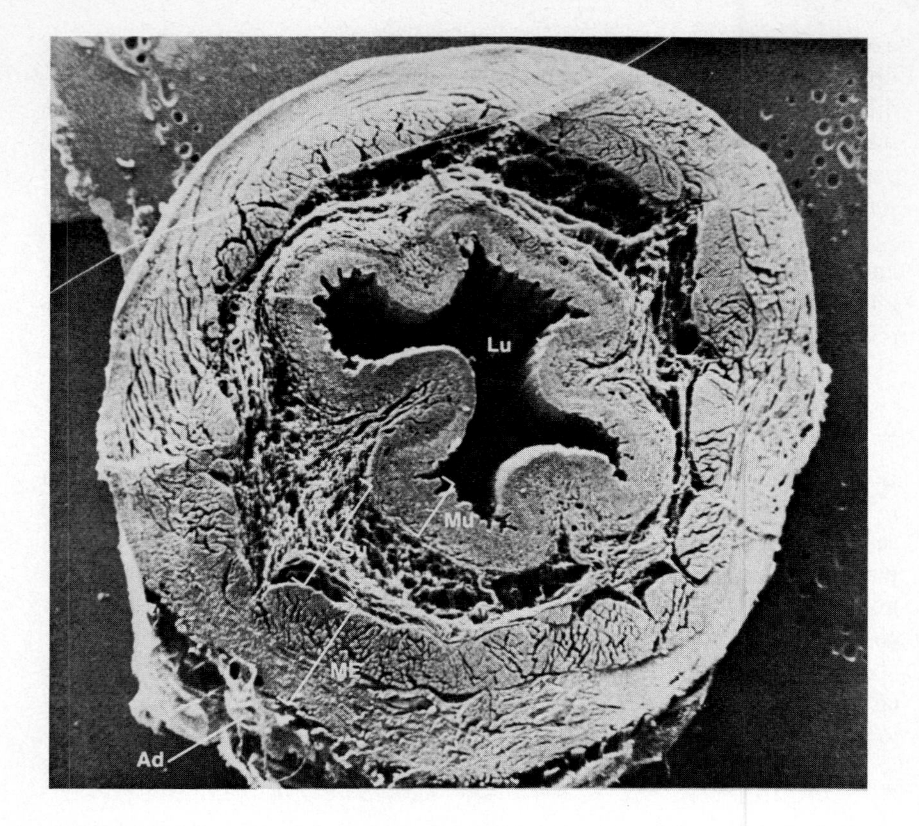

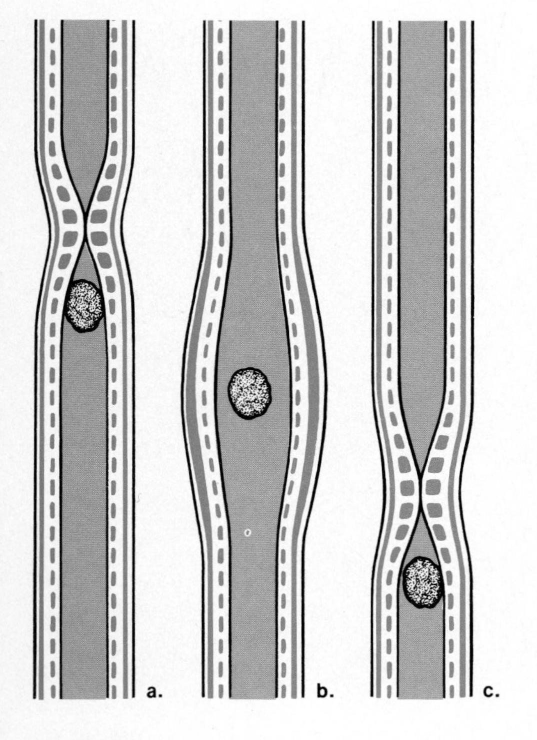

Figure 9.6
Peristalsis in the digestive tract. Rhythmic waves of muscle contraction move material along the digestive tract. The three drawings show how a peristaltic wave moves through a single section of gut over time (a. *to* c.)

Esophagus

After swallowing occurs, the **esophagus** conducts the bolus through the thoracic cavity. The wall (fig. 9.5) of the esophagus is representative of the gut in general. A *mucous membrane* lines the **lumen** (space within the tube); this is followed by a *submucosal layer* of connective tissue that contains nerve and blood vessels; a *smooth muscle layer* having both longitudinal and circular muscles; and finally a *serous membrane layer.*

A rhythmical contraction of the esophageal wall, called **peristalsis,** (fig. 9.6) pushes the food along. Occasionally peristalsis begins even though there is no food in the esophagus. This produces the sensation of a lump in the throat.

The food bolus enters the stomach by way of a sphincter. **Sphincters** are muscles that encircle tubes and act as valves; tubes close when sphincters contract, and they open when sphincters relax. Normally the sphincter prevents food from moving up out of the stomach but when vomiting occurs, a reverse peristaltic wave causes the sphincter to relax and the contents of the stomach are propelled upwards toward the esophagus.

Stomach

The stomach is a thick-walled, J-shaped organ that lies on the left side of the body beneath the diaphragm. It is an enlarged portion of the gut which can stretch to hold about a half gallon of liquids and solids. The walls contain three layers of muscle instead of two layers, and contraction of these muscles causes the stomach to churn and mix its contents. Hunger pangs are felt when an empty stomach churns.

The mucosal lining of the stomach contains millions of microscopic digestive glands called **gastric glands** (the term *gastric* always refers to the stomach). The gastric glands (fig. 9.7) produce a gastric juice. Gastric juice

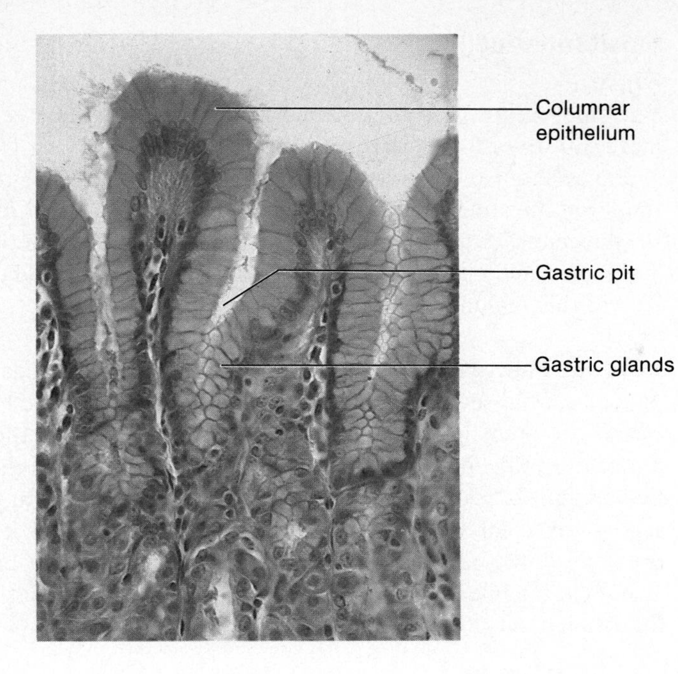

Figure 9.7
Photomicrograph of the mucosa of the stomach. Gastric glands produce gastric juice rich in pepsin, an enzyme that digests protein to peptides.

contains pepsinogen and hydrochloric acid (HCl). When *pepsinogen* is exposed to hydrochloric acid within the stomach, it becomes the digestive enzyme pepsin. **Pepsin** is a hydrolytic enzyme that acts on protein to produce peptides:

$$\text{protein} + H_2O \xrightarrow{\text{pepsin}} \text{peptides}$$

Peptides vary in length, but always consist of a number of amino acids joined together. Peptides are too large to be absorbed by the gut lining. However, they are later broken down to amino acids in another part of the digestive tract.

The presence of hydrochloric acid causes the contents of the stomach to have an acid pH of about 3. This low pH is beneficial in that it kills most bacteria present in food. Although HCl is not an enzyme, it still has a corrosive action on food and, at times, on the stomach wall itself. Normally, the wall is protected by a thick layer of mucus secreted by mucosal cells. If by chance HCl does penetrate this mucus, pepsin starts to digest the stomach lining and an ulcer results. An *ulcer* (fig. 9.8) is an open sore in the wall caused by the gradual disintegration of tissues. It is believed that the most frequent cause of an ulcer is oversecretion of gastric juice due to too much nervous stimulation. Persons under stress tend to have a greater incidence of ulcers.

Normally, the stomach empties in about two to six hours. By this time, the bolus of food has become a semiliquid food mass called *acid chyme*. Acid chyme leaves the stomach and enters the small intestine by way of a sphincter.

Figure 9.8
Peptic ulcer in the stomach. Normally the digestive tract produces enough mucus to protect itself from digestive juices. An ulcer often begins when an excessive amount of gastric juices is produced due to an increased amount of nervous stimulation.

Gastric glands produce HCl and a precursor of pepsin, an enzyme that breaks down protein to peptides.

Small Intestine

Digestion

The small intestine gets its name from its small diameter, compared to the large intestine. But perhaps it should be called the long intestine because it averages about 6.0 m (20 ft.) in length compared to about 1.5 m (5 ft.) for the large intestine. The first 25 cm (10 in.) of the small intestine are called the **duodenum.** Duodenal ulcers sometimes occur because the acid and pepsin within the acid chyme from the stomach may corrode and digest the internal wall in this region.

Liver and Pancreas Two very important accessory glands, the liver and the pancreas, send secretions to the duodenum (fig. 9.1). The liver produces **bile,** which is stored in the **gallbladder,** and sent by way of the bile duct to the duodenum. Bile looks green because it contains pigments that are products of hemoglobin breakdown. This green color is familiar to anyone who has observed the color changes of a bruise. Within a bruise, the hemoglobin is breaking down into the same type of pigments found in bile. However, bile also contains bile salts, which are emulsifying agents that break up fat into fat droplets:

$$\text{fat} \xrightarrow{\text{bile salts}} \text{fat droplets}$$

Following emulsification (fig. 1.30), the resulting fat droplets are ready for chemical digestion. Almost all fat digestion occurs in the small intestine as discussed following.

The pancreas sends *pancreatic juice* into the duodenum by way of the pancreatic duct (fig. 9.1). You may be more familiar with the pancreas as the source of the hormone insulin. But some other pancreatic cells produce a juice that contains digestive enzymes and *sodium bicarbonate* ($NaHCO_3$). The latter makes pancreatic juice basic, which is also called alkaline. The alkaline pancreatic juice of about pH 8.5 neutralizes the acid of the acid chyme and causes the pH of the small intestine to be basic. The enzymes found in the small intestine prefer a basic pH and only function to their optimum at this pH.

Pancreatic juice contains enzymes that act on every major component of food. There is a pancreatic enzyme called **pancreatic amylase** that digests starch:

$$\text{starch} + H_2O \xrightarrow[\text{amylase}]{\text{pancreatic}} \text{maltose}$$

Trypsin is an example of a pancreatic enzyme that digests protein:

$$\text{protein} + H_2O \xrightarrow{\text{trypsin}} \text{peptides}$$

Trypsin is secreted as trypsinogen and changes to trypsin in the gut.

Lipase digests the fat droplets:

$$\text{fat droplets} + H_2O \xrightarrow{\text{lipase}} \text{glycerol} + \text{fatty acids}$$

The digestion of fat is now complete and the molecules of glycerol and fatty acids are small enough to be absorbed by the lining of the small intestine.

In the small intestine, fat is emulsified by bile salts into fat droplets before being acted upon by pancreatic lipase. Protein is digested by pancreatic trypsin, and starch is digested by pancreatic amylase.

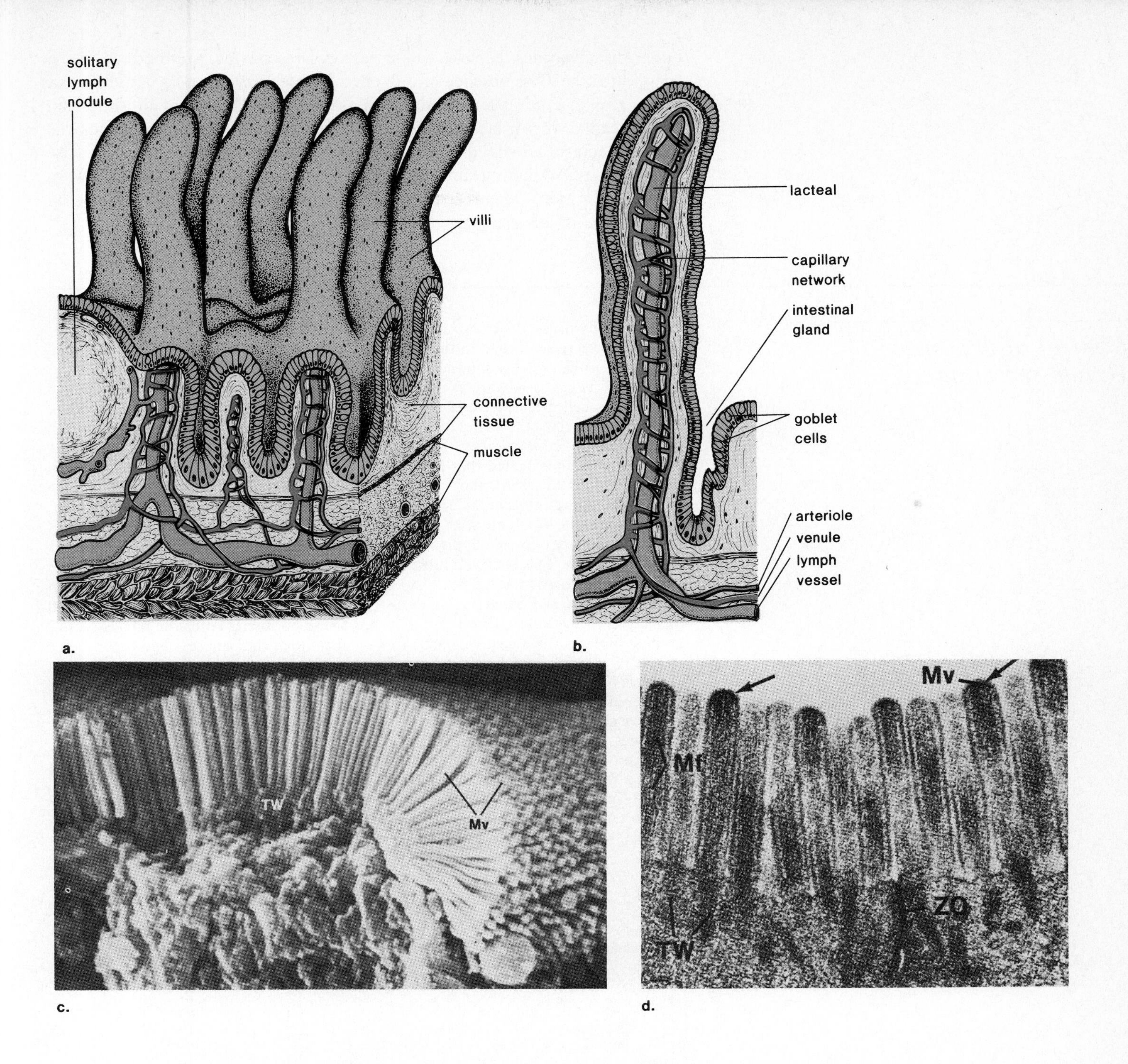

Intestinal Glands The wall of the small intestine contains millions of intestinal glands (fig. 9.9) that produce a digestive juice. The enzymes in this juice complete the digestion of protein and carbohydrates. Instead of being deposited in the lumen, these enzymes remain attached to the cell membranes of adjacent microvilli with their active sites exposed.

Peptides, which result from the first step in protein digestion, are digested to amino acids:

$$\text{peptides} + H_2O \xrightarrow{\text{peptidases}} \text{amino acids}$$

Maltose, which results from the first step in starch digestion, is digested to glucose:

$$\text{maltose} + H_2O \xrightarrow{\text{maltase}} \text{glucose}$$

Figure 9.9
Anatomy of intestinal lining. a. *The products of digestion are absorbed by villi, fingerlike projections of the intestinal wall,* b. *each of which contains blood vessels and a lacteal.* c. *The scanning electron micrograph shows that the villi themselves are covered with microvilli* (Mv). d. *A transmission electron micrograph shows that the microvilli contain microfilaments* (Mf). *These allow limited motion of the microvilli.*

(c. and d.) From: Tissues and Organs: A Text-Atlas of Scanning Electron Microscopy by R. G. Kessel and R. Kardon. W. H. Freeman and Company, © 1979.

Other disaccharides, each of which has its own enzyme, are digested in the small intestine. The absence of any one of these enzymes may cause illness. For example, many people, including as many as 75 percent of American blacks, cannot digest lactose, the sugar found in milk, because they lack the enzyme lactase, an enzyme that converts lactose to its components, glucose and galactose. Drinking untreated milk often gives these individuals severe diarrhea. In most areas it is possible to purchase milk made lactose-free by the addition of lactase.

Control of Digestive Gland Secretion

The study of the control of digestive gland secretion began in the late 1800s, when Ivan Pavlov showed that dogs would begin to salivate at the ringing of a bell, because they had learned to associate the sound of the bell with being fed. Pavlov's experiments demonstrated that even the thought of food can bring about the secretion of digestive juices. Certainly if food is present in the mouth, stomach, and small intestine, digestive secretion occurs. This is attributable to a simple reflex occurrence. The presence of food sets off nerve impulses that travel to the brain. Thereafter, the brain stimulates the digestive glands to secrete.

In this century, investigators have discovered that specific control of digestive secretions is achieved by hormones. A *hormone* is a substance that is produced by one set of cells but affects a different set of cells, the so-called target cells. Hormones are transported by the bloodstream. For example, when a person has eaten a meal particularly rich in protein, the hormone **gastrin** produced by the lower part of the stomach, enters the bloodstream and soon reaches the upper part of the stomach, where it causes the gastric glands to secrete more pepsinogen.

Experimental evidence has also shown that the duodenal wall produces hormones, the most important of which are **secretin** and **CCK (cholecystokinin).** Acid, especially HCl present in the acid chyme, stimulates the release of secretin, while partially digested protein and fat stimulate the release of CCK. These hormones enter the bloodstream and signal the pancreas and the gallbladder to send secretions to the duodenum.

Another hormone has recently been discovered. **GIP** (gastric inhibitory peptide), produced by the small intestine, apparently works opposite to gastrin because it inhibits gastric acid secretion. This is not surprising since very often the body has hormones that have opposite effects.

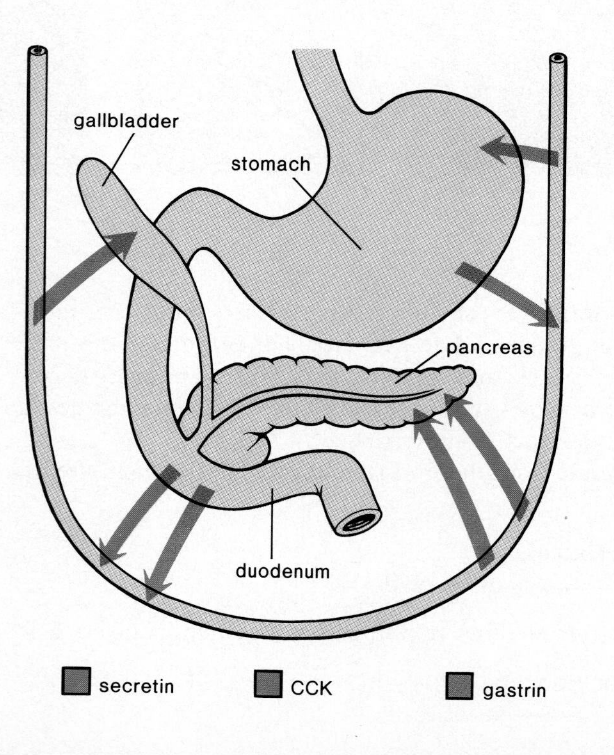

Hormonal control of digestive gland secretions. Especially after eating a protein-rich meal, gastrin produced by the lower part of the stomach enters the bloodstream and thereafter stimulates the upper part of the stomach to produce more digestive juices. Acid chyme from the stomach causes the duodenum to release secretin and CCK. Both of these hormones stimulate the pancreas to secrete its digestive juices and CCK alone stimulates the gallbladder to release bile.

Absorption

The small intestine is specialized for absorption. First, it is quite long with convoluted walls. Secondly, the absorptive surface is increased by the presence of fingerlike projections called **villi,** and the villi (fig. 9.9) themselves have tiny microvilli. The huge number of villi that cover the entire surface of the small intestine give it a soft, velvety appearance. Each villus has an outer layer of columnar cells and contains blood vessels and a small lymph vessel called a **lacteal.** The lymphatic system is an adjunct to the circulatory system and returns fluid to the veins.

Absorption takes place across the wall of each villus, continuing until all small molecules have been absorbed. Thus absorption is an active process involving active transport of molecules across cell membranes and requiring an expenditure of cellular energy (p. 69). Sugars and amino acids cross the columnar cells to enter the blood, but glycerol and fatty acids enter the lacteals.[1]

Intestinal glands at the base of the villi secrete enzymes that hydrolyze disaccharides, polypeptides, and other substances. Following digestion, amino acids and glucose enter the blood; reformed glycerol and fatty acids enter the lymph.

Liver

Blood vessels from the villi merge to form the **hepatic portal vein,** which leads to the liver (fig. 9.10). The liver acts in some ways as the gatekeeper to the blood; it removes poisonous substances from the blood and works to keep the contents of the blood constant. In particular, we may note that the glucose level of the blood is always about 0.1 percent even though we eat intermittently. Any excess glucose present in the hepatic portal vein is removed and stored by the liver as glycogen:

$$\text{glucose} \longrightarrow \text{glycogen} + H_2O$$

[1]For the fate of these molecules see also pages 196–202.

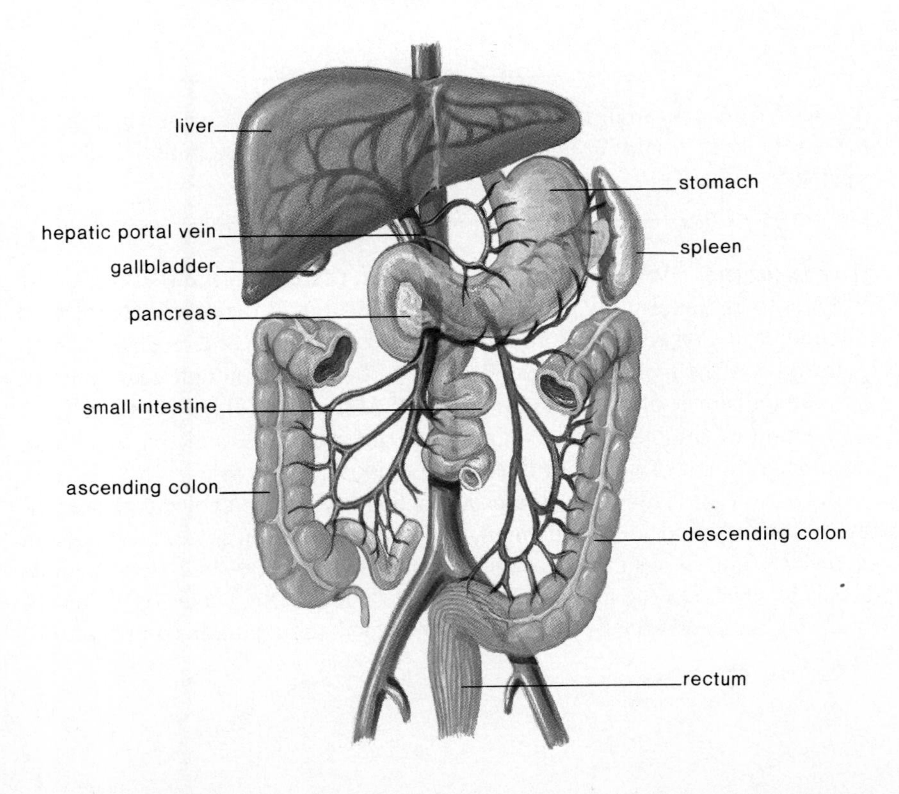

Figure 9.10
The hepatic portal vein takes the products of digestion from the digestive system to the liver where they are processed before entering the circulatory system proper.

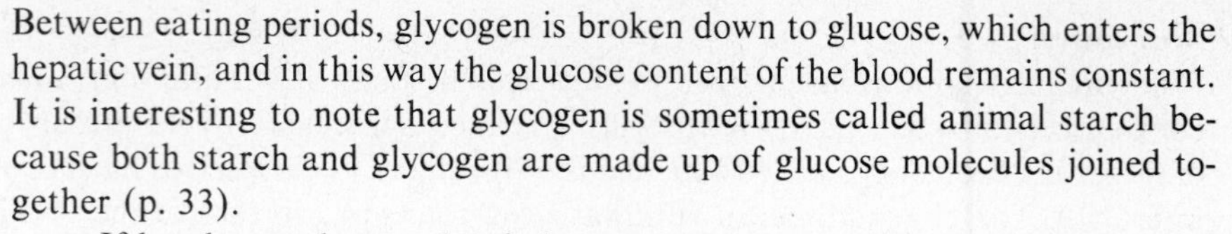

Between eating periods, glycogen is broken down to glucose, which enters the hepatic vein, and in this way the glucose content of the blood remains constant. It is interesting to note that glycogen is sometimes called animal starch because both starch and glycogen are made up of glucose molecules joined together (p. 33).

If by chance the supply of glycogen or glucose runs short, the liver will convert amino acids to glucose molecules:

$$\text{amino acids} \longrightarrow \text{glucose} + \text{amino groups}$$

You will recall that amino acids contain nitrogen in the form of amino groups, whereas glucose contains only carbon, oxygen, and hydrogen. Therefore, before amino acids can be converted to glucose molecules, **deamination,** or the removal of amino groups from the amino acids, must take place. By an involved metabolic pathway, the liver converts these amino groups to urea:

$$H_2N - \overset{\overset{\displaystyle O}{\|}}{C} - NH_2$$

Urea is the common nitrogen waste product of humans; and after its formation in the liver, it is transported to the kidneys for excretion.

The liver also makes blood proteins from amino acids. These proteins are not used as food for cells; rather, they serve important functions within the blood itself.

Altogether we have mentioned the following functions of the liver:

1. Destroys old red blood cells and converts hemoglobin to the breakdown products (bilirubin and biliverdin) excreted along with bile salts in bile.
2. Produces bile, which is stored in the gallbladder before entering the small intestine where it emulsifies fats.
3. Stores glucose as glycogen after eating and breaks down glycogen to glucose to maintain the glucose concentration of the blood between eating.
4. Produces urea from the breakdown of amino acids.
5. Makes the blood proteins.
6. Detoxifies the blood by removing poisonous substances and metabolizing them.

Blood from the small intestine enters the hepatic portal vein which goes to the liver, a vital organ that has numerous important functions listed previously.

Liver Disorders When a person is *jaundiced,* there is a yellowish tint to the skin due to an abnormally large amount of bilirubin in the blood. In one type of jaundice, called *hemolytic jaundice,* red blood cells are broken down in such quantity that the liver cannot excrete the bilirubin fast enough, and an extra amount spills over into the bloodstream. In *obstructive jaundice,* there is an obstruction of the bile duct or damage to the liver cells, and this causes an increased amount of bilirubin to enter the bloodstream.

Obstructive jaundice often occurs when crystals of cholesterol precipitate out of bile and form **gallstones,** which on occasion also contain calcium carbonate. The stones may be so numerous that passage of bile along the bile duct is blocked and the gallbladder must be removed (fig. 9.11). In the meantime, the bile leaves the liver by way of the blood and a jaundiced appearance results.

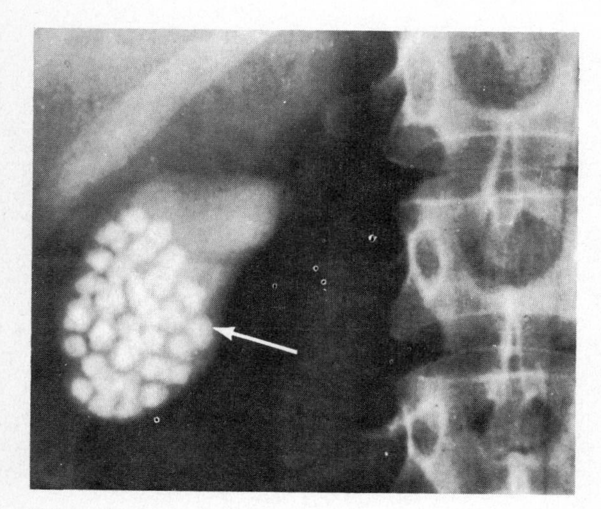

a.

b.

Figure 9.11
a. *An X ray of a gallbladder filled with gallstones.*
b. *After removal, this gallbladder was cut open to show its contents—numerous gallstones. The dime was added later merely to indicate the size of the stones.*

Jaundice is also frequently caused by liver damage due to *viral hepatitis,* a term that includes two separate but similar diseases. *Type A* virus causes *infectious hepatitis* transmitted by unsanitary food or water. In recent years, persons have been known to acquire the disease after eating shellfish from polluted waters, for example. *Type B* virus causes *serum hepatitis,* commonly spread by means of blood transfusions, kidney dialysis, and injection with inadequately sterilized needles. However, there is evidence that both types of hepatitis can be sexually transmitted. To recover from hepatitis, a long recuperation period is commonly required, during which time the patient is in a very weakened condition. To prevent the possibility of passing on the disease, a person who has had serum hepatitis cannot give blood.

Cirrhosis is a chronic disease of the liver in which the organ first becomes fatty and then liver tissue is replaced by inactive fibrous scar tissue. This condition, common among alcoholics, is most likely caused by the need for the liver to break down excessive amounts of alcohol. When alcohol, a two-carbon compound, is metabolized, active acetate results, and these molecules can be synthesized to fatty acids. To accomplish this synthesis, smooth endoplasmic reticulum increases dramatically in the liver—this may be the first step toward cirrhosis.

The liver is a very critical organ and any malfunctioning is a matter of considerable concern.

Large Intestine

The large intestine includes the colon and rectum. The **colon** has three parts: the ascending colon goes up the right side of the body to the level of the liver; the transverse colon crosses the abdominal cavity just below the liver and stomach; and the descending colon passes down the left side of the body to the rectum. The **rectum** is the last approximately 20 cm (7.5 in.) of the intestinal tract. The opening of the rectum to the exterior is called the **anus.** Most of the water in our food and drink is reabsorbed by the cells that line the colon. If too little water is absorbed, diarrhea results, and if too much water is absorbed, constipation occurs. These conditions are discussed shortly.

Most of the indigestible remains fill the colon and the reflex defecation arises from stimuli generated in the colon and rectum (fig. 9.12). In addition to indigestible remains, **feces** also contains certain excretory substances such as bile pigments and heavy metals, and large quantities of the noninfectious bacterium *E. coli.*

The large intestine normally contains a large population of bacteria that live off any substances that were not digested earlier. When they break this material down, they give off odorous molecules that cause the characteristic odor of feces. Some of the vitamins, amino acids, and other growth factors produced by these bacteria spill out into the gut and are absorbed by the gut lining. In this way *E. coli* performs a service for us.

Water is considered unsafe for swimming when the *E. coli* count reaches a certain level. This is not because *E. coli* normally causes disease, but because a high count is an indication of the amount of fecal material that has entered the water. The more fecal material present, the greater the possibility that pathogenic, or disease-causing, organisms are also present.

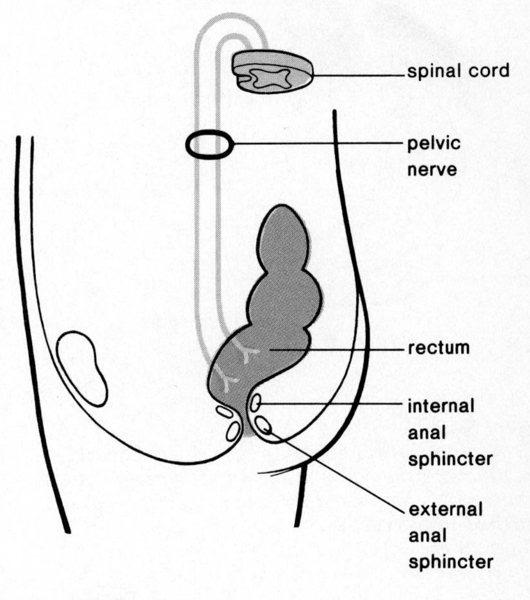

Figure 9.12
Defecation reflex. The accumulation of feces in the rectum causes it to stretch, which initiates a reflex action resulting in rectal contraction and expulsion of the fecal material.

Diarrhea and Constipation

Two common everyday complaints associated with the large intestine are diarrhea and constipation.

The major causes of *diarrhea* are infection of the lower tract and nervous stimulation. In the case of infection, such as food poisoning caused by eating contaminated food, the intestinal wall becomes irritated and peristalsis increases. Lack of absorption of water is a protective measure, and the diarrhea that results serves to rid the body of the infectious organisms. In nervous diarrhea, the nervous system stimulates the intestinal wall and diarrhea results. Loss of water due to diarrhea may lead to dehydration, a serious condition in which the body tissues lose their normal water content.

When a person is *constipated,* the feces are dry and hard. One cause of this condition is that socialized persons have learned to inhibit defecation to the point that the normal reflexes are often ignored. Two components of the diet can help to prevent constipation: water and roughage. A proper intake of water prevents the drying out of the feces. Dietary inclusion of roughage (fiber) or nondigestible plant substances, provides the bulk needed for elimination and may even help protect one from colon cancer. It is believed that the less time that feces are in contact with the membrane of the colon, the better. Even so the frequent use of laxatives is certainly discouraged, but if it should be necessary to take a laxative, a bulk laxative is the most natural because, like roughage, it produces a soft mass of cellulose in the colon. Lubricants like mineral oil make the colon slippery and saline laxatives like milk of magnesia act osmotically; they prevent water from exiting or even cause water to enter the colon depending on the dosage. Some laxatives are irritants; they increase peristalsis to the degree that the contents of the colon are expelled.

Chronic constipation is associated with the development of hemorrhoids, a condition that is discussed on page 226.

The large intestine does not produce digestive enzymes; it does absorb water. In diarrhea, too little water has been absorbed; in constipation, too much water has been absorbed.

Appendicitis and Colostomy

Two other, usually more serious medical conditions, associated with the large intestine are appendicitis and colostomy.

Appendicitis The small intestine joins the large intestine in such a way that there is a blind end to one side (fig. 9.13). This blind sac, or cecum, has a small projection about the size of the little finger, called the **appendix.** In humans, the appendix is vestigial, meaning that the organ is underdeveloped. But in other animals it is developed and serves as a location for bacterial digestion of cellulose. Unfortunately, the appendix can become infected. When this happens, the individual has *appendicitis,* a very painful condition. It is possible for the fluid content of the appendix to rise to the point that it bursts. It is better to have the appendix removed before this occurs, because it may lead to a generalized infection of the serous membranes of the abdominal cavity.

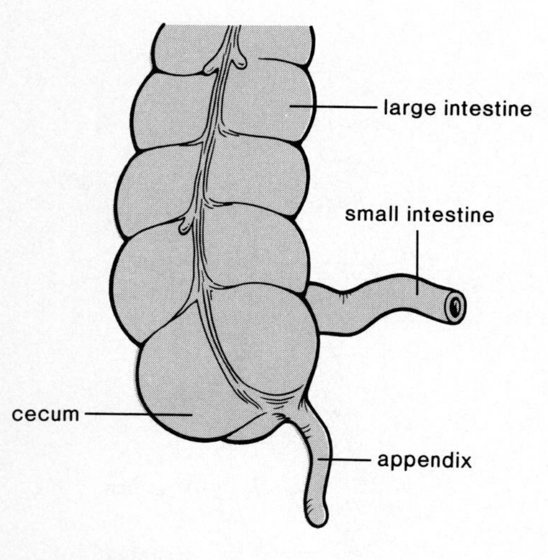

Figure 9.13
The anatomical relationship between the small intestine and the colon is shown. The cecum is the blind end of the ascending colon. The appendix is attached to the cecum.

Colostomy The colon is subject to the development of *polyps,* small growths that generally appear on epithelial tissue like those that line the digestive tract. Polyps, whether they are benign or cancerous, can be individually removed, along with a portion of the colon if necessary. Should it be necessary to remove the rectum and the anal canal, then the intestine is sometimes attached to the abdominal wall and the digestive remains are collected in a plastic bag fastened around the opening. Recently, the use of metal staples has permitted surgeons to join the colon to a piece of rectum that formerly would have been considered too short.

Two serious conditions associated with the large intestine require surgical removal of its parts. An infection of the appendix (appendicitis) requires removal of the appendix and cancer of the colon requires removal of cancerous tissue.

Digestive Enzymes

The digestive process in each part of the digestive tract has been described on the preceding pages. However, it is also possible to take each type of food—protein, carbohydrate, and fat—and discuss the digestion of each. Table 9.2 lists the enzymes needed for each of these major components of food. As you can see, there are two enzymes for the digestion of starch to maltose: salivary amylase and pancreatic amylase. Once starch has been converted to maltose, an intestinal enzyme breaks down maltose to glucose. Glucose is absorbed into the blood by the intestinal villi.

Protein is broken down to peptides by both pepsin, an enzyme produced by the gastric glands of the stomach, and trypsin, an enzyme produced by the pancreas. Peptides must be further digested by peptidases, produced by the intestinal glands. After digestion, the released amino acids can be absorbed into the blood of the intestinal villi.

Fats are first emulsified by bile to fat droplets and then these are digested by pancreatic lipase to glycerol and fatty acids. Glycerol and fatty acids are absorbed into the lacteals of the intestinal villi.

Digestive enzymes are hydrolytic enzymes that catalyze degradation by the introduction of water at specific bonds. Digestive enzymes are no different from any other enzyme of the body. For example, they are proteins with a particular shape that fits their substrate. Enzymes have a preferred pH because this pH maintains their shape in order that they may speed the reaction for which they are specific. Table 9.3 lists the enzymes and their preferred pH.

Table 9.2 *Digestive Enzymes*

Reaction	Enzyme	Gland	Site of Occurrence
starch + H_2O ⟶ maltose	Salivary amylase	Salivary	Mouth
	Pancreatic amylase	Pancreas	Small intestine
maltose + H_2O ⟶ glucose*	Maltase	Intestinal	Small intestine
protein + H_2O ⟶ peptides	Pepsin	Gastric	Stomach
	Trypsin	Pancreas	Small intestine
peptides + H_2O ⟶ amino acids*	Peptidases	Intestinal	Small intestine
fat + H_2O ⟶ glycerol + fatty acids*	Lipase	Pancreas	Small intestine
nucleic acids + H_2O ⟶ nucleotides	Nucleases	Intestinal	Small intestine

*Absorbed by villi.

Food is largely made up of carbohydrate (starch), protein, and fat. These very large macromolecules are broken down by digestive enzymes to small molecules that can be absorbed by intestinal villi. This chart indicates the steps needed for carbohydrate digestion (starch and maltose), protein digestion (protein and peptides), and fat digestion (fat) and shows that they are all hydrolytic reactions.

Table 9.3 *Comparison of Enzymes*

Enzyme	Source	Optimum pH	Type of Food Digested	Product
Salivary amylase	Saliva	Neutral	Starch	Maltose
Pepsin	Stomach	Acid	Protein	Peptides
Pancreatic amylase	Pancreas	Alkaline	Starch	Maltose
Lipase	Pancreas	Alkaline	Fat	Glycerol; fatty acids
Trypsin	Pancreas	Alkaline	Protein	Peptides
Nucleases	Pancreas	Alkaline	RNA, DNA	Nucleotides
Peptidases	Intestine	Alkaline	Peptides	Amino acids
Maltase	Intestine	Alkaline	Maltose	Glucose

All enzymes have a preferred pH that maintains their proper shape to do their job. This table indicates the pH for each of the enzymes in table 9.2.

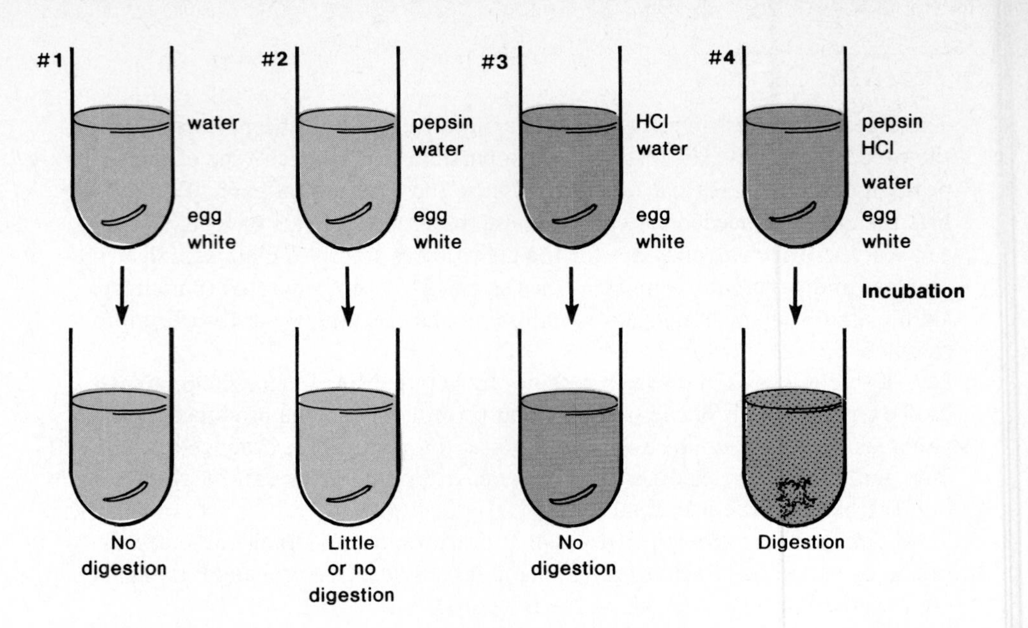

Figure 9.14
An experiment to demonstrate that enzymes digest food when the environmental conditions are correct. #1 lacks the enzyme pepsin and no digestion occurs; #2 has too high a pH and little or no digestion occurs; #3 has the proper pH because of the presence of HCl, but still no digestion occurs because the enzyme is missing; #4 contains enzyme and the environmental conditions are correct for digestion.

Simple laboratory experiments (fig. 9.14) can be done to show that it is enzymes that bring about the breakdown of food and not some other substance, such as hydrochloric acid (HCl) or bile. For example, the following four test tubes may be prepared in which egg white, or the protein called albumin, is to be tested for digestion:

1. Water + a small sliver of egg white
2. Pepsin + water + a small sliver of egg white
3. HCl + water + a small sliver of egg white
4. Pepsin + HCl + water + a small sliver of egg white

All tubes are now placed in an incubator at body temperature for at least an hour. At the end of this time, we can predict that tube 4 will show the best digestive action. Tube 3 does not contain the enzyme and tube 2 has too high a pH, so these two tubes will show little or no digestion. Tube 1 is a control tube and there will be no digestion in it.

This experiment can be expanded and three test tubes similar to test tube 4 can be prepared. One is kept in the cold, one at body temperature, and the last is boiled. The body temperature tube will, of course, show the best action. Like any chemical reaction, enzymatic reactions speed up if warmed; however, boiling denatures enzymes and so the boiled tube will show no digestion.

Nutrition

The four major classes of nutrients are proteins, carbohydrates, lipids, and vitamins and minerals. Nutrients supply us with the building blocks and energy we need to maintain our bodies.

Energy

Humans need energy primarily for basal metabolism but also for physical activities. Each individual's *basal metabolic rate (BMR)* is the amount of energy measured in **calories**[2] needed to maintain the body at rest. BMR is best measured 14 hours after the last meal (because absorption of digestive products requires energy) with the subject lying down at complete physical and mental rest. BMR is usually lower for women than for men and, in general, is affected by size, shape, weight, age, activity of endocrine glands, and so forth. Any

[2]Calorie (capital C) is the amount of heat required to raise the temperature of 1,000 grams of water one degree centigrade, while one calorie (lowercase c) is the amount needed to raise the temperature of one gram of water one degree.

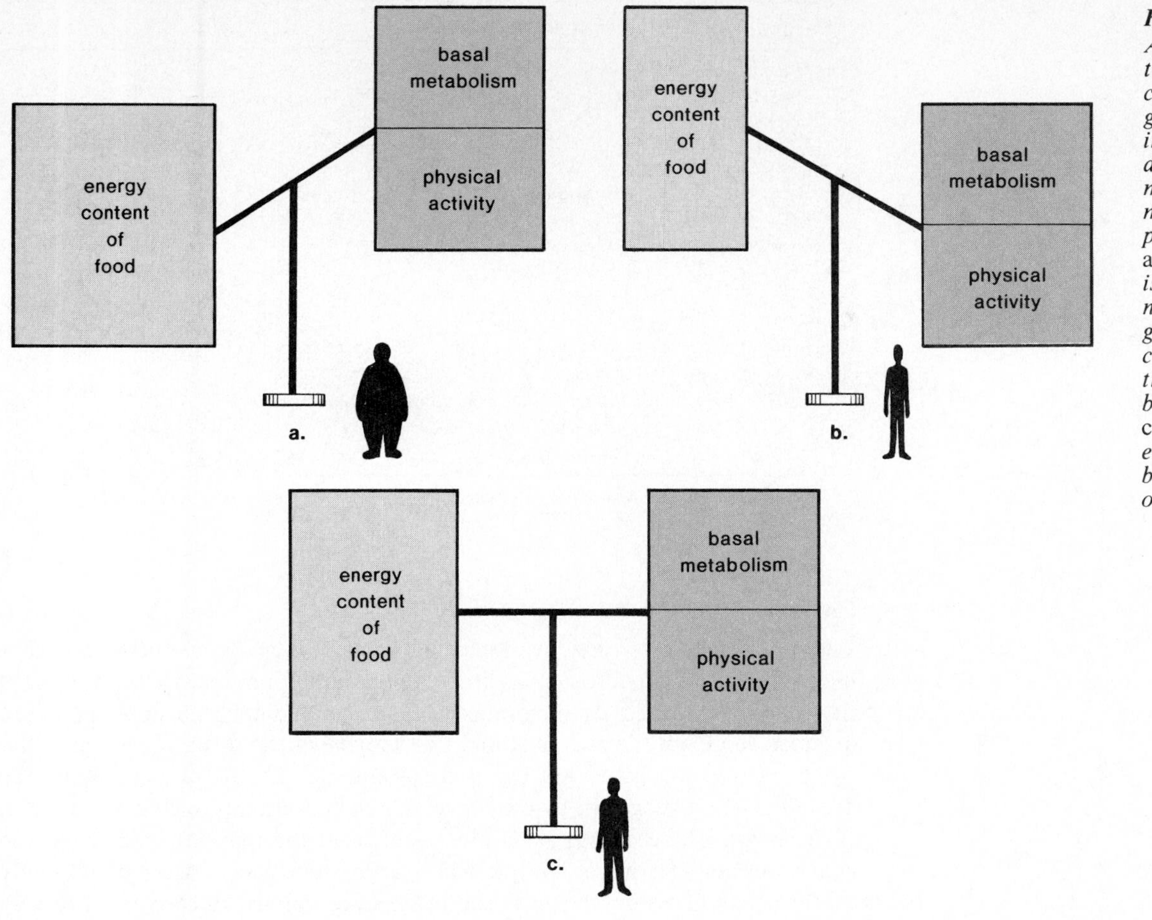

Figure 9.15
A diagram illustrating the relationship between caloric intake and weight gain or loss. In each instance, energy needs are divided between basal metabolism (basal metabolic rate) and physical activity.
a. *Energy content of food is greater than energy needs of body—weight gain occurs.* b. *Energy content of food is less than energy needs of body—weight loss occurs.* c. *Energy content of food equals energy needs of body—no weight change occurs.*

type of exercise requires more energy than the BMR as shown by the values in table 9.4 The recommended daily intake of calories for a woman 19 to 22 years of age, height 64 inches, who is only doing light exercise each day, is 2,100 C. The recommended daily intake of calories for a man 19 to 22 years of age, height 70 inches, who is only doing light exercise each day, is 2,900 C.

Table 9.4 Number of Calories Utilized by Various Actions

Kinds of Activity	Calories (per hour)*
Walking up stairs	1100
Running (a jog)	570
Swimming	500
Vigorous exercise	450
Slow walking	200
Dressing and undressing	118
Sitting at rest	100

*Includes BMR.

Dieting

Figure 9.15 indicates that a loss of weight will occur if the caloric intake is reduced while still maintaining the same level of activity. However, as discussed in the reading on page 200, dieting and/or increased exercise does not always immediately result in weight loss. Of particular interest is the theory that the body has a *"set point"* for its usual amount of fat. If the amount of stored fat falls below this point, the fat cells are believed to signal the brain by the release of a chemical substance. In response, the metabolic rate is lowered so that fewer calories (less food) are needed to stay at the same weight. At the same time, the hunger drive disproportionately increases so that the dieter feels even more inclined to eat. Also, because of the set point, a person tends immediately to regain any weight that has been lost after dieting is given up.

It is most reasonable, then, to make dieting and exercise a way of life that is always maintained. The reading makes suggestions as to the proper choice of foods. In any case, it is wise to pick a variety of foods from the four recognized food groups (table 9.5); for reasons that are discussed following.

Table 9.5 The Four Major Food Groups

1. Fruits and vegetables
2. Whole-grain and enriched breads, cereals, and other products made from grains
3. Milk, cheese, yogurt, and other products made from milk
4. Meats, poultry, fish, eggs, and legumes (dry peas and beans)

Reducing the amount of caloric intake and/or increasing the amount of exercise will eventually result in weight loss.

Table 9.6 Dietary Recommendations
The less-fat recommendations:
1. Choose as protein foods lean meat, poultry, fish, dry beans, and peas. Trim fat off before you eat.
2. Eat eggs and such organ meats as liver in moderation. (Actually, these are high in cholesterol rather than fat.)
3. Broil, boil, or bake, rather than fry.
4. Limit your intake of butter, cream, hydrogenated oils, shortenings, and coconut oil.
The less-salt recommendations:
1. Learn to enjoy unsalted food flavors.
2. Add little or no salt to foods at the table and add only small amounts of salt when you cook.
3. Limit your intake of salty prepared foods such as pickles, pretzels, and potato chips.
The less-sugar recommendations:
1. Eat less sweets such as candy, soft drinks, ice cream, and pastry.
2. Eat fresh fruit or canned fruit without heavy syrup.
3. Use less sugar—white, brown, raw—and less honey and syrups.

Sources: American Dietetic Association based on *Dietary Guidelines for Americans* 1980, U.S. Department of Agriculture, and Department of Health, Education, and Welfare.

Carbohydrates

The quickest, most easily available source of energy for the body is carbohydrates. Starchy foods, such as bread and potatoes, provide the largest quantity of carbohydrates. As mentioned previously, all dietary carbohydrates are digested to glucose, which is stored by the liver in the form of glycogen. Between eating, the blood glucose is maintained at about 0.1 percent by the breakdown of glycogen or by the conversion of amino acids to glucose (p. 192). If necessary, these amino acids are taken from the muscles, even from the heart muscle. This is how people who starve waste away. A constant supply of glucose is necessary because the brain utilizes only glucose as an energy source. Other organs can metabolize fatty acids for energy, but unfortunately this results in *acidosis,* an acid blood pH. In order to avoid this situation, it is suggested that the diet contain at least 100 grams of carbohydrate daily.[3]

A limited amount of carbohydrate is needed in the diet because the brain requires a constant supply of glucose.

Even so, foods such as candy, ice cream, sugar-coated cereals, soft drinks, and alcohol, which are rich in simple sugars, are labeled "empty calories" by some because they contribute to energy needs and weight gain without supplying any other nutritional requirements. Government agencies charged with advising the public about their dietary needs suggest that we limit our sugar intake (table 9.6).

Fats

Fats are present not only in butter and margarine but also in meat, eggs, milk, nuts, and a variety of vegetable oils. Fats from an animal origin tend to have saturated fatty acids and those from plants tend to have unsaturated fatty acids. An increase in the amount of fats in the diet can greatly increase the

[3]A slice of bread contains approximately 14 grams of carbohydrate.

number of calories consumed. The pat of butter or margarine on a potato contains almost as many calories as the potato. This is understandable when you compare the amount of calories derived from a gram of fat to the amount derived from a gram of carbohydrate (table 9.7). After being absorbed, the products of fat digestion are transported by the lymph and blood to the tissues. The liver can alter ingested fats to suit the body's needs, except it is unable to produce the fatty acids linolenic and linoleic acids. Since these are required for phospholipid production, they are considered the **essential fatty acids.** Essential molecules must be present in our food because the body is unable to manufacture them.

As discussed in the reading on page 224, dietary lipids, especially saturated fatty acids and cholesterol, have been found to cause circulatory difficulties, such as hypertension and heart attack due to hardening of the arteries. Not only does the American Heart Association and the governmental agencies mentioned earlier recommend that we limit our fat intake (table 9.6), they also suggest that in doing so we may be protecting ourselves from certain types of cancer. Statistical studies have shown a strong correlation between fat intake and the occurrence of breast, colon, and prostate cancer.

Table 9.7 Caloric Energy Release

	Calories/Gram
Carbohydrate	4.1
Fat	9.3
Protein	4.1

Fats have the highest caloric content but they cannot be avoided entirely because of the essential fatty acids.

Proteins

Foods rich in protein include meat, fish, poultry, cheese, nuts, milk, eggs, and cereals. Various legumes such as beans and peas also contain lesser amounts of protein. Following digestion of protein, amino acids enter the blood and are transported to the tissues. Some are incorporated into structural proteins and some are used to synthesize such proteins as hemoglobin, plasma proteins, enzymes, and hormones.

You will recall that protein formation requires twenty different types of amino acids. Of these, nine are required in the diet because the body is unable to produce them. These are termed the *essential amino acids.* The body produces the other amino acids by simply transforming one type into another type. Some protein sources, such as meat, are complete in the sense that they provide all the different types of amino acids. Vegetables do supply us with amino acids but they are incomplete sources because at least one of the essential amino acids is absent. However, it is possible to combine foods in order to acquire all of the essential amino acids.

Although meat is the best source of dietary protein, the method of fattening conventional cattle in feedlots where they are grain fed results in a high cholesterol, fatty meat that most certainly is not beneficial to one's health. For this reason innovative cattle ranchers are beginning to experiment with new breeds of beef-producing cattle and/or feeding techniques that produce a meat much lower in fat. Brae, Beefalo, and Chianina are among the substitute kinds of beef that are now available in certain places.

A complete source of protein is absolutely necessary to ensure a sufficient supply of the essential amino acids.

Dieting: The Losing Game

Dieting in particular causes people to purchase all sorts of specially prepared foods. Nutritionists advise that this is not necessary—simply eat a variety of wholesome foods in small quantities and exercise. Dieting should be considered a life-long project rather than a short-term one.

The U.S. population already ranks as the world's fattest. "Europeans visit Disneyland and go back home thinking we all weigh 250 lbs.," observes Cardiologist John Farquhar, of the Stanford University Medical Center. Since the Civil War, when statistics on inductees were first gathered, U.S. adults have been growing not only taller but fatter. Youngsters too have been getting heavier. About 10% to 15% of children and 20% of teens are judged to be overweight. Remarks Dr. George Bray, of the Los Angeles County—University of Southern California Medical Center: "We seem to be winning the race in the wrong direction."

Theoretically, heading the opposite way is easy. To lose a pound a week, for example, a person needs to burn 3,500 more calories than are taken in over that period. That can be done by eating less, by exercising more, or both. Says Dr. Edward Horton, of the University of Vermont College of Medicine: "For the vast majority of people, what we aim for is a negative caloric balance of 500 calories a day." Those 500 can be found without much difficulty: eliminate that 350-calorie vanilla milkshake each day and add a 150-calorie brisk half-hour walk. After one week, there should be a loss of 1 lb. It sounds simple, except that it is not, as millions attest. Researchers still cannot fully explain why there is such difficulty, but they have made dramatic progress over the past ten years puzzling out some of the mystery. Being overweight, they have found, is as much a result of physiology and heredity as it is of behavior and environment. . . .

Doctors now urge Americans to exercise regularly. That is a major change from a few decades ago, when physical activity was rarely mentioned, but the discouraging truth is that exercise works off calories very slowly. "Fat is such a concentrated source of energy that it takes a long time to burn," notes Vermont's Horton. Brisk walking uses up a mere five calories a minute, easy jogging only 15. Orlando Pizzolato of Italy won the grueling New York City Marathon last October, running 26 miles 385 yds. in 2 hr. 11 min. 34 sec. He would have lost about three-quarters of a pound of fat, based on tables devised by the President's Council on Physical Fitness and Sports.

If exercise alone is an inefficient way to lose weight, dieting by itself has proved almost equally unsuccessful. . . .

Researchers today increasingly accept the relatively new theory that the body "defends" a certain weight at certain times. These plateaus are often referred to as set points or settling points. Most experts believe that set points are the result of a number of hereditary and cultural forces. "It's becoming apparent," says Psychologist Richard Keesey of the University of Wisconsin at Madison, "that displacing someone from a set point, which is what dieting does, initiates a whole series of physical forces that fight that displacement." Furthermore, notes New York's Van Itallie, "it's very clear that if you lose weight, adaptations come into play that make it possible to regain that weight more efficiently." In short, a person's body may be the most formidable foe in losing weight. . . .

A metabolic slowdown also seems to account for at least part of the weight gained by people who stop smoking. Added pounds have popularly been assumed to be the result of increased eating. Last week, though, European researchers reported in the *New England Journal of Medicine* that when diet and exercise were stable, people burned 10% more calories in the course of a day when they smoked 24 cigarettes than when they abstained. People who quit smoking will have to eat less to maintain their weight, suggests the study. If they want to lose, they will have to cut back even more. No one proposes that smokers keep puffing to control weight, however. Says Bray: "A person would have to be 50 or 60 lbs. heavier to have the same health risks from weight that you would get from being a normal-weight smoker." . . .

Behavior, the experts insist, can override biology. Set points and genes and metabolism may make losing weight difficult, but they do not make it impossible. "In order to become successful in losing weight you have to become a thinner person in your thinking," says George Bray. "That means you have to modify your view about exercise and how you use it and about food and how you use it.". . .

Behavior-modification programs seek to do just that. To change eating habits, participants record when and what they eat, the amount of food and the calorie count, learn to avoid eating stimuli like food commercials on television, and subdue self-defeating thoughts and emotions associated with food. . . .

All reputable programs now reject fad diets, with their list of banned foods and their customary emphasis on certain food groups and exclusion of others. Instead, the experts stress sound, nutritionally balanced eating. . . .

Dieters are taught to make useful substitutions—more chicken and fish in place of beef and pork, low-sugar cereals for breakfast instead of butter-laden muffins or cheese-stuffed Danish, vegetables like cabbage and summer squash rather than lima beans and avocados. They are urged to use herbs and spices instead of salt for seasoning, lemon juice and vinegar in place of creamy salad dressings, to rely on low-fat milk, cottage cheese and yogurt instead of whole-milk products, and to drink seltzer and mineral water rather than cocktails and wine. People learn that some foods normally thought of as fattening, namely potatoes, pasta, rice and bread, are acceptable as long as they are not fried, or doused in butter or cream. Counsels Nutritionist Nestle: "The best diets—losing no more than one pound a week—take a long time to do and occur with healthy eating, a reasonable amount of calories, a great deal of exercise and a program of behavior modification. It's got to be slow and long and for a lifetime." . . .

In losing weight, mind over matter or, more to the point, mind over body, is very much within the realm of possibility. Ready, forget that set point, and go for it.

"Dieting: The Losing Game." *Time*. January 20, 1986.

Table 9.8 Vitamins and Minerals—Their Best Food Sources

Fat-soluble vitamins	Best food sources
A	Liver, milk and milk products, sweet potatoes, carrots, spinach, cantaloupe, squash, broccoli, apricots.
D	Sun on skin, fish-liver oils, sardines, salmon, milk and milk products, egg yolk.
E	Vegetable oils, margarine.
K	Green vegetables, tomato.
Water-soluble vitamins	
C	Citrus, strawberry, and other fruits, broccoli, potatoes.
Thiamin (B-1)	Liver, pork products, peas, legumes, whole grain and enriched breads and cereals.
Riboflavin (B-2)	Liver, milk, beef, pork, chicken, eggs, cheese, broccoli, salmon, whole grain and enriched breads and cereals.
Niacin (B-3)	Enriched flour, red meat, poultry, peanut butter, whole grain and enriched breads and cereals.
Pyridoxine (B-6)	Liver, pork, red meat, whole grains, vegetables.
Folacin	Liver, kidney, yeast, green leafy vegetables.
B-12	Organ meats, eggs, fish, milk.
Biotin	Yeast, organ meats, egg yolk, whole grains.
Pantothenic acid	Organ meats, red meat, eggs.
Inositol	Liver, yeast, cereals, fruit.
Choline	Egg yolk, cereals, vegetables.
Minerals	
Calcium	Milk and milk products, salmon, legumes, broccoli, oranges, sweet potatoes, lettuce.
Phosphorus	Milk and milk products, whole grains.
Magnesium	Seeds, nuts, whole grains, milk, eggs, fish, red meat.
Iron	Liver, red meat, chicken, legumes, spinach, peas, prune, apricot, tomato juice, whole grain enriched breads and cereals.
Zinc	Meat, liver, eggs, oysters.
Iodine	Seafood, iodized salt.
Copper	Oysters, nuts, organ meats, corn oil, legumes, raisins.
Manganese	Nuts, whole grains.
Fluoride	Drinking water, fish, legumes.
Chromium	Yeast, red meat, cheese, whole grains.
Selenium	Seafood, organ meats, red meat.
Molybdenum	Meat, grain, legumes.
Sodium	Salt, cured meats, and vegetables.
Potassium	Bananas, citrus fruits.
Chloride	Salt.
Sulfur	Protein foods.
Cobalt	Red meat.

Adapted from a *Time* Magazine Special Advertising section on "Eating Well, Looking Fit, Feeling Better."

Vitamins and Minerals

Vitamins are organic compounds (other than carbohydrates, lipids, and proteins) that the body is unable to produce and therefore must be present in the diet. Table 9.8 lists some of the more important vitamins and minerals and the best sources of food for each one. Various symptoms develop when vitamins are lacking in the diet (fig. 9.16).

Although vitamins are an important part of a balanced diet, they are required only in very small amounts. As was discussed in chapter 2, many vitamins are portions of *coenzymes,* or enzyme helpers. For example, niacin is part of the coenzyme NAD, and riboflavin is part of another dehydrogenase, FAD. Coenzymes are needed in only small amounts because each one can be used over and over again. This means that the daily requirement for vitamins is relatively low and a properly balanced diet usually provides the amount needed. For this reason it is important to include fresh fruits and vegetables in the diet. The National Academy of Sciences suggests that the consumption of oranges, broccoli, and tomatoes which are high in vitamin C, and squash,

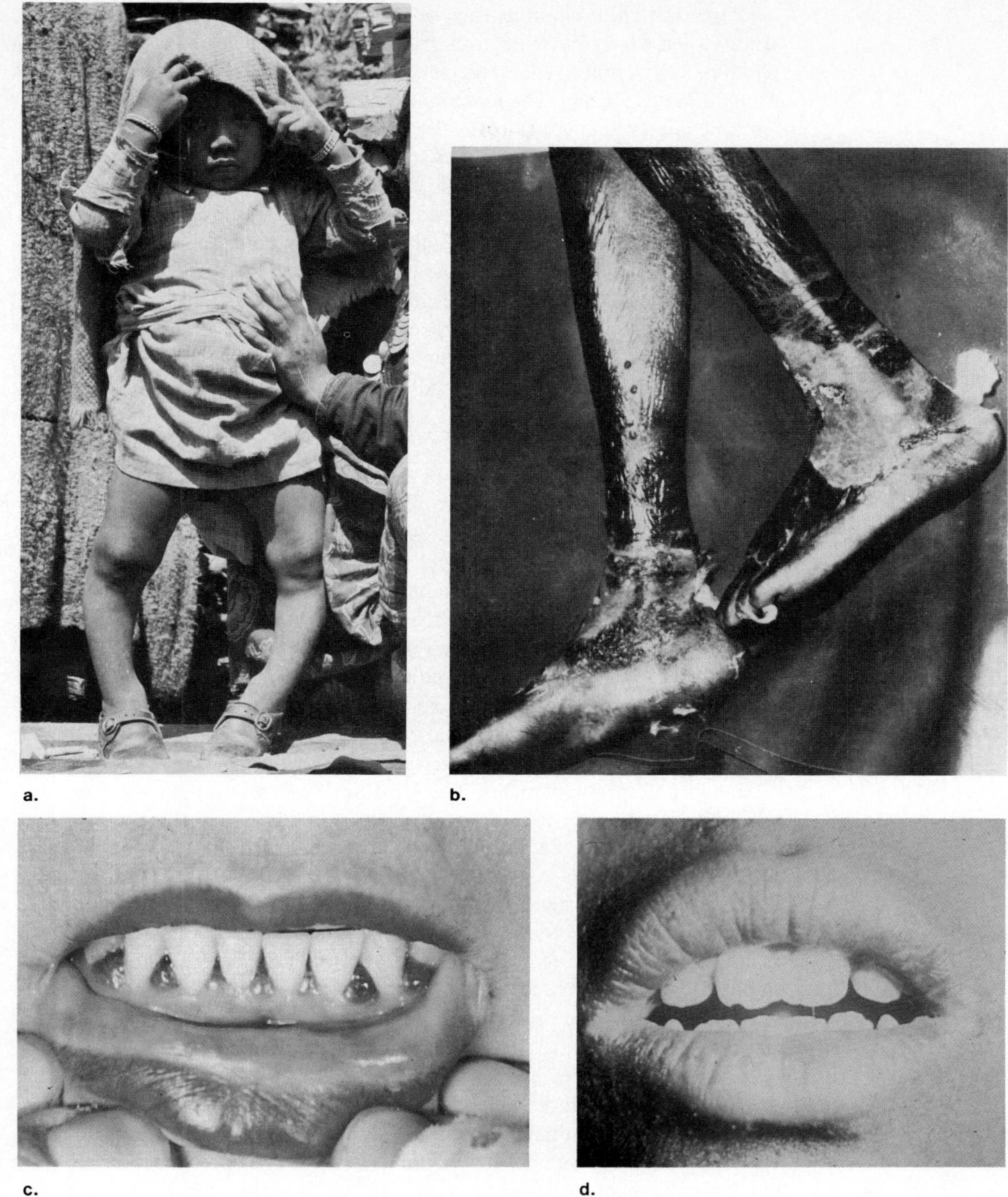

carrots, and other yellow and green vegetables which are high in vitamin A may even be a protection against the development of cancer. Nevertheless they discourage the intake of excess vitamins because it can possibly lead to illness. For example, excess vitamin C is converted to a product, oxalic acid, that is toxic to the body. Vitamin A taken in excess over long periods can cause loss of hair, bone and joint pains, and loss of appetite. Excessive vitamin D can cause an overload of calcium in the blood, which in children leads to loss of appetite and retarded growth. Megavitamin therapy should always be supervised by a physician.

Figure 9.16
Illnesses due to vitamin deficiency. a. *Bowing of bones (rickets) due to vitamin D deficiency.* b. *Dermatitis of areas exposed to light (pellagra) due to niacin deficiency.* c. *Bleeding of gums (scurvy) due to vitamin C deficiency.* d. *Fissures of lips (cheilosis) due to riboflavin deficiency.*

A properly balanced diet includes all the vitamins needed to maintain health.

In addition to vitamins, various **minerals** are also required by the body. Minerals are divided into macronutrients which are needed in gram amounts per day and micronutrients (trace elements) which are needed in only microgram amounts per day. The macronutrients sodium, magnesium, phosphorus, chlorine, potassium, and calcium serve as constituents of cells and body fluids and as structural components of tissues. For example, calcium is needed for the construction of bones and teeth and also for nerve conduction and muscle contraction. The micronutrients seem to have very specific functions. For example, iron is needed for the production of hemoglobin, and iodine is used in the production of thyroxin, a hormone produced by the thyroid gland. As research continues, more and more elements have been added to the list of those considered essential. During the past three decades, molybdenum, selenium, chromium, nickel, vanadium, silicon, and even arsenic have been found to be essential to good health in very small amounts.

Food Preparation

As the American public has come to depend more and more on processed and convenience foods, the use of *food additives* has increased until now it is estimated that 26,000 different chemicals are deliberately added to foods. Some of these are natural substances, such as salt and spices, but many others are artificial, such as synthetically produced preservatives, sweeteners, and dyes. All additives must be approved by the Food and Drug Administration (FDA), which is responsible for determining the safety and effectiveness of such products. The FDA has taken some additives off the market because they have been found to produce cancer in animals. About half of the food additives banned by the FDA have been food colorings made from coal-tar dye.

It has been very difficult to establish a standardized method by which to deal with suspect chemicals. The Food Safety Council, a private industry/consumer group, after meeting for six years, finally issued some suggestions in 1982. The council felt that tests should be run for each chemical in order to establish a dose that was most likely safe for that chemical; then a safety factor should be subtracted from this in order to determine the maximum dose considered acceptable for human exposure. The council acknowledged that it would sometimes be necessary to make a risk-benefit determination. For example, there is evidence to suggest that some preservatives are not altogether safe and yet their use protects the public from very dangerous bacterial infections. In such instances it may be more beneficial in the long run to continue to use the preservative.

There is one additive which people themselves add to food that they might be well advised to eliminate or reduce. This additive is salt, or sodium chloride. The nutritional requirement for sodium is in the range of .1–.2 grams per day, which is equivalent to .25–.5 grams of salt per day. The average intake of salt is twenty times the nutritional requirement. There is evidence to suggest that excess salt in the diet promotes high blood pressure in susceptible individuals. Hypertension is absent in some nonindustrialized populations, such as those of the Solomon Islands, the Amazon basin, and the Coco Islands of Polynesia where the salt intake is about 2 grams per day. Americans receive about 3 grams of salt per day even if they add no salt at all to their food either during cooking or at the table; this is due to modern methods of commercial food processing. It is recommended that we try to eliminate salty foods from our diets (table 9.6).

Summary

In the mouth, food is chewed and starch is acted upon by salivary amylase. After swallowing, peristaltic action moves the food along the esophagus to the stomach. Here pepsin, in the presence of HCl, acts on protein. In contrast, the small intestine has an alkaline environment. Here, fat is emulsified by bile salts to fat droplets before being acted upon by pancreatic lipase. Protein is digested by pancreatic trypsin, and starch is digested by pancreatic amylase. The intestinal wall contains digestive glands that secrete intestinal enzymes to finish the digestion of proteins and carbohydrates. The action of digestive enzymes is summarized in tables 9.2 and 9.3. Only nondigestible material passes from the small intestine to the large intestine. The large intestine absorbs water from this material and contains a large population of bacteria that can use it as food. In the process, the bacteria produce vitamins that can be absorbed and used by our bodies.

The walls of the small intestine have fingerlike projections called villi within which are blood capillaries and a lymphatic lacteal. Amino acids and glucose enter the blood; glycerol and fatty acids enter the lymph. The blood from the small intestine moves into the hepatic portal vein, which goes to the liver, an organ that monitors and contributes to blood composition.

A balanced diet is required for good health. Food should provide us with all necessary vitamins, amino acids, fatty acids, and an adequate amount of energy. If the caloric value of food consumed is greater than that needed for bodily functions and activity, weight gain will occur.

Objective Questions

1. In the mouth, salivary __________ digests starch to __________ .
2. When swallowing, the __________ covers the opening to the larynx.
3. The __________ takes food to the stomach where __________ is primarily digested.
4. The gallbladder stores __________ , a substance that __________ fat.
5. The pancreas sends digestive juices to the __________ , the first part of the small intestine.
6. Pancreatic juice contains __________ for digesting protein, __________ for digesting starch, and __________ for digesting fat.
7. Whereas pepsin prefers a __________ pH, the enzymes found in pancreatic juice prefer a __________ pH.
8. The products of digestion are absorbed into the cells of the __________ , fingerlike projections of the intestinal wall.
9. After eating, the liver stores glucose as __________ .
10. The diet should include a complete protein source, one that includes all the __________ .

Answers to Objective Questions

1. amylase, maltose 2. epiglottis 3. esophagus, protein 4. bile, emulsifies 5. duodenum 6. trypsin, pancreatic amylase, lipase 7. strongly acid, slightly basic 8. villi 9. glycogen 10. essential amino acids

Study Questions

1. List the parts of the digestive tract, anatomically describe them, and state the contribution of each to the digestive process. (pp. 182–191)
2. List the accessory glands and describe the part that they play in the digestion of food. (p. 188)
3. Discuss the absorption of the products of digestion into the circulatory system. (p. 191)
4. List six functions of the liver. How does the liver maintain a constant glucose level in the blood? (p. 192)
5. What is jaundice? Cirrhosis of the liver? (pp. 192–193)
6. What is the common intestinal bacterium? What do these bacteria do for us? (p. 193)
7. What are gastrin, secretin, and CCK? Where are they produced? What are their functions? (p. 190)
8. Discuss the digestion of starch, protein, and fat, listing all the steps that occur to bring about digestion of each of these. (p. 195)
9. What factors determine how many calories should be ingested? (pp. 196–197)
10. Give reasons why carbohydrates, fats, proteins, vitamins, and minerals are all necessary to good nutrition. (pp. 198–204)

Key Terms

amylase (am'i-lās) a starch-digesting enzyme secreted by the salivary glands (salivary amylase) and the pancreas (pancreatic amylase). *185*

CCK cholecystokinin; hormone produced by the duodenum that stimulates release of bile from gallbladder. *190*

colon (ko'lon) the large intestine of vertebrates. *193*

epiglottis (ep''ĭ-glot'is) a structure that covers the glottis during the process of swallowing. *185*

esophagus (ĕ-sof'ah-gus) a tube that transports food from the mouth to the stomach. *186*

gastric gland (gas'trik gland) gland within the stomach wall that secretes gastric juice. *186*

gastrin (gas'trin) a hormone secreted by stomach cells to regulate the release of pepsin by the stomach wall. *190*

glottis (glot'is) slitlike opening between the vocal cords. *185*

hard palate (hard pal'at) anterior portion of the roof of the mouth which contains several bones. *184*

hydrolytic enzyme (hi-dro-lit'ik en'zīm) an enzyme that catalyzes a reaction in which the substrate is broken down with the addition of water. *185*

lipase (li'-pās) a fat-digesting enzyme secreted by the pancreas. *188*

lumen (lu'men) the cavity inside any tubular structure, such as the lumen of the gut. *186*

pepsin (pep'sin) a protein-digesting enzyme secreted by gastric glands. *187*

peristalsis (per''ĭ-stal'sis) a rhythmical contraction that serves to move the contents along in tubular organs such as the digestive tract. *186*

pharynx (far'ingks) a common passageway (throat) for both food intake and air movement. *185*

salivary gland (sal'ĭ-ver-e gland) a gland associated with the mouth that secretes saliva. *184*

secretin (se-kre'tin) hormone secreted by the small intestine that stimulates the release of pancreatic juice. *190*

soft palate (soft pal'at) entirely muscular posterior portion of the roof of the mouth. *184*

sphincter (sfingk'ter) a muscle that surrounds a tube and closes or opens the tube by contracting and relaxing. *186*

trypsin (trip'sin) a protein-digesting enzyme secreted by the pancreas. *188*

villi (vil'i) fingerlike projections that line the small intestine and function in absorption. *191*

vitamin (vi'tah-min) essential requirement in the diet, needed in small amounts, that is often a part of coenzymes. *202*

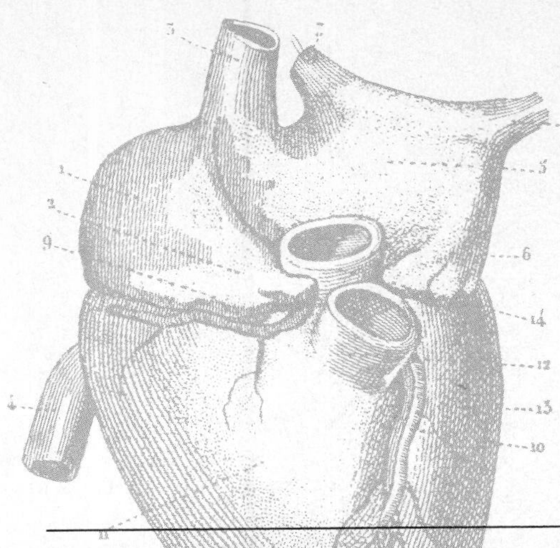

10

Circulation

Chapter Concepts

1 In human beings, the blood, kept in motion by the pumping of the heart, circulates through a series of vessels.

2 The heart is actually a double pump: the right side pumps blood to the lungs and the left to the rest of the body.

3 Although the circulatory system is very efficient, it is still subject to various degenerative illnesses.

4 The lymph vessels form a one-way lymphatic system that transports lymph from the tissues to certain cardiovascular veins.

Chapter Outline

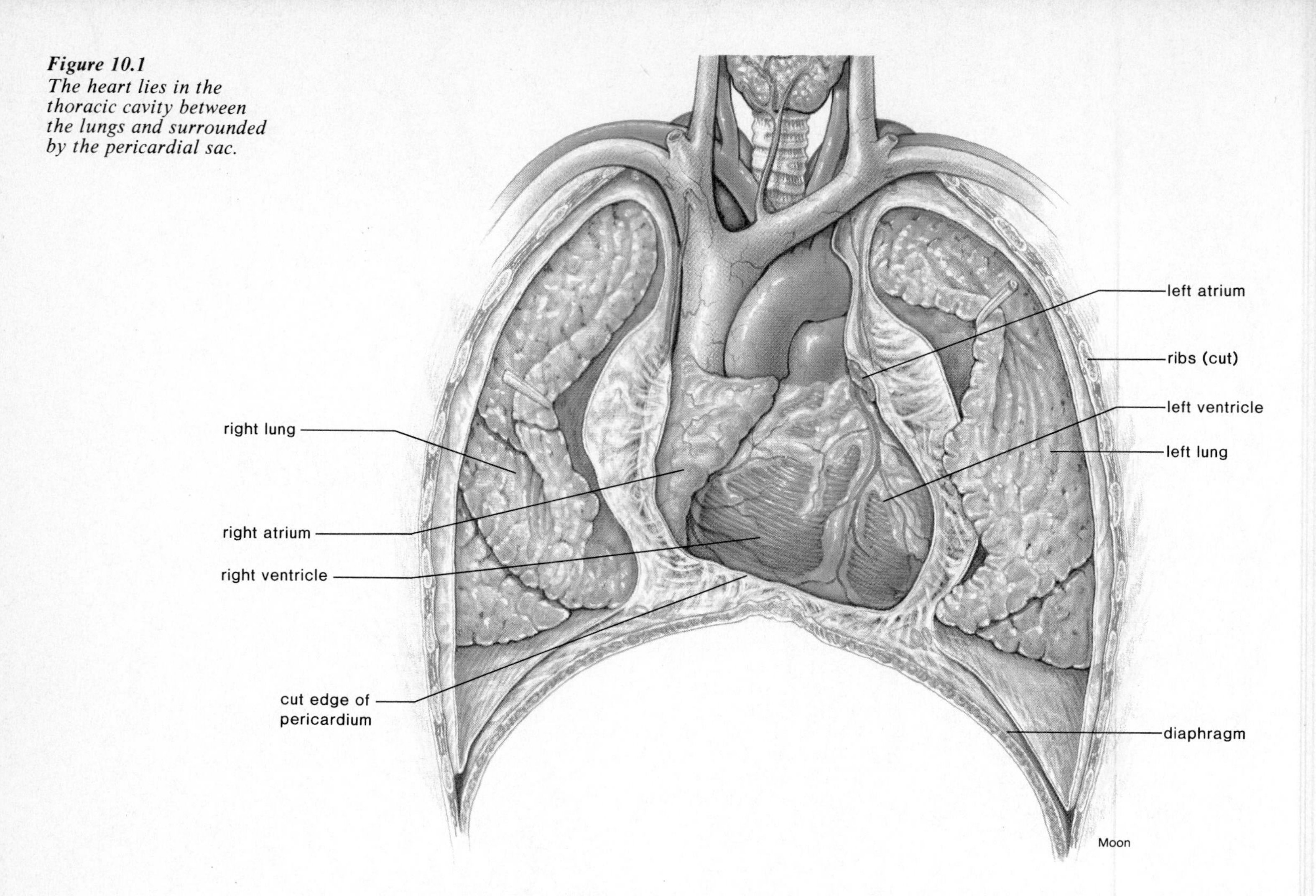

Figure 10.1
The heart lies in the thoracic cavity between the lungs and surrounded by the pericardial sac.

Single-celled organisms do not have need of a circulatory system. Their watery environment brings them their food and removes their wastes. But most of our trillions of cells are far removed from the external environment and need to be serviced. It is the circulatory system that brings them their daily supply of nutrients, such as amino acids and glucose, and takes away their wastes, such as carbon dioxide and ammonia. At the center of the system is the heart (fig. 10.1) which keeps the blood moving along its predetermined circular path. Circulation of the blood is so important that if the heart discontinues beating for only a few minutes, death will result.

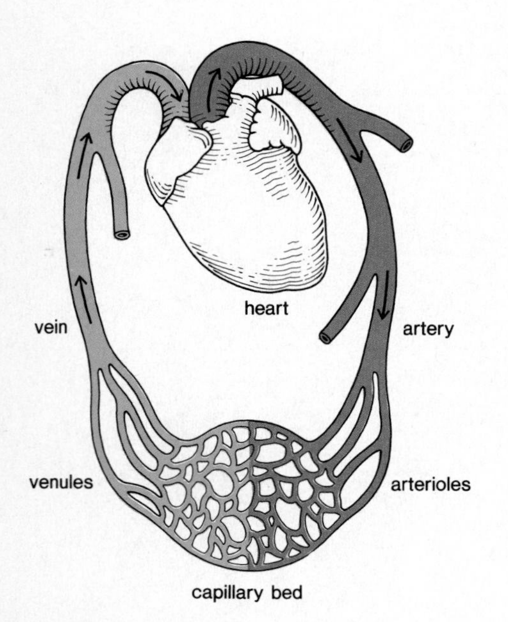

Figure 10.2
Diagram illustrating the path of blood. Blood leaving the heart moves from an artery to arterioles to capillaries to venules and then returns to the heart by way of a vein. Therefore, arteries are vessels that take blood away from the heart and veins are vessels that return blood to the heart.

Circulatory System

Blood Vessels

The blood vessels are arranged so that they continually carry blood from the heart to the tissues and then return it from the tissues to the heart. Blood vessels (fig. 10.2) are of three types: the **arteries** (and **arterioles**) carry blood away from the heart; the **capillaries** exchange material with the tissues; and the **veins** (and **venules**) return blood to the heart.

Arteries and Arterioles

Arteries have thick walls (fig. 10.3) because, in addition to an inner endothelium layer and an outer connective tissue layer, they have a thick middle layer of elastic and muscle fibers. The elastic fibers enable an artery to expand and accommodate the sudden increase in blood volume that results after each heartbeat. Arterial walls are so thick that the walls themselves are supplied with blood vessels. The *arterioles* are small arteries just visible to the naked

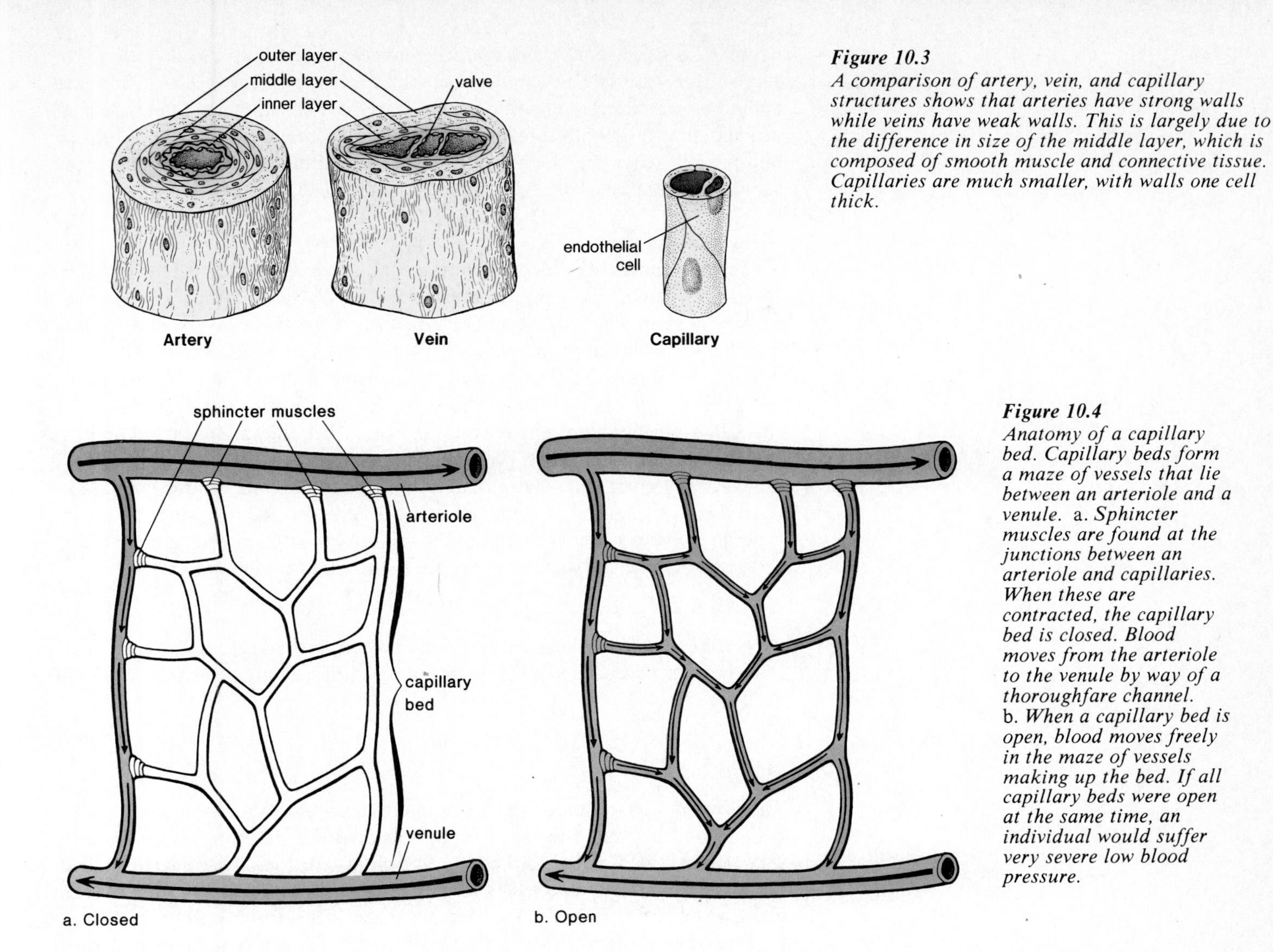

Figure 10.3
A comparison of artery, vein, and capillary structures shows that arteries have strong walls while veins have weak walls. This is largely due to the difference in size of the middle layer, which is composed of smooth muscle and connective tissue. Capillaries are much smaller, with walls one cell thick.

Figure 10.4
Anatomy of a capillary bed. Capillary beds form a maze of vessels that lie between an arteriole and a venule. a. *Sphincter muscles are found at the junctions between an arteriole and capillaries. When these are contracted, the capillary bed is closed. Blood moves from the arteriole to the venule by way of a thoroughfare channel.* b. *When a capillary bed is open, blood moves freely in the maze of vessels making up the bed. If all capillary beds were open at the same time, an individual would suffer very severe low blood pressure.*

eye. The middle layer of these vessels has some elastic tissue but is composed mostly of smooth muscle whose fibers encircle the arteriole. The contraction of the smooth muscle cells is under the control of an involuntary portion of the autonomic nervous system. If the muscle fibers contract, the bore of the arteriole gets smaller; if the fibers relax, the bore of the arteriole enlarges. Whether arterioles are constricted or dilated affects blood pressure. The greater the number of vessels dilated, the lower the blood pressure.

Capillaries

Arterioles branch into small vessels called capillaries. Each one is an extremely narrow, microscopic tube with a wall composed of only one layer of endothelial cells (fig. 10.3). *Capillary beds* (a network of many capillaries) are present in all regions of the body; consequently, a cut to any body tissue draws blood. The capillaries are the most important part of a closed circulatory system because an exchange of nutrient and waste molecules takes place across their thin walls. Oxygen and glucose diffuse out of a capillary into the tissue fluid that surrounds cells, and carbon dioxide and ammonia diffuse into the capillary (fig. 11.14). Since it is the capillaries that serve the needs of the cells, the heart and other vessels of the circulatory system can be thought of as a means by which blood is conducted to and from the capillaries.

Not all capillary beds (fig. 10.4) are open or in use at the same time. After eating, the capillary beds of the digestive tract are usually open; during muscular exercise, the capillary beds of the skeletal muscles are open. Most capillary beds have a "thoroughfare channel" that allows blood to move directly from arteriole to venule when the capillary bed is closed. There are sphincter muscles that encircle the entrance to each capillary. These are constricted, preventing blood from entering the capillaries, when the bed is closed and are relaxed when the bed is open. As would be expected, the larger the number of capillary beds open, the lower the blood pressure.

Veins and Venules

Veins and venules take blood from the capillary beds to the heart (fig. 10.2). First, the *venules* drain the blood from the capillaries and then join together to form a vein. The wall of a venule has the same three layers as an artery but the wall is much thinner than that of an artery because the middle layer of muscle and elastic fibers is poorly developed (fig. 10.3). Within some veins, especially in the major veins of the arms and legs, there are **valves** (fig. 10.3) that allow blood to flow only toward the heart when they are open and prevent the backward flow of blood when they are closed.

At any given time, more than half of the total blood volume is found in the veins and venules. If a loss of blood occurs, for example due to hemorrhaging, nervous stimulation causes the veins to constrict, providing more blood to the rest of the body. In this way, the veins act as a blood reservoir.

---

*A*rteries and arterioles carry blood away from the heart; veins and venules carry blood to the heart and capillaries join arterioles to venules.

Heart

The **heart** is a cone-shaped (fig. 10.5) muscular organ, about the size of a fist. It is located between the lungs, directly behind the sternum, and is tilted so that the apex is directed to the left. The major portion of the heart is called the **myocardium,** which consists largely of cardiac muscle tissue. The muscle fibers within the myocardium are branched and joined to one another so tightly that, prior to studies with the electron microscope, it was thought that they formed one continuous muscle. Now it is known that there are individual fibers. The inner surface of the heart is lined with endothelial tissue called *endocardium,* which resembles squamous epithelium. The outside of the heart is covered with an epithelial and fibrous tissue called *pericardium,* which forms a sac called the pericardial sac, within which the heart is located. Normally, this sac contains a small quantity of liquid to lubricate the heart.

Internally (fig. 10.6), the heart has a right and left side, separated by the **septum.** The heart has four chambers: two upper, thin-walled **atria** (singular, atrium), sometimes called auricles, and two lower, thick-walled **ventricles.** The atria are much smaller than the strong, muscular ventricles.

The heart also has valves that direct the flow of blood and prevent a backflow. The valves that lie between the atria and ventricles are called the **atrioventricular valves.** The valves are supported by strong fibrous strings called chordae tendineae. These cords, which are attached to muscular projections of the ventricular walls, support the valves and prevent them from inverting. The atrioventricular valve on the right side is called the tricuspid valve because it has three cusps, or flaps; and the valve on the left side is called the bicuspid, or mitral, because it has two flaps. There are also **semilunar valves,** which resemble half moons, between the ventricles and their attached vessels.

---

*H*umans have a four-chambered heart (two atria and two ventricles) in which the right side is separated from the left by a septum.

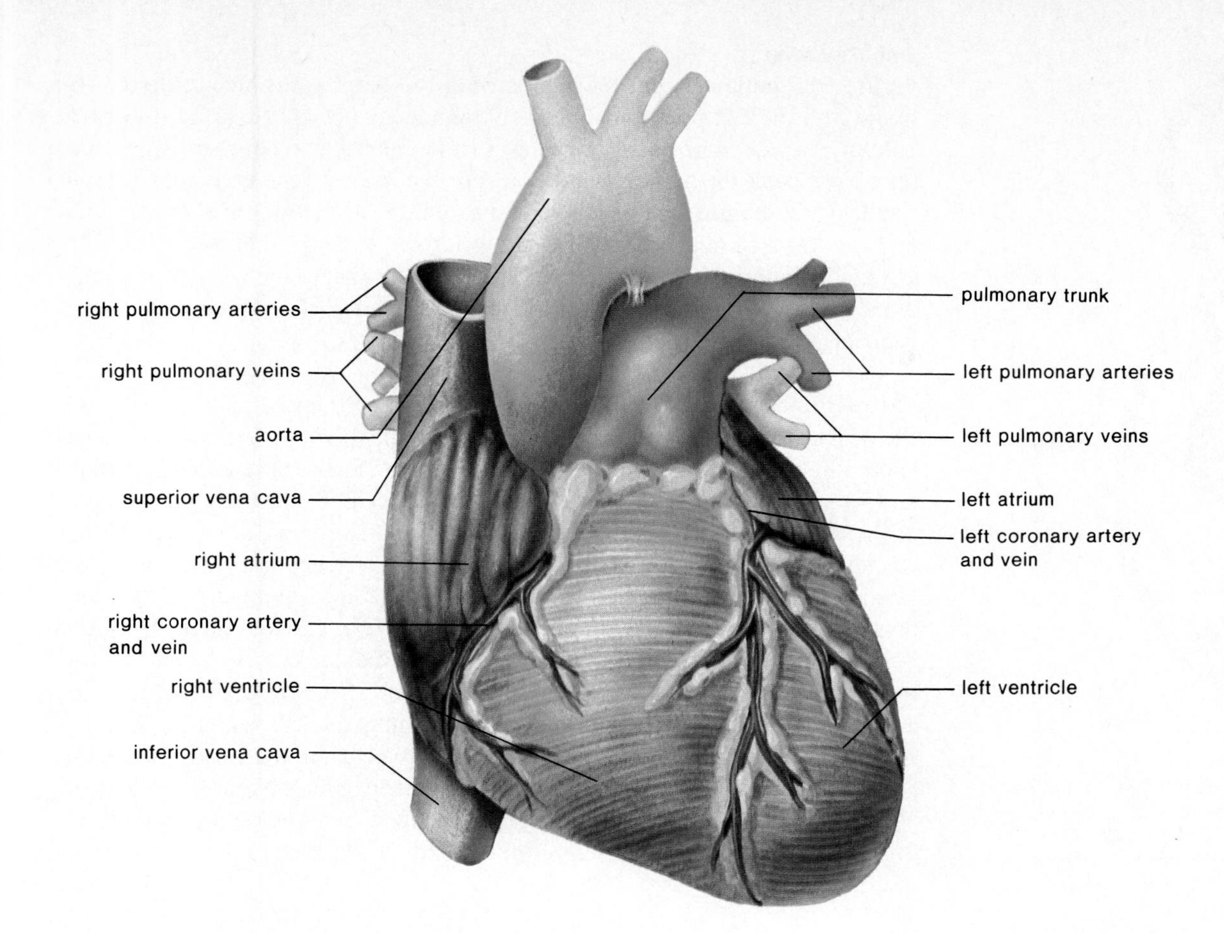

Figure 10.5
External heart anatomy. The coronary arteries bring oxygen and nutrients to the heart muscle. The individual suffers a heart attack should they fail to do so.

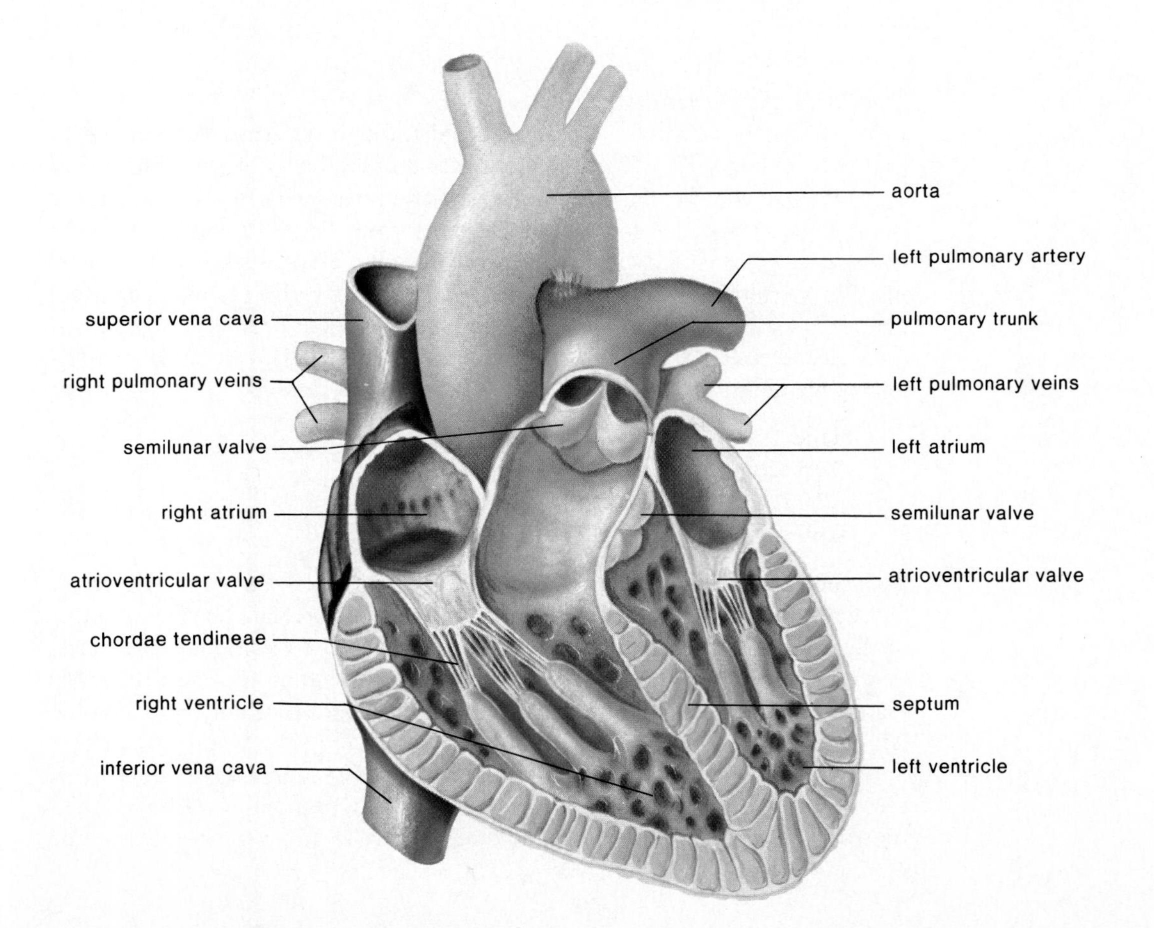

Figure 10.6
Internal view of heart. The chordae tendineae support the atrioventricular valves and keep them from inverting during ventricle contraction.

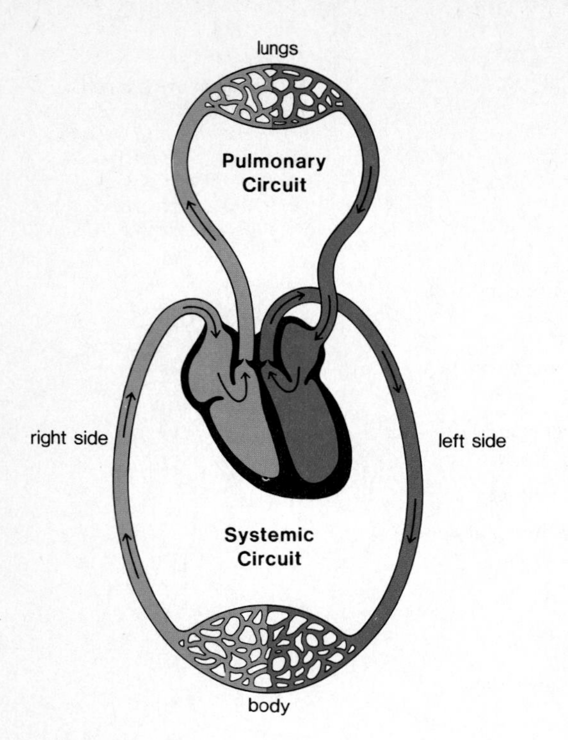

Figure 10.7
Diagram of pulmonary and systemic circuits. The blue-colored vessels carry deoxygenated blood, while the red-colored vessels carry oxygenated blood. Notice that the blood cannot move from the right side of the heart to the left side without passing through the lungs.

Double Pump

Figure 10.7 indicates that the right side of the heart sends blood through the lungs, and the left side sends blood throughout the body. Therefore there are actually two circular paths (circuits) of the blood: (1) from the heart to the lungs and back to the heart and (2) from the heart to the body and back to the heart. The right side of the heart is a pump for the first of these circuits, and the left side of the heart is a pump for the second. Therefore, the heart is a double pump. Since the left ventricle has the harder job because it pumps blood to the entire body, its walls are much thicker than those of the right ventricle.

Path of Blood in the Heart It is possible to trace the path of blood through the heart in the following manner (figs. 10.6 and 10.7). Blood low in oxygen and high in carbon dioxide enters the right atrium from the **superior** (anterior) and **inferior** (posterior) **venae cavae,** the largest veins in the body. Contraction of the right atrium forces the blood through an atrioventricular valve, the tricuspid valve, to the right ventricle. The right ventricle pumps it through the pulmonary semilunar valve into the pulmonary trunk. The pulmonary trunk divides into the **pulmonary arteries,** which take blood to the lungs. From the lungs, blood high in oxygen and low in carbon dioxide enters the left atrium from the **pulmonary veins.** Contraction of the left atrium forces blood through an atrioventricular valve, the bicuspid valve, into the left ventricle. The left ventricle pumps it through the aortic semilunar valve into the **aorta,** the largest artery in the body. The aorta sends blood to all body tissues. Notice that oxygen-poor blood never mixes with oxygen-rich blood and that blood must pass through the lungs before entering the left side of the heart.

The right side of the heart pumps blood to the lungs and the left side pumps blood to the tissues.

Heartbeat and Heart Sounds

From this description of the path of blood through the heart, it might seem that the right and left side of the heart beat independently of one another, but actually they contract together. First, the two atria contract simultaneously; then the two ventricles contract at the same time. The word **systole** refers to contraction of heart muscle, and the word **diastole** refers to relaxation of heart muscle; therefore, atrial systole is followed by ventricular systole. The heart contracts, or beats, about 70 times a minute and each heartbeat lasts about 0.85 second. Each heartbeat, or *cardiac cycle* (fig. 10.8), consists of the following elements:

Time	Atria	Ventricles
0.15 sec.	Systole	Diastole
0.30 sec.	Diastole	Systole
0.40 sec.	Diastole	Diastole

This shows that while the atria contract, the ventricles relax, and vice versa, and that all chambers rest at the same time for 0.40 second. The short systole of the atria is appropriate since the atria send blood only into the ventricles. It is the muscular ventricles that actually pump blood out into the circulatory system proper. When the word systole is used alone, it usually refers to the left ventricular systole.

When the heart beats, the familiar lub-DUPP sound may be heard as the valves of the heart close. The lub is caused by vibrations of the heart when the atrioventricular valves close, and the DUPP is heard when vibrations occur

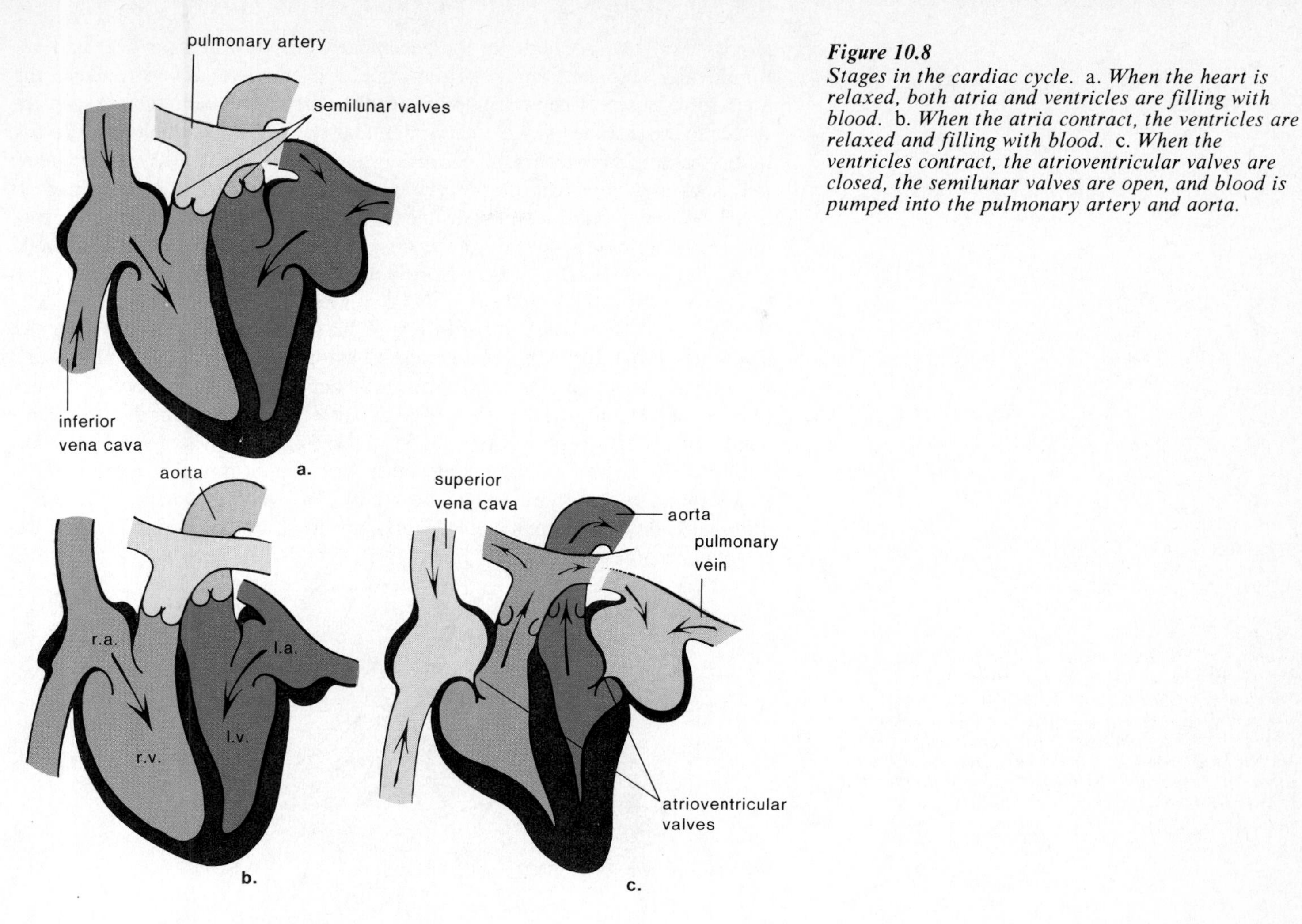

Figure 10.8
Stages in the cardiac cycle. a. *When the heart is relaxed, both atria and ventricles are filling with blood.* b. *When the atria contract, the ventricles are relaxed and filling with blood.* c. *When the ventricles contract, the atrioventricular valves are closed, the semilunar valves are open, and blood is pumped into the pulmonary artery and aorta.*

due to the closing of the semilunar valves (fig. 10.9). Heart murmurs, or a slight slush sound after the lub, are often due to ineffective valves that allow blood to pass back into the atria after the atrioventricular valves have closed. Rheumatic fever resulting from a strep infection is one cause of a faulty valve, particularly the bicuspid valve. If operative procedures are unable to restructure the valve, it may be replaced by an artificial valve.

The heartbeat is divided into two phases, first the atria contract and then the ventricles contract. When the atria are in systole the ventricles are in diastole and vice versa. The heart sounds are due to the closing of the heart valves.

Figure 10.9
Close-up view of closed semilunar valves. Refer to figure 10.6 for their anatomical position in the heart.

The beat of the heart is *intrinsic,* meaning the heart will beat independently of any nervous stimulation. In fact, it is possible to remove the heart of a small animal, such as a frog's heart, and watch it undergo contraction in a petri dish. The reason for this lies in the fact that there is a unique type of tissue called nodal tissue, with both muscular and nervous characteristics, located in two regions of the heart. The first of these, the **SA (sinoatrial) node,** is found in the upper dorsal wall of the right atrium; the other, the **AV (atrioventricular) node,** is found in the base of the right atrium very near the septum (fig. 10.10). The SA node, or the **pacemaker,** initiates the heartbeat and automatically sends out an excitation impulse every 0.85 second to cause the atria to contract. When the impulse reaches the AV node, it signals the ventricles to contract by way of specialized fibers called Purkinje fibers. The SA node is called the pacemaker because it usually keeps the heartbeat regular. If the SA node fails to work properly, the heart will still beat, but irregularly. To correct this condition, it is possible to implant in the body an artificial pacemaker that automatically gives an electric shock to the heart every 0.85 second. This causes the heart to beat regularly again.

The rate of the heartbeat is also under nervous control. There is a heart-rate center in the medulla oblongata (p. 322) of the brain which can alter the beat of the heart by way of the autonomic nervous system (p. 319). This system is made up of two divisions: the parasympathetic, which promotes those functions we tend to associate with normal activities, and the sympathetic system, which brings about those responses we associate with times of stress. For example, the parasympathetic system causes the heartbeat to slow down and the sympathetic system increases the heartbeat. Various factors, such as the relative need for oxygen or the blood pressure level, determine which of these systems becomes activated.

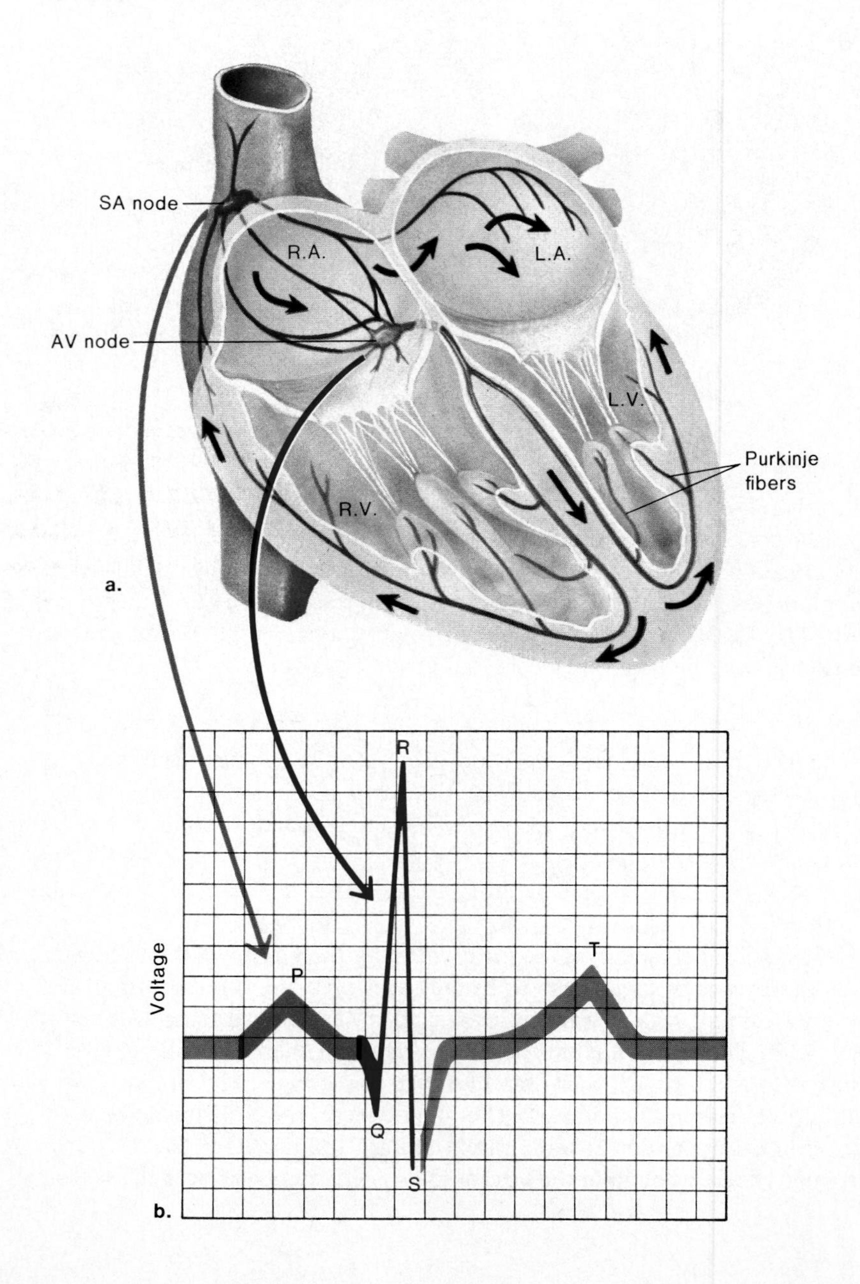

Figure 10.10
Control of the heart cycle. a. *The* SA *node sends out a stimulus that causes the atria to contract. When this stimulus reaches the* AV *node it signals the ventricles to contract by way of the Purkinje fibers.* b. *A normal EKG indicates that the heart is functioning properly. The* P *wave indicates that the atria have contracted; the* QRS *wave indicates that the ventricles have contracted; and the* T *wave indicates that the ventricles are recovering from contraction.*

The SA node is the natural pacemaker that keeps the heart beating regularly.

Vascular Pathways

The cardiovascular system, which is represented in figure 10.11, includes two circuits: the **pulmonary circuit,** which circulates blood through the lungs, and the **systemic circuit** which serves the needs of the body's tissues.

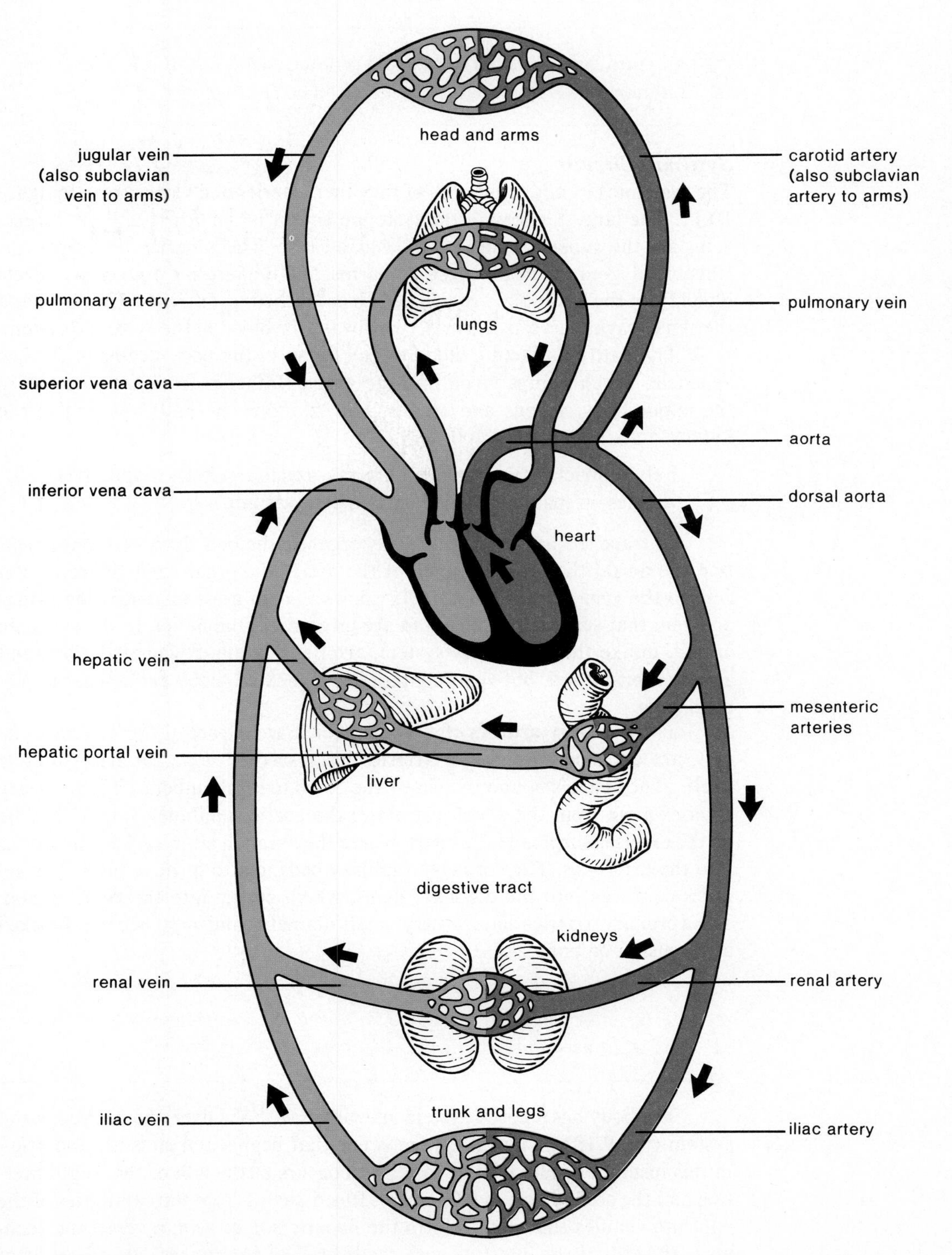

Figure 10.11
Blood vessels in the pulmonary and systemic circuits. The blue-colored vessels carry deoxygenated blood and the red-colored vessels carry oxygenated blood; the arrows indicate the flow of blood. Compare this diagram, useful for learning to trace the path of blood, to figure 10.12 in order to realize that both arteries and veins go to all parts of the body.

Pulmonary Circuit

The path of blood through the lungs can be traced as follows. Blood from all regions of the body first collects in the right atrium and then passes into the right ventricle, which pumps it into the pulmonary trunk. The pulmonary trunk divides into the *pulmonary arteries,* which divide up into the arterioles of the lungs. The arterioles take blood to the pulmonary capillaries, where carbon dioxide and oxygen are exchanged. The blood then enters the pulmonary venules that lead back through the *pulmonary veins* to the left atrium. Since the blood in the pulmonary arteries is low in oxygen but the blood in the pulmonary veins is high in oxygen, it is not correct to say that all arteries carry blood high in oxygen and all veins carry blood low in oxygen. It is just the reverse in the pulmonary circuit.

The pulmonary arteries take deoxygenated blood to the lungs, and the pulmonary veins return oxygenated blood to the heart.

Systemic Circuit

The systemic circuit includes all of the other arteries and veins shown in figure 10.11. The largest artery in the systemic circuit is the *aorta,* and the largest veins are the *superior* and *inferior* venae cavae. The superior vena cava collects blood from the head, chest, and arms, and the inferior vena cava collects blood from the lower body regions. Both enter the right atrium. The aorta and the venae cavae serve as the major pathways for blood in the systemic system.

The path of systemic blood to any organ in the body begins in the left ventricle, which pumps blood into the aorta. Branches from the aorta go to the major body regions and organs. For example, the path of blood to the kidneys may be traced as follows:

> Left ventricle—aorta—renal artery—renal arterioles, capillaries, venules—renal vein—vena cava—right atrium.

To trace the path of blood to any organ in the body, you need only mention the aorta, the proper branch of the aorta, the organ, and the returning vein to the vena cava. Figure 10.11 shows that in most instances the artery and vein that serve the same organ are given the same name. In the systemic circuit, unlike the pulmonary system, arteries contain oxygenated blood and appear a bright red, but veins contain deoxygenated blood and appear a purplish color.

The **coronary arteries** (fig. 10.5), which are a part of the systemic circuit, are extremely important arteries because they serve the heart muscle itself. (The heart is not nourished by the blood in its chambers.) The coronary arteries arise from the aorta just above the aortic semilunar valve. They lie on the exterior surface of the heart, where they branch off in various directions into the arterioles. The coronary capillary beds join to form venules. The venules converge into the coronary veins, which empty into the right atrium. The coronary arteries have a very small diameter and may become blocked as discussed on page 223.

The systemic circuit takes blood from the left ventricle of the heart to the right atrium of the heart. It serves the body proper.

The body has a portal system associated with the liver, the hepatic portal system (fig. 9.10). A portal system is one that begins and ends in capillaries; in this instance the first set of capillaries occurs at the villi of the small intestine and the second occurs in the liver. Blood passes from the capillaries of the villi into venules that join to form the *hepatic portal vein,* a vessel that connects the villi of the intestine with the liver. The *hepatic vein* leaves the liver and enters into the inferior vena cava.

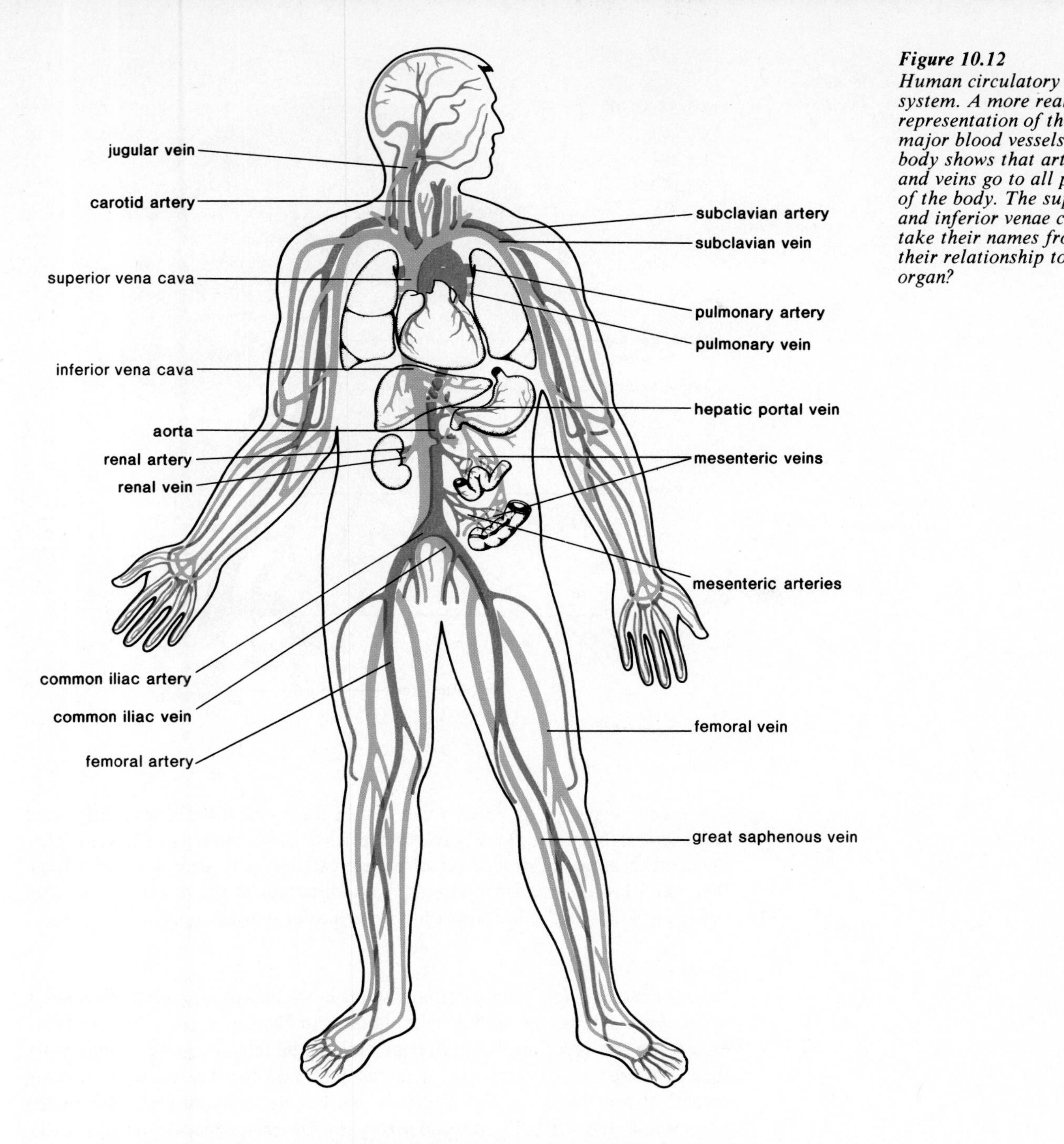

Figure 10.12
Human circulatory system. A more realistic representation of the major blood vessels in the body shows that arteries and veins go to all parts of the body. The superior and inferior venae cavae take their names from their relationship to what organ?

While figure 10.11 is helpful in tracing the path of the blood, it must be remembered that all parts of the body receive both arteries and veins, as illustrated in figure 10.12.

Cardiovascular Measurement Procedures

Knowledge of the cardiac cycle is helpful in understanding three medical procedures.

Electrocardiogram (EKG)

With the contraction of any muscle, including the myocardium, ionic changes occur that can be detected by electrical recording devices. Therefore it is possible to study the heartbeat by recording voltage changes that occur when the heart contracts. (Voltage, which in this case is measured in millivolts, is the difference in polarity between two electrodes attached to the body.) The record that results is called an **electrocardiogram** (fig. 10.10b) which clearly shows an atrial phase and a ventricular phase. The first wave in the electrocardiogram, called the P wave, represents the excitation and contraction of the atria.

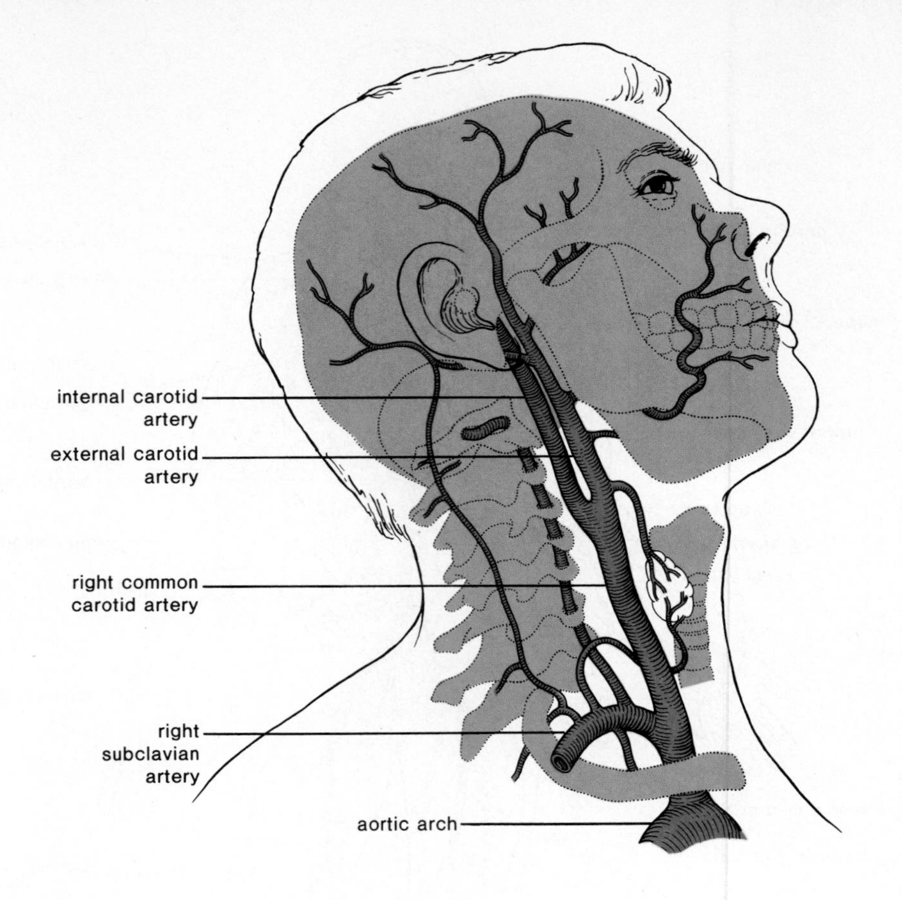

Figure 10.13
The common carotid artery is located in the neck region and can be used to take the pulse. The pulse indicates how rapidly the heart is beating.

The second wave, or the QRS wave, occurs during ventricular excitation and contraction. The third, or T wave, is caused by the recovery of the ventricles. During this period of time, an electrochemical change is recorded even though the ventricles are not contracting. An examination of the electrocardiogram indicates whether the heartbeat has a normal or irregular pattern.

Pulse

When the left ventricle contracts and sends blood out into the aorta, the elastic walls of the arteries swell, but then almost immediately recoil. This alternate expanding and recoiling of an arterial wall can be felt as a **pulse** in any artery that runs close to the surface. It is customary to feel the pulse by placing several fingers on the radial artery, which lies near the outer border of the palm side of the wrist. The carotid artery is another good location (fig. 10.13). Normally the pulse rate indicates the rate of the heartbeat because the arterial walls pulse whenever the left ventricle contracts.

Blood Pressure

Blood pressure, the pressure of the blood against the wall of a vessel, is created by the pumping action of the heart.

To measure blood pressure (fig. 10.14), a sphygmomanometer is used. This consists of a hollow cuff attached to a pressure gauge. The cuff is placed about the upper arm over the brachial artery and inflated with air until there is no pulse felt in the wrist. At this point, no blood is flowing in the brachial

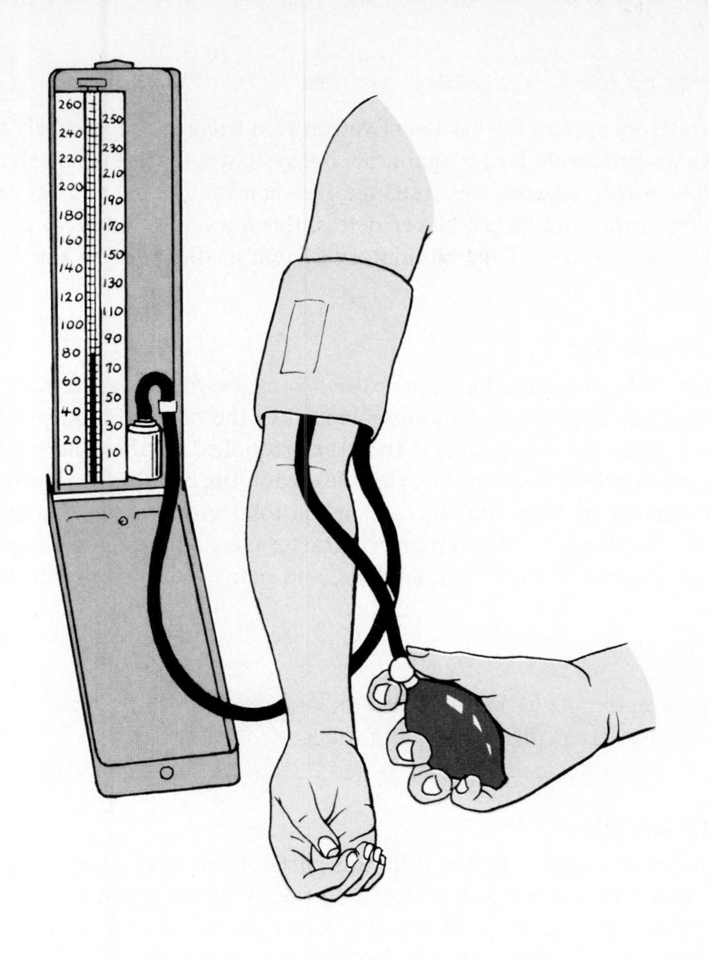

Figure 10.14
Determination of blood pressure by the use of a sphygmomanometer. The technician inflates the cuff with air and then as he or she gradually reduces the pressure, he or she listens by means of a stethoscope for the sounds that indicate the blood is moving past the cuff. A pressure gauge on the cuff is used to tell the systolic and diastolic blood pressure.

artery. Air is slowly released and the cuff is deflated until the first tapping sounds caused by arterial vibrations as the blood begins to flow intermittently can be detected through a stethoscope placed on the arm just beneath the cuff. The examiner glances at the manometer, or pressure gauge, and notes the pressure at this point. This is the value to be assigned to **systolic blood pressure,** the highest arterial pressure, reached during ejection of blood from the heart. The systolic pressure has overcome the pressure exerted by the cuff and has caused the blood to flow in the artery. The cuff is further deflated while the examiner continues to listen. The tapping sounds become louder as the pressure is lowered. Finally, the sounds become dull and muffled, just before there are no sounds at all. Now the examiner again notes the pressure. This is **diastolic blood pressure,** the lowest arterial pressure. Diastolic pressure occurs while the heart ventricles are relaxing.

Normal resting blood pressure for a young adult is said to be 120 millimeters of mercury (Hg) over 80 millimeters, or simply 120/80. Actually, this is the expected blood pressure in the brachial artery of the arm; blood pressure varies in different parts of the body (fig. 10.15). When the resting blood pressure reading is higher than expected, the person is said to have hypertension, and when the reading is lower than expected the person is said to have hypotension. Hypertension, in particular, is often associated with cardiovascular disease, p. 221.

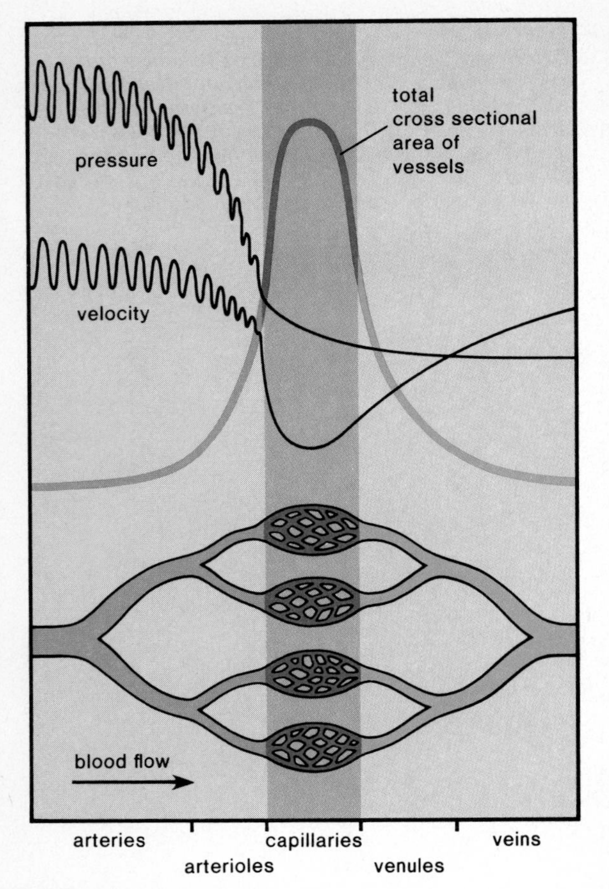

Figure 10.15
Diagram illustrating how velocity and blood pressure are related to the total cross-sectional area of blood vessels. Capillaries have the greatest cross-sectional area and the least pressure and velocity. Skeletal muscle contraction, not blood pressure, accounts for velocity of blood in the veins.

Features of the Circulatory System

The circulatory system is a system of vessels that become progressively smaller and then progressively larger again. Such a system is desirable because it permits a large surface area for exchange between the tissues and the blood in the region of the capillaries. Nevertheless, this does have an effect on certain physiological features of the circulatory system as discussed in the following sections.

Blood Pressure

As figure 10.15 indicates, blood pressure decreases with distance from the left ventricle, the chamber that pumps blood into the aorta. Blood pressure is, therefore, higher in the arteries than in the arterioles. Further, there is a sharp drop in blood pressure when the arterioles reach the capillaries. The decrease may be correlated with the increase in the total cross-sectional area of the vessels as blood moves through arteries, arterioles, and then into capillaries. There are more arterioles than arteries, and many more capillaries than arterioles.

The pumping of the heart creates blood pressure, which steadily decreases from the aorta to the veins.

Velocity and Blood Flow

The velocity of blood varies in different parts of the circulatory system (fig. 10.15). Blood pressure accounts for the velocity of the blood flow in the arterial system and, therefore, as blood pressure decreases due to the increased cross-sectional area of the arterial system, so does velocity. The blood moves much slower through the capillaries than it does in the aorta. This is important because the slow progress allows time for the exchange of molecules between the blood and the tissues.

Blood pressure cannot account for the movement of blood through the venules and veins, since they lie on the other side of the capillaries. Instead, movement of the blood through the venous system is due to skeletal muscle contraction. When the skeletal muscles contract, they press against the weak walls of the veins and this causes the blood to move past a *valve* (fig. 10.16). Once past the valve, the blood will not fall back. The importance of muscle contraction in moving blood in the venous system may be demonstrated by forcing a person to stand rigidly still for a number of hours. Frequently, fainting will occur because the blood collects in the limbs, robbing the brain of oxygen. In this case, fainting is beneficial because the resulting horizontal position aids in getting blood to the head.

Blood flow gradually increases in the venous system (fig. 10.15) due to a progressive reduction in the cross-sectional area as small venules join to form veins. The two venae cavae together have a cross-sectional area of only about double that of the aorta. Blood pressure is lowered in the chest cavity whenever the chest expands during inspiration. This also aids the flow of venous blood into the chest area because blood flows in the direction of reduced pressure.

Blood pressure accounts for the flow of blood in the arteries and arterioles; skeletal muscle contraction accounts for the flow of blood in the venules and veins.

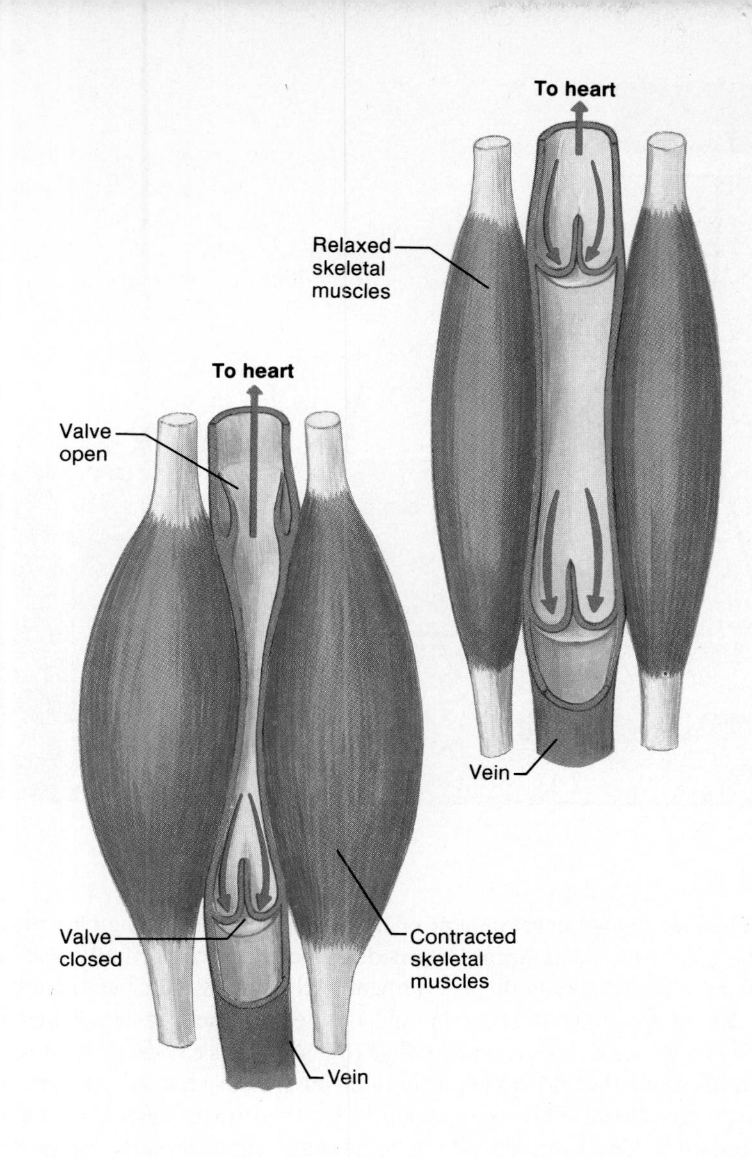

Figure 10.16
Skeletal muscle contraction moves blood in veins. (below) *Muscle contraction exerts pressure against the vein and blood moves past the valve.* (above) *Once blood has moved past the valve, it cannot slip back.*

Circulatory Disorders

During the past 30 years, the number of deaths due to cardiovascular disease has declined more than 30 percent. Even so, more than 50 percent of all deaths in the United States are still attributable to cardiovascular disease. The number of deaths due to hypertension, stroke, and heart attack are more than those due to cancer and accidents combined.

---

*C*ardiovascular disease is the number one killer in the United States.

Hypertension

It is estimated that about 20 percent of all Americans suffer from *hypertension,* or high blood pressure. Even though hypertension is easily detected by blood pressure readings (p. 219) it is believed that at least one-third of these people are unaware they have this condition that can lead to failure of the cardiovascular system.

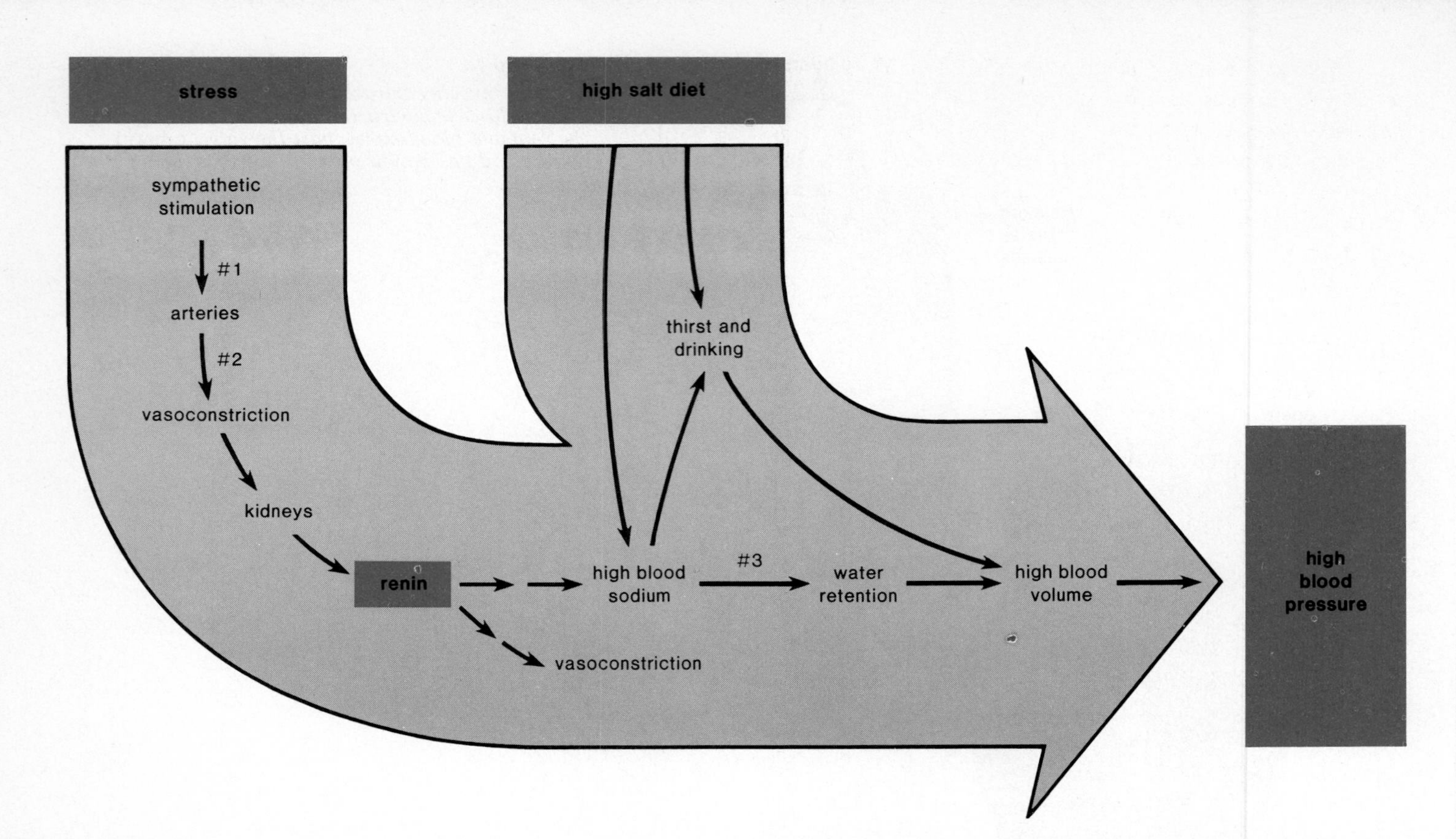

Figure 10.17
A scheme that explains the development of high blood pressure due to either stress or a high salt diet. The numbers (#1–3) indicate the site of action of hypertensive drugs as discussed by the text.

The reasons for development of hypertension are various. One possible scenario (fig. 10.17) has been described. Blood pressure normally rises with excitement or alarm due to the involvement of the sympathetic nervous system, which causes arterioles to constrict and the heart to beat faster. When arterioles are constricted, reduced blood flow to the kidneys causes them to release renin, a hormone that brings about high blood pressure partially by retaining sodium in the blood. This excess sodium leads to water retention and high blood pressure. The same effect can be brought about directly by an excess intake of salt (sodium chloride) in the diet.

Medical treatment for hypertension is based on this sequence of events. Sympathetic-blocking agents act at arrow #1 and prevent action of the sympathetic nervous system. Vasodilators act at arrow #2 and prevent the arteries from constricting. Diuretics act at arrow #3 and cause the kidneys to excrete excess salts and fluids.

Diet, stress, and kidney involvement are implicated in the development of hypertension in some persons.

Hypertension is also seen in individuals who have *atherosclerosis,* an accumulation of soft masses of fatty materials, particularly cholesterol, beneath the inner linings of arteries. Such deposits are called *plaque,* and as they develop, they tend to protrude into the vessel and interfere with the flow of blood. As described in the reading on page 224, it may be beneficial to reduce the amount of cholesterol in the diet by selecting certain foods.

The occurrence of plaque can cause a clot to form on the irregular arterial wall. As long as the clot stays stationary it is called a *thrombus,* but if and when it dislodges and moves along with the blood, it is called an *embolus.* Thrombi reduce the flow of blood, particularly in small arteries. An embolus causes embolism when it comes to a standstill and entirely blocks the flow of blood in a small vessel. If an embolism is not treated, the tissue fed by the vessel dies from lack of blood.

Plaque development often occurs in the arteries of individuals with hypertension.

Stroke and Heart Attack

A stroke occurs when a portion of the brain dies due to a lack of oxygen, and a heart attack occurs when a portion of the heart muscle dies due to a lack of oxygen. A *stroke,* characterized by paralysis or death, often results when a small arteriole bursts or becomes blocked by an embolism.

When a person has hypertension, cardiac muscles require more oxygen because they must work harder to pump the blood against the increased arterial pressure. However, the coronary arteries may be lined by plaque and unable to provide an adequate blood supply. A thrombus may be present, termed coronary thrombosis, or an embolus may have moved into a coronary arteriole. At first, the individual may suffer *angina pectoris,* characterized by a radiating pain in the left arm. Then if circulation to a significant portion of the heart is blocked entirely, a *heart attack* occurs.

Stroke and heart attack are both associated with hypertension.

Surgical Treatments

Surgical treatments are now available for blocked coronary arteries.

Thrombolytic Therapy

Recently, new technologies have been developed that do away with the obstruction without the need for a major operation. In both of these, a plastic tube is threaded into an artery of an arm or leg and guided through a major blood vessel toward the heart. In one procedure, when the tube reaches the blockage, a balloon attached to the end of the tube inflates and breaks up the obstruction. In the other procedure, a drug called streptokinase is injected to dissolve a clot. Then the artery opens and the blood begins to flow again. The latter procedure has been done even while a person was suffering a heart attack. If the clot is removed within six hours after a heart attack begins, there is usually no damage to the heart muscle.

There are procedures to clear blocked arteries that do not require major surgery.

Hypertension

High blood pressure, or hypertension, is not often listed as the cause of death on death certificates, but it often precedes and accompanies death due to heart attack, stroke, and thromboembolism. Altogether, cardiovascular diseases account for more deaths in the United States than do cancer and accidents combined.

Patients with hypertension frequently have atherosclerosis and vice versa. This common condition involves the formation of plaques within arterial linings. Plaques contain large quantities of cholesterol. Cholesterol is a substance found within the cell membrane of all cells and is also modified into hormones by certain glands. Cholesterol is carried in the blood by lipoprotein molecules of two types: LDLs (low-density lipoproteins) and HDLs (high-density lipoproteins).

The cholesterol-LDL molecular combination is the one that triggers the atherosclerotic accumulation. When these molecules adhere to arterial walls, white blood cells called monocytes (figure a.) begin to invade and damage the blood vessel lining. These cells are transformed into macrophages (large phagocytic cells) that absorb the cholesterol and burst, leaving behind fatty streaks. Platelets are also naturally attracted to the damaged area. They release a growth factor that causes the muscle cells of the arterial wall to divide. This accumulating mass constitutes a plaque that can eventually grow so large that it hinders or even stops blood flow through the artery. Eventually calcium collects in the cells of the plaque and hardens it.

There has been a great deal of attention given to the role of cholesterol in the development of atherosclerotic plaque. A diet high in saturated fatty acids and cholesterol leads to high blood levels of cholesterol carried by LDL (table A). Recently, investigators discovered that omega-3 fatty acids—those in which there is a double bond in the fatty acid chain on the third carbon from the end (omega being the last letter of the Greek alphabet)—reduces the level of LDL in the blood. Omega-3 fatty acids are prevalent in fish oil. Thus a diet high in fish may account for the lower incidence of heart disease among Eskimos.

A significant clinical study showed that lowering the cholesterol level can help prevent heart disease. In 1984, the National Heart, Lung, and Blood Institute in the United States reported the results of a ten-year study involving nearly a half-million middle-aged men who had high blood levels of cholesterol. Half

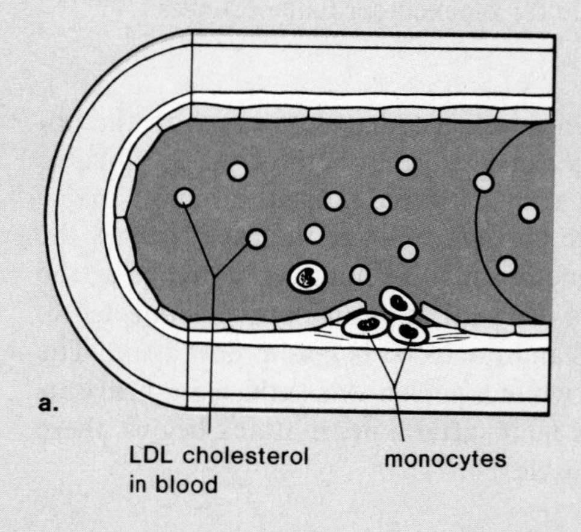

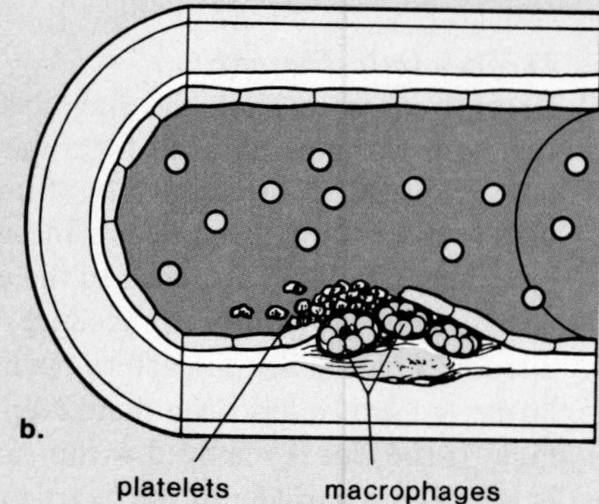

Plaque formation. a. *Cholesterol in combination with LDL attracts monocytes that invade the arterial lining.* b. *Monocytes become cholesterol-ingesting macrophages that burst, leaving behind fatty streaks. Release of growth factor by platelets causes muscle cells to divide.* c. 1. *Plaque that obstructs blood flow consists of all components mentioned.* c.2. *Micrograph of plaque (yellow) in the coronary artery of a heart patient.*

Table A Dietary Reduction of Blood Cholesterol Levels	
Eat Less	**Eat More**
Fatty meats	Fish, poultry, lean cuts of meat
Organ meats (liver, kidney, brain, pancreas), shrimp	
Sausage, bacon, processed meats	Fruit, vegetables, cereals, starches
Whole milk	Skim or low-fat milk
Butter, hard margarine	Soft margarine
High-fat cheese (bleu, cheddar)	Low-fat cheese
Ice cream and cream	Yogurt
Egg yolks	Egg whites
Foods fried in animal fats	Vegetable fats for frying and salad dressing
Commercial baked goods	

From Rifkind, Basil M., and S. E. Epstein, "Heart and blood vessels" in *1986 Medical and Health Annual*. Copyright © 1985 Encyclopaedia Brittanica, Inc., Chicago, Illinois. Reprinted by permission.

of the men followed a moderate cholesterol-lowering diet and also took a drug that reduces blood cholesterol levels. The other half followed the diet and took a placebo. The group that received the drug had fewer cardiovascular incidences than did the other group. It was concluded that those subjects with a 25 percent lower blood-cholesterol level had their risk of heart attack cut in half.

Heart attacks and strokes are common causes of death in individuals with atherosclerosis. The heart muscle requires a large and continuous supply of oxygen. Coronary arteries narrowed by atherosclerosic plaques cannot supply enough oxygen. First, angina (chest pain) and then heart attack (death of heart muscle due to lack of oxygen) may result. Then, too, blood clots often form in vessels damaged by plaque. The clot could develop in the coronary arteries or develop in a larger vessel some distance away and be brought to the coronary arteries by circulating blood. A thrombus is a stationary blood clot, while an embolus is a moving clot; therefore, thromboembolism (blood clots) sometimes causes heart attacks. Thromboembolism is also a cause of stroke when a blood clot occludes (clogs up) an artery in the brain.

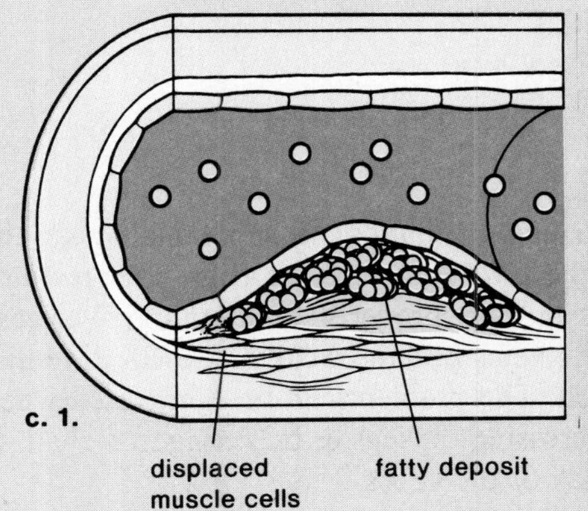

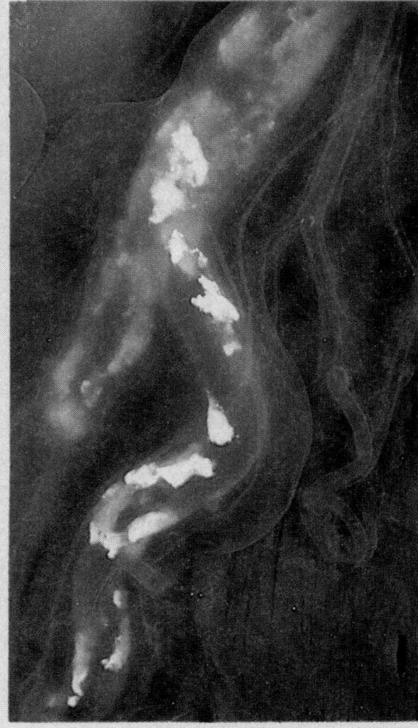
c.2.

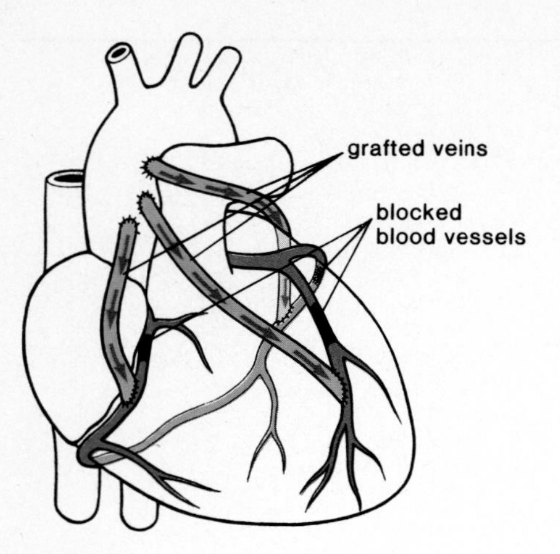

Figure 10.18
Coronary bypass operation. During this operation, the surgeon grafts segments of a leg vein between the aorta and the coronary vessels, bypassing areas of blockage. Patients who are ill enough to require surgery often receive two or three bypasses in a single operation.

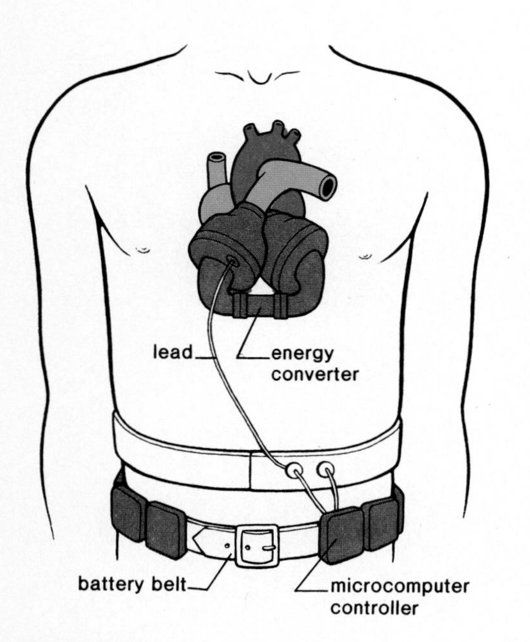

Figure 10.19
Artificial heart. It is hoped that some day the artificial heart will be powered by batteries the patient can wear about the waist. A microcomputer would control the signals that enable the artificial heart to pump blood.

Coronary Bypass Operations

As many as 100,000 persons a year have *coronary bypass* surgery. During this operation, surgeons take segments from another blood vessel, often a large vein in the leg, and stitch one end to the aorta and the other end to a coronary artery past the point of obstruction (fig. 10.18). Between 75 and 90 percent of those who have had bypass surgery say their angina pain has been relieved.

Once the heart is exposed, some physicians have used lasers to open up clogged coronary vessels. Presently the technique is used in conjunction with coronary bypass operations but eventually it may be possible to use lasers without the need to open the thoracic cavity.

Donor Heart Transplants and Artificial Heart Implants

Persons with weakened hearts may eventually suffer from *congestive heart failure,* meaning that the heart is no longer able to pump blood adequately. These individuals are candidates for a donor heart transplant, or even implantation of an artificial heart. The difficulty with a donor heart transplant is, first, one of availability and, second, the tendency of the body to reject foreign organs. It would be helpful to find ways to repair the heart instead of replacing it. In one recent procedure, a surgeon took a back muscle and wrapped it around a heart weakened by removal of myocardial tissue. Later he implanted an artificial pacemaker that caused the muscle to contract regularly and help pump the blood.

On December 2, 1982, Barney Clark was the first person to receive an artificial heart. The heart's two polyurethane ventricles were attached to Clark's own atria and blood vessels by way of dacron fittings. Two long tubes stretched between the artificial heart and an external machine that periodically sent bursts of air into the ventricles, forcing the blood out into the aorta and pulmonary trunk. It is hoped that eventually an artificial heart can be powered by batteries so that the patient will be completely mobile (fig. 10.19). However it will be necessary first to solve a major problem; the components of the artificial heart apparently cause the development of blood clots leading to the occurrence of strokes.

Coronary bypass, donor heart transplants, and artificial heart implantation all require major surgery.

Varicose Veins and Phlebitis

Varicose veins are abnormal and irregular dilations in superficial (near the surface) veins, particularly those in the lower legs. Varicose veins in the rectum, however, are commonly called piles, or more properly, *hemorrhoids.* Varicose veins develop when the valves of the veins become weak and ineffective due to a backward pressure of the blood. The problem can be aggravated when venous blood flow is obstructed by crossing the legs or by sitting in a chair so that its edge presses against the back of the knees.

Phlebitis, or inflammation of a vein, is a more serious condition, particularly when a deep vein is involved. Blood in the inflamed vessel may clot, in which case *thromboembolism* occurs. An embolus that originates in a systemic vein may eventually come to rest in a pulmonary arteriole, blocking circulation through the lungs. This condition, termed *pulmonary embolism,* can result in death.

Veins have weak walls, and this occasionally leads to medical disorders.

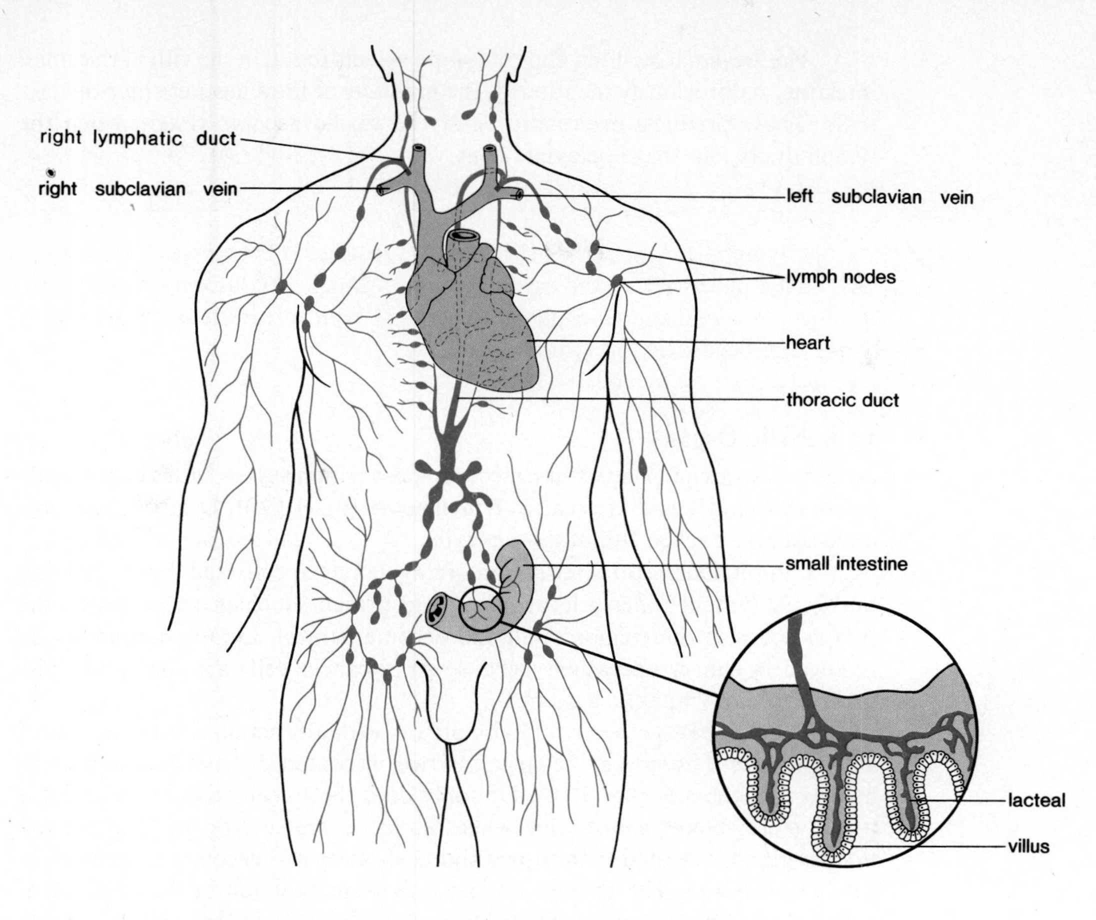

Figure 10.20
Lymphatic system. The lymphatic system drains excess fluid from the tissues and returns it to the cardiovascular system. The thoracic duct is one of the major lymph vessels. The blowup shows the lymph vessels, called lacteals, which are present in the intestinal villi. Lacteals are involved in the absorption of the products of fat digestion from the intestine.

Lymphatic System

The **lymphatic system** is closely associated with the cardiovascular system because it consists of vessels that take up excess tissue fluid and transport it to the bloodstream. **Tissue fluid** is the fluid that surrounds cells. Localized swelling due to excess tissue fluid not collected by the lymphatic system is called **edema.**

The lymphatic system is a one-way system that starts in the tissues and empties into the cardiovascular system.

Lymphatic Vessels

Lymph vessels consist of *lymph capillaries* and *lymph veins.* The latter have a construction similar to cardiovascular veins, including the presence of valves (fig. 10.16).

The lymphatic system (fig. 10.20) is a one-way system rather than a circulatory system. The system begins with lymph capillaries that lie near blood capillaries and take up fluid that has diffused from the capillaries and has not been reabsorbed by them. Once tissue fluid enters the lymph vessels, it is called **lymph.** Lymph also contains *lymphocytes,* a type of white blood cell (p. 240). Some lymphocytes produce antibodies, proteins that are capable of combining with foreign proteins called antigens. At times, the foreign proteins are associated with disease-causing bacteria and viruses, and therefore the lymphatic system helps fight this source of infection.

Lymph is collected in vessels that join to form two main trunks: the right lymphatic duct, which drains the upper right portion of the body, and the thoracic duct, which drains the rest of the body. The former empties into the right subclavian vein and the latter into the left subclavian vein.

The *lacteals* are blind ends of lymph vessels found in the villi of the small intestine. As previously mentioned, the products of fat digestion enter the lacteals. These products eventually enter the cardiovascular system when the lymph ducts join the subclavian veins.

The lymphatic system has three main functions: (1) transport of excess tissue fluid back to the cardiovascular system; (2) absorption of fat from the intestine and transport to blood; (3) fight infection by cleansing lymph and production of lymphocytes.

Lymphatic Organs

At certain strategic points along medium-sized lymph vessels, there occur small ovoid, or round, structures called **lymph nodes** (fig. 10.20). Lymph nodes produce and are packed full of lymphocytes.

Lymph nodes also filter lymph of any damaged cells and debris, helping purify the lymph. When a local infection is present, such as a sore throat, the lymph nodes in that region swell and become painful. Lymph nodes may be removed in cancer operations because stray cancer cells are sometimes dispersed by the lymphatic system.

There are two other lymphatic organs with a function similar to that of a lymph node. The **spleen,** the largest of two, is located in the abdominal cavity behind the stomach (fig. 9.10). Not only does the spleen contain white cells, it also stores blood, contracting when the blood pressure drops. The bilobed **thymus** gland is located in the upper thoracic cavity and becomes progressively smaller with age. The thymus has an important function in the production and maturation of some lymphocytes, and its decrease in size may be important in the aging process.

Summary

Movement of blood in the circulatory system is dependent on the beat of the heart. During the cardiac cycle, the SA node (pacemaker) initiates the beat and causes the atria to contract. The AV node picks up the stimulus and initiates contraction of the ventricles. The heart sounds, lub-DUPP, are due to the closing of the atrioventricular valves followed by the closing of the semilunar valves.

The circulatory system is divided into the pulmonary and systemic circuits. In the pulmonary circuit the pulmonary artery takes blood from the right ventricle to the lungs and the pulmonary veins return it to the left atrium. To trace the path of blood in the systemic circuit, start with the aorta from the left ventricle. Follow its path until it branches to an artery going to a specific organ. It may be assumed that the artery will divide into arterioles and capillaries, and that the capillaries will lead to venules. The vein that takes blood to the vena cava most likely has the same name as the artery.

Lymph vessels, or veins, are constructed similarly to cardiovascular veins and contain valves to keep lymph moving from the tissues to the veins. The lymphatic system is a one-way system taking excess tissue fluid to the subclavian veins. The lymphatic organs, such as lymph nodes and the thymus, are involved in the infection-fighting capacity of the body.

Objective Questions

1. Arteries are blood vessels that take blood ______________ from the heart.
2. When the left ventricle contracts blood enters the ______________ .
3. The right side of the heart pumps blood to the ______________ .
4. The ______________ node is known as the pacemaker.
5. The blood vessels that serve the heart are the ______________ arteries and veins.
6. The pressure of blood against the walls of a vessel is termed ______________ .
7. Blood moves in arteries due to ________ and in veins due to ______________ .
8. Reducing the amount of ______________ in the diet reduces the chances of plaque buildup in arteries.
9. Varicose veins develop when __________ become weak and ineffective.
10. There are only two types of lymph vessels, the lymph ______________ and the lymph ______________ .

Answers to Objective Questions

1. away 2. aorta 3. lungs 4. SA 5. coronary 6. blood pressure 7. blood pressure, skeletal muscle contraction 8. cholesterol 9. valves 10. capillaries, veins

Study Questions

1. What types of blood vessels are there? Discuss their structure and function. (pp. 208–210)
2. Trace the path of blood in the pulmonary circuit as it travels from and returns to the heart. (p. 212)
3. Describe the cardiac cycle (using the terms systole and diastole) and explain the heart sounds. (p. 212)
4. Trace the path of blood from the mesenteric arteries to the aorta, indicating which of the vessels are in the systemic circuit and which are in the pulmonary circuit. (pp. 215–216)
5. Describe an EKG and tell how its components are related to the cardiac cycle. (pp. 214, 217)
6. What is blood pressure, and why is the average normal arterial blood pressure said to be 120/80? (p. 219)
7. In which type of vessel is blood pressure highest? Lowest? Velocity is lowest in which type vessel and why is it lowest? Why is this beneficial? What factors assist venous return of the blood? (p. 220)
8. What is atherosclerosis? (p. 222) Name two illnesses associated with hypertension and thromboembolism. (p. 223) Discuss treatment of cardiovascular disease. (p. 226)
9. What is a lymph vessel? (p. 227) Give three functions of the lymphatic system and tell how these functions are carried out. (pp. 227–228)

Key Terms

aorta (a-or'tah) major systemic artery that receives blood from the left ventricle. *212*

arterioles (ar-te're-ōlz) vessels that take blood from arteries to capillaries. *208*

arteries (ar'ter-ēz) vessels that take blood away from the heart; characteristically possessing thick elastic walls. *208*

atria (a'tre-ah) chambers; particularly the upper chambers of the heart that lie above the ventricles. (*sing.* atrium) *210*

AV node (a-ve nōd) a small region of neuromuscular tissue that transmits impulses received from the SA node to the ventricular walls. *213*

capillaries (kap'i-lar''ēz) microscopic vessels connecting arterioles to venules through whose thin walls molecules either exit or enter the blood. *208*

coronary arteries (kor'ŏ-na-re ar'ter-ēz) arteries that supply blood to the wall of the heart. *216*

diastole (di-as'to-le) relaxation of the heart chambers. *212*

lymph (limf) fluid having the same composition as tissue fluid and carried in lymph vessels. *227*

lymphatic system (lim-fat'ik sis'tem) vascular system which takes up excess tissue fluid and transports it to the bloodstream. *227*

pulmonary circuit (pul'mo-ner''e ser'kit) that part of the circulatory system that takes deoxygenated blood to and oxygenated blood away from the lungs. *215*

SA node (es a nōd) small region of neuromuscular tissue that initiates the heartbeat. Also called the pacemaker. *213*

systemic circuit (sis-tem'ik ser'kit) that part of the circulatory system that serves body parts other than the gas-exchanging surfaces in the lungs. *215*

systole (sis'to-le) contraction of the heart chambers. *212*

tissue fluid (tish'u floo'id) fluid found about tissue cells containing molecules that enter from and exit to the capillaries. *227*

valves (valvz) openings that open and close, insuring one-way flow; common to the systemic veins, the lymphatic veins, and the heart. *210*

veins (vānz) vessels that take blood to the heart; characteristically having nonelastic walls. *208*

venae cavae (ve'nah ka'vah) large systemic veins that return blood to the right atrium of the heart. *212*

ventricles (ven'trĭ-k'lz) cavities in an organ such as the lower chambers of the heart. *210*

venules (ven'ūlz) vessels that take blood from capillaries to veins. *208*

11

Blood

Chapter Concepts

1 Blood, which is composed of cells and a fluid containing many inorganic and organic molecules, has three primary functions: transport, clotting, and fighting infection.

2 Exchange of molecules between blood and tissue fluid takes place across capillary walls.

3 Blood is typed according to the antigens present on the red blood cells.

4 All of the functions of blood may be correlated with the ability of the body to maintain a constant internal environment.

Chapter Outline

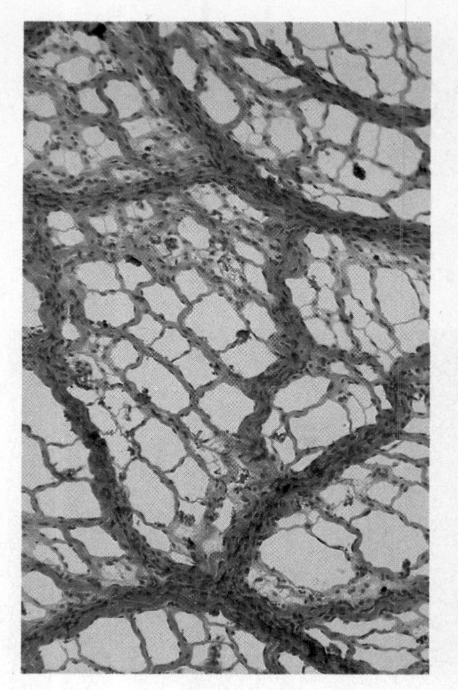

a.

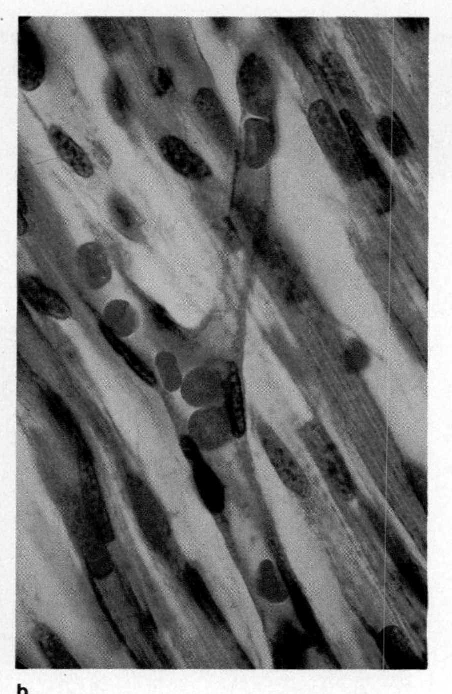

b.

Figure 11.1
a. *Low power micrograph of blood coursing through blood capillaries. The many red blood cells are just visible.* b. *High power micrograph of a capillary shows the red blood cells more clearly.*

It is a curious fact that more than half the body is water; the total quantity of water is around 70 percent of the body's weight. By far, most of this water is found within the cells. A much smaller amount lies outside the cells. The water outside the cells is found in (1) the *tissue fluid* that surrounds the cells, (2) *lymph* contained within lymph vessels, and (3) in *blood* vessels (fig. 11.1).

If blood is transferred from a person's vein to a test tube and prevented from clotting, it separates into two layers (fig. 11.2). The lower layer consists of red blood cells (erythrocytes), white blood cells (leukocytes), and blood platelets (thrombocytes). Collectively, these are called the **formed elements** (fig. 11.3) that take up about 45 percent of the volume of whole blood. The upper layer, called *plasma,* contains a variety of inorganic and organic substances dissolved or suspended in water. Plasma accounts for about 55 percent of the volume of whole blood. Table 11.1 lists the components of blood, which we will discuss in terms of three functions: transport, clotting, and infection fighting. All of these can be related to blood's primary function of maintaining a constant internal environment, or homeostasis.

plasma constitutes 55% total volume

cells constitute 45% total volume

Figure 11.2
Volume relationship of plasma and formed elements (cells) in blood. Red cells are by far the most prevalent blood cell and this accounts for the color of blood.

Blood is a liquid tissue. The liquid portion is termed plasma and the solid portions are the formed elements.

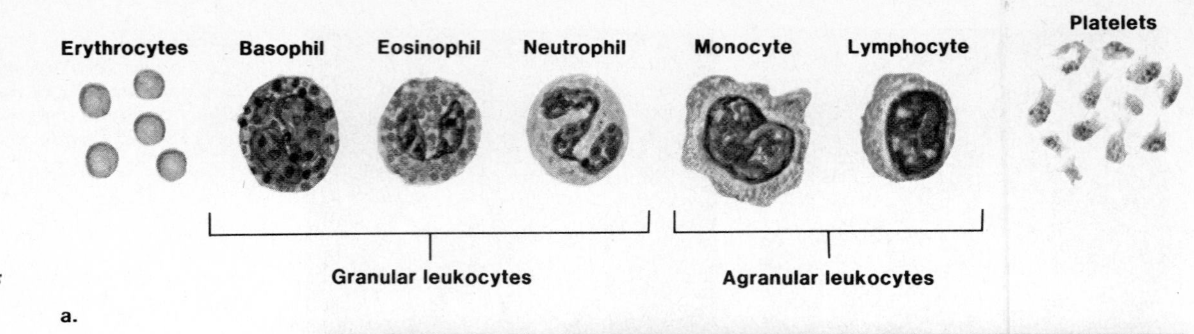

Figure 11.3
Formed elements in blood are the cells and platelets which are fragments of cells. a. *Diagram of formed elements. Erythrocytes are red blood cells. Leukocytes (both granular and agranular) are white blood cells. Thrombocytes are platelets.* b. *More realistic view of formed elements as they would appear within a blood vessel.*

Table 11.1 Components of Blood

Blood	Function	Source
I. Formed Elements		
Red cells	Transport oxygen	Bone marrow
Platelets	Clotting	Bone marrow
White cells	Fight infection	Bone marrow and lymphoid tissue
II. Plasma*		
Water	Maintains blood volume and transports molecules	Absorbed from intestine
Plasma proteins	All maintain blood osmotic pressure and pH	
Albumin	Transport	Liver
Fibrinogen	Clotting	Liver
Globulins	Fight infection	Lymphocytes
Gases		
Oxygen	Cellular respiration	Lungs
Carbon dioxide	End product of metabolism	Tissues
Nutrients		
Fats, glucose, amino acids, etc.	Food for cells	Absorbed from intestinal villi
Salts	Maintain blood osmotic pressure and pH; aid metabolism	Absorbed from intestinal villi
Wastes		
Urea and ammonia	End products of metabolism	Tissues
Hormones, vitamins, etc.	Aid metabolism	Varied

*Plasma is 90–92 percent water, 7–8 percent plasma proteins, not quite 1 percent salts, and all other components are present in even smaller amounts.

Functions of Blood

Transport

The transport function of the blood helps maintain the constancy of tissue fluid. The blood transports oxygen from the lungs and nutrients from the intestine to the capillaries where they enter tissue fluid. Here, it also takes up carbon dioxide and nitrogen waste (i.e., ammonia) given off by the cells and transports them away. Carbon dioxide exits the blood at the lungs, and ammonia exits at the liver where it is converted to urea, a substance that later travels by way of the bloodstream to the kidneys and is excreted. Figure 11.4 diagrams the major transport functions of blood, indicating the manner in which this function helps keep the internal environment relatively constant.

---

*H*omeostasis is only possible because blood brings nutrients to the cells and removes their wastes.

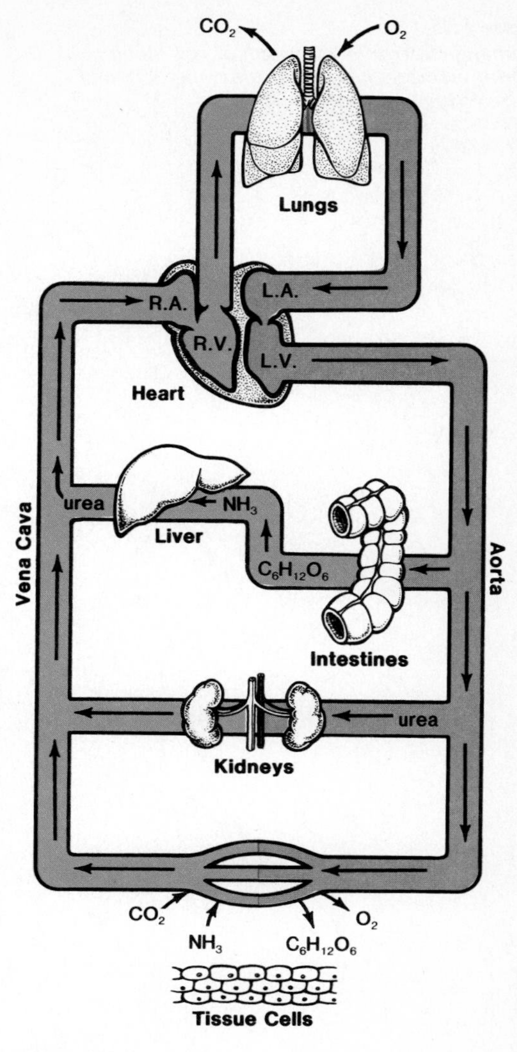

Figure 11.4
Diagram illustrating the transport function of blood. Oxygen (O_2) is transported from the lungs to the tissues, and carbon dioxide (CO_2) is transported from the tissues to the lungs. Ammonia (NH_3) is transported from the tissues to the liver where it is converted to urea, a molecule excreted by the kidneys. Glucose ($C_6H_{12}O_6$) is absorbed by the gut and may be temporarily stored in the liver as glycogen before it is transported to the tissues.

Blood Proteins

Although small organic molecules such as glucose and urea simply dissolve in plasma, large organic molecules such as hormones, vitamins, and lipids combine with proteins for transport. For example, cholesterol is a lipid that will not dissolve in blood plasma and therefore must be carried in combination with protein. As we discussed on page 224, a combination of cholesterol with HDL (high-density lipoprotein) is far more desirable than a combination with LDL (low-density lipoprotein). It is the latter combination that leads to atherosclerosis.

You will recall that blood pressure accounts for the movement of blood in the arteries. But blood needs a certain *viscosity,* or thickness, in order to exert pressure, and this property of blood is largely dependent on the presence of plasma proteins and on red blood cells. Maintenance of viscosity is therefore another way that proteins contribute to transport of molecules.

Blood also needs a certain *volume* in order to exert a pressure. Plasma proteins, together with salts, create an osmotic pressure that maintains the water volume of the blood. You will recall that water moves across cell membranes from the area of greater concentration to the area of lesser concentration of water. Since proteins are too large to pass through or across a capillary wall, the fluid within the capillaries is always the area of lesser concentration of water, and water will therefore pass into the capillaries. The blood proteins we will be discussing in this chapter are listed in table 11.2. All of these except albumin have a special function. Albumin, globulin, and fibrinogen are present in the plasma, but hemoglobin is present in the red blood cells. Hemoglobin is a conjugated protein because it is composed not only of the protein globin but also the nonprotein group, heme.

---

*B*lood proteins assist in the transport function of blood. They serve as carriers for some molecules, and they help maintain the viscosity and volume of the blood.

***Table 11.2** Blood Proteins*

Name	Location	Special Function
Albumin	Plasma	———
Globulin	Plasma	Antibodies to fight infection
Fibrinogen	Plasma	Blood clotting
Hemoglobin	Red cells	Carries gases (oxygen and carbon dioxide)

Figure 11.5
Scanning electron micrograph of red blood cells, the formed element that contains hemoglobin and carries oxygen within the blood.

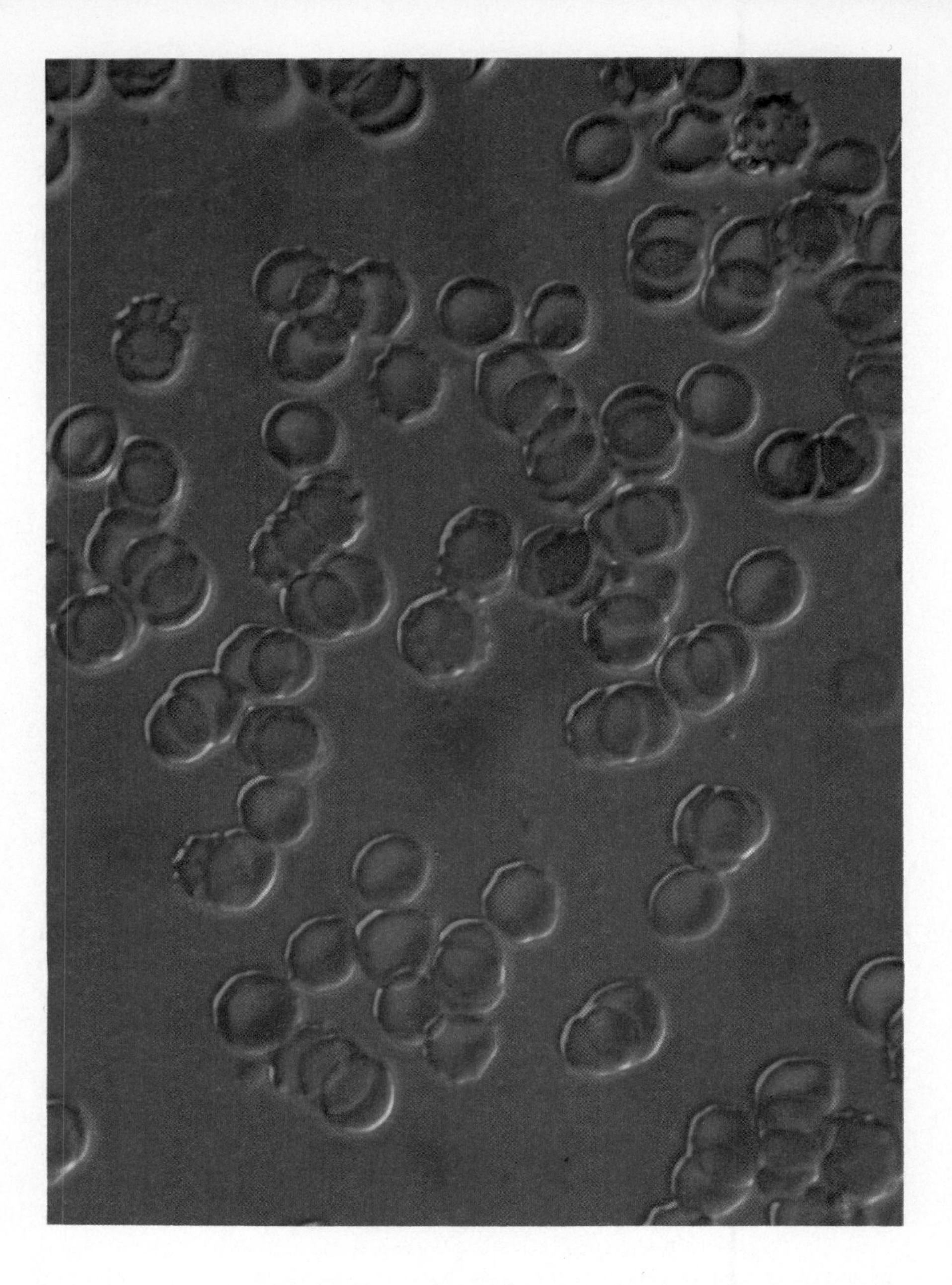

Figure 11.6
The hemoglobin molecule is a protein that contains four polypeptide chains, two of which are alpha (α) and two of which are beta chains (β). The plane in the center of each chain represents an iron-containing heme group. When hemoglobin carries oxygen, it combines with the iron.

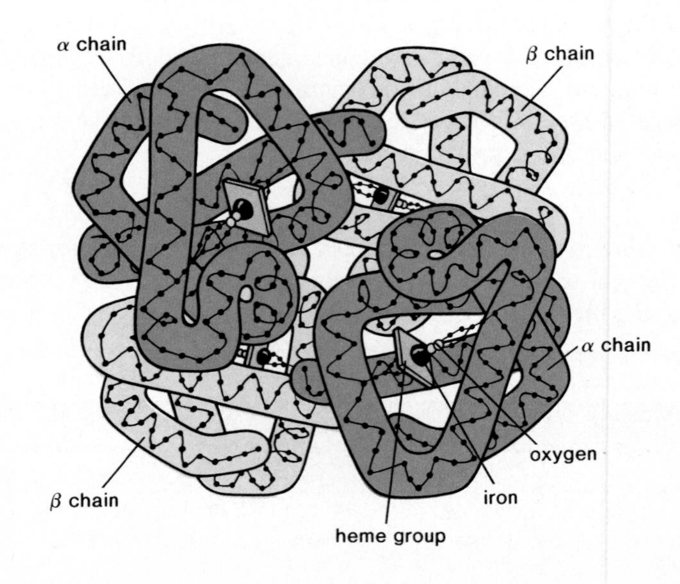

Oxygen

Since humans are active warm-blooded animals, primarily the brain and muscle cells require much oxygen within a short period of time. The use of the respiratory pigment ***hemoglobin*** allows the blood to carry much more oxygen than would otherwise be possible. Plasma carries only about 0.3 milliliters of oxygen per 100 milliliters, but whole blood carries 20 milliliters of oxygen per 100 milliliters. This shows that the presence of hemoglobin increases the carrying capacity of blood 60 times.

Although the iron portion of hemoglobin carries oxygen, the equation for oxygenation of hemoglobin is usually written as:

$$Hb + O_2 \underset{\text{tissues}}{\overset{\text{lungs}}{\rightleftharpoons}} HbO_2$$

The hemoglobin on the right, which is combined with oxygen, is called **oxyhemoglobin.** Oxyhemoglobin forms in the lungs and is a bright red color. The hemoglobin on the left, which has given up oxygen to tissue fluid, is called **reduced hemoglobin** and is a dark purple color.

Hemoglobin is an excellent carrier for oxygen because it forms a loose association with oxygen in the cool, neutral conditions of the lungs and readily gives it up under the warm and more acidic conditions of the tissues.

Carbon monoxide, present in automobile exhaust, combines with hemoglobin more readily than does oxygen, and it stays combined for several hours, regardless of the environmental conditions. Accidental death or suicide from carbon monoxide poisoning occurs because the hemoglobin of the blood is not available for oxygen transport. This transport function of blood is so important that life can be temporarily sustained by giving the patient a hemoglobin substitute transfusion when whole blood is not available or cannot be given. The reading on page 237 discusses the possible benefits of this "artificial blood."

Hemoglobin does not float free within the plasma; it is enclosed within cells. Since hemoglobin is a red pigment, the cells appear red and their color also makes the blood red. There are 5 million red cells (fig. 11.5) per cubic millimeter of whole blood, and each of these cells contains about 200 million hemoglobin molecules. If this much hemoglobin were suspended within the plasma rather than being enclosed within the cells, the blood would be so thick the heart would have difficulty pumping it.

Each hemoglobin molecule (fig. 11.6) contains four polypeptide chains that make up the protein globin. Each of the polypeptide chains is joined to **heme,** a complex iron-containing structure. It is the iron that forms a loose association with oxygen allowing red cells to carry oxygen in the blood.

Oxygen is transported to the tissues in combination with hemoglobin, a pigment found in red blood cells.

Red Blood Cells (Erythrocytes)

The red cells in humans are small biconcave, disk-shaped cells without nuclei (fig. 11.5). They pass through several developmental stages during which time they lose the nucleus and acquire hemoglobin (fig. 11.7).

Red cells are continuously manufactured in the red bone marrow of the skull, ribs, vertebrae, and ends of the long bones (fig. 16.6). Evidence indicates that the oxygen tension of arterial blood serves to regulate red cell formation and, consequently, also hemoglobin production. At high altitudes, where the oxygen tension is low, the red cell count increases. It is believed that low oxygen

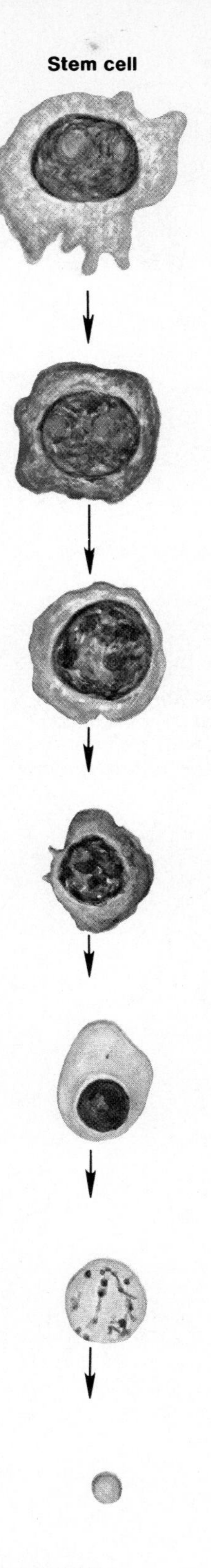

Figure 11.7
Maturation of a red blood cell. Red blood cells are made in red bone marrow where stem cells continuously divide. During the maturation process, a red blood cell loses its nucleus and gets much smaller.

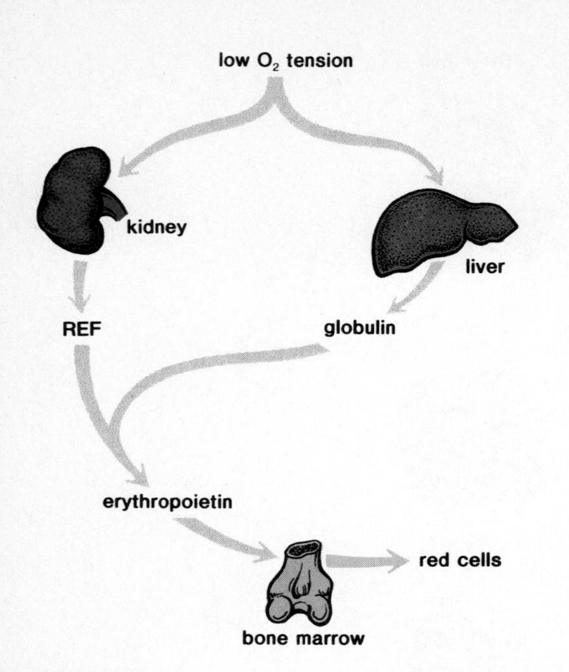

Figure 11.8
Control of red blood cell formation. When oxygen tension is low, the kidneys produce a chemical (REF) that, after combination with globulin from the liver, causes the bone marrow to produce more red cells.

low oxygen tension in the blood causes the kidneys to produce a substance called *renal erythropoietic factor* (REF). This joins with liver globulin to give a combination that stimulates the red bone marrow to produce more red cells (fig. 11.8).

Red cells live only about 120 days and are destroyed chiefly in the liver and spleen, where they are engulfed by large phagocytic cells. When red cells are broken down, the hemoglobin is released. The iron is recovered and returned to the red bone marrow for reuse. The heme portion of the molecule undergoes chemical degradation and is excreted by the liver in the bile as bile pigments. The bile pigments are primarily responsible for the color of feces.

Red blood cells are made in the bone marrow and are broken down in the liver and spleen.

Anemia When there is an insufficient number of red cells or the cells do not have enough hemoglobin, the individual suffers from *anemia*[1] and has a tired, run-down feeling. In iron-deficiency anemia, the hemoglobin count is low. It may be that the diet does not contain enough iron. Certain foods, such as spinach, raisins, and liver, are rich in iron and the inclusion of these in the diet can help prevent this type of anemia.

In another type of anemia, called pernicious anemia, the digestive tract is unable to absorb enough vitamin B_{12}. This vitamin is essential to the proper formation of red cells and without it, immature red cells tend to accumulate in the bone marrow in large quantities. A special diet and administration of vitamin B_{12} by injection is an effective treatment for pernicious anemia.

Illness (anemia) results when the blood has too few red cells and/or not enough hemoglobin.

Carbon Dioxide

Red cells that have given up oxygen to tissue fluid are now ready to take part in the transport of carbon dioxide. Reduced hemoglobin will combine with carbon dioxide to form *carbaminohemoglobin:*

$$Hb + CO_2 \underset{\text{lungs}}{\overset{\text{tissues}}{\rightleftharpoons}} HbCO_2$$

However, such a combination with hemoglobin actually represents only a small portion of the carbon dioxide in the blood. Most of the carbon dioxide is transported as the *bicarbonate ion,* HCO_3^-. This ion is formed after carbon dioxide has combined with water. Carbon dioxide combined with water forms carbonic acid; this dissociates (breaks down) to a hydrogen ion and a bicarbonate ion:

$$CO_2 + H_2O \underset{\text{lungs}}{\overset{\text{tissues}}{\rightleftharpoons}} H_2CO_3 \underset{\text{lungs}}{\overset{\text{tissues}}{\rightleftharpoons}} H^+ + HCO_3^-$$

There is an enzyme within red cells, called *carbonic anhydrase,* that speeds up this reaction. The released hydrogen ions, which could drastically change the pH, are absorbed by the globin portions of hemoglobin, and the bicarbonate ions diffuse out of the red cells to be carried in the plasma. Reduced hemoglobin, which combines with a hydrogen ion, may be symbolized as HHb. This combination plays a vital role in maintaining the pH of the blood.

[1]Sickle-cell anemia and Cooley's anemia are discussed in chapter 21.

Artificial Blood

There are risks to having a blood transfusion (fig. a). The donor's red cells may carry an antigen that leads to an immune reaction or the donor's plasma may contain a disease-causing agent. A cross-matching test between the donor's blood and the recipient's blood usually detects whether an antigen-antibody reaction is likely. Blood is also screened for the presence of two types of viruses that are especially troublesome. They are the hepatitis B virus and the AIDS virus. Blood donors are questioned carefully and their blood is tested for the presence of these viruses. Despite the care that is taken to avoid immune reactions and transference of disease, it would be advantageous to develop an artificial blood that would have neither of these risks.

Investigation is proceeding in two directions. There is an emulsion of perfluorocarbon (PFC) that can be transfused and this substance will carry oxygen much like hemoglobin does. The emulsion can serve as a blood substitute for humans in emergency situations where only the oxygen-carrying function of blood is required. FDA approval has been sought by a Japanese corporation to market PFC under the trade name of Fluosol-DA but thus far the FDA has denied permission on the grounds that clinical trials have not been successful enough.

Recently, artificial red blood cells called neohemocytes (NHCs) have been developed (fig. b). Purified human hemoglobin, taken from outdated donor blood, is encapsulated in a lipid bilayer membrane. The artificial cells are much smaller than normal human red blood cells and they do not contain as much hemoglobin. However, rats which received transfusions did survive until they were sacrificed for gross toxicity studies. The investigators believe that the tests are successful enough to warrant further study. The stability and vascular retention time of the cells needs to be improved since they are removed from the bloodstream and broken down at a faster rate than normal cells. Anthony Hunt, representing the researchers at the University of California at San Francisco who are developing the NHCs says, "Neohemocytes are one step along the road of constructing biological systems from scratch."

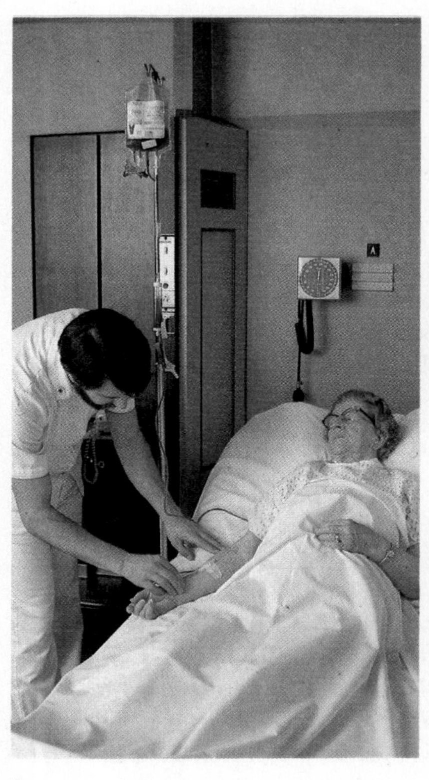

a.

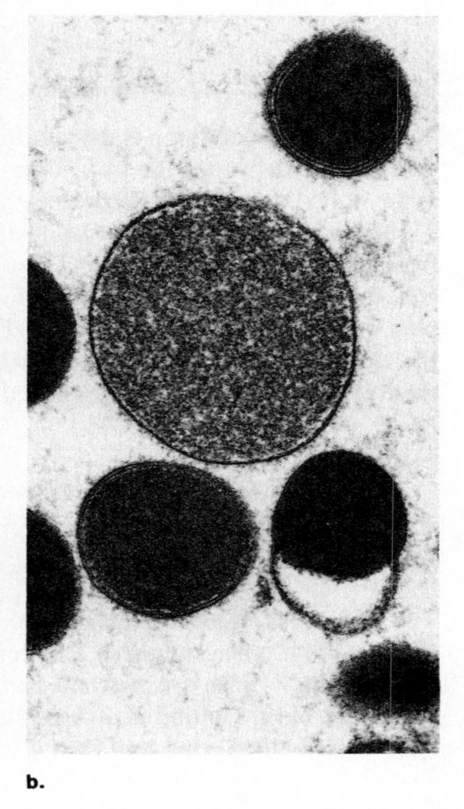

b.

a. *If blood were not available for this woman's transfusion, she might receive artificial blood.*
b. *Artificial blood cells called neohemocytes.*

Table 11.3 Hemoglobin	
Heme	**Globin**
Nonprotein	Protein
Contains iron	
Carries oxygen	Carries carbon dioxide; acts as a buffer by absorbing H^+
Becomes bile pigments	May be reused

Once systemic venous blood has reached the lungs, the reaction just described takes place in the reverse: the bicarbonate ion joins with a hydrogen ion to form carbonic acid and this splits into carbon dioxide and water. The carbon dioxide diffuses out of the blood into the lungs for expiration. Now hemoglobin is ready again to transport oxygen. Table 11.3 summarizes the structure and function of hemoglobin.

Hemoglobin also participates in the transport of carbon dioxide in the blood.

Blood Clotting

When an injury occurs to a blood vessel, **clotting,** or coagulation, of the blood takes place. This is obviously a protective mechanism to prevent excessive blood loss. As such, blood clotting is another mechanism by which blood components maintain homeostasis.

Portions of the blood that have been identified as necessary for clotting are (1) platelets, (2) prothrombin, a globulin protein, and (3) fibrinogen. **Platelets** (fig. 11.3) result from fragmentation of certain large cells, called megakaryocytes, in the red bone marrow. They are produced at a rate of 200 billion a day and the bloodstream possesses more than a trillion. **Fibrinogen** and **prothrombin** are manufactured and deposited in the blood by the liver. Vitamin K is necessary to the production of prothrombin, and if by chance this vitamin is missing from the diet, hemorrhagic disorders develop.

The steps necessary for blood clotting are quite complex but may be summarized in this simplified manner:

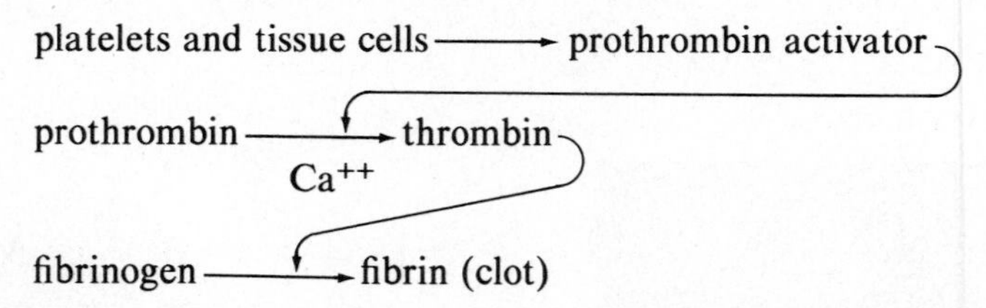

When a blood vessel is damaged, platelets clump at the site of the puncture and partially seal the leak. They and the injured tissues release an enzyme, called prothrombin activator, that converts prothrombin to thrombin. **Thrombin,** in turn, acts as an enzyme that severs two short amino acid chains from each fibrinogen molecule. These activated fragments then join end to end, forming long threads of **fibrin.** Fibrin threads wind around the platelet plug in the damaged area of the blood vessel and provide the framework for the clot. Red cells are also trapped within the fibrin threads (fig. 11.9) and their presence makes a clot appear red.

A blood clot consists of platelets and red cells entangled within fibrin threads.

If blood is placed in a test tube, blood clotting can be prevented by adding citrate or any other substance that combines with calcium. This is because calcium (Ca^{++}) ions are required for the blood clotting reactions to occur. Also since blood clotting is an enzymatic process, clotting takes place at a faster rate if blood is warmed than if it is kept cool. If the blood is allowed to

Figure 11.9
Scanning electron micrograph showing an erythrocyte caught in the fibrin threads of a clot. Fibrin threads form from activated fibrinogen, a normal component of blood plasma.

clot, a yellowish fluid comes to lie above the clotted material (fig. 11.10). This fluid is called **serum,** and it contains all the components of plasma except fibrinogen. Since we have now used a number of different terms to refer to portions of the blood, table 11.4 reviews these terms for you.

A fibrin clot is only a temporary way to repair the blood vessel. Eventually an enzyme called plasmin destroys the fibrin network and restores the fluidity of plasma. This is a protective measure because a blood clot can act as a thrombus or an embolus. In either case it interferes with circulation and may even cause the death of tissues in the area (p. 223).

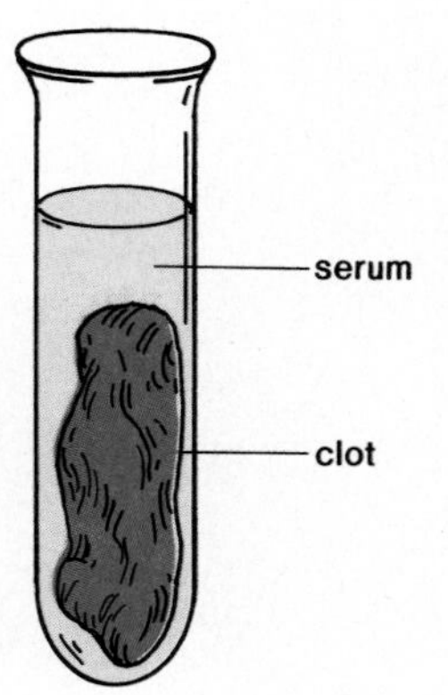

Figure 11.10
When blood clots, serum is squeezed out as a solid plug is formed. In a blood vessel this plug helps prevent further blood loss.

Infection Fighting

The body defends itself against parasites, such as bacteria and viruses, in several ways. The so-called first line of defense is the outer covering (skin and mucous membranes), which resists invasion by parasites. The second line of defense is dependent on two components of blood: white blood cells and gamma globulins. Infection fighting is the third of the three ways in which the blood components contribute to homeostasis.

White blood cells fight infection. They attack bacteria and viruses that have invaded the body.

Table 11.4 Body Fluids

Name	Composition
Blood	Formed elements and plasma
Plasma	Liquid portion of blood
Serum	Plasma minus fibrinogen
Tissue fluid	Plasma minus proteins
Lymph	Tissue fluid within lymph vessels

Table 11.5 White Cells (Leukocytes)

	Granulocytes (Polymorphonuclear)	
	Size	Granules stain
Neutrophils	9–12 μm	Lavender
Eosinophils	9–12 μm	Red
Basophils	9–12 μm	Deep blue
	Agranulocytes	
	Size	Type of nucleus
Lymphocytes	8–10 μm	Circular
Monocytes	12–20 μm	Indented

White Blood Cells (Leukocytes)

White cells may be distinguished from red cells in that they are usually larger; have a nucleus; and without staining, would appear to be white in color. With staining, white cells characteristically appear a bluish shade in color. White cells are less numerous than red cells, with only 7,000 to 8,000 cells per cubic millimeter.

Table 11.5 lists the different types of white cells and figure 11.11 shows two in detail. On the basis of structure, it is possible to divide white cells into the granulocytes and the agranulocytes. The **granulocytes** have granules in the cytoplasm and a many-lobed nucleus joined by nuclear threads; therefore, they are called polymorphonuclear. Granulocytes like red blood cells and platelets are formed in the red bone marrow. The **agranulocytes** do not have granules and have a circular, or indented, nucleus. They are produced in lymphoid tissue found in the spleen, lymph nodes, and tonsils.[2]

The granulocytes (e.g., neutrophils) are made in bone marrow; the agranulocytes (e.g., lymphocytes) are made in lymphoid tissue.

[2]Stem cells in the bone marrow produce specialized cells; certain ones seed the thymus, spleen, and lymph nodes and produce lymphocytes there.

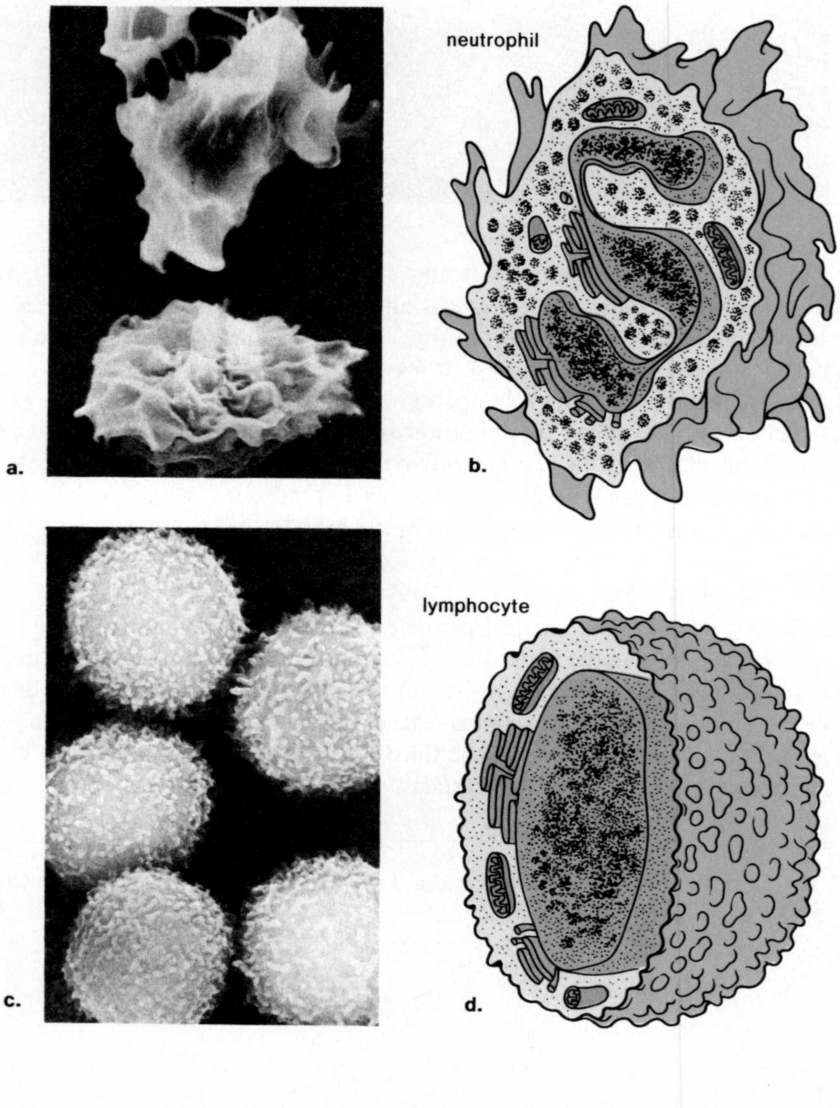

Figure 11.11
Two leukocytes of interest. a. and b. *Scanning electron micrograph of a neutrophil and artist's representation.* c. and d. *Scanning electron micrograph of a lymphocyte and artist's representation.*

Infection fighting by white cells is primarily dependent on the neutrophils, which comprise 60 to 70 percent of all leukocytes, and the lymphocytes which make up 25 to 30 percent of the leukocytes. Neutrophils are phagocytic; they destroy bacteria and viruses by traveling to the site of invasion and engulfing the foe. Like neutrophils, monocytes and eosinophils are phagocytic. Lymphocytes secrete gamma globulins[3] called immunoglobulins, or antibodies, that combine with foreign substances to inactivate them. Neutrophils and lymphocytes may be contrasted in the following manner.

Neutrophils	Lymphocytes
Granules in cytoplasm	No granules in cytoplasm
Polymorphonuclear	Mononuclear
Produced in bone marrow	Produced in lymphoid tissue
Phagocytic	Make antibodies

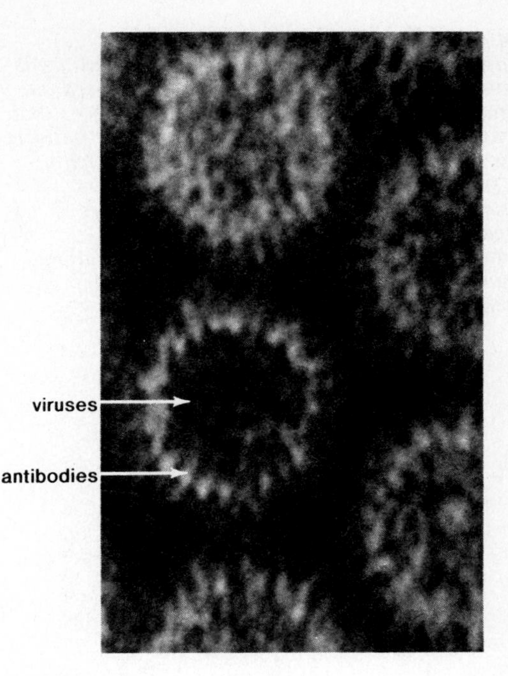

Figure 11.12
An electron micrograph that shows antibodies (light areas) *attached to viruses* (dark areas). *Each antibody will combine with only one type of antigen, usually a foreign protein such as those that occur in the coat of a virus.*

Antibodies

Parasites and their toxins cause lymphocytes to produce antibodies. Each lymphocyte produces one type of antibody that is specific for one type of antigen. **Antigens,** which are most often proteins, sometimes polysaccharides, are apt to be found in the outer covering of a parasite or present in its toxin. **Antibodies** combine with their antigens (fig. 11.12) in such a way that the antigens are rendered harmless. Sometimes the antibodies cause precipitation of the antigen, or agglutination (clumping) of the antigen, or simply prepare it for phagocytosis. In any case, it is well to keep in mind that the antigen is the foreigner, and the antibody is the substance prepared by the body. The antigen-antibody reaction may be symbolized as follows:

antigen	+	antibody	⟶ inactive complex
(foreign substance)		(globulin protein)	

The antigen-antibody reaction is a lock-and-key reaction in which the two molecules fit together like a lock and key. This seems surprising at first because it has been shown that all antibodies have the same overall shape. Even so, each type of antibody has a variable region, a unique sequence of amino acids that results in a receptor site capable of combining with one type antigen. In other words, this particular sequence of amino acids shapes a site where the antibody fits the antigen.

Immunity An individual is actively immune when the body has antibodies that can react to a specific disease-causing antigen. The blood in these individuals contains lymphocytes that are long capable of producing the necessary antibodies. Exposure to the antigen, either naturally or by way of a vaccine, can cause active immunity to develop. Chapter 12 deals with immunity and explores the topic in detail.

*L*ymphocytes are responsible for immunity and produce antibodies that specifically combine with disease-causing antigens.

[3]The gamma globulins get their name because it was observed that if globulins underwent electrophoresis (were put in an electrical field), they separated into three major components called alpha globulin, beta globulin, and gamma globulin. Almost all circulating antibodies were found in the gamma globulin fraction and, as a result, this term is used for circulating antibodies.

Figure 11.13
Inflammatory reaction. (above) *When a capillary is ruptured due to injury, substances are released into the tissues. Among these are certain precursors that quickly become bradykinin, a chemical that initiates nerve impulses, resulting in the sensation of pain, and stimulates mast cells to release histamine. Histamine causes a capillary to become more permeable.* (below) *Now neutrophils (and monocytes) squeeze through the capillary wall and phagocytize bacteria that have already been attacked by antibodies released by lymphocytes.*

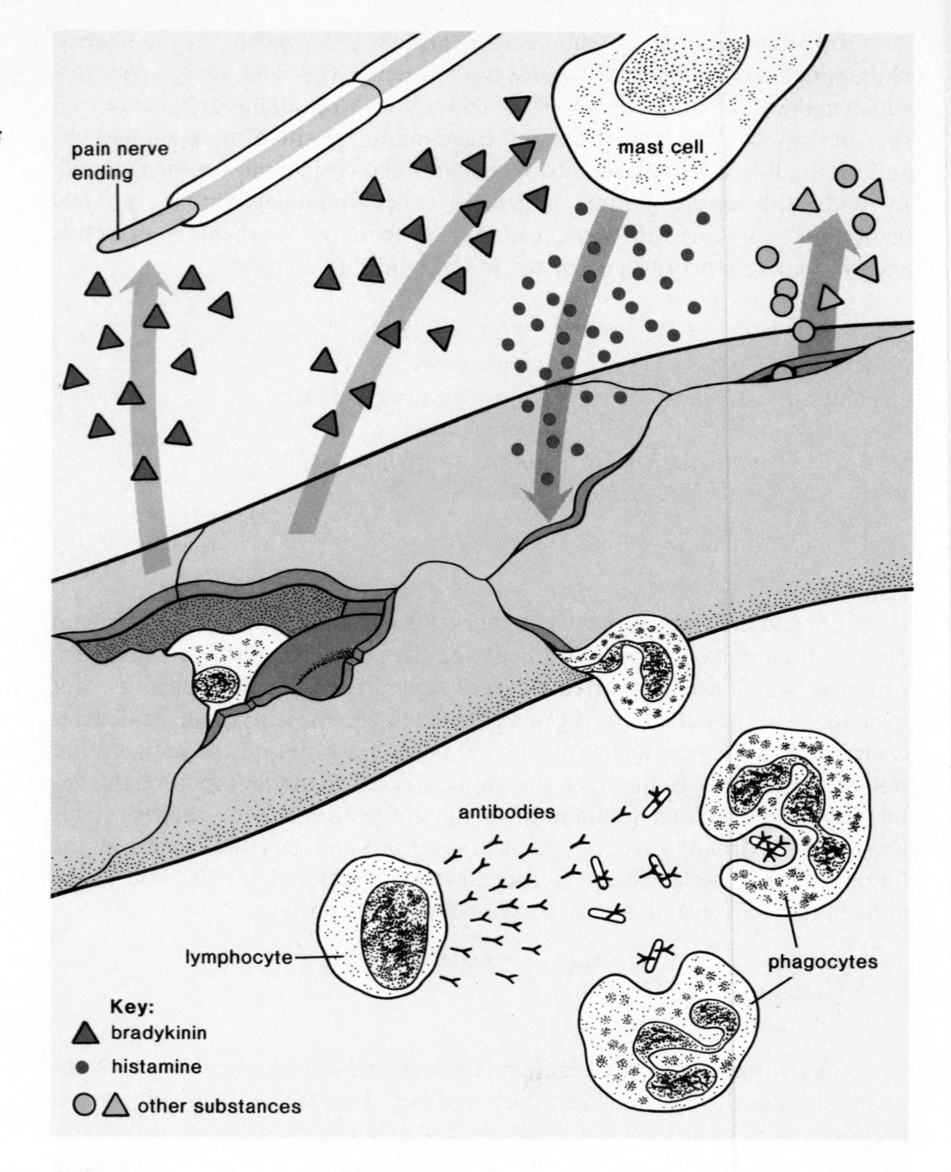

Diseases

Often illnesses cause an increase in a particular type of white cell. For this reason, a differential white cell count, involving the microscopic examination of a blood sample and the counting of each type of white cell, may be done as part of the diagnostic procedure. For example, the characteristic finding in the viral disease **mononucleosis** is a great number of lymphocytes that are larger than mature lymphocytes and stain more darkly. This condition takes its name from the fact that lymphocytes are mononuclear.

Leukemia **Leukemia** is a form of cancer characterized by uncontrolled production of abnormal white blood cells. These cells accumulate in the bone marrow, lymph nodes, spleen, and liver so that these organs are unable to function properly. Patients with leukemia are subject to severe anemia due to a low red cell count, clotting difficulties due to a low platelet count, and development of infections due to the presence of nonfunctioning white cells.

Since the initiation of "total therapy," the combined use of anticancer medication with radiation therapy, a large proportion of patients are able to remain symptom-free for years.

Inflammatory Reaction

Whenever the skin is broken due to a minor injury, a series of events occur that are known as the **inflammatory reaction** because there is swelling and reddening at the site of the injury. Figure 11.13 illustrates the participants in

Anchored by small microextensions, this macrophage has an irregular surface signifying it is ready to engulf invading bacteria in a fraction of a second.

the inflammatory reaction. One participant, the mast cells are derived from basophils, a type of white blood cell that takes up residence in the tissues.

When an injury occurs, a capillary and several tissue cells are apt to rupture and release certain precursors that lead to the presence of **bradykinin,** a chemical that (1) initiates nerve impulses resulting in the sensation of pain, (2) stimulates mast cells to release **histamine,** another chemical that together with bradykinin (3) causes the capillary to become enlarged and more permeable. The enlarged capillary causes the skin to redden and its increased permeability allows proteins and fluids to escape so that swelling results.

Any breakage of the skin will allow bacteria and viruses to enter the body. In figure 11.13, a lymphocyte is releasing antibodies that attack the bacteria (or viruses), preparing them for **phagocytosis.** When a neutrophil or monocyte phagocytizes a foreign substance, an intracellular vacuole is formed. Now the engulfed material is destroyed or neutralized by hydrolytic enzymes when the vacuole combines with a lysosome. (The granules of a neutrophil are in fact lysosomes.)

Once monocytes have arrived on the scene, they swell to five to ten times their original size and become **macrophages,** large phagocytic cells that are able to devour a hundred invaders and still survive. Some tissues, particularly connective tissues, have resident macrophages that routinely act as scavengers devouring old blood cells, bits of dead tissue, and other debris. Such macrophages are also capable of bringing about an explosive increase in the number of leukocytes, by liberating a substance that passes, by way of the blood, to the bone marrow, where it stimulates the production and release of white cells, usually neutrophils.

As the infection is being overcome, some neutrophils die and these, along with dead tissue, cells, bacteria, and living white cells, form **pus,** a thick yellowish fluid. The presence of pus indicates that the body is trying to overcome the infection.

The inflammatory reaction is a "call to arms"—it marshalls phagocytic white cells and antibody producing lymphocytes at a site of invasion by bacteria and viruses.

Capillary Exchange within the Tissues

Arterial Side

When arterial blood enters the tissue capillaries (fig. 11.14) it is bright red because the red cells are carrying oxygen. It is also rich in nutrients that are dissolved in the plasma. At this end of the capillary, blood pressure (25 mm Hg) is higher than the osmotic pressure of the blood (15 mm Hg). Blood pressure, you will recall, is created by the pumping of the heart; the osmotic pressure is caused by the presence of salts and also, in particular, by the plasma proteins that are too large to pass through the wall of the capillary. Since the blood pressure is higher than the osmotic pressure, fluid, together with oxygen and nutrients (glucose and amino acids), will exit from the capillary. This is a *filtration* process because large substances, such as red cells and plasma proteins, remain but small substances such as water and nutrient molecules leave the capillaries. Tissue fluid, created by this process, consists of all the components of plasma except the proteins.

Midsection

Along the length of the capillary, molecules will follow their concentration gradient as diffusion occurs. Diffusion, you will recall, is the movement of molecules from an area of greater concentration to an area of lesser concentration. The area of greater concentration for nutrients is always the blood, because after these molecules have passed into the tissue fluid they are taken up and metabolized by the tissue cells. The cells use glucose and oxygen in the process of cellular respiration, and they use amino acids for protein synthesis. Following cellular respiration, the cells give off carbon dioxide and water. Whenever the cells break down amino acids, they remove the amino group, which is released as ammonia. Carbon dioxide and ammonia, being waste products of metabolism, leave the cell by diffusion. Since tissue fluid is always the area of greater concentration for these waste materials, they diffuse into the capillary.

Oxygen and nutrient molecules (e.g., glucose and amino acids) exit a capillary near the arterial end; waste molecules (e.g., carbon dioxide and ammonia) enter a capillary near the venous end.

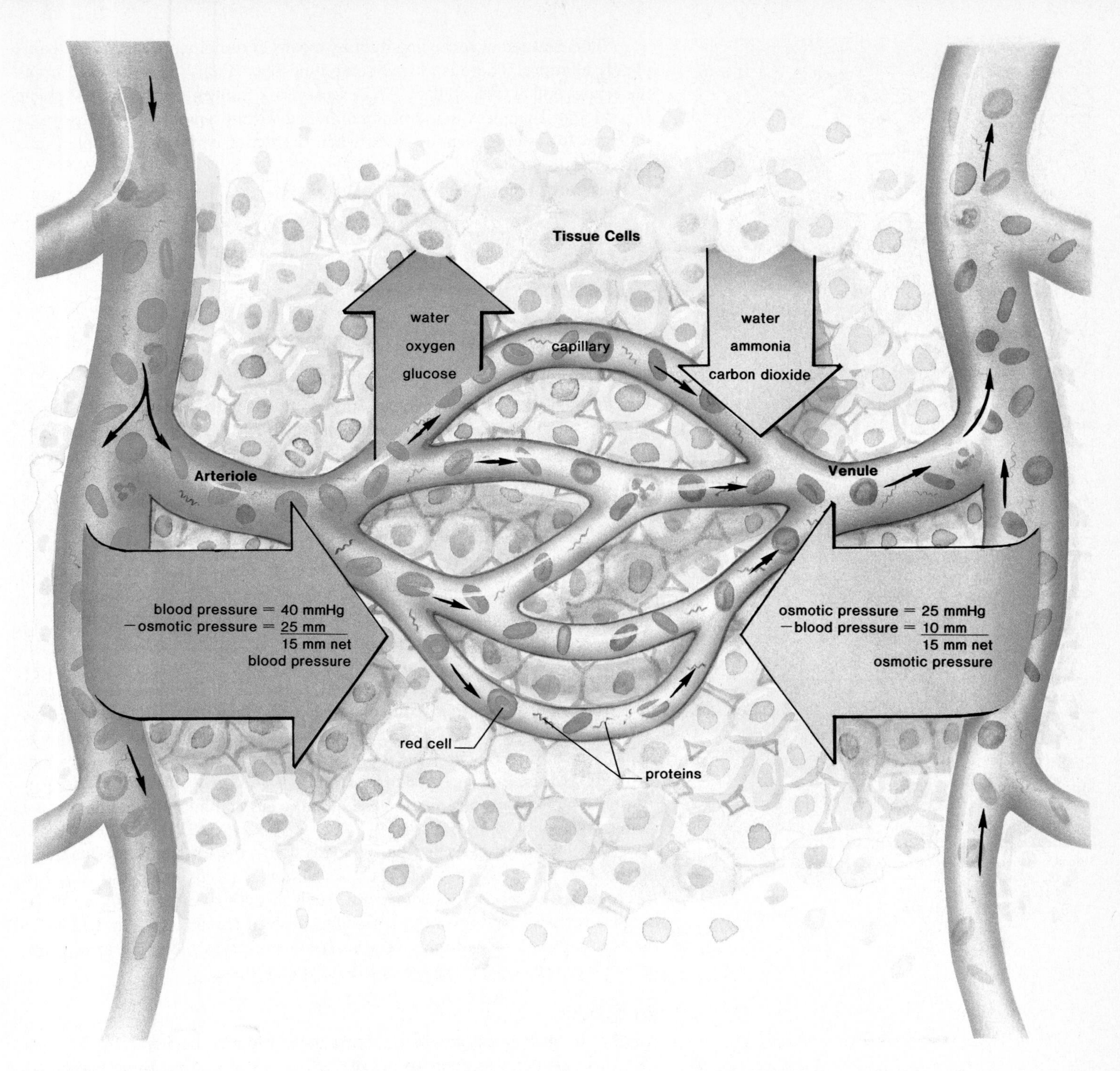

Venous Side

At the venous end of the capillary, blood pressure is much reduced (10 mm Hg), as can be verified by reviewing figure 10.15 in the previous chapter. However, there is no reduction in osmotic pressure (25 mm Hg), which tends to successfully force fluid into the capillary. When it does, it brings with it additional amounts of waste molecules (carbon dioxide and ammonia). As the blood leaves the capillaries, it is deep purple in color because the red cells contain reduced hemoglobin. Carbon dioxide is carried as the bicarbonate ion, and this, along with ammonia, is dissolved in the plasma.

Figure 11.14
Diagram of a capillary illustrating the exchanges that take place and the forces that aid the process. At the arterial end of a capillary, the blood pressure is higher than the osmotic pressure and therefore water, oxygen, and glucose tend to leave the bloodstream. At the venous end of a capillary, the osmotic pressure is higher than the blood pressure and, therefore, water, ammonia, and carbon dioxide tend to enter the bloodstream. Notice that the red cells and plasma proteins are too large to exit from a capillary.

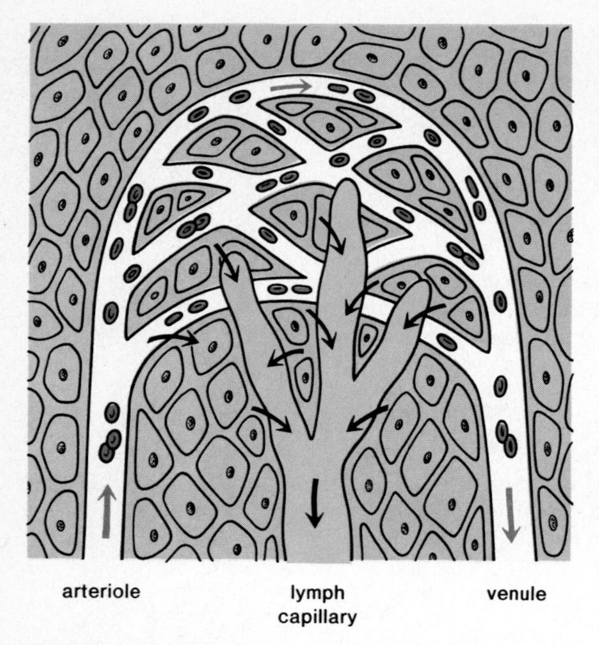

Figure 11.15
Lymph capillaries lie close to cardiovascular capillaries. Arrows indicate that tissue fluid is formed as water leaves blood capillaries and that thereafter some of it is taken up by lymph capillaries.

Table 11.6 Blood Groups				
Type	**Antigen**	**Antibody**	**% U.S.* Black**	**% U.S.* Caucasian**
A	A	b	25	41
B	B	a	20	7
AB	A,B	none	4	2
O	none	a,b	51	50

*Blood type frequency for other races is not available.

This method of retrieving fluid by means of osmotic pressure is not completely effective. There is always some fluid that is left and not picked up at the venous end of the capillary. This excess tissue fluid enters the lymph vessels (fig. 11.15). Lymph is tissue fluid contained within lymph vessels. Lymph is returned to systemic venous blood when the major lymph vessels enter subclavian veins (p. 227).

Lymphatic capillaries lie in close proximity to cardiovascular capillaries where they collect excess tissue fluid.

Blood Typing

ABO Grouping

Two antigens that may be present on the red cells have been designated as A and B. As table 11.6 shows, an individual may have one of these antigens (i.e., type A or type B), or both (type AB), or neither (type O). Therefore, blood type is dependent on which antigens are present on the red cells. As you can see, type O blood is most common in the United States.

In the ABO blood grouping system there are four types of blood: A, B, AB, and O. Type O blood has no antigens on the red cells; the other types of blood are designated by the antigens present on the red cells.

Within the plasma of an individual, there are antibodies to the antigens that are *not* present on that individual's red cells. Therefore, for example, type A blood has antibody b. Type AB blood has no antibodies because both antigens are on the red cells. This is reasonable because if the same antigen and antibody are present, **agglutination** or clumping of red cells will occur. Agglutination of red cells can cause the blood to stop circulating and death will follow.

For a recipient to receive blood from a donor, the recipient's plasma must not have an antibody that would cause the donor's cells to agglutinate. For this reason, it is important to determine each person's blood type. Figure 11.16 demonstrates a way to use the antibodies derived from plasma to determine the type of blood. If clumping occurs after a sample of blood is exposed to a particular antibody, the person has that type of blood.

Rh System

Another important antigen in matching blood types is the **Rh factor.** Persons with this particular antigen on the red cells are Rh positive; those without it are Rh negative. Rh negative individuals do not normally make antibodies to the Rh factor but they will make them when exposed to the Rh factor. It is possible to extract these antibodies and use them for blood type testing. When Rh positive blood is mixed with Rh antibodies, agglutination occurs.

The designation of blood type usually also includes whether the person has the Rh factor (Rh^+) or does not have the Rh factor (Rh^-) on the red cells.

The Rh factor is particularly important during pregnancy (fig. 11.17). If the mother is Rh negative and the father is Rh positive, the child may be Rh positive. The Rh positive red cells may begin leaking across into the mother's circulatory system as placental tissues normally break down before and

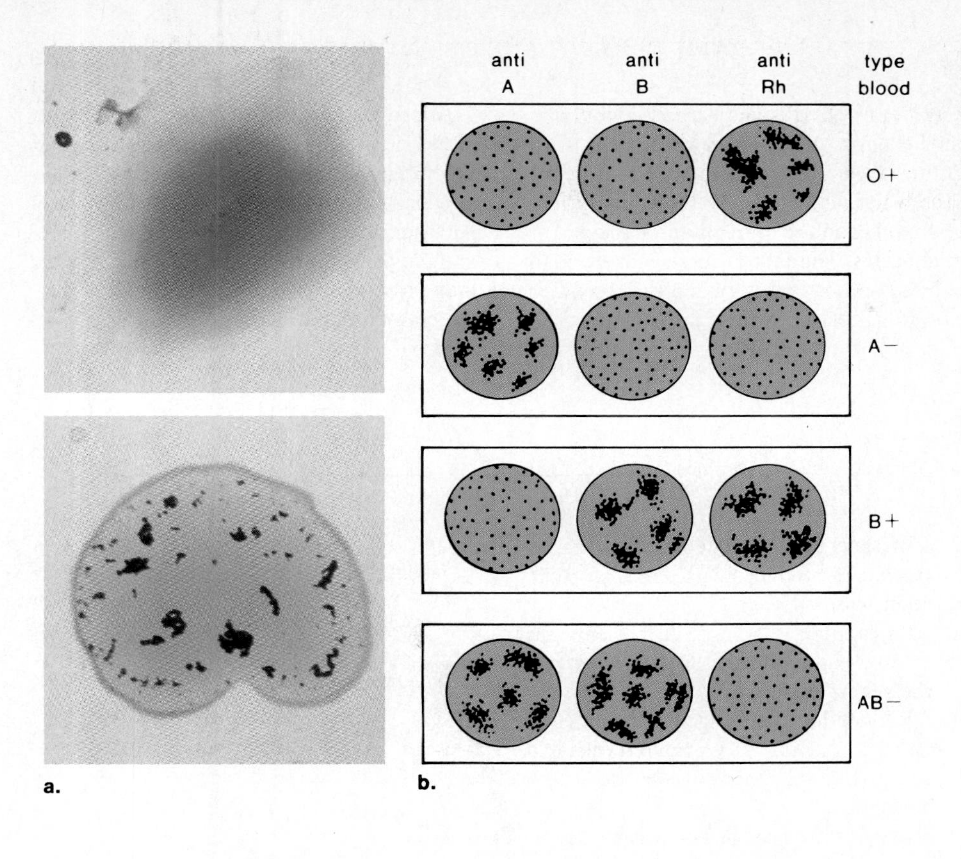

Figure 11.16
The standard test to determine ABO and Rh blood type consists of putting a drop of Anti-A antibodies, Anti-B antibodies, and Anti-Rh antibodies on a slide. To each of these a drop of the person's blood is added. a. *If agglutination occurs, as seen at lower left, the person has this antigen on the red cells.* b. *Several possible results.*

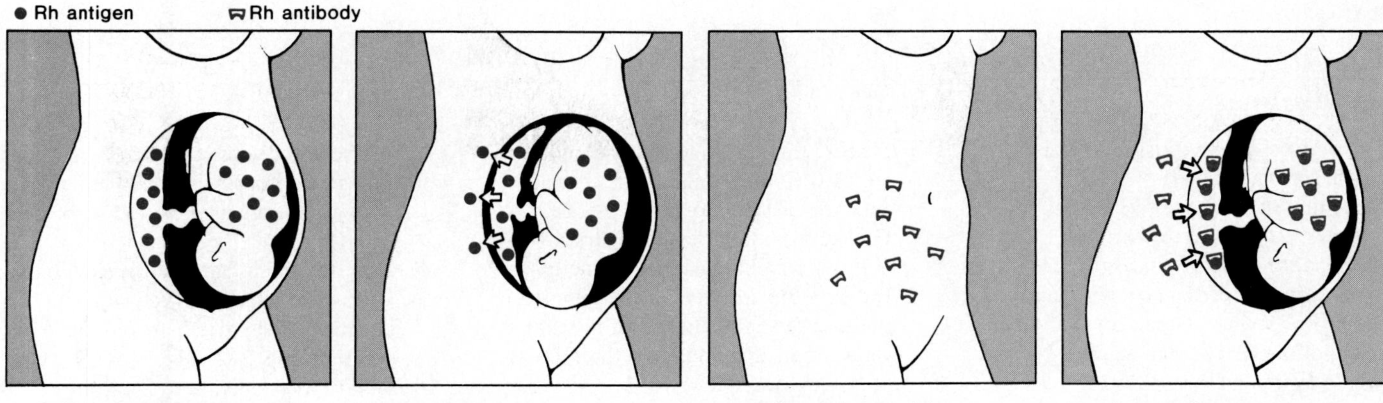

Figure 11.17
Diagram describing the development of erythroblastosis. a. *Baby's red blood cells carry Rh antigen.* b. *Some of these cells escape into mother's system.* c. *Mother begins manufacturing Rh antibodies.* d. *During a subsequent pregnancy, mother's Rh antibodies cross placenta to destroy the new baby's red cells.*

at birth. This causes the mother to produce Rh antibodies. If the mother becomes pregnant with another Rh positive baby, Rh antibodies (but not antibodies a and b discussed earlier) may cross the placenta and cause destruction of the child's red cells. This is called **fetal erythroblastosis**.

This problem has been solved by giving Rh negative women an Rh immune globulin injection called RhoGAM just after the birth of any Rh positive child. This injection contains Rh antibodies that attack the baby's red cells before these cells can stimulate the mother to produce her own antibodies. The RhoGAM injection is not beneficial if the woman has already begun to produce antibodies; therefore, the timing of the injection is most important.

The possibility of fetal erythroblastosis exists when the mother is Rh^- and the father is Rh^+.

Summary

Nutrients and wastes are transported in plasma but oxygen is combined with hemoglobin within red blood cells. The end result of transport is capillary exchange with tissue fluid, regulated by blood pressure and osmotic pressure. Blood clotting requires a series of enzymatic reactions involving platelets, prothrombin, and fibrinogen. In the final reaction fibrinogen becomes fibrin threads entrapping red cells.

White blood cells and gamma globulin proteins are required in the process of fighting infections. The two most prevalent of the white blood cells are the phagocytic neutrophils and the antibody-producing lymphocytes. The first of these is involved in the inflammatory reaction and the second is involved in development of immunity.

Blood transfusions require that the types of blood be compatible. Of consideration are the antigens (A and/or B) on the red cells and antibodies (a and/or b) in the plasma. Another important antigen is the Rh antigen particularly because an Rh^- mother may possess antibodies that will attack the red cells of an Rh^+ fetus.

Objective Questions

1. The liquid part of blood is called ________.
2. Red blood cells carry ________ and white blood cells ________.
3. Hemoglobin that is carrying oxygen is called ________.
4. Human red blood cells lack a ________ and only live about ________ days.
5. A blood clot occurs when fibrinogen becomes ________ threads.
6. The most common granulocyte is the ________, a phygocytic white blood cell.
7. Lymphocytes are made in ________ tissue and produce ________ that react with antigens.
8. At a capillary, ________ leave the arterial end and ________ enter the venous end.
9. AB blood has the antigens ________ and ________ on the red cells and ________ antibodies in the plasma.
10. Fetal erythroblastosis can occur when the mother is ________ and the father is ________.

Answers to Objective Questions

1. plasma 2. oxygen, fight infection 3. oxyhemoglobin 4. nucleus, 120 5. fibrin 6. neutrophil 7. lymphoid, antibodies 8. oxygen, amino acids, glucose; carbon dioxide, ammonia 9. A and B, no 10. Rh^-, Rh^+

Study Questions

1. Define blood, plasma, tissue fluid, lymph, serum. (p. 239)
2. Name three functions of blood and tell how they are related to the maintenance of homeostasis. (p. 231)
3. State the major components of plasma. Name the plasma proteins and tell their common function as well as their specific functions. (pp. 232–233)
4. Give the equation for the oxygenation of reduced hemoglobin. Where does this reaction occur? Where does the reverse reaction occur? (p. 235)
5. Discuss the life cycle of red blood cells. (p. 235) Compare the structure and function of heme to globin. (p. 238)
6. Give an equation that indicates how CO_2 is commonly carried in the blood. Indicate the direction of the reaction in the tissues and in the lungs. In what ways does hemoglobin aid the process of transporting CO_2? (p. 236)
7. Name the steps that take place when blood clots. Which substances are present in the blood at all times and which appear during the clotting process? (p. 238)
8. Name and discuss two ways that blood fights infection. Associate each of these with a particular type of white blood cell. (p. 239)
9. Describe the inflammatory reaction and give a role for each type cell and chemical that participates in the reaction. (pp. 242–244)
10. What forces operate to facilitate exchange of molecules across the capillary wall? (pp. 244–245)
11. What are the four ABO blood types in humans? (p. 246) For each, state the antigen(s) on the red blood cells and the antibody(ies) in the plasma. (p. 246)
12. Problems can arise during childbearing if the mother is which Rh type and the father is which Rh type? Explain why this is so. (pp. 246–247)

Key Terms

agglutination (ah-gloo″tĭ-na′shun) clumping of cells, particularly in reference to red blood cells involved in an antigen-antibody reaction. *246*

agranulocytes (ah-gran′u-lo-sīts″) white blood cells that do not contain distinctive granules. *240*

antibody (an′tĭ-bod″e) a protein produced in response to the presence of some foreign substance in the blood or tissues. *241*

antigen (an′tĭ-jen) a foreign substance, usually a protein, that stimulates the immune system to produce antibodies. *241*

clotting (klot′ing) process of blood coagulation, usually when injury occurs. *238*

erythroblastosis (ĕ-rith″ro-blas-to′sis) destruction of Rh^+ fetal red cells due to antibodies produced by a mother who is Rh^-. *247*

fibrinogen (fi-brin′o-jen) plasma protein that is converted into fibrin threads during blood clotting. *238*

formed element (form′d el′ĕ-ment) a constituent of blood that is either cellular (red cells and white cells) or at least cellular in origin (platelets). *231*

granulocytes (gran′u-lo-sīts) white blood cells that contain distinctive granules. *240*

inflammatory reaction (in-flam′ah-to″re re-ak′shun) a tissue response to injury that is characterized by dilation of blood vessels and an accumulation of fluid in the affected region. *242*

macrophage (mak′ro-fāj) an enlarged monocyte that ingests foreign material and cellular debris. *243*

phagocytosis (fag″o-si-to′sis) the taking in of bacteria and/or debris by engulfing. *243*

platelet (plāt′let) a formed element that is necessary to blood clotting. *238*

prothrombin (pro-throm′bin) plasma protein that is converted to thrombin during the process of blood clotting. *238*

pus (pus) thick yellowish fluid composed of dead phagocytes, dead tissue, and bacteria. *244*

serum (se′rum) light-yellow liquid left after clotting of the blood. *239*

thrombin (throm′bin) an enzyme that converts fibrinogen to fibrin threads during blood clotting. *238*

12

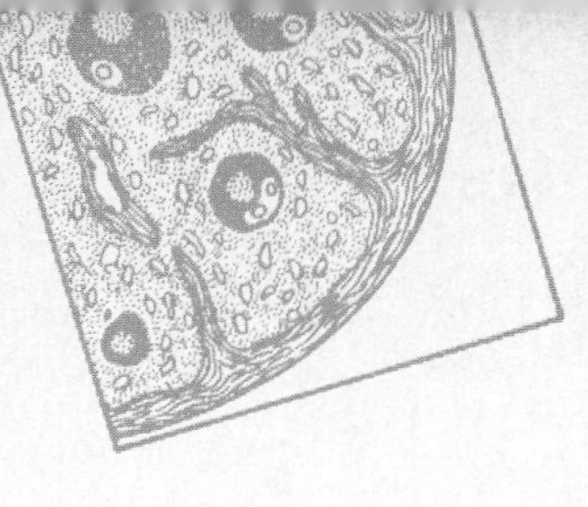

Immunity

Chapter Concepts

1 The immune system consists of lymphocytes and the structures that produce them and within which they reside.

2 One type of lymphocyte, the B cell, is responsible for antibody-mediated immunity and another type, the T cell, is responsible for cell-mediated immunity.

3 While the immune system preserves our existence, it is also responsible for certain undesirable effects, such as tissue rejection, allergies, and autoimmune diseases.

4 Active immunity, the ability of the body to produce specific antibodies, can be promoted by immunization.

5 Passive immunity, acquired when antibodies are received from an outside source, has gained importance due to the production of specific human antibodies in the laboratory.

Chapter Outline

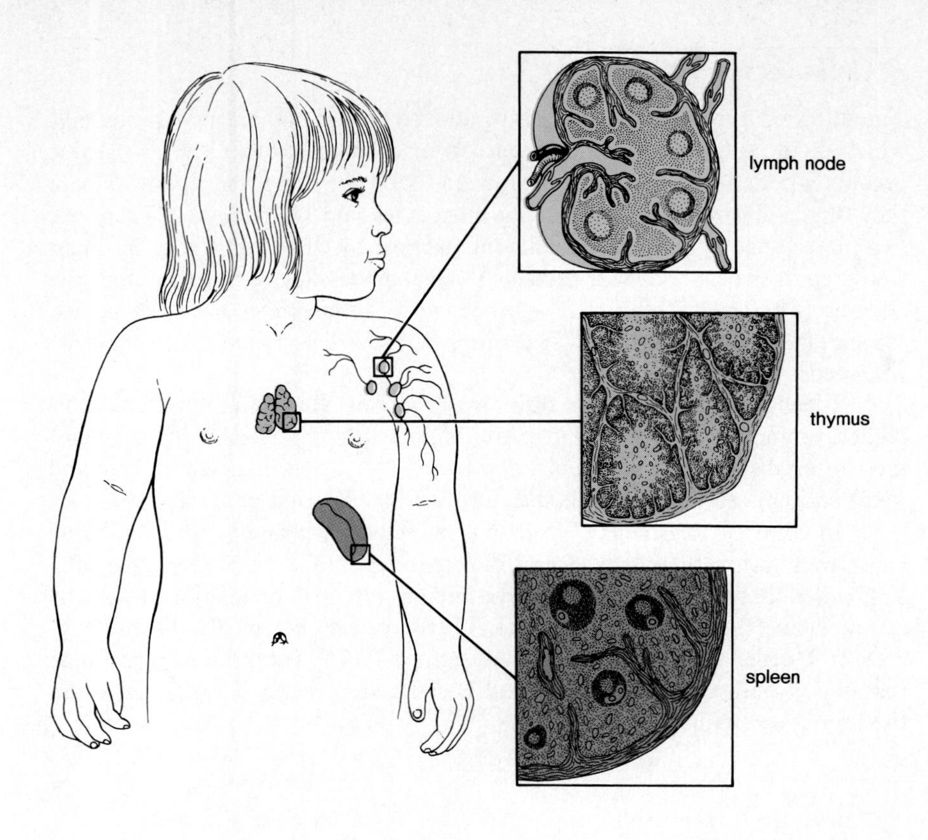

Figure 12.1
The immune system includes all those organs in which lymphocytes are produced and found, such as lymph nodes, thymus, and spleen. The thymus is important to the maturation of T cells, a particular type of lymphocyte. The lymphocytes and monocytes within lymph nodes help purify lymph and those within the spleen help purify the blood.

The immune system (fig. 12.1) consists of about a trillion agranulocytes (lymphocytes and monocytes); the spleen, thymus, and lymph nodes (p. 228) where they are produced, and the lymph vessels where these cells are originally found. Following maturation, agranulocytes are also found in the bloodstream.

The immune system protects us from disease. For example, it destroys bacteria and viruses and the cells they have infected. It also does away with foreign cells and cancerous cells. If functioning properly, it does not attack the body's normal cells.

The immune system includes the lymphoid organs and the agranulocytes.

Immunity is dependent on the proper functioning of two types of lymphocytes, the B cells and the T cells.

B Cells versus T Cells

The different types of blood cells are derived from bone marrow stem cells, which divide to give a specific precursor for each cell type. Certain of these precursor cells leave the bone marrow and seed the lymphoid organs, where they divide and produce functioning monocytes and lymphocytes. There are two main types of lymphocytes, distinguishable by their maturation process. The precursor cells that will become **T lymphocytes** have passed through the *thymus* (fig. 12.1). Those that will become **B lymphocytes** have not passed through the thymus and therefore are derived more directly from *bone marrow* stem cells.

T cells and B cells have different functions. B cells (1) produce antibodies, proteins that are capable of combining with and inactivating antigens, most often disease-causing agents and their toxins (poisons) (tables 12.4 and 12.5) and (2) release the antibodies into the bloodstream or lymph.

In contrast to B cells, T cells do not release antibodies into the blood and lymph. Also, there are several different types of T cells. One type of T cell, called a *cytotoxic T cell* destroys foreign, infected, or malignant cells by lysing them (causing them to burst). There are also helper T cells and suppressor T cells. As their name implies, *helper T cells* potentiate the immune reaction of both cytotoxic T cells and B cells. *Suppressor T cells* shut down the immune response.

There are two types of lymphocytes: the B cells are derived more directly from bone marrow stem cells. The T cells are derived from cells that have passed through the thymus. Both B and T cells are produced in lymphoid organs.

B Lymphocytes

Each B cell (and T cell) will respond to only one type antigen. It is estimated that during our lifetime we encounter a million different antigens and that we need the same number of different lymphocytes to protect ourselves against these antigens. It is remarkable to think that so much diversification occurs during the maturation process that in the end there is a different type cell for each possible antigen. None of these cells, though, are supposed to attack the body's normal cells. It is believed that if by chance a lymphocyte develops that could respond to the body's own proteins, it is normally suppressed and develops no further.

When an antigen is present in the bloodstream (fig. 12.2) an appropriate B cell recognizes its presence and begins to divide, producing **plasma cells** that actively secrete antibodies. All the plasma cells derived from one parent lymphocyte are called a clone, and they all produce the same type antibody. Notice that the particular antigen determines which lymphocyte will be stimulated to produce antibodies. This is called the *clonal selection theory* because the antigen has selected which B cells will produce a clone of plasma cells.

Once antibody production is high enough, suppressor T cells prevent further antibody production; the antigen disappears from the system and development of plasma cells ceases. Some members of the clone do not participate in the current antibody production; instead they remain in the bloodstream as

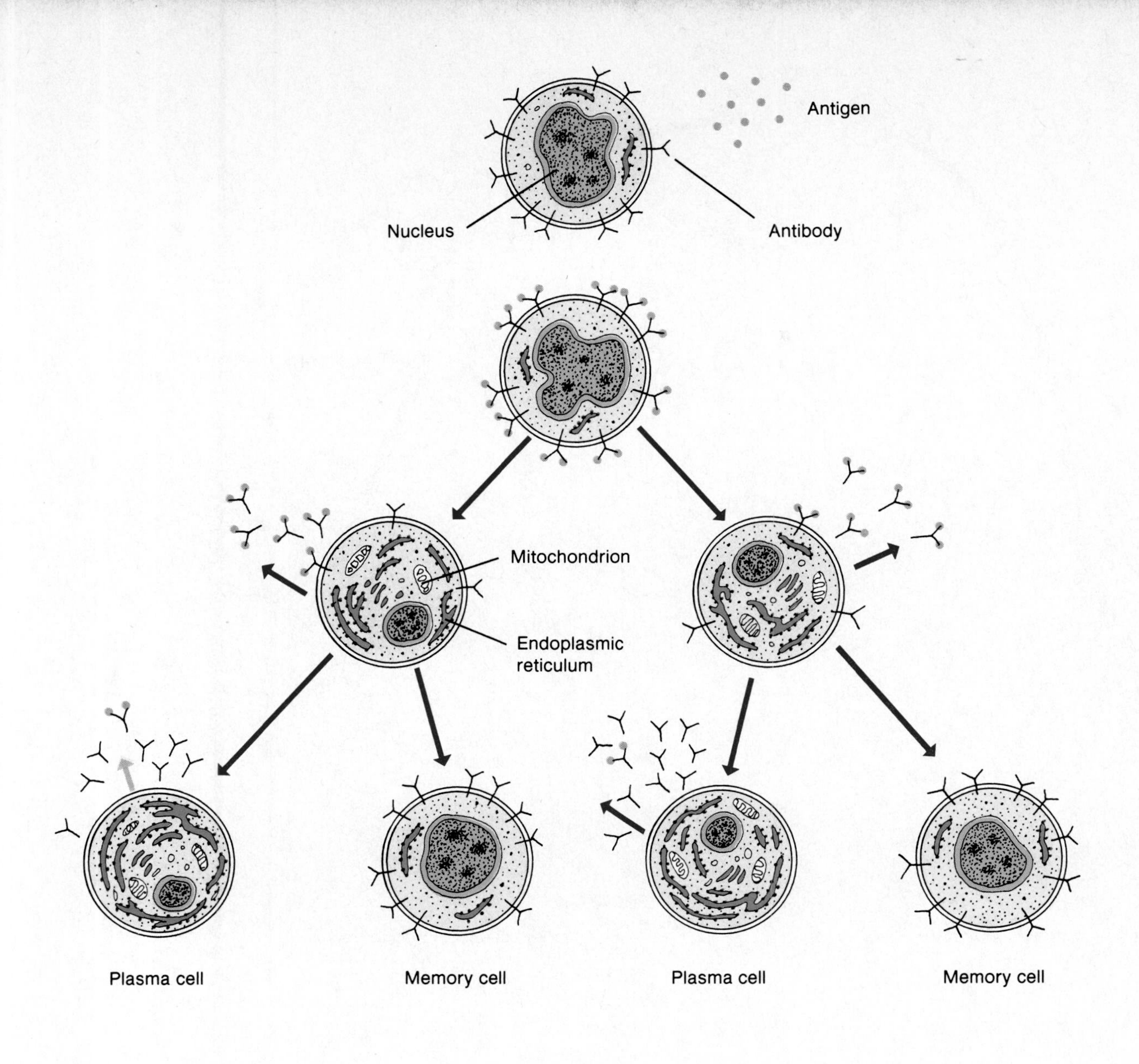

memory cells capable for some time of producing the antibody specific to a particular antigen. As long as this is the case, the individual is said to be *actively immune,* because a certain number of antibodies are always present and also because memory cells can produce more plasma cells if the same antigen invades the system again.

Defense by B cells is called *antibody-mediated immunity* because B cells produce antibodies (fig. 12.3 and table 12.1). Sometimes it is also called humoral immunity because these antibodies are present in the bloodstream.

Figure 12.2
Clonal selection theory. Each type B cell has membrane-bound antibodies that combine with only a particular antigen. The antigen finds or selects this B cell and it clones, producing, by the fifth day, many mature plasma cells, which actively secrete antibodies, and memory cells that retain the ability to secrete these antibodies at a future time.

B cells are responsible for antibody-mediated immunity.

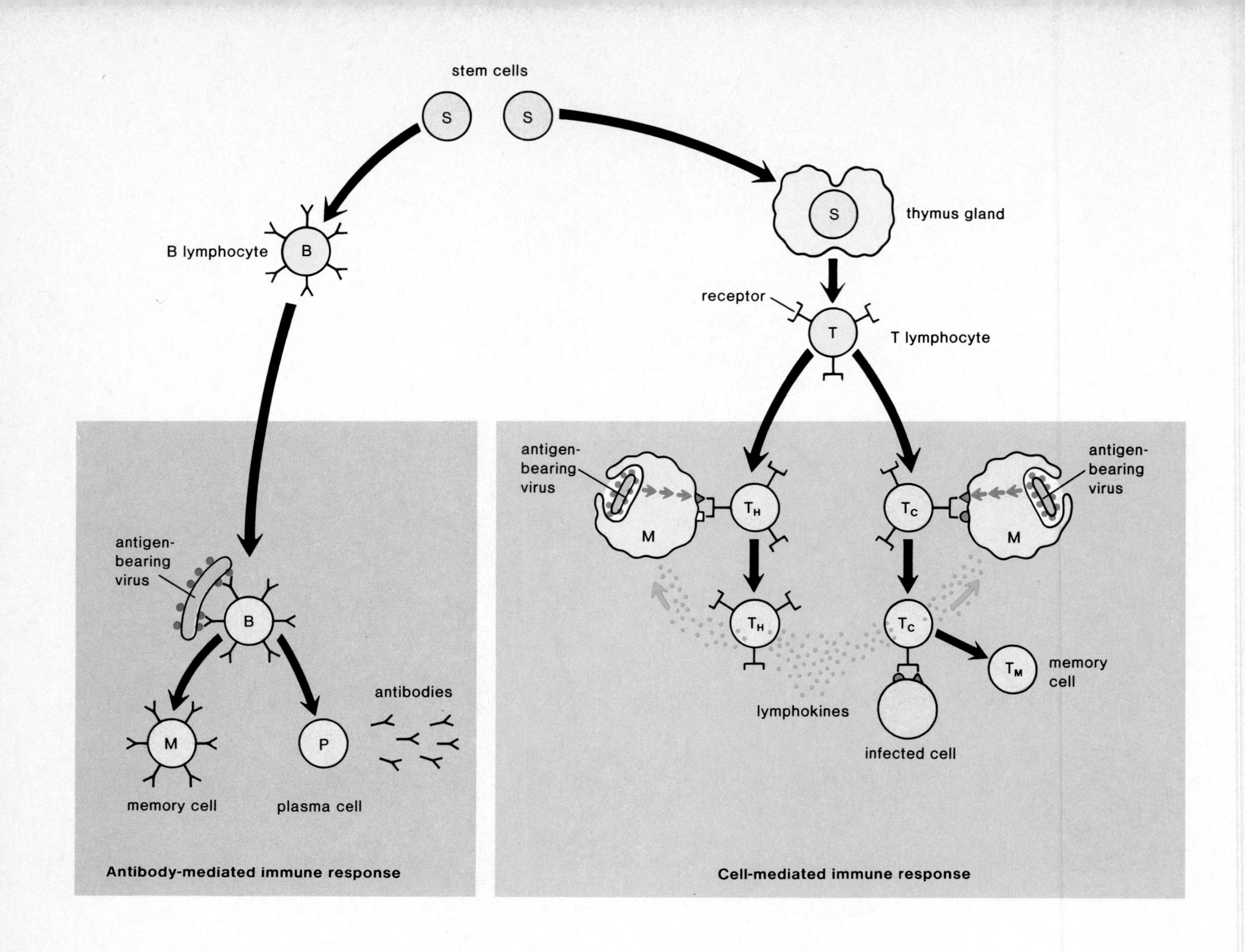

Figure 12.3
Bone marrow stem cells produce precursor cells that take up residence in the lymphoid organs where they produce the B cells and T cells. Only if the precursor cells are in or have passed through the thymus do they produce T cells. When a B cell recognizes an antigen, it divides to give antibody secreting plasma cells and memory cells. In this way B cells are responsible for antibody-mediated immunity. When helper T cells (T_H cells) recognize an antigen, because it has been presented to them by a macrophage (M), they produce lymphokines and otherwise stimulate other immune cells, including B cells and cytotoxic T cells. T_C cells are responsible for cell-mediated immunity.

***Table 12.1** Some Properties of B Cells and T Cells*

Property	B Cells	T Cells
Antigen recognition	Direct recognition	Must be presented by macrophages
Early response	Produce plasma cells	Enlarge, multiply, and secrete lymphokines
Later response	Antibody production	Helper cells stimulate other cells; cytotoxic cells attack antigen-bearing cells
Final response	Memory cells	Memory cells; suppressor cells shut down immune response

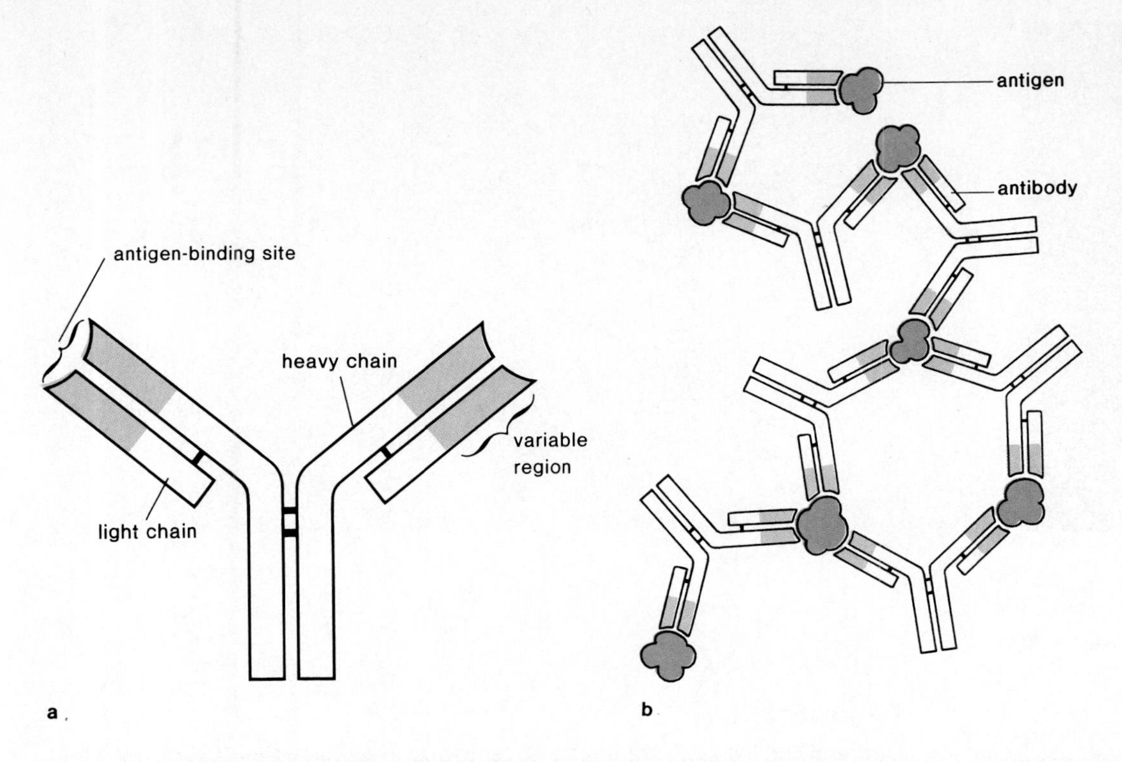

Figure 12.4
Antigen-antibody reaction. a. *An antibody contains two heavy* (long) *amino acid chains and two light* (short) *amino acid chains arranged to give two variable regions where a particular antigen is capable of binding with the antibody in a lock-and-key manner.* b. *Quite often the antigen-antibody reaction produces complexes of antigens combined with antibodies.*

Table 12.2 Reactions of Antibodies

Types of Antibodies	Reactions
Antitoxins	Neutralize toxins of infective agents
Agglutinins	Cause clumping of certain infective agents
Opsonins	Make certain infective agents more susceptible to work of phagocytes
Lysins	Dissolve certain infective agents
Precipitins	Bring about a precipitation or flocculation of extracts of infective agents

From *Living: Health, Behavior, and Environment*, 5th ed. by Fred V. Hein, Dana L. Farnsworth, and Charles E. Richardson. © 1970 Scott, Foresman and Company. Reprinted by permission of the author.

Antigen-Antibody Complexes

An antibody is a Y-shaped molecule having two long "heavy" chains and two short "light" chains of amino acids. At the end of each arm of every antibody is a tiny segment, called the variable portion, that binds to an antigen in a lock-and-key manner. In figure 12.4, the variable region of each antibody is shown in color.

The antigen-antibody reaction can take several forms as indicated in table 12.2. Many times, the antigen-antibody complex, sometimes called the **immune complex,** marks the antigen for destruction by other forces. For example, the complex may be engulfed by neutrophils or macrophages, or it may activate a portion of blood serum called the complement system. **Complement** refers to twenty different proteins that stimulate phagocytic cells and also bring about lysis of bacterial cells in the following manner. Nine of the proteins form a channel that allows water and salts to enter the cell until it bursts (fig. 12.5).

An antibody combines with its antigen in a lock-and-key manner. The antigen-antibody reaction can lead to complexes that contain several antibodies and antigens.

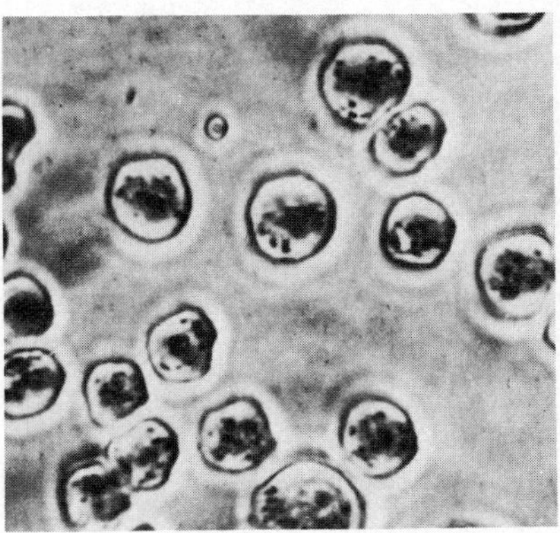

a.

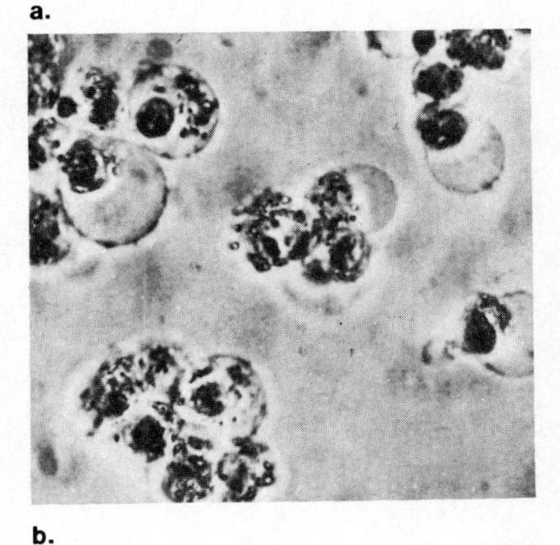

b.

Figure 12.5
Photomicrographs of tumor cells. a. *Normal appearance.* b. *After treatment with antibody and complement, the cells swell and burst. Complement is a series of about nine proteins that are converted to enzymes when activated by immune complexes.*

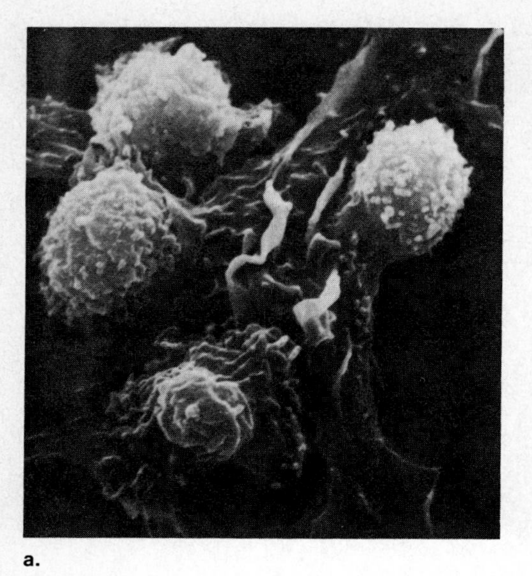

a.

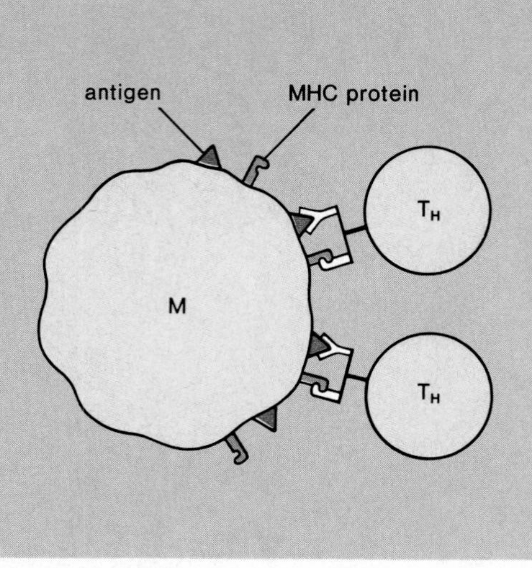

b.

T_H B

c.

Figure 12.6
a. *Scanning electron micrograph of macrophage with adhering T cells.* b. *T cells have membrane receptors that combine with only a particular antigen. Only when a macrophage presents a processed antigen along with an MHC protein can a T cell recognize an antigen. After a helper T cell recognizes an antigen it can stimulate a B cell by direct contact.*

T Lymphocytes

Unlike B cells, T cells are unable to recognize an antigen that is simply present in lymph or blood. Instead, the antigen must be presented to them by a (monocyte-derived) macrophage. When a *macrophage* engulfs a bacteria or virus it enzymatically breaks it down and displays only a protein fragment on the cell membrane together with a **MHC** (major histocompatible complex) **protein** (fig. 12.6). The importance of these proteins was first recognized when it was discovered that these cell membrane proteins contribute to the specificity of tissues and make it difficult to transplant a tissue from one person to another. In other words, the donor and the recipient must be histo- (tissue) compatible (the same or nearly so) for a transplant to be successful without the administration of immunosuppressive drugs.

In order for a T cell to recognize an antigen, the antigen must be presented along with a MHC protein to the T cell by a macrophage.

Once T cells have recognized the antigen, they enlarge and secrete lymphokines including interferons and interleukins.

Lymphokines are stimulatory chemicals for all types of immune cells which now begin to proliferate. Among these are **helper T cells** (T_H cells) which directly combine with B cells (fig. 7.6c) stimulating them to divide and produce plasma and memory cells. Because the AIDS (acquired immune deficiency syndrome) virus attacks helper T cells, it also inactivates the antibody producing capability of the immune system. This is discussed in the reading on page 258.

T cells secrete lymphokines, chemicals that powerfully stimulate the immune response.

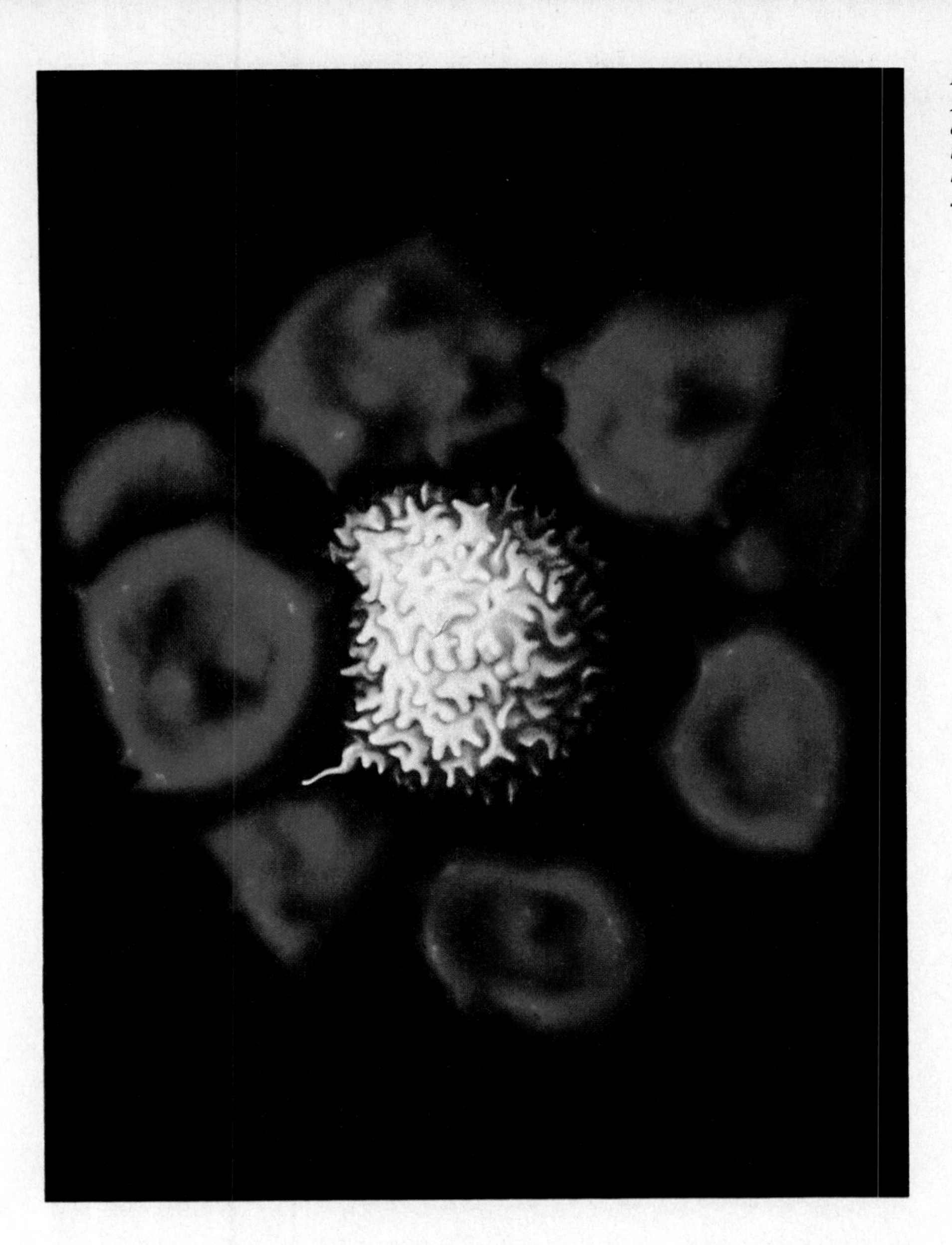

Figure 12.7
Activated T lymphocyte surrounded by red blood cells. T cells are responsible for cell-mediated immunity and attack all cells to which they have been synthetized. In this instance T cells have been sensitized to the red cells of another individual.

Cytotoxic T cells (T_C cells) actively defend the body by attacking and lysing cells (fig. 12.7) that display the antigen they recognize. Cytotoxic T cells have a more obvious function than helper or suppressor T cells—they are responsible for *cell-mediated immunity* (fig. 12.3).

Suppressor T cells (T_S cells) increase in number more slowly than the other two types of T cells just discussed. Once there is sufficient number, however, the immune response ceases. Following suppression, a population of **memory T cells** (T_M cells) also persists, perhaps for life.

Cytotoxic T cells are responsible for cell-mediated immunity. T_H cells promote the immune response, and T_S cells suppress the immune response.

Treatment for AIDS

AIDS (acquired immune deficiency syndrome) is characterized by the destruction of the immune system, which leads to various infectious diseases and a virulent type of skin cancer (Kaposi's sarcoma). Many victims also show early brain impairment with ever-increasing neuromuscular and psychological consequences.

Since the number of new cases of AIDS will be at least thirty thousand this year and victims rarely live longer than four years, there is great interest in developing a suitable treatment and a vaccine to prevent infection in the first place.

One name given to the virus is HTLV-III (human T cell lymphotropic virus) because it primarily attacks

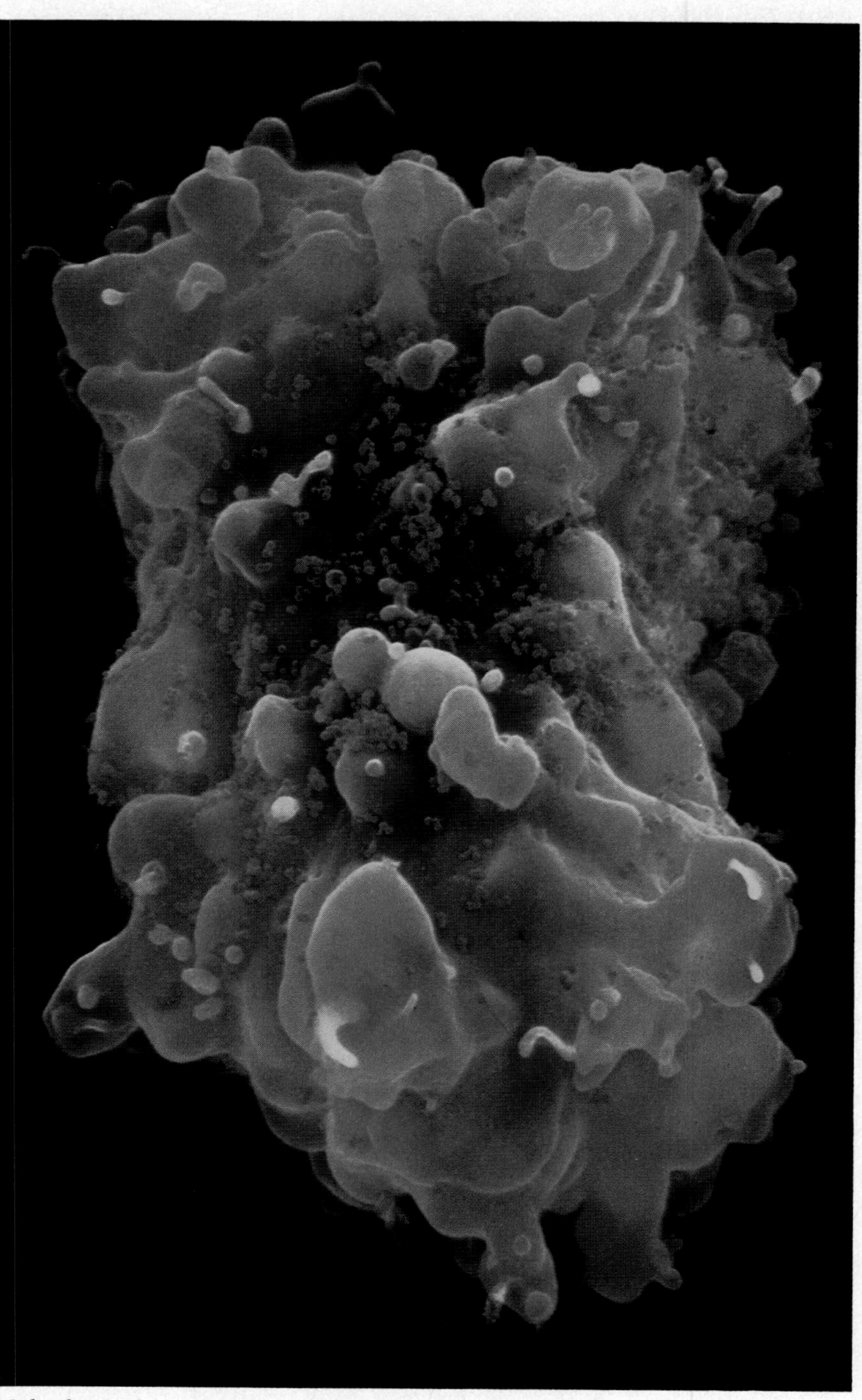

Multiple numbers of the AIDS virus are seen escaping from this infected T cell.

helper T cells and slows their growth. (It is possible that T cells are most susceptible after they have proliferated due to a previous infection with the hepatitis B or the Epstein Barr virus.) Once the AIDS virus has taken hold, very few helper T cells can be found in the bloodstream. The virus also attacks those immunoglobulin cells that interact with T cells, such as macrophages and B cells. Secondarily, it is found in skin cells, in endothelial cells lining the blood and lymph vessel, and in cells of the nervous system.

Since antibodies to the virus are found in the bloodstream, it would seem that at first the body is able to mount an offensive to the invader. Eventually, however, the body succumbs. Various explanations have been given: (1) Perhaps the virus spreads by causing immunoglobulin cells to fuse with one another and therefore it need never return to the bloodstream where antibodies are located. (2) There is evidence that the AIDS virus alters the MHC proteins that appear on the surfaces of infected cells. This would make T cells incapable of recognizing and destroying infected cells. (3) The AIDS virus mutates so rapidly that antibody production lags behind and never catches up enough to be useful. Then, too, the virus invades the very cells that are sent to destroy it and this may cause antibody production to never be adequate.

Many viruses consist of double-stranded DNA surrounded by a capsid (protein coat). When they invade a cell, the viral DNA takes over the machinery of the host cell and orders it, by way of RNA, to produce copies of the virus, which either bud from the host cell or burst forth to infect other cells. The AIDS virus is a retrovirus, so named because it contains single-stranded RNA which serves as a template for the production of DNA—the reverse of the normal sequence of steps. One drug that has shown some promise this past year, AZT (azidothymidine), interferes with the ability of the virus to perform this RNA—DNA step because the drug is used in place of thymidine, one of the usual bases in DNA. The result is nonfunctioning DNA that cannot lead to the production of new viruses. A possible drawback to the use of this drug is that it would inhibit production of all DNA and not just the viral DNA. Therefore, the drug itself might lead to a decrease in the number of T cells in the blood!

Other investigators are working on the development of a vaccine that could be used to make people immune to the AIDS virus. A vaccine has to be able to promote antibody production without actually causing the symptoms of the disease. Perhaps it would be possible to alter the AIDS virus so that the virus itself could be used as a vaccine. The AIDS virus is believed to have evolved from a virus that commonly infects African green monkeys, apparently causing them no harm. It may have spread to humans in Africa either by their being bitten or by their eating monkey meat. Thereafter it mutated to its present deadly form.

Investigators have found a related nonvirulent virus in the people of Senegal, Africa. There is hope that this harmless relative will help determine the difference between virulent and nonvirulent strains of the virus. Perhaps this knowledge can be used to help develop a harmless viral vaccine for AIDS.

A concern is that any vaccine would have to enable the immune system to recognize the many variants of the AIDS virus. Or it may be possible to find a portion of the virus that does stay constant in the various strains and could be used as an effective vaccine. Recently, investigators placed only a portion of the genetic material from the AIDS virus in vaccinia, a live virus that serves as a smallpox vaccine. They were gratified to see that the recipient, himself an investigator, developed antibodies to the AIDS virus. In laboratory tests these antibodies were effective against more than one strain.

While there is much to be learned about AIDS, we do at least have knowledge of its means of transmission. We know that AIDS is transmitted by blood-to-blood contact. This puts intravenous drug abusers at great risk. Also, infected semen (and possibly saliva) passes the disease to a sexual partner when it comes in contact with the partner's bloodstream. This can happen if a women is menstruating, if the partner has an open abrasion, or if the sexual act itself damages the mucosal lining.

There is little danger of contracting AIDS by casual encounters between individuals, such as shaking hands, touching the same utensils, breathing the same air, etc. However, the public is advised to avoid promiscuity. At present, individuals can best protect themselves by practicing monogamy (always the same sexual partner) with someone who is free of the disease.

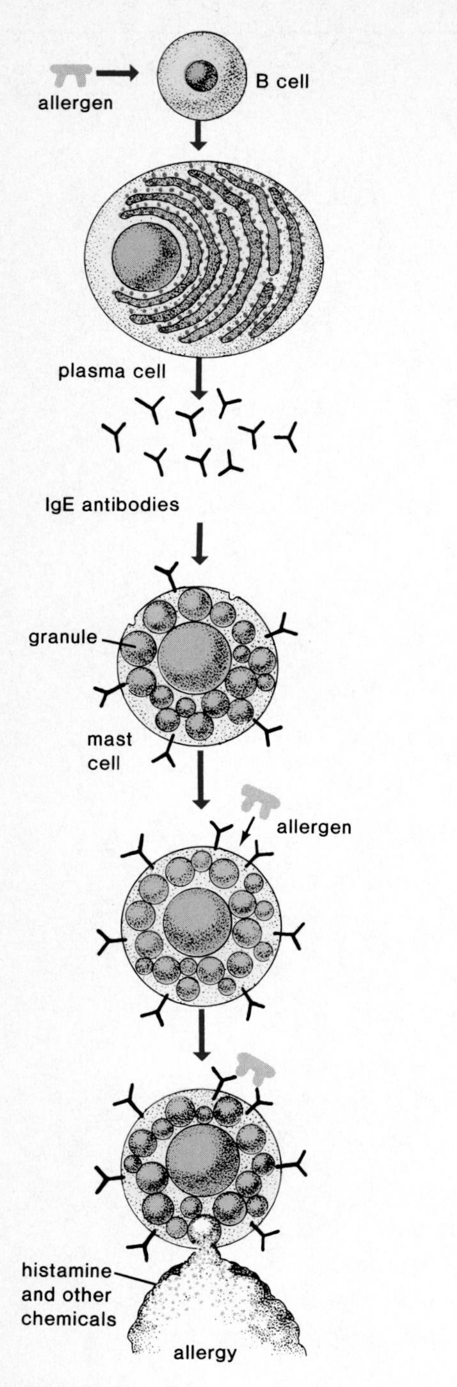

Figure 12.8
Allergic response. An allergen causes a plasma cell to secrete IgE antibodies. Some of these bind to mast cells. Then, when the individual is exposed again, the allergen attaches to these IgE antibodies and the mast cell releases histamine and other chemicals that cause the allergic response.

***Table 12.3** The Immunoglobulins*

Immunoglobulin	Examples of Functions
IgG	Main form of antibodies in circulation; production increased after immunization
IgA	Main antibody type in external secretions, such as saliva and mother's milk
IgE	Responsible for allergic symptoms in immediate hypersensitivity reactions
IgM	Present on lymphocyte surface prior to immunization; secreted during primary response
IgD	Present on lymphocyte surface prior to immunization; other functions unknown

Immunological Side Effects and Illnesses

The immune system protects us from disease because it can tell self from nonself. Sometimes, however, the immune system is underprotective as when an individual develops cancer or is overprotective, as when an individual has allergies.

Allergies

Allergies are caused by an overactive immune system that forms antibodies to substances that are not usually recognized as being foreign substances. Unfortunately, allergies are usually accompanied by coldlike symptoms or, even at times, severe systemic reactions such as a sudden low blood pressure, termed shock.

Among the five varieties (table 12.3) of antibodies—immunoglobulin A (IgA), IgD, IgE, IgG, and IgM—it is IgE that causes allergies. IgE antibodies are found in the bloodstream but they, unlike the other types of antibodies, also reside in the membrane of mast cells found in the tissues. Some investigators contend that mast cells are basophils that have left the bloodstream and have taken up residence in the tissues. In any case, when the *allergen,* an antigen that provokes an allergic reaction, attaches to the IgE antibodies on mast cells, these cells release histamine, and other substances (fig. 12.8) that cause secretion of mucus and constriction of the airways, resulting in the characteristic wheezing and labored breathing of someone with asthma. On occasion, basophils and other white cells release these chemicals into the bloodstream. The increased capillary permeability that results from this can lead to fluid loss and shock.

Allergy shots sometimes prevent the occurrence of allergic symptoms. Injections of the allergen cause the body to build up high quantities of IgG antibodies, and these combine with allergens received from the environment before they have a chance to reach the IgE antibodies located in the membrane of mast cells.

---

*A*llergic symptoms are caused by the release of histamine and other substances from mast cells.

Tissue Rejection

Certain organs such as skin, the heart, and the kidneys could easily be transplanted from one person to another if the body did not attempt to reject them. It is obvious that the transplanted organ is foreign to the individual, and for this reason the immune system reacts to it. At first T cells appear on the scene, and later antibodies bring about a disintegration of the foreign tissue.

Organ rejection can be controlled in two ways: careful selection of the organ to be transplanted and the administration of *immunosuppressive drugs*. In the first instance, it is best if the organ is made up of cells having the same type MHC proteins (p. 256) as those on the cells of the prospective recipient. It is these that the T cell recognizes as foreign to the recipient. In regard to immunosuppression, there is now available a drug called Cyclosporine, which suppresses cell-mediated immunity as long as it is administered. After the body has adjusted to the new organ, the drug can be withdrawn and immunity will return. This withdrawal is desirable because a person is more apt to catch an infectious disease when immunity is suppressed.

When an organ is rejected, the immune system is attacking cells that bear different MHC proteins from those of the individual.

Autoimmune Diseases

Certain human illnesses are believed to be due to the production of antibodies that act against an individual's own tissues. In myasthenia gravis, autoantibodies attack the neuromuscular junctions so that the muscles do not obey nervous stimuli. Muscular weakness results. In multiple sclerosis (MS), antibodies attack the myelin sheath of nerve fibers, causing various neuromuscular disorders. A person with systemic lupus erythematosus (SLE) forms various antibodies to different constituents of the body, including the DNA of the cell nucleus. The disease sometimes results in death, usually due to kidney damage. In rheumatoid arthritis it is the joints that are affected. When an autoimmune disease occurs, a viral infection of tissues has often set off an immune reaction to the body's own tissues in an attempt to attack the virus. There is evidence to suggest diabetes is also the result of this sequence of events.

Autoimmune diseases seem to be preceded by a viral infection that fools the immune system into attacking the body's own tissues.

Aging

The thymus (fig. 12.1) is large in relation to the rest of the body during fetal development and childhood; however, it stops growing by puberty and then begins to atrophy and get progressively smaller. This has led some researchers to suggest that aging may be associated with a general decline in the immune system. Certain diseases, such as cancer, are more prevalent with age.

The thymus gland becomes smaller with age and perhaps this occurrence contributes to the aging process.

Table 12.4 Infectious Diseases Caused by Viruses

Respiratory Tract	Nervous System
Common colds	Encephalitis
*Flu	*Polio
Viral pneumonia	*Rabies
Skin Reactions	**Liver**
*Measles	*Yellow fever
*German measles	Infectious hepatitis
Chicken pox	**Other**
*Smallpox	AIDS
Warts	*Mumps
	Herpes
	Cancer

*Vaccines available. Yellow fever, rabies, flu, and smallpox vaccines are given if the situation requires them. Others are routinely given.

Table 12.5 Infectious Diseases Caused by Bacteria

Respiratory Tract	Nervous System
Strep throat (sometimes causing rheumatic and scarlet fever)	*Tetanus
Pneumonia	Botulism
*Whooping cough	Meningitis
*Diphtheria	**Digestive Tract**
*Tuberculosis	Food poisoning (salmonella, botulism, and staph)
Skin Reactions	*Typhoid fever
Staph (pimples and boils)	*Cholera
*Gas gangrene (wound infections)	**Sexually Transmitted Diseases**
	Chlamydia
	Gonorrhea
	Syphilis

*Vaccines are available. Tuberculosis vaccine is not used in this country. Typhoid fever, cholera, and gas gangrene vaccines are given if the situation requires it. Others are routinely given.

Immunotherapy

The immune system can be manipulated to help people avoid or recover from diseases. Some of these techniques have been utilized for quite some time and some are relatively new.

Active Immunity

Active immunity that provides long-lasting protection against a disease-causing agent develops after an individual becomes infected with a virus or bacterium (tables 12.4 and 12.5). Today, however, it is not necessary to suffer an illness to become immune because it is possible to be medically immunized against a disease. One possible recommended immunization schedule for children is given in figure 12.9. Immunization requires the use of **vaccines,** that traditionally are bacteria and viruses that have been treated so that they are no longer virulent (able to cause disease). New methods of producing vaccines are being developed. For example, it is possible to alter the DNA of a bacterium or virus and use the recombinant DNA technique (p. 492) to mass produce an altered protein that is still antigenic and can be used as a vaccine.

After a vaccine is injected, it is possible to determine the amount of antibody present in the bloodstream. This is called the *antibody titer.* After the first injection, a primary response occurs. There is a period of several days during which no antibodies are present; then there is a slow rise in the titer, which is followed by a gradual decline (fig. 12.10). After a second injection, a secondary response occurs. The titer rises rapidly to a level much greater than before. This second injection is often called the *"booster shot"* since it boosts the antibody titer to a high level. The antibody titer is now high enough to prevent disease symptoms even if the individual is exposed to the disease. For some time thereafter the individual is immune to that particular disease.

Primary and secondary responses occur whenever an individual is exposed twice to the same disease-causing agent. The difference in the responses may be related to the number of plasma and memory cells. Upon the second exposure, these cells are already present and antibodies can be rapidly produced.

Vaccines can be used to make people actively immune.

Age	Shots
2 months old	1 DTP[a] immunization 1 polio immunization
4 months old	2 DTP immunizations 2 polio immunizations
6 months old	3 DTP immunizations 2 polio immunizations[b]
15 months old	3 DTP immunizations 2 polio immunizations[b] 1 measles immunization[c] 1 rubella immunization[c,d] 1 mumps immunization[c]
18 months and older	4 DTP immunizations 3 polio immunizations[b]
4 to 6 years, before starting to school	A DTP booster (the 5th immunization) A polio booster (the 4th immunization)[b]
Thereafter	Tetanus-diphtheria (Td) booster should be given every 10 years or following a dirty wound if a booster has not been given in the preceding 5 years.

[a]D = diphtheria, T = tetanus, P = pertussis (whooping cough)
[b]Some doctors give one additional dose when the child is 6 months old.
[c]Only one shot needed. Some doctors combine these vaccines in a single injection.
[d]rubella (German measles)

a.

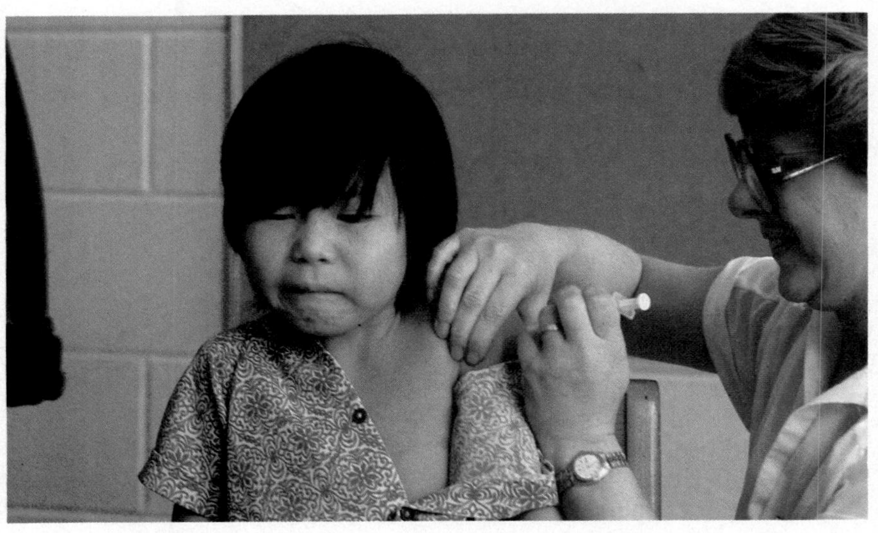

b.

Figure 12.9
a. *Suggested immunization schedule for infants and young children.* b. *Preschool shots are a necessity. Children who are not immunized are subject to diseases that can cause serious health consequences.*
(a.) Source: DHRS Immunization Program, State of Florida.

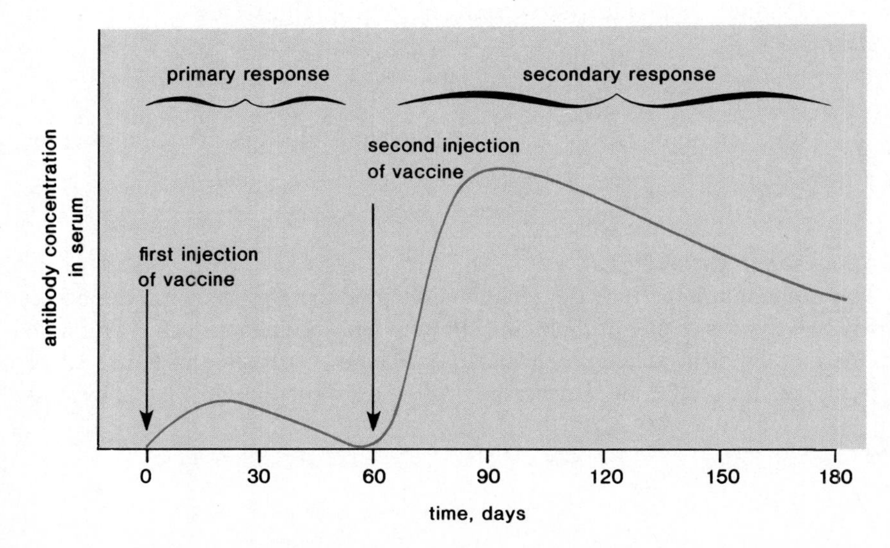

Figure 12.10
Immunization responses. The primary response after the first injection of a vaccine is minimal, but the secondary response after the second injection shows a dramatic rise in the amount of antibody present in serum.

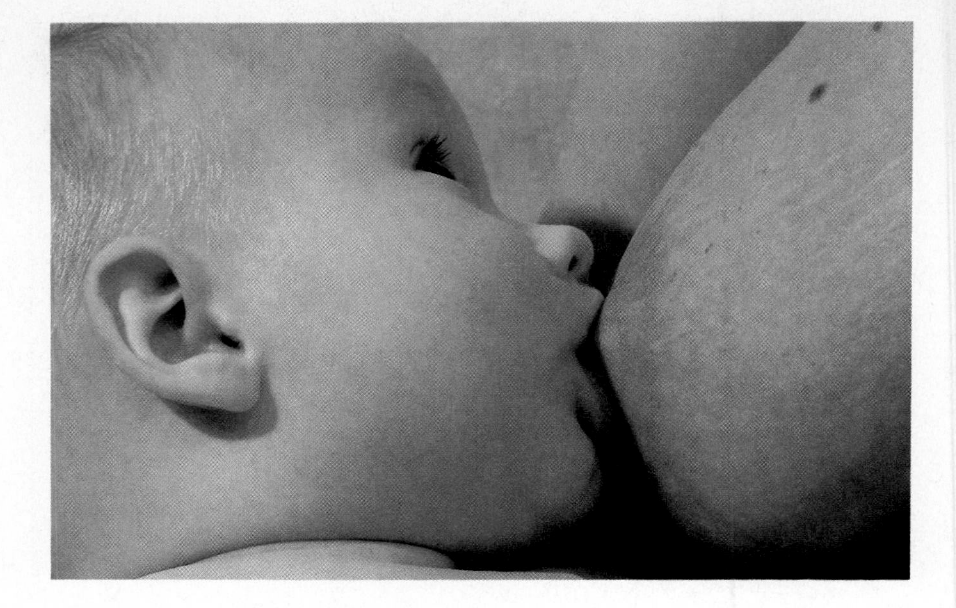

Figure 12.11
Example of passive immunity. Breast feeding is believed to provide a newborn with antibodies during the period of time when the infant is not yet producing antibodies.

Passive Immunity

Passive immunity occurs when an individual is given antibodies to combat a disease. Since these antibodies are not produced by the individual's B cells, passive immunity is short-lived. For example, newborn infants possess passive immunity because antibodies have crossed the placenta from their mother's blood. These antibodies soon disappear however, so that within a few months, infants become more susceptible to infections. Breast feeding (fig. 12.11) prolongs the passive immunity an infant receives from its mother because there are antibodies in the mother's milk.

Even though passive immunity is not lasting, it is sometimes used to prevent illness in a patient who has been unexpectedly exposed to an infectious disease. Usually, the person receives an injection of a serum containing antibodies. This may have been taken from donors who have recovered from the illness. In other instances, horses have been immunized and serum taken from them to provide the needed antibodies. Horses have been used to produce antibodies against diphtheria, botulism, and tetanus. Occasionally a patient who receives these antibodies becomes ill because the serum contains proteins that the individual's immune system recognizes as foreign. This is called serum sickness.

Passive immunity is short-lived because the antibodies are administered to and not made by the individual.

Monoclonal Antibodies

There is much hope that the technique of producing monoclonal antibodies will one day make plentiful amounts of pure and specific antibodies available. One method of producing monoclonal antibodies is depicted in figure 12.12. Lymphocytes are removed from the body and exposed in vitro (in laboratory glassware) to a particular antigen. Then they are fused with a cancer cell because cancerous cells, unlike normal cells, will divide an unlimited number of times. The fused cells are called **hybridomas;** hybrid because they are a fusion of two different cells and oma because one of the cells is a cancer cell.

The antibodies produced by hybridoma cells are called **monoclonal antibodies** because all of them are the same type (mono) and because they are produced by cells derived from the same type parent cell (clone). At present monoclonal antibodies are being used to allow quick and certain diagnosis of various conditions. For example, a certain hormone is present in the urine when a woman is pregnant (p. 405). A monoclonal antibody can be used to detect the hormone and indicate that the woman is pregnant. Soon, monoclonal antibodies may also be available to detect and fight certain types of cancer within the individual. It may be possible to have them carry chemicals that will help destroy the cancerous cells.

Monoclonal antibodies are produced in pure patches—they are specific against just one antigen.

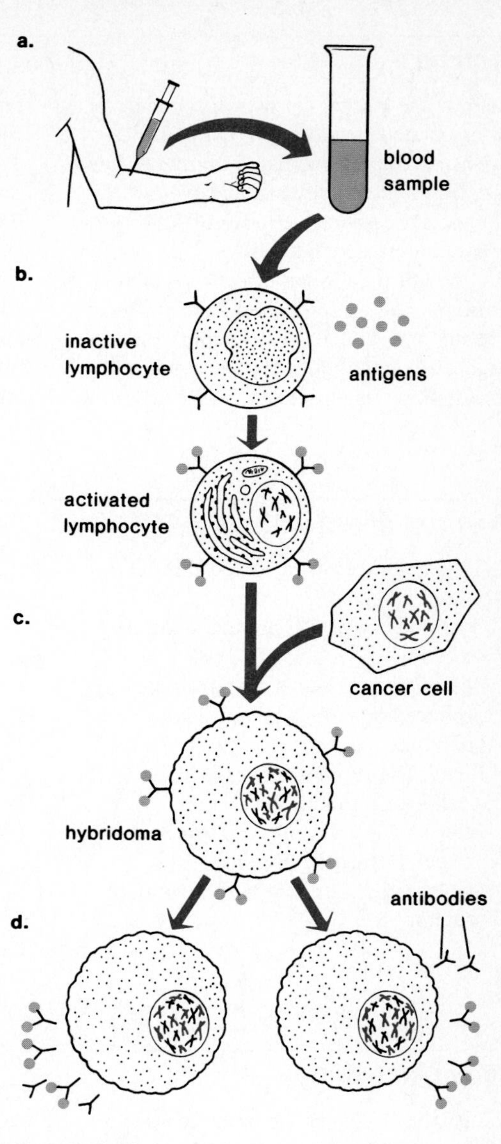

Figure 12.12
One possible method for producing human monoclonal antibodies. a. *Blood sample is taken from patient.* b. *Inactive lymphocytes from sample are exposed to antigen.* c. *Activated lymphocytes are fused with cancer cells.* d. *Resulting hybridomas divide repeatedly, giving many cells that produce monoclonal antibodies.*

Lymphokines

Both interferon and various interleukins have been used as immunotherapeutical drugs particularly to potentiate the ability of the individual's own T cells (and possibly B cells) to fight cancer.

Interferon is a substance produced by leukocytes, fibroblasts, and probably most cells, in response to a viral infection. When it is produced by T cells it is called a *lymphokine.* Interferon enters noninfected cells and in some way interferes with viral replication in these cells. Interferon is *species specific,* meaning, for example, that only human interferon is effective in humans. In the past, supplies of interferon had to be extracted from white cells and the amount available was so scarce that treatment cost $20,000 to $30,000 per patient. Now, however, interferon is being commercially produced by the recombinant DNA technique and the cost of treatment has been reduced to $300 per patient. Interferon is still being investigated as a possible cancer drug, but thus far it has proven to be effective only in certain patients and the exact reasons cannot as yet be discerned. For example, interferon has been found to be effective in up to 90 percent of patients with a type of leukemia known as hairy-cell leukemia, so named because of the hairy appearance of the malignant cells.

When and if cancer cells carry an altered protein on their cell surface, by all rights they should be attacked and destroyed by T_C cells. Whenever cancer develops, it is possible that the T_C cells have not been activated. Then the use of lymphokines might awaken the immune system and lead to destruction of the cancer. In one technique being investigated, researchers first withdraw T cells from the patient and activate the cells by culturing them in the presence of an interleukin. The cells are then injected back into the patient, who is given doses of interleukin to maintain the killer activity of the T cells.

Lymphokine therapy shows some promise of potentiating the individual's own immune system.

Summary

The immune system consists of lymphocytes and associated structures. T lymphocytes responsible for cell-mediated immunity have passed through the thymus and B lymphocytes responsible for antibody-mediated immunity have not.

Although B cells can directly recognize an antigen and thereafter produce antibody-secreting plasma cells, T cells must have the antigen, along with an MHC protein, presented by a macrophage. There are three types of T cells: T_C cells kill cells on contact; T_H cells stimulate B cells and other T cells; and T_S cells suppress the immune response. Following an immune response, memory T and B cells remain in the body providing long-lasting immunity.

Immunity has certain unwanted side effects. Allergies are due to an overactive immune system that forms antibodies to substances not normally recognized as foreign. T_C cells attack transplanted organs, but immunosuppressive drugs are available. Autoimmune illnesses occur when antibodies form against the body's own cells.

Immunity can be fostered by immunotherapy. Vaccines are available to promote active immunity and monoclonal antibodies are sometimes available to provide an individual with short-term passive immunity. Lymphokines, notably interferon and the interleukins, are used to promote the body's ability to recover from cancer.

Objective Questions

1. T lymphocytes have passed through the ______________ .
2. A stimulated B cell produces antibody-secreting ______________ cells and ______________ cells that are ready to produce the same type antibody at a later time.
3. B cells are responsible for ______________ -mediated immunity.
4. In order for a T cell to recognize an antigen it must be presented by a ______________ along with a MHC protein.
5. T cells produce ______________ that are stimulatory chemicals for all types of immune cells.
6. Cytotoxic T cells are responsible for ______________-mediated immunity.
7. Allergic reactions are associated with the release of ______________ from mast cells.
8. The body recognizes foreign cells because they bear different ______________ proteins than the body's cells.
9. Immunization with ______________ brings about active immunity.
10. Hybridomas produce ______________ antibodies.

Answers to Objective Questions

1. thymus 2. plasma, memory 3. antibody 4. macrophage 5. lymphokines 6. cell 7. histamine 8. MHC 9. vaccines 10. monoclonal

Study Questions

1. List the organs of the immune system and give their functions. (p. 251)
2. What is the clonal selection theory? (p. 252)
3. B cells are said to be responsible for what type of immunity? (p. 253)
4. Explain the process that allows a T cell to recognize an antigen. (p. 256)
5. List the three types of T cells and state their function. (pp. 256–257)
6. What are lymphokines and how are they used in immunotherapy? (pp. 256, 265)
7. Cytotoxic T cells are said to be responsible for what type of immunity? (p. 257)
8. Discuss allergies, tissue rejection, autoimmune disease, and aging as they relate to the immune system. (pp. 260–261)
9. Relate active immunity to the presence of plasma cells and memory cells. (p. 262)
10. How is active immunity achieved? passive immunity achieved? (p. 264)

Key Terms

B lymphocytes (lim'fo-sīts) lymphocytes that react against foreign substances in the body by producing and secreting antibodies. *252*

complement (kom'plĕ-ment) a group of proteins in plasma that produce a variety of effects once an antigen-antibody reaction has occurred. *255*

cytotoxic T cells (si''to-tok'sik te selz) T lymphocytes which attack cells bearing foreign bodies. *257*

helper T cells (hel'per te selz) T lymphocytes that stimulate certain other T and B lymphocytes to perform their respective functions. *256*

hybridomas (hi''brid-o'mahz) fused lymphocytes and cancer cells used in the manufacture of monoclonal antibodies. *265*

immune complex (ĭ-mūn' kom'pleks) the product of an antigen–antibody reaction. *255*

interferon (in''ter-fēr'on) a protein formed by a cell infected with a virus that can increase the resistance of other cells to the virus. *265*

lymphokines (lim'fo-kīnz); chemicals secreted by T cells that have the ability to affect the characteristics of lymphocytes and monocytes. *256*

memory B cells (mem'o-re be selz) from B cells that remain within the body producing a specific antibody and accounting for the development of active immunity. *257*

MHC protein (em āch se pro'te-in) major histocompatibility protein; a surface molecule that serves as a genetic marker. *256*

monoclonal antibodies (mon''-o-klōn'al an'tĭ-bod''ēz) antibodies of one type that are produced by cells that are derived from a lymphocyte that has fused with a cancer cell. *265*

plasma cells (plaz'mah selz) cells derived from B cell lymphocytes that are specialized to mass produce antibodies. *252*

suppressor T cells (su-pres' or te selz) T lymphocytes that suppress certain other T and B lymphocytes from continuing to divide and perform their respective functions. *257*

T lymphocytes (lim'fo-sīts); lymphocytes that interact directly with antigen-bearing cells and are responsible for cell-mediated immunity. *252*

vaccine (vak'sēn) antigens prepared in such a way that they can promote active immunity without causing disease. *262*

13

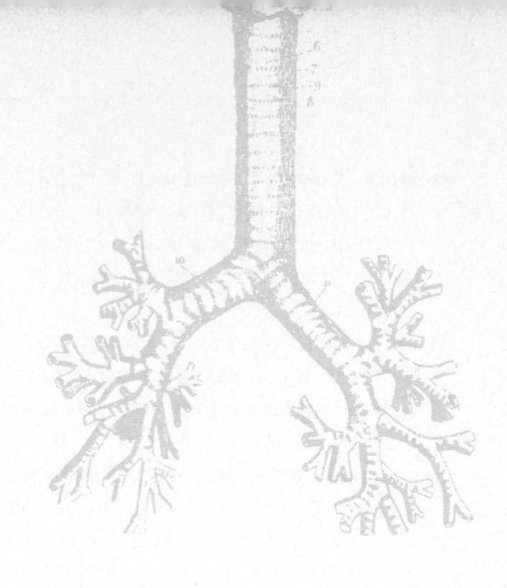

Respiration

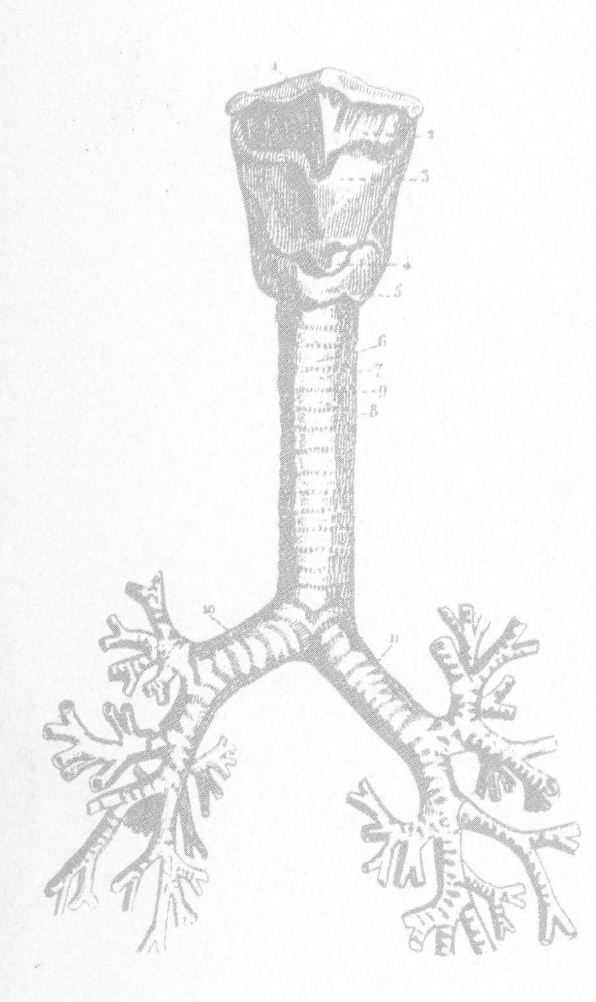

Chapter Concepts

1 As air passes along the respiratory tract, it is filtered, warmed, and saturated with water before gas exchange takes place across a very extensive moist surface.

2 Breathing brings in oxygen needed by the cells for cellular respiration, and rids the body of carbon dioxide, a by-product of cellular respiration.

3 The respiratory pigment, hemoglobin, combines with oxygen in the lungs and releases oxygen in the tissues. It also aids in the transport of carbon dioxide largely by its ability to buffer.

4 The respiratory tract is especially subject to disease because it serves as an entrance for infectious agents. Polluted air contributes to two major lung disorders—emphysema and cancer.

Chapter Outline

Figure 13.1
Exercising increases the body's need for oxygen because cellular respiration is sped up in order to provide ATP needed for muscle contraction. The heart pumps faster to deliver oxygen to the tissues in a more timely manner. Most likely, for this reason, regular and moderate exercising has been found to prolong the life span.

Breathing is more eminently necessary than eating. While it is possible to stop eating altogether for several days, it is not possible to remain alive for longer than several minutes without breathing. Breathing supplies the body with the oxygen needed for cellular respiration, as indicated in the following equation.[1]

$$38\ ADP + 38Ⓟ \longrightarrow 38\ ATP$$

$$C_6H_{12}O_6 + 6\ O_2 \longrightarrow 6\ H_2O + 6\ CO_2$$

This equation indicates that the body requires oxygen to convert the energy within glucose to phosphate-bond energy. Therefore, the more energy expended, the greater the need for oxygen (fig. 13.1). The minimum amount of oxygen a person consumes at complete rest, without eating previously, is related to the basal metabolic rate (p. 196). The average young adult male utilizes about 250 milliliters of oxygen per minute in a basal, or restful, state. Exercise and digestion of food raise the need for oxygen. The average amount of oxygen needed with mild exercise is 500 milliliters of oxygen per minute. The equation for cellular respiration also indicates that cells produce carbon dioxide. This metabolic end product must be eliminated from the body by the breathing process.

B reathing is necessary to supply the body with oxygen so that ATP can be formed by cellular respiration.

---

[1]The body also requires oxygen for the respiration of fats and amino acids in addition to glucose.

Altogether, respiration may be used to refer to the complete process of getting oxygen to body cells for cellular respiration and the reverse process of ridding the body of carbon dioxide given off by cells. Respiration may be said to include the following:

1. **Breathing:** entrance and exit of air into and out of the lungs
2. **External respiration:** exchange of gases (O_2 + CO_2) between air and blood
3. **Internal respiration:** exchange of gases between blood and tissue fluid
4. **Cellular respiration:** production of ATP in cells. In this chapter, we are studying the first three portions of the respiratory process. Cellular respiration was discussed in chapter 5.

Breathing

The normal breathing rate is about 14 to 20 times per minute. Breathing consists of taking air in, **inspiration,** and forcing air out, **expiration.** Expired air contains less oxygen and more carbon dioxide than inspired air, indicating that the body takes in oxygen and gives off carbon dioxide (table 13.1).

Table 13.1 *Composition of Inspired and Expired Air*

Component of Air	Inspired Air (%/vol)	Expired Air (%/vol)
N_2	79.00	79.60
O_2	20.96	16.02
CO_2	0.04	4.38

Passage of Air

During inspiration and expiration, air is conducted toward or away from the lungs by a series of cavities, tubes, and openings, listed in order in table 13.2 and illustrated in figure 13.2.

Table 13.2 *Path of Air*

Structure	Function
Nasal cavities	Filters, warms, and moistens
Nasopharynx	Passage of air from nose to throat
Pharynx (throat)	Connection to surrounding regions
Glottis	Passage of air
Larynx (voice box)	Sound production
Trachea (windpipe)	Passage of air to thoracic cavity
Bronchi	Passage of air to each lung
Bronchioles	Passage of air to each alveolus
Alveoli	Air sacs for gas exchange

As air moves in along the air passages, it is filtered, warmed, and moistened. The filtering process is accomplished by coarse hairs and cilia in the region of the nostrils and by cilia alone in the rest of the nose and windpipe. In the nose, the hairs and cilia act as a screening device. In the trachea, cilia beat upward, carrying mucus, dust, and occasional bits of food that "went the wrong way" into the pharynx, where the accumulation may be swallowed or expectorated. The air is warmed by heat given off by the blood vessels lying close to the surface of the lining of the air passages, and it is moistened by the wet surface of these passages. By the time the inspired air reaches the lower end of the trachea, it is about 99.5 percent saturated with water.

On the other hand, as air moves out during expiration, it becomes progressively cooler and loses its moisture. As the gas cools, it deposits its moisture on the lining of the windpipe and nose, and the nose may even drip as a result of this condensation. But the air still retains so much moisture that upon expiration on a cold day it condenses and forms a small cloud.

Air is warmed, filtered, and moistened as it moves from the nose toward the lungs.

Each portion of the air passage also has its own unique structure and function, as described in the sections that follow.

Nose

The nose contains two nasal cavities, narrow canals with convoluted lateral walls that are separated from one another by a septum. Up in the narrow recesses of the nasal cavities are special ciliated cells (fig. 17.5) that act as odor receptors. Nerves lead from these cells and go to the brain, where the impulses are interpreted as smell.

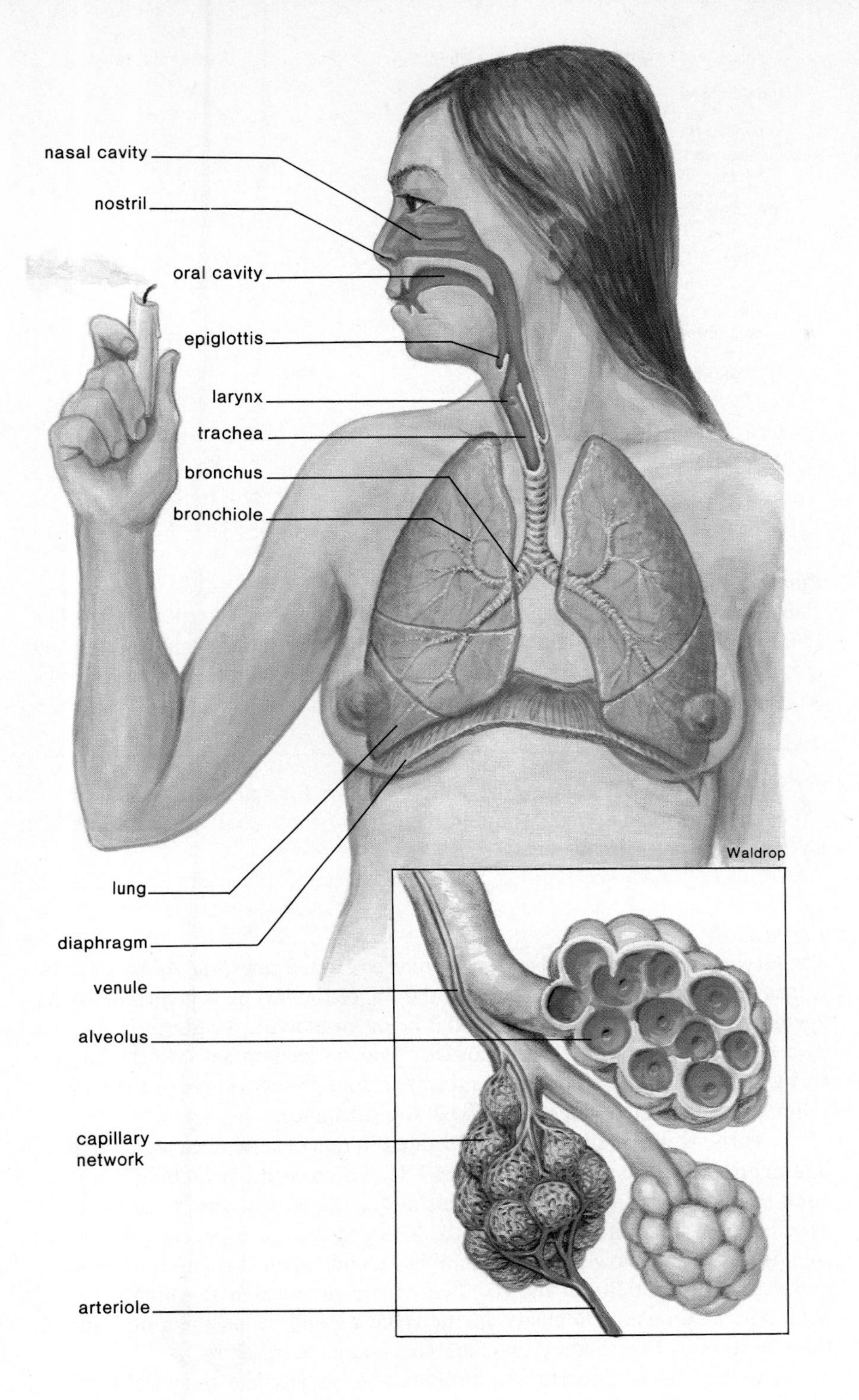

Figure 13.2
Diagram of human respiratory tract, with internal structure of one lung revealed in an enlargement of a section of this lung. Gas exchange occurs in the alveoli, which are surrounded by a capillary network.

The nasal cavities have a number of openings. Tears produced by tear (lacrimal) glands drain into the nasal cavities by way of tear ducts, producing a runny nose. The nasal cavities open into the cranial sinuses, air-filled spaces in the skull, and empty into the nasopharynx, a chamber just beyond the soft palate. The *eustachian tubes* lead into the nasopharynx from the middle ears (fig. 17.16).

Figure 13.3
Pathway of air. When we breathe, the glottis is open, and when we swallow, the epiglottis covers the glottis.

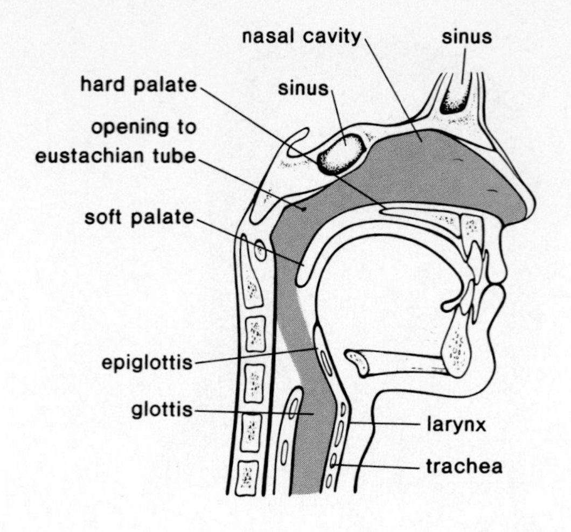

Pharynx

The air and food channels cross in the pharynx (fig. 13.3) so that the trachea (windpipe) lies in front of the esophagus, which normally opens only during the process of swallowing food. Just below the pharynx lies the larynx, or voice box.

The nasal cavities contain the sense receptors for smell and open into the pharynx. The pharynx is in the back of the throat where the air passage and food passage cross.

Larynx

The larynx may be imagined as a triangular box whose apex, the Adam's apple, is located at the front of the neck. At the top of the larynx is a variable-sized opening called the glottis. When food is being swallowed, the glottis is covered by a flap of tissue called the epiglottis so that no food passes into the larynx. If, by chance, food or some other substance does gain entrance to the larynx, reflex coughing usually occurs to expel the substance.

At the edges of the glottis, embedded in mucous membrane, are elastic ligaments called the **vocal cords** (fig. 13.4). These cords, stretching from the back to the front of the larynx at the sides of the glottis, vibrate when air is expelled past them through the glottis. Vibration of the vocal cords produces sound. The high or low pitch of the voice depends upon the length, thickness, and degree of elasticity of the vocal cords and the tension at which they are held. The loudness, or intensity, of the voice depends upon the amplitude of the vibrations, or the degree to which vocal cords vibrate.

At the time of puberty, the growth of the larynx and the vocal cords is much more rapid and accentuated in the male than in the female, causing the male to have a more prominent Adam's apple and a deeper voice. The voice "breaks" in the young male due to his inability to control the longer vocal cords.

The larynx is the voice box because it contains the vocal cords at the sides of the glottis, an opening sometimes covered by the epiglottis.

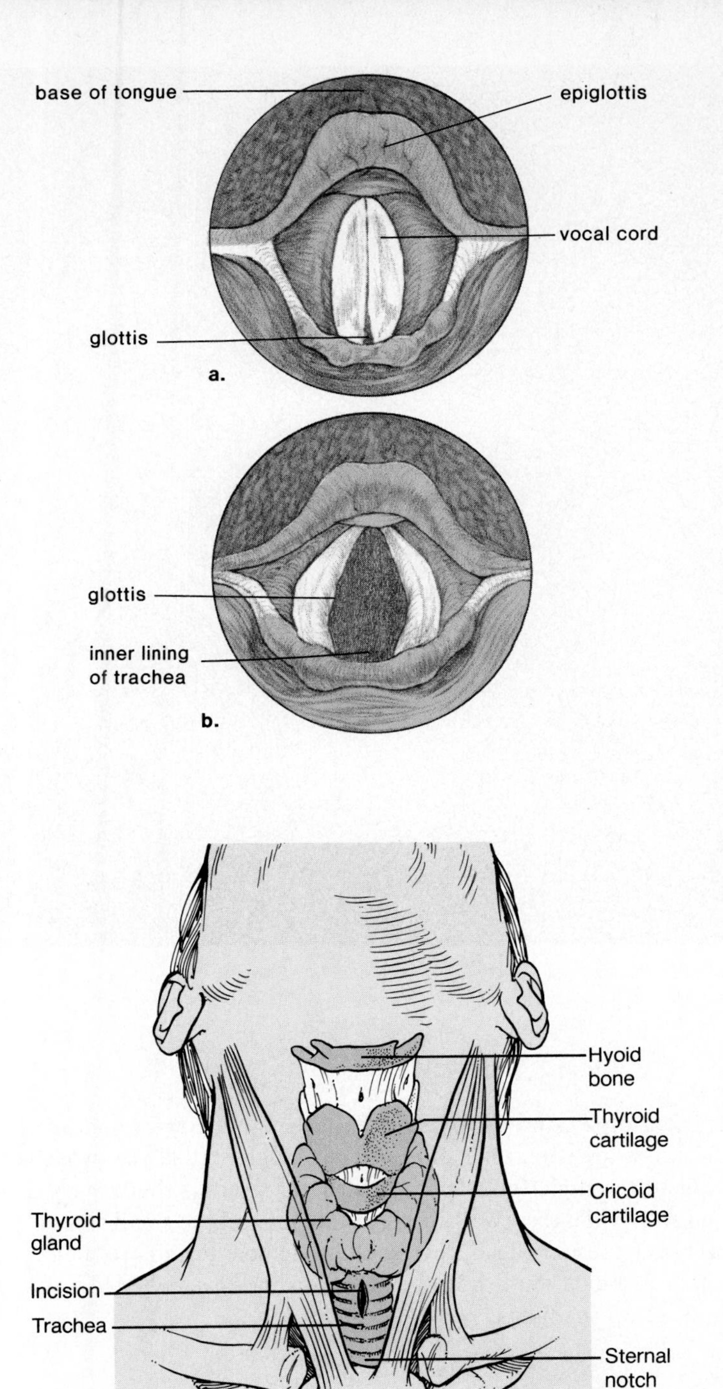

Figure 13.4
Vocal cords lie at the edge of the glottis. When air is expelled from the larynx, the cords vibrate, producing the voice. (a. *closed position* b. *open position.*)

Figure 13.5
Site of tracheostomy to allow air to enter the trachea beneath the location of an obstruction.

Trachea

The larynx, continuous with the **trachea,** is held open by c-shaped cartilaginous rings. Ciliated mucous membrane (fig. 8.4) also lines the trachea, and normally these cilia keep the windpipe free of debris. Smoking is known to destroy the cilia, and consequently the soot in cigarette smoke collects in the lungs. Smoking will be discussed more fully at the end of this chapter.

If the trachea is blocked because of illness or accidental swallowing of a foreign object, it is possible to insert a tube by way of an incision made in the trachea; this tube acts as an artificial air intake and exhaust duct. The operation is called a *tracheostomy* (fig. 13.5).

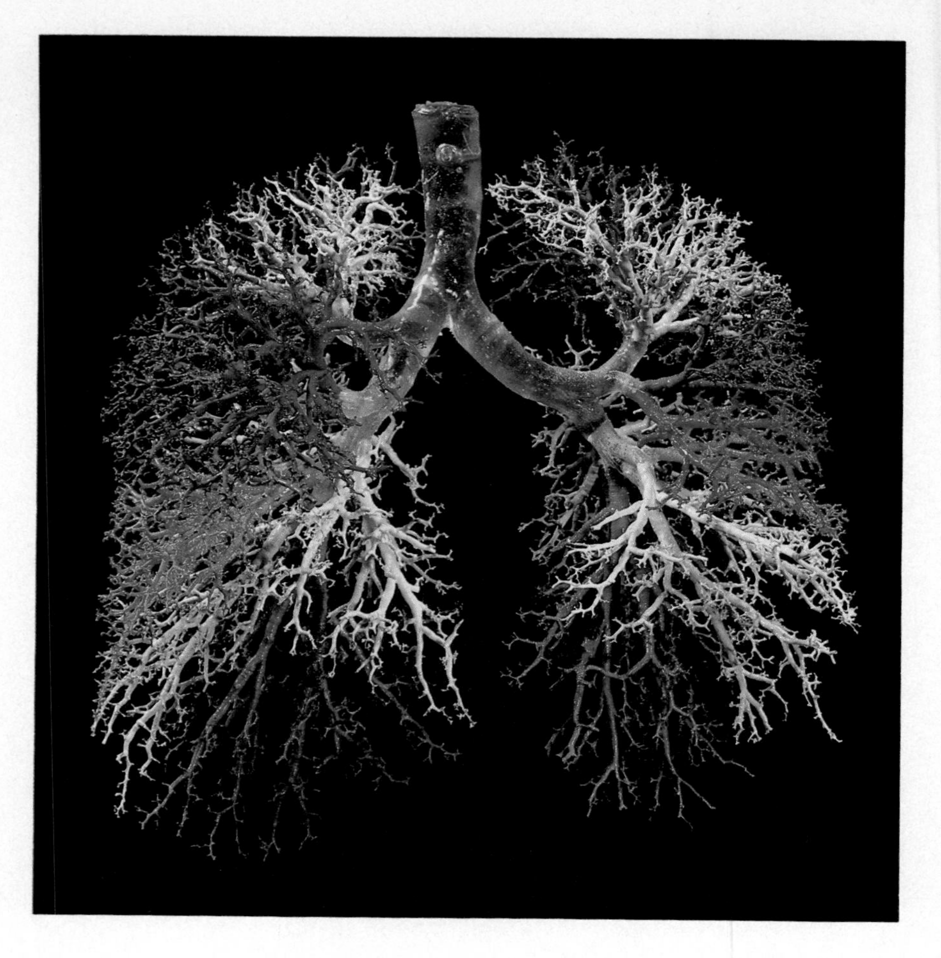

Figure 13.6
Cast of lungs showing the number of airways. The bronchi branch into the bronchioles that branch and rebranch until termination in the alveoli. Each lobe of the lung is in a different color in this cast.

Bronchi

The trachea divides into two **bronchi** that enter the right and left lungs and branch into a great number of smaller passages called the **bronchioles.** The two bronchi resemble the trachea in structure, but as the bronchial tubes divide and subdivide, their walls become thinner and the small rings of cartilage are no longer present (fig. 13.6). Each bronchiole terminates in an elongated space enclosed by a multitude of air pockets, or sacs, called **alveoli** (fig. 13.2) which make up the lungs.

Lungs

Within the lungs each alveolar sac is only one layer of squamous epithelium surrounded by blood capillaries. Gas exchange occurs between the air in the alveoli and the blood in the capillaries (fig. 13.2).

A film of lipoprotein that lines the alveoli of mammalian lungs lowers the surface tension and prevents them from closing up. The lungs collapse in some newborn babies, especially premature infants, who lack this film. This condition, called infant respiratory distress syndrome, often results in death.

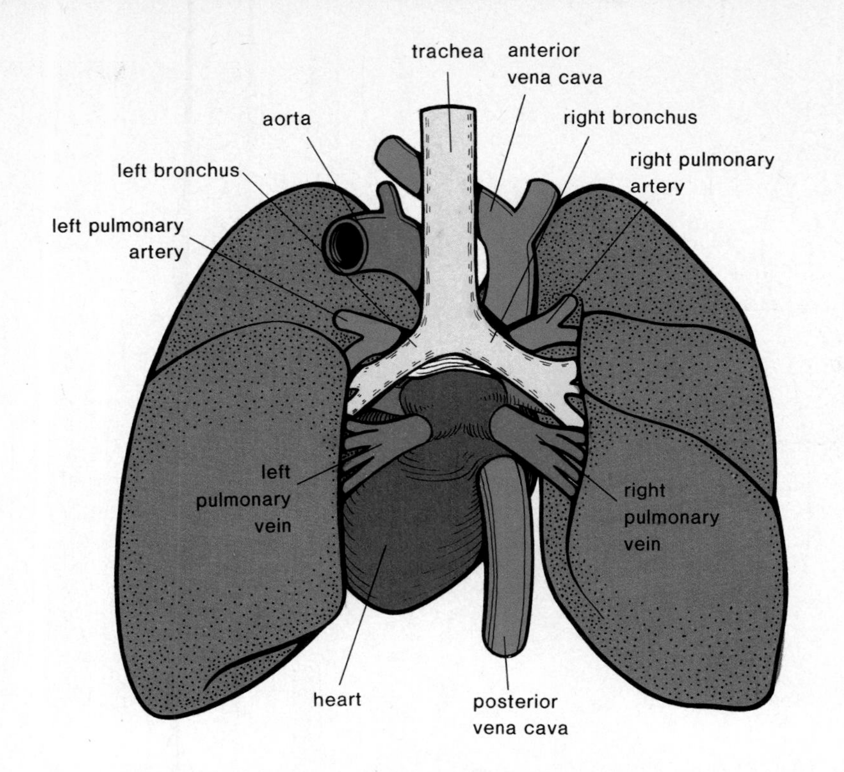

Figure 13.7
Posterior view of heart and lungs shows the relationship of the pulmonary vessels to the trachea and bronchial tubes. Trace the path of air to the left lung and trace the path of blood from the heart to the left lung and return.

There may be as many as 700,000,000 alveoli in the human lung. This is the equivalent of 75 square yards, or 100 times the surface of the skin! Because of their many air spaces, the lungs are very light; normally, a piece of lung tissue dropped in a glass of water will float.

Externally, the lungs are cone-shaped organs that lie on both sides of the heart in the thoracic cavity. The branches of the pulmonary artery accompany the bronchial tubes and form a mass of capillaries around the alveoli.[2] The pulmonary veins collect blood from these capillaries and empty into the left atrium of the heart. Figure 13.7 shows the relationship of the pulmonary vessels to the trachea and bronchial tubes.

Air moves from the trachea, held open by cartilaginous rings, into the two bronchi, one for each of the lungs. The lungs are composed of air sacs, called alveoli.

Mechanism of Breathing

In order to understand **ventilation,** the manner in which air is drawn into and expelled out of the lungs, it is necessary to remember first that when one is breathing there is a continuous column of air from the pharynx to the alveoli of the lungs (the air passages are open).

[2]These capillaries are used for gas exchange and do not furnish the lungs with oxygen. The bronchial artery supplies the lungs with their oxygen needs.

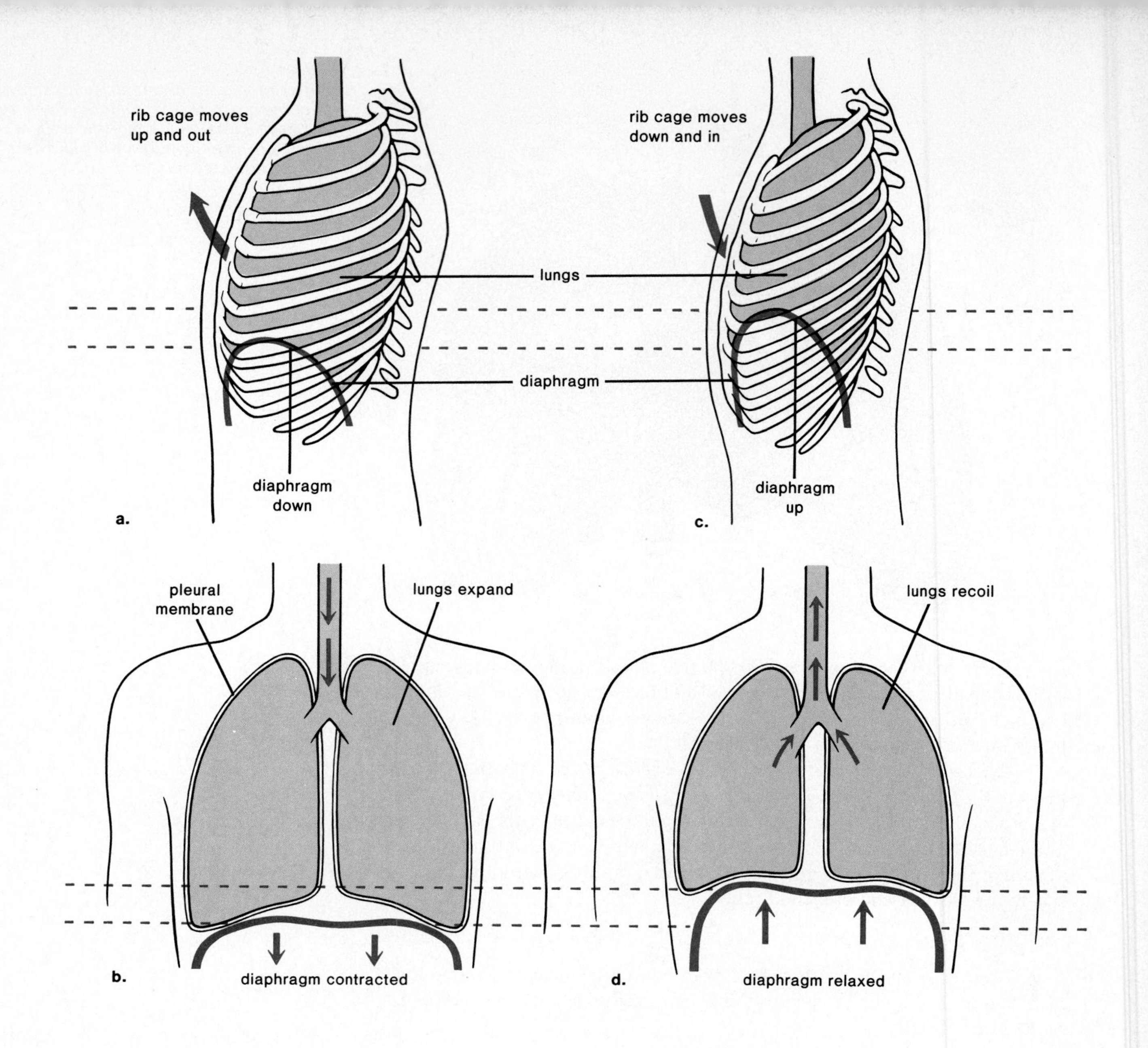

Figure 13.8
Inspiration versus expiration. a. *When the rib cage lifts up and outward and the diaphragm lowers, the lungs expand so that* b. *air is drawn in. This sequence of events is only possible because the pressure within the intrapleural space, containing a thin film of water, is less than atmospheric pressure.* c. *When the rib cage lowers and the diaphragm rises* d. *the lungs recoil so that air is forced out.*

Secondly, we may note that the lungs lie within the sealed-off chest (thoracic) cavity. The **ribs,** hinged to the vertebral column at the back and to the sternum (breastbone) at the front, along with the muscles that lie between them, make up the top and sides of the chest cavity. The **diaphragm,** a dome-shaped horizontal muscle, forms the floor of the chest cavity. The lungs themselves are enclosed by the **pleural membranes** (fig. 13.8). The outer pleural membrane adheres closely to the walls of the chest and diaphragm, and the inner is fused to the lungs. The two pleural layers lie very close to one another, being separated only by a thin film of fluid. Normally, the intrapleural pressure is less than atmospheric pressure. The importance of this reduced pressure is demonstrated when, by design or accident, air enters the intrapleural space. Now the lungs collapse and inspiration is impossible.

The lungs are completely enclosed and, by way of the pleural membranes, adhere to the chest cavity walls.

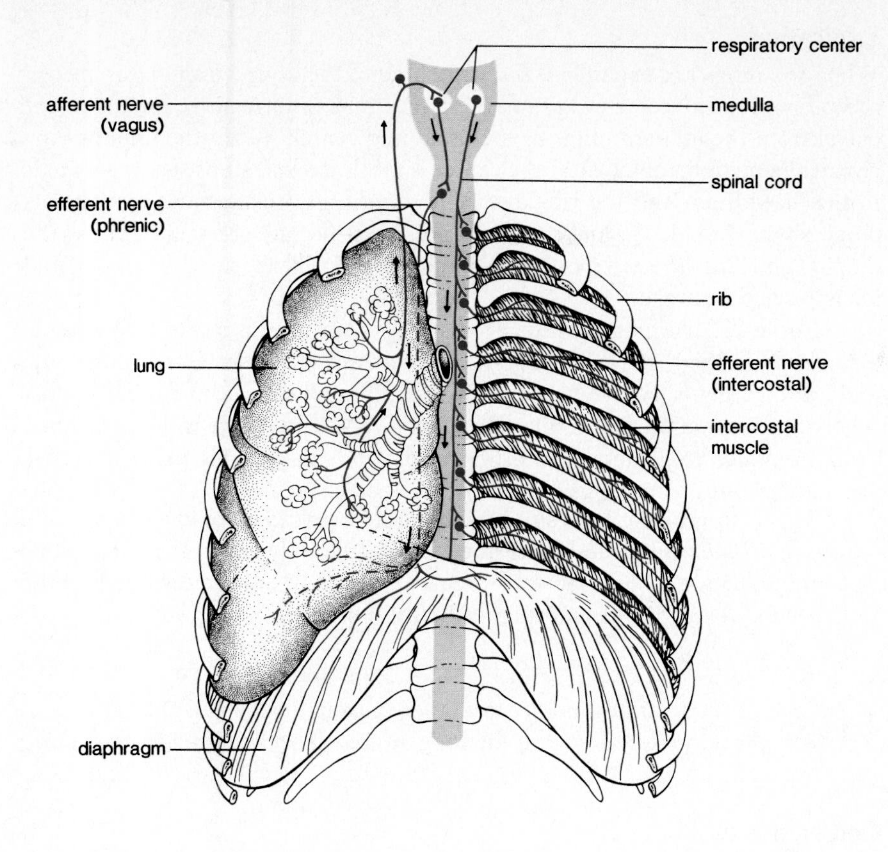

Figure 13.9
Nervous control of breathing. During inspiration, the respiratory center stimulates the rib (intercostal) muscles and the diaphragm to contract by way of the efferent (phrenic) nerve. Nerve impulses from the expanded lungs by way of the afferent (vagus) nerve then inhibit the respiratory center. Lack of stimulation causes the rib muscles and diaphragm to relax and expiration follows.

Inspiration

It can be shown that carbon dioxide and hydrogen ions are the primary stimuli that cause us to breathe. When the concentration of CO_2 and subsequently H^+ reach a certain level in the blood, the *breathing center* in the medulla oblongata—the stem portion of the brain—is stimulated. This center is not affected by low oxygen levels, but there are chemoreceptors in the *carotid bodies* (located in the carotid arteries) and in the *aortic bodies* (located in the aorta) that do respond to low blood oxygen in addition to carbon dioxide and hydrogen ion concentration.

As diagrammed in figure 13.9, when the breathing center is stimulated, a nerve impulse goes out by way of nerves to the diaphragm and rib cage. In its relaxed state, the *diaphragm* is dome-shaped, but upon stimulation it contracts and lowers. When the rib muscles contract, the *rib cage* moves upward and outward. Both of these contractions serve to increase the size of the chest cavity. As the chest cavity increases in size, the lungs expand. When the lungs expand, air pressure within the enlarged alveoli lowers and is immediately rebalanced by air rushing in through the nose or mouth.

Inspiration (fig. 13.8a and b) is the active phase of breathing. It is during this time that the diaphragm contracts, the rib muscles contract, and the lungs are pulled open, and air comes rushing in. Note that the air comes in because the lungs have already opened up; the air does not force the lungs open. This is why it is sometimes said that *humans breathe by negative pressure.* It is the creation of a partial vacuum that sucks air into the lungs.

Stimulated by nervous impulses, the rib cage lifts up and out and the diaphragm lowers to expand the chest cavity and lungs. Air flows in as inspiration occurs.

Table 13.3 Breathing Process

Inspiration	Expiration
Medulla sends stimulatory message to diaphragm and rib muscles.	Stretch receptors in lungs send inhibitory message to medulla.
Diaphragm contracts and flattens.	Diaphragm relaxes and resumes dome position.
Rib cage moves up and out.	Rib cage moves down and in.
Lungs expand.	Lungs recoil.
Negative pressure in lungs.	Positive pressure in lungs.
Air is pulled in.	Air is forced out.

Expiration

When the lungs are expanded, the stretching of the alveoli stimulates special receptors in the alveolar walls, and these receptors initiate nerve impulses that travel from the inflated lungs to the breathing center. When the impulses arrive at the medulla oblongata, the center is inhibited and stops sending signals to the diaphragm and the rib cage. The *diaphragm* relaxes and resumes its dome shape (fig. 13.8c and d). The abdominal organs press up against the diaphragm. The *rib cage* moves down and inward. The elastic lungs recoil and air is pushed outward.

Table 13.3 summarizes the events that occur during inspiration and expiration. It is clear that while inspiration is an active phase of breathing, normally expiration is passive because the breathing muscles automatically relax following contraction. But it is possible, with deeper and more rapid breathing, for both phases to be active. In other words, you can forcibly bring more air into and/or out of the lungs.

It is also possible for you to deliberately breathe more rapidly or more slowly or to even hold the breath for a short time. However, it is impossible to commit suicide by holding your breath; eventually, carbon dioxide buildup in the blood forces the resumption of breathing.

When nervous stimulation ceases, the rib cage lowers and the diaphragm rises, allowing the lungs to recoil and expiration to occur.

Consequences

There are certain physiological consequences that result from the manner in which we breathe. First of all, we may note that breathing is initiated, and continues, because of carbon dioxide in the blood. Therefore, when necessary, it is better to give a person oxygen gas containing carbon dioxide rather than pure oxygen alone. The mixture of gases stimulates the resumption of breathing, whereas pure oxygen does not.

In humans, the breathing rate is regulated by the amount of carbon dioxide in the blood.

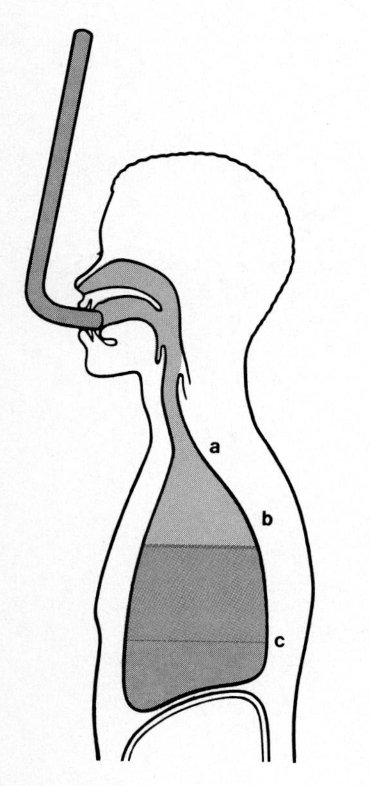

Figure 13.10
Distribution of air in lungs. The air between a *and* b *does not immediately reach the alveoli; therefore this is called dead space. The air below* c *represents the amount of residual air that has not left the lungs. Only the air between* b *and* c *brings with it additional oxygen to respiration.*

Also, students who have followed this discussion carefully will realize that the respiratory tract contains a certain amount of dead space, or space that contains air not used in gas exchange. If we diagram the tract as illustrated in figure 13.10, we see that the dead space can be considered to extend from the top *a* of the tube (pharynx) to a certain level within the lungs *b*. Not all this fresh air immediately reaches the respiratory surface of the alveoli. Also there is a certain amount of air that never leaves the lungs. This is called residual air and is indicated below *c* in the diagram. This means that the air between *b* and *c* is the portion that immediately supplies additional oxygen. Normally we inhale about 500 cubic centimeters of air; of this amount, about 350 cubic centimeters immediately reaches the alveoli. With deep breathing we can inhale a maximum of 1,650 cubic centimeters. Even so, it may be readily seen that humans cannot breathe through a very long tube. Any device that increases the amount of dead space beyond maximal inhaling capacity spells death to the individual because the air inhaled would never reach the alveoli.

Some of the inspired air never reaches the lungs and there is always a mixture of used and new air in the lungs.

Figure 13.11
Mountain climbers should ascend slowly and rest often, giving their bodies time to adjust to the progressively lower oxygen tension. Above 10,000 feet it is wise to hyperventilate by taking numerous deep breaths in rapid succession or to carry oxygen equipment as a safeguard. Mountain sickness in its mildest form includes headache, fatigue, dizziness, palpitations, loss of appetite, nausea, and insomnia. High-altitude pulmonary edema (HAPE) is more serious because the lung alveoli fill up with fluid preventing gas exchange from occurring. HAPE can be recognized by severe shortness of breath, coughing, bubbling noises in the chest, weakness, and stupor. Even more dangerous is cerebral edema, in which the brain tissues swell and the person begins to hallucinate. Should either of these edemas occur the person should be transported down immediately.

External and Internal Respiration

External Respiration

The term *external respiration* refers to the exchange of gases between the air in the alveoli and the blood within the pulmonary capillaries. The wall of an alveolus consists of a thin, single layer of cells, and the wall of a blood capillary also consists of such a layer. Since neither wall offers resistance to the passage of gases, *diffusion* alone governs the exchange of oxygen and carbon dioxide between alveolar air and the blood. Active cellular absorption and secretion do not appear to play a role. Rather, the direction in which the gases move is determined by the pressure or tension gradients between blood and inspired air.

Atmospheric air contains little carbon dioxide, but blood flowing into the lung capillaries is almost saturated with the gas. Therefore, *carbon dioxide diffuses out of the blood into the alveoli.* The pressure pattern is the reverse for oxygen. Blood coming into the pulmonary capillaries is oxygen poor and alveolar air is oxygen rich; therefore, *oxygen diffuses into the capillary.* Breathing at high altitudes is less effective than at low altitudes because the air pressure is less, making the concentration of oxygen (and other gases) lower than normal; therefore less oxygen diffuses into the blood (fig. 13.11). Breathing problems do not occur in airplanes because the cabin is pressurized to maintain an appropriate pressure. Emergency oxygen is available in case the pressure should, for one reason or another, be reduced.

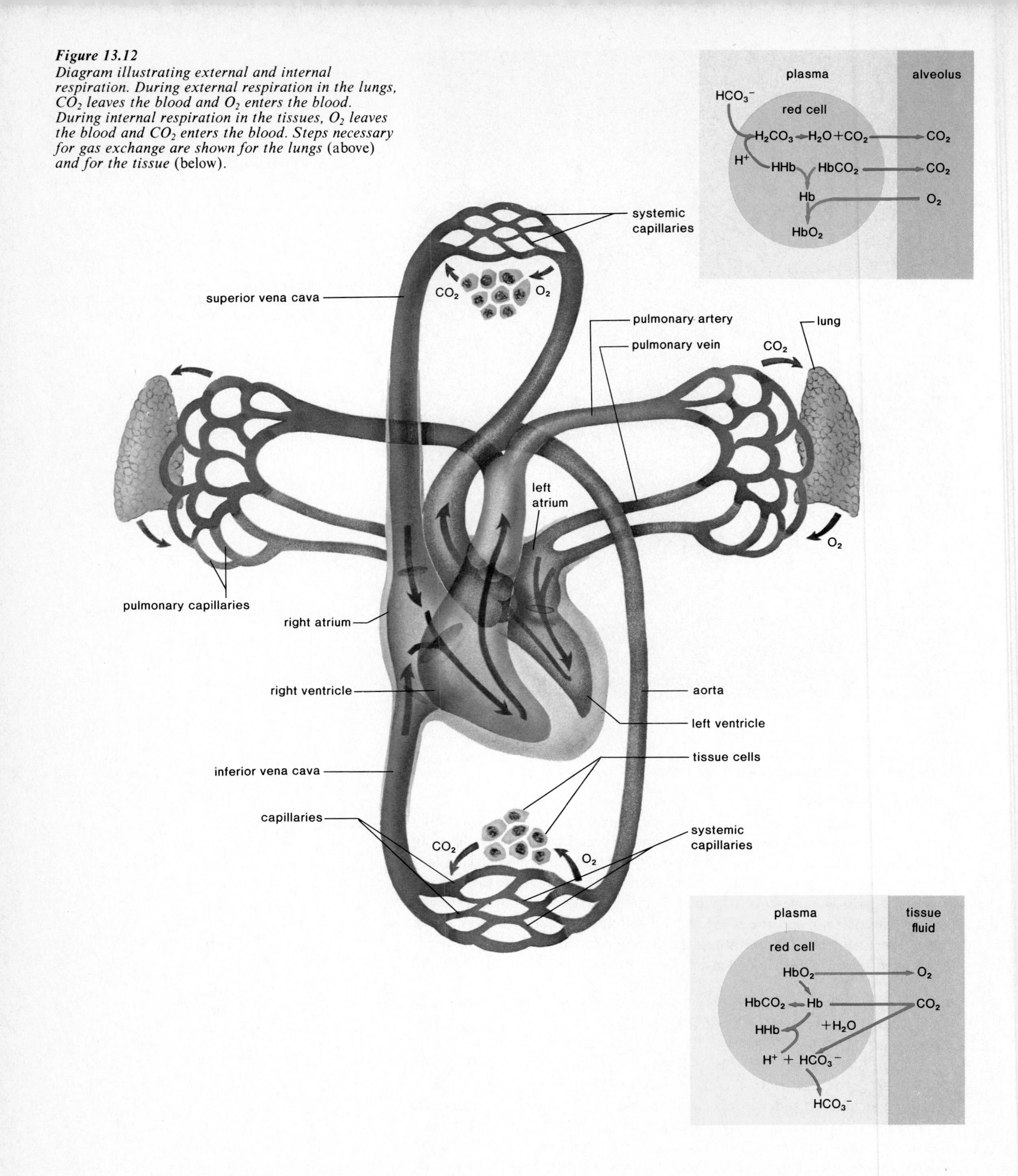

Figure 13.12
Diagram illustrating external and internal respiration. During external respiration in the lungs, CO_2 leaves the blood and O_2 enters the blood. During internal respiration in the tissues, O_2 leaves the blood and CO_2 enters the blood. Steps necessary for gas exchange are shown for the lungs (above) *and for the tissue* (below).

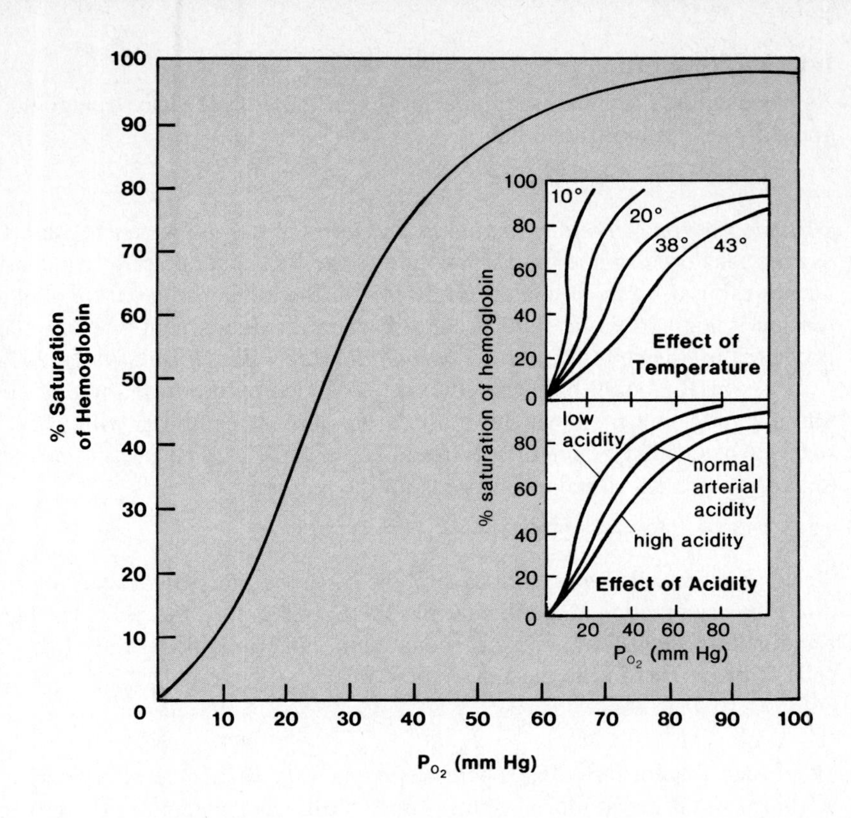

Figure 13.13
Properties of hemoglobin as revealed by the hemoglobin dissociation curve. The large curve shows the percentage of saturation of hemoglobin at 38° and normal arterial blood acidity. As the partial pressure of oxygen (PO_2) decreases, hemoglobin gives up its oxygen and this effect is also promoted by the higher temperature and higher acidity of the tissues.

As blood enters the pulmonary capillaries (fig. 13.12), most of the carbon dioxide is being carried as the bicarbonate ion, HCO_3^-. As the little free carbon dioxide remaining begins to diffuse out, the following reaction is driven to the right:

$$H^+ + HCO_3^- \longrightarrow H_2CO_3 \longrightarrow H_2O + CO_2$$

The enzyme carbonic anhydrase (p. 236), present in red cells, speeds up the reaction. As the reaction proceeds, **hemoglobin** (p. 234) gives up the hydrogen ions it has been carrying; HHb becomes Hb.

Now hemoglobin more readily takes up oxygen and becomes oxyhemoglobin.

$$Hb + O_2 \longrightarrow HbO_2$$

It is a remarkable fact that at the oxygen tension in inspired air (150 mm Hg), hemoglobin is about 100 percent saturated. Figure 13.13 indicates that hemoglobin takes up oxygen in increasing amounts as the oxygen tension increases, until about 100 mm Hg when the uptake levels off. This means that hemoglobin easily retains oxygen at the oxygen tension in the lungs and tends to release it at the oxygen tension in the tissues. Further, another remarkable fact about hemoglobin is that it more readily takes up oxygen in the neutral pH and cool temperature of the lungs. On the other hand, it will give up oxygen more readily at the more acid pH and warmer temperature of the tissues.[3]

External respiration, the exchange of oxygen for carbon dioxide between air within the alveoli and the blood in pulmonary capillaries, is dependent on the process of diffusion.

---

[3]	*pH*	*Temperature*
Lungs	7.40	37° C (98.6°F)
Body	7.38	38° C (100.4° F)

Internal Respiration

As blood enters the systemic capillaries (fig. 13.12), oxygen leaves hemoglobin and diffuses out into tissue fluid:

$$HbO_2 \longrightarrow Hb + O_2$$

Diffusion of oxygen out of the blood and into the tissues occurs because the oxygen tension in tissue fluid is low due to the fact that the cells are continuously using it up in cellular respiration. On the other hand, carbon dioxide tension is high in tissue fluid because carbon dioxide is continuously being produced by the cells. Therefore, *carbon dioxide will diffuse into the blood.*

A small amount of carbon dioxide is carried by the formation of carbaminohemoglobin (p. 236). But most carbon dioxide combines with water to form carbonic acid, which dissociates to $H^+ + HCO_3^-$. The enzyme carbonic anhydrase present in red cells speeds up the reaction:

$$CO_2 + H_2O \longrightarrow H_2CO_3 \longrightarrow H^+ + HCO_3^-$$

The globin portion of hemoglobin combines with the excess hydrogen ions produced by the reaction, and Hb becomes HHb. In this way, the pH of the blood remains fairly constant. The bicarbonate ion, HCO_3^-, diffuses out of the red cells to be carried in the plasma.

Internal respiration, the exchange of oxygen for carbon dioxide between blood in tissue capillaries and in tissue fluid, is dependent on the process of diffusion.

Respiration and Health

We have seen that the full length of the respiratory tract is lined with a warm, wet mucosa lining that is constantly exposed to environmental air. The quality of this air, determined by the pollutants and germs it contains, can affect the health of the individual.

Germs frequently spread from one individual to another by way of the respiratory tract. Droplets from one single sneeze may be loaded with billions of bacteria or viruses. The mucous membranes are protected by the production of mucus and by the constant beating of the cilia; but if the number of infective agents is large and/or the resistance of the individual is reduced, an upper respiratory infection may result.

Upper Respiratory Infections

An upper respiratory infection (URI) is one that affects only the nasal cavities, throat, trachea, bronchi, and associated organs.[4] Vaccines for these infections are not in wide use and if they are viral in nature, antibiotics are not helpful. Since viruses take over the machinery of the cell when they reproduce, it is difficult to develop drugs that will affect the virus without affecting the cell itself.

Common Cold

A cold is a viral infection that usually begins as a scratchy sore throat, followed by a watery mucus discharge from the nasal cavities. There is rarely a fever and symptoms are usually mild, requiring little or no medication. While colds have a short duration, immunity is also brief. Since there are estimated to be over 150 cold-causing viruses (fig. 13.14) it is very difficult to isolate oneself in order to avoid infection. A nasal spray containing interferon has shown some promise in helping prevent colds but the cost remains prohibitive for most people.

[4]Allergies, including asthma, are discussed on pages 260–261.

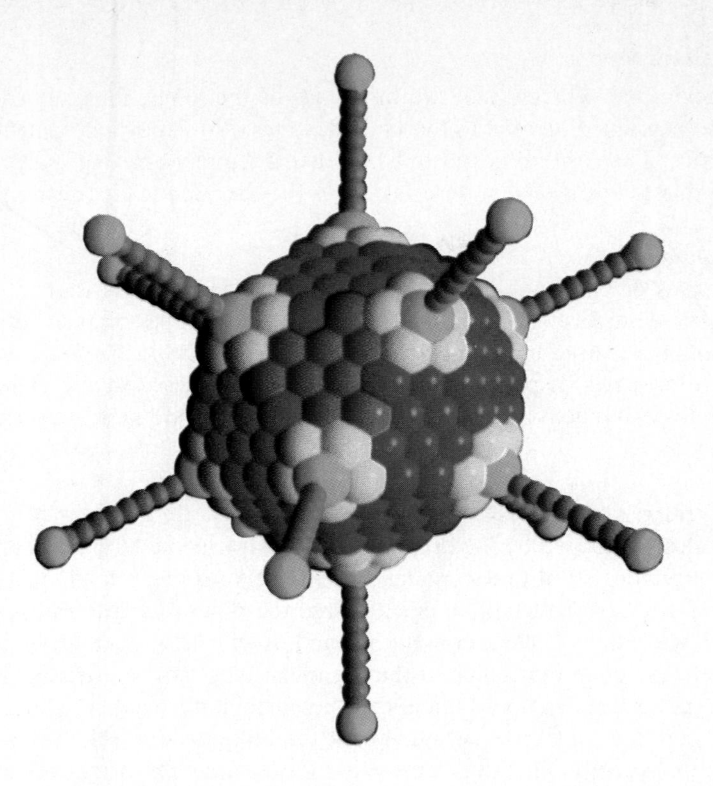

Figure 13.14
Computer graphics allows us to see the structure of a cold virus. These germs are spread through the air when a person with a cold sneezes or coughs.

Influenza

"Flu" is also a viral infection; but although it begins as an upper respiratory infection, it spreads to other parts of the body, causing aches and pains in the joints. There is usually a fever and the illness lasts for a longer length of time than a cold. Immunity is possible, but only the vaccine developed for the particular virus prevalent that season can be successful in protecting the individual during a flu epidemic. Since flu viruses constantly mutate, there can be no buildup in immunity and a new viral illness rapidly spreads from person to person and from place to place. Pandemics, in which a newly mutated flu virus spreads about the world, have occurred on occasion.

Bronchitis

Viral infections can spread from the nasal cavities to the sinuses (sinusitis), to the middle ears (otitis media), to the larynx (laryngitis), and to the bronchi (bronchitis). Acute bronchitis is usually caused by a secondary bacterial infection of the bronchi, resulting in a heavy mucoid discharge with much coughing. Acute bronchitis usually responds to antibiotic therapy. Chronic bronchitis is not necessarily due to infection. It is often caused by a constant irritation of the lining of the bronchi which, as a result, undergoes degenerative changes including the loss of cilia and their normal cleansing action. There is frequent coughing and the individual is more susceptible to upper respiratory infections. Chronic bronchitis most often affects cigarette smokers.

Strep Throat

This is a very severe throat infection caused by the bacterium *Streptococcus*. Swallowing is very difficult and there is a fever. Unlike a viral infection, strep throat should be treated with antibiotics. If not treated, it may lead to complications, such as rheumatic fever in which the heart valves may be permanently affected.

Upper respiratory infections due to viral infections are not treatable by antibiotics but bacterial ones are.

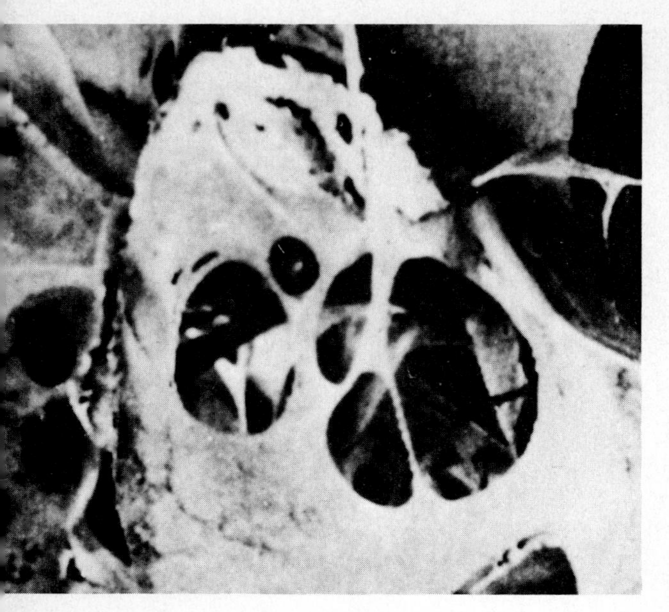

Figure 13.15
Scanning electron micrograph of the lungs of a person with emphysema. There are large cavities in the lungs due to the breakdown of alveoli.

Lung Disorders

Pneumonia and tuberculosis, two infections of the lungs, formerly caused a large percentage of deaths in the United States. Now they are controlled by antibiotics. Two other illnesses to be discussed, emphysema and lung cancer, are not due to infections; in most instances they are due to cigarette smoking.

Pneumonia

Most forms of pneumonia are caused by bacteria or viruses that infect the lungs. The demise of AIDS patients is usually due to a particularly rare form of pneumonia caused by a protozoan *Pneumocystis carinii.* Sometimes pneumonia is localized in specific lobes of the lungs and these become inoperative as they fill with mucus and pus. The more lobes involved, the more serious the infection.

Tuberculosis

Tuberculosis is caused by the tubercle bacillus. It is possible to tell if a person has ever been exposed to tuberculosis by use of a skin test in which a highly diluted extract of the bacilli is injected into the skin of the patient. A person who has never been in contact with the bacillus will show no reaction, but one who has developed immunity to the organism will show an area of inflammation that peaks in about 48 hours. If these bacilli do invade the lung tissue, the cells build a protective capsule about the foreigners to isolate them from the rest of the body. This tiny capsule is called a *tubercle.* If the resistance of the body is high, the imprisoned organisms may die, but if the resistance is low the organisms may eventually be liberated. If a chest X ray detects the presence of tubercles, the individual is put on appropriate drug therapy to ensure the localization of the disease and eventual destruction of any live bacterial organisms.

Emphysema

Emphysema refers to the destruction of lung tissue, with accompanying ballooning or inflation of the lungs due to trapped air. The trouble stems from the destruction and collapsing of the bronchioles. When this occurs, the alveoli are cut off from renewed oxygen supply and the air within them is trapped. The trapped air very often causes rupturing of alveolar walls (fig. 13.15) and fibrous thickening of associated blood vessel walls. In any case, the victim is breathless and may have a cough. Since the surface area for gas exchange is reduced, not enough oxygen reaches the heart and brain. Even so, the heart works furiously to force more blood through the lungs and this may lead to a heart condition. Lack of oxygen for the brain may make the person feel depressed, sluggish, and irritable.

Chronic bronchitis and emphysema, two conditions most often caused by smoking, together are called chronic obstructive pulmonary disease (COPD).

Pulmonary Fibrosis

Inhaling particles, such as silica (sand), coal dust, and asbestos (fig. 13.16) can lead to pulmonary fibrosis in which fibrous connective tissue builds up in the lungs. Breathing capacity can be seriously impaired and the development of cancer is common. Since asbestos has been so widely used as a fireproofing and insulating agent, unwarranted exposure has occurred.

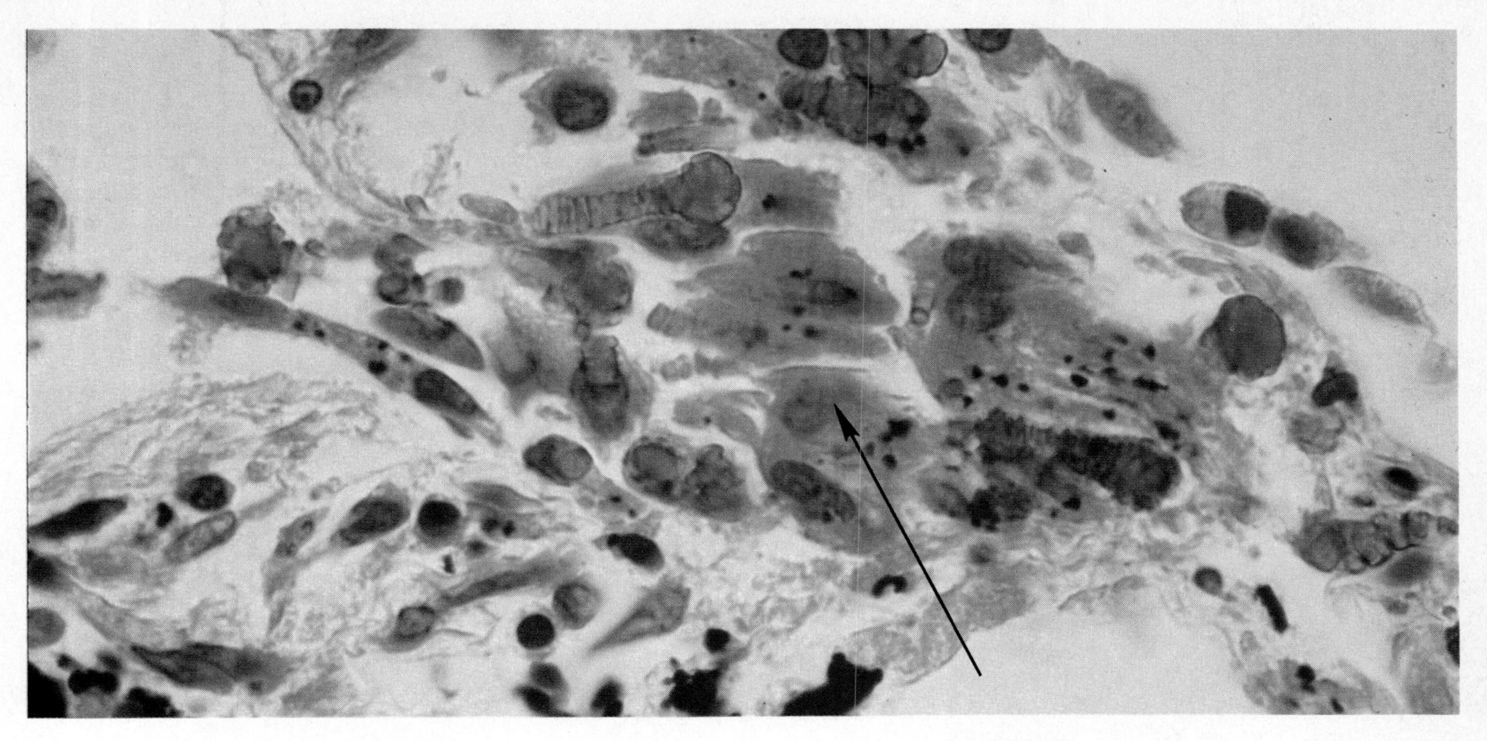

Figure 13.16
Asbestos fibers get caught in the lungs (arrow). *Fibrous tissue develops and eventually there may be cancer.*

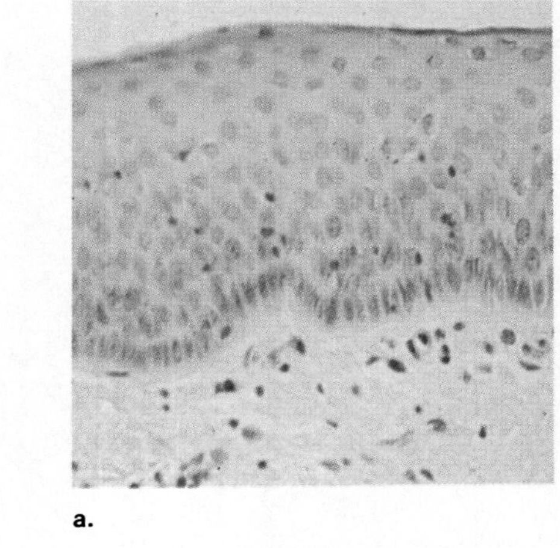

a.

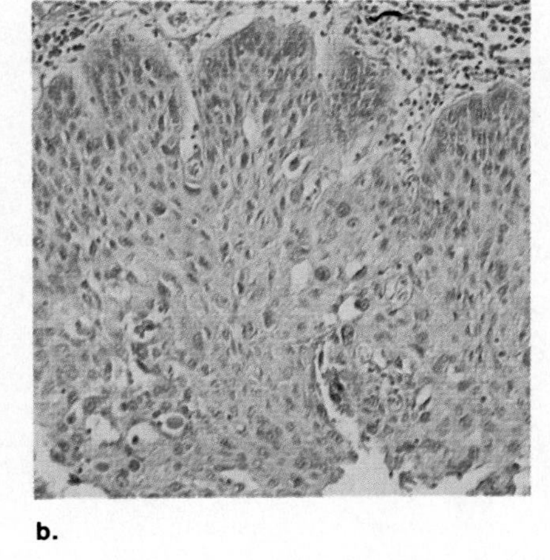

b.

Figure 13.17
Stages in the development of bronchial cancer.
a. *Cilia have disappeared and epithelial cells have increased in number.* b. *Precancerous cells continue to invade other tissue layers.*

Lung Cancer

Lung cancer use to be more prevalent in men than women, but recently lung cancer has surpassed breast cancer as a cause of death in women. This can be linked to an increase in the number of women who smoke today. Autopsies on smokers have revealed the progressive steps by which the most common form of lung cancer develops (fig. 13.17). The first event appears to be a thickening of the cells that line the bronchi. (Callusing occurs whenever cells are exposed to irritants.) Then there is a loss of cilia so that it is impossible to prevent dust and dirt from settling in the lungs. Following this, cells with atypical nuclei appear in the thickened lining. A disordered collection of cells with atypical nuclei may be considered to be cancer in situ (at one location). The final step occurs when some cells break loose and penetrate the other tissues, a process called metastasis. This is true cancer. The tumor may grow until the bronchus is blocked, cutting off the supply of air to that lung. The lung then collapses, and the secretions trapped in the lung spaces become infected, with a resulting pneumonia or the formation of a lung abscess. The only treatment that offers a possibility of cure, before secondary growths have had time to form, is to remove the lung completely. This operation is called *pneumonectomy*.

Risks of Smoking versus Benefits of Quitting

Based on available statistics, the American Cancer Society informs us of the risks of smoking versus the benefits of quitting:

Risks of Smoking	**Benefits of Quitting**
Shortened life expectancy. 25-year-old 2-pack a day smokers have life expectancy 8.3 years shorter than nonsmoking contemporaries. Other smoking levels: proportional risk.	*Reduced risk of premature death* cumulatively. After 10–15 years, ex-smokers' risk approaches that of those who have never smoked.
Lung cancer. Smoking cigarettes "major cause in both men and women."	Gradual decrease in risk. *After 10–15 years, risk approaches that of those who never smoked.*
Larynx cancer. In all smokers (including pipe and cigar) it is 2.9 to 17.7 times that of nonsmokers.	*Gradual reduction of risk* after smoking cessation. *Reaches normal after 10 years.*
Mouth cancer. Cigarette smokers have 3 to 10 times as many oral cancers as nonsmokers. Pipes, cigars, chewing tobacco also major risk factors. Alcohol seems synergistic carcinogen with smoking.	Reducing or eliminating smoking/drinking reduces risk in first few years; *risk drops to level of nonsmokers in 10–15 years.*
Cancer of esophagus. Cigarettes, pipes, and cigars increase risk of dying of esophageal cancer about 2 to 9 times. Synergistic relationship between smoking and alcohol.	Since risks are dose related, reducing or eliminating smoking/drinking *should have risk-reducing effect.*
Cancer of bladder. Cigarette smokers have 7 to 10 times risk of bladder cancer as nonsmokers. Also synergistic with certain exposed occupations: dye-stuffs, etc.	Risk decreases gradually to that of nonsmokers over 7 years.
Cancer of pancreas. Cigarette smokers have 2 to 5 times risk of dying of pancreatic cancer as nonsmokers.	Since there is evidence of dose-related risk, reducing or eliminating smoking should have risk-reducing effect.

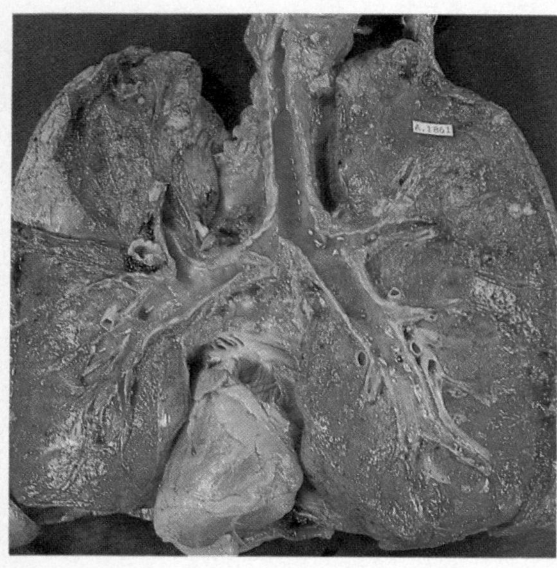

a.

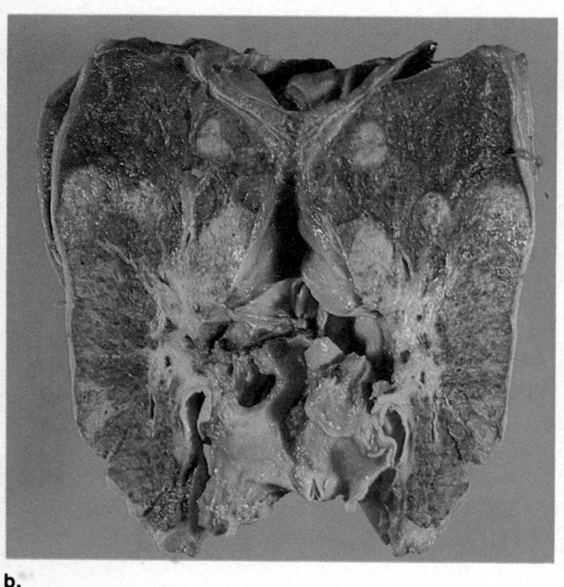

b.

a. *Normal lung with heart in place. Notice the healthy red color.* b. *Lungs of a heavy smoker. Notice how black the lungs are except where cancerous tumors have formed.*

The incidence of lung cancer is much higher in individuals who smoke compared to those who do not smoke.

A recent finding of late has been that *involuntary smoking,* simply breathing in air filled with cigarette smoke, can also cause lung cancer and other illnesses associated with smoking. The reading on page 286 discusses these various illnesses including other types of cancer. If a person does stop both voluntary and involuntary smoking and if the body tissues are not already cancerous, they return to normal.

Risks of Smoking	**Benefits of Quitting**
Coronary heart disease. Cigarette smoking is major factor; responsible for 120,000 excess U.S. deaths from coronary heart disease (CHD) each year.	*Sharply decreases risk after one year.* After 10 years ex-smokers' risk is same as that of those who never smoked.
Chronic bronchitis and pulmonary emphysema. Cigarette smokers have 4–25 times risk of death from these diseases as nonsmokers. Damage seen in lungs of even young smokers.	*Cough and sputum disappear* during first few weeks. *Lung function may improve* and rate of deterioration slow down.
Stillbirth and low birthweight. Smoking mothers have more stillbirths and babies of low birthweight—more vulnerable to disease and death.	Women who stop smoking before 4th month of pregnancy *eliminate risk of stillbirth and low birthweight* caused by smoking.
Children of smoking mothers smaller, underdeveloped physically and socially, seven years after birth.	Since children of nonsmoking mothers are bigger and more advanced socially, inference is that *not smoking during pregnancy might avoid such underdeveloped children.*
Peptic ulcer. Cigarette smokers get more peptic ulcers and die more often of them; cure is more difficult in smokers.	Ex-smokers get ulcers but these are *more likely to heal rapidly and completely* than those of smokers.
Allergy and impairment of immune system.	Since these are direct, immediate effects of smoking, they are obviously *avoidable by not smoking.*
Alters pharmacologic effects of many medicines, diagnostic tests and greatly increases risk of thrombosis with oral contraceptives.	*Majority of blood components elevated by smoking return to normal after cessation.* Nonsmokers on the Pill have much lower risks of thrombosis.

Reproduced by permission of the American Cancer Society, Inc.

Summary

Air enters and exits the lungs by way of the respiratory tract (table 13.2). Inspiration begins when the breathing center in the medulla oblongata sends excitatory nerve impulses to the diaphragm and rib cage. As they contract, the diaphragm lowers and the rib cage moves up and out; the lungs expand, creating a partial vacuum that causes air to rush in. Nerves within the expanded lungs then send inhibitory impulses to the breathing center. As the diaphragm relaxes, it resumes its dome shape and as the rib cage retracts, air is pushed out of the lungs during expiration.

External respiration occurs when carbon dioxide leaves the blood and oxygen enters the blood at the alveoli. Oxygen is transported to the tissues in combination with hemoglobin. Internal respiration occurs when oxygen leaves the blood and carbon dioxide enters the blood at the tissues. Carbon dioxide is carried to the lungs in the form of the bicarbonate ion.

There are a number of illnesses associated with the respiratory tract. In addition to colds and flu, pneumonia and tuberculosis are serious lung infections. Two illnesses that have been attributed to breathing polluted air are emphysema and lung cancer.

Objective Questions

1. In tracing the path of air, the ________ immediately follows the pharynx.
2. The lungs contain air sacs called ________.
3. The breathing rate is primarily regulated by the amount of ________ and ________ in the blood.
4. Air enters the lungs after they have ________.
5. Carbon dioxide is carried in the blood as the ________ ion.
6. The H^+ ions given off when carbonic acid dissociates are carried by ________.
7. Gas exchange is dependent on the physical process of ________.
8. Reduced hemoglobin becomes oxyhemoglobin in the ________.
9. The most likely cause of emphysema and chronic bronchitis is ________.
10. Most cases of lung cancer actually begin in the ________.

Answers to Objective Questions

1. larynx 2. alveoli 3. CO_2, H^+ 4. expanded 5. HCO_3^- 6. hemoglobin 7. diffusion 8. lungs 9. smoking cigarettes 10. bronchi

Study Questions

1. What are the four parts of respiration? In which of these is oxygen actually used up and carbon dioxide produced? (p. 270)
2. List the parts of the respiratory tract. What are the special functions of the nasal cavity, larynx, and alveoli? (p. 270)
3. What are the steps in inspiration and expiration? How is breathing controlled? (pp. 275–278)
4. Why can't we breathe through a very long tube? (p. 278)
5. What physical process is believed to explain gas exchange? (p. 279)
6. What two equations are needed to explain external respiration? (p. 281)
7. How is hemoglobin remarkably suited to its job? (p. 281)
8. What two equations are needed to explain internal respiration? (p. 282)
9. Name some infections of the respiratory tract. (pp. 282–284)
10. What are emphysema and pulmonary fibrosis, and how do they affect one's health? (p. 284)
11. By what steps is cancer believed to develop in the person who smokes? (p. 285)

Key Terms

alveoli (al-ve'o-lī) saclike structures that are the air sacs of a lung. *274*

bronchi (brong'ki) the two major divisions of the trachea leading to the lungs. *274*

bronchiole (brong'ke-ōl) the smaller air passages in the lungs of mammals. *274*

diaphragm (di'ah-fram) a sheet of muscle that separates the chest cavity from the abdominal cavity. *276*

expiration (eks''pi-ra'shun) process of expelling air from the lungs; exhalation. *270*

external respiration (eks-ter'nal res''pi-ra'shun) exchange between blood and alveoli of carbon dioxide and oxygen. *270*

hemoglobin (he''mo-glo'bin) a red iron-containing pigment in blood that combines with and transports oxygen. *281*

inspiration (in''spi-ra'shun) the act of breathing in. *270*

internal respiration (in-ter'nal res''pi-ra'shun) exchange between blood and tissue fluid of oxygen and carbon dioxide. *270*

ribs (ribz) bones hinged to the vertebral column and sternum which, with muscle, define the top and sides of the chest cavity. *276*

trachea (tra'ke-ah) a tube that is supported by C-shaped cartilagenous rings that lies between the larynx and the bronchi; the windpipe. *273*

ventilation (ven''tī-la'shun) breathing; the process of moving air into and out of the lungs. *275*

vocal cords (vo'kal kordz) folds of tissue within the larynx that create vocal sounds when they vibrate. *272*

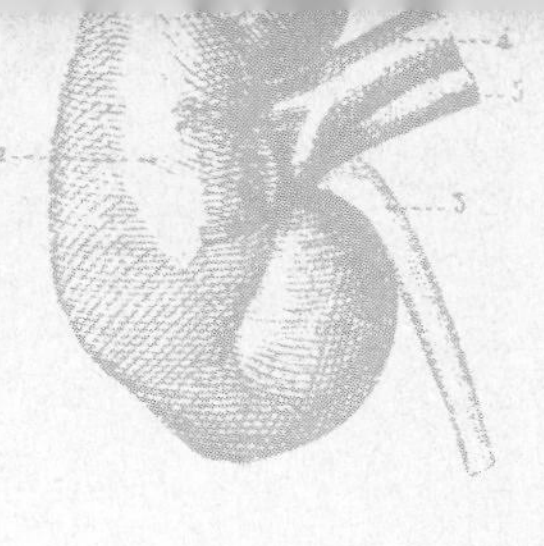

14

Excretion

Chapter Concepts

1 Excretion rids the body of unwanted substances, particularly end products of metabolism.

2 Several organs assist in the excretion process, but the kidneys, which are a part of the urinary system, are the primary organs of excretion.

3 The formation of urine by the more than one million nephrons present in each kidney serves not only to rid the body of nitrogenous wastes but also to regulate the water content, the salt levels, and the pH of the blood.

4 The kidneys, whose malfunction brings illness and may cause death, are important organs of homeostasis.

Chapter Outline

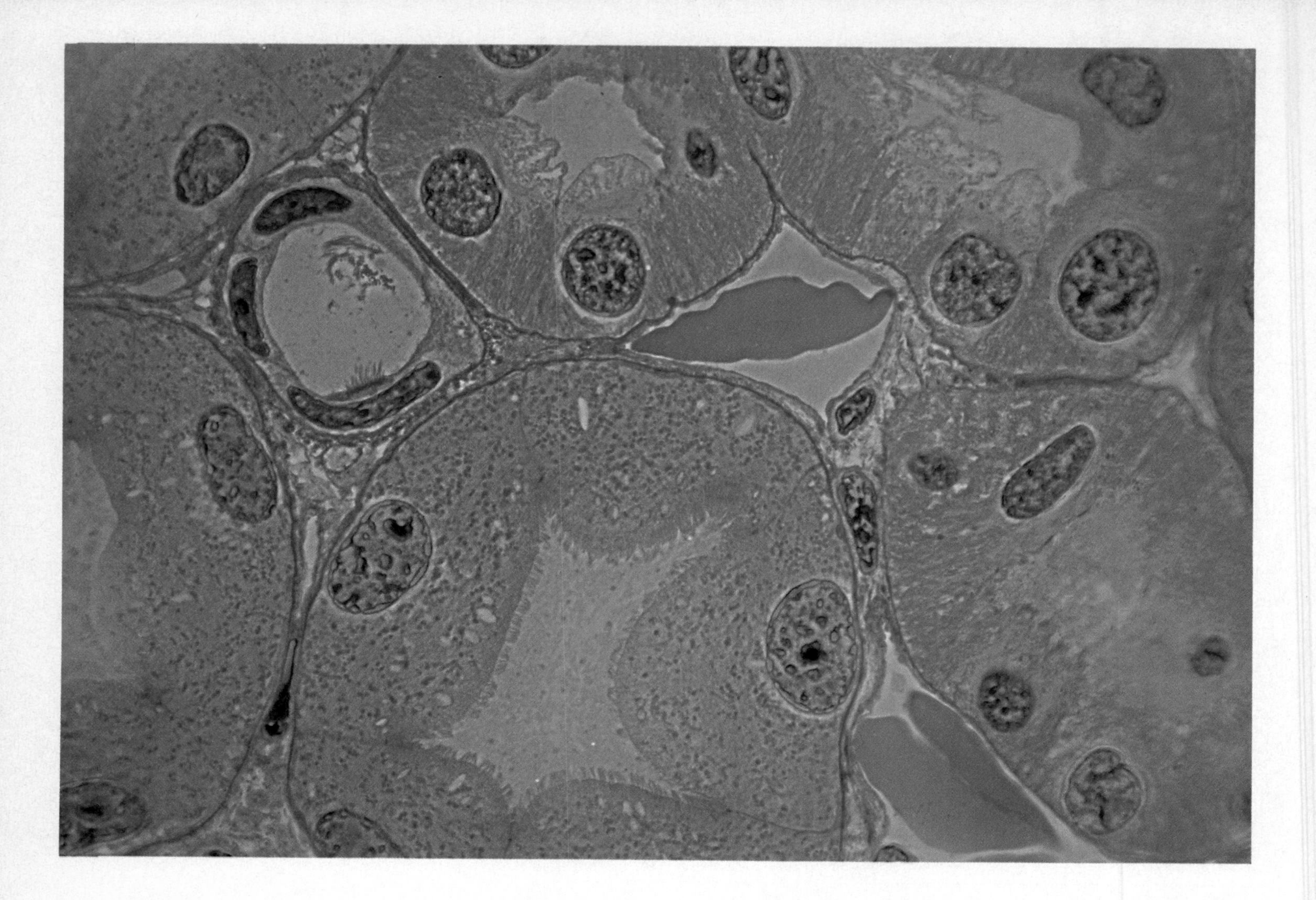

Figure 14.1
Cross section through kidney tubules, the structures that rid the body of nitrogenous wastes. The microvilli that form a "brush border" absorb substances needed by the body, leaving behind the wastes that are to be excreted.

The composition of blood serving the tissues remains relatively constant due to both the continual addition of substances needed by cells and the continual removal of substances not needed by cells (fig. 14.1). In previous chapters we have discussed how the digestive tract and lungs add nutrients and oxygen to the blood. In this chapter, we will discuss how the organs of excretion (fig. 14.2) remove substances from the blood and thereby help maintain homeostasis.

Excretory Substances

Excretion rids the body of metabolic wastes. Among these are the toxic products listed in table 14.1. In addition, salts and water are constantly being excreted.

Several of the end products excreted by humans are related to nitrogen metabolism since amino acids, nucleotides, and creatine all contain nitrogen.

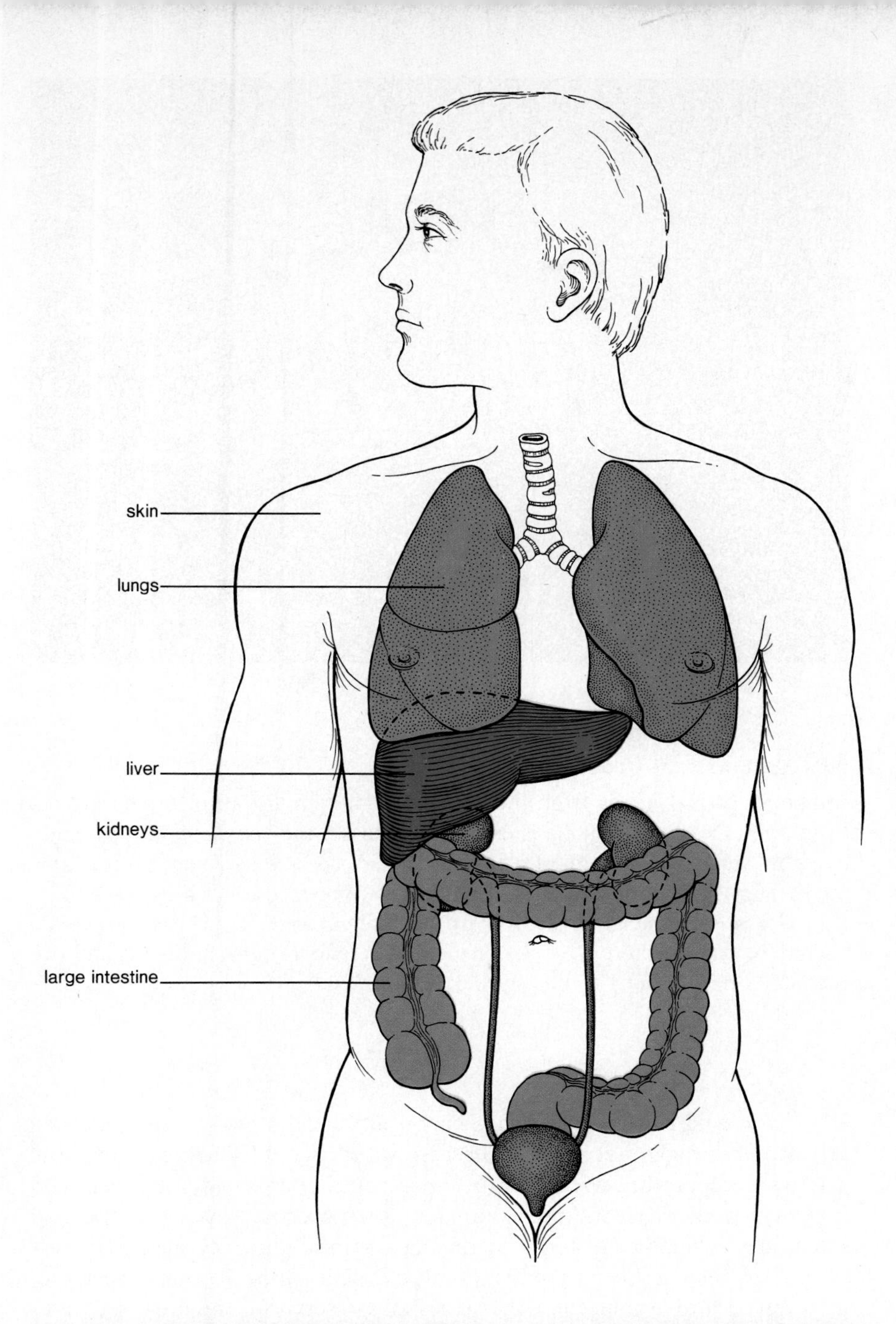

Figure 14.2
The organs of excretion include not only the kidneys but also the skin, lungs, liver, and large intestine. The lungs excrete carbon dioxide; the liver excretes hemoglobin breakdown products, and the intestine excretes certain heavy metals. Excretion, ridding the body of metabolic wastes, should not be confused with defecation, ridding the body of nondigestable remains.

***Table 14.1** Some Metabolic End Products*

Name	End Product of	Primarily Excreted by
Nitrogenous Wastes		
Ammonia	Amino acid metabolism	Kidneys
Urea	Ammonia metabolism	Kidneys
Uric acid	Nucleotide metabolism	Kidneys
Creatinine	Creatine phosphate metabolism	Kidneys
Other		
Bile pigments	Hemoglobin metabolism	Liver
Carbon dioxide	Cellular respiration	Lungs

Figure 14.3
Fishes can excrete ammonia as their nitrogenous waste. They do not need to conserve water since they live in an aquatic environment.

Figure 14.4
Birds excrete uric acid, a solid material, as their nitrogenous waste. It is mixed with fecal material in a common repository for the urinary, digestive, and reproductive systems. Sea birds congregate in such numbers that their droppings build up to give a nitrogen-rich substance called guano. At one time guano was harvested as a natural fertilizer.

Nitrogenous End Products

Ammonia (NH_3) arises from the deamination, or removal, of amino groups from amino acids. Ammonia is extremely toxic to the body, and only animals living in water, who continually flush out their bodies with water, excrete ammonia (fig. 14.3). In our bodies, ammonia is converted to urea by the liver.

Urea is produced in the liver by a complicated series of reactions called the urea cycle. In this cycle, carrier molecules take up carbon dioxide and two molecules of ammonia to finally release urea:

$$H_2N - \overset{\overset{\displaystyle O}{\|}}{C} - NH_2$$

Uric acid is excreted as the general nitrogenous end product of many terrestrial animals that need to conserve water (fig. 14.4.) In humans, uric acid only occurs when nucleotides are metabolically broken down; if uric acid is present in excess, it will precipitate out of the plasma. Crystals of uric acid sometimes collect in the joints, producing a painful ailment called gout.

Creatinine is an end product of muscle metabolism. It results when creatine phosphate, a molecule that serves as a reservoir of high energy phosphate, breaks down.

Other Excretory Substances

Other excretory substances are bile pigments, carbon dioxide, ions (salts), and water.

Bile Pigments

Bile pigments are derived from the heme portion of hemoglobin and are incorporated into bile within the liver (fig. 14.5). Although the liver produces bile, it is stored in the gallbladder before passing into the small intestine by way of ducts. If for any reason the bile duct is blocked, bile spills out into the blood, producing a discoloration of the skin called jaundice (p. 192).

Carbon Dioxide

The lungs are the major organs of *carbon dioxide* excretion, although the kidneys are also important. The kidneys excrete bicarbonate ions, the form in which carbon dioxide is carried in the blood.

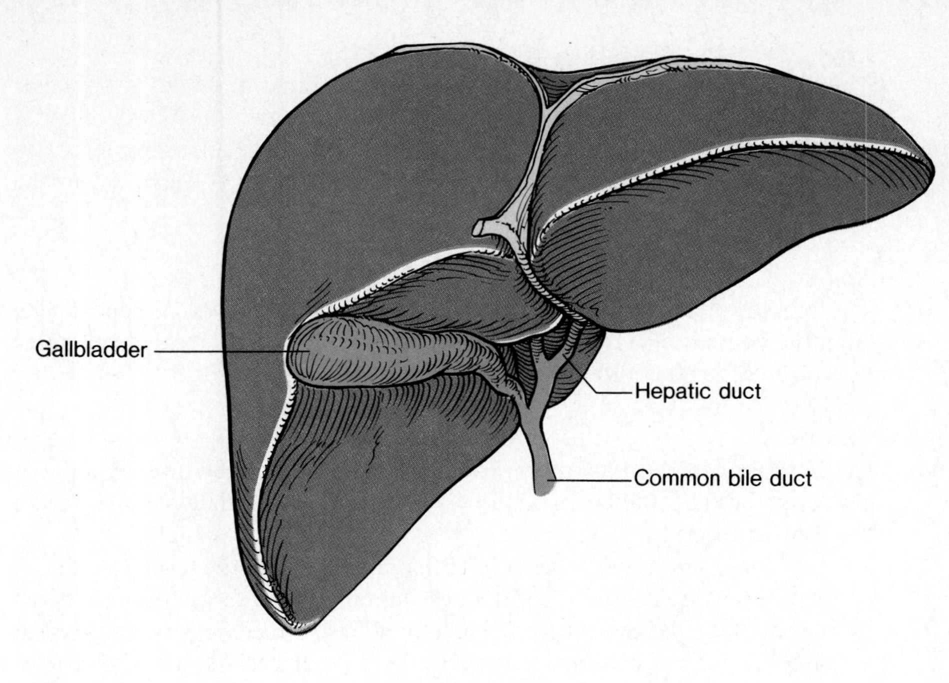

Figure 14.5
The liver is an organ of excretion because it breaks down hemoglobin and the products become the bile pigments. The liver makes bile, which is stored in the gall bladder, before being sent to the duodenum by way of ducts.

Ions

Ions (salts) are excreted, not because they are end products of metabolism, but because their proper concentration in the blood is so important to the pH, osmotic pressure, and electrolyte balance of the blood. The balance of potassium (K^+) and sodium (Na^+) is important to nerve conduction. The level of calcium (Ca^{++}) in the blood affects muscle contraction; iron (Fe^{++}) takes part in hemoglobin metabolism; magnesium (Mg^{++}) helps many enzymes function properly.

Water

Water is an end product of metabolism; it is also taken into the body when food and liquids are consumed. The amount of fluid in the blood helps determine blood pressure. Treatment of hypertension sometimes includes the administration of a diuretic drug that increases the excretion of sodium and water by the kidneys.

Urea, salts, and water are the primary constituents of human urine. Carbon dioxide is excreted as a gas in the lungs and as the bicarbonate ion in the kidneys.

Organs of Excretion

The kidneys are the primary excretory organs, but there are other organs that also function in excretion (fig. 14.2), such as those described in the discussion that follows.

Skin

The sweat glands in the skin (fig. 8.13) excrete perspiration, a solution of water, salt, and some urea. The sweat glands are made up of a coiled tubule portion in the dermis and a narrow, straight duct that exits from the epidermis. Although perspiration is an excretion, we perspire not so much to rid the body of waste as to cool the body. The body cools because heat is lost as perspiration evaporates. Sweating keeps the body temperature within normal range during muscular exercise or when the outside temperature rises. In times of renal failure, more urea than usual may be excreted by the sweat glands, to the extent that a so-called urea frost is observed on the skin.

Liver

The liver excretes bile pigments that are incorporated into bile, a substance stored in the gallbladder before it passes into the small intestine by way of ducts (fig. 14.5). The yellow pigment found in urine, called urochrome, is also derived from the breakdown of heme, but this pigment is deposited in the blood and is subsequently excreted by the kidneys.

Lungs

The process of expiration (breathing out) not only removes carbon dioxide from the body; it also results in the loss of water. The air we exhale contains moisture, as demonstrated by blowing onto a cool mirror.

Table 14.2 *Composition of Urine*

Water	95%
Solids	5%
Organic wastes (per 1500 ml of urine)	
urea	30 g
creatinine	1–2 g
ammonia	1–2 g
uric acid	1g
Ions (Salts)	25 g
Positive	*Negative*
sodium	chlorides
potassium	sulfates
magnesium	phosphates
calcium	

Intestine

Certain salts, such as those of iron and calcium, are excreted directly into the cavity of the intestine by the epithelial cells lining it. These salts leave the body in the feces.

At this point, it might be helpful to remember that the term *defecation,* and not *excretion,* is used to refer to the elimination of feces from the body. Substances that are excreted are waste products of metabolism. Undigested food and bacteria, which make up feces, have never been a part of the functioning of the body, but salts that are passed into the gut are excretory substances because they were once metabolites in the body.

Kidneys

The kidneys excrete urine, which contains a combination of the end products of metabolism (table 14.2). The kidneys are a part of the urinary system.

There are various organs that excrete metabolic wastes but only the kidneys consistently rid the body of urea.

Urinary System

The urinary system includes the structures illustrated in figure 14.6 and listed in table 14.3. The organs are listed in order according to the path of urine.

The **kidneys** are reddish-brown organs about 4 inches long, 2 inches wide, and 1 inch thick. They lie on either side of the midline against the dorsal body wall where they are anchored by connective tissue. The renal artery, renal vein, nerves, and ureters join the kidney on the concave side toward the midline.

The **ureters** are muscular tubes that convey the urine toward the bladder by peristaltic contractions. Urine enters the bladder in jets that occur at the rate of one to five per minute.

The **urinary bladder,** which can hold up to 600 milliliters of urine, is a hollow muscular organ that gradually expands as urine enters. In the male, the bladder lies ventral to the rectum, seminal vesicles, and ductus deferens. In the female, it is ventral to the uterus and upper vagina.

The **urethra,** extending from the urinary bladder to an external opening, differs in length in females and males. In females, the urethra lies ventral to the vagina and is only about 1 inch long. The short length of the female urethra facilitates bacterial invasion and explains why females are more prone to urethral infections. In males, the urethra averages 6 inches when the penis is flaccid. As the urethra leaves the bladder, it is encircled by the prostate gland (fig. 14.7). In older men, enlargement of the prostate gland may prevent urination, a condition that can usually be corrected by surgery.

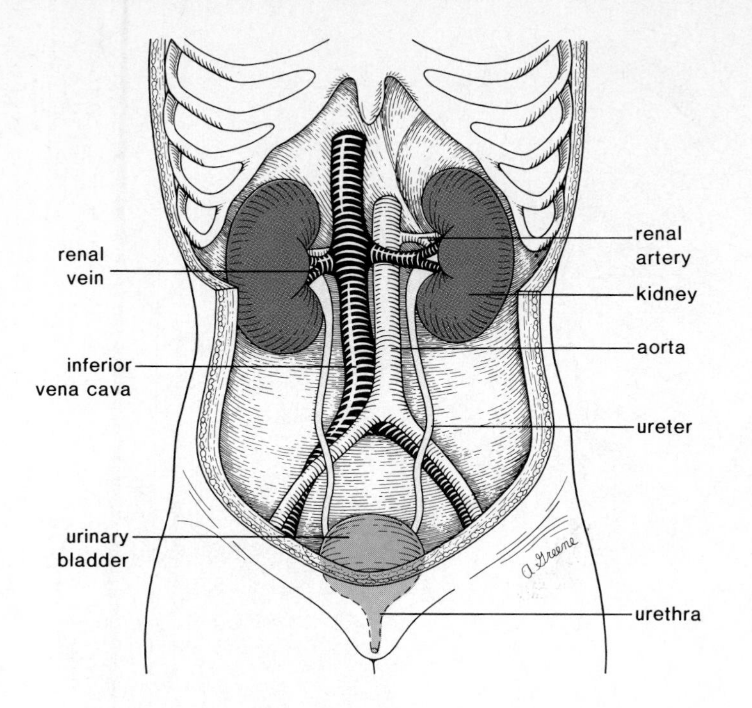

Figure 14.6
The urinary system. Urine is found only within the kidneys, ureters, bladder, and urethra.

Table 14.3 Urinary System	
Organ	**Function**
Kidneys	Produce urine
Ureters	Transport urine
Bladder	Storage of urine
Urethra	Elimination of urine

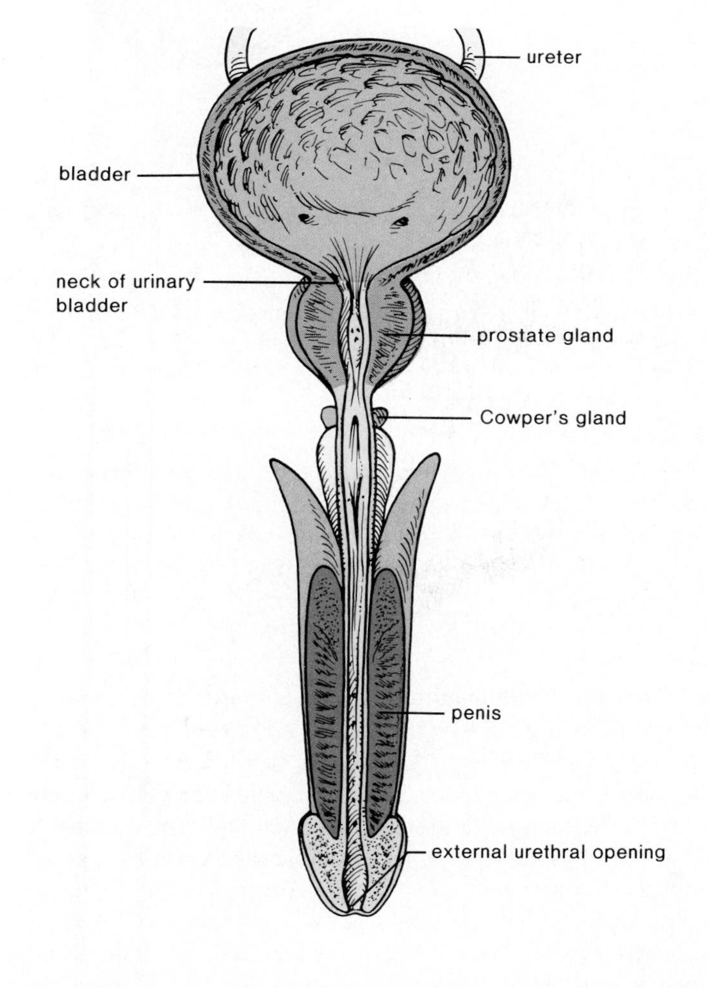

Figure 14.7
Longitudinal section of male urethra leaving the bladder. Note the position of the prostate gland, which can enlarge to obstruct the flow of urine.

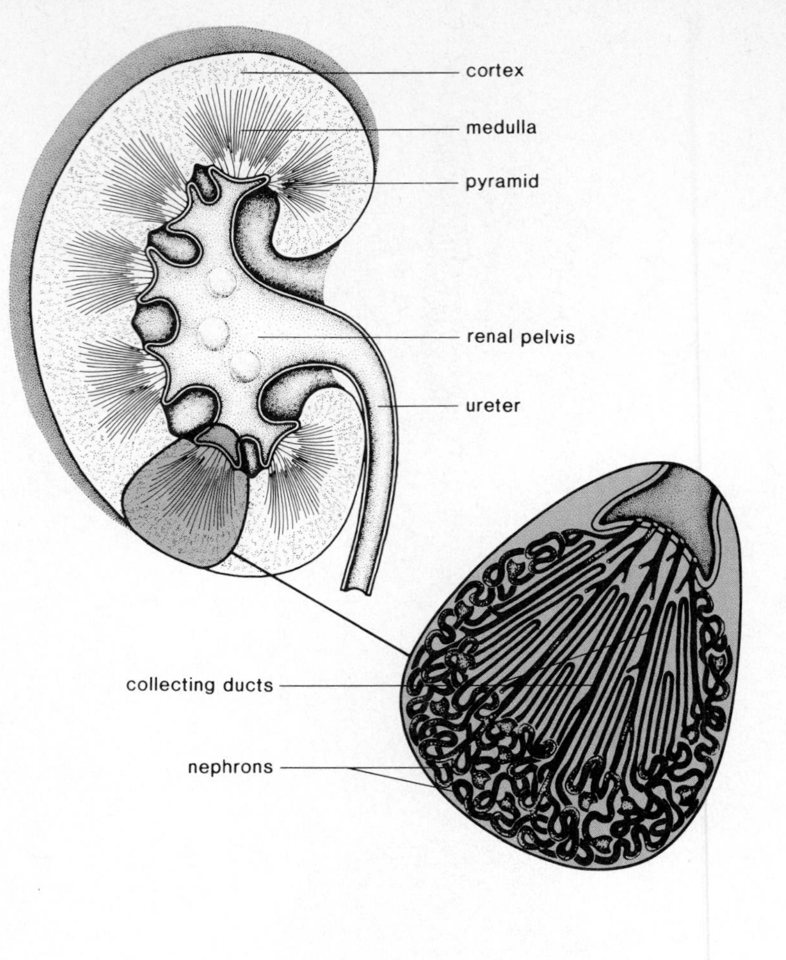

Figure 14.8
Macroscopic anatomy of the kidney with an enlargement of a pyramid. The structure of a pyramid is dependent on the structure of kidney tubules, called nephrons.

Notice that there is no connection between the genital (reproductive) and urinary systems in females, but there is a connection in males. When urinating, the urethra in the male carries urine, and during sexual orgasm the urethra transports semen. This double function does not alter the path of urine, and it is important to realize that urine is found only in those structures listed in table 14.3.

Urination

When the bladder fills with urine, stretch receptors send nerve impulses to the spinal cord; nerve impulses leaving the cord then cause the bladder to contract and the sphincters to relax so that urination may take place. In older children and adults, it is possible for the brain to control this reflex, delaying urination until a suitable time.

Kidneys

When a kidney is sliced longitudinally, it is shown to be composed of three macroscopic parts (fig. 14.8): (1) an outer granulated layer called the **cortex,** which dips down in between (2) a radially striated, or lined, layer called the **medulla** and (3) an inner space, or cavity, called the **pelvis,** where the urine collects before entering the ureters. Occasionally **kidney stones** form in the pelvis. They are formed from fairly insoluble substances such as uric acid and calcium salts that precipitate out of the urine instead of remaining in the solution. Previously, either the stones passed naturally or they were surgically removed. Now new methods of treatment are being developed. Sometimes the stones can be crushed by soundwave generation within a special device called a lithotripter. Laser light has also been tried and found to be successful in shattering the stones.

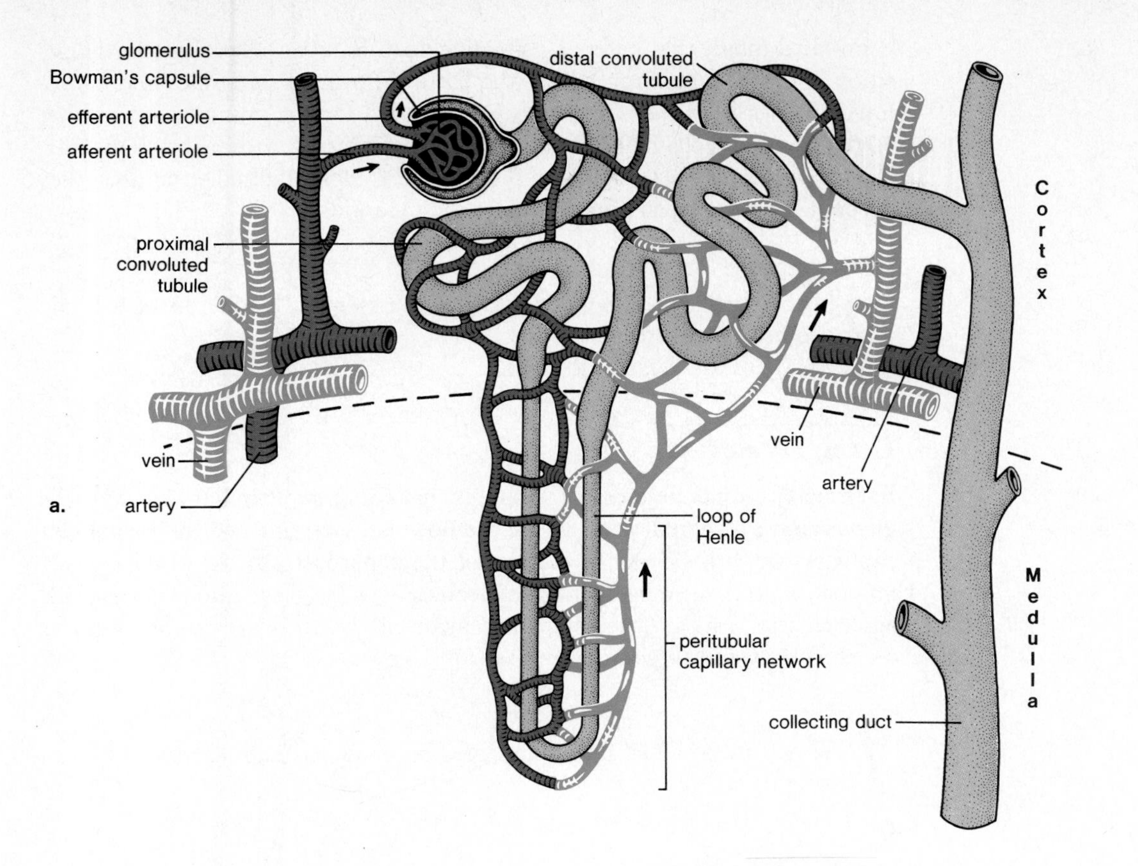

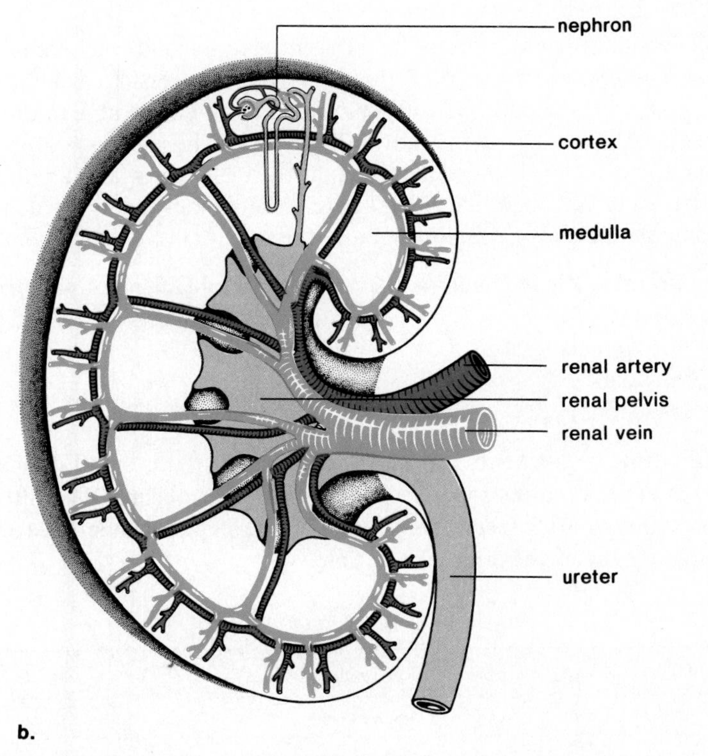

Figure 14.9
a. *Diagram of nephron (kidney tubule) gross anatomy. You may trace the path of blood about the nephron by following the arrows. Note that the dotted line indicates which portions of the nephron are in the cortex and which portions are in the medulla of the kidney.* b. *Each kidney receives a renal artery that divides into arterioles within the kidney. Venules leaving the kidney join to form the renal vein. This drawing also shows how one nephron is placed in the kidney. Parts of each nephron are in the cortex and other parts are in the medulla.*

Microscopically, the kidney is composed of over 1 million **nephrons,** sometimes called renal tubules. Each nephron is made up of several parts (fig. 14.9). The blind end of the tubule is pushed in on itself to form a cup-like structure called **Bowman's capsule,** within which there is a capillary tuft called the *glomerulus.* Next, there is a **proximal** (meaning near the Bowman's capsule) **convoluted tubule** which makes a U-turn to form the portion of the tubule called the loop of Henle. This leads to the **distal** (far from Bowman's capsule)

convoluted tubule that enters a **collecting duct.** Figure 14.9a indicates the position of a single nephron within the kidney. Bowman's capsules and convoluted tubules lie within the cortex and account for the granular appearance of the cortex. Loops of Henle and collecting ducts lie within the triangular-shaped *pyramids* of the medulla. Since these are longitudinal structures, they account for the striped appearance of the pyramids.

Only the urinary system consisting of the kidneys, bladder, ureters, and urethra ever hold urine.

Urine Formation

Each nephron has its own blood supply, including two capillary regions; the **glomerulus** is a capillary tuft inside the Bowman's capsule and the **peritubular capillary network** surrounds the rest of the nephron (table 14.4). Urine formation requires the movement of molecules between these capillaries and the nephron (fig. 14.9). Three steps are involved: *pressure filtration, selective reabsorption,* and *tubular excretion* (fig. 14.15).

The pattern of blood flow about the nephron is critical to urine formation.

Pressure Filtration

Whole blood, of course, enters the afferent arteriole and the glomerulus (fig. 14.10). Under the influence of glomerular blood pressure, which is usually about 60 mm Hg, small molecules move from the glomerulus to the inside of Bowman's capsule across the thin walls of each. The process is a **pressure filtration** because large molecules and formed elements are unable to pass through. In effect, then, blood that enters the glomerulus is divided into two portions: the filterable components and the nonfilterable components.

Filterable Blood Components	Nonfilterable Blood Components
Water	Formed elements (blood cells and platelets)
Nitrogenous wastes	Proteins
Nutrients	
Salts (ions)	

The filterable components form a **filtrate** that contains small dissolved molecules in approximately the same concentration as plasma. The filtrate stays inside of Bowman's capsule and the nonfilterable components leave the glomerulus by way of the efferent arteriole.

***Table 14.4** Circulation about a Nephron**

Name of Structure	Comment
Afferent arteriole	Brings arteriolar blood toward Bowman's capsule
Glomerulus	Capillary tuft enveloped by Bowman's capsule
Efferent arteriole	Takes arteriolar blood away from Bowman's capsule
Peritubular capillary network	Capillary bed that envelops the rest of the tubule
Venule	Takes venous blood away from the tubule

*Compare to figure 14.9.

In order for filtration to occur, there must be adequate blood pressure within the glomerulus. Blood pressure is constantly monitored by a special region of the afferent arteriole called the *juxtaglomerular apparatus,* or polar cushion (fig. 14.11). If necessary, *renin,* a substance that brings about increased blood pressure is released in order to increase the pressure. Renin seems to be continually released by persons with kidney disease and this accounts for the hypertension usually present in these patients.

A consideration of the preceding filterable substances leads one to conclude that if the composition of urine was the same as that of glomerular filtrate, the body would continually lose nutrients, water, and salts. Obviously, death from dehydration, starvation, and low blood pressure would quickly follow. Therefore, we can assume that the composition of the filtrate must be altered as this fluid passes through the remainder of the tubule.

---

*D*uring pressure filtration, water, nutrient molecules, and waste molecules move from the glomerulus to the inside of Bowman's capsule. The filtered substances are called the glomerular filtrate.

Figure 14.10
Scanning electron micrograph of a glomerulus (the outer layer of Bowman's capsule has been removed). Renal failure is due to glomeruli that no longer function properly.

Selective Reabsorption

Both passive and active reabsorption of molecules from the tubule to the blood occur as the filtrate moves along the proximal convoluted tubule. Active transport provides **selective reabsorption** for reasons discussed following.

Passive reabsorption involves particularly the movement of water molecules from the area of greater concentration in the filtrate to the area of lesser concentration in the blood. Two factors aid the process. The nonfilterable proteins remain in the blood, where they exert an osmotic pressure that pulls water back into the bloodstream. Following the active reabsorption of sodium (Na^+), which is discussed later, chlorine (Cl^-) follows passively because, being a negative ion, it is attracted to the positive charge of sodium.

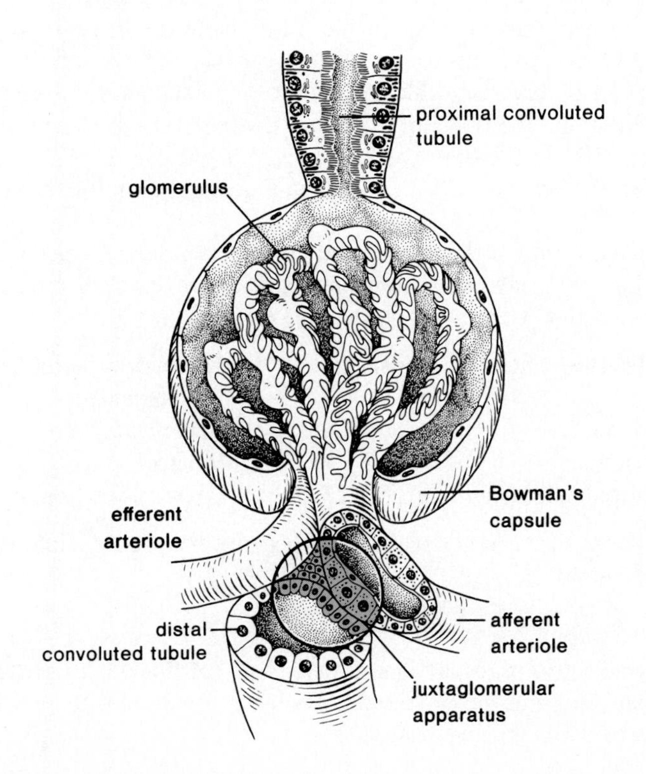

Figure 14.11
Drawing of glomerulus and adjacent distal convoluted tubule. The juxtaglomerular apparatus (circled) *is sensitive to the fluid pressure within the distal convoluted tubule and releases renin if this pressure falls below normal.*

Figure 14.12
Nutrient molecules and sodium are actively reabsorbed from a kidney tubule in the manner illustrated. These molecules move passively into the tubule cell but are then actively transported out of the tubule cell into the blood. Active transport requires the participation of carrier molecules.

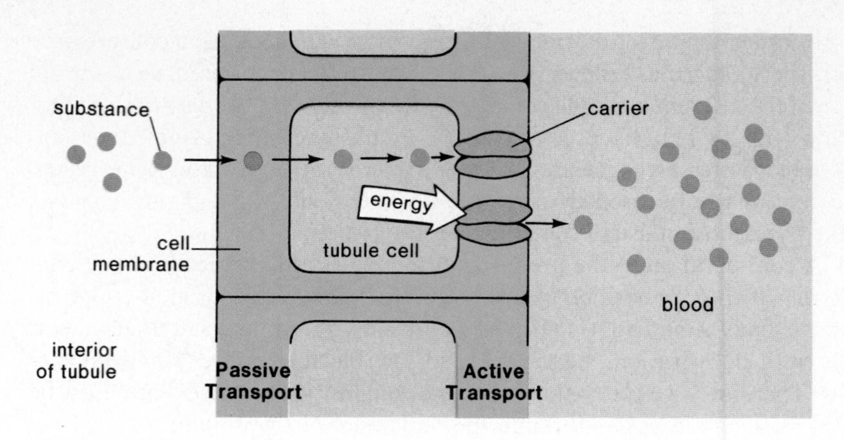

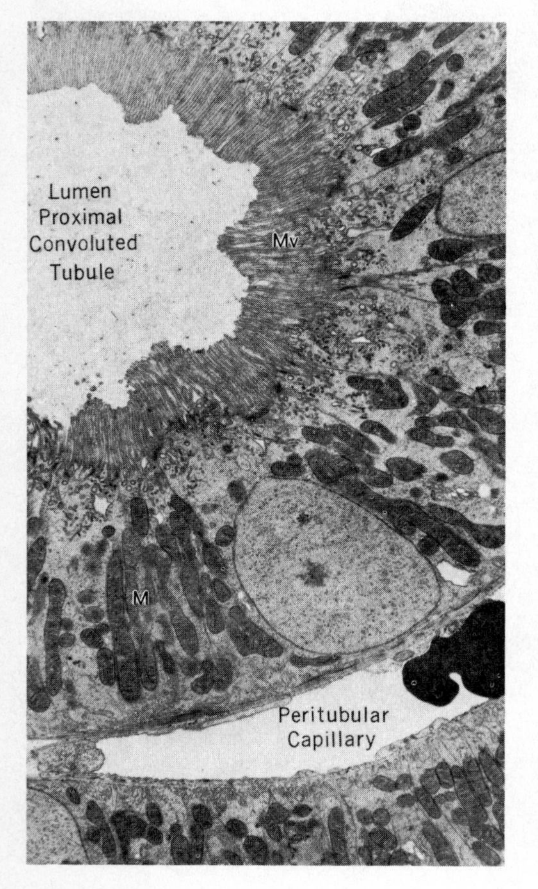

Figure 14.13
Electron micrograph of cells that line the lumen (inside) of proximal convoluted tubule where selective reabsorption takes place. The cells have a brush border composed of microvilli that greatly increase the surface area exposed to the lumen. Each cell has many mitochondria that supply the energy needed for active transport. A red blood cell (lower right) *is seen in the peritubular capillary. (Mv = microvilli; M = mitochondria)*

Active reabsorption involves active transport and this accounts for the large energy needs of the kidney. In figure 14.12 two membranes are shown; the first of these exists between the tubule cavity and the tubule cell, and the second lies between the tubule cell and the blood. A molecule, glucose, for example, diffuses passively into the tubule cell but is then actively transported from the cell into the blood, requiring the use of a carrier molecule. *Reabsorption* by active transport is *selective* since only molecules recognized by carrier molecules move across the membrane.

The structure of the cells that line the proximal convoluted tubule is anatomically adapted for absorption (fig. 14.13). These cells have numerous microvilli, about one micron in length, that increase the surface area for reabsorption. In addition, the cells contain numerous mitochondria that produce the energy necessary for active transport. Reabsorption occurs until the threshold level of a substance is obtained. Thereafter the substance will appear in the urine. For example, the threshold level for glucose is 0.15 grams glucose per 100 milliliters of blood. After this amount is reabsorbed, any excess present in the filtrate will appear in the urine. In diabetes mellitus (sugar diabetes, p. 382) the filtrate contains excess glucose because the liver fails to store glucose as glycogen. The real difficulty is that the pancreas is failing to produce insulin, a hormone that promotes glucose uptake in all cells.

In contrast to the high threshold level of glucose, urea has a very low threshold level that is quickly reached, so that nearly all urea remains in the urine.

We have seen that the filtrate that enters the proximal convoluted tubule is divided into two portions—components that are reabsorbed and components that are not reabsorbed.

Reabsorbed Filtrate Components	**Nonreabsorbed Filtrate Components**
Most water	Some water
Nutrients	Wastes
Required salts (ions; e.g., Na^+, Cl^-)	Excess salts (ions)

The substances that are not reabsorbed become the tubular fluid that enters the loop of Henle.

During selective reabsorption, nutrient and salt molecules are actively reabsorbed from the proximal convoluted tubule into the peritubular capillary, and water follows passively.

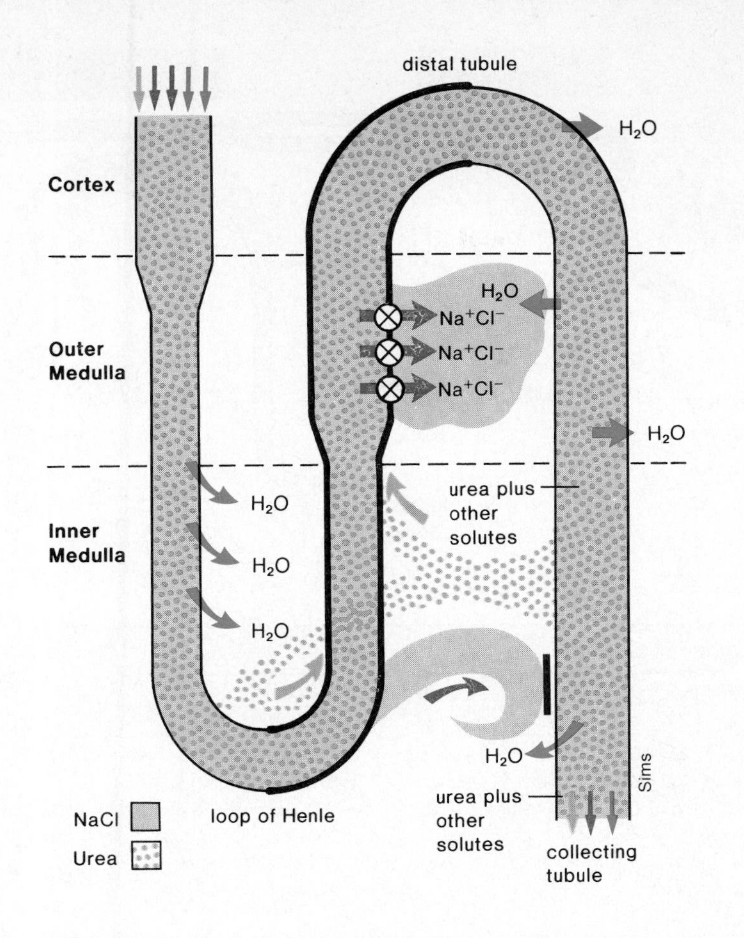

Figure 14.14
The presence of a loop of Henle allows the nephron to concentrate the urine. Salt (Na^+Cl^-) diffuses and is extruded by the ascending limb into the medulla; also, the collecting duct is believed to extrude urea into the tissues of the medulla. This produces a hypertonic environment that draws water out of the descending limb and the collecting duct. This water is returned to the circulatory system.

Loop of Henle

The presence of a loop of Henle (fig. 14.14) allows humans to excrete a *hypertonic urine* (i.e., a larger amount of metabolic wastes per volume than that of blood plasma). The loop of Henle is made up of a *descending* (going down) and an *ascending* (going up) *limb.*

Salt (Na^+Cl^-) passively diffuses out of the lower portion of the ascending limb, but the upper portion is believed to extrude salt actively into the tissue of the medulla. Therefore, the medulla is hypertonic to the fluid within the descending limb, and water passively diffuses out of the descending limb to be carried away by the blood. Some believe that the salt leaving the ascending limb subsequently diffuses into the descending limb, and they refer to this occurrence as a *countercurrent exchange,* because as the fluid moves in the opposite direction within the two limbs, there is an exchange of salt between them.

During the past few years, an increasing number of researchers have begun to believe that presence of urea also contributes to the high osmolarity of the medullary tissue. They point out that the deeper regions of the medulla are very hypertonic even though no salt is being actively extruded here by the ascending limb. Apparently there may be a recycling of urea between the collecting duct and the loop of Henle as depicted in figure 14.14.

It is the hypertonicity of medulla tissues that causes water to leave the descending limb. This water cannot turn around and diffuse into the ascending limb because the ascending limb is impermeable to water.

Unique features of the loop of Henle allow humans to excrete a hypertonic urine.

Figure 14.15
Diagram of nephron showing steps in urine formation: filtration, reabsorption, and tubular excretion. Note also that water enters the tissues at the loop of Henle and collecting duct.

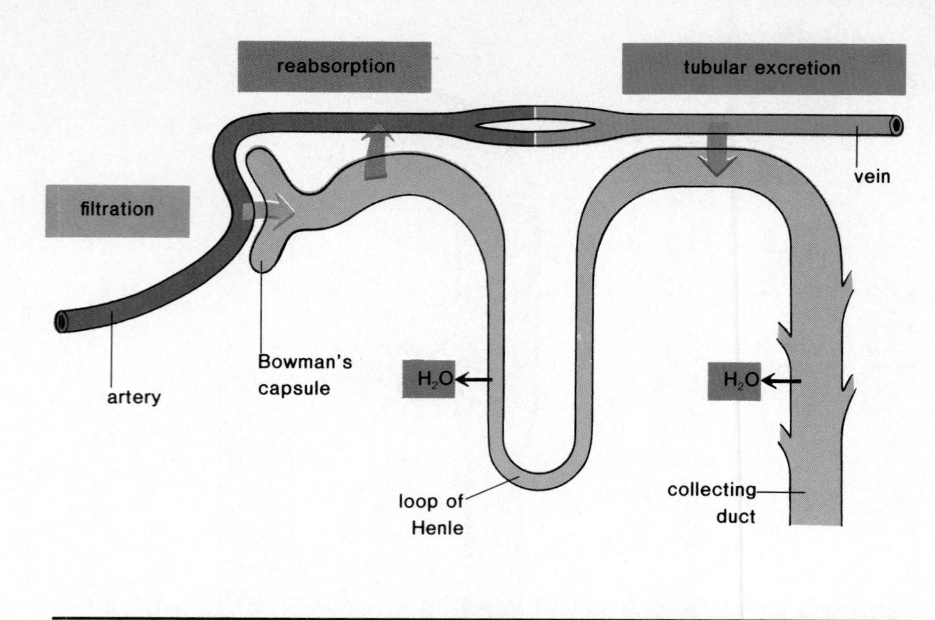

Table 14.5 Urine Formation		
Name of Part	**Location in Kidney**	**Function**
Bowman's capsule	Cortex	Forms filtrate
Proximal convoluted tubule	Cortex	Selective reabsorption
Loop of Henle	Medulla	Extrusion of sodium and reabsorption of water
Distal convoluted tubule	Cortex	Tubular excretion
Collecting duct	Cortex and medulla	Reabsorption of water

Tubular Excretion

The distal convoluted tubule continues the work of the proximal convoluted tubule in that sodium and water are both reabsorbed. As before, sodium is actively reabsorbed into the blood capillary, and thereafter water follows passively. In this region of the tubule also, substances may be added to the urine by a process called tubular excretion, or augmentation. The cells that line this portion of the tubule have numerous mitochondria. **Tubular excretion** (fig. 14.15) is an active process just like selective reabsorption, but the molecules are moving in the opposite direction. Histamine and penicillin are actively excreted, as are hydrogen ions and ammonia (p. 304).

During tubular excretion, certain substances, penicillin and histamine, for example, are actively secreted from the peritubular capillary into the fluid of the tubule.

Collecting Duct

The fluid that enters the collecting duct is isotonic to the cells of the cortex. This means that to this point the net effect of the reabsorption of water and sodium has been to produce a fluid in which the proportion of water to sodium is the same as in most tissues. However, the medulla is hypertonic to the contents of the collecting duct, due to the extrusion of salt by the ascending limb of the loop of Henle (and possibly also due to the presence of urea). Therefore, water diffuses out of the collecting duct into the medulla, and the urine within the collecting duct becomes hypertonic to blood plasma. *Urine* (table 14.2) now passes from the collecting duct to the pelvis of the kidney. Table 14.5 summarizes urine formation.

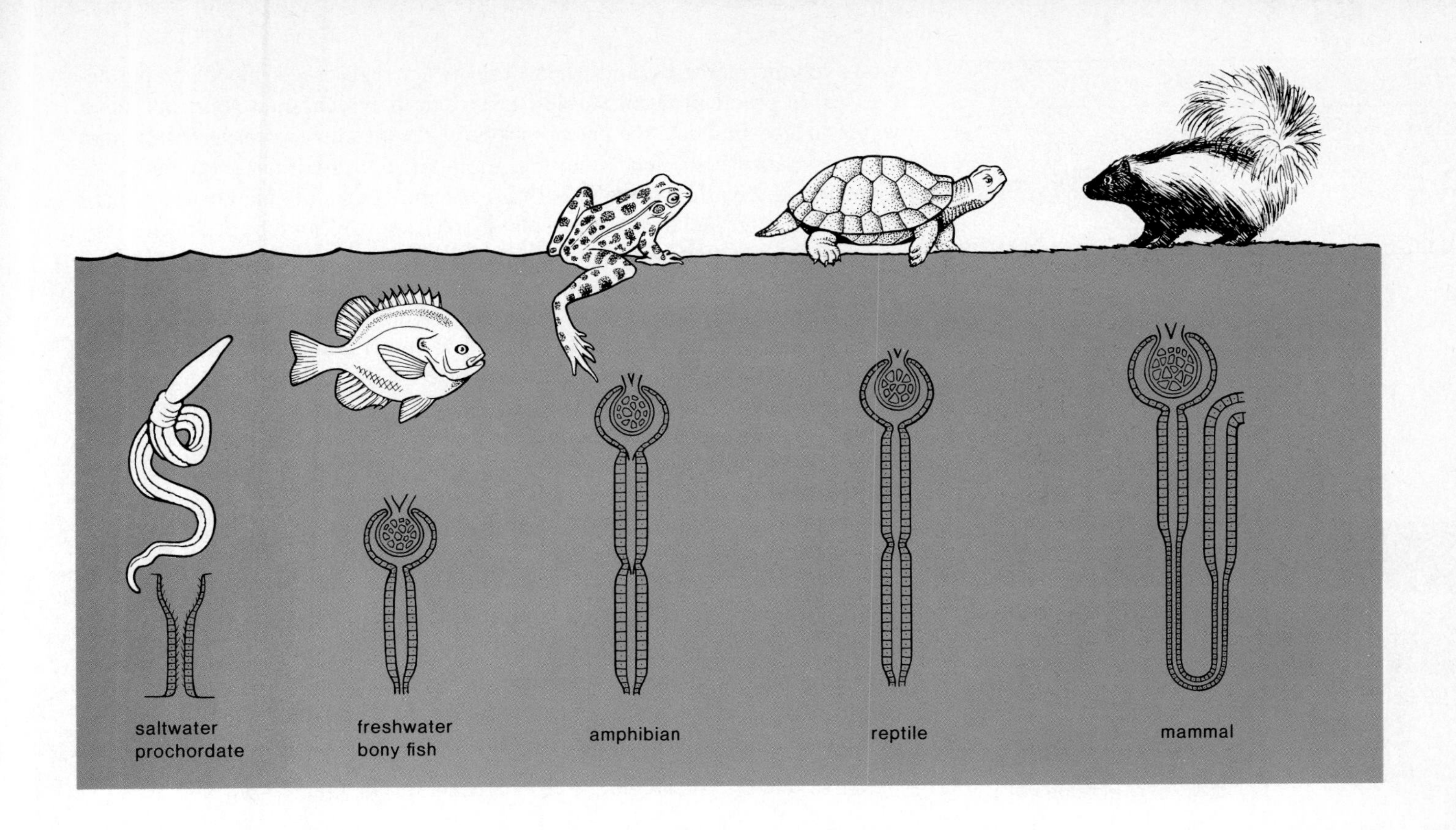

Figure 14.16
Presence or absence of loop of Henle in groups of animals. The internal fluids of marine protochordates (p. 611) are isotonic to sea water and there is no loop of Henle. Freshwater bony fishes and amphibians do not need to conserve water and there is no loop of Henle. Reptiles are terrestrial and conserve water by excreting a solid waste, uric acid, rather than by reabsorbing water. Only mammals are solely dependent on excreting a hypertonic urine to conserve water. They have a well-developed loop of Henle.

Water diffuses from the collecting duct and the urine becomes increasingly hypertonic as it nears the pelvis of the kidney.

In conclusion, then, human beings, along with other mammals, excrete a concentrated, or hypertonic, urine, not because water fails to enter the nephron but because the water that enters is reabsorbed. The consistent presence of a loop of Henle in all nephrons makes it possible for efficient reabsorption to occur. In lower vertebrates (fig. 14.16) up to and including the reptiles, the nephrons lack a loop of Henle and the kidney is incapable of producing a hypertonic urine. In birds, some nephrons do have a loop of Henle, but the arrangement is not as efficient as in mammals. Reptiles and birds, unlike mammals, conserve water by excreting uric acid, a solid nitrogenous waste. This is possible because the urinary system empties into the cloaca, a common repository for both the urinary and digestive systems.

Regulatory Functions of the Kidneys

Blood Volume

Reabsorption of water is under the control of a hormone called **ADH (antidiuretic hormone),** which is released by the posterior lobe of the pituitary gland (p. 374). ADH increases the permeability of the distal convoluted tubule and collecting duct so that more water can be reabsorbed. In order to understand the function of this hormone, consider its name. Diuresis means increased excretion of urine, and antidiuresis means suppression of urinary excretion. When ADH is present, more water is reabsorbed and a decreased amount of urine results. This hormone is secreted according to whether blood volume needs to be increased or decreased. When water is reabsorbed at the collecting duct,

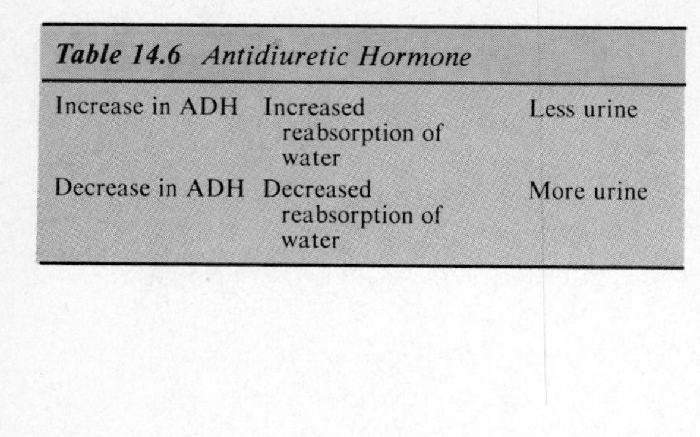

Table 14.6 Antidiuretic Hormone

Increase in ADH	Increased reabsorption of water	Less urine
Decrease in ADH	Decreased reabsorption of water	More urine

blood volume increases, and when water is not reabsorbed, blood volume decreases. In practical terms (table 14.6), if an individual does not drink much water on a certain day, the posterior lobe of the pituitary releases ADH; more water is reabsorbed; blood volume is maintained at a normal level; and, consequently, there is less urine. On the other hand, if an individual drinks a large amount of water and does not perspire much, the posterior lobe of the pituitary does not release ADH; more water is excreted; blood volume is maintained at a normal level; and a greater amount of urine is formed.

Drinking alcohol causes diuresis because it inhibits the secretion of ADH. The dehydration that follows is believed to contribute to the symptoms of a "hangover." Drugs, called diuretics, are often prescribed for high blood pressure. The drugs cause increased urinary excretion and thus reduce blood volume and blood pressure. Concomitantly, any edema present is also reduced.

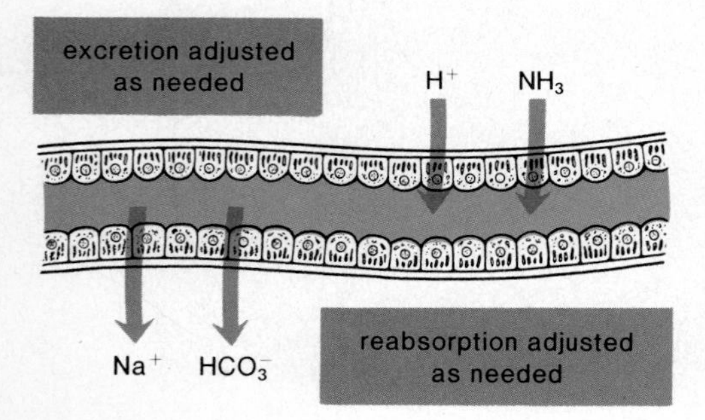

Figure 14.17
The kidneys maintain homeostasis, in part, by adjusting the ionic composition of the blood. In order to maintain the pH fairly constant, the excretion of H^+ and reabsorption of Na^+ and HCO_3^- is adjusted as needed. If the blood is acid, hydrogen ions are excreted in conjunction with ammonia. If the blood is alkaline, reabsorption of sodium bicarbonate is minimized.

Adjustment of pH

The kidneys aid in maintaining a constant pH of the blood, and the whole nephron takes part in this process. Figure 14.17 indicates that the excretion of hydrogen ions and ammonia ($H^+ + NH_3$), together with the reabsorption of bicarbonate ions and sodium, is adjusted in order to keep the pH within normal limits. If the blood is acid, hydrogen ions are excreted in combination with ammonia, and sodium bicarbonate is reabsorbed. This will restore alkalinity because the sodium bicarbonate can react with water to give hydroxide ions, which are balanced by the sodium ions:

$$Na^+HCO_3^- + HOH \longrightarrow H_2CO_3 + Na^+OH^-$$

If the blood is alkaline, fewer hydrogen ions are excreted and fewer sodium and bicarbonate ions are reabsorbed. The reabsorption and/or excretion of ions (salts) by the kidneys illustrates their homeostatic ability to maintain not only the pH of the blood but also the osmolarity of the blood. Osmolarity increases as salts are reabsorbed. Reabsorption of ions, such as K^+ and Mg^{++}, also maintains the proper electrolyte balance of the blood, as discussed on page 293.

The kidneys contribute to homeostasis by excreting urea. They also regulate the volume and pH of the blood, two very important functions.

Problems with Kidney Function

Because of the great importance of the kidney to the maintenance of body fluid homeostasis, renal failure is a life-theatening event. There are many types of illnesses that bring on progressive renal disease and renal failure.

Infections of the urinary tract themselves are a fairly common occurrence, particularly in the female since the urethra is considerably shorter than that of the male. If the infection is localized in the urethra, it is called *urethritis*. If it invades the bladder, it is called *cystitis*. And finally, if the kidneys are affected, it is called *nephritis*. Glomerular damage sometimes leads to blockage of the glomeruli so that no fluid moves into the tubules. Or it can cause the glomeruli to become more permeable than usual. This is detected when a **urinalysis** is done. If the glomeruli are too permeable, albumin, white cells, or even red cells may appear in the urine. Trace amounts of protein in the urine is not a matter of concern, however.

When glomerular damage is so extensive that more than two-thirds of the nephrons are incapacitated, waste substances accumulate in the bood. This condition is called *uremia* because urea begins to accumulate in the blood.

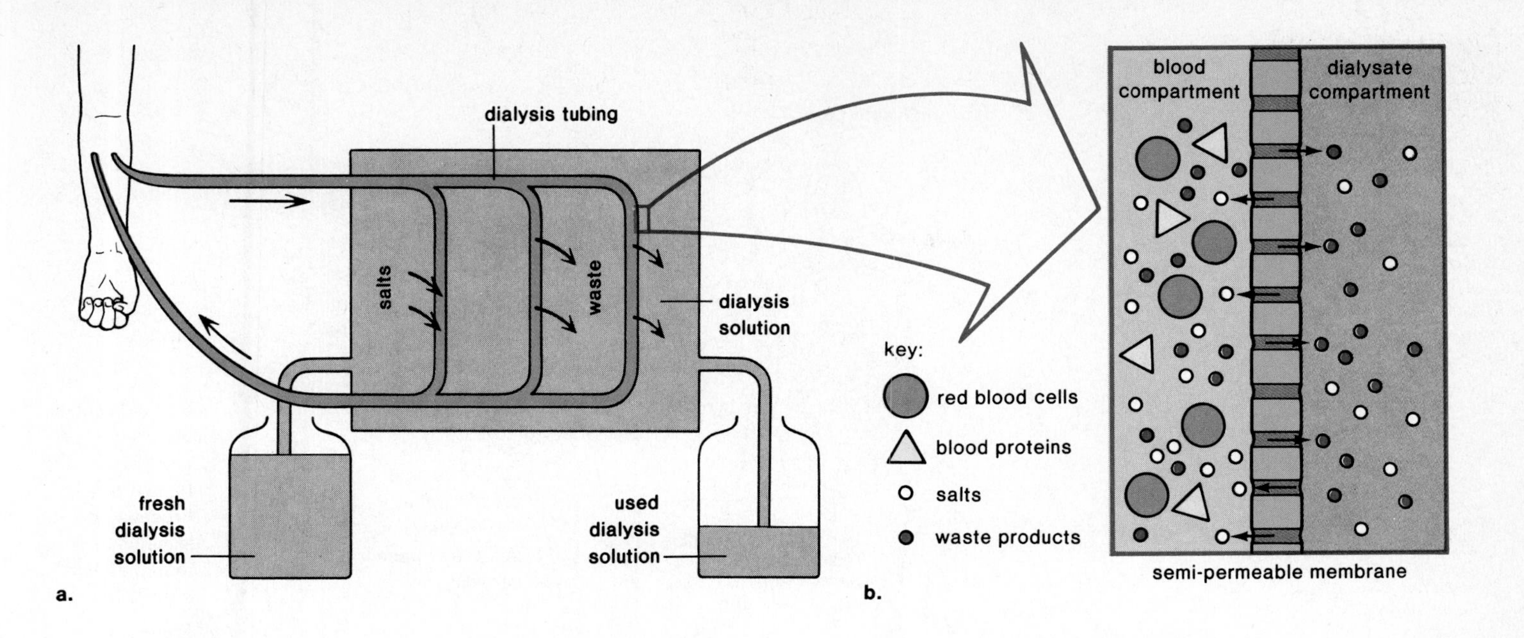

Figure 14.18
Diagram of an artificial kidney. a. *As the patient's blood circulates through dialysis tubing, it is exposed to a solution.* b. *Wastes exit from the blood into the solution because of preestablished concentration gradient. In this way the blood is not only cleansed, the pH can also be adjusted.*

Although the presence of nitrogenous wastes can cause serious damage, the retention of water and salts is of even more concern. The latter causes edema, fluid accumulation in the body tissues. Imbalance in the ionic composition of body fluids can even lead to loss of consciousness and heart failure.

Kidney Replacements

Kidney Transplant

Patients with renal failure can sometimes undergo kidney transplant operations during which they receive a functioning kidney from a donor. As with all organ transplants, there is the possibility of organ rejection, so that receiving a kidney from a close relative has the highest chance of success. The current one-year survival rate is 97 percent if the kidney is received from a relative and 90 percent if it is received from a nonrelative. Recently, investigators have discovered that the drug Cyclosporine is most helpful in preventing organ rejection while at the same time allowing the patient to fight infections. Others are hopeful that monoclonal antibodies that react against cytotoxic T cells will be available soon.

Kidney transplants and dialysis are available procedures for persons who have suffered renal failure.

Dialysis

If a satisfactory donor cannot be found for a kidney transplant, which is frequently the case, the patient may undergo dialysis treatments, utilizing either a kidney machine or continuous ambulatory peritoneal (abdominal) dialysis, CAPD. Dialysis is defined as the diffusion of dissolved molecules through a semipermeable membrane. These molecules will, of course, move across a membrane from the area of greater concentration to one of lesser concentration.

In the case of the kidney machine (fig. 14.18), the patient's blood is passed through a semipermeable membranous tube that is in contact with a balanced salt (dialysis) solution. Substances more concentrated in the blood diffuse into the dialysis solution, also called the dialysate. Conversely, substances more concentrated in the dialysate diffuse into the blood. Accordingly, the artificial

Figure 14.19
CAPD (continuous ambulatory peritoneal dialysis). a. *Dialysate fluid is introduced into the abdominal cavity by way of a plastic bag.* b. *After bag is securely placed at waist, patient can move freely about.* c. *After four to eight hours, the old fluid is removed before the procedure is repeated.*

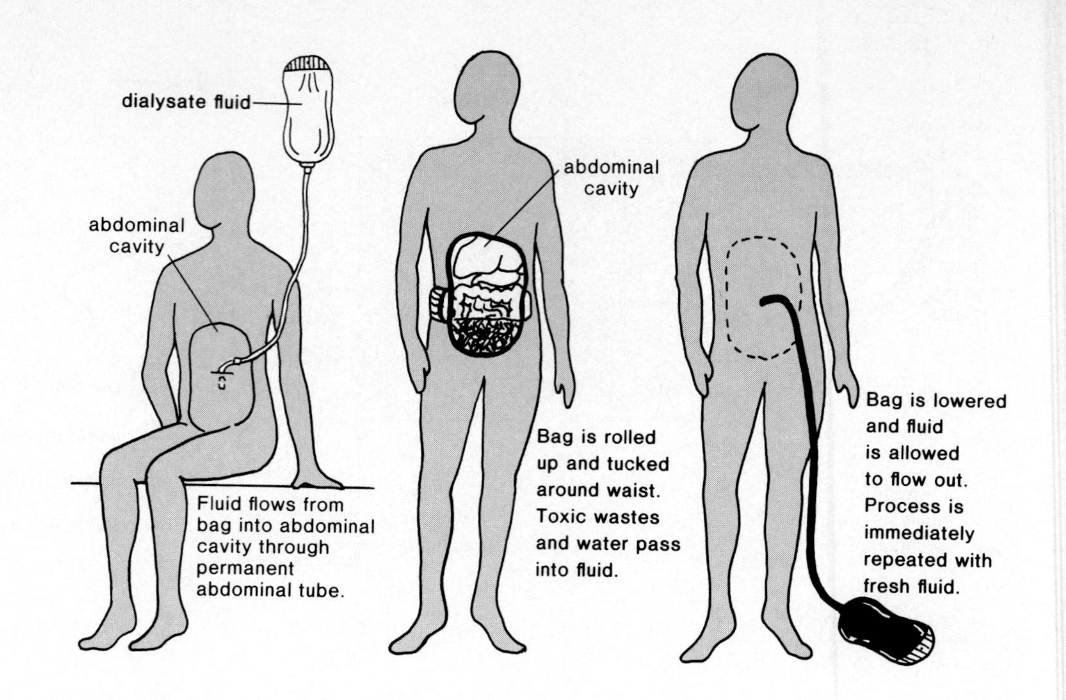

kidney can be utilized either to extract substances from the blood, including waste products or toxic chemicals and drugs, or to add substances to the blood, for example, bicarbonate ions if the blood is acid. In the course of a six-hour dialysis, from 50 to 250 grams of urea can be removed from a patient, which greatly exceeds the urea clearance of normal kidneys. Therefore, a patient need undergo treatment only about twice a week.

In the case of CAPD, a fresh amount of dialysate is introduced directly into the abdominal cavity from a bag attached to a permanently implanted plastic tube. Waste and water molecules pass into the fluid from the surrounding organs before the fluid is collected four or eight hours later (fig. 14.19).

Summary

The end products of metabolism are, for the most part, nitrogenous wastes, such as ammonia, urea, uric acid, and creatinine, all of which are excreted primarily by the urinary system. The urinary system contains the kidneys, whose macroscopic anatomy is dependent on the presence of nephrons. Urine formation requires three steps: during pressure filtration, small components of plasma pass into Bowman's capsule from the glomerulus, due to blood pressure; during selective reabsorption, nutrients and sodium are actively reabsorbed from the proximal convoluted tubule back into the blood; during tubular excretion, a few types of substances are actively secreted into the distal convoluted tubule from the blood.

Water is reabsorbed along the length of the tubule but especially from the loop of Henle and collecting duct. ADH, a hormone produced by the posterior pituitary, controls the reabsorption of water. The whole tubule participates in maintaining the pH of the blood by regulating the pH of urine. In practice, hydrogen ions are excreted, and sodium bicarbonate is reabsorbed to maintain the pH.

Various types of problems can lead to kidney failure, necessitating that the person either receives a kidney from a donor or undergoes dialysis treatments by means of the kidney machine or CAPD.

Objective Questions

1. The primary nitrogenous end product of humans is ____________ .
2. The intestines are an organ of excretion because they rid the body of ________ .
3. Urine leaves the bladder in the ____________ .
4. The capillary tuft inside Bowman's capsule is called the ____________ .
5. ____________ is a substance that is found in the filtrate, is reabsorbed, and still is in urine.
6. ____________ is a substance that is found in filtrate, is not reabsorbed, and is concentrated in urine.
7. Tubular excretion takes place at the ____________, a portion of the nephron.
8. Reabsorption of water from the collecting duct is regulated by the hormone ____________ .
9. In addition to excreting nitrogenous wastes, the kidneys adjust the ________ and ____________ of blood.
10. Persons who have nonfunctioning kidneys are often on ____________ machines.

Answers to Objective Questions

1. urea 2. certain salts, e.g., calcium and iron 3. urethra 4. glomerulus 5. water 6. urea 7. distal convoluted tubule 8. ADH 9. volume and pH 10. dialysis

Study Questions

1. Name four nitrogenous end products and explain how each is formed in the body. (pp. 291–293)
2. Name several excretory organs and the substances they excrete. (pp. 291, 293–294)
3. What is the composition of urine? (p. 294)
4. Give the path of urine. (pp. 294–295)
5. Name the parts of the kidney tubule, or nephron. (p. 297)
6. Trace the path of blood about the tubule. (pp. 297–298)
7. Describe how urine is made by telling what happens at each part of the tubule. (pp. 298–303)
8. Explain these terms: pressure filtration, active reabsorption, and countercurrent exchange. (pp. 298–301)
9. How does the nephron regulate the pH of the blood? (p. 304)
10. Explain how the artificial kidney machine and CAPD work. (pp. 305–306)

Key Terms

antidiuretic hormone (an''tī-di''u-ret'ik hōr'mōn) ADH; sometimes called vasopressin, a hormone secreted by the posterior pituitary that controls the rate at which water is reabsorbed by the kidneys. *303*

Bowman's capsule (bo'manz kap'sūl) a double-walled cup that surrounds the glomerulus at the beginning of the kidney tubule. *297*

collecting duct (kō-lekt'ing dukt) a tube that receives urine from several distal convoluted tubules. *298*

distal convoluted tubule (dis'tal kon'vo-lūt-ed tu'būl) highly coiled region of a nephron that is distant from Bowman's capsule. *297*

excretion (eks-kre'shun) removal of metabolic wastes. *290*

filtrate (fil'trāt) the filtered portion of blood that is contained within Bowman's capsule. *298*

glomerulus (glo-mer'u-lus) a cluster; for example, the cluster of capillaries surrounded by Bowman's capsule in a kidney tubule. *298*

kidneys (kid'nēz) organs in the urinary system which form, concentrate, and excrete urine. *294*

nephron (nef'ron) the anatomical and functional unit of the vertebrate kidney; kidney tubule. *297*

pelvis (pel'vis) a hollow chamber in the kidney that lies inside the medulla and receives freshly prepared urine from the collecting ducts. *296*

peritubular capillary (per''ī-tu'bu-lar kap'i-lar''e) capillary that surrounds a nephron and functions in reabsorption during urine formation. *298*

pressure filtration (presh'ur fil-tra'shun) the movement of small molecules from the glomerulus into Bowman's capsule due to the action of blood pressure. *298*

proximal convoluted tubule (prok'sī-mal kon'vo-lūt-ed tu'būl) highly coiled region of a nephron near Bowman's capsule. *297*

selective reabsorption (sē-lek'tiv re''ab-sorp'shun) the movement of nutrient molecules, as opposed to waste molecules from the contents of the kidney tubule into the blood at the proximal convoluted tubule. *299*

tubular excretion (tu'bu-lar eks-kre'shun) the movement of certain molecules from the blood into the distal convoluted tubule so that they are added to urine. *302*

urea (u-re'ah) primary nitrogenous waste of mammals derived from amino acid breakdown. *292*

uric acid (u'rik as'id) waste product of nucleotide breakdown. *292*

ureters (u-re'terz) tubes that take urine from the kidneys to the bladder. *294*

urethra (u-re'thrah) tube that takes urine from bladder to outside. *294*

urinary bladder (u'rī-ner''e blad'der) an organ where urine is stored before being discharged by way of the urethra. *294*

15

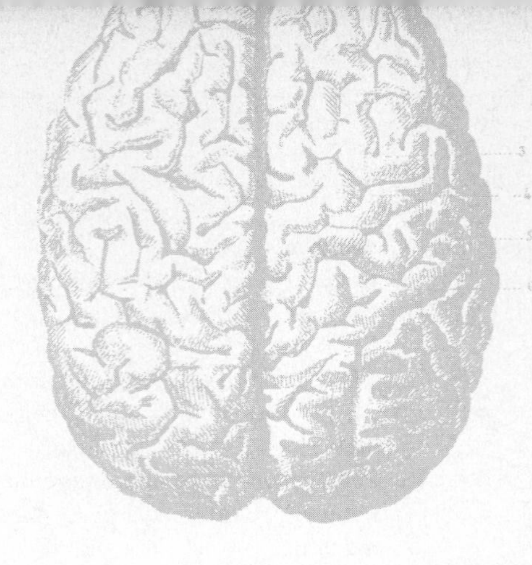

Nervous System

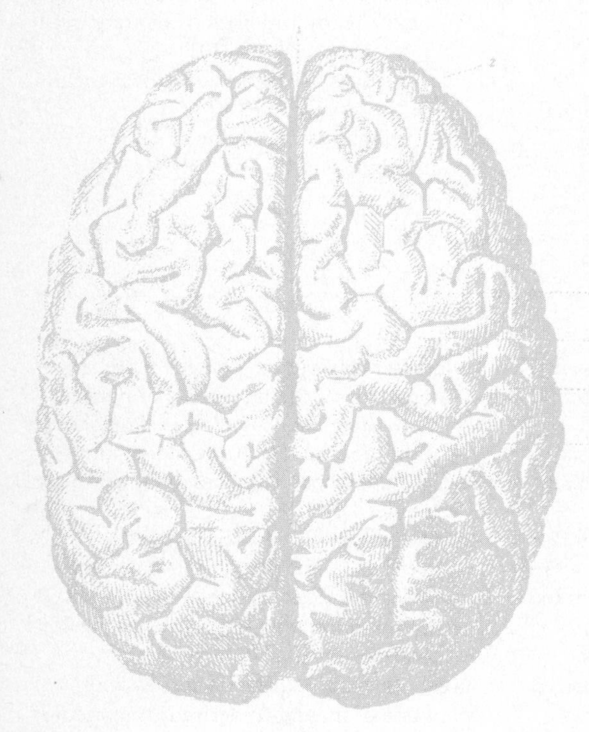

Chapter Concepts

1 The nervous system is made up of neurons that are specialized to carry nerve impulses. A nerve impulse is an electrochemical change.

2 Transmission between neurons is accomplished by means of chemicals called neurotransmitter substances. Mood-altering drugs affect the transmission of these neurotransmitters.

3 The nervous system consists of the central and peripheral nervous systems. The two systems are joined when a reflex occurs.

4 The central nervous system, made up of the spinal cord and brain, is highly organized. Consciousness is a function only of the cerebrum, which is most highly developed in humans.

Chapter Outline

Figure 15.1
The nervous system contains nerve cells that are specialized for conducting and passing on nerve impulses from one to the other, much as a baton is passed from one runner to another in a relay race.

The nervous system tells us that we exist and, along with the muscles, accounts for our distinctly animal characteristics of mobility and quick reaction to environmental stimuli. It is the one system we associate most clearly with what we take to be our very essence or being. Yet the nervous system is composed simply of nerve cells called **neurons,** which are specialized to carry nerve impulses (fig. 15.1).

As figure 15.2 indicates, the nervous system has two major divisions: the central and peripheral nervous systems. The **central nervous system** (**CNS**) includes the brain and spinal cord (nerve cord), which lie in the midline of the body where the brain is protected by the skull and the spinal cord is protected by the vertebrae. The **peripheral nervous system** (**PNS**) which is further divided into the somatic division and the autonomic division includes all the cranial and spinal nerves. These nerves project out from the central nervous system; thus the name peripheral nervous system. Figure 15.11 illustrates what is meant by the central nervous system and the peripheral nervous system. The division is arbitrary; the two systems work together and are connected to one another.

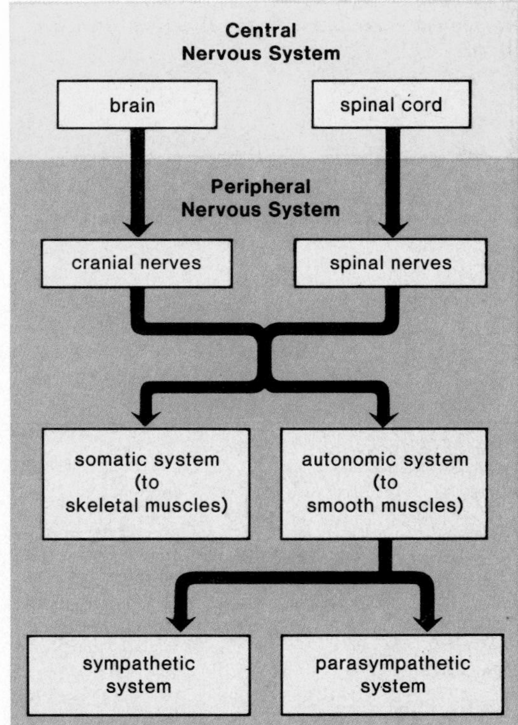

Figure 15.2
Overall organization of the nervous system in human beings. The central nervous system is at the top of the diagram and the peripheral nervous system is below. These portions of the nervous system take their names from their locations in the body.

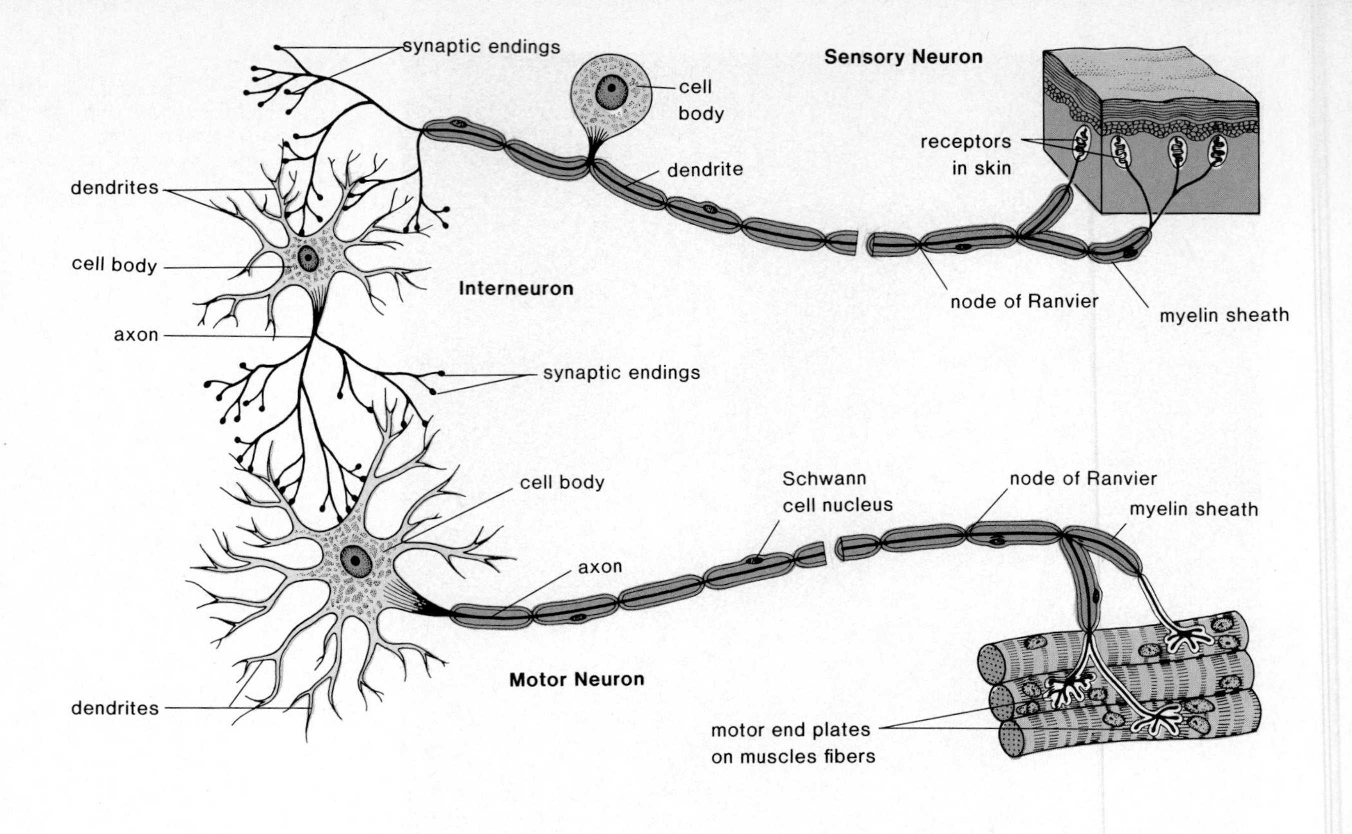

Figure 15.3
The three types of neurons—sensory, interneuron, and motor—are drawn here to show their arrangement in the body. How does this arrangement correlate with the function of each neuron?

Table 15.1 Neurons		
Neuron	**Structure**	**Function**
Sensory (afferent)	Long dendrite, short axon	Carry nerve impulses (messages) from periphery to CNS*
Motor (efferent)	Short dendrites, long axon	Carry nerve impulses (messages) from CNS to periphery
Interneuron	Short dendrites, long or short axon	Carry nerve impulses (messages) within CNS

*CNS = central nervous system.

Nerve Impulse

Structure of Neuron

All neurons (fig. 15.3) have three parts: dendrite(s), cell body, and axon. A **dendrite** conducts nerve impulses (messages) toward the **cell body,** and an **axon** conducts nerve impulses away from the cell body. There are three types of neurons: sensory, motor, and interneuron. A **sensory neuron** takes a message from a sense organ to the central nervous system and has a long dendrite and short axon, while a **motor neuron** takes a message away from the central nervous system to a muscle fiber or gland and has short dendrites and a long axon. Because motor neurons cause muscle fibers and glands to react, they are said to **innervate** these structures. Sometimes a sensory neuron is referred to as the *afferent neuron,* and the motor neuron is called the *efferent neuron.* These words, which are derived from Latin, mean running to and running away from, respectively. Obviously, they refer to the relationship of these neurons to the central nervous system.

An **interneuron** (also called association neuron, or connector neuron) is always found completely within the central nervous system and conveys messages between parts of the system. An interneuron can have short dendrites and a long axon or short dendrites and a short axon. Table 15.1 summarizes the three types of neurons that are also illustrated in figure 15.3.

The dendrites and axons of neurons are sometimes called **fibers** or processes. Most long fibers, whether dendrites or axons, are covered by a **myelin sheath** (fig. 15.4), formed by tightly packed spirals of the cell membrane of Schwann cells. This sheath, which gives nerves their white appearance, is interrupted by gaps at intervals called the nodes of Ranvier.

Although all neurons have the same three parts, each is specialized as to its structure and function. Specialization is dependent on the location of the neuron in relation to the CNS.

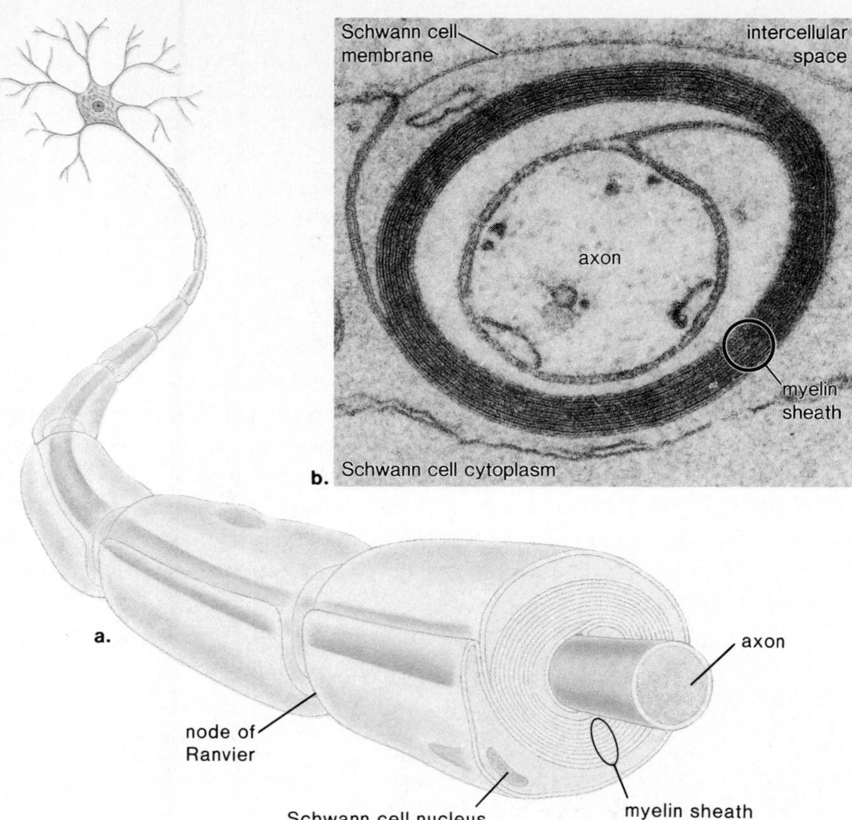

Figure 15.4
Myelin sheath a. *Axon of motor neuron ending in a cross section of the myelin sheath that encloses the long fibers of all neurons. The myelin sheath is composed of Schwann cell membrane and has a white glistening appearance in the body.* b. *Electron micrograph of a cross section of the myelin sheath.*

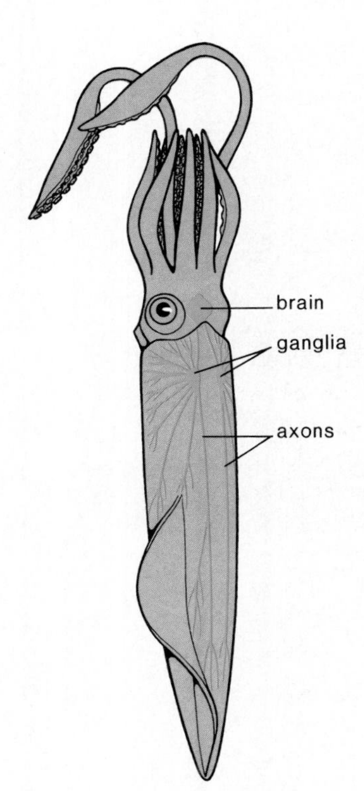

Figure 15.5
The squid axons shown here produce rapid muscular contraction so that the squid can move quickly. These neurons are so large that a microelectrode can be inserted in the axon to study the nature of the nerve impulse and also inserted in the cell bodies within ganglia to study the nature of transmission between neurons.

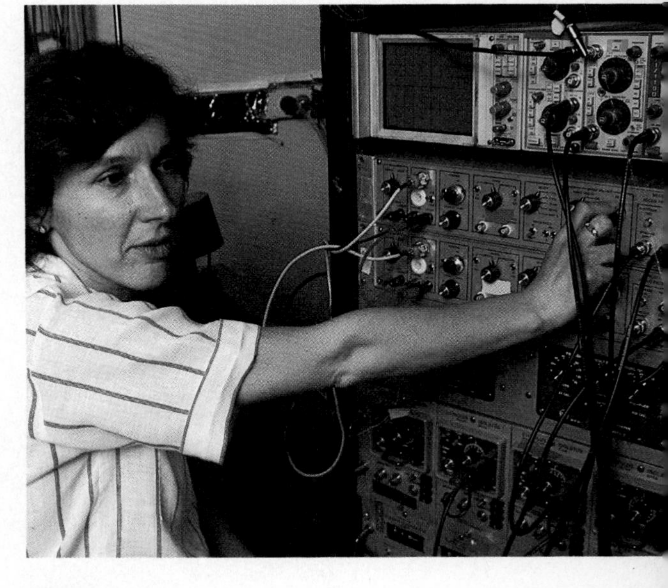

Figure 15.6
Scientist working at an oscilloscope, the electrical recording device that measures changes in voltage wherever an electrode is placed on or inserted in a neuron.

Conduction along Neuron

A neuron is specialized to conduct nerve impulses. The nature of a **nerve impulse** has been studied by using giant axons in the squid (fig. 15.5) and a type of voltmeter called an oscilloscope, which shows a trace or pattern indicating changes in voltage as time elapses (fig. 15.6). Voltage is a measurement of the potential difference between two points. When a potential difference exists, we can say that a plus and a minus pole exist; therefore, an oscilloscope indicates the existence of polarity and records polarity changes.

Resting Potential

In the experimental setup shown in figure 15.7a, an oscilloscope is wired to two electrodes, one of which is an internal recording electrode that records from inside a giant axon of the squid. The axon, being a process that extends from a cell body, is essentially a membranous tube filled with cytoplasm, or in this case, axoplasm. When the axon is not conducting an impulse, the oscilloscope records a potential difference across the membrane equal to −60 millivolts (mV). This is the **resting potential** because the axon is not conducting an impulse.

Such polarity is not unexpected because it is known that there is a difference in ion distribution on either side of the membrane. As figure 15.7b shows, there is a concentration of sodium ions (Na^+) outside the axon and a concentration of potassium ions (K^+) inside the axoplasm. Also, there are large organic negative ions in the axoplasm, which cause the resting fiber to be negative inside. These organic ions are held inside due to the differentially permeable nature of the axomembrane. The distribution of sodium and potassium is maintained by a form of active transport called the sodium/potassium pump (p. 69), which requires energy and is believed to function whenever the neuron is not conducting an impulse.

Figure 15.7
The resting and action potential. a. *Resting potential. The oscilloscope reads a resting potential of −60 millivolts due to* (b) *the presence of large negative organic ions inside the axoplasm. Note also the unequal distribution of Na^+ and K^+ across the membrane.* c. *Action potential. The action potential is a change in polarity that may be explained by* (d) *first the movement of Na^+ to the inside and second, by the movement of K^+ to the outside of the axon.*

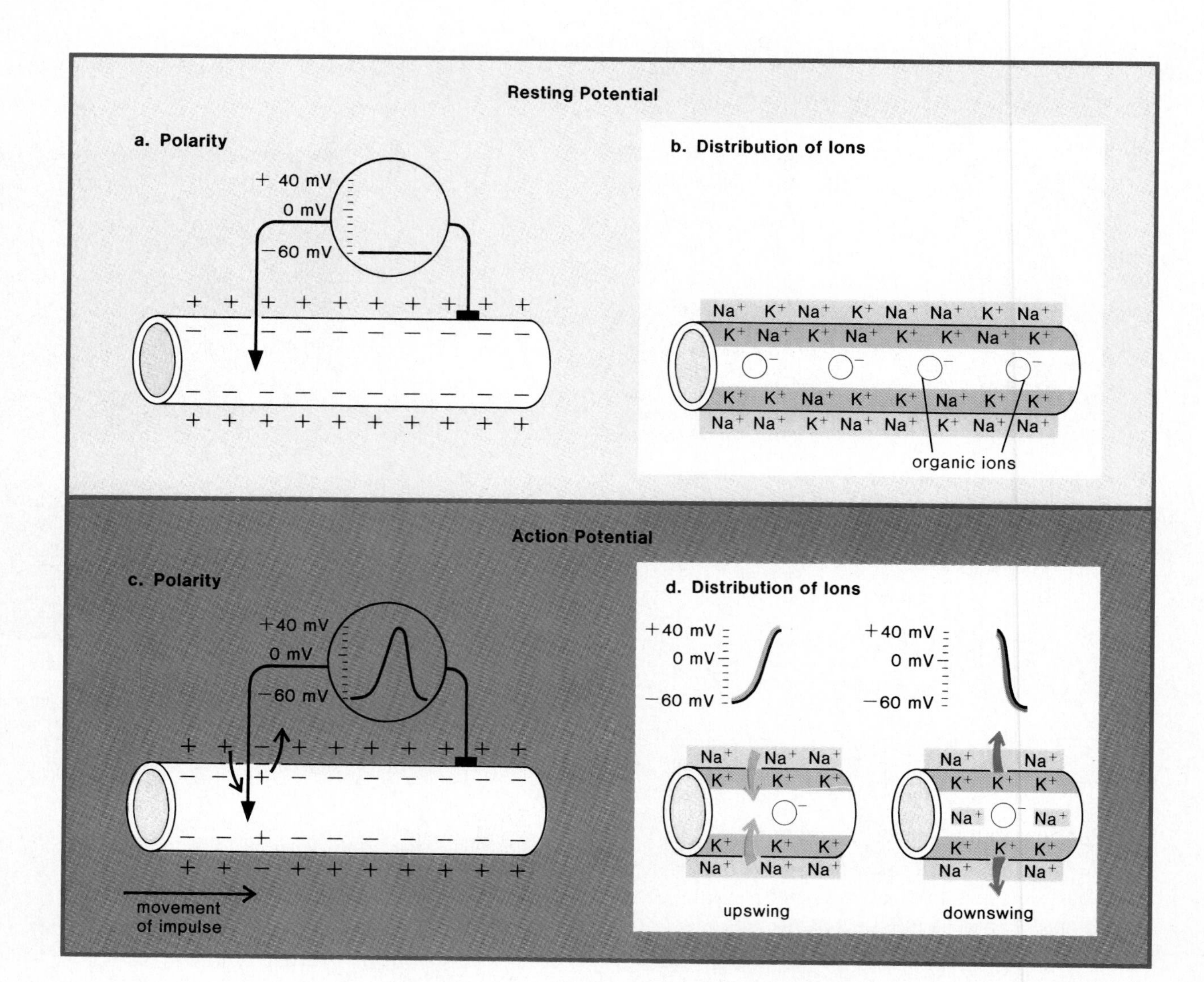

Action Potential

If the axon is stimulated to conduct a nerve impulse by an electric shock, by a sudden difference in pH, or by a pinch, a trace appears on the oscilloscope screen. This pattern caused by rapid polarity changes, called the **action potential,** has an upswing and a downswing (figs. 15.7c and 15.7d).

The Upswing (from −60 mV to +40 mV) Sophisticated experiments indicate that as the action potential goes to +40 mV, sodium ions are rapidly moving to the inside of the axon. Evidence shows that stimulation of the axon causes sodium gates, or channels, to open, allowing sodium to flow into the axon. This sudden permeability of the membrane causes the oscilloscope to record a **depolarization:** the inside of the fiber goes from negative to positive as sodium ions enter.

The Downswing (from +40 mV to −60 mV) It is now known that the restoration of the resting potential (or the return to −60 mV) is caused by the exit of potassium ions from the axoplasm. The membrane has suddenly become permeable to potassium because the potassium gates, or channels, have opened. The oscilloscope records a **repolarization** as the inside of the axon returns to negative again.

Table 15.2 summarizes the events that occur during transmission of a nerve impulse.

Table 15.2 Summary of Nerve Impulse

Resting Potential	Action Potential or Nerve Impulse
−60 mV (inside negative). Sodium/potassium pump at work. Large organic ions cause negativity inside.	a. −60 mV to + 40 mV Sodium gates are open and sodium moves to inside. Inside becomes positive compared to outside. b. + 40 mV to −60 mV Potassium gates are open and potassium moves to outside. Inside returns to negative again.

Recovery Phase A fiber can conduct a volley of nerve impulses because only a small number of ions are exchanged with each impulse. When the fiber rests, however, there is a recovery phase during which the sodium/potassium pump (p. 70) returns the sodium to the outside and the potassium to the inside.

Speed of Conduction

Although the oscilloscope records from only one area of the axon, the nerve impulse actually travels along the length of an axon. In nonmyelinated fibers, the action potential at one point generates the action potential at the very next point so that the nerve impulse is propagated along the entire length of a fiber. Conduction is hundreds of times faster (200 meters per second compared to 0.5 meter per second) in myelinated fibers because depolarization occurs only at the nodes of Ranvier (fig. 15.8). Thus the action potential jumps from node to node rather than traveling from point to point.

All neurons, whether sensory or motor, transmit the same type nerve impulse—an electrochemical change that is propagated along the fiber(s).

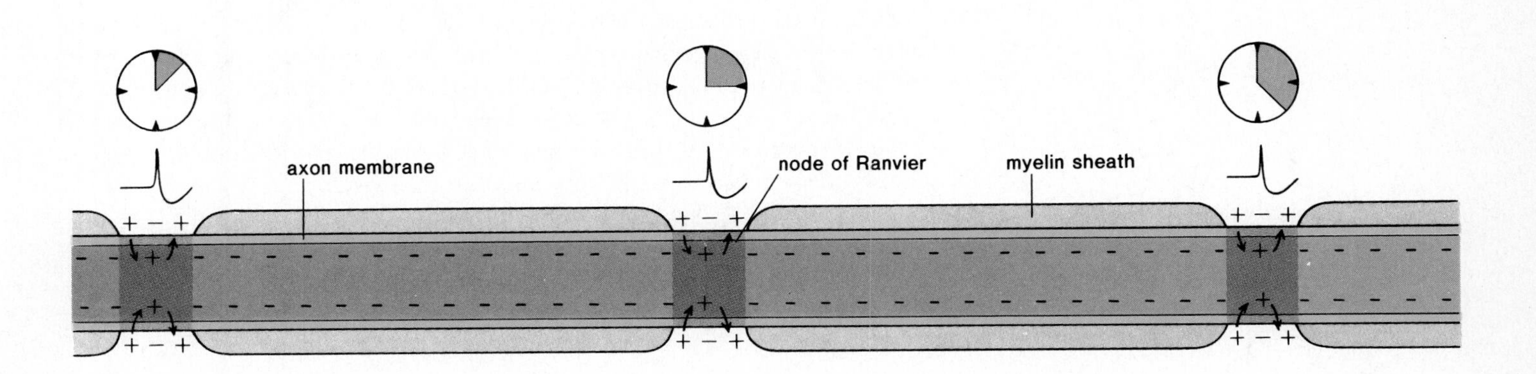

Figure 15.8
Longitudinal section of a vertebrate axon illustrates the manner by which the nerve impulse travels down a long nerve fiber. The speed of the impulse is due to the fact that it jumps from one node of Ranvier to the next.

Transmission across a Synapse

The mechanism by which an action potential passes from one neuron to another is not the same as the mechanism by which an action potential is conducted along a neuron. Each axon branches into many fine terminal branches, each of which is tipped by a small swelling, or terminal knob (fig. 15.9). Each knob lies very close to the dendrite (or cell body) of another neuron. This region is called a **synapse** and the knob is called a **synaptic ending.** The membrane of the knob is called the **presynaptic membrane,** and the membrane of the next neuron just beyond the knob is called the **postsynaptic membrane.** The small gap between is the **synaptic cleft.**

Transmission of nerve impulses across a synaptic cleft is carried out by chemicals called **neurotransmitter substances.** Figure 15.9b shows the location of synaptic vesicles at the end of an axon (fig. 15.9c is an actual electron micrograph); the transmitter substance is stored in these vesicles before it is released. When nerve impulses traveling along an axon reach a synaptic ending, they modify the membrane in such a way that calcium ions flow into the ending. These ions appear to interact with contractile proteins to pull the synaptic vesicles up to the inner surface of the presynaptic membrane. When the vesicles merge with this membrane, a neurotransmitter substance is discharged into the cleft. The neurotransmitter molecules diffuse across the cleft to the postsynaptic membrane where they bind with **receptor sites** in a lock-and-key manner. This reception alters the potential of the postsynaptic membrane in either an excitatory or inhibitory direction.

Neurotransmitter Substances

Acetylcholine (ACh) and **noradrenalin** (NA) are well-known excitatory transmitters active in both the peripheral and central nervous systems. Examples of inhibitory substances, so far discovered only in the central nervous system, are given on page 327.

Once a transmitter substance has been released into a synaptic cleft, it has only a short time to act. In some synapses, the cleft contains enzymes that rapidly inactivate the neurotransmitter. For example, the enzyme **acetylcholinesterase** (AChE), or simply cholinesterase, breaks down acetylcholine. In other synapses, the synaptic ending rapidly absorbs the transmitter substance, possibly for repackaging in synaptic vesicles or for chemical breakdown. The enzyme monoamine oxidase breaks down noradrenalin after it is absorbed. The short existence of neurotransmitters in the synapse prevents continuous stimulation (or inhibition) of postsynaptic membranes.

Summation and Integration

A neuron is on the receiving end of many synapses (fig. 15.9a). Whether a neuron fires or not depends on the summary effect of all the excitatory and/or inhibitory neurotransmitters received. If the amount of excitatory neurotransmitters received is sufficient to overcome the amount of inhibitory neurotransmitters received, the neuron fires. If the amount of excitatory neurotransmitters received is not sufficient, only **local excitation** occurs. It can be seen then that synapses are regions where a "summing up" occurs and, therefore, are also regions of **integration,** where the nervous system can fine tune its response to the environment. The structure and function of synapses allows them to carry on this very important activity.

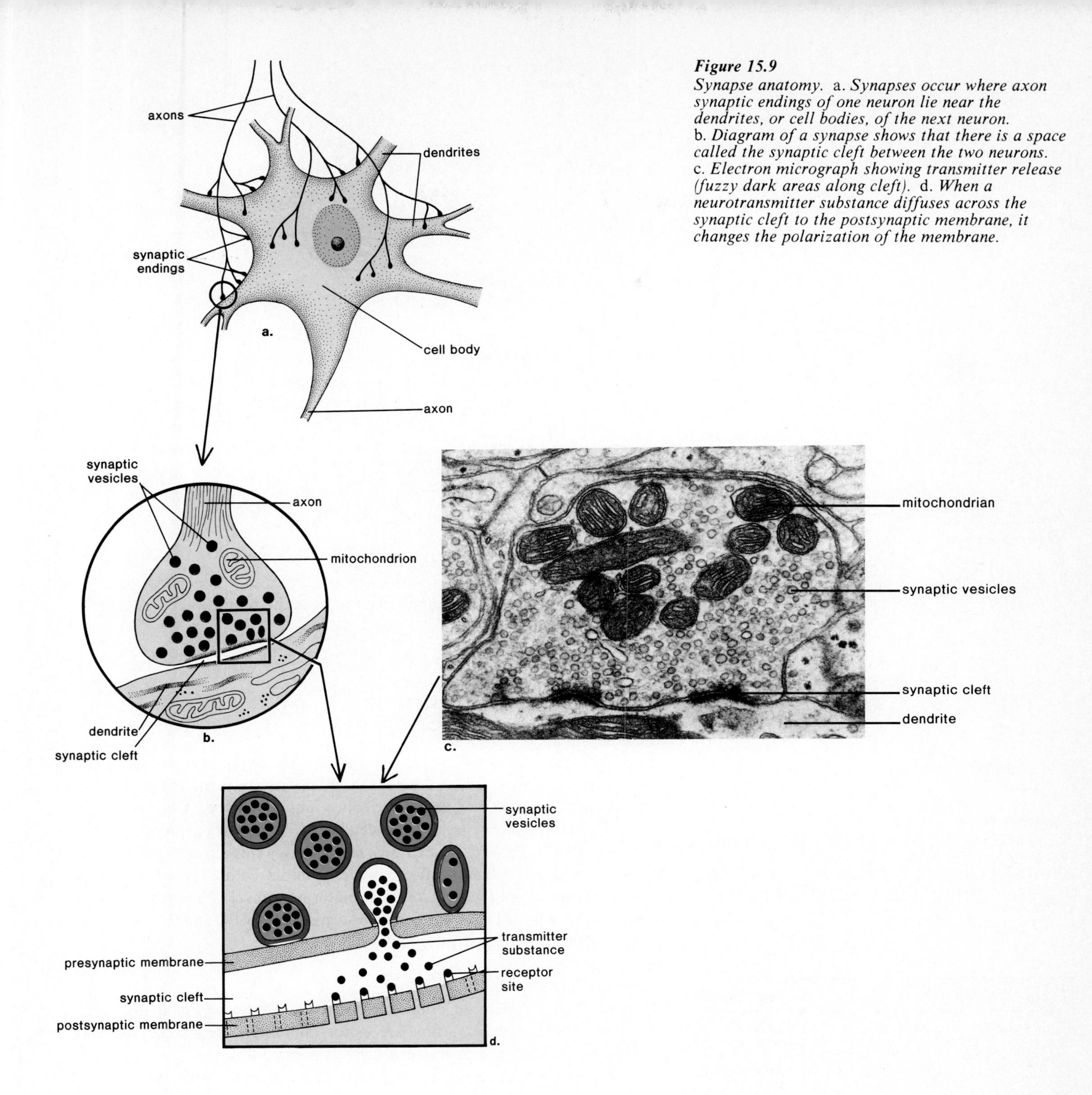

Figure 15.9
Synapse anatomy. a. *Synapses occur where axon synaptic endings of one neuron lie near the dendrites, or cell bodies, of the next neuron.* b. *Diagram of a synapse shows that there is a space called the synaptic cleft between the two neurons.* c. *Electron micrograph showing transmitter release (fuzzy dark areas along cleft).* d. *When a neurotransmitter substance diffuses across the synaptic cleft to the postsynaptic membrane, it changes the polarization of the membrane.*

Figure 15.10
Diagram of a cross section of a nerve, with one axon extended to show that each fiber is enclosed by a myelin sheath. Because nerves contain so many fibers it has been difficult to successfully rejoin them after they are severed in an accident. Scientists have now found that if they hold well-cut pieces together, and then surround them by a solution that resembles cytoplasm, the nerve will repair itself and be functional.

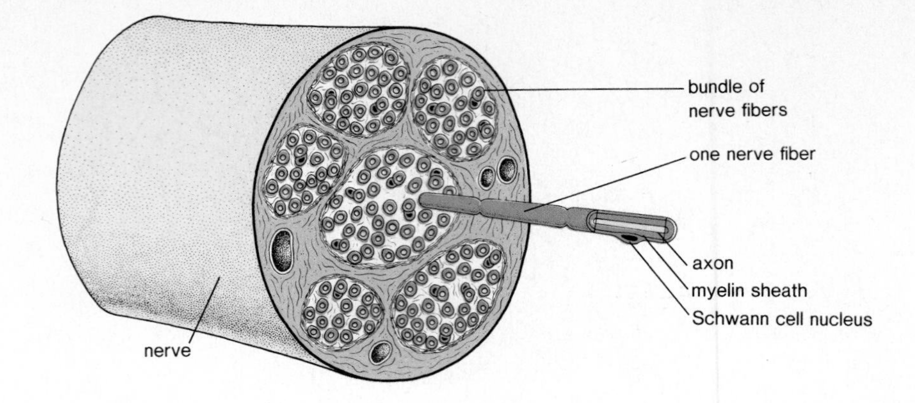

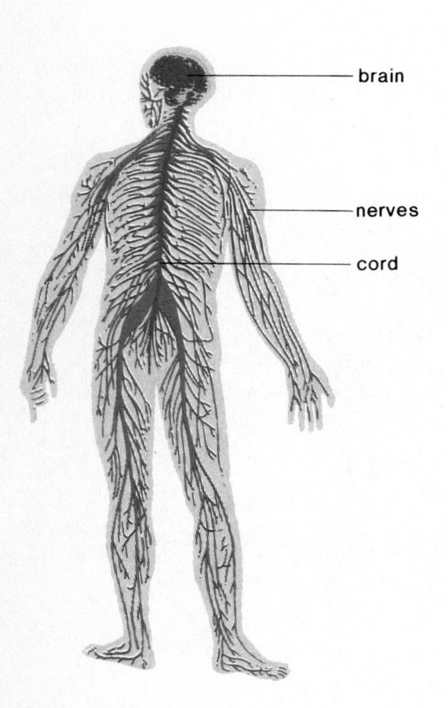

Figure 15.11
Location of central nervous system (brain and cord) and peripheral nervous system (nerves). The central nervous system (CNS) lies in the center of the body and the peripheral nervous system (PNS) lies to either side.

***Table 15.3** Nerves**

Type of Nerve	Consists of	Function
Sensory nerves	Long dendrites only of sensory neurons	Carry message from receptors to CNS
Motor nerves	Long axons only of motor neurons	Carry message from CNS to effectors
Mixed nerves	Both long dendrites of sensory neurons and long axons of motor neurons	Carry message in dendrites to CNS and away from CNS in axons

*Compare to table 15.1.

One-way Propagation

Transmission across a synapse is one-way because only the ends of axons have synaptic vesicles that are able to release neurotransmitter substances to affect the potential of the next neuron. Also, neurons obey the *all-or-none law,* meaning that a neuron either fires maximally or it does not fire at all. A nerve does not obey the all-or-none law, because a nerve contains many fibers (fig. 15.10), any number of which may be carrying nerve impulses. Thus a nerve may have degrees of performance.

Transmission of nerve impulses across a synapse is dependent on a neurotransmitter substance that changes the permeability of the postsynaptic membrane.

Peripheral Nervous System

Nerves

The PNS (fig. 15.11) consists of nerves that contain only long dendrites and/or long axons. This is so because neuron cell bodies are found only in the brain, spinal cord, and ganglia. **Ganglia** are collections of cell bodies within the PNS.

There are three types of nerves (table 15.3). **Sensory nerves** contain only the long dendrites of sensory neurons; **motor nerves** contain only the long axons of motor neurons; **mixed nerves,** however, contain both the long dendrites of sensory neurons and the long axons of motor neurons. Each nerve fiber within a nerve is surrounded by a white myelin sheath (fig. 15.10) and therefore nerves have a white, shiny, glistening appearance.

Cranial Nerves

Humans have 12 pairs of **cranial nerves** attached to the brain (fig. 15.12). Some of these are sensory, some are motor, and others are mixed. Notice that although the brain is a part of the CNS, the cranial nerves are a part of the PNS. All cranial nerves, except the vagus, are concerned with the head, neck, and face regions of the body; but the vagus nerve has many branches to serve the internal organs.

Spinal Nerves

Each **spinal nerve** emerges from the cord (fig. 15.13) by two short branches, or *roots,* which lie within the vertebral column. The dorsal root can be identified by the presence of an enlargement called the **dorsal root ganglion.** This ganglion contains the cell bodies of the sensory neurons whose dendrites conduct impulses toward the cord. The ventral root of each spinal nerve contains axons of motor neurons that conduct impulses away from the cord. These two roots join just before the spinal nerve leaves the vertebral column. Therefore all spinal nerves are mixed nerves that contain many sensory dendrites and motor axons.

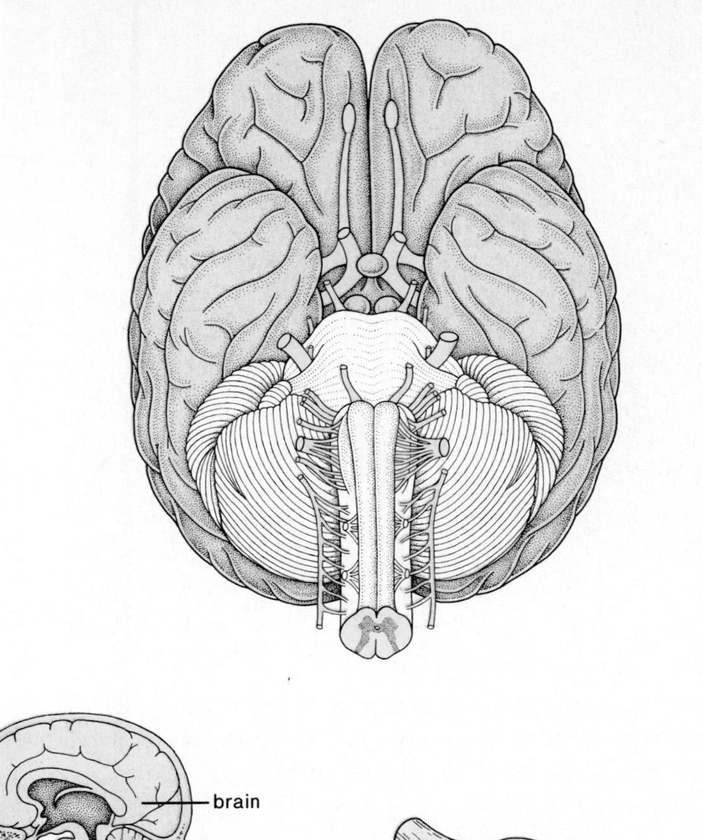

Figure 15.12
Underside of the brain showing the origins of the cranial nerves. Many cranial nerves are either sensory nerves that receive impulses from the sense organs or motor nerves that control muscles of the face and neck.

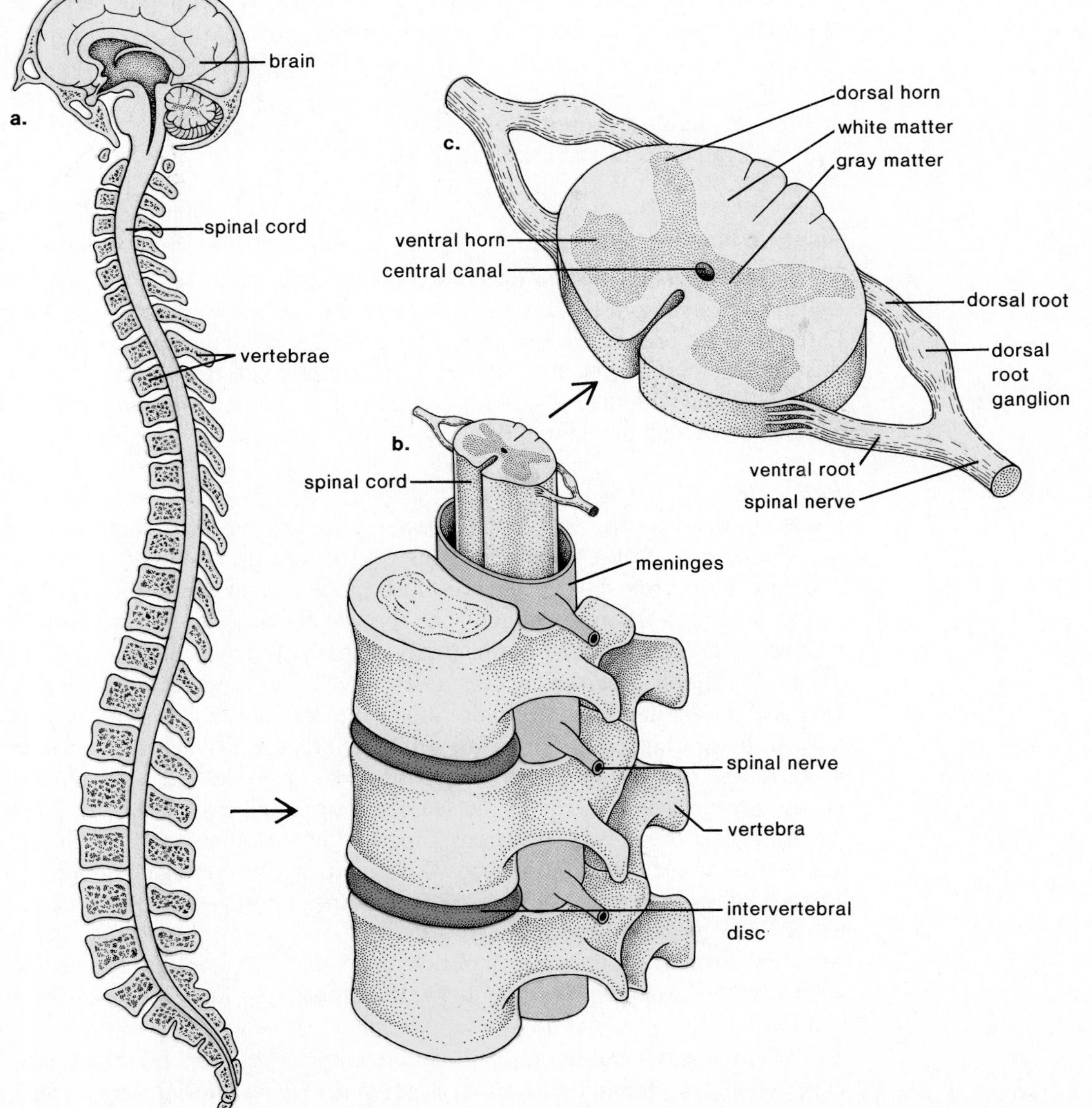

Figure 15.13
Central nervous system. a. *The brain and spinal cord are protected by bone.* b. *Anatomy of the spinal cord and spinal nerve. A spinal nerve arises after the dorsal and ventral roots join.* c. *The spinal cord is protected by vertebrae, and the spinal nerves are not apparent until they project from between the vertebrae.*

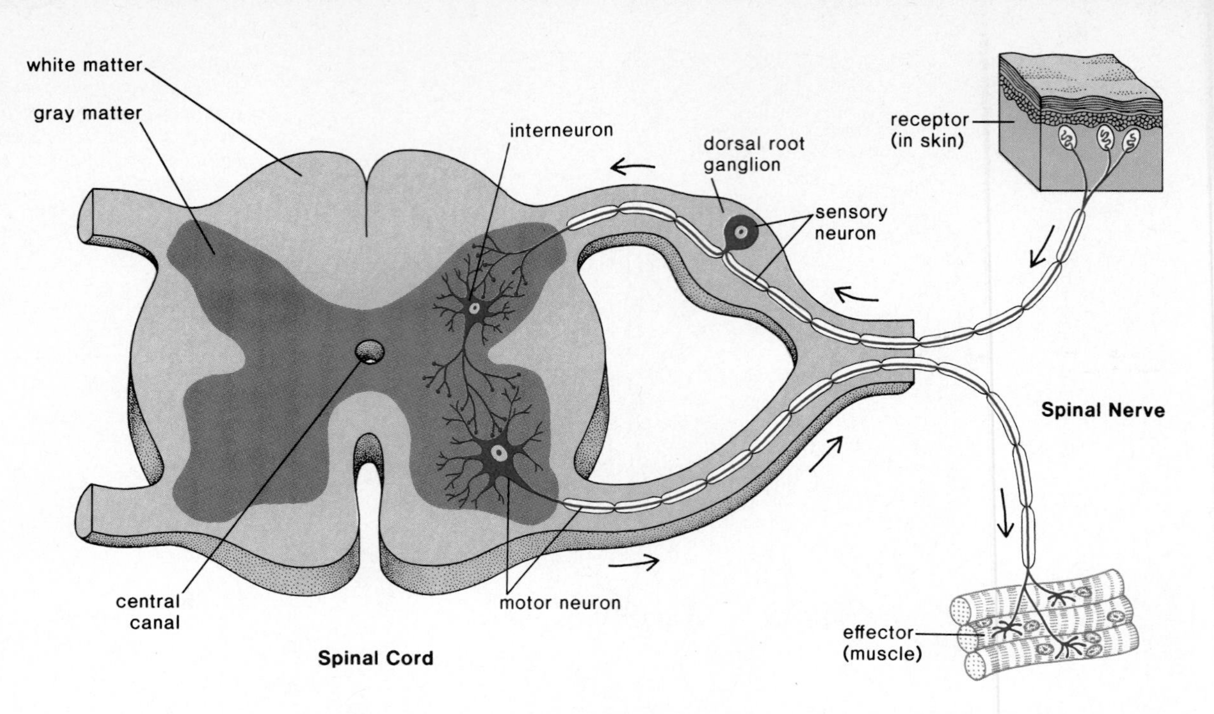

Figure 15.14
Diagram of a reflex arc, the functional unit of the nervous system. Trace the path of a reflex by following the black arrows. Name the three types of neurons that are required for a simple reflex, such as the rapid response to touching a hot object with the hand.

Human beings have 31 pairs of spinal nerves, which give evidence that humans are segmented animals, especially since the spinal nerves serve the particular region of the body where they are located.

Cranial nerves take impulses to and/or from the brain. Spinal nerves take impulses to and from the spinal cord.

Somatic Nervous System

The **somatic nervous system** includes all those nerves that serve the musculoskeletal system and the exterior sense organs, including those in the skin. Exterior sense organs are **receptors** that receive environmental stimuli and then initiate nerve impulses. Muscle fibers are **effectors** that bring about a reaction to the stimulus. Receptors are studied in chapter 17 and muscle effectors are studied in chapter 16.

Reflex Arc

Reflexes are automatic, involuntary responses to changes occurring inside or outside the body. In the somatic nervous system, outside stimuli often initiate a reflex action. Some reflexes, such as blinking the eye, involve the brain, while others, such as withdrawing the hand from a hot object, do not necessarily involve the brain. Figure 15.14 illustrates the path of the second type of reflex action. Whenever a person touches a very hot object, a receptor in the skin generates nerve impulses that move along the dendrite of a sensory neuron toward the cell body and CNS. The cell body of a sensory neuron is located in the dorsal root ganglion just outside the cord. From the cell body, the impulses travel along the axon of the sensory neuron and enter the cord; there they may pass to many interneurons, one of which lies completely within the gray matter of the cord and connects with a motor neuron. The short dendrites and cell body of the motor neuron are in the ventral region (horn) of the gray matter, and the axon leaves the cord by way of the ventral root. The nerve impulses travel along the axon to muscle fibers that then contract so that the hand is withdrawn from the hot object. (See table 15.4 for a listing of these events.)

Various other reactions usually accompany a reflex response; the person may look in the direction of the object, jump back, and utter appropriate exclamations. This whole series of responses is explained by the fact that the

Table 15.4 Path of a Simple Reflex

1. Receptor (formulates message)*	Generates nerve impulses
2. Sensory neuron (takes message to central nervous system)	Impulses move along dendrite (spinal nerve)†—and proceed to cell body (dorsal root ganglia) and then go from cell body to axon (gray matter of cord)
3. Interneuron (passes message to motor neuron)	Impulses picked up by dendrites and pass through cell body to axon (completely within gray matter)
4. Motor neuron (takes message away from central nervous system)	Impulses travel through short dendrites and cell body (gray matter of cord) to axon (spinal nerve)
5. Effector (receives message)	Receives nerve impulses and reacts: glands secrete and muscles contract

*Phrases within parentheses state overall function.
†Words within parentheses indicate location of structure.

sensory neuron stimulates several interneurons, which take impulses to all parts of the central nervous system, including the cerebrum, which, in turn, makes the person conscious of the stimulus and his or her reaction to it.

The reflex arc is the main functional unit of the nervous system. It allows us to react to internal and external stimuli.

Autonomic Nervous System

The autonomic nervous system, a part of the PNS, is made up of motor neurons that control the internal organs automatically and usually without the need for conscious intervention. There are two divisions to the autonomic nervous system: the sympathetic and parasympathetic systems. Both of these (1) function automatically and usually subconsciously in an involuntary manner; (2) innervate all internal organs; and (3) utilize two motor neurons and one ganglion for each impulse. The first neuron has a cell body within the central nervous system and a **preganglionic axon.** The second neuron has a cell body within the ganglion and a **postganglionic axon.**

The autonomic nervous system controls the functioning of internal organs without need of conscious control.

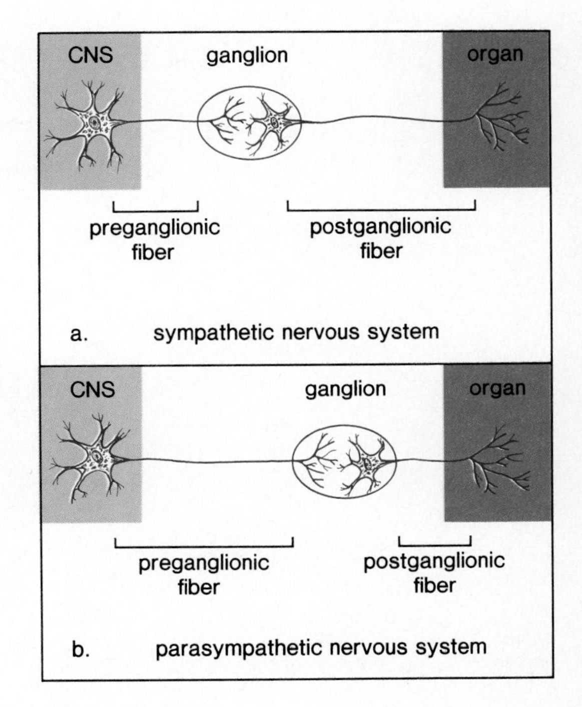

Figure 15.15
Location of ganglia in the sympathetic and parasympathetic nervous systems. a. *In the sympathetic nervous system, each ganglion lies close to the spinal cord (CNS) and therefore the preganglionic fiber is short and the postganglionic fiber is long.* b. *In the parasympathetic nervous system, each ganglion lies close to the organ being innervated and therefore the preganglionic fiber is long and the postganglionic fiber is short.*

Sympathetic System

The preganglionic fibers of the **sympathetic nervous system** arise from the middle or *thoracic-lumbar portion* of the cord and almost immediately terminate in ganglia that lie near the cord. Therefore, in this system, the preganglionic fiber is short, but the postganglionic fiber that makes contact with an organ is long (fig. 15.15).

The sympathetic nervous system is especially important during emergency situations and is associated with "fight or flight." For example, it inhibits the digestive tract, but dilates the pupil, accelerates the heartbeat, and increases the breathing rate. It is not surprising, then, that the neurotransmitter released by the postganglionic axon is noradrenalin, a chemical close in structure to adrenalin, a well-known heart stimulant.

The sympathetic nervous system brings about those responses we associate with "fight or flight."

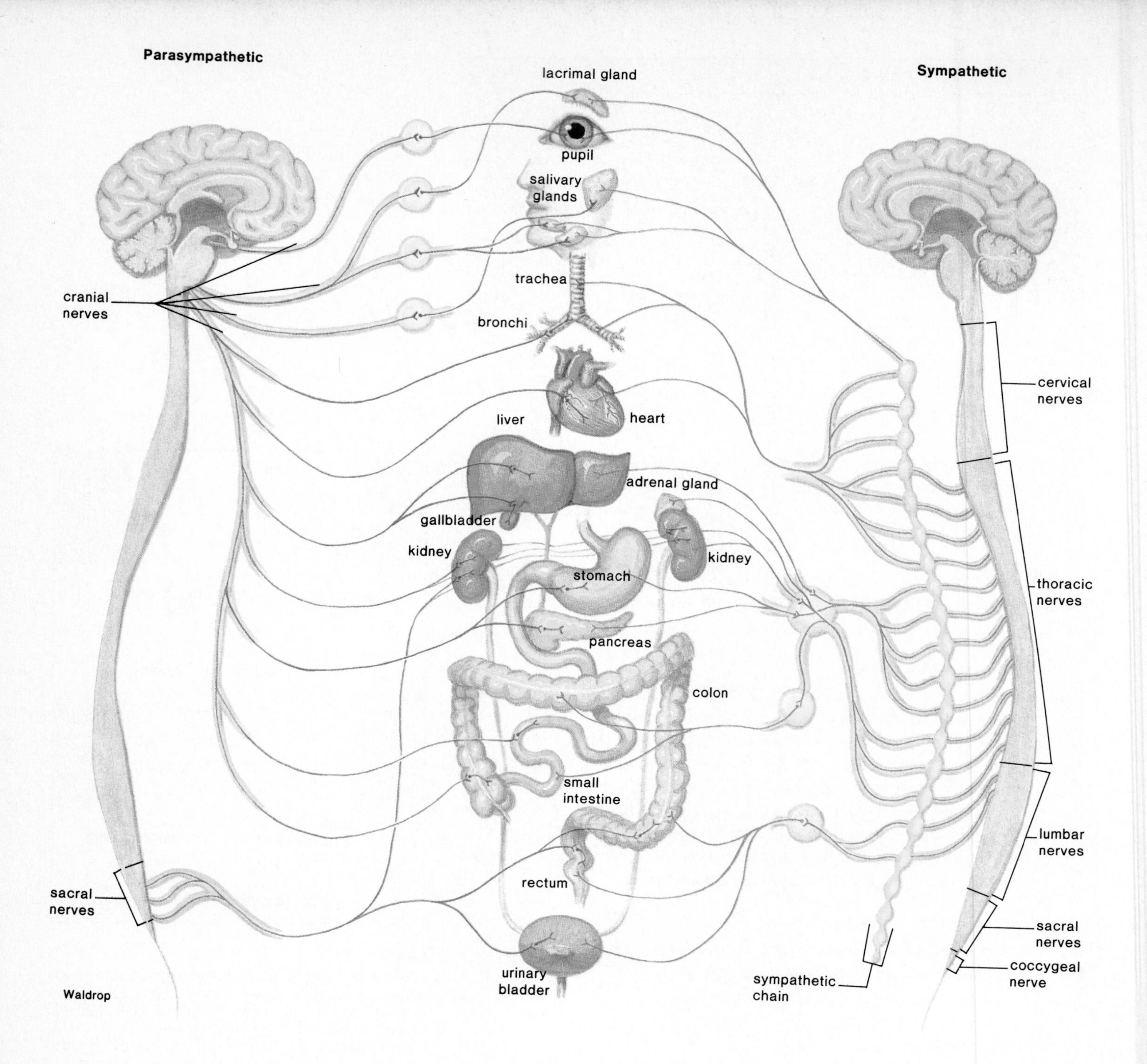

Figure 15.16
Structure and function of the autonomic nervous system. The sympathetic fibers arise from the thoracic and lumbar portion of the cord; the parasympathetic fibers arise from the brain and sacral portion of the cord. Each system innervates the same organs but have contrary effects. For example, the sympathetic system speeds up and the parasympathetic system slows down the beat of the heart.

***Table 15.5** Sympathetic versus Parasympathetic System*

Sympathetic	Parasympathetic
Fight or flight	Normal activity
Noradrenalin is neurotransmitter	Acetylcholine is neurotransmitter
Postganglionic fiber is longer than preganglionic	Preganglionic fiber is longer than postganglionic
Preganglionic fiber arises from middle portion of cord	Preganglionic fiber arises from brain and lower portion of cord

Parasympathetic System
The vagus nerve and fibers that arise from the bottom portion of the cord form the **parasympathetic nervous system.** Therefore this system is often referred to as the *craniosacral portion* of the autonomic nervous system. In the parasympathetic nervous system, the preganglionic fiber is long and the postganglionic fiber is short because the ganglia lie near or within the organ (fig. 15.15). The parasympathetic system promotes all those internal responses we associate with a relaxed state; for example, it causes the pupil of the eye to contract, promotes digestion of food, and retards the heartbeat. The neurotransmitter utilized by the parasympathetic system is acetylcholine.

Figure 15.16 contrasts the sympathetic and parasympathetic systems, and table 15.5 lists all the differences we have noted between these two systems.

---

*T*he parasympathetic nervous system brings about those responses we associate with normally restful activities.

Central Nervous System

The CNS consists of the spinal cord and brain. As figures 15.13 and 15.17 illustrate, the CNS is protected by bone: the brain is enclosed within the skull and the spinal cord is surrounded by vertebrae. Also, both the brain and spinal cord are wrapped in three protective membranes known as **meninges;** spinal meningitis is a well-known infection of these coverings. The spaces between the meninges are filled with **cerebrospinal fluid,** which cushions and protects

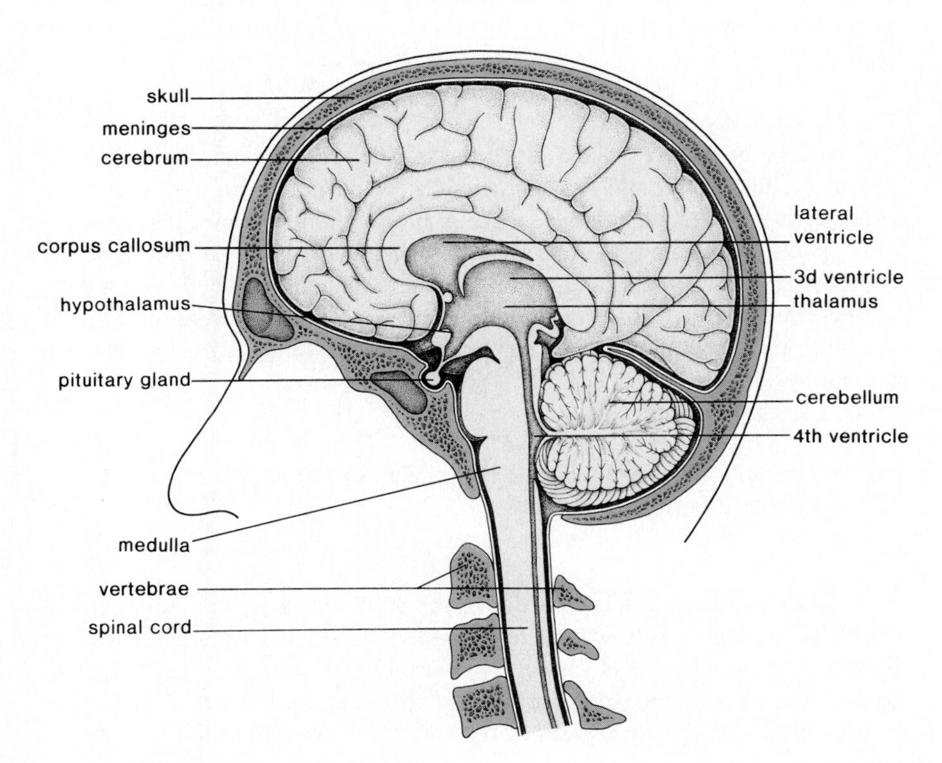

Figure 15.17
Anatomy of the human brain. The cerebrum, the highest and largest part of the human brain, is responsible for consciousness. The medulla, the last part of the brain before the spinal cord, controls various internal organs.

the CNS. A small amount of this fluid is sometimes withdrawn for laboratory testing when a spinal tap is done (fig. 15.18). Cerebrospinal fluid is also contained within the **central canal** of the spinal cord and the **ventricles** of the brain. The latter are interconnecting spaces that produce and serve as a reservoir for cerebrospinal fluid.

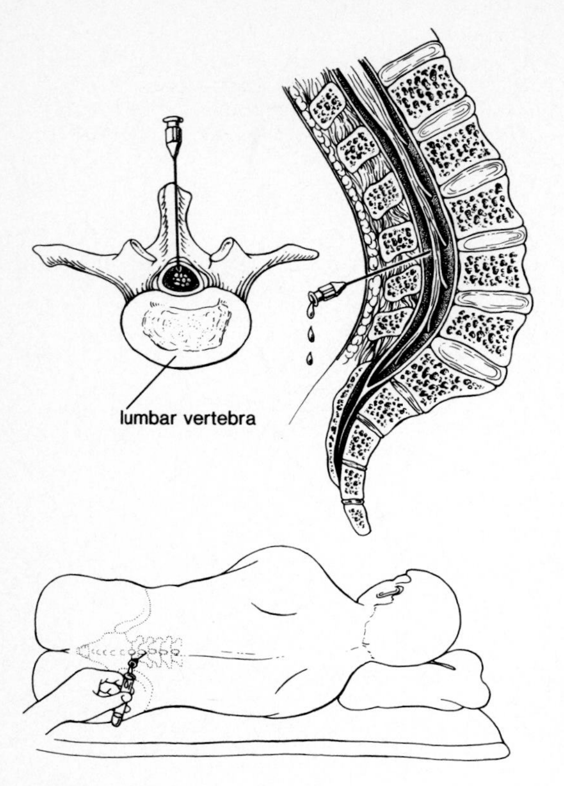

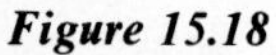

Figure 15.18
The central nervous system is protected not only by bone but also by three layers of meninges, or membranes. The spaces between the meninges are filled with cerebrospinal fluid. Since the spinal cord ends at the upper portion of the lumbar region, it is possible to make a puncture at this level in order to obtain cerebrospinal fluid for examination purposes.

Spinal Cord

The cord (fig. 15.13 and fig. 15.14) contains (1) the central canal filled with cerebrospinal fluid, (2) gray matter containing cell bodies and short fibers, and (3) white matter containing long fibers of interneurons that run together in bundles called **tracts.** These tracts connect the cord to the brain.

In the gray matter, dorsal cell bodies function primarily in receiving sensory information, and ventral cell bodies send along primarily motor information. Within the white matter of the spinal cord, ascending tracts of dorsal axons take information to the brain, and descending tracts in the ventral part of the cord primarily carry information down from the brain. Because the tracts cross over, the left side of the brain controls the right side of the body and the right side of the brain controls the left side of the body.

The central nervous system lies in the midline of the body and consists of the brain and spinal cord where sensory information is received and motor control is initiated.

Brain

The largest and most prominent portion of the human brain (fig. 15.17) is the cerebrum. Consciousness resides only in the cerebrum; the rest of the brain functions below the level of consciousness. We will consider significant portions of the "unconscious brain" before discussing the "conscious brain." In addition to the portions mentioned, it should be remembered that the unconscious brain contains many tracts that relay messages to and from the spinal cord.

The Unconscious Brain

The **medulla oblongata** lies closest to the spinal cord and contains centers for heartbeat, breathing, and vasoconstriction (blood pressure), and also reflex centers for vomiting, coughing, sneezing, hiccoughing, and swallowing.

The **hypothalamus** is concerned with homeostasis, or the constancy of the internal environment, and contains centers for hunger, sleep, thirst, body temperature, water balance, and blood pressure. The hypothalamus controls the pituitary gland and thereby serves as a link between the nervous and endocrine systems.

The medulla oblongata and the hypothalamus are both concerned with control of the internal organs.

The **thalamus** is the last portion of the brain for sensory input before the cerebrum. It serves as a central relay station for sensory impulses traveling upward from other parts of the cord and brain to the cerebrum. It receives all sensory impulses (except those associated with the sense of smell) and channels them to appropriate regions of the cortex for interpretation.

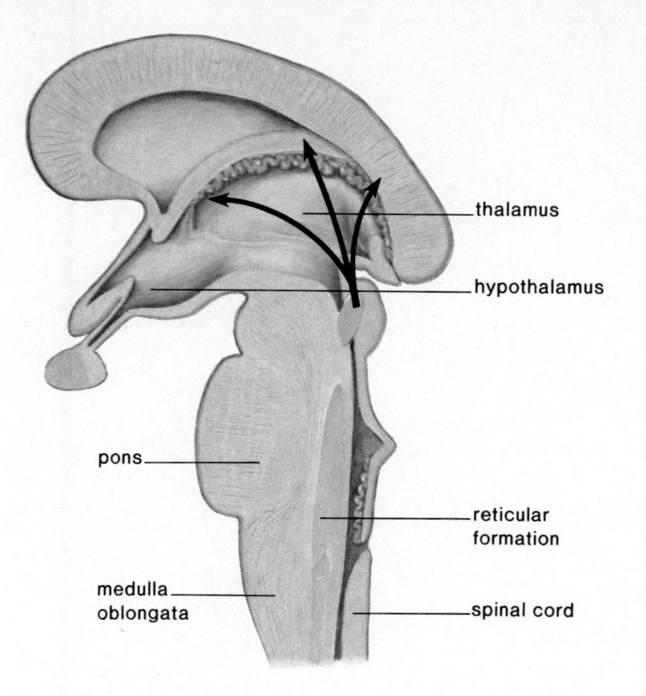

Figure 15.19
The reticular formation is a part of the ascending reticular activating system. The sensory information received by the reticular formation is sorted out by the thalamus before nerve impulses are sent to other parts of the brain.

The thalamus has connections to various parts of the brain by way of the diffuse thalamic projection system, an extension of the reticular formation, (fig. 15.19) a complex network of cell bodies and fibers that extends from the medulla to the thalamus. Together they form the **ARAS, the ascending reticular activating system,** which is believed to sort out incoming stimuli, passing on only those that require immediate attention. For this reason, the thalamus is sometimes called the gatekeeper to the cerebrum because it alerts the cerebrum to only certain sensory input. In this way it may allow you to concentrate on your homework while the television is on.

The thalamus receives sensory impulses from all parts of the body and channels them to the cerebrum.

The **cerebellum,** a bilobed structure that resembles a butterfly, is the second largest portion of the brain. It functions in muscle coordination (fig. 15.20) integrating impulses received from higher centers to ensure that all the skeletal muscles work together to produce smooth and graceful motions. The cerebellum is also responsible for maintaining normal muscle tone and transmitting impulses that maintain posture. It receives information from the inner ear indicating the position of the body, and sends impulses to those muscles whose contraction maintains or restores balance.

The cerebellum controls balance and complex muscular movements.

Figure 15.20
Gymnastic skills require muscular coordination and balance that is controlled by the cerebellum.

Figure 15.21
The convoluted cortex of the cerebrum is divided into four lobes: frontal, temporal, parietal, and occipital. Further, it is possible to map the cerebrum since each particular area has a particular function.

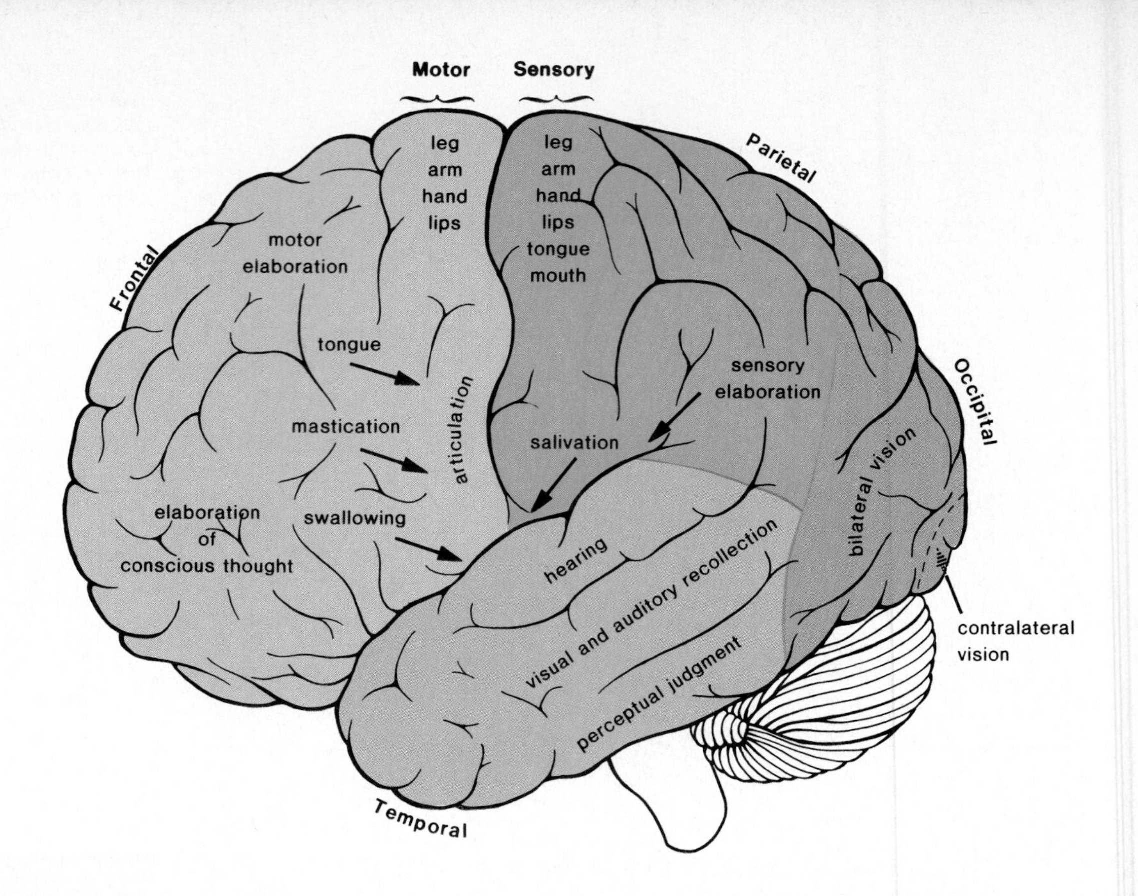

Table 15.6 Functions of the Cerebral Lobes

Lobe	Functions
Frontal lobes	Motor areas control movements of voluntary skeletal muscles. Association areas carry on higher intellectual processes such as those required for concentration, planning, complex problem solving, and judging the consequences of behavior.
Parietal lobes	Sensory areas are responsible for the sensations of temperature, touch, pressure, and pain from the skin. Association areas function in the understanding of speech and in using words to express thoughts and feelings.
Temporal lobes	Sensory areas are responsible for hearing and smelling. Association areas are used in the interpretation of sensory experiences and in the memory of visual scenes, music, and other complex sensory patterns.
Occipital lobes	Sensory areas are responsible for vision. Association areas function in combining visual images with other sensory experiences.

From Hole, John W., Jr. *Human Anatomy and Physiology*, 4th ed.

The Conscious Brain

The **cerebrum,** the only area of the brain responsible for consciousness, is the largest portion of the brain in humans. The outer layer of the cerebrum, called the cortex, is gray in color and contains cell bodies and short fibers. The cerebrum is divided into halves known as the right and left **cerebral hemispheres.** Each half contains four types of lobes: **frontal, parietal, temporal,** and **occipital** (fig. 15.21). The cerebrum can be mapped according to the particular functions of each of the lobes (table 15.6).

Association areas are believed to be areas for intellect, artistic and creative ability, learning, and memory. Sensory areas receive nerve impulses from the sense organs and produce what we call sensations. The particular sensation produced is the prerogative of the area of the brain that is stimulated, since the nerve impulse itself always has the same nature (as described previously). Motor areas of the cerebrum initiate nerve impulses that control muscle fibers. A momentary lack of oxygen during birth can damage the motor areas of the cerebral cortex so that the individual develops the symptoms of cerebral palsy, a condition characterized by a spastic weakness of the arms and legs.

Consciousness is the province of the cerebrum, the most highly developed portion of the human brain responsible for higher mental processes, including the interpretation of sensory input and the initiation of voluntary muscular movements.

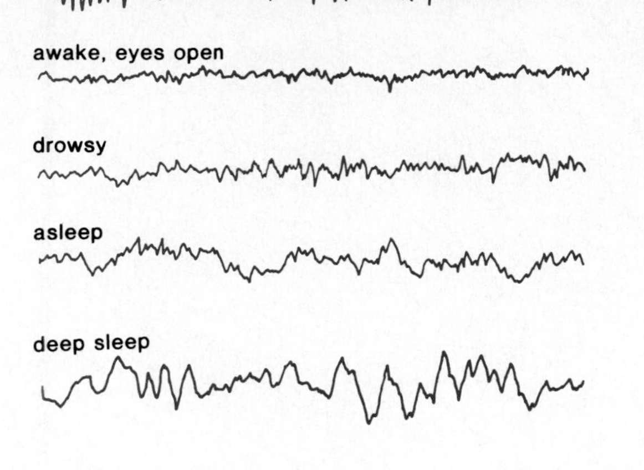

Figure 15.22
Encephalograms are recordings of the electrical activity of the brain. The alpha waves, which appear when the subject is awake with eyes closed, are the most common. Second most common are the beta waves recorded when the subject is awake with eyes open. Sleep has various stages, as indicated.

There has been a great deal of testing to determine whether the right and left half of the cerebrum serve different functions. These studies have tended to suggest that the left half of the brain is the verbal (word) half and the right half of the brain is the visual (spatial relation) and artistic half. However, other results indicate that such a strict dichotomy does not always exist between the two halves. In any case the two cerebral hemispheres normally share information because they are connected by a horizontal tract called the **corpus callosum** (fig. 15.17).

Severing the corpus callosum can control severe epileptic seizures but results in a person with two brains, each with its own memories and thoughts. Today, use of the laser permits more precise treatment without these side effects. *Epilepsy* is caused by a disturbance of the normal communication between the ARAS and the cortex. In a grand mal seizure, the cerebrum becomes extremely excited. Due to a reverberation of signals within the ARAS and cerebrum, the individual loses consciousness, even while convulsions are occurring. Finally the neurons become fatigued and the signals cease. Following an attack, the brain is so fatigued the person must sleep for a while.

The electrical activity of the brain can be recorded in the form of an **electroencephalogram (EEG).** Electrodes are taped to different parts of the scalp, and an instrument called the electroencephalograph records the so-called brain waves (fig. 15.22).

When the subject is awake, two types of waves are usual: *alpha waves,* with a frequency of about 6 to 13 per second and a potential of about 45 microvolts, predominate when the eyes are closed, and *beta waves,* with higher frequencies but lower voltage, appear when the eyes are open.

During an eight-hour sleep there are usually five times when the brain waves become slower and larger than alpha waves. During each of these times, there are irregular flurries as the eyes move back and forth rapidly. When subjects are awakened during the latter, called **REM** (rapid eye movement) **sleep,** they always report that they were dreaming. The significance of REM sleep is still being debated, but some studies indicate that REM sleep is needed for memory to occur.

The EEG is a good diagnostic tool; for example, an irregular pattern can signify epilepsy or a brain tumor. A flat EEG signifies lack of electrical activity of the brain, or brain death, and thus it may be used to determine the precise time of death.

Figure 15.23
The extrapyramidal and limbic systems. The extrapyramidal region, which includes portions of cerebrum, cerebellum, and pons, controls body movement and posture. The limbic system, which includes portions of the cerebrum, thalamus, and hypothalamus, is concerned mainly with emotion and memory.

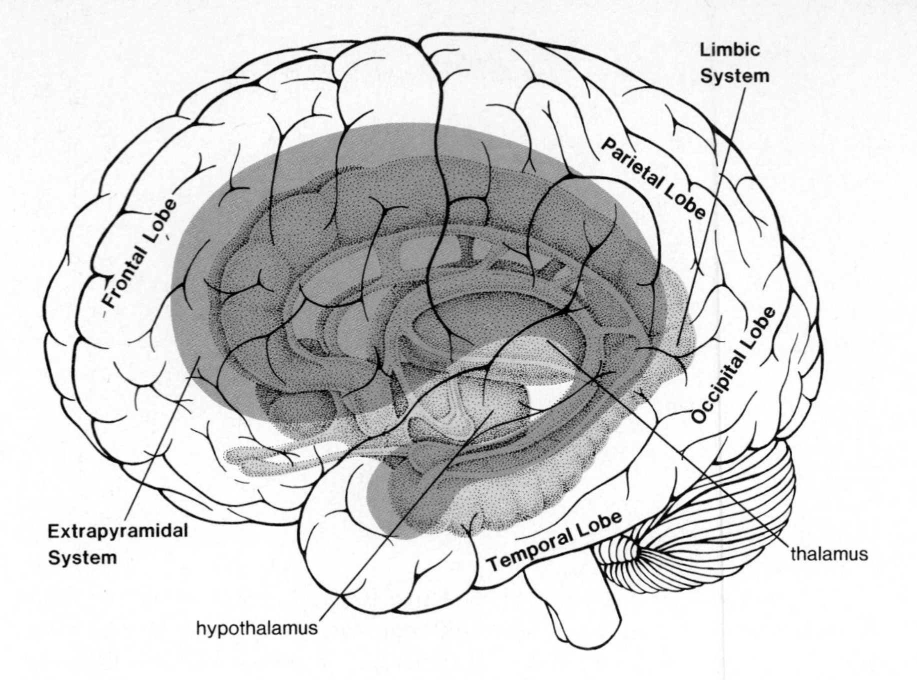

Figure 15.24
Individual nerve cells in a snail, Hermissenda, *are being stimulated by microelectrodes, simulating the signals that scientists had previously recorded when a snail learns to avoid light. When this snail is freed, it automatically avoids the light and does not need to be taught like other snails. To teach snails to avoid light, they are placed on a table that rotates every time they venture toward light.*

Extrapyramidal and Limbic Systems

Just beneath the cerebral cortex are masses of white matter that belong to the ascending and descending tracts. Some of the latter, called the extrapyramidal system (fig. 15.23), pass into the basal nuclei, several masses of gray matter that lie deep within each hemisphere of the cerebrum. The basal nuclei are a part of the **limbic system,** an interior loop that connects portions of the frontal lobes, temporal lobes, thalamus, and hypothalamus. Stimulation of different areas of the limbic system causes the subject to experience rage, pain, pleasure, or sorrow. By causing pleasant or unpleasant feelings about experiences, the limbic system apparently guides the individual into behavior that is likely to increase the chance of survival.

Learning and Memory

Learning requires memory, but just what permits memory to occur is not yet known. Investigators have been working with invertebrates such as slugs and snails because their nervous systems are very simple and yet they can be taught to perform a particular behavior. In order to determine how learning takes place, it has been possible to insert electrodes into individual cells to alter or record the electrochemical responses of these cells (fig. 15.24). This work suggests that learning results from changes in nerve transmission at existing synapses and not from the creation of new synapses.

Thus far, learning requiring only short-term memory has been studied in this manner. A good example of *short-term memory* in humans is the ability to recall a telephone number long enough to dial the number. *Long-term memory,* such as the ability to recall the events of the day, has been shown to involve the limbic system and protein synthesis. Even so, memories appear to be stored throughout the association areas; when stimulated by an experimenter, no particular region is richer in memories than another.

Certain portions of the brain (e.g., limbic system) are particularly involved in the emotions and in memory and learning.

Neurotransmitters in the Brain

As discussed previously, p. 314, neurotransmitters take nerve impulses across synapses. Proper functioning of the brain seems to depend on the proper balance of excitatory and inhibitory synaptic transmitters. The *excitatory transmitters* include acetylcholine (ACh), noradrenalin (NA), serotonin, and dopamine. The *inhibitory transmitters* include gamma aminobutyric acid (GABA), and glycine. It has been discovered that several neurological illnesses are due to an imbalance in these transmitters. Parkinson's disease and Huntington's chorea result from malfunctions of the extrapyramidal system. Parkinson's disease is a condition characterized by a wide-eyed, unblinking expression, an involuntary tremor of the fingers and thumbs, muscular rigidity, and a shuffling gait. All these symptoms are due to dopamine deficiencies. Huntington's chorea is characterized by a progressive deterioration of the individual's nervous system that eventually leads to constant thrashing and writhing movements leading to insanity and death. The problem is believed to be GABA malfunctions. Most recently it has been discovered that Alzheimer's disease, a severe form of senility with marked memory loss found in 5 to 10 percent of all people over age 65, seems to be due to acetylcholine deficiencies. Alzheimer's disease is termed a disorder of the limbic system because the limbic system is concerned not only with emotion but also memory. Treatment of individuals with brain disorders has thus far been directed toward restoring the proper balance of neurotransmitter substances. However, it now appears that it might be possible one day to implant new cells to replace the deficient cells in brains.

The *enkephalins* and *endorphins* are two neurotransmitters of interest because they are involved in the transmission and perception of pain as discussed in the reading on page 328.

Both inhibitory and excitatory neurotransmitters are active in the CNS.

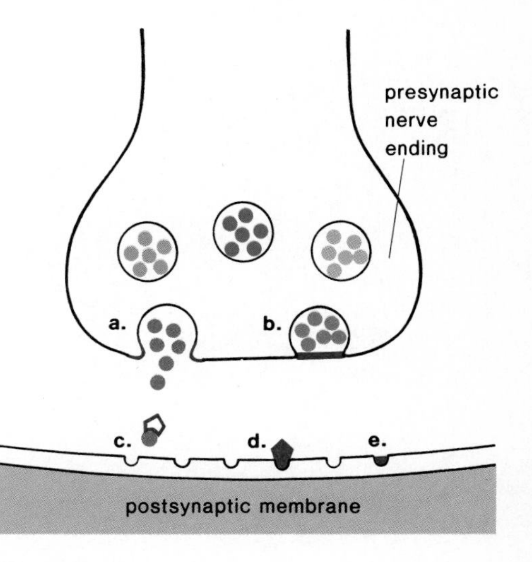

Figure 15.25
Drug action at synapses. a. *Drug stimulates release of neurotransmitter.* b. *Drug blocks release of neurotransmitter.* c. *Drug combines with neurotransmitter preventing its breakdown.* d. *Drug mimics neurotransmitter.* e. *Drug blocks receptor so that neurotransmitter cannot be received.*

Drug Action

There is a wide variety of neurological drugs that are used to alter the mood and/or emotional state. Nevertheless, two general principles of drug action have been discovered: (1) mood-altering drugs particularly affect the ARAS (p. 323) and limbic system, and (2) they either promote or decrease the action of a particular neurotransmitter. There are a number of different ways in which drugs can influence transmission of neurotransmitters and some of these are described in figure 15.25. It is clear, as outlined in table 15.7, that stimulants can either enhance the action of an excitatory transmitter or block the action of an inhibitory transmitter. Also, depressants can either enhance the action of an inhibitory transmitter or block the action of an excitatory transmitter.

Table 15.7 Drug Action

Drug Action	Neurotransmitter	Result
Blocks	Excitatory	Depression
Enhances	Excitatory	Stimulation
Blocks	Inhibitory	Stimulation
Enhances	Inhibitory	Depression

Types of Drugs

Domestic Drugs

People may not realize that they daily imbibe drugs that act on the CNS. For example, caffeine in coffee and theophylline in tea are stimulants that block the action of adenosine, a chemical that inhibits the release of neurotransmitters. Nicotine, the addicting substance in cigarettes, enhances the action of ACh.

Pain

A physiological picture is emerging to explain the phenomenon of pain. When a portion of the body is traumatized, potent chemicals are released including bradykinin and prostaglandins. These are believed to initiate nerve impulses that travel, perhaps by special small diameter fibers, to the cord where they stimulate SG (substantia gelatinosa) cells (fig. a). Ordinarily the activity of the SG cells is kept at bay by messages received from large diameter fibers that also synapse with these cells. The perception of pain depends on which of these fibers (small diameter or large diameter) is most active.

This explanation can be reconciled with the gate control theory of pain, which simply states that pain perception is controlled by a gate mechanism. If the gate is open, information regarding pain passes through and if it is closed this information cannot pass through. Consider that pain can be alleviated by simply rubbing the site of injury. In other words, competition from other sensations prevents the pain stimulus from passing through the gate. Or to phrase it differently, pain is alleviated when both large and small diameter fibers are carrying nerve impulses to SG cells.

SG cells release the neurotransmitter responsible for sending on the message that injury has occurred and pain is present. This transmitter substance has been labeled substance P (for pain). On occasion people have undergone serious bodily injury and yet report that they felt no pain. The potent pain reliever heroin has provided an answer to this circumstance. It turns out that neurons (and presumably SG cells also) have receptors for heroin because the body produces opioids, substances that act like opium, termed the endorphins and enkephalins. When these are received, the message that results in the sensation of pain is not transmitted. In other words, opioids prevent the release of substance P.

Endorphin and enkephalin research gives us an explanation for heroin addiction. Narcotic intake causes the CNS to stop producing its own natural opiates. Therefore the addict must take more and more of the drug in order to get the same effect. Furthermore, when the intake of heroin is stopped, release of neurotransmitters are no longer opposed, causing withdrawal symptoms. As the body gradually begins to produce opioids again, the individual returns to a normal state.

This explanation for the sensation of pain has proven to be fruitful. Aspirin has long been found to alleviate pain but now it is known that it interferes with prostaglandin production in tissues. Tylenol and other popular analgesics work in much the same way. Steroids also prevent prostaglandin production but at a later junction. This action has been misused—athletes who "break the pain barrier" by cortisone injections have damaged their muscles without feeling a thing.

The gate theory of pain has provided a basis for the use of TENS, transcutaneous electrical nerve stimulation. Stimulating electrodes are placed above the painful area and a mild current is applied. This produces a sensation of tingling or warmth that may relieve pain by increasing large fiber activity and closing the gate. Acupuncture, the technique of placing fine needles into the skin, most likely works similarly.

Poisons

Some poisons, such as strychnine, block inhibitory synapses in the spinal cord and brain stem. Other poisons, such as insecticides and nerve gas, inactivate AChE. The result is the same—increased excitability, convulsions, and possibly death. There are also poisons that have exactly the opposite effect. Botulism toxin decreases the amount of ACh in the synaptic cleft. Death from paralysis follows since, as discussed in the next chapter, ACh causes muscle contraction.

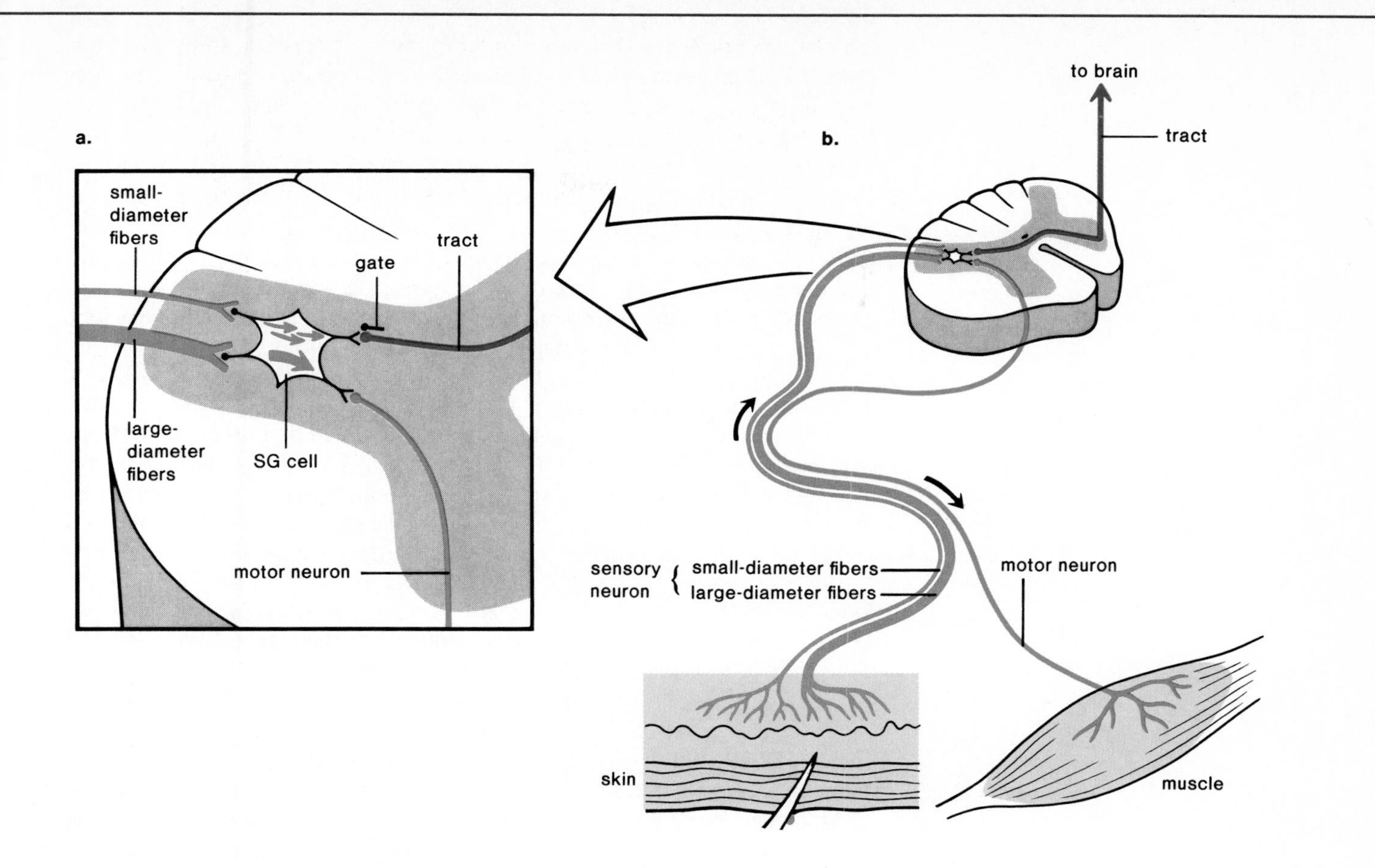

Some physicians believe that pain should be stopped at the second link in the chain—heroin should be made available for medicinal purposes. The Passionate Relief Act which is now before Congress would permit this use which is now prohibited. Physicians testify, however, that there are synthetic opiates that are just as effective as heroin.

No doubt pain has its psychological aspects and, not surprisingly, antidepressant drugs have been found to alleviate pain. These drugs most likely block the transmission of certain impulses and in this way promote the transmission of others; perhaps, the very ones that promote endorphin and/or enkephalin release. Those who engage in athletic activities report an elevation of mood and this has now been linked to the stimulation of fibers that bring about the release of the natural opiates.

How the sensation of pain begins. Figure a. *is an enlargement of a portion of figure* b. *Small diameter fibers are stimulated by bradykinin and prostaglandins if an injury occurs. The fibers lead to special SG cells which pass on the stimuli to tracts that travel to the cerebrum if a "gate" is open. When heroin or the natural opioids, endorphins and enkephalins, are received by SG cells, the "gate" closes and stimuli that lead to the sensation of pain are blocked.*

Depressants

Sedatives, including barbiturates, depress all nervous functions, acting first on the cortex and then on the rest of the brain, depending on the dose taken. They induce a sleeplike EEG. *Tranquilizers,* such as Librium and Valium, and more recently, *alcohol,* have been shown to enhance the action of the inhibitory transmitter GABA. Dependency develops when the body begins to produce less GABA. Consumption of a large amount of alcohol within a short period of time can cause death because of its depressing effect on brain functions. A habitual use of excess alcohol can cause damage to several areas of the brain, particularly the hippocampus, which results in memory impairment. The breakdown of alcohol in the liver leads to cirrhosis of the liver (p. 193) and other physiological side effects, eliminating alcohol as a drug of choice in treating anxiety.

Stimulants

Amphetamines have a structure similar to the excitatory transmitter NA and are believed to promote the synaptic release of dopamine and NA, thereby increasing the amount received by the postsynaptic membrane. Unpleasant side effects of amphetamines, including hallucinations, may be due to stimulant-induced insomnia. Cocaine blocks the uptake of NA and thereby increases the length of time it is present in the synaptic cleft, and this may account for some of its psychological effects. One study found, however, that cocaine users who took the drug under controlled laboratory conditions could not distinguish its effects from other drugs or even from a placebo. These investigators concluded that environmental factors may play a role in the sensation of euphoria produced by the drug.

There are two types of *antidepressants*. One type, represented by Elavil, is believed, like cocaine, to prevent the reabsorption of NA and the other excitatory neurotransmitters. Another type, represented by Parnate, inhibit the enzyme monoamine oxidase, which breaks down NA. In any case, both types of antidepressants increase the amount of NA in the synaptic cleft and thereby relieve depression. Recently it has been found that both types of antidepressants also block receptors for the neurotransmitter histamine.

Antipsychotics

There is a delicate balance of neurotransmitters at the cerebral synapses, and it logically follows that mental illness might be caused by an imbalance. (This does not mean that all mental illnesses are caused by neurotransmitter imbalance.) Drugs that can restore the normal balance alleviate the symptoms of mental illness. For example, it is possible that lithium is effective in treating manic-depressive symptoms because it blocks the release of NA from the presynaptic membrane and thereby controls the manic (euphoric) phase. When the manic phase is prevented from developing, the depression phase does not follow. Drugs, such as Thorazine, relieve the symptoms of schizophrenia, apparently because they bind to dopamine receptors, interfering with its normal action.

Hallucinogens

LSD (lysergic acid diethylamide) and mescaline, a chemical derived from the peyote plant, affect the action of serotonin on certain ARAS cells involved in vision and emotion. This explains their ability to produce hallucinations. "Bad trips" may be due to a concomitant (simultaneous) effect on dopamine.

The active ingredient in marijuana (tetrahydrocannabinol, or THC) causes hallucinations only when taken in large doses. In low doses, THC is like a mild sedative, acting as a hypnotic drug whose effect resembles the effect of alcohol and tranquilizers. The mode of action of marijuana is not yet fully understood although it does impair short-term memory and slows learning.

Narcotics

Opium and heroin bind to receptors meant for the body's opioids, the endorphins and enkephalins. As discussed in the reading on page 328, the opioids are believed to alleviate pain by preventing the release of a neurotransmitter, termed substance P, from certain sensory neurons in the region of the spinal cord. When substance P is released, pain is felt, and when substance P is not released, pain is not felt. Evidence also indicates that there are opioid receptors in neurons that travel from the spinal cord to the limbic system, and that stimulation of these can cause a feeling of pleasure. This explains why opium and heroin not only kill pain but also produce a feeling of tranquility.

Drugs that affect the alertness or mood of the individual are most often either potentiating or depressing the effects of neurotransmitters at CNS synapses.

Drug Tolerance

Even when a drug does not gradually replace the natural substances produced by the body, an individual may find that ever-increasing consumption is needed to achieve the same effect. This may be related to drug breakdown. The liver produces enzymes to detoxify and break down drugs, preparing them for excretion. As more drugs, including alcohol, are consumed, more and more of these enzymes are produced. This explains, for example, the ability of a heavy drinker to consume more alcohol than others.

Drug Interaction

It is dangerous to take concomitantly two or more kinds of drugs that have the same effect on the CNS. At any one time, the liver contains only a certain quantity of detoxification enzymes. Therefore, if two drugs are taken in the usual dosage at the same time, the liver requires twice as long to detoxify them. In the meantime, the drugs cause a compound effect on the central nervous system. The consequences may be dire. For example, if alcohol and barbiturates are taken together, the total depressive effect on the CNS can lead to coma and death.

Summary

The cell bodies of nerve cells are found in the CNS and ganglia. Axons and dendrites make up nerves. The nerve impulse is a change in permeability of the membrane so that sodium ions move to the inside of a neuron and potassium ions move to the outside. The nerve impulse is transmitted across the synapse by neurotransmitter substances.

During a spinal reflex, a sensory neuron transmits nerve impulses from a receptor to an interneuron, which in turn transmits impulses to a motor neuron, which conducts them to an effector. Reflexes are automatic and some do not require involvement of the brain.

Long fibers of sensory and/or motor neurons make up cranial and spinal nerves of the somatic and autonomic divisions of the PNS. While the somatic division controls skeletal muscle the autonomic controls smooth muscle and internal organs.

The CNS consists of the spinal cord and brain. Only the cerebrum is responsible for consciousness and the other portions of the brain have their own specific functions. The cerebrum can be mapped, and each lobe seems to also have particular functions. Neurological drugs, although quite varied, have been found to affect the ARAS and limbic system by either promoting or preventing the action of neurotransmitters.

Objective Questions

1. An ______________ carries nerve impulses away from the cell body.
2. During the upswing of the action potential, ______________ ions are moving to the ______________ of the nerve fiber.
3. The space between the axon of one neuron and the dendrite of another is called the ______________ .
4. ACh is broken down by the enzyme ______________ after it has altered the permeability of the postsynaptic membrane.
5. Motor nerves innervate ______________ .
6. The vagus nerve is a ______________ nerve that controls ______________ .
7. In a reflex arc only the ______________ is completely within the CNS.
8. The brain and cord are covered by protective layers called ______________ .
9. The ______________ is that part of the brain that allows us to be conscious.
10. The ______________ is the part of the brain responsible for coordination of body movements.

Answers to Objective Questions

1. axon 2. sodium, inside 3. synaptic cleft 4. acetylcholinesterase (AChE) 5. muscles 6. either cranial, motor or parasympathetic, internal organs 7. interneuron 8. meninges 9. cerebrum 10. cerebellum

Study Questions

1. What are the two main divisions of the nervous system? How are these divisions subdivided? (p. 309)
2. What are the three types of neurons? How are they similar and how are they different? (p. 310)
3. What does *resting potential* mean and how is it brought about? (p. 312) Describe the two parts of an action potential and the changes that may be associated with each part. (p. 313)
4. What is the sodium/potassium pump and when is it active? (pp. 69 and 312)
5. What is a neurotransmitter substance; where is it stored; how does it function; and how is it destroyed? Name two well-known neurotransmitters. (p. 314) What is *summation?* (p. 314)
6. What are the three types of nerves and how are they anatomically different? Functionally different? Distinguish between cranial and spinal nerves. (p. 316)
7. Trace the path of a reflex action after discussing the structure and function of the spinal cord and spinal nerve. (p. 318)
8. What is the autonomic nervous system and what are its two major divisions? (p. 319) Give several similarities and differences between these divisions. (pp. 319–321)
9. Name the major parts of the brain and give a function for each. (pp. 322–324)
10. Describe the EEG and discuss its importance. (p. 325)
11. Name the various categories of drugs that affect the CNS. How does each type of drug affect transmission across the synapse? (pp. 327–331)

Key Terms

axon (ak'son) process of a neuron that conducts nerve impulses away from the cell body. *310*

cell body (sel bod'e) portion of a nerve cell that includes a cytoplasmic mass and a nucleus and from which the nerve fibers extend. *310*

central nervous system (sen'tral ner'vus sis'tem) CNS; the brain and spinal cord in vertebrate animals. *309*

cerebral hemisphere (ser'ĕ-bral hem'ĭ-sfēr) one of the large, paired structures that together constitute the cerebrum of the brain. *324*

dendrite (den'drīt) process of a neuron, typically branched, that conducts nerve impulses toward the cell body. *310*

effector (ē-fek'tor) a structure such as the muscles and glands that allows an organism to respond to environmental stimuli. *318*

ganglion (gang'gle-on) a collection of neuron cell bodies outside the central nervous system. *316*

innervate (in'er-vāt) to activate an organ, muscle, or gland by motor neuron stimulation. *310*

interneuron (in''ter-nu'ron) a neuron that is found within the central nervous system and takes nerve impulses from one portion of the system to another. *310*

motor neuron (mo'tor nu'ron) a neuron that takes nerve impulses from the central nervous system to the effectors. *310*

myelin sheath (mi'ĕ-lin shēth) the fatty cell membranes that cover long neuron fibers and give them a white, glistening appearance. *310*

nerve impulse (nerv im'puls) an electrochemical change due to increased membrane permeability that is propagated along a neuron from the dendrite to the axon following excitation. *311*

neurotransmitter substance (nu''ro-trans-mit'er sub'stans) a chemical made at the ends of axons that is responsible for transmission across a synapse. *314*

parasympathetic nervous system (par''ah-sim''pah-thet'ik ner'vus sis'tem) that part of the autonomic nervous system that usually promotes those activities associated with a normal state. *321*

peripheral nervous system (pĕ-rif'er-al ner'vus sis'tem) PNS; nerves and ganglia that lie outside the central nervous system. *309*

receptor (re-sep'tor) a sense organ specialized to receive information from the environment. *318*

sensory neuron (sen'so-re nu'ron) a neuron that takes nerve impulses to the central nervous system; afferent neuron. *310*

somatic nervous system (so-mat'ik ner'vus sis'tem) that portion of the PNS containing motor neurons that control skeletal muscles. *318*

sympathetic nervous system (sim''pah-thet'ik ner'vus sis'tem) that part of the autonomic nervous system that usually causes effects associated with emergency situations. *319*

synapse (sin'aps) the region between two nerve cells where the nerve impulse is transmitted from one to the other; usually from axon to dendrite. *314*

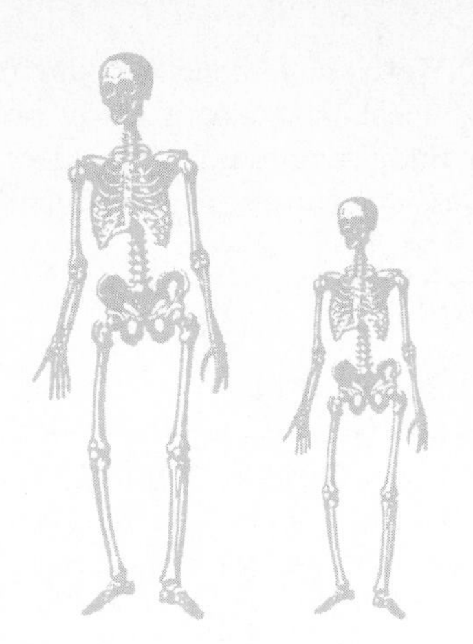

16

Musculoskeletal System

Chapter Concepts

1 The skeleton, which contributes greatly to our general appearance, has various functions and is divided into the axial and appendicular skeletons.

2 Macroscopically, skeletal muscles work in antagonistic pairs and exhibit certain physiological characteristics due to the composition of their muscle fiber components.

3 Microscopically, muscle fiber contraction is dependent on actin and myosin filaments and a ready supply of Ca^{++} and ATP.

Chapter Outline

Skeleton
- Functions
- Structure

Skeletal Muscles
- Macroscopic Anatomy and Physiology
- Microscopic Anatomy and Physiology

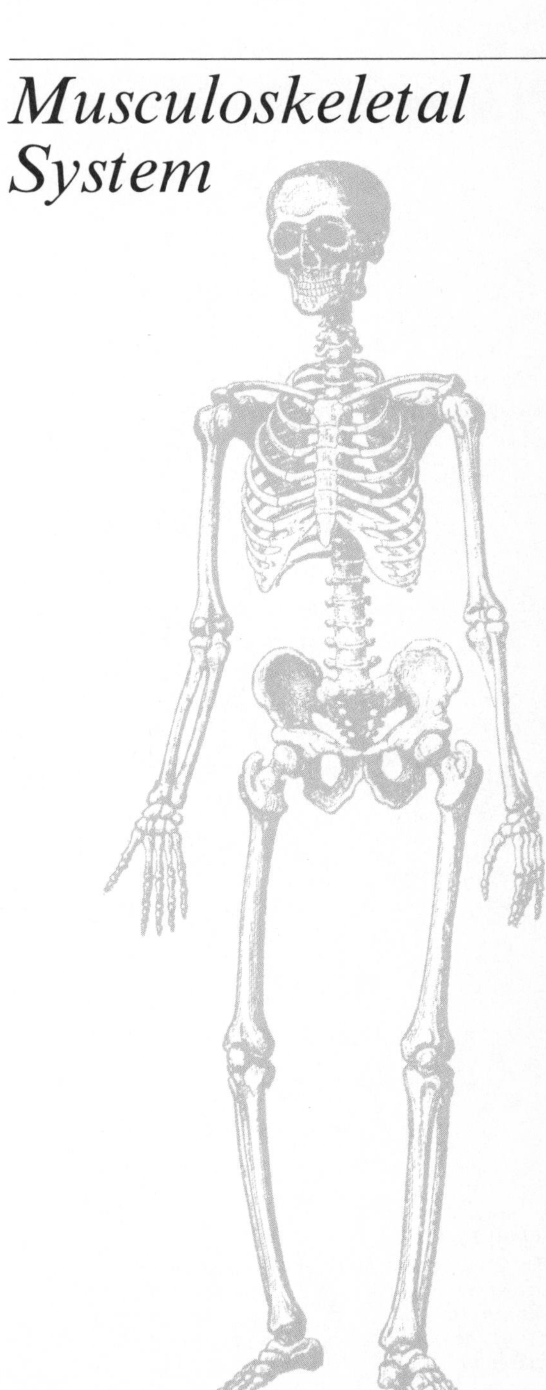

Figure 16.1
When muscles contract due to nervous stimulation, the bones move, producing even the graceful motions of ballet dancers.

Muscles and bones working together allow humans to perform many mechanical tasks, some of which require grace and agility (fig. 16.1). Body weight and appearance are largely accounted for by these organs whose structure suits their functions as discussed in the following.

Skeleton

Functions

The skeleton (fig. 16.2), notably the large heavy bones of the legs, supports the body against the pull of gravity. The skeleton also protects soft body parts. For example, the skull forms a protective encasement for the brain, as does the rib cage for the heart and lungs. Flat bones, such as those of the skull, ribs, and breastbone, produce red blood cells in both adults and children. All bones are a storage area for inorganic calcium and phosphorus salts. Bones also provide sites for muscle attachment. The long bones, particularly those of the legs and arms, permit flexible body movement.

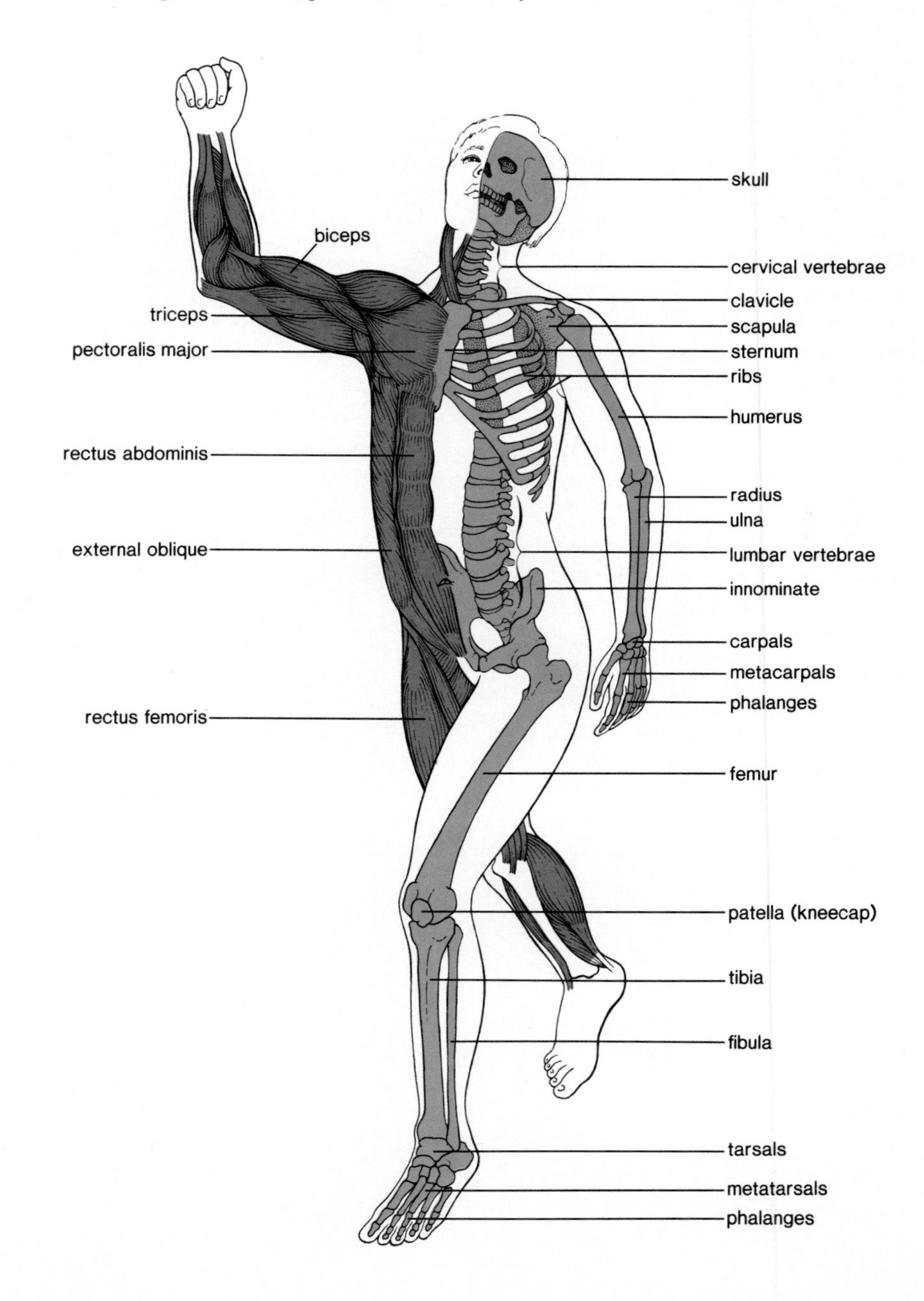

Figure 16.2
Major bones and muscles of the human body. The axial skeleton, composed of the skull, vertebral column, sternum, and ribs, lies in the midline; the rest of the bones belong to the appendicular skeleton.

The skeleton not only permits flexible movement, it also supports and protects the body, produces red blood cells, and serves as a storehouse for certain inorganic salts.

Structure

The skeleton may be divided into two parts: the **axial skeleton** and the **appendicular skeleton.**

Axial Skeleton

The **skull,** or cranium, is composed of many bones fitted tightly together in adults. In newborns, certain bones are not completely formed and instead are joined by membranous regions called **fontanels,** all of which usually close by the age of 16 months. The bones of the skull contain the **sinuses,** air spaces lined by mucous membrane. Two of these, called the mastoid sinuses, drain into the middle ear. Mastoiditis, a condition that can lead to deafness, is an inflammation of these sinuses. Whereas the skull protects the brain, the several bones of the face join together to support and protect the special sense organs and to form the jawbones.

The **vertebral column** extends from the skull to the pelvis and forms a dorsal backbone that protects the spinal cord (fig. 15.13). Normally, the vertebral column has four curvatures that provide more resiliency and strength than a straight column could. It is composed of many parts, called **vertebrae,** that are held together by bony facets, muscles, and strong ligaments. The vertebrae are named according to their location in the body (fig. 16.3).

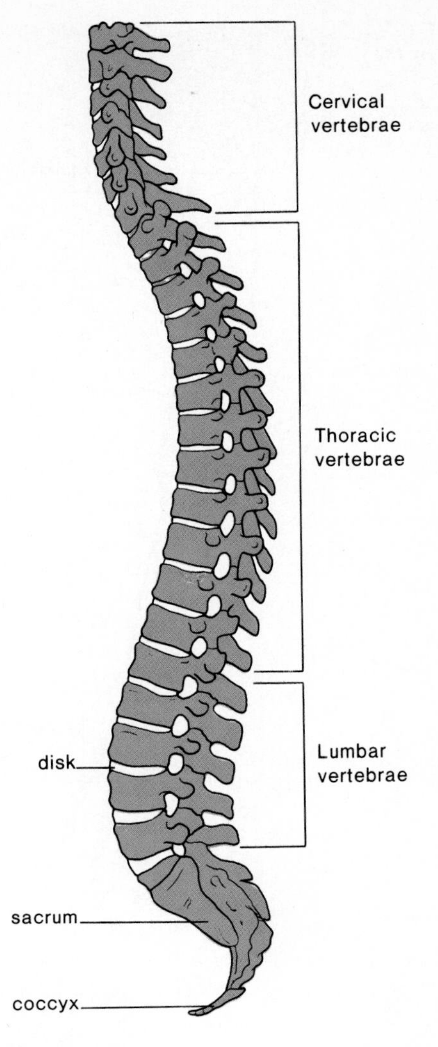

Figure 16.3
The vertebral column. The vertebrae are named according to their location in the column, which is flexible due to the presence of disks between the vertebrae. Note the presence of the coccyx, the vestigial "tailbone."

There are disks between the vertebrae that act as a kind of padding. They prevent the vertebrae from grinding against one another and also absorb shock caused by movements such as running, jumping, and even walking. Unfortunately, these disks become weakened with age and can slip, or even rupture. This causes pain when the damaged disk presses up against the spinal cord and/or spinal nerves. The body may heal itself, or else the disk can be removed surgically. If the latter occurs, the vertebrae can be fused together but this will limit the flexibility of the body. The presence of the disks allows motion between the vertebrae so that we can bend forward, backward, and from side to side.

The vertebral column, directly or indirectly, serves as an anchor for all the other bones of the skeleton (fig. 16.2). All the 12 pairs of **ribs** connect directly to the thoracic vertebrae in the back, and all but two pairs connect either directly or indirectly via shafts of cartilage to the **sternum** (breastbone) in the front. The lower two pairs of ribs are called "floating ribs" because they do not attach to the sternum.

Appendicular Skeleton

The appendicular skeleton consists of the bones within the pectoral and pelvic girdles and the attached appendages. The pectoral (shoulder) girdle and appendages (arms and hands) are specialized for flexibility, but the pelvic girdle (hip bones) and appendages (legs and feet) are specialized for strength.

The components of the **pectoral girdle** (fig. 16.4) are loosely linked together by ligaments rather than firm joints. Each **clavicle** (collar bone) connects with the sternum in front and the **scapula** (shoulder blade) behind, but the scapula is freely movable and held in place by muscles. This allows it to freely follow the movements of the arm. The single long bone in the upper arm (fig. 16.4), the **humerus,** has a smoothly rounded head that fits into a socket of the scapula. The socket, however, is very shallow and much smaller than the head. Although this means that the arm can move in almost any

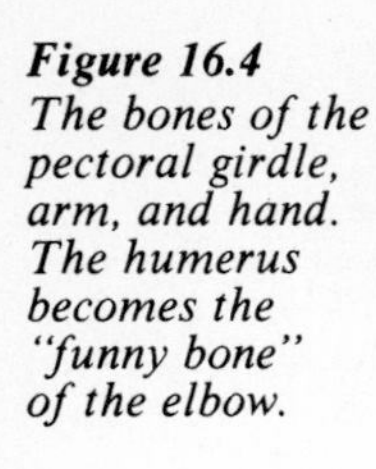

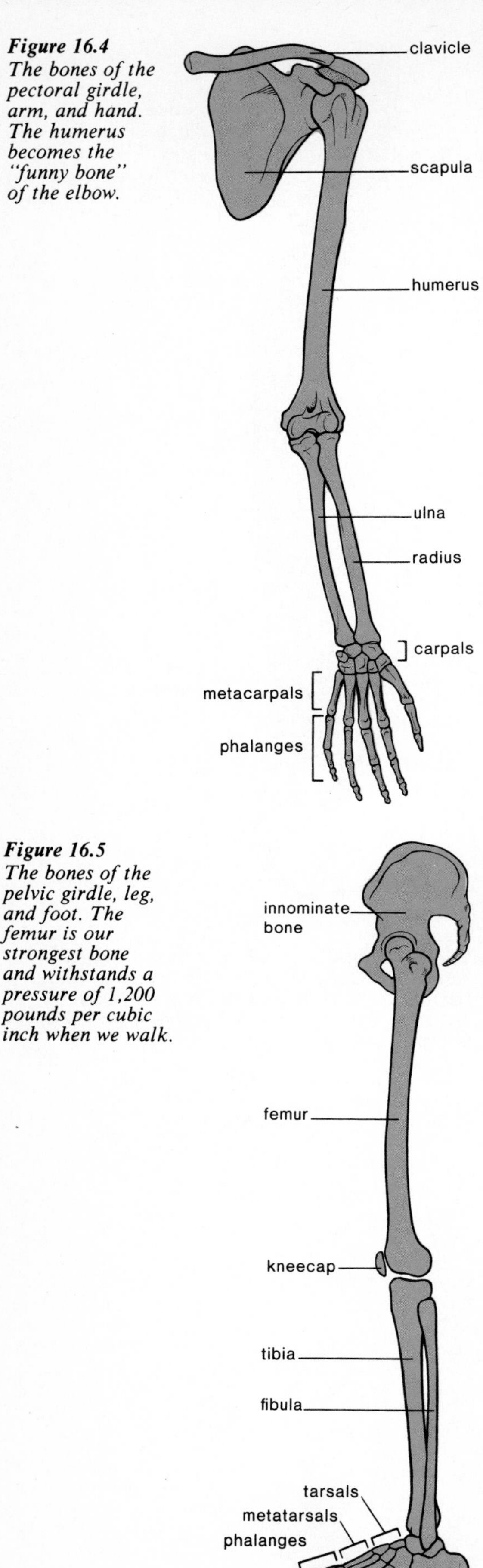

Figure 16.4
The bones of the pectoral girdle, arm, and hand. The humerus becomes the "funny bone" of the elbow.

Figure 16.5
The bones of the pelvic girdle, leg, and foot. The femur is our strongest bone and withstands a pressure of 1,200 pounds per cubic inch when we walk.

direction, there is little stability. Therefore this is the joint that is most apt to dislocate. The opposite end of the humerus meets the two bones of the lower arm, the **ulna** and the **radius,** at the elbow. (The prominent bone in the elbow is the topmost part of the ulna.) When the arm is held so that the palm is turned frontward, the radius and ulna are about parallel to one another. When the arm is turned so that the palm is next to the body, the radius crosses in front of the ulna, a feature that contributes to the easy twisting motion of the lower arm.

The many bones of the hand increase its flexibility. The wrist has eight **carpal** bones which look like small pebbles. From these, five **metacarpal** bones fan out to form a framework for the palm. The metacarpal bone that leads to the thumb is placed in such a way that the thumb can reach out and touch the other digits. (**Digits** is a term that refers to either fingers or toes.) Beyond the metacarpals are the **phalanges,** the bones of the fingers and the thumb. The phalanges of the hand are long, slender, and lightweight.

The **pelvic girdle** (fig. 16.5) consists of two heavy, large **innominate** hip-bones. The innominate bones are anchored to the sacrum and together these bones form a hollow cavity, the pelvis. The weight of the body is transmitted through the pelvis to the legs and then onto the ground. The largest bone in the body is the **femur,** or thigh bone. Although the femur is a strong bone, it is doubtful that the femurs of a fairy-tale giant could support the increase in weight. If a giant were 10 times taller than an ordinary human being, he would also be about 10 times wider and thicker, making him weigh about one thousand times as much. This amount of weight would break even giant-size femurs.

In the lower leg the larger of the two bones, the **tibia** (fig. 16.5), has a ridge we call the shin. Both of the bones of the lower leg have a prominence that contributes to the ankle—the tibia on the inside of the ankle and the **fibula** on the outside of the ankle. Although there are seven **tarsal** bones in the ankle, only one receives the weight and passes it on to the heel and the ball of the foot. If one wears high-heel shoes the weight is thrown even further toward the front of the foot. The **metatarsal** bones form the arches of the foot. There is a longitudinal arch from the heel to the toes and a transverse arch across the foot. These provide a stable, springy base for the body. If the tissues that bind the metatarsals together become weakened, flatfeet are apt to result. The bones of the toes are called *phalanges,* just like those of the fingers, but in the foot the phalanges are stout and extremely sturdy.

The axial and appendicular skeletons contain the bones that are listed in table 16.1.

Long Bones

A long bone, such as the femur, illustrates principles of bone anatomy. When the bone is split open, as in figure 16.6, the longitudinal section shows that it is not solid but has a cavity bounded at the sides by compact bone and at the ends by spongy bone. Beyond the spongy bone there is a thin shell of compact bone and finally a layer of cartilage.

Compact bone, as discussed on page 170, contains bone cells in tiny chambers called lacunae, and are arranged in concentric circles around Haversian canals, containing blood vessels and nerves. The lacunae are separated by a matrix that contains protein fibers of collagen and mineral deposits, primarily calcium and phosphorus salts.

Spongy bone contains numerous bony bars and plates separated by irregular spaces. Although lighter than compact bone, spongy bone is still designed for strength. Just as braces are used for support in buildings, the solid

Table 16.1 Bones of Skeleton

Part	Bones
Axial skeleton	Skull, Vertebral Column, Sternum, Ribs
Appendicular skeleton	
Pectoral girdle	Clavicle, Scapula
Arm	Humerus, Ulna, Radius
Hand	Carpals, Metacarpals, Phalanges
Pelvic girdle	Innominate
Leg	Femur, Tibia, Fibula
Foot	Tarsals, Metatarsals, Phalanges

portions of spongy bone follow lines of stress. The spaces in spongy bone are often filled with **red marrow,** a specialized tissue that produces blood cells. The cavity of a long bone usually contains **yellow marrow,** which is a fat-storage tissue.

Bones are not inert. They contain cells and perform various functions, including production of red blood cells.

Growth and Development of Bones Most of the bones of the skeleton are cartilaginous during prenatal development. Later, bone-forming cells known as **osteoblasts** replace cartilage with bone. At first, there is only a primary ossification center at the middle of a long bone but later, secondary centers form at the ends of the bones. There remains a *cartilaginous disk* between the primary ossification center and each secondary center. The length of a bone is dependent on how long the cartilage cells within the disk continue to divide. Eventually, though, the disks disappear and the bone stops growing as the individual attains adult height.

In the adult, bone is continually being broken down and then built up again. Bone-absorbing cells, called **osteoclasts,** are derived from cells carried in the bloodstream. As they break down bone, they remove worn cells and deposit calcium in the blood. Apparently, after about three weeks, they disappear. The destruction caused by the work of osteoclasts is repaired by **osteoblasts.** As they form new bone, they take calcium from the blood. Eventually some of these cells get caught in the matrix they secrete and are converted to **osteocytes,** the cells found within Haversian systems, page 170.

Because of continual renewal, the thickness of bones can change, depending on the amount of physical activity or a change in certain hormone balances (see chapter 18). In most adults, the bones become weaker due to a loss of mineral content. Strange as it may seem, adults seem to require more calcium in the diet than children in order to promote the work of osteoblasts. Many older women, due to a lack of estrogen, suffer from *osteoporosis,* a condition in which weak and thin bones cause aches and pains and fracture easily. A tendency toward osteoporosis may also be augmented by lack of exercise and too little calcium in the diet.

one is a living tissue and it is always being rejuvenated.

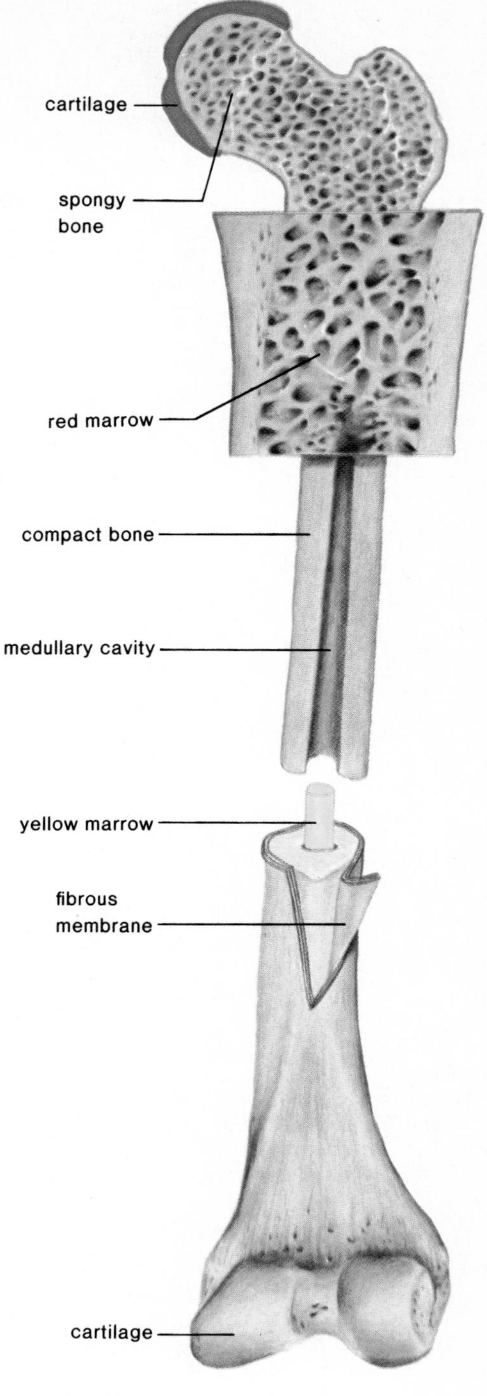

Figure 16.6
Anatomy of a long bone. A long bone is encased by fibrous membrane except where it is covered by articular cartilage at the ends. The central shaft is composed of compact bone, but the ends are spongy bone, which can contain red marrow. A central medullary cavity contains yellow marrow.

Joints

Bones are linked at the joints, which are often classified according to the amount of movement they allow. Some bones, such as those that make up the cranium, are sutured together and are *immovable.* Other joints are *slightly movable,* such as the joints between the vertebrae. The vertebrae are separated by disks, described earlier, that increase their flexibility. Similarly, the two hipbones are slightly movable where they are ventrally joined by cartilage. Owing to hormonal changes, this joint becomes more flexible during late pregnancy, which allows the pelvis to expand during childbirth. Most joints are *freely movable* **synovial joints** in which the two bones are separated by a cavity. **Ligaments** are composed of fibrous connective tissue that binds the two bones to one another, holding them in place as they form a capsule. In a "double-jointed" individual, the ligaments are unusually loose. The joint capsule is lined by synovial membrane which produces *synovial fluid,* a lubricant for the joint. The knee is an example of a synovial joint (fig. 16.7). In the knee, as in other freely movable joints, the bones are capped by cartilage, but in the knee there are also crescent-shaped pieces of cartilage between the bones, called **menisci.** These give added stability, helping to support the weight placed on the knee joint. Unfortunately, as discussed in the reading on page 340, athletes often suffer injury of the menisci, known as torn cartilage. The knee joint also contains 13 fluid-filled sacs called bursae which ease friction between tendons and ligaments and between tendons and bones. Inflammation of bursae is called bursitis. Tennis elbow is a form of bursitis.

There are different types of movable joints. The knee and elbow joints are *hinge joints* because, like a hinged door, they largely permit movement in one direction only. More movable are the ball-and-socket joints; for example, the ball of the femur fits into a socket on the hipbone. *Ball-and-socket joints* allow movement in all planes and even a rotational movement.

Synovial joints are subject to *arthritis.* In rheumatoid arthritis, the synovial membrane becomes inflamed and grows thicker. Degenerative changes take place that make the joint almost immovable and painful to use. There is

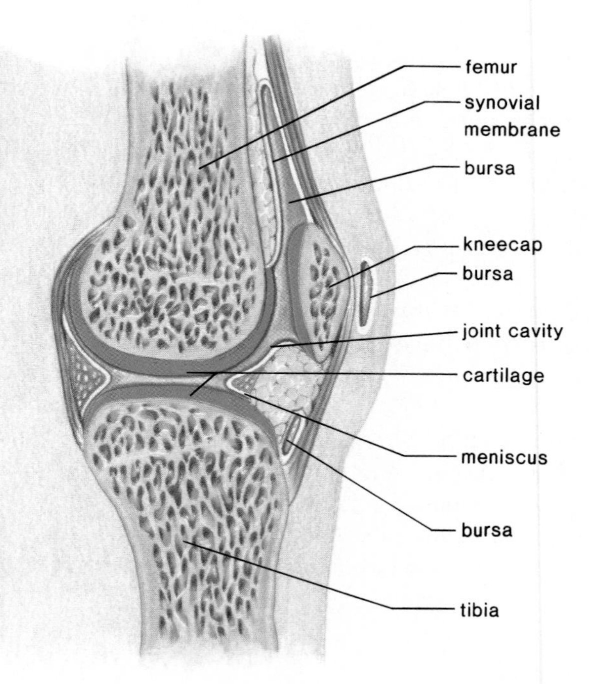

Figure 16.7
The knee joint, an example of a freely movable synovial joint. Notice that there is a cavity between the bones that is encased by ligaments and lined by synovial membrane. The kneecap protects the joint.

evidence that these effects are brought on by an autoimmune reaction. In old-age arthritis, or osteoarthritis, the cartilage at the ends of the bones disintegrates so that the two bones become rough and irregular. This type of arthritis is apt to affect the joints that have received the greatest use over the years.

---

*J*oints are classified according to the degree of movement. Some joints are immovable, some are slightly movable, and some are movable.

Skeletal Muscles

Muscles are effectors, which enable the organism to respond to a stimulus (p. 318). Skeletal muscles are attached to the skeleton, and their contraction accounts for voluntary movements. Involuntary muscles, both smooth and cardiac, were discussed previously on page 172. Here we will divide our discussion of skeletal muscle into macroscopic anatomy and physiology and microscopic anatomy and physiology.

Macroscopic Anatomy and Physiology

Muscles are typically attached to bone by **tendons** made of fibrous connective tissue. Tendons most often attach muscles to the far side of a joint so that the muscle extends across the joint (fig. 16.8). When the central portion of the muscle, called the belly, contracts, one bone remains fairly stationary and the other one moves. The **origin** of the muscle is on the stationary bone, and the **insertion** of the muscle is on the bone that moves.

When a muscle contracts, it shortens. Therefore, muscles can only pull; they cannot push. Because we need to both extend and flex at a joint, muscles generally work in *antagonistic pairs*. For example, the biceps and triceps are a pair of muscles that move the lower arm up and down (fig. 16.8). When the *biceps* contracts, the lower arm flexes and when the *triceps* contracts, the lower arm extends.

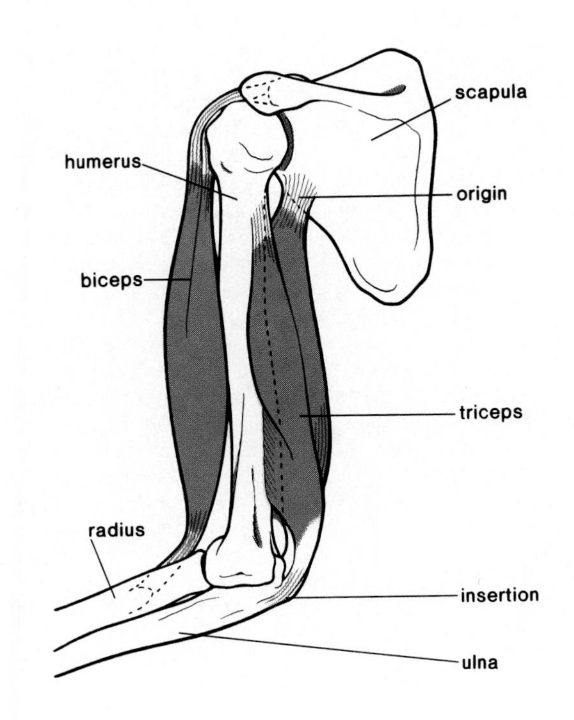

Figure 16.8
Attachment of skeletal muscles as exemplified by the biceps and triceps. The origin of a muscle remains stationary while the insertion moves. These muscles are antagonistic. When the biceps contracts, the lower arm is raised and when the triceps contracts, the lower arm is lowered.

The Knee

Every year surgeons remove some 52,000 pieces of cartilage from knees that were previously propelling someone down the gridiron. "Football isn't just a contact sport, it's a collision sport," says Edward Percy, an orthopedic surgeon and head of the University of Arizona's Sports Medicine program. Even so, less than half the knee injuries in football are caused by a direct blow to the knee. Most arise during the tangling of bodies that occurs in tackling or blocking. Twenty percent come about with no contact at all. . . .

The knee's vulnerability to sports injuries is not the result of some intrinsic flaw in its design. In fact, the knee is a nice compromise of mobility and stability. It can bend 150 degrees, swing from side to side, and twist on itself. And it can absorb the force, equal to nearly seven times an athlete's weight, that occurs as a tight end makes the final cut of a down-and-out pass pattern.

The knee is the meeting place of the two major leg bones: the thighbone, called the femur, and the shinbone, called the tibia. The bottom end of the thighbone is shaped like a baby's behind and coated with cartilage. The top of the lower leg bone has two shallow scoops hollowed out of it and a ridge along the middle. The femur rides along this ridge like a cowboy in a saddle. Wedges of cartilage line the top of the tibia, acting as shock absorbers and keeping the femur from rocking too much from side to side. . . .

Cartilage can be torn if it's twisted too tight. This can occur as a result of a collision or fall, but it can also happen if an athlete simply combines a lot of pressure with a twist. For example, a tennis player may lunge to return a volley, bending her knee to the inside, as if she were knock-kneed. This puts pressure on the cartilage between the outer parts of the femur and tibia. The action is not unlike that of a mortar and pestle: The cartilage is ground between the two bones.

Because cartilage is the only tissue in the human body that has no blood supply of its own, it often cannot get enough nourishment to repair itself after it is damaged. Once it's torn, it stays torn. Most doctors recommend removing all or at least the damaged part of it.

Five years ago, this type of operation meant a stay in the hospital, five weeks on crutches, weeks of therapy, and a scar along the inside or outside of the knee that impressed friends. A new instrument has become widely used, however, that shortens both recovery time and scars. Called an arthroscope, it enables some patients to be out of the hospital in as little as a day, spend only a few days on crutches, and begin running after two weeks of therapy.

Essentially composed of a tiny lens and light mounted on a pencil-thin shaft (and usually hooked up to a television camera), the arthroscope allows a surgeon to peer into the knee through a small slit in the skin. It also enables a surgeon to use similarly shaped knives, scissors, and other instruments to remove cartilage while leaving only a tiny scar. . . .

There is one injury to the knee, however, that can end a career if not immediately repaired: tearing a ligament. The knee would be nothing but a balancing act if not for the tough, fibrous ligaments that grip the bones like leather straps on a hinge and hold it steady. . . .

There are five ligaments around the knee and two more buried deep within the joint. Most knee sprains involve the ligaments at the inside of the knee because collisions usually occur at the outside of the leg. If the blow is hard enough, the ligament will tear. . . .

The sooner the severed ligament is repaired the better. The fluid in the knee will break down tissue, and if the injury is not repaired within 10 days, says Tab Blackburn, a physical therapist and trainer at the Rehabilitation Services of Columbus, Inc., in Georgia, "the fluids in the knee will turn the ligament to mush."

The front-most ligament, which runs from the front of the shinbone to the back of the thighbone, is particularly prone to injury. Doctors are not quite sure why, but this ligament, called the anterior cruciate, is stretched or torn in nearly 70 percent of all serious knee injuries. Unfortunately, the ligament also has a poor blood supply and usually does not heal, even if it is sewn back together.

Some athletes can continue to play without this ligament. It depends on the job the knee has to do. "A football player who has built up the muscles and other ligaments around his knee might not have any difficulty coming back," says Blackburn, "but a wiry gymnast with a lot of flexibility in her knees may have trouble." If a knee remains unstable, one solution is to jerry-rig a knee with new "ligaments" fashioned out of pieces of other tissue. A piece of tendon from the thigh, for example, is sometimes cut out and attached to the bone around the knee with staples. Percy and his colleagues are experimenting with using collagen, a fibrous substance extracted from cattle hide and woven into a band to help bind the bones together.

Knee specialists blame overtraining for the majority of knee problems afflicting amateur athletes, particularly those who are just beginning an exercise program. An estimated 40 percent of all women runners, for example, have knee problems severe enough to require a doctor within the first three months of training. One problem—actually a hodgepodge of muscle, tendon, and bone irritation sometimes called runner's knee—affects nearly a third of the estimated 15 million joggers in the United States. During running each foot hits the ground 1,500 times a mile. Unlike walking, where both feet are on the ground 30 percent of the time, each knee must alternately bear the full load of the body hitting the pavement during running.

William F. Allman, "The Knee." Reprinted by permission from the November issue of Science 83. *Copyright © 1983 by the American Association for the Advancement of Science.*

Playing football often results in knee injuries.

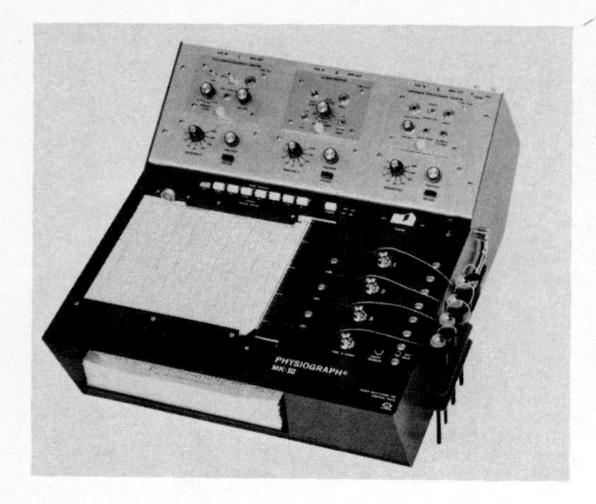

Figure 16.9
A physiograph. This apparatus can be used to record a myogram, a visual representation of the contraction of a muscle that has been dissected from an animal.

In the laboratory, it is possible to study the contraction of individual whole muscles. Customarily, a calf muscle is removed from a frog and fibers are mounted in an apparatus called a **physiograph** (fig. 16.9). The muscle is stimulated electrically, and when it contracts, it pulls on a lever. The lever's movement is recorded, and the resulting pattern is called a **myogram.**

All-or-None Response

Either a *single* muscle fiber (muscle cell, p. 172) responds to a stimulus and contracts, or it does not. At first, the stimulus may be so weak that no contraction occurs, but as soon as the strength of the stimulus reaches the *threshold stimulus,* the muscle fiber will contract completely. Therefore, a muscle fiber behaves in an **all-or-none manner.**

Contrary to that of an individual fiber, the strength of contraction of a whole muscle can increase according to the degree of stimulus beyond the *threshold stimulus.* A whole muscle contains many fibers, and the degree of contraction is dependent on the total number of fibers contracting. The maximum stimulus is the one beyond which the degree of contraction does not increase.

Muscle fibers obey the all-or-none law but whole muscles do not obey this law.

Muscle Twitch

If a muscle is placed in a physiograph and given a maximum stimulus, it will contract and then relax. This action—a single contraction that lasts only a fraction of a second—is called a muscle **twitch.** Figure 16.10a is a myogram of a twitch, which is customarily divided into a *latent period,* or the period of time between stimulation and initiation of contraction; the period of *contraction;* and the period of *relaxation.*

If a muscle is exposed to two maximum stimuli in quick succession, it will respond to the first but not to the second stimulus. This is because it takes an instant following a contraction for the muscle fibers to recover in order to respond to the next stimulus. The very brief moment following stimulation, during which a muscle remains unresponsive, is called the refractory period.

Summation and Tetanus

If a muscle is given a rapid series of threshold stimuli, it may respond to the next stimulus without relaxing completely. In this way, muscle tension **summates** until maximal sustained **tetanic contraction** is achieved (fig. 16.10b). The myogram no longer shows individual twitches; rather, they are completely fused and blended into a straight line. Tetanic contraction continues until the muscle fatigues, due to depletion of energy reserves. **Fatigue** is apparent when a muscle relaxes, even though stimulation is continued.

Tetanic contractions occur whenever skeletal muscles are being actively used. Ordinarily, however, only a portion of any particular muscle is involved—while some fibers are contracting others are relaxing. Because of this, intact muscles rarely fatigue completely.

Muscle twitch, summation, and tetanus are related to the frequency with which a muscle is stimulated.

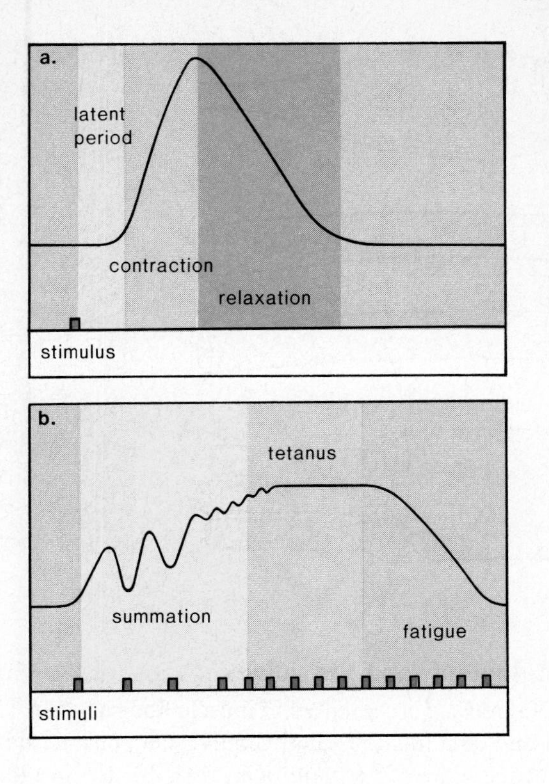

Figure 16.10
Physiology of muscle contraction. a. *Simple muscle twitch is composed of three periods: latent, contraction, and relaxation.* b. *Summation and tetanic contraction. When a muscle is not allowed to relax completely between stimuli, the contractions increase in size and then the muscle remains contracted until it fatigues.*

Muscle Tone

Intact skeletal muscles also have **tone,** a condition in which there are always some fibers contracted. Muscle tone is particularly important in maintaining posture. If the muscles of the neck, trunk, and legs suddenly become relaxed, the body collapses.

The maintenance of the right amount of tone requires the use of special sense receptors called **muscle spindles** (fig. 16.11). A muscle spindle consists of a bundle of modified muscle fibers with sensory nerve fibers wrapped around a short, specialized region somewhere near the middle of their length. A spindle contracts along with muscle fibers, but thereafter it sends stimuli to the CNS that enable it to regulate muscle contraction so that tone is maintained.

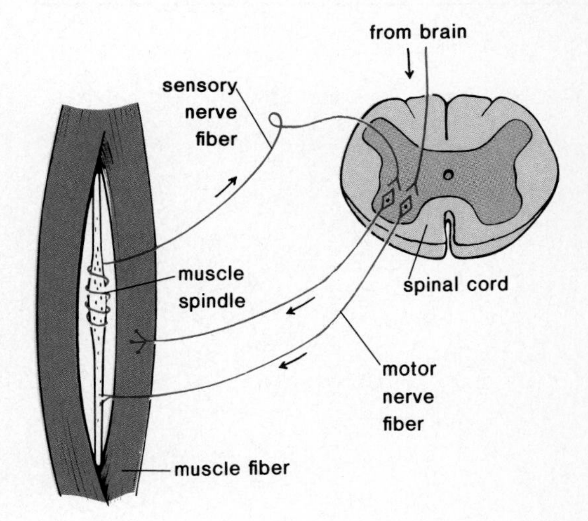

Figure 16.11
Muscle spindle structure and function. After associated muscle fibers contract, spindle fibers aid in the coordination of muscular contraction by sending sensory impulses to the central nervous system, which then directs only certain other muscle fibers to contract.

Effect of Contraction on Size of Muscle

As mentioned previously, forceful muscular activity over a prolonged period of time causes muscles to increase in size. This increase, called *hypertrophy,* occurs only if the muscle contracts to at least 75 percent of its maximum tension. However, only a few minutes of forceful exercise a day are required for hypertrophy to occur. The number of muscle fibers do not increase. Instead the size of each fiber increases. The fibers show a gain in metabolic potential as well as in the number of myofibrils. This means that the muscle can work longer before it gets tired. Some athletes take steroids, either testosterone or related chemicals, to promote muscle growth.

When muscles are not used or are used for only very weak contractions, they decrease in size, or atrophy. Atrophy can occur when a limb is placed in a cast or when the nerve serving a muscle is damaged. If nerve stimulation is not restored, the muscle fibers will gradually be replaced by fat and fibrous tissue. Unfortunately, atrophy causes the fibers to shorten progressively, leaving body parts in contorted positions.

Figure 16.12
Anatomy of a muscle as revealed by the light microscope. a. *Whole muscle.* b. *Muscle fibers.* c. *A single muscle fiber with a few exposed myofibrils.* d. *Light micrograph shows the striations characteristic of skeletal muscle. The striations are due to the light and dark banding pattern of myofibrils.*

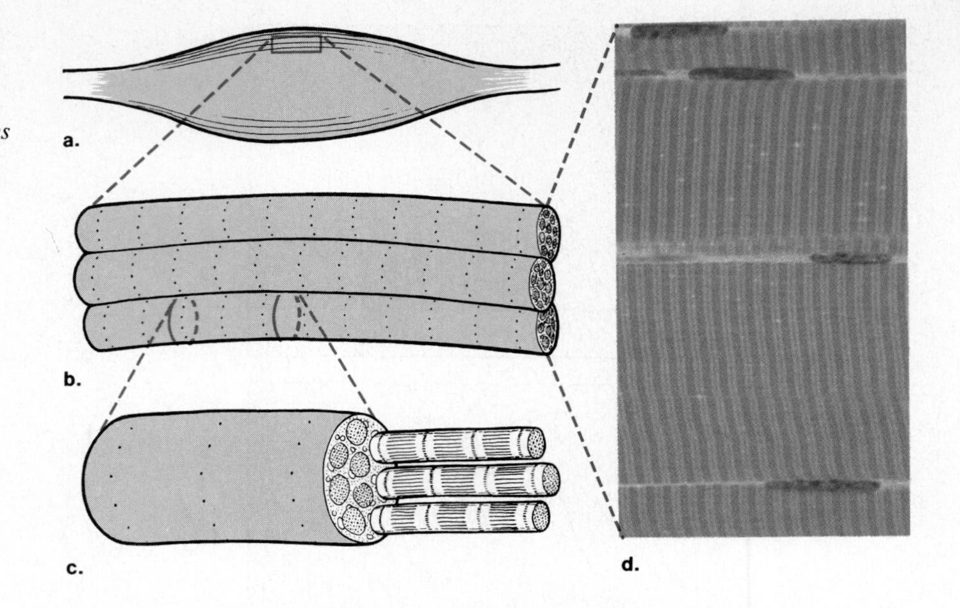

Table 16.2 Muscle Cells	
Component	**Term**
Cell membrane	Sarcolemma
Cytoplasm	Sarcoplasm
Endoplasmic reticulum	Sarcoplasmic reticulum

Microscopic Anatomy and Physiology

A whole skeletal muscle is composed of muscle fibers (fig. 16.12a and b). Each fiber is a cell and contains the usual cellular components, but special terminology has been assigned to certain ones, as indicated in table 16.2. Also, a muscle fiber has some unique anatomical characteristics. For one thing, the **sarcolemma,** or cell membrane, forms tubules that penetrate or dip down into the cell so that they come into contact with, but do not fuse with, expanded portions of modified ER, which is termed the **sarcoplasmic reticulum** in muscle cells. The tubules comprise the *T* (for transverse) *system* (fig. 16.13). The expanded portions of the sarcoplasmic reticulum, called *calcium-storage sacs,* contain calcium, Ca^{++}, an element that is essential to muscle contraction. The sarcoplasmic reticulum encases hundreds, and even sometimes thousands, of **myofibrils,** which are the contractile portions of the fibers. Myofibrils are cylindrical in shape and run the length of the fiber. The light microscope shows that the myofibrils have light and dark bands called striations (fig. 16.12c and 16.12d). It is the banding pattern of myofibrils that cause skeletal muscle to be striated. Electron micrographs (figs. 16.13 and 16.14a) have revealed that there are even areas of light and dark within the bands themselves. These areas can be studied in relation to a unit of a myofibril called a sarcomere.

Myofibrils are the contractile portions of muscle fibers.

Sarcomere Structure and Function

A sarcomere extends between two dark lines called the *Z lines* (fig. 16.14b). The *I band* is a light region that takes in the sides of two sarcomeres; therefore, an I band includes a Z line. The center dark region of a sarcomere is called the *A band.* The A band is interrupted by a light center, the *H zone,* and a fine, dark stripe called the *M line* cuts through the H zone.

These bands and zones relate to the placement of filaments within each sarcomere. A sarcomere contains two types of filaments, *thin filaments* and *thick filaments* (fig. 16.14b). The thin filaments are attached to a Z line, and the thick filaments are anchored by an M line. The I band is light because it contains only thin filaments; the A band is dark because it contains both thin and thick filaments, except at the center where, in the lighter H zone, only thick filaments are found.

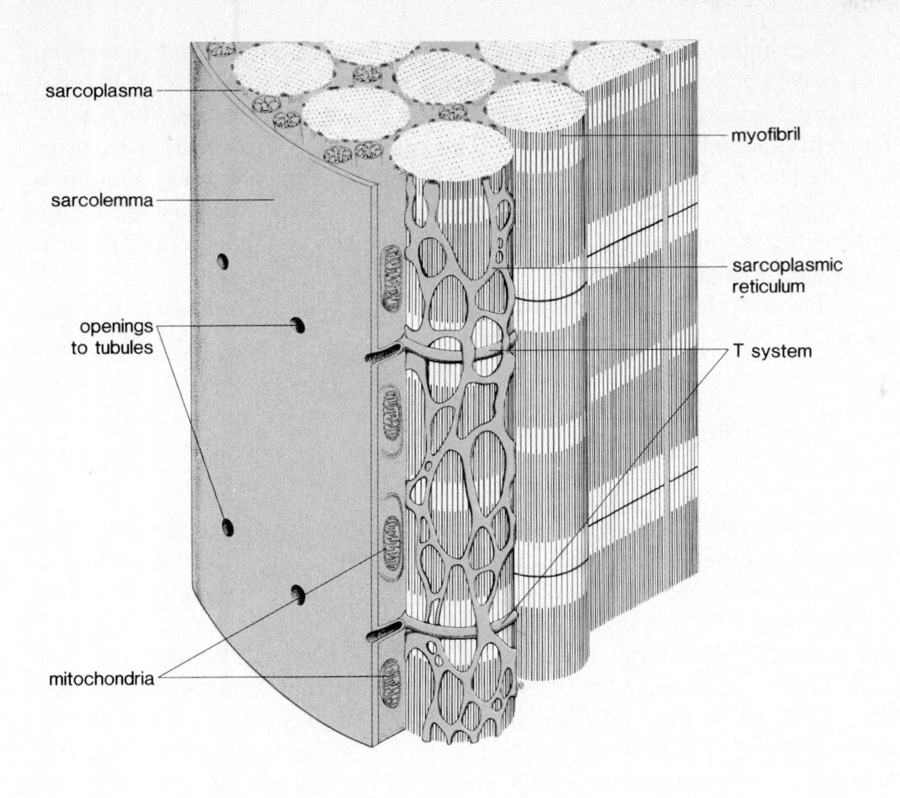

Figure 16.13
Anatomy of a muscle fiber as revealed by the electron microscope. A muscle fiber contains numerous myofibrils, each of which is enclosed by sarcoplasmic reticulum. The sarcolemma forms tubules that dip down and come in contact with the sarcoplasmic reticulum.

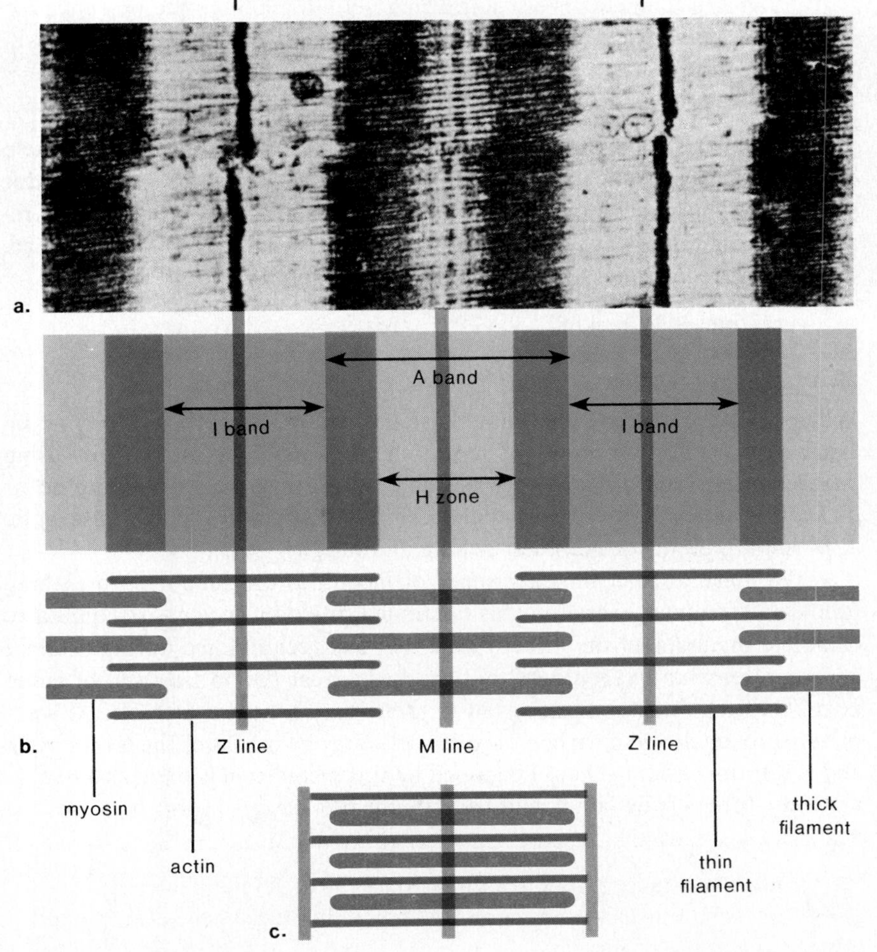

Figure 16.14
a. *Electron micrograph of a sarcomere showing the typical striations of skeletal muscle.* b. *The striations contain various bands and dark lines. The I band contains the Z line and thin filaments. The A band contains both thin and thick filaments except at the center where the H zone has only thick filaments anchored by the M line.* c. *Notice that the I band has decreased in size and H zone has disappeared in the contracted sarcomere because the thin filaments have moved to the center.*

Table 16.3 Contractile Elements	
Component	**Definition**
Myofibril	Muscle cell contractile subunit
Sarcomere	Functional unit of myofibril
Myosin	Thick filament
Actin	Thin filament

Sarcomere contraction is dependent on two proteins, **actin** and **myosin,** that make up the thin and thick filaments respectively (table 16.3). When a sarcomere contracts (fig. 16.14c), the actin filaments slide past the myosin filaments and approach one another. This causes the I band to become shorter and the H zone almost or completely to disappear. The movement of actin in relation to myosin is called the **sliding filament theory** of muscle contraction. During the sliding process, the sarcomere shortens, even though the filaments themselves remain the same length (fig. 16.15).

The overall formula for muscle contraction can be represented as follows:

$$\text{actin} + \text{myosin} \xrightarrow[\text{Ca}^{++}]{\text{ATP} \rightarrow \text{ADP} + \text{Ⓟ}} \text{actomyosin}$$

The participants in this reaction have the functions listed in table 16.4. Even though the actin filaments slide past the myosin filaments, it is the latter that does the work. In the presence of Ca^{++}, portions of the myosin filaments, called *cross bridges* (fig. 16.15b), reach out and attach to the actin filaments, pulling them along. Following attachment, ATP is broken down as detachment occurs. Myosin brings about ATP breakdown and therefore it is not only a structural protein, it is also an ATPase enzyme. The cross bridges attach and detach some 50 to 100 times as the thin filaments are pulled to the center of a sarcomere. If, by chance, more ATP molecules are not available, detachment cannot occur. This explains *rigor mortis,* permanent muscle contraction after death.

The sliding filament theory states that actin filaments slide past myosin filaments because myosin has cross bridges that pull the actin filaments inward.

It is obvious from our discussion that ATP provides the energy for muscle contraction. In order to assure a ready supply, muscle fibers contain **creatine phosphate** (phosphocreatine), a storage form of high-energy phosphate. Creatine phosphate does not participate directly in muscle contraction. Instead, it is used to regenerate ATP by the following reaction:

$$\text{creatine} \sim \text{P} + \text{ADP} \rightarrow \text{ATP} + \text{creatine}$$

Oxygen Debt

When all of the creatine phosphate has been depleted and there is no oxygen available for aerobic respiration, a muscle fiber can generate ATP by using anaerobic respiration (p. 108). Anaerobic respiration, which is apt to occur during strenuous exercise, can supply ATP for only a short time because lactic acid buildup produces muscular aching and fatigue.

We have all had the experience of having to continue deep breathing following strenuous exercise. This continued intake of oxygen is required to complete the metabolism of lactic acid that has accumulated during exercise and represents an **oxygen debt** that the body must pay to rid itself of lactic acid. The lactic acid is transported to the liver, where one-fifth of it is completely broken down to carbon dioxide and water by means of the Krebs cycle and respiratory chain. The ATP gained by this respiration is then used to convert four-fifths of the lactic acid back to glucose.

Muscle contraction requires a ready supply of ATP. Creatine phosphate is used to generate ATP rapidly. If oxygen is in limited supply, anaerobic respiration produces ATP and results in oxygen debt.

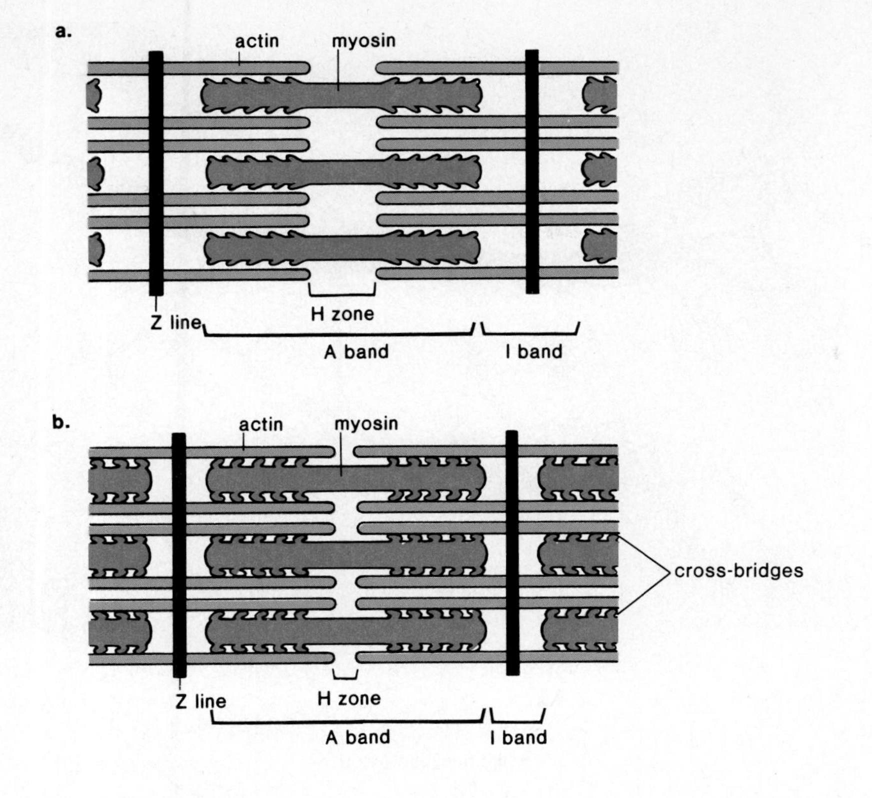

Figure 16.15
Sliding filament theory. a. *Relaxed sarcomere.* b. *Contracted sarcomere. Note that during contraction, the I band and H zone decrease in size. This indicates that the thin filaments slide past the thick filaments. Even so, the thick filaments do the work by pulling the thin filaments by means of cross bridges.*

***Table 16.4** Muscle Contraction*

Name	Function
Actin filaments	Slide past myosin causing contraction
Ca^{++}	Needed for actin to bind to myosin
Myosin filaments	a. Enzyme that splits ATP b. Pulls actin by means of cross bridges
ATP	Supplies energy for bonding between actin and myosin = actomyosin

Muscle Fiber Contraction and Relaxation

Nerves innervate muscles, and nerve impulses cause muscles to contract. A motor axon within a nerve sends branches to several muscle fibers, which collectively are termed a *motor unit*. Each branch has terminal knobs that contain synaptic vesicles filled with the neuromuscular transmitter acetylcholine (ACh). A terminal end knob lies in close proximity to the sarcolemma of the muscle fiber. This region, called a **neuromuscular junction** (fig. 16.16) has the same components as a synapse: a presynaptic membrane, a synaptic cleft, and a postsynaptic membrane. Only in this case, the postsynaptic membrane is a portion of the sarcolemma of a muscle fiber. The sarcolemma, just like a neural membrane, is polarized; the inside is negatively charged and the outside is positively charged.

Nerve impulses cause synaptic vesicles to merge with the presynaptic membrane and release ACh into the synaptic cleft. When ACh reaches the sarcolemma, it is depolarized. The result is a **muscle action potential** that spreads over the sarcolemma and down the T system (fig. 16.13) to where calcium ions are stored in calcium-storage sacs of the sarcoplasmic reticulum. When the action potential reaches a sac, Ca^{++} ions are released and diffuse into the sarcoplasm, after which they attach to the thin filaments. A thin filament is a twisted double strand of globular actin molecules. Associated with the actin filaments are two other proteins: *tropomyosin* forms threads that twist about the actin filaments, and *troponin* is located at intervals along the tropomyosin. The calcium combines with the troponin and this causes the tropomyosin threads to shift in position so that myosin binding sites are exposed (fig. 16.17).

Figure 16.16
Muscle fiber innervation. a. *Micrograph of a motor unit showing several terminal end knobs.* b. *A neuromuscular junction occurs where a terminal end knob comes in close proximity to a muscle fiber.* c. *The knob contains synaptic vesicles filled with ACh. When these vesicles fuse with the presynaptic membrane, ACh diffuses across the synaptic cleft to initiate a muscle action potential.*

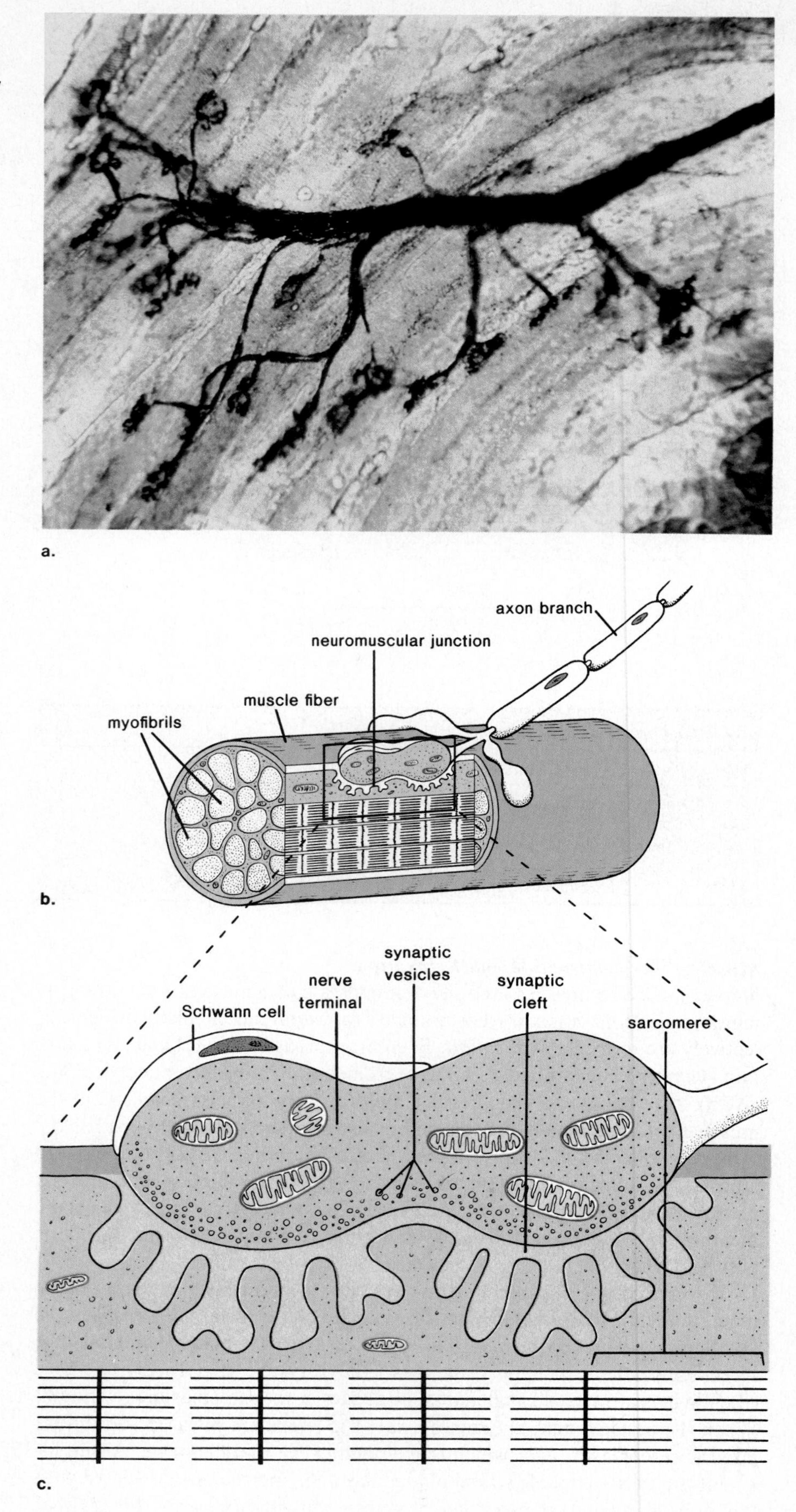

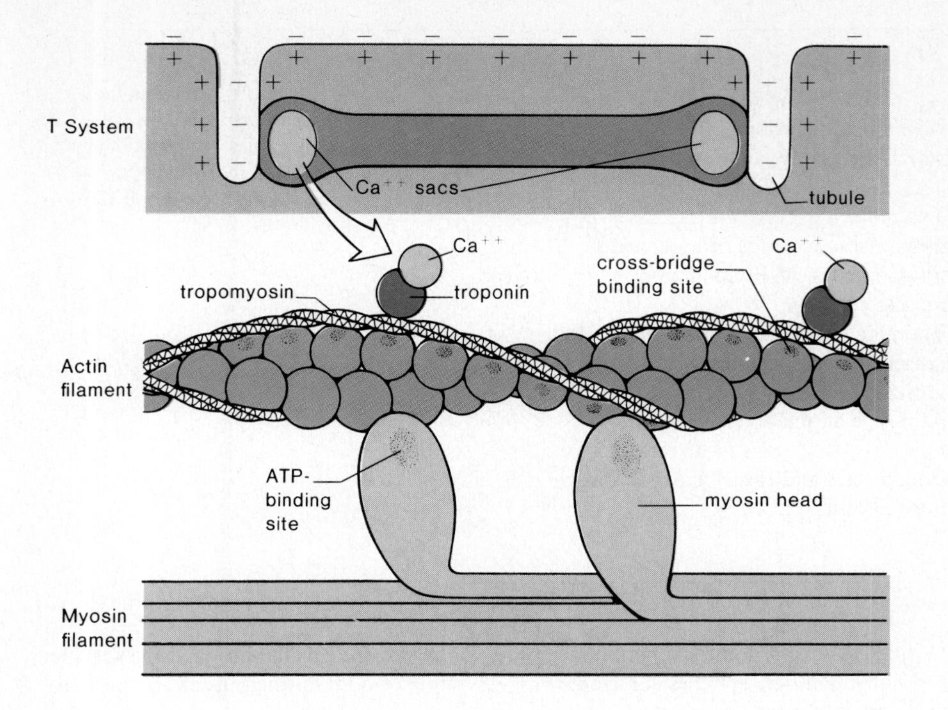

Figure 16.17
Detailed structure and function of sarcomere contraction. After calcium, Ca^{++} is released from its storage sacs, it combines with troponin, a protein that occurs periodically along tropomyosin threads. This causes the tropomyosin threads to shift their position so that cross-bridge binding sites are revealed along the actin filaments. The myosin filament extends its globular heads, forming cross bridges that bind to these sites. The breakdown of ATP by myosin causes the cross bridges to detach and reattach farther along the actin. In this way, the actin filaments are pulled along past the myosin filaments.

The thick myosin filament is a bundle of myosin molecules, each having a globular head. These heads form the cross bridges that attach to the binding sites on actin. Also, as they break down ATP, detachment and reattachment to a site farther along occurs. In this way actin filaments are pulled along the myosin filament.

Contraction discontinues when nerve impulses no longer stimulate the muscle fiber. With the cessation of a muscle action potential, Ca^{++} ions are pumped back into their storage sacs by active transport. Relaxation now occurs.

A neuromuscular junction functions similar to a synapse except that a muscle action potential causes calcium to be released from calcium-storage sacs, and thereafter muscle contraction occurs.

Summary

The skeleton aids movement of the body while it also supports and protects the body. Bones serve as deposits for inorganic salts, and some bones are sites for blood-cell production. The skeleton is divided into two parts: (1) the axial skeleton, which is made up of the skull, ribs, and vertebrae; and (2) the appendicular skeleton, which is composed of the appendages and their girdles. Joints are regions where bones are linked together.

Whole skeletal muscles work in antagonistic pairs and have degrees of contraction. Muscle fibers obey the all-or-none law; it is possible to study a single contraction (muscle twitch) and sustained contraction (summation and tetanus) by using a physiograph.

Muscle fibers are cells that contain myofibrils in addition to the usual components of cells. Longitudinally myofibrils are divided up into sarcomeres, where it is possible to note the arrangement of actin and myosin filaments. When a sarcomere contracts, the actin filaments slide past the myosin filaments. Myosin has cross bridges that attach to and pull the actin filaments along. ATP breakdown by myosin is necessary for detachment to occur.

Nerve innervation of a muscle fiber begins at a neuromuscular junction. Here synaptic vesicles release ACh into the synaptic cleft. When the sarcolemma receives the ACh, a muscle action potential moves down the T system to calcium-storage sacs. When calcium is released, contraction occurs. When calcium is actively transported back into the storage sacs, muscle relaxation occurs.

Objective Questions

1. The skull, ribs, and sternum are all in the ____________ skeleton.
2. The vertebral column protects the ____________ cord.
3. The two bones of the lower arm are the ____________ and ____________ .
4. Most joints are freely movable ________ joints in which the two bones are separated by a cavity.
5. Muscles work in ____________ pairs, the biceps flexes and the triceps extends the lower arm.
6. Maximal sustained contraction of a muscle is called ____________ .
7. Actin and myosin filaments are found within cell inclusions called ____________ , which are divided into units called ____________ .
8. The molecule ____________ serves as an immediate source of high energy phosphate for ATP production in muscle cells.
9. The juncture between axon ending and muscle cell sarcolemma is called a ____________ junction.
10. A muscle action potential causes ____________ ions to be released from storage sacs and this signals the muscle fiber to contract.

Answers to Objective Questions

1. axial 2. spinal 3. radius, ulna 4. synovial 5. antagonistic 6. tetanus 7. myofibrils, sarcomeres 8. creatine phosphate 9. neuromuscular 10. calcium

Study Questions

1. Distinguish between the axial and appendicular skeletons. (p. 335)
2. List the bones that form the pectoral and the pelvic girdles. (p. 335)
3. Describe the anatomy of a long bone (p. 336); of a freely movable joint. (p. 338)
4. Describe how muscles are attached to bones. Why do muscles act in antagonistic pairs? (p. 339)
5. The study of whole muscle physiology often includes observing both threshold and maximal stimulus, muscle twitch, summation, and tetanic contraction. Describe the significance of each of these. (pp. 342–343)
6. How is the tone of a muscle maintained and how do muscle spindles contribute to the maintenance of tone? (p. 343)
7. Discuss the microscopic anatomy of a muscle fiber and the structure of a sarcomere. What is the sliding filament theory? (pp. 344–346)
8. Give the function of each participant in the following reaction:

$$\text{actin} + \text{myosin} \xrightarrow[\text{Ca}^{++}]{\text{ATP} \rightarrow \text{ADP} + \text{Ⓟ}} \text{actomyosin}$$

(p. 346)

9. Discuss the availability and the specific role of ATP during muscle contraction. (pp. 346; 349)
10. What is oxygen debt and how is it repaid? (p. 346)
11. What causes a muscle action potential? How does the muscle action potential bring about sarcomere and muscle fiber contraction? (pp. 347–348)

Key Terms

actin (ak'tin) one of the two major proteins of muscle; makes up thin filaments in myofibrils of muscle fibers. *See* myosin. *346*

appendicular skeleton (ap''en-dik'u-lar skel'ĕ-ton) portion of the skeleton forming the upper extremities, pectoral girdle, lower extremities, and pelvic girdle. *335*

axial skeleton (ak'se-al skel'ĕ-ton) portion of the skeleton that supports and protects the organs of the head, neck, and trunk. *335*

compact bone (kom-pakt' bōn) bone composed of osteocytes located in lacunae that are arranged concentrically around Haversian canals. *336*

creatine phosphate (kre'ah-tin fos'fāt) a compound unique to muscles that contains a high-energy phosphate bond. *346*

insertion (in-ser'shun) the end of a muscle that is attached to a movable bone. *339*

muscle action potential (mus'el ak'shun po-ten'shal) an electrochemical change, due to increased sarcolemma permeability, that is propagated down the T system and results in muscle contraction. *347*

myofibrils (mi''o-fi'brilz) the contractile portions of muscle fibers. *344*

myosin (mi'o-sin) one of two major proteins of muscle; makes up thick filaments in myofibrils and is capable of breaking down ATP. *See* actin. *346*

neuromuscular junction (nu''ro-mus'ku-lar jungk'shun) the point of contact between a nerve cell and a muscle fiber. *347*

origin (or'i-jin) end of a muscle that is attached to a relatively immovable bone. *339*

osteocyte (os'te-o-sīt) a mature bone cell. *337*

oxygen debt (ok'sĭ-jen det) oxygen that is needed to metabolize lactic acid, a compound that accumulates during vigorous exercise. *346*

red marrow (red mar'o) blood-cell-forming tissue located in spaces within bones. *337*

synovial joint (si-no've-al joint) a freely movable joint. *338*

tetanus (tet'ah-nus) sustained muscle contraction without relaxation. *342*

tone (tōn) the continuous partial contraction of muscle; due to contraction of a small number of muscle fibers at all times. *343*

17

Senses

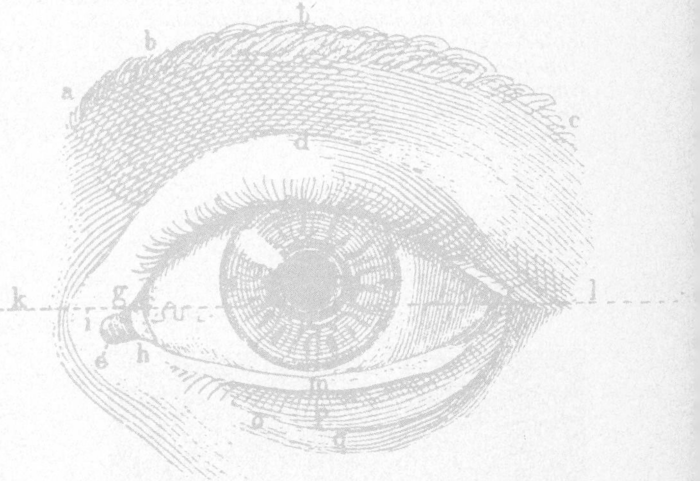

Chapter Concepts

1 Sense organs are sensitive to environmental stimuli and are therefore termed receptors. Each receptor responds to one type of stimulus, but they all initiate nerve impulses.

2 The sensation realized is the prerogative of the region of the cerebrum receiving nerve impulses.

3 There are sense receptors that respond to mechanical stimuli, chemical stimuli, and light energy. Our knowledge of the outside world is dependent on these stimuli.

Chapter Outline

Figure 17.1
Do you see a vase in this picture, or a man and woman looking at each other? Our sense organs are dependent on the stimuli received and they can be fooled. This vase was especially designed to commemorate the 25th year of the Queen of England's reign. The man and woman are Prince Philip and Queen Elizabeth.

Table 17.1 Receptors

Receptors	Sense	Stimulus
General		
Temperature	Hot-cold	Heat flow*
Touch	Touch	Mechanical displacement of tissue†
Pressure	Pressure	Mechanical displacement of tissue†
Pain	Pain	Tissue damage‡
Proprioceptors	Limb placement	Mechanical displacement†
Special		
Eye	Sight	Light*
Ear	Hearing	Sound waves†
	Balance	Mechanical displacement†
Taste buds	Taste	Chemicals‡
Olfactory cells	Smell	Chemicals‡

*Radioreceptors
†Mechanoreceptors
‡Chemoreceptors

Sense organs receive external and internal stimuli (fig. 17.1); therefore they are called **receptors.** Each type of receptor is sensitive to only one type of stimulus. Table 17.1 lists the receptors discussed in this chapter and the stimulus to which each reacts.

Receptors are the first components of the reflex arc described in chapter 15. When a receptor is stimulated, it generates nerve impulses that are transmitted to the spinal cord and/or brain; but we are conscious of a sensation only if the impulses reach the cerebrum. The sensory portion of the cerebrum can be mapped according to the parts of the body and the type of sensation realized at different loci (fig. 15.21).

General Receptors

Microscopic receptors (table 17.1) are present in the skin, in the visceral organs, and in the muscles and joints. They are all specialized nerve endings for the detection of touch, pressure, pain, temperature (hot and cold), and proprioception. Proprioception refers to the sense of knowing the position of the limbs; for example, if you close your eyes and move your arm about slowly, you still have a sense of where your arm is located.

Skin

The skin (fig. 17.2) contains receptors for touch, pressure, pain, and temperature. It is a mosaic of these tiny receptors, as you can determine by passing a metal probe slowly over the skin. At certain points there will be a feeling of pressure; at others, a feeling of hot or cold (depending on the temperature of the probe). Certain parts of the skin contain more receptors for a particular sensation; for example, the fingertips have an abundance of touch receptors.

A simple experiment suggests that temperature receptors are sensitive to the flow of heat. Fill three bowls with water—one cold, one warm, and one hot. Put your left hand in the cold water and your right in the hot water for a few moments. Your hands will adjust, or adapt, to these temperatures so that when you put both hands in warm water, each hand will indicate a different water temperature. Therefore, it seems that when the outside temperature is higher than the temperature to which we have adjusted, we detect a sensation of warm or hot as heat flows into the skin. When the outside temperature is low enough that heat flows out from the skin, we detect coolness or cold.

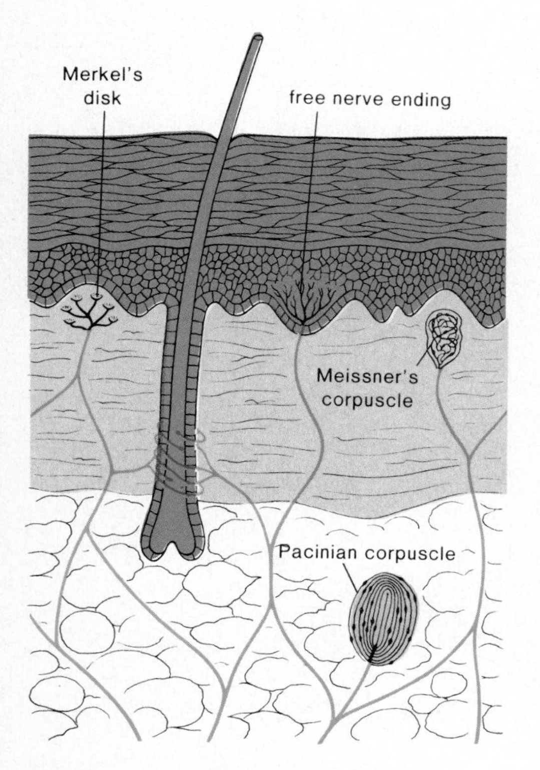

Figure 17.2
Receptors in human skin. Free nerve endings are pain receptors. Pacinian corpuscles are pressure receptors; Merkel's disks and Meissner's corpuscles are touch receptors, as are the nerve endings surrounding the hair follicle.

Other skin receptors beside those for temperature also demonstrate adaptation. **Adaptation** occurs when the receptor becomes so accustomed to stimuli that it stops generating impulses even though the stimulus is still present. The touch receptors are of this type. They can quickly adapt to the clothing we put on so that we are not constantly aware of the feel of clothes against our skin.

The receptors of the skin can be used to illustrate that sensation actually occurs in the brain and not in the sense organ itself. If the nerve fiber from the sense organ is cut, there is no sensation. Also, since a nerve impulse is always the same electrochemical charge, the particular sensation realized does not have to do with the nerve impulse. It is the brain that is responsible for the type of sensation felt and the localization of the sensation. For example, if we connected a pain receptor in the foot to a nerve normally receiving impulses from a heat receptor in the hand, and then proceeded to stick the pain receptor in the foot, the subject would report the feeling of warmth in the hand. The brain indicates the sensation and the localization. This realization is mildly disturbing, because it makes us aware of how dependent we are on the anatomical wholeness of the body in order to be properly aware of our surroundings.

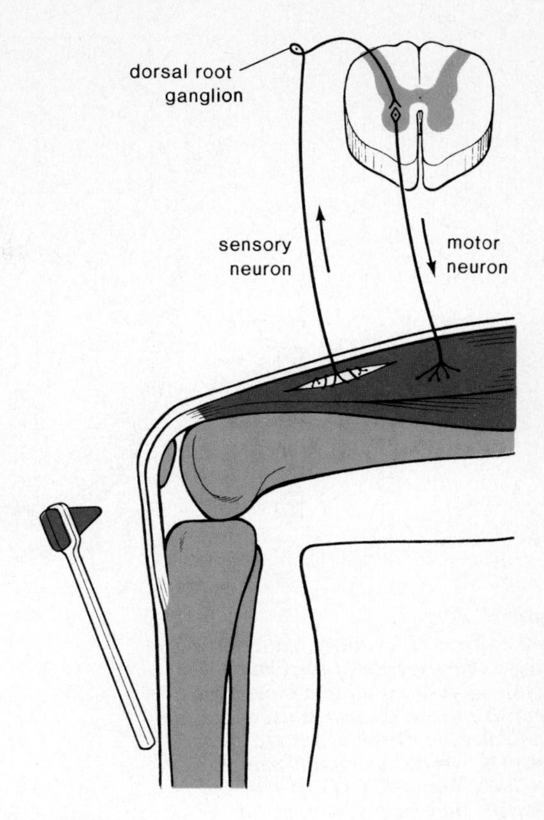

Figure 17.3
Proprioceptors in muscles and joints. Muscle spindle fibers are stretch receptors as can be demonstrated by the knee reflex.

Specialized receptors in the human skin respond to temperature, touch, pressure, and pain.

Muscles and Joints

The sense of position and movement of limbs (i.e., proprioception) is dependent upon receptors termed **proprioceptors.** Muscle spindles discussed in chapter 16 are sometimes considered to be proprioceptors. Stretching of associated muscle fibers causes muscle spindles to increase the rate at which they fire and for this reason, they are sometimes called **stretch receptors.** The *knee jerk* is a common example of the manner in which muscle spindles act as stretch receptors (fig. 17.3). When the legs are crossed at the knee and the tendon at the knee is tapped, both the tendon and muscles in the thigh are stretched. Stimulated by the stretching, muscle spindles transmit impulses to the spinal cord and thereafter the thigh muscles contract. This causes the lower leg to jerk upward in a kicking motion.

There are proprioceptors located in the joints and associated ligaments and tendons that respond to stretching, pressure, and pain. Nerve endings from these receptors are integrated with those received from other types of receptors so that the person knows the position of body parts.

Sense receptors in the muscles and joints, called proprioceptors, give us a sense of how our body parts are positioned.

Special Senses

The special senses include the chemoreceptors for taste and smell, the light receptors for sight, and the mechanoreceptors for hearing and balance.

Chemoreceptors

Taste and smell are called the *chemical senses* because these receptors are sensitive to certain chemical substances in the food we eat and the air we breathe.

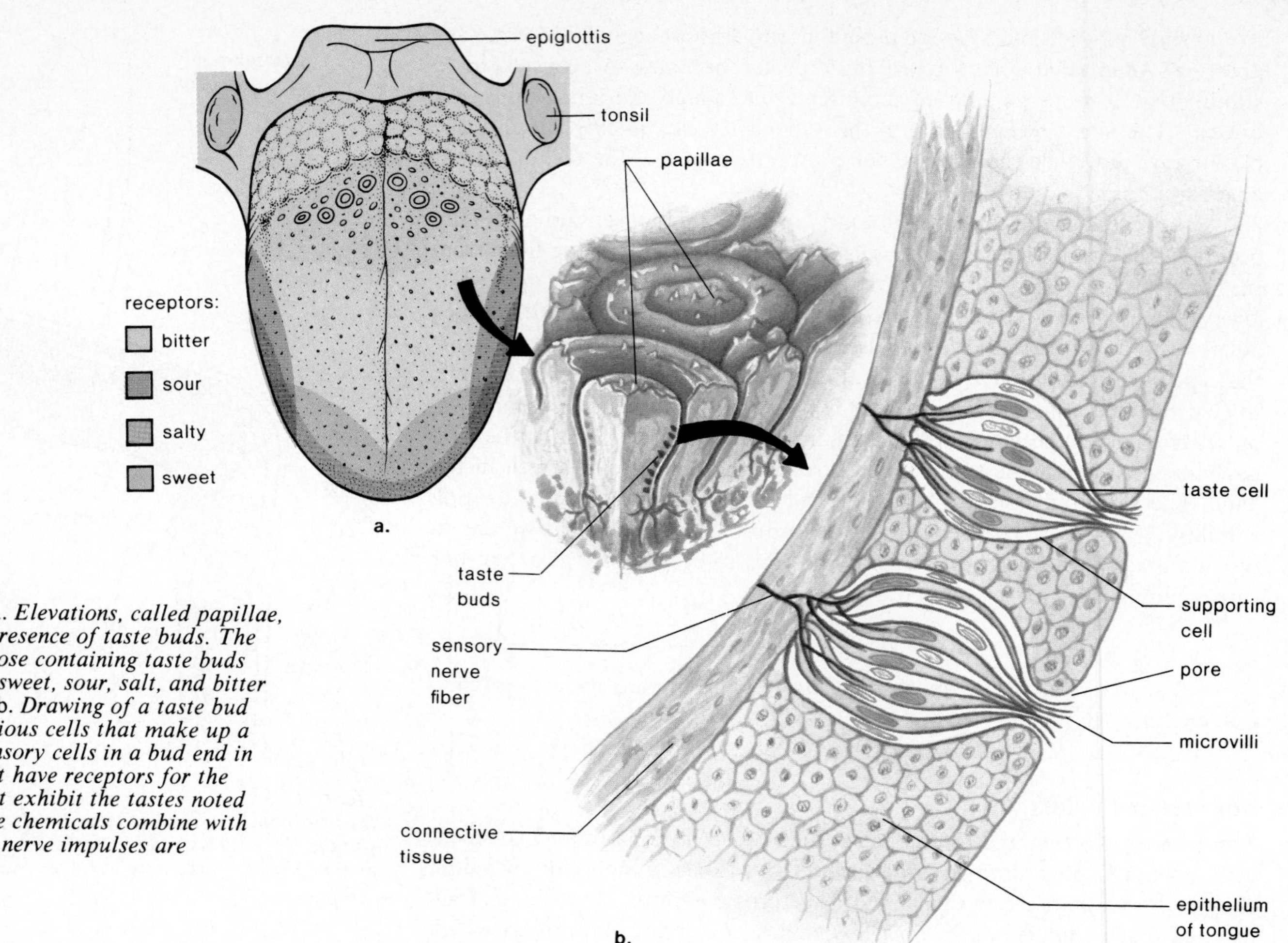

Figure 17.4
Taste buds. a. *Elevations, called papillae, indicate the presence of taste buds. The location of those containing taste buds responsive to sweet, sour, salt, and bitter is indicated.* b. *Drawing of a taste bud shows the various cells that make up a taste bud. Sensory cells in a bud end in microvilli that have receptors for the chemicals that exhibit the tastes noted in* a. *When the chemicals combine with the receptors, nerve impulses are generated.*

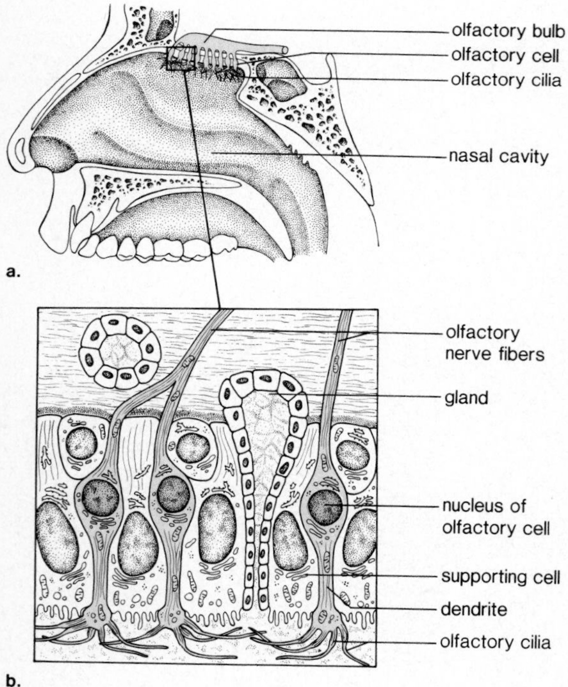

Figure 17.5
a. *Position of olfactory epithelium in a nasal passageway.* b. *The olfactory receptor cells, which have cilia projecting into the nasal cavity, are supported by columnar epithelial cells. When these cells are stimulated by chemicals in the air, nerve impulses begin and are conducted to the brain by olfactory nerve fibers.*

Taste buds are located on the tongue. Many lie along the walls of the papillae (fig. 17.4), the small elevations visible to the naked eye. Isolated ones are also present on the palate, pharynx, and epiglottis.

Taste buds (fig. 17.4b) are pockets of cells that extend through the tongue epithelium and open at a taste pore. Within the oval pocket are supporting cells and a number of elongated cells that end in microvilli. These cells, which have associated nerve fibers, are sensitive to chemicals. Nerve impulses are most probably generated when the chemicals bind to receptor sites found on the microvilli.

The sense of taste has been shown to be genetically inherited, and foods taste differently to various people. This might very well account for the fact that some persons dislike a food that is preferred by others.

It is believed that there are four types of tastes (bitter, sour, salty, sweet) and that taste buds for each are concentrated on the tongue in particular regions. Sweet receptors are most plentiful near the tip of the tongue. Sour receptors occur primarily along the margins of the tongue. Salt receptors are most common on the tip and the upper front portion of the tongue. And bitter receptors are located toward the back of the tongue.

The **olfactory cells** (fig. 17.5) are located high in the roof of the nasal cavity. These cells, which are specialized endings of the fibers that make up the olfactory nerve, lie among supporting epithelial cells. Each cell ends in a tuft of about five cilia which bear receptor sites for various chemicals. Research, resulting in the stereochemical theory of smell, suggests that different types of smell are related to the various shapes of molecules rather than to the atoms that make up the molecules. When chemicals combine with the receptor sites, nerve impulses are generated. The olfactory receptors, like the touch and temperature receptors, also adapt to outside stimuli. In other words, we can become accustomed to a smell and no longer take notice of it.

Figure 17.6
If you are on a diet, it is safer to eat the same monotonous food each day instead of tempting the palate. A variety of interesting tastes and smells can cause you to eat more than is warranted.

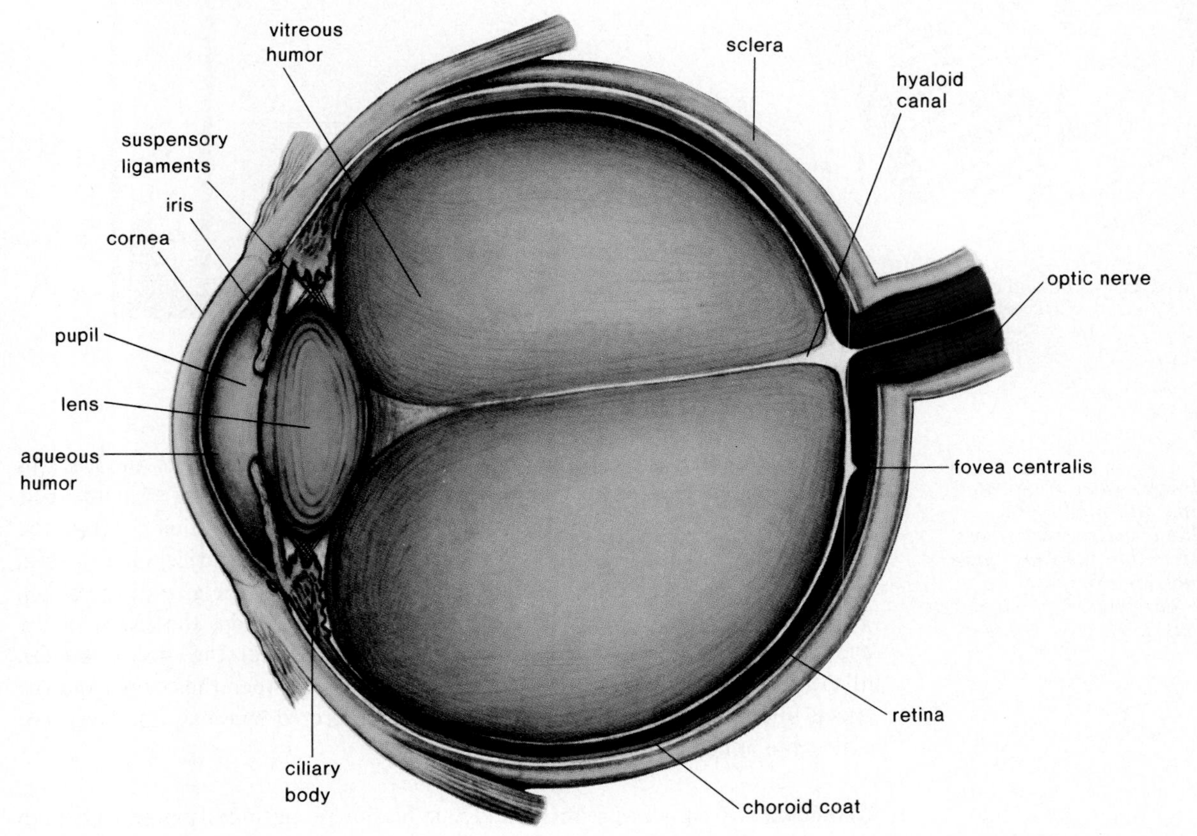

Figure 17.7
Anatomy of the human eye. Notice that the sclera becomes the cornea, the window of the eye; the choroid becomes the ciliary body and iris. The ciliary body is thrown into 70 to 80 radiating folds that contain the ciliary muscle and ligaments that hold and adjust the shape of the lens. The retina contains the receptors for sight and vision is most acute in the fovea centralis where there are only cones. A blind spot occurs when the optic nerve leaves the retina and there are no receptors for sight. The hyaloid canal, which contains an artery only in the fetus, has no known function in the adult.

The sense of taste and the sense of smell supplement each other, creating a combined effect when interpreted by the cerebral cortex. For example, when we have a cold, we think that food has lost its taste but actually we have lost the ability to sense its smell. This may work in reverse also. When we smell something, some of the molecules move from the nose down into the mouth region and stimulate the taste buds there. Therefore, part of what we refer to as smell may actually be taste (fig. 17.6).

The receptors for taste (taste buds) and the receptors for smell (olfactory microvilli) work together to give us our sense of taste and smell.

Photoreceptor—the Eye

The eye (table 17.2 and fig. 17.7), an elongated sphere about 1 inch in diameter, has three layers or coats. The outer **sclera** is a white fibrous layer except for the transparent cornea, the window of the eye. The middle, thin,

Table 17.2 *Name and Function of Parts of the Eye*

Part	Function
Lens	Refraction and focusing
Iris	Regulates light entrance
Pupil	Hole in iris
Choroid	Absorbs stray light
Sclera	Protection
Cornea	Refraction of light
Humors	Refraction of light
Ciliary body	Holds lens in place
Retina	Contains receptors
Rods	Black-and-white vision
Cones	Color vision
Optic nerve	Transmits impulse
Fovea	Region of cones in retina
Ciliary muscle	Accommodation

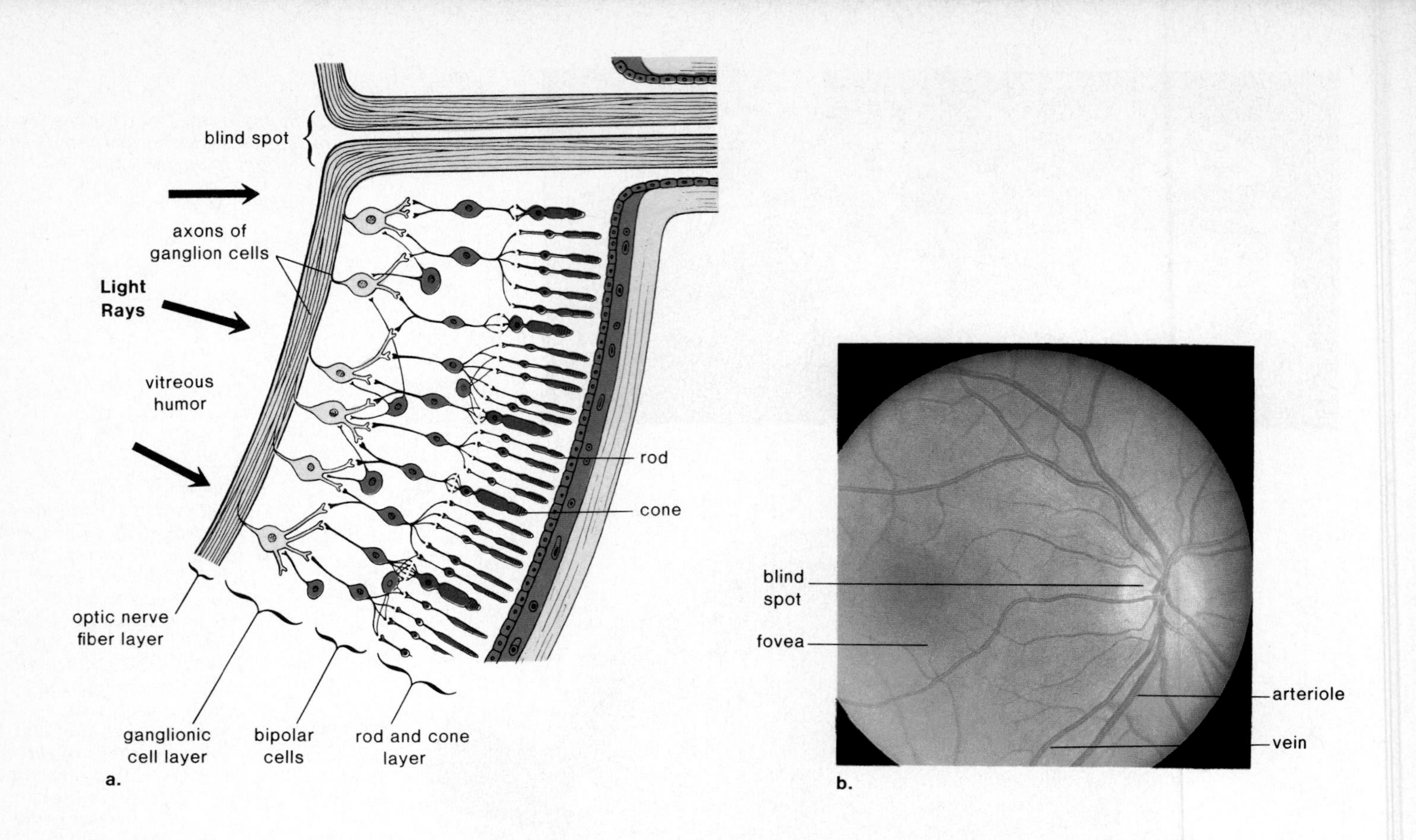

Figure 17.8
a. *Structure of retina. Rods and cones are located toward the back of the retina, followed by the bipolar cells and the ganglionic cells whose fibers become the optic nerve. Notice that the rods share bipolar cells but cones do not. Cones, therefore, distinguish more detail.* b. *The blind spot where the optic nerve pierces the eyeball is clearly visible in eye examinations.*

dark brown layer, the **choroid coat,** contains many blood vessels and absorbs stray light rays. Toward the front, the choroid thickens and forms a ring-shaped structure, the ciliary body, containing the **ciliary muscle,** which controls the shape of the lens for near and far vision. Finally the choroid becomes a thin, circular, muscular diaphragm, the iris, which regulates the size of the pupil. The **lens,** attached to the ciliary body by ligaments, divides the cavity of the eye into two chambers. A viscous and gelatinous material, the **vitreous humor,** fills the large cavity behind the lens. The chamber between the cornea and the lens is filled with an alkaline, watery solution, secreted by the ciliary body and called the **aqueous humor.**

Glaucoma A small amount of aqueous humor is continually produced each day. Normally it leaves the anterior chamber by way of tiny ducts that are located where the iris meets the cornea. If these drainage ducts are blocked, pressure rises and compresses the retinal arteries whose capillaries feed nerve fibers located in the retina. With the passage of time, some of these fibers die and the result is partial or total blindness.

Retina

The inner layer of the eye, the **retina,** has three layers of cells (fig. 17.8a). The layer closest to the choroid contains the sense receptors for sight, the **rods** and **cones** (fig. 17.8b); the middle layer contains bipolar cells; and the innermost layer contains ganglion cells whose fibers become the **optic nerve.** Only the rods and cones contain light-sensitive pigments and therefore light must penetrate to the back of the retina before nerve impulses are generated. Nerve impulses initiated by the rods and cones are passed to the bipolar cells, which in turn pass them to the ganglion cells whose fibers pass in front of the retina, forming the optic nerve, which turns to pierce the layers of the eye. The presence of these three layers of nerve cells permits a certain amount of integration

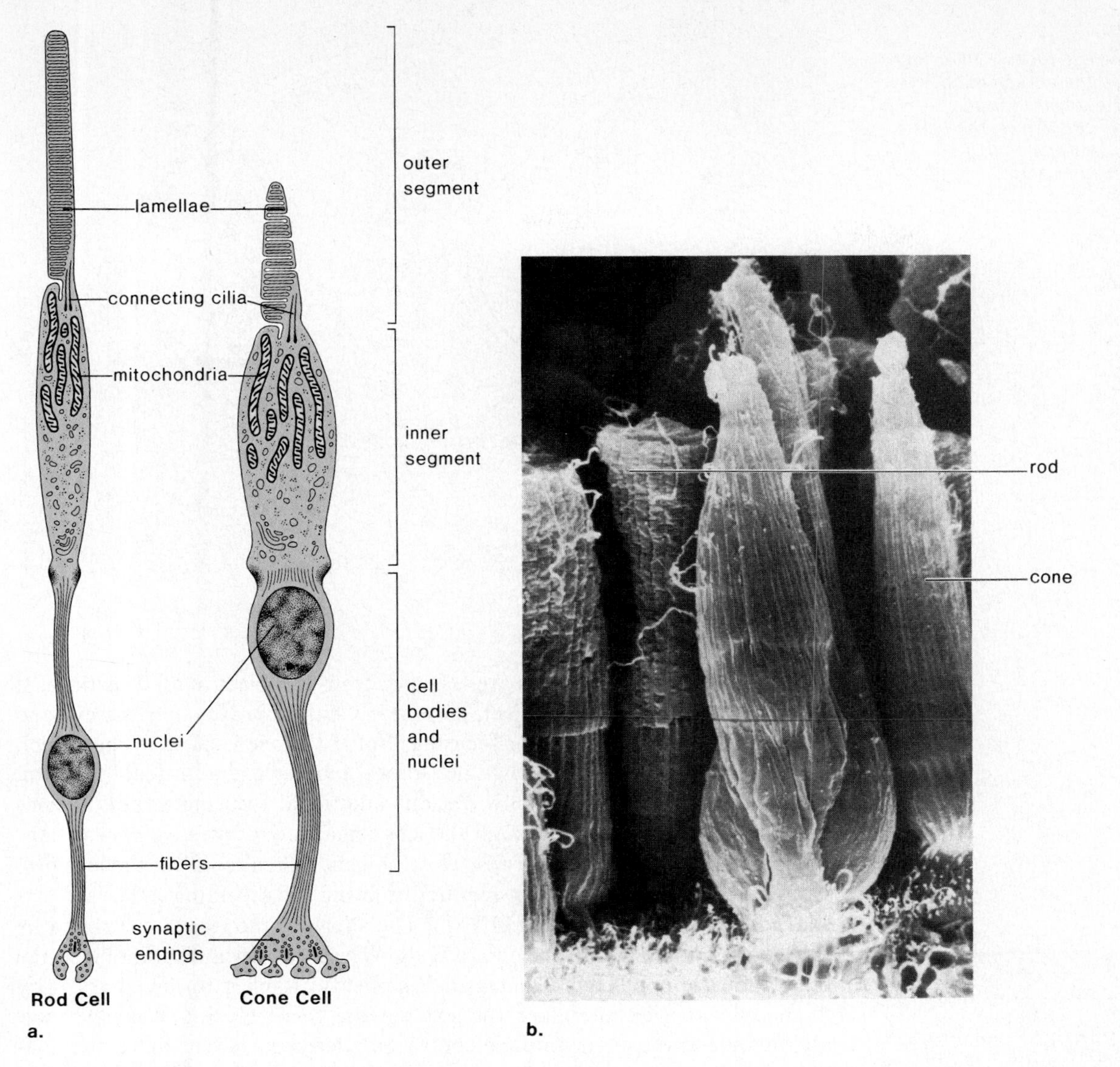

Figure 17.9
Receptors for sight.
a. *Drawing of rod and cone. The photosensitive pigment is located in the lamellae of the outer segment, which apparently is a modified cilium. A cilium-like stalk with nine sets of microtubules connects the outer segment to the inner segment.* b. *Scanning electron micrograph of rods and cones. The cones in the foreground are responsible for color vision and the rods in the background are responsible for night vision.*

before nerve impulses are sent to the brain. There are no rods and cones where the optic nerve passes through the retina; therefore, this is a **blind spot** (fig. 17.8b) where vision is impossible.

The retina contains a very special region called the **fovea centralis** (fig. 17.7), an oval yellowish area with a depression where there are only cone cells. Vision is most acute in the fovea centralis.

The eye has three layers: the outer sclera, the middle choroid, and the inner retina. Only the retina contains receptors for sight.

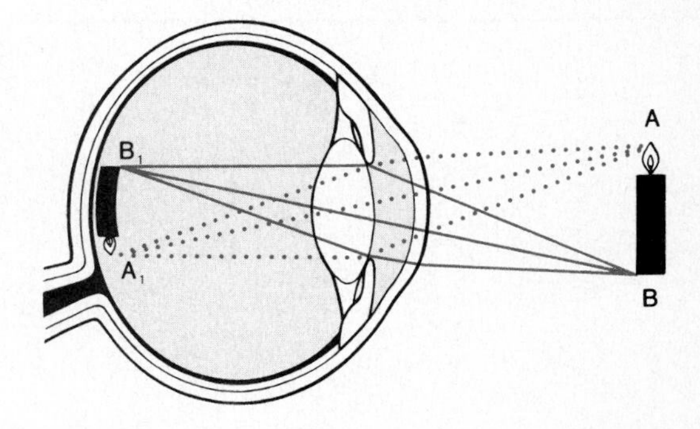

Figure 17.10
Focusing. Light rays from each point on an object are bent by the cornea in such a way that they are directed to a single point after emerging from the lens. By this process an inverted image of the object forms on the retina.

Physiology

Focusing When we look at an object, light rays are **focused** on the retina (fig. 17.10). In this way, an *image* of the object appears on the retina. The image on the retina occurs when the rods and cones in a particular region are excited. Obviously, the image is much smaller than the object. In order to produce this small image, light rays must be bent (refracted) and brought to a focus. They are bent as they pass through the cornea. Further bending occurs as the rays pass through the lens and humors.

Figure 17.11
Accommodation. a. *When the eye focuses on a far object, the lens is flat because the ciliary muscle is relaxed and the suspensory ligament is taut.*
b. *When the eye focuses on a near object, the lens rounds up because the ciliary muscle contracts, causing the suspensory ligament to relax.*

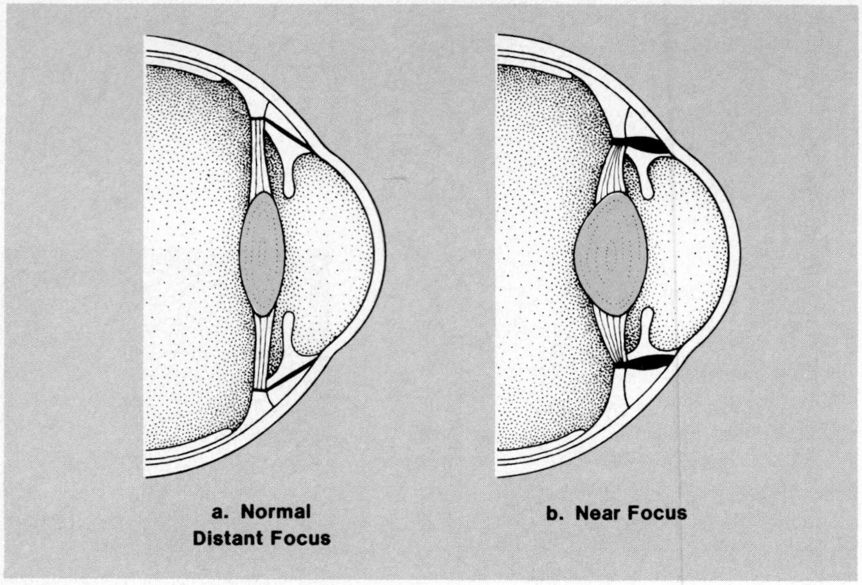

Accommodation Light rays are reflected from an object in all directions. If the eye is distant from an object, only nearly parallel rays enter the eye and the cornea alone is needed for focusing. But if the eye is close to the object, many of the rays are at sharp angles to one another and additional focusing is required. When the lens provides this additional focusing power, **accommodation** has occurred. Although the lens remains flat when we view distant objects, it rounds up when we view close objects. When rounded, the lens provides the additional refraction required to bring the diverging light rays to a sharp focus on the retina (fig. 17.11). The shape of the lens is controlled by the ciliary muscle within the ciliary body. When we view a distant object, the ciliary muscle is relaxed, causing the ligaments attached to the ciliary body to be under tension; therefore, the lens remains relatively flat. When we view a close object, the ciliary muscle contracts, releasing the tension on the ligaments and, therefore, the lens rounds up due to its natural elasticity (table 17.3). Since close work requires contraction of the ciliary muscle, it very often causes "eye strain."

Table 17.3 *Accommodation*

Object	Ciliary Muscle	Lens
Near object	Ciliary muscle contracts, ligaments relax	Lens becomes round
Far object	Ciliary muscle relaxes, ligaments under tension	Lens is flattened

With aging, the lens loses some of its elasticity and is unable to accommodate and bring close objects into focus. This usually necessitates the wearing of glasses, as is discussed on page 360. The lens is also subject to *cataracts;* it can become cloudy and opaque and unable to transmit rays of light. Special cells within the interior of the lens contain proteins called crystallin. Recent research suggests that cataracts are brought on when these proteins become oxidized, causing their three-dimensional shape to change. If so, researchers believe that they may eventually be able to find ways to restore the normal configuration of crystallin so that cataracts can be treated medically instead of surgically. First, a surgeon opens the eye near the rim of the cornea and may use Zonulysin, an enzyme, to digest away the ligaments holding the lens in place. Most then use a cryoprobe that freezes the lens for easy removal. An intraocular lens, attached to the iris, can be implanted in the eye, so that the patient need not wear thick glasses or contact lenses.

The lens, assisted by the cornea and humors, focuses images on the retina.

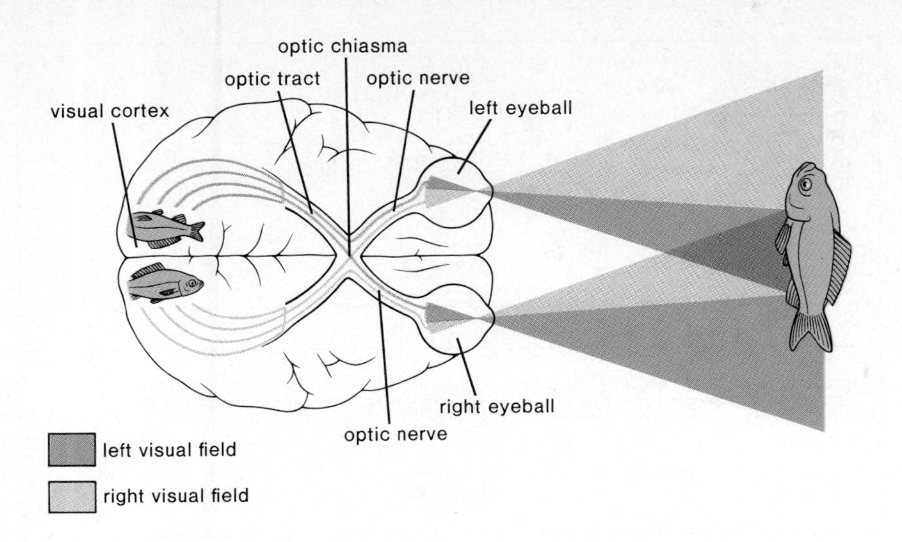

Figure 17.12
Both eyes "see" the entire object but information from the right half of each retina goes to the right visual cortex and information from the left half of the retina goes to the left visual cortex. When the information is pooled the brain "sees" the entire object and in depth.

Inverted Image The image on the retina is upside down (fig. 17.10) and it is thought that perhaps this image is righted in the brain by experience. In one experiment, scientists wore glasses that inverted the field. At first, they had difficulty adjusting to the placement of the objects, but then they soon became accustomed to their inverted world. Experiments such as this suggest that if we see the world upside down, the brain learns to see it right side up.

Stereoscopic Vision We can see well with either eye alone, but the two eyes functioning together provide us with **stereoscopic vision.** Normally, the two eyes are directed by the eye muscles toward the same object and therefore the object is focused on corresponding points of the two retinas. But each eye sends its own information to the brain about the placement of the object because each forms an image from a slightly different angle. These data are pooled to produce depth perception by a two-step process. First, because the optic nerves cross at the optic chiasm (fig. 17.12), one-half of the brain receives information from both eyes about the same part of an object. Later, the two halves of the brain communicate to arrive at a complete three-dimensional interpretation of the whole object.

The anatomy and physiology of the brain allow us to see the world right side up and in three dimension.

Biochemistry In *dim light,* the pupils enlarge so that more rays of light can enter the eyes. As the rays of light enter, they strike the rods and cones, but only the 160 million rods located in the periphery, or sides, of the eyes are sensitive enough to be stimulated by this faint light. The rods do not detect fine detail or color, so at night, for example, all objects appear to be blurred and have a shade of gray. Rods do detect even the slightest motion, however, because of their abundance and position in the eyes.

The rods contain **rhodopsin,** a pigment called visual purple. When light strikes rhodopsin it breaks down into a protein, **scotopsin** (opsin), and the pigment portion of the molecule, **retinene** (retinal). This leads to a depolarization of the rod cells and a release of transmitter substance from the rod cells so that nerve impulses are generated. When the eye is exposed to a flash of light, the stimulus generated lasts one-tenth of a second. This is why we continue to see an image if we close our eyes immediately after looking at an object. It also allows us to see motion if still frames are presented at a rapid rate, as in "movies."

a.

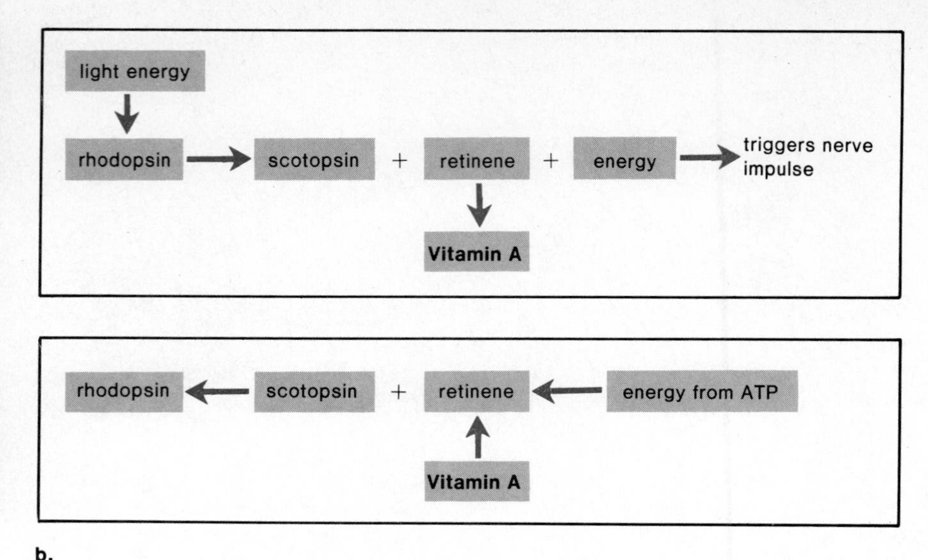

b.

Figure 17.13
a. *Vitamin A, available from fresh fruits and vegetables, especially carrots, is involved in biochemistry of vision.* b. (top) *In the presence of light, rhodopsin breaks down to a protein and retinene and this reaction triggers nerve impulses. Retinene becomes vitamin A.* (bottom) *In the dark, ATP energy and vitamin A are used to reform rhodopsin.*

The more rhodopsin present in the rods, the more sensitive are our eyes to dim light. Therefore, during the time required for adaptation (adjustment) to dim light when we find it difficult to see, rhodopsin is being formed in the rods. As figure 17.13 shows, retinene eventually breaks down to vitamin A. Vitamin A is abundant in carrots, so the notion that we should eat carrots for good vision is not without foundation. Actually, most of the vitamin A is transformed back to retinene, which then combines in the dark with scotopsin to form rhodopsin again.

In *bright light* the pupils get smaller, so that less light enters the eyes. The cones located primarily in the fovea, are active and detect the fine detail and color of an object. In order to perceive depth, as well as to see color, we turn our eyes so that reflected light from the object strikes the fovea. Color vision has been shown to depend on three kinds of cones that contain pigments sensitive to either blue, green, or red light. The nerve impulses generated from one type cone not only stimulate certain cells in the visual cortex of the brain, they also inhibit the reception of impulses from other type cones. For example, when we see red, certain cells in the brain are prohibited from receiving impulses from green cones. Similarly, impulses sent through blue cones tend to oppose the combination of signals sent by red and green cones—which together produce yellow. This process assists integration and enables the brain to tell the location of various colors in the environment (fig. 17.14). Complete color blindness is extremely rare. In most instances a particular type of cone is lacking or deficient in number. The lack of red or green cones is the most common, affecting about 5 percent of the American population. If the eye lacks red cones, the green colors become accentuated, and vice versa.

The sense receptors for sight are the rods and cones. The rods are responsible for vision in dim light and the cones are responsible for vision in bright light and color vision. When either is stimulated nerve impulses begin and are transmitted in the optic nerve to the brain.

Corrective Lenses

The majority of people can see what is designated as a size "20" letter 20 feet away and are said to have 20/20 vision. Persons who can see close objects but cannot see the letters from this distance are said to be nearsighted. Nearsighted persons can see near better than they can see far. These individuals often have an elongated eyeball and when they attempt to look at a far object, the image is brought to focus in front of the retina (fig. 17.15). They can see

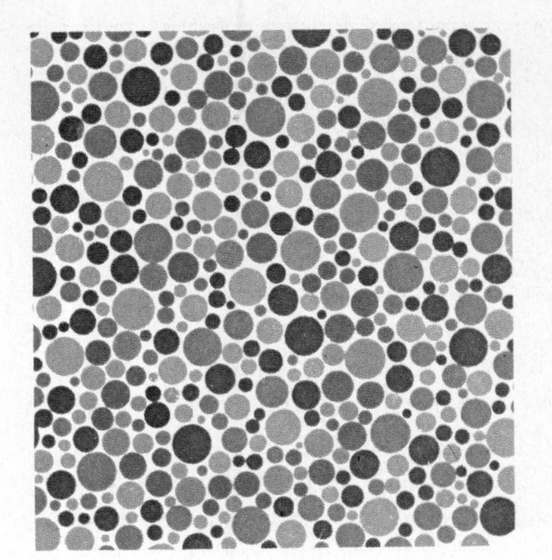

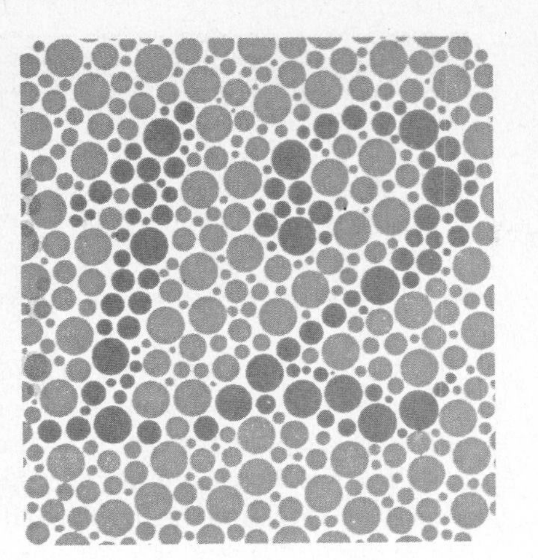

Figure 17.14
Test plates for color blindness. When looking at the plate on the left, the person with normal color vision will see the number 8 and when looking at the plate on the right, the person with normal color vision will see the number 12. The most common form of color blindness involves an inability to distinguish reds and greens.

These plates have been reproduced from Ishihara's Tests for Colour Blindness published by Kanehara & Co., Ltd., Tokyo, Japan, but tests for color blindness cannot be conducted with this material. For accurate testing, the original plate should be used.

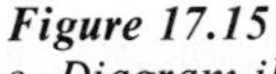

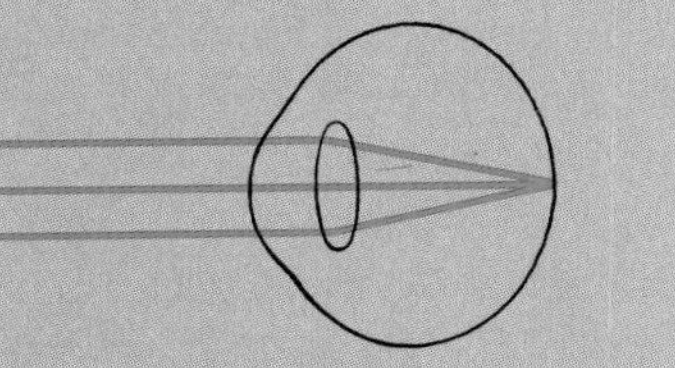

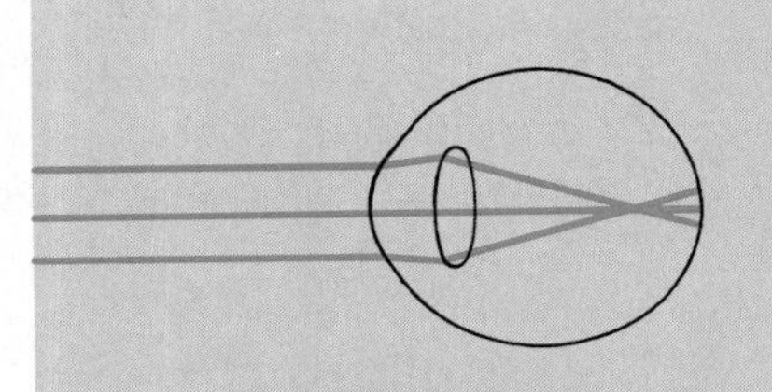

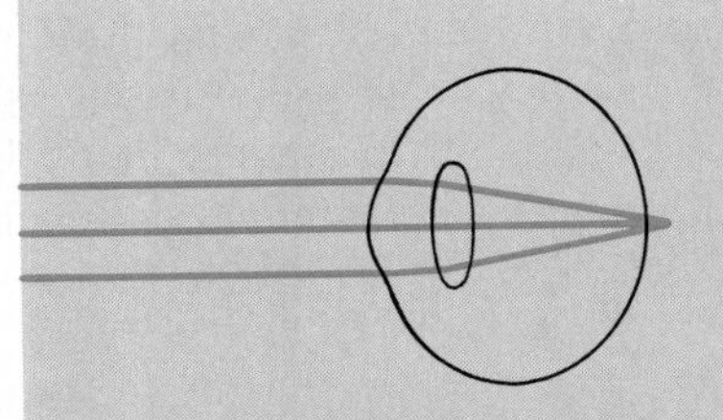

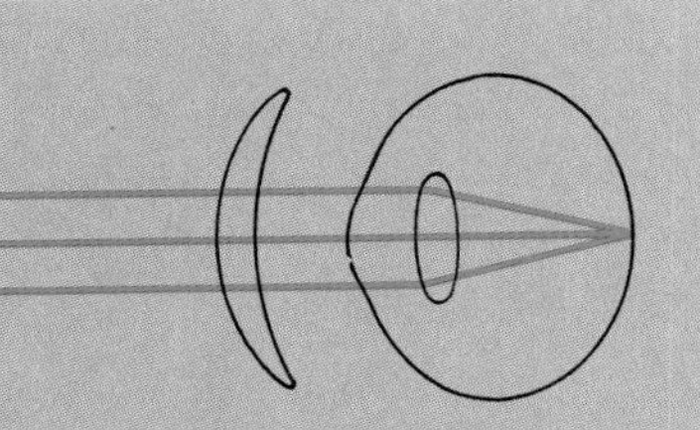

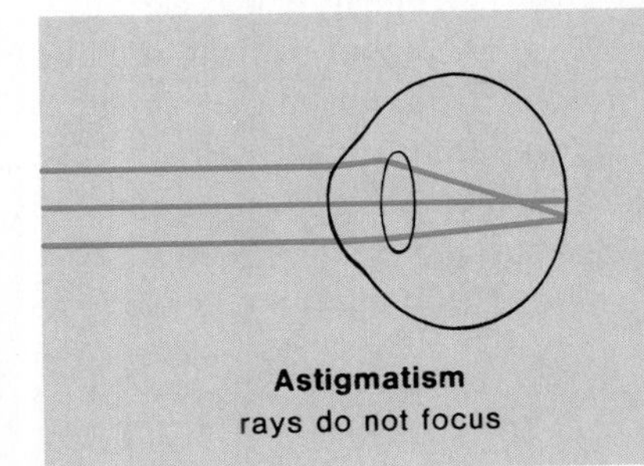

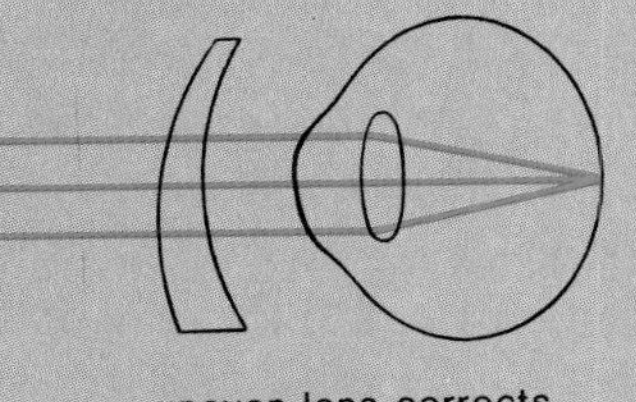

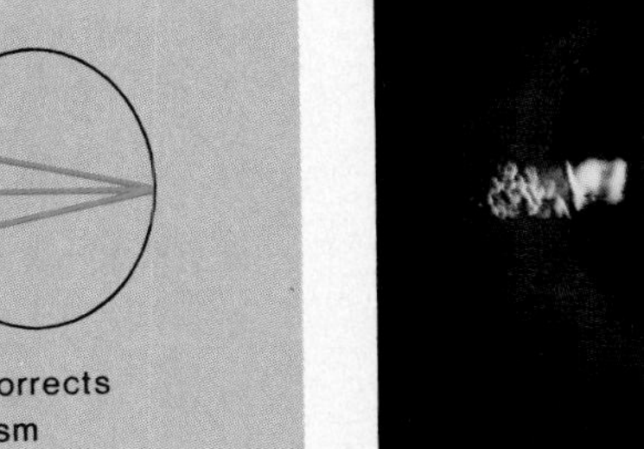

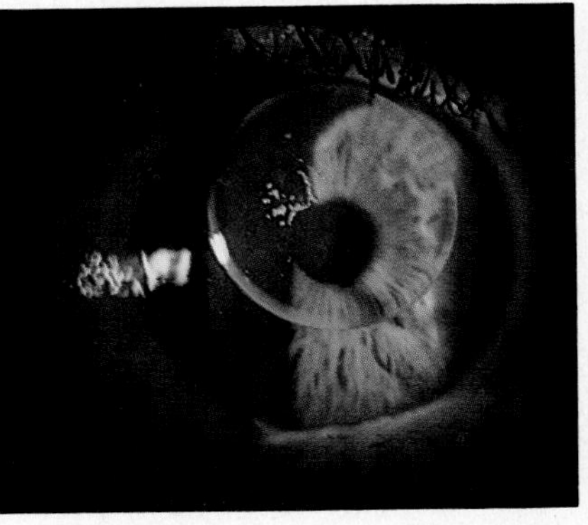

Figure 17.15
a. *Diagram illustrating common abnormalities of the eye. Both the cornea and the lens function in bringing light rays* (lines) *to focus, but sometimes they are unable to compensate for the shape of the eyeball or for an uneven cornea.* b. *Corrective lenses can be provided to allow the individual to see normally.* c. *Opthalmologists can examine the fit of a contact lens by using a narrow beam of light from a "slit lamp" while looking through a biomicroscope.*

Table 17.4 *Common Abnormalities of the Eye*

Name	Effect	Fault	Result	Correction
Nearsighted	Can't see far, can see near	Long eyeball	Image focused in front of retina	Concave lens
Farsighted	Can't see near, can see far	Short eyeball	Image focused behind retina	Convex lens
Astigmatism	Can't focus	Irregular eyeball	Image not focused	Irregular lens

near because they can adjust the lens to allow the image to focus on the retina; but to see far, these people must wear concave lenses that diverge the light rays so that the image can be focused on the retina.

Persons who can easily see the optometrist's chart but cannot see close objects well are farsighted; these individuals can see far away better than they can see near. They often have a shortened eyeball, and when they try to see near, the image is focused behind the retina. When the object is far away, the lens can compensate for the short eyeball, but when the object is close, these persons must wear a convex lens to increase the bending of light rays so that the image will be focused on the retina.

When the cornea or lens is uneven, the image is fuzzy because the light rays cannot be focused evenly on the retina. This fault can be corrected by an unevenly ground lens to compensate for the uneven cornea. Table 17.4 summarizes these conditions and their corrections.

Bifocals As mentioned earlier, with normal aging, the lens loses some of its ability to change shape in order to focus on close objects. Since nearsighted individuals still have difficulty seeing objects clearly in the distance, they must wear bifocals, which means that the upper part of the lens is for distant vision and the remainder for near vision.

Shape of the eyeball determines the need for corrective lenses; the inability of the lens to accommodate as we age also requires corrective lenses for close vision.

Mechanoreceptor—the Ear

The ear accomplishes two sensory functions: balance and hearing. The sense cells for both of these are located in the inner ear and consist of hair cells with cilia that respond to mechanical stimulation. Each hair cell has from 30 to 150 extensions called cilia, despite the fact that they contain tightly packed filaments rather than microtubules. When the cilia of any particular hair cell are displaced in a certain direction, the cell generates nerve impulses which are sent along a cranial nerve to the brain.

Anatomy

Table 17.5 lists the parts of the ear, and figure 17.16 is a drawing of the ear. The ear has three divisions: outer, middle, and inner. The **outer ear** consists of the **pinna** (external flap) and **auditory canal.** The opening of the auditory canal is lined with fine hairs and sweat glands. In the upper wall are modified sweat glands that secrete earwax, a substance that helps guard the ear against the entrance of foreign materials such as air pollutants.

The **middle ear** begins at the **tympanic membrane** (eardrum) and ends at a bony wall containing two small openings covered by membranes. These openings are called the **oval** and **round windows.** Three small bones are found

Table 17.5 The Ear				
	Outer Ear	**Middle Ear**	**Inner Ear**	
			Cochlea	**Sacs plus semicircular canals**
Function	Directs sound waves to tympanic membrane	Picks up and amplifies sound waves	Hearing	Maintains equilibrium
Anatomy	Pinna Auditory canal	Tympanic membrane Ossicles	Contains organ of Corti Auditory nerve starts here	Saccule and utricle Semicircular canals
Media	Air	Air (eustachian tube)	Fluid	Fluid

Path of vibration: Sound waves—vibration of tympanic membrane—vibration of hammer, anvil, and stirrup—vibration of oval window—fluid pressure waves in canals of inner ear lead to stimulation of hair cells—bulging of round window.

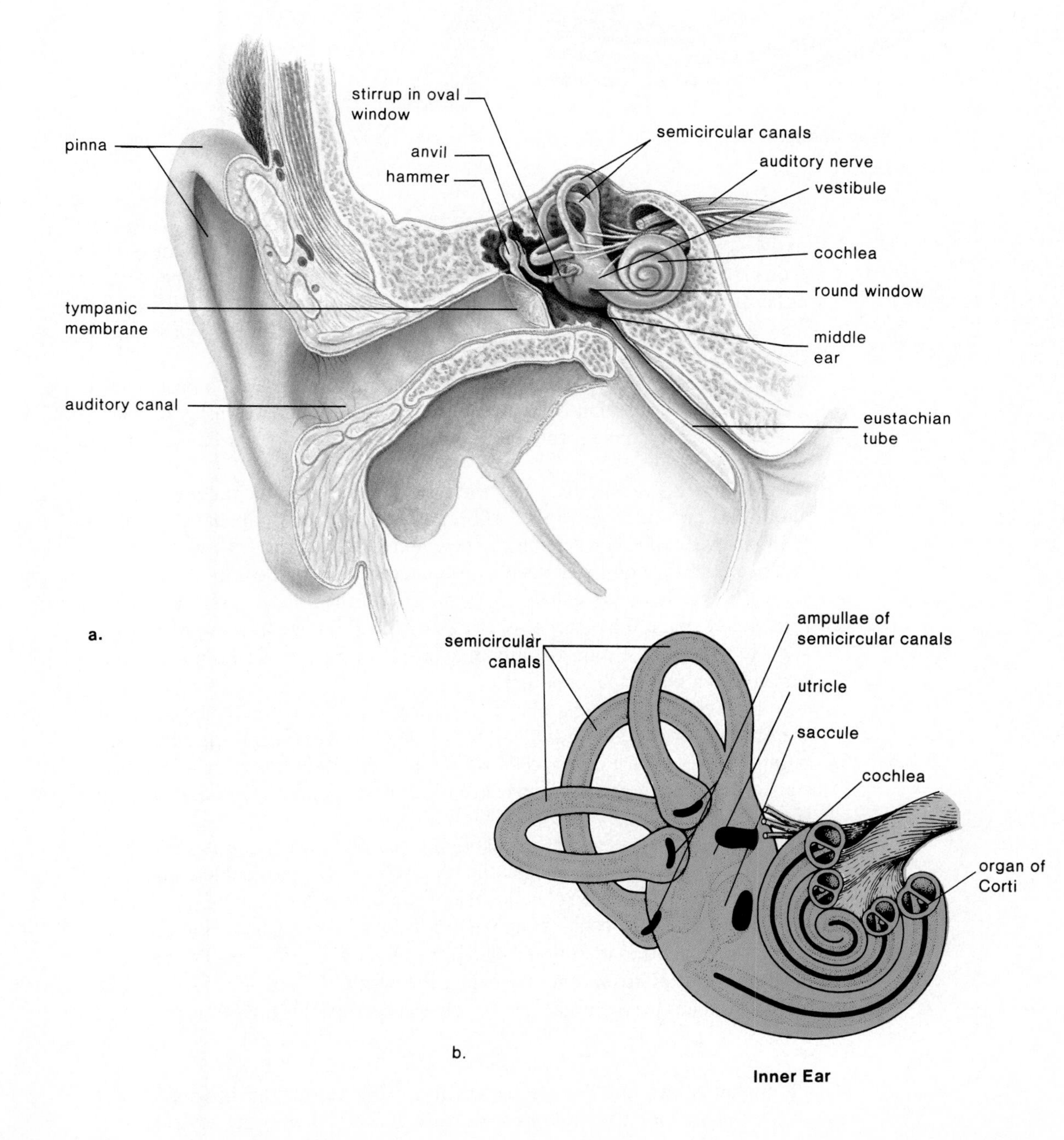

Figure 17.16
Anatomy of the human ear. a. *In the middle ear, the hammer, anvil, and stirrup amplify sound waves. Otosclerosis is a condition in which the stirrup becomes attached to the inner ear and is unable to carry out its normal function. It can be replaced by a plastic piston and thereafter the individual hears normally because sound waves are transmitted as usual to the cochlea that contains the receptors for hearing.* b. *Inner ear. The sense organs for balance are in the inner ear: the vestibule contains the utricle and saccule, and the ampullae are at the bases of the semicircular canals. The receptors for hearing are also in the inner ear: the cochlea has been cut to show the location of the organ of Corti.*

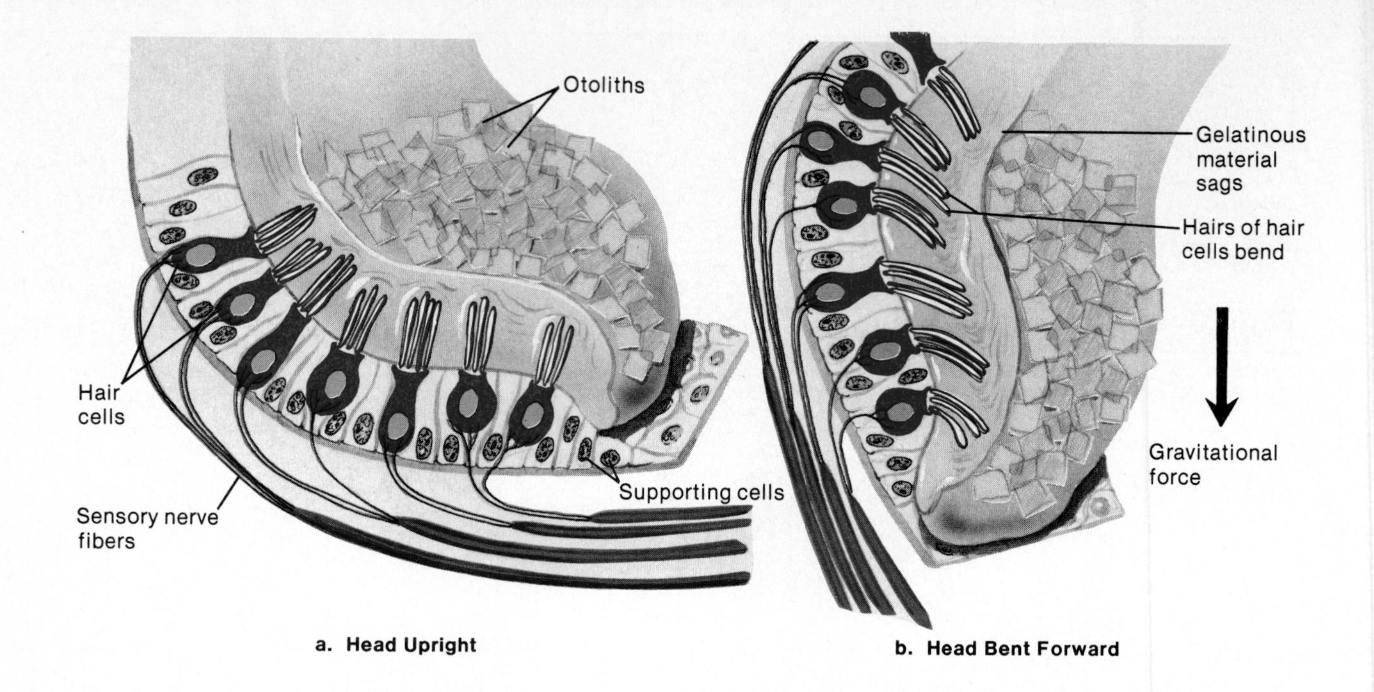

Figure 17.17
Receptor hair cells in the utricle and saccule are involved in our sense of static equilibrium: responsiveness to movement sideways or up and down. a. *When the head is upright otoliths are balanced directly on the cilia of hair cells.* b. *When the head is bent forward the otoliths shift and the cilia are bent causing nerve impulses to begin.*

between the tympanic membrane and the oval window. Collectively called the **ossicles,** individually they are the **hammer** (malleus), **anvil** (incus), and **stirrup** (stapes) (fig. 17.16) because their shapes resemble these objects. The hammer adheres to the tympanic membrane, and the stirrup touches the oval window. The posterior wall also has an opening that leads to many air spaces within the mastoid process.

The eustachian tubes extend from the middle ear to the nasopharynx and permit equalization of air pressure. Chewing gum, yawning, and swallowing in elevators and airplanes helps move air through the eustachian tubes upon ascent and descent.

Whereas the outer ear and middle ear contain air, the inner ear is filled with fluid. The **inner ear** (fig. 17.16b), anatomically speaking, has three areas: the first two, called the vestibule and semicircular canals, are concerned with balance; and the third, the cochlea, is concerned with hearing.

The **semicircular canals** are arranged so that there is one in each dimension of space. The base of each canal, called the **ampulla** (fig. 17.16b), is slightly enlarged. Within the ampullae are little hair cells whose cilia are inserted into a gelatinous medium.

The **vestibule** is a chamber that lies between the semicircular canals and the cochlea. It contains two small sacs called the **utricle** and **saccule.** Within both of these are little hair cells whose cilia protrude into a gelatinous substance. Resting on this substance are calcium carbonate granules, or **otoliths** (fig. 17.17).

The **cochlea** resembles the shell of a snail because it spirals. Within the tubular cochlea are three canals: the vestibular canal, the **cochlear canal,** and the tympanic canal. Along the length of the basilar membrane, which forms the lower wall of the cochlear canal, are little hair cells whose cilia come into contact with another membrane called the tectorial membrane. The hair cells, plus the **tectorial membrane,** are called the **organ of Corti** (fig. 17.19). When this organ sends nerve impulses to the cerebral cortex, it is interpreted as sound.

The outer ear, middle ear, and cochlea are necessary for hearing. The vestibule and semicircular canals are concerned with the sense of balance.

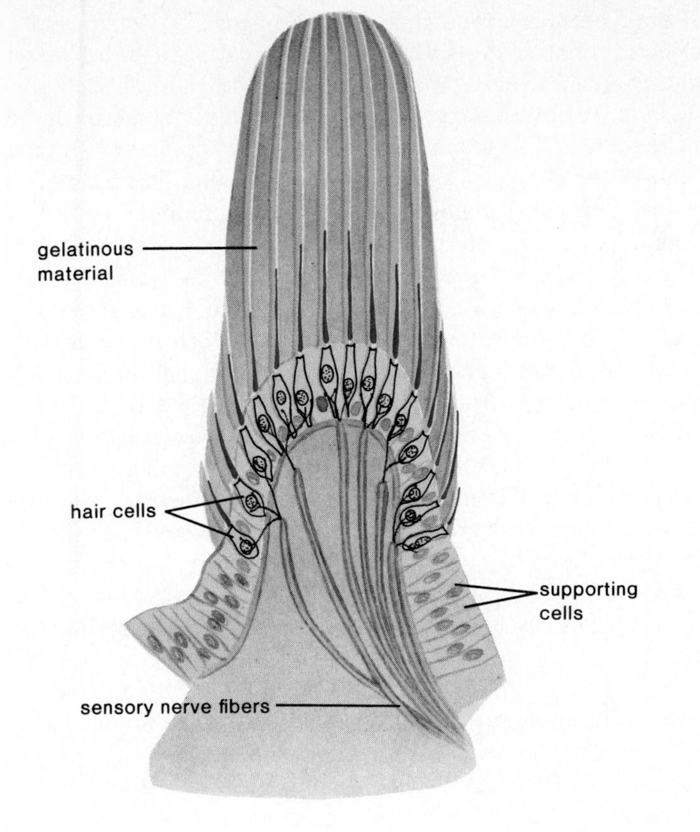

Figure 17.18
Receptor hair cells in an ampulla are involved in dynamic equilibrium. Within each ampulla hair cells are surrounded by a gelatinous material. When the body rotates, the fluid within the semicircular canals is displaced causing the material to move and the cilia to bend.

Physiology

Balance (Equilibrium) The sense of balance has been divided into two senses: *static equilibrium,* referring to knowledge of movement in one plane, either vertical or horizontal, and *dynamic equilibrium,* referring to knowledge of angular and/or rotational movement.

When the body is still, the otoliths in the utricle (fig. 17.17) and saccule rest on the hair cells. When the head and/or body moves horizontally or vertically, the granules in the utricle and saccule are displaced. Displacement causes the cilia to bend slightly so that the cell generates nerve impulses that travel by way of a cranial nerve to the brain.

When the body is moving about, the fluid within the semicircular canals moves back and forth. This causes bending of the cilia attached to hair cells within the ampullae (fig. 17.18) and they initiate nerve impulses that travel to the brain. Continuous movement of the fluid in the semicircular canals causes one form of motion sickness.

Movement of the otoliths within the utricle and saccule are important for static equilibrium. Movement of fluid within the semicircular canals contributes to our sense of dynamic equilibrium.

Hearing The process of hearing begins when sound waves enter the auditory canal. Just as ripples travel across the surface of a pond, sound travels by the successive vibrations of molecules. Ordinarily, sound waves do not carry much energy, but when a large number of waves strike the eardrum, it moves back and forth (vibrates) ever so slightly. The hammer then takes the pressure from the inner surface of the eardrum and passes it by way of the anvil to the stirrup in such a way that the pressure is multiplied about 20 times as it moves from the eardrum to the stirrup. The stirrup strikes the oval window, causing it to vibrate and in this way the pressure is passed to the fluid within the inner ear.

If the cochlea is unwound, as shown in figure 17.19b, it can be seen that the vestibular canal connects with the tympanic canal and that pressure waves will move from one canal to the other toward the round window, a membrane that can bulge to absorb the pressure. As a result of the movement of the fluid within the cochlea, the basilar membrane moves up and down, and the cilia of the hair cells rub against the tectorial membrane. This bending of the cilia initiates nerve impulses that pass by way of the auditory nerve to the brain, where the impulses are interpreted as a sound.

The organ of Corti is narrow at its base but widens as it approaches the tip of the cochlear canal. Each part of the organ is sensitive to different wave frequencies, or pitch. Near the tip, the organ of Corti responds to low pitches such as a tuba, and near the base it responds to higher pitches such as a bell or whistle. The neurons from each region along the length of the cochlea lead to slightly different areas in the brain. The pitch sensation we experience depends upon which of these areas of the brain is stimulated. Volume is a function of the amplitude of sound waves. Loud noises cause the fluid of the cochlea to oscillate to a greater degree, and this, in turn, causes the basilar membrane to move up and down to a great extent. The resulting increased stimulation is interpreted by the brain as loudness. It is believed that tone is an interpretation of the brain based on the distribution of hair cells stimulated.

The sense receptors for sound are hair cells on the basilar membrane (the organ of Corti). When the basilar membrane vibrates, the delicate hairs touch the tectorial membrane and nerve impulses begin and are transmitted in the auditory nerve to the brain.

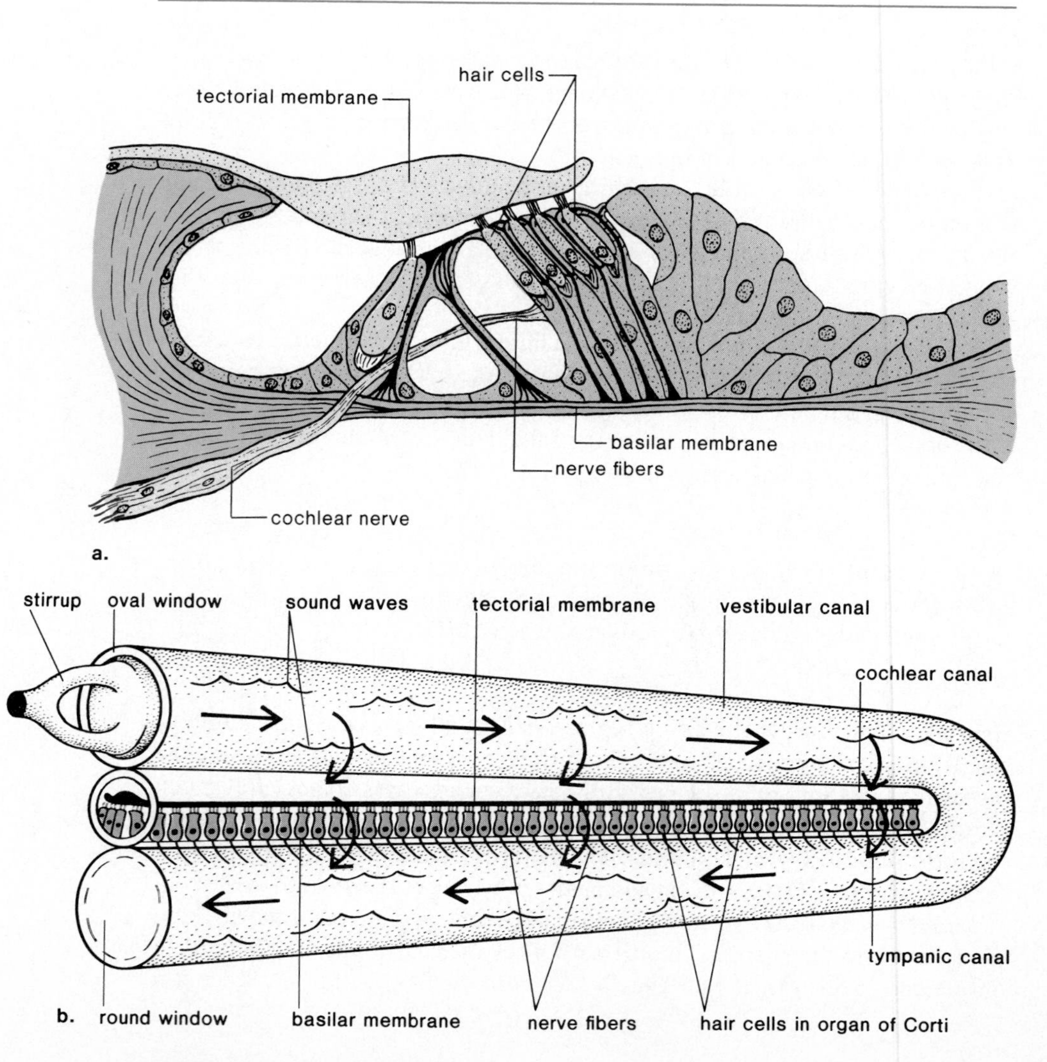

Figure 17.19
Organ of Corti.
a. *Enlarged cross section through the organ of Corti showing the receptor hair cells in side view.* b. *Cochlea unwound showing the placement of the organ of Corti along its length. The arrows represent the pressure waves that move from the oval window to the round window. These cause the basilar membrane to vibrate and the cilia of at least a portion of the 15,000 hair cells to bend against the tectorial membrane. The resulting nerve impulses result in hearing.*

Deafness There are two major types of deafness: *conduction* and *nerve deafness*. Conduction deafness can be due to a congenital defect, as those that occur when a pregnant woman contracts German measles during the first trimester of pregnancy. (For this reason every female should be sure to be immunized against rubella before the childbearing years.) Or, conduction deafness can be due to infections that have caused the ossicles to fuse together, restricting the ability to magnify sound waves. As mentioned in chapter 13, respiratory infections can spread to the ear by way of the eustachian tubes; therefore, every cold and ear infection should be taken seriously.

Nerve deafness most often occurs when cilia on the sense receptors within the cochlea have worn away. Since this may happen with normal aging, old people are more likely to have trouble hearing; however, nerve deafness also occurs when young people listen to loud music amplified to 130 decibels. Because the usual types of hearing aids are not helpful for nerve deafness, it is wise to avoid subjecting the ears to any type of continuous loud noise. Costly cochlear implants that directly stimulate the auditory nerve are available. Those who have these electronic devices report that the speech they hear is like that of a robot.

Picturing the Effects of Noise

"We have an idea of what noise does to the ear," David Lipscomb [of the University of Tennessee Noise Laboratory] says. "There's a pretty clear cause-effect relationship." And these photomicrographs of the cochlea's tiny structures graphically document noise trauma to the inner ear.

Hair cells transmit the mechanical energy of sound waves into those neural impulses that the brain interprets as sound. Loud noise can damage or destroy hair cells as these scanning electron micrographs illustrate.

Hair cells come in two varieties: a single row of inner cells and a triple row of outer ones. "Outer cells degenerate before inner cells," notes Clifton Springs, N.Y. otolaryngologist Stephen Falk. The most subtle change wrought by noise is a development of vesicles, or blister-like protrusions along the walls of the hair cells' stereocilia. Continued assault by noise will lead to a rupturing of the vesicles and damage. In addition, the "cuticular plate"—base tissue supporting the stereocilia—may soften, followed by a swelling and ultimate degeneration of hair cells.

But sensory hair cells are not the only structures at risk. Adjacent inner-ear cells . . . may undergo vacuolation—development of degenerative empty spaces in cells. Even nerve fibers synapsing at the hair cells' roots may die. In the final phase of noise-induced cochlear damage, the organ of Corti—of which hair cells and supporting cells are a part—is completely denuded and covered by a layer of scar tissue.

From *Science News*, 5 June 1982. Used by permission.

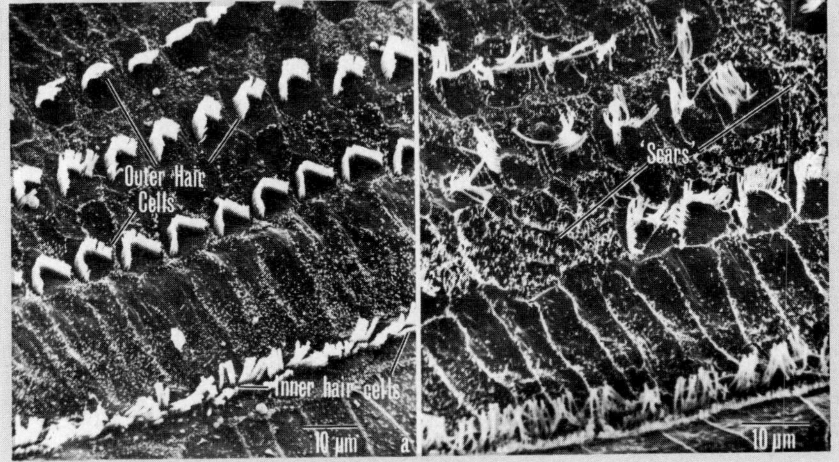

Damage to organ of Corti due to loud noise a. *Normal organ of Corti* b. *Organ of Corti after twenty-four-hour exposure to noise level typical of rock music. Note scars where cilia have worn away.*

Summary

All receptors are the first part of a reflex arc and they initiate nerve impulses that eventually reach the cerebrum where sensation occurs. Among the general receptors are those located in the skin and the chemoreceptors for taste and smell.

Vision is a specialized sense dependent on the eye, optic nerve, and occipital lobe of the brain. The rods, receptors for vision in dim light, and the cones, receptors that depend on bright light and provide color and detailed vision, are both located in the retina, the inner layer of the eyeball. The cornea, humors, and especially the lens bring the light rays to focus on the retina. To see a close object, accommodation occurs as the lens rounds up. Due to the optic chiasm, both sides of the brain must function together to give us three-D vision.

Hearing is a specialized sense dependent on the ear, auditory nerve, and the temporal lobe of the brain. The outer and middle portions of the ear simply convey and magnify the sound waves which strike the oval window. Its vibrations set up pressure waves within the cochlea which contains the organ of Corti, consisting of hair cells with the tectorial membrane above. When the hair cells strike this membrane, nerve impulses are initiated that finally result in hearing.

The ear also contains receptors for our sense of balance. Static equilibrium relies on the stimulation of hair cells by otoliths within the utricle and saccule. Dynamic equilibrium is dependent on the stimulation of hair cells within the ampullae of the semicircular canals.

Objective Questions

1. The sense organs for position and movement are called ____________ .
2. Taste buds and olfactory receptors are termed ____________ because they are sensitive to chemicals in the air and food.
3. The receptors for sight, the ____________ and ____________ , are located in the ____________ , the inner layer of the eye.
4. The cones give us ____________ vision and work best in ____________ light.
5. The lens ____________ for viewing close objects.
6. People who are nearsighted cannot see objects that are ____________ . A ____________ lens will restore this ability.
7. The ossicles are the ____________ , ____________ , and ____________ .
8. The semicircular canals are involved in our sense of dynamic ____________ .
9. The organ of Corti is located in the ____________ canal of the ____________ .
10. Vision, hearing, taste, and smell do not occur unless nerve impulses reach the proper portion of the ____________ .

Answers to Objective Questions

1. proprioceptors 2. chemoreceptors 3. rods, cones, retina 4. color, bright (day) 5. rounds up 6. distant, concave 7. hammer, anvil, stirrup 8. equilibrium 9. cochlear, cochlea 10. brain

Study Questions

1. Name three factors that all receptors have in common. (p. 352)
2. What type receptors are categorized as general and what type are categorized as special receptors? (p. 352)
3. Discuss the receptors of the skin, viscera, and joints. (p. 352)
4. Discuss the chemoreceptors. (pp. 353–354)
5. Describe the anatomy of the eye (pp. 355–357) and explain focusing and accommodation. (pp. 357–358)
6. Describe sight in dim light. What chemical reaction is responsible for vision in dim light? (p. 359) Discuss color vision. (p. 360)
7. Relate the need for corrective lenses to three possible shapes of the eye. (pp. 360–361) Discuss bifocals. (p. 362)
8. Describe the anatomy of the ear (pp. 362–366) and how we hear.
9. Describe the role of the utricle, saccule, and semicircular canals in balance. (p. 365)
10. Discuss the two causes of deafness, including why young people frequently suffer loss of hearing. (p. 367)

Key Terms

accommodation (ah-kom″o-da′shun) lens adjustment in order to see close objects. *358*

choroid (ko′roid) the vascular, pigmented middle layer of the wall of the eye. *356*

ciliary muscle (sil′e-er″e mus′el) a muscle that controls the curvature of the lens of the eye. *356*

cochlea (kok′le-ah) that portion of the inner ear that resembles a snail's shell and contains the organ of Corti, the sense organ for hearing. *364*

cones (kōnz) bright-light receptors in the retina of the eye that detect color and provide visual acuity. *356*

fovea centralis (fo′ve-ah sen-tral′is) region of the retina, consisting of densely packed cones, which is responsible for the greatest visual acuity. *357*

lens (lenz) a clear membranelike structure found in the eye behind the iris. The lens brings objects into focus. *356*

organ of Corti (or′gan uv kor′ti) a portion of the inner ear that contains the receptors for hearing. *364*

otoliths (o′to-liths) granules associated with ciliated cells in the utricle and saccule. *364*

proprioceptor (pro″pre-o-sep′tor) receptor that assists the brain in knowing the position of the limbs. *353*

retina (ret′ĭ-nah) the innermost layer of the eyeball that contains the rods and cones. *356*

rhodopsin (ro-dop′sin) visual purple, a pigment found in the rods. *359*

rods (rodz) dim-light receptors in the retina of the eye that detect motion but no color. *356*

saccule (sak′ūl) a saclike cavity that makes up part of the membranous labyrinth of the inner ear; contains receptors for static equilibrium. *364*

sclera (skle′rah) white fibrous outer layer of the eyeball. *355*

semicircular canals (sem″e-ser′ku-lar kah-nal′) tubular structures within the inner ear that contain the receptors responsible for the sense of dynamic equilibrium. *364*

tympanic membrane (tim-pan′ik mem′brān) membrane located between outer and middle ear that receives sound waves; the eardrum. *362*

utricle (u′tre-k′l) saclike cavity that makes up part of the membranous labyrinth of the inner ear; contains receptors for static equilibrium. *364*

18

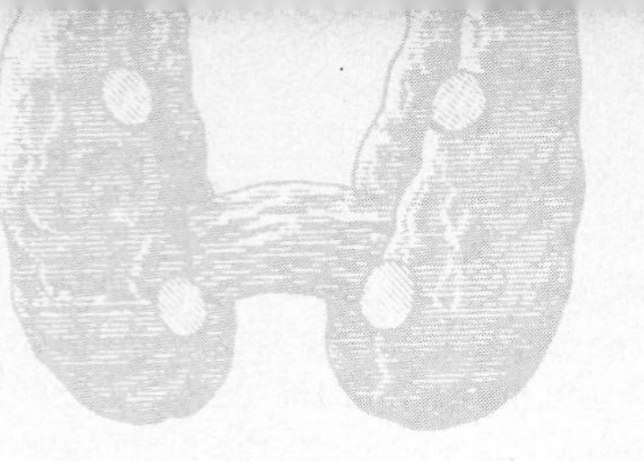

Hormones

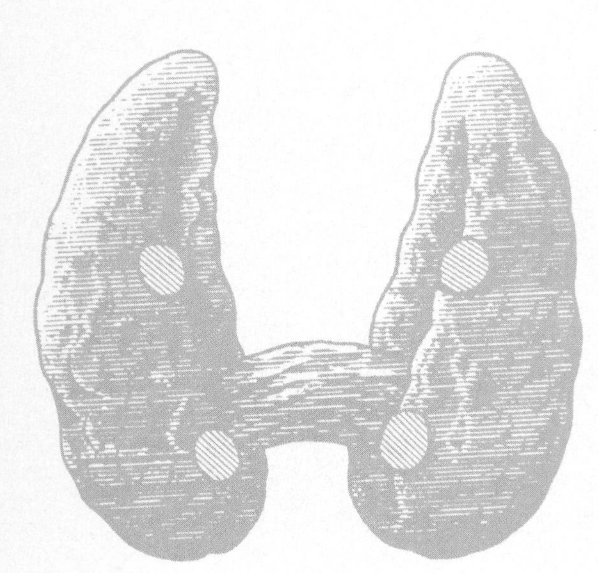

Chapter Concepts

1 The hormonal glands are usually ductless glands that secrete directly into the bloodstream.

2 Hormonal secretions coordinate the biochemical functioning of the body by acting on target organs.

3 In general, hormonal glands are controlled by a negative feedback mechanism.

4 Malfunctioning of hormonal glands can bring about a dramatic change in appearance and cause early death.

Chapter Outline

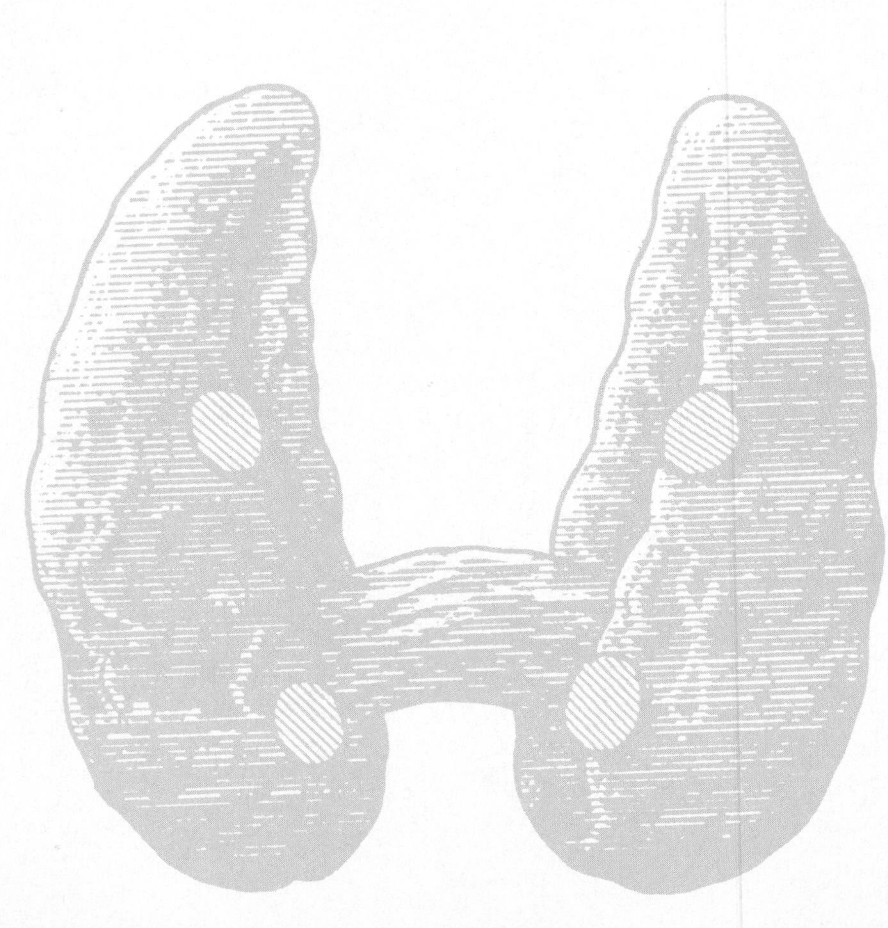

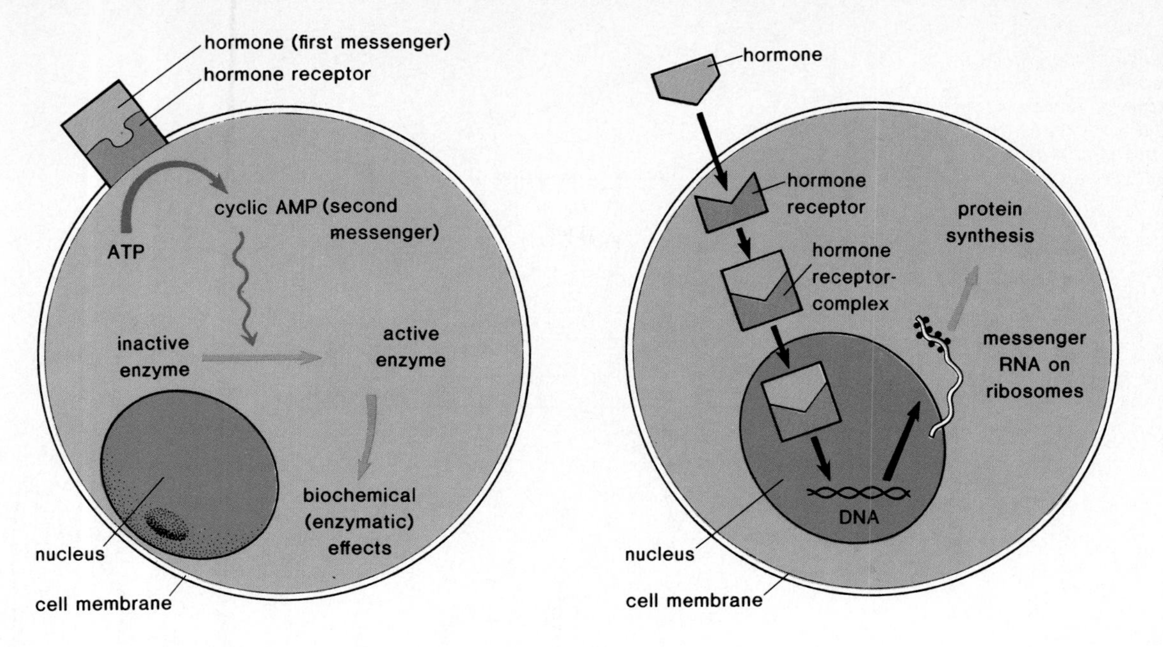

Figure 18.1
Cellular activity of hormones. a. *Peptide hormones combine with receptors located in the cell membrane. This promotes the production of cAMP, which in turn leads to activation of a particular enzyme.*
b. *Steroid hormones pass through the membrane to combine with receptors, and the complex is believed to activate certain genes, leading to protein synthesis.*

Along with the nervous system, hormones coordinate the functioning of body parts. Their presence or absence affects our metabolism, our appearance, and our behavior. It is now known that hormones are chemical regulators of cellular activity.

Hormones are organic substances that fall into two basic categories: (1) amino acids, polypeptides, or proteins and (2) steroid hormones. Steroids are complex rings of carbon and hydrogen atoms (fig. 1.32). The difference between steroids is due to the atoms attached to these rings.

When hormones of the first type are received by a cell, they bind to specific receptor sites (fig. 18.1a) in the membrane. This hormone-receptor complex activates an enzyme that produces for example cAMP (cyclic adenosine monophosphate). Cyclic AMP is a compound made from ATP, but it contains only one phosphate group attached to adenine at two locations. The cAMP now activates the enzymes of the cell to carry out their normal functions.

Unlike peptide hormones, steroid hormones pass through the cell membrane with no difficulty because they are relatively small and lipid soluble; there are receptor molecules present in the cytoplasm (fig. 18.1b). After the hormone has combined with the receptor, the hormone-receptor complex moves into the nucleus, where it binds with chromatin at a location that promotes activation of a particular gene. Protein synthesis follows. In this manner, steroid hormones lead to protein synthesis.

---

*H*ormones are chemical messengers that influence the metabolism of the cell either directly or indirectly, depending on the hormone type.

Figure 18.2
Anatomical location of major endocrine glands in the body. The hypothalamus controls the pituitary, which in turn controls the hormonal secretions of the thyroid, adrenal cortex, and sex organs. Both sets of sex organs are shown; ordinarily an individual has only one set of these.

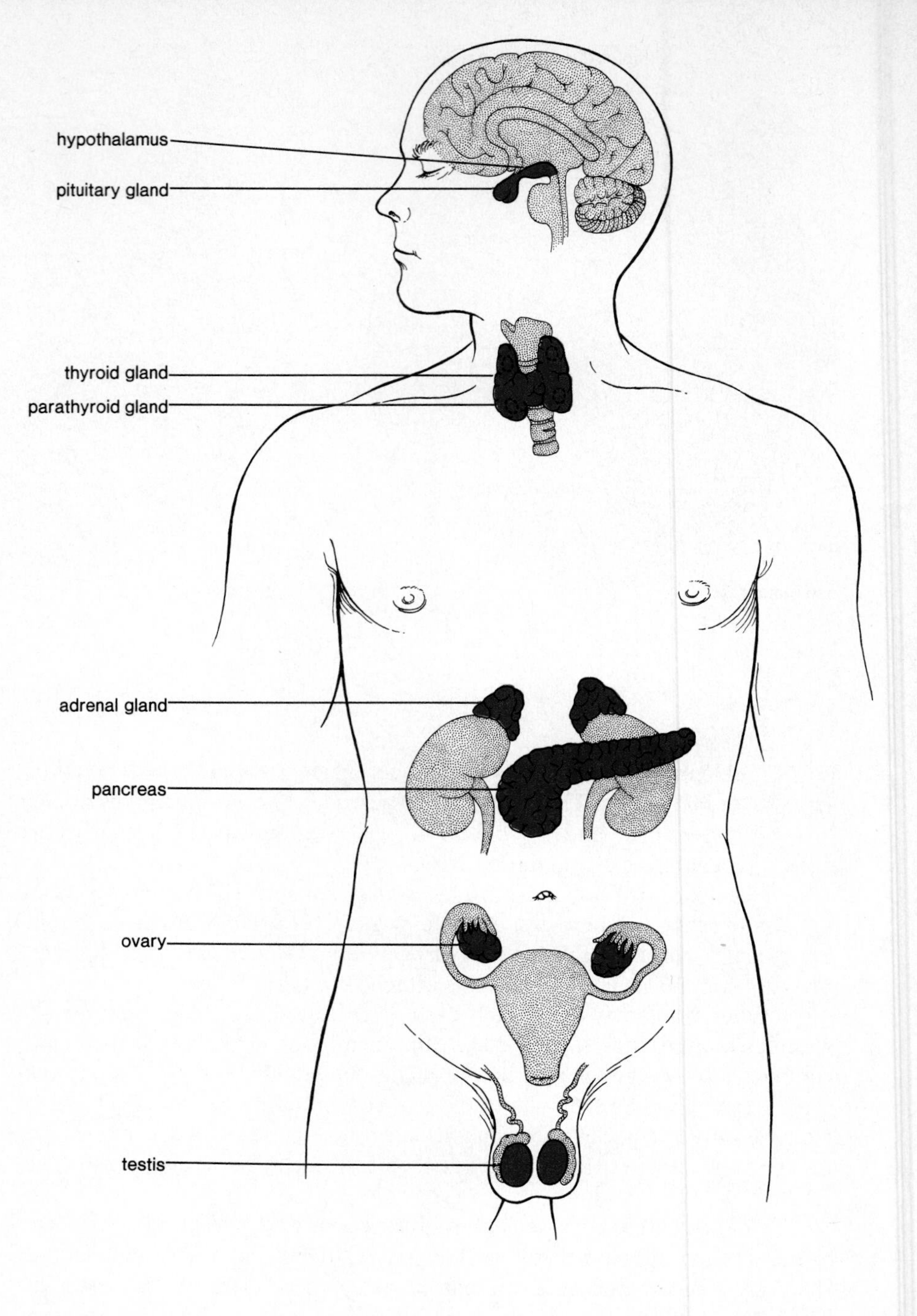

Hormones are produced by glands (fig. 18.2) called **endocrine glands** that secrete their products internally, placing them directly in the blood. Since these glands do not have ducts for the transport of their secretions, they are sometimes called *ductless glands.* All hormones are carried throughout the body by the blood, but each one affects only a specific part or parts, appropriately termed *target organs.*

Table 18.1 lists the major endocrine glands in humans, the hormones produced by each, and the associated disorders that occur when there is an abnormal level of the hormones—either too much or too little. The adrenal cortex and sex glands produce steroid hormones; the other glands produce hormones that are either amino acids, polypeptides, or proteins.

Table 18.1 The Principal Endocrine Glands and their Hormones

Gland	Hormones	Chief Functions	Disorders Too Much/Too Little
Hypothalamus	Releasing hormones	Regulate anterior pituitary hormone secretion	*See* anterior pituitary
Anterior pituitary	Thyroid-stimulating (TSH, thyrotropic)	Stimulates thyroid	*See* thyroid
	Adrenocorticotropic (ACTH)	Stimulates adrenal cortex	*See* adrenal cortex
	Gonadotropic	Stimulates gonads	*See* testis and ovary
	Follicle-stimulating (FSH)	Egg and sperm	
	Luteinizing (LH)	Sex hormones	
	Lactogenic (LTH, prolactin)	Milk production	
	Growth (GH, somatotropic)	Growth	Giant, acromegaly/midget
Posterior pituitary	Antidiuretic (ADH, vasopressin)	Water retention by kidneys	Diverse*/diabetes insipidus
	Oxytocin	Uterine contraction	
Thyroid	Thyroxin	Increases metabolic rate (cellular respiration)	Exophthalmic goiter/simple goiter myxedema, cretinism
	Calcitonin	Plasma level of calcium	Tetany/weak bones
Parathyroid	Parathormone (PTH)	Plasma levels of calcium and phosphorus	Weak bones/tetany
Adrenal cortex	Glucocorticoids (cortisol)	Gluconeogenesis	
	Mineralocorticoids (aldosterone)	Sodium retention; potassium excretion by kidneys	Cushing's syndrome/Addison's disease
	Sex hormones	Sex characteristics	
Adrenal medulla	Adrenalin (epinephrine)	Fight or flight	
Pancreas	Insulin	Lowers blood sugar	Shock/diabetes mellitus
	Glucagon	Raises blood sugar	
Testis	Androgens (testosterone)	Secondary male characteristics	Diverse/eunuch
Ovary	Estrogen (by follicle)	Secondary female characteristics	Diverse/masculinization
	Progesterone (by corpus luteum)		

*The word *diverse* in this table means that the symptoms have not been described as a syndrome in the medical literature.

The endocrine glands secrete their hormones into the bloodstream for transport to target organs.

Pituitary Gland

The pituitary gland, which has two portions called the **anterior pituitary** and the **posterior pituitary,** is a small gland about 1 centimeter in diameter that lies at the base of the brain.

Posterior Pituitary

The posterior pituitary is connected by means of a stalk to the hypothalamus, the portion of the brain concerned with homeostasis. The hormones released by the posterior pituitary are made in nerve cell bodies in the hypothalamus after which they migrate through axons that terminate in the posterior pituitary (fig. 18.3).

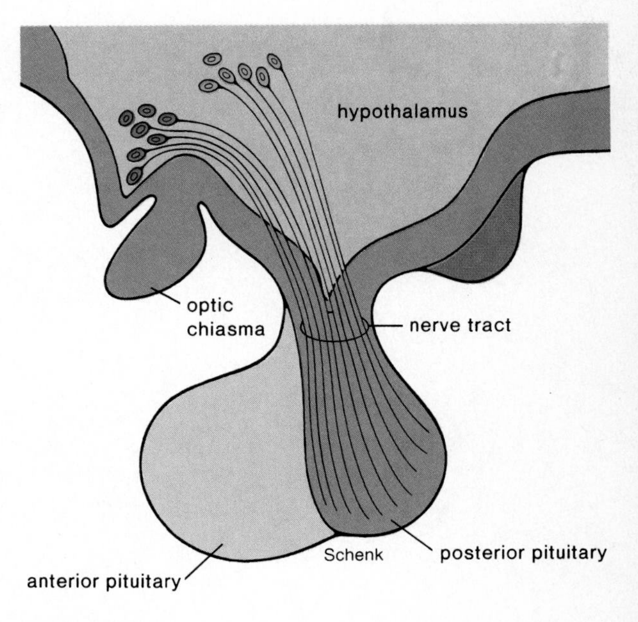

Figure 18.3
The posterior pituitary is connected to the hypothalamus by a stalk. The hypothalamus produces the hormones (ADH and oxytocin) that are secreted by the posterior pituitary. These are sent to the posterior pituitary by way of nerve fibers.

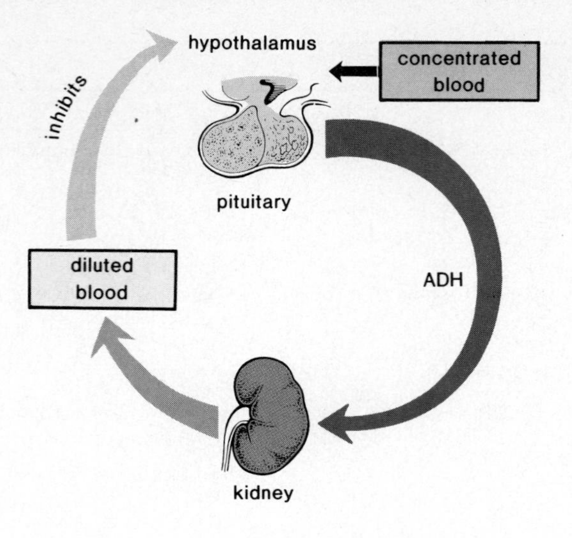

Figure 18.4
Regulation of ADH secretion. When the blood is concentrated, the hypothalamus produces ADH that is released by the posterior pituitary. This acts on the kidneys to retain more water so that the blood is diluted. Thereafter the hypothalamus does not produce ADH.

The posterior pituitary releases **antidiuretic hormone (ADH),** sometimes called **vasopressin.** ADH, as discussed in chapter 14, promotes the reabsorption of water from the collecting duct, a portion of the kidney tubules. It is believed that the hypothalamus contains osmoreceptors, or cells that are sensitive to the amount of water in the blood. When these cells detect that the blood lacks sufficient water, ADH is produced by hypothalamic neurons and is transported by their fibers to the posterior pituitary, where it is released (fig. 18.4). As the blood becomes more dilute, the hormone ceases to be produced and released. Figure 18.4 illustrates how the level of this hormone is controlled by a circular pattern in which the effect of the hormone (diluted blood) acts to shut down the production and release of the hormone. This is an example of control by negative feedback. Negative feedback mechanisms regulate the activities of most hormonal glands.

Inability to produce ADH causes **diabetes insipidus** (watery urine) in which a person produces copious amounts of urine with a resultant loss of salts from the blood. The condition can be corrected by the administration of ADH.

Oxytocin, is another hormone released by the posterior pituitary that is made in the hypothalamus. Oxytocin causes the uterus to contract and may be used to artificially induce labor. It also stimulates the release of milk from the breast when a baby is nursing.

The hormones of the posterior pituitary, ADH and oxytocin, are produced in the hypothalamus.

Anterior Pituitary

A portal system composed of tiny blood vessels connects the anterior pituitary to the hypothalamus. The hypothalamus controls the anterior pituitary by producing hypothalamic-releasing hormones that are transported to the anterior pituitary by the blood within the portal system connecting the two organs. Each type of releasing hormone causes the anterior pituitary either to secrete or to stop secreting a specific hormone. The anterior pituitary produces at least six different types of hormones (fig. 18.5), each by a distinct cell type.

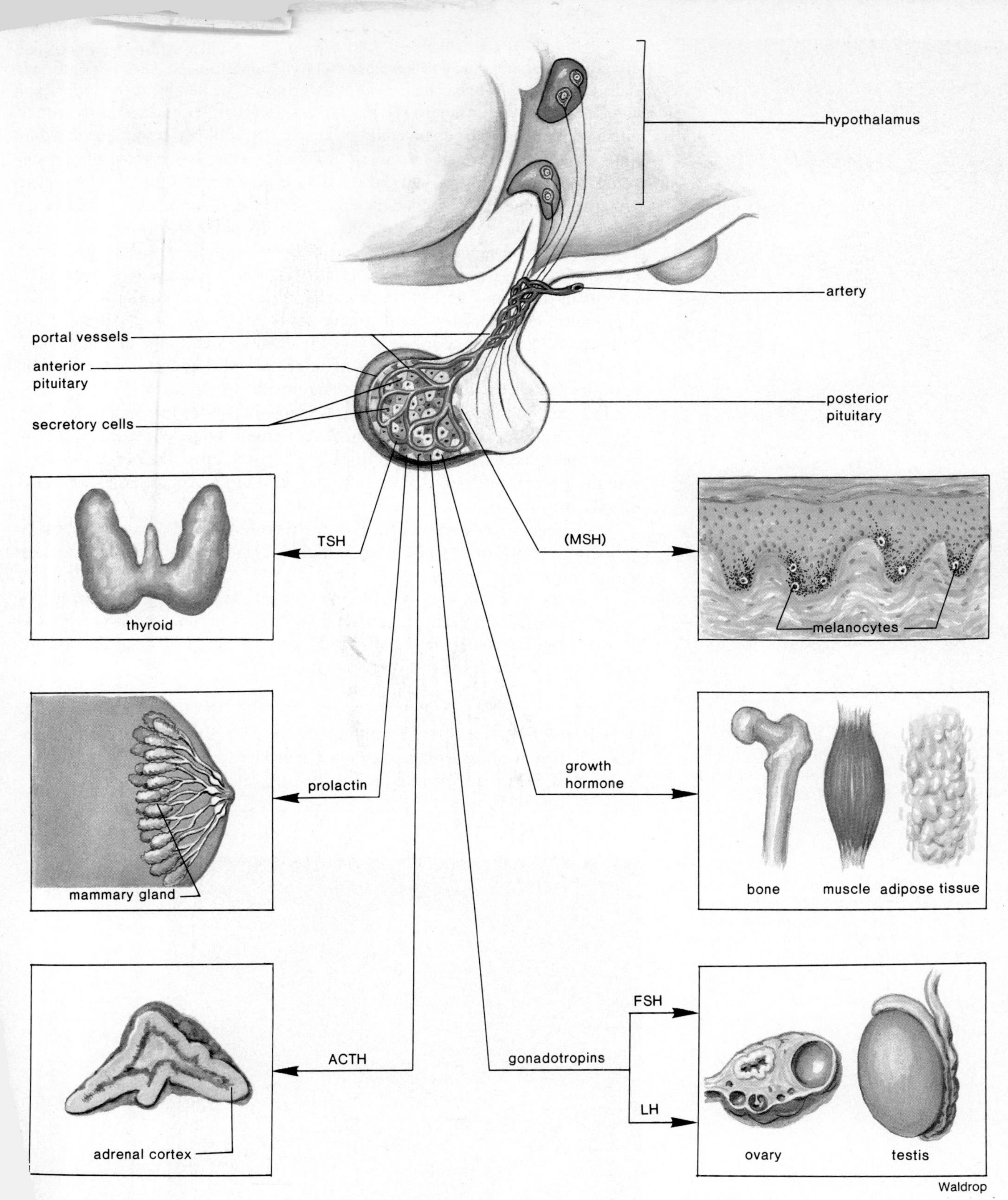

Figure 18.5
The anterior pituitary is connected to the hypothalamus only by a portal system. The hypothalamus sends releasing hormones to the anterior pituitary by this circulatory route. The releasing hormones specifically promote or inhibit the secretion of anterior pituitary hormones.

Figure 18.6
Manute Bol of the Washington Bullets. His height is 7′6¾″ due to a higher than usual amount of growth hormone produced by the anterior pituitary.

Three of the hormones produced by the anterior pituitary have a direct effect on the body. **Growth hormone (GH),** or somatotropin, dramatically affects the physical appearance since it determines the height of the individual (fig. 18.6). If little or no growth hormone is secreted by the anterior pituitary during childhood, the person could become a dwarf of perfect proportions but quite small in stature. If too much growth hormone is secreted, the person could become a giant. Giants usually have poor health, primarily because growth hormone has a secondary effect on blood sugar level, promoting an illness called diabetes (sugar) mellitus, to be discussed following.

Growth hormone promotes cell division, protein synthesis, and bone growth. It stimulates the transport of amino acids into cells and increases the activity of ribosomes, both of which are essential to protein synthesis. In bones, it promotes growth of the cartilaginous plates and causes osteoblasts to form bone (p. 171). Evidence suggests that the effects on cartilage and bone may actually be due to hormones called somatomedins, released by the liver. Growth hormone causes the liver to release somatomedins.

If the production of GH increases in an adult after full height has been obtained, only certain bones respond. These are the bones of the jaw, eyebrow ridges, nose, fingers, and toes. When these begin to grow, the person takes on a slightly grotesque look with huge fingers and toes, a condition called **acromegaly** (fig. 18.7).

Lactogenic hormone (LTH), also called **prolactin,** is produced in quantity only after childbirth. It causes the mammary glands in the breasts to develop and produce milk.

Melanocyte-stimulating hormone (MSH) causes skin color changes in lower vertebrates—but no one knows what it does in humans. However, it is derived from a molecule that is also the precursor for ACTH and the opioids, endorphins and enkephalins, discussed following.

GH and LTH are two hormones of the anterior pituitary. GH influences the height of children and brings about a condition called acromegaly in adults. LTH promotes milk production after childbirth.

Figure 18.7
Acromegaly is caused by the production of growth hormone in the adult. It is characterized by an enlargement of the bones in the face and fingers of an adult. Compare the normal-size fingers (hand at left) to those of the patient.

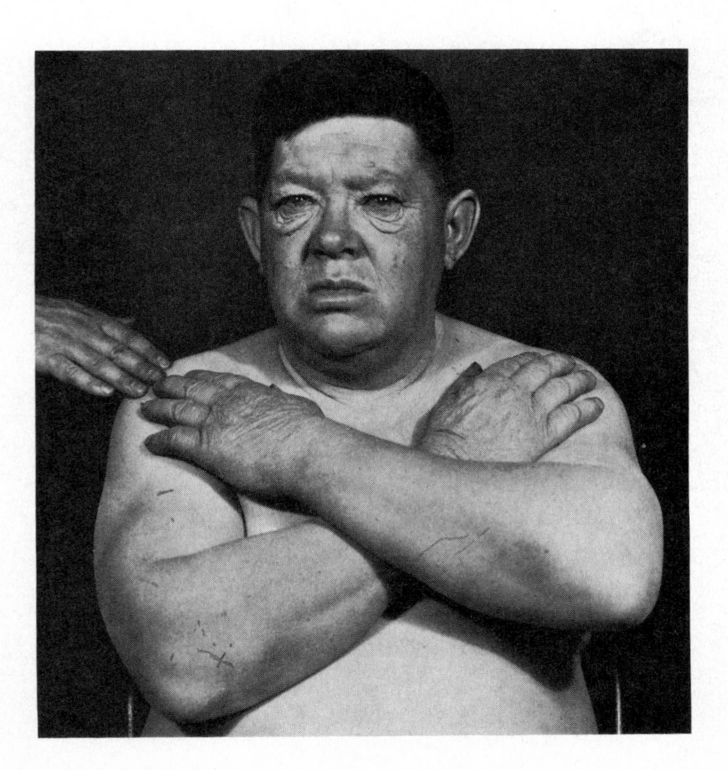

Master Gland

The anterior pituitary is called the master gland because it controls the secretion of other endocrine glands (fig. 18.5). As indicated in table 18.1, the anterior pituitary secretes the following hormones, which have an effect on other glands:

1. **TSH,** thyroid-stimulating hormone
2. **ACTH,** a hormone that stimulates the adrenal cortex
3. **Gonadotropic hormones** (FSH and LH) that stimulate the gonads, the testes in males and the ovaries in females

TSH causes the thyroid to produce thyroxin; ACTH causes the adrenal cortex to produce cortisol; and gonadotropic hormones cause the gonads to secrete sex hormones. Notice that it is now possible to indicate a three-tiered relationship between the hypothalamus, pituitary, and other endocrine glands. The hypothalamus produces releasing hormones that control the anterior pituitary, and the anterior pituitary produces hormones that control the thyroid, adrenal cortex, and gonads. Figure 18.8 illustrates the feedback mechanism that controls the activity of these glands.

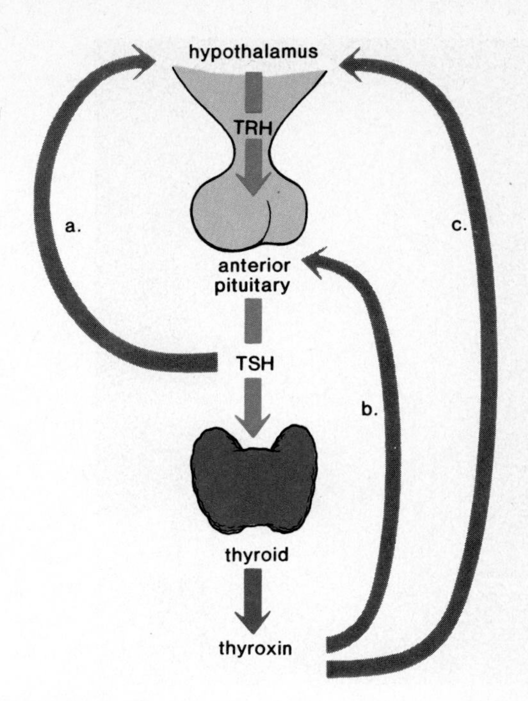

Figure 18.8
Control of hormone secretion. The level of thyroxin in the body is controlled in three ways, as shown: a. *the level of TSH exerts feedback control over the hypothalamus;* b. *the level of thyroxin exerts feedback control over the anterior pituitary, and* c. *over the hypothalamus. In this way thyroxin controls its own secretion. Substitution of the appropriate terms would also allow this diagram to illustrate control of cortisol and sex hormone levels.*

The hypothalamus, anterior pituitary, and the other endocrine glands controlled by the anterior pituitary are all involved in a self-regulating feedback loop.

Thyroid and Parathyroid Glands

Thyroid Gland

The thyroid gland (fig. 18.2) is located in the neck and is attached to the trachea just below the larynx. Internally (fig. 18.9), the gland is composed of a large number of follicles filled with thyroglobulin, the storage form of thyroxin. The production of both of these requires iodine. Iodine is actively transported into the thyroid gland, where the concentration may become as much

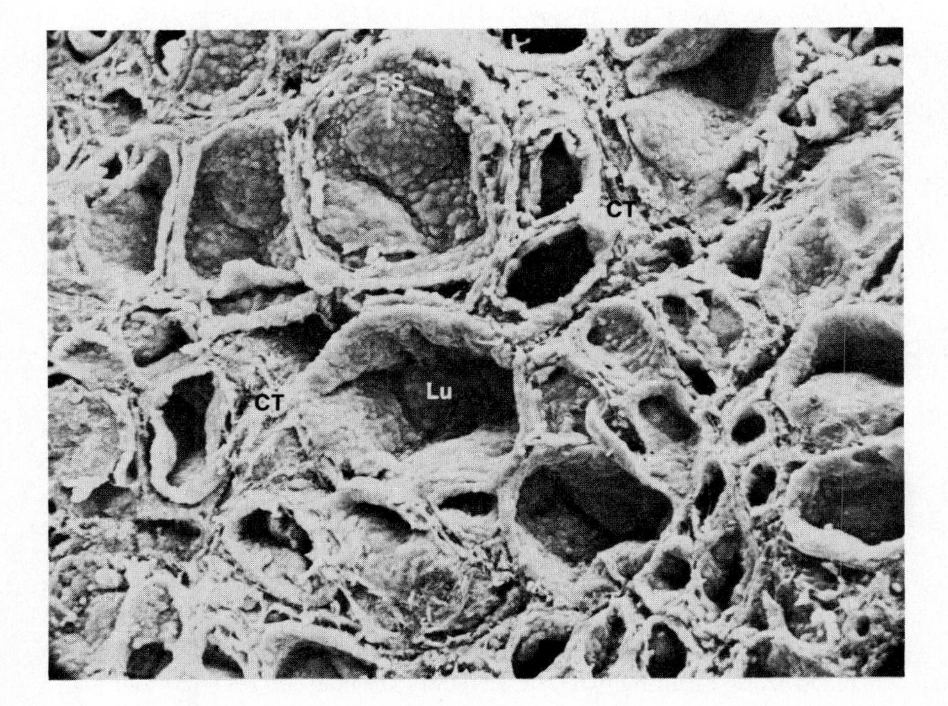

Figure 18.9
Thyroid gland. The thyroid is composed of many follicles, lined by epithelial cells, that secrete a precursor to thyroxin. ES = *epithelial cell surface;* Lu = *lumen of follicle;* CT = *connective tissue.*

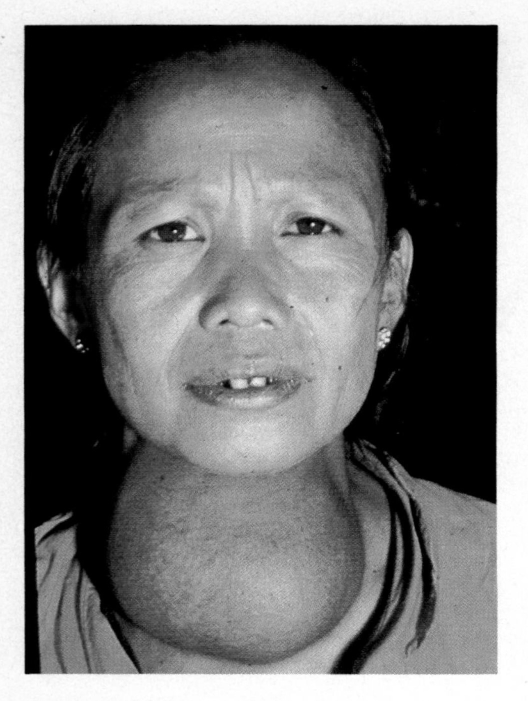

Figure 18.10
Simple goiter. An enlarged thyroid gland is often caused by a lack of iodine in the diet. Without iodine the thyroid is unable to produce thyroxin and continued anterior pituitary stimulation causes the gland to enlarge.

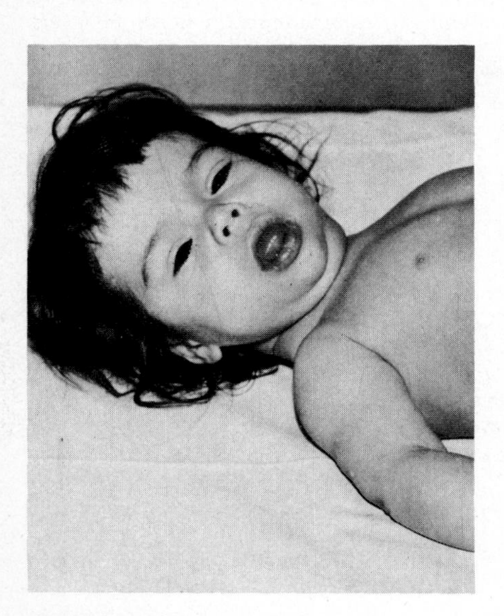

Figure 18.11
Cretinism. Cretins are individuals who have suffered from thyroxin insufficiency since birth or early childhood. Skeletal growth is usually inhibited to a greater extent than soft tissue growth; therefore, the child appears short and stocky. Sometimes the tongue becomes so large that it obstructs swallowing and breathing.

as 25 times that of the blood. If iodine is lacking in the diet, the thyroid gland enlarges, producing a goiter (fig. 18.10). The cause of this enlargement becomes clear if we refer to figure 18.8. When there is a low level of thyroxin in the blood, a condition called hypothyroidism, the anterior pituitary is stimulated to produce *TSH*. TSH causes the thyroid to increase in size so that enough **thyroxin** usually is produced. In this case, enlargement continues because enough thyroxin is never produced. An enlarged thyroid that produces some thyroxin is called a **simple goiter.**

Thyroxin

Thyroxin increases the metabolic rate. It does not have a target organ; instead, it stimulates most of the cells of the body to metabolize at a faster rate. The number of respiratory enzymes in the cell increases, as does oxygen uptake.

If the thyroid fails to develop properly, a condition called **cretinism** results. Cretins (fig. 18.11) are short, stocky persons who have had extreme hypothyroidism since infancy and/or childhood. Thyroid therapy can initiate growth, but unless treatment is begun within the first two months, mental retardation results. The occurrence of hypothyroidism in adults produces the condition known as **myxedema** (fig. 18.12), which is characterized by lethargy, weight gain, loss of hair, slower pulse rate, decreased body temperature, and thickness and puffiness of the skin. The administration of adequate doses of the thyroid hormone restores normal function and appearance.

In the case of hyperthyroidism (too much thyroxin), the thyroid gland is enlarged and overactive, causing a goiter to form and the eyes to protrude for some unknown reason. This type of goiter is called **exophthalmic goiter** (fig. 18.13). The patient usually becomes hyperactive, nervous, irritable, and suffers from insomnia. Removal or destruction of a portion of the thyroid by means of radioactive iodine is sometimes effective in curing the condition.

Calcitonin

In addition to thyroxin, the thyroid gland also produces the hormone **calcitonin.** This hormone helps regulate the calcium level in the blood and opposes the action of parathyroid hormone. The interaction of these two hormones is discussed following.

The anterior pituitary produces TSH, a hormone that promotes the production of thyroxin by the thyroid, a gland subject to goiters. Thyroxin, which speeds up metabolism, can affect the body as a whole, as exemplified by cretinism and myxedema.

Parathyroid Glands

The parathyroid glands are embedded in the posterior surface of the thyroid gland, as shown in figure 18.14. Many years ago, these four small glands were sometimes removed by mistake during thyroid surgery. Under the influence of **parathyroid hormone (PTH),** also called parathormone, the calcium (Ca^{++}) level in the blood increases and the phosphate (PO_4^{-3}) level decreases. The hormone stimulates the absorption of calcium from the gut, the retention of calcium by the kidneys, and the demineralization of bone. In other words, PTH promotes the activity of osteoclasts, the bone-resorbing cells. Although this also raises the level of phosphate in the blood, PTH acts on the kidneys to excrete phosphate in the urine. When a woman stops producing the female sex hormone estrogen following menopause, she is more likely to suffer from osteoporosis, characterized by a thinning of the bones. It is therefore reasoned that estrogen makes bones less sensitive to PTH.

If insufficient parathyroid hormone is produced, the level of calcium in the blood drops, resulting in **tetany.** In tetany, the body shakes from continuous

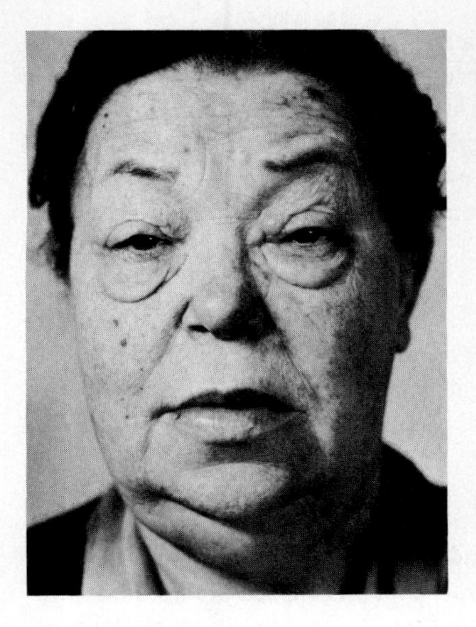

Figure 18.12
Myxedema is caused by thyroid insufficiency in the older adult. An unusual type of edema leads to swelling of the face and bagginess under the eyes.

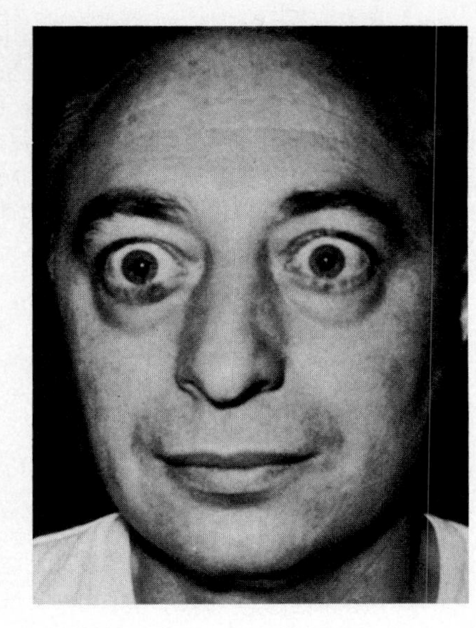

Figure 18.13
Exophthalmic goiter. Protruding eyes occur when an active thyroid gland enlarges.

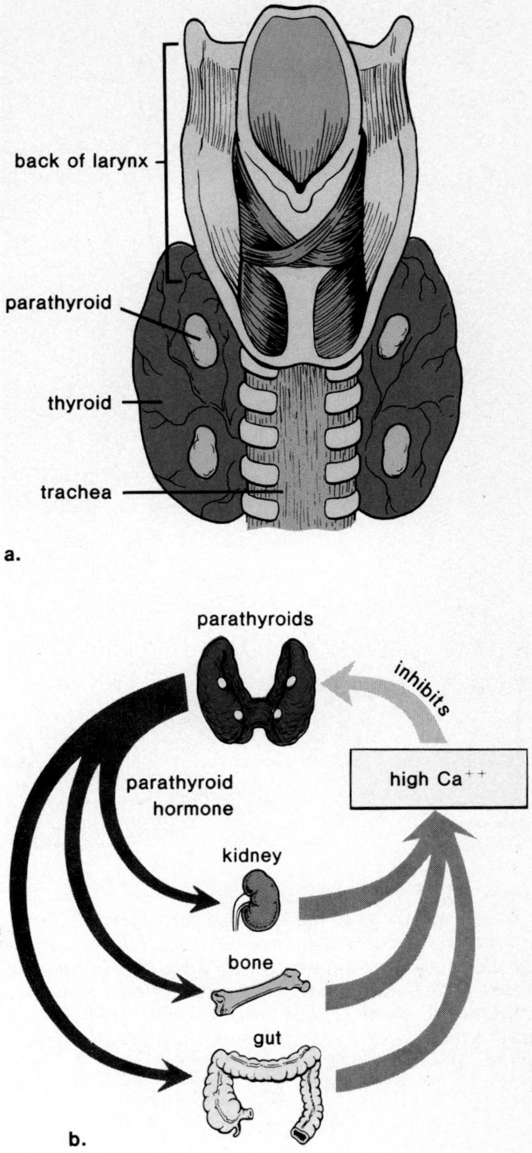

muscle contraction. The effect is really brought about by increased excitability of the nerves, which fire spontaneously and without rest. Calcium plays an important role in both nervous conduction and muscle contraction.

The level of PTH secretion is controlled by a feedback mechanism involving calcium (fig. 18.14b). When the calcium level rises, PTH secretion is inhibited, and when the calcium level lowers, PTH secretion is stimulated.

As mentioned previously, the thyroid secretes calcitonin, which also influences blood calcium level. Although calcitonin has the opposite effect of PTH, particularly on the bones, its action is not believed to be as significant. Still, the two hormones function together to regulate the level of calcium in the blood.

Parathyroid hormone maintains a high blood level of calcium by promoting its absorption in the gut, its reabsorption by the kidneys, and demineralization of bone. These actions are opposed by calcitonin produced by the thyroid.

Adrenal Glands

The adrenal glands, as their name implies (*ad* = near; *renal* = kidneys), lie atop the kidneys (fig. 18.2). Each consists of an outer portion, called the *cortex,* and an inner portion, called the *medulla.* These portions, like the anterior and posterior pituitary, have no connection with one another.

The hypothalamus exerts control over the activity of both portions of the adrenal glands. The hypothalamus can initiate nerve impulses that travel by way of the brain stem, nerve cord, and sympathetic nerve fibers (fig. 15.16) to the adrenal medulla, which then secretes its hormones. The hypothalamus, by means of ACTH-releasing hormone, controls the anterior pituitary's secretion of ACTH, which in turn stimulates the adrenal cortex. Stress of all types, including both emotional and physical trauma, prompts the hypothalamus to stimulate the adrenal glands.

Figure 18.14
Parathyroid glands. a. *These small glands are embedded in the posterior surface of the thyroid gland. Yet the parathyroids and thyroid glands have no anatomical or physiological connection with one another.* b. *Regulation of parathyroid hormone secretion. A low blood level of calcium causes the parathyroids to secrete parathyroid hormone, which causes the kidneys and gut to retain calcium and osteoclasts to break down bone. The end result is an increased level of calcium in the blood. A high blood level of calcium inhibits hormonal secretion of parathyroid hormone.*

Figure 18.15
It is now known that the anterior pituitary produces beta-endorphins, internal opioids that, like morphine, can produce a feeling of euphoria and a higher threshold of pain. It is possible that physical activity causes the release of endorphins and can elevate the mood. This would account for what is called "the runner's high."

Scientists were very interested to learn that ACTH and the opioids, endorphins, and enkephalins, are all chemically related. Although enkephalins are found in other parts of the nervous system, they are not predominant in the pituitary, as are ACTH and beta-endorphins. It is possible that not only ACTH but also beta-endorphins are released as a response to stress. The "runner's high" (fig. 18.15) has been attributed to the release of endorphin from the anterior pituitary.

The adrenal glands have two parts, an outer cortex and an inner medulla. The adrenal medulla is under nervous control and the cortex is under the hormonal control of ACTH, an anterior pituitary hormone.

Adrenal Medulla

The adrenal medulla (18.16a) secretes **adrenalin** and **noradrenalin.** The postganglionic fibers of the sympathetic nervous system also secrete noradrenalin. In fact, the adrenal medulla is often considered to be an adjunct to the sympathetic nervous system.

Adrenalin and noradrenalin are involved in the body's immediate response to stress. They bring about all those effects that occur when an individual reacts to an emergency. Blood glucose level rises, the metabolic rate increases, as does breathing and the heart rate. The blood vessels in the intestine constrict, but those in the muscles dilate. Increased circulation to the muscles causes them to have more strength than usual. The individual has a wide-eyed look and is extremely alert. Adrenalin has such a profound effect on the heart that it is often injected directly into a heart that has stopped beating in an attempt to stimulate its contraction.

The adrenal medulla releases adrenalin and noradrenalin into the bloodstream helping us cope with situations that seem to threaten our survival.

Figure 18.16
Adrenal glands. a. *The adrenal glands lie atop the kidneys and consist of the cortex and the medulla.* b. *Reaction to stress. Stress causes the hypothalamus to produce a releasing hormone that stimulates the anterior pituitary to produce ACTH. This hormone causes the adrenal cortex to produce cortisol, a hormone that brings about gluconeogenesis, which is thought to relieve stress.*

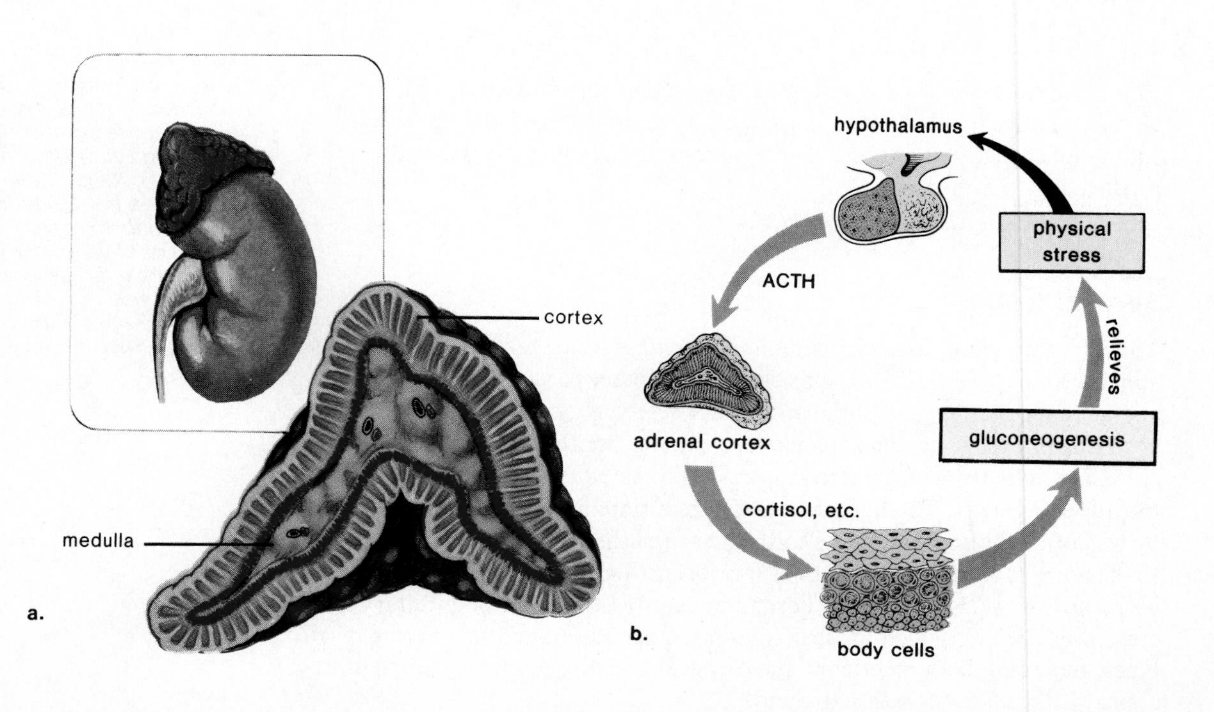

Adrenal Cortex

Although the adrenal medulla may be removed with no ill effects, the adrenal cortex is absolutely necessary to life. The two major types of hormones made by the adrenal cortex are the glucocorticoids and the mineralocorticoids. It also secretes a small amount of male and female sex hormones. All of these hormones are steroids.

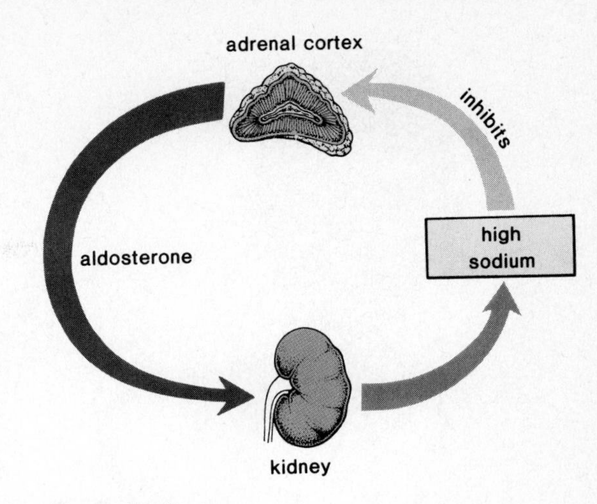

Figure 18.17
Regulation of aldosterone secretion. The adrenal cortex produces aldosterone, which acts on the kidneys to retain sodium. Once the sodium level in the blood is sufficient, the adrenal cortex no longer produces aldosterone.

Glucocorticoids

Of the various glucocorticoids, the hormone responsible for the greatest amount of activity, is cortisol. The secretion of cortisol helps an individual recover from stress (fig. 18.16b) such as an infection or trauma. **Cortisol** raises the level of amino acids in the blood, which, in turn, leads to an increased level of glucose when the liver converts these amino acids into glucose. It is said that the adrenal cortex brings about gluconeogenesis, or the production of glucose from nonglucose substances. Gluconeogenesis aids recovery because it allows an individual to maintain cellular respiration, especially in the brain, even when the body is not being supplied with dietary glucose. The amino acids not converted to glucose can be used for tissue repair should injury occur.

Cortisol also counteracts the inflammatory response (p. 242). During the inflammatory response, capillaries become more permeable and fluid leaks out, causing swelling in surrounding tissues. This causes the pain and swelling of joints that accompany arthritis and bursitis. The administration of cortisol aids these conditions because it reduces inflammation.

Mineralocorticoids

The secretion of mineralocorticoids, the most important of which is **aldosterone,** is not believed to be under the control of the anterior pituitary. These hormones regulate the level of sodium (Na^+) and potassium (K^+) in the blood, their primary target organ being the kidney where they promote renal reabsorption of sodium and renal excretion of potassium. The level of sodium in the blood seems to regulate the secretion of aldosterone (fig. 18.17).

The concentration of sodium in the blood affects the reabsorption of water by the kidneys. If the level of sodium falls too low, too little water is reabsorbed, blood volume falls, and hypotension results. If the level of sodium rises too high, too much water is reabsorbed, blood volume increases, and hypertension results. Sodium and potassium levels are also critical for nerve conduction and muscle contraction; in fact, cardiac failure may result from too low a level of potassium.

Cortisol leading to gluconeogenesis and aldosterone leading to sodium retention are two hormones secreted by the adrenal cortex.

Sex Hormones

The adrenal cortex produces a small amount of both male and female sex hormones. In males, the cortex is a source of female sex hormones and in females, it is a source of male hormones. A tumor in the adrenal cortex can cause the production of a large amount of sex hormones, which can lead to feminization in males and masculinization in females.

Disorders

Addison's Disease When there is a low level of adrenal cortex hormones in the body, the person begins to suffer from Addison's disease. Because of the lack of cortisol, the patient is unable to maintain the glucose level of the blood, tissue repair is suppressed, and there is a high susceptibility to any kind of stress. Even a mild infection can cause death. Due to the lack of aldosterone, the blood sodium level is low and the person experiences low blood pressure along with acidosis and low pH. In addition, the patient has a peculiar bronzing of the skin (fig. 18.18).

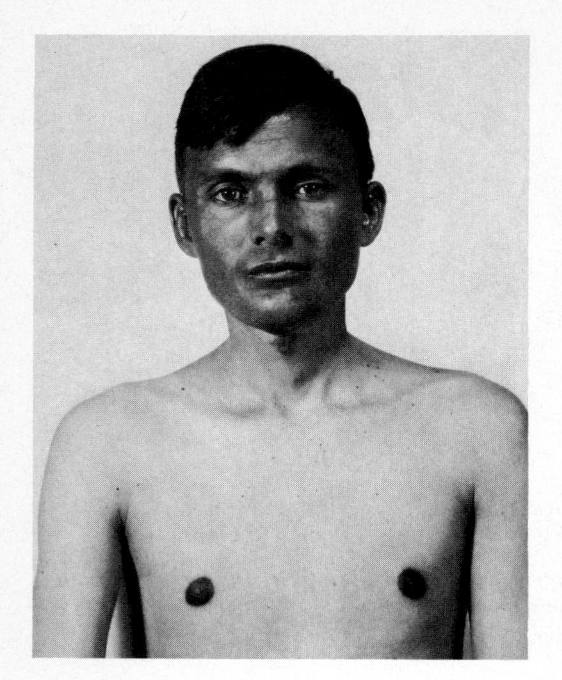

Figure 18.18
Addison's disease. This condition is characterized by a peculiar bronzing of the skin, as seen in the face and the thin skin of the nipples of this patient.

Cushing's Syndrome When there is a high level of adrenal cortex hormones in the body, the person suffers from Cushing's syndrome (fig. 18.19). Cortisol causes a tendency toward diabetes mellitus, a decrease in muscular protein, and an increase in subcutaneous fat. Because of these effects, the person usually develops thin arms and legs and an enlarged trunk. Due to the high level of sodium in the blood, the patient has alkaline blood, hypertension, and edema of the face, which gives the face a moon shape.

Addison's disease is due to adrenal cortex hyposecretion and Cushing's syndrome is due to adrenal cortex hypersecretion.

Pancreas

The pancreas is a long, soft organ that lies transversely in the abdomen (fig. 18.20) between the kidneys and near the duodenum of the small intestine. It is composed of two types of tissues; one of these produces and secretes the digestive juices that go by way of the pancreatic duct to the small intestine, and the other type, called the **islets of Langerhans,** produces and secretes the hormones insulin and glucagon directly into the blood. **Insulin** and glucagon are hormones that affect the blood glucose level in opposite directions—insulin decreases the level and **glucagon** increases the level of glucose.

Insulin is secreted when there is a high level of glucose in the blood, which usually occurs just after eating. Once the glucose level returns to normal, insulin is not secreted, as illustrated in figure 18.21. Insulin is believed to cause all the cells of the body to take up glucose. When the liver and muscles take up glucose, they convert to glycogen any glucose not needed immediately. Therefore, insulin promotes the storage of glucose as glycogen.

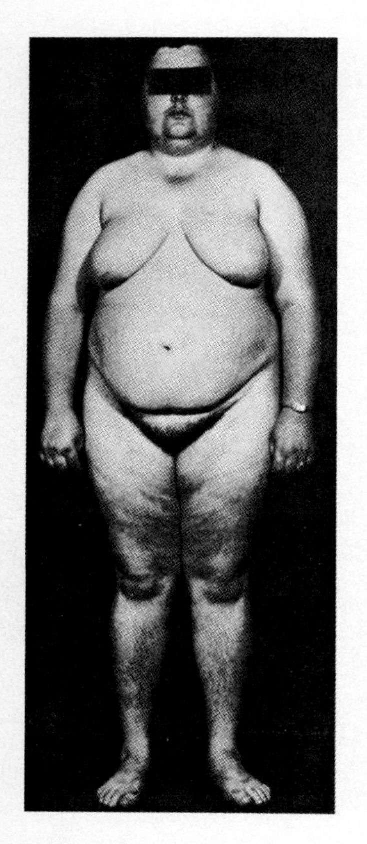

Figure 18.19
Cushing's syndrome. Persons with this condition tend to have an enlarged trunk and moonlike face. Masculinization may occur in women due to the excessive male sex hormones in the body.

Diabetes Mellitus

The symptoms of **diabetes mellitus** (sugar diabetes) include the following:

Sugar in the urine
Frequent, copious urination
Abnormal thirst
Rapid loss of weight
General weakness
Drowsiness and fatigue
Itching of the genitals and skin
Visual disturbances, blurring
Skin disorders, such as boils, carbuncles, and infection

Many of these symptoms develop because sugar is not being metabolized by the cells. The liver fails to store glucose as glycogen, and all the cells fail to utilize glucose as an energy source. This means that the blood glucose level rises very high after eating, causing glucose to be excreted in the urine. More water than usual is therefore excreted so that the diabetic is extremely thirsty.

Since carbohydrates are not being metabolized, the body turns to the breakdown of proteins and fat for energy. Unfortunately, the breakdown of these molecules leads to the buildup of acids in the blood (acidosis) and respiratory distress. It is the latter that can eventually cause coma and death of the diabetic. The symptoms that lead to coma (table 18.2) develop slowly.

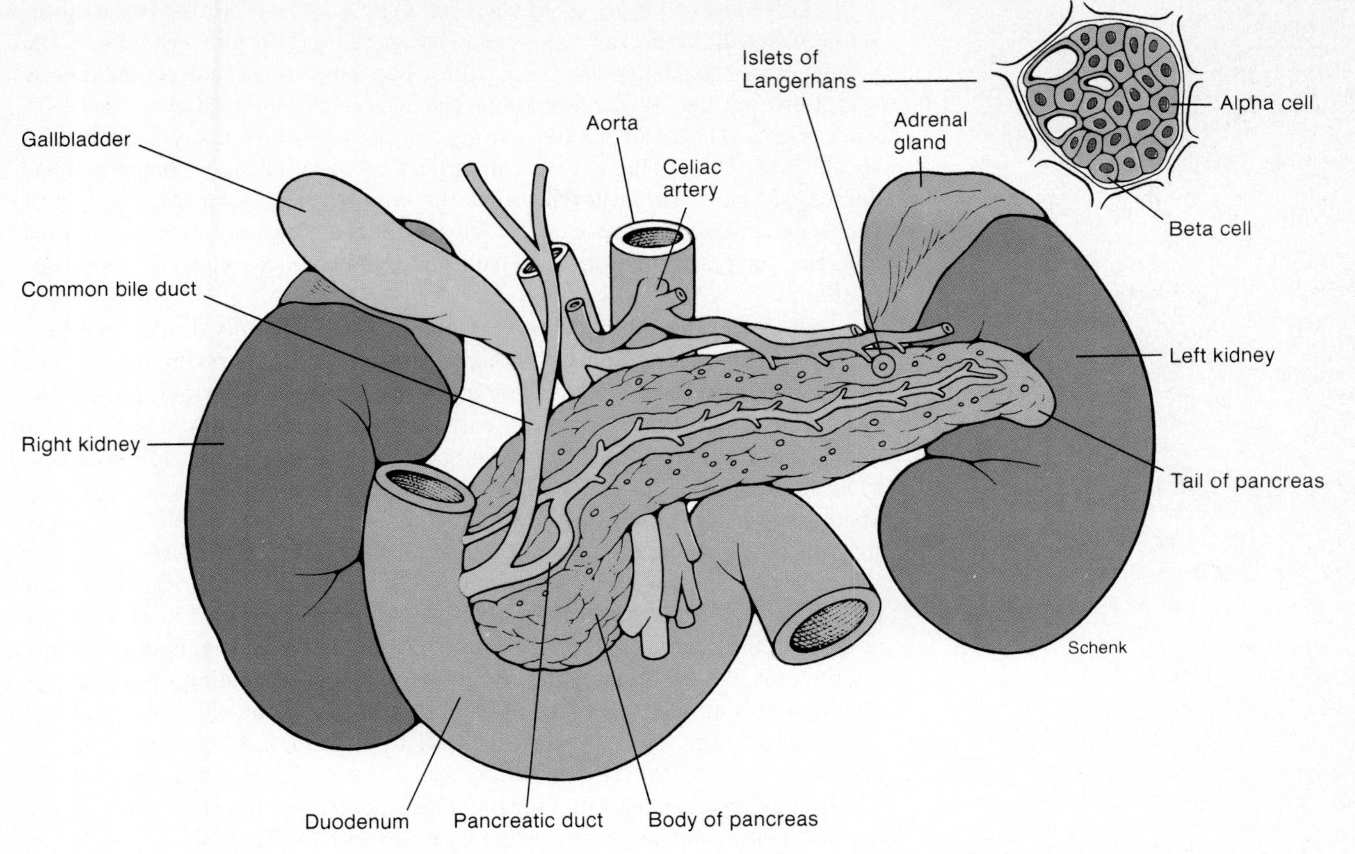

Figure 18.20
Gross and microscopic anatomy of the pancreas. The pancreas lies in the abdomen between the kidneys near the duodenum. As an exocrine gland it secretes digestive enzymes that enter the duodenum by the common bile duct. As an endocrine gland it secretes insulin and glucagon into the bloodstream. (Top right) The "alpha" cells of the islets of Langerhans produce glucagon and the "beta" cells produce insulin.

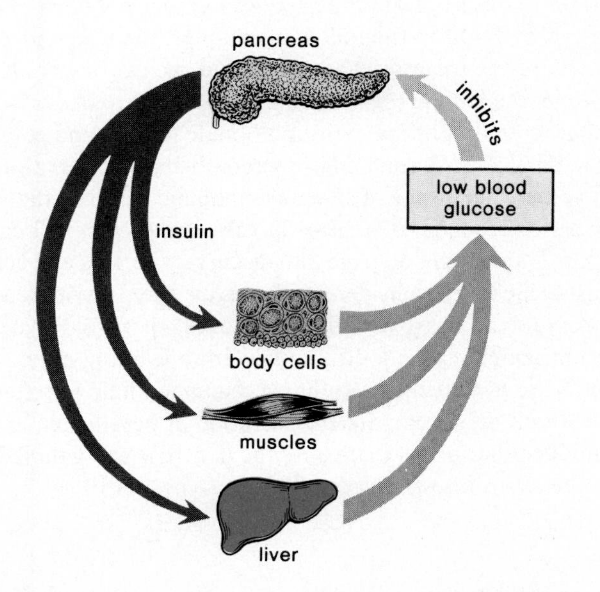

Figure 18.21
Regulation of insulin secretion. In response to a high blood sugar level, the pancreas secretes insulin, which promotes the uptake of glucose in body cells, muscles, and the liver. As a result of a low blood glucose level, the pancreas stops secreting insulin.

Table 18.2 *Symptoms of Insulin Shock and Diabetic Coma*

Insulin Shock	Diabetic Coma
Sudden onset	Slow, gradual onset
Perspiration, pale skin	Dry, hot skin
Dizziness	No dizziness
Palpitation	No palpitation
Hunger	No hunger
Normal urination	Excessive urination
Normal thirst	Excessive thirst
Shallow breathing	Deep, labored breathing
Normal breath odor	Fruity breath odor
Confusion, disorientation, strange behavior	Drowsiness and great lethargy leading to stupor
Urinary sugar absent or slight	Large amounts of urinary sugar
No acetone in urine	Acetone present in urine

From Dolger, Henry and Bernard Seeman, *How to Live with Diabetes*. Copyright © 1972, 1965, 1958 W. W. Norton & Company, Inc., New York. Reprinted by permission.

There are two types of diabetes. In *Type I diabetes,* formerly called juvenile onset diabetes, the pancreas is not producing insulin. Therefore, the patient must have daily insulin injections. These injections control the diabetic symptoms but may still cause inconveniences since either an overdose of insulin or the absence of regular eating can bring on the symptoms of insulin shock (table 18.2). These symptoms appear because the blood sugar level has decreased below normal levels. Since the brain requires a constant supply of sugar, unconsciousness results. The cure is quite simple: an immediate source of sugar, such as a sugar cube or fruit juice, can counteract insulin shock immediately.

Obviously, insulin injections are not the same as a fully functioning pancreas that responds on demand to high glucose level by supplying insulin. For this reason, some doctors advocate a pancreas transplant, which does, however, require that the recipient take immunosuppressive drugs.

Of the over six million people that now have diabetes in the United States, at least five million have *Type II diabetes* formerly called maturity-onset diabetes. In this type diabetes, now known to occur in obese people of any age, the pancreas is producing insulin, but the cells do not respond to it. At first the cells lack the receptors necessary to detect the presence of insulin, and then later the cells are even incapable of taking up glucose. If Type II is left untreated, the results can be as serious as Type I diabetes. Diabetics are prone to blindness, kidney disease, and circulatory disorders including strokes. Pregnancy carries an increased risk of diabetic coma, and the child of a diabetic is somewhat more likely to be stillborn or to die shortly after birth. It is important, therefore, that Type II diabetes be prevented or at least controlled. The best defense is a nonfattening diet and regular exercise. If that fails, there are oral drugs that make the cells more sensitive to the effects of insulin or stimulate the pancreas to make more of it.

The most common illness due to hormonal imbalance is diabetes mellitus, caused by a lack of insulin or insensitivity of cells to insulin. Insulin lowers blood glucose levels by causing the cells to take it up and the liver to convert it to glycogen.

Other Hormones

It is appropriate to discuss certain hormones in other chapters of the text. As will be discussed in detail in the following chapter, the testes produce the androgens, which are the male sex hormones, and the ovaries produce estrogen and progesterone, the female sex hormones. The sex hormones control the secondary sex characteristics of the male and female (p. 398 and p. 406). Among other traits, males have greater muscle strength than do females. Generally, athletes believe that the intake of so-called anabolic steroids, that is, the male sex hormone testosterone or synthetically related steroids, will cause greater muscle stength. The reading on page 386 discusses the pros and cons of taking these steroids, which are considered illegal by the International Olympic Committee. Any Olympic athlete whose urine tests positive for steroids at the time of an event is immediately disqualified from winning a medal.

Previously we have mentioned that the stomach lining produces the hormone *gastrin* which helps regulate the secretion of pepsinogen and the intestinal lining produces hormones that cause the pancreas and gallbladder to send secretions to the small intestine (p. 190).

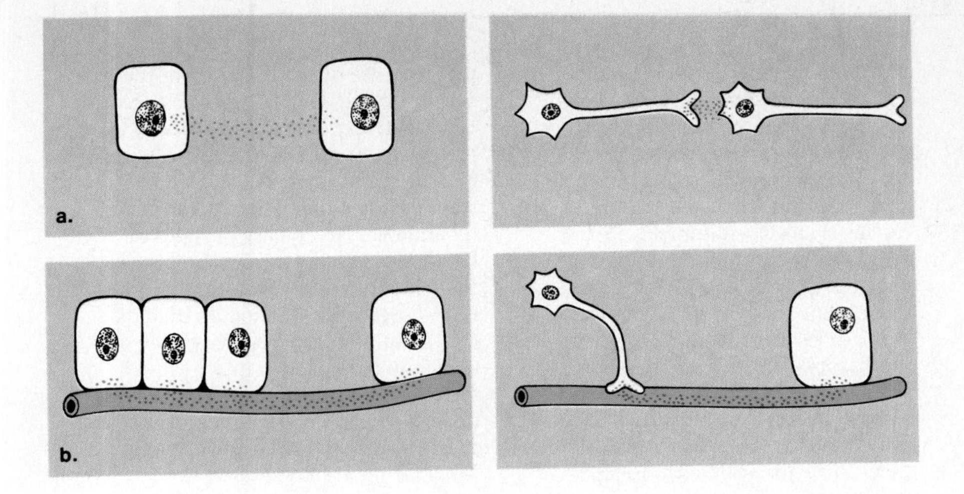

Figure 18.22
Local hormones versus nonlocal hormones. a. *Some hormones are produced and effective in the immediate environment and are therefore called local hormones. It would be possible to include neurotransmitters in this category.* b. *Some hormones are secreted into the bloodstream and transported by the blood to target organs. It would be possible to include hypothalamic releasing hormones in this category.*

It should be mentioned also that the thymus gland secretes *thymosins,* a family of hormones that assure the ability of the immune system to combat disease.

Local Hormones

Aside from hormones produced by endocrine glands and carried in the bloodstream, there are hormones, called *autocoids,* that act on tissues or cells in their immediate vicinity. Such hormones could even be said to include the well-known neurotransmitters (fig. 18.22) and the newly discovered *opioids,* the endorphins and enkephalins.

Other autocoids we have mentioned previously are bradykinin, histamine, and prostaglandins. All of these are involved in the inflammatory reaction (p. 242) and the origination of stimuli that result in the sensation of pain.

Prostaglandins

Prostaglandins (PG) are derived from cell membrane phospholipids. First discovered in semen, they were named after the prostate gland from which they were thought to originate. The prostaglandins in semen cause the uterus to contract during sexual intercourse which is believed to assist the movement of sperm as they traverse this organ. Because of their effect on the uterus, they are sometimes used to abort a fetus. Prostaglandins are also being considered for possible use with a self-administered, once-a-month means of preventing pregnancy (p. 408). Prostaglandins are involved in many other aspects of reproduction and development.

Prostaglandins also function in nonreproductive organs. They are being considered as possible treatment for ulcers because they reduce gastric secretion; as treatment of hypertension because they lower blood pressure; and as prevention of thrombosis because they inhibit platelet aggregation.

Sometimes prostaglandins have contrary effects. For example, one type helps prevent blood clots but another helps blood clots to form. Also a large dose of PG may have an effect that is opposite to that of a small dose. Therefore, it has been very difficult to standardize PG therapy and, in most instances, prostaglandin drug therapy is still considered experimental.

The Toughest Test for Athletes

The Olympic Committee added anabolic steroids to its list of banned substances in 1973. Since then, the drugs have become ever more widely used as men and women seek to push their bodies to still higher levels of attainment. In the U.S., where synthetic steroids were developed about half a century ago, their use is thought to extend from world-class athletes to high school football players. Soviet and East European trainers are widely believed to have been giving the drugs to their athletes since the 1950s. Nor will it be possible for athletes to escape detection simply by stopping use of steroids immediately before the Los Angeles Games: their presence in the body can be detected as much as six months after the drugs have been taken. Indeed, some doctors suspect that fear of detection may have contributed to the Soviet decision to boycott the Los Angeles Games.

Being a steroid user may cost an athlete far more than his or her Olympic medal: a growing body of medical evidence indicates that athletes who take steroids have experienced problems ranging from sterility to loss of libido, and the drug has been implicated in the deaths of young athletes from liver cancer and a type of kidney tumor. Steroid use has also been linked to heart disease. "Athletes who take steroids are playing with dynamite," says Robert Goldman, 29, a former wrestler and weight lifter who is now a research fellow in sports medicine at Chicago Osteopathic Medical Center and who has just published a book on steroid abuse, *Death in the Locker Room* (Icarus; $19.95). "Any jock who uses these drugs is taking chances not just with his health but with his life."

Anabolic steroids are essentially the male hormone testosterone and its synthetic derivatives. They were developed to alleviate strictly medical problems: correcting delayed puberty and preventing the withering of muscle tissue in people undergoing prolonged recovery from surgery, starvation or other traumas. Curiously, U.S. athletes were indirectly introduced to the drugs by Soviet athletes. In 1956 the late Dr. John Ziegler attended a world weight-lifting championship in Vienna and was told that the drugs were greatly improving the performance of the lifters from the Soviet Union. Ziegler, believing that U.S. athletes could also be helped by the drugs, worked with CIBA Pharmaceutical Co. to develop a steroid drug called Dianabol for use by athletes. He quickly abandoned his research, however, when he saw that the drug was being abused. CIBA ceased production of Dianabol for the same reason, although the company continues to make these steroids for medical use. These drugs, doctors say, are being brought into the U.S. illegally from Europe and Mexico and are being used by athletes without a doctor's prescription. Said a disillusioned Ziegler, shortly before he died last year: "I wish I had never heard the word steroid."

The great majority of physicians say the drugs upset the body's natural hormonal balance, particularly that involving testosterone, which is present, though in different amounts, in both men and women. Normally, the hypothalamus, the part of the brain that regulates many of the body's functions, "tastes" the testosterone levels; if it finds them too low, it signals the pituitary gland to trigger increased production. When the hypothalamus finds the testosterone levels too high, as it does in the case of steroid abusers, it signals the pituitary to stop production. Problems can also arise in some cases after athletes stop taking the drugs and the hypothalamus fails to get the system started again.

The results can be traumatic. Many men experience atrophy, or shrinking, of the testicles, falling sperm counts, temporary infertility, and a lessening of sexual desire; some men grow breasts, while others may develop enlargement of the prostate gland, a painful condition not usually found in men under 50. Women who take too many steroids can develop male sexual characteristics. Some grow hair on their chests and faces and lose hair from their heads; many experience abnormal enlargement of the clitoris. Some cease to ovulate and menstruate, sometimes permanently.

There are several other health risks. Steroids can cause the body to retain fluid, which results in rising blood pressure. This often tempts

Athletes whose urine tests positive for anabolic steroids are barred from receiving medals at the Olympic games.

users to fight "steroid bloat" by taking large doses of diuretics. A postmortem on a young California weight lifter who had a fatal heart attack after using steroids within the past year showed that by taking diuretics he had purged himself of electrolytes, chemicals that help regulate the heart. Convincing athletes of the dangers of steroids is far from easy. Earlier this year, Author Goldman asked this hypothetical question of 198 world-class athletes: would they take a pill that would guarantee them a gold medal even if they knew that it would kill them in five years? One hundred and three said that they would.

Summary

Hormones are chemical messengers having a metabolic effect on cells. The hypothalamus produces the hormones, ADH and oxytocin, released by the posterior pituitary and also controls the anterior pituitary by means of releasing hormones. In addition to LTH and GH which affect the body directly, the anterior pituitary secretes hormones that control other endocrine glands: TSH stimulates the thyroid to release thyroxin; ACTH stimulates the adrenal cortex to release glucocorticoids and mineralocorticoids; gonadotropins stimulate the gonads to release the sex hormones. The secretion of hormones is controlled by negative feedback. In the case of the hormones just mentioned, this mechanism involves the hypothalamus and the anterior pituitary, in addition to the hormonal gland in question.

The most common illness due to hormonal imbalance is diabetes mellitus (sugar diabetes). This condition occurs when the islets of Langerhans within the pancreas fail to produce insulin. Insulin promotes the uptake of glucose by the cells and the conversion of glucose to glycogen, thereby lowering the blood glucose levels. Without the production of insulin, the blood sugar level rises and some of it spills over into the urine. The real problem in diabetes mellitus, however, is acidosis, which may cause the death of the diabetic if therapy is not begun.

In addition to hormones produced by endocrine glands, it has become evident there are many hormones that work locally.

Objective Questions

1. The hypothalamus ____________ the hormones ____________ and ____________ released by the posterior pituitary.
2. The ____________ secreted by the hypothalamus controls the anterior pituitary.
3. Generally hormone production is self-regulated by a ____________ mechanism.
4. Growth hormone is produced by the ____________ pituitary.
5. Simple goiter occurs when the thyroid is producing ____________ (too much or too little) ____________ .
6. ACTH, produced by the anterior pituitary, stimulates the ____________ of the adrenal glands.
7. An overproductive adrenal cortex results in the condition called ____________ .
8. Parathyroid hormone increases the level of ____________ in the blood.
9. Type I diabetes mellitus is due to a malfunctioning ____________ while Type II diabetes is due to limited uptake of insulin by ____________ .
10. Prostaglandins are not carried in the ____________ as are hormones secreted by the endocrine glands.

Answers to Objective Questions

1. produces, ADH, oxytocin 2. releasing hormones 3. feedback 4. anterior 5. too little, thyroxin 6. cortex 7. Cushing's syndrome 8. calcium 9. pancreas, cells 10. blood

Study Questions

1. Hormones fall into what two groups, chemically speaking? (p. 371) Name some hormones in each group. (p. 372)
2. What mechanisms of action have been suggested to explain how hormones work? (p. 371)
3. Define the endocrine gland and target organ. (p. 372)
4. How does the hypothalamus control the posterior pituitary; the anterior pituitary? (pp. 373; 374)
5. Discuss two hormones secreted by the anterior pituitary that have an effect on the body proper rather than on other glands. (pp. 374–376) Why is the anterior pituitary called the master gland? (p. 377)
6. For each of the following endocrine glands name the hormone(s) secreted, the effect of the hormone(s), and the medical illnesses, if any, that result from too much or too little of each hormone: posterior pituitary, thyroid, parathyroids, adrenal cortex, adrenal medulla, pancreas. (p. 373)
7. Give the anatomical location of each of the endocrine glands listed in #6. (p. 372)
8. Draw a diagram to describe the action and control of ADH, thyroxin, glucocorticoids (e.g., cortisol), aldosterone, parathyroid hormone, and insulin. (pp. 374–383)

Key Terms

acromegaly (ak''ro-meg'ah-le) a condition resulting from an increase in growth hormone production after adult height has been achieved. *376*

anterior pituitary (an-te're-or pi-tu'i-tar''e) a portion of the pituitary gland which produces six hormones and is controlled by a hypothalamic releasi hormones. *373*

cretinism (kre'tin-izm) a condit resulting from a lack of thyro in an infant. *378*

diabetes insipidus (di''ah-be dus) condition charac abnormally large pro rine due to a deficiency of a hormone. *374*

dia' i''ah-be'tez mē-li'- racterized by a high nd the appearance rine due to a sulin production or ells. *382*

and (en'do-krin gland) a that secretes hormones directly o the blood or body fluids. *372*

ophthalmic goiter (ek''sof-thal'mik goi'ter) an enlargement of the thyroid gland accompanied by an abnormal protrusion of the eyes. *378*

islets of Langerhans (i'lets uv lahng'er-hanz) distinctive groups of cells within the pancreas that secrete insulin and glucagon. *382*

myxedema (mik''sĕ-de'mah) a condition resulting from a deficiency of thyroid hormone in an adult. *378*

posterior pituitary (pos-tēr'e-or pi-tu'i-tar''e) portion of the pituitary gland connected by a stalk to the hypothalamus. *373*

simple goiter (sim'p'l goi'ter) condition in which an enlarged thyroid produces low levels of thyroxin. *378*

Further R ngs for Part Three

Barr, M. 3. *The human nervous system, an anat cal viewpoint.* 4th ed. New York: H er and Row Publishers.

Bl , F. E. 1981. Neuropeptides. *Scientific American* 245(4):148.

uisseret, P. D. 1982. Allergy. *Scientific American* 247(2):86.

Doolittle, R. F. 1981. Fibrinogen and fibrin. *Scientific American* 245(6):126.

Eckert, R., and D. Randall. 1983. *Animal physiology.* 2d ed. San Francisco: W. H. Freeman.

Feder, M. E. and W. W. Burggren. 1985. Skin breathing in vertebrates. *Scientific American* 253(5):126.

Fincher, J. 1981. *The brain: mystery of matter and mind.* Washington, D.C.: U.S. News Books.

Fine, A. 1986. Transplantation in the central nervous system. *Scientific American* 255(2):52.

Goldstein, G. W. and A. L. Betz. 1986. The blood-brain barrier. *Scientific American* 255(3):74.

Guyton, A. C. 1984. *Physiology of the human body.* 6th ed. Philadelphia: W. B. Saunders.

Hickman, C. P., et al. 1982. *Biology of animals.* 3d ed. St. Louis: C. V. Mosby.

Hill, R. W. 1976. *Comparative physiology of animals: an environmental approach.* New York: Harper & Row, Publishers, Inc.

Hole, J. W. 1987. *Human anatomy and physiology.* 4th ed. Dubuque, Ia: Wm. C. Brown Publishers.

Hudspeth, A. J. 1983. The hair cells of the inner ear. *Scientific American* 248(1):54.

Human Nutrition: Readings from Scientific American. 1978. San Francisco: W. H. Freeman.

Jarvik, R. K. 1981. The total artificial heart. *Scientific American* 244(1):74.

Kennedy, R. C. 1986. Anti-idiotypes and immunity. *Scientific American* 255(1):48.

Kessel, R. G., and R. H. Kardon. 1979. *Tissues and organs: a text-atlas of scanning electron microscopy.* San Francisco: W. H. Freeman.

Keynes, R. D. 1979. Ion channels in the nerve cell membrane. *Scientific American* 240(3):126.

Llinas, R. R. 1982. Calcium in synaptic transmission. *Scientific American* 247(4):56.

Loeb, G. E. 1985. The functional replacement of the ear. *Scientific American* 252(2):104.

Masland, R. H. 1986. The functional architecture of the retina. *Scientific American* 255(6):102.

Moog, F. 1981. The lining of the small intestine. *Scientific American* 245(5):154.

Morell, P., and W. T. Norton. 1980. Myelin. *Scientific American* 242(5):88.

Nauta, W. J. H., and M. Feirtag. 1979. The organization of the brain. *Scientific American* 241(3):88.

Norman, D. A. 1982. *Learning and memory.* San Francisco: W. H. Freeman.

Perutz, M. F. 1978. Hemoglobin structure and respiratory transport. *Scientific American* 239(6):92.

Robinson, T. F., et al. 1986. The heart as a suction pump. *Scientific American* 254(6):84.

Rose, N. R. 1981. Autoimmune diseases. *Scientific American* 244(2):80.

Rubenstein, E. 1980. Diseases caused by impaired communication among cells. *Scientific American* 242(3):102.

Schwartz, J. H. 1980. The transport of substances in nerve cells. *Scientific American* 242(4):152.

Scrimshaw, N. S., and V. R. Young. 1976. The requirements of human nutrition. *Scientific American* 235(3):50.

Selim, R. D. 1985. *Muscles: The magic of motion.* Washington, D.C.: U.S. News Books.

Shashoua, V. E. 1985. The role of extracellular proteins in learning and memory. *American Scientist* 73:364.

Snyder, S. H. 1985. The molecular basis of communication between cells. *Scientific American* 253(4):132.

Stevens, C. F. 1979. The neuron. *Scientific American* 241(3):54.

Tonegawa, S. 1985. The molecules of the immune system. *Scientific American* 253(4):122.

Vander, A. J. 1985. *Human physiology: The mechanisms of body function.* 4th ed. New York: McGraw-Hill.

Wertenbaker, L. 1981. *The eye: Window to the world.* Washington, D.C.: U.S. News Books.

Wurtman, R. J. 1982. Nutrients that modify brain function. *Scientific American* 246(4):50.

Zucker, M. B. 1980. The functioning of blood platelets. *Scientific American* 242(6):86.

Zwislocki, J. J. 1981. Sound analysis in the ear: A history of discoveries. *American Scientist* 69:184.

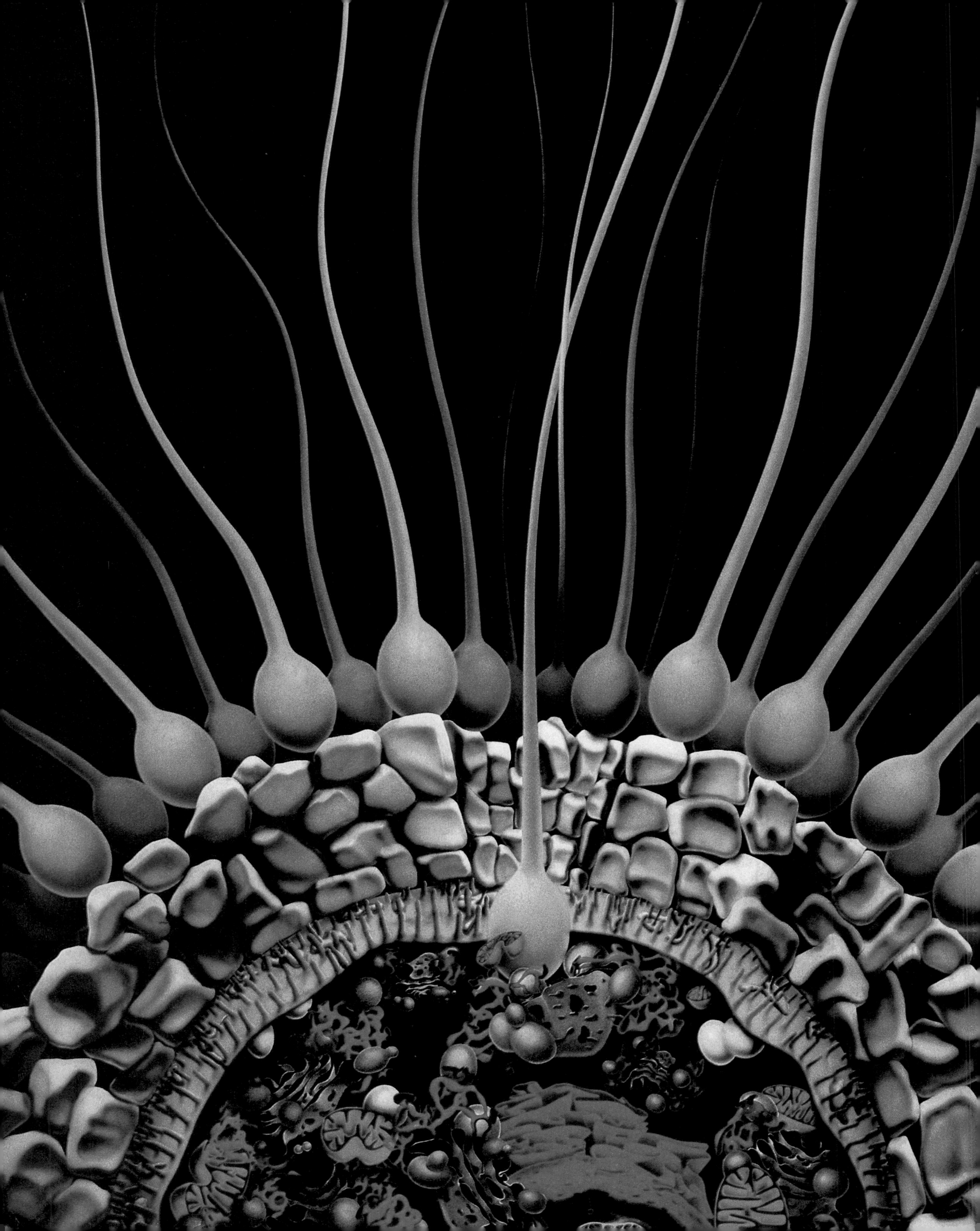

Part Four

Human Reproduction, Development, and Inheritance

The anatomy of the human male and female serve to bring the sperm to the egg, resulting in fertilization, followed by the gradual steps of development. Sexual reproduction results in a recombination of genes and therefore produces offspring that are, at the same time, both similar to and different from the parents. Knowing the genes carried by the parents sometimes makes it possible to predict certain features of the offspring. However, this simple relationship can be influenced by the interaction of genes during development and by the prenatal environment.

Biochemical knowledge of the makeup of the hereditary material and how it operates has forged a biological revolution. It is now possible to manipulate the genes, an advance that may make it possible to cure genetic diseases and cancer.

Many sperm approach the egg but only one enters to contribute chromosomes and initiate development of the new individual. This artist's representation clearly shows that the egg has a complex outer coating now known to prohibit entrance of other sperm.

19

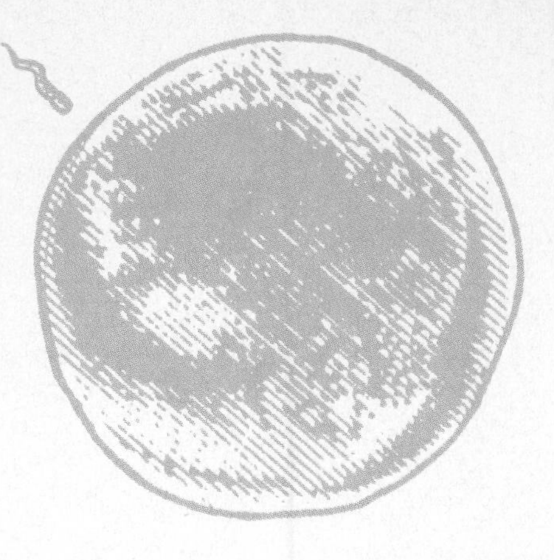

Reproductive System

Chapter Concepts

1 The male reproductive system is designed for the continuous production of a large number of sperm within a fluid medium.

2 The female reproductive system is designed for the monthly production of an egg and preparation of the uterus for possible implantation of the fertilized egg.

3 Hormones control the reproductive process and the sex characteristics of the individual.

4 Birth control measures vary in effectiveness from those that are very effective to those that are minimally effective.

5 There are alternative methods of reproduction today, including in vitro fertilization followed by artificial implantation.

6 There are several serious and prevalent sexually transmitted diseases.

Chapter Outline

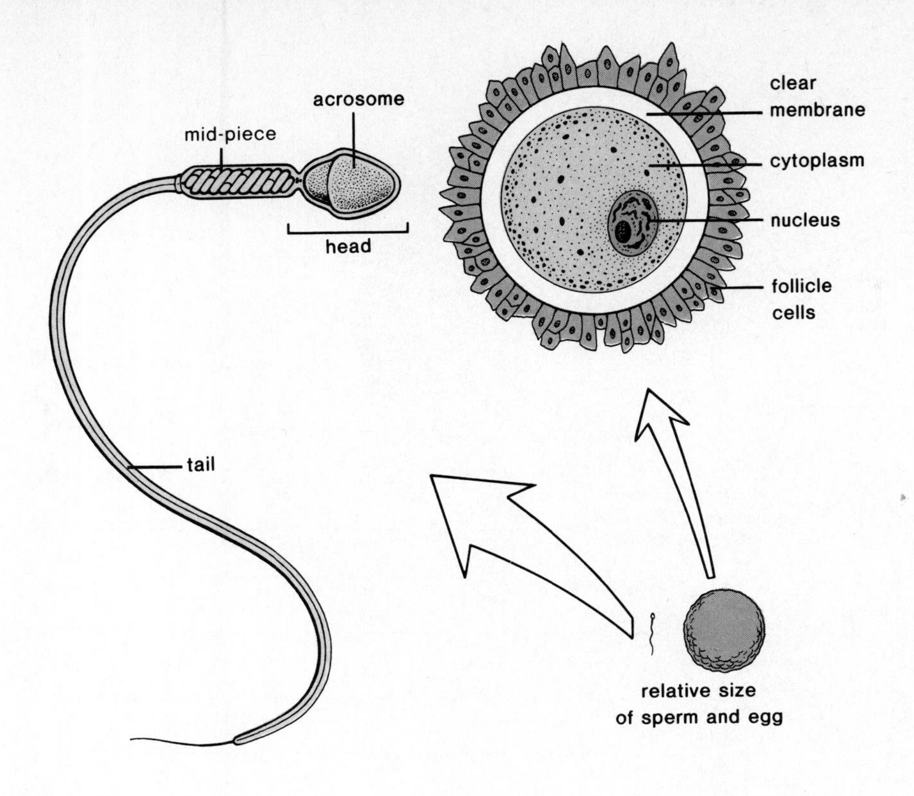

Figure 19.1
Microscopic anatomy of sperm and egg. Note their relative sizes. The head of sperm contains the chromosomes and little else; the larger egg contributes most of the cytoplasm to the zygote.

In advanced forms of sexual reproduction there are two types of gametes (sex cells) both of which contribute the same number of chromosomes to the new individual (fig. 4.4). The sperm are small and swim to the stationary egg, a much larger cell that contains food for the developing embryo (fig. 19.1). It seems reasonable that there should be a large number of sperm to ensure that a few will find the egg. In humans, the male continually produces sperm, which are temporarily stored before being released.

Male Reproductive System

Figure 19.2 shows the reproductive system of the male and table 19.1 lists the anatomical parts of this system.

Testes

The **testes** lie outside the abdominal cavity of the male within the **scrotum.** The testes begin their development inside the abdominal cavity but descend into the scrotal sacs during the last two months of fetal development. If by chance the testes do not descend and the male is not treated or operated on to place the testes in the scrotum, sterility—the inability to produce offspring—usually follows. This is because the internal temperature of the body is too high to produce viable sperm.

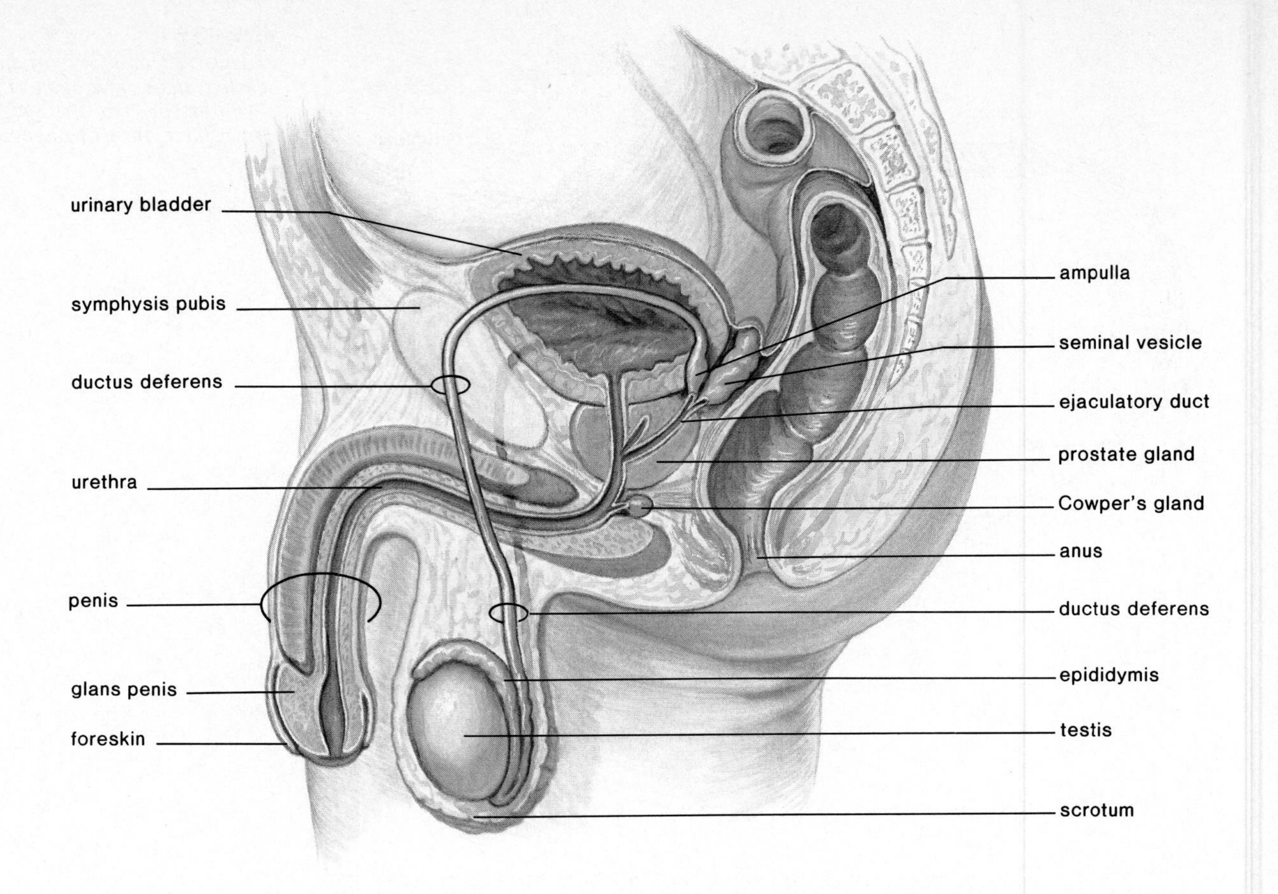

Figure 19.2
Side view of male reproductive system. Trace the path of the genital tract from a testis to the exterior. The seminal vesicles, Cowper's gland, and prostate gland produce seminal fluid and do not contain sperm. Notice that the penis in this drawing is not circumcised, since the foreskin is present.

Table 19.1 *Male Reproductive System*

Organ	Function
Testes	Produce sperm and sex hormones
Epididymis	Maturation and some storage of sperm
Ductus deferens	Conducts and stores sperm
Seminal vesicles	Contributes to seminal fluid
Prostate gland	Contributes to seminal fluid
Urethra	Conducts sperm
Cowper's glands	Contribute to seminal fluid
Penis	Organ of copulation

Seminiferous Tubules

Connective tissue forms the wall of each testis and divides it into lobules (fig. 19.3). Each lobule contains one to three tightly coiled **seminiferous tubules** that have a combined length of approximately 250 m. A microscopic cross section through a tubule shows it is packed with cells undergoing spermatogenesis. These cells are derived from undifferentiated germ cells, called spermatogonia (singular, spermatogonium), that lie just inside the outer wall and divide mitotically, always producing new spermatogonia. Some spermatogonia move away from the outer wall to increase in size and become primary spermatocytes that undergo meiosis, a type of cell division described in chapter 4. Although these cells have forty-six chromosomes, they divide to give secondary spermatocytes, each with twenty-three chromosomes. Secondary spermatocytes divide to give spermatids, also with twenty-three chromosomes, but single-stranded. Spermatids then differentiate into spermatozoa, or mature sperm.

Sperm The mature sperm, or spermatozoan (fig. 19.1), has three distinct parts: a head, a mid-piece, and a tail. The *tail* contains the 9 + 2 pattern of microtubules typical of cilia and flagella (fig. 2.15) and the *mid-piece* contains energy-producing mitochondria. The *head* contains the 23 chromosomes within a nucleus. The tip of the nucleus is covered by a cap called the **acrosome,** which is believed to contain enzymes needed for fertilization. The human egg is surrounded by several layers of cells and a mucoprotein substance. The acrosome enzymes are believed to aid the sperm in reaching the surface of the egg and allowing a single sperm to penetrate the egg. It is hypothesized that each acrosome contains such a minute amount of enzyme that it requires the action of many sperm to allow just one to actually penetrate the egg. This may explain why so many sperm are required for the process of fertilization. A normal human male usually produces several hundred million sperm per day, an adequate number for fertilization. Sperm are continually produced throughout a male's reproductive life.

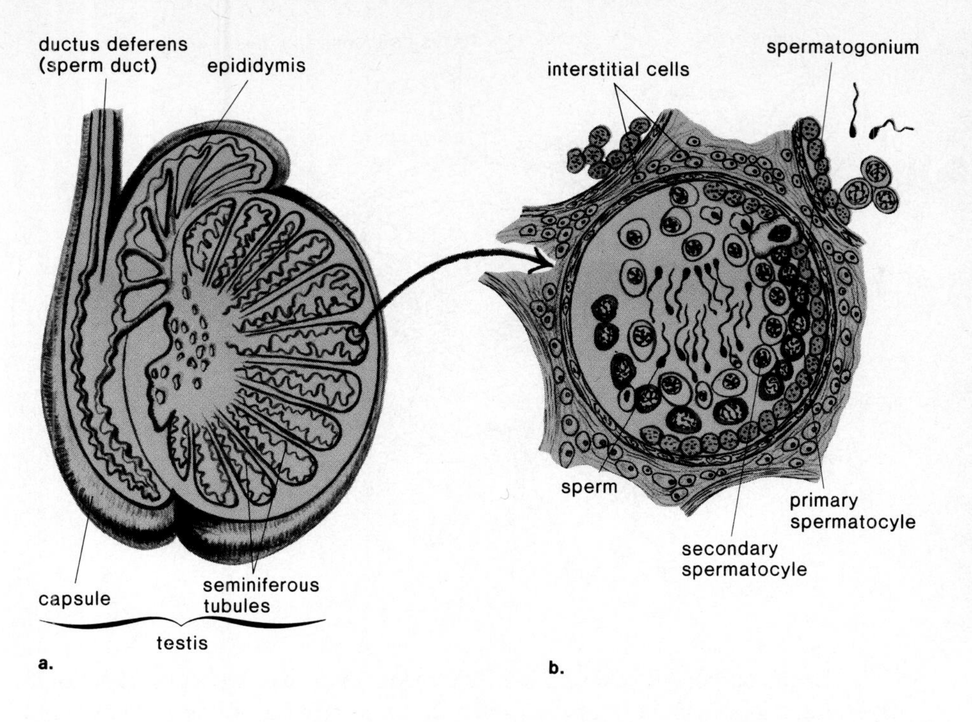

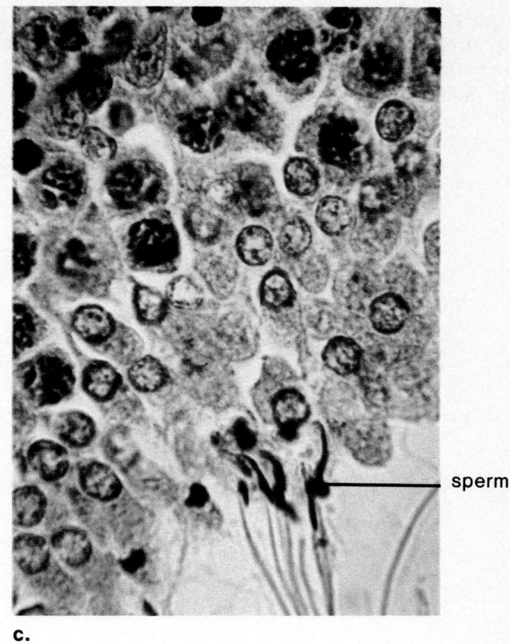

Figure 19.3
Sections through a human testis. a. *Longitudinal section showing lobules containing seminiferous tubules.* b. *Cross section of a tubule showing germ cells in various stages of spermatogenesis.* c. *Micrograph of* b.

Interstitial Cells

The male sex hormones, the androgens, are secreted by cells that lie between the seminiferous tubules and are therefore called **interstitial cells.** The most important of the androgens is testosterone, whose functions are discussed on page 398.

Genital Tract

Sperm are produced in the testes, but they mature in the **epididymis** (fig. 19.2), a tightly coiled tubule about 20 feet in length that lies just outside each testis. During the two-to-four day maturation period, the sperm develop their characteristic swimming ability. Also, it is possible that during this time defective sperm are removed from the epididymis. Each epididymis joins with a **ductus (vas) deferens,** which ascends through a canal called the *inguinal canal* and enters the abdomen where it curves around the bladder and empties into the urethra. Sperm are stored in the first part of the ductus deferens. They pass from each ductus into the urethra only when ejaculation (p. 396) is imminent.

Spermatic Cords The testes are suspended in the scrotum by the *spermatic cords,* each of which consists of connective tissue and muscle fibers that enclose the ductus deferens, blood vessels, and nerves. The region of the inguinal canal, where the spermatic cord passes into the abdomen, remains a weak point in the abdominal wall. As such, it is frequently the site of hernias. A **hernia** is an opening or separation of some part of the abdominal wall through which a portion of an internal organ, usually the intestine, protrudes.

Seminal Fluid

At the time of ejaculation (p. 396), sperm leave the penis in a fluid called **seminal fluid.** This fluid is produced by three types of glands—the seminal vesicles, the prostate gland, and Cowper's glands. The **seminal vesicles** lie at the base of the bladder, and each has a duct that joins with a ductus deferens. The **prostate gland** is a single doughnut-shaped gland that surrounds the upper portion of the urethra just below the bladder. In older men, the prostate may enlarge and cut off the urethra, making urination painful and difficult. This condition may be treated medically or surgically. **Cowper's glands** are pea-sized organs that lie posterior to the prostate on either side of the urethra.

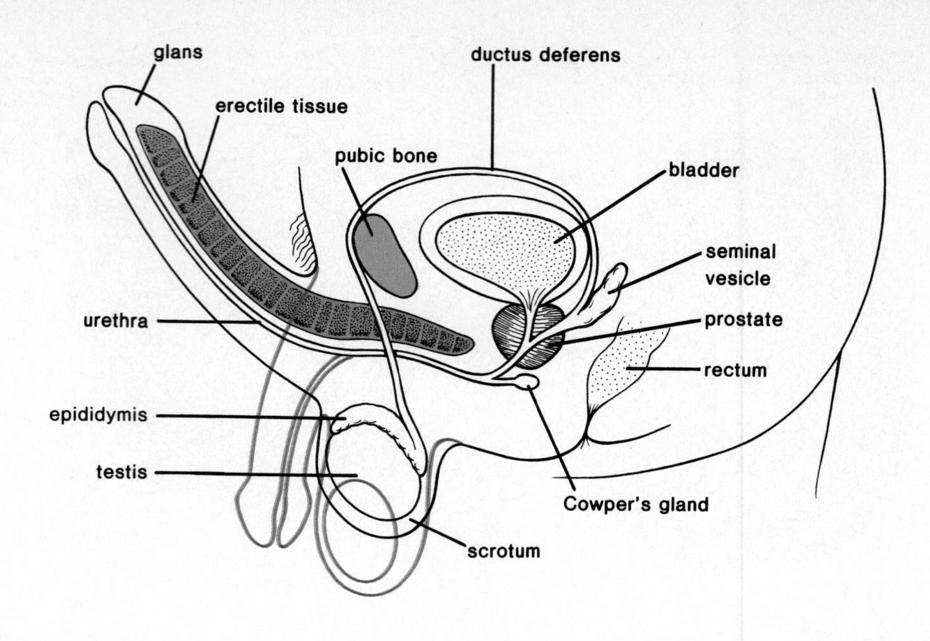

Figure 19.4
Erection contrasted to a flaccid penis and relaxed scrotum. The erectile tissue fills with blood when the penis becomes erect. Note that the penis lacks a foreskin, due to circumcision.

Each component of a seminal fluid seems to have a particular function. Sperm are more viable in a basic solution; seminal fluid, which is white and milky in appearance, has a slightly basic pH (about 7.5). Swimming sperm require energy, seminal fluid contains the sugar fructose, which presumably serves as an energy source. Seminal fluid also contains prostaglandins, chemicals that cause the uterus to contract. Some investigators now believe that uterine contraction is necessary to propel the sperm and that the sperm swim only when they are in the vicinity of the egg.

Penis

The **penis** has a long shaft and enlarged tip called the glans penis. At birth the glans penis is covered by a layer of skin called the **foreskin** or prepuce. Gradually over a period of five to ten years, the foreskin becomes separated from the glans and may be retracted. During this time there is a natural shedding of cells between the foreskin and glans. These cells, along with an oil secretion that begins at puberty, is called smegma. In the child no special cleansing method is needed to wash away smegma but in the adult the foreskin can be retracted to do so. **Circumcision** is the surgical removal of the foreskin soon after birth.

The penis is the copulatory organ of males. When the male is sexually aroused, the penis becomes erect and ready for intercourse (fig. 19.4). **Erection** is achieved because blood sinuses within the erectile tissue of the penis become filled with blood. Parasympathetic impulses dilate the arteries of the penis while the veins are passively compressed so that blood flows into the erectile tissue under pressure. If the penis fails to become erect, the condition is called **impotency.** Although it was formerly believed that almost all cases of impotency were due to psychological reasons, it has recently been reported that some cases may be due to hormonal imbalances. Treatment consists of finding the precise imbalance and restoring the proper level of testosterone.

Ejaculation

As sexual stimulation becomes intense, sperm enter the urethra from each ductus deferens and the glands secrete seminal fluid. Sperm and seminal fluid together are called **semen.** Once semen is in the urethra, rhythmical muscle contractions cause it to be expelled from the penis in spurts. During ejaculation, a sphincter closes off the bladder so that no urine enters the urethra. (Notice that the urethra carries either urine or semen at different times.)

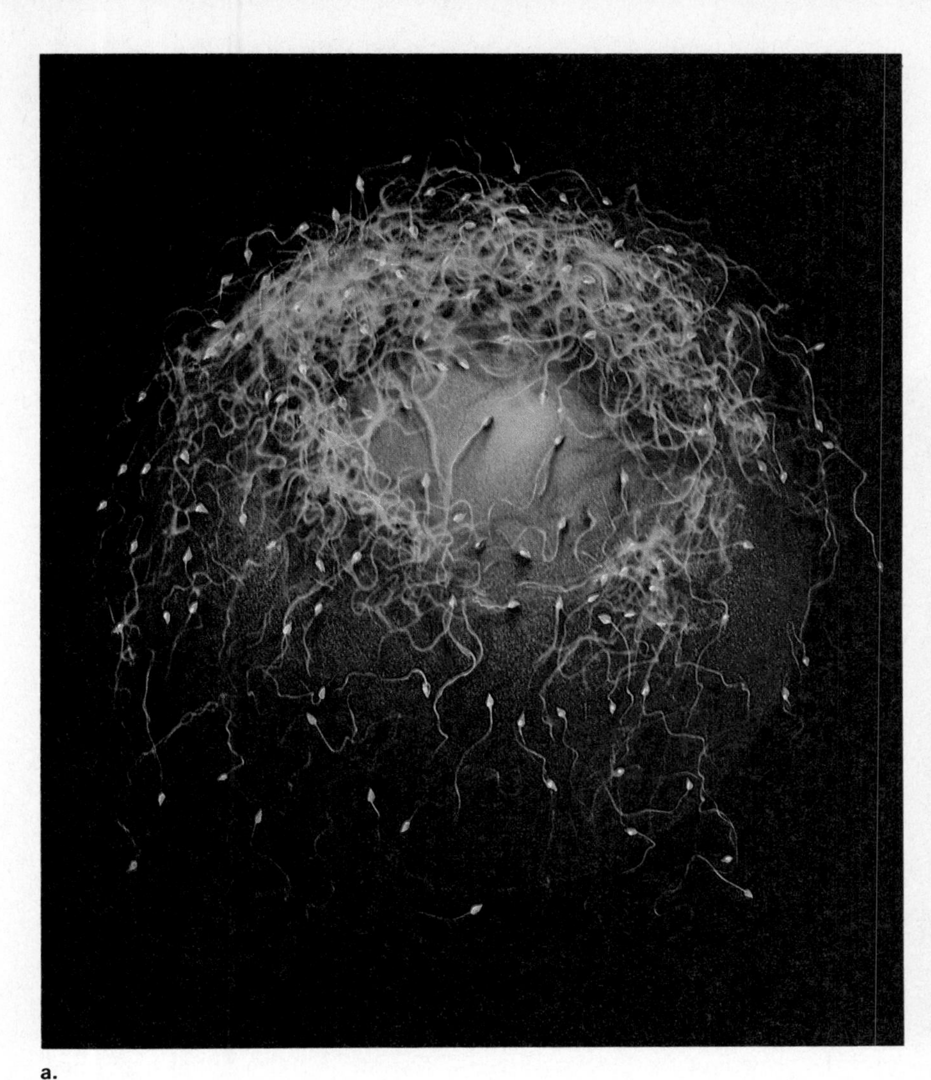

a.

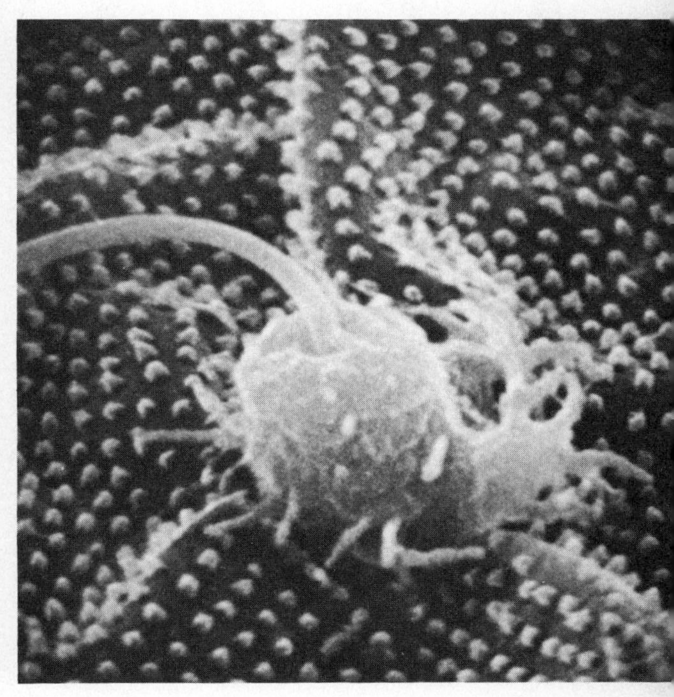

b.

Figure 19.5
Fertilization. a. *Scanning electron micrographs of an echinoderm egg surrounded by a large number of sperm.* b. *Only one sperm penetrates the egg to achieve fertilization.*

The contractions that expel semen from the penis are a part of male **orgasm,** the physiological and psychological sensations that occur at the climax of sexual stimulation. The psychological sensation of pleasure is centered in the brain, but the physiological reactions involve the genital (reproductive) organs and associated muscles as well as the entire body. Marked muscle tension is followed by contraction and relaxation.

Following ejaculation and/or loss of sexual arousal, the penis returns to its normal flaccid state. After ejaculation, a male typically experiences a period of time, called the refractory period, during which stimulation does not bring about an erection.

There may be in excess of 400 million sperm in 3.5 milliliters of semen expelled during ejaculation. The sperm count can be much lower than this, however, and fertilization (fig. 19.5) will still take place.

Sperm mature in the epididymis and are stored in the ductus deferens before entering the urethra just prior to ejaculation. The accessory glands (seminal vesicles, prostate gland, and Cowper's gland) produce seminal fluid. Semen, which contains sperm and seminal fluid, leaves the penis during ejaculation.

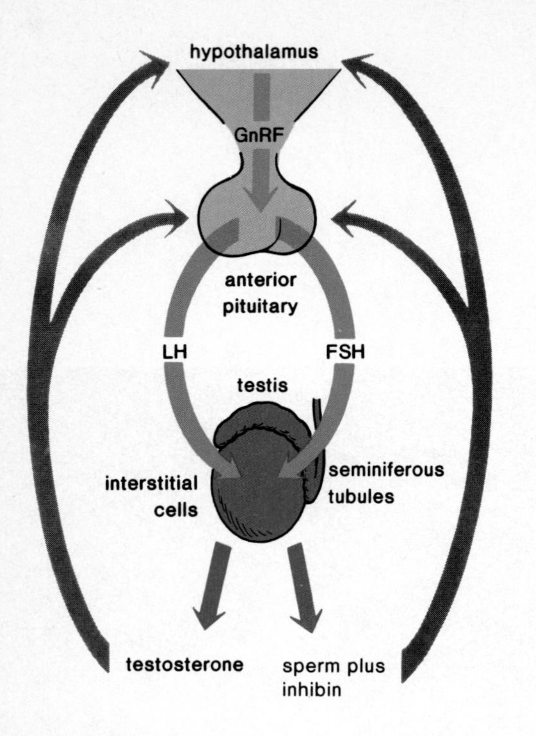

Figure 19.6
Hypothalamic-pituitary-gonad system as it functions in the male. GnRH is a hypothalamic-releasing hormone that stimulates the anterior pituitary to secrete LH and FSH. These gonadotropic hormones act on the testes. LH promotes the production of testosterone, and FSH promotes spermatogenesis. Negative feedback controls the level of all hormones involved.

Hormonal Regulation in the Male

The hypothalamus has ultimate control of the testes' sexual functions because it secretes a releasing hormone that stimulates the anterior pituitary to produce the gonadotropic hormones. Two gonadotropic hormones, **FSH (follicle-stimulating hormone)** and **LH (luteinizing hormone),** are named for their function in females but exist in both sexes, stimulating the appropriate gonads in each. It is believed that FSH promotes spermatogenesis in the seminiferous tubules and that LH promotes the production of testosterone in the interstitial cells. Sometimes LH in males is given the name interstitial cell-stimulating hormone (ICSH).

The hormones mentioned are involved in a feedback process (fig. 19.6) that maintains the production of testosterone at a fairly constant level. For example, when the amount of testosterone in the blood rises to a certain level, it causes the hypothalamus to decrease its secretion of releasing hormone, which causes the anterior pituitary to decrease its secretion of LH. As the level of testosterone begins to fall, the hypothalamus increases secretion of the releasing hormone and the anterior pituitary increases its secretion of LH, and stimulation of the interstitial cells reoccurs. It should be emphasized that only minor fluctuations of testosterone level occur in the male and that the feedback mechanism in this case acts to maintain testosterone at a normal level. It has long been suspected that the seminiferous tubules produce a hormone that blocks FSH secretion. This substance, termed *inhibin,* has recently been isolated.

Testosterone

The male sex hormone, **testosterone,** has many functions. It is essential for the normal development and functioning of the primary sex organs, those structures we have just been discussing. It is also necessary for the maturation of sperm, probably after diffusion from the interstitial cells into the seminiferous tubules.

Greatly increased testosterone secretion at the time of puberty stimulates the growth of the penis and testes. Testosterone also brings about and maintains the secondary sex characteristics in males that develop at the time of puberty. Testosterone causes growth of a beard, axillary (underarm) hair, and pubic hair. It prompts the larynx and vocal cords to enlarge, causing the voice to change. It is responsible for the greater muscle strength of males and this is the reason why some athletes take supplemental amounts of *anabolic steroids,* which are either testosterone or related chemicals. The pros and cons of taking anabolic steroids are discussed in the reading on page 386. Testosterone also causes oil and sweat glands in the skin to secrete; therefore, it is largely responsible for acne and body odor. Another side effect of testosterone activity is baldness. Genes for baldness are probably inherited by both sexes, but baldness is seen more often in males because of the presence of testosterone.

Testosterone is believed to be largely responsible for the sex drive and may even contribute to the supposed aggressiveness of males.

In males, FSH promotes spermatogenesis and LH promotes testosterone production within the testes. Testosterone stimulates growth of the male genitals during puberty and is necessary for maturation of sperm and development of the secondary sex characteristics.

Female Reproductive System

Table 19.2 lists the anatomical parts of this system, and figure 19.7 shows the reproductive system of the female.

Ovaries

The **ovaries** lie in shallow depressions, one on each side of the upper pelvic cavity. A longitudinal section through an ovary shows that it is made up of an outer cortex and an inner medulla. The cortex contains ovarian **follicles** at various stages of maturation. A female is born with a large number of follicles (400,000) in both ovaries, each containing a potential egg. In contrast to the male, the female produces no new gametes after she is born. Only a small number of eggs (about 400) ever mature because a female produces only one egg per month during her reproductive years. Since eggs are present at birth, they age as the woman ages. This is one possible reason why older women are more likely to produce children with genetic defects.

***Table 19.2** Female Reproductive System*

Organ	Function
Ovaries	Produce egg and sex hormones
Fallopian tubes (oviducts)	Conduct egg
Uterus (womb)	Location of developing fetus
Cervix	Contains opening to uterus
Vagina	Organ of copulation and birth canal

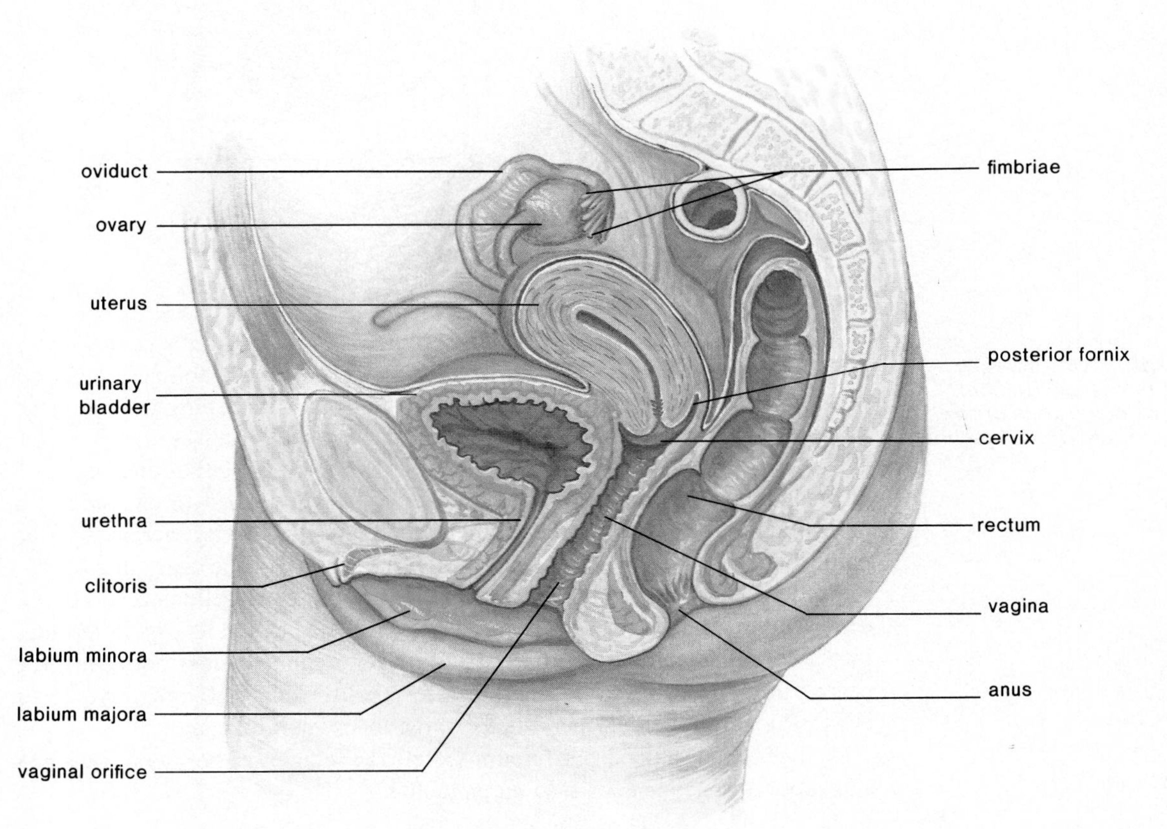

Figure 19.7
Side view of female reproductive system. The ovaries produce one egg a month; fertilization occurs in the oviduct and development occurs in the uterus. The vagina is the birth canal and organ of copulation.

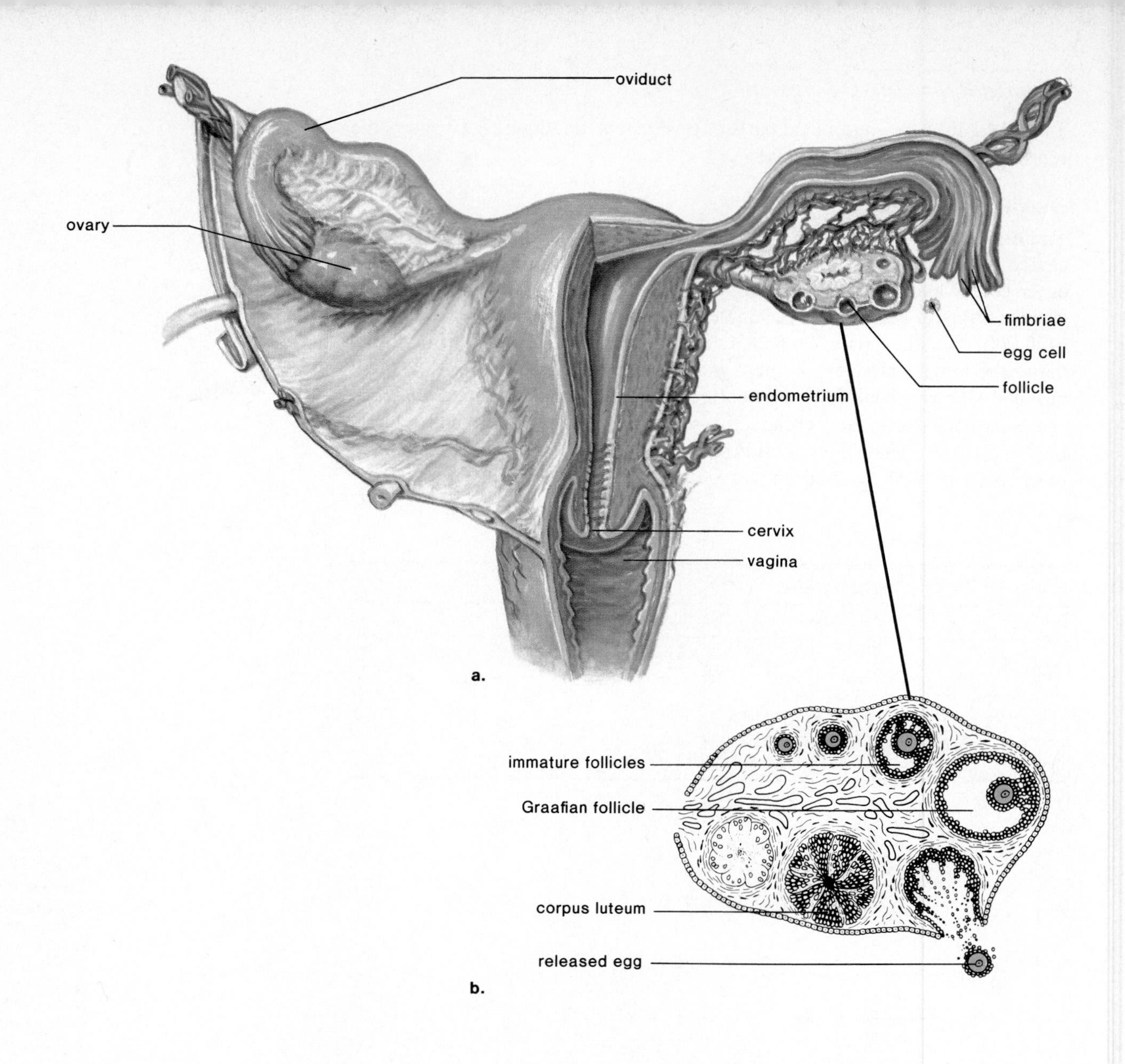

Figure 19.8
a. *Reproductive organs of female, front view. Left side is surface view and right side is longitudinal view.* b. *The enlargement shows maturation of the follicle, release of egg, and resulting corpus luteum.*

As the follicle undergoes maturation, it develops from a primary to a secondary to a **Graafian follicle** (fig. 19.8). In a primary follicle an oocyte divides meiotically into two cells, each having twenty-three chromosomes (fig. 4.21). One of these cells, termed the secondary oocyte, receives almost all the cytoplasm nutrients and enzymes. The other is a polar body that disintegrates. A secondary follicle contains the secondary oocyte pushed to one side of a fluid-filled cavity. In a Graafian follicle, the fluid-filled cavity increases to the point that the follicle wall balloons out on the surface of the ovary and bursts, releasing the secondary oocyte surrounded by the zona pellucida and a few cells. This is referred to as **ovulation.** Once a follicle has lost its egg, it develops into a **corpus luteum,** a glandlike structure. If pregnancy does not occur, the corpus luteum begins to degenerate after about 10 days. If pregnancy does occur, the corpus luteum persists for three to six months.

The follicle and corpus luteum secrete the female sex hormones estrogen and progesterone, as discussed on page 403.

In females, oogenesis occurs within the ovaries where one follicle reaches maturity each month. This follicle balloons out of the ovary and bursts to release the egg. The ruptured follicle develops into a corpus luteum. The follicle and corpus luteum produce the female sex hormones estrogen and progesterone.

Genital Tract

The female genital tract includes the oviducts, uterus, and vagina.

Oviducts

The oviducts, also called uterine or fallopian tubes, extend from the uterus to the ovaries. The oviducts are not attached to the ovaries but instead have fingerlike projections called **fimbriae** that sweep over the ovary at the time of ovulation. When the egg bursts (fig. 19.8) from the ovary during ovulation, it is usually swept up into an oviduct by the combined action of the fimbriae and the beating of cilia that line the tubes.

Since the egg must traverse a small space before entering an oviduct, it is possible for the egg to get lost and instead enter the abdominal cavity. Such eggs usually disintegrate but in some rare cases have been fertilized in the abdominal cavity and have implanted themselves in the wall of an abdominal organ. Very rarely, such embryos have come to term, the child being delivered by surgery.

Once in the oviduct, the egg is propelled slowly by cilia movement and tubular muscular contraction toward the uterus. Fertilization usually occurs in an oviduct because the egg only lives approximately 6 to 24 hours. The developing zygote normally arrives at the uterus after several days and then embeds, or implants, itself in the uterine lining, which has been prepared to receive it. Occasionally, the zygote becomes embedded in the wall of an oviduct, where it begins to develop. Tubular pregnancies cannot succeed because the tubes are not anatomically capable of allowing full development to occur.

Uterus

The **uterus** is a thick-walled, muscular organ about the size and shape of an inverted pear. Normally it lies above and is tipped over the urinary bladder. The oviducts join the uterus anteriorly, while posteriorly, the cervix enters into the vagina nearly at a right angle. A small opening in the cervix leads to the vaginal canal. Development of the embryo takes place in the uterus. This organ, sometimes called the womb, is approximately 2 inches wide in its usual state but is capable of stretching to over 12 inches to accommodate the growing baby. The lining of the uterus, called the **endometrium,** participates in the formation of the placenta (p. 434), which supplies nutrients needed for embryonic and fetal development. The endometrium has two layers: a basal layer and an inner functional layer. In the nonpregnant female, the functional layer of the endometrium varies in its thickness according to a monthly reproductive cycle, called the uterine cycle (p. 403).

Cancer of the cervix is a common form of cancer in women. Early detection is possible by means of a **Pap test,** which requires that the doctor remove a few cells from the region of the cervix for microscopic examination. If the cells are cancerous, a hysterectomy may be recommended. A hysterectomy is the removal of the uterus. Removal of the ovaries in addition to the uterus is termed an ovariohysterectomy. Since the vagina remains, the woman may still engage in sexual intercourse.

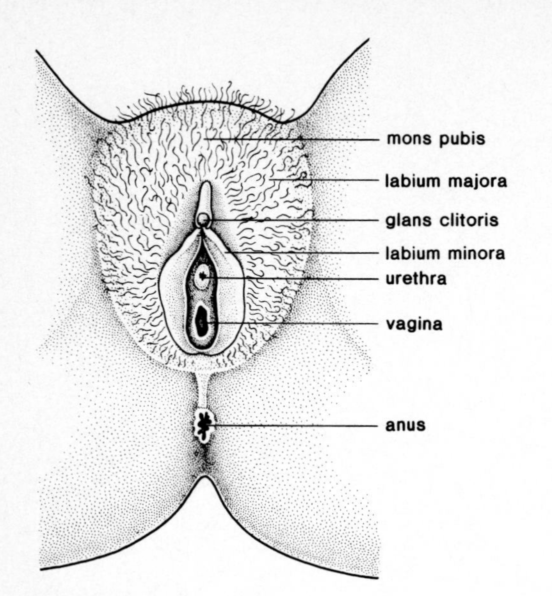

Figure 19.9
External genitalia of female. At birth, the opening of the vagina is partially occluded by a membrane called the hymen. Physical activities and sexual intercourse disrupt the hymen.

Vagina

The **vagina** is a tube that makes a 45-degree angle with the small of the back. The mucosal lining of the vagina lies in folds that extend as the fibromuscular wall stretches. This capacity to extend is especially important when the vagina serves as the birth canal, and it may also facilitate intercourse when the vagina receives the penis during copulation.

External Genitalia

The external genital organs of the female (fig. 19.9) are known collectively as the **vulva.** The vulva includes two large, hair-covered folds of skin called the **labia majora.** They extend backward from the *mons pubis,* a fatty prominence underlying the pubic hair. The **labia minora** are two small folds lying just inside the labia majora. They extend forward from the vaginal opening to encircle and form a foreskin for the *clitoris,* an organ that is homologous to the penis. Although quite small, the clitoris has a shaft of erectile tissue and is capped by a pea-shaped glans. The glans clitoris also has sense receptors that allow it to function as a sexually sensitive organ.

The *vestibule,* a cleft between the labia minora, contains the openings of the urethra and the vagina. The vagina may be partially closed by a ring of tissue called the hymen. The hymen is ordinarily ruptured by initial sexual intercourse; however, it can also be disrupted by other types of physical activities. If the hymen persists after sexual intercourse, it can be surgically ruptured.

Notice that the urinary and reproductive systems in the female are entirely separate. For example, the urethra carries only urine and the vagina serves only as the birth canal and the organ for sexual intercourse.

The egg enters the oviducts, which lead to the uterus followed by the vagina. The vagina opens into the vestibule, the location of female external genitalia.

Orgasm

Sexual response in the female may be more subtle than in the male, as discussed in the reading on page 406, but there are certain corollaries. The clitoris is believed to be an especially sensitive organ for initiating sexual sensations. It is possible for the clitoris to become ever so slightly erect as its erectile tissues become engorged with blood. But vasocongestion is more obvious in the labia minora, which expand and deepen in color. Erectile tissue within the vaginal wall also expands with blood, and the added pressure in these blood vessels causes small droplets of fluid to squeeze through the vessel walls and lubricate the vagina.

Release from muscle tension occurs in females, especially in the region of the vulva and vagina but also throughout the entire body. Increased uterine motility may assist the transport of sperm toward the oviducts. Since female orgasm is not signaled by ejaculation, there is a wide range in normality regarding sexual response.

Hormonal Regulation in the Female

Hormonal regulation in the female is quite complex, so we will begin with a simplified presentation and follow with a more in-depth presentation for those who wish to study the matter in more detail. The following glands and hormones are involved in hormonal regulation:

Hypothalamus: secretes *GnRH* (gonadotropic-releasing hormone)
Anterior pituitary: secretes *FSH* (follicle-stimulating hormone) and *LH* (luteinizing hormone), the gonadotropic hormones
Ovaries: secrete estrogen and progesterone, the female sex hormones

Table 19.3 Ovarian and Uterine Cycles (Simplified)

Ovarian Cycle Phases	Events	Uterine Cycle Phases	Events
Follicular (Days 1–13)	FSH secretion by pituitary	Menstruation Days 1–5	Endometrium breaks down
	Follicle maturation and secretion of estrogen	Proliferative Days 6–13	Endometrium rebuilds
*Ovulation Day 14**			
Luteal (Days 15–28)	LH secretion by pituitary	Secretory Days 15–28	Endometrium thickens and glands are secretory
	Corpus luteum formation and secretion of progesterone		

*Assuming a 28-day cycle.

Hormonal Regulation (Simplified)

Ovarian Cycle The gonadotropic and sex hormones are not present in constant amounts in the female and instead are secreted at different rates during a monthly ovarian cycle, which lasts an average of 28 days but may vary widely in specific individuals. For simplicity's sake it is convenient to emphasize that during the first half of a 28-day cycle (days 1 to 13, table 19.3) FSH from the anterior pituitary is promoting the development of a follicle in the ovary and that this follicle is secreting estrogen. As the blood estrogen level rises, it exerts feedback control over the anterior pituitary secretion of FSH so that this follicular phase comes to an end (fig. 19.10). The end of the follicular phase is marked by ovulation on the fourteenth day of the 28-day cycle. Similarly it may be emphasized that during the last half of the ovarian cycle (days 15 to 28, table 19.3) anterior pituitary production of LH is promoting the development of a corpus luteum, which is secreting progesterone. As the blood progesterone level rises, it exerts feedback control over anterior pituitary secretion of LH so that the corpus luteum begins to degenerate. As the luteal phase comes to an end, menstruation occurs.

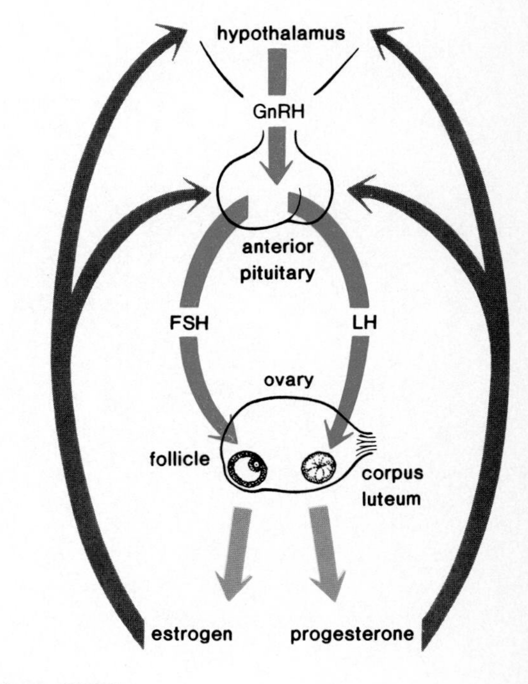

Figure 19.10
Hypothalamic-pituitary-gonad system (simplified) as it functions in the female. GnRH is a hypothalamic-releasing hormone that stimulates the anterior pituitary to secrete LH and FSH. These gonadotropic hormones act on the ovaries. FSH promotes the development of the follicle that later, under the influence of LH, becomes the corpus luteum. Negative feedback controls the level of all hormones involved.

Uterine Cycle The female sex hormones estrogen and progesterone have numerous functions, one of which is discussed here. The effect these hormones have on the endometrium of the uterus causes the uterus to undergo a cyclical series of events known as the **uterine cycle** (table 19.3). Cycles that last 28 days are divided as follows:

During *days 1 to 5* there is a low level of female sex hormones in the body, causing the uterine lining to disintegrate and its blood vessels to rupture. A flow of blood, known as the *menses,* passes out of the vagina during a period of **menstruation,** also known as the menstrual period.

During *days 6 to 13* increased production of estrogen by an ovarian follicle causes the endometrium to thicken and become vascular and glandular. This is called the proliferative phase of the uterine cycle.

Ovulation usually occurs on the fourteenth day of the 28-day cycle.

During *days 15 to 28* increased production of progesterone by the corpus luteum causes the endometrium to double in thickness and the uterine glands to become mature, producing a thick mucoid secretion. This is called the secretory phase of the uterine cycle. The endometrium is now prepared to receive the developing zygote, but if pregnancy does not occur, the corpus luteum degenerates and the low level of sex hormones in the female body causes the uterine lining to break down. This is evident, due to the menstrual discharge that begins at this time. Even while menstruation is occurring, the anterior pituitary begins to increase its production of FSH and a new follicle begins maturation. Table 19.3 indicates how the ovarian cycle controls the uterine cycle.

Figure 19.11
Ovarian and uterine cycles. Above shows that the plasma levels of FSH and LH control the ovarian cycle, which consists of a follicular and luteal phase. Notice how ovulation is associated with a sudden spurt of these hormones. Below shows that the plasma levels of estrogen and progesterone control the uterine cycle, which consists of menstruation, proliferative, and secretory phase.

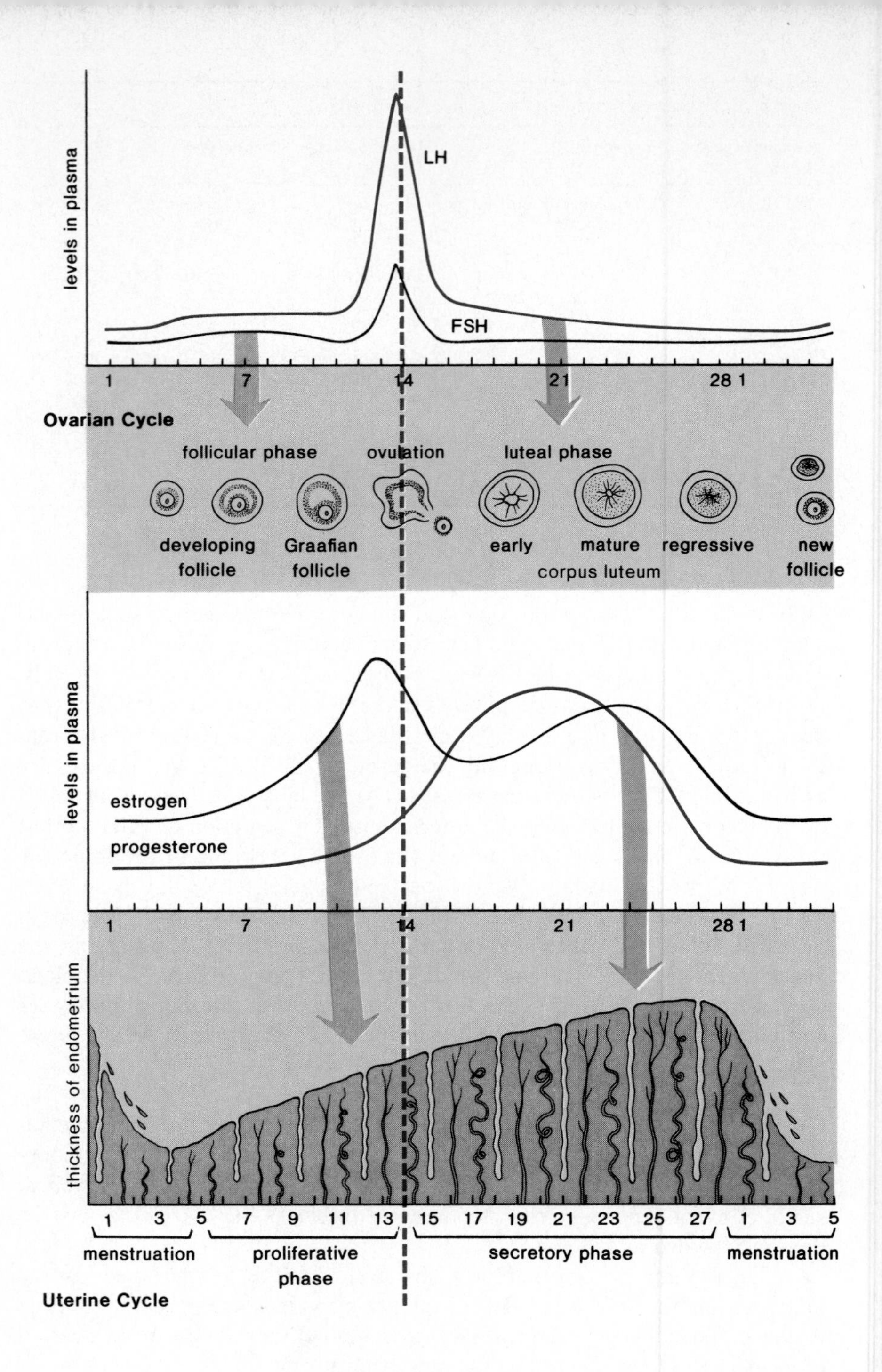

Hormonal Regulation (in Detail)

Figure 19.11 shows the changes in blood concentration of all four hormones participating in the ovarian and uterine cycles. Notice that all four of these hormones (FSH, LH, estrogen, and progesterone) are present during the entire 28 days of the cycle. Therefore, in actuality, both FSH and LH *are* present during the follicular phase and both are needed for follicle development and maturation of the egg. The follicle secretes primarily estrogen and a very minimal amount of progesterone. Similarly, both LH and FSH are present in decreased amounts during the luteal phase. LH may be primarily responsible for corpus luteum formation, but the corpus luteum secretes both progesterone and estrogen. The effect that these hormones have on the endometrium has already been stated. Estrogen stimulates growth of the endometrium and readies it for reception of progesterone, which causes it to thicken and become secretory.

Feedback Control It has been frequently mentioned that a hormone can exert feedback inhibition. Therefore it comes as no surprise to find that as the estrogen level increases during the first part of the follicular phase, FSH secretion begins to decrease. However, toward the end of the follicular phase there is a sharp increase in FSH and LH secretion at the point when the estrogen level is the highest. Regarding this phenomenon, it is believed that the high level of estrogen exerts *positive feedback on the hypothalamus,* causing it to secrete gonadotropic-releasing hormone, after which the pituitary momentarily produces an unusually large amount of FSH and LH. It is the surge of LH that is believed to promote ovulation. During the luteal phase, estrogen and progesterone bring about feedback inhibition as expected and the level of both LH and FSH declines steadily. In this way, all four hormones eventually reach their lowest levels, causing menstruation to occur. It still is not known what causes the corpus luteum to degenerate if pregnancy does not occur. In some mammals there is evidence to suggest that prostaglandins (p. 385) cause degeneration, but this is not believed to be the case in humans.

During the first half of the ovarian cycle, FSH from the anterior pituitary causes maturation of a follicle, which secretes estrogen. After ovulation, and during the second half of the cycle, LH from the anterior pituitary converts the follicle into the corpus luteum, which produces progesterone. Estrogen and progesterone regulate the uterine cycle in which the endometrium builds up and then is shed during menstruation.

Pregnancy

If pregnancy occurs, menstruation does not occur. Instead, the developing zygote embeds itself in the endometrium lining several days following fertilization. This process, called **implantation,** is what causes the female to become *pregnant.* During implantation, an outer layer of cells surrounding the zygote produces a gonadotropic hormone called **h**uman **c**horionic **g**onadotropic hormone (HCG) that prevents degeneration of the corpus luteum and instead causes it to secrete even larger quantities of progesterone. The corpus luteum may be maintained for as much as six months, even after the placenta is fully developed.

The **placenta** (fig. 20.19) originates from both maternal and fetal tissue and is the region of exchange of molecules between fetal and maternal blood, although there is no mixing of the two types of blood. After its formation, the placenta continues production of HCG and begins production of progesterone and estrogen. The latter hormones have two effects: they shut down the anterior pituitary so that no new follicles mature, and they maintain the lining of the uterus so that the corpus luteum is not needed. There is no menstruation during the period of pregnancy.

Pregnancy Tests Pregnancy tests, which are readily available in hospitals, clinics, and, now, even drug and grocery stores, are based on the fact that HCG is present in the blood and urine of a pregnant woman.

Before the advent of monoclonal antibodies, only a blood test requiring the use of radioactive material in the hospital was available to detect pregnancy before the first missed menstrual period. Now there is a monoclonal antibody (p. 265) test for the detection of pregnancy 10 days after conception. This test can be done on a urine sample in a doctor's office and the results are available within the hour.

The physical signs that might prompt a woman to have a pregnancy test are cessation of menstruation, increased frequency of urination, morning sickness, and increase in the size and fullness of the breasts, as well as darkening of the areolae (fig. 19.12).

The Reasons for Orgasm

Why does orgasm exist? A glance at the reproduction of primitive forms of animal life shows that such a fancy thing far exceeds what's necessary to pass genes from one generation to another, which even in mammals can be attained by a no-frills male ejaculation and mere passive receptivity on the part of the female. Yet it's unlikely that nature, as economical as it is, would create something so baroque without good reason.

The simpler the creature, the more difficult it is to determine whether its sexual relations are pleasurable, much less orgasmic. However, all mammals show a marked interest in sex, and will even work for it—a good indication that it's rewarding. Scientists prefer low-key terms like "ejaculatory reflex" and "estrus behavior" when discussing animal sexuality; they're reluctant to apply the term "orgasm," because it can't be verified in animals. The best gauge of orgasm is uniquely human: Did you or didn't you?

Male sexual responses, particularly erection and ejaculation, have been better studied than female processes. Even the humble rat, that scrupulously observed mammal, exhibits the basic criteria associated with male orgasm: ejaculation followed by refraction, accompanied by characteristic movements. "Males of most mammalian species have what at least looks like a precursor to human orgasm, demonstrated by skeletal and facial patterns," says Benjamin Sachs, a reproductive behavior researcher at the University of Connecticut. "You may as well call it orgasm. I'd be more cautious about females." Moreover, in addition to appearances, males have an excellent motive for ejaculation: progeny.

Alas for researchers, human females don't always show characteristic muscular movements during orgasm. Even more vexing, they need not have orgasms to conceive—or even to seek and enjoy mating. The best evidence for female orgasm is subjective and anecdotal—the yea or nay of the woman in question. Until there's a way to acquire such information from animals, subhuman female orgasm can only be inferred. Does this mean that for female animals mating is simply a selfless matter of lie still and think of England?* Ronald Nadler, a primatologist at the Yerkes Regional Primate Research Center in Atlanta,

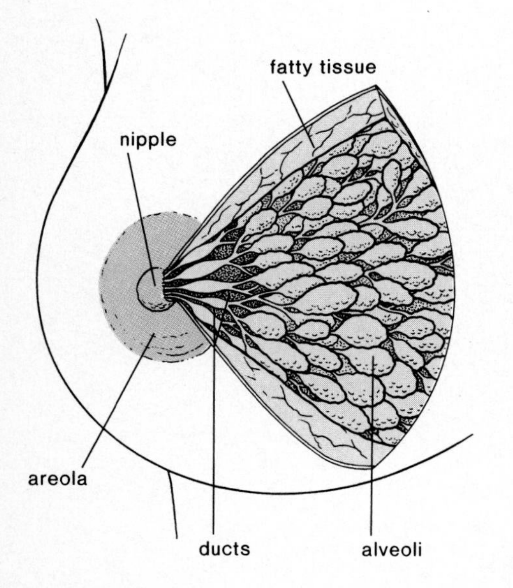

Figure 19.12
Anatomy of breast. The female breast contains lobules consisting of ducts and alveoli. The alveoli are lined by milk-producing cells in the lactating (milk-producing) breast.

Female Sex Hormones

The female sex hormones, estrogen and progesterone, have many effects on the body. In particular, estrogen secreted at the time of puberty stimulates the growth of the uterus and vagina. Estrogen is necessary for egg maturation and is largely responsible for the secondary sex characteristics in females. For example, it is responsible for the onset of the uterine cycle, as well as female body hair and fat distribution. In general, females have a more rounded appearance than males because of a greater accumulation of fat beneath the skin. Also, the pelvic girdle enlarges in females so that the pelvic cavity has a larger relative size compared to males; this means that females have wider hips. Both estrogen and progesterone are also required for breast development.

Breasts A female breast contains 15 to 25 lobules (fig. 19.12), each with its own milk duct which begins at the nipple and divides into numerous other ducts that end in blind sacs called *alveoli.* In a nonlactating (nonmilk-producing) breast, the ducts far outnumber the alveoli because alveoli are made up of cells that can produce milk.

Milk is not produced during pregnancy. *Lactogenic hormone* (prolactin) is needed for lactation (milk production) to begin, and the production of this hormone is suppressed because of the feedback inhibition estrogen and progesterone have on the pituitary during pregnancy. It takes a couple of days after delivery for milk production to begin and, in the meantime, the breasts believes it's unlikely that women are alone among female primates in having orgasms, but the real controversy is not whether, but why, females have orgasms at all.

There are two schools of thought on this. The first maintains that the female has orgasms because the male does, though hers serve no essential reproductive function (an analogy is the male nipple). The female genitals are poorly designed for easy stimulation to orgasm, particularly during intercourse, and if orgasm were really important to procreation, anatomy would have evolved differently to facilitate it. This idea of female orgasm as vestigial is supported by the fact that the inability to have orgasms is common in women yet rare in men, and that orgasm sometimes requires considerable education and practice for women to achieve.

The second school says that female orgasm is designed for a reason or reasons. Sarah Blaffer Hardy, a primatologist at the University of California at Davis, notes that a female could accumulate sufficient stimulation for orgasm as well as increase her chances of conception via the repeated matings typical of primates during estrus. Furthermore, if the motions of the vagina and uterus during orgasm can be proved to enhance the mobility of sperm, as some researchers speculate, orgasm might be a way to enable a female to be somewhat selective about when and even by whom she becomes pregnant. Although neither camp can prove its thesis, a majority holds that female orgasm has a purpose—without knowing what the purpose is.

While scholars debate, couples unwittingly contend with evolutionary traits that have more practical application. As sex therapists are wont to put it, "Men heat up like light bulbs and women like irons." Trying as this difference in timing can be, it makes sociobiological sense: a female motivated to seek repeated sexual encounters increases her opportunities for pregnancy and assures male benignity to a greater extent than if she were quickly aroused and satisfied. On the other hand, a male, always on the defensive, can't afford lengthy dalliances with his back to potential enemies.

Though many theories about the evolution of orgasm are specific to males or females, the response also makes sense in social, as well as reproductive, terms for both. While the sex lives of other mammals are regulated by hormones—totally so in non-primates like rats, partially so in nonhuman primates—humans appear to be freed from such controls, and even castrated men and women can enjoy sex and have orgasms. There are reliable reports of children seven years old having orgasms, and there's no limit at the other end of the age spectrum—further indications that there's more to sex than reproduction. Roger Short, a reproductive biologist at Monash University in Melbourne, Australia, thinks that because human females, unlike those of other species, will accept intercourse at any time, whether or not they are in a fertile period, most couplings are performed for social purposes such as strengthening pair-bonding and releasing sexual tension, so that the more mundane business of life can be attended to.

*Traditional advice to Victorian brides on their wedding nights.
From Winifred Gallagher, "The Reasons for Orgasm May be More Social than Sexual."

produce a watery, yellowish white fluid called **colostrum,** which differs from milk in that it contains more protein and less fat. The continued production of milk requires continued breastfeeding. When a breast is suckled, the nerve endings in the areola are stimulated and nerve impulses travel to the hypothalamus, which causes oxytocin to be released by the posterior pituitary. When this hormone arrives at the breasts, it causes contraction of the lobules so that milk flows into the ducts (called milk letdown).

Menopause

Menopause, the period in a woman's life during which the ovarian and uterine cycles cease, is likely to occur between ages 45 and 55. The ovaries are no longer responsive to the gonadotropic hormones produced by the anterior pituitary, and the ovaries no longer secrete estrogen or progesterone. At the onset of menopause, the uterine cycle becomes irregular, but as long as menstruation occurs it is still possible for a woman to conceive and become pregnant. Therefore, a woman is usually not considered to have completed menopause until there has been no menstruation for a year. The hormonal changes during menopause often produce physical symptoms, such as "hot flashes" that are caused by circulatory irregularities, dizziness, headaches, insomnia, sleepiness, and depression. Again, there is a great variation among women and any of these symptoms may be absent altogether.

Women sometimes report an increased sex drive following menopause, and it has been suggested that this may be due to androgen production by the adrenal cortex.

Estrogen and, to some extent, progesterone affect the female genitals, promote development of the egg, and maintain the secondary sex characteristics. Lactogenic hormone causes the breasts to begin to secrete milk after delivery while another hormone, oxytocin, is responsible for milk letdown. When menopause occurs, FSH and LH are still produced by the anterior pituitary, but the ovaries are no longer able to respond.

Control of Reproduction

Birth Control

Several means of birth control have been available for quite some time (table 19.4). The use of these contraceptive methods decreases the probability of pregnancy. A common way to discuss pregnancy rate is to indicate the number of pregnancies expected per 100 women per year. For example, it is expected that 80 out of 100 young women, or 80 percent, who are regularly engaging in unprotected intercourse will be pregnant within a year. Another way to discuss birth control methods is to indicate their effectiveness, in which case the emphasis is placed on the number of women who will not get pregnant. For example, with the least effective method given in table 19.4, we expect that 70 out of 100, or 70 percent, sexually active women will not get pregnant while 30 women will get pregnant within a year.

Future Means of Birth Control

There are three areas in which birth control investigations have been directed. There is a need for a morning-after medication, a long-lasting method, and a medication that is specifically for males.

In this country, DES, a synthetic estrogen, which affects the uterine lining, making implantation difficult, is sometimes given following intercourse. Since large doses are required, causing nausea and vomiting, DES is usually given only for incest or rape. Investigators in France have recently announced promising results with a substance known as *RU 486*. RU 486 is a synthetic steroid that prevents progesterone from acting on the uterine lining because it has a high affinity for progesterone receptors. In clinical tests the uterine lining sloughed off within four days in 85 percent of women who were less than a month pregnant. To improve the success rate, investigators are thinking of combining RU 486 with small quantities of prostaglandins that cause contraction of the uterus and disintegration of the corpus luteum. The promoters of this treatment are using the term "contragestation" to describe its effects; however, it should be recognized that this medication rather than preventing implantation, brings on an **abortion,** the loss of an implanted fetus. It is expected that the medication will be used by many women who are experiencing delayed menstruation without knowing whether they are actually pregnant.

Depo-Provera is an injectable contraceptive that is commercially available in many countries outside the United States. The injection contains crystals that gradually dissolve over a period of three months. The crystals contain a chemical related to progesterone and, like the pill, it suppresses ovulation. The drug has not been approved for use in the United States because cancer developed in some test animals receiving the injections. More animal studies are now underway. An even more potent progesterone-like molecule is now being tested for implantation under the skin. The implant consists of narrow tubes that slowly release the drug over a period of five years.

Table 19.4 Common Birth Control Methods

Name	Procedure	Methodology	Effectiveness*	Action Needed	Risk
Vasectomy	Ductus deferentia are cut and tied	No sperm in semen	Almost 100%	Sexual freedom	Irreversible sterility
Tubal ligation	Oviducts are cut and tied	No eggs in oviduct	Almost 100%	Sexual freedom	Irreversible sterility
Pill	Must take medication daily	Shuts down pituitary	Almost 100%	Sexual freedom	Thrombo-embolism
IUD	Must be inserted into uterus by physician	Prevents implantation	More than 90%	Sexual freedom	Infection
Diaphragm	Plastic cup inserted into vagina to cover cervix	Blocks entrance of sperm into uterus	With jelly about 90%	Must be inserted each time before intercourse	—
Condom	Sheath that fits over erect penis	Traps sperm	About 85%	Must be placed on penis at time of intercourse	—
Coitus interruptus (Withdrawal)	Male withdraws penis before ejaculation	Prevents sperm from entering vagina	About 80%	Intercourse must be interrupted before ejaculation	—
Jellies, creams, foams	Contain spermicidal chemicals	Kill a large number of sperm	About 75%	Must be inserted before intercourse	—
Rhythm method	Determine day of ovulation by record keeping; testing by various methods	Avoid day of ovulation	About 70%	Limits sexual activity	—
Douche	Cleanses vagina and uterus after intercourse	Washes out sperm	Less than 70%	Must be done *immediately* after intercourse	—

*Effectiveness is the average percentage of women who did not become pregnant in a population of 100 sexually active women using the technique for one year.

Various possibilities exist for a *"male pill."* Scientists have made analogs of gonadotropic-releasing hormones that interfere with the action of this hormone and prevent it from stimulating the pituitary. Experiments in both animal and human subjects suggest that one of these might possibly inhibit spermatogenesis in males (and ovulation in females) without affecting the secondary sex characteristics. The seminiferous tubules produce a horomone termed inhibin that inhibits FSH production by the pituitary (p. 398). It is hoped that this chemical may someday be produced commercially and made available in pill form for males. Testosterone and/or related chemicals can be used to inhibit spermatogenesis in males, but there are usually feminizing side effects because the excess is changed to estrogen by the body.

There are numerous well-known birth control methods and devices available to those who wish to prevent pregnancy. These differ as to their effectiveness. In addition, new methods are expected to be developed.

Infertility

Sometimes couples do not need to prevent pregnancy; conception or fertilization does not occur despite frequent intercourse. The American Medical Association estimates that 15 percent of all couples in this country are unable to have any children and are therefore properly termed sterile; another 10 percent have fewer children than they wish and are therefore termed *infertile.*

Infertility can be due to a number of factors. It is possible that fertilization takes place but the embryo dies before implantation takes place. One area of concern is that radiation, chemical mutagens, and the use of psychoactive drugs can contribute to sterility, possibly by causing chromosomal mutations that prevent development from proceeding normally. The lack of progesterone can also prevent implantation and therefore the proper administration of this hormone is sometimes helpful.

It is also possible that fertilization never takes place. There may be a congenital malformation of the reproductive tract or there may be an obstruction of the oviduct or ductus deferens. Sometimes these physical defects can be corrected surgically. If no obstruction is apparent, it is possible to give females a substance rich in FSH and LH that is extracted from the urine of postmenopausal women. This treatment causes multiple ovulations and sometimes multiple pregnancies, however.

Endometriosis

Endometriosis, an often unrecognized disease that afflicts anywhere from 4 million to 10 million American women, is a major cause of infertility. The condition is caused by the spread and growth of tissue from the lining of the uterus (or endometrium) beyond the uterine walls. These endometrial cells form bandlike patches and scars throughout the pelvis and around the ovaries and Fallopian tubes, resulting in a variety of symptoms and degrees of discomfort. Because endometriosis has been associated with delayed childbearing, it is sometimes called the "career woman's disease." But recent studies have shown that the disorder strikes women of all socioeconomic groups and even teenagers, though those with heavier, longer or more frequent periods may be especially susceptible. Says Dr. Donald Chatman of Chicago's Michael Reese Hospital: "Endometriosis is an equal-opportunity disease."

How the disease begins is something of a mystery. One theory ascribes it to "retrograde menstruation." Instead of flowing down through the cervix and vagina, some menstrual blood and tissue back up through the Fallopian tubes and spill out into the pelvic cavity (*see box figure*). Normally this errant flow is harmlessly absorbed, but in some cases the stray tissue may implant itself outside the uterus and continue to grow. A second theory suggests that the disease arises from misplaced embryonic cells that have lain scattered around the abdominal cavity since birth. When the monthly hormonal cycles begin at puberty, says Dr. Howard Judd, director of gynecological endocrinology at UCLA Medical Center, "some of these cells get stirred up and could be a major cause of endometriosis."

If anything about endometriosis is clear, it is that once the disease has begun, it will probably get worse. Stimulated by the release of estrogen, the implanted tissue grows and spreads. Cells from the growths break away and are ferried by lymphatic fluid throughout the body, sometimes, although rarely, forming islands in the lungs, kidneys, bowel or even the nasal passages. There they respond to the menstrual cycle, causing monthly bleeding from the rectum or wherever else they have settled.

The most common symptom of endometriosis is pain, which can occur during menstruation, urination, and sexual intercourse. Unfortunately, these warnings are often overlooked by women and their doctors.

To confirm that a patient has endometriosis, doctors look for the telltale tissue by peering into the pelvic cavity with a fiberoptic instrument called a laparoscope. After diagnosis, a number of treatments can be prescribed. One is pregnancy—if it is still feasible; the nine-month interruption of menstruation can help shrink misplaced endometrial tissue. Taking birth-control pills may also help, but more effective is a drug called danazol, a synthetic male hormone that stops ovulation and causes endometrial tissue to shrivel. But it can also produce acne, facial-hair growth, weight gain and other side effects.

A new experimental treatment with perhaps fewer ill effects involves a synthetic substance called nafarelin, similar to gonadotropin-releasing hormone. Normally GnRH is released in bursts by the hypothalamus gland, eventually triggering the process of ovulation. But "if the GnRH stimulation is given continuously instead of in pulses," explains Dr. Robert Jaffe of the University of California, San Francisco, "the whole [ovulatory] system shuts off," and the endometrial implants "virtually melt away."

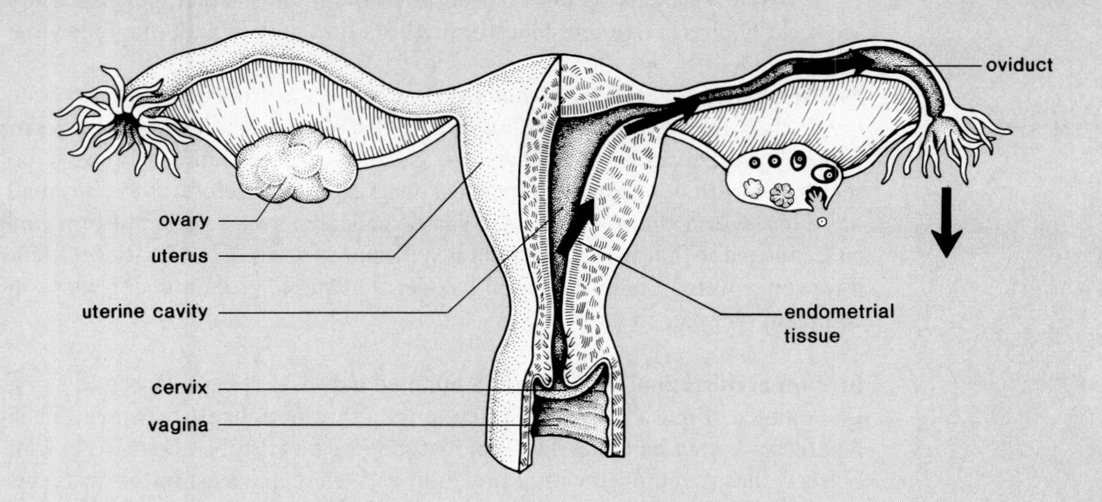

It is speculated that endometriosis could be caused by a backward menstrual flow represented by the arrows in this diagram. This would allow endometrial cells to enter the abdominal cavity where they take up residence and respond to the monthly cyclic changes in hormonal levels, including those that result in menstruation.

For severe cases of endometriosis, surgical removal of the ovaries and uterus may be the only solution. But less extreme surgery can often help. At Atlanta's Northside Hospital, Dr. Camran Nezhat has had success with a high-tech procedure called videolaseroscopy, which employs a laparoscope rigged with a tiny video camera and a laser. The camera images, enlarged on a video screen, enable Nezhat to zero in on endometrial tissue and vaporize it with the laser. In a study of 102 previously infertile patients, Nezhat found that 60.7% were able to conceive within two years of videolaseroscopy treatment.

Like many other doctors who see the unfortunate consequences of endometriosis, Nezhat is concerned that a "lot of women do not seek help for this problem." Any serious pain, he notes, needs investigating. Agrees Cheri Bates (a victim); "If a doctor tells you that suffering is a woman's lot in life, get another doctor."

Figure 19.13
Sometimes couples utilize alternative methods of reproduction in order to experience the joys of parenthood.

When reproduction does not occur in the usual manner, couples today are seeking alternative reproductive methods that may include the following:

Artificial Insemination by Donor (AID) Since the 1960s, there have been hundreds of thousands of births following artificial insemination in which sperm is placed in the vagina by a physician. Sometimes a woman is artificially inseminated with her husband's sperm. This is especially helpful if the husband has a low sperm count—the sperm can be collected over a period of time and concentrated so that the sperm count is sufficient to result in fertilization. Often, however, a woman is inseminated by sperm acquired from a donor who is a complete stranger to her.

In Vitro Fertilization (IVF) Over a hundred babies have been conceived using this method. First a woman is given appropriate hormonal treatment. Then laparoscopy may be done. The laparoscope is a metal tube about the size of a pencil that is equipped with a tiny light and telescopic lens. In this instance, it is also fitted with a tube for retrieving eggs. After insertion through a small incision near the woman's naval, the physician guides the laparoscope to the ovaries where they are sucked up into the tube. Alternately, it is possible to place a needle through the vaginal wall and guide it, by the use of ultrasound, to the ovaries where the needle is used to retrieve the eggs. This method is called transvaginal retrieval.

Concentrated sperm from the male is placed in a solution that approximates the conditions of the female genital tract. When the eggs are introduced, fertilization occurs. The resultant zygotes begin development and after about two to four days they are inserted into the uterus of the woman, who is now in the secretory phase of her menstrual cycle. If implantation is successful, development is normal and continues to term.

Gamete Intrafallopian Transfer (GIFT) This method was devised as a means to overcome the low success rate (15 to 20 percent) of in vitro fertilization. The method is exactly the same as in vitro fertilization except the eggs and sperm are immediately placed in the oviducts after they have been brought together. This procedure would be helpful in couples whose eggs and sperm never make it to the oviducts; sometimes the egg gets lost between the ovary and the oviducts, and sometimes the sperm never reach the oviducts. GIFT has an advantage in that it is a one-step procedure for the woman—the eggs are removed and reintroduced all in the same time period. For this reason it is less expensive, $1,500, compared with $3,000 and up for in vitro fertilization.

Surrogate Mothers Over a hundred babies have been born to women paid to have them by other individuals who have contributed sperm (or egg) to the fertilization process. If all the alternative methods discussed above are considered it is possible to imagine that a baby could have five parents: (1) sperm donor, (2) egg donor, (3) surrogate mother, and (4) and (5) adoptive mother and father.

Some couples are infertile. There may be a hormonal imbalance or a blockage of the oviducts. When corrective medical procedures fail, it is possible today to consider an alternative method of reproduction.

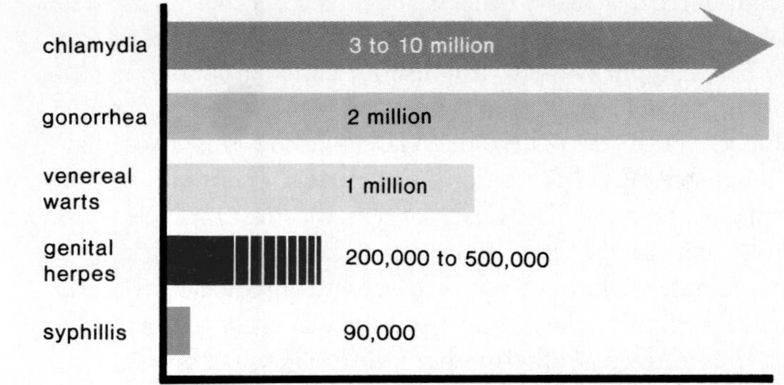

Figure 19.14
Statistics for the most common sexually transmitted diseases show that chlamydia, gonorrhea, and veneral warts all are much more common than herpes and syphilis. Chlamydia, gonorrhea, and syphilis are all curable with antibiotic therapy.

Sexually Transmitted Diseases

There are many diseases that are transmitted from person to person by sexual contact. We will consider only five of the most troublesome, AIDS (acquired immune deficiency syndrome), genital herpes, gonorrhea, chlamydia, and syphilis (fig. 19.14). The first two are viral diseases and, therefore, are difficult to treat since the traditional antibiotics are not helpful (p. 543). The drug AZT (azidothymidine) is now being used in patients with AIDS (p. 259) and both ointment and oral ACV (acyclovir) is available for herpes. Although gonorrhea and chlamydia are treatable with appropriate antibiotic therapy they are not always promptly diagnosed. Unfortunately, as yet, there are no vaccines available for these infections.

AIDS

The organism that causes *AIDS* (acquired immune deficiency syndrome) is a retrovirus that attacks helper T cells (p. 256). The virus also attacks those cells that interact with helper T cells, such as macrophages and B lymphocytes. The scientific community has settled on the name HID (human immuno-deficiency) virus.

It is estimated that there will be 270,000 people with AIDS in the United States by 1991. Figure 19.15 indicates the three stages of infection. During the first stage, the individual usually has detectable antibodies in the bloodstream and may exhibit swollen lymph nodes. During the second stage, called ARC (AIDS-related complex), symptoms also include weight loss, night sweats, fatigue, fever, and diarrhea. Finally, the person may develop full-blown AIDS, especially characterized by development of pneumonia, skin cancer, and also neuromuscular and psychological disturbances. Death is usually within a few years. As figure 19.15 implies, most people do not develop AIDS in the first two years of infection but in the next 3 to 4 years the rate is fairly constant, and new evidence indicates that rates get even higher after the fifth year of infection.

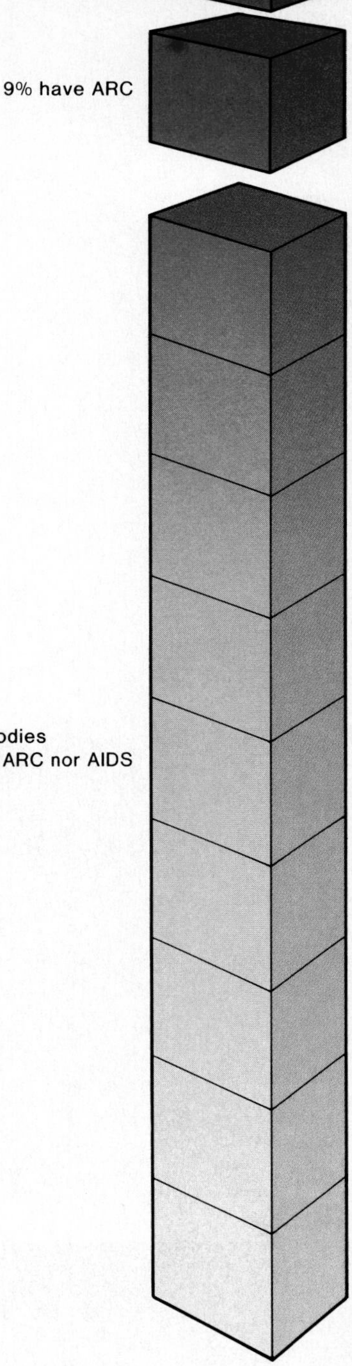

Figure 19.15
Out of every 100 people who test positive for AIDS antibodies in their bloodstream, 90 will have no outward symptoms of the disease except swollen lymph nodes; 9 will have ARC with numerous symptoms; and only 1 will have AIDS itself. An unknown portion of the 90 people will ever go on to develop ARC and an unknown portion of those with ARC will ever develop AIDS itself.

Transmission

Although AIDS is considered a sexually transmitted disease, the virus is spread by blood-to-blood contact. It is for this reason that unsterilized needles can transmit the disease between intravenous drug users, and hemophiliacs in particular have contracted the disease from infected donor blood. Subsequently, today all donated blood that tests positive for AIDS antibodies is discarded.

The greatest incidence of AIDS is among homosexual males but the proportion of heterosexual cases in the United States is increasing at a faster rate. This means that both anal and vaginal intercourse can pass the virus to a partner. Apparently, females who have intercourse with infected men are at a greater risk than vice versa. Because the HID virus can cross the placenta, a steady increase in the number of children born with AIDS is expected.

Health officials emphasize that unprotected intercourse with multiple partners or a single infected partner increases the chance of transmission. The use of a condom reduces the risk but the very best preventive measure at this time is monogamy with a sexual partner who is free of the disease. Casual contact with someone who is infected, such as shaking hands, eating at the same table, or swimming in the same pool, is not a mode of transmission.

Genital Herpes

Genital herpes is caused by herpes simplex virus (fig. 19.16) of which there are two types: type 1 usually causes cold sores and fever blisters while type 2 more often causes genital herpes.

Genital herpes is one of the more prevalent sexually transmitted diseases today (fig. 19.14); an estimated 20.5 million persons in the United States have it, with an estimated 500,000 new cases appearing each year. Immediately after infection there are no symptoms, but the individual may experience a tingling or itching sensation before blisters appear at the infected site within 2 to 20 days. Once the blisters rupture, they leave painful ulcers that may take as long as three weeks or as little as five days to heal. These symptoms may be accompanied by fever, pain upon urination, and swollen lymph nodes.

After the ulcers heal, the disease is dormant. But blisters can reoccur repeatedly at variable intervals. Sunlight, sex, menstruation, and stress seem to cause the symptoms of genital herpes to reoccur. While the virus is latent, it resides in nerve cells. Type 1 occasionally travels via a nerve fiber to the eye and causes an eye infection that can lead to blindness. Type 2 has been known to cause a form of meningitis and also has been associated with a form of cervical cancer.

Infection of the newborn can occur if the child comes in contact with a lesion in the birth canal. In 1 to 3 weeks the infant is gravely ill and may become blind, or have neurological disorders including brain damage, or may die. Birth by caesarean section prevents these occurrences.

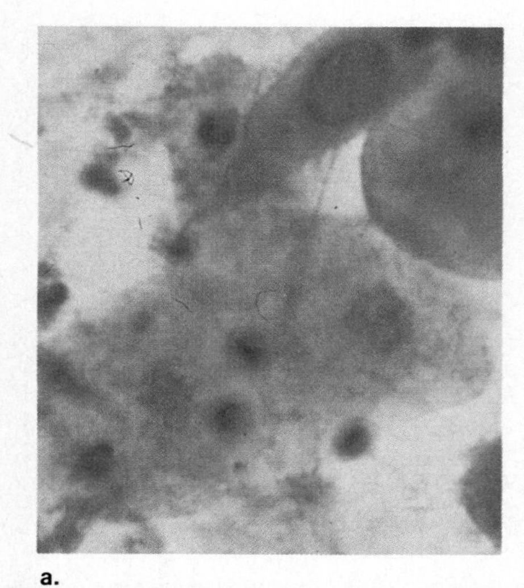

a.

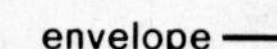

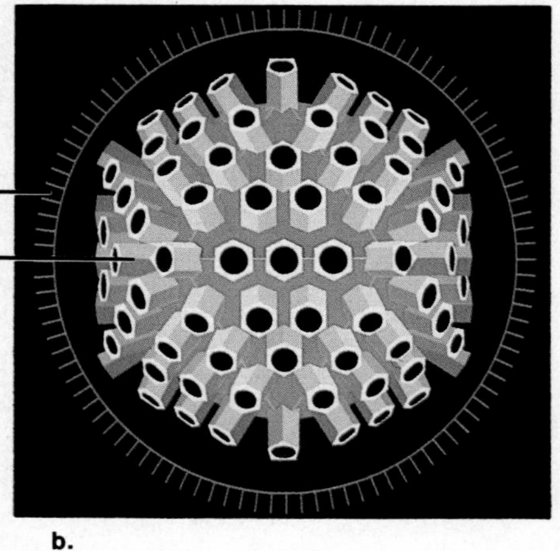

b.

Figure 19.16
a. *Cell infected with herpes viruses.* b. *Enlarged model of herpes virus.*

Gonorrhea

Gonorrhea is caused by the bacterium *Neisseria gonorrheae* which is a nonmotile (lacks flagella), and nonspore-forming diplococcus, meaning that two cells generally stay together (fig. 19.17).

The diagnosis of gonorrhea in the male is not difficult as long as he displays typical symptoms (as many as 40 percent of males may be asymptomatic). The patient complains of pain on urination and has a thick, greenish-yellow urethral discharge three to five days after contact. In the female, the bacteria may first settle within the urethra or near the cervix, from which they may spread to the oviducts, causing **pelvic inflammatory disease (PID).** As the inflamed tubes heal, they may become partially or completely blocked by scar tissue. As a result, the female is sterile or, at best, subject to **ectopic pregnancy,** a pregnancy that begins at a location other than the uterus (p. 401). Unfortunately, 60 to 80 percent of females are asymptomatic until they develop severe pains in the abdominal region due to PID.

Homosexual males develop gonorrhea proctitis, or infection of the anus, with symptoms including pain in the anus and blood or pus in the feces. Oral sex can cause infection of the throat and tonsillitis. Gonorrhea may also spread to other parts of the body, causing heart damage or arthritis. If, by chance, the person touches infected genitals and then his or her eyes, a severe eye infection can result.

Eye infection leading to blindness can occur as a baby passes through the birth canal. Because of this, all newborn infants receive eye drops containing antibacterial agents, such as silver nitrate, tetracycline, or penicillin, as a protective measure.

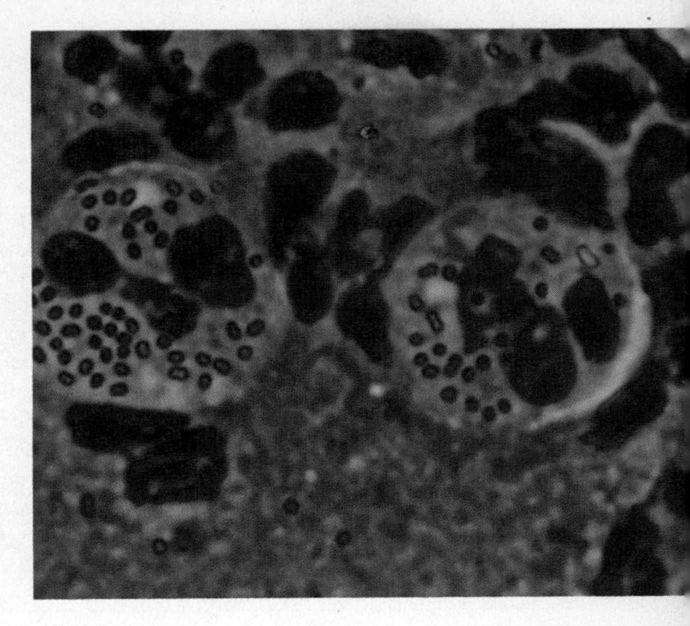

Figure 19.17
Gonorrhea bacteria in the bloodstream. If you look carefully you will notice that these round bacteria occur in twos; for this reason they are called diplococci.

Chlamydia

This sexually transmitted disease is named for the tiny bacterium that causes it (*Chlamydia trachomatis*). New chlamydiae infections occur at even a faster rate than gonorrhea infections (fig. 19.14). They are the most common cause of NGU, nongonococcal urethritis. About 8 to 21 days after exposure, men experience a mild burning sensation upon urinating and a mucoid discharge. Women may have a vaginal discharge along with the symptoms of a urinary tract infection. Unfortunately, a physician may mistakenly diagnose a gonorrheal or urinary infection and prescribe the wrong type antibiotic. Or the person may never seek medical help. In either case the infection may eventually cause PID and sterility or ectopic pregnancy.

If a newborn comes in contact with chlamydia during delivery, inflammation of the eyes or pneumonia may result. There are also those who believe that chlamydia infections increase the possibility of premature and stillborn births.

Syphilis

Syphilis is caused by the bacteria *Treponema pallidum.* Syphilis has three stages, which may be separated by latent stages in which the bacteria are resting before multiplying again. During the *primary stage,* a hard chancre (ulcerated sore with hard edges) indicates the site of infection. The chancre may go unnoticed, especially since it usually heals spontaneously leaving little scarring. During the *secondary stage,* proof that bacteria have invaded and spread throughout the body is evident when the victim breaks out in a rash. Curiously, the rash does not itch and is seen even on the palms of the hands and soles of the feet.

There may be hair loss and gray patches on the mucous membranes, including the mouth. These symptoms disappear of their own accord. During a *tertiary stage,* which lasts until the patient dies, gummas (large, destructive ulcers) appear on the skin and within the internal organs, especially the small arteries (cardiovascular syphilis) and brain (neurosyphilis). Because syphilis can be cured even if it reaches the tertiary stage, a person should never feel that it is too late for treatment. However, organ damage cannot be reversed.

Congenital syphilis is caused by syphilis bacteria crossing the placenta. The newborn is born blind and/or with numerous anatomical malformations. Penicillin has been used as an effective antibiotic to cure syphilis.

The sexually transmitted diseases, AIDS, herpes, gonorrhea, chlamydia, and syphilis are most troublesome at this time. AIDS often results in death within a few years, recurrent attacks of herpes occur throughout life, and gonorrhea and chlamydia often lead to sterility. The best preventive measure against these diseases is monogamy with a partner who is free of them.

Summary

In males, spermatogenesis occurs within the seminiferous tubules of the testes, which also produce testosterone within the interstitial cells. Sperm mature in the epididymis and are stored in the ductus deferens before entering the urethra, along with seminal fluid prior to ejaculation. Hormonal regulation involving secretions from the hypothalamus, anterior pituitary, and the testes in the male maintains testosterone at a fairly constant level.

In females, oogenesis occurs within the ovaries where one follicle produces an egg each month. Fertilization, if it occurs, takes place in the oviducts and the resulting embryo travels to the uterus where it embeds itself in the uterine lining. In the nonpregnant female, hormonal regulation in the female involves the ovarian and uterine cycle, dependent upon the hypothalamus, anterior pituitary, and the female sex hormones, estrogen and progesterone.

Numerous birth control methods and devices are available for those who wish to prevent pregnancy. Infertile couples are increasingly resorting to alternative methods of reproduction.

Sexually transmitted diseases are of concern to all. AIDS, genital herpes, and chlamydia are presently of the greatest concern but still prevalent are gonorrhea and syphilis.

Objective Questions

1. If one were tracing the path of sperm, the structure that follows the epididymis is the ______________ .
2. The prostate gland, Cowper's glands, and the ______________ all contribute to seminal fluid.
3. The primary male sex hormone is ______________ .
4. An erection is caused by the entrance of ______________ into sinuses within the penis.
5. In the female reproductive system, the uterus lies between the oviducts and the ______________ .
6. In the ovarian cycle, once each month a ______________ produces an egg. In the uterine cycle, the ______________ lining of the uterus is prepared to receive the zygote.
7. The female sex hormones are ________ and ______________ .
8. Pregnancy in the female is detected by the presence of ______________ in the blood or urine.
9. In vitro fertilization occurs in ________ .
10. Although a sexually transmitted disease, AIDS is actually spread by ______________ contact.
11. Herpes simplex virus type 1 causes ______________ and type 2 causes ______________ .
12. The most prevalent sexually transmitted disease today is ________________ .

Answers to Objective Questions

1. ductus deferens 2. seminal vesicles 3. testosterone 4. blood 5. vagina 6. follicle, endometrial 7. estrogen, progesterone 8. HCG 9. laboratory glassware 10. blood 11. cold sores, genital herpes 12. chlamydia

Study Questions

1. Discuss the anatomy and physiology of the testes. (p. 393) Describe the structure of sperm. (p. 394)
2. Give the path of sperm. (p. 394)
3. What glands produce seminal fluid? (p. 395)
4. Discuss the anatomy and physiology of the penis. (p. 396) Describe ejaculation. (p. 396)
5. Discuss hormonal regulation in the male. Name three functions for testosterone. (p. 398)
6. Discuss the anatomy and physiology of the ovaries. (p. 399) Describe ovulation. (p. 400)
7. Give the path of the egg. Where does fertilization and implantation occur? Name two functions for the vagina. (pp. 399–401)
8. Describe the external genitalia in females. (p. 402)
9. Compare male and female orgasms. (pp. 396, 402)
10. Discuss hormonal regulation in the female, either simplified and/or in detail. (pp. 403, 404) Give the events of the uterine cycle and relate them to the ovarian cycle. (pp. 403, 404) In what way is menstruation prevented if pregnancy occurs? (p. 405)
11. Name four functions of the female sex hormones. (p. 406) Describe the anatomy and physiology of the breast. (p. 406)
12. Discuss the various means of birth control and their relative effectiveness. (p. 408)
13. Describe the most common types of sexually transmitted diseases. (pp. 413–415)

Key Terms

endometrium (en″do-me′tre-um) the lining of the uterus that becomes thickened and vascular during the uterine cycle. *401*

erection (ĕ-rek′shun) referring to a structure such as the penis, that is turgid and erect as opposed to being flaccid or lacking turgidity. *396*

Graafian follicle (graf′e-an fol′lĭ-k′l) mature follicle within the ovaries which houses a developing egg. *400*

implantation (im″plan-ta′shun) the attachment of the embryo to the lining (endometrium) of the uterus. *405*

interstitial cells (in″ter-stish′al selz) hormone-secreting cells located between the seminiferous tubules of the testes. *395*

menopause (men′o-pawz) termination of the ovarian and uterine cycles in older women. *407*

menstruation (men″stroo a′shun) loss of blood and tissue from the uterus at the end of a uterine cycle. *403*

ovarian cycle (o-va′re-an si′k′l) monthly occurring changes in the ovary that affect the level of sex hormones in the blood. *403*

ovaries (o′var-ez) the female gonads, the organs that produce eggs, and estrogen and progesterone. *399*

semen (se′men) the sperm-containing secretion of males; seminal fluid plus sperm. *396*

seminiferous tubules (sem″i-nif′er-us tu′būlz) highly coiled ducts within the male testes that produce and transport sperm. *394*

testes (tes′tēz) the male gonads, the organs that produce sperm and testosterone. *393*

uterine cycle (u′ter-in si′k′l) monthly occurring changes in the characteristics of the uterine lining. *403*

20

Development

Chapter Concepts

1 The first stages of human development, which lead to the establishment of the embryonic germ layers, are the same as those of other animal embryos.

2 Induction, or the ability of one tissue to influence the development of another, can explain differentiation, or specialization of parts, and can account for the orderliness of development.

3 Human embryos have the same extraembryonic membranes as the embryos of reptiles and birds but their function has been altered to suit internal development.

4 It is possible to outline in a precise manner the steps in human development from the fertilized egg to the birth of a child.

Chapter Outline

Chordate Development
- Developmental Processes
- Early Developmental Stages
- Vertebrate Cross Section
- Extraembryonic Membranes

Human Development
- Embryonic Development
- Fetal Development

Birth of Human
- Stages
- Prepared Childbirth

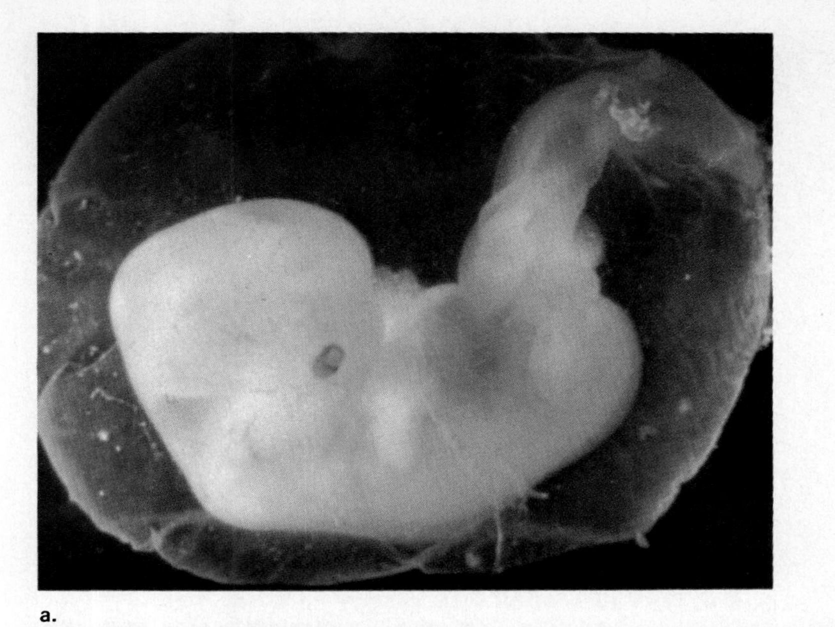

a.

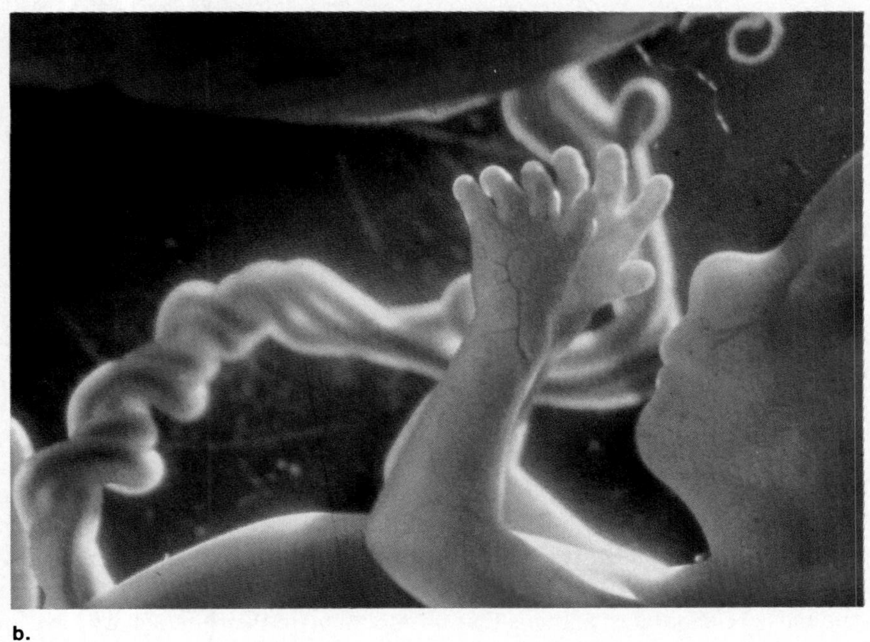

b.

Figure 20.1
Human development is divided into the embryonic period (first two months) and fetal development (third through ninth month). a. *Embryo is not recognizably human.* b. *Fetus is recognizably human.*

Development encompasses the time from *conception* (fertilization) to *birth* (parturition). In humans the *gestation period,* or length of pregnancy, is approximately nine months. It is customary to calculate the time of birth by adding 280 days to the start of the last menstruation because this date is usually known, whereas the day of fertilization is usually unknown. Because the time of birth is influenced by so many variables, only about 5 percent of babies actually arrive on the forecasted date.

Human development is very often divided into *embryonic development* (first two months) and *fetal development* (three through nine months). The embryonic period consists of early development, during which all the major organs form, and fetal development consists of a refinement of these structures. The fetus, not the embryo, is recognizable as a human being (fig. 20.1).

Figure 20.2
Amphioxus development. Since the egg has little yolk, cleavage is complete (first row), *gastrulation occurs by invagination* (second row), *and the coelom develops by outpocketing* (third row). *Presumptive notochord induces the formation of the neural tube.*

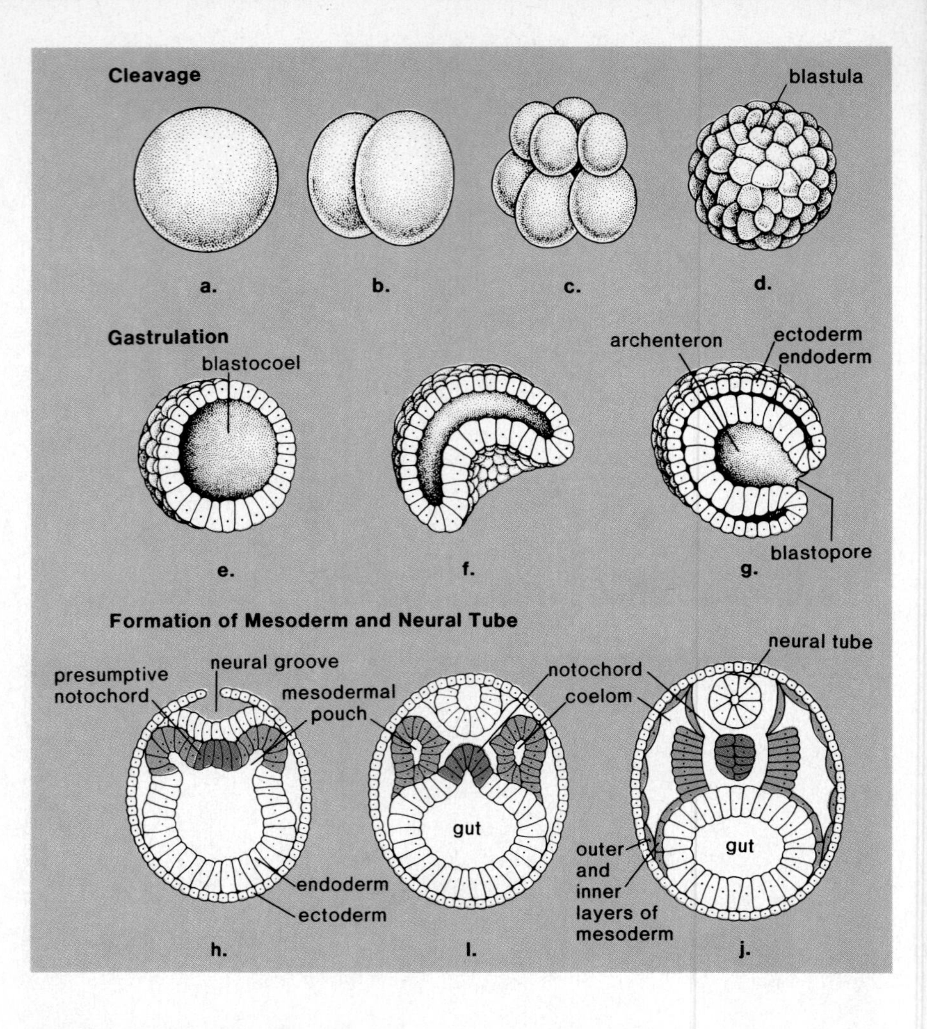

Table 20.1 Early Developmental Stages		
Stage	**Process**	**Result**
Cleavage	Cell division without growth	Many-celled morula
Blastula	Morphogenesis and growth	Hollow ball of cells
Gastrula	Morphogenesis and growth	An embryo with three germ layers
Neurula	Differentiation by induction	Nervous system development

Chordate Development

Much has been learned about human embryonic development by studying the early development of other animals whose eggs are more accessible, easier to see, and may be freely subjected to experimentation. As we discuss human embryonic development, we will have occasion to refer to the development of amphioxus (fig. 20.2), the frog, and chick in addition to humans. All these animals are chordates, animals that at some time in their life history have an elastic supporting rod known as a *notochord.* In vertebrates, this rod is replaced by the vertebral column. All the animals mentioned except amphioxus are vertebrates.

Developmental Processes

All animal embryos develop by means of the following processes, which are also listed in table 20.1.

Cleavage Immediately after fertilization, the zygote begins to divide so that at first there are 2, then 4, 8, 16, and 32 cells, and so forth. Since increase in size does not accompany these divisions, the embryo is at first no larger than the zygote was. Cell division during **cleavage** is mitotic, and each cell receives a full complement of chromosomes and genes.

Growth Later cell division is accompanied by an increase in size of the daughter cells, and **growth** in the true sense of the term takes place.

Morphogenesis **Morphogenesis** refers to the shaping of the embryo and is first evident when certain cells are seen to move, or migrate, in relation to other cells. By these movements, the embryo begins to assume various shapes.

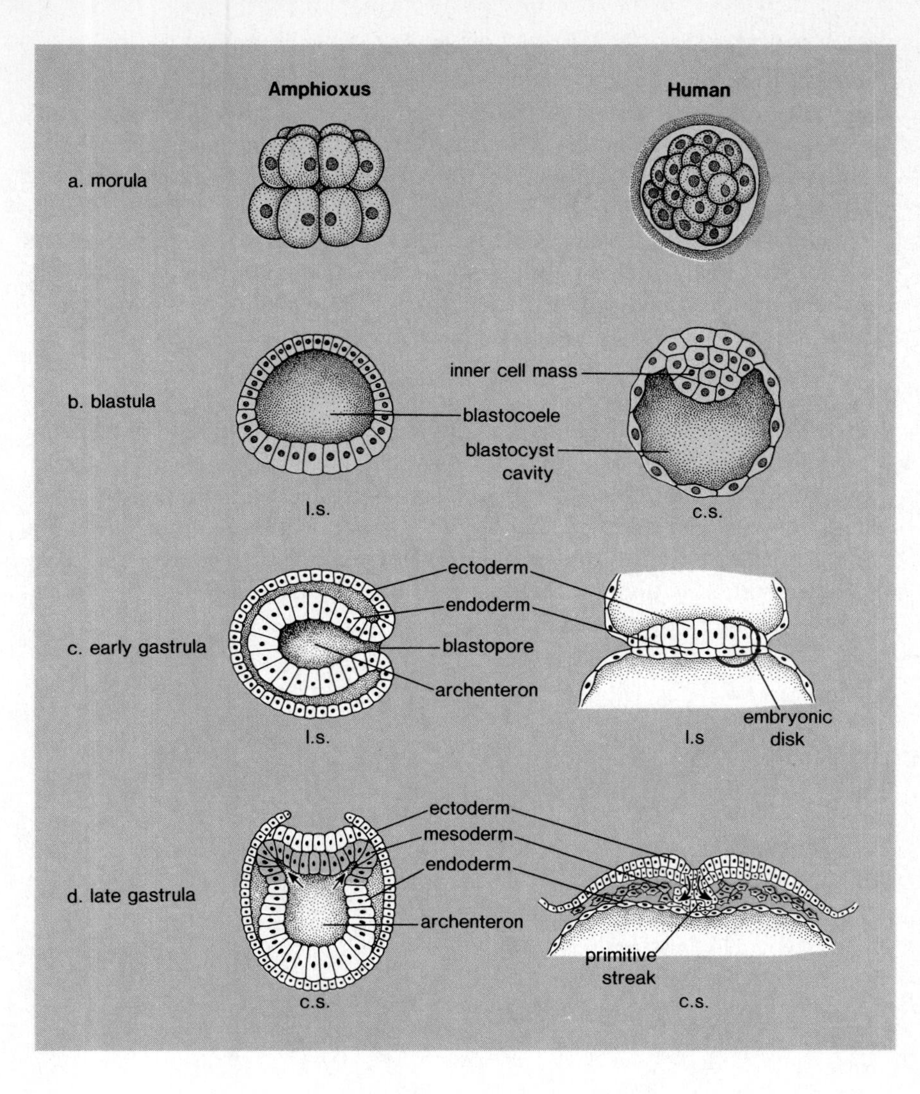

Figure 20.3
Comparison of amphioxus development and human development. a. *Comparative morula stages. Cleavage is complete in both.* b. *Comparative blastula stages. The observed cavity is the blastocoele in amphioxus, but the blastocyst in humans.* c. *Comparative gastrula stages. The gastrula in amphioxus is spherical; the gastrula in humans is flattened.* d. *Comparative late gastrula stages. Outpocketing produces mesoderm in amphioxus, while invagination between ecotderm and endoderm produces mesoderm in humans.*

Differentiation When cells take on a specific structure and function, **differentiation** occurs. The first system to become visibly differentiated is the nervous system.

Early Developmental Stages

All embryos of higher animals go through the same early stages of development, as listed in table 20.1. Each stage can be identified by the events or the results of that stage.

Cleavage

This first stage is cell division without growth. It is best observed in an embryo such as amphioxus, which has little yolk, a rich nutrient material. (The yellow portion of a chick egg is the yolk.) Since amphioxus has little yolk, cell division is about equal and the cells are of a fairly uniform size (fig. 20.2a–d). Cleavage continues until there is a solid ball of cells called the *morula.*

Blastula

The second stage occurs when the cells of the morula more or less position themselves to create a space. In amphioxus a completely hollow ball, the *blastula,* results and the space within the ball is called the *blastocoel.* The human blastula, termed the *blastocyst,* consists of a hollow ball with a mass of cells—the *inner cell mass*—at one end. Figure 20.3 compares the appearance of a human embryo to that of amphioxus during the first stages of development.

Inner Cell Mass Each cell within the inner cell mass has the genetic capability of becoming a complete individual. Scientists have been able to demonstrate the genetic potential of inner cell mass cells within the laboratory (fig. 20.4). Using mice embryos, scientists transplanted nuclei from these cells to newly enucleated eggs of a different mouse. The mice that developed had the genetic characteristics of the original embryo. Mice developing from the same inner cell mass are clones because they have exactly the same genes and are propagated by an asexual means. Cloning of human embryos has not as yet been achieved and cloning of an adult human, while frequently imagined, is considered to be very much in the future.

Gastrula

Gastrulation is evident in amphioxus when certain cells begin to push, or invaginate, into the blastocoel (fig. 20.2e–g). This creates a double layer of cells. The outer layer of cells is now called the **ectoderm** and the inner layer is called the **endoderm.** The space created by invagination will become the gut and is called either the primitive gut or the **archenteron.** The pore or hole created by the invagination is the *blastopore* and in amphioxus, as well as other vertebrates, this pore becomes the anus.

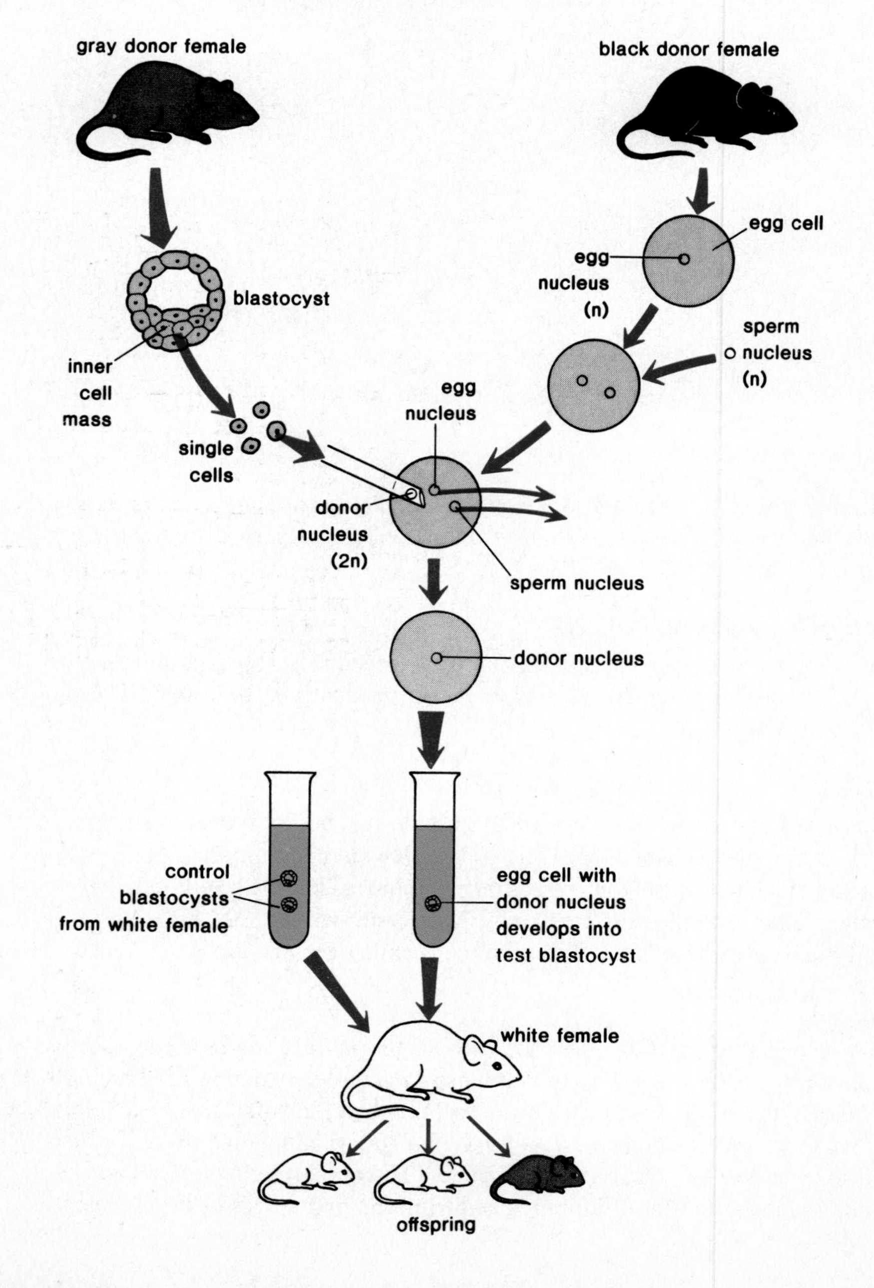

Figure 20.4
A cloning (producing exact copies) experiment that was performed in recent years. At the upper left, nuclei are removed from an embryo taken from a gray mouse. At upper right, an egg cell from a black mouse is prepared to receive the donated nucleus. At lower half, development proceeds normally and a gray mouse is born to a white mouse. This experiment shows that the nuclei taken from the differentiated cells of a gray mouse embryo contained a complete set of genes. This is a cloning experiment because all the gray mice have exactly the same genotype and phenotype.

Gastrulation is not complete until a third middle layer of cells, the **mesoderm,** has been formed. In amphioxus this layer begins as outpocketings from the primitive gut; these outpocketings grow in size until they meet and fuse. In effect, then, two layers of mesoderm are formed and the space between them is called the coelom. A *coelom* is defined as a body cavity lined by mesoderm and within which the internal organs form.

Figure 20.3 compares human gastrulation to that of amphioxus. In humans a space called the amniotic cavity appears within the inner cell mass. The portion of the mass below this cavity is the embryonic disc, which elongates to form the primitive streak (fig. 20.5). Some of the upper cells within the primitive streak invaginate and spread out between the lower layer, now called endoderm, and the remaining cells of the upper layer, now called ectoderm. The invaginating cells are the mesoderm layer. Mesoderm later forms blocklike portions called *somites* in the posterior half of the embryo, and these become muscle tissue.

It is interesting to note that human development resembles chick development. In the chick there is a primitive streak rather than a spherical gastrula because the yolk does not participate in the early stages of development. The human egg contains very little yolk, yet early development resembles that of the chick. The evolutionary history of these two animals can provide an answer for this amazing resemblance. Both birds (e.g., chicks) and mammals (e.g., humans) are related to the reptiles, and this evolutionary relationship manifests itself in the manner in which development proceeds.

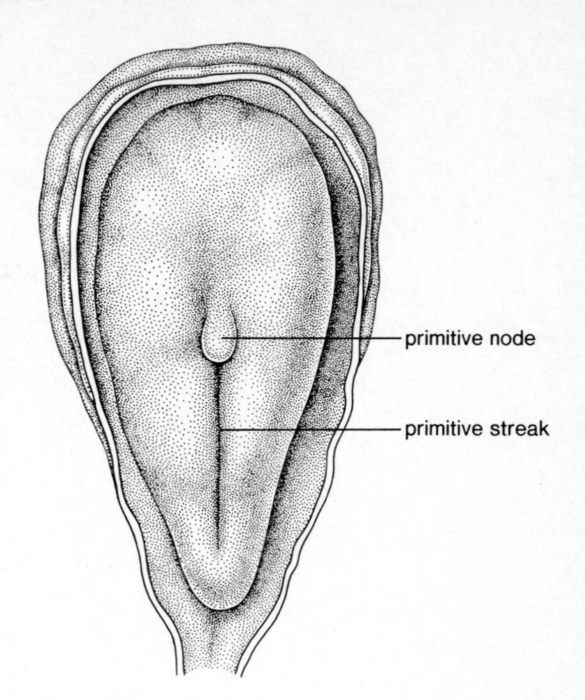

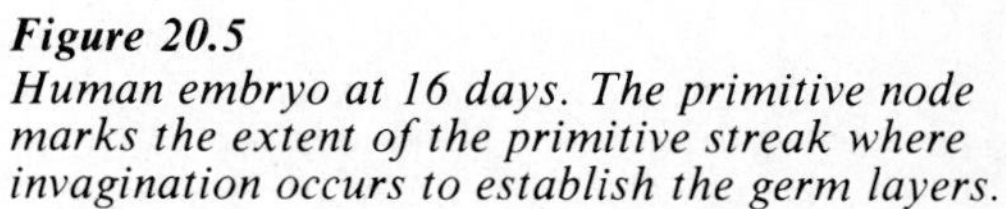
Figure 20.5
Human embryo at 16 days. The primitive node marks the extent of the primitive streak where invagination occurs to establish the germ layers.

Germ Layers Ectoderm, mesoderm, and endoderm are called the primary *germ layers* of the embryo, and no matter how gastrulation takes place the end result is the same: three germ layers are formed. It is possible to relate the development of future organs to these germ layers, as is done in table 20.2.

The processes of development are cleavage, growth, morphogenesis, and differentiation. Cleavage results in many cells that become first a morula, then a blastula, and finally a gastrula. The result of gastrulation is the establishment of three germ layers: ectoderm, endoderm, and mesoderm. Later development can be related to these germ layers.

Neurula

During the **neurula** stage of development *differentiation,* or specialization, of cells first becomes apparent. Differentiation cannot be explained by a parceling out of genes to the various cells since each and every cell of the animal's body receives a full complement of genes. Rather, genes must be controlled in such a way that only certain ones are active in certain cells. Recent investigative studies have suggested that the cytoplasm may contain either inhibitors or stimulators that combine with the chromosome and inactivate or

Table 20.2 *Organs Developed from the Three Primary Germ Layers*

Ectoderm	Mesoderm	Endoderm
Skin epidermis including hair, nails, and sweat glands	All muscles	Lining of digestive tract, trachea, bronchi, lungs, gallbladder, and urethra
Nervous system including brain, spinal cord, ganglia, nerves	Dermis of skin	Liver
Retina, lens, and cornea of eye	All connective tissue including bone, cartilage, and blood	Pancreas
Inner ear	Blood vessels	Thyroid, parathyroid, and thymus glands
Lining of nose, mouth, and anus	Kidneys	Urinary bladder
Teeth enamel	Reproductive organs	

activate certain genes. Support for the importance of cytoplasmic control of genes comes from a study of frog development. The frog's egg contains a special section of cytoplasm called the *gray crescent*. If an experimenter ties the fertilized egg so that each half has a portion of the gray crescent, both halves develop successfully into a complete embryo. If the experimenter ties an egg so that only one-half has the gray crescent, only that half develops successfully. The other half becomes a mass of nondifferentiated cells (fig. 20.6).

Early investigators referred to the gray crescent as the *primary organizer* for the embryo. An organizer is believed to be a group of cells that can influence the development of other cells. The concept of an organizer was further developed when it was found that mesoderm tissue located at the upper lip of the blastopore brings about or induces the formation of the nervous system in animals.

Induction In vertebrate animals a central portion of the mesoderm is the future *notochord*, which will later be replaced by the vertebral column. The nervous system develops from ectoderm located just above the notochord. At first, a thickening of cells called the neural plate is seen along the dorsal surface of the embryo. Then neural ridges develop on either side of a neural groove that becomes the *neural tube* when the ridges fuse. Figure 20.7 shows cross

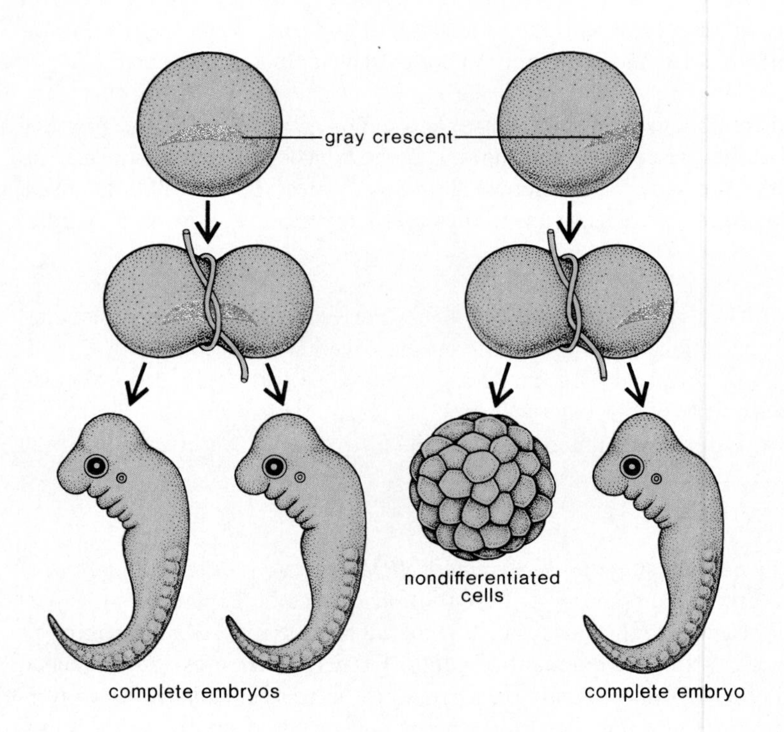

Figure 20.6
Gray crescent and development. If a frog's egg is divided so that each half receives an ample share of gray crescent, then each half develops into a complete embryo. On the other hand, if a frog's egg is divided so that only one-half receives the gray crescent, only that half develops into a complete embryo.

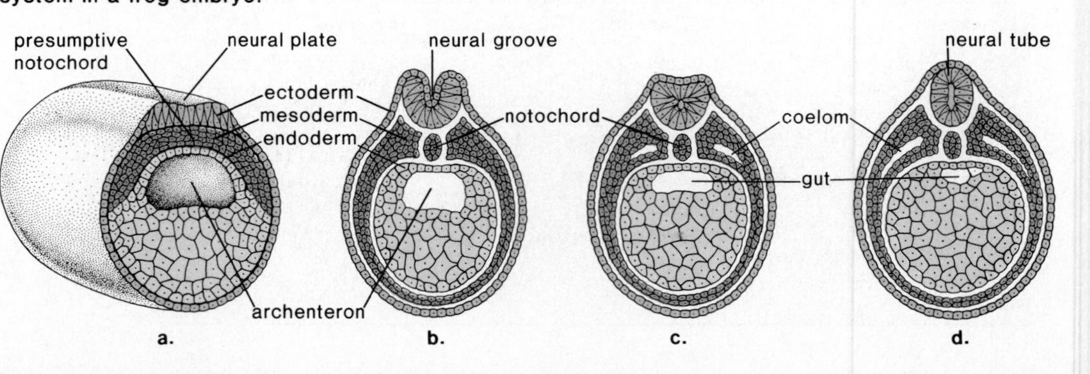

Figure 20.7
Induction. Presumptive notochord induces the formation of the neural tube. This series of drawings shows how the neural tube forms by closure of the neural folds and how the coelom forms by a splitting of mesoderm.

sections of frog development, which illustrate the internal development of the nervous system. Figure 20.8 shows an intact human embryo, allowing you to see the external appearance of the developing nervous system.

Experiments have been performed with frogs to show that if the presumptive (potential) nervous system, lying just above the notochord, is cut out and transplanted to another region of the embryo, it will not form a neural tube. On the other hand, if the presumptive notochord is cut out and transplanted beneath what would be belly ectoderm, this ectoderm now differentiates into neural tissue (fig. 20.9). These experiments indicate that notochord mesoderm brings about the formation of the nervous system, and it is said that this mesoderm *induces* the formation of the neural tube.

The process of **induction** can largely explain the orderly development of the embryo. One tissue is induced by another and this, in turn, induces another tissue, and so forth, until development is complete. Thus, induction can account for the stepwise and timewise progression of development.

This theory of orderly development is supported by the fact that once the closure of the neural tube is complete, the embryonic brain induces the formation of the lens of the eyes. First, the sides of the brain bulge out and widen just beneath overlying ectoderm. This seems to trigger a thickening of these ectoderm cells. Then the bulge dips in to form a cuplike structure that will be the future eyeball, while the overlying thickened ectoderm grows into a ball of cells to form the lens of the eye.

Experimentation has suggested that RNA formation and diffusion may be responsible for the process of induction. For example, if ectoderm is placed in a solution that formerly contained notochord mesoderm, the ectoderm differentiates into nervous tissue. This indicates that the mesoderm must have left some chemicals behind in the solution and these chemicals carry out the inductive process. An analysis of such a solution, following mesoderm removal, shows that nucleic acids have been added to the solution by the mesoderm tissue. The nucleic acid most likely to appear is RNA.

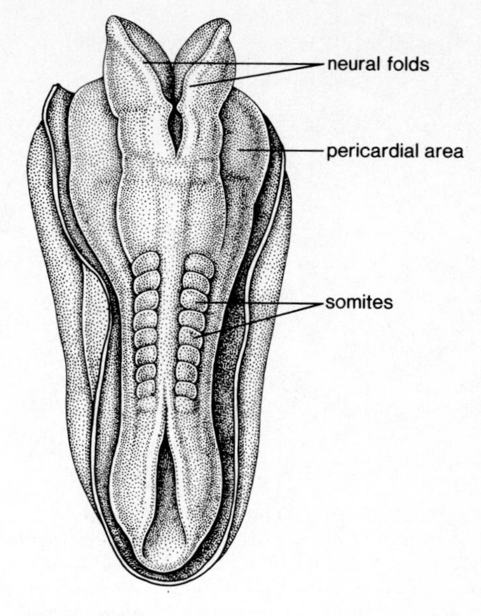

Figure 20.8
Human embryo at 21 days. The neural folds still need to close at the anterior and posterior of the embryo. The pericardial area contains the primitive heart, and the somites are the precursors of the muscles.

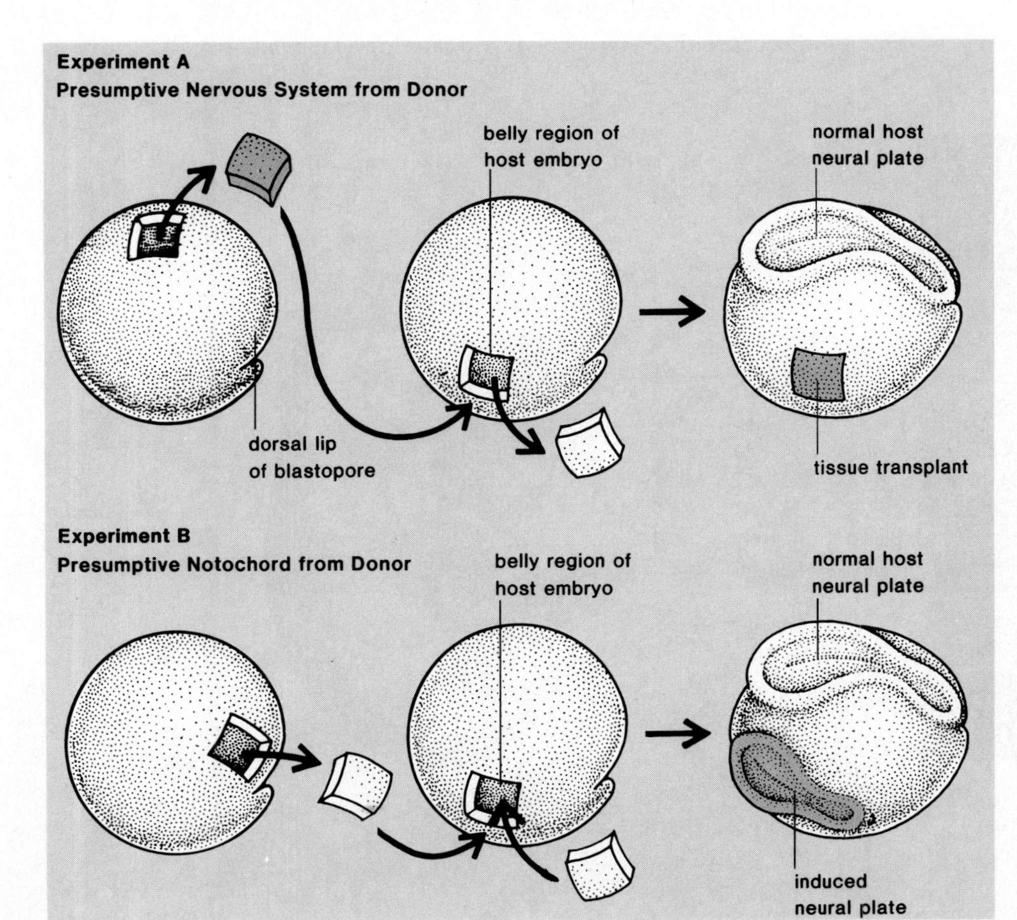

Figure 20.9
Experiments proving importance of presumptive notochord. In experiment A, *presumptive nervous system tissue does not complete its development when moved from its location above the notochord. On the other hand, in experiment* B, *presumptive notochord can cause even presumptive belly ectoderm to develop into a nervous system.*

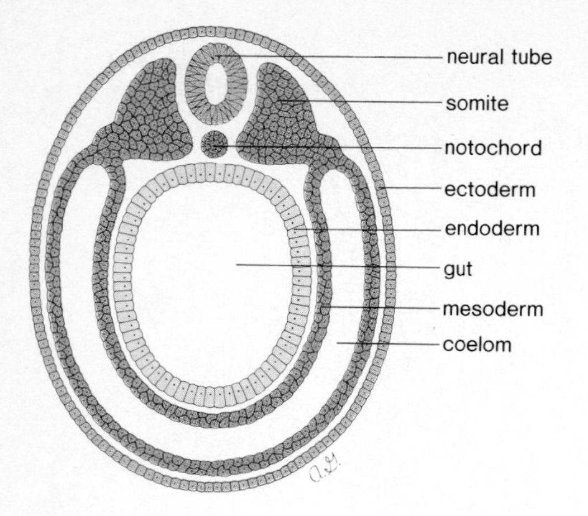

Figure 20.10
Typical vertebrate embryonic cross section. The nervous system and outer covering are derived from the ectoderm; somites and notochord are derived from the mesoderm, which also lines the coelom; the digestive tract is derived from the endoderm. The consistency of these derivations supports the germ layer theory.

The control of differentiation, whereby tissues take on specific structures and function, may be accounted for by the process of induction. Good examples of this are the induction of the nervous system by the notochord and the induction of the lens by the embryonic brain.

Vertebrate Cross Section

With the formation of the nervous system, it is possible to show a generalized diagram (fig. 20.10) of a vertebrate embryo to illustrate placement of parts. Correlation of figure 20.10 with table 20.2 will help you relate the formation of vertebrate structures and organs to the three embryonic layers of cells: the ectoderm, mesoderm, and endoderm. Thus, the skin and nervous system develop from the ectoderm; muscles, skeleton, kidneys, circulatory system, and gonads develop from the mesoderm; and the digestive tract, lungs, liver, and pancreas develop from the endoderm.

The diagram illustrates that embryonic vertebrates have a notochord and a dorsal hollow nerve cord called the spinal cord once it is enveloped by vertebrae. Another characteristic is the presence of gill pouches or slits supported by gill arches (fig. 20.11). Obviously, only the lower vertebrates (fishes and amphibian larvae) have actual use for gill slits as functioning structures. The fact that higher forms go through this embryonic stage of the lower forms indicates a relationship between them. The phrase *ontogeny* (development) *recapitulates* (repeats) *phylogeny* (evolutionary history) was coined some years ago as a dramatic way to suggest that all animals share the same embryonic stages. This theory has been modified today since embryos proceed only through those stages that are necessary to their later development. For example, in higher vertebrates actual gill slits never form; instead, the first gill pouch becomes the cavity of the middle ear and eustachian tube. The second pouch becomes the tonsils, while the third and fourth pouches become the thymus and parathyroids, respectively. The fifth pouch disappears. Thus, gill pouches develop because they are necessary to later development.

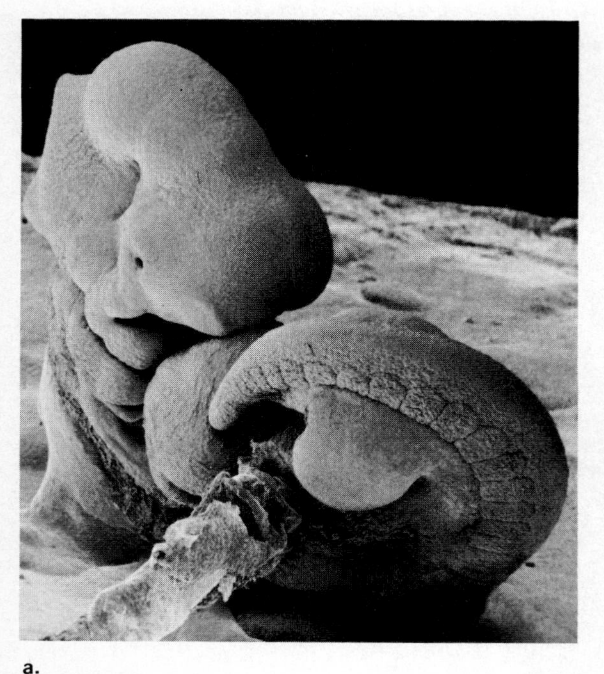
a.

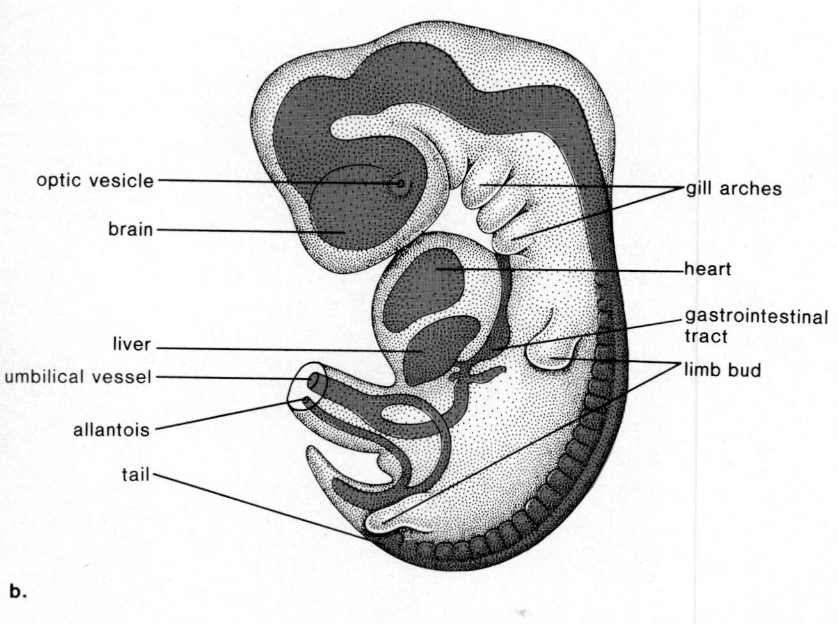

b.

Figure 20.11
Human embryo at beginning of fifth week.
a. *Scanning electron micrograph.* b. *Drawing. The embryo is curled so that the head touches the heart, two organs whose development is further along than the rest of the body. The organs of the gastrointestinal tract are forming. The presence of the tail is an evolutionary remnant; its bones will regress and become those of the coccyx. The arms and legs will develop from the bulges that are called limb buds.*

All vertebrates at some time in their development portray a similar cross section that displays typical vertebrate embryonic characteristics: dorsal hollow nerve cord, notochord, and a coelom completely surrounded by mesoderm. Also, at some time in their embryonic history, vertebrates have gill pouches or slits.

Extraembryonic Membranes

Before we consider human development chronologically, we must understand the placement of extraembryonic membranes. Extraembryonic membranes are best understood by considering their function in reptiles and birds. The formation of these membranes in reptiles first made development on land possible. If an embryo develops in the water, the water supplies oxygen for the embryo and takes away waste products. The surrounding water prevents drying out and provides a protective cushion. For an embryo that develops on land, all these functions are performed by the **extraembryonic membranes.** Figure 20.12 shows the chick within its hard shell surrounded by the membranes. The **chorion** lies next to the shell and carries on gas exchange. The **yolk sac** surrounds the remaining yolk. The **allantois** collects nitrogenous waste and the **amnion** contains the amniotic fluid that bathes the developing embryo.

As figure 20.12 indicates, humans (and other mammals as well) also have these extraembryonic membranes. The chorion develops into the fetal half of the placenta; the yolk sac is present but lacks yolk and is largely nonfunctional; the allantoic blood vessels become the umbilical blood vessels; and the amnion contains fluid to cushion and protect the fetus. Thus, the function of the membranes has been modified to suit internal development, but their very presence indicates our relationship to birds and reptiles. It is interesting to note that all animals develop in water, either directly or within amniotic fluid.

The presence of extraembryonic membranes in reptiles made development on land possible. Humans also have these membranes, but their function has been modified for internal development.

Figure 20.12
Extraembryonic membranes are not part of the embryo. These membranes are also found during the development of chicks and humans, where each has a specific function. a. *In the chick the chorion lies just beneath the shell and performs gas exchange. The allantois collects nitrogenous wastes. The yolk sac provides nourishment, and the amnion provides a watery environment.* b. *In humans, only the chorion and amnion have comparable functions. The chorion forms the fetal half of the placenta, where exchange occurs with mother's blood, and the amnion provides a watery environment. The yolk sac and allantois are vestigial and disappear as the umbilical cord forms.*

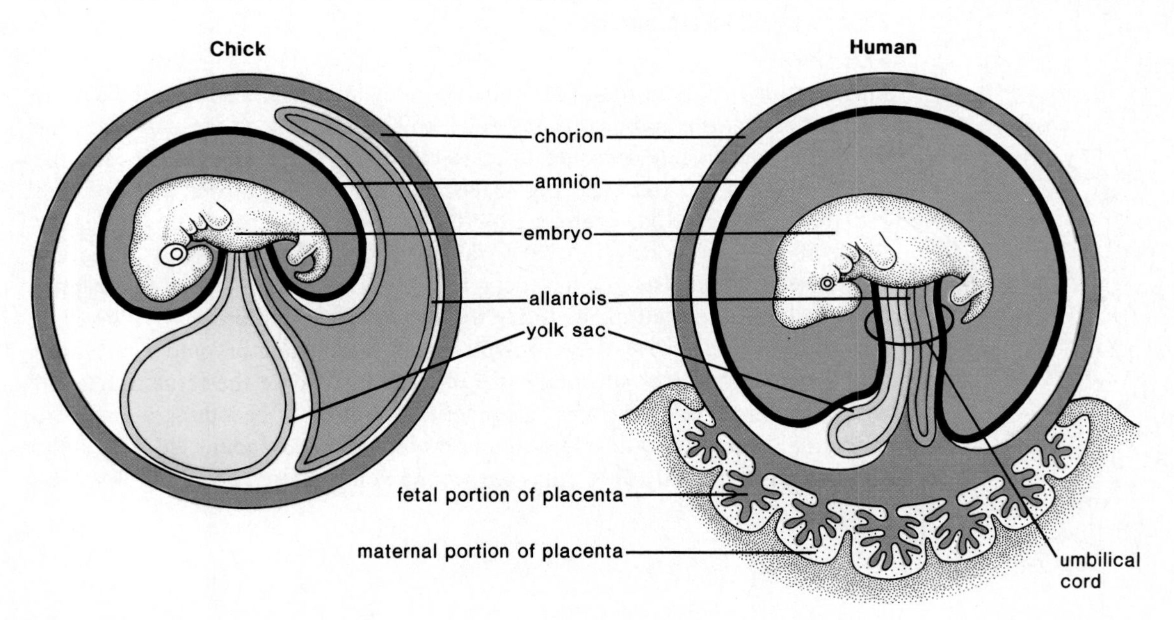

Figure 20.13
Fertilization and implantation of human embryo. After the egg is fertilized, it begins cleavage as it moves toward the uterus. At the time of implantation, the embryo is in the gastrula stage of development.

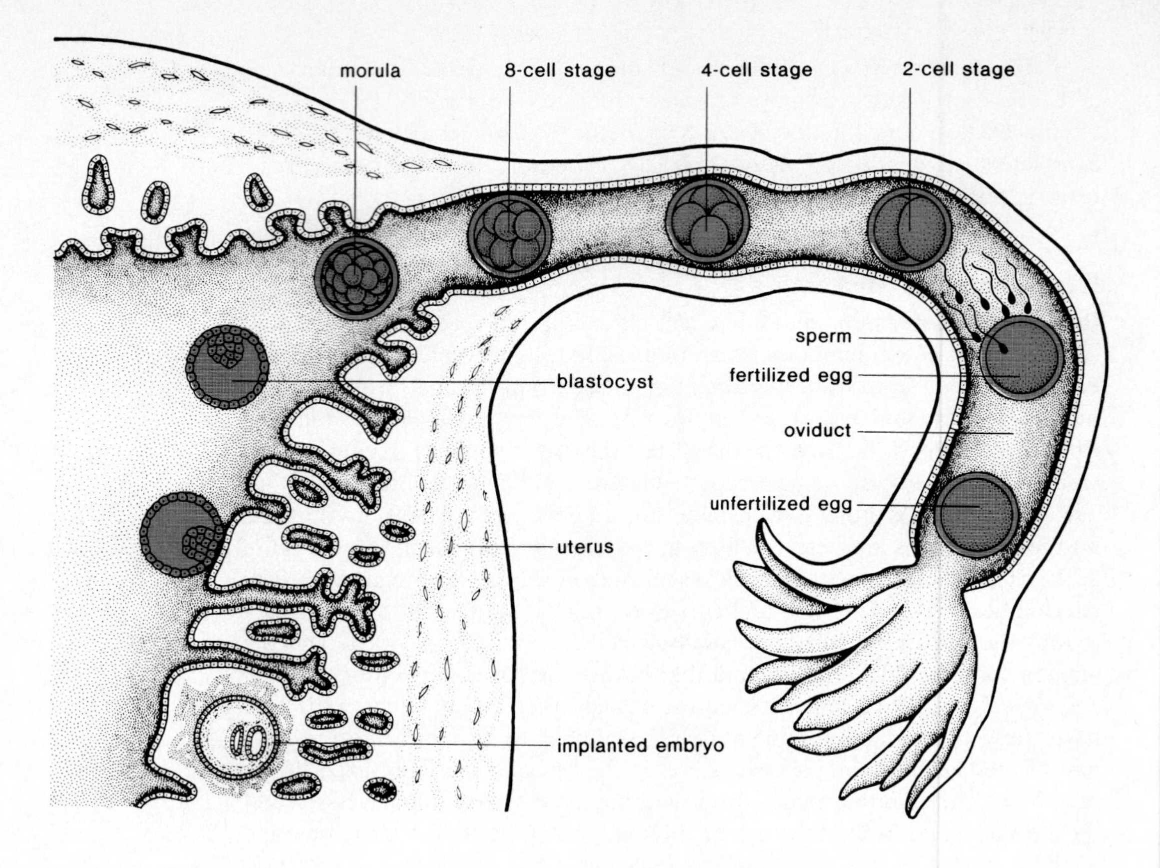

Human Development

We are now prepared to consider development in a stepwise manner. The human egg has little yolk but this is not a detriment because the extraembryonic membranes and placenta develop very early during development. The placenta fulfills the needs of the developing embryo.

Embryonic Development

First Week

Fertilization occurs in the upper third of an oviduct (fig. 20.13) and cleavage begins even as the embryo passes down this tube to the uterus. By the time the embryo has reached the uterus on the third day it is a *morula.* The morula is not much larger than the zygote because although multiple cell divisions have occurred there has been no growth of these newly formed cells. By about the fifth day, the morula has been transformed into the *blastocyst,* a hollow ball of cells. The blastocyst has a single layer of outer cells called the **trophoblast** and an inner cell mass. Later the trophoblast, reinforced by a layer of mesoderm, will give rise to the *chorion,* one of the extraembryonic membranes (fig. 20.12). The *inner cell mass* will eventually become the fetus. Each cell within the inner cell mass has the genetic capability of becoming a complete individual. Sometimes during human development, the inner cell mass splits and two embryos start developing rather than one. These two embryos will be

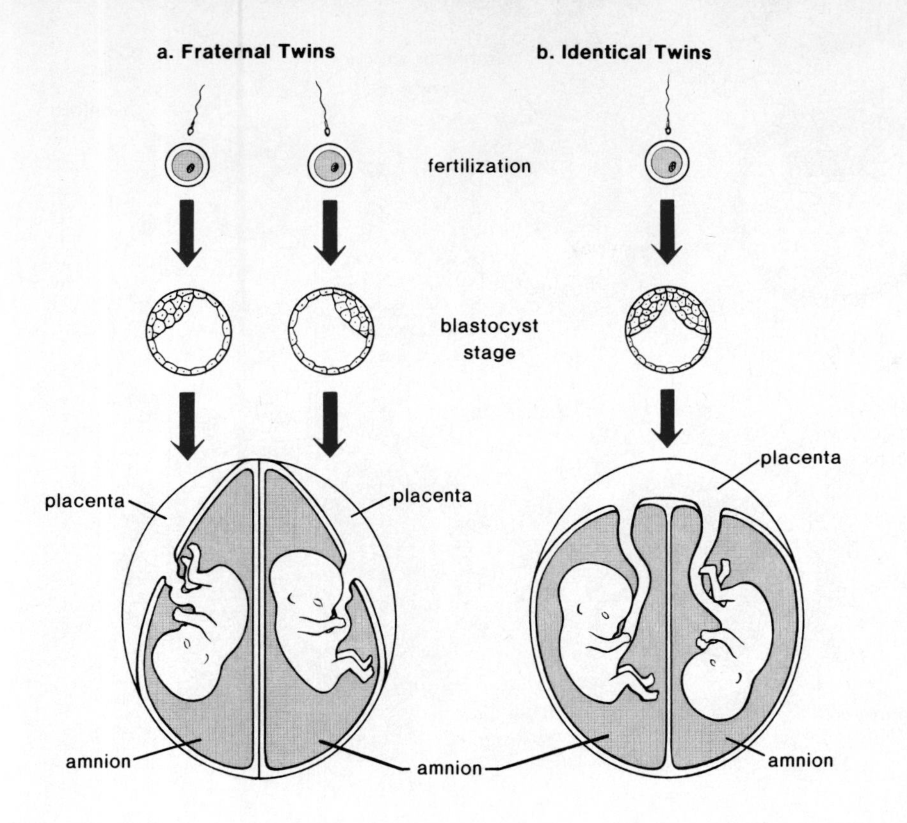

Figure 20.14
Conception of fraternal versus identical twins.
a. *Fraternal twins are formed when two eggs are released and fertilized. Fraternal twins receive a different genetic inheritance from both the mother and father. They can even have separate fathers.*
b. *Identical twins occur when an embryo breaks in two during an early stage of development. Identical twins have the exact same genetic inheritance from both the mother and father.*

identical twins (fig. 20.14) because they have inherited exactly the same chromosomes. Fraternal twins, which arise when two different eggs are fertilized by two different sperm, do not have identical chromosomes. It has even been known to happen that these "twins" have different fathers.

---

*D*uring the first week, the human embryo undergoes cleavage and then the morula becomes the blastocyst, having two main parts, the outer trophoblast (to become the chorion) and the inner cell mass (to become the fetus).

Second Week

At the end of the first week, the embryo begins the process of *implanting* itself in the wall of the uterus as the trophoblast secretes enzymes to digest away some of the tissue and blood vessels of the uterine wall (fig. 20.13). The embryo is now about the size of the period at the end of this sentence.

The trophoblast begins to secrete *HCG* (human chorionic gonadtrophin) the hormone that is the basis for the pregnancy test, and which serves to maintain the corpus luteum past the time it would normally disintegrate. Because of this the endometrium is maintained and menstruation does not occur.

---

*D*uring the second week, the embryo embeds itself in the uterine lining and the trophoblast begins to produce HCG.

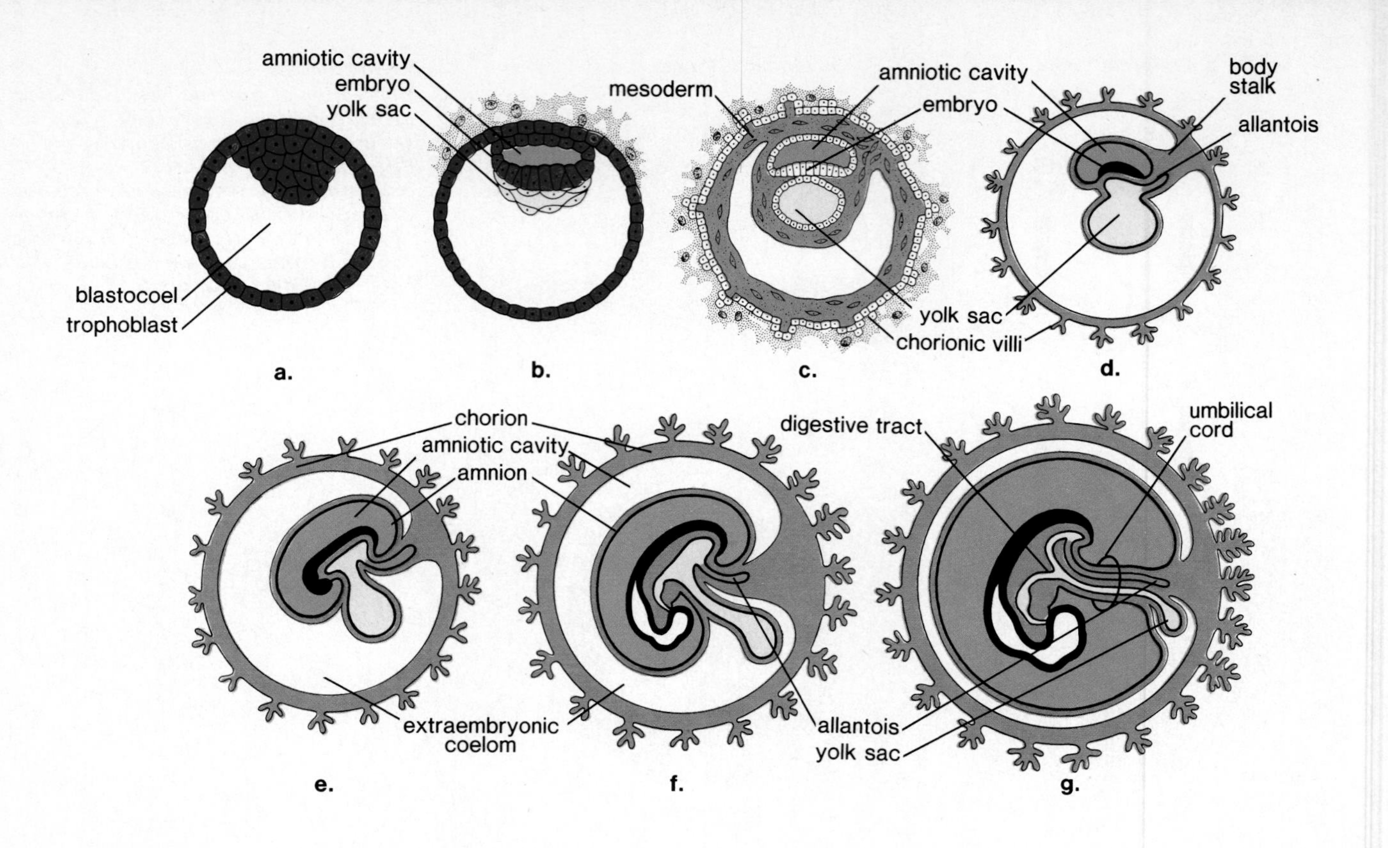

Figure 20.15
Early development of fetus and umbilical cord. a. *The blastocyst with its inner cell mass and surrounding trophoblast.* b. *Amniotic cavity and yolk sac appear.* c. *Chorionic villi first appear.* d. *Embryo is connected to chorion by a body stalk.* e.–g. *Embryo becomes more differentiated as the umbilical cord forms.*

Third Week

During the third week of development, the inner cell mass detaches itself from the trophoblast, which is then reinforced by mesoderm and becomes the chorion. Two more extraembryonic membranes (fig. 20.15) form. The *yolk sac,* which forms below the embryo, is vestigial in the human and has no nutritive function, as it does in other animals. However, the *amnion* and its cavity are where the embryo and then the fetus will develop. The amniotic fluid acts as an insulator against cold and heat and also absorbs any shock, such as a blow to the mother's abdomen.

Gastrulation occurs during the third week. The inner cell mass has now flattened into the *embryonic disc* composed of two layers of cells: the *ectoderm* above and the *endoderm* below. Once the embryonic disc elongates to become the *primitive streak* (fig. 20.5) the third germ layer, the mesoderm, forms by invagination of cells along the streak. The mesoderm lies between the ectoderm and the endoderm where it forms blocks of tissue along the midline called somites.

It is possible to relate the development of future organs to these germ layers (table 20.2). In general, the ectoderm will become the nervous system, skin, hair, and nails; the endoderm will produce the inner linings of the digestive, respiratory, and urinary tracts; the mesoderm produces the muscles, skeleton, and circulatory systems (fig. 20.16).

The germ layers (ectoderm, endoderm, and mesoderm) are laid down during the third week of development. Development of each of the organs can be related to certain of the germ layers.

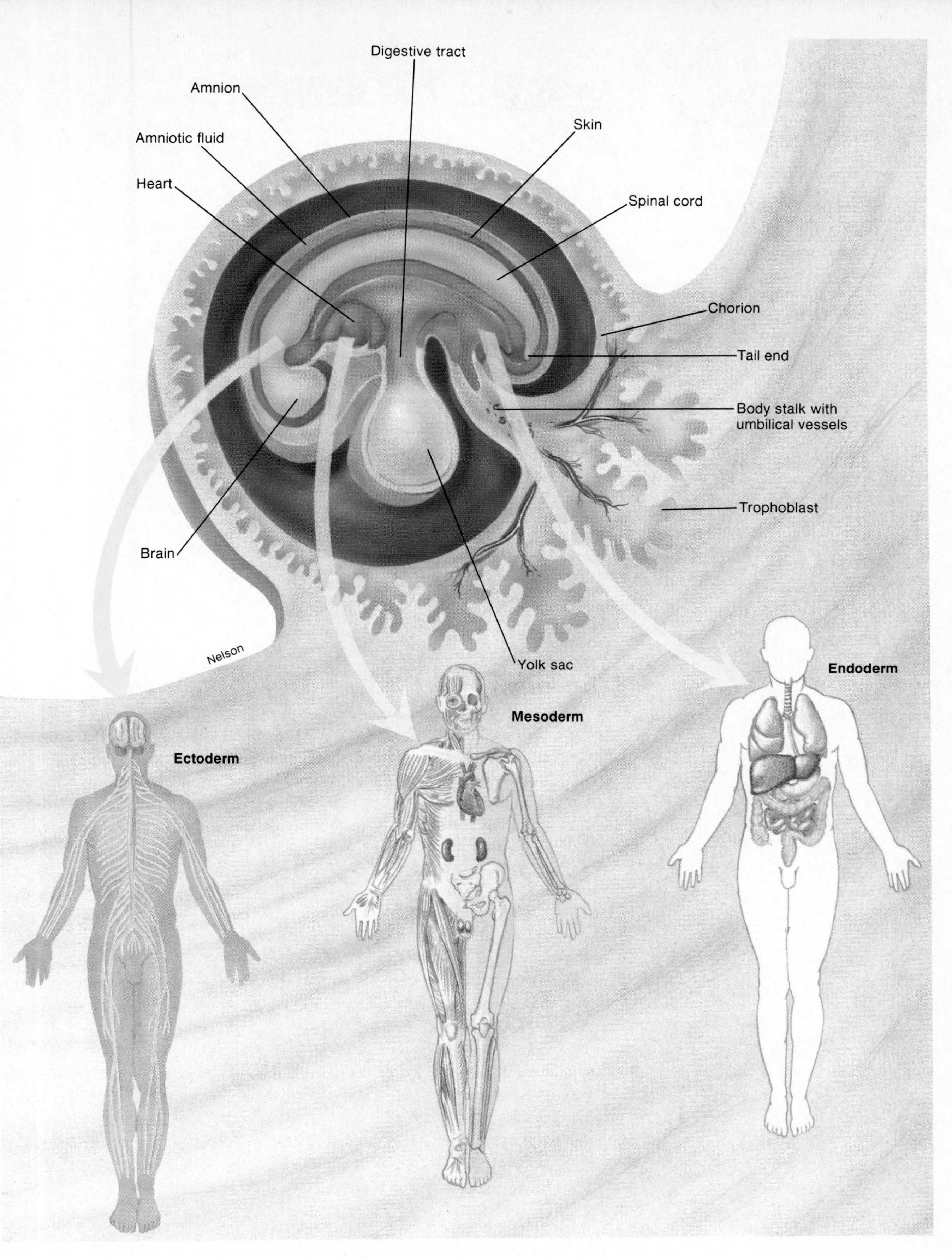

Figure 20.16
The organs of the body develop from one of the three germ layers as indicated.

Figure 20.17
Neurulation. a. *Neural plate is an ectoderm layer that will give rise to neural tube.* b. *As microfilaments contract, cells begin to invaginate.* c. *Continued constriction causes the neural tube to pinch off from the outer ectoderm layer.*

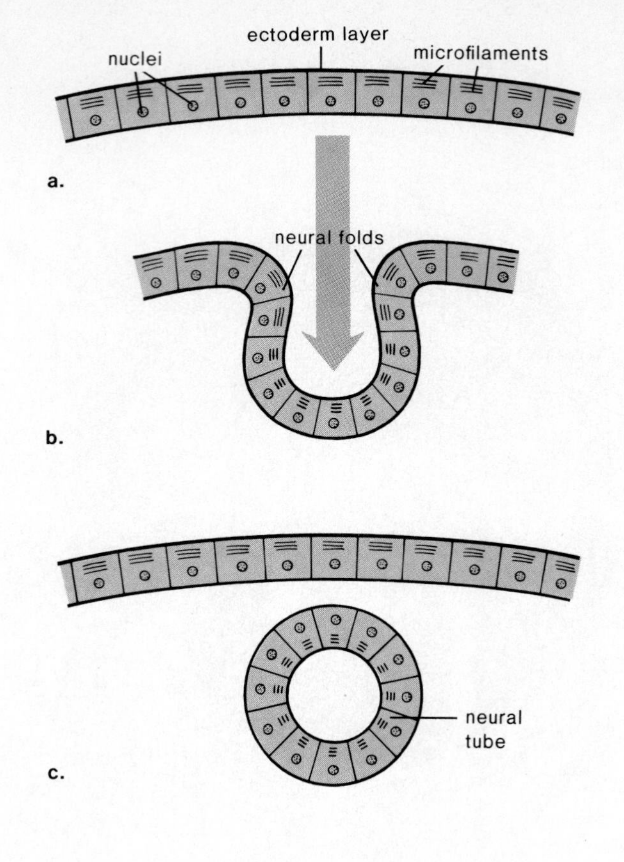

Fourth Week

At the fourth week, there is a bridge of mesoderm called the *body stalk* connecting the caudal end of the embryo with the chorion. The fourth extraembryonic membrane, the *allantois* (fig. 20.12) is contained within this stalk, and its blood vessels become the umbilical blood vessels. After the head and tail lift up and the body stalk moves anteriorly by a constriction (fig. 20.15e–g), the **umbilical cord** which connects the developing embryo to the placenta is fully formed.

Neurulation The *nervous system* is the first organ system to be visually evident. At first a thickening appears along the entire dorsal length of the embryo and then invagination occurs as neural folds appear. When the neural folds meet (fig. 20.17) at the midline, the neural tube which later develops into the brain and nerve cord is formed. Figure 20.8 shows a human embryo during the fourth week when neurulation is in progress. The notochord will later be replaced by the vertebral column.

Development of the Heart Development of the *heart* begins in the third week and continues into the fourth week. At first there are right and left heart tubes; when these fuse the heart begins pumping blood even though the chambers of the heart are not yet fully formed. The veins enter posteriorly and the arteries exit anteriorly from this largely tubular heart but later the heart will twist so that all major blood vessels are located anteriorly.

During the fourth week major organs like the central nervous system and the heart make their appearance.

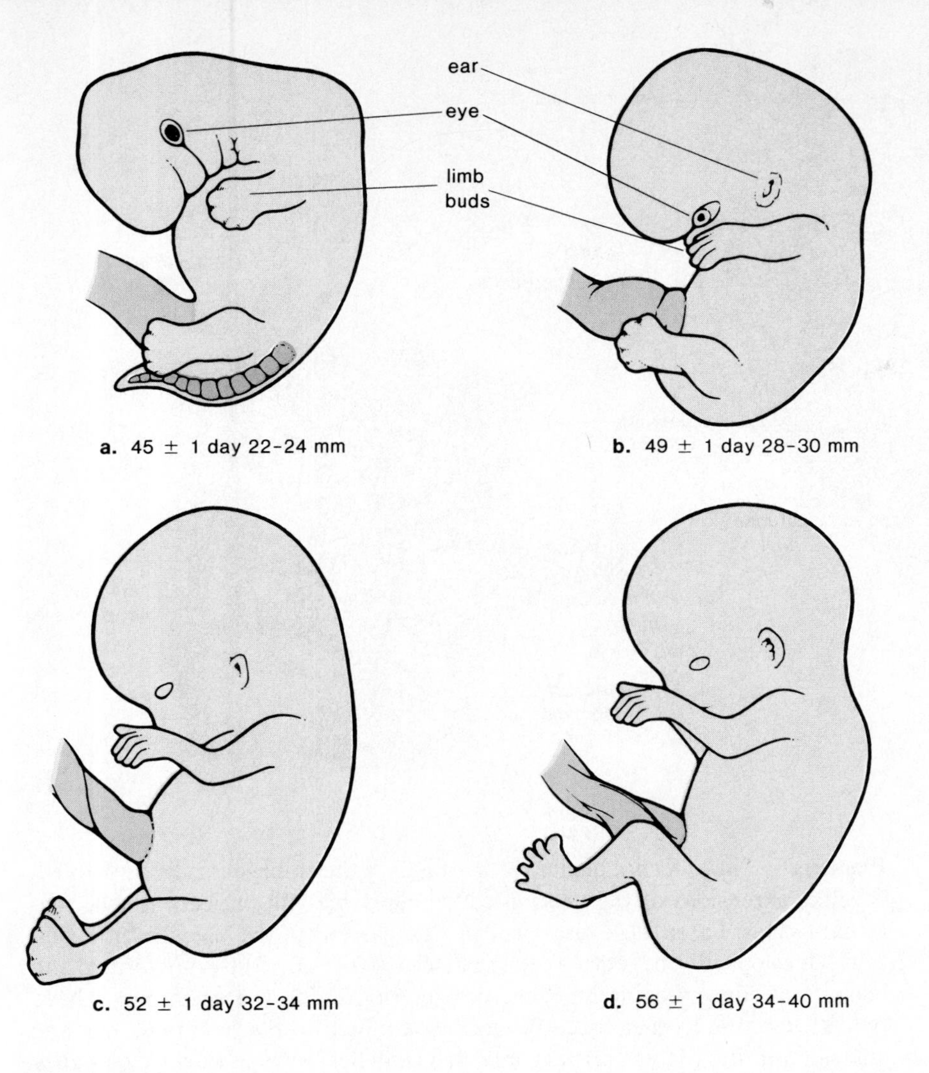

Figure 20.18
Human embryo at the days indicated and at sizes noted. Ten millimeters (mm) equals 0.4 inches.

Fifth Week

At five weeks the embryo (fig. 20.11) is barely larger than the height of this print. Little flippers called **limb buds** have appeared and are developing into arms and legs from which hands and feet, fingers and toes will soon form. Notice in figure 20.18 that within the limb buds the *cartilaginous skeleton* is also growing and differentiating. A cartilaginous skull is evident as the head enlarges and facial features are becoming obvious. The mouth breaks through when ectoderm invaginates to meet the newly closed digestive tube which begins to have various parts including the accessory digestive organs.

During the fifth week, human features like the head, arms, and legs begin to make their appearance.

Sixth to Eighth Week

There is a remarkable change in external appearance during these weeks (fig. 20.18) from a form that would be difficult to recognize as human to one that is easily recognizable as human. Concurrent with brain development the head achieves its normal relationship with the body as a neck region develops. The nervous system is well enough developed to permit reflex actions, such as a startle response to being touched. At the end of this period, the embryo is about one and one half inches (38 mm) long and weighs no more than an aspirin tablet, even though all organ systems have been established.

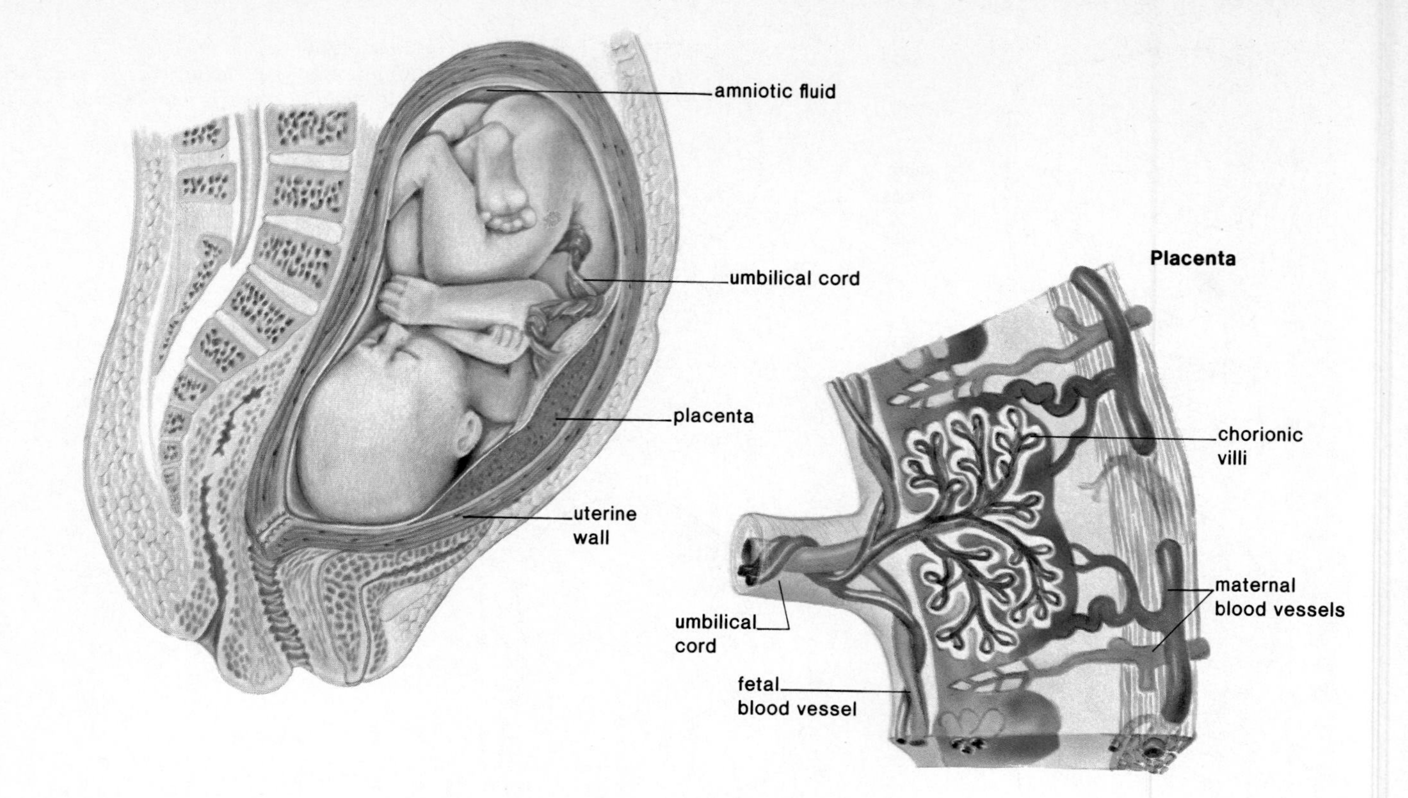

Figure 20.19
Anatomy of the placenta. The placenta is composed of both fetal and maternal tissues. Chorionic villi penetrate the uterine lining and are surrounded by maternal blood. Exchange of molecules between fetal and maternal blood takes place across the walls of the villi.

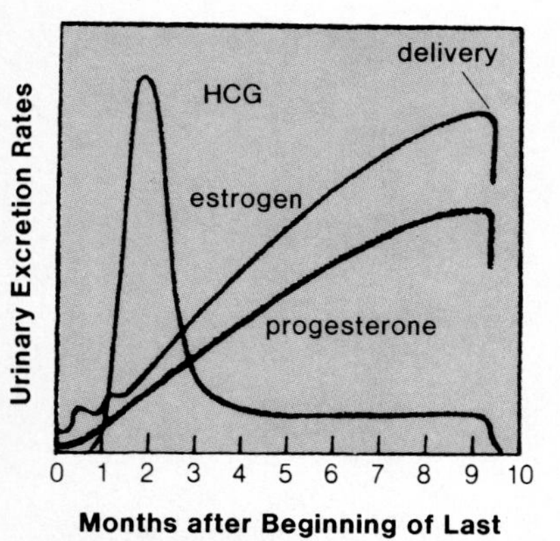

Figure 20.20
Urinary excretion of HCG (human chorionic gonadotrophin), estrogen, and progesterone during pregnancy. Urinary excretion rates are an indication of blood concentrations of these hormones.

Placenta The *placenta* begins formation once the embryo is fully implanted. Treelike extensions of the chorion called **chorionic villi** project into the maternal tissues. Later these disappear in all areas except the one where the placenta develops. By the tenth week the placenta (fig. 20.19) is fully formed and begins to produce progesterone and estrogen (fig. 20.20). These hormones have two effects; due to their negative feedback effect on the hypothalamus and anterior pituitary, they prevent any new follicles from maturing, and they maintain the lining of the uterus so that the corpus luteum is not needed. There is no menstruation during the period of pregnancy.

The placenta has a fetal side contributed by the chorion and a maternal side consisting of uterine tissues. Notice how the chorionic villi are surrounded by maternal blood sinuses; yet the blood of the mother and fetus never mix since exchange always takes place across cell membranes. Carbon dioxide and other wastes move from the fetal to the maternal side, and nutrients and oxygen move from the maternal to the fetal side of the placenta. The umbilical cord stretches between the placenta and the fetus. Although it may seem that the umbilical cord travels from the placenta to the intestine, actually the umbilical cord is simply taking fetal blood to and from the placenta. The umbilical cord is the lifeline of the fetus because it contains the umbilical arteries and vein, which transport waste molecules (carbon dioxide and urea) to the placenta for disposal and take oxygen and nutrient molecules from the placenta to the rest of the fetal circulatory system.

Harmful chemicals can also cross the placenta and this is of particular concern during the embryonic period when various structures are first forming. Each organ or part seems to have a sensitive period, during which a substance can alter its normal formation. The reading on page 436 concerns the origination of birth defects.

At the end of the embryonic period all organ systems have been established and there is a mature and fully functioning placenta. The embryo is only about one and one half inches (38 mm) long.

Figure 20.21
It's a boy. The sex of the fetus can be determined at about the third month of development.

Fetal Development

Third and Fourth Months

At the beginning of the third month, the head is still very large, the nose is flat, the eyes are far apart, and the ears are distinctively present. Head growth will now begin to slow down as the rest of the body increases in length. Epidermal refinements such as eyelashes, eyebrows, hair on head, fingernails, and nipples appear.

Cartilage is replaced by *bone* as ossification centers appear in most of the bones. Cartilage remains at the ends of the long bones, and ossification will not be complete until age eighteen or twenty. The skull has six large membranous areas called *fontanels* that permit a certain amount of flexibility as the head passes through the birth canal and allow rapid growth of the brain during infancy. The fontanels disappear by two years of age.

At the third month it is possible to distinguish males from females (fig. 20.21). Apparently the Y chromosome brings about the production of a protein called the H-Y antigen (because females form antibodies to it) that triggers the differentiation of gonads into testes. Once the testes differentiate, they produce androgens, the male sex hormones. It is the androgens, especially testosterone, that stimulates the growth of the male external genitalia. In the absence of androgens, female genitalia form. The ovaries have no need to produce estrogen because there is plenty of it circulating in the mother's bloodstream.

At this time, both testes and ovaries are located within the abdominal cavity but later, in the last trimester of fetal development, the testes will descend into the scrotal sacs (scrotum). Sometimes the testes fail to descend and in that case an operation may later be done, to place them in their proper location.

Birth Defects

It is believed that at least 1 in 16 newborns has a birth defect, either minor or serious, and the actual percentage may be even higher. Most likely only 20 percent of all birth defects are due to heredity. Those that are can sometimes be detected before birth by subjecting embryonic and/or fetal cells to various tests following chorionic villi sampling or amniocentesis.

Treatment of the fetus in the womb is a rapidly developing area of medical expertise. Biochemical defects can sometimes be treated by giving the mother appropriate medicines. For example, if a baby is unable to synthesize vitamin B and/or is unable to use biotin efficiently, the mother can take these substances in doses large enough to prevent any untoward effects. Structural defects can sometimes be corrected by surgery. For example, if the fetus has water on the brain or is unable to pass urine, tubes that temporarily allow the fluid to pass out into the amniotic fluid can be inserted even while the fetus is still in the womb. Physicians are hopeful that eventually all sorts of structural defects can be corrected by lifting the fetus from the womb long enough for corrective surgery to be done.

Birth defects not due to heredity, called congenital defects, cannot be passed on to the next generation. These are often caused by microbes or substances that have crossed the placenta and altered normal development. Already there have been 300 babies who have contracted AIDS while in their mother's womb and the number is expected to increase to more than 3,000 by 1991.

Medications can also sometimes cause problems. When the synthetic hormone DES was given to pregnant women to prevent miscarriage, their daughters showed various abnormalities of the reproductive organs and an increased tendency toward cervical cancer. Other sex hormones, including birth control pills, can possibly cause abnormal fetal development, including abnormalities of the sex organs.

Drugs of all types should be avoided. Most people are aware that women taking the tranquilizer thalidomide produce children with deformed arms and legs. Also, aspirin, caffeine (present in coffee, tea, and cola), and alcohol should be severely limited. Mood-altering drugs most likely should not be taken at all. It is not unusual for babies of drug addicts and alcoholics to display withdrawal symptoms and have various abnormalities. Babies born to women who have about 45 drinks a month and as many as 5 drinks on one occasion, are apt to have FAS (fetal alcohol syndrome) with decreased weight, height, and head size, malformation of the head and face, and mental retardation (see figure). A recent new threat is the use of cocaine during pregnancy; "cocaine babies" now make up 60 percent of drug-affected babies. Severe fluctuations in blood pressure accompany the use of cocaine and these temporarily deprive a fetal brain of oxygen. Cocaine babies have visual problems, lack coordination, and are mentally retarded. Other so-called fetotoxic chemicals should also be avoided. These include pesticides and many organic industrial chemicals. Cigarette smoke includes some of these very same chemicals so that babies born to smokers are often underweight and subject to convulsions.

It is recommended that all women who are capable of reproduction and sexually active take precautions to protect developing embryos. An Rh negative woman who has an Rh positive child should receive a RhoGam injection to prevent the production of Rh antibodies, which can cause birth defects, including nervous system and heart defects during a subsequent pregnancy. Also, women should be immunized for German measles before the childbearing years. German measle viruses also can cause defects, such as deafness.

X-ray diagnostic therapy should be avoided during pregnancy because X rays are mutagenic to a developing embryo or fetus. Children born to women who have received X-ray treatment are apt to have birth defects and/or develop leukemia later on. As discussed in chapter 18, children born to women with herpes, gonorrhea, and chlamydia are subject to blindness and other mental and physical defects when they become infected after passing through the birth canal. Birth by caesarean section could prevent these occurrences.

Now that physicians and lay people are aware of the various ways in which birth defects can be prevented, it is hoped that all types of birth defects, both genetic and congenital, will decrease dramatically.

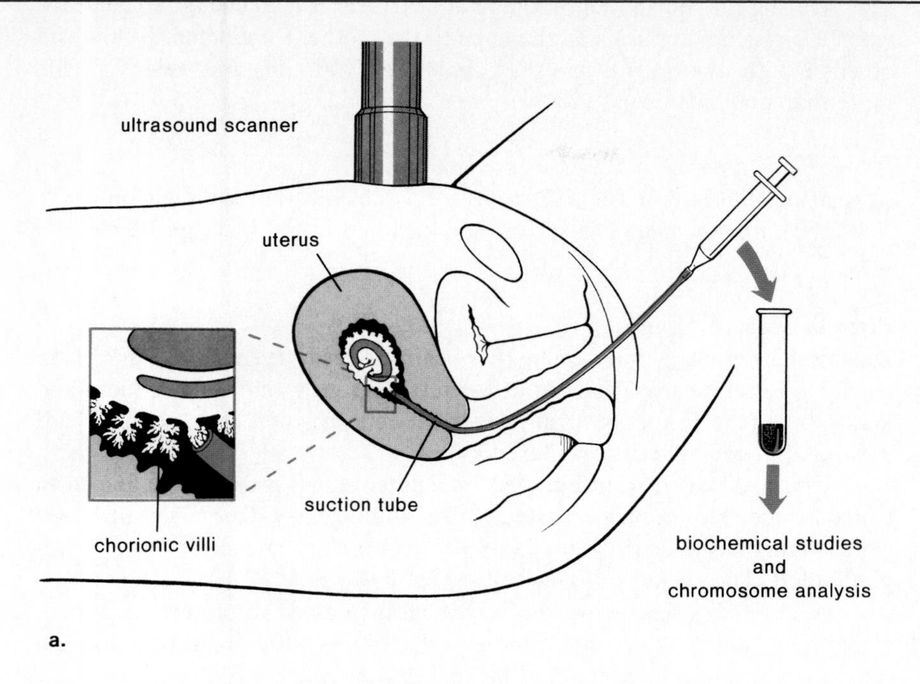

fetal
blood
endoscope
amniotic cavity
amniotic
fluid
cells
centrifuge
needle
vein
cell culture
biochemical studies
and
chromosome analysis
b.

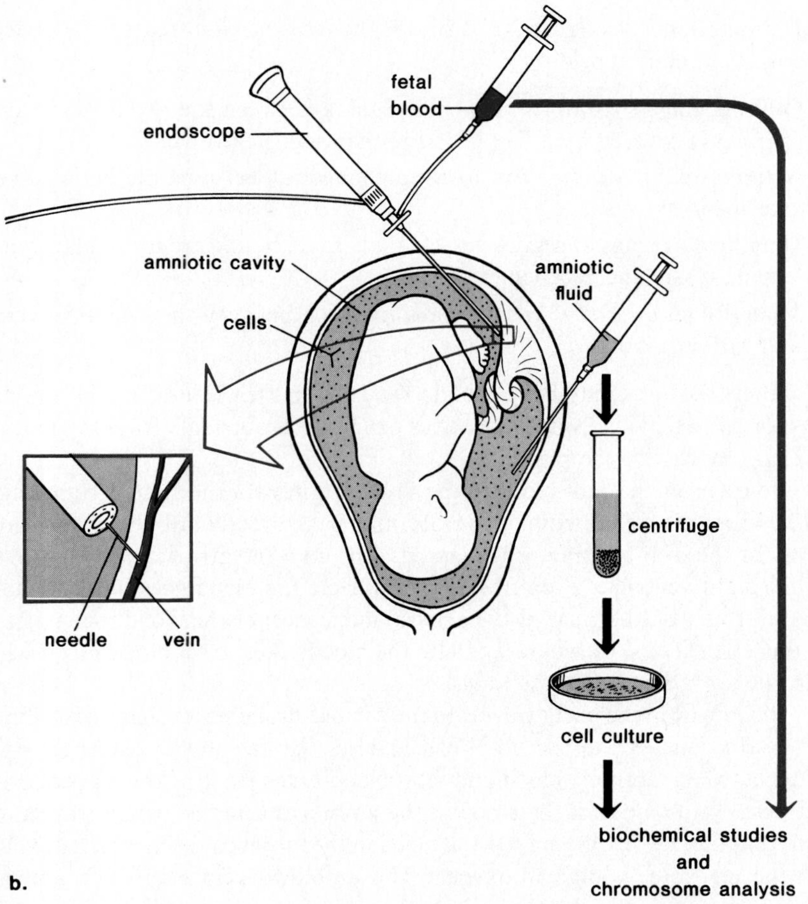

Two methods of acquiring fetal cells for chromosomal and genetic defects are in use.
a. *Chorionic villi sampling allows physicians to collect embryonic cells as early as the fifth week. The doctor inserts a long thin tube through the vagina into the uterus. With the help of ultrasound, which gives a picture of the uterine contents, the tube is placed between the lining of the uterus and the chorion. Then suction is used to remove a sampling of the chorionic villi cells. Chromosomal analysis and biochemical tests for several different genetic defects can be done immediately on these cells.*
b. *Physicians have to wait until about the sixteenth week of pregnancy to perform amniocentesis. In amniocentesis, a long needle is passed through the abdominal wall to withdraw a small amount of amniotic fluid along with fetal cells. Since there are only a few cells in the amniotic fluid, testing must be delayed for four weeks until cell culture produces enough cells for testing purposes. In fetoscopy, another possible procedure, the physician uses an endoscope to view the fetus so that blood can be withdrawn for prenatal diagnosis.*

During the fourth month, the fetal heartbeat is loud enough to be heard when a physician applies a stethoscope to the mother's abdomen. By the end of this month, the fetus is less than six inches (140 mm) and weighs a little more than one-half pound (200 g).

During the third and fourth months, it is obvious that the skeleton is becoming ossified. The sex of the individual is now distinguishable.

Fifth to Seventh Months

During this period of time, the mother begins to feel movement. At first there is only a fluttering sensation but as the fetal legs grow and develop kicks and jabs will be felt. The fetus, though, is in the fetal position with the head bent down and in contact with the flexed knees.

The wrinkled, translucent, pinkish-colored skin is covered by a fine down called **lanugo.** This in turn is coated with a white, greasy, cheeselike substance called **vernix caseosa** which probably protects the delicate skin from the amniotic fluid. The eyelids are now fully open, however.

At the end of this period, the weight has increased to almost three pounds (1,350 gm) and the length to almost twelve inches (300 mm). It is possible that, if born now, the baby might be able to survive.

Fetal Circulation As figure 20.22 shows, the fetus has four features that are not present in adult circulation:

1. **Oval opening** or *foramen ovale:* an opening between the two atria. This opening is covered by a flap of tissue that acts as a valve.
2. **Arterial duct** or *ductus arteriosus:* a connection between the pulmonary artery and the aorta.
3. **Umbilical arteries** and **vein:** vessels that travel to and from the placenta, leaving waste and receiving nutrients.
4. **Venous duct** or *ductus venosus:* a connection between the umbilical vein and the vena cava.

All of these features may be related to the fact that the fetus does not use its lungs for gas exchange, since it receives oxygen and nutrients from the mother's blood by way of the placenta.

To trace the path of blood in the fetus, begin with the right atrium (fig. 20.22). From the right atrium, the blood may pass directly into the left atrium by way of the oval opening or it may pass through the atrioventricular valve into the right ventricle. From the right ventricle the blood goes into the pulmonary artery, but because of the arterial duct, most of the blood then passes into the aorta. Thus, by whatever route the blood takes, most of the blood will reach the aorta instead of the lungs.

Blood within the aorta travels to the various branches, including the iliac arteries that connect to the umbilical arteries leading to the placenta. Exchange between mother's blood and fetal blood takes place at the placenta. It is interesting to note that the blood in the umbilical arteries, which travels to the placenta, is low in oxygen, but the blood in the umbilical vein, which travels from the placenta, is high in oxygen. The umbilical vein enters the venous duct, which passes directly through the liver. The venous duct then joins with the inferior vena cava, a vessel that contains deoxygenated blood. The vena cava returns this "mixed blood" to the heart.

At birth, the oval opening usually closes. With the tying of the cord and the expansion of the lungs, blood enters the lungs in quantity. Return of this blood to the left side of the heart usually causes a flap to cover over the opening.

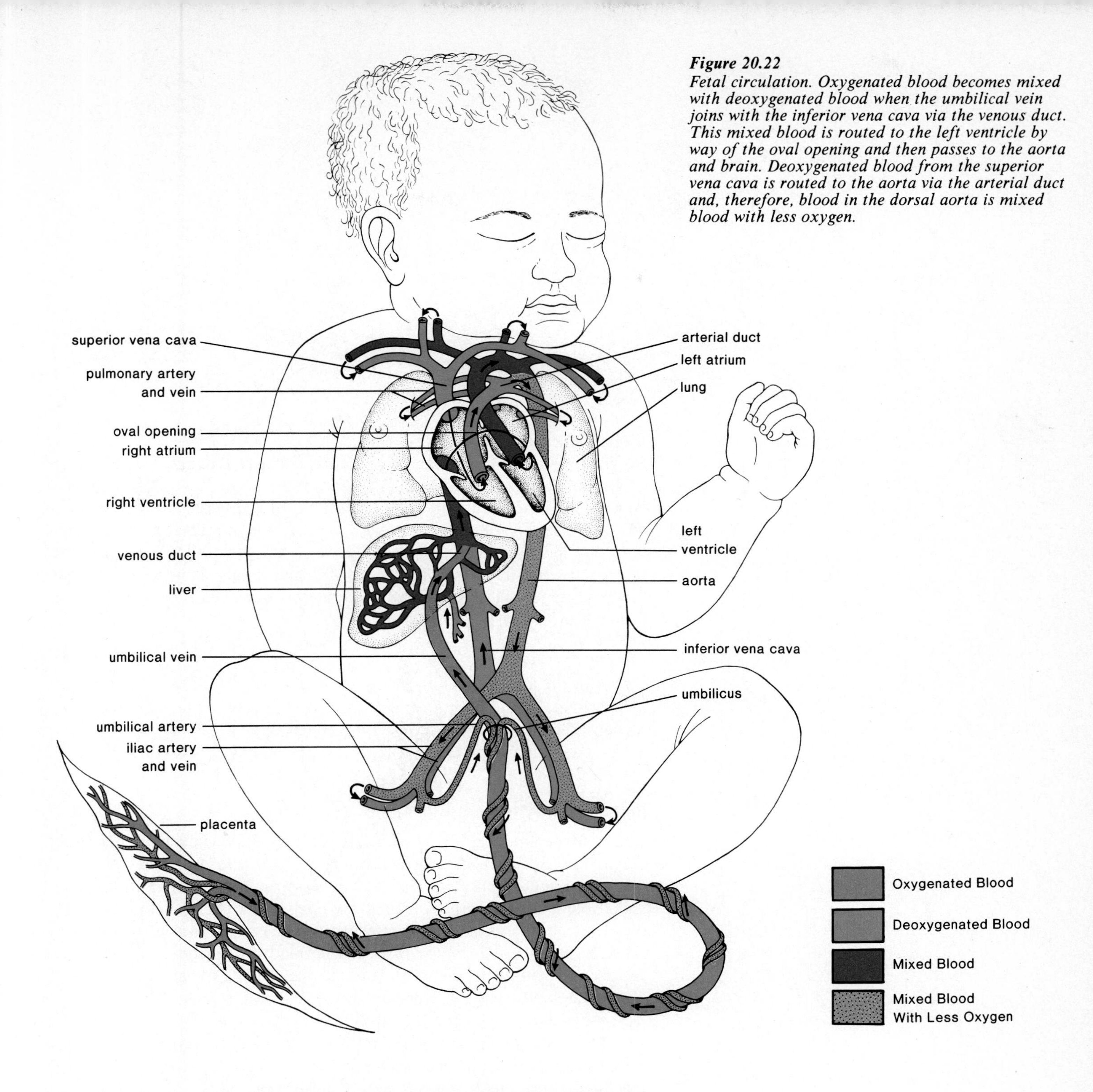

Figure 20.22
Fetal circulation. Oxygenated blood becomes mixed with deoxygenated blood when the umbilical vein joins with the inferior vena cava via the venous duct. This mixed blood is routed to the left ventricle by way of the oval opening and then passes to the aorta and brain. Deoxygenated blood from the superior vena cava is routed to the aorta via the arterial duct and, therefore, blood in the dorsal aorta is mixed blood with less oxygen.

However, the most common of all cardiac defects in the newborn is a persistence of the oval opening. Incomplete closure occurs in nearly one out of four individuals; but even so, passage of the blood from the right to the left atrium rarely occurs because either the opening is small or it closes when the atria contract. In a small number of cases, the passage of impure blood from the right to the left side of the heart is sufficient to cause a "blue baby." Such a condition may now be corrected by open-heart surgery.

The arterial duct closes because endothelial cells divide and block off the duct. Remains of the arterial duct and parts of the umbilical arteries and vein are later transformed into connective tissue.

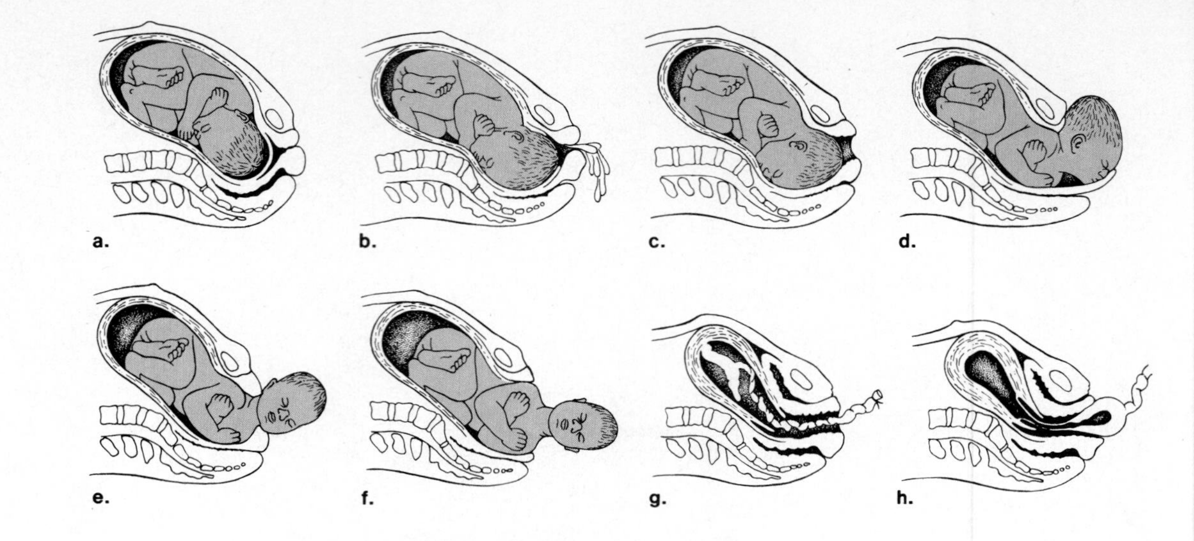

Figure 20.23
Process of birth.
a. *Dilation of cervix.*
b. *Amnion bursts.*
c. *Fetus descends into pelvis.* d. *Head appears.*
e. *Rotation.* f. *Delivery of shoulders.* g–h. *Expulsion of afterbirth.*

Eighth and Ninth Months

As the time of birth approaches the fetus rotates so that the head is pointed toward the cervix (fig. 20.23a). If the fetus does not turn, then the likelihood of a breech birth (rump first) may call for a caesarean section. It is very difficult for the cervix to expand enough to accommodate this form of birth and asphyxiation of the baby is more likely. At the end of this time period the fetus is about twenty-one inches (530 mm) long and weighs about seven and one-half pounds (3,400 gm). Weight gain is largely due to an accumulation of fat beneath the skin.

From the fifth to the ninth month, the fetus continues to grow and gain weight. Babies born after 6–7 months may survive but are subject to various illnesses that may have lasting effects or cause an early death.

Birth of Human

The uterus characteristically contracts throughout pregnancy. At first, light, often indiscernible, contractions last about 20 to 30 seconds and occur every 15 to 20 minutes, but near the end of pregnancy they become stronger and more frequent so that the woman may falsely think that she is in labor. The onset of true labor is marked by uterine contractions that occur regularly every 15 to 20 minutes and last for 40 seconds or more. **Parturition,** which includes labor and expulsion of the fetus, is usually considered to have three stages.

The events that cause parturition are still not entirely known but there is now evidence suggesting the involvement of prostaglandins. It may be too that the prostaglandins cause the release of oxytocin from the maternal posterior pituitary. Both prostaglandins and oxytocin do cause the uterus to contract and either can be given to induce parturition.

Stages

During the *first stage,* the cervix dilates; during the *second,* the baby is born; and during the *third,* the afterbirth is expelled.

Stage 1

Prior to the first stage of parturition or concomitant with it, there may be a "bloody show" caused by the expulsion of a mucus plug from the cervical canal. This plug prevents bacteria and sperm from entering the uterus during pregnancy.

Uterine contractions during the first stage of labor occur in such a way that the cervical canal slowly disappears (fig. 20.23a–c) as the lower part of the uterus is pulled upward toward the baby's head. This process is called *effacement*, or "taking up the cervix." With further contractions, the baby's head acts as a wedge to assist cervical dilation. The baby's head usually has a diameter of about 4 inches and therefore the cervix has to dilate to this diameter in order to allow the head to pass through. If it has not occurred already, the amniotic membrane is apt to rupture now, releasing the amniotic fluid, which escapes out the vagina. The first stage of labor ends once the cervix is completely dilated.

Stage 2

During the second stage, the uterine contractions occur every one to two minutes and last about one minute each. They are accompanied by a desire to push or bear down. As the baby's head gradually descends into the vagina, the desire to push becomes greater. When the baby's head reaches the exterior, it turns so that the back of the head is uppermost (fig. 20.23d). Since the vagina may not expand enough to allow passage of the head without tearing, an *episiotomy* is often performed. This incision, which enlarges the opening, is stitched later and will heal more perfectly than a tear would. As soon as the head is delivered, the baby's shoulders rotate so that the baby faces either to the right or left. The physician may at this time hold the head and guide it downward while one shoulder and then the other emerges (fig. 20.23e–f). The rest of the baby follows easily.

Once the baby is breathing normally, the umbilical cord is cut and tied, severing the child from the placenta. The stump of the cord shrivels and leaves a scar, which is the navel.

Stage 3

The placenta, or *afterbirth*, is delivered during the third stage of labor (fig. 20.23g–h). About 15 minutes after delivery of the baby, uterine muscular contractions shrink the uterus and dislodge the placenta. The placenta is then expelled into the vagina. As soon as the placenta and its membranes are delivered, the third stage of labor is complete.

Prepared Childbirth

Some doctors and expectant couples feel that nervous system depressants may be harmful not only to the expectant mother but to the baby as well. This sentiment, together with a desire to enjoy and share the process of giving birth, has given impetus to the prepared childbirth movement. Usually couples who wish to practice prepared childbirth use the methods espoused by Dr. Fernand LaMaze and attend several teaching sessions in which they learn about the events of labor and delivery, the phenomenon of conditioned pain, and suggestions for behavior during labor and delivery.

It is believed that the woman may help prevent discomfort during labor by concentrating on mild, shallow breathing at the time of contractions. This breathing method prevents the diaphragm from exerting pressure on the abdominal organs and guarantees an adequate supply of oxygen for uterine contraction. When delivery begins and the woman feels a great need to push, her partner coaches her to use deep inhalation, along with a controlled type of pushing at the time of each strong contraction. Advocates of the Lamaze method of prepared childbirth feel that this active participation on the part of the couple will not only help the woman overlook discomfort but will also give them the reward of seeing their baby born.

Summary

All animals undergo the same processes of development: cleavage, growth, morphogenesis, and differentiation. Cleavage results in many cells that become first a morula, then a blastula, and finally a gastrula. The result of gastrulation is the establishment of three germ layers: ectoderm, endoderm, and mesoderm. Later development can be related to these germ layers and induction is believed to be important to origination of organs from the germ layers.

All vertebrates at some time in their development portray a similar cross section that displays typical vertebrate embryonic characteristics; dorsal hollow nerve cord, notochord, and a coelom completely lined by mesoderm. The presence of extraembryonic membranes in reptiles made development on land possible. These same membranes are found in humans but their function has been modified. For example, the chorion becomes the fetal part of the placenta.

During the first two months of development all major organs are formed and the embryo takes on a human appearance. After this, we refer to the developing new life as the fetus. Various features of the fetus are refined during the next several months, while weight gain characterizes the last few months.

The process of birth requires three stages. During dilation, the cervix is increasing in size; during delivery, the baby is born; and finally the afterbirth is expelled.

Objective Questions

1. When cells take on a specific structure and function, ____________ occurs.
2. The morula becomes the ____________, a structure that contains the inner cell mass.
3. The ____________ membranes include the chorion, ____________, yolk sac, and allantois.
4. The blastocyst ____________ itself in the uterine lining.
5. The notochord ____________ the formation of the nervous system.
6. During embryonic and fetal development, gas exchange occurs at the ____________.
7. During development, there is a connection between the pulmonary artery and the aorta called the ____________.
8. Fetal development begins with the ____________ month.
9. The fontanels are commonly called ____________.
10. If delivery is normal, the ____________ appears before the rest of the body.

Answers to Objective Questions
1. differentiation 2. blastocyst
3. extraembryonic, amnion 4. implants
5. induces 6. placenta 7. arterial duct
8. third 9. soft spots 10. head

Study Questions

1. List the processes of development for any organism. (pp. 420–421)
2. List the early stages of development and describe what occurs during each stage. (pp. 421–426) Compare the appearance of amphioxus and the human embryo during these stages. (p. 421)
3. Explain why it is believed that the cytoplasm controls the genetic potential of cells. (pp. 423–424) How do experiments with frogs support this belief? (p. 424)
4. What are the three germ layers? What structures are associated with each germ layer? (pp. 423; 431)
5. What is induction and what experiments have been done to show that induction takes place? (pp. 424–425)
6. Draw a generalized cross section of a vertebrate embryo and label the parts. (p. 426)
7. What are the extraembryonic membranes for the chick? For the human? And what are their respective functions? (p. 427)
8. Describe in general the happenings during embryonic development of the human; during fetal development of the human. (pp. 428–440)
9. Trace the path of blood in the fetus from the umbilical vein to the aorta using two different routes. (p. 438)
10. Describe the three stages of parturition. (pp. 440–441)

Key Terms

allantois (ah-lan'to-is) one of the extraembryonic membranes; in reptiles and birds a pouch serving as a repository for nitrogenous waste; in mammals a source of blood vessels to and from the placenta. *427*

amnion (am'ne-on) an extraembryonic membrane; a sac around the embryo containing fluid. *427*

chorion (ko're-on) an extraembryonic membrane that forms an outer covering around the embryo; in reptiles and birds it functions in gas exchange, in mammals it contributes to the formation of the placenta. *427*

differentiation (dif''er-en''she-a'shun) the process and developmental stages by which a cell becomes specialized for a particular function. *421*

ectoderm (ek'to-derm) the outer germ layer of the embryonic gastrula; it gives rise to the skin and nervous system. *422*

endoderm (en'do-derm) an inner layer of cells that line the primitive gut of the gastrula. It becomes the lining of the digestive tract and associated organs. *422*

extraembryonic membranes (eks''trah-em''bre-on'ik mem'brānz) membranes that are not a part of the embryo but are necessary to the continued existence and health of the embryo. *427*

induction (in-duk'shun) a process by which one tissue controls the development of another, as when the embryonic notochord induces the formation of the neural tube. *425*

lanugo (lah-nu'go) downy hair with which a fetus is born; fetal hair. *438*

mesoderm (mes'o-derm) the middle germ layer of an animal embryo that gives rise to the muscles, connective tissue, and circulatory system. *423*

morphogenesis (mor''fo-jen'ĭ-sis) the movement of cells and tissues to establish the shape and structure of an organism. *420*

parturition (par''tu-rish'un) the processes that lead to and include the birth of a human, and the expulsion of the extraembryonic membranes through the terminal portion of the female reproductive tract. *440*

trophoblast (trof'o-blast) the outer membrane that surrounds the human embryo and, when thickened by a layer of mesoderm, becomes the chorion, an extraembryonic membrane. *428*

umbilical cord (um-bil'ĭ-kal kord) cord connecting the fetus to the placenta through which blood vessels pass. *432*

vernix caseosa (ver'niks ka''se-o'sah) cheeselike substance covering the skin of the fetus. *438*

yolk sac (yōk sak) one of the extraembryonic membranes within which yolk is found. *427*

21

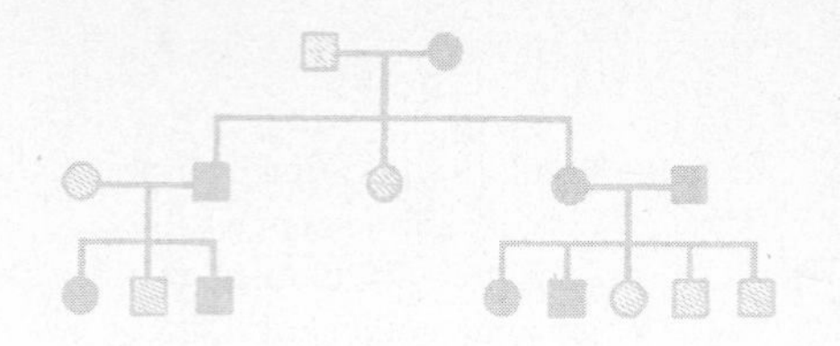

Inheritance

Chapter Concepts

1 Genes, located on chromosomes, are passed from one generation to the next.

2 The Mendelian laws of genetics relate the genotype (inherited genes) to the phenotype (physical characteristics).

3 Exceptions to Mendel's laws apply to traits controlled by more than one gene and to genes located on the same homologous pair of chromosomes.

4 The karyotype (chromosomal inheritance) is also related to the phenotype. For example, persons receiving two X chromosomes are females and persons receiving an X and a Y chromosome are males.

5 Humans are subject to many disorders due to inheritance of faulty genes.

Chapter Outline

Figure 21.1
Genes are passed from one generation to the next; therefore, a child resembles its parents and other family members.

When a sperm fertilizes an egg, a new individual with the diploid number of chromosomes begins development. These chromosomes determine what the individual will be like; even if the zygote develops in a surrogate mother, the individual will still resemble the original parents (fig. 21.1).

Today we say that the chromosomes located within the nuclei of cells contain the genes. By this we mean that it is possible to imagine that the chromosomes can be divided up into sections and that each section controls a particular trait of the individual. We will use the word *trait* to mean some aspect of the individual, such as height. In figure 21.2, the rectangles stand for a pair of homologous chromosomes and the letters stand for genes that control particular traits. Genes, like the letters in the rectangles, are in a particular sequence and are at particular spots, or loci, on the chromosomes. Alternate forms of a gene having the same position on a pair of chromosomes and affecting the same trait are called alleles. In our example, *A* is an *allele* of *a,* and vice versa. Also, *F* is an allele of *f,* and vice versa. *F* could never be an allele for *A* because *F* and *A* are at different loci. Each allelic pair controls some particular trait of the individual, such as color of hair, type of fingers, length of nose.

r	R
S	s
T	t
u	U
V	v
W	w

Figure 21.2
Diagrammatic representation of a homologous pair of chromosomes. The letters rR, Ss, and so forth stand for alleles.

Mendel's Laws

The first person to conduct a successful study of genetic or particulate inheritance was Gregor Mendel, a Catholic priest who grew peas in a small garden plot in 1860. Mendel knew little about cell structure, but his studies described in the reading led him to conclude that inheritance is governed by *factors* that exist within the individual and are passed on to offspring. Mendel said *that every* **trait,** for example, height, *is controlled by two factors, or a pair of factors.* We now call these factors **alleles.** He also observed that one of the factors controlling the same trait can be **dominant** over the other, which is **recessive.** The individual may show the dominant characteristic, for example, tallness[1], while the recessive factor for shortness, although present, is not expressed.

[1]Tallness is dominant in peas, but not in humans.

Mendel's Results

Mendel's use of pea plants as his experimental material was a good choice because pea plants are easy to cultivate, have a short generation time, and can be self-pollinated or cross-pollinated at will. Mendel selected certain traits for study and before beginning his experiments made sure his parental (P_1 generation) plants bred true. He observed that when these plants self-pollinated, the offspring were like one another and like the parent plant. For example, a parent with yellow seeds always had offspring with yellow seeds; a plant with green seeds always had offspring with green seeds. Following that observation Mendel cross-pollinated the plants by dusting the pollen of plants with yellow seeds on the stigma of plants with green seeds whose own anthers had been removed, and vice versa. Either way the offspring (called F_1, or first filial generation) resembled the parents with yellow seeds. These results caused Mendel to allow the F_1 plants to self-pollinate. Once he had obtained an F_2 generation, he observed the color of the peas produced. As table 21.A (p. 447) indicates, he counted over 8,000 plants and found an approximate 3:1 ratio (about three plants with yellow seeds for every plant with green seeds) in the F_2 generation.

Mendel realized that these results were explainable, assuming (a) there are two factors for every trait; (b) one of the factors can be dominant over the other that is recessive; (c) the factors separate when the gametes are formed. He assigned letters to these factors and displayed his results similar to this:

P_1	yellow YY	$\times$	green yy
F_1		all yellow	
$F_1 \times F_1$	yellow Yy	$\times$	yellow Yy
F_2		3 yellow : 1 green	

He believed that the F_2 plants with yellow seeds carried a dominant factor because his results could be related to the binomial equation, $a^2 + 2ab + b^2$, in this manner: $a^2 = YY$; $2ab = 2Yy$, and $b^2 = yy$. Thus the plants with yellow seeds would be YY or Yy, and there would be three plants with yellow seeds for every plant with green seeds.

As a test to determine if the F_1 generation was indeed Yy, Mendel back-crossed it with the recessive parent, yy. His results of 1:1 indicated that he had reasoned correctly. Today when a testcross is done, a suspected heterozygote is crossed with the recessive phenotype because this cross gives the best chance of producing the recessive phenotype.

Mendel performed a second series of experiments in which he crossed true-breeding plants that differed in two traits. For example, he crossed plants with yellow, round peas by plants with green, wrinkled peas. The F_1 generation always had both dominant characteristics and therefore he allowed the F_1 plants to self-pollinate. Among the F_2 generation he achieved an almost perfect ratio of 9:3:3:1 (table 21.A). For example, for every plant that had green, wrinkled seeds he had approximately nine that had yellow, round seeds, and so forth. Mendel saw that these results were explainable if pairs of factors separate independently from one another when the gametes form, allowing all possible combinations of factors to occur in the gametes. This would mean that the probability of achieving any two factors together in the F_2 offspring was the product of their chances of occurring separately. Thus, since the chance of yellow peas was 3/4 (in a one-trait cross) and the chance of round peas was 3/4 (in a one-trait cross), the chance of their occurring together was 9/16, and so forth.

Mendel achieved his success in genetics by studying large numbers of offspring, keeping careful records, and treating his data quantitatively. He showed that the application of mathematics to biology was extremely helpful in producing testable hypotheses.

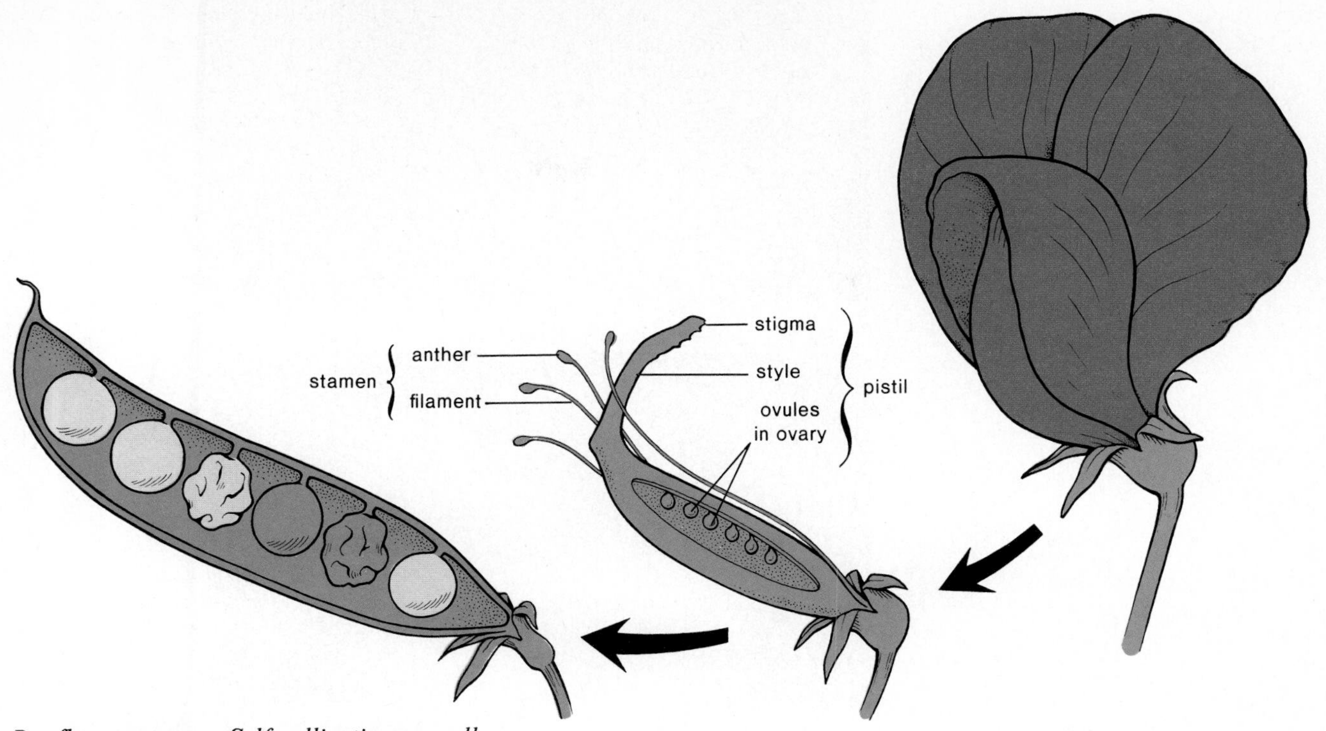

Pea flower anatomy. Self-pollination normally occurs within the flowers of the pea plant. In order to cross the dominant by the recessive, Mendel removed the anthers of the plant with the dominant trait and used pollen from the plant with the recessive trait to bring about cross-fertilization. He also performed the reverse cross in the same manner. The cross under discussion pertains to the shape and color of peas in a pod. All possible shapes and colors are illustrated.

***Table 21.A** Mendel's Results*

Single-trait Cross	F_1	F_2	Actual F_2 Ratio
yellow × green	all yellow	6,022 yellow 2,001 green	3.01:1
Two-trait Cross	**F_1**	**F_2**	**Actual F_2 Ratio**
yellow, round × green, wrinkled	all round yellow	315 yellow, round 101 yellow, wrinkled 108 green, round 32 green, wrinkled	9.8:2.9:3.11:1.0

Figure 21.3
In humans, widow's peak a. *is dominant over continuous hairline* b.

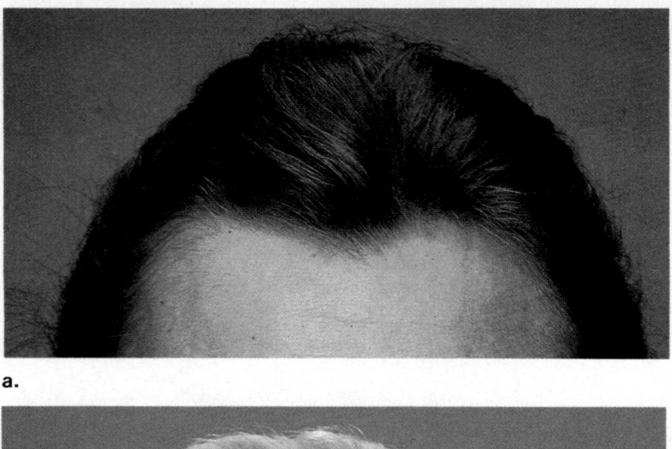
a.

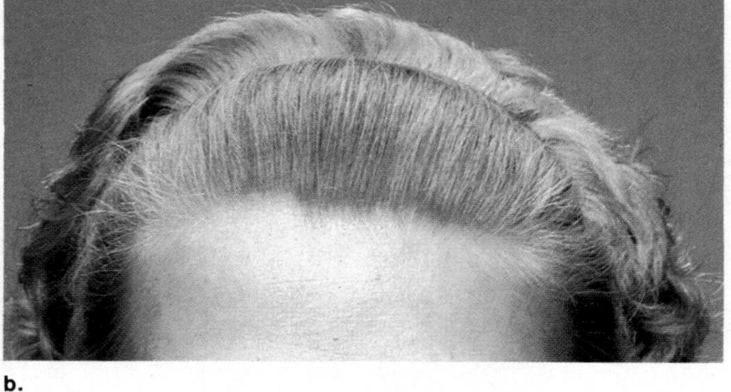
b.

Mendel's experimental crosses made him realize that it is possible for a tall pea plant to pass on a factor for shortness. Therefore he concluded that while the individual has two factors for each trait, the gametes contained only one factor for each trait. This is often called Mendel's law of segregation.

Law of segregation: The factors separate when the gametes are formed, and only one factor of each pair is present in each gamete.

Inheritance of a Single Trait

Mendel suggested that letters be used to indicate factors so that *crosses* (gamete union resulting in offspring) might be more easily described. A capital letter indicates a dominant factor, and a lowercase letter indicates a recessive factor. The same procedure is used today, only the letters are now said to represent alleles. Also, Mendel's procedure and laws are applicable not only to peas but to all diploid individuals. Therefore, we will take as our example not peas but human beings. Figure 21.3 illustrates the difference between a widow's peak and a continuous hairline. In doing a problem concerning hairline, the *key* would be represented as:

W = Widow's peak (dominant allele)

w = Continuous hairline (recessive allele)

Table 21.1 *Genotype versus Phenotype*

Genotype	Genotype	Phenotype
WW	Homozygous (pure) dominant	Widow's peak
Ww	Heterozygous (hybrid)	Widow's peak
ww	Homozygous (pure) recessive	Continuous hairline

The key simply tells us what letter of the alphabet to use for the gene in a particular problem and tells which allele is dominant, a capital letter signifying dominance.

Genotype and Phenotype

When we indicate the genes of a particular individual, two letters must be used for each trait mentioned. This is called the **genotype** of the individual. The genotype may be expressed not only by using letters but also by a short descriptive phrase, as table 21.1 shows. Thus the word **pure,** or **homozygous,** means that the two members of the allelic pair in the zygote (*zygo*) are the same (*homo*); genotype *WW* is called *homozygous dominant* and *ww* is called *homozygous recessive.* The word **heterozygous** means that the members of the allelic pair are different (*hetero*); only *Ww* is heterozygous. Another term, **hybrid,** is sometimes used in place of heterozygous and means that the genotype is not pure.

As table 21.1 also indicates, the word **phenotype** refers to the physical characteristics of the individual. What the individual actually looks like is the phenotype. (Also included in the phenotype are the microscopic and metabolic characteristics of the individual.) Notice that both homozygous dominant and heterozygous show the dominant phenotype.

Gamete Formation

Whereas the genotype has two alleles for each trait, the gametes have only one allele for each trait. This, of course, is related to the process of meiosis. The alleles are present on a homologous pair of chromosomes and these chromosomes separate during meiosis (fig. 21.6). Therefore, the members of each allelic pair separate during meiosis, and there is only one allele for each trait in the gametes. When doing genetic problems it should be kept in mind that no two letters in a gamete may be the same. For this reason *Ww* would represent a possible genotype and the gametes for this individual could contain either a *W* or *w*.

When doing genetics problems, the same alphabet letter is used for both the dominant and recessive alleles; a capital letter indicates the dominant and a lowercase letter indicates the recessive. A homozygous dominant individual is indicated by two capital letters, and a homozygous recessive individual is indicated by two lowercase letters. The genotype of a heterozygous individual is indicated by a capital and a lowercase letter. Contrary to the individual, gametes have one letter of each type, either capital or lowercase, as appropriate. All possible combinations of letters indicate all possible gametes.

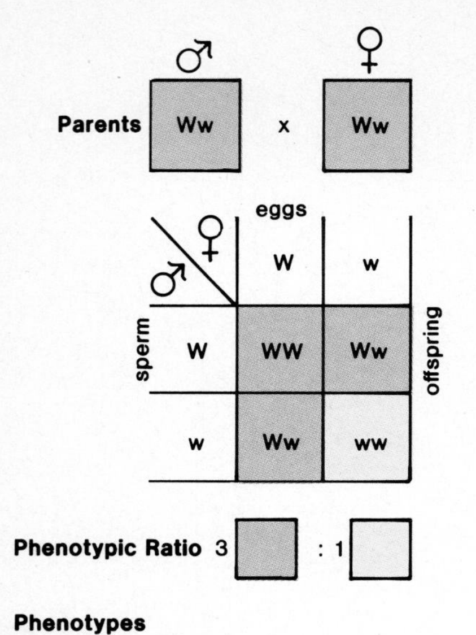

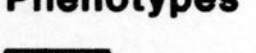

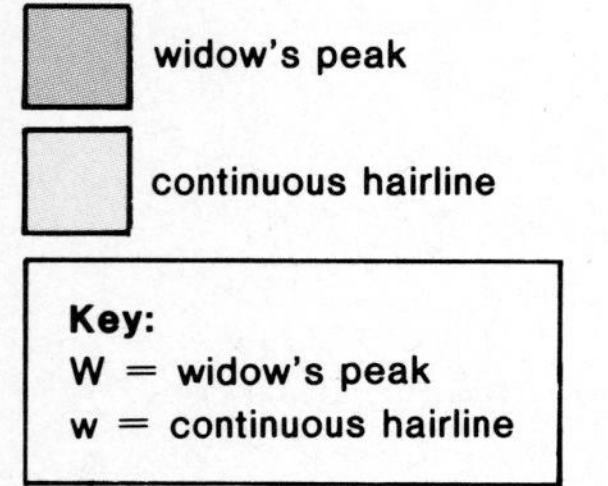

Figure 21.4
Monohybrid cross. In this cross, the parents are heterozygous for widow's peak. The chances of any child having a continuous hairline are one out of four, or 25 percent.

Practice Problems 1

1. For each of the following genotypes, give all possible gametes.
   a. *WW* d. *Ttgg*
   b. *WWSs* e. *AaBb*
   c. *Tt*
2. For each of the following state whether it represents a genotype or a gamete.
   a. *D* c. *Pw*
   b. *Ll* d. *LlGg*

*Answers to problems are on page 472.

Crosses

It is now possible for us to consider a particular cross. If a homozygous man with a widow's peak (fig. 21.3a) marries a woman with a continuous hairline (fig. 21.3b), what kind of hairline will their children have?

In solving the problem, we must indicate the genotype of each parent by using letters, determine what the gametes are, and what the genotypes of the children are after reproduction. In the format that follows, P_1 stands for the parental generation, and the letters in this row are the genotypes of the parents. The second row shows that each parent has only one type of gamete in regard to hairline, and therefore all the children (F_1 = first filial generation) will have a similar genotype, that is, heterozygous. Heterozygotes show the dominant characteristic, and so all the children will have a widow's peak.

P_1	Widow's peak	×	Continuous hairline
	WW		*ww*
Gametes:	*W*		*w*
F_1		Widow's peak	
		Ww	

These individuals are **monohybrids** because they are heterozygous for only one pair of alleles. If they marry someone else with the same genotype, what type of hairline will their children have?

P_1	Widow's peak	×	Widow's peak
	Ww		*Ww*
Gametes:	*W, w*		*W, w*

In this problem, each parent has two possible types of gametes. In calculating F_1, it is assumed that either type of sperm has an equal chance to fertilize either type of egg. One way to assure that we have accounted for this is to use a **Punnett square** (fig. 21.4) in which all possible types of sperm are lined up vertically and all possible types of eggs are lined up horizontally (or vice versa), and every possible fertilization is considered.

When this is done, the results show a 3:1 phenotypic ratio; that is, three with widow's peak to one with continuous hairline. Such a ratio will actually be observed only if a large number of crosses of the same type take place and a large number of offspring result. Only then will all possible sperm have an equal chance to fertilize all possible eggs. It is obvious that we do not routinely observe hundreds of offspring from a single type cross in humans, and so it is customary to merely state that each child has three chances out of four to have a widow's peak, or one chance out of four to have a continuous hairline. It is important to realize that *chance has no memory;* for example, if two heterozygous parents have already had three children with a widow's peak and are expecting a fourth child, this child still has three chances out of four to have a widow's peak and only one chance out of four of not having one. Each individual child has the same chances.

Probability

Another way to calculate the possible results of a cross is to realize that the chance, *or probability of receiving a particular combination of alleles, is simply the product of the individual probabilities.* In the cross just considered:

Ww × *Ww*

the offspring have an equal chance of receiving *W* or *w* from each parent. Therefore:

Probability of *W* = ½
Probability of *w* = ½

and

Probability of *WW* = ½ × ½ = ¼
Probability of *Ww* = ½ × ½ = ¼
Probability of *wW* = ½ × ½ = ¼
¾ = Widow's peak
Probability of *ww* = ½ × ½ = ¼
¼ = Continuous hairline

Testcross

If a plant, animal, or person has the dominant phenotype, it is not possible to tell by inspection if the organism is homozygous dominant or heterozygous. However, if the plant or animal is crossed with a homozygous recessive, the results may indicate what the original genotype was. For example, figure 21.5 shows the different results if a man with a widow's peak is homozygous dominant, or if he is heterozygous, and married to a woman with a continuous hairline. (She must be homozygous recessive or she would not have a continuous hairline.) In the first case, the man can only sire children with widow's peaks, and in the second the chances are 2:2 or 1:1 that a child will or will not have one. Thus, the cross of a possible heterozygote with an individual having the recessive phenotype gives the best chance of producing the recessive phenotype among the offspring. Therefore, this type cross is called the **testcross.**

In doing an actual cross, it is assumed that all possible types of sperm fertilize all possible types of eggs. The results may be expressed as a probable phenotype ratio; it is also possible to state the chances of an offspring showing a particular phenotype.

Practice Problems 2

1. Both a man and a woman are heterozygous for freckles. Freckles are dominant over no freckles. What are the chances that their children will have freckles?
2. A woman is homozygous dominant for short fingers. Short fingers are dominant over long fingers. Will any of her children have long fingers?
3. Both you and your sister or brother have attached earlobes, yet your parents have unattached ones. Unattached earlobes are dominant over attached. What are the genotypes of your parents?
4. A father has dimples, the mother does not have dimples, all the children have dimples. Dimples are dominant over no dimples. Give the probable genotype of all persons concerned.

*Answers to problems are on page 472.

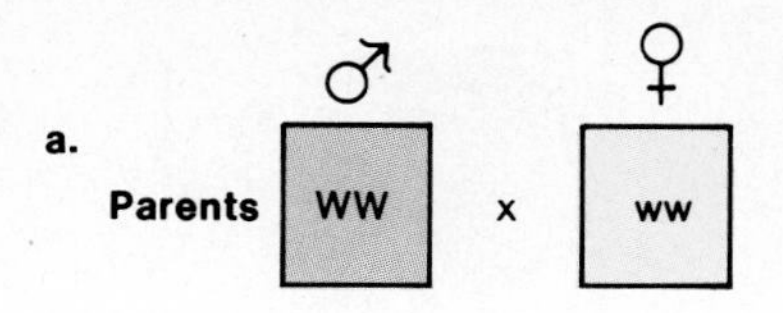

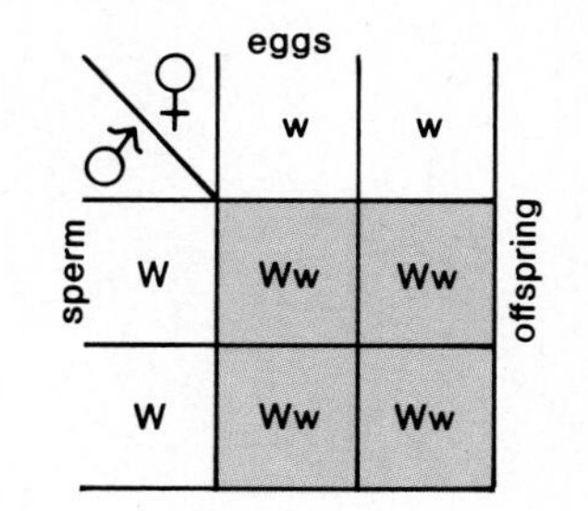

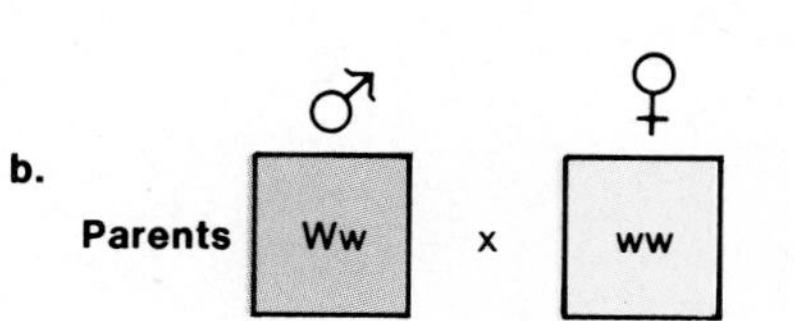

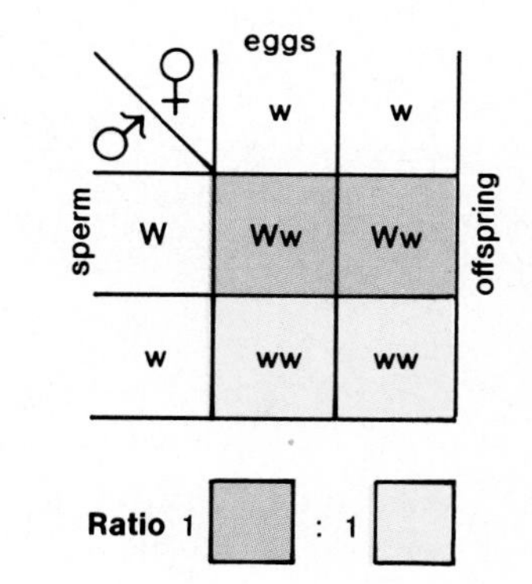

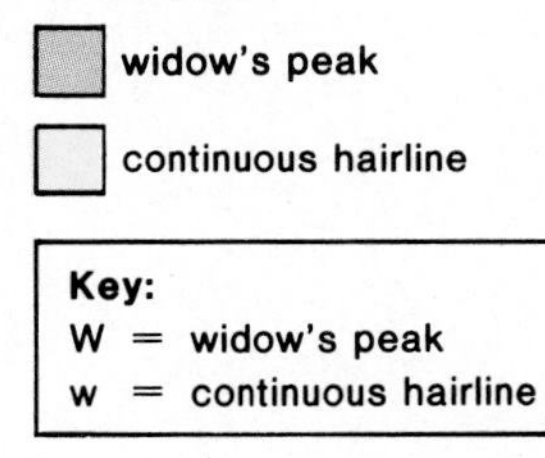

Figure 21.5
Testcross. In this example, it is impossible to tell if the male parent is homozygous or heterozygous by inspection. The results of reproduction with a homozygous recessive, however, may help determine his genotype. If he is homozygous, none of the offspring will show the recessive characteristic, but if he is heterozygous there is a 50-50 chance that any offspring will show the recessive characteristic.

Inheritance of Multitraits

Two Traits (Unlinked)

Although it is possible to consider the inheritance of just one trait, actually each individual passes on to his or her offspring many genes for many traits. In order to arrive at a general understanding of multitrait inheritance, we will consider the inheritance of two traits. The same principles will apply to as many traits as we might wish to consider.

When Mendel performed two-trait crosses, he formulated his second law:

Law of independent assortment: Pairs of factors separate independently of one another to form gametes, and therefore all possible combinations of factors may occur in the gametes.

Figure 21.6 illustrates that the law of segregation and the law of independent assortment hold because of the manner in which meiosis occurs. The law of segregation is dependent on the separation of members of homologous pairs of chromosomes; and the law of independent assortment is dependent on the random arrangement of homologous pairs with respect to one another during metaphase I prior to the separation process.

Crosses

When doing a two-trait cross, we realize that the genotypes of the parents require four letters because there is an allelic pair for each trait. Second, the gametes of the parents contain one letter of each kind in every possible combination, as predicted by Mendel's law of independent assortment. Finally, in order to produce the probable ratio of phenotypes among the offspring, all possible matings are presumed to occur.

To give an example, let us cross a person homozygous for widow's peak and short fingers with a person who has a continuous hairline and long fingers.

Key:
W = Widow's peak   *S* = Short fingers
w = Continuous hairline   *s* = Long fingers

P_1	Widow's peak, Short fingers *WWSS*	×	Continuous hairline Long fingers *wwss*
Gametes:	*WS*		*ws*
F_1		Widow's peak, Short fingers *WwSs*	

In this particular cross, only one type of gamete is possible for each parent; therefore, all of the F_1 will have the same genotype (*WwSs*) and the same phenotype (widow's peak with short fingers). This genotype is called a **dihybrid** because the individual is heterozygous in two regards: hairline and fingers.

When a dihybrid reproduces with a dihybrid, each parent has four possible types of gametes:

WwSs × *WwSs*

Gametes:	*WS*	*WS*
	Ws	*Ws*
	wS	*wS*
	ws	*ws*

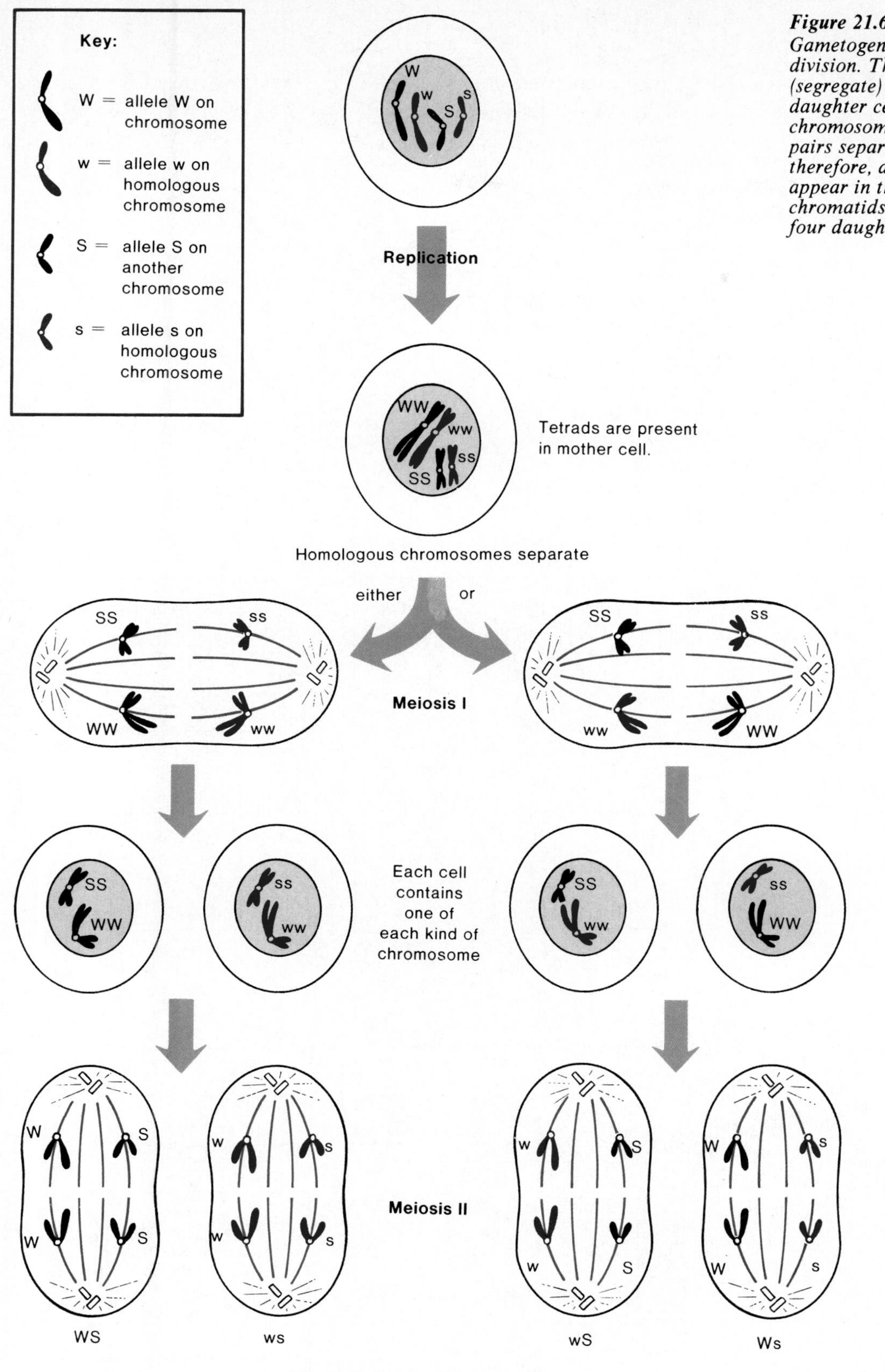

Figure 21.6
Gametogenesis includes meiosis, a type of cell division. The homologous chromosomes separate (segregate) during meiosis I; therefore, each daughter cell has only one of each kind of chromosome and one of each type of gene. Also, the pairs separate independently of one another; therefore, all possible combinations of alleles appear in the gametes. During meiosis II, the chromatids separate and therefore meiosis produces four daughter cells.

Figure 21.7
Dihybrid cross. A dihybrid results when an individual homozygous dominant in two regards reproduces with an individual homozygous recessive in two regards. When a dihybrid reproduces with a dihybrid, there are four possible phenotypes among the offspring and the phenotypic ratio is 9:3:3:1 as indicated.

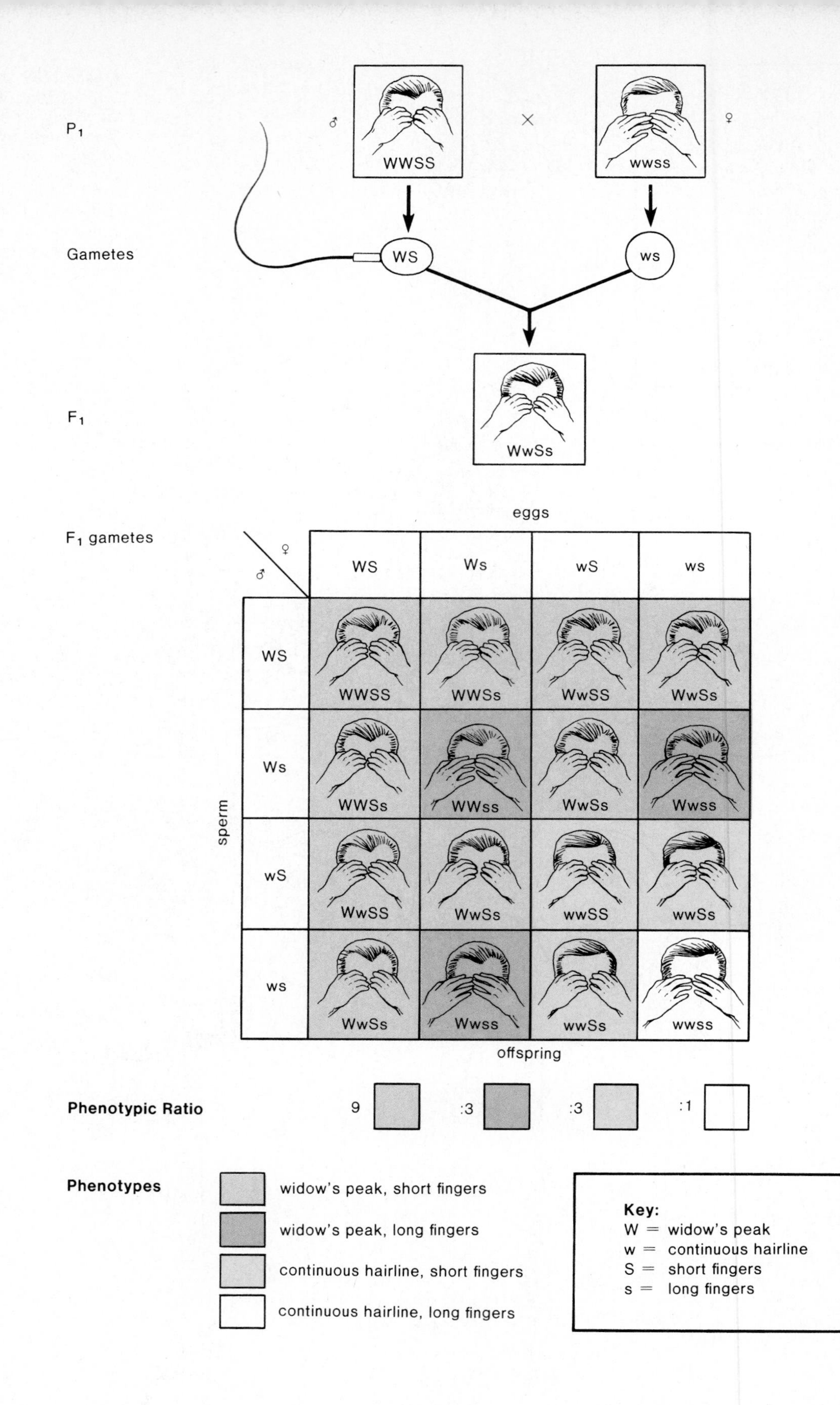

The Punnett square (fig. 21.7) for such a cross shows the expected genotypes among 16 offspring if all possible sperm fertilize all possible eggs. An inspection of the various genotypes in the square shows that among the offspring, *nine* will have a widow's peak and short fingers, *three* will have a widow's peak and long fingers, *three* will have a continuous hairline and short fingers, and *one* will have a continuous hairline and long fingers. This is called a 9:3:3:1 phenotype ratio, and this ratio always results when a dihybrid is mated with a dihybrid and simple dominance is present.

Probability

We can use the previous ratio to predict the chances of each child receiving a certain phenotype. For example, the possibility of getting the two dominant phenotypes together is 9 out of 16 (9+3+3+1 = 16), and that of getting the two recessive phenotypes together is 1 out of 16.

We can also calculate the chance, or probability, of these various phenotypes occurring by knowing that the *probability of combinations of independent events is the product of the probabilities of each of the events.* Thus:

Probability of widow's peak = ¾
Probability of short fingers = ¾
Probability of continuous hairline = ¼
Probability of long fingers = ¼

Therefore

Probability of widow's peak and short fingers = ¾ × ¾ = 9/16
Probability of widow's peak and long fingers = ¾ × ¼ = 3/16
Probability of continuous hairline and short fingers = ¼ × ¾ = 3/16
Probability of continuous hairline and long fingers = ¼ × ¼ = 1/16

Testcross

A plant or animal that shows the dominant traits can be tested for the dihybrid genotype by a mating with the recessive in both traits.

P_1 *WwSs* × *wwss*

Gametes: *WS*, *Ws*, *wS*, *ws*   *ws*

The Punnett square (fig. 21.8) shows that the resulting ratio is 1 widow's peak with short fingers : 1 widow's peak with long fingers : 1 continuous hairline with short fingers : 1 continuous hairline with long fingers, or 1:1:1:1.

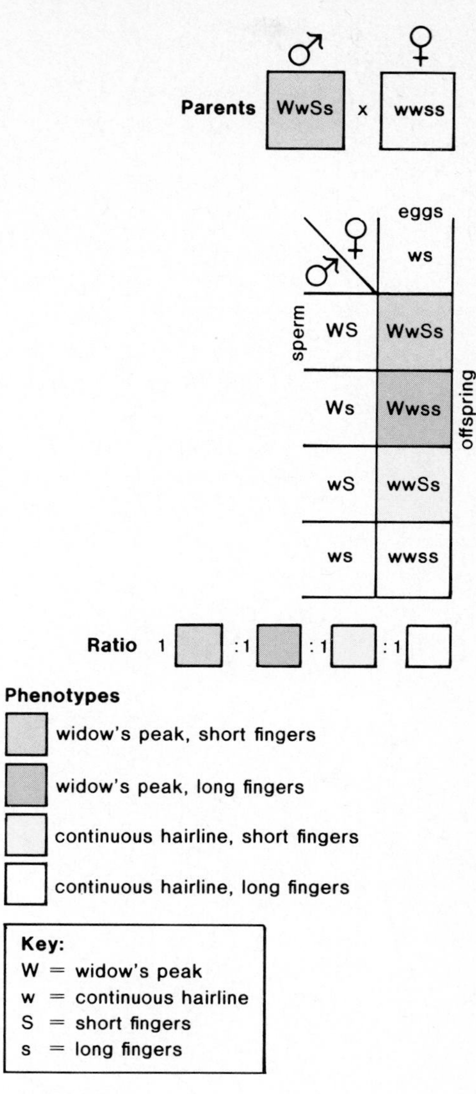

Figure 21.8
Testcross. In this example it is impossible to tell by inspection if the male parent is homozygous dominant or if he is heterozygous for both traits. However, reproduction with a female who is recessive for both traits is likely to show which he is. If he is heterozygous, there is a 25 percent chance that the offsprings will show both recessive characteristics and a 50 percent chance that they will show one or the other of the recessive characteristics.

Table 21.2 lists all of the crosses we have studied thus far, which show a frequently observed ratio. When these types of crosses are done, these ratios are observed.

***Table 21.2** Phenotypic Ratios of Common Crosses*

Genotypes:	Phenotypes:
Monohybrid × monohybrid	3:1 (dominant to recessive)
*Monohybrid × recessive	1:1 (dominant to recessive)
Dihybrid × dihybrid	9:3:3:1 (9 both dominant, 3 one dominant, 3 other dominant, 1 both recessive)
*Dihybrid × recessive	1:1:1:1 (all possible combinations in equal number)

*Called a backcross because it is as if the F_1 were mated back to the recessive parent. Also called a testcross because it can be used to test if the individual showing the dominant gene is homozygous or heterozygous. For a definition of all terms, see the end-of-chapter key terms.

Practice Problems 3

Using the information provided in Practice Problems 2 solve these problems.

1. What is the genotype of the offspring if a man homozygous recessive for type of earlobes and homozygous dominant for type of hairline is married to a woman who is homozygous dominant for earlobes and homozygous recessive for hairline.
2. If the offspring of this cross marries someone of the same genotype, then what are the chances that this couple will have a child with a continuous hairline and attached earlobes?
3. A person who has dimples and freckles marries someone who does not. This couple produces a child that does not have dimples nor freckles. What is the genotype of all persons concerned?

*Answers to problems are on page 472.

Beyond Mendel's Laws

Although the study of Mendel's laws is helpful, we know today that they are an oversimplification. There are many exceptions, such as those discussed following.

Polygenic Inheritance

Two or more genes may affect the same trait in an additive fashion. When a black person has children by a white person, the children are mulatto, but two mulattoes can produce children who range in skin color from black to white. This can be explained if we assume that there are two genes that control skin color and that only capital letters in the example contribute equally to skin color.

Black = *AABB*

Dark = *AABb* or *aABB*

Mulatto = *AaBb* or *AAbb* or *aaBB*

Light = *Aabb* or aabB

White = *aabb*

Polygenic inheritance can cause the distribution of human traits according to a bell-shaped curve, with most individuals exhibiting the average phenotype. The more genes that control the trait, the more continuous the distribution will be. Just how many genes control skin color and height, another possible example of polygenic inheritance, is not known.

Figure 21.9
Inheritance of skin color. This white husband (aabb) and his mulatto wife (AaBb) had fraternal twins, one of which was white and one of which was mulatto.

Multiple Alleles

ABO Blood Type

Three alleles for the same gene control the inheritance of A-B-O blood types. These alleles determine the presence or absence of antigens on the red blood cells. Therefore, *I* standing for immunogen (antigen) is used to signify the gene, and a superscript letter is used to signify the particular allele:

I^A = type A antigen on red cells

I^B= type B antigen on red cells

i^o = no antigens on the red cells

Each person has only two of the three possible alleles, and both I^A and I^B are dominant over i^o. Therefore as table 21.3 shows, there are two possible genotypes for type A blood and two possible genotypes for type B blood. On the other hand, I^A and I^B are fully expressed in the presence of the other. Therefore, if a person inherits one of each of these alleles, that person will have type AB blood. Type O blood can only result from the inheritance of two i^o alleles.

Table 21.3 Blood Groups

Phenotype	Genotypes
A	I^AI^A, I^Ai^o
B	I^BI^B, I^Bi^o
AB	I^AI^B
O	i^oi^o

I = immunogen gene

An examination of possible matings between different blood types sometimes produces surprising results; for example,

P_1 $I^Ai^o \times I^Bi^o$

F_1 I^AI^B, i^oi^o, I^Ai^o, I^Bi^o

Thus from this particular mating every possible phenotype (AB, O, A, or B blood type) is possible.

Blood typing can sometimes aid in paternity suits. However, a blood test of a supposed father can only suggest that he *might* be the father, not that he definitely *is* the father. For example, it is possible, but not definite, that a man with blood type A (having genotype I^Ai^o) is the father of a child with blood type O. On the other hand, a blood test can sometimes definitely prove that a man is not the father. For example, a man with blood type AB could not possibly be the father of a child with blood type O. Therefore, blood tests may legally be used only to exclude a man from possible paternity.

Rh Blood Factor

It might be noted here that the blood factor called Rh is inherited separately from A, B, AB, or O type blood. In each instance it is possible to be Rh^+ or Rh^-, meaning in the first case that an Rh factor is present on the red cells and in the second that an Rh factor is not present. It may be assumed that Rh is controlled by a single allelic pair in which simple dominance prevails: Rh^+ is dominant over Rh^-. Complications arise when an Rh^- woman marries an Rh^+ man and the child in the womb is Rh^+. With the birth of the first child of this phenotype, the mother may begin to build up antibodies to the factor and in later pregnancies these antibodies may cross the placenta to destroy the baby's blood cells (p. 247).

Incomplete Dominance and Codominance

In incomplete dominance, neither member of an allelic pair is dominant over the other and the phenotype is intermediate between the two. For example, when a curly-haired Caucasian person reproduces with a straight-haired Caucasian person, their children will have wavy hair. And two wavy-haired individuals can produce all possible phenotypes: straight hair, curly hair, and wavy hair. To signify that neither allele is dominant, one allele is designated as *H* and the other as *H'*, as is done in figure 21.10.

In codominance, each member of an allelic pair is dominant and the phenotype exhibits both characteristics. For example, an individual with the genotype I^AI^B has the blood type AB.

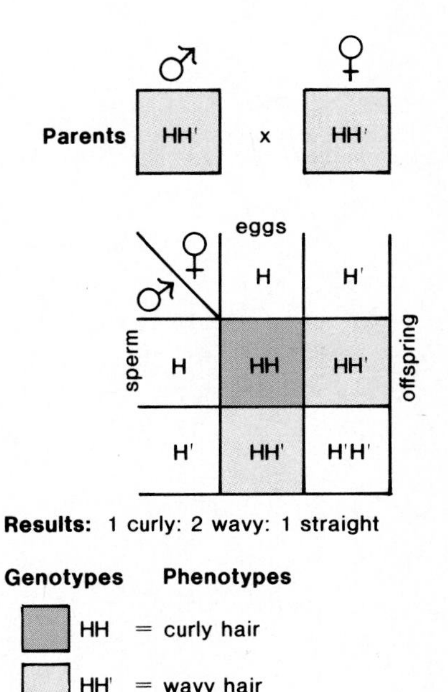

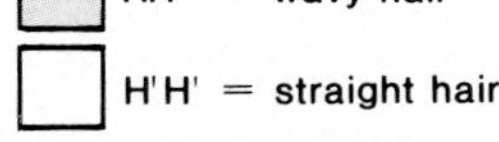

Figure 21.10
Incomplete dominance. Among Caucasians, neither straight nor curly hair is dominant. When two wavy-haired individuals reproduce, the offspring has a 25 percent chance of having either straight or curly hair and a 50 percent chance of having wavy hair, the intermediate phenotype.

Figure 21.11
Complete linkage (hypothetical). In this example, the genes for dimples and type of fingers are linked. Even though the parents are dihybrids, the offspring show only two possible phenotypes.

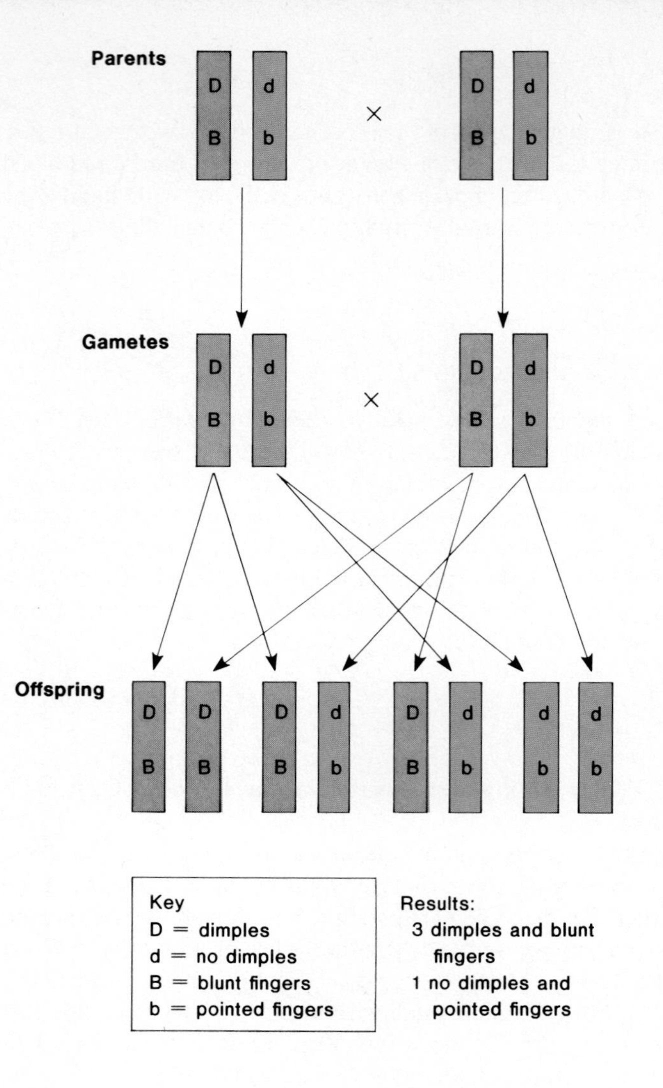

Linkage

As illustrated in figure 21.2 a chromosome pair has a series of genes. Genes on the same chromosome are said to form a **linkage** group. Mendel's law of independent assortment cannot hold for linked genes since they tend to appear together in the same gamete. Thus, traits controlled by linked genes tend to be inherited together.

To take a hypothetical example, let us remember that dimples are dominant over no dimples and blunt fingers are dominant over pointed fingers. If a dihybrid were married to a dihybrid, you would expect four possible phenotypes among the offspring. But as figure 21.11 shows, only two phenotypes would appear if the genes were absolutely linked in the manner illustrated. When doing linkage problems, it is better to use the method illustrated in figure 21.11 rather than a Punnett square so that the genes can be shown on a single chromosome.

When a dihybrid is crossed with a recessive, you normally would expect all possible phenotypes among the offspring. If linkage is present, however, the number of possible phenotypes could possibly be reduced to two types. To take an actual example, it has been reported that A-B-O blood type gene and the gene for a very unusual dominant condition called nail-patella syndrome (NPS) are on the same chromosome. A person with NPS has fingernails and toenails that are reduced or absent and a kneecap (patella) that is small. In

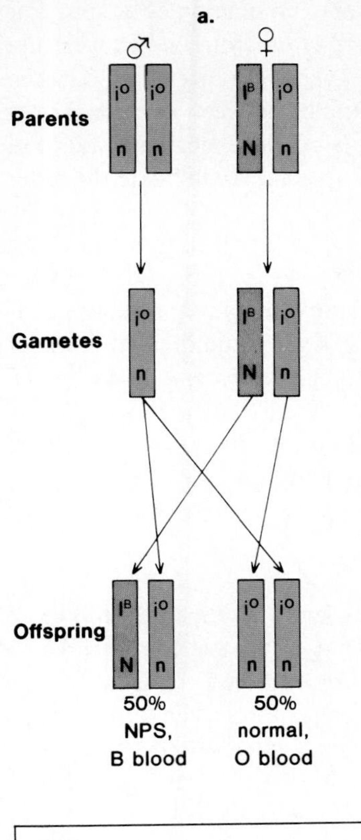

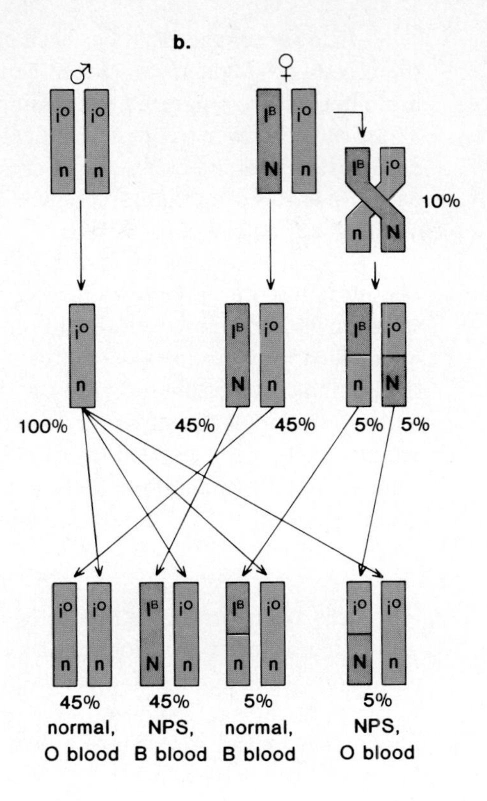

Figure 21.12
Linkage (in practice).
a. *If linkage is complete, then only two phenotypes would appear among the offspring resulting from this cross.* b. *In practice, linkage is not complete because of crossing over. In this actual example, 10 percent of the offspring had recombinant characteristics because crossing over had occurred.*

Key:
I^Bi^O = type B blood
i^Oi^O = type O blood
N = nail-patella syndrome (NPS)
n = normal

one family, the female parent had the genotype I^Bi^o for blood type and the genotype *Nn* for NPS; furthermore, it could be established that the allele I^B was on the same chromosome as N and that the allele i^o was on the same chromosome as *n*. Notice in figure 21.12a that if linkage holds, this individual would form only two possible gametes.

The male parent in this example had the recessive genotype for both traits and therefore could form only one type of gamete carrying the recessive alleles of each gene as illustrated in figure 21.12a. Therefore, assuming linkage, the children of this couple should have only two possible phenotypes: blood type B with NPS and blood type O without NPS.

However, at least 10 percent of all offspring showed a phenotype in which blood type B is found without NPS and blood type O is found with NPS. This indicates that crossing over occurred (fig. 21.12b).

Crossing Over

When tetrads form during meiosis, the chromatids may exchange portions by a process of breaking and then reassociating (fig. 4.18). The gametes that receive recombined chromosomes are called recombinant gametes. Recombinant gametes indicate that the linkage between two genes has been broken by crossing over. Figure 21.12 shows how recombinant gametes produced unexpected phenotypes in our example.

In lower organisms, it has been possible to map chromosomes by studying the crossover frequency of linked genes. Genes distant from one another are more likely to be separated by crossing over than genes that are close together. Thus, the crossover frequency indicates the distance between two genes on a chromosome. Each percentage of crossing over is taken to mean a distance of one map unit. Using these frequencies, then, it is possible to indicate the order of the genes on the chromosome.

Human disorders, such as NPS (fig. 21.12), have helped to map human chromosomes, but laboratory investigations involving the chromosomes themselves have been more helpful. In one type of study, human and mouse cells were fused together in tissue cultures. As the cells grew and divided, some of the human chromosomes were lost and eventually the daughter cells contained only a few human chromosomes. Analysis of the proteins[2] made by the various human-mouse cells enabled investigators to determine which genes should be associated with which human chromosomes. Today, it is possible to use DNA probes (p. 493) to help determine which genes are on which chromosomes.

There are many exceptions to Mendel's laws and these include polygenic inheritance (skin color), multiple alleles (blood type), incomplete dominance (curly hair), and codominance (AB blood type). Also, the presence of linkage groups reduces the number of different genotypes and phenotypes among the offspring. But recombinant gametes do occur by the process of crossing over, and the frequency of crossing over can be used to map chromosomes.

Practice Problems 4

1. What is the genotype of a person with straight hair? Could this individual ever have a child with curly hair?
2. What is the darkest child that could result from a mating between a light individual and a white individual?
3. What is the lightest child that could result from a mating between two mulatto individuals?
4. From the following blood types, determine which baby belongs to which parents:

Mrs. Doe	Type A
Mr. Doe	Type A
Mrs. Jones	Type A
Mr. Jones	Type AB
Baby 1	Type O
Baby 2	Type B

5. Prove that a child does not have to have the blood type of either parent by indicating what blood types *might* be possible when a person with type A blood reproduces with a person with type B blood.
6. Imagine that ability to curl the tongue is dominant and that this characteristic is linked to a rare form of mental retardation, which is also dominant. The parents are both dihybrids, with the two dominant alleles on one chromosome and the two recessive alleles on the other. What phenotypic ratio is possible among the offspring if crossing over does not occur?

*Answers to problems are on page 472.

---

[2]Gene control of protein synthesis is discussed in the next chapter.

Sex-Linked Inheritance

The genes which determine the development of the sexual organs are on the sex chromosomes (p. 76). The fetus is unisexual and will automatically develop into a female unless a Y chromosome is present. There is evidence for a gene on the Y chromosome which directs the nondifferentiated gonad to produce testosterone and thereafter the fetus begins to develop into a male. Several other genes are necessary to complete the process of normal male sexual development. Genes are also required to complete normal female maturation.

Some genes on the sex chromosomes have nothing to do with sexual development and are instead concerned with other body traits. These genes are said to be **sex linked** because they are on the sex chromosomes. A few sex-linked genes are on the Y chromosome, but the most important ones discovered so far are on only the X chromosome.

X-Linked Genes

X-linked genes have alleles on the X chromosome but determine body traits unrelated to sex. Since there are no alleles for these genes on the Y chromosome, any recessive allele present on the X chromosome in males will be expressed. In X-linked trait problems, the females are indicated by XX and the males by XY. An X-linked allele appears as a letter attached to the X chromosome. For example, in human beings color blindness is controlled by an X-linked recessive allele and therefore the key is:

X^C = normal vision

X^c = color blindness

The possible genotypes and phenotypes in both males and females are:

X^CX^C = a female with normal color vision

X^CX^c = a carrier female with normal vision

X^cX^c = a female who is color-blind

X^CY = a male with normal vision

X^cY = a male who is color-blind

Note that the second genotype is called a carrier female because although a female with this genotype appears normal, she is capable of passing on an allele for color blindness. Color-blind females are rare because they must receive the allele from both parents, but color-blind males are more common since they need only one recessive allele in order to be color-blind. The allele for color blindness had to have been inherited from their mother because it is on the X chromosome; males only inherit the Y chromosome from their father.

Cross

If a heterozygous woman is married to a man with normal vision, what are their chances of having a color-blind daughter? A color-blind son?

Parents $X^CX^c \times X^CY$

Inspection indicates that all daughters will have normal vision because they will all receive an X^C from their father. The sons, however, have a 50:50 chance of being color-blind, depending on whether they receive an X^C or X^c from their mother. The inheritance of a Y from their father cannot offset the inheritance of an X^c from their mother. Figure 21.13 illustrates the use of the Punnett square in doing sex-linked problems.

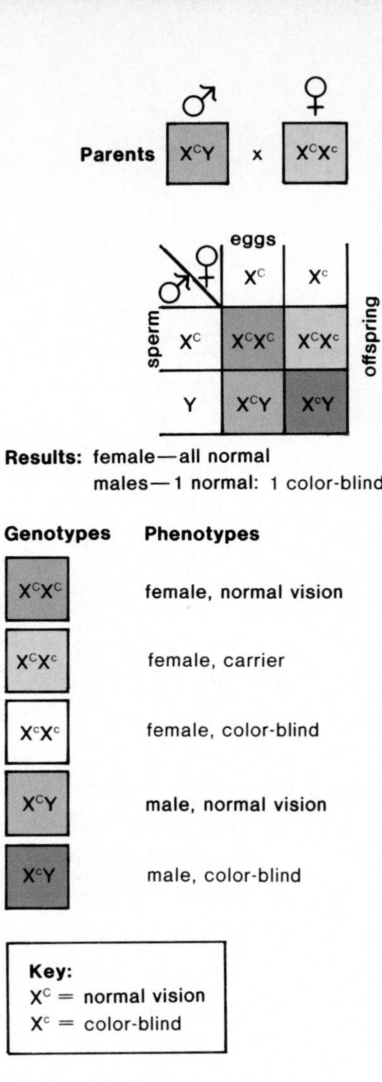

Figure 21.13
Cross involving X-linked genes. The male parent is normal, but the female parent is a carrier; an allele for color blindness is located on one of her chromosomes. Therefore each son stands a 50 to 50 chance of being color-blind.

X-Linked Traits

Some of the ways in which it is possible to recognize X-linked traits are:

1. More males than females are afflicted.
2. In order for a female to have the characteristic, her father must also have it. Her mother must have it or be a carrier.
3. The characteristic often skips a generation from the grandfather to the grandson (fig. 21.23).
4. If a woman has the characteristic, all of her sons will have it.

Most sex-linked genes are carried on the X chromosome and the Y is blank. There are four ways, noted above, to recognize X-linked inheritance which also can be used to help solve X-linked genetics problems.

Practice Problems 5

1. Both the mother and father of a hemophilic son appear to be normal. From whom did the son inherit the gene for hemophilia? What is the genotype of the mother, the father, and the son?
2. A woman is color-blind. What are the chances that her sons will be color-blind? If she is married to a man with normal vision, what are the chances that her daughters will be color-blind? Will be carriers?
3. Both parents are right-handed (R = right-handed, r = left-handed) and have normal vision. Their son is left-handed and color-blind. Give the genotype of all persons involved.
4. Both the husband and wife have normal vision. A woman has a color-blind daughter. What can you deduce about the girl's father?

*Answers to problems are on page 472.

Sex-Influenced Traits

Sex-influenced traits are characteristics that often appear in one sex but only rarely appear in the other. It is believed that these traits are governed by genes that are turned on or off by hormones. For example, the secondary sex characteristics such as the beard of a male and the breasts of a female probably are indirectly controlled by hormone balance.

Baldness is believed to be caused by the male sex hormone, testosterone, because males who take the hormone to increase masculinity begin to lose their hair. A more detailed explanation has been suggested by some investigators. It has been reasoned that due to the effect of hormones, males require only one gene for the trait to appear, whereas females require two genes. In other words, the gene acts as a dominant in males but as a recessive in females. This means that males born to a bald father and a mother with hair at *best* would have a 50 percent chance of going bald. Females born to a bald father and a mother with hair at *worst* would have a 25 percent chance of going bald.

Another trait of interest is the length of the index finger. In women the index finger is at least equal to if not longer than the fourth finger. In males the index finger is shorter than the fourth finger.

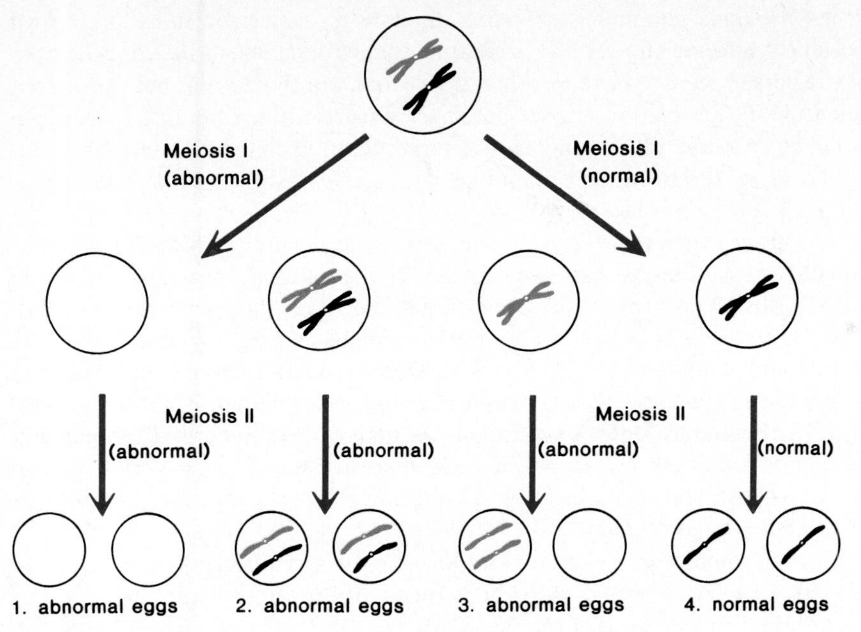

Figure 21.14
Nondisjunction during oogenesis. Nondisjunction can occur during meiosis I if the chromosome pairs fail to separate and during meiosis II if the chromatids fail to separate completely. In either case, the abnormal eggs carry an extra chromosome. Nondisjunction of the number 21-chromosome leads to Down's syndrome.

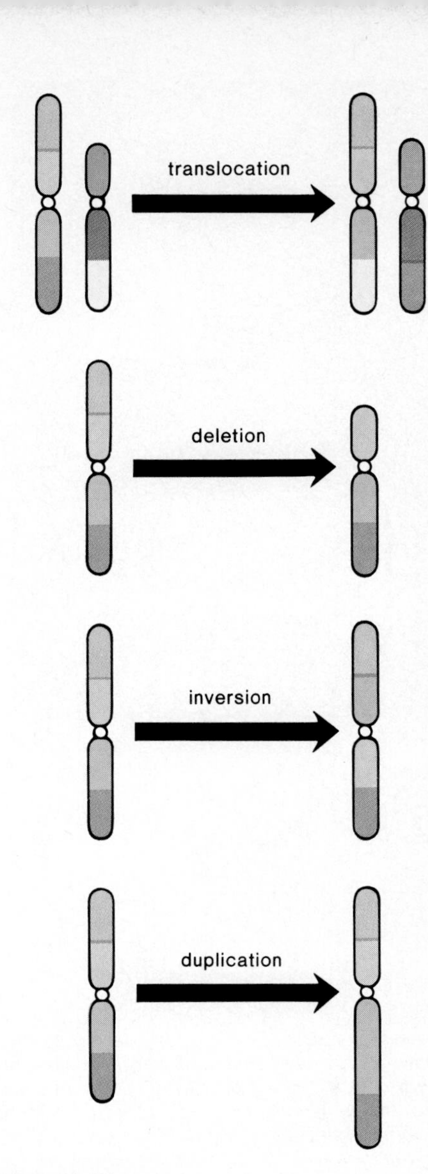

Figure 21.15
Types of chromosome mutations. a. *Translocation is the exchange of chromosome pieces between nonhomologous pairs.* b. *Deletion is the loss of a chromosome piece.* c. *Inversion is when a piece of chromosome breaks loose and then rejoins in the reversed direction.* d. *Duplication occurs when the same piece is repeated within the chromosome.*

Human Genetic Disorders

Abnormal Chromosome Inheritance

Abnormal Autosomal Chromosome Inheritance

Sometimes individuals are born with either too many or too few autosomal chromosomes due most likely to nondisjunction of chromosomes or sister chromatids during meiosis (fig. 21.14). It is possible also that even though there is the correct number of chromosomes, one chromosome may be defective in some way because of a chromosomal mutation (fig. 21.15). Chromosomal mutations are known to occur after chromosomes are broken due to exposure to radiation, addictive drugs, or pesticides, for example. When the chromosomes reform, the pieces may be rearranged. An *inversion* results when a piece of a chromosome has turned in the opposite direction. A *deletion* occurs when a piece of a chromosome is lost and the chromosome is shorter. In contrast, a *duplication* is the presence of a chromosome piece more than once in the same chromosome. Deletion and duplication are apt to occur between homologous chromosomes. When nonhomologous chromosomes exchange pieces a *translocation* has taken place. The presence of a mutated chromosome can cause the individual to have reproductive problems because the abnormal chromosome can result in nonviable zygotes or a child with birth defects.

a.

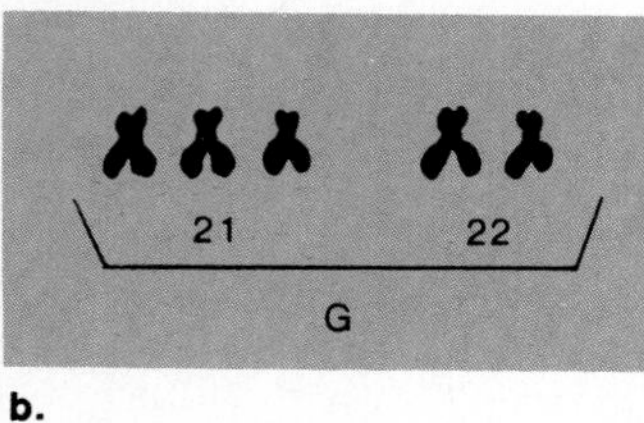

b.

Figure 21.16
a. *Down's syndrome. Common characteristics include a wide, rounded face and a fold of the upper eyelids that roughly resembles that of the Mongolian races. Mental retardation, along with an enlarged tongue, makes it difficult for persons with Down's syndrome to learn to speak coherently.*
b. *Karyotype of individual with Down's syndrome has an extra 21st chromosome in the G set.*

The most common autosomal abnormality is seen in individuals with **Down's syndrome** (fig. 21.16), sometimes called mongolism because the eyes of the person seem to have an oriental-like fold, but this term is not considered scientific. Other characteristics are short stature; stubby fingers; a wide gap between the first and second toes; a large, fissured tongue; a round head; a palm crease, the so-called simian line; and, unfortunately, mental retardation that can sometimes be severe.

Persons with Down's syndrome usually have three number 21 chromosomes because the egg had two number 21 chromosomes instead of one (fig. 21.14). (In 23 percent of the cases studied, however, the sperm had the extra 21st chromosome.) It would appear that **nondisjunction** is most apt to occur in the older female since children with Down's syndrome are usually born to women over age forty. If a woman wishes to know whether or not her unborn child is affected by Down's syndrome, she may elect to undergo chorionic villi testing or amniocentesis (p. 437). Following this procedure, a karyotype can reveal whether the child has Down's syndrome. If so, she may elect to continue or to abort the pregnancy.

A chromosomal deletion is responsible for a syndrome known as *cri du chat* (cat's cry). Affected individuals meow like a kitten when they cry, but more important perhaps is the fact that they tend to have a small head with malformations of the face and body, and mental defectiveness that usually causes retarded development. Chromosomal analysis shows that a portion of chromosome number 5 is missing (deleted), while the other number 5 chromosome is normal.

Down's syndrome is most often due to the inheritance of an extra 21st chromosome and cri du chat syndrome is due to the inheritance of a defective 5th chromosome.

***Table 21.4** Incidence of Selected Chromosomal Abnormalities*

Name	Frequency/100,000 Live Births
Down's syndrome general	140
Down's syndrome (Mothers over 40)	1,000
Turner's syndrome	8
Superfemale	50
Klinefelter's syndrome	80
XYY	100

Abnormal Sexual Chromosome Inheritance

Due to nondisjunction of the sex chromosomes during oogenesis, an egg may be produced that has either no X chromosome or two X chromosomes. When the first of these is fertilized by X-bearing sperm, a female with **Turner's syndrome** may be born (fig. 21.17b). These XO individuals have only one sex chromosome, an X; the O signifies the absence of the second sex chromosome. Because the ovaries never become functional these females do not undergo puberty or menstruate, and there is a lack of breast development. Generally, these individuals have a stocky build and a webbed neck. They also have difficulty recognizing various spatial patterns.

When an egg having two X chromosomes is fertilized by an X-bearing sperm, a **superfemale** having three X chromosomes results. While it might be supposed that the XXX female with 47 chromosomes would be especially feminine, this is not the case. Although there is a tendency toward learning disabilities, most superfemales have no apparent physical abnormalities and many are fertile and have children with a normal chromosome count.

When an egg having two X chromosomes is fertilized by a Y-bearing sperm, a male with **Klinefelter's syndrome** results. This individual is male in general appearance, but the testes are underdeveloped and the breasts may be enlarged (fig. 21.17a). The limbs of these XXY males tend to be longer than average, body hair is sparse, and many have learning disabilities.

XYY males also occur possibly due to nondisjunction during spermatogenesis. Afflicted males are usually taller than average, suffer from persistent acne, and tend to have barely normal intelligence. At one time it was suggested that these men were likely to be criminally aggressive, but it has been shown that the incidence of such behavior is no greater than that among normal XY males.

From an examination of these abnormal sex chromosome constituencies, it can be deduced that at least one X chromosome is required for human survival. There are no YO males. Also, the presence of a Y signifies a male regardless of the number of X chromosomes.

Individuals are sometimes born with the sex chromosomes XO (Turner's syndrome), XXX (superfemale), XXY (Klinefelter's syndrome), and XYY. Individuals with a Y are always male no matter how many X chromosomes there may be; however, at least one X chromosome is needed for survival.

Autosomal Genetic Diseases

Dominant Genetic Diseases

We have already mentioned nail-patella syndrome and the reading on page 466 describes *Marfan's syndrome,* which is recognized by skeletal, eye, and cardiovascular defects. All of these are due to the inability to produce normal connective tissue. Abraham Lincoln is believed to have suffered from this disease, caused by a dominant allele. More common genetic diseases due to the inheritance of a single dominant allele are the following:

Achondroplasia: a form of dwarfism
Chronic simple glaucoma (some forms): a major cause of blindness if untreated
Huntington's chorea: progressive nervous system degeneration
Hypercholesterolemia: high blood cholesterol levels, propensity to heart disease
Polydactyly: extra fingers or toes[3]

Most often one parent is heterozygous for the characteristic and the other is recessive and the child therefore has a 50:50 chance of receiving the characteristic or escaping it (fig. 21.18).

Notice that since these characteristics are dominant, the heterozygous parent must necessarily show it. But in the case of *Huntington's chorea* (Huntington's disease) the characteristic does not appear until the thirties or early forties. There is progressive deterioration of the individual's nervous system that eventually leads to constant thrashing and writhing movements until insanity precedes death. A screening test, discussed in the next chapter, p. 493, has now been developed for Huntington's disease, and so it is possible to know early in life whether one has the gene or not.

[3]National Foundation/March of Dimes.

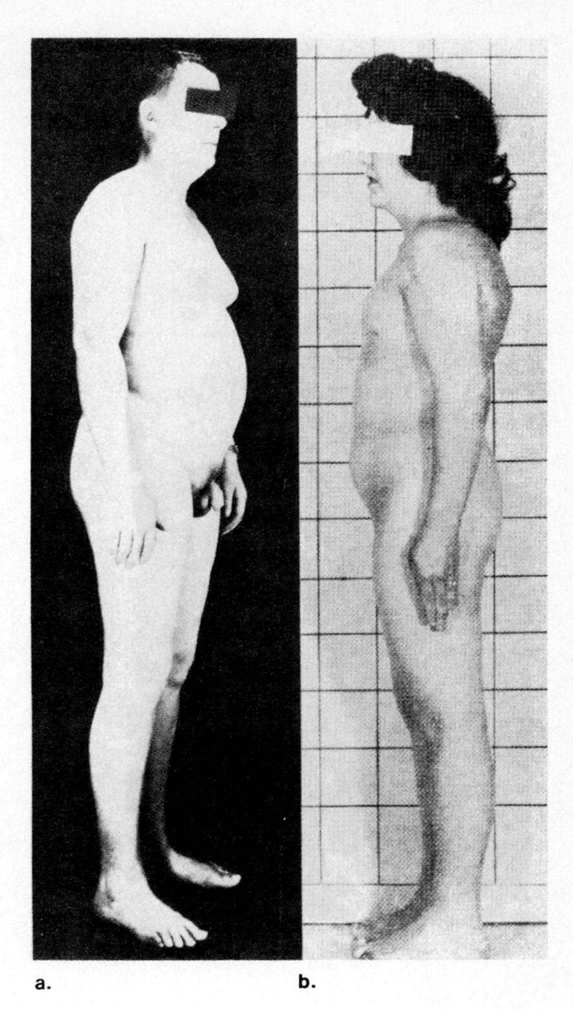

Figure 21.17
Abnormal sex chromosome inheritance. a. *A male with Klinefelter's (XXY) syndrome which is marked by immature sex organs and development of the breasts.* b. *Female with Turner's (XO) syndrome which includes a bull neck, short stature, and immature sexual features.*

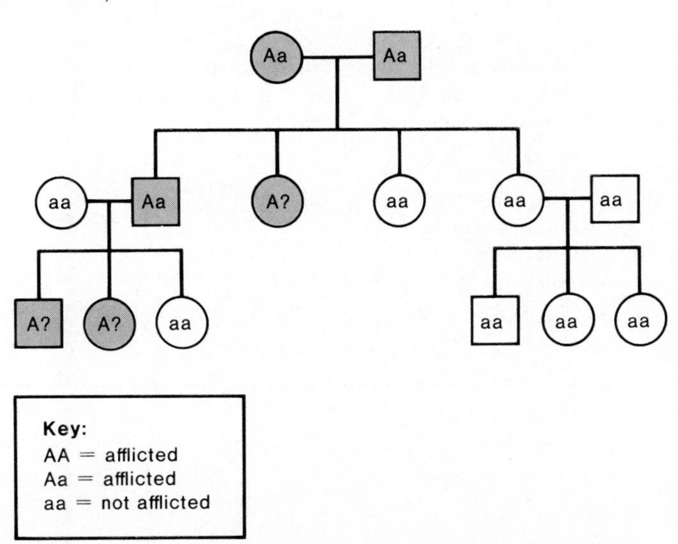

Figure 21.18
Pedigree chart for a dominant characteristic. In a pedigree chart, males are designated by squares and females by circles; shaded squares and circles are afflicted individuals. A line between a square and a circle represents a marriage. A vertical line going downward leads to the children, who are placed off a horizontal line. Notice that when the affliction is dominant, a child without the affliction can be born to parents that have the affliction.

Abe's Malady

The blurred left foot may be evidence that our sixteenth president suffered from Marfan's syndrome.

To Dr. Harold Schwartz, the signs left little doubt. The seven-year-old boy visiting his Huntington Park, Calif., office in 1959 had Marfan's syndrome, a genetic disorder of the connective tissue that can cause heart and eye problems, affect skeletal growth and occasionally be fatal. A few months later, the boy's grandmother dropped in to inquire about his condition and revealed that her husband had died of Marfan's. The grandmother's married name was Lincoln.

Says Schwartz: "I call that my 'burning bush' moment. I had read Carl Sandburg's biography of Abraham Lincoln, which contains a great deal about Lincoln's physical characteristics." Suddenly everything connected. The Great Emancipator, Schwartz realized, was probably afflicted by Marfan's syndrome.

Since then, Schwartz, now 60, has traced the Lincoln Marfan gene back to 16th century England and now is more certain than ever about his theory. In the *Western Journal of Medicine,* he strongly suggests that had John Wilkes Booth not fired the fatal shot on April 14, 1865, Lincoln would have died within a year from complications of Marfan's syndrome—for which there is still no cure.

Schwartz points to the well-documented fact that Lincoln had disproportionately long arms, legs, hands and feet, even for a man of his height. While watching a regiment of Maine lumbermen during the Civil War, the President himself noted: "I don't believe that there is a man in that regiment with longer arms than mine." In 1907 a sculptor working with Lincoln casts observed that "the first phalanx of the middle finger is nearly half an inch longer than that of an ordinary hand." The President sometimes squinted with his left eye. All of these characteristics, according to Schwartz, are typical of Marfan's syndrome. In fact, Lincoln's "spider-like legs," a phrase used by one of the President's contemporaries, was the very simile used in 1896 by French Physician Bernard-Jean Antonin Marfan when he described the syndrome that was named for him.

Schwartz has also presented an ingenious bit of evidence that Lincoln had a specific cardiovascular problem also associated with Marfan's syndrome: imperfect closure of the valves of the aorta, the large artery that carries blood from the heart. The clue appeared in a picture of the President taken in 1863. Lincoln had his legs crossed, and in an otherwise sharp photo, the left foot—suspended in the air—is blurred. When viewing the print, Lincoln asked why the foot was fuzzy. A friend familiar with physiology suggested that the throbbing arteries in the leg might have caused some movement. Lincoln promptly crossed his legs and watched. "That's it!" he exclaimed. "Now that's very curious, isn't it?" Not to Schwartz. The Marfan-caused defect, he points out, results in "aortic regurgitation," which causes pulses of blood strong enough to shake the lower leg.

Schwartz has also found in the President's own words what he believes to be good evidence that before Lincoln was shot he was "in a state of early congestive heart failure"—brought on by his aortic condition. About seven weeks before Lincoln's assassination, for example, he told his friend Joshua Speed: "My feet and hands of late seem to be always cold, and I ought perhaps to be in bed." Though he was only 56 in 1865, Abe was also easily fatigued toward the end. "There is only one word that can express my condition," he said, "and that is 'flabbiness.' " Once, shortly before his death, he tried to get out of bed but fell back, too weak to rise. Only a day before Lincoln was shot, his wife Mary wrote of the President's "severe headache" and indisposition. Concludes Schwartz: the faulty aortic valves resulted in "a decompensating left ventricle which was the undiagnosed or concealed cause of the President's failing health."

Recessive Genetic Diseases

Recessive conditions are not expressed unless the individual inherits a pair of alleles for the characteristic. *Albinism* is the inability to produce melanin and the individual lacks pigment in all parts of the body (fig. 21.19). More common are the following, which are also controlled by one pair of alleles:

Cystic fibrosis: disorder affecting function of mucous and sweat glands
Galactosemia: inability to metabolize milk sugar
Phenylketonuria (PKU): essential liver enzyme deficiency
Thalassemia (Cooley's anemia): blood disorder primarily affecting persons of Mediterranean ancestry
Tay-Sachs disease: fatal brain damage primarily affecting infants of East European Jewish ancestry[4]

In inheritance of these conditions, both parents may appear to be normal but are carriers of the defective allele. A carrier is a heterozygous person who does not show the characteristic but who can pass on the allele to an offspring. If both parents carry the allele, each child will run a 25 percent risk of manifesting the disease. Each child has a 25 percent chance of being homozygous normal but a 50:50 chance of receiving a single defective allele and being a carrier (fig. 21.20).

Cystic fibrosis, characterized by abnormal mucous-secreting tissues, is now one of the most commonly inherited disorders among Caucasian children. At first the infant may have difficulty regaining the birth weight despite good appetite and vigor. A cough associated with a rapid respiratory rate but no fever indicates lung involvement. Large, frequent, and foul-smelling stools are due to abnormal pancreatic secretions. As discussed on page 72, there is reason to believe that all these problems are due to defective cell membrane permeability. These children previously died in infancy due to infections, but now they often survive because of antibiotic therapy.

Figure 21.19
Albinism. Albinos are unable to produce the pigment melanin and therefore any genes received for coloring cannot be expressed. This is rock singer Johnny Winter who is an albino.

[4]National Foundation/March of Dimes.

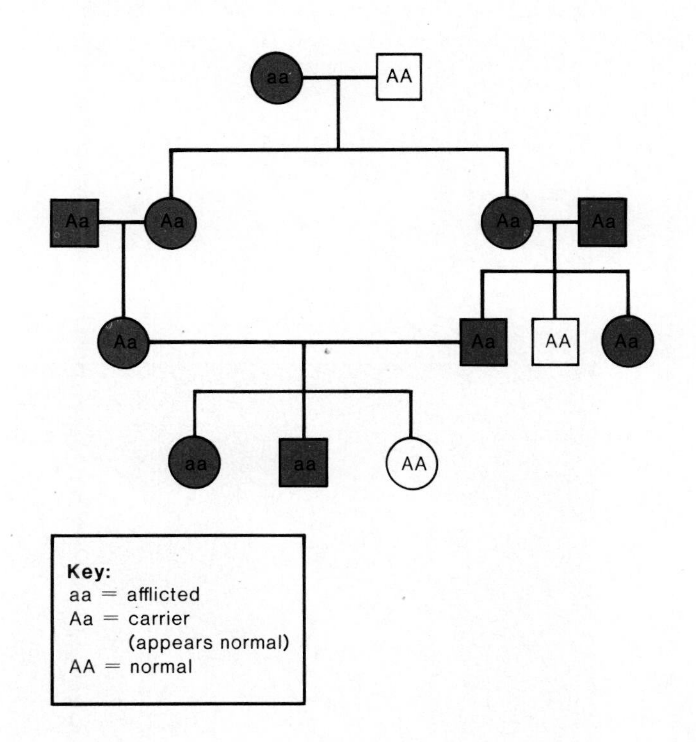

Figure 21.20
Pedigree chart for a recessive characteristic in which afflicted individuals are shaded more heavily than carriers. Notice that an afflicted child can be born to parents (carriers) that are not afflicted. A marriage between cousins can bring out recessive characteristics.

Phenylketonuria (PKU) is characterized by severe mental retardation due to an abnormal accumulation of the common amino acid phenylalanine within cells, including neurons. The disorder takes its name from the presence of a breakdown product, phenylketone, in the urine and blood. Newborn babies are routinely tested at the hospital and, if necessary, are placed on a diet low in phenylalanine. This diet improves the mental capabilities of affected individuals.

Tay-Sachs disease is caused by the inability to break down a certain type of fat molecule that accumulates in nerve cells until they are destroyed. Afflicted newborns appear normal and healthy at birth, but they do not develop normally. At first, they may learn to sit up and stand, but later they regress and become mentally retarded, blind, and paralyzed. Death usually occurs between ages three and four.

Codominant

Sickle-cell anemia occurs when the individual has the genotype Hb^SHb^S. In these individuals the red blood cells are sickle-shaped (fig. 21.21) because the abnormal hemoglobin molecule is less soluble than the normal hemoglobin, Hb^A. Sickle-shaped cells have a limited ability to transport oxygen. Inheritance involves codominance in the following manner: Individuals with the genotype Hb^AHb^A are normal; those with Hb^SHb^S have sickle-cell anemia; and those with Hb^AHb^S have *sickle-cell trait,* a condition in which the cells are sometimes sickle-shaped, as described in the paragraphs that follow. Two individuals with sickle-cell trait can produce children with all three phenotypes, as indicated in figure 21.21.

Sickle-cell anemia is prevalent among members of the black race because the shape of the cells seems to give protection against the malaria parasite, which utilizes red cells during its life cycle. Although infants with sickle-cell anemia often die, those with the trait are protected from malaria, especially during ages two to four. This means that in Africa these children survived and grew up to reproduce and pass on the allele to their offspring. As many as 60 percent of some tribes in malaria-infected regions of Africa have the allele. In the United States, about 10 percent of the black population carry it.

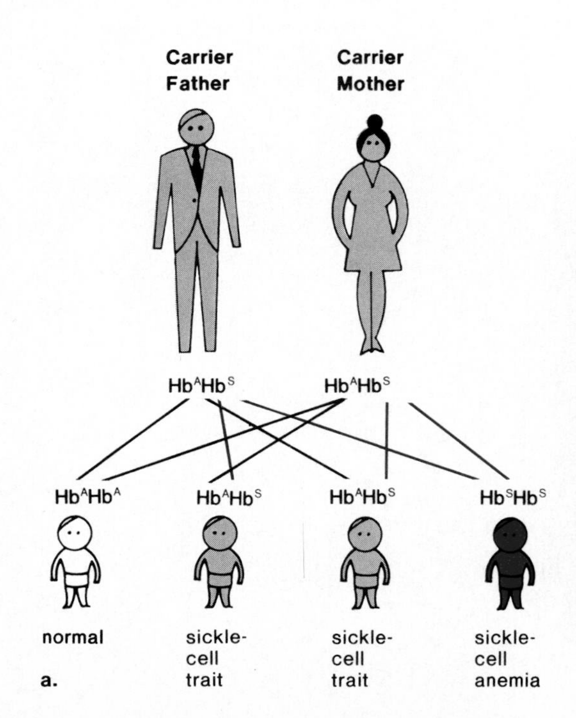

Figure 21.21
a. *Inheritance of sickle-cell anemia. In this example, both parents have sickle-cell trait and are therefore carriers. Therefore, each child has a 25 percent chance of having sickle-cell anemia or of being perfectly normal and a 50 percent chance of having sickle-cell trait. Courtesy National March of Dimes.*
b. *Sickled cells. Individuals with sickle-cell anemia have sickled red blood cells that tend to clump as illustrated here.*

The blood cells in persons with sickle-cell anemia cannot easily pass along small blood vessels. The sickle-shaped cells either break down or they clog blood vessels. Thus the individual suffers from poor circulation, anemia, and sometimes internal hemorrhaging. Jaundice, episodic pain of the abdomen and joints, poor resistance to infection, and damage to internal organs are all symptoms of sickle-cell anemia. Few patients live beyond age 40.

Persons with the sickle-cell trait do not usually have any difficulties unless they are exposed to air that is low in oxygen. At such times, the cells become sickle-shaped, with accompanying disturbances in circulation.

Polygenic Genetic Diseases

A number of serious genetic diseases, such as cleft lip or palate (fig. 21.22), club foot, congenital dislocation of the hip, and certain spine conditions, occur only when several different genes interact. Because a combination of genes brings about these conditions, it is difficult to predict the chances that any couple will have such a child.

However, if a couple is concerned about the birth of a child with a neural tube defect, an analysis of the amniotic fluid, following amniocentesis (p. 437) can reveal if there has been a leakage of neural tube substance into the fluid. If such a leakage has taken place, then it is known that the unborn child is not developing normally.

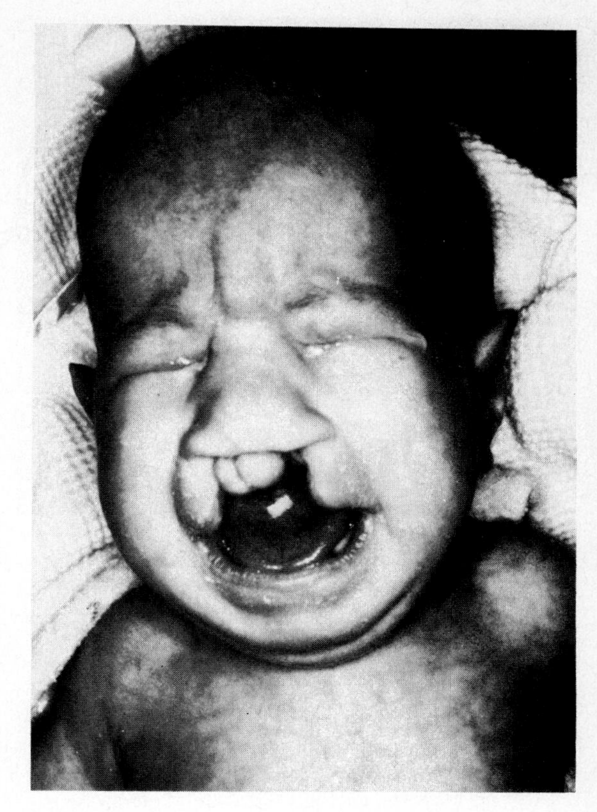

Figure 21.22
Child with cleft palate and harelip.

X-Linked Genetic Diseases

Among 150 human disorders transmitted by a gene or genes on the X chromosome are the following:[5]

Agammaglobulinemia: lack of immunity to infections
Color blindness: inability to distinguish certain colors
Hemophilia: defect in blood-clotting mechanisms
Muscular dystrophy (some forms): progressive wasting of muscles
Spinal ataxia (some forms): spinal cord degeneration

X-linked recessive genetic diseases are more common in males than females. Since a male receives a Y from his father, the recessive X-linked gene was received from his mother. Such genes often pass from a maternal grandfather to a grandson by way of a carrier female (fig. 21.23).

[5]National Foundation/March of Dimes.

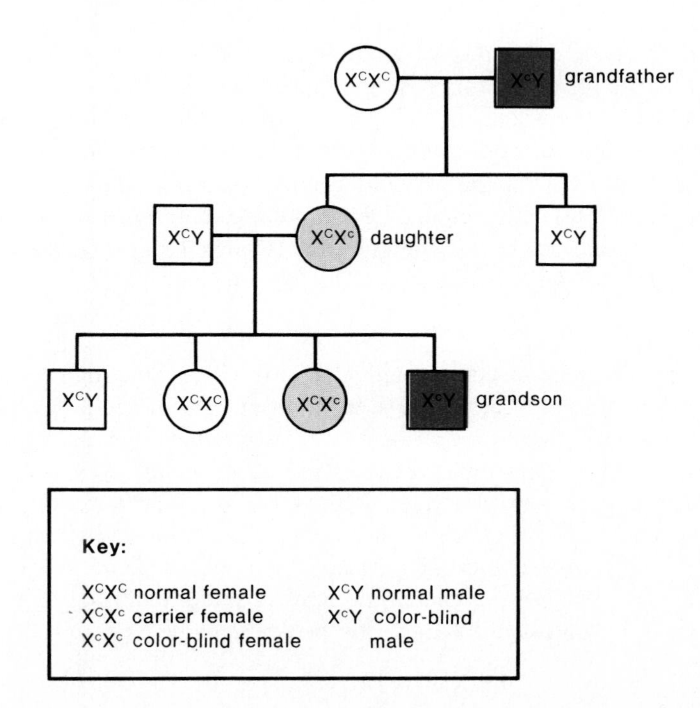

Figure 21.23
Pedigree chart for an X-linked recessive characteristic in which afflicted individuals are shaded more heavily than carriers. Notice that X-linked recessive traits often skip a generation and pass from grandfather to grandson by way of a carrier female.

Figure 21.24
The boy pictured here lived in a germ-free bubble for 12 years because his body failed to produce antibodies to fight infections. It was hoped that a bone marrow transplant would give him the capability to produce antibodies so that he could leave his bubble home. A bone marrow transplant was given by his sister but unfortunately it was infected with a virus that caused him to develop cancer. The virus hadn't caused this difficulty in his sister because she already had a functioning immune system when exposed.

The most common type of *hemophilia,* hemophilia A, is due to the absence, or minimal presence, of a particular clotting factor called Factor VIII. Hemophilia is called the bleeder's disease because the afflicted person's blood is unable to clot. Although hemophiliacs do bleed externally after an injury, they also suffer from internal bleeding, particularly around joints. Years ago, a hemophiliac received blood plasma to stop the bleeding. Today, concentrated Factor VIII is available and may be self-injected to stop the bleeding.

Muscular dystrophy, as the name implies, is characterized by the wasting away of muscles. The most common form, Duchenne type, is X linked; other types are not. Symptoms such as waddling gait, toe-walking, frequent falls, and difficulty in rising may appear as soon as the child starts to walk. Muscle weakness intensifies until the individual is confined to a wheelchair. Death usually occurs during the teenage years. Recently the gene for muscular dystrophy has been isolated by a technique discussed in the next chapter. It is hoped that this breakthrough will lead to a cure for the disease.

Persons with the genetic disease *agammaglobulinemia* have no functional T nor B lymphocytes and, therefore, they have no immunity. Because of this they are susceptible to repeated infections and an early death. Their only hope is to receive a bone marrow transplant from a close relative. The bubble boy (fig. 21.24) lived for twelve years in a protective bubble until the development of a new technique allowed him to receive a bone marrow transplant from his sister, whose tissue was previously not considered a close enough match. Unfortunately, the bone marrow transplant carried a latent virus (p. 540) that became active in his body and caused him to contract a cancer that ran rampant throughout his body in a short time.

Genetic Counseling

Now that persons are becoming aware that many illnesses are caused by faulty genes, more couples are seeking genetic counseling. The counselor studies the background of the couple and tries to determine if any immediate ancestor may have had a genetic disease. Then the counselor studies the couple themselves. As much as possible, laboratory tests are performed on all persons involved.

Carrier tests are now available for a large number of potential genetic diseases. Blood tests can identify carriers of thalassemia and sickle-cell anemia. By measuring enzyme levels in blood, tears, or skin cells, carriers of enzyme defects can now be identified for some inborn metabolic errors, such as Tay-Sachs disease, in which a defect of a single enzyme interferes with a vital metabolic process. From this information, the counselor can sometimes predict the chances of the couple having a defective child.

Whenever the woman is pregnant, chorionic villi sampling can be done early and amniocentesis can be done later in the pregnancy. These procedures, which were discussed in the previous chapter (p. 437), allow the testing of embryonic and fetal cells, respectively, in order to determine if the child has a genetic disease. Most of the time the baby is normal, but should a defect be discovered the couple may decide to abort the pregnancy.

Pedigree Charts

Pedigree charts are often constructed to show the inheritance of a certain condition within a family. Such charts are a great help in deciding whether a phenotype is controlled by a dominant or recessive allele. For example, if only one parent shows the trait and yet all or several of the children show it, then it is most probably dominant (fig. 21.18). Or if two individuals do not show the characteristic but their children do, then the characteristic must be determined by a recessive allele (fig. 21.20). On the other hand, if the characteristic appears primarily in males and passes from grandfather to grandson, then it must be controlled by an X-linked recessive allele (fig. 21.23).

Summary

In keeping with Mendel's laws of inheritance, it is customary to use letters to indicate the genotype and gametes of individuals. Homozygous dominant is indicated by two capital letters, and homozygous recessive is indicated by two lowercase letters. Heterozygous is indicated by a capital and lowercase letter. Contrary to the individual, gametes have one letter of each type, either capital or lowercase as appropriate. All possible combinations of letters can occur in the gametes (except if the genes are linked). In doing an actual cross, it is assumed that all possible types of sperm fertilize all possible types of eggs. The results of some crosses may be determined by simple inspection, but certain others that commonly recur are given in table 21.2.

There are many exceptions to Mendel's laws and these include polygenic inheritance (skin color), multiple alleles (ABO blood type), and incomplete dominance (curly hair).

Genes that appear on the same chromosomes are linked to one another and their alleles tend to go into the same gamete together. Linkage reduces the number of different genotypes and phenotypes among the offspring. However, recombinant gametes do occur by the process of crossing over and the frequency of crossing over can be used to map chromosomes.

Some genes are sex-linked and most of these occur on the X chromosome while the Y is blank. More males than females have X-linked genetic disorders and they often skip a generation passing from grandfather to grandson. Studies of human genetics have shown that there are also many autosomal genetic diseases.

Objective Questions

1. Whereas an individual has two genes for every trait, the gametes have ______ gene for every trait.
2. The recessive allele for the dominant gene *W* is ____________ .
3. Mary has a widow's peak, and John has a continuous hairline. This would be a description of their ____________ .
4. *W* = widow's peak and *w* = continuous hairline; therefore, only the phenotype ____________ could be heterozygous.
5. Two heterozygotes, each having a widow's peak, already have a child with a continuous hairline. The next child has what chance of having a continuous hairline? ____________
6. In a testcross, an individual having the dominant phenotype is crossed with an individual having the ____________ phenotype.
7. How many letters are required to designate the genotype of a dihybrid individual? ____________
8. If a dihybrid is crossed with a dihybrid, how many offspring out of sixteen are expected to have the dominant phenotype for both traits? ____________
9. How many different phenotypes among the offspring are possible when a dihybrid is crossed with a dihybrid? ____________
10. According to Mendel's Law of Independent Assortment, a dihybrid can produce how many types of gametes having different combinations of genes? ____________
11. Do sex-linked genes determine the sex of the individual? ____________
12. If a male is color-blind, he inherited the allele for color blindness from his ____________ .
13. What is the genotype of a female who has a color-blind father but a homozygous normal mother? ________
14. In a pedigree chart, it is observed that although the children have a characteristic, neither parent has it. The characteristic must be inherited as a ____________ gene.

Answers to Objective Questions

1. one 2. *w* 3. phenotype 4. widow's peak 5. 25 percent 6. recessive 7. four 8. nine 9. four 10. four 11. no 12. mother 13. X^CX^c 14. recessive

Study Questions

1. Explain why there is a pair of alleles for every trait except for sex-linked traits in males. (p. 445)
2. Relate Mendel's laws of inheritance to one-trait and two-trait crosses. (pp. 450, 452)
3. What is the difference between genotype and phenotype? (p. 449)
4. What are the expected results from these crosses:
   heterozygous × heterozygous;
   heterozygous × recessive;
   dihybrid × dihybrid;
   dihybrid × double recessive? (p. 455)
5. What does the phrase "chance has no memory" mean? (p. 450)
6. Give four examples of exceptions to Mendel's laws. (pp. 456–460)
7. How would linkage (assuming dominant alleles on one chromosome and recessive on the other) affect the results of the last two crosses mentioned in question 4? (p. 458)
8. What is sex linkage? Give all possible genotypes for an X-linked trait and discuss each. (p. 461)
9. Name four types of chromosomal mutations and give examples. (p. 463)
10. What is nondisjunction and how does it occur? (p. 463) What is the most common autosomal chromosomal abnormality? (p. 464) Name four sex chromosomal abnormalities. (p. 464)
11. How could you determine whether a pedigree chart is depicting the inheritance of a dominant, recessive, and X-linked characteristic? (p. 470)

Additional Genetic Problems

1. A woman heterozygous for polydactyly (dominant) is married to a normal man. What are the chances that their children will have six fingers and toes? (p. 450)
2. John cannot curl his tongue (recessive) but both his parents can curl their tongues. Give the genotypes of all persons involved. (p. 450)
3. Parents who do not have Tay-Sachs (recessive) produce a child who has Tay-Sachs. What are the chances that each child born to this couple will have Tay-Sachs? (p. 450)
4. A man with widow's peak (dominant) who cannot curl his tongue (recessive) is married to a woman with a continuous hairline who can curl her tongue. They have a child who has a continuous hairline and cannot curl the tongue. Give the genotype of all persons involved. (p. 452)
5. Both Mr. and Mrs. Smith have freckles (dominant) and attached earlobes (recessive). Some of the children do not have freckles. What are the chances that the next child will have freckles and attached earlobes? (p. 452)
6. Mary has wavy hair (incomplete dominance) and marries a man with wavy hair. They have a child with straight hair. Give the genotype of all persons involved. (p. 457)
7. A man has type AB blood. What is his genotype? Could this man be the father of a child with type B blood? If so, what blood types could the child's mother have? (p. 457)
8. A woman with white skin has mulatto parents. If this woman married a light man, what is the darkest skin color possible for their children? the lightest? (p. 456)
9. What is the genotype of a man who is color-blind (X-linked recessive) and has a continuous hairline? If this man has children by a woman who is homozygous dominant for normal color vision and widow's peak, what will be the genotype and phenotype of the children? (p. 461)
10. Is the characteristic represented by the darkened individuals inherited as a dominant, recessive, or X-linked recessive? (pp. 456, 467, 469)

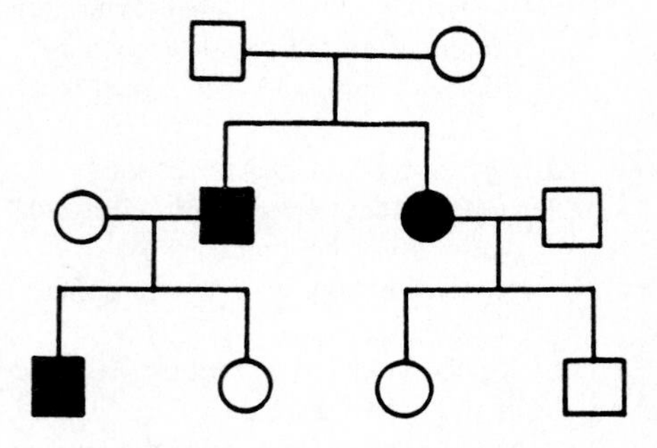

11. Fill in this pedigree chart to give the probable genotypes of the twins pictured in figure 21.9. (p. 456)

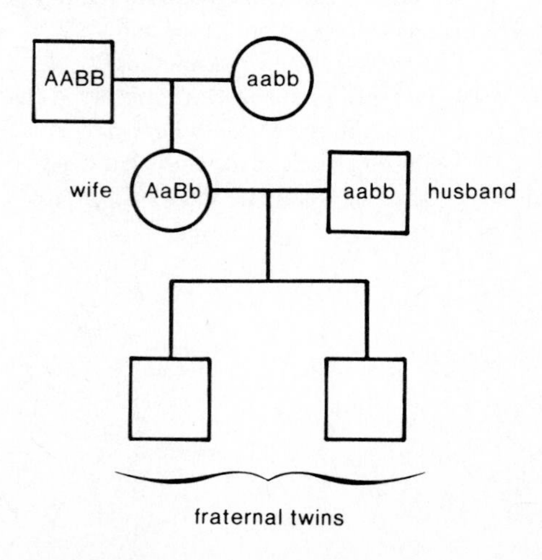

Answers to Additional Genetics Problems

1. 50%
2. John = tt; parents = Tt
3. 25%
4. man = $Wwtt$; woman = $wwTt$; child = $wwtt$
5. 75%
6. Mary and husband = HH'; child = $H'H'$
7. I^AI^B; yes; A, B, O, AB
8. light; white
9. X^cYww; girls = X^CX^cWw; boys = X^CYWw; both = normal vision and widow's peak
10. Autosomal recessive
11. AaBb and aabb

Answers to Practice Problems

Practice Problems 1

1. a. W b. WS, Ws c. T, t d. Tg, tg e. AB, Ab, aB, ab
2. a. gamete b. genotype c. gamete d. genotype

Practice Problems 2

1. 75%
2. No
3. Heterozygous
4. $DD \times dd$; Dd

Practice Problems 3

1. Dihybrid
2. 1/16
3. $DdFf \times ddff$; $ddff$

Practice Problems 4

1. $H'H'$; No
2. Light
3. White
4. Baby 1 = Doe; Baby 2 = Jones
5. AB, O, A, B
6. 3:1

Practice Problems 5

1. his mother; X^HX^h, X^HY, X^hY
2. 100%; None; 100%
3. $RrX^CX^c \times RrX^CY$; rrX^cY
4. The husband is not father.

Key Terms

allele (ah-lēl') an alternative form of a gene that occurs at a given chromosomal site (locus). *445*

dihybrid (di-hi'brid) the offspring of parents who differ in two ways: shows the phenotype governed by the dominant alleles but carries the recessive alleles. *452*

dominant (dom'ĭ-nant) hereditary factor that expresses itself or a characteristic that is present even when the genotype is heterozygous. *445*

Down's syndrome (downz sin'drōm) human congenital disorder associated with an extra 21st chromosome. *464*

genotype (jen'o-tīp) the genetic makeup of any individual. *449*

heterozygous (het"er-o-zi'gus) having two different alleles (as *Aa*) for a given trait. *449*

homozygous (ho"mo-zi'gus) having identical alleles (as *AA* or *aa*) for a given trait; pure breeding. *449*

Klinefelter's syndrome (klīn'fel-terz sin'drōm) a condition caused by the inheritance of a chromosome abnormality in number; an XXY individual. *464*

linkage (lingk'ij) alleles on the same chromosome are linked in the sense that they tend to move together to the same gamete; crossing over interferes with linkage. *458*

monohybrid (mon"o-hi'brid) the offspring of parents who differ in one way only; shows the phenotype of the dominant allele but carries the recessive allele. *450*

nondisjunction (non"dis-jungk'shun) the failure of homologous chromosomes or sister chromatids to separate during the formation of gametes. *464*

phenotype (fe'no-tīp) the outward appearance of an organism caused by the genotype and environmental influences. *449*

Punnett square (pun'et skwār) a gridlike device that enables one to calculate the results of simple genetic crosses by lining gametic genotypes of two parents on the outside margin and their recombination in boxes inside the grid. *450*

pure (pūr) *see* homozygous. *449*

recessive (re-ses'iv) hereditary factor that expresses itself or a characteristic that is present only when the genotype is homozygous. *445*

sex linked (seks lingkt) alleles located on sex chromosomes which determine traits unrelated to sex. *461*

superfemale (su"per-fe'māl) a female that has three X chromosomes. *464*

testcross (test kros) the crossing of a heterozygote with an organism homozygous recessive for the characteristic(s) in question in order to determine the genotype. *451*

trait (trāt) specific term for a distinguishing feature studied in heredity. *445*

Turner's syndrome (tur'nerz sin'drōm) a condition caused by the inheritance of an abnormality in chromosome number; an X chromosome lacks a homologous counterpart-XO. *464*

X-linked (eks lingkt) an allele located on X chromosome that determines a characteristic unrelated to sex. *461*

XYY male (eks wi wi māl) a male that has an extra Y chromosome. *464*

22

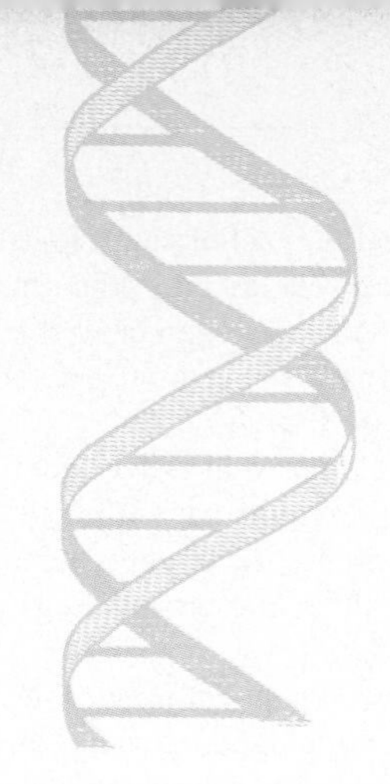

Molecular Basis of Inheritance

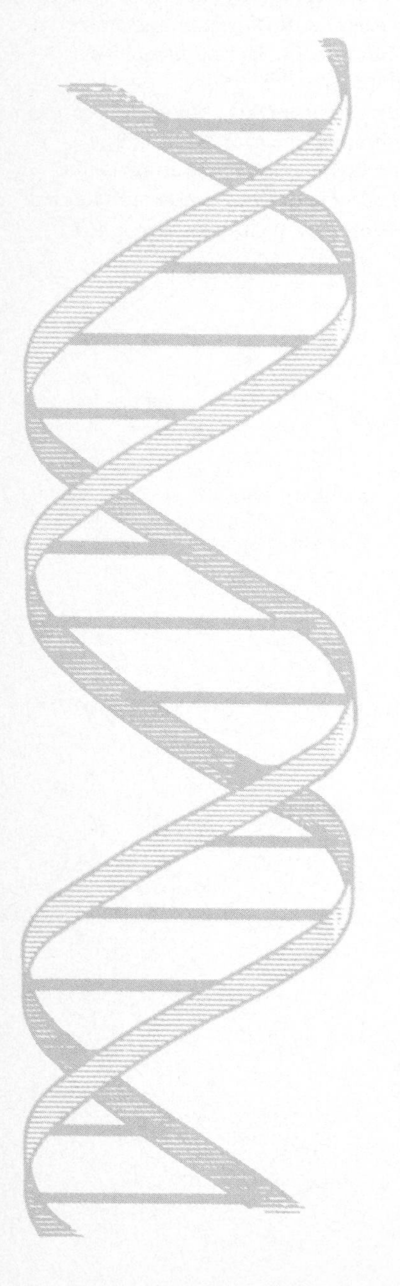

Chapter Concepts

1 DNA is the genetic material and therefore its structure and function constitute the molecular basis of inheritance.

2 DNA is able to replicate, mutate, and control the phenotype.

3 DNA controls the phenotype by controlling protein synthesis, a process that also requires the participation of RNA.

4 The manner in which gene action is regulated in eukaryotes is an area of important research today.

5 Genetic engineering, especially recombinant DNA research, is being used to help treat disorders including genetic diseases.

6 Mutations, accounting for the origin of genetic diseases and cancer, can affect both structural and regulatory genes.

Chapter Outline

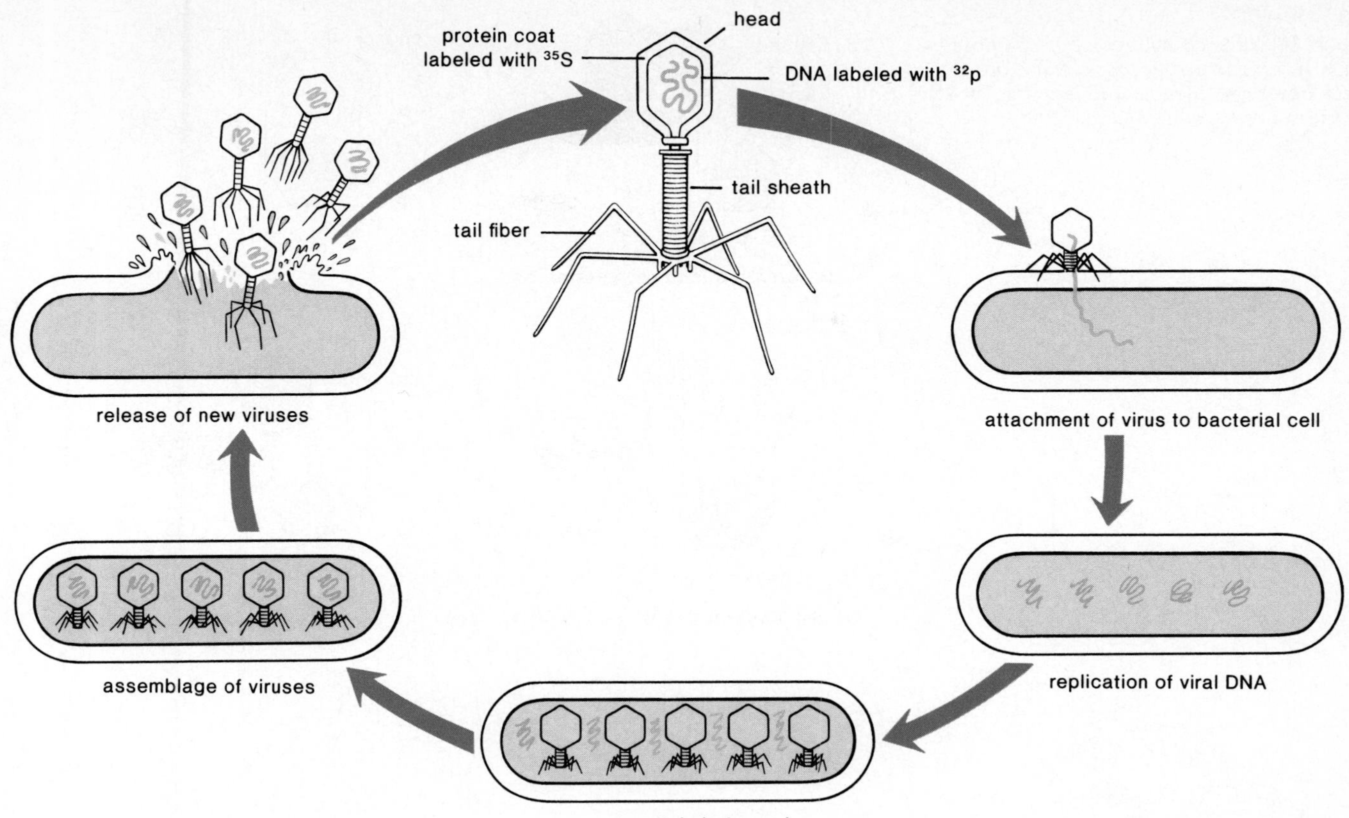

Figure 22.1
Life cycle of a T virus. A T virus is a complex virus with a head and a tail. Even so, it is composed of just a protein coat and inner core of DNA. Experimenters labeled the coat with ^{35}S and the DNA with ^{32}P and allowed the viruses to attack bacteria. Later, they found only ^{32}P in the cell and yet the cell produced many new viruses. From this they knew that DNA was the genetic material.

When the sperm fertilizes the egg, a new individual comes into being. Each of the gametes contributes genes that direct not only the development but the continued functioning of the individual from birth to death. In the previous chapter we saw how the inheritance of various genes affected the phenotype, but of what are the genes composed?

DNA

In the mid-1900s it was known that the genes are on the chromosomes and that the chromosomes contain both DNA and protein, but it was uncertain which of these was the genetic material. Scientists turned to experiments with viruses to determine which of these is the genetic material because they knew that viruses are tiny particles having just two parts: an outer coat of protein and an inner core of nucleic acid, most often DNA.

They chose to work with a virus called a T virus (the T simply means "type") that infects bacteria. They wished to determine which part of a T virus, the outer protein coat or the inner DNA core, enters a bacterium, taking over its machinery so that it produces more viruses. They began by culturing bacteria and viruses in radioactive sulfur and phosphorus until they had a batch with ^{35}S, labeled protein coats and ^{32}P, labeled DNA.[1] Then they allowed these viruses to attack new bacteria (fig. 22.1) and determined whether the labeled protein or the labeled DNA entered the cell. They found that only DNA enters the cell to take over the metabolism of the cell so that more viral particles are made. In other words, DNA is the genetic material.

[1]In actuality, two different labeled batches were needed, but for simplicity's sake they are described as one batch here.

Figure 22.2
Nucleotides in DNA. Each nucleotide is composed of phosphate, the sugar deoxyribose, and a base. a. *The purine bases are adenine and guanine.* b. *The pyrimidine bases are thymine and cytosine.*

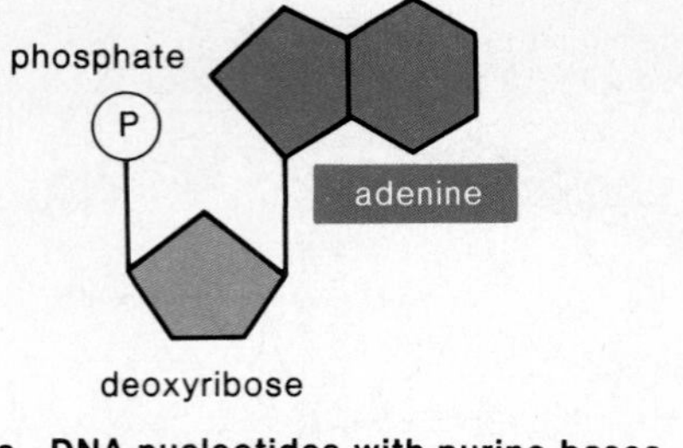

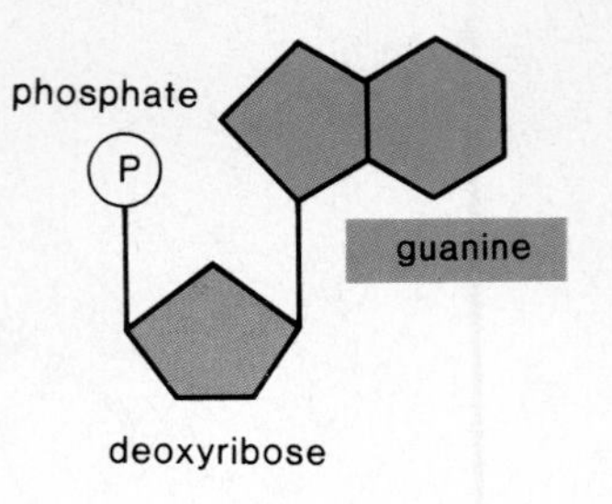

a. DNA nucleotides with purine bases

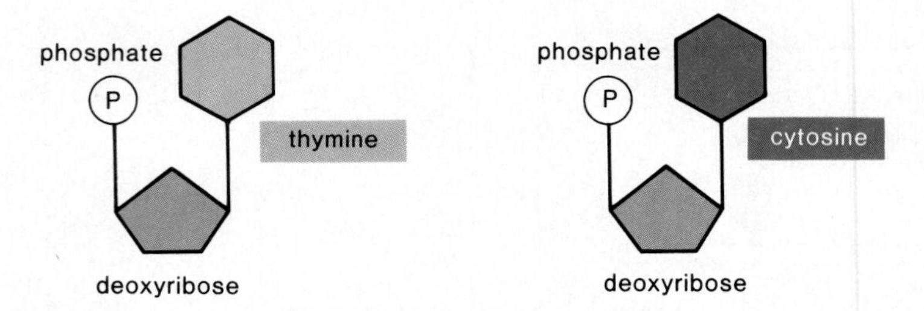

b. DNA nucleotides with pyrimidine bases

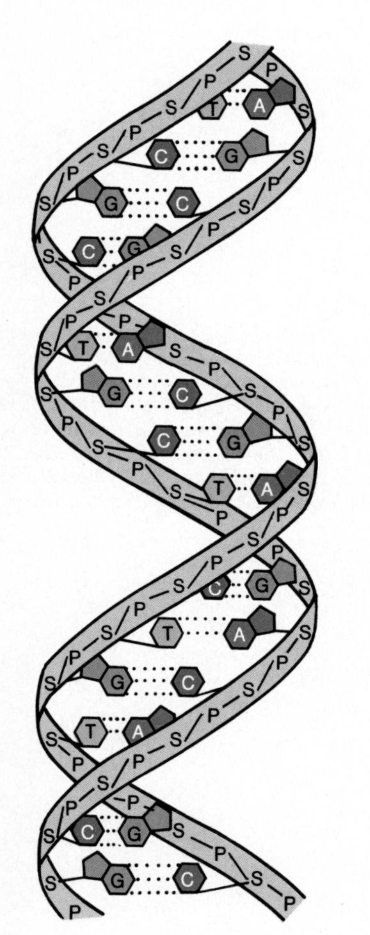

a.

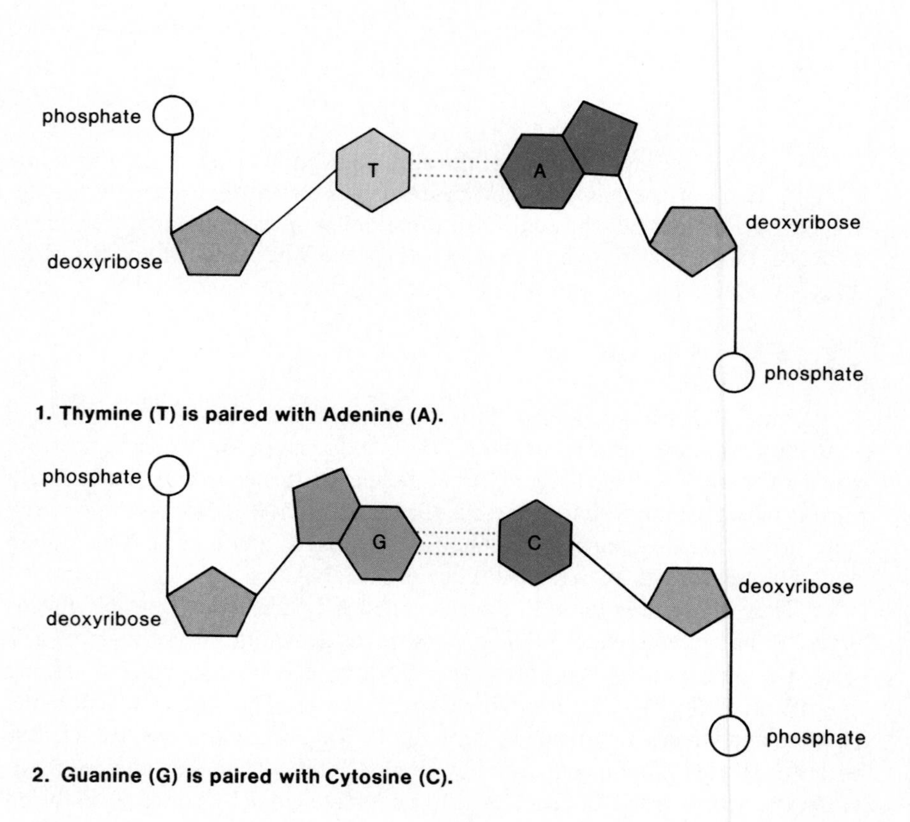

b. **2. Guanine (G) is paired with Cytosine (C).**

Figure 22.3
Structure of DNA. a. *DNA is a double helix. The backbone of each strand is composed of phosphate and sugar molecules, and the bases project to the side.* b. *Complementary pairing between the bases in which a purine is always paired with a pyrimidine. Specifically, (1) Thymine (T) is paired with Adenine (A) and (2) Guanine (G) is paired with Cytosine (C).*

Structure of DNA

DNA is a nucleic acid that contains multiple copies of just four nucleotides (fig. 22.2). Notice that each nucleotide is a complex of three united subunits: phosphoric acid (phosphate), a pentose sugar (deoxyribose), and a nitrogen base. The bases can be either the **purines,** adenine or guanine, which have a double ring, or the **pyrimidines,** thymine or cytosine, which have a single ring. These structures are called bases because they have basic characteristics that raise the pH of a solution.

When nucleotides join together, they form a polymer, or *strand,* in which the backbone is made up of phosphate-sugar-phosphate-sugar—with the bases to one side of the backbone (fig. 1.34). DNA contains two such strands and therefore it is *double stranded.* The two strands of DNA twist about one another in the form of a **double helix** (fig. 22.3a). The two strands are held together by hydrogen bonds between purine and pyrimidine bases. Thymine (T) is always paired with adenine (A), and guanine (G) is always paired with cytosine (C) (fig. 22.3b). This is called **complementary base pairing.**

If we unwind the DNA helix, it resembles a ladder (fig. 22.4). The sides of the ladder are made entirely of phosphate and sugar molecules, and the rungs of the ladder are made only of the complementary paired bases. The bases can be in any order but A is always paired with T and G is always paired with C, and vice versa. Therefore no matter what the order or the quantity of any particular base pair, the number of purine bases *always equals* the number of pyrimidine bases.

The structure of the DNA molecule was first determined by two young scientists, James Watson and Francis Crick. The data that was available to them and the way they used it to deduce DNA's structure is reviewed in the reading.

Figure 22.4
Overview of DNA structure. a. *The double helix structure is a* (b) *twisted ladder.* c. *DNA unwound shows that the sides of the ladder are composed of sugar and phosphate molecules and the rungs are complementary paired bases.*

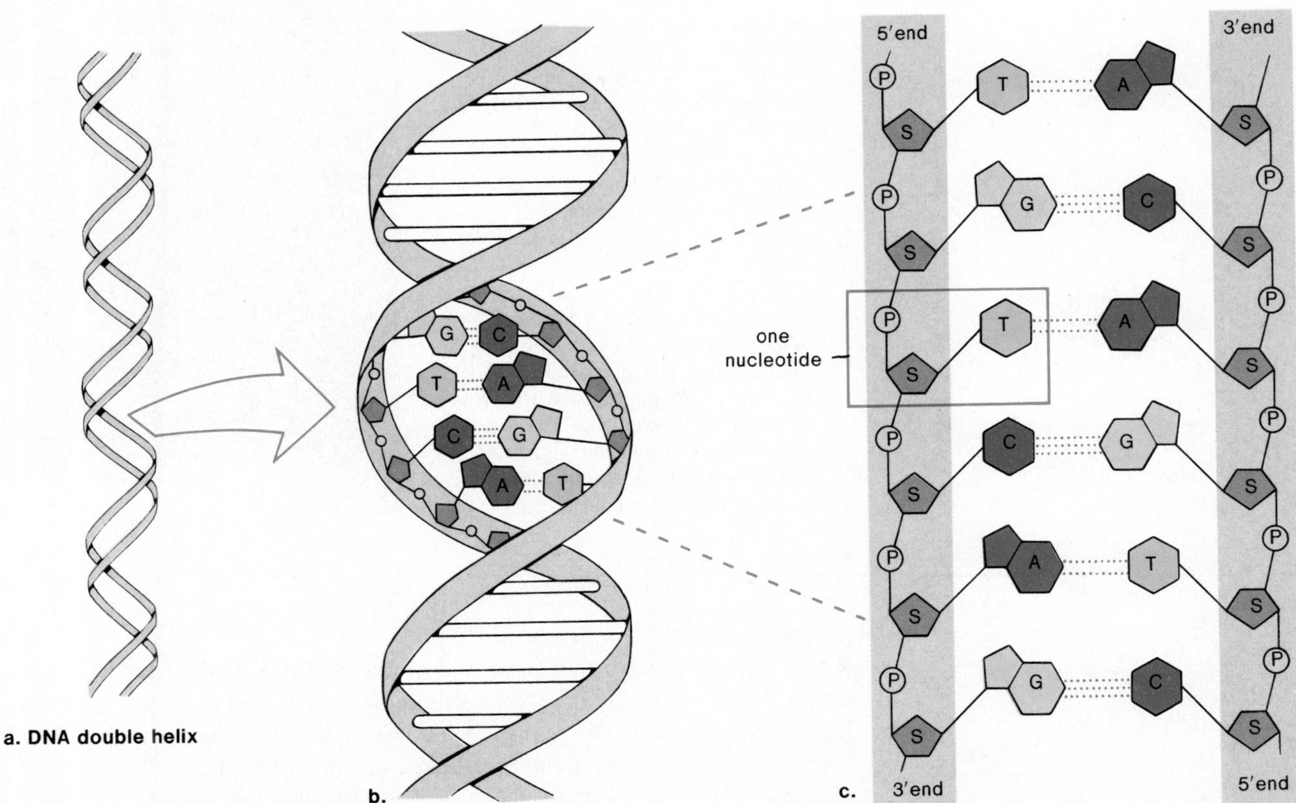

Solving the Puzzle

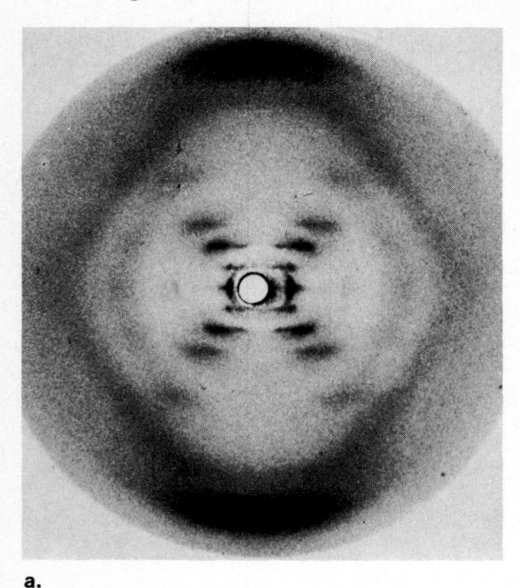

a.

a. *X-ray diffraction photograph of DNA taken by Rosalind Franklin. The crossing pattern of dark spots in the center of the picture indicated that DNA was helical. The dark regions at the top and bottom of the photograph showed that base pairs were stacked on top of one another.*
b. *A portion of the actual wire and tin model constructed by Watson and Crick.*

In 1951, James Watson, an American biologist, began an internship at the University of Cambridge, England. There he met Francis Crick, an English physicist, who was interested in molecular structures. They began to try to build a model that would show the molecular structure of DNA, the known hereditary material. They knew that their model should explain the manner in which DNA can vary from species to species and even from individual to individual, and also that it should show how it is possible for DNA to replicate (make a copy of itself) so that copies of the genetic material can be passed on to daughter cells and from the parent to the offspring.

Bits and pieces of data were available to Watson and Crick, and they undertook to solve the puzzle by putting the pieces together. This is what they knew from research done by others:

1. DNA is a polymer of nucleotides, each one having a phosphate group, the sugar deoxyribose, and a nitrogenous base. There are four different nucleotides that differ according to the base: adenine (A) and guanine (G) are purines while cytosine (C) and thymine (T) are pyrimidines.
2. A chemist, Erwin Chargaff, had determined in the late 1940s that, regardless of the species under consideration, the number of purines in DNA always equals the number of pyrimidines and that the amount of adenine equals the amount of thymine, that is, [A] = [T], and the amount of guanine equals the amount of cytosine, that is, [G] = [C]. These findings came to be known as *Chargaff's rules.*
3. Rosalind Franklin and Maurice Wilkins, working at King's College, London, had just prepared an X-ray diffraction photograph (fig. a.) of DNA. It showed that DNA is a double helix of constant diameter and that the bases are regularly stacked above one another.

Using this data, Watson and Crick deduced that DNA has a twisted ladder-type structure: the sugar-phosphate molecules make up the sides of the ladder and the bases make up the rungs of the ladder. Further, they determined that if A was always hydrogen-bonded with T and G was always hydrogen-bonded with C (in keeping with Chargaff's rules), then the rungs would always have a constant width (as required by the X-ray photograph).

Watson and Crick built an actual model of DNA out of wire and tin (fig. b.). This double-helix model does indeed allow for differences in DNA structure between species because, while A must always pair with T and G must always pair with C, there is no set order in the sequencing of these pairs. Also, the model provides a means by which DNA can replicate, as Watson and Crick alluded to in their original paper; "It has not escaped our notice that the specific pairing we have postulated immediately suggests a possible copying mechanism for the genetic material."

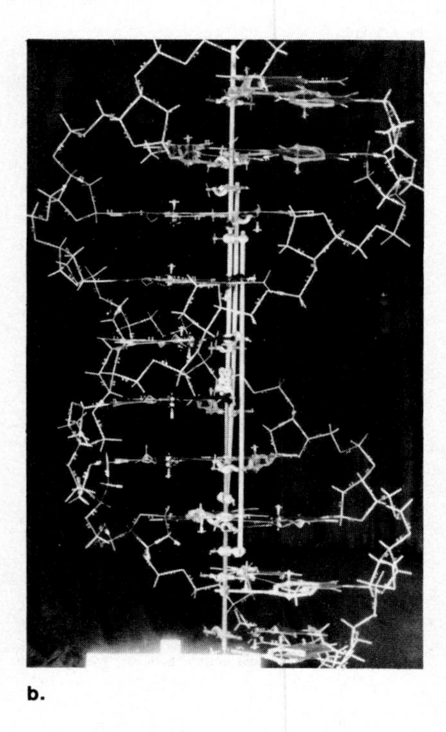

b.

Structure of DNA Compared to RNA

RNA is also a nucleic acid composed of multiple copies of nucleotides. However, the pentose sugar in RNA is ribose—not deoxyribose. Also, the pyrimidine thymine does not appear in RNA; it is replaced by the pyrimidine uracil. Therefore RNA contains the four bases, adenine (A), guanine (G), cytosine (C), and uracil (U). Finally RNA is *single stranded.* Figure 22.5 shows the structure of RNA and table 22.1 summarizes the differences between DNA and RNA structure.

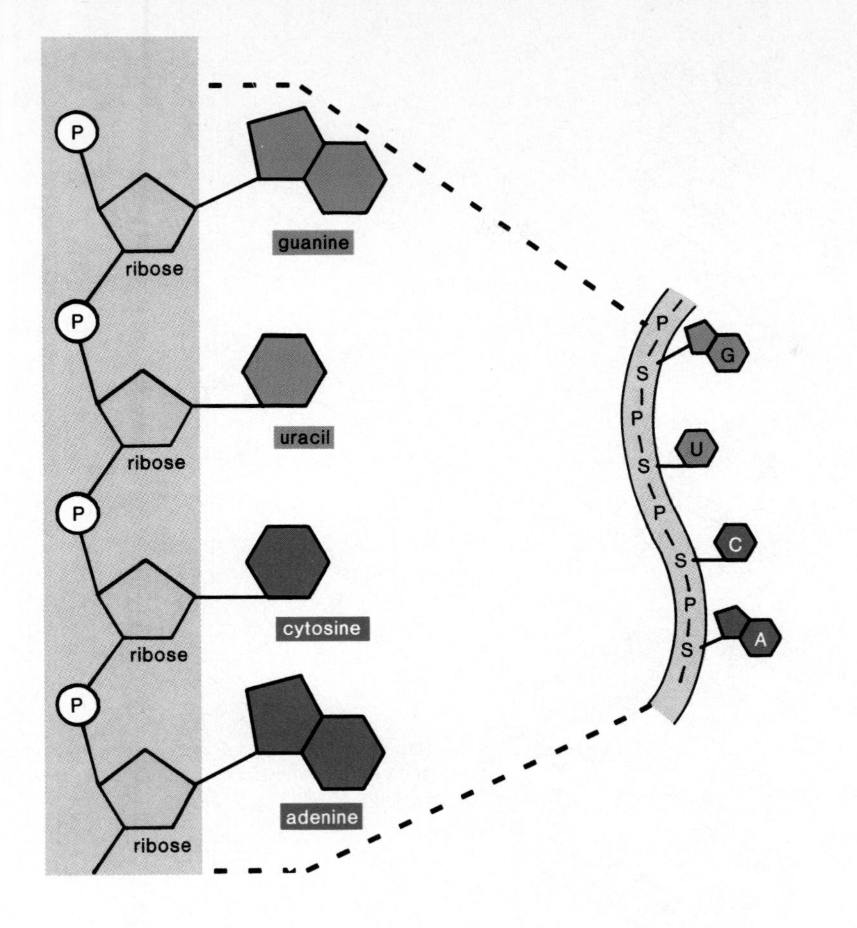

Figure 22.5
Structure of RNA. RNA is single stranded. The backbone contains the sugar ribose instead of deoxyribose. The bases are guanine, uracil, cytosine, and adenine.

DNA has a structure like a twisted ladder: sugar-phosphate backbones make up the sides of the ladder; hydrogen-bonded bases make up the rungs of the ladder. The base A is always paired with the base T, and the base C is always paired with the base G. RNA differs from DNA in several respects (table 22.1).

Table 22.1 *DNA Structure Compared to RNA Structure*

	DNA	RNA
Sugar	Deoxyribose	Ribose
Bases	Adenine, guanine, thymine, cytosine	Adenine, guanine, uracil, cytosine
Strands	Double stranded with base pairing	Single stranded
Helix	Yes	No

Functions of DNA

Any hereditary material will have at least three functions. The hereditary material must be able to:

1. replicate, make copies of itself, that may be passed on from cell to cell and from generation to generation;
2. control the activities of the cell; thereby producing the phenotypic characteristics of the individual and the species;
3. undergo *mutations*—permanent genetic changes passed on to the offspring—in order to account for the evolutionary history of life.

We now wish to explore the manner in which DNA carries out these functions.

Replication

The double-stranded structure of DNA lends itself to replication because each strand can serve as a template for the formation of a complementary strand. A **template** is most often a mold, used to produce a shape opposite to itself. In this case, the word *template* is appropriate because each new strand of DNA has a sequence of bases complementary to the bases of the old strand of DNA.

Figure 22.6
DNA replication. Replication is called semiconservative because each new double helix is composed of an old parental strand and a new daughter strand.

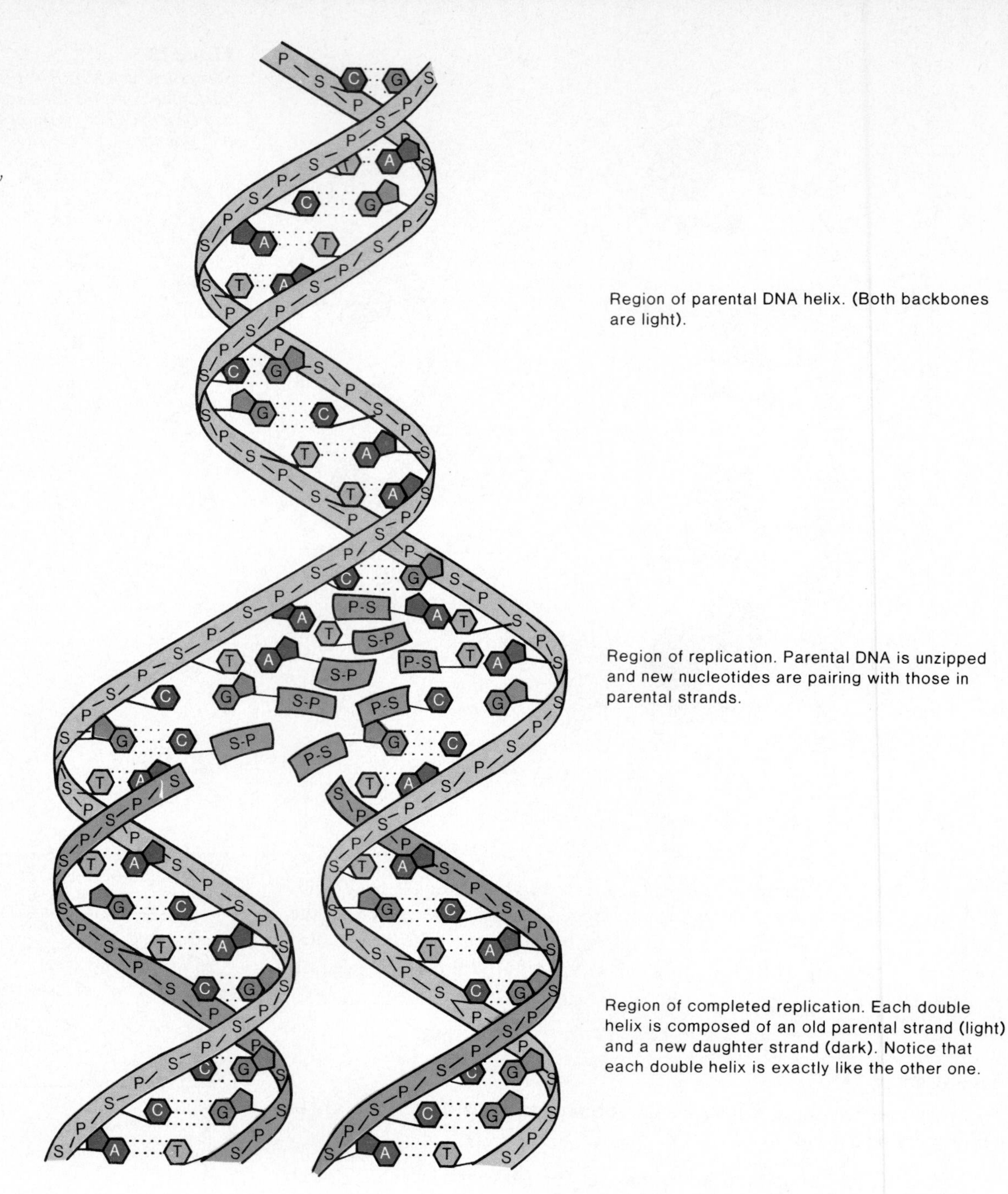

Replication requires the following steps (fig. 22.6):

1. The two strands that make up DNA become "unzipped" (i.e., the weak hydrogen bonds between the paired bases are broken).
2. New complementary nucleotides, always present in the nucleus, move into place by the process of complementary base pairing.
3. The adjacent nucleotides, through their sugar-phosphate components, become joined together along the newly forming chain.
4. When the process is finished, two complete DNA molecules are present, identical to each other and to the original molecule.

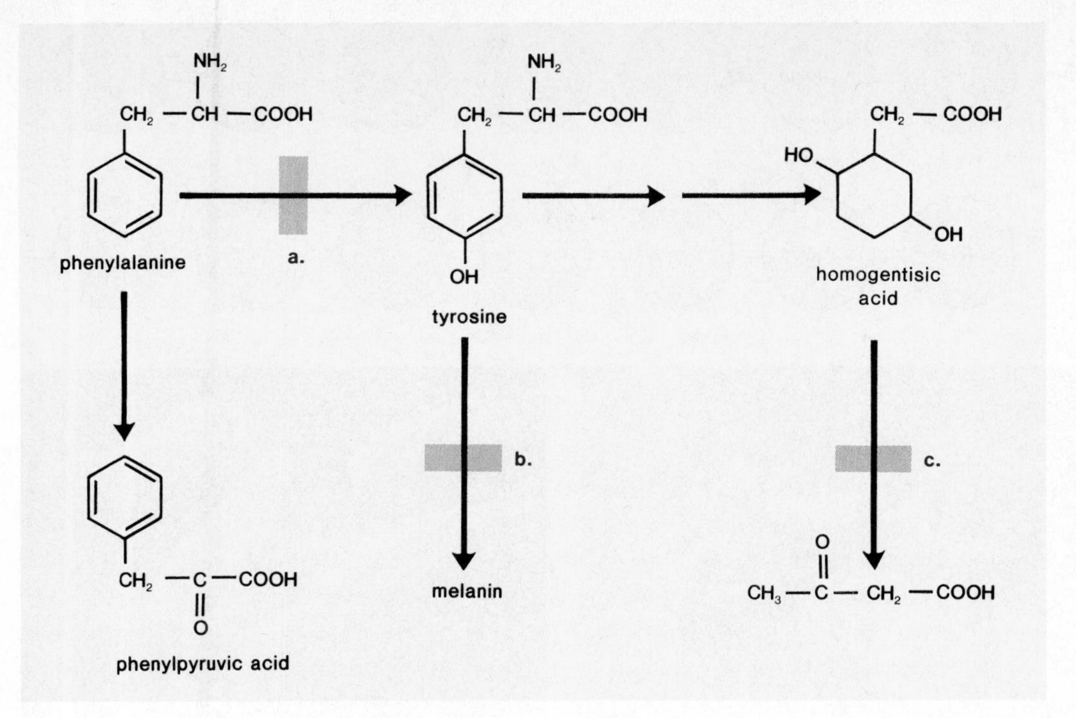

Figure 22.7 *Metabolic pathway by which phenylalanine is converted to other metabolites.* a. *If the enzyme that converts phenylalanine to tyrosine is defective, tyrosine is converted to phenylpyruvic acid instead, and the accumulation of this substance leads to PKU (phenylketonuria).* b. *If the enzyme that converts tyrosine to melanin is defective, albinism results.* c. *If homogentisic acid cannot be metabolized, alkaptonuria, meaning that the urine turns a dark color, results.*

This replication process is described as *semiconservative* because each double strand of DNA contains one old strand and one new strand. Although DNA replication can be easily explained, it is in actuality an extremely complicated process involving many steps and enzymes. There are enzymes that assist the unwinding process, that join together the nucleotides, and that assist the rewinding process, just to mention a few. On occasion, errors are made that cause a change in the DNA and, in this way, a mutation can arise.

---

*D*uring replication, DNA becomes "unzipped," and a complementary strand forms opposite to each original strand. This is called semiconservative replication.

Genes and Enzymes

Many novel experiments and years of research allowed scientists to conclude that there is a relationship between genetic inheritance and the structure of enzymes and other proteins. For example, the metabolic pathway outlined in figure 22.7 was discovered in the early 1900s. In this pathway, three genetic diseases are known. In the disease known as *phenylketonuria (PKU),* phenylpyruvic acid accumulates in the body and spills over into the urine because the enzyme needed to convert phenylalanine to tyrosine is missing. If the condition is not treated, the continued accumulation of phenylpyruvic acid can cause mental retardation. Albinism (fig. 21.19) results because tyrosine cannot be converted to melanin, the natural pigment in human skin. The genetic disease alkaptonuria results if the enzyme needed to metabolize homogentisic acid is missing.

At first these conditions were simply called *inborn errors of metabolism,* and only later was it confirmed that the genetic fault lay in the absence of particular enzymes. This gave rise to the suggestion that genes in some way controlled the presence of enzymes in the individual. This was called the *one gene–one enzyme* theory.

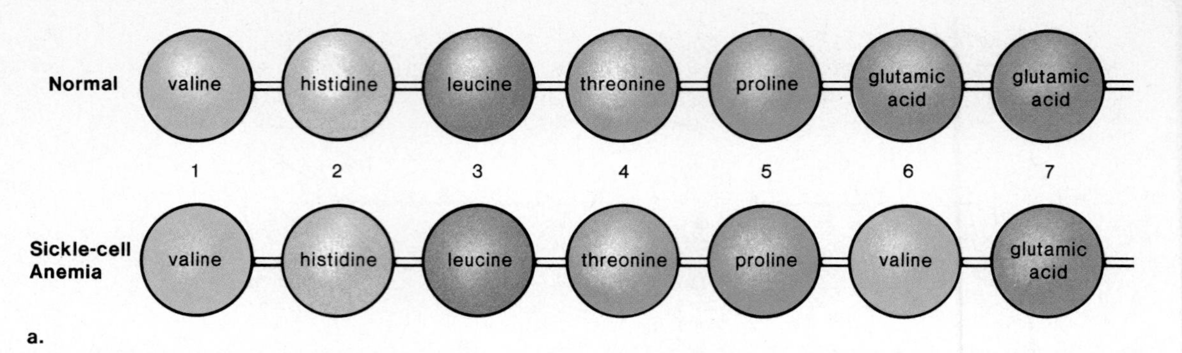

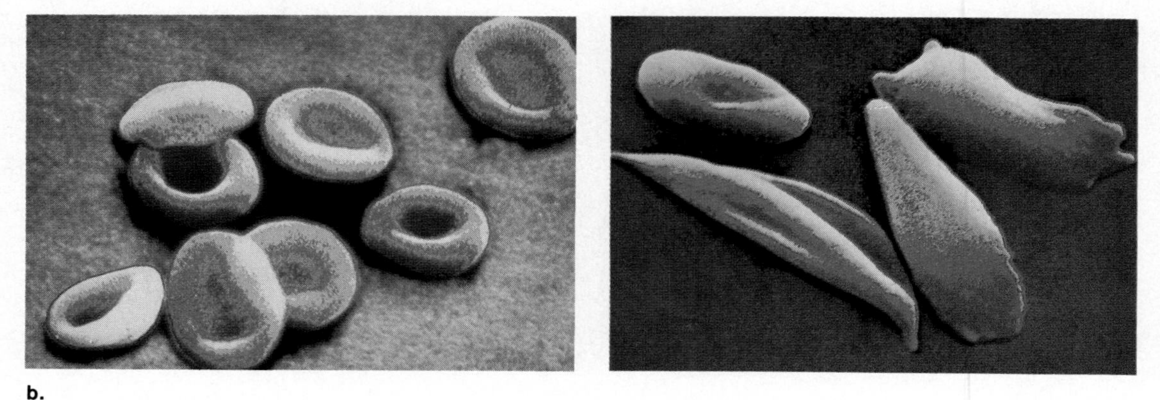

Figure 22.8
Sickle-cell anemia in humans. a. *The first seven amino acids found in the normal and in the abnormal β chain. The substitution of a single amino acid (valine substituted for glutamic acid) at the sixth position results in sickle-cell anemia.*
b. *Photomicrographs of normal (left) and sickled (right) red blood cells.*

Figure 22.9
Eukaryotic cell structure. Ribosomal RNA is produced in the nucleolus. Chromatin contains DNA. Protein synthesis occurs at the ribosomes. Proteins that are synthesized at ribosomes attached to the endoplasmic reticulum are usually for export. Those that are synthesized at polysomes are usually for use inside the cell.

Since enzymes are proteins, the concept was soon broadened to be the one gene–one protein theory. However, then, some proteins like hemoglobin have more than one type of polypeptide chain (fig. 11.6). In persons with sickle-cell anemia, it is only the β polypeptide chain that has an altered sequence of amino acids (fig. 22.8) compared to the normal β chain. Therefore it may be more appropriate to state that a gene controls the sequence of amino acids in a polypeptide. Today we define a gene as a section of a DNA molecule that determines the sequence of amino acids in a single polypeptide chain of a protein.

It was a breakthrough to discover that there is a relationship between genetic inheritance and the primary structure of proteins in an individual.

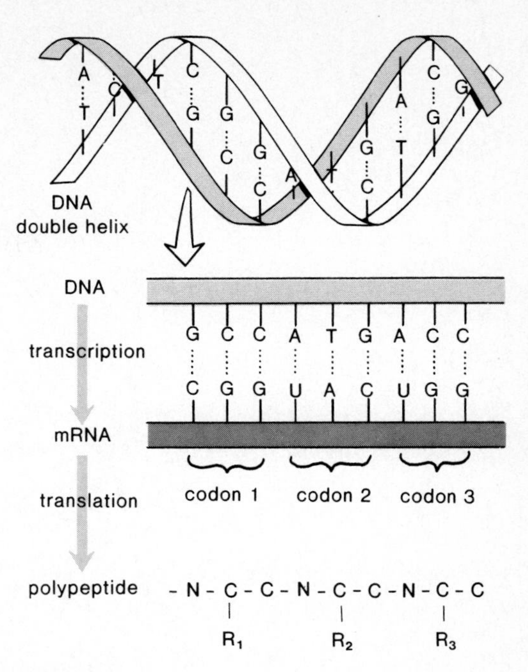

Figure 22.10
Transcription and translation in protein synthesis. Transcription occurs when DNA acts as a template for mRNA synthesis. Translation occurs when sequence of the codons found in mRNA determines the sequence of the amino acids in a polypeptide.

Protein Synthesis

The fact that DNA controls the production of proteins may at first seem surprising when we consider that genes are located in the nucleus of higher cells but proteins are synthesized at the ribosomes in the cytoplasm. However, although DNA is found only in the nucleus, RNA exists in both the nucleus and the cytoplasm (fig. 22.9).

Biochemical genetic research indicates that a type of RNA, called **messenger RNA (mRNA),** serves as a go-between for DNA in the nucleus and the ribosomes in the cytoplasm. (This is possible because a strand of DNA can serve as a template for the production of a complementary strand of RNA, as well as a template for another strand of DNA. The RNA molecule contains a sequence of nucleotides that are complementary to those of a single gene.) The mRNA is processed (p. 494) and then moves from the nucleus to the ribosomes in the cytoplasm where it dictates the sequence of amino acids in a polypeptide. This concept is often called the central dogma of modern genetics and can be diagrammed as follows:

DNA ——transcription——→ mRNA ——translation——→ protein

The diagram indicates that the control of protein synthesis requires transcription and translation. During the process of **transcription,** complementary mRNA is formed in the nucleus, and during **translation,** its message is used in the cytoplasm to produce the correct order of amino acids in a polypeptide (fig. 22.10).

Code of Heredity

DNA provides mRNA with a message that directs the order of amino acids during protein synthesis, but what is the nature of the message? The message cannot be contained in the sugar-phosphate backbone because it is constant in every DNA molecule. However, the order of the bases in DNA and mRNA can and does change. Therefore, it must be the bases that contain the message. The order of the bases in DNA must code for the order of the amino acids in a polypeptide. Can four bases provide enough combinations to code for 20 amino acids? If the code were a doublet one (any two bases stand for one amino acid), it would not be possible to code for 20 amino acids (table 22.2). But if the code were a triplet code, then the four bases would be able to supply 64 different triplets, far more than are needed to code for 20 different amino acids. It should come as no surprise, then, to learn that the code is a triplet code.

Table 22.2 Number of Bases in Code

Number of Bases in Code	Number of Amino Acids Coded for
1	4
2	16
3	64

***Table 22.3** Three-Letter Codons of Messenger RNA, and the Amino Acids Specified by the Codons*

Codon	Amino Acid	Codon	Amino Acid	Codon	Amino Acid	Codon	Amino Acid
AAU	Asparagine	CAU	Histidine	GAU	Asparatic acid	UAU	Tyrosine
AAC		CAC		GAC		UAC	
AAA	Lysine	CAA	Glutamine	GAA	Glutamic acid	UAA	(Stop)*
AAG		CAG		GAG		UAG	
ACU	Threonine	CCU	Proline	GCU	Alanine	UCU	Serine
ACC		CCC		GCC		UCC	
ACA		CCA		GCA		UCA	
ACG		CCG		GCG		UCG	
AGU	Serine	CGU	Arginine	GGU	Glycine	UGU	Cysteine
AGC		CGC		GGC		UGC	
AGA	Arginine	CGA		GGA		UGA	(Stop)*
AGG		CGG		GGG		UGG	Tryptophan
AUU	Isoleucine	CUU	Leucine	GUU	Valine	UUU	Phenylalanine
AUC		CUC		GUC		UUC	
AUA		CUA		GUA		UUA	Leucine
AUG	Methionine	CUG		GUG		UUG	

*Stop codons signal the end of the formation of a polypeptide chain.

From Volpe, E. Peter, *Biology and Human Concerns*, 3d ed. © 1975, 1979, 1983 Wm. C. Brown Publishers, Dubuque, Iowa. All Rights Reserved. Reprinted by permission.

To crack the code, artificial RNA was added to a medium containing bacterial ribosomes and a mixture of amino acids. Comparison of the bases in the RNA with the resulting polypeptide allowed investigators to decipher the code. Each three-letter unit of a messenger RNA is called a **codon.** All 64 codons have been determined (table 22.3). Sixty-one triplets correspond to a particular amino acid; the remaining three code for chain termination. The one codon that stands for the amino acid methionine also signals polypeptide initiation.

The Genetic Code Is Universal Research indicates that the code is essentially universal. The same codons stand for the same amino acids in all living things, including bacteria, plants, and animals. This illustrates the remarkable biochemical unity of living things and suggests that all living things have a common evolutionary ancestor.

DNA controls the phenotype because it controls protein synthesis. DNA and several forms of RNA participate in protein synthesis. DNA, which always stays within the nucleus, contains a triplet code: a series of three bases codes for one particular amino acid.

Transcription

During transcription, the DNA code is passed to mRNA, and thus the code is "transcribed" or rewritten.

Messenger RNA

Transcription allows the formation of a mRNA that contains a sequence of bases complementary to DNA. A segment of the DNA helix unravels; complementary RNA nucleotides pair with DNA nucleotides of one of the strands. After an enzyme joins the nucleotides by way of their sugar-phosphate components, the resulting mRNA molecule carries a sequence of bases that are triplet codons complementary to the DNA triplet code (fig. 22.11).

The mRNA strand is processed (p. 494) and then passes from the cell nucleus into the cytoplasm, carrying the transcribed DNA code.

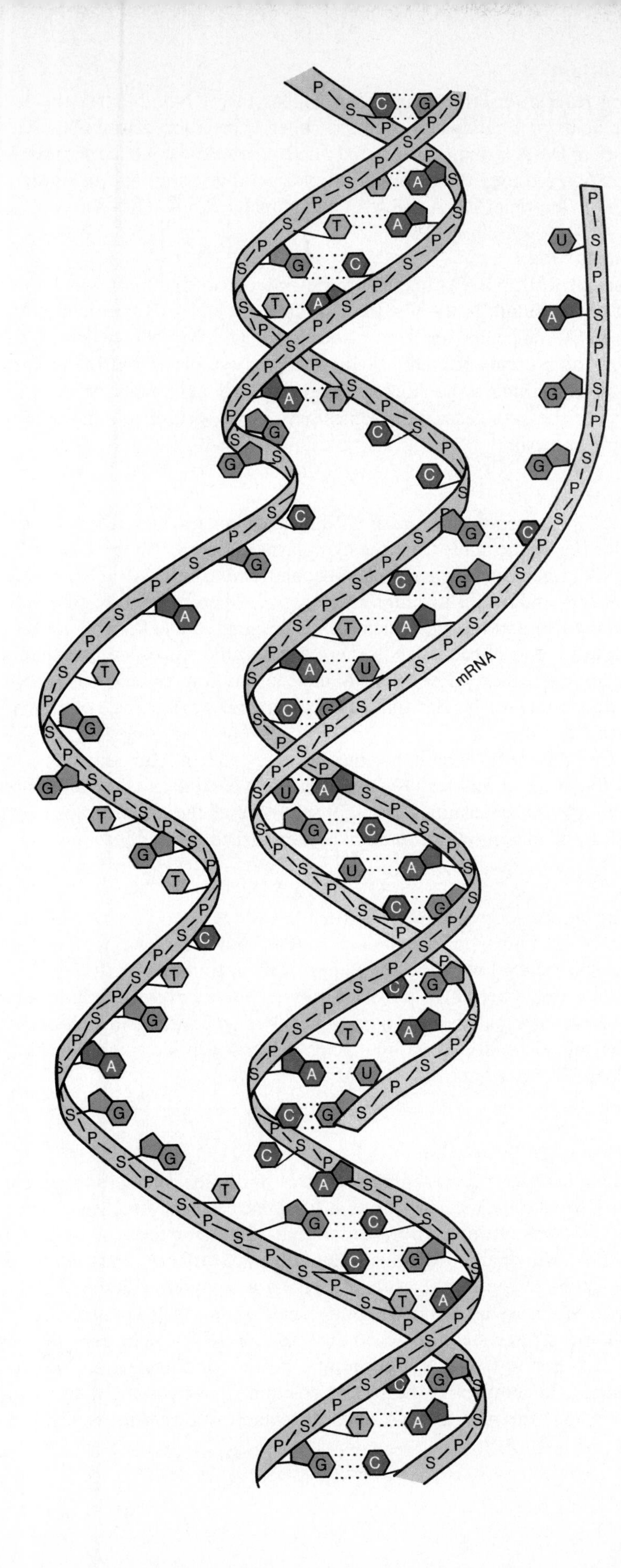

Figure 22.11
Transcription. When mRNA is formed, its bases are complementary to those found in one strand of DNA.

Translation

During translation, the sequence of codons in mRNA dictates the order of amino acids in a polypeptide. This is called translation because the sequence of bases in DNA is finally translated into a particular sequence of amino acids. Translation requires the involvement of several enzymes and two other types of RNA: ribosomal RNA (rRNA) and transfer RNA (tRNA).

Ribosomal RNA

Ribosomal RNA (rRNA) makes up the ribosomes (fig. 2.6c) which are composed of two subunits, each with characteristic RNA and protein molecules. The rRNA molecules are transcribed from DNA in the region of the nucleolus; the proteins are manufactured in the cytoplasm but then migrate to the nucleolus, where the ribosomal subunits are assembled before they migrate into the cytoplasm. Ribosomes play an important role in coordinating protein synthesis.

Transfer RNA

Located in the cytoplasm are small molecules of **transfer RNA (tRNA)** that transfer the amino acids from the cytoplasm to the ribosomes. Each molecule of tRNA attaches at one end to a particular amino acid. Attachment requires ATP energy and results in a high energy bond. Therefore this bond is indicated by a wavy line and the entire complex is designated by *tRNA ~ amino acid.* At the other end of each tRNA there is a specific **anticodon** complementary to an mRNA codon. (Each tRNA molecule is transcribed from DNA and then, due to intramolecular binding of complementary bases, the anticodon is exposed.)

Complementary base pairing between codons and anticodons determines the order in which tRNA ~ amino acid complexes come to a ribosome, and this in turn determines the final sequence of the amino acids in a polypeptide. The making of a protein is accomplished codon by codon.

During transcription, mRNA is made complementary to one of the DNA strands. It then contains codons, moves to the cytoplasm, and becomes associated with the ribosomes. During translation, tRNA molecules, each carrying a particular amino acid, travel to the mRNA, and through complementary base pairing between anticodon and codon, the tRNAs and, therefore, the amino acids in a polypeptide chain become sequenced in a predetermined order.

The Process of Translation

Protein synthesis requires three processes: initiation, elongation, and termination. During the initiation process, a ribosome becomes attached to a mRNA. Initiation always begins with a codon that stands for the amino acid methionine. First, the smaller ribosomal subunit binds to mRNA, and then the larger subunit joins to the smaller subunit, giving a complete ribosomal structure. *Elongation* occurs as the polypeptide chain grows in length (fig. 22.12). A ribosome is large enough to accommodate two tRNA molecules; the peptide chain attached to the tRNA in the first position is transferred to the tRNA ~ amino acid complex in the second position. The ribosome then moves laterally so that the next mRNA codon becomes available to receive the next

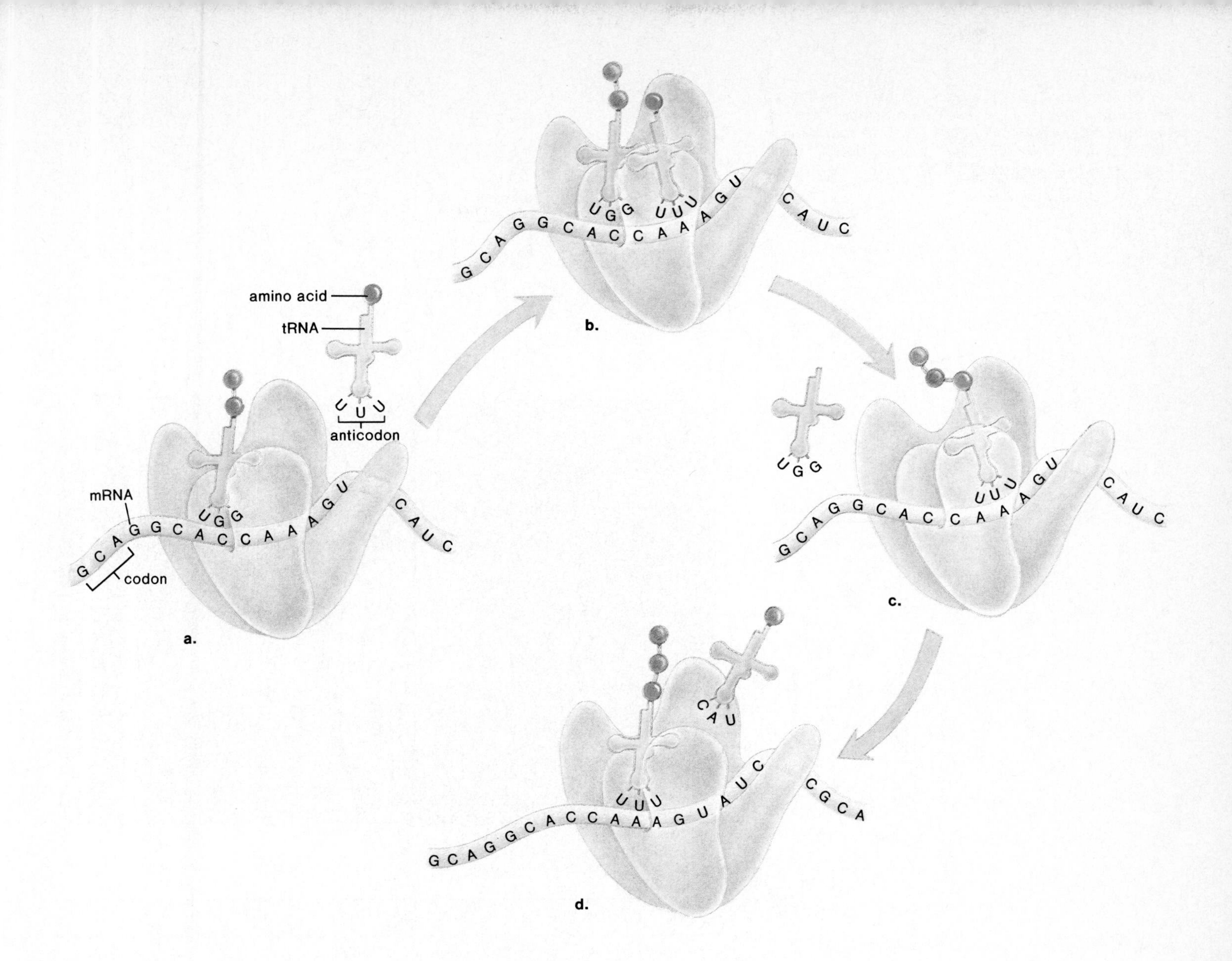

tRNA ~ amino acid complex. In this manner, the peptide chain grows and the primary structure of a protein comes about. The secondary and tertiary structures occur after termination, as the predetermined sequence of amino acids within the polypeptide chain interact with one another.

Termination of protein synthesis occurs at a specific nucleotide sequence on the mRNA where the last tRNA and completed polypeptide are liberated from the ribosomal complex. The ribosome dissociates into its two subunits and falls off the messenger. Several ribosomes, collectively called a **polysome,** may move along one mRNA at a time; therefore, several of the same type proteins may be synthesized at once (fig. 22.13). After the translation process is completed, the mRNA disintegrates.

Figure 22.12
During translation, a ribosome moves along a mRNA. A codon on the mRNA attracts an anticodon of a tRNA. When a codon pairs with an anticodon, the tRNA brings an amino acid to the ribosome. a. *In this diagram, translation is already in progress and a single tRNA molecule is in place on a ribosome as another approaches.* b. *Both tRNA molecules are positioned at the ribosome. The first of these with the anticodon UGG bears a peptide chain, while the second having the anticodon UUU bears an amino acid.* c. *In this diagram the first tRNA has passed the peptide chain to the second tRNA before departing.* d. *The ribosome has moved to the right so that the tRNA with the anticodon UUU is now in the first position on the ribosome. Another tRNA ~ amino acid complex with the anticodon CAU is approaching the ribosome. The same sequence of events* (a–d) *will now reoccur.*

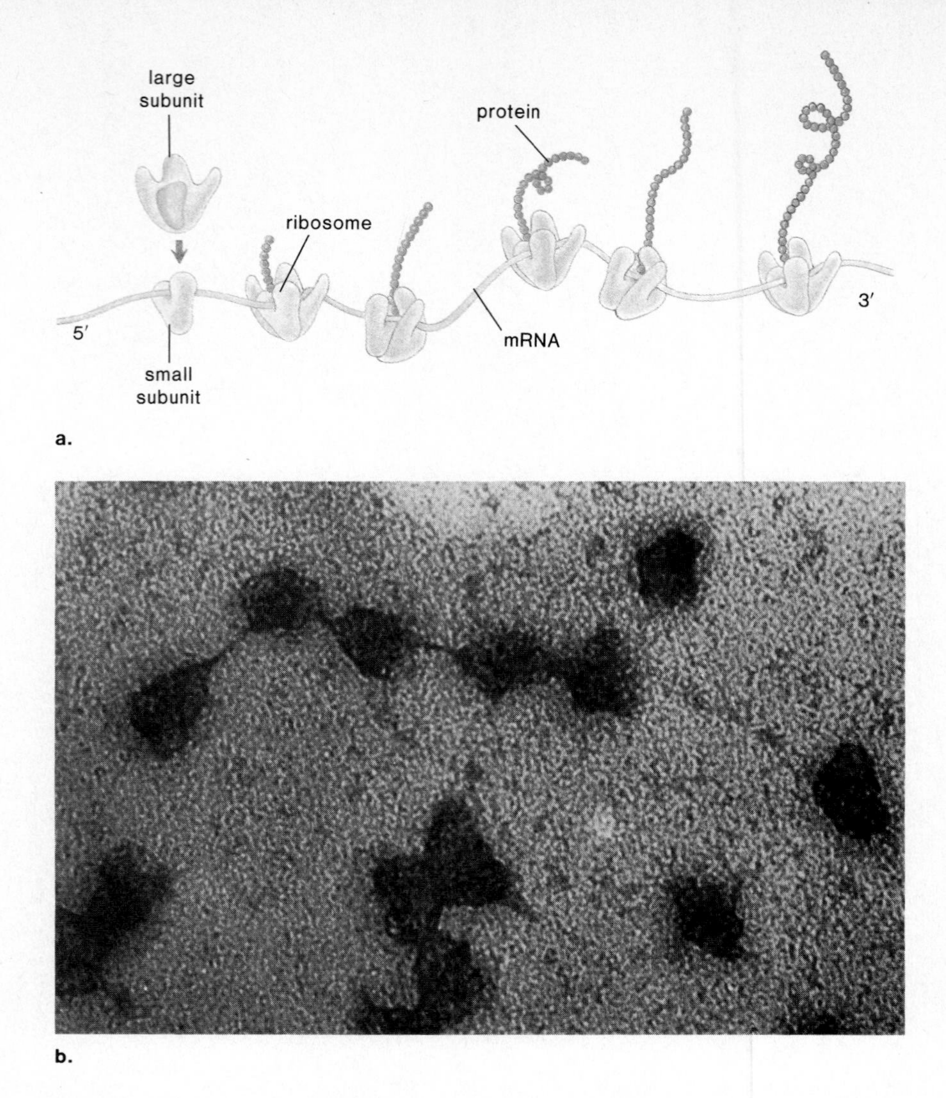

Figure 22.13
a. *Polysome structure. Several ribosomes, called a polysome, move along a mRNA at a time. They function independently of each other so that several polypeptides can be made at the same time.*
b. *Electron micrograph of several polysomes.*

Table 22.4 *Steps in Protein Synthesis*

Name of Molecule	Special Significance	Definition
DNA	Code	Sequence of bases in threes
mRNA	Codon	Complementary sequence of bases in threes
tRNA	Anticodon	Sequence of three bases complementary to a codon
Amino acids	Building blocks	Transported to ribosomes by tRNAs
Protein	Enzyme	Amino acids joined in a predetermined order

Summary of Protein Synthesis

The following list, along with table 22.4, provides a brief summary of the steps involved in protein synthesis.

1. DNA, which always remains in the nucleus, contains a series of bases that serve as a *triplet code* (every three bases codes for an amino acid).
2. During transcription, one strand of DNA serves as a template for the formation of messenger RNA (mRNA), which contains *triplet codons* (sequences of three bases complementary to DNA code).

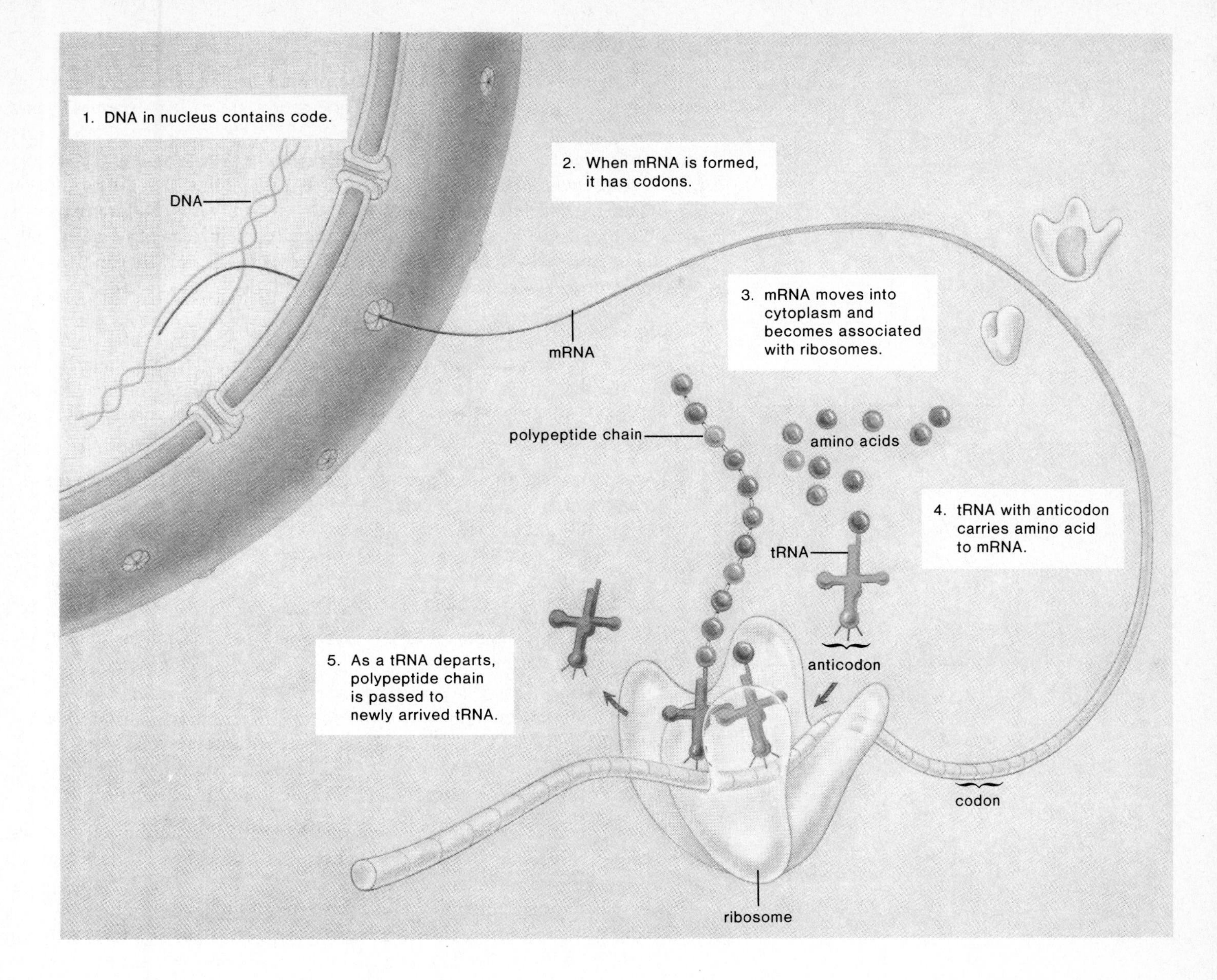

Figure 22.14
Summary of protein synthesis. Transcription occurs in the nucleus and translation occurs in the cytoplasm (blue). During translation, the ribosome moves along the mRNA. In the diagram, as the ribosome moves to the right, a tRNA bearing an amino acid comes to the ribosome. Thereafter the polypeptide chain will be passed to this tRNA ~ amino acid complex. Each time the ribosome moves, a tRNA departs.

3. Messenger RNA goes into the cytoplasm and becomes associated with the *ribosomes,* which are composed of ribosomal RNA (rRNA) and proteins.
4. Transfer RNA (tRNA) molecules, each of which is bonded to a particular amino acid, have *anticodons* that pair complementarily to the codons in mRNA.
5. As the ribosome moves along mRNA, a newly arrived tRNA ~ amino acid complex receives the growing polypeptide chain from a tRNA molecule, that leaves the ribosome to pick up another amino acid. During translation, therefore, the linear sequence of codons determines the order in which the tRNA molecules arrive at the ribosomes and thus determines the *primary structure* of a protein (i.e., the order of its amino acids).

The transcription-translation process is illustrated in figure 22.14.

Regulatory Genes

Thus far, genes that provide coded information for the synthesis of proteins (polypeptides) have been discussed. These genes are called **structural genes** because they determine protein structure. Evidence indicates that another class of genes, called regulatory genes, exists. **Regulatory genes** regulate the activity of structural genes. For example, we previously mentioned that differentiation occurs during development of the human embryo and fetus. Differentiation is possible because although each cell receives a full complement of genes, only certain ones are active in each cell type. First, however, we will consider regulation in prokaryotes before we consider regulation in eukaryotes.

Prokaryotes

Research with the bacterium *E. coli* has resulted in at least two models that explain the regulation of gene transcription and, therefore, protein synthesis.

Prokaryotic regulatory models have the following components (fig. 22.15):

Table 22.5 Participants in Regulatory Models

Participants	Action
Operon	Genes that code for enzymes in a metabolic pathway
Operator	An on/off switch for transcription of the operon
Regulatory gene	A gene that codes for a repressor
Inducer	A metabolite that inactivates a repressor
Corepressor	A metabolite that activates a repressor

1. An *operon* is a group of genes, called structural genes, that code for enzymes active in a particular metabolic pathway, such as enzymes 1, 2, 3 in this pathway:

$$A \xrightarrow{1} B \xrightarrow{2} C \xrightarrow{3} D$$

2. An *operator* is a segment of DNA that acts as an on/off switch for transcription of the operon.
3. A *regulatory gene* is a gene that codes for a protein that either
   a. immediately combines with the operator preventing transcription or
   b. must first join with a metabolite before it combines with the operator. The two models (fig. 22.15) are dependent on whether (a) or (b) controls transcription of a particular operon.

In figure 22.15a, the operon is normally inactive because the regulatory gene codes for a protein, called a *repressor,* that combines with the operator, preventing transcription. The operon becomes active when the repressor joins with an inducer molecule, and the complex is unable to bind with the operator. The inducer, so named because it induces protein synthesis, is a metabolite in a metabolic pathway. For example, *A* in the pathway above could be an inducer. In this **inducible operon model,** then, the first metabolite indicates the need for particular enzymes.

In figure 22.15b, the operon is normally active because the regulatory gene codes for an inactive repressor that must join with the corepressor before the complex combines with the operator. A *corepressor,* so named because it prevents protein synthesis, is a metabolite in a metabolic pathway. For example, *D* in the pathway above could be a corepressor. In this **repressible operon model,** the presence of the end product indicates that particular enzymes are no longer needed.

Notice that the inducible model accounts for the fact that some structural genes are normally inactive and the repressible model accounts for the fact that some structural genes are normally active. Therefore, some genes could normally be turned on, while others could normally be turned off.

Eukaryotes

Most likely eukaryotes have regulatory mechanisms that are similar to the inducible and repressible operon models that have been formulated on the basis of experimentation with prokaryotes. There is also the possiblity that

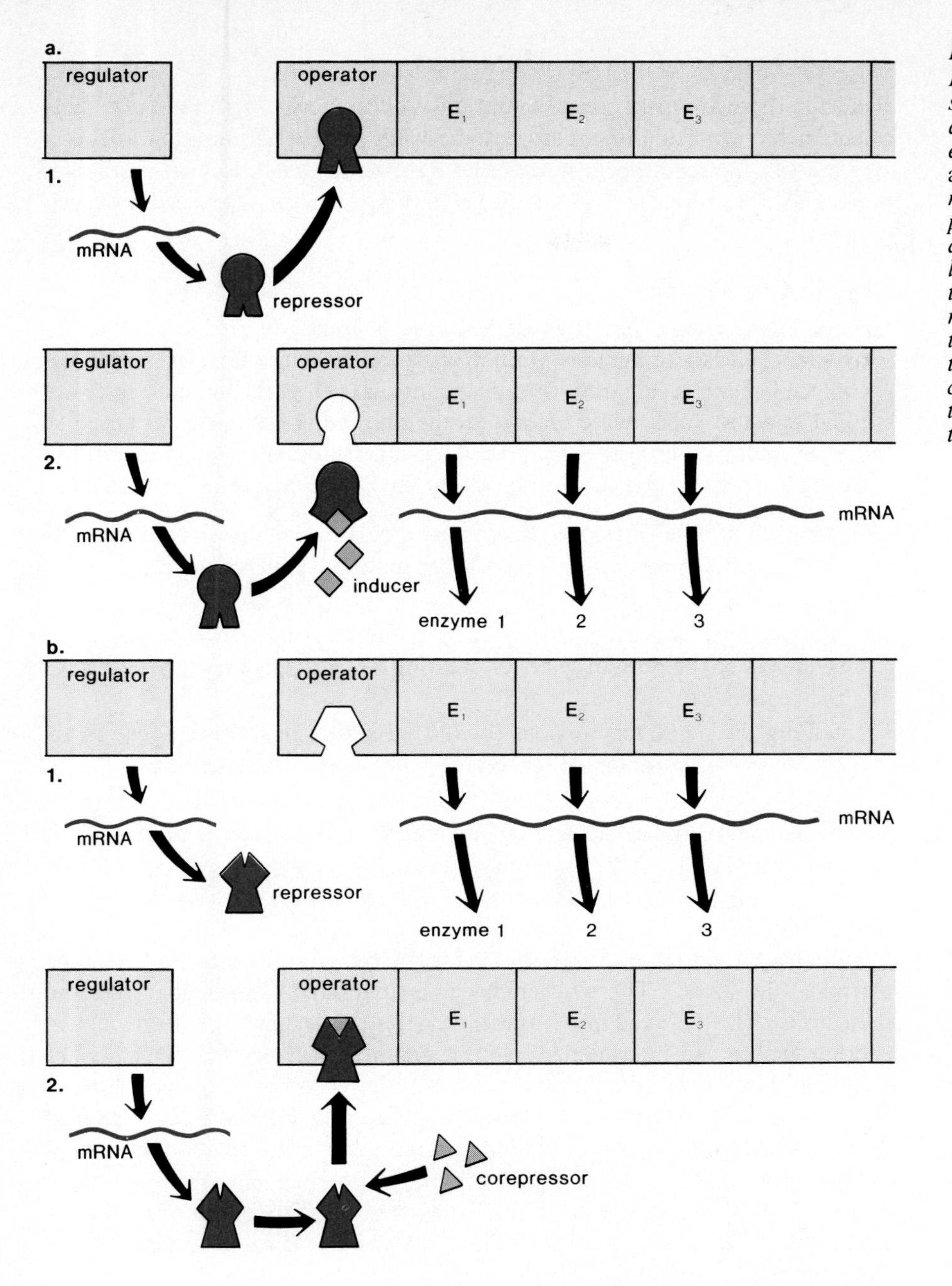

Figure 22.15
Prokaryote models for the regulation of protein synthesis. Notice that the first model explains how an operon can get turned on and the second model explains how an operon can get turned off.
a. *Inducible models: (1) A regulator gene codes for a repressor that can immediately bind to the operator preventing transcription from taking place. (2) When an inducer combines with the repressor it is no longer capable of binding to the operator and therefore transcription takes place.* b. *Repressible models: (1) The regulator gene codes for a repressor that is unable to bind to the operator and therefore transcription takes place. (2) When a corepressor is combined with the repressor, it is now able to bind to the operator and therefore transcription cannot take place.*

eukaryotes have additional mechanisms especially because (1) the eukaryote chromosome differs from the prokaryote chromosome (p. 495) and because (2) mRNA is processed before it leaves the nucleus in eukaryotes (p. 494). Even after mRNA leaves the nucleus it still may be translated or not translated depending on regulatory proteins present in the cytoplasm.

Knowledge about regulatory genes is extremely important; genes can best be manipulated when we know how to turn them on and off. Mutations of regulatory genes probably account for some genetic diseases and/or the development of cancer.

Structural genes code for proteins that function in the cytoplasm. Regulatory genes control the expression of structural genes.

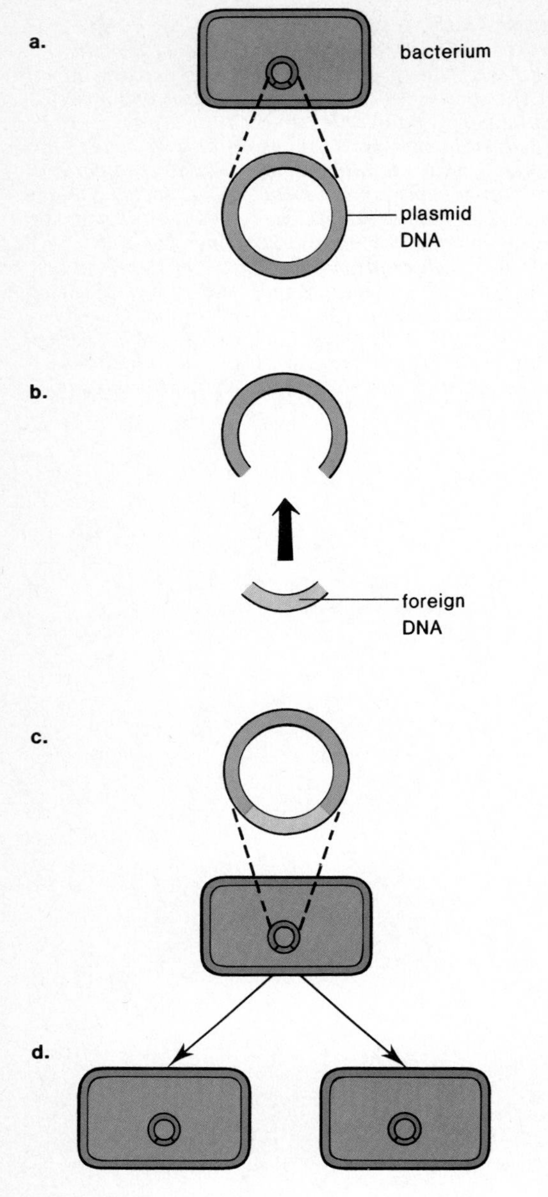

Figure 22.16
Recombinant DNA experimentation. a. *Plasmid DNA is removed from bacterium.* b. *Foreign DNA is incorporated into the plasmid.* c. *Plasmid is reintroduced into the bacterium.* d. *The incorporated gene is cloned when the bacterium reproduces and, if all goes well, will also direct protein synthesis.*

Mechanisms for Genetic Mutations

Basically there are two types of mutations. A chromosomal mutation (p. 463) affects a large portion of a chromosome while a genetic mutation affects a single gene. In recent years it has been possible to profoundly affect the genetic makeup of the individual by performing genetic engineering experiments.

Genetic Engineering

Genetic engineering refers to procedures that manipulate the genes of organisms—from viruses to humans. Both in vitro and in vivo experiments are commonplace in genetic research. *In vitro* means that cell parts and molecules are studied in a test tube, while *in vivo* means that living organisms are used in the experiments. The following procedures are common in experiments involving both prokaryotic DNA and eukaryotic DNA, including human DNA.

1. Isolation of a gene; that is, removal of a particular portion of DNA from a cell. Thereafter it is possible to determine the sequence of nucleotides in the gene.
2. Manufacture of a gene; that is, joining nucleotides together in the sequence of the normal gene, or creating a mutated gene by altering the sequence.
3. Joining the regulatory regions (p. 490) of a viral or bacterial gene to an isolated or machine-made structural gene so that transcription is assured.
4. Placement of the constructed gene in another cell where it undergoes replication and directs protein synthesis.

Gene Cloning

Certain bacteria, such as *E. coli,* have rings of extrachromosomal DNA called **plasmids** (fig. 22.16). These plasmids can be extracted from the bacteria and then enzymatically sliced into fragments. After this is done, DNA taken from another source can be attached to these fragments before the plasmids are reformed. Notice that the plasmids now contain recombined, or **recombinant, DNA** molecules and that such plasmids will be taken up by other bacterial cells (fig. 22.16). Once a plasmid has entered a bacterial cell, the cell is said to be **transformed** because it can now make a protein product it was unable to make before. Further, whenever a bacterium reproduces, the plasmid, including the foreign gene, is copied. Since exact multiple copies of the plasmid and gene are now available, they are said to have been **cloned.**

Applications of Genetic Engineering

More and more applications have been found for genetic engineering procedures. We will discuss a few of these.

Protein Products as Medicines All sorts of proteins have been made by *E. coli.* Human insulin and growth hormone are now being marketed and other such products are currently undergoing clinical testing. The recombinant DNA procedure can also help us fight disease because *E. coli* can be engineered to make both antibodies, and proteins that can serve as vaccines. Vaccines for hepatitis B and childhood meningitis are already available.

Cloned plasmids can be removed from *E. coli* and introduced into other types of cells. For example, yeast cells have been engineered to make human interferon (p. 265), a chemical useful in immunity and cancer research. Since yeast is used in the production of beer, technology already exists for mass production of these cells. Perhaps, then, interferon will soon be available in bulk.

Figure 22.17
After studying the chromosomes of hundreds of persons with Huntington's disease, many of whom were members of related families in Venezuela, investigators determined that the gene(s) causing this genetic disease are located on chromosome 4. Using recombinant DNA techniques, a test has been developed that enables investigators to analyze this chromosome in order to predict whether a person will eventually develop the adult-onset disease.

Protein Engineering Before a naturally occurring gene is spliced into plasmids, its sequence of bases can be altered. These so-called synthetic mutations cause *E. coli* to produce proteins not normally produced in cells. It is believed that these mutations may eventually tell us more about human genetic diseases. Also, industrial chemists are hopeful that synthetically mutated genes will gives us enzymes superior to those now being used as catalysts in the production of important and useful chemicals.

DNA Probes Cloned DNA fragments are helping physicians diagnose infectious and genetic diseases. These fragments are called *DNA probes* because they will search out and bind to complementary DNA sequences. For example, a portion of DNA can be removed from an infectious organism and cloned. Blood from a patient is exposed to these fragments and if binding occurs, it can be detected by radioactive or fluorescent techniques. Binding indicates that the infectious organism is present in the patient's blood. Or, the abnormal portion of a gene from a person with a genetic disease can be cloned. If this probe binds to the DNA of a patient's cells, it indicates that the patient has the genetic disease (fig. 22.17).

Alteration of Genetic Inheritance Genes placed in specially prepared plant cells (fig. 22.18) and in animal eggs remain active as the organism grows and matures. This offers hope that one day it will be possible to engineer plants with genes that promote resistance to pests and/or that fix aerial nitrogen, thereby reducing the need for pesticide and fertilizer use. It may also be possible one day to cure human genetic diseases.

Figure 22.18
a. *Photomicrograph of protoplasts, plant cells that lack cell walls. Whereas plant cells with cell walls cannot take up recombinant plasmids, protoplasts will do so. (These plasmids are from the bacterium* A. tumefaciens.*) Therefore, it is possible to transfer genes for desired characteristics to protoplasts. Given proper treatment, protoplasts will develop into whole plants that display the characteristics. Therefore, scientists are hopeful that this technique will allow agricultural plants to be remodeled within the next five or ten years.* b. *Plants grown from protoplasts.*

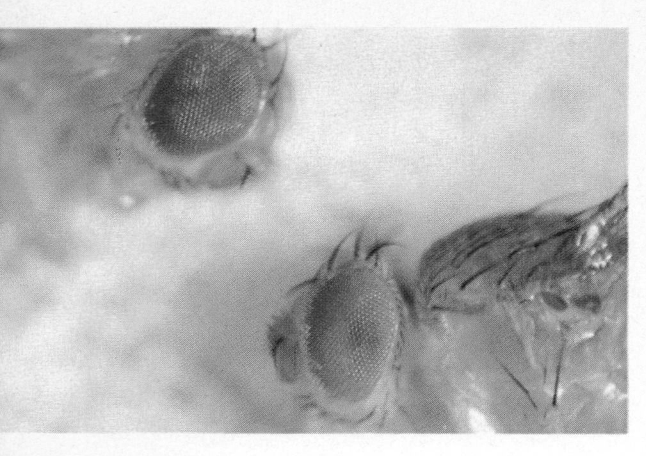

Figure 22.19
Transposons bearing a gene for red eyes were injected into the cells of this brown-eyed Drosophilia (left) *when it was an embryo. Subsequently its offspring* (right) *had red eyes, indicating that transposons had successfully incorporated the normal gene into the chromosomes of the parent.*

As discussed in the reading on page 496, researchers have developed the means to attempt to cure human genetic diseases. They have chosen a viral DNA, rather than a plasmid, to serve as the *vector* (carrier) for normal human genes. This particular type of virus has the capability of inserting foreign DNA into host DNA.[2] The virus will be extensively altered. The harmful viral genes will be replaced by the normal human gene needed by the patient. The researchers then plan to infect blood stem cells from the patient's bone marrow. Blood stem cells have been chosen as recipients for the normal genes because they can be removed, treated, and then reinjected into the bone marrow. Hopefully, the active normal genes present in these cells will overcome the detrimental effects of the defective genes in the rest of the patient's body.

Recombinant DNA techniques allow DNA to be particularly placed in the bacterium *E. coli,* where it usually replicates and functions normally. If so, this cloned DNA can be removed and put into other types of cells. There have been many applications of this technique.

Natural Mechanisms

A surprising finding of late has been that cells have built-in mechanisms for producing genetic mutations. In eukaryotes, somatic mutations occur in body cells rather than the gametes. Germinal mutations are passed on from one generation to the next, such as those that were studied in the previous chapter.

Transposons

Transposons are specific DNA sequences that have the ability to move within and out of chromosomes. Their movement to a new location sometimes alters neighboring genes, particularly by increasing or decreasing their expression. Although "movable elements" in corn were described forty years ago, their significance was only recently realized. So-called "jumping genes" have now been discovered in bacteria, fruit flies, and humans, and it is likely that all organisms have such elements. Some investigators have suggested that transposons tend to become active during times of environmental stress and in that way increase the likelihood of mutations that can aid survival. In modern genetics laboratories, transposons have been used as vectors to carry selected genes into new hosts (fig. 22.19).

Split Genes

Eukaryotic structural genes (fig. 22.20) are now known to be interrupted by sections of DNA that are not part of the gene. These portions are called *introns* because they are *intra*gene segments. The other portions of the gene are called *exons* because they are ultimately *ex*pressed. When DNA is transcribed, the mRNA contains bases that are complementary to both exons and introns. But before the mRNA exits from the nucleus, it is *processed*—the nucleotides complementary to the introns are enzymatically removed. Some evidence suggests that the presence of introns facilitates the occurrence of routine mutations. Just as sectional furniture can be arranged differently, it has been shown that exons are like modules that can be selected and arranged to give different sequences of DNA (fig. 22.21). Each rearrangement appears to be a mutation.

Figure 22.20 shows that the eukaryotic chromosome also contains *repetitive DNA*—the same short-to-intermediate length DNA sequence is repeated over and over. The exact function of repetitive DNA has not been

[2]The virus is a retrovirus. Cancer-causing retroviruses are discussed on page 542.

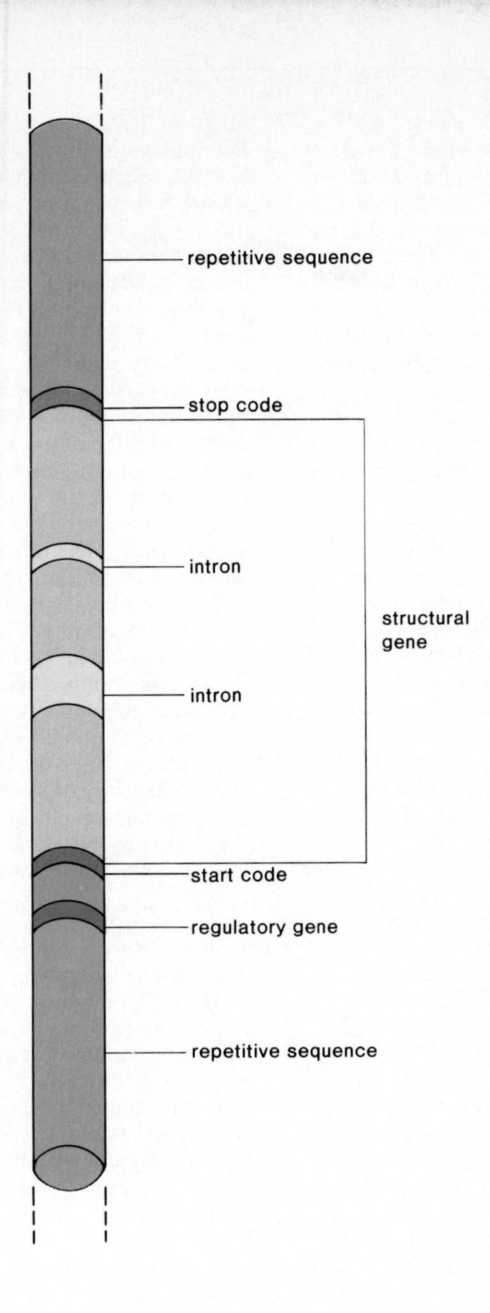

Figure 22.20
The chromosomes of higher organisms contain segments that do not code for polypeptides. Repetitive DNA is noncoding and consists of sequences of base pairs repeated many times, one after another. Structural genes, themselves, are interrupted by intervening sequences or introns, which do not code for polypeptides. During transcription, the entire gene, including these segments, are copied into mRNA. Before the mRNA leaves the nucleus, these segments are spliced out. Structural genes are flanked at one end by regulatory genes that control whether transcription takes place or not. There is a start code at the beginning and a stop code at the end of a structural gene.

determined, but perhaps these sequences are transposons that are now acting as introns. In other words, movement of transposons sometimes causes a structural gene to be interrupted by portions that are not part of the gene. Another view is that repetitive DNA has no function and simply is getting a free ride as it is duplicated and passed from generation to generation.

View of DNA

In the past, most biologists thought of DNA as fixed and static, but the recent findings discussed previously have changed this view. Indeed, DNA has been shown to be labile; it is constantly changing as portions move here or there or are joined in new and different ways prior to transcription.

The human chromosome has built-in mechanisms for mutations. Transposons can move to new sites and the chromosome contains introns that allow exons to be mixed and matched to give different products.

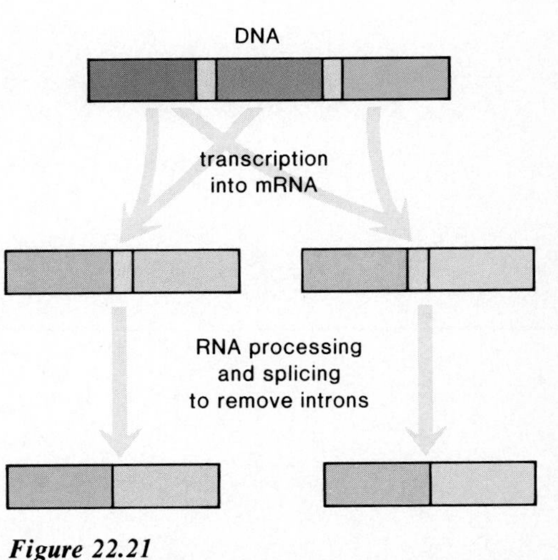

Figure 22.21
What is the function of introns? Perhaps they permit mutations by allowing different sections of DNA to be selected and subsequently pieced together to give various genes.

Beating Nature's Odds

Nature's lottery is never more tragically played out than in genetic disease. An estimated 25 percent of all hospital beds hold patients suffering from some degree of genetic abnormality. There are approximately 3,000 known genetic diseases. Symptoms can be alleviated in a number of cases: diet therapy in phenylketonuria, blood transfusions in Cooley's anemia, factor VIII replacement in hemophilia, insulin administration in diabetes. But these are just stopgaps. Presently no genetic disease can be cured.

The treatment of disease, genetic and otherwise, may soon shift dramatically. Such a revolution occurred in the 19th century, Lewis Thomas observed (*Science 84,* November), when physicians learned to stop bleeding, purging, and blistering sick patients. Medicine's second revolution, Thomas said, started mid-century when it became possible to cure some infectious diseases with antibiotics. Medicine's third revolution hinges on the applications of molecular biology: the mapping, cloning, and study of thousands of human genes to understand the body's normal functions at the molecular level. The practice of medicine will then become vastly more precise.

Gene therapy is one of the most exciting ramifications in this third revolution. Physicians should be able to treat many disorders by inserting a normal gene into the cells of a patient. If a child is born with a defective gene—for example, a hemoglobin gene that prevents the manufacture of blood, as in Cooley's anemia—a normal gene would be inserted into the appropriate cells so that the disease can be cured. Not just treated with blood transfusions, but cured.

Is this rosy picture really possible in the next 15 years? Yes, probably, for a number of diseases. It is possible now to insert foreign genes into animals and get these genes to function. The first successful gene therapy in a mammal was reported in 1984, when researchers injected a growth hormone gene into a fertilized mouse egg. The mouse would have developed into a dwarf because of a genetic deficiency of growth hormone. The growth hormone deficiency was, in fact, overcorrected in these first experiments: the mouse grew to nearly twice normal size. But the technical ability to cure a genetic disease was demonstrated.

A type of virus known as retrovirus, an RNA tumor virus, can be used to carry genes into an animal's body. The retrovirus is altered so that it functions only as a delivery system: it can no longer cause an infection, but it can carry a functional gene into cells (see figure). Researchers remove bone marrow cells from an animal, "infect" the marrow cells with the disabled retrovirus carrying the gene of choice, and then reinject the marrow cells into the animal. In this way they have engineered an active foreign gene into the blood cells of mice. . . .

The most promising premier candidate for gene therapy is a disease called adenosine deaminase deficiency. A missing ADA enzyme in immune cells can result in severe combined immune deficiency disease—infants who have little or no immunity and who die from simple childhood infections. They are the "bubble babies" who often cannot survive except in the germ-free environment of a sterile tent or bubble. Inserting a normal ADA gene into the bone marrow cells of these patients should produce normal resistance to infection in them.

Another early candidate will be Lesch-Nyhan disease—a severe neurological disease that results in uncontrollable self-mutilation. Victims bite off their own lips and fingers. The normal version of the defective gene that produces the disease has been isolated. It has been inserted into the bone marrow of mice using a disabled retroviral vector and the human enzyme has been produced in the animals. In addition, the gene has been put into human bone marrow cells growing in culture that have been isolated from patients with Lesch-Nyhan disease. Partial correction of the enzyme deficiency in these cells has been achieved.

Ultimately such genes should be simply packaged and injected into the bloodstream like any other common medication. The packaged gene would have signals directing it to target cells in the patient and, if necessary, into the correct place in the genome of the cell. Thus, the technology should be available for any patient, anywhere in the world, with a genetic disease.

Certainly there will be problems. There is no evidence yet that the technique will even work in human beings. And if it does, there is no assurance that the procedure or the gene itself might not produce other problems. Might the disabled retroviral carrier in some way be reactivated and cause its own disease? Might the insertion of a foreign gene interfere with normal cell function? These questions still must be examined in animal studies before the first attempts to treat humans should be carried out.

Then there is the further question: If we begin changing genes, are we tampering with the essence of our humanness? Should we ever attempt to alter germ line cells so that the patient's offspring would also be corrected? Gene therapy offers enormous hope for the alleviation of human suffering, but we must go forward carefully as we develop greater power to alter the lottery of nature.

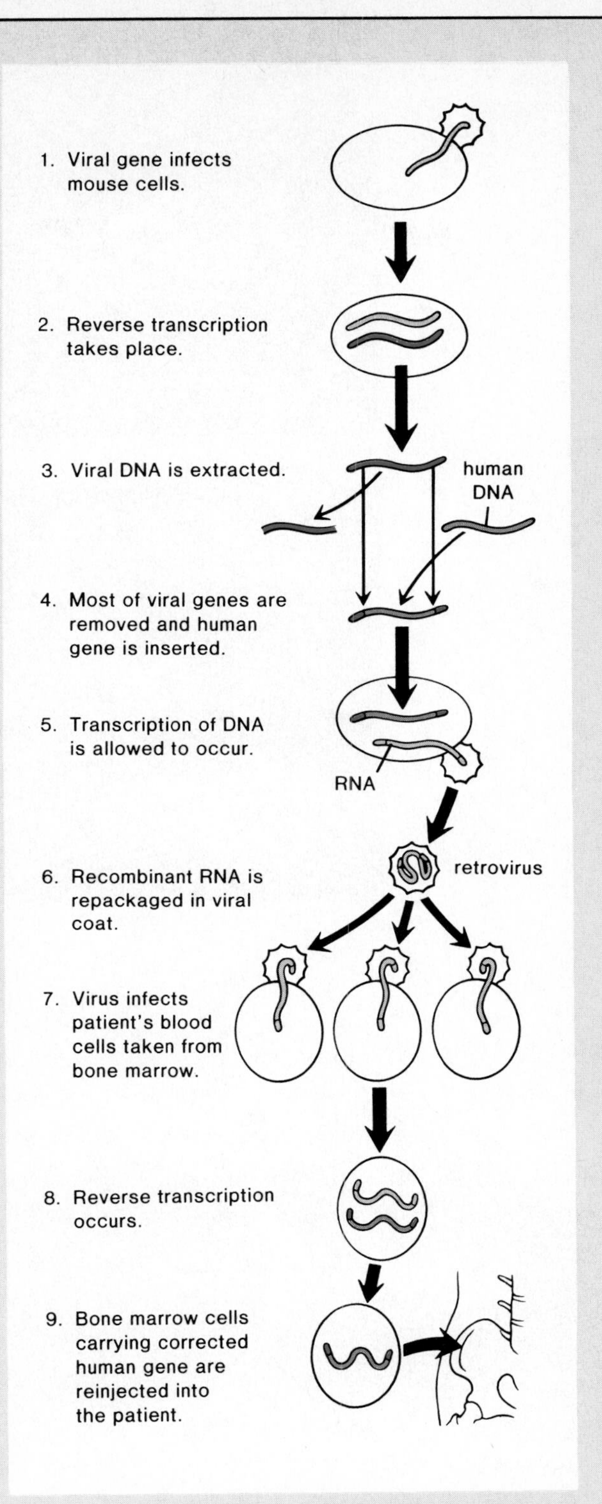

The virus selected for gene therapy is a retrovirus. Retroviruses have an RNA chromosome. When this chromosome enters a cell, reverse transcription must take place before reproduction of the virus can begin. You can see that this complicates preparing this virus to serve as a vector for the purpose of gene therapy.

Figure 22.22
Transformation from normal to cancerous cells.
a. *Normal fibroblasts are flat and extended.*
b. *After being infected with Rouse sarcoma virus, the cells become round and cluster together in piles. The virus carries an oncogene.*

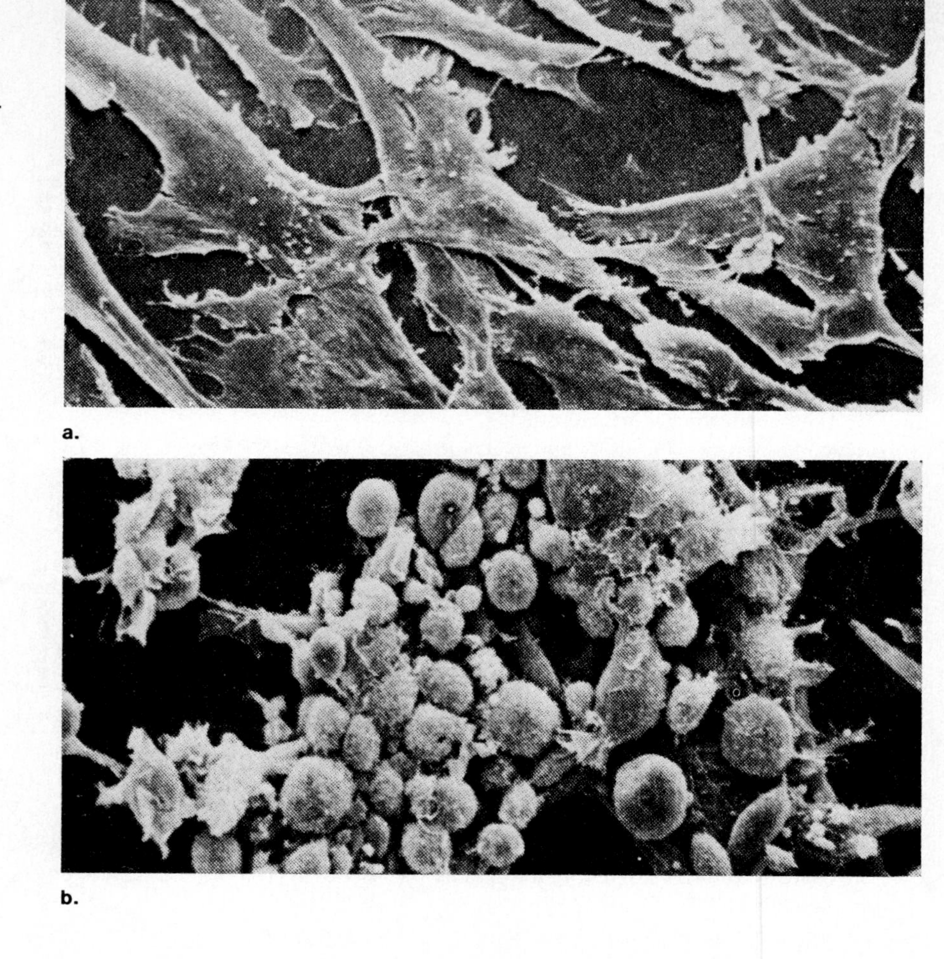

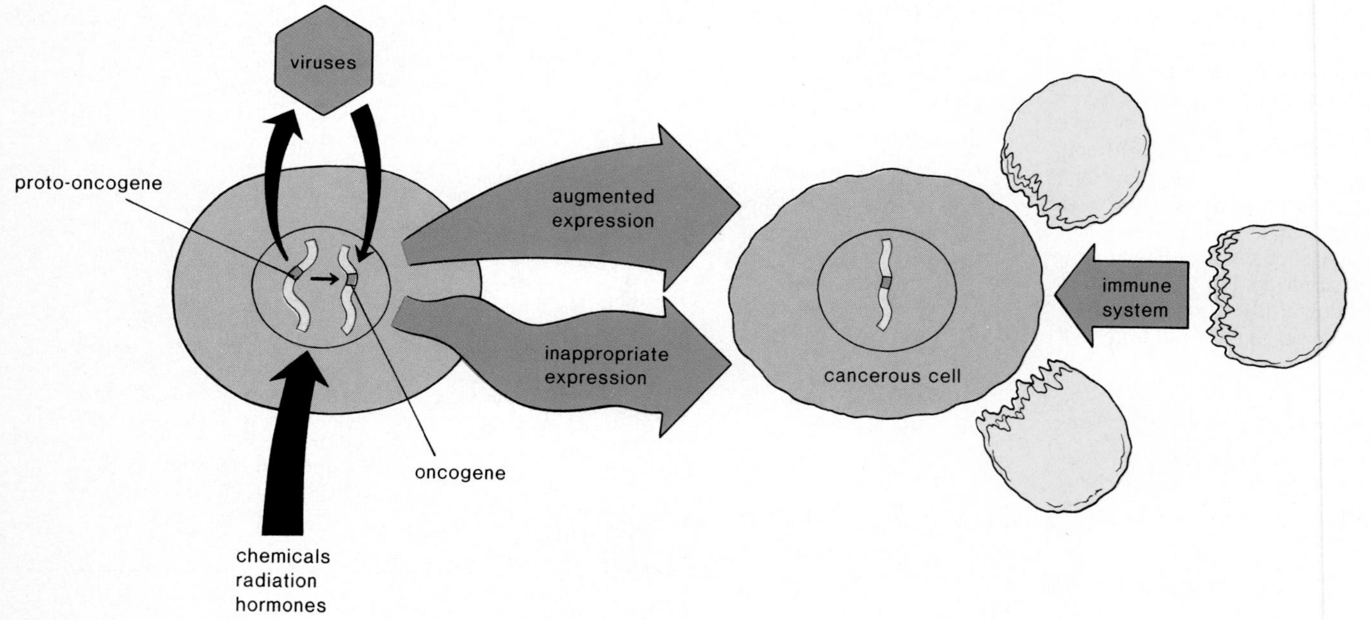

Figure 22.23
Summary of the development of cancer. A virus can pass an oncogene to a cell. A normal gene called a proto-oncogene, can become an oncogene due to a mutation caused by a chemical or radiation. The oncogene either expresses itself to a greater degree than normal or else expresses itself inappropriately. Thereafter the cell becomes cancerous. Cancer cells are usually destroyed by the immune system and the individual only develops cancer when the immune system fails to perform this function.

Cancer

Cancer is characterized by irregular (fig. 22.22) and uncontrolled growth of cells that do not stay in the organ where they arose. For a time, the cancer remains at the site of origin, but eventually cancer cells invade underlying tissues, become detached, and are carried by the lymphatic and circulatory systems to other parts of the body where new cancer growth may begin. The process by which cancer spreads to other parts of the body is called *metastasis,* and it is this tendency of malignant cells to metastasize widely that usually results in the death of the patient.

It has long been believed that cancer begins with a change in the DNA, but the exact nature of the change was unknown. Recently, investigators have been able to determine that cells contain genes, called *proto-oncogenes* (*proto* = before; *onco* = tumor), that can become **oncogenes,** cancer-causing genes (fig. 22.23). These genes are not alien to the cell; they are normal, essential genes that have undergone a mutation. By using recombinant DNA techniques, for example, investigators have shown that an oncogene that causes both lung cancer and bladder cancer differs from a normal gene by a change in only one nucleotide. It is now believed that almost any type of mutation can convert a proto-oncogene into an oncogene. For instance, a chromosomal translocation may place a normally dormant structural gene next to an active regulatory gene. If this structural gene is a proto-oncogene, the translocation may cause it to become an oncogene. Or the movement of a transposon might suddenly cause the transformation of a proto-oncogene into an oncogene. Also, environmental **mutagens** (any factor that increases the chances of a mutation), such as chemicals and X rays, can cause cancer when they bring about the conversion of a proto-oncogene to an oncogene. On the other hand, cancer-causing viruses most likely bring active oncogenes into a cell.

All cancer-causing viruses discovered so far are retroviruses (p. 542). *Retroviruses* contain a core of RNA rather than a core of DNA and contain an enzyme that carries out the transcription of RNA to DNA. This is the *reverse* of normal transcription, and contradicts the central dogma discussed on page 483. Retroviruses normally insert their newly transcribed DNA into host DNA, where it replicates along with host DNA. Only after a time does it cause the production of new viruses. The fact that retroviruses insert their DNA in host DNA allows them to plant a foreign oncogene into host DNA.

Sometimes it seems that a cell becomes cancerous when it contains one oncogene, and sometimes several oncogenes are required. Perhaps this difference can be attributed to the influence of the environment on the cell. Oncogene-bearing cells surrounded by normal neighbors remain normal, but if these normal cells are removed, the presence of a single oncogene seems to induce uncontrollable growth.

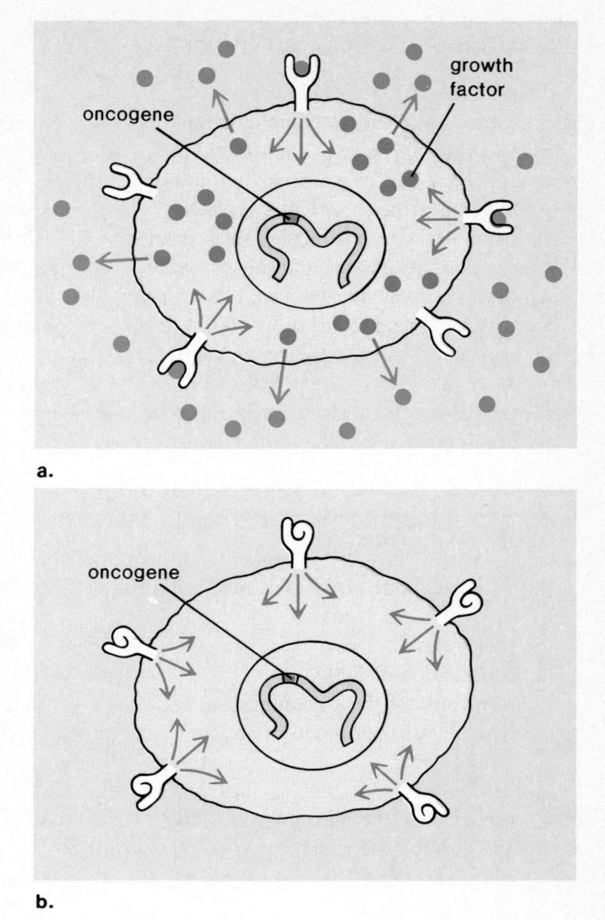

Figure 22.24
Function of oncogenes. a. *Oncogenes sometimes code for a growth factor (colored dots) that exits the cell where it is received by cell membrane receptors. Then the cell begins to grow and divide.* b. *Other times, oncogenes code for a growth-factor receptor that, due to a malformation, is capable of bringing about stimulation even when no growth factor is present.*

Function of Oncogenes

Cells transformed (fig. 22.22) by oncogenes often require few or no growth factors in the medium in order to grow and divide without limit. *Growth factors* are proteins that can stimulate a cell to grow and divide once it is received by a particular receptor on the cell's membrane. As figure 22.24 shows, it now appears that any gene that codes for a growth factor and any one that codes for a growth-factor receptor can be considered a proto-oncogene. Growth-factor genes typically become relatively inactive as the cell matures, which accounts for the fact that most cells are capable of only a few divisions during their life span. A mutation that suddenly causes a growth-factor gene to become active or more active than usual also accounts for the gene's conversion from proto-oncogene to oncogene.

Mutations can cause cancer to develop. Apparently they activate oncogenes and all those thus far discovered affect the amount of growth factor available to the cell or the function of a growth-factor receptor.

Summary

During replication, DNA becomes "unzipped" and then a complementary strand forms opposite to each original strand. DNA controls protein synthesis. During transcription, mRNA is made complementary to one of the DNA strands. It then contains codons and moves to the cytoplasm and becomes associated with the ribosomes. During translation, tRNA molecules, attached to their own particular amino acid, travel to the mRNA and through complementary base pairing, the tRNAs and therefore the amino acids in a polypeptide chain become sequenced in a predetermined way. This sequence of events pertains to structural genes.

The recombinant DNA technique allows eukaryotic DNA to be placed in the bacterium *E. coli,* where it usually replicates and functions normally. The many uses of the recombinant DNA technique are explored.

Regulatory control of structural genes in prokaryotes and eukaryotes is an important part of research today. A large portion of the eukaryotic chromosome has no known function. Other parts are movable elements or introns. After mRNA is formed, the sequences corresponding to the introns are excised and the exons may be mixed and matched to give different products.

These special features of the eukaryotic chromosome provide a means for mutations to occur. While germinal mutations often lead to nonfunctional enzymes, somatic mutations involving the transformation of proto-oncogenes to oncogens often accompany cancer. Oncogenes are involved in production or reception of growth factors.

Objective Questions

1. The backbone of DNA is made up of ____________ and ____________ molecules.
2. Replication of DNA is semiconservative, meaning that each new helix is composed of an ________ strand and a ____________ strand.
3. The base ____________ in DNA is replaced by the base uracil in RNA.
4. The DNA code is a ____________ code, meaning that every three bases stands for an ____________ .
5. The three types of RNA that are necessary to protein synthesis are ____________ , ____________ , and ____________ .
6. Which of these types carries amino acids to the ribosomes? ____________
7. Plasmids that are carrying a foreign gene contain ____________ DNA.
8. *E. coli,* transformed by recombinant DNA, multiplies and makes many copies of a foreign gene. This gene is said to have been ____________ .
9. Another name for transposons is ____________ .
10. When mRNA is processed within the eukaryotic nucleus, the portions complementary to ____________ in DNA are removed.

Study Questions

1. Describe the experiment that designated DNA rather than protein as the genetic material. (p. 475)
2. Describe DNA and RNA structure. (pp. 477–479)
3. Explain how DNA replicates. (pp. 479–480)
4. Various genetic diseases indicate that DNA controls the formation of proteins. Name and discuss some of these diseases. (p. 481)
5. If the code is TTA'TGC'TCC'TAA, what are the codons and what is the sequence of amino acids? (p. 484)
6. List the five steps involved in protein synthesis. (pp. 488–489)
7. Define operon, operator, and regulatory gene. Describe the two prokaryote models for control of gene transcription. (p. 490)
8. You are a scientist who has decided to "clone a gene." Tell precisely how you would proceed. (p. 492)
9. With reference to figure 22.20, discuss the recent findings in regard to eukaryotic chromosomes. (p. 494)
10. With reference to figure 22.23, explain the current findings regarding the development of cancer. (p. 498)

Answers to Objective Questions

1. sugar (deoxyribose), phosphate 2. old, new 3. thymine 4. triplet, amino acid 5. messenger RNA (mRNA), ribosomal RNA (rRNA), transfer RNA (tRNA) 6. tRNA 7. recombinant 8. cloned 9. movable elements or "jumping genes" 10. introns

Key Terms

anticodon (an″tĭ-ko′don) a "triplet" of three nucleotides in transfer RNA that pairs with a complementary triplet (codon) in messenger RNA. *486*

cloned (klōnd) DNA fragments from an external source that have been reproduced by *E. coli*. *492*

codon (ko′don) a "triplet" of three nucleotides in messenger RNA that directs the placement of a particular amino acid into a polypeptide chain. *484*

complementary base pairing (kom″plĕ-men′tă-re bās pār′ing) pairing of bases between nucleic acid strands; adenine is always paired with either thymine (DNA) or uracil (RNA) and cytosine is always paired with guanine. *477*

double helix (dŭ′b′l he′liks) a double spiral often used to describe the three-dimensional shape of DNA. *477*

messenger RNA (mes′en-jer) mRNA; a nucleic acid (ribonucleic acid) complementary to genetic DNA and bearing a message to direct cell protein synthesis at the ribosome. *483*

oncogene (ong′ko-jēn) a gene that contributes to the transformation of a cell into a cancerous cell. *499*

plasmid (plaz′mid) a circular DNA segment that is present in bacterial cells but is not part of the bacterial chromosome. *492*

polysome (pol′e-sōm) a cluster of ribosomes all attached to the same mRNA molecule and thus all participating in the synthesis of the same polypeptide. *487*

purines (pu′rinz) nitrogenous bases found in DNA and RNA that have two interlocking rings as in adenine and guanine. *477*

pyrimidines (pi-rim′ĭ-dinz) nitrogenous bases found in DNA and RNA that have just one ring as in cytosine and thymine. *477*

recombinant DNA (re-kom′bĭ-nant) DNA having genes from two different organisms often produced in the laboratory by introducing foreign genes into a bacterial plasmid. *492*

regulatory genes (reg′u-lah-tor″e jēnz) genes which code for proteins that are involved in regulating the activity of structural genes. *490*

replication (re″plĭ-ka′shun) the duplication of DNA; occurs when the cell is not dividing. *480*

ribosomal RNA (ri′bo-sōm″al) rRNA; RNA occurring in ribosomes, structures involved in protein synthesis. *486*

structural genes (struk′tūr-al jēnz) genes which direct the synthesis of enzymes and also structural proteins in the cell. *490*

template (tem′plāt) a pattern that serves as a mold for the production of an oppositely shaped structure; one strand of DNA is a template for the complementary strand. *479*

transcription (trans-krip′shun) the process that results in the production of a strand of mRNA that is complementary to a segment of DNA. *483*

transfer RNA (trans′fer) tRNA; molecule of RNA that carries an amino acid to a ribosome engaged in the process of protein synthesis. *486*

translation (trans-la′shun) the process involving mRNA, ribosomes, and tRNA molecules that results in a synthesis of a polypeptide having an amino acid sequence dictated by the sequence of codons in mRNA. *483*

Further Readings for Part Four

Ayala, F., and J. A. Kiger. 1984. *Modern genetics.* 2d ed. Menlo Park, CA: Benjamin/Cummings.

Baconsfield, P., G. Birdwood, and R. Baconsfield. 1980. The placenta. *Scientific American* 243(2):94.

Bishop, J. M. 1982. Oncogenes. *Scientific American* 246(3):80.

Cech, T. R. 1986. RNA as an enzyme. *Scientific American* 255(5):64.

Chambon, P. 1981. Split genes. *Scientific American* 244(5):60.

Chilton, M. 1983. A vector for introducing new genes into plants. *Scientific American* 248(6):50.

Cohen, S. N., and J. A. Shapiro. 1980. Transposable genetic elements. *Scientific American* 242(2):40.

Darnell, J. E. 1983. The processing of RNA. *Scientific American* 249(4):90.

———. 1985. RNA. *Scientific American* 253(4):68.

DeRobertis, E. M., and J. B. Gurdon. 1979. Gene transplantation and the analysis of development. *Scientific American* 241(6):74.

Dickerson, R. E. 1983. The DNA helix and how it is read. *Scientific American* 249(6):94.

Doolittle, R. F. 1985. Proteins. *Scientific American* 253(4):88.

Felsenfield, G. 1985. DNA. *Scientific American* 253(4):58.

Gilbert, W., and L. Villa-Komaroff. 1980. Useful proteins from recombinant bacteria. *Scientific American* 242(4):74.

Goldberg, Susan, and Barbara DeVitto. 1983. *Born too soon: Preterm birth and early development.* San Francisco: W. H. Freeman and Company.

Guttmacher, Alan F. 1986. *Pregnancy, birth, and family planning.* New York: New American Library.

Lake, J. A. 1981. The ribosome. *Scientific American* 245(2):84.

Leach, Penelope. 1980. *Your baby and child from birth to age five.* New York: Alfred A. Knopf.

Lein, A. 1979. *The cycling female: Her menstrual rhythm.* San Francisco: W. H. Freeman.

Mader, S. S. 1980. *Human reproductive biology.* Dubuque, IA: Wm. C. Brown Publishers.

Nilsson, Lennart. 1977. *A child is born,* rev. ed. New York: Delacorte Press.

Nomura, M. 1984. The control of ribosome synthesis. *Scientific American* 205(1):102.

Novick, R. P. 1980. Plasmids. *Scientific American* 243(6):102.

Ptashne, M. 1982. A genetic switch in a bacterial virus. *Scientific American* 247(5):128.

Rugh, R., et al. 1971. *From conception to birth: The drama of life's beginnings.* New York: Harper & Row, Publishers, Inc.

Shepard, J. F. 1982. The regeneration of potato plants from leaf-cell protoplasts. *Scientific American* 246(5):154.

Stahl, F. W. 1987. Genetic recombination. *Scientific American.* 256(2):90

Vander, A. J., and J. Sherman. 1985. *Human physiology.* 4th ed. New York: McGraw-Hill.

Volpe, E. P. 1983. *Biology and human concerns.* 3d ed. Dubuque, IA: Wm. C. Brown Publishers.

Weinberg, R. A. 1983. A molecular basis of cancer. *Scientific American* 249(5):126.

Part Five

Evolution and Diversity

Evolution depends on the retention of genetic changes that have been tested by the environment. This process, termed natural selection, results in adaptation to both the abiotic and biotic environment.

A gradual increase in chemical complexity produced the first cell(s) and this (these) evolved into all the forms of life we see about us. Taxonomists try to classify living things according to their evolutionary relationship; therefore, when we study taxonomy we are also studying evolutionary history. This text recognizes five kingdoms: Monera (bacteria), Protista (protozoans and unicellular algae), Fungi (molds and mushrooms), Plants (multicellular algae and terrestrial plants), and Animals.

Humans are primates, animals adapted to living in trees. They share a common ancestor with apes, some of whom still live in trees. The first human ancestor may have left the trees when grasslands replaced trees in Africa. Walking erect could have been an adaptation to this change of habitat. Later, tool use and intelligence evolved together, and allowed humans to eventually take up a hunting way of life.

Culture, which began with tool use, soon also included art, science, and religion. Unfortunately, twentieth-century culture tends to make humans unaware of their natural place in the biosphere.

Some frogs are adapted to reproducing on land. The female leaf frog lays eggs enveloped by leaves and repeatedly empties her bladder over them to keep them moist. Notice the circular disks that act like adhesive pads enabling this frog to climb up the smooth surfaces of the leaves.

23

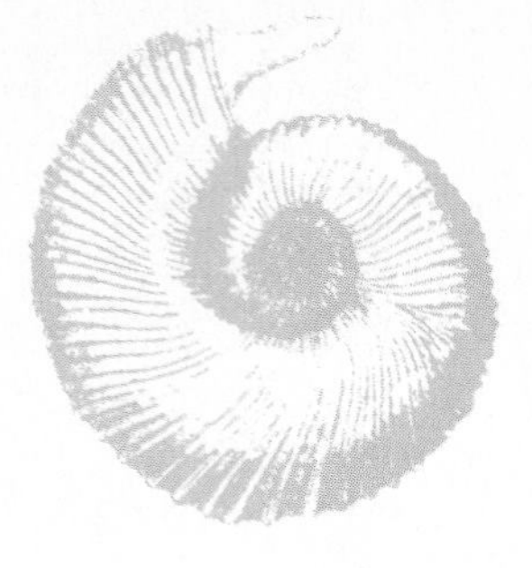

Evolution

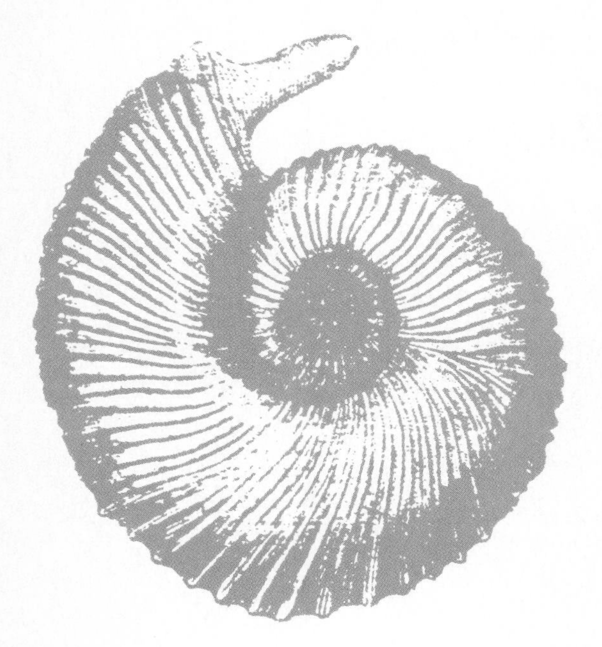

Chapter Concepts

1 Life evolved from the first cell(s) into all the forms of life, present or extinct.

2 The fossil record, comparative anatomy, embryology, and biochemistry all provide evidences of evolution.

3 Evolution, defined as a change in frequency of genes in the gene pool of a population, results in adaptation to the environment as a result of natural selection.

4 Natural selection occurs when the better adapted members of a population reproduce to a greater degree than the less well adapted members.

5 New species come about when a population is at first geographically isolated and then later reproductively isolated from other similar populations.

Chapter Outline

a.

b.

Figure 23.1
Fossils. a. *Impression of fern leaves on a rock.* b. *Small dinosaur fossil bones are being exposed by chipping away rocks holding them.*

Evolution is the process that explains the history and diversity of life. Data from various fields of biology give us evidence that evolution produced the myriad of organisms now present on earth.

Evidences for Evolution

Fossil Record

Our knowledge of the history of life is based primarily on the fossil record. **Fossils** (fig. 23.1) are the remains or evidence of some organism that lived long ago. Most fossils are formed when an organism is buried in mud or sand before the hard mineralized parts have decayed. A fossil may be the remains of this part, or it may be the impression or mold that the part made in the rock developing about it. Sometimes, fossils are formed by a replacement of the original organic material by a durable mineral, such as silica.

The fossil record depicted in figure 23.2 shows that life began with the simplest of organisms and progressed with ever increasing complexity. A reasonable explanation is that the more primitive forms gave rise to more complex forms of life. The fossil record has provided evidence of the presence of intermediate forms between groups of organisms. *Archaeopteryx* (fig. 23.3a) appears to be a flying reptile except that it clearly had the feathers and beak of a bird. A living example of the same principle is *Peripatus* (fig. 23.3b), a two-inch long animal that looks like a caterpillar and has characteristics of both the annelids (segmented worms) and arthropods (insects, crustaceans).

Figure 23.2
The fossil record provides a history of life. The record indicates that primitive, single-celled organisms were the first life forms to appear during the Proterozoic era. In time, more complex cellular forms appeared, followed by multicellular plants and animals. After that, there was an increase in complexity.

Era	Period	Epoch	Years from Start of Period to Present
Cenozoic	Quaternary	Recent	10,000
		Pleistocene (Ice Age)	3 million
	Tertiary		63 million
Mesozoic "Age of Reptiles"	Cretaceous		135 million
	Jurassic		181 million
	Triassic		230 million
Paleozoic	Permian		280 million
	Carboniferous		345 million
	Devonian		405 million
	Silurian		425 million
	Ordovician		500 million
	Cambrian		600 million
Proterozoic			1.5 billion
			2.5 billion
Archeozoic			4.5 billion

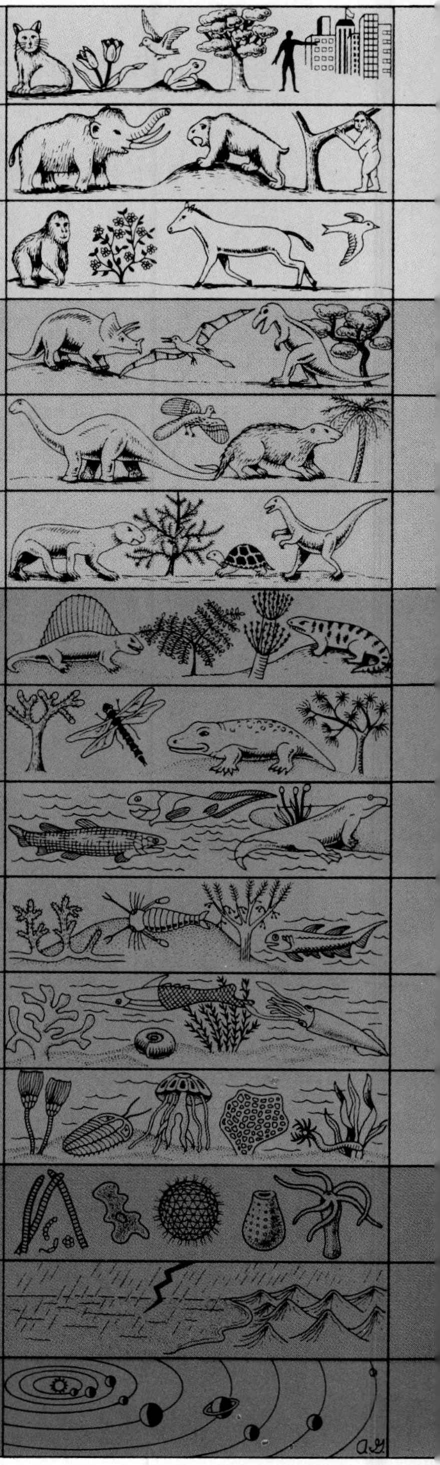

Plant Life	Animal Life
Increase in the number of herbaceous plants	Age of human civilization
Extinction of many species of plants	Great mammals such as woolly mammoth and saber-toothed tiger became extinct First human social life
Dominance of land by angiosperms	Dominance of land by mammals, birds, insects Mammalian radiation First humans
Angiosperms prevalent, gymnosperms decline Trees resembling modern-day maples, oaks, and palms flourish	Dinosaurs reach peak, then become extinct Second great radiation of insects First primates
Gymnosperms such as cycads and conifers still prevalent	Dinosaurs large, specialized, more abundant First mammals appear First birds appear
Dominance of land by gymnosperms and ferns Decline of club mosses and horsetails	First dinosaurs appear Mammallike reptiles evolve
Gymnosperm and angiosperm(?) evolve	Expansion of reptiles Decline of amphibians
Age of great coal forests including club mosses, horsetails, and ferns	"Age of Amphibians" First great radiation of insects First reptiles appear
Expansion of land plants; first forests of club mosses, horsetails, and ferns	"Age of Fishes" First land vertebrates, the amphibians, appear
First vascular plants, modern groups of algae and fungi	First air-breathing land animals, such as land scorpion, appear Rise of fishes
Invasion of land by plants(?)	Diverse marine invertebrates, coral and nautaloid common First vertebrates appear as fish
Marine algae common	Diverse primitive marine invertebrates, trilobites common Animals with skeletons appear
Multicellular acoelomate and coelomate animals evolve Eukaryotic protists and fungi evolve	
Prokaryotes abundant	
Anaerobic and photosynthetic bacteria evolve Formation of earth and rest of solar system	

Figure 23.3
a. *An artist's conception of* Archaeopteryx. *This fossil had features common to both reptiles and birds. Note the indication of feathers, a birdlike feature, and the long bony tail, a reptilian feature.*
b. Peripatus. *This animal has features common to both annelids and arthropods. It is obviously a segmented animal; its excretory, reproductive, and nervous systems are similar to those of the annelids, while its circulatory and respiratory systems are similar to those of the arthropods.*

Table 23.1	*The Classification of Modern Humans*
Kingdom	Animalia (animals)
Phylum	Chordata (chordates)
Class	Mammalia (mammals)
Order	Primates (primates)
Suborder	Anthropoidea (anthropoids)
Superfamily	Hominoidea (hominoids)
Family	Hominidae (hominids)
Genus	*Homo* (humans)
Species	*Sapiens* (modern humans)

These two animals may not be those of the precise species that respectively gave rise to birds or arthropods, but they do indicate a relationship between reptiles and birds and a relationship between annelids and arthropods.

Comparative Anatomy

A comparative study of the anatomy of groups of organisms has shown that each has a *unity of plan.* For example, the reproductive organs of all flowering plants are basically similar, and all vertebrate animals have essentially the same type of skeleton. Unity of plan allows organisms to be classified into various groups. Organisms most similar to one another are placed in the same **species,** similar species are placed in a **genus,** similar genera in a family; therefore, we proceed from **family** to **order** to **class** to **phylum** (animals) or **division** (plants) to **kingdom.** The classification of any particular organism (table 23.1) indicates to what kingdom, phylum, class, order, family, genus, and species the organism belongs. According to the **binomial system** of naming organisms, each organism is given a two-part name, which consists of the genus and species to which it belongs. For example, a human is *Homo sapiens* and the domesticated cat is *Felis domestica.* **Taxonomy** is the branch of biology that is concerned with classification, and biologists who specialize in classifying organisms are called taxonomists.[1]

A *unity of plan* is explainable by descent from the same **common ancestor.** Species that share a recent common ancestor will share a large number of the same genes and will be quite similar to each other and to this ancestor. Species that share a more distant common ancestor will have fewer genes in common and will be less similar to each other and to this ancestor, because differences arise as organisms continue on their own evolutionary pathways. This principle allows biologists to construct **evolutionary trees,** diagrams that tell how various organisms are believed to be related to one another. All evolutionary trees have a branchlike pattern (fig. 23.7), indicating that evolution does not proceed in a single steplike manner; rather, evolution proceeds by way of common ancestors that often give rise to two different groups of organisms. For example, reptiles are believed to have produced both birds and mammals.

Even after related organisms have become adapted to different ways of life, they may continue to show similarities of structure. For example, the forelimb of all vertebrates contains the same fundamental bone structure (fig. 23.4) despite their specific specializations. Similarities in structure that have arisen through descent from a common ancestor are called **homologous structures.** Homologous structures indicate that organisms are related. Sometimes two groups of organisms have structures that function similarly but are constructed differently. In contrast to homologous structures, **analogous structures,** such as an insect wing and a bird wing, have similar functions but differ in their anatomy, and therefore we know they evolved independently of one another.

Comparative Embryology

Some groups of organisms share the same type of embryonic stages. For example, the larva of certain lower chordates (the phylum containing vertebrates) is strikingly similar to that of certain echinoderms (e.g., starfish). As we would expect, the embryonic stages of all vertebrates are also similar (fig. 23.5). Therefore, during development a human embryo at one point has gill pouches, even though it will never breathe by means of gills as do fishes, and a rudimentary tail, even though it will never have a long tail as do some four-legged vertebrates. In this way, embryological observations indicate evolutionary relationships.

[1]The classification system utilized in this text is given in Appendix C.

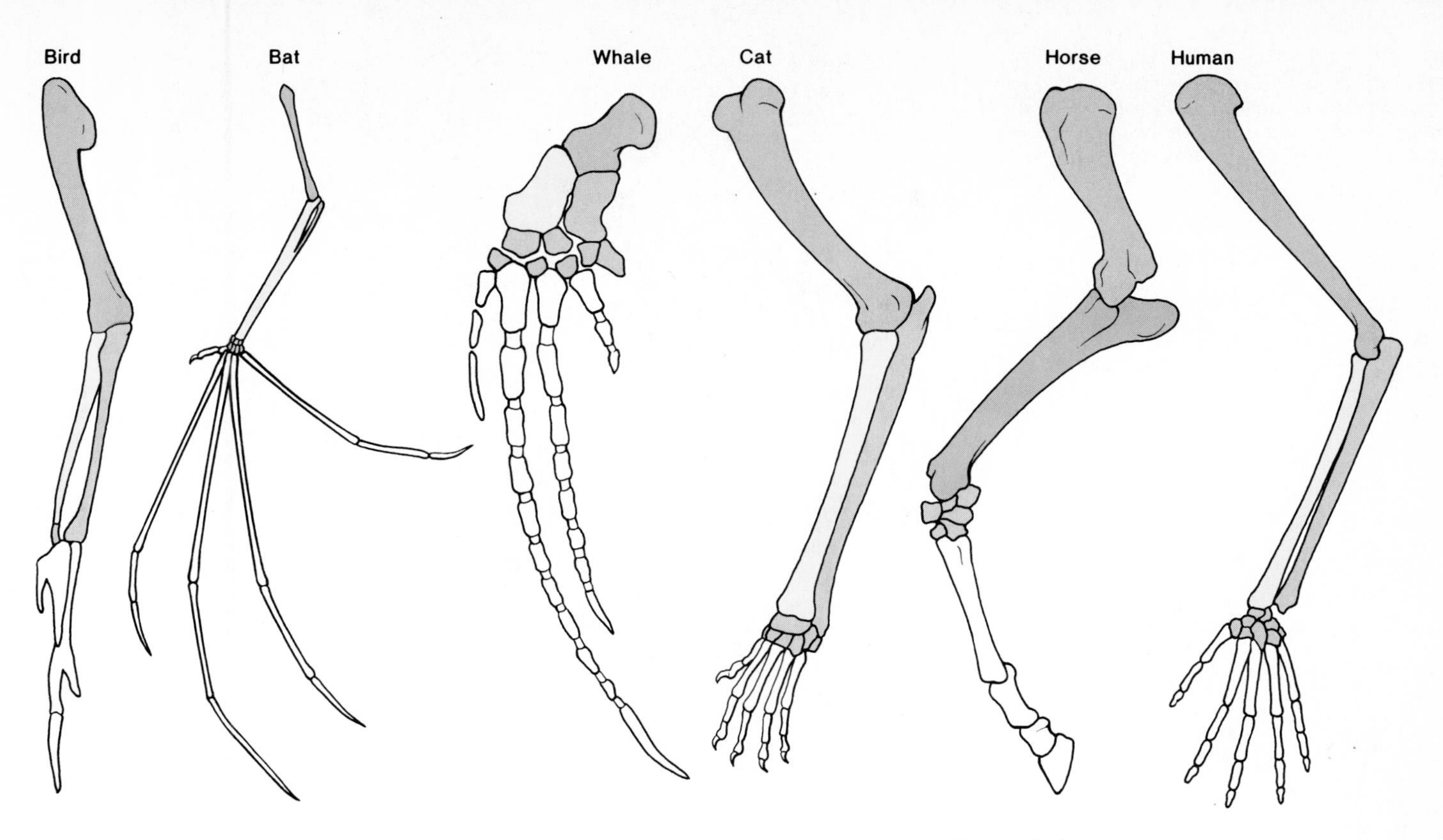

Figure 23.4
Homologous structures. The bones are coded so that you may note the similarity in the bones of the forelimbs of these vertebrates. This similarity is to be expected since all vertebrates trace their ancestry back to a common ancestor.

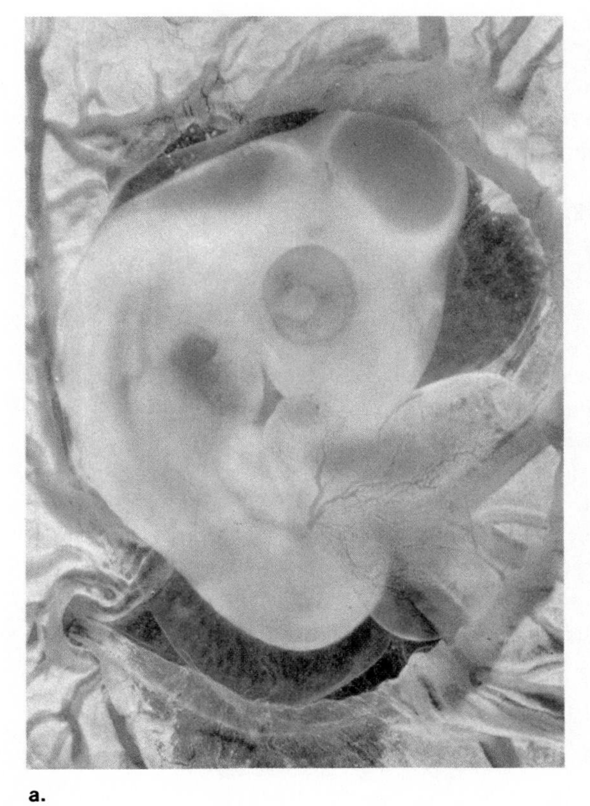
a.

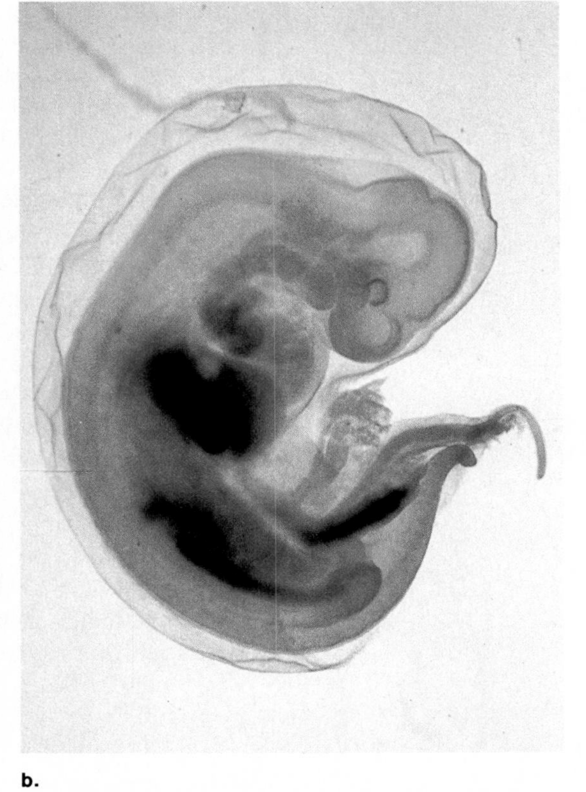
b.

Figure 23.5
A chick and a pig embryo at comparable stages have many features in common. a. *Chick embryo.* b. *Pig embryo.*

Figure 23.6
Vestigial structures. Human beings have various vestigial structures such as those shown. These show our relationships to animals in which these structures are fully developed and functional.

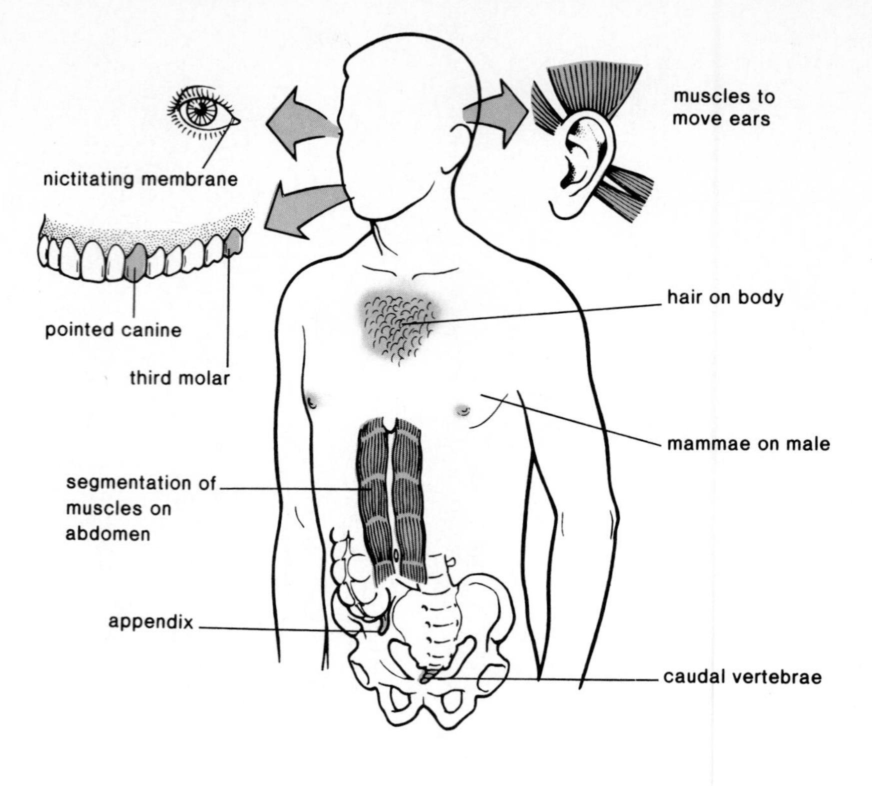

Vestigial Structures

An organism may have structures that are underdeveloped and seemingly useless, and yet these same structures may be fully developed and functional in related organisms. Such structures are called **vestigial.** Figure 23.6 illustrates numerous vestigial structures in humans. The presence of these structures is understandable when we realize that related organisms share genes in common.

Genes that code for vestigial structures are retained for various genetic reasons, some of which are discussed in this chapter, even though they do not contribute to the organism's adaptation to the environment. Some retained genes are not even expressed. Modern chickens do not have teeth, but they have been found to have genes for the production of tooth dentin.

Comparative Biochemistry

Almost all living organisms use the same basic biochemical molecules, including DNA, ATP, and many identical or nearly identical enzymes. It would seem that these molecules evolved very early in the evolution of life and have been passed on ever since.

Analyses of amino acid sequences in certain proteins like hemoglobin and cytochrome C have been done in various animals in order to determine how distantly related they are. The rationale is that the number of differences will reflect how long ago the two species shared a common ancestor. Also, analyses of DNA nucleotide differences of the genome (all of the genes) have been done for the same purpose. Figure 23.7 shows the results of one such study. Investigators have been gratified to find that evolutionary trees based on biochemical data are quite similar to those based on anatomical data. Whenever the same conclusions are drawn from independent data, they substantiate scientific theory, in this case organic evolution, even more than usual.

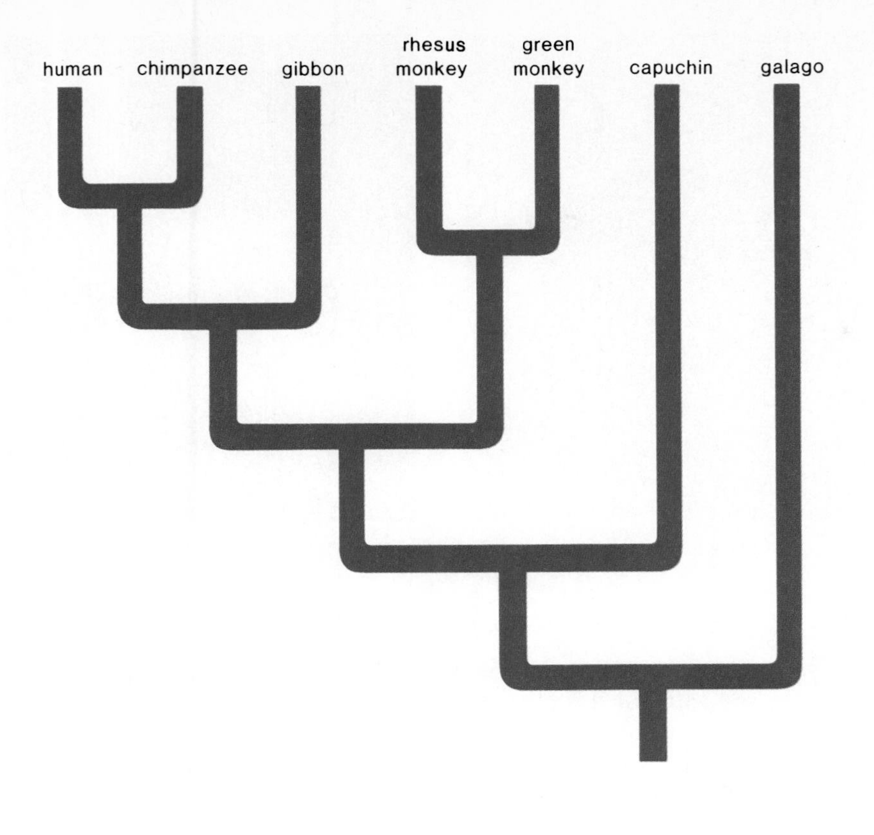

Figure 23.7
Evolutionary tree of primate species based on a biochemical study of their genomes. The length of the branches indicates the approximate number of nucleotide pair differences that were found between groups.

Figure 23.8
The quagga was striped on the front end like a zebra, but had a solid, chestnut-colored coat like a horse on the hind end. Although it became extinct a century ago, scientists have been able to clone a portion of its DNA chemically extracted from tissue removed from museum hides. Comparative biochemical studies indicate that the quagga is a zebra. Such a study emphasizes only the physical and chemical aspects of life and does not in the least suggest that biochemists can bring whole organisms back from the dead.

Where biochemists formerly restricted their studies to living organisms, they have recently begun to study extinct organisms and even fossils. For example, investigators have extracted proteins and DNA from a scrap of muscle on the pelt of a 140-year-old museum specimen of a quagga (fig. 23.8), an animal that became extinct a century ago. Cloning provided enough DNA (see page 492) to establish that a quagga is a zebra, not a true horse. Protein studies showed that a quagga is most closely related to a Burchell's zebra (see cover of this text). Such studies are only possible because all living things share the same types of chemical molecules.

Figure 23.9
The African ostrich (a) *and the South American rhea* (b) *are not closely related. They look and behave similarly because they have been exposed to the same type of environment.*

Biogeography

In chapter 32, we will have an opportunity to study the geographic distribution of plants and animals. It is observed that similar but geographically separate environments have different plants and animals that are similarly adapted. For example, plants of the cactus family are found in the deserts of southwestern North America, while members of the spurge family are found in the deserts of Africa. Both types of plants have made similar adaptations to arid habitats. Also, the African ostrich and the South American rhea (fig. 23.9) are not closely related. Yet they look and behave similarly because they have been exposed to the same type of environment. These observances show that the evolutionary process causes organisms to be adapted to their environments.

Evolution explains the history and diversity of life. Evidences for evolution can be taken from the fossil record, comparative anatomy, embryology, and biochemistry. Also, vestigial structures and biogeography support the occurrence of evolution.

Evolutionary Process

In order to study the mechanism of evolution, it is best to consider the evolution of populations rather than the evolution of major groups of organisms. Since this is evolution in miniature, it is called microevolution.

For the sake of discussion, a **population** will be defined as a group of interbreeding individuals living in a particular area. The various alleles and their frequencies constitute the genetic makeup of the population, which is very often referred to as the **gene pool.** If there were nothing to upset the equilibrium, we would expect the gene pool to remain constant generation after generation. For example, suppose it is known that one-fourth of all persons in a human population are homozygous dominant for widow's peak, one-half are heterozygous, and one-fourth are homozygous recessive for continuous hairline. What will be the ratio of genotypes in the next generation?

Using the key given in the previous chapter, W = widow's peak and w = continuous hairline, we can describe the population in this manner:

$$\tfrac{1}{4}\, WW + \tfrac{1}{2}\, Ww + \tfrac{1}{4}\, ww$$

Necessarily, the homozygous dominant individuals will produce one-fourth of all the gametes of the population, and these gametes will all carry the dominant allele, *W;* the heterozygotes will produce one-half of all the gametes, but one-fourth will be *W* and one-fourth will be *w;* the homozygous recessive will produce one-fourth of all the gametes and they will be *w*. Therefore, in summary, one-half of the gametes will be *W* and one-half will be *w*.

Assuming that all possible gametes have an equal chance to combine with one another, then, as the Punnett square shows, the next generation will have exactly the same ratio of genotypes as the previous generation.

	½ *W*	½ *w*
½ *W*	¼ *WW*	¼ *Ww*
½ *w*	¼ *Ww*	¼ *ww*

Results:

¼ *WW* + ½ *Ww* + ¼ *ww*

To take another example, let's suppose that 64 percent of the population is homozygous dominant and does not have PKU; 32 percent are heterozygous; and 4 percent are homozygous recessive and have PKU. Figure 23.10a shows that the genetic makeup of the next population will be exactly the same as the parental generation. The same results would be obtained no matter what alleles we consider and no matter how many generations were included. This means that (*a*) dominant alleles do not tend to take the place of recessive alleles and that recessive alleles do not tend to disappear and (*b*) sexual reproduction in and of itself cannot bring about a change in the allele frequency of the population.

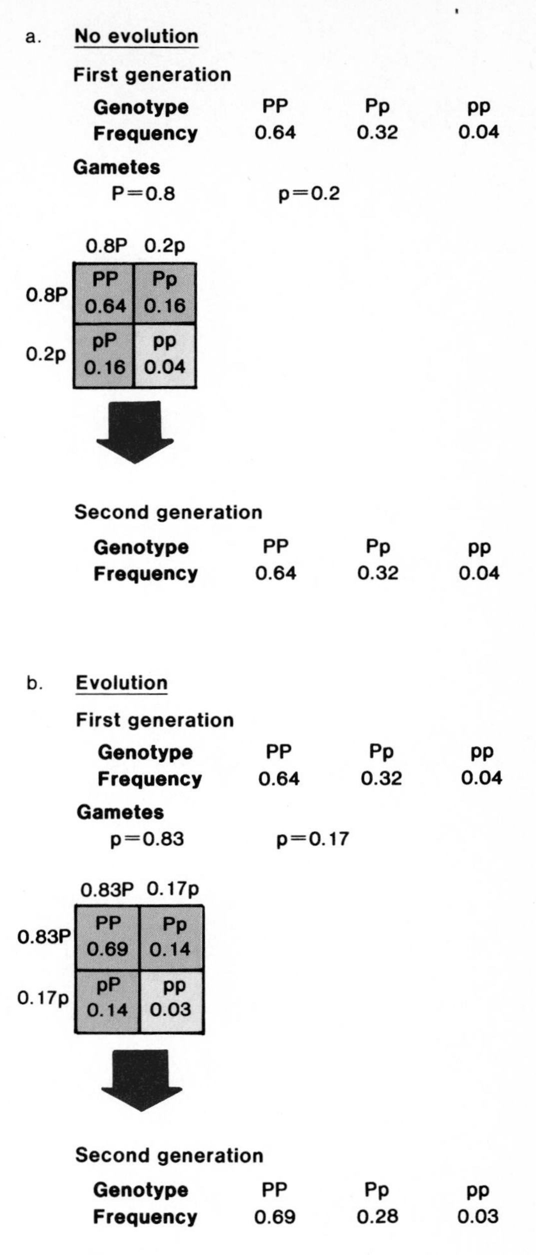

Figure 23.10
Effect on gene pool when evolution does not occur and when it does occur. a. *Gene pool frequencies are constant generation after generation if evolution does not occur.* b. *Gene pool frequencies change when evolution occurs.*

Hardy-Weinberg Law

The Hardy-Weinberg law states that, all things being equal, the gene pool stays constant and may be described by means of the quadratic equation:

$$p^2 + 2pq + q^2 = 1.00$$

In this case p represents the frequency of the dominant allele and q represents the frequency of the recessive allele. Therefore:

p^2 = homozygous dominant individuals

q^2 = homozygous recessive individuals

$2pq$ = heterozygous individuals

The real value of this mathematical approach to population genetics is that by observation or inspection it is possible to determine the percentage of individuals who are recessive, and from this it is possible to calculate the frequencies of the alleles and genotypes. Frequencies are given in decimals rather than fractions or percentages. For our example, then,

$$\text{if } q^2 = 0.25,\quad p^2 = 0.25,\quad 2pq = \underline{0.50}\ \ 1.00 \qquad q = 0.50,\quad p = \underline{0.50}\ \ 1.00$$

Notice that $p + q$ *(frequencies of the two alleles) must equal 1.00 and* $q^2 + p^2 + 2pq$ *(frequencies of the various genotypes) must also equal 1.00.*

To take another example, suppose by inspection we determine that 1 percent of the population has dimples. Therefore, 99 percent of the population does not have dimples. Of these, how many are homozygous dominant? How many are heterozygous?

To answer these questions, first convert 1 percent to a decimal. Then we know that $q^2 = 0.01$ and that therefore $q = 0.1$. Since $p + q = 1.0$, then we know that $p = 0.9$ and that therefore p^2 (frequency of the population that is homozygous dominant) $= 0.81$. To determine the frequency of the heterozygote, we simply realize that so far we have accounted for only 0.82 of the population, and that therefore $0.18 =$ heterozygous. Or if you prefer, calculate that $2pq = 0.18$. In summary we have found that

Homozygous recessive	= 0.01	= 1 percent do have dimples
Homozygous dominant	= 0.81	= 99 percent do not have dimples
Heterozygous	= 0.18	

Practice Problems

1. A student places 600 fruit flies with the genotype *Ll* and 400 with the genotype *ll* in a culture bottle. Assuming that evolution does not occur, what will be the genotype frequencies in the next generation and each generation thereafter?
2. Four percent of the members of a population of pea plants are short. What is the frequency of the recessive allele and the dominant allele? What are the genotype frequencies in this population?
3. Twenty-one percent of a population is homozygous dominant, 49 percent are heterozygous, and 29 percent are recessive. What percentage of the next generation is predicted to be recessive?

*Answers to problems are on page 526.

Theoretically, it would be possible for the gene pool of a population to remain constant generation after generation. In other words, sexual recombination, in and of itself, cannot alter gene frequencies in large populations. But, in fact, the gene pool rarely if ever remains constant and the Hardy-Weinberg law recognizes this by adding the qualification: The gene pool stays constant only if (1) the population is large and mating is random, (2) no mutations occur, (3) there is no gene flow, and (4) there is no natural selection. *When the gene pool does not stay constant, then evolution has occurred* (fig. 23.10b).

Evolution is believed to take place by means of a gradual change in the frequency of genes in the gene pool of a population. Such change continuing for a vast period of time (millions of years) will produce large changes.

Synthetic Theory

The explanation of evolution in terms of modern genetic principles is a synthesis; it takes data and hypotheses from all sources and blends them into one whole. According to the synthetic theory, the evolutionary process requires two steps (table 23.2):

1. Production of genotype and therefore phenotype variations.
2. Sorting out of these variations through successive generations.

***Table 23.2** Mechanism of Evolution*

Produce Variation	Reduce Variation
Mutations	Genetic drift
Gene flow	Natural selection
Recombination	

Production of Variations

It is readily apparent that members of a human population (fig. 23.11) vary, but just as humans differ one from the other, so do members of other populations. The daisies on the hill and the earthworms in your backyard are not genotypically or phenotypically the same. Metabolic, structural, and behavioral differences exist between them. Genetic variations in the gene pool of sexually reproducing diploid organisms have three sources: mutations, gene flow, and recombinations.

Mutations are the raw material for evolution and the ultimate source of all variations found in natural populations. Many times, observed mutations, such as those that cause human genetic diseases, seem to harm rather than benefit an individual. This may be because members of a population are so adapted (suited) to an environment that only nonbeneficial changes are apparent. Nevertheless, recessive nonobserved mutations may be occurring that could be beneficial should the environment change.

In organisms that lack sexual reproduction, such as bacteria, genotype variability is dependent entirely on mutations. This is sufficient because the short generation time of these organisms allows new mutations to be immediately tested by the environment.

Gene flow, which occurs when individuals immigrate and emigrate between populations, brings new genes into the pool of each population. The sharing of genes can cause the two populations to become similarly adapted but can also keep each one from becoming very closely adapted to a local environment.

Recombination of genes occurs during meiosis and fertilization. During meiosis, crossing over and independent assortment produces unlike gametes. During fertilization, gamete union also brings about a genotype unlike those of the parents.

Recombination is an important source of variability in sexually reproducing organisms. It allows different combinations of genes to be tested by the environment. After all, it is the entire genotype, represented by the combination of genes, not an individual gene, that determines whether the individual is suited to the environment. The phrase *unity of the genotype* means that the genotype should be viewed not as a composite of individual genes but as a cohesive whole. Only then is it possible to take into account gene interactions (fig. 23.12), such as *pleiotropy* (one gene can affect several different characteristics), *polygeny* (one characteristic can be controlled by several genes), and regulatory genes, which modify the action of other genes.

Genetic variation in a population of sexually reproducing diploid organisms has three sources: mutation (both chromosomal and gene), gene flow (emigration or immigration), and recombination (at the time of meiosis or fertilization).

Reduction in Variations

Genetic drift and *natural selection* both act in such a way that variations in a population are sorted out and reduced. But only natural selection consistently results in adaptation. Adaptations may be structural (land animals breathe by means of lungs), physiological (desert animals make do with metabolic water), or behavioral (some animals forage at night and others forage in the daytime).

Figure 23.11
Variations among individuals of a population. It is easy for us to note that humans vary one from the other. However, the same is true for populations of any organism.

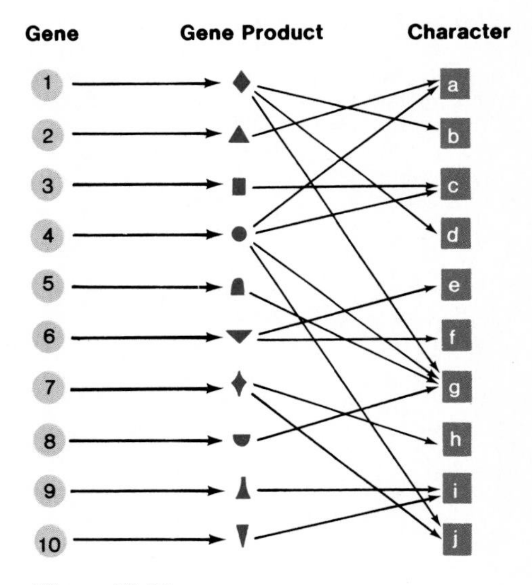

Figure 23.12
Gene interactions. This diagram illustrates that one gene can affect many characteristics of the individual (pleiotropy) and one characteristic can be controlled by several genes (polygeny).

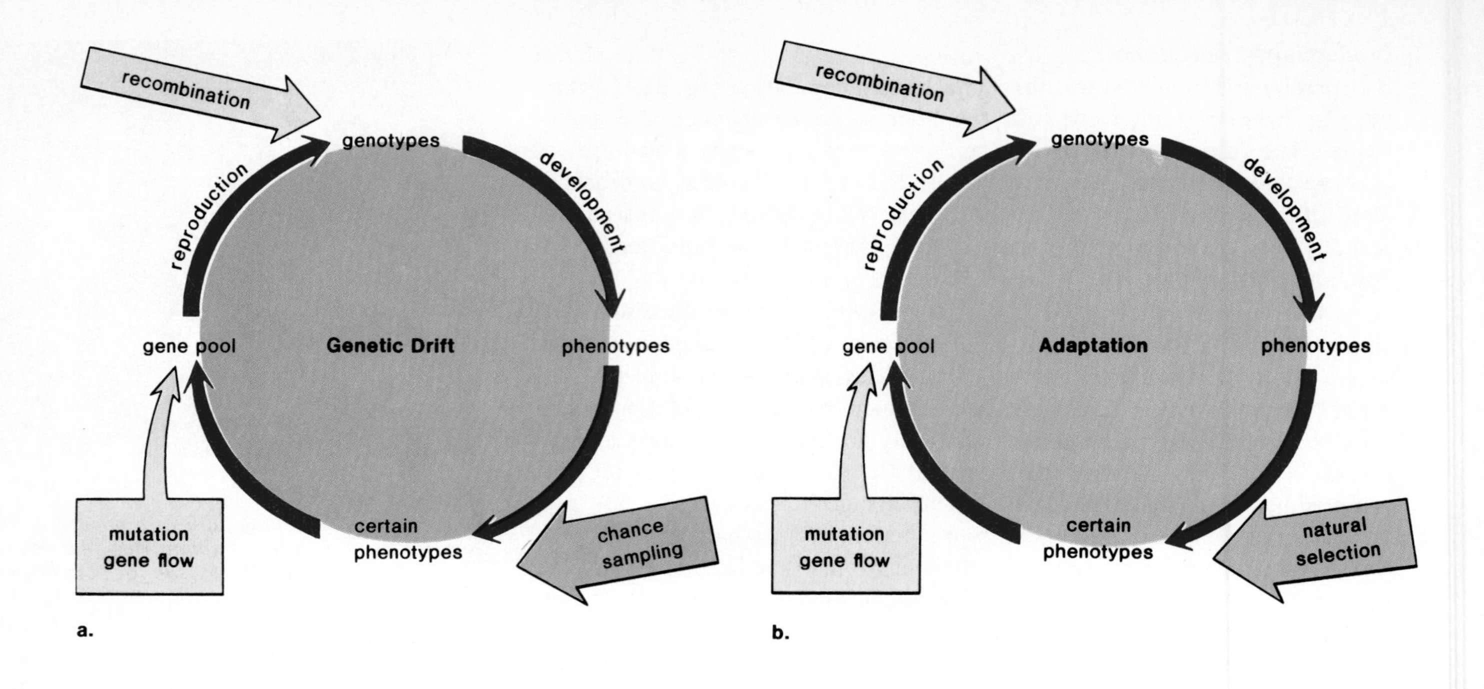

Figure 23.13
Genetic drift versus adaptation. In both cases, mutation, gene flow, and recombination are sources of genetic variation. The various genotypes develop into various phenotypes. a. *A chance sampling of these phenotypes can lead to genetic drift.*
b. *Selection of adapted phenotypes can lead to adaptation to the environment.*

Genetic drift, as diagrammed in figure 23.13a, is a reduction in gene pool variation that occurs purely due to chance. It should be visualized as a chance drifting toward certain genes so that others are eliminated. In the diagram, *chance sampling* means that only a few individuals among all the various phenotypes available produce offspring.

Genetic drift operates in both large and small populations, but it is significant only in small populations. Imagine that by chance a small number of individuals, representing a fraction of the gene pool, found a colony, and then during gamete formation, only certain of their genes were passed on to the next generation. As a result, a severe reduction in genetic variation compared to the original population has taken place and genetic drift has occurred. This combination of circumstances has been historically observed and is called the *founder principle.* For example, an investigation of a small religious group, called the Dunkers, showed that the blood type was 60 percent blood group A. Since the frequency of this blood group is 40 percent in the United States and 45 percent in West Germany (the country from which the Dunkers are derived), the high occurrence of blood group A among the Dunkers can only be explained by drift.

Genetic drift is *not* expected to produce adaptation to the environment because phenotypes are not selected for reproduction; rather, chance alone determines who will reproduce. We can imagine that a natural disaster, for example, would severely reduce a large population so that only a few individuals would remain to reproduce. Only those genes that happened to be passed on to the next generation would then be available in the gene pool. As gene pool variation decreases, the possibility of fixation of a few genes and therefore genetic drift increases.

Natural selection, as diagrammed in figure 23.13b, is the process by which populations become better adapted to their environment. Individuals that are more suited to the environment are the very ones that are more likely to survive and reproduce. In this way **adaptation** of the species to the environment tends to increase with each generation. Adaptations to various environments explains the diversity of life.

Natural selection operates by means of both biotic and abiotic factors that influence survival and reproduction. Biotic factors involve other organisms in the environment and abiotic factors involve physical conditions. So, for example, an organism that is better able to escape a predator would be selected, that is, would be more likely to have offspring. Similarly, an organism better able to withstand the climate would be selected and again would be more likely to have offspring. "Survival of the fittest," a phrase often used to refer to natural selection, means, according to modern thinking, that a better adapted genotype has a greater *probability* of producing offspring than a less adapted one.

Figure 23.10b illustrates that natural selection brings about a change in frequency of genes in the gene pool. This occurs when better adapted individuals increase their contribution to the gene pool and less well adapted individuals decrease their contribution. As this process continues generation after generation, the gene pool naturally tends toward stabilization.

Figure 23.14 indicates that natural selection has two common effects: (*a*) to stabilize variations and (*b*) to direct variations. **Stabilizing selection** tends to eliminate atypical phenotypes and enhances adaptation of the population to current environmental circumstances, but **directional selection** selects an extreme phenotype better adapted to a new environmental circumstance. Directional selection occurs during a time when the environment is changing rapidly or when members of a population are adapting to a new environmental situation.

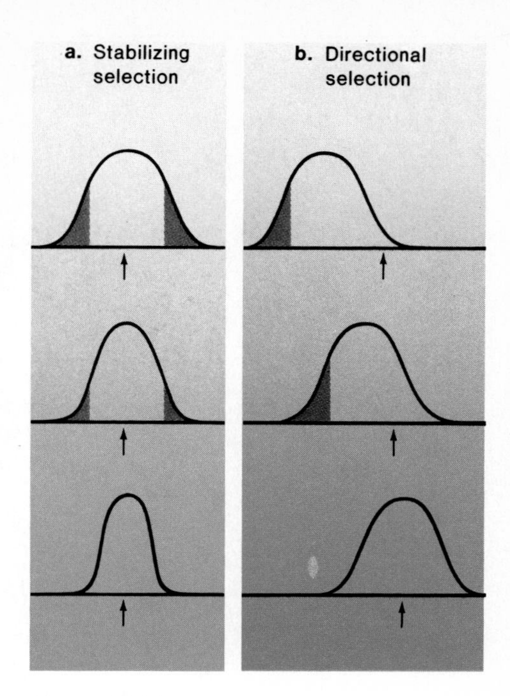

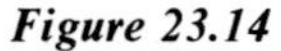

Figure 23.14
Two types of selection. a. *If stabilizing selection occurs, variation is decreased as the extreme phenotypes are eliminated.* b. *If directional selection occurs, an extreme phenotype is favored over other phenotypes.*

Maintenance of Variations

Even though stabilization of the gene pool is an expected result of the evolutionary process, genetic and phenotypic variation may still be retained for reasons such as those listed in table 23.3.

Diploidy helps maintain variation because recessive genes may remain hidden in the gene pool, serving as a potential source of future phenotypic variations. Pleiotropic genes may produce variations aside from those that adapt the organism to a current environment and when a characteristic is controlled by several genes (polygeny) a change in one of these can produce an unexpected variation. Such phenotypic variations may be neutral or harmful to the organism.

The environment may actually promote the maintenance of two distinctly different phenotypes due to opposing selection pressures. In Africa, persons with normal red blood cells carry oxygen more efficiently but are more susceptible to malaria, while those with sickle-shaped cells escape malaria but die from sickle-cell anemia. Clearly the heterozygote is best adapted, but in order to perpetuate it, both homozygotes must be maintained. Cases such as this are called *balanced polymorphism* (*poly* = many; *morphism* = shape).

The environment itself may vary and therefore may call for different adaptations at different times, such as seasonal changes within certain environments. Members of a population may become generally adapted, never specializing for any particular season, or they may become *polymorphic,* having a different phenotype for each season. Polymorphic adaptation is demonstrated by the arctic hare, which has a white coat in the winter and a brown coat in the summer.

Table 23.3 Maintenance of Variation

Genetic Causes	Ecological Factors
Diploidy	Opposing ecological pressures
Heterozygote superiority (heterosis)	Geographic differences
Polygeny and pleiotropy	Changing environment

Genetic drift and natural selection act in such a way that gene pool variations are reduced, but only natural selection consistently results in adaptation. Natural selection has two possible primary effects: (1) to stabilize population variations and (2) to direct population variations. Eventually, stabilization always occurs, except that genetic variations are maintained because of genetic and ecological reasons (table 23.3).

Charles Darwin and the Theory of Natural Selection

At the age of twenty-two, Charles Darwin signed on as a naturalist with the HMS *Beagle*, a ship that took a five-year trip around the world in the latter half of the nineteenth century. Because the ship sailed in the southern hemisphere where life is most abundant and varied, Darwin encountered forms of life very different from those in his native England.

Even though it was not his original intent, Darwin began to realize, and gather evidence, that life forms change over time and from place to place. He read a book by Charles Lyell, a geologist, who suggested the world is very old and has been undergoing gradual changes for many, many years. He found the remains of a giant ground sloth and an armadillo on the east coast of South America and wondered if these extinct forms were related to the living forms of these animals. When he compared the animals of Africa to those of South America he noted that the African ostrich and the South American rhea, although similar in appearance, were actually different animals. He reasoned that they had a different line of descent because they were on different continents. When Darwin arrived at the Galápagos Islands he began to study the 13 species of finches (fig. 23.18) whose adaptations could best be explained by assuming they had diverged from a common ancestor. With this type of evidence, Darwin concluded that species do *evolve* (change) with time.

When Darwin arrived home, he spent the next twenty years gathering data to support the principle of organic evolution. His most significant contribution to this principle was his theory of *natural selection,* which explains how a species becomes adapted to its environment. Before formulating the theory, he read an essay on human population growth written by Thomas Malthus. Malthus observed that although the reproductive potential of humans is great, there were many environmental factors, such as availability of food and living space, which tend to keep the human population within bounds. Darwin applied these ideas to all populations of organisms. For example, he calculated that a single pair of elephants could have 19 million descendants in 750 years. He realized that other organisms have even greater reproductive potential than this pair of elephants; yet usually the number of each type organism remains about the same. Darwin decided there is a constant *struggle for existence,* and only a few survive to reproduce. The ones that do survive and contribute to the evolutionary future of the species are, by and large, the better adapted individuals. This so-called "survival of the fittest" causes the next generation to be better adapted than the previous generation.

Darwin's theory of natural selection was nonteleological. Organisms do not strive to adapt themselves to the environment. Rather, the environment acts on them to select those individuals that are best adapted. These are the ones that have been "naturally selected" to pass on their characteristics to the next generation. In order to emphasize the nonteleological nature of Darwin's theory, it is often contrasted to that of Lamarck's, another nineteenth century naturalist (fig. a). The Lamarckian explanation for the long neck of the giraffe was based on the assumption that the ancestors of the modern giraffe were trying to reach into the trees to browse on high-growing vegetation. Continual stretching of the neck caused it to become longer, and this *acquired characteristic* was passed on to the next generation. Lamarck's theory is teleological because, according to him, a species does shape its own future. This type of explanation has not stood the test of time, but Darwin's theory of evolution by natural selection has been fully substantiated by later investigations.

In summary, these are the critical elements in Darwin's theory:

Variations. Individual members of a species vary in their physical characteristics. Physical variations can be passed on from generation to generation. (Darwin was never aware of genes but we know today that the inheritance of the genotype determines the phenotype.)

Struggle for existence. The members of all species compete with each other for limited resources. Certain members are better able to capture these resources than others.

Survival of the fittest. Just as humans have artificial breeding programs and select which plants and animals will reproduce, so there is a natural selection by the environment of which organisms will survive and reproduce. While Darwin emphasized the importance of survival, modern evolutionists emphasize the importance of unequal reproduction. In any case, however, the selection process is not teleological. Certain members of the

a. Lamarck's Theory

Early giraffes probably had short necks which they stretched to reach food.

Their offspring had longer necks which they stretched to reach food.

Eventually the continued stretching of the neck resulted in today's giraffe.

Existing data do not support this theory.

b. Darwin's Theory

Early giraffes probably had necks of various lengths.

Competition and natural selection led to survival of the longer-necked giraffes and their offspring.

Eventually only long-necked giraffes survived the competition.

Existing data support this theory.

population are selected to produce more offspring simply because they *happen* to have a variation that makes them more suited to the environment.

Adaptation. Natural selection causes a population of organisms, and ultimately a species, to become adapted to the environment. The process is slow but each subsequent generation will include more individuals that are adapted to the environment.

Can natural selection account for the origin of new species and for the great diversity of life? Yes, if we are aware that life has been evolving for a very long time and that variously adapted populations can arise from a common ancestor.

Darwin was only prompted to publish his findings after he received a letter from another naturalist, Alfred Russel Wallace, who had come to the exact same conclusions about evolution as had Darwin. Although both scientists subsequently presented their ideas at the same meeting of the famed Royal Society in London in 1858, only Darwin later gathered together detailed evidence in support of his ideas. He described his experiments and reasonings at great length in his book, *The Origin of Species by Means of Natural Selection,* a book still studied by all biologists today.

a.

b.

Figure 23.15
Industrial melanism. In industrial areas, black-colored moths are protected from predation because they are not seen against the trees darkened with pollutants. Gradually the populations of moths in the area acquire more and more colored members.

Examples of Natural Selection

While natural selection usually takes many hundreds, or even thousands, of years to produce a noticeable change in the phenotype, there are a few examples of rapid adaptation.

Industrial Melanism

Before the industrial revolution in England, collectors of a moth called the peppered moth (fig. 23.15) noted that most moths were light-colored, although occasionally a dark-colored moth was captured. Several decades after the industrial revolution, however, the black moths made up 99 percent of the moth population in polluted areas. An explanation for this rapid change can be found in natural selection. The color of the moths, dark or light, is caused by their genetic makeup; black is a mutation that occurs with some regularity. Moths rest on the trunks of trees during the day; if they are seen there by predatory birds, they are eaten. As long as the trees in the environment are light in color, the light-colored moths live to reproduce. But once the trees turn black due to industrial pollutants, natural selection enables the dark moths to avoid being eaten and to survive and reproduce; therefore, the black phenotype becomes the more frequent one in the population. This explanation has been supported by experiments in which both dark- and light-colored moths were released into industrial and nonindustrial areas. In the industrial areas, the light moths suffered more attrition; in the nonindustrial areas, the dark moths did not survive. This shows that the phenotype most adapted to the environment is the one that is preserved in nature. Industrial melanism has also been noted in the United States; around major cities the insects have taken on a darker color than in the nonpolluted countryside.

Resistance

Indiscriminate use of antibiotics and pesticides has caused pathogenic organisms to become resistant to these chemicals. The mutation enabling them to survive the unfavorable environment is already present before exposure; the chemicals are merely acting as selective agents. For example, when bacteria are grown on a medium containing streptomycin, a few survive. These few can grow on a medium both with and without streptomycin and, therefore, this new generation of bacteria is now resistant to the antibiotic.

A more recent example is the human struggle against malaria, a disease caused by an infection of the liver and red blood cells (fig. 25.15). The dread *Anopheles* mosquito transfers the disease-causing protozoan *Plasmodium* from person to person. In the early 1960s, international health authorities thought that malaria would soon be eradicated. The administration of new drug, chloroquine, was more effective than quinine and DDT spraying killed the mosquitoes. But in the mid-1960s, *Plasmodium* showed signs of chloroquine resistance, and worse yet, mosquitoes were becoming resistant to DDT. A few drug-resistant parasites and a few DDT-resistant mosquitoes had survived and multiplied, making the fight against malaria more difficult than ever. New tactics have to be devised. Recombinant DNA procedures have enabled researchers to identify proteins that will hopefully be effective as vaccines in humans.

It has been possible to witness the process of natural selection in several recent instances: the adaptation of moths to polluted areas, the adaptation of bacteria to modern drugs, and the adaptation of insects to pesticides.

Figure 23.16
Song sparrows of the species Melospiza melódia. *This species has such a wide range that these members show numerous variations.*

From G. Ledyard Stebbins, PROCESSES IN ORGANIC EVOLUTION, 3rd ed. © 1977, p. 134. Reprinted by permission of Prentice-Hall, Inc., Englewood Cliffs, N.J.

***Table 23.4** Reproductive Isolating Mechanisms*

Isolating Mechanisms	Example
Premating	
Habitat	Species at same locale occupy different habitats
Temporal	Species mate at different seasons or different times of day
Behavioral	In animals, courtship behavior differs or they respond to different songs, calls, pheromones, or other signals
Mechanical	Genitalia unsuitable to one another
Postmating	
Gametic mortality	Sperm cannot reach or fertilize egg
Zygote mortality	Hybrid dies before maturity
Hybrid sterility	Hybrid survives but is sterile and cannot reproduce

Speciation

For our present discussion, we are defining a *species* as a group of interbreeding populations that share a gene pool reproductively isolated from other species. The populations that belong to the same species are spread over a certain geographic range. If the geographic range of a species is large, more phenotypic variations between populations are apt to be seen than if the geographic range is small. For example, differences in the song sparrows pictured in figure 23.16 can be accounted for in that this species, *Melospiza melódia,* has a range that extends across the United States from the east to the west coast. Not only do the birds differ anatomically, they also sing a slightly different song. However, as long as there is gene flow, or the movement of genes from one population to another as a consequence of the immigration of individuals from one population to another, it is possible to regard these sparrows as belonging to one species. If interbreeding between populations is possible but rarely occurs, taxonomists sometimes assign subspecies or race designations to populations by giving them a third name in addition to their normal binomial name.

The populations of the same species exchange genes, but different species do not exchange genes. Reproductive isolation of the gene pools of similar species is accomplished by such mechanisms as those listed in table 23.4. **Premating isolating mechanisms** are those that prevent mating from ever taking

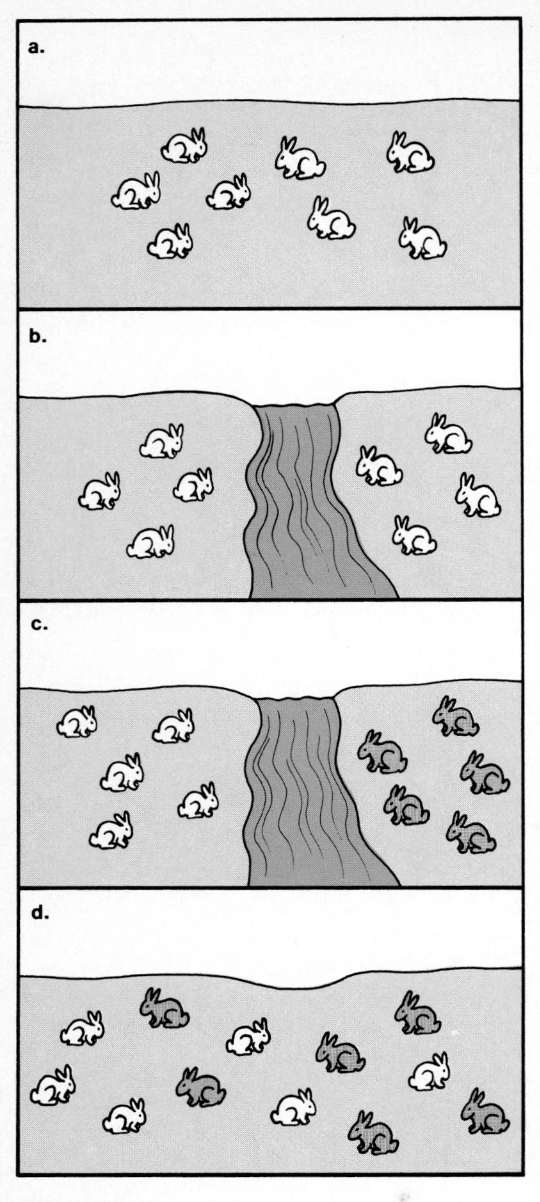

Figure 23.17
Allopatric speciation. a. *A hypothetical species has many populations, each one represented by a single rabbit.* b. *Geographic isolation occurs, and gene flow between all populations is not possible.* c. *Divergent evolution occurs because the populations represented by white rabbits and those represented by black rabbits are exposed to different selective pressures.* d. *When geographic isolation has ended, the populations represented by black rabbits are not able to reproduce successfully with those represented by white rabbits. This shows we are now dealing with two different species.*

place, and **postmating isolating mechanisms** are those that prevent hybrid offspring from developing or breeding if mating has occurred. In evolutionary terms, a hybrid is an offspring of members of populations that do not normally reproduce with one another.

Process of Speciation

One species can give rise to two different species and when this happens, speciation has occurred. There are two criteria by which we can recognize that speciation has occurred: the presence of structural differences and the failure to reproduce successfully in the same natural setting. It is now generally accepted that in most instances *speciation is a two-step process: geographic isolation is later followed by reproductive isolation.*[2]

The two-step process for speciation is illustrated in figure 23.17. Initially, all populations of the species live in the same environment and share a common gene pool. In our example, this is exemplified by giving all populations, each represented by a rabbit, the same color. Then let us suppose that a canal is dug to divert water from a nearby source and that the canal separates our populations of rabbits into two groups so that some populations are now geographically isolated from other populations (fig. 23.17b). With this, the first step of speciation has occurred.

Once geographic isolation has occurred, the two sets of populations will begin to undergo divergent evolution. This is represented in our example by coloring the two sets of populations different colors (fig. 23.17c). From this time on we may speak of these two sets of populations as subspecies. Why has divergent evolution taken place? There are three reasons for this: (1) The subspecies have different gene pools. This is obvious when we consider that the original populations had slight differences. Since gene flow between certain populations is now prevented, the genes for these structural differences have a greater chance of being passed on to the next generation. (2) Each new gene pool will now separately be subject to the normal increase in variation caused by mutation and recombination of genes during gamete formation. (3) The subspecies have different environments and therefore experience different selective pressures. Consequently, natural selection will cause the subspecies to diverge further genetically and phenotypically from one another. Given enough time, this divergence will eventually result in reproductive isolation so that even if the geographic barrier is removed, the two population systems cannot successfully reproduce with one another. In our example, we can imagine that if the canal were filled in gene flow between the two subspecies would still not be observed (fig. 23.17d). The second step of speciation has now occurred; what was formerly one species has become two species.

Speciation is the origin of species and this usually requires geographic isolation followed by reproductive isolation.

Adaptive Radiation

One of the best examples of speciation is provided by the finches on the Galápagos Islands, which are very often called Darwin's finches because Darwin (p. 518) first realized their significance as an example of how evolution works. The Galápagos Islands (fig. 23.18a) located 600 miles west of Ecuador, South America, are volcanic but do have forest regions at higher elevations. The 13 species of finches (fig. 23.18b), placed in three genera, are believed to be descended from mainland finches that migrated to one of the islands some years ago. Therefore, Darwin's finches are an example of *adaptive radiation,* or the proliferation of a species by adaptation to different ways of life. We can imagine

[2]Sometimes termed allopatric speciation.

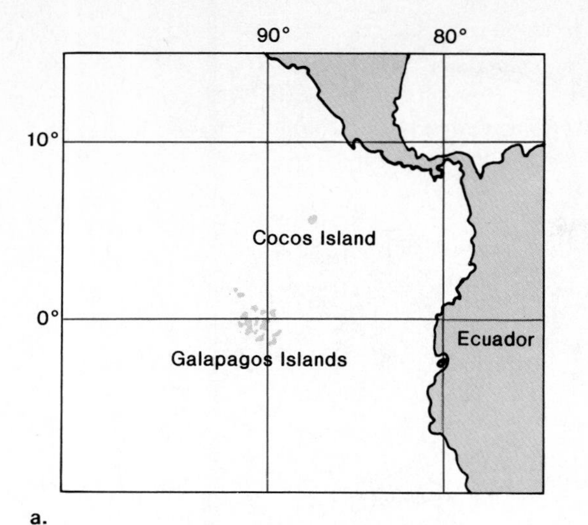

a.

Figure 23.18

a. *Location of the Galápagos Islands. These islands are close enough to South America that finches could have come there from the mainland. Once they arrived, the original population on one island is presumed to have spread out to the other islands where varying selective pressures would have caused divergent evolution to occur. With time, the various populations of finches were unable to reproduce with one another.* b. *The ancestral form from which the Galápagos finches are descended most likely had a strong beak capable of crushing seeds. Each of the present-day thirteen species of finches has a bill adapted to a particular way of life. For example,* (1) *the large tree finch grinds fruit and insects with a parrotlike bill. The small ground finch* (2) *has a pointed bill and eats tiny seeds and ticks picked from iguanas. The woodpecker finch* (3) *has a stout, straight bill that chisels through tree bark to uncover insects, but because it lacks a woodpecker's long tongue it uses a tool—usually a cactus spine or a small twig—to ferret insects out.*

b.1.

b.2.

b.3

that after the original population of a single island increased, some individuals dispersed to other islands. The islands are ecologically different enough to have promoted divergent feeding habits. This is apparent because, although the birds physically resemble each other in many respects, they have different bills, each of which is adapted to a particular food-gathering method. There are seed-eating ground finches with bills appropriate to cracking small-, medium-, or large-size seeds; insect-eating tree finches, also with different size bills; and a warbler-type finch with a bill adapted to nectar gathering. Among the tree finches there is a woodpecker type, which lacks the long tongue of a true woodpecker but makes up for this by using a cactus spine or twig to ferret out insects.

Most likely as the populations on the various islands began to assume their own evolutionary pathways, they developed postmating isolating mechanisms. When secondary contact occurred between the original and the derived populations, premating mechanisms would subsequently have been selected (see table 23.4).

One frequently cited example of speciation is the evolution of several species of finches on the Galápagos Islands. This is also an example of adaptive radiation into unfilled habitats, as was the evolution of many types of reptiles and mammals on a grander scale.

Figure 23.19
Adaptive radiation among reptiles and mammals. Both reptiles and mammals evolved into the types of forms shown. However, the mammals did not begin their adaptive radiation until the reptiles had declined. Therefore, each group shows similar adaptations because they were subjected to the same selective pressures.

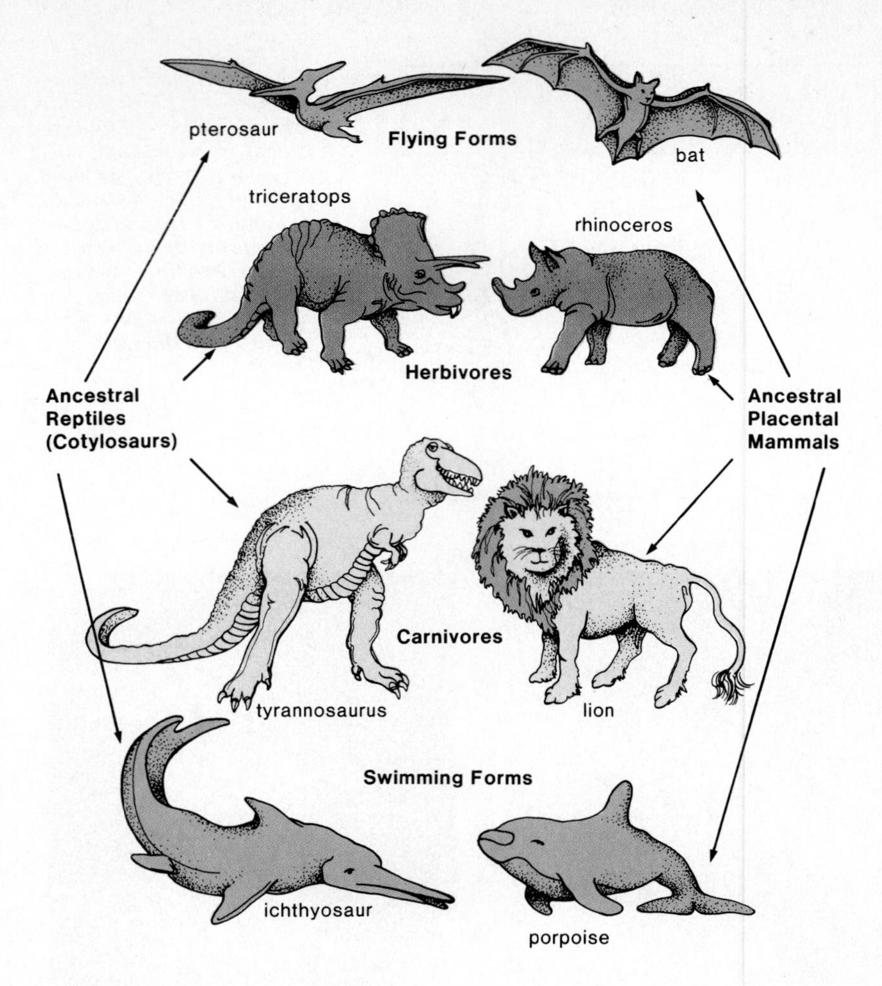

Higher Taxonomic Categories

Adaptive radiation is observed not only among species but among higher taxonomic categories also. Both reptiles and mammals underwent large-scale adaptive radiation as they became adapted to different ways of life, as illustrated in figure 23.19. Mammals began their radiation after many reptiles, including the dinosaurs, had become extinct and therefore mammals were free to evolve in the same directions as the reptiles had evolved.

Extinction It is possible that extinction of intermediate forms (p. 505) leads to our recognition of higher taxonomic categories among fossil and/or living forms. As related species become extinct, wider and wider gaps are observed until it becomes obvious that the ones remaining should be placed in different taxonomic categories.

The fossil record indicates that on the average two to five families, each containing many species, have become extinct per every million years. These extinctions, termed *background extinctions* are explained on the basis of inability to adapt to changing environmental conditions. The fossil record also gives evidence of periods of *mass extinctions* when about 20 families have become extinct per million years. It was during one of these periods that the dinosaurs became extinct and, as the reading on page 525 discusses, there has been much speculation to account for this and other mass extinctions.

The extinction of intermediate species can help explain the wide separation we now observe between major groups of animals.

Extinction of the Dinosaurs

Dinosaurs dominated the earth for 135 million years and then suddenly, at the end of the Cretaceous period, they and much of the rest of life on earth vanished. Geologists have found that there is an abnormally high level of iridium in clay of the late Cretaceous period and have speculated that it could have been caused by a worldwide fallout of radioactive material created by a comet or asteroid impact at the time of the dinosaur disappearance. While this may seem to be an absurd idea, evidence suggests that it may very well be the case: (1) paleontologists (biologists who study the fossil record) have now found evidence of other such *mass extinctions* every 26 million years or so and (2) astronomers tell us that the movement of our solar system in the Milky Way could possibly cause the earth to be subjected to such regular but infrequent bombardments.

A comet bombardment can be likened to a worldwide atomic bomb explosion. A cloud of dust would mushroom into the atmosphere and shade out the sun, causing plants to freeze and die. Mass extinction of animals will follow. Once the cosmic winter is over, plant seeds would germinate. This new growth would serve as food for the few remaining animals. Lack of competition would permit adaptive radiation on a grand scale.

Others do not look to the heavens for the causes of mass extinctions. They point out that the continents have not always been as they are now. At one time there was only one huge continent called Pangaea. Pangaea eventually broke up into two major continents and during the Cretaceous period, these fragmented into the seven present-day continents. Even now these continents are slowly drifting further apart. It's possible that climatic changes due to continental drift are sufficient to explain mass extinctions.

The continents drift because they are on mammoth pieces of crust that are formed at midoceanic ridges and disappear at oceanic trenches.

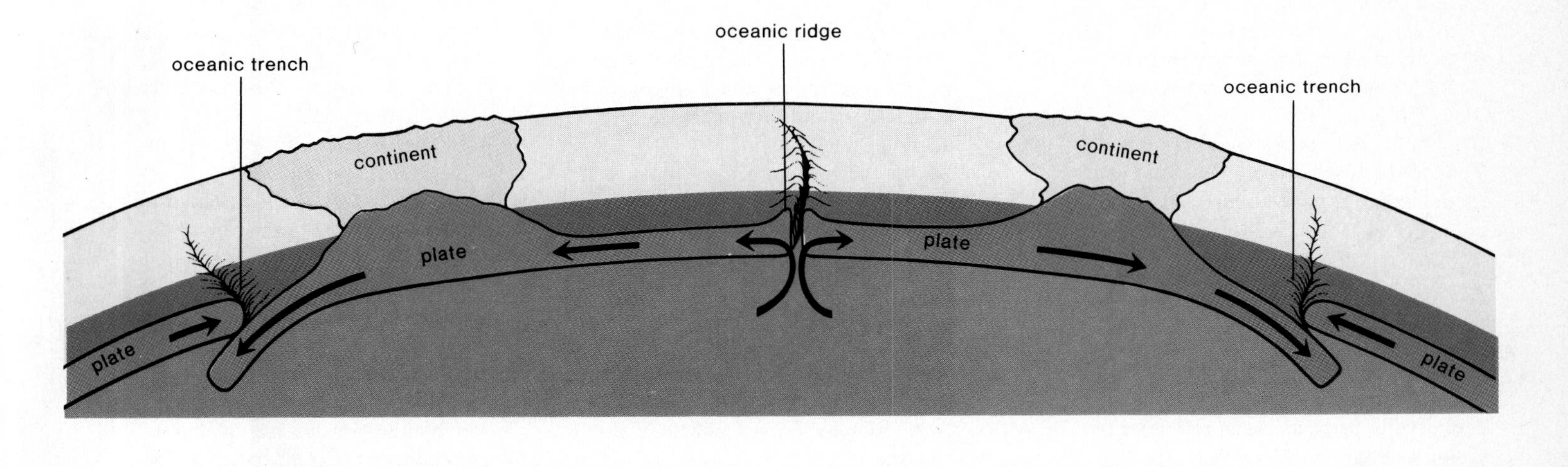

Summary

Evidences of macroevolution can be found in various fields of biology but the process of evolution is believed to involve a gradual change in gene frequencies within the gene pool of a population. The Hardy-Weinberg law states that the gene pool will remain constant only under certain prescribed conditions and therefore evolution is expected to occur.

The process of evolution is presented as a two-step process requiring the production of genotypic variations (by mutation, gene flow, and recombination) and a sorting out of these variations (by genetic drift and natural selection). Several examples of natural selection at work are given.

Populations that share the same gene pool are all part of the same species. Similar types of species are in the same genus and thereafter we proceed from family to order to class to phylum (animals) or division (plants) and finally kingdom, the largest of the categories of classification.

Speciation is the origin of species and this usually requires geographic isolation, followed by reproductive isolation. One frequently cited example of speciation is the evolution of several species of finches on the Galápagos Islands. This is an example of adaptive radiation, as was the evolution of many types of reptiles and mammals on a grander scale. It is possible that extinction of intermediate species led to the wide separation we now observe between major groups of organisms.

Objective Questions

1. If the Hardy-Weinberg law holds, evolution will ____________ (choose always or not) occur.
2. Twenty-one percent of a population is homozygous recessive. If the Hardy-Weinberg law holds, what percentage is expected to be homozygous recessive in the next generation? ____________
3. Can sexual reproduction in and of itself cause evolution to occur? ____________
4. Recombination of genes, mutations, and ____________ are expected to increase the amount of variation among individuals of a population.
5. Natural selection and ____________ are expected to reduce the amount of variation among individuals of a population.
6. The fossil record shows that the brain size of humans has steadily increased since they evolved. This is an example of ____________ selection.
7. There are genetic reasons for the maintenance of variation among individuals of a population, but ________ factors are also important.
8. During the first stage of speciation, populations become ____________ .
9. During the second stage of speciation, populations become ____________ .
10. Two species of butterflies have different courtship behavioral patterns. This is an example of a ____________ ____________ isolating mechanism of the ____________ type.

Answers to Objective Questions
1. not 2. 21 percent 3. no 4. gene flow 5. genetic drift 6. stabilizing 7. ecological 8. geographically separated 9. reproductively isolated 10. premating reproductive, behavioral

Answers to Practice Problems
1. 60 percent *Ll* and 40 percent *ll*
2. recessive allele = 0.2 and dominant allele = 0.8; homozygous dominant = 0.64; homozygous recessive = 0.04; heterozygous = 0.32
3. 29 percent

Study Questions

1. Show that the fossil record, comparative anatomy, comparative embryology, comparative biochemistry and biogeography all give evidence that evolution has occurred. (pp. 505–512)
2. What is an evolutionary tree and how are evolutionary trees constructed? (p. 508)
3. What is the Hardy-Weinberg law? (p. 513) If genotype *gg* is found in 16 percent of a population, what is the frequency of the *g* allele? The *G* allele? What proportion of the next generation will be *gg* if the law holds? (p. 513)
4. What factors prevent the Hardy-Weinberg law from operating? (p. 514)
5. Name and describe the sources of gene pool variations in a population made up of diploid sexually reproducing individuals. (p. 515)
6. Discuss the concept of the "unity of the genotype." (p. 515)
7. Name and contrast two processes that reduce gene pool variations. (pp. 515–516)
8. Name several reasons for the maintenance of variation in a gene pool. (p. 517)
9. Give two modern examples of natural selection. (p. 520)
10. Define a species. How do new species originate? (pp. 521–522)
11. When is adaptive radiation apt to take place? (pp. 522–523)

Key Terms

analogous structure (ah-nal'o-gus) similar in function but not in structure; particularly in reference to similar adaptations. *508*

binomial system (bi-no'me-al sis'tem) the assignment of two names to each organism, the first of which designates the genus and the second of which designates the species. *508*

class (klas) in taxonomy, the category below phylum and above order. *508*

common ancestor (kŏ'mun an'ses-tor) an ancestor to two or more branches of evolution. *508*

evolutionary trees (ev''o-lu'shun-ar-e trēz) diagrams describing the evolutionary relationship of groups of organisms. *508*

family (fam'ĭ-le) in taxonomy, the category above genus and below order. *508*

fossils (fos''lz) any remains of an organism that have been preserved in the earth's crust. *505*

gene flow (jēn flo) the movement of genes from one population to another via reproduction between members of the populations. *515*

gene pool (jēn po͞ol) the total of all the genes of all the individuals in a population. *512*

genetic drift (jĕ-net'ik drift) evolution by chance processes alone. *516*

genus (je'nus) in taxonomy, the category above species and below family. *508*

homologous structures (ho-mol'o-gus struk'tūrz) similar in structure but not necessarily function; homologous structures in animals share a common ancestry. *508*

kingdom (king'dum) the largest taxonomic category into which organisms are placed: Monera, Protista, Fungi, Plants, and Animals. *508*

natural selection (nat'u-ral sĕ-lek'shun) the process by which better adapted organisms are favored to reproduce to a greater degree and pass on their genes to the next generation. *516*

order (or'der) in taxonomy, the category below class and above family. *508*

phylum (fi'lum) in taxonomy, the category applied to animals that is below kingdom and above class. *508*

population (pop''u-la'shun) all the organisms of the same species in one place (area/space). *512*

species (spe'shēz) a group of similarly constructed organisms that are capable of interbreeding and producing fertile offspring; organisms that share a common gene pool. *508*

taxonomy (tak-son'o-me) the science of naming and classifying organisms. *508*

vestigial (ves-tij'e-al) the remains of a structure that was functional in some ancestor but is no longer functional in the organism in question. *510*

24

Origin of Life

Chapter Concepts

1 The first cell or cells most likely arose by a slow process of chemical evolution.

2 It is generally accepted that the first cell was a heterotroph that utilized anaerobic respiration.

3 Autotrophic nutrition made life on land possible by releasing free oxygen into the atmosphere.

4 Although life may have originated by chemical evolution, this process does not occur today. Today, life comes only from life.

Chapter Outline

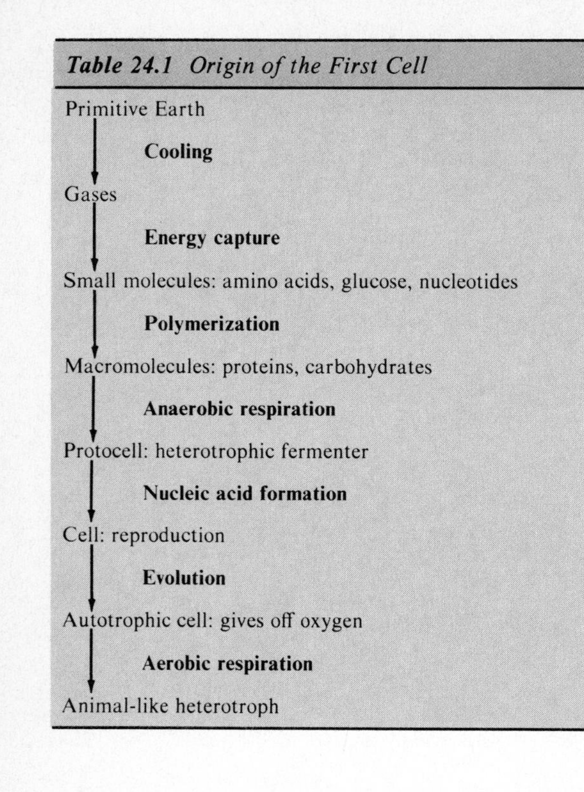

Table 24.1 Origin of the First Cell

Primitive Earth
↓ **Cooling**
Gases
↓ **Energy capture**
Small molecules: amino acids, glucose, nucleotides
↓ **Polymerization**
Macromolecules: proteins, carbohydrates
↓ **Anaerobic respiration**
Protocell: heterotrophic fermenter
↓ **Nucleic acid formation**
Cell: reproduction
↓ **Evolution**
Autotrophic cell: gives off oxygen
↓ **Aerobic respiration**
Animal-like heterotroph

Today we do not believe that life arises spontaneously from nonlife, and we say that "life comes only from life." But if this is so, how did the first form of life come about? First, we could assume that the first form of life was very simple; for example, a single cell (or cells). As soon as this cell could grow, reproduce, and mutate, it could be said to be alive. Since it was the very first living thing, it had to come from nonliving chemicals. In fact, it is possible that slow progression of chemicals from the simple to the complex finally resulted in a live cell. In other words, a **chemical evolution** produced the first form of life. Table 24.1 lists the steps that could have occurred to produce life. The evidence for these steps is based on our knowledge of the primitive earth (fig. 24.1) and on experiments that have been performed in the laboratory.

Chemical Evolution

When the earth was first formed about 5 billion years ago, it was a glowing mass of free atoms (fig. 24.6a), which sorted themselves out according to weight. The heavy ones, such as iron and nickel, sank toward the center of the earth; the lighter atoms, such as silicon and aluminum, formed the middle shell; and the very lightest atoms, hydrogen, nitrogen, oxygen, and carbon, may have collected on the outside. The temperature was so hot that atoms could not permanently bind together; whenever bonds formed, they were quickly broken.

As the earth cooled, the heavy atoms tended to liquefy and solidify, but the intense heat at the center prevented complete solidification, and even today the earth contains a hot, thickly flowing molten core. In the middle shell, the lighter atoms congealed and formed the outer surface of the earth, the so-called crust. Cooling may also have allowed the first atmosphere to form.

Figure 24.1
The primitive earth was devoid of living things.

Primitive Atmosphere

There are two current hypotheses about the origin of the **primitive atmosphere.** One hypothesis suggests that the gases of the primitive atmosphere came about when cooling allowed the lightest of the atoms to react with one another. Since hydrogen was the most abundant of these atoms, it combined with itself and with carbon, oxygen, and nitrogen to form hydrogen gas (H_2), methane gas (CH_4), water vapor (H_2O), and ammonia vapor (NH_3). An abundance of hydrogen atoms would have caused the first atmosphere to be a highly **reducing atmosphere** (p. 23).

The second hypothesis is also compatible with the steps listed in table 24.1. This hypothesis suggests that the gases of the primitive atmosphere were released from volcanic eruptions. Further, it is believed that while the atmosphere may have been mildly reducing, it probably also contained carbon dioxide (CO_2) and nitrogen gas (N_2).

Both hypotheses support the contention that the first atmosphere contained little or no free oxygen. Therefore, the first atmosphere is believed to have lacked O_2.

The earth began as a mass of hot, glowing individual atoms. The primitive atmosphere may have formed as the lightest of these atoms cooled and joined together to form hydrogen gas, methane, water vapor, and ammonia vapor. Another hypothesis proposes that volcanic eruptions produced a primitive atmosphere much like today's atmosphere, except that it lacked free oxygen.

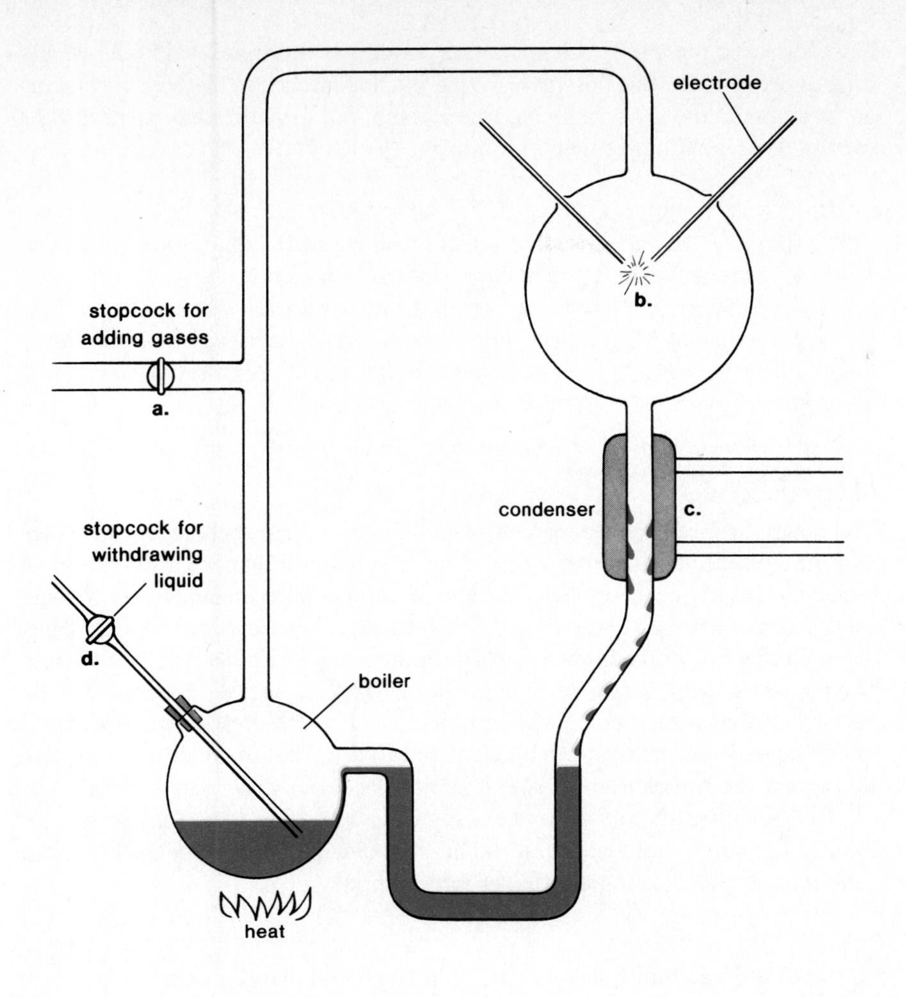

Figure 24.2
In Miller's experiment, gases are admitted to the apparatus (a), *circulated past an energy source* (b), *and cooled* (c) *to produce a liquid that can be withdrawn for testing* (d).

Simple Organic Molecules

Water, present at first as vapor in the atmosphere, formed dense, thick clouds, but cooling eventually caused the vapor to condense to liquid, and rain began to fall. This rain was in such quantity that it produced the oceans of the world (fig. 24.1). The gases, dissolved in the rain, were carried down into the newly forming oceans (fig. 24.6b).

The remaining steps shown in table 24.1 took place in the sea, where life arose. The dissolved gases, although relatively inert, are believed to have reacted together to form simple organic compounds when they were exposed to the strong outside **energy sources** present on the primitive earth (fig. 24.6c). These energy sources included heat from volcanoes and meteorites, radioactivity from the earth's crust, powerful electric discharges in lightning, and solar radiation, especially ultraviolet radiation. In a classic experiment (fig. 24.2), Stanley Miller showed that an atmosphere containing methane and ammonia could have produced organic molecules. These gases were dissolved in water and circulated in a closed container past an electric spark. After a week's run, he analyzed the contents of the reaction mixture and found, among other organic compounds, amino acids and nucleotides. Other investigators have achieved the same results by utilizing carbon monoxide and nitrogen gas dissolved in water.

These experiments indicate that the primitive gases not only could have but probably did react with one another to produce simple organic compounds that accumulated in the ancient seas. Neither oxidation (there was no free oxygen in the ancient atmosphere) nor decay (there were no bacteria) would have destroyed these molecules, and they would have accumulated in the oceans for hundreds of millions of years. With the accumulation of these simple organic compounds, the oceans became a thick, hot **organic soup,** containing a variety of organic molecules.

Cooling caused water vapor to turn to rain, and the quantity of this rain was great enough to produce the oceans in which the first atmospheric gases were dissolved. Here in the ocean, as Miller and more recent investigators have shown, these gases could have reacted with one another under the influence of an outside energy source, such as lightning or ultraviolet radiation, to produce simple organic molecules.

Macromolecules

The newly formed organic molecules combined to form still larger molecules and macromolecules. Perhaps this came about by a chance combination of molecules in the ocean, or perhaps some of the smaller molecules were washed ashore where dry heat would have encouraged **polymerization** (fig. 24.6d).

Sidney Fox of the University of Miami supports the idea that amino acid polymers[1] were the first macromolecules to form. In various experiments, he has shown that amino acids will combine in a preferred order when exposed to dry heat. Presumably, amino acids collected in shallow puddles along the shore, and the sun caused polymerization to occur as drying took place.

Other investigators believe that nucleic acid may have formed first in exactly the same manner. These scientists point out that since DNA is the genetic material, it is logical that it formed first.

Small organic molecules polymerized to produce large organic macromolecules similar to proteins or nucleic acids.

Biological Evolution

Biological evolution, the evolution of life forms, began with the origination of rudimentary cells termed protocells. Protocells presumably arose from the newly formed organic macromolecules.

Protocells

Fox has shown that when amino acid polymers are exposed to water they form **proteinoid microspheres** (fig. 24.3), which have many properties similar to today's cells (table 24.2).

He feels that the **protocells** could have been these proteinoid microspheres and that they could have evolved to contain other macromolecules characteristic of true cells. He calls this a **cell-first hypothesis,** meaning that the protocell came first—before true proteins and nucleic acids.

Fox has proposed a cell-first hypothesis, which is dependent on the observation that amino acid polymers can form spheres resembling cell-like bodies called proteinoids. Proteinoids have many features in common with cells.

---

[1]Fox suggests that these polymers should not be called proteins because they do not have all the characteristics of true cellular proteins.

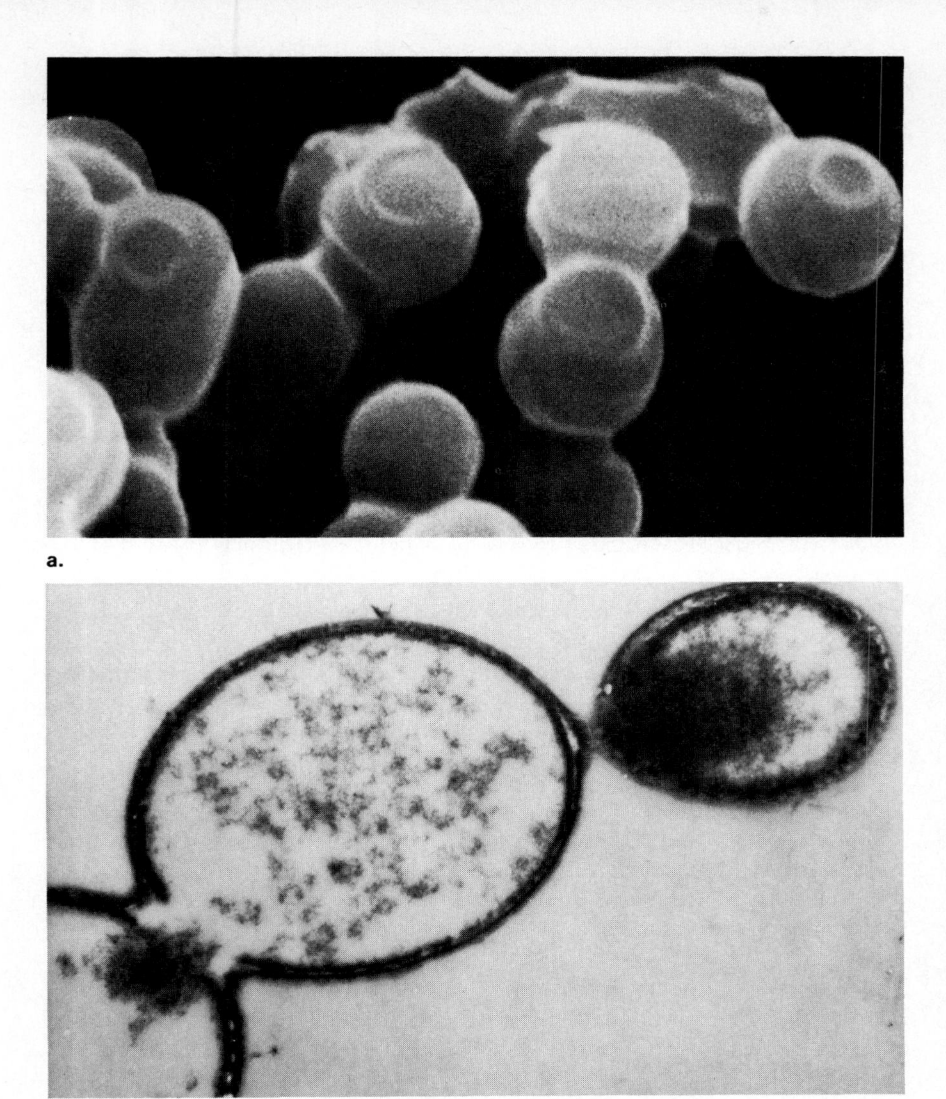

Figure 24.3
Microspheres. a. *Scanning electron micrograph of the exterior of proteinoid microspheres, which may be similar to the structure of the protocells.*
b. *Electromicrograph of the interior of protein microspheres.*

Table 24.2 Proteinoid Microparticles Possess Many Properties Similar to Contemporary Cells
Stability (to standing, centrifugation, sectioning)
Microscopic size
Variability in shape but uniformity in size
Numerousness
Stainability
Ultrastructure (electron microscope)
Double-layered boundary
Selective passage of molecules through boundary
Catalytic activities
Patterns of association
Propagation by "budding" and fission
Growth by accretion
Motility
Propensity to form junctions and to communicate

From Fox, Sidney W., "Chemical origins of cells" in *Chemical & Engineering News, 49,* 50, Dec. 6, 1971. Copyright © 1971 American Chemical Society. Reprinted by permission.

Other investigators believe that protocell formation may have required the presence of various macromolecules aside from amino acid polymers. Some researchers support the theoretical work of Oparin, who wrote a definitive book on the topic in 1938. Oparin pointed out that, under appropriate conditions of temperature, ionic composition, and pH, concentrated mixtures of macromolecules tend to give rise to complex units called **coacervate** droplets (fig. 24.4). Coacervate droplets have a tendency to absorb and incorporate various substances from the surrounding solution. Eventually, a semipermeable-type boundary may form about the droplet.

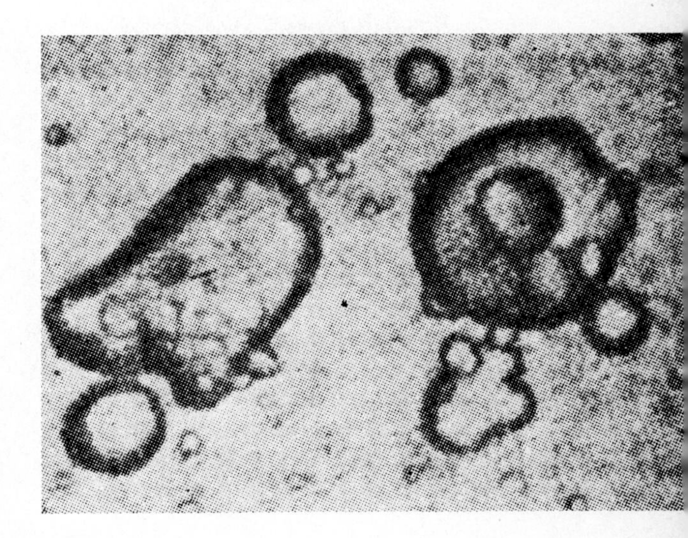

Figure 24.4
Coacervates are polymer-rich colloidal droplets. It is theorized by some that coacervates could have been protocells.

Heterotroph Hypothesis We might now ask, How did the protocell carry on nutrition and respiration? Nutrition would have been no problem because the protocell lived in the "organic soup," which contained simple organic molecules that would have been its food. Therefore, the protocell would have been a **heterotroph,** an organism that takes in preformed food. Notice, too, that heterotrophs are believed to have preceded **autotrophs,** organisms that make their own organic food.

In regard to respiration, the protocell must have carried on **anaerobic respiration,** or *fermentation,* which does not require free oxygen:

glucose $\rightarrow$ alcohol and/or acids + CO_2 + energy

The protocell would have been capable of growing since it carried on nutrition and respiration; but without genes, it could not have mutated and evolved. However, we can theoretically assume that the protocell may have incorporated nucleotides or nucleic acids that eventually formed genes. With this development, the protocell would have become a *cell* capable of passing on genetic traits by means of cell division.

True Cells

The first cells would have been anaerobic heterotrophs just like the protocells. There are anaerobic bacteria living today that may very closely resemble the first true cells. Other evolutionary lines, however, produced the autotrophs.

The first cell-like structure, called the protocells, would have carried on anaerobic respiration and heterotrophic nutrition. Once the protocells acquired genes they became true cells, which allowed other types of cells to evolve.

Autotrophs The oldest microfossils (microscopic fossils) are believed to be those of autotrophs. They are the remains of giant colonies of bacteria found in a rock dated 3.5 billion years ago (see reading on this page). The rock was found during a geological expedition in Western Australia (fig. 24.5).

Why Is the Earth Neither Too Hot Nor Too Cold?

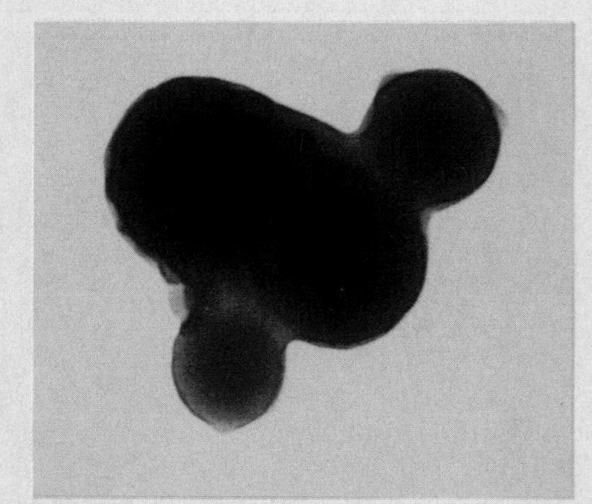

Electron micrograph of organic matter found in 3.5 billion-year-old sedimentary rock. It is possible that these forms are microorganisms, in which case they are the oldest ever pictured. The picture was taken by geologist Miryam Glickson.

The Earth is quite literally poised between fire and ice. Consider, for example, what would happen if we somehow moved the Earth slightly closer to the sun.

As the oceans grew warmer, more and more water vapor would begin to steam into the atmosphere. Once there, the vapor would begin to act like glass in a greenhouse, preventing heat from radiating back into space. So the Earth would grow warmer still, until oceans began to dry and the carbonate rocks—the limestones and dolomites—began to bake in the heat and release carbon dioxide.

The greenhouse effect caused by this gas is famous: By burning fossil fuels we are already releasing enough carbon dioxide to warm the climate measurably. The carbonate rocks, however, contain billions upon billions of tons of it, enough to trigger a "runaway greenhouse." In the end our planet would become a twin of unfortunate Venus, the next planet inward to the sun: a gaseous, dry, searing hell, its surface covered with clouds, oppressed by a massive atmosphere of carbon dioxide, and hot enough to melt lead.

Suppose, on the other hand, we moved the Earth further out from the sun. As the planet grew colder, glaciers would grind southward over Canada, Europe, and Siberia, while sea ice crept northward from Antarctica. The ice would reflect more sunlight back into space, cooling the planet even more. Step by step, the ice would extend toward the equator. In the end, the Earth would gleam brilliantly—but its oceans would be frozen solid.

Thus, the climate is balanced precariously indeed—so precariously that many geologists now believe that tiny, cyclic variations in the Earth's orbit, known as the Milankovitch cycles, were enough to have triggered the ice ages.

But geologists also have found fossilized marine microbes in rocks more than 3.5 billion years old (see accompanying figure) and they assure us that the oceans of the

Figure 24.5
Field researchers examine rocks near North Pole, Australia, for evidence of the earth's earliest forms of life.

Earth have remained warm and liquid throughout its 4.6-billion-year history.

Perhaps this is a lucky accident—after all, if the Earth had not formed at just the right distance from the sun to have liquid oceans, we would not be here to worry about it. But the astrophysicists point out that things aren't quite that simple.

The sun, they say, is a quiet and stable star. But like others of its type, it is inexorably getting hotter with age. In fact, it is about 40 percent brighter now than when the Earth was born. So how could the climate possibly stay constant? If the Earth is comfortable now, then billions of years ago, under a colder sun, the oceans must have been frozen solid. But they were not. On the other hand, if the oceans were liquid then, why has the sun not broiled us into a second Venus by now?

One theory, advocated by a number of biologists, is that the early Earth started out with a good deal more carbon dioxide in the air than it has now, which gave enough of a greenhouse effect to keep the planet warm even when the sun was cool. If nothing had changed, the greenhouse eventually would have "run away" as it did on Venus. (There is some evidence that Venus started out with oceans much like ours.) But fortunately for us, about three billion years ago, certain blue-green algae [cyanobacteria] devised a way of taking carbon dioxide out of the air and turning it into organic carbon compounds. We call that process photosynthesis. In the eons since, as the algae and their descendants, the plants, have evolved and multiplied, the decreasing levels of atmospheric carbon dioxide have just about kept pace with the warming sun. Thus, say the biologists, it was life that saved the world for life.

Excerpted from W. Mitchell Waldrop, "Why Is the Earth Neither Too Hot Nor Too Cold?". Reprinted by permission from the July/August issue of *Science '83*.

When bacteria photosynthesize, they do not give off oxygen. Therefore further evolution was required before there were autotrophs similar to today's algae. These autotrophs would have possessed chlorophyll *a* and would have photosynthesized and made their own food by using carbon dioxide and water:

$$\begin{array}{c} \text{carbon dioxide} \\ + \\ \text{water} \end{array} \xrightarrow[\text{from the sun}]{\text{energy}} \begin{array}{c} \text{glucose} \\ + \\ \text{oxygen} \end{array}$$

The evolution of autotrophs was critical because the preformed organic molecules in the ocean no doubt would have been running out. The newly evolved autotrophs would have provided a much needed continual source of food. These autotrophs also made aerobic respiration possible because they put oxygen gas into the atmosphere (fig. 24.6e).

Aerobic Respiration The presence of oxygen in the atmosphere permitted **aerobic** respiration:

$$\text{oxygen} + \text{glucose} \longrightarrow \text{carbon dioxide} + \text{water} + \text{energy}$$

It also permitted plants and animals to invade the land because oxygen in the upper atmosphere forms ozone (O_3), which filters out the ultraviolet rays of the sun. Before the formation of this so-called **ozone shield,** the amount of radiation would have destroyed land-dwelling organisms. There is concern today that the ozone shield may be in danger of breaking down due to air pollutants, particularly nitric oxides and chlorine, that are capable of reacting with ozone. Nitric oxides are given off by jet planes, and chlorine is released from an aerosol propellant called freon. Although freon is no longer used in this country, it is still used in other countries around the world. It has been suggested that a 10 percent annual increase in the use of aerosols could reduce the ozone layer by 10 percent in 20 years and by 40 percent by the year 2014.

Eventually autotrophs arose that not only supplied food for themselves and other organisms but also released oxygen into the atmosphere. This oxygen allowed the evolution of eukaryotic plants and animals and formed an ozone shield that permits these organisms to live on the land.

Origin of Life Today

We have shown that life could have come into existence by means of a chemical evolution (fig. 24.6). However, we do not believe that this same chemical evolution is occurring today for the following reasons:

1. Appropriate energy sources, particularly ultraviolet radiation, are unavailable. The ozone layer now acts as a shield to prevent these rays from reaching the earth in quantity.
2. While the first atmosphere was believed to be a reducing one, that promoted the buildup of organic molecules, today's atmosphere is an **oxidizing atmosphere,** that tends to break down organic molecules.
3. Living organisms, already present, would use any newly formed organic molecules for food.

Today, life is believed to come only from life because an appropriate energy source is lacking to cause the formation of organic molecules, and if they did form, they would be oxidized by oxygen or eaten by preexisting life.

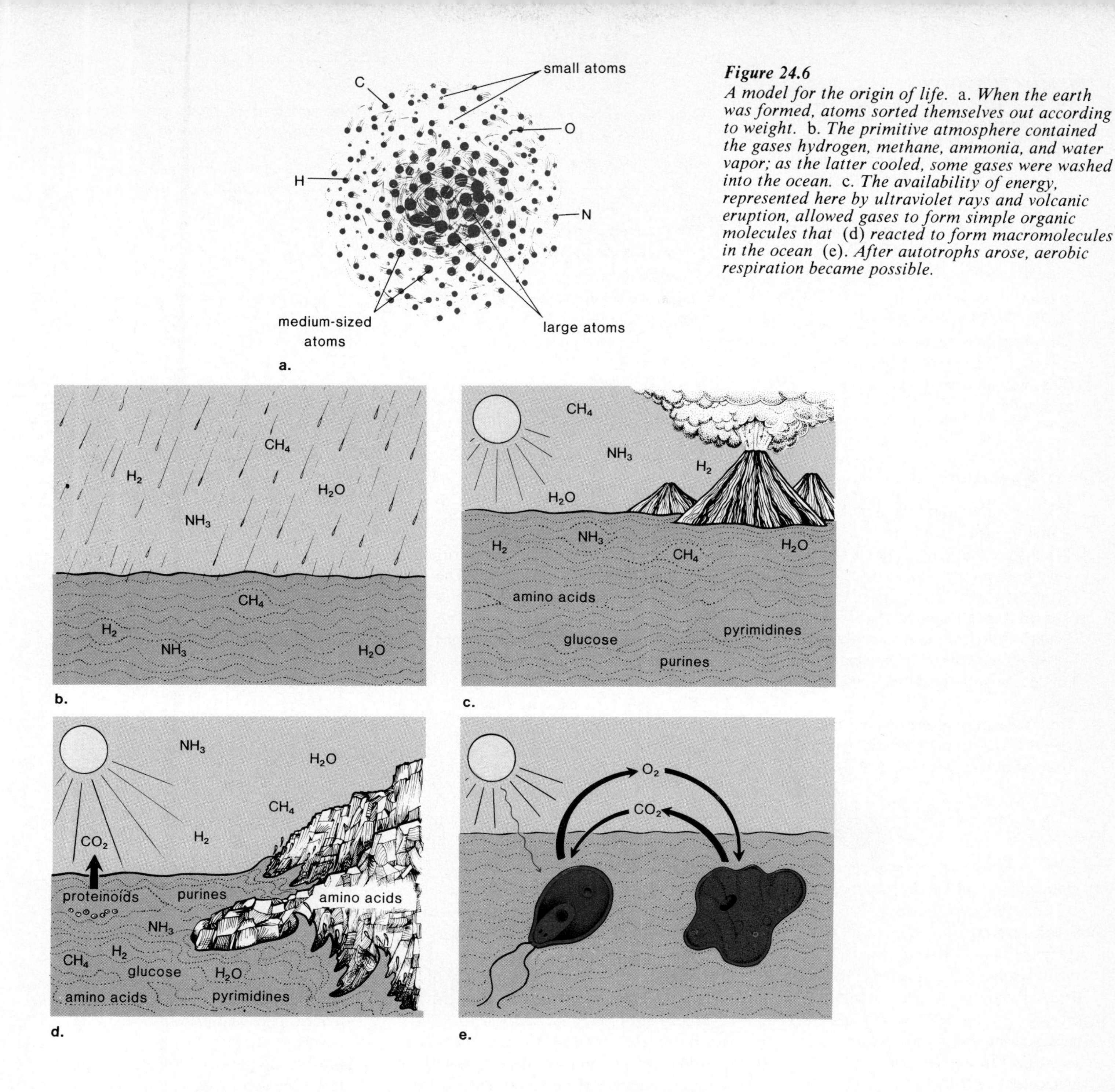

Figure 24.6
A model for the origin of life. a. *When the earth was formed, atoms sorted themselves out according to weight.* b. *The primitive atmosphere contained the gases hydrogen, methane, ammonia, and water vapor; as the latter cooled, some gases were washed into the ocean.* c. *The availability of energy, represented here by ultraviolet rays and volcanic eruption, allowed gases to form simple organic molecules that* (d) *reacted to form macromolecules in the ocean* (e). *After autotrophs arose, aerobic respiration became possible.*

Summary

The gases of the primitive atmosphere probably included hydrogen gas, methane, water vapor, and ammonia. There was no free oxygen. As the newly formed earth cooled, water vapor-turned-to-rain produced the oceans in which the gases were dissolved. In the ocean, gases could have reacted with one another under the influence of an outside energy source (lightning or ultraviolet radiation) to give small organic molecules. These polymerized to produce large organic molecules similar to proteins or nucleic acids. In the laboratory, amino acid polymers form spheres resembling cell-like bodies called proteinoids. Proteinoids have many features in common with cells. The first cell-like structures, called the protocells, would have carried on anaerobic respiration and heterotrophic nutrition. Once the protocells acquired genes they became true cells, which allowed other types of cells to evolve. Eventually autotrophs arose that not only provided a source of food for other organisms but also released oxygen into the atmosphere. This oxygen allowed the evolution of plants and animals and formed an ozone shield that permits these organisms to live on the land.

Today, life is believed to come only from life because an appropriate energy source is lacking to cause the formation of organic molecules, and if they did form, they would be oxidized by oxygen or eaten by preexisting life.

Objective Questions

1. A ____________ evolution is believed to have preceded an organic evolution.
2. Miller's experiment showed that the first gases could have reacted together to form ____________.
3. Amino acid polymerization could have produced cell-like bodies called ______.
4. The protocell would have carried on ____________ nutrition and fed on the organic material in the ocean.
5. The protocell would have carried on ____________ respiration because there was no free oxygen in the atmosphere.
6. Once the protocell could have ________ it would have been a true cell.
7. The earliest fossils found so far have been ____________, organisms capable of making their own food.
8. The presence of oxygen in the atmosphere permitted ____________ respiration to evolve.
9. Today, we believe that life comes only from ____________.

Answers to Objective Questions

1. not 2. 21 percent 3. no 4. gene flow 5. genetic drift 6. stabilizing 7. ecological 8. geographically separated 9. reproductively isolated 10. premating reproductive, behavioral

Study Questions

1. What was the primitive earth like when it first formed? (p. 528)
2. What were the lightest atoms that may have been present outside the new planet? (p. 528) What gases may have formed from these atoms? (p. 528)
3. Under what conditions was it possible for gases to react with one another to produce small organic molecules? (p. 529)
4. Describe a type of experiment that shows that organic molecules can form from primitive gases. (p. 529)
5. What is the cell-first hypothesis? (p. 530)
6. What type of respiration and nutrition did the protocell have? (p. 531) The protocell became a true cell when it could do what? (p. 532)
7. How did the evolution of autotrophs change the primitive atmosphere? (p. 534)
8. Why doesn't life arise by chemical evolution today? (p. 534)

Key Terms

aerobic (a″er-ōb′ik) growing or metabolizing only in the presence of oxygen as in aerobic respiration. *534*

anaerobic respiration (an-a″er-ōb′ik res″pĭ-ra′shun) growing or metabolizing in the absence of molecular oxygen. *531*

autotroph (aw′to-trōf) an organism that is capable of making its food (organic molecules) from inorganic molecules. *531*

biological evolution (bi-o-loj′ĕ-kal ev″o-lu′shun) changes that have occurred in life forms from the origination of the first cell or cells to the many diverse forms in existence today. *530*

cell-first hypothesis (sel ferst hi-poth′ĕ-sis) a suggestion that only the first cell or cells contained true proteins and nucleic acids. *530*

chemical evolution (kem′ĭ-kal ev″o-lu′shun) a gradual increase in the complexity of chemical compounds that is believed to have brought about the origination of the first cell or cells. *528*

coacervate (ko-as′er-vāt) a mixture of polymers that may have preceded the origination of the first cell or cells. *531*

energy sources (en′er-je sor′sez) ways by which energy can be made available to organisms from the environment. *529*

heterotroph (het′er-o-trof″) an organism that takes in preformed food. *531*

organic soup (or-gan′ik sōōp) an expression used to refer to the ocean before the origin of life when it contained newly formed organic compounds. *530*

oxidizing atmosphere (ok′sĭ-dīz-ing at′mos-fēr) an atmosphere that contains oxidizing molecules such as O_2 rather than reducing molecules such as H_2. *534*

ozone shield (o′zōn shēld) a layer of O_3 present in the upper atmosphere which protects the earth from damaging U.V. light. *534*

polymerization (pol″ĭ-mer″ĭ-za′shun) the joining together of unit molecules to form large organic molecules such as nucleic acids and protein molecules. 530

primitive atmosphere (prim′ĭ-tiv at′mos-fēr) the gases that were found in the atmosphere when the earth first arose. *528*

proteinoid microspheres (pro′te-in-oid mi-kro-sfērz′) a phase, consisting of polypeptides only, during the chemical evolution of the first cell or cells. *530*

protocells (pro′to-selz) the structures that preceded the true cell in the history of life. *530*

reducing atmosphere (re-dūs′ing at′mos-fēr) an atmosphere that contains reducing molecules such as H_2 rather than oxidizing ones such as O_2. *528*

25

Viruses and Kingdoms Monera, Protista, and Fungi

Chapter Concepts

1 Viruses are noncellular; whether they should be considered living organisms is questionable.

2 The monerans are prokaryotes, while the protistans and fungi are eukaryotes.

3 The kingdom Monera includes bacteria which despite their small size are important organisms.

4 The kingdom Protista contains unicellular organisms that may resemble the first eukaryotic cells.

5 The kingdom Fungi contains the most complex organisms to rely on saprophytic nutrition.

Chapter Outline

It is our aim to discuss living organisms from the simple to the complex and from the primitive (earliest evolved) to the most advanced (latest evolved). The designation *primitive* means only that these organisms have changed less with time (fig. 23.2) than have the advanced organisms. It is also well to remember that no living group of organisms is the direct ancestor of another living group of organisms, although it is possible for two living groups to have shared a common ancestor. The great variety of living organisms within any particular group is the result of adaptive radiation from a common ancestor, which may at times be determined from the fossil record.

It is curious that we must begin our discussion with viruses when they are not even included in the classification table found in Appendix C. We begin with viruses only because they are on the borderline between living and nonliving things.

Viruses

Viruses are considered noncellular because they do not have a cellular type of organization. They are tiny particles (5–200 nm) composed of at least two parts: *an outer coat of protein* and *an inner core of nucleic acid.* The coat, called a **capsid,** is sometimes surrounded by a membranous outer envelope.

Figure 25.1
Viruses; micrographs below and drawings above a. *The tobacco mosaic virus is an RNA virus that attacks tobacco. The drawing shows how the nucleic-acid core circles within a capsid composed of individual proteinaceous units.* b. *The adenovirus is a DNA virus that causes respiratory and intestinal infections in humans.* c. *A T virus is a bacteriophage with a complex structure. The DNA within the head passes through the tail to enter the host* E. coli *cells.* d. *The influenza virus is an RNA virus whose protein capsid is enclosed by a membranous envelope.*

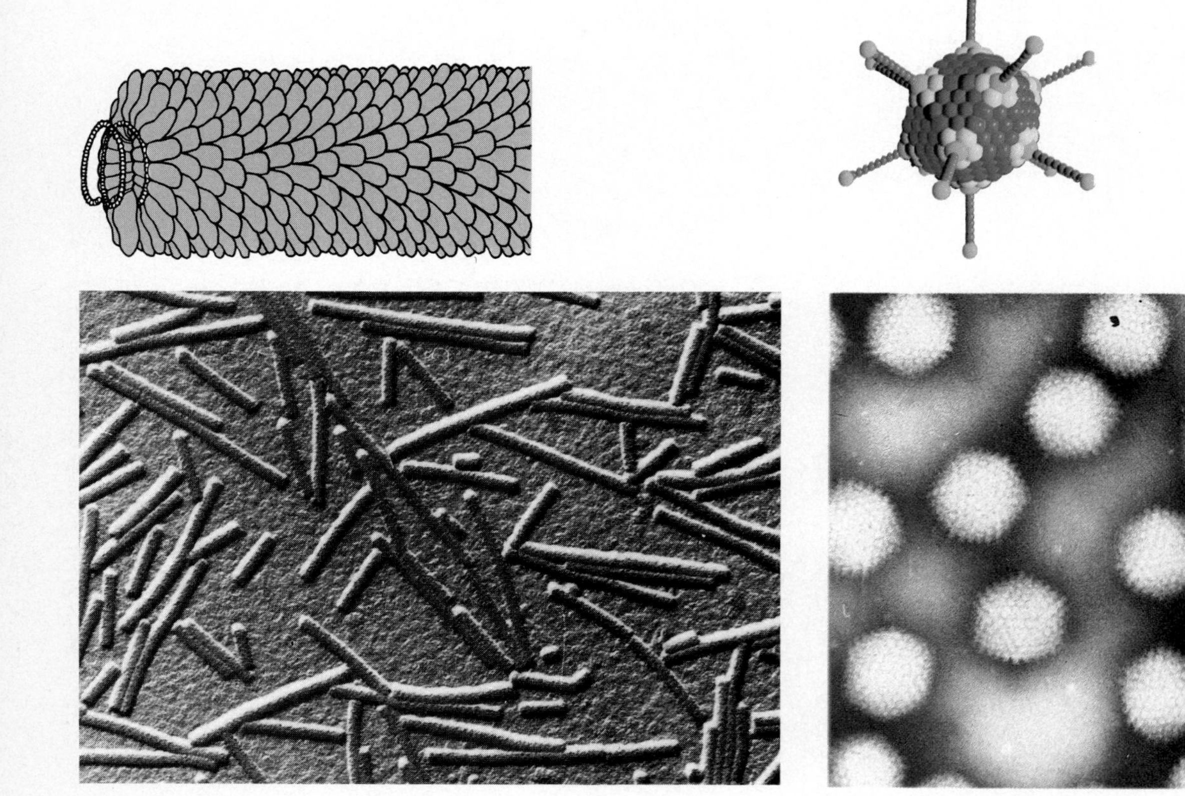

The core of nucleic acid may be either DNA or RNA. Although viruses cannot be seen with the light microscope, the electron microscope has permitted a study of their structure (fig. 25.1). Notice how the capsid is made up of repeating protein subunits.

Viruses are capable of reproduction, but only within living cells; therefore, they are called *obligate parasites*. In the laboratory, active animal viruses are maintained by injecting them into live chick embryos (fig. 25.2). Outside living cells, viruses are nonliving and can be stored just as chemicals are stored. Therefore, it is proper to ask if viruses should be considered alive.

Typically viruses have a specific host range. Certain ones attack only plants; others attack only animals, either birds or mammals; and the viruses called **bacteriophages** attack only bacteria. The human disease-causing viruses (table 12.3) even attack only certain type cells. It is now known that there must be a match between a protein in the outer coat or envelope of the virus and the cell's outer surface before a virus can gain entry.

Viruses are noncellular obligate parasites that always have a coat of protein and a nucleic acid core.

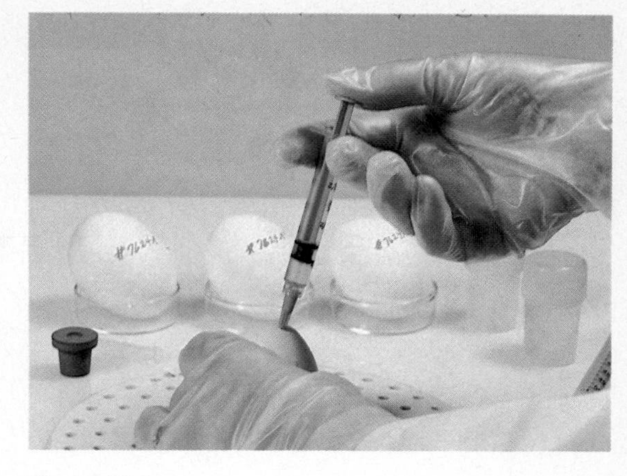

Figure 25.2
Inoculation of live chick eggs with virus particles. A virus only reproduces inside a living cell, not because it uses the cell as nutrients but rather because it takes over the machinery of the cell.

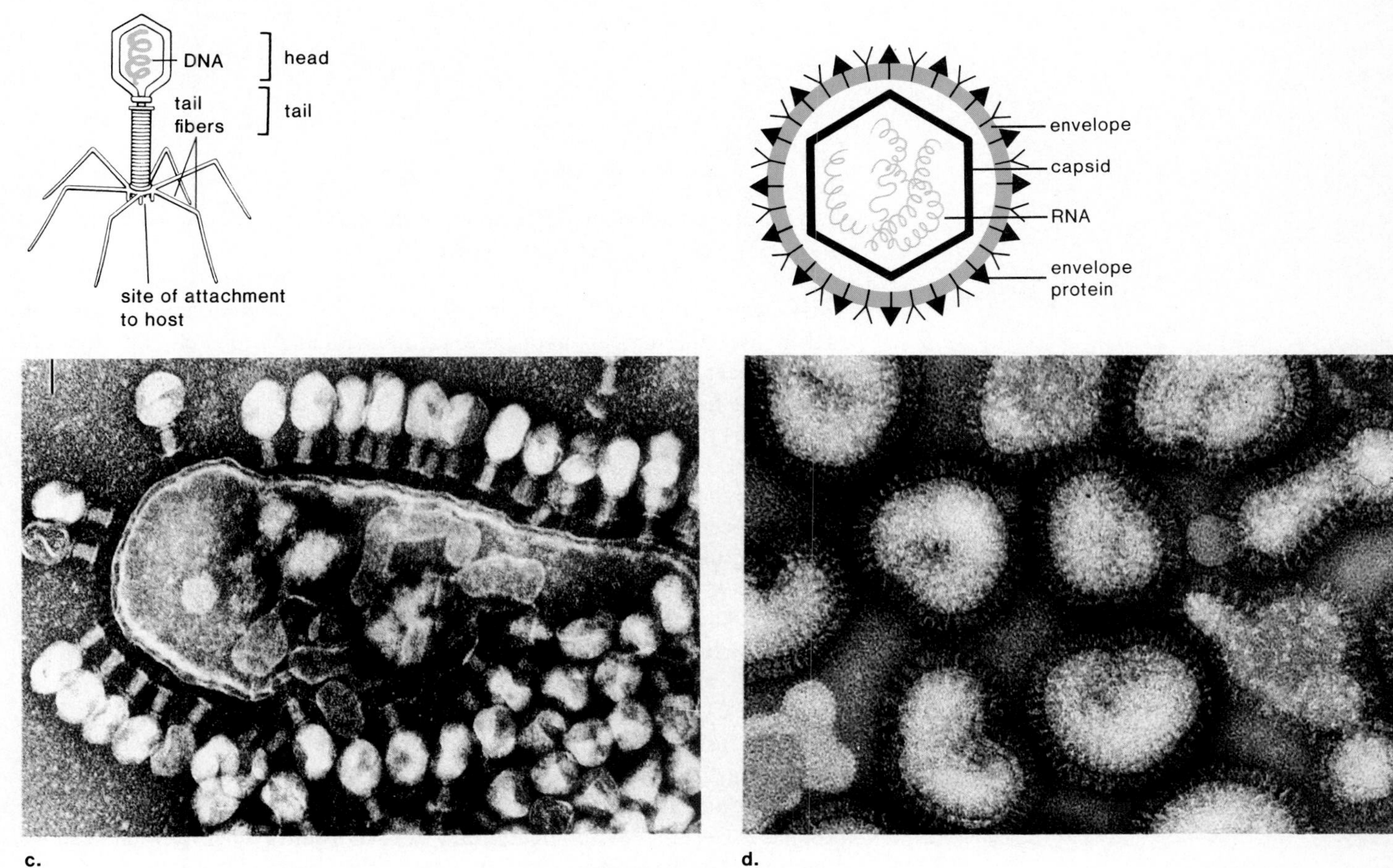

Figure 25.3
Lytic versus lysogenic life cycle. In the lytic cycle, a bacteriophage reproduces within the host bacterial cell and then the cell is lysed (broken open), allowing the viral particles to escape. In the lysogenic cycle, the DNA of a bacteriophage integrates into the host DNA and becomes a prophage. Thereafter, the prophage is replicated along with the bacterial chromosome and passed to all the daughter cells. When and if the viral DNA leaves the chromosomes, the lysogenic cycle can be followed by the lytic cycle.

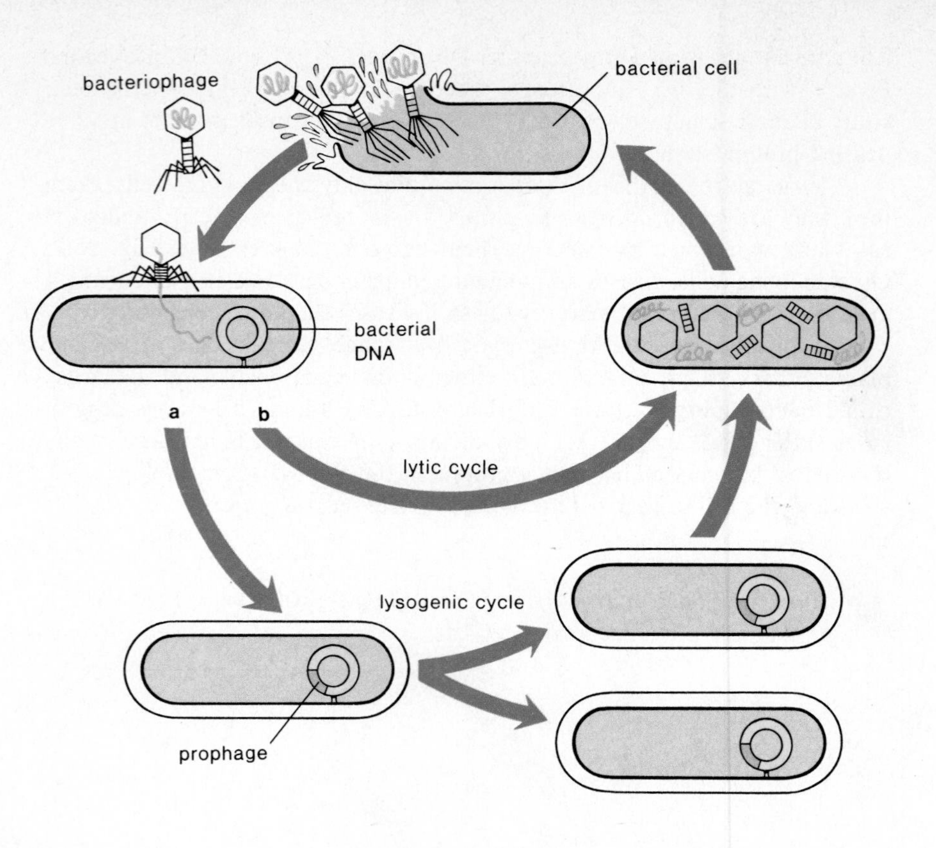

Life Cycles

Bacteriophages

Two types of bacteriophage life cycles, termed the lytic and the lysogenic cycle (fig. 25.3) have been carefully studied.

Lytic Cycle When a T-even virus (meaning that it is designated by an even rather than odd number) happens to collide with an *E. coli* cell, it attaches to specific receptors by means of its protein tail fibers. An enzyme digests away part of the bacterial cell wall and the viral DNA enters the bacterial cell by way of the tail. Once inside, the viral DNA brings about disintegration of host DNA and takes over the operation of the cell. Viral DNA replication, utilizing the nucleotides within the host cell, produces many copies of viral DNA. Transcription occurs and mRNA molecules utilize host ribosomes to bring about the production of multiple copies of coat proteins. Viral DNA and capsids are *assembled* to produce about 100 viral particles. In the meantime, viral DNA has directed the synthesis of lysozyme, an enzyme that disrupts the cell wall and the particles are released.

Lysogenic Cycle Some bacteriophages do not immediately undergo a lytic life cycle. Instead, the viral DNA becomes integrated into the bacterial DNA. In this stage it is called a *prophage*. The prophage is replicated along with the host DNA and all subsequent cells, called *lysogenic cells,* carry a copy of the prophage. Certain environmental factors such as ultraviolet radiation can induce the lytic cycle. The prophage leaves the bacterial chromosome; replication of viral DNA, production of capsids, assemblage, and lysis follow.

During the lytic cycle of a bacteriophage, the bacterial cell dies when the viral particles burst from the cell. During the lysogenic cycle, viral DNA integrates itself into bacterial DNA for an indefinite period of time.

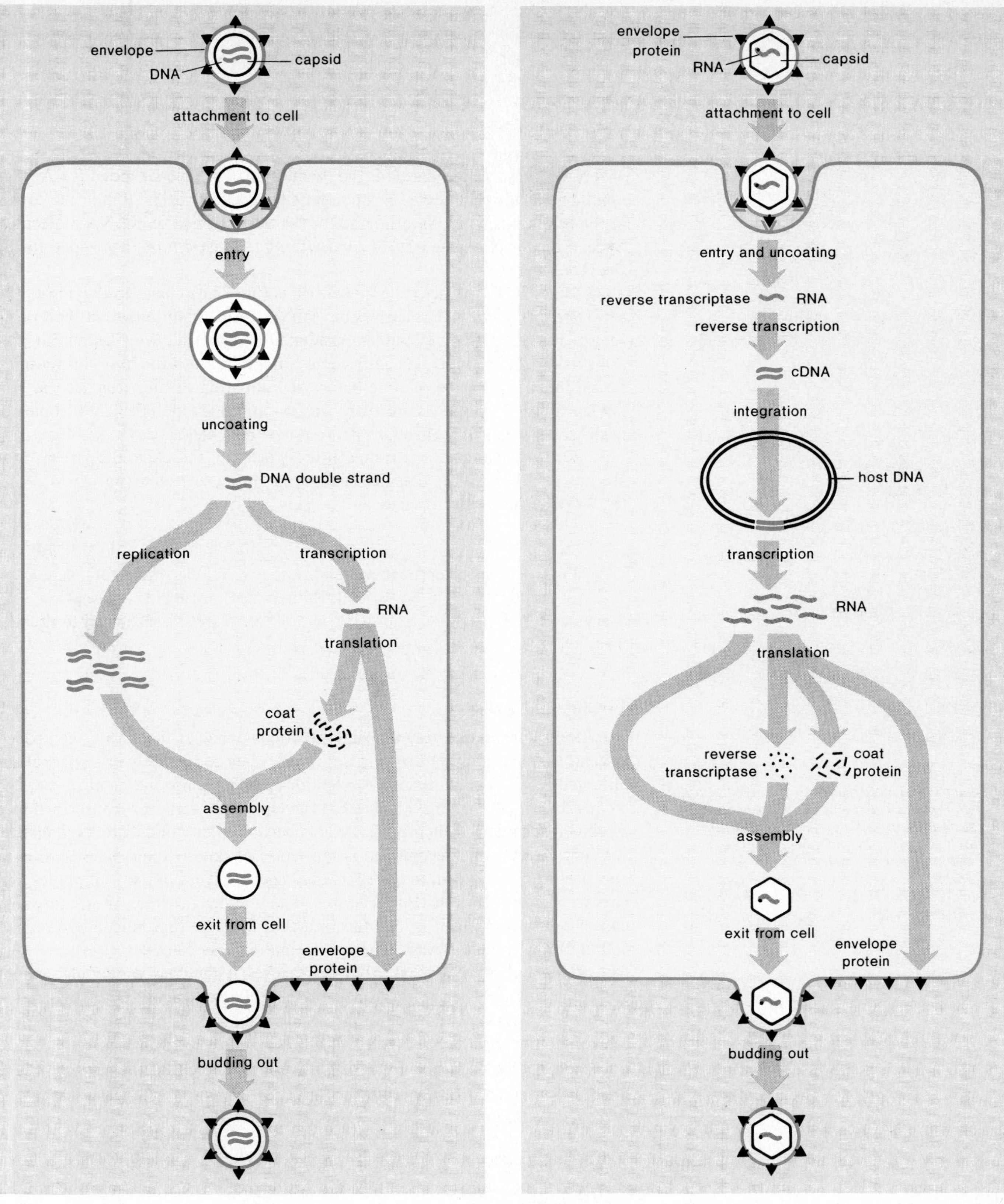

Animal Viruses

Animal viruses typically have a membranous outer *envelope* along with a different mode of entry and exit from the host cell (fig. 25.4a). The entire virus enters by endocytosis and *uncoating,* which releases viral nucleic acid from the envelope and capsid, occurs inside the cell. Following assemblage, the virus exits from the cell by exocytosis, also called *budding*. The process of budding,

Figure 25.4
Life cycle of an animal virus. After entering by endocytosis, the virus becomes uncoated. The nucleic acid then codes for proteins, some of which are capsid proteins and some of which are envelope proteins. Assemblage follows replication of the nucleic acid core. When the virus exits by budding, it becomes enclosed by an envelope made up of host-cell membrane lipids and viral envelope proteins.

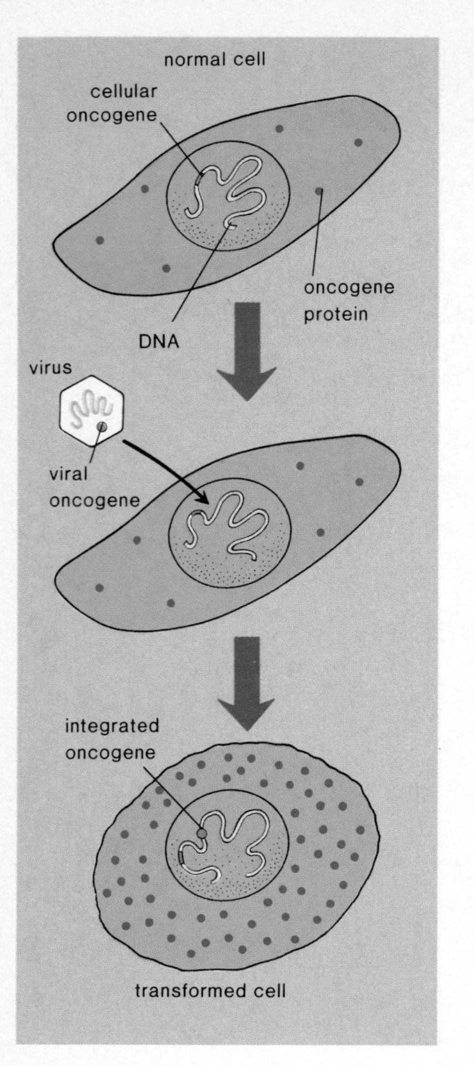

Figure 25.5
How a retrovirus may contribute to the development of cancer. The normal cell contains one oncogene but is not yet cancerous. The virus also contains an oncogene which it passes to the cell. Now the presence of two oncogenes causes the cell to be transformed and it becomes cancerous because it produces a protein, most likely a growth hormone, that causes the cell to become abnormal.

which does not necessarily kill the host cell, provides the new viral particles with the membranous envelope. This envelope often contains glycoproteins, coded for by viral DNA, that interact with the next host cell's membrane permitting entry of the virus into this cell.

RNA Animal Viruses Some animal viruses have RNA genes, that is, only RNA is enclosed within the capsid. In most of these viruses, the single strand of RNA serves as a template for the production of double-stranded RNA. This unique molecule then serves as a template for the replication of multiple copies of the original genetic material and for the transcription of mRNA molecules. Special enzymes, termed *RNA replicase and transcriptase,* are coded for by the RNA genes.

Other RNA viruses called **retroviruses** (fig. 25.4b) have an enzyme called *reverse transcriptase* that carries out RNA → DNA transcription. Following replication, the resulting double-stranded DNA (called *cDNA* because it contains a copy of the viral RNA) becomes integrated into the host chromosome for an indefinite period of time before reproduction of the virus occurs. Following integration, newly formed viruses sometimes include RNA copies of host genes which they then carry into a new host cell.

Retroviruses are of extreme interest because these are the viruses that are apt to bring cancer-causing oncogenes, into a host cell (fig. 25.5). Also, the AIDS virus is a retrovirus.

Animal viruses often have a membranous envelope which they acquire by budding from the host cell. Retroviruses are RNA viruses that carry out RNA → DNA transcription and integrate this cDNA into the host cell.

Viroids and Prions

Viroids and prions are very unusual infectious particles that can be compared to viruses. **Viroids** differ from viruses in that they consist only of a short chain of naked RNA. The number of nucleotides present is not sufficient to code for a capsid and there is no evidence that the viroid RNA is ever transcribed into a protein. We know of their presence only because they cause diseases in plants; the first viroid to be recognized is the cause of potato spindle tuber disease and they are also known to have attacked coconut trees in the Philippines and chrysanthemums in the United States. The manner in which they cause disease is not known but it is hypothesized that they are reproduced by the host cell and may interfere with gene regulation in these cells.

Prions also differ markedly from viruses. They consist only of a glycoprotein having only one polypeptide of about 250 amino acids. DNA that codes for the polypeptide is not found in the host cell so it is not known how they manage to be reproduced. Like viroids, these particles are only known because they have been associated with a disease. It is possible that they are the cause of nervous system diseases previously termed slow-virus diseases because it takes a long time for symptoms to gradually worsen.

Viral Infections

Viruses are best known for causing infectious diseases in animals and plants. In plants, infectious diseases can only be controlled by destroying those plants that show symptoms of disease. In animals, especially humans, they are controlled by administering vaccines and, only recently, by the administration of antiviral drugs, as discussed in the reading on page 543. Some well-studied viral diseases in humans are flu, mumps, measles, polio, rabies, and infectious hepatitis (table 12.4). As mentioned above, retroviruses have been implicated in the development of cancer.

Antibiotics and Antiviral Drugs

Penicillium chrysogenum.

An antibiotic is a chemical that selectively kills bacteria when it is taken into the body as a medicine. There has been a dramatic reduction in the number of deaths due to pneumonia, tuberculosis, and other infections since 1900 and this can, in part, be attributed to the increasing use of antibiotic therapy.

Most antibiotics are produced naturally by soil microorganisms. Penicillin is made by the fungus *Penicillium;* streptomycin, tetracycline, and erythromycine are all produced by a bacterium, *Streptomyces.* Sulfa, an analog of a bacterial growth factor, can be produced in the laboratory.

Antibiotics are metabolic inhibitors specific for bacterial enzymes. This means that they poison bacterial enzymes without harming host enzymes. Penicillin blocks the synthesis of the bacterial cell wall, streptomycin, tetracycline, and erythromycine block protein synthesis, and sulfa prevents the production of a coenzyme.

There are problems associated with antibiotic therapy. Some patients are allergic to antibiotics and their reaction to them may even be fatal. Antibiotics not only kill off disease-causing bacteria, they also reduce the number of beneficial bacteria in the intestinal tract. The latter may have held in check a pathogen that now is free to multiply and invade the body. The use of antibiotics sometimes prevents natural immunity from occurring, leading to the necessity for recurring antibiotic therapy. Most important, perhaps, is the growing resistance of certain strains of bacteria. While penicillin used to be 100 percent effective against hospital strains of *Staphylococci aureus,* today it is far less effective. Tetracycline and penicillin, long used to cure gonorrhea, now have a failure rate of more than 20 percent against certain strains of *Gonococcus.* Most physicians believe that antibiotics should only be administered when absolutely necessary and some believe that if this is not done then resistant strains of bacteria will completely replace present strains and antibiotic therapy will no longer be effective at all. They are very much opposed to the current practice of adding antibiotics to livestock feed in order to make animals grow fatter because resistant bacteria are easily transferred from animals to humans.

The development of antiviral drugs has lagged far behind the development of antibiotics. Viruses lack most enzymes and instead utilize the metabolic machinery of the host cell. Rarely has it been possible to find a drug that successfully interferes with viral reproduction without also interfering with host metabolism. One such drug, however, called Vidarabine was approved in 1978 for treatment of viral encephalitis, an infection of the nervous system. Acyclovir (ACV) seems to be helpful in treating genital herpes, and the drug AZT (azidothymidine) is being used in AIDS patients.

Viruses cause diseases of bacteria, plants, and animals. They have also been implicated in the development of cancer.

Kingdom Monera

Kingdom Monera contains only prokaryotes, the first types of cells to evolve. The prokaryotes are divided into the archaebacteria and the eubacteria which now include the cyanobacteria (formerly called blue-green algae). Prokaryotic cells do not have the cytoplasmic organelles found in eukaryotic cells, except ribosomes (table 2.2). They do have DNA, but it is not contained within a nuclear envelope; therefore they are said to lack a nucleus. They have respiratory enzymes, but no mitochondria; and if they possess chlorophyll, it may be found within thylakoids, but there are no chloroplasts.

Prokaryotic cells are small but are generally larger than viruses; they range in size from 1 to 10 μm (micrometers) in length and from 0.2 to 0.3 μm in width. Since they are microscopic, it is not always obvious that they are abundant in the air, water, soil, and on most objects. It has even been suggested that the combined weight of all bacteria would exceed that of any other type organism on earth.

The kingdom Monera includes the archaebacteria and the eubacteria, which are prokaryotic, meaning that their cells lack the organelles found in eukaryotic cells.

Characteristics of Bacteria

Most of our knowledge about prokaryotes comes from the study of eubacteria. These bacteria have a cell wall that contains unique amino sugars cross-linked by peptide chains. Its composition is entirely different from the cell wall of any eukaryotic cell. Parasitic bacteria are also surrounded by a polysaccharide or polypeptide capsule that enhances their virulence (ability to cause disease). Cyanobacteria have a sheath of gelatinous material outside their cell walls.

Bacteria occur in three basic shapes (fig. 25.6): *rod* (bacillus), *round,* or spherical (coccus), and *spiral* (a curved shape called a spirillum). The rods and bacilli may form chains of a length typical of the particular bacterium.

Figure 25.6
Scanning electron micrographs of bacteria. a. *Spherical-shaped bacteria.* b. *Rod-shaped bacteria.* c. *Spiral-shaped bacteria. See Figure 2.17 for generalized drawings of bacteria.*

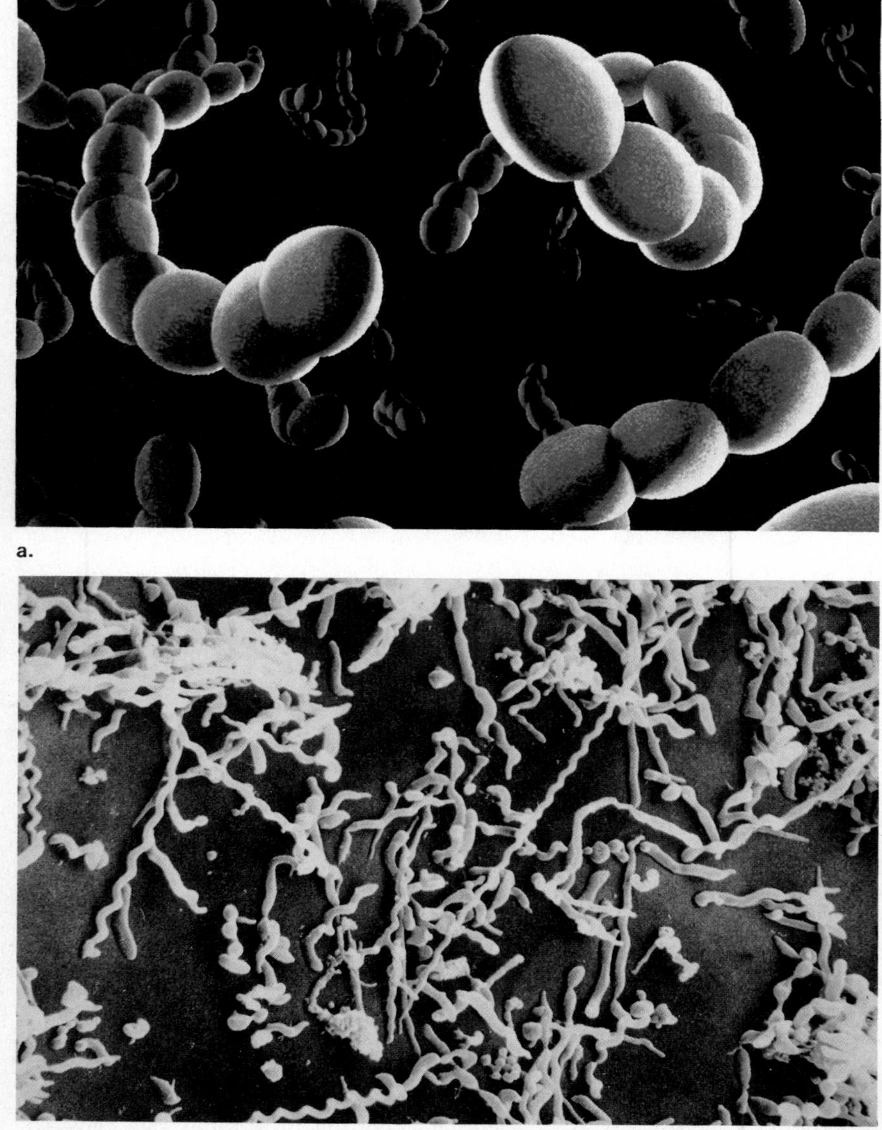

a.

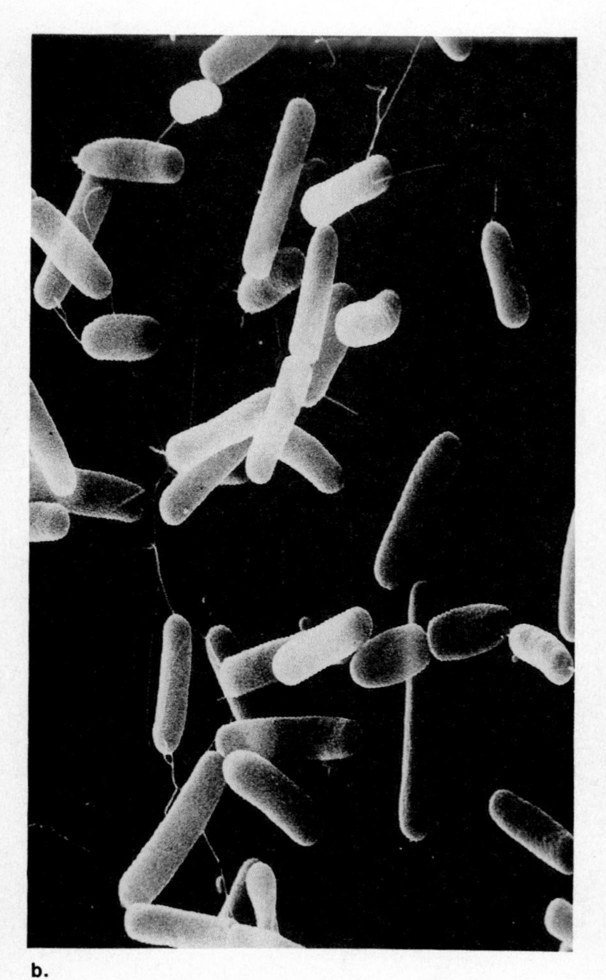

b.

c.

Some bacteria can locomote by means of flagella, which are composed only of a protein called flagellin, and some can adhere to surfaces by means of pili, projections also composed of protein.

Reproduction

Bacteria reproduce asexually by **binary fission.** First, the single chromosome duplicates and then the two chromosomes move apart into separate areas; then the cell membrane grows inward and partitions the cell into two daughter cells, each of which now has its own chromosome (fig. 25.7).

Sexual exchange has been observed between two bacteria when the so-called male passes DNA to the female by way of a pilus called a conjugation pilus. Recombination of the genetic material occurs within the female, which then divides. Bacterial cells can also pick up fragments of DNA released into the medium from dead cells. This is called **transformation** because the cell receiving the DNA is transformed into a cell with the genotype and phenotype of the absorbed DNA. Bacteriophages provide a third means by which it is possible to recombine bacterial DNA. In this process, called **transduction,** the phages carry portions of bacterial DNA from one cell to another.

When faced with unfavorable environmental conditions, some bacteria can form endospores. During spore formation, the cell shrinks, rounds up within the former cell membrane, and secretes a new, thicker wall inside the old one (fig. 25.8). *Endospores* are amazingly resistant to extreme temperatures, drying out, and harsh chemicals, including acids and bases. When conditions are again suitable for growth, the spore absorbs water, breaks out of the inner shell, and becomes a typical bacterial cell.

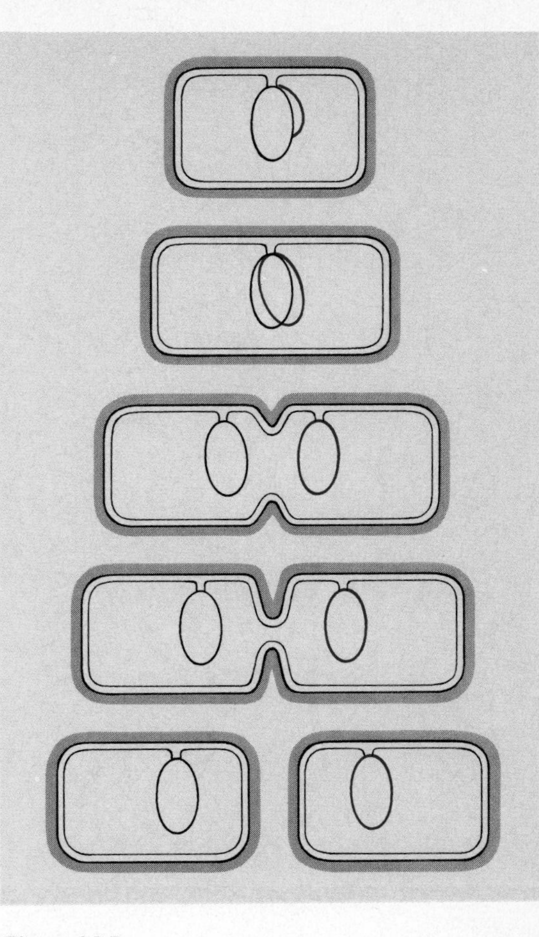

Figure 25.7
Reproduction in bacteria. Above, the single chromosome is seen to be attached to the cell membrane where it is replicating. As the cell membrane lengthens, the two chromosomes separate. Once fission has taken place, each bacterium has its own chromosome.

Metabolism

Some living bacteria are obligate anaerobes and are unable to live in the presence of oxygen. A few serious illnesses, such as botulism, gas gangrene, and tetanus, are caused by anaerobic bacteria. Some bacteria, called *facultative anaerobes,* are indifferent to oxygen and can survive whether or not it is present. Most bacteria, however, are aerobic and, like animals, require a constant supply of oxygen to carry out complete cellular respiration.

Every type of nutrition is found among bacteria except holozoism (eating of whole food). A few bacteria are autotropic, being either chemosynthetic or photosynthetic. The *chemosynthetic bacteria* oxidize inorganic compounds to obtain the necessary energy to produce their own food. Among the organic

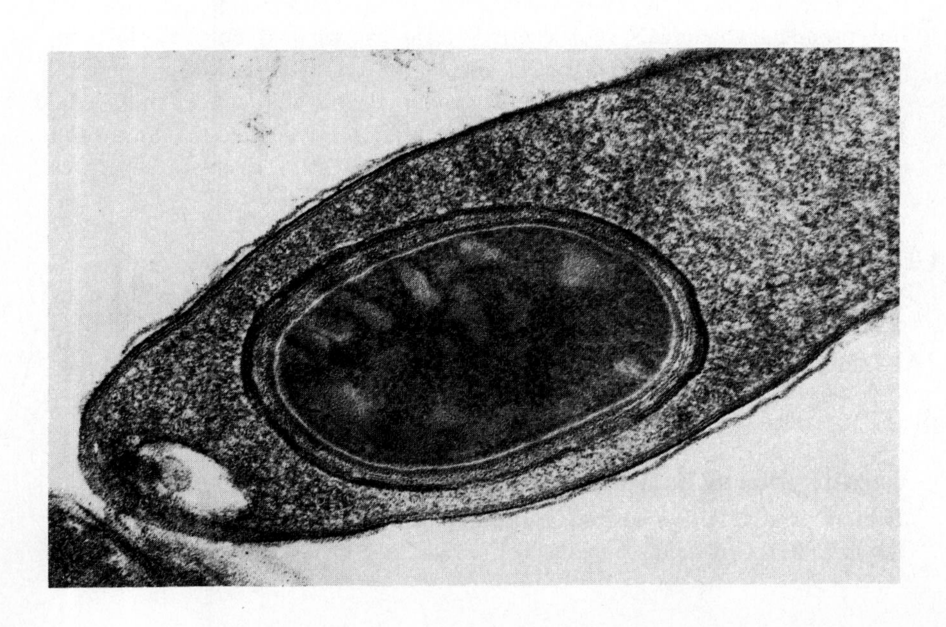

Figure 25.8
Spore formation. This bacterium contains a spore, the dark oval at the lower end of the cell. A spore protects the organism's DNA from exposure to environmental conditions that could destroy it.

compounds oxidized by specific bacteria are ammonia, nitrite, sulfur, hydrogen, and ferrous iron. *Photosynthetic bacteria* contain chlorophyll and can obtain energy from the sun to produce their own food but some are more complex than others (p. 547). Most bacteria are heterotrophic **saprophytes,** which send out digestive enzymes into the environment to break down large molecules into small molecules that can be absorbed across the cell membrane.

Nitrogen fixation occurs among bacteria and when it does, these bacteria are able to reduce aerial nitrogen (N_2) to ammonia which can be incorporated into organic compounds. This extremely important function not only makes nitrogen available to themselves but for all living things.

Bacteria may be rod-shaped, or have a spherical or spiral shape. They are usually aerobic saprophytes and reproduce by binary fission.

Importance of Bacteria

Within ecosystems, the cyanobacteria are important producers of food for other aquatic organisms. Most bacteria, however, are decomposers (p. 690) and as such they play a role in the carbon cycle. In the nitrogen cycle, nitrogen-fixing bacteria in the soil or in the nodules of legumes (fig. 31.11) provide a usable source of nitrogen for terrestrial plants. Then, too, the nitrifying bacteria oxidize ammonia, an excretion product of animals, to nitrate, a nutrient for plants.

Bacteria often live in association with other organisms. The nitrogen-fixing bacteria in the nodules of legumes are mutualistic, as are the bacteria, mainly *E. coli,* within our own intestinal tract. We provide the bacteria with a home and they provide us with certain vitamins. Commensalistic bacteria reside on our skin where they usually cause no problems. Because of their presence, parasitic bacteria and fungi may have difficulty establishing residence.

The parasitic bacteria cause numerous diseases, such as those listed in table 12.5. General cleanliness is the first step toward preventing the spread of these diseases. Disinfectants and antiseptics also help reduce the number of infectious bacteria. Typhoid is kept in check by adding chlorine to water; diphtheria, typhoid, and tuberculosis are also partially controlled by the pasteurization (heating to 145° F. for 30 minutes) of milk. Sterilization, a process that kills all living things, even endospores, is used whenever all bacteria and viruses must be killed. Sterilization can be achieved by use of an autoclave (fig. 25.9), a container that maintains steam under pressure.

Figure 25.9
Sterilization by autoclaving. Hospital employees are closing the door and operating a large sterilizer. Sterilization permits surgical procedures to be done with reduced fear of subsequent infection. During autoclaving, steam under pressure kills bacterial cells and endospores.

Bacteria have long been used by humans to commercially produce various products. Chemicals, such as ethyl alcohol, acetic acid, butyl alcohol, and acetones, are produced by bacteria. Bacterial action is involved in the production of butter, cheese, sauerkraut, rubber, cotton, silk, coffee, and cocoa. By means of gene splicing, bacteria are now used to produce human insulin and interferon, as well as other types of proteins (p. 492). Even antibiotics, as discussed in the reading, are produced by bacteria.

Bacteria play important roles in nature's cycles, and are used by humans to produce all sorts of chemical products, some even as a result of genetic engineering. They often live in association with other organisms and, as such, sometimes cause diseases, including human diseases.

Classification of Bacteria

Bacteria are classified into the archaebacteria (division Archaebacteria) and the eubacteria (division Eubacteria).

Table 25.1 Major Groups of Eubacteria

Group	Characteristics
Purple bacteria	Some photosynthesizers (e.g., purple sulfur) but many nonphotosynthesizers such as *E. coli* (symbiotic in our gut) and *Rhizobium* (nitrogen-fixing in legumes) and some parasitic (e.g., the plague bacterium)
Green sulfur bacteria	Photosynthesizers that use H_2 or H_2S as hydrogen source
Cyanobacteria	Mostly photosynthetic, possessing chlorophyll a, and using water as the hydrogen source resulting in O_2 release; some carry out nitrogen fixation in heterocysts
Gram-positive bacteria	Have thick walls that take up a stain called the gram stain; some causing diseases (e.g., botulism, food poisoning, diptheria, strep throat)
Spirochetes	Spiral or curved bacteria moving by means of flagella such as *Treponema pallidum* (cause of syphilis)
Rickettsias and chlamydias	Parasitic, very small bacteria Rickettsias cause Rocky Mountain spotted fever and chlamydias cause NGU (nongonococcal urethritis)

Archaebacteria

Most likely these were the earliest of the prokaryotes and their cell wall and cell membrane do not have the same composition as the eubacteria. Some believe that they should even be placed in a different kingdom.

The **archaebacteria** are able to live in the most extreme of environments, perhaps representing the kinds of habitats that were available when the earth first formed. The methanogens are anaerobic and live in swamps and marshes producing the methane known as marsh gas. They also live in the guts of organisms, including humans. The halophiles live where it is salty, such as the Great Salt Lake in Utah. Curiously a type of rhodopsin pigment (related to the one found in our own eyes) allows them to carry on a primitive type of photophosphorylation for ATP production. The thermoacidophiles live where it is both hot and acidic. Those that live in the hot sulfur springs of Yellowstone National Park obtain energy by oxidizing sulfur.

Eubacteria

There are many different types of eubacteria, and space does not permit us to even list all of these, but table 25.1 points out those that have been referred to in this section and other sections of the text. In addition, we will discuss the photosynthetic bacteria here.

Photosynthetic Bacteria The most prevalent of the photosynthetic bacteria are the **cyanobacteria** (fig. 25.10) which used to be called blue-green algae because they contain a blue pigment in addition to the green pigment chlorophyll. Actually, however, many have additional pigments and may appear black, brown, yellow, or red.

Cyanobacteria may be *unicellular, filamentous,* or *colonial.* The filaments and colonies are not considered multicellular because each cell is independent of the others. Cyanobacteria lack any visible means of locomotion although some oscillate (sway back and forth).

Cyanobacteria have a form of photosynthesis that is more complex than the other photosynthetic bacteria. The other types of photosynthetic bacteria most often have a unique form of chlorophyll, possess only Photosystem I, and do not give off oxygen when they photosynthesize. They lack this feature because instead of using water as a hydrogen source, they use either molecular hydrogen, hydrogen sulfide, or organic compounds. None of these molecules gives off oxygen when broken down. In contrast, the cyanobacteria carry out photosynthesis, as do plants. They have a type of chlorophyll found in plants, utilize both Photosystems I and II, and do give off oxygen because they use water as a hydrogen source.

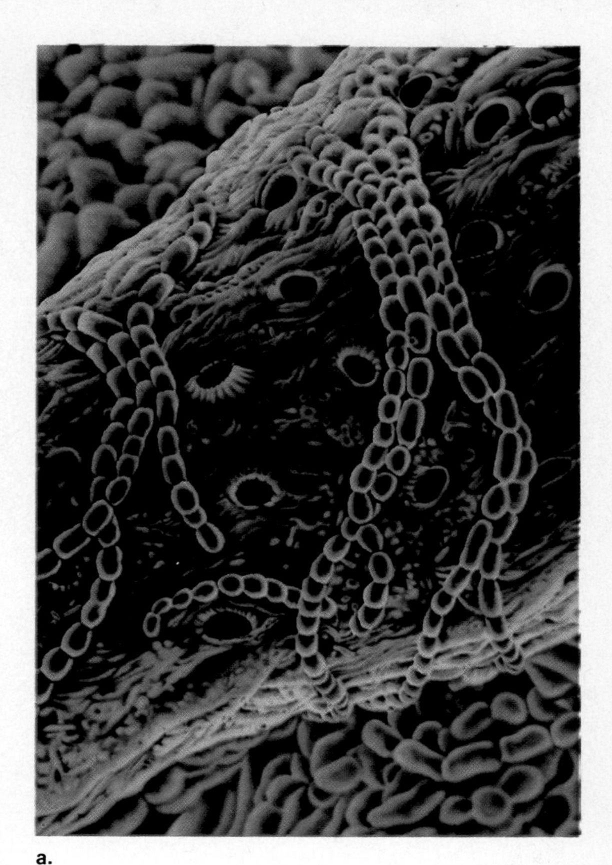

a.

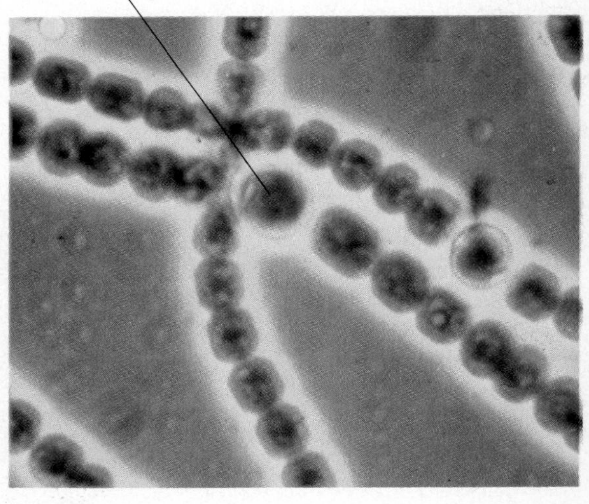

b.

Figure 25.10
a. *Filamentous cyanobacterium.* b. *Heterocysts develop when the only source of nitrogen is aerial nitrogen. Nitrogen fixation occurs in these enlarged cells.*

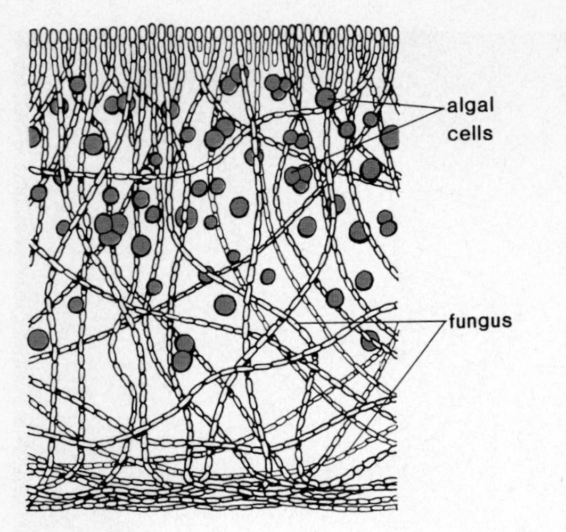

Figure 25.11
Lichen anatomy. Lichens are made up of two types of organisms, an alga and a fungus. In this drawing, the algal cells are represented by circles and the fungus is represented by the filaments. Biologists are still debating whether this is a completely mutualistic relationship or whether the fungus is at least somewhat parasitic on the alga.

Cyanobacteria are common in fresh water, soil, and moist surfaces. They are also found in habitats devoid of other types of life, such as hot springs. With the fungi, they form lichens that can grow on rocks. A **lichen** (figs. 25.11 and 25.21) is a mutualistic relationship in which the cyanobacteria provides organic nutrients to the fungus, while the latter possibly protects and furnishes inorganic nutrients to its partner. Some cyanobacteria have a special advantage because they can fix aerial nitrogen and therefore their nutrient requirements are minimal. Lichens help transform rocks to soil so that other forms of life may follow.

In fresh water, cyanobacteria are sometimes responsible for the bloom associated with cultural eutrophication (p. 709). The occasional red color of bodies of fresh water and rivers is due to a species of blue-green that contains a red pigment.

Cyanobacteria are photosynthesizers that sometimes can also fix aerial nitrogen. They are often responsible for the algal blooms of polluted waters and together with fungi, they form lichens, important soil formers.

Kingdom Protista

The protists are unicellular eukaryotes. They have all the organelles with which we are familiar: a nucleus, mitochondria, endoplasmic reticulum, and Golgi apparatus, for example.

We will include in the kingdom Protista unicellular (or colonial) organisms whose relationships to either plants or animals has not clearly been established. (Unicellular organisms whose relationship is clearer are included as members of these other kingdoms.) While unicellular organisms must cope with the environment without benefit of tissues and organ systems, they are not simple; their complexity resides in the organization of their cells.

The kingdom Protista contains unicellular eukaryotic organisms whose relationship to multicellular forms is not well established.

Table 25.2 *Types of Protozoans*

Protozoans	Locomotor Organelles	Example
Amoeboid	Pseudopodia	*Amoeba*
Ciliate	Cilia	*Paramecium*
Flagellate	Flagella	*Trypanosoma*
Sporozoa	No locomotor organelle	*Plasmodium*

Protozoans

Protozoans are small (2 μm to 1,000 μm), usually colorless, unicellular organisms that lack a cell wall. Like animals they tend to have special structures for food gathering and locomotion; excretion and respiration are carried out across the cell membrane. Although sexual exchange is sometimes observed, reproduction occurs by cell division. Protozoans are classified according to their type of locomotor organelle (table 25.2).

Amoeboids

Amoeboids (phylum Sarcodina), such as *Amoeba proteus* (fig. 25.12), are a small mass of cytoplasm without any definite shape. They move about and feed by means of cytoplasmic extensions called pseudopodia, or false feet. A pseudopod forms when the cytoplasm streams forward in a particular direction.

The organelles within an amoeba include food, or digestive vacuoles, and contractile vacuoles. *Food vacuoles* are characteristic of holozoic protozoans and are formed within an amoeba when a morsel of food is surrounded by pseudopodia. This is a form of phagocytosis that produces a vacuole that later becomes a digestive vacuole. *Contractile vacuoles* first collect excess water from the cytoplasm and then appear to "contract," releasing the water through a temporary opening in the cell membrane. Contractile vacuoles are most often seen in freshwater protozoans.

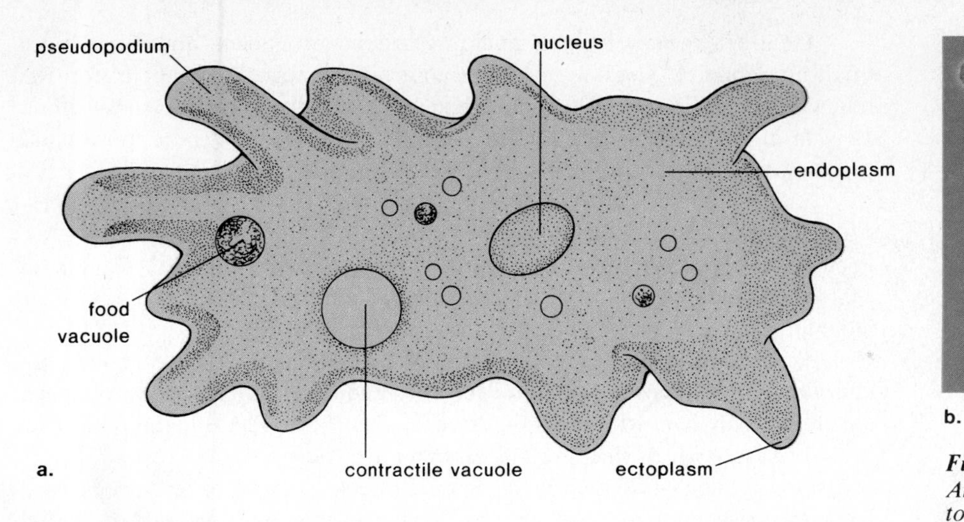

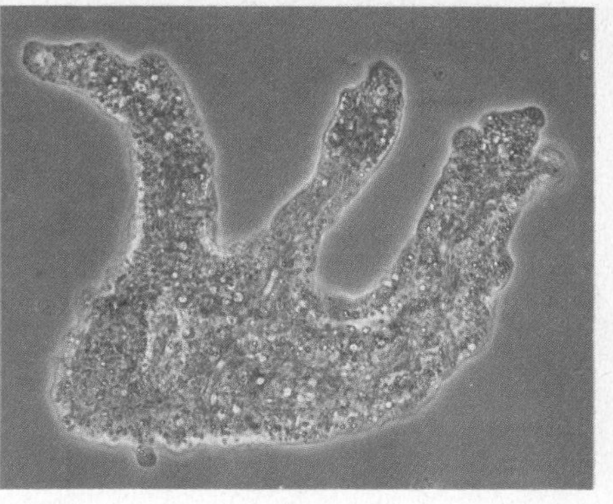

Figure 25.12
Amoeba anatomy. a. *Cytoplasmic streaming occurs to form pseudopodia. It has been observed that the gel-like ectoplasm must become more fluid before streaming of the sol-like endoplasm is possible. Specialized vacuoles carry on digestive and excretory functions. At one time it was thought that the contractile vacuoles had the power to contract, but most likely they are simply squeezed by the cytoplasm.* b. *Micrograph showing how the shape of the organism can vary.*

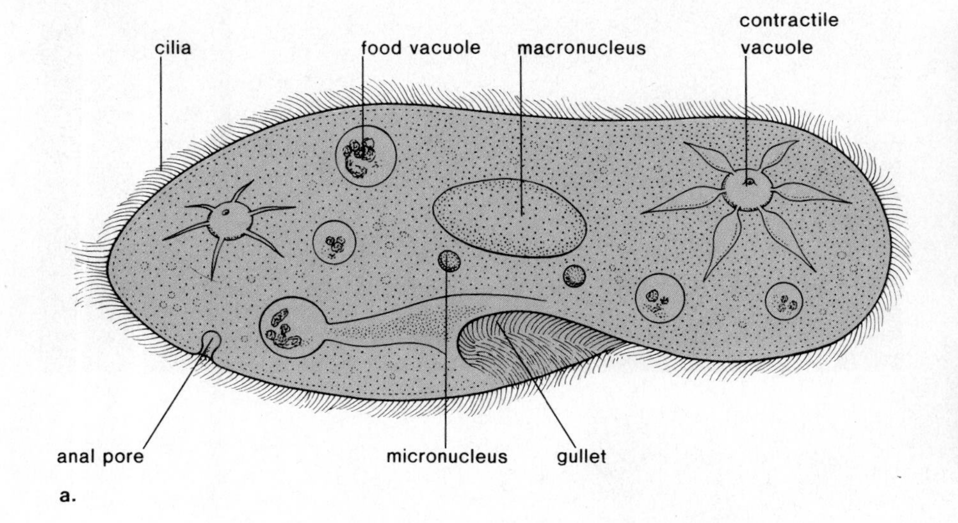

Figure 25.13
Paramecium *anatomy.* a. *This single cell is the most complex of the protozoans. Food moves from the gullet to a vacuole to the anal pore. An extensive network of fibers (not shown) keeps the cilia beating in one direction to propel the organism. The macronucleus controls the vegetative cell; the micronuclei only serve to allow sexual exhange of DNA which is occurring in* (b), *a process termed conjugation.* b. *The scanning electron micrograph not only illustrates conjugation, it also shows how the organism is covered by cilia.*
From: SCANNING ELECTRON MICROSCOPY IN BIOLOGY by R. G. Kessel and C. Y. Shih. © Springer-Verlag, Berlin.

Two forms of marine amoeba have shells. *Foraminifera* have a chalky, sometimes many-chambered shell; these organisms were so numerous at one time that their remains built the White Cliffs of Dover. *Radiolaria* secrete a beautiful skeleton of silica that becomes the bottom ooze in deeper parts of the ocean. One form of amoeba, *Entamoeba histolytica,* causes amoebic dysentery in humans.

Ciliates

The **ciliates** (phylum Ciliophora), such as those in the genus *Paramecium* (fig. 25.13), are the more complex of the protozoans. Hundreds of cilia project through tiny holes in the outer covering, or pellicle. Lying in the ectoplasm just beneath the pellicle are numerous oval capsules that contain **trichocysts.** The contents of the trichocysts may be discharged as long threads, which are used for defense. When a *Paramecium* feeds, food is swept down a gullet to the cell mouth, below which food vacuoles form. Following digestion, the soluble nutrients are absorbed by the cytoplasm and the indigestible residue is eliminated at the anal pore.

Ciliates have two types of nuclei: a large macronucleus and one or more small micronuclei. The macronucleus controls the normal metabolism of the cell, while the micronuclei are concerned with reproduction. Following meiotic division of the micronuclei, two *Paramecia* may exchange these in a sexual process called *conjugation* (fig. 25.13b).

Flagellates

Protozoans that move by means of flagella are sometimes called **zooflagellates** (phylum Zoomastigina) to distinguish them from unicellular algae that have flagellates.

Many zooflagellates enter symbiotic relationships (p. 680). *Trichonympha collaris* lives in the gut of termites and enzymatically converts wood to soluble carbohydrates easily digested by the insect. The trypanosomes (fig. 25.14) cause African sleeping sickness and are transmitted to vertebrates by the tsetse fly. The tsetse fly, which becomes infected when it takes a blood meal from a diseased animal, passes on the disease when it feeds on another victim. The white cells in an infected animal accumulate around the blood vessels leading to the brain and cut off the circulation. The lethargy characteristic of the disease is caused by an inadequate supply of oxygen for normal brain alertness.

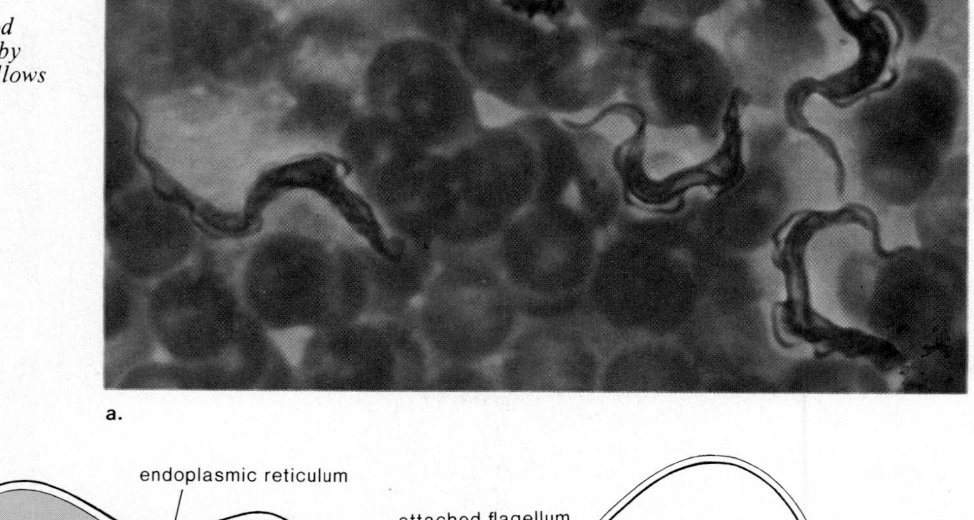

a.

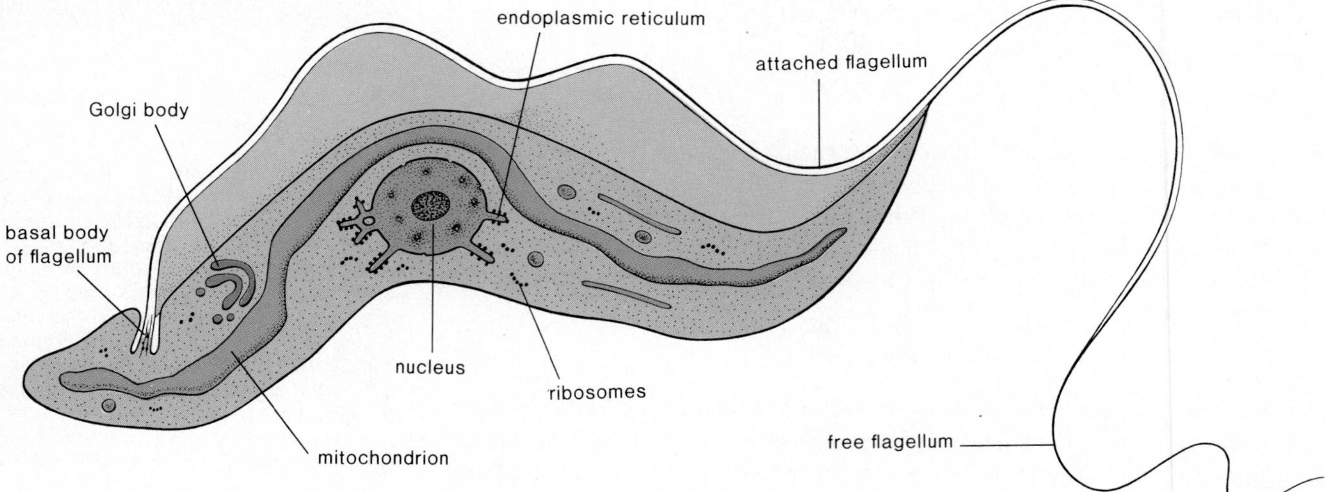

Figure 25.14
Trypanosome infection. a. *A stained blood smear from a patient suffering with African sleeping sickness showing trypanosomes among the blood cells.* b. *Structure of trypanosome as revealed by the electron microscope. The flexible pellicle allows undulating movement of the cell.*

Sporozoa

The **sporozoa** (phylum Sporozoa) are nonmotile parasites with a complicated life cycle that always involves the formation of infective spores. The most important human parasite among the sporozoa is *Plasmodium vivax* (fig. 25.15), the causative agent of one type of malaria. When a human being is bitten by an infected female *Anopheles* mosquito, the parasite eventually invades the blood cells. The chills and fever of malaria occur when the infected cells burst and release toxic substances into the blood.

The eradication of malaria has centered on the destruction of the mosquito, since without this host the disease cannot be transmitted from one human being to another. However, the use of pesticides has caused the development of resistant strains of mosquitoes. It is hoped that genetic engineering techniques will soon allow the production of vaccine (p. 492).

The protozoans are animal-like in that they are heterotrophic and motile. They are classified according to the type of locomotor organelle employed (table 25.2).

Unicellular Algae

The unicellular algae placed in the kingdom Protista all have chloroplasts and carry out photosynthesis in a manner similar to plants. They do not represent forms believed to be direct ancestors of plants, however.

Euglenoids

Organisms belonging to the genus *Euglena* (fig. 25.16) are freshwater organisms having both animal-like and plantlike characteristics. Their animal-like characteristics include motility and a flexible body wall. **Euglenoids** (phylum Euglenophyta) move by means of flagellae, typically having one much longer than another, that project anteriorly. Because they are bounded by a flexible pellicle instead of a rigid cell wall, they can assume different shapes as the underlying cytoplasm undulates or contracts.

Their plantlike characteristics include the possession of chloroplasts. A light-sensitive swelling at the base of one flagellum is shaded by a pigment spot and this allows *Euglena* to judge the direction of the light. *Euglena* moves toward light so that photosynthesis can take place.

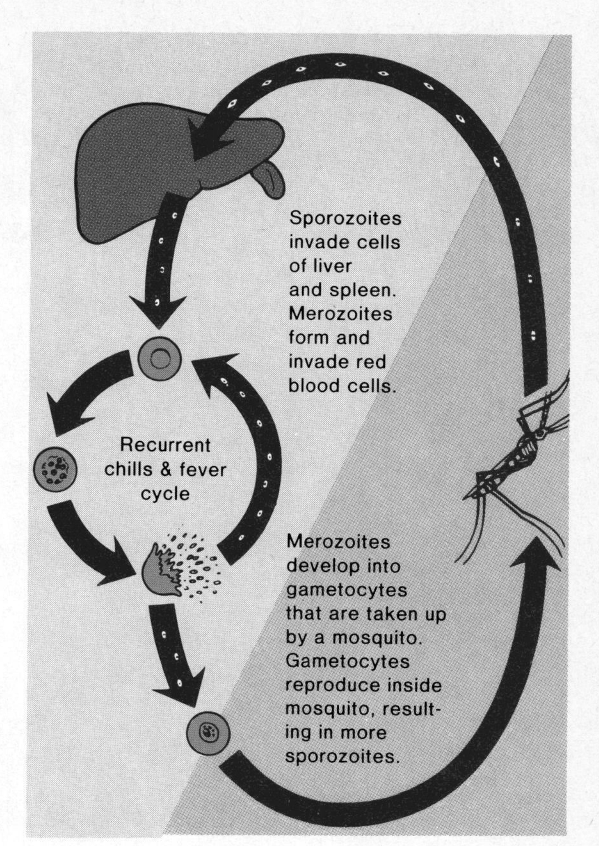

Figure 25.15
Life cycle of Plasmodium vivax. *Asexual reproduction occurs in humans, while the sexual life cycle takes place within the* Anopheles *mosquito.*

Figure 25.16
Euglena *anatomy.* a. Euglena *is typical of those protozoans that have both animal-like and plantlike characteristics. Movement is by means of a very long flagellum. A photoreceptor allows* Euglena *to seek the light where photosynthesis occurs in the numerous chloroplasts.* b. *Micrograph of* Euglena.

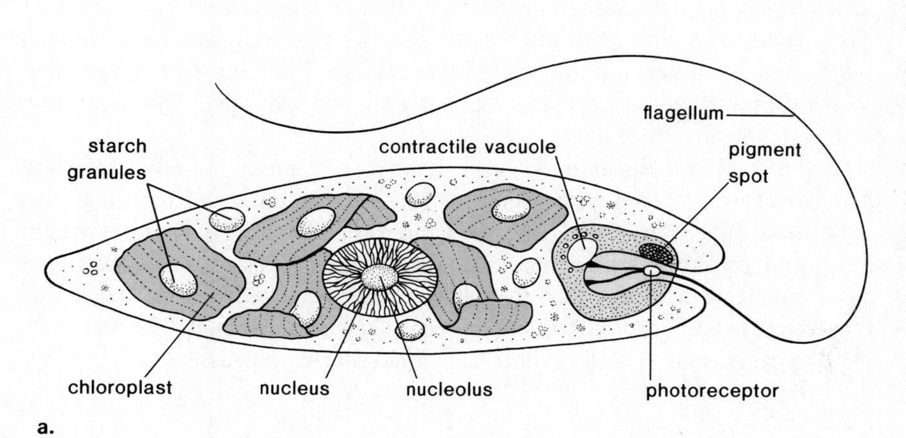

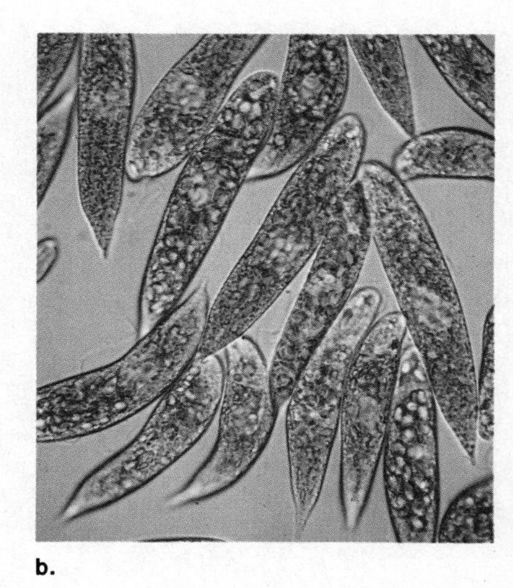

Figure 25.17
Scanning electron micrograph of a dinoflagellate envelope. The cell can be either naked or covered by a cellulose envelope, sometimes divided into plates like this. There are two flagella, one circles the cell, while the other extends posteriorly.

Dinoflagellates

Dinoflagellates (phylum Pyrrophyta) (fig. 25.17) have two external grooves or furrows, each containing a single flagellum. One furrow is transverse and completely encircles the cell; the other is longitudinal and extends along one side only. The beating of these flagella causes the organism to spin like a top. The cell wall, when present, is frequently divided into polygonal plates of cellulose closely joined together. At times there are so many of these organisms in the ocean that they cause a "red tide." The toxins they give off cause widespread fish kill and can cause a paralysis in humans if they eat shellfish that have fed on the dinoflagellates.

Usually the dinoflagellates are an important source of food for small animals in the ocean and they live as symbiotes within the bodies of some invertebrates. For example, corals usually contain large numbers of these organisms and this allows them to grow much faster than otherwise.

Diatoms

Diatoms (phylum Chrysophyta) (fig. 25.18) are golden brown in color due to an accessory pigment in their chloroplasts that masks the color of chlorophyll. The structure of a diatom is often compared to a box because the cell wall has two halves, or valves, and the larger valve acts as a "lid" for the smaller valve. When diatoms reproduce, each receives only one old valve. The new valve always fits inside the old one.

The cell wall has an outer layer of silica, a common ingredient of glass. The valves are covered with a great variety of striations and markings that form beautiful patterns when observed under the microscope. Diatoms are the most numerous of all unicellular organisms in the oceans. As such, they serve as an important source of food for other organisms. Their remains, called diatomaceous earth, accumulate on the ocean floor and are mined for use in filtering agents, soundproofing materials, and scouring powders.

The protistan unicellular algae include the euglenoids, dinoflagellates, and diatoms. They all have chloroplasts and carry on photosynthesis in the same manner as plants.

a.

b.

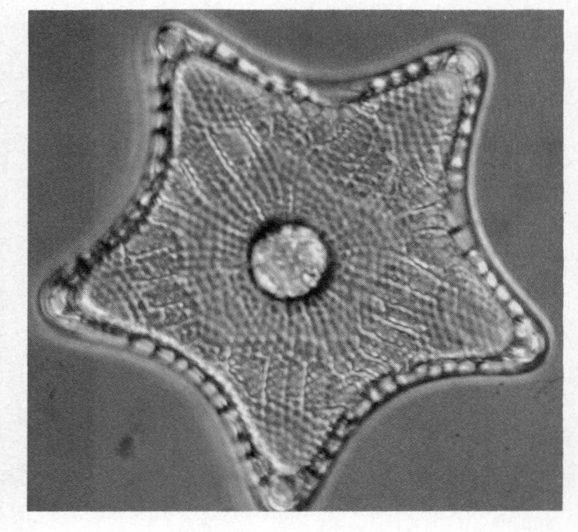

c.

Figure 25.18
Diatom anatomy. a. *Scanning electron micrograph shows how the structure of the organism can be compared to a box, such as here to a "pill box," because one half acts as a lid to the other half of the organism. These unicellular algae have a unique golden-brown pigment in addition to chlorophyll, but actually may be various colors as shown in* (b) *and* (c). *These beautiful patterns are markings on their silica-embedded walls.*

Kingdom Fungi

Fungi, being saprophytic like bacteria, are also decomposers. Since they are the most complex of the saprophytic organisms, fungi are placed in a separate kingdom in the classification system used by this text.

The body of all fungi, except unicellular yeast, is made up of filaments called hyphae. A **hypha** consists of an elongated cylinder, containing a mass of cytoplasm and hundreds of haploid nuclei, which may or may not be separated by cross walls. A collection of hyphae is called a **mycelium.** Most fungi reproduce both asexually and sexually. They are adapted to reproduction on land and produce windblown spores during both these aspects of their life cycle (fig. 25.19). Classification emphasizes mode of sexual reproduction.

Fungi are saprophytic heterotrophic eukaryotes composed of hyphae filaments (a mycelium). Fungi produce spores during both sexual and asexual reproduction and the major groups of fungi are distinguishable on the basis of sexual reproduction.

Figure 25.19
Life cycle of black bread mold Rhizopus *including both asexual (lower left) and sexual cycles. Notice that the sexual cycle requires two mating strains.*

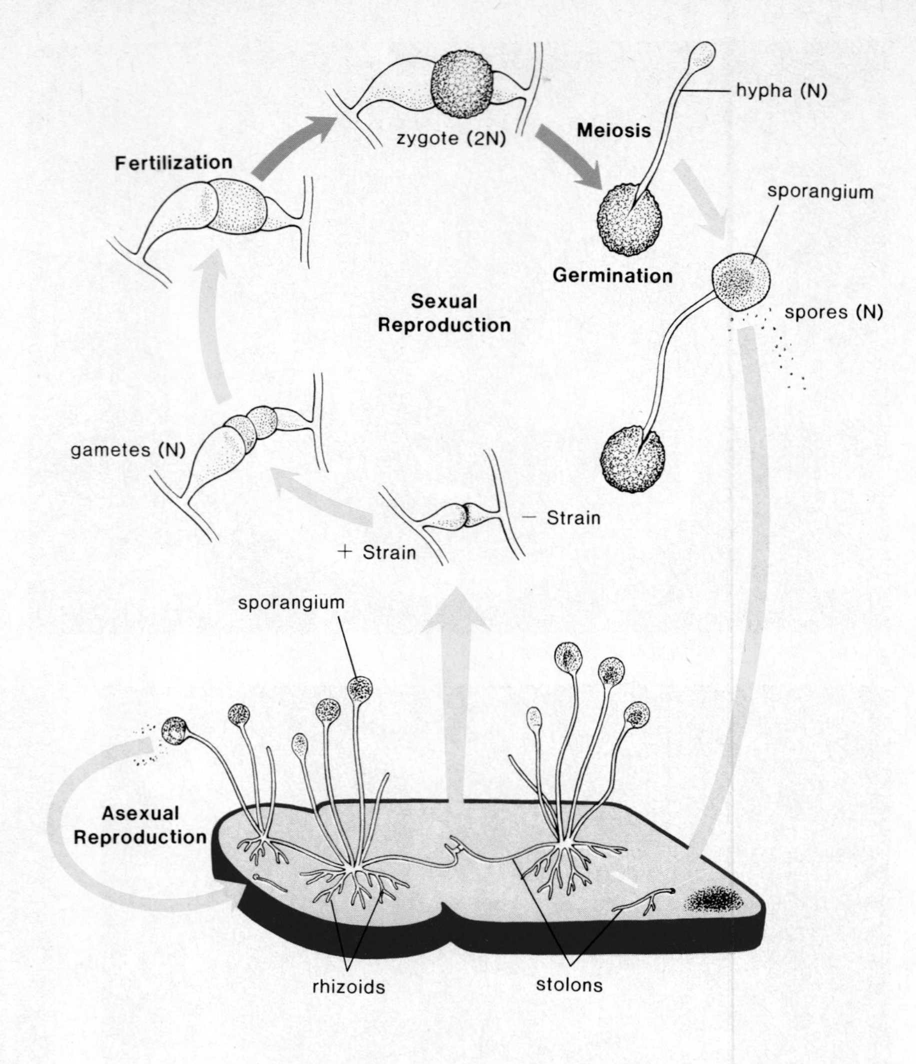

Black Bread Molds

Black bread molds (division Zygomycota) belonging to the genus *Rhizopus* are often used as an example of this group. These molds exist as a whitish or grayish haploid mycelium on bread or fruit (fig. 25.19). During asexual reproduction, some hyphae grow upright and bear a spherical **sporangium** within which thousands of spores are formed.

During sexual reproduction, hyphae of two different strains (usually called plus and minus) reach out to one another and form a diploid zygote which darkens as it enlarges into a zygospore. After remaining dormant for several months, the zygospore germinates. Now the nucleus of the zygospore undergoes meiosis to produce a short haploid hypha, which immediately forms a sporangium, with the subsequent release of windblown spores.

Sac Fungi

There are many different types of sac fungi (division Ascomycota) (fig. 25.20), most of which produce asexual spores, called **conidia.** During sexual reproduction, sac fungi form spores called ascospores within saclike cells called **asci.** In most species, the asci are supported within **fruiting bodies,** a collection of specialized hyphae. In cup fungi, the largest group of sac fungi, the fruiting body takes the shape of a cup.

a.

b.

c.

d.

Yeasts are sac fungi that do not form fruiting bodies. In fact, yeasts are different from all other fungi in that they are unicellular and most often reproduce asexually by budding. Yeasts, as you know, carry out fermentation as follows:

glucose ⟶ carbon dioxide + alcohol

In baking, the carbon dioxide from this reaction makes bread rise. In the production of wines and beers, it is the alcohol that is desired, and the carbon dioxide is allowed to escape into the air.

Figure 25.20
Sac and club fungi. a. *Bracket fungi growing on a tree limb.* b. *Colorful cup fungi.* c. *Morel, a common edible fungus.* d. *Mushrooms of the genus* mycena *have bell-shaped caps.*

Blue-green molds, notably *Penicillium,* are sac fungi. This mold grows on many different organic substances, such as bread, fabrics, leather, and wood. It is purposefully used by humans to provide the characteristic flavor of Camembert and Roquefort cheese; more important, it produces the antibiotic penicillin. Most natural penicillin today, however, is synthetically altered to increase its effectiveness. Another **mold,** the red bread mold *Neurospora,* was used in the experiments that helped decipher the function of genes.

Unfortunately, sac fungi also cause many diseases of plants; chestnut tree blight and Dutch elm disease are caused by sac fungi, as are powdery mildews, apple scab, and ergot. Ergot, a disease of cultivated cereals, contains LSD; persons who eat infected grain are likely to experience an LSD trip.

The fungus portion of *lichens* (fig. 25.21) is usually a sac fungus, while the alga may be cyanobacteria or green algae. Lichens can live on bare rock or poor soil and are able to survive great extremes of heat, cold, and dryness in all regions of the world. Reindeer moss is a lichen that is an important food source for arctic animals.

Figure 25.21
Lichens. Lichens can be either low lying or leafy or branchlike. All types are composed of algal cells and fungi filaments. The grow on a. *fallen trees and* b. *rocks.*

Club Fungi

Among the club fungi (division Basidiomycota) (fig. 25.20), asexual reproduction may also be accomplished by formation of conidial spores. As a result of sexual reproduction, members of this group form club-shaped structures called **basidia,** often within fruiting bodies. The mushroom, puffball, and bracket (shelf) fungi are club fungi; the visible portions of these are actually fruiting bodies, and the mycelia lie beneath the surface. On the underside of a mushroom cap, the basidia project from the gills. Within each basidium, a diploid nucleus undergoes meiosis to produce windblown spores (fig. 25.22).

Club fungi are economically important. Mushrooms are commercially raised and sold as a delicacy. Rusts and smuts are parasitic club fungi that attack grains, resulting in great economic loss and necessitating expensive control measures. They do not have a conspicuous fruiting body and consist of vegetative hyphae, together with spores of various kinds. On the other hand, the mycelia of club fungi that lie beneath the soil often form beneficial symbiotic relationships with plants, notably pine trees. These so-called "*fungus roots*" (mycorrhizae) help the trees garner nutrients from the soil and grow at a faster rate. Therefore, foresters make sure the tree roots are exposed to the fungi before they are planted (fig. 25.23).

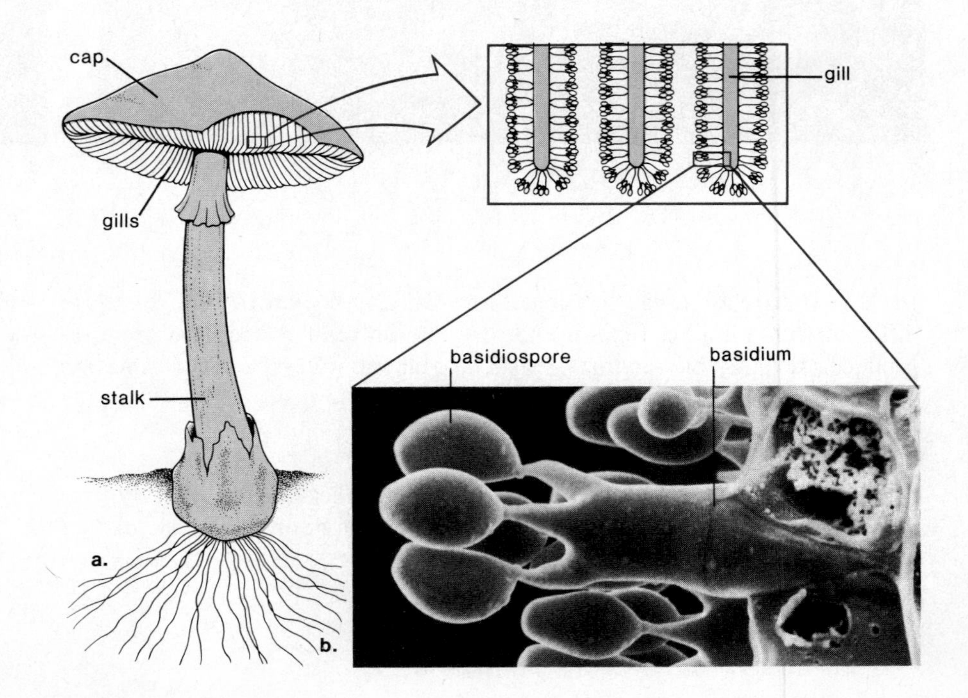

Figure 25.22
A mushroom is a fruiting body. The gills on the underside of the cap are lined with basidia (club-shaped structures). As a result of sexual reproduction, spores called basidiospores are produced here. a. *Entire mushroom.* b. *Scanning electron micrograph of basidia and basidiospores.*

Other

Some fungi (division Deuteromycota) cannot be assigned to a definite group because the sexual portion of the life cycle has not been observed. For this reason these fungi are sometimes called *imperfect fungi.* Ringworm and athlete's foot (fig. 25.24) belong to this group, as does *Candida albicans,* which causes moniliasis, a fairly common vaginal infection in females who take the birth control pill.

The black bread molds produce spores in sporangia; the sac fungi produce spores in saclike cells; and the club fungi produce spores in club-shaped structures. The sac and club fungi typically have fruiting bodies. The sexual life cycle for the fungi which parasitize humans is unknown.

a.

b.

Figure 25.23
Fungus roots. a. *One-year-old loblolly pine seedlings with the normally low levels of naturally occurring mycorrhizae.* b. *One-year-old loblolly seedlings after being inoculated with mycorrhizae.*

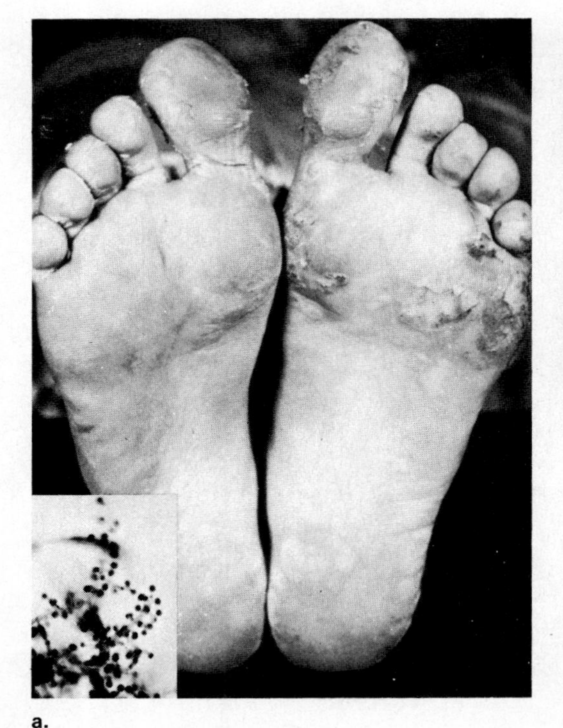

a.

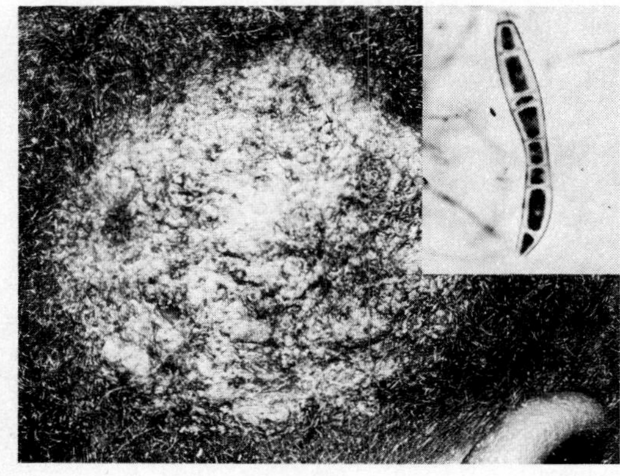

b.

Figure 25.24
Human diseases caused by fungi. a. *Athlete's foot can be caused by* Trichophyton mentagrophytes, *which reproduces by means of spores (shown in insert).* b. *Ringworm can be caused by* Microsporum audouini, *which also reproduces by means of spores (insert).*

Summary

Viruses are noncellular obligate parasites that have a capsid and nucleic acid core. A virus must enter a host cell before reproduction is possible. The kingdom Monera includes prokaryotic, usually single-celled organisms, the archaebacteria and eubacteria. The archaebacteria are adapted to living in extreme habitats. Most information relates to the eubacteria. Reproduction is by binary fission, but sexual exchange does occasionally take place. Some bacteria form endospores that can survive the harshest of treatment except sterilization. Usually bacteria are aerobic, but some are facultative anaerobes or even obligate anaerobes. All types of nutrition are found except holozoism. Many bacteria are saprophytes but the cyanobacteria are important photosynthesizers.

The kingdom Protista contains protozoans and unicellular eukaryotic algae whose relationship to plants is not well established. Protozoans are animal-like in that they are heterotrophic and motile. They are classified according to the type of locomotor organelle employed (table 25.2). The protistan unicellular algae include the euglenoids, dinoflagellates, and diatoms. They all have chloroplasts and carry on photosynthesis in the same manner as plants.

Fungi are saprophytic heterotrophic eukaryotes composed of hyphae filaments that form a mycelium. Along with heterotrophic bacteria, they are organisms of decay. The fungi produce spores during both sexual and asexual reproduction. The major groups of fungi are distinguishable on the basis of their mechanism of sexual reproduction.

Objective Questions

1. Viruses always have a ____________ core and ____________ coat called a capsid.
2. The ____________ are RNA viruses that carry an enzyme for RNA → DNA transcription.
3. The two types of monerans are the ____________ and the ____________ .
4. Most bacteria are saprophytic decomposers, meaning that they ____________ .
5. In contrast, cyanobacteria are ________ , using the energy of the sun to make their own ____________ .
6. Amoeba move by means of __________ and ciliates move by means of ________ .
7. Dinoflagellates are classified as what type of protista? ____________
8. The body of a fungus is a ____________ that contains filamentous ____________ .
9. The ____________ and the ____________ fungi both have fruiting bodies.
10. A lichen is a symbiotic relationship between a ____________ and an ____________ .

Answers to Objective Questions
1. nucleic acid, protein 2. retroviruses 3. archaebacteria and eubacteria 4. break down dead organic matter 5. photosynthetic, food 6. pseudopodia, cilia 7. unicellular algae 8. mycelium, hyphae 9. sac, club 10. fungus, algal cell

Study Questions

1. Compare and contrast the bacteriophage lytic and lysogenic cycles. (p. 540)
2. Describe the life cycle of a DNA animal virus and an RNA retrovirus. (p. 541)
3. Describe the prokaryote characteristics shared by most bacteria. (p. 544)
4. What are the three shapes of bacteria? (p. 544) How do bacteria reproduce? (p. 545) What are endospores? (p. 545)
5. Discuss the importance of bacteria including cyanobacteria. (pp. 546–547)
6. What two major groups of organisms are included in the kingdom Protista? (p. 548)
7. Give an example for each type of protozoan studied. Describe the anatomy of those that are free living and the life cycle of the parasitic ones. (pp. 548–551)
8. Discuss the significance of the euglenoids, dinoflagellates, and diatoms. (pp. 551–552)
9. Describe the structure of black bread mold and its life cycle. (p. 554)
10. Define a fruiting body and name two groups of fungi that typically have fruiting bodies. (pp. 554–556)
11. Name several human diseases caused by fungi (p. 557)
12. Describe the makeup of a lichen. (pp. 548, 556)

Key Terms

bacteriophage (bak-te′re-o-fāj″) a virus that infects a bacterial cell. *539*

bacteria (bak-te′re-ah) prokaryotes that lack the organelles of eukaryotic cells; archaebacteria and eubacteria. *544*

binary fission (bi′na-re fish′un) reproduction by division into two equal parts by a process that does not involve a mitotic spindle. *545*

cyanobacteria (si″ah-no-bak-te′re-ah) photosynthetic prokaryotes that contain chlorophyll and release O_2; formerly called blue-green algae. *547*

diatoms (di′ah-tomz) a large group of fresh and marine unicellular algae having a cell wall consisting of two silica impregnated valves that fit together as in a pill box. *552*

dinoflagellates (di″no-flaj′ĕ lāts) a large group of marine unicellular algae that have two flagella; one circles the body while the other projects posteriorly. *552*

euglenoids (u-gle′noidz) a small group of unicellular algae that are bounded by a flexible pellicle and move by flagella. *551*

fungus (fung′gus) an organism, usually composed of strands called hyphae, that lives chiefly on decaying matter; e.g., mushroom and mold. *553*

hypha (hi′fah) one filament of a mycelium that constitutes the body of a fungus. *553*

lichen (li′ken) fungi and algae coexisting in a mutualistic relationship. *548*

mycelium (mi-se′le-um) a mass of hyphae that make up the body of a fungus. *553*

protozoans (pro″to-zo′anz) animal-like protists that are classified according to means of locomotion: amoebas, flagellates, ciliates. *548*

retroviruses (ret″ro-vi′rus-ez) viruses that contain only RNA and carry out RNA to DNA transcription. *542*

saprophytes (sap′ro-fits) heterotrophic organisms such as bacteria and fungi that externally break down dead organic matter before absorbing the products. *546*

sporangium (spo-ran′je-um) a structure within which spores are produced. *554*

transduction (trans-duk′shun) the transfer by bacteriophages of DNA from one cell to another. *545*

transformation (trans″for-ma′shun) a change in phenotype of a cell due to an introduction of genes from the environment. *545*

26

Plant Kingdom

Chapter Concepts

1 Some plants are adapted to an aquatic existence and some are adapted to a terrestrial existence.

2 Full adaptation is seen in the seed plants that utilize pollen to transport the sperm to the egg. The resulting seed protects the embryo from drying out.

3 A study of evolutionary trends among plants shows that full adaptation to life on land developed slowly.

Chapter Outline

Plants (fig. 26.1) can be defined as living organisms that carry on photosynthesis and cannot move about voluntarily. We will include in this kingdom multicellular photosynthetic forms and those unicellular ones that appear to be very closely related to the multicellular forms.

Most plants have a life cycle, termed **alternation of generations,** that includes two phases (fig. 26.2)—the sporophyte generation and the gametophyte generation:

1. The *diploid* **sporophyte generation** produces spores by meiosis.
2. The *haploid* **gametophyte generation** produces gametes.

The relative prominence of these generations varies from plant group to plant group, as we will be discussing in this chapter.

Plants first appeared in the seas but eventually became adapted to a land existence. As we survey present-day plants in this chapter, we will attempt to trace the steps by which adaptation to a land existence may have occurred.

Figure 26.1
Plants dominate the environment. They provide beauty, purify the air, and produce food not only for themselves but also for all other living things.

Algae

Green algae, brown algae, and red algae are classified as members of the plant kingdom. The term *algae,* as explained previously, refers to aquatic photosynthesizing organisms. The term is old but not very exact. In this text, the blue-green algae, now termed cyanobacteria, are members of the kingdom Monera; euglenoids, dinoflagellates, and diatoms, being exclusively unicellular, are classified in the kingdom Protista. The algae placed in the plant kingdom all have multicellular representatives.

The plant kingdom includes multicellular photosynthetic organisms and closely related unicellular ones.

Green Algae

Green algae (division Chlorophyta) are believed to be ancestral to the first terrestrial plants because both of these groups possess chlorophylls *a* and *b,* both store reserve food as starch, and both have cell walls that contain cellulose. The green algae are a diverse group that range from simple to complex in regard to both structure and sexual reproduction. Some green algae are unicellular, but many types are multicellular.

Figure 26.2
Alternation of generations life cycle. In this life cycle, the zygote does not undergo meiosis and instead develops into the sporophyte generation. The sporophyte generation produces haploid spores that develop into the gametophyte generation and the gametophyte generation produces gametes.

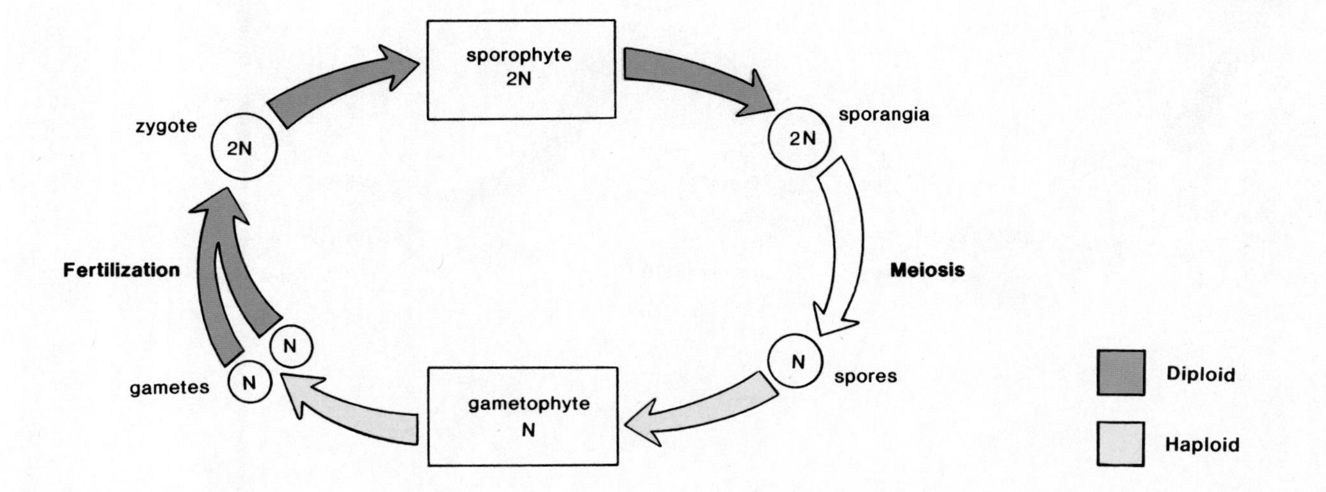

Flagellated Green Algae

The genus *Chlamydomonas* contains unicellular green algae that move by means of a pair of flagella. The single large, cup-shaped chloroplast contains a conspicuous pyrenoid, which is the site of starch production. The stigma, or "eyespot," is a portion of the chloroplast that aids the organism in moving toward the light. There are two small "contractile vacuoles" at the base of the flagella that rhythmically discharge water.

In these organisms, the adult is haploid and the only diploid structure is the zygote. During the asexual portion of the life cycle (fig. 26.3), the interior of the cell divides one or more times to form two to eight zoospores (motile spores) that resemble the parent cell. With the disintegration of the cell wall, the zoospores are released into the water and soon become adult cells like the parent. Sexual reproduction begins when the adult divides repeatedly to produce **isogametes** (gametes that look alike). When gametes from two different strains come into contact, the contents of the two cells join to form a zygote. A heavy wall forms around the zygote, and it becomes a resistant body able to survive until conditions are favorable, at which time the zygote germinates and produces four zoospores by meiosis. Notice that *Chlamydomonas* produces zoospores, or flagellated spores, in both the asexual and sexual life cycles.

Figure 26.3
The structure and life cycle of Chlamydomonas, *a motile green alga. During asexual reproduction, all structures are haploid; during sexual reproduction only the zygote is diploid.*

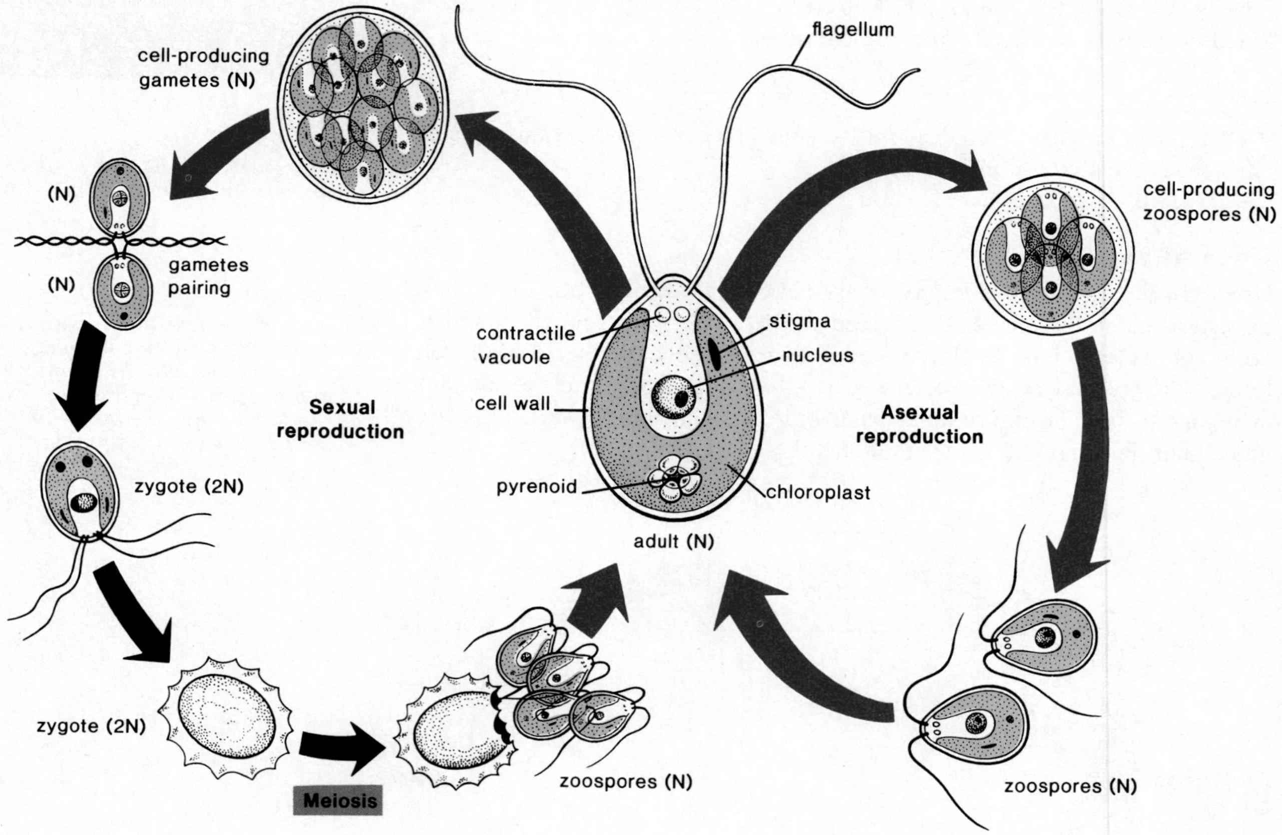

The life cycle of *Chlamydomonas* illustrates the primary differences between asexual and sexual reproduction (table 26.1). *Sexual reproduction* is simply reproduction that involves the use of gametes. Distinct and separate sexes are not required and heterogametes (unlike gametes such as egg and sperm) are not required. Sexual reproduction aids the process of evolution because concurrent recombination of genes may produce a product more suited to the environment than either parent.

A number of colonial (loose association of cells) forms occur among the flagellated green algae. *Volvox* is considered the most complex genus among these colonial green algae. A ***Volvox* colony** is a hollow sphere with thousands of cells arranged in a single layer surrounding a watery interior. The cells of a *Volvox* colony, each one of which resembles a *Chlamydomonas* cell, cooperate in that the flagella beat in a coordinated fashion. Some cells are specialized for reproduction, and each of these can divide asexually to form a new daughter colony (fig. 26.4 left). This daughter colony resides for a time within the parental colony. A daughter colony leaves the parental colony by releasing an enzyme that dissolves away a portion of the matrix of the parental colony, allowing it to escape. During sexual reproduction there are heterogametes—large nonmotile eggs and small flagellated sperm (fig. 26.4 right).

***Table 26.1** Asexual versus Sexual Reproduction*

	Asexual	Sexual
Number of parents	One	Two
Gametes	None	Gametes
Recombination of genes	Does not occur	Does occur

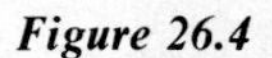

Figure 26.4
A typical green algal colony from the genus Volvox. Left: *the colony produces a definite egg and sperm during sexual reproduction.* Right: *the zygotes develop into daughter colonies that are retained within the mother colony for a while.*
(left) © Carolina Biological Supply Company.

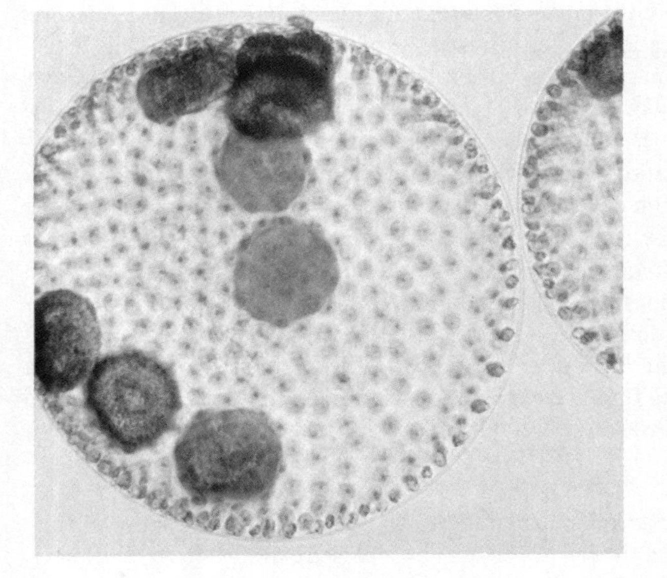

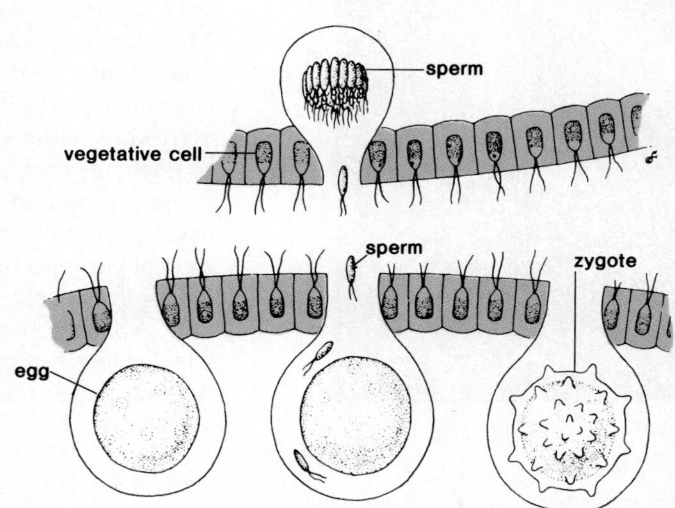

Filamentous Green Algae

Filaments are end-to-end chains of cells that form after cell division occurs in only one plane.

Spirogyra (fig. 26.5), found in green masses on the surface of ponds and streams, is a genus whose members are filamentous. In *Spirogyra,* the chloroplasts are ribbonlike and spiral within the cell. Asexual reproduction occurs when a filament breaks up and each piece begins to produce new cells. Sexual reproduction occurs when two filaments line up next to one another, and **conjugation** *tubes* form between their respective cells. The contents of the cells of one filament move into the cells of the other filament, forming 2N zygotes. These zygotes may survive the winter and in the spring undergo meiosis to produce new haploid filaments.

Among the filamentous algae, there are plants that use heterogametes during sexual reproduction. For example, the genus *Oedogonium* contains filamentous algae in which the cells have cylindrical and netlike chloroplasts. Sexual reproduction (fig. 26.6) occurs when an enlarged specialized cell produces an egg, and other short, disklike cells each produce two sperm. The sperm, which look like small zoospores, escape and swim to an egg, after which the zygote is released and enters a period of dormancy. Upon germination, the zygote produces four zoospores, each of which may grow into a filament. Also, any vegetative cell may produce a zoospore asexually and this may develop directly into a filament.

It appears that the development of sexual reproduction was advantageous because it provided these green algae with (*a*) a means to survive unfavorable conditions and (*b*) offspring that had a new combination of genes. The protective coat of the zygote allows it to overwinter. Upon germination, the offspring has a new combination of genes that may be more suited to different environmental conditions. We also note that while some forms have isogametes (*Chlamydomonas*), others have heterogametes (*Volvox* and *Oedogonium*).

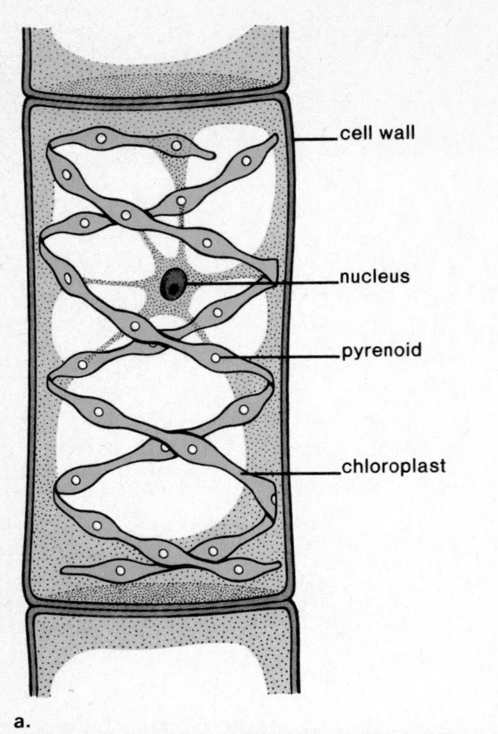

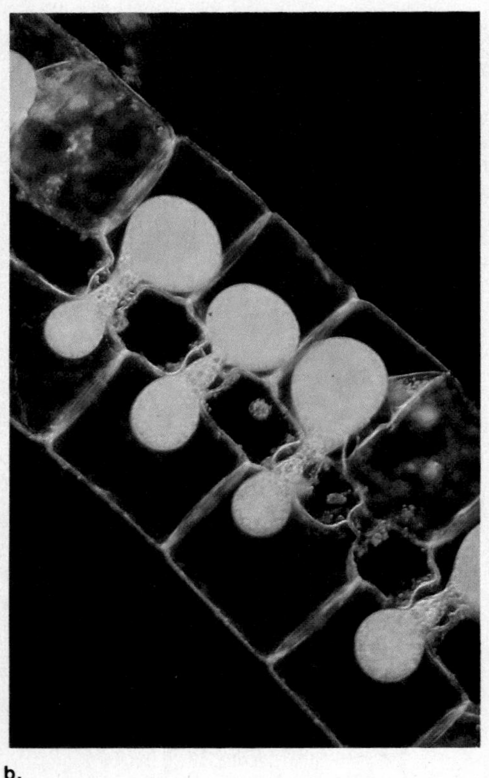

Figure 26.5
Members of the genus Spirogyra *are filaments of cells.* a. *Anatomy of one cell.* b. *Micrograph depicting conjugation between two filaments.*

Multicellular Sheets

The genus *Ulva* contains green algae that are found in the sea close to shore. *Ulva* is commonly called sea lettuce because of a leafy appearance (fig. 26.7). Members of this genus may resemble a form ancestral to later plants. *Ulva* is multicellular; and like primitive plants adapted to land, it has a noticeable sporophyte and gametophyte generation. However, both generations look exactly alike. Observation verifies that each haploid zoospore produced by the sporophyte generation develops directly into a gametophyte generation, while gametes produced by the gametophyte generation fuse to give a zygote that develops into the diploid sporophyte generation. While the gametophyte and sporophyte generations are codominant in *Ulva,* in terrestrial plants, one generation is typically dominant over (longer lasting than) the other.

Notice that the life cycles of both *Chlamydomonas* and *Ulva* contain flagellated zoospores that are capable of swimming. This is an adaptation to reproduction in the water.

The green algae are plants adapted to an aquatic existence. They have biochemical characteristics that show a close relationship to terrestrial plants. The green alga *Ulva* has a life cycle that has two distinct generations like that of terrestrial plants.

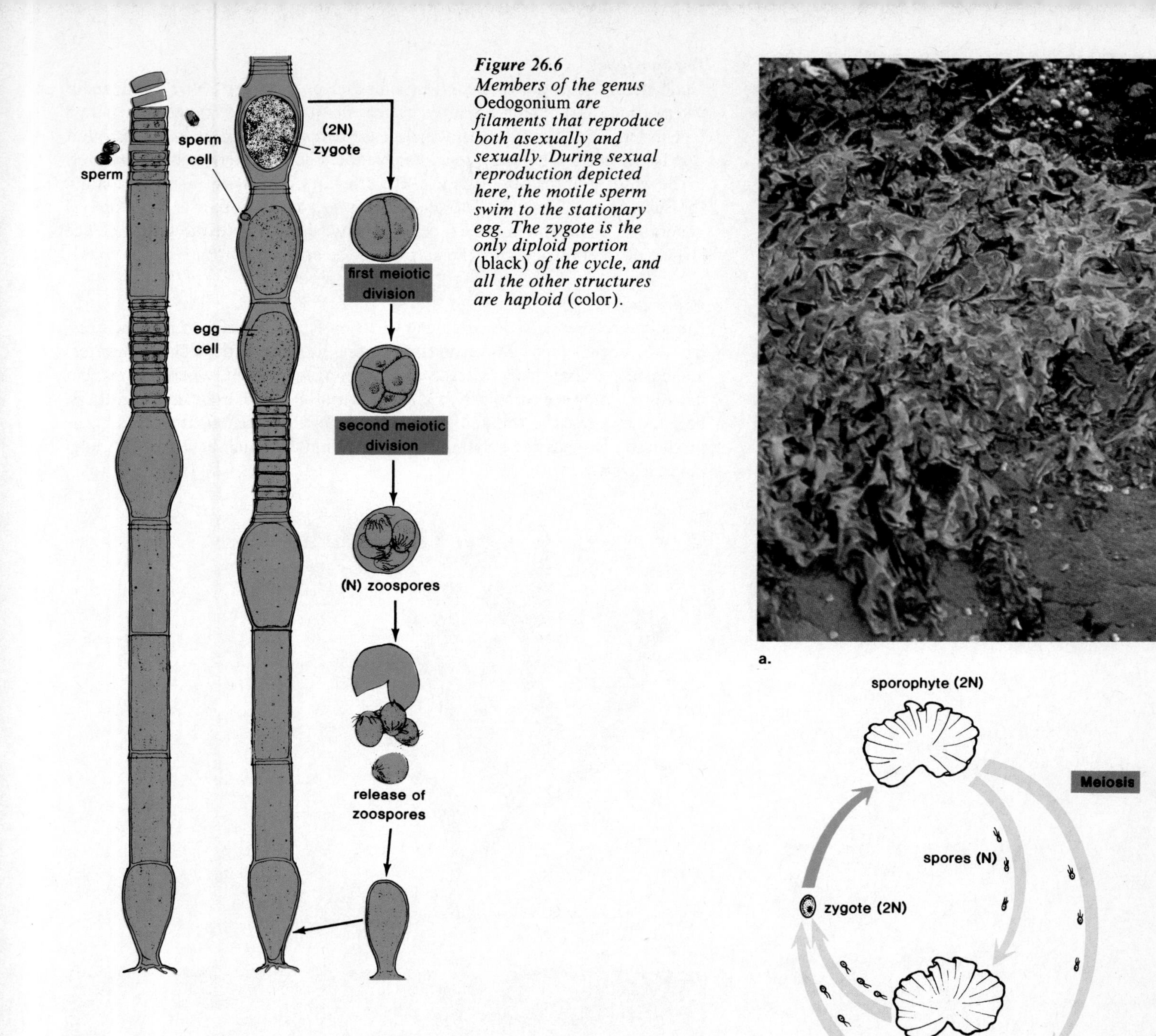

Figure 26.6
Members of the genus Oedogonium *are filaments that reproduce both asexually and sexually. During sexual reproduction depicted here, the motile sperm swim to the stationary egg. The zygote is the only diploid portion* (black) *of the cycle, and all the other structures are haploid* (color).

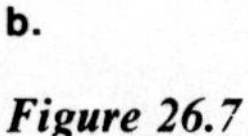

Figure 26.7
Members of the genus Ulva *have a life cycle known as alternation of generations. In these plants the sporophyte generation and the gametophyte generation have the same appearance.*
a. *Micrograph.* b. *Life cycle.*

Seaweeds

Multicellular green algae (such as *Ulva*), red algae, and brown algae are all seaweeds. Their color is dependent on the pigments they contain, which in red and brown algae mask the green color of their chlorophyll. These pigments enable the algae to collect the light they need for photosynthesis. When light strikes seawater, the various wave lengths penetrate to different depths. The accessory pigments in red algae absorb those wavelengths that penetrate deepest; those in brown algae absorb wavelengths that are quickly filtered out of water. Therefore it is possible to predict the depth at which these two forms of algae could be found.

Brown Algae

Good examples of brown algae (division Phaeophyta) are found in the genus *Fucus* (fig. 26.8), which are often called *rockweeds*. They range from 1 to 3 feet in length and may be attached in great masses to rocks exposed at low tide in the north temperate zone. Air bladders hold the forked branches aloft in the water, and the tips of some of the branches, which are enlarged, contain the sex organs. The life cycle of *Fucus* is so advanced that a gametophyte generation is not seen in some species; instead, the adult is always diploid. The largest of the brown algae, the kelps, may be as long as 100 feet.

Red Algae

Like the brown algae, the red algae (division Rhodophyta) are multicellular, but they occur chiefly in warmer seawaters, growing both in shallow waters and as deep as light penetrates. Some forms of red algae are filamentous, but more often they are complexly branched, with the branches having a feathery, flat, and expanded or ribbonlike appearance. Notably, red algae is used in the production of agar, the gelatinous solid medium on which bacteria and fungi may be grown.

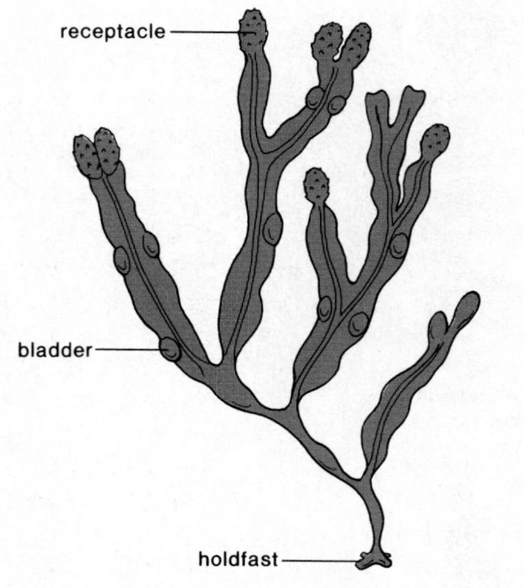

Figure 26.8
A common representative of the genus Fucus. a. *General anatomy.* b. *Micrograph of receptacles (reproductive structures).*

Seaweeds include multicellular green, red, and brown algae. Green algae and brown algae are typically found in cool waters along rocky coasts. Red algae are found farther from the shore in warm tropical waters.

Terrestrial Plants

A land existence offers some advantages to plants. One advantage is the greater availability of light for photosynthesis since water, even if clear, filters out light. Another advantage is that carbon dioxide and oxygen are present in higher concentrations and diffuse more readily in air than in water. Many adaptations, however, are required to live successfully on land (table 26.2). Plants have become adapted in ways that are quite different from those of humans and at the end of this chapter, we will compare the adaptation of humans to that of trees.

There are two main groups of plants adapted to living on land. These two groups—the bryophytes and tracheophytes—may be compared in these ways:

Bryophytes	**Tracheophytes**
Gametophyte dominant	Sporophyte dominant
No vascular tissue	Vascular tissue

The *dominant generation* is the conspicuous generation in the life cycle of a higher plant, the one that lasts longer and is usually considered *the* plant by the layperson.

Bryophytes

The bryophytes (division Bryophyta) include the liverworts and mosses. Liverworts (fig. 26.9) that have flattened lobed bodies, although widespread and well-known, represent but a small fraction of the total number of species. Most species of liverworts are "leafy" and look like mosses, but examination shows that they have a distinct top and bottom surface, with numerous rhizoids (rootlike hairs) projecting into the soil. A moss plant is composed of a stemlike structure with radially arranged leaflike structures. Rhizoids anchor the plant and absorb minerals and water from the soil. Since bryophytes do not have vascular tissue, they *lack true roots, stems, and leaves.* Instead, they are said to have rhizoids, stemlike structures, and leaflike structures.

***Table 26.2** Comparison of Water Environment with Land Environment*

Water	Land
1. The surrounding water prevents the organism from drying out, that is, prevents desiccation.	1. In order to prevent desiccation, the organism must obtain water, provide it to all body parts, and possess a covering that prevents evaporation.
2. The surrounding water buoys up the organism and keeps it afloat.	2. An internal structure is required for a large body to oppose the pull of gravity.
3. The water prevents desiccation and allows easy transport of reproductive units such as zoospores and swimming sperm.	3. The organism may provide a water environment for swimming reproductive units. Alternately, the reproductive units must be adapted to transport by wind currents or by motile animals.
4. The surrounding water prevents the fertilized egg (zygote) from drying out.	4. The developing zygote must be protected from possible desiccation.
5. The water maintains a relatively constant environment in regard to temperature, pressure, and moisture.	5. The organism must be capable of withstanding extreme external fluctuations in temperature, humidity, and wind.

Figure 26.9
Marchantia *is a liverwort that can reproduce asexually by means of gemmae—minute bodies that give rise to new plants. As shown here, the gemmae are located in cuplike structures called gemma cups.*

a.

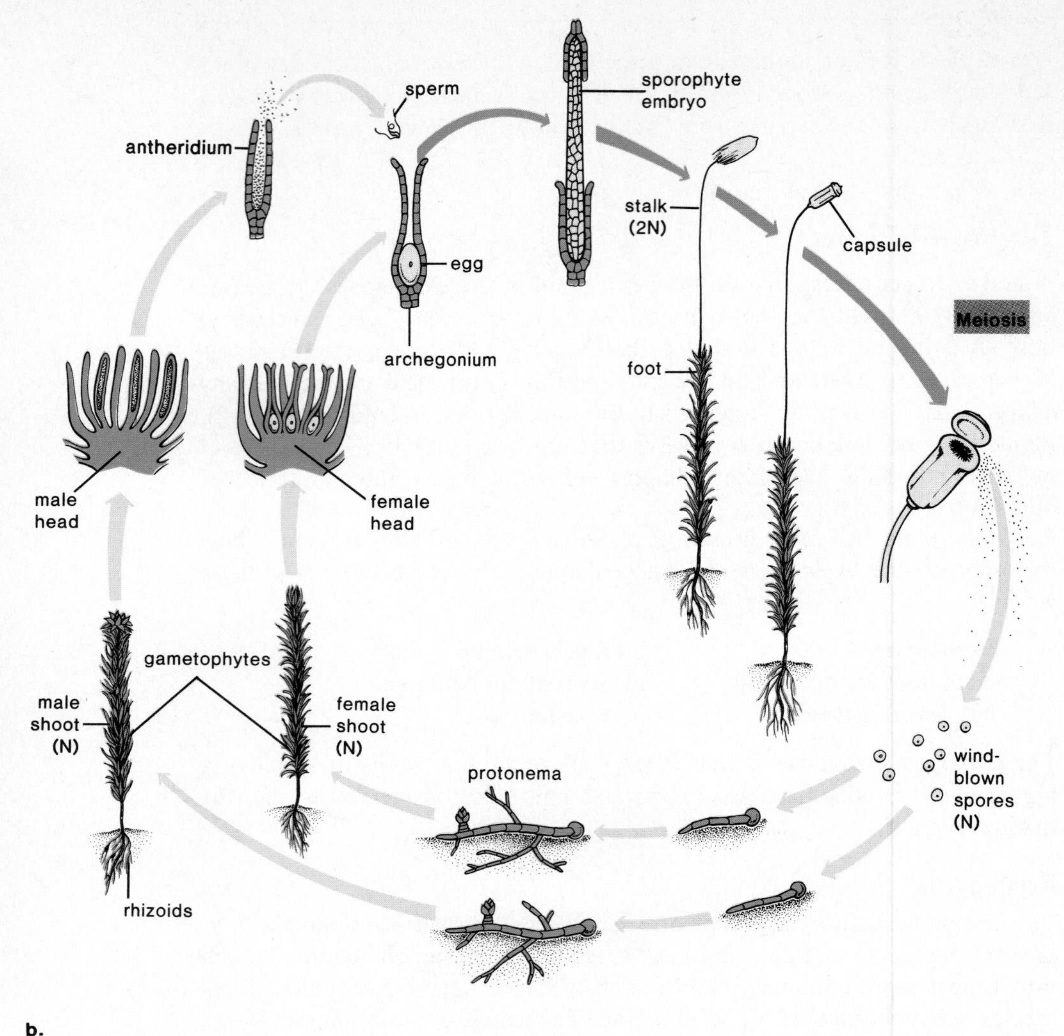

b.

Figure 26.10
a. *Haircup moss with both gametophyte generation* (below) *and sporophyte generation* (above). *Spores are produced in the capsules held aloft by stalks.*
b. *Life cycle in moss. Spores germinate to give gametophyte generation. The sporophyte generation occurs after fertilization.*

Moss Life Cycle

The moss plant just described is the dominant gametophyte generation that produces the gametes. This is the more permanent and longer-lasting generation in bryophytes. In some mosses there are separate male and female gametophytes (fig. 26.10). At the tip of a male gametophyte are **antheridia** in which swimming sperm are produced. After a rain or heavy dew, the sperm swim to the tip of a female gametophyte where structures called **archegonia** are found and within which eggs are produced. Antheridia and archegonia are both multicellular sex organs, and each has an outer layer of jacket cells that help protect the enclosed gametes from desiccation. After an egg is fertilized, it is retained within the archegonium and begins development as the sporophyte generation. The *sporophyte generation,* which is parasitic on the gametophyte, consists of a *foot* that grows down into the gametophyte tissue, a stalk, and upper capsule, or *sporangium,* where meiosis occurs and haploid spores are produced. The spores are windblown and released in dry weather.

When conditions are favorable, a spore germinates into a branching filamentous structure called a **protonema.** The leafy shoots we recognize as the gametophyte generation arise from buds that form on the protonema.

Figure 26.11
The growth of moss on rocks contributes to succession, the process by which rocks are eventually converted to fertile soil.

Adaptation
Bryophytes are incompletely adapted to life on land. To prevent water loss, the entire body is covered by a waxy cuticle. The zygote and embryo are also protected from drying out by remaining within the archegonium. The organism is dispersed or distributed to new locations by windblown spores. However, the bryophytes are limited in their adaptations. They lack vascular tissue, and the sperm have to swim in external moisture to reach the egg. It is for these reasons that bryophytes are restricted to moist locations.

Importance
Certain bryophytes and lichens colonize rocks (fig. 26.11) and slowly convert them to soil that can be used for the growth of other organisms.

Sphagnum, bog or peat moss, has commercial importance. This moss has special nonliving cells that can absorb moisture, which is why peat moss is often used in gardening to improve the water-holding capacity of the dirt. In some areas, like bogs, where the ground is wet and acid, dead mosses, especially sphagnum, accumulate and do not decay; this accumulation, called peat, can be used as a fuel.

The bryophytes include the inconspicuous liverworts and mosses, plants that have the gametophyte generation dominant. These plants are not well adapted to life on land because they lack vascular tissue, and fertilization requires an outside source of moisture. Windblown spores do disperse the species.

Nonseed Tracheophytes

The tracheophytes (division Tracheophyta), with a dominant sporophyte generation, include those plants that are best adapted to a land existence. One advantage of having the sporophyte dominant is that this is the diploid generation; and if a faulty gene is present, it may be masked by a functional gene. In addition, possession of vascular (transport) tissue is an important adaptation to the land environment.

The tracheophytes have two types of vascular tissue. *Xylem* conducts water and minerals up from the soil, and *phloem* transports organic nutrients from one part of the body to another. Because they have vascular tissue, the specialized body parts of tracheophytes can properly be called roots, stems, and leaves. Further, the strong-walled xylem cells support the body of the plant against the pull of gravity. The tallest organisms in the world are tracheophytes—the redwood trees of California.

There are both nonseed plants and seed plants among the tracheophytes. The first tracheophytes to evolve, including the ferns, are nonseed plants; but later the seed plants, that is, the gymnosperms and angiosperms, evolved.

Primitive Tracheophytes

Among the first tracheophytes to evolve are the psilopsids (division Psilophyta), the club mosses (division Lycophyta), and the horsetails (division Sphenophyta) (fig. 26.12). The psilopsids are of particular interest because they may resemble the common ancestral form of the other tracheophytes. Formerly *Psilotum* was considered to be a psilopsid. Although this assumption is now being questioned and *Psilotum* is considered by some to be a fern, it is discussed here because it is a primitive genus of living tracheophytes. The sporophyte body consists of stems with scalelike structures but no leaves. In these living fossils, there is a horizontal stem (lacking roots) from which rhizoids extend on the underside, while green, photosynthetic upright branches, with tiny scalelike structures, grow upward. This is the sporophyte generation, and sporangia are located on the branches. The gametophyte generation is separate, smaller than the sporophyte, and water dependent. In fact, the life cycle of *Psilotum* is very close to that of the fern, which is discussed next.

Ferns

Ferns (division Pterophyta) vary in appearance. Most are only a few feet tall, but in tropical rain forests some are very tall, resembling palm trees. The common temperate zone ferns (e.g., *Pteridium*) often have a horizontal stem

a.

b.

c.

Figure 26.12
Representative lower tracheophytes. These plants were widespread when the tracheophytes first evolved. a. Equisetum *(horsetail).* b. Psilotum c. Lycopodium *(club moss).*

(rhizome) from which hairlike roots project beneath and large leaves, or **fronds** (fig. 26.13), project above. The fronds are often subdivided into a large number of leaflets. All parts of the plant contain vascular tissue and therefore a fern has true roots, stems, and leaves.

The plant referred to as the fern is the sporophyte generation; sori (sing. **sorus**), collections of sporangia, are present on the underside of some leaflets (fig. 26.14). Within the sporangia, meiosis occurs and spores are produced. A band of thickened cells breaks open and expels the mature spores that are carried by the wind.

The gametophyte generation is a small, heart-shaped structure called a **prothallus.** Typically, archegonia develop at the notch and at the tip there are antheridia. Swimming sperm are released by the antheridia and swim to the eggs in the archegonia. The zygote begins development within the archegonia but soon it sends forth its first leaf above the prothallus and anchors itself in the soil below. Now it is the independent sporophyte generation and begins to carry on photosynthesis.

Adaptation Ferns are incompletely adapted to life on land due to the water-dependent gametophyte generation. This generation lacks vascular tissue and is separate from the sporophyte generation. Swimming sperm require an outside source of water in which to swim to the eggs in the archegonia. Ferns are most often found in moist environments where water is available for the gametophyte generation.

Figure 26.13
Representative fern. The sporophyte generation of a bracken fern. The undersurfaces of the large fronds are covered with sori, clusters of tiny sporangia.

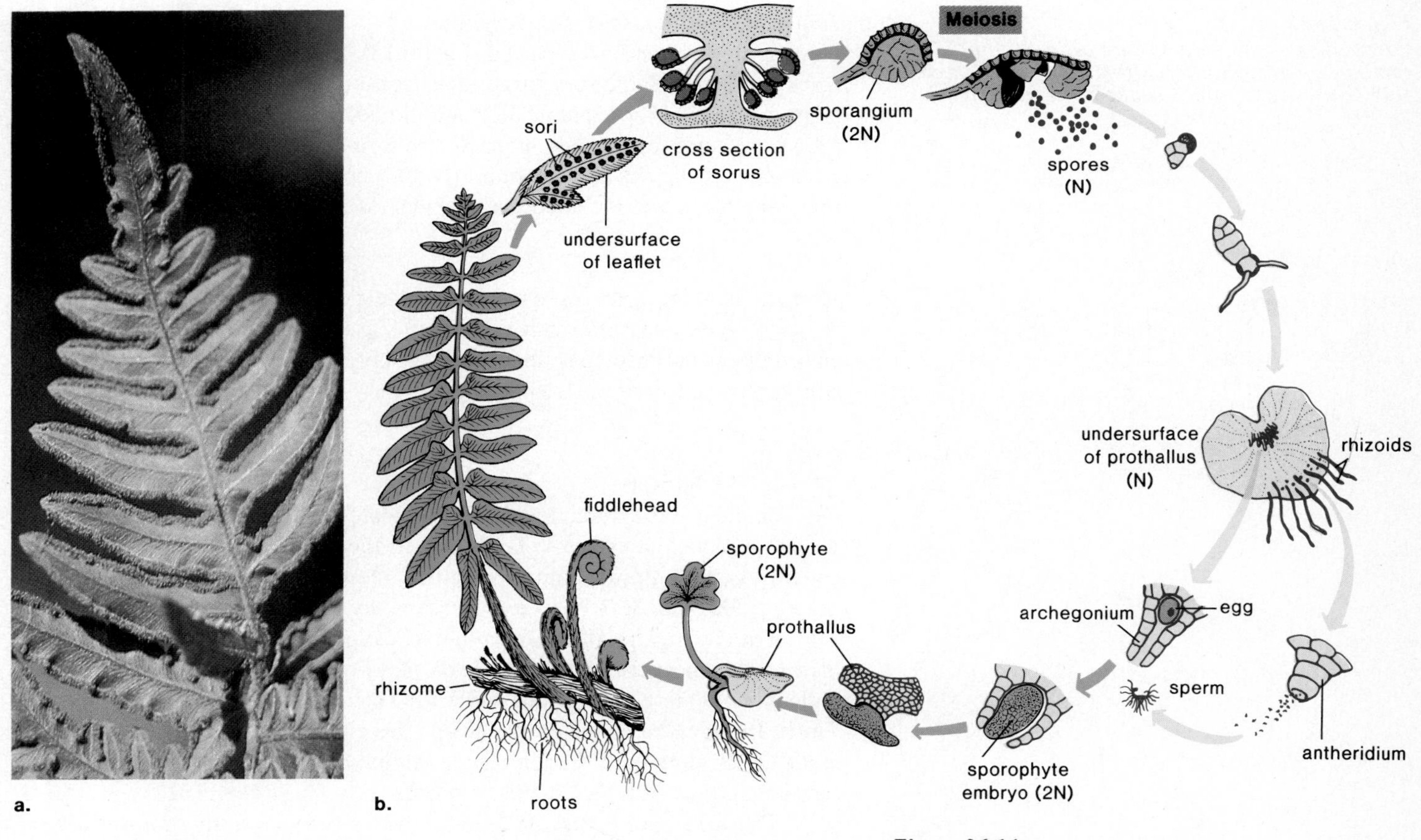

Figure 26.14
Life cycle of a fern. a. *Photo shows location of sori on the underside of the leaflets of a bracken fern.* b. *Spores released from sori germinate to give a prothallus, the gametophyte generation. Swimming sperm move from an antheridium to an archegonium where fertilization takes place. The zygote develops into the sporophyte generation.*

Figure 26.15
Carboniferous coal swamps are believed to have contained plants with fernlike foliage (left), *treelike club mosses* (left), *and treelike horsetails* (right).

Importance of Nonseed Tracheophytes
During the Carboniferous period (fig. 26.15), the horsetails, club mosses, and ferns were abundant, very large, and treelike. For some unknown reason a large quantity of these plants died and did not decompose completely. Instead they were compressed and compacted to form the coal that we still mine and burn today. (Oil was formed similarly, but most likely formed in marine sedimentary rocks and included animal remains.)

---

*I*n the nonseed tracheophytes, such as ferns, there is a dominant vascular sporophyte generation but these plants are restricted to a moist environment because of an independent nonvascular gametophyte generation that produces swimming sperm.

Seed Tracheophytes

Seed plants are fully adapted to a land existence. The primary weakness observed in plant life cycles so far has been the water-dependent gametophyte generation with its swimming sperm that require outside moisture to swim to the egg. This difficulty has been overcome by the seed plants.

These plants produce **heterospores,** called microspores and megaspores, instead of homospores, or identical spores. Microspores develop into immature male gametophyte generations while still retained within a microsporangium. These are released as **pollen grains** and are either carried by the wind or animals to the vicinity of the female gametophyte. Later, a mature pollen grain contains sperm—the *pollen grain replaces swimming sperm* in the seed plants.

The megasporangium is found within an ovule. Here a megaspore develops into the female gametophyte generation. While still within the ovule, the female gametophyte generation produces an egg that is fertilized by a

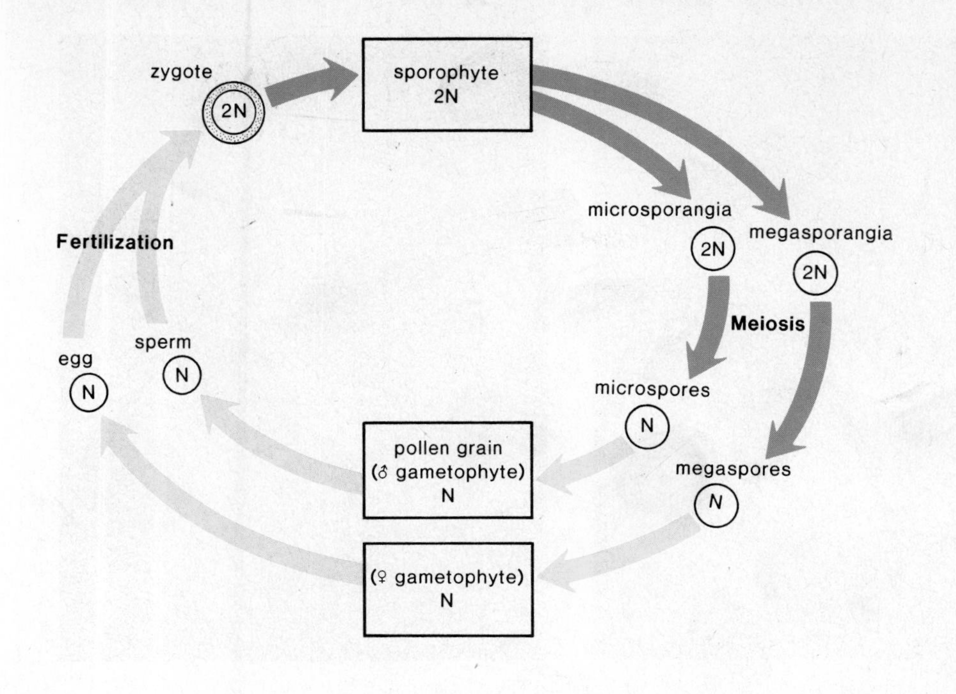

Figure 26.16
Life cycle in seed plants. Notice there is a separate male and female gametophyte generation.

sperm. The zygote becomes an embryonic plant enclosed within a seed—the *seed contains the embryonic plant* of the next sporophyte generation plus stored food.

Figure 26.16 diagrams the life cycle of seed plants. In this life cycle:

1. The dominant generation is a diploid sporophyte.
2. Meiosis produces microspores within microsporangia and megaspores are produced within megasporangia. Each megasporangium is found within an ovule.
3. The mature male gametophyte generation is the pollen grain and this structure contains a sperm that fertilizes the egg. The female gametophyte generation is retained within the ovule, and within the ovule it produces an egg.
4. Fertilization results in an embryo that lies inside the original ovule. Upon maturation the ovule becomes the released seed.
5. The seed contains the new sporophyte generation, along with stored food, usually within several protective layers.

While nonseed plants are dispersed by spores, in seed plants the seeds distribute the species.

Gymnosperms

There are three major groups of gymnosperms (subdivisions Cycadophyta, Ginkgophyta, and Coniferophyta). All are woody plants but the most common are those that bear cones, the conifers. Pine trees (fig. 26.17a) are well-known conifers. They often inhabit colder regions of the globe and therefore their needlelike leaves are modified to withstand low temperatures and harsh winds. The mesophyll lacks air spaces and the stomata are sunken. Thick-walled cells occur beneath the epidermis and the leaf veins are surrounded by an endodermis.

a.1

a.2

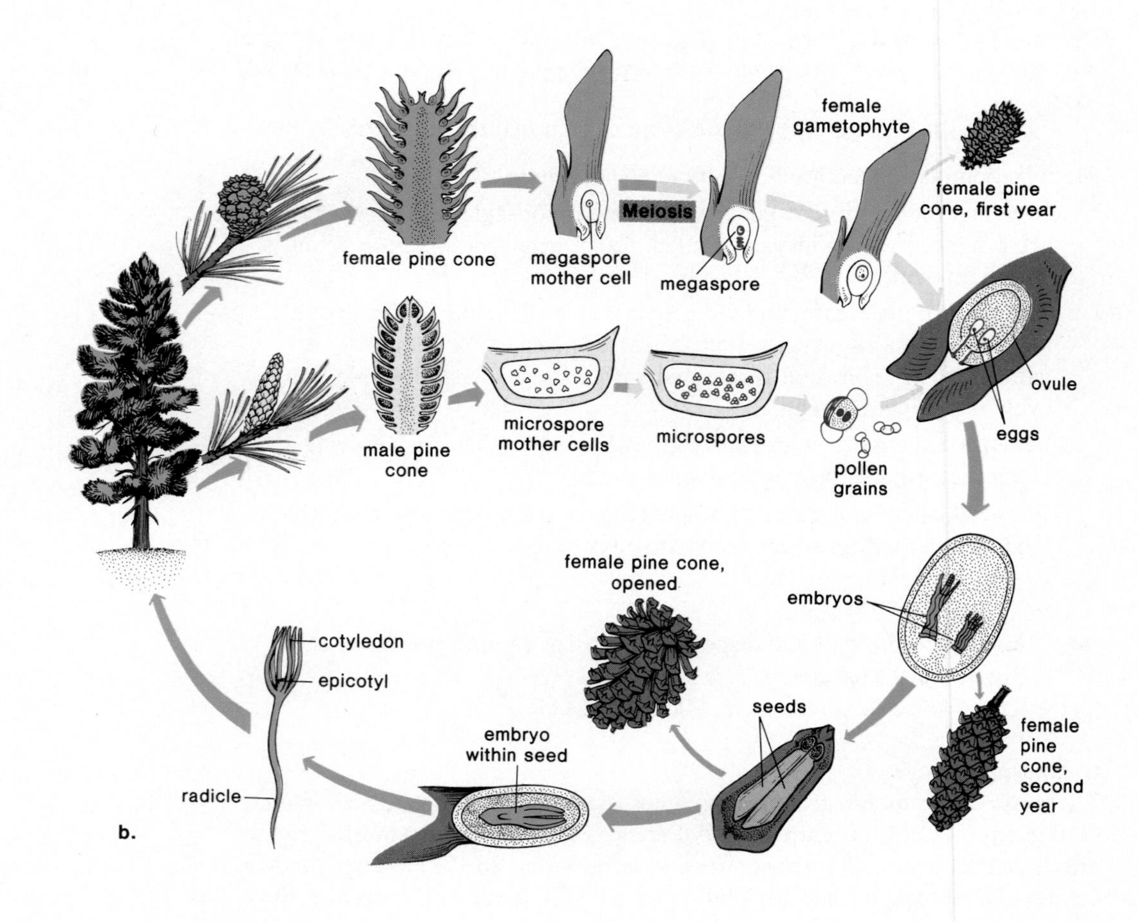

Figure 26.17
a. *Pitch pine; (1) entire sporophyte generation, (2) close-up of branches showing cones.* b. *Life cycle of pine. The mature sporophyte (pine tree) has female pine cones that produce megaspores that develop into female gametophyte generations and male pine cones that produce microspores that develop into male gametophyte generations (mature pollen grains). Following fertilization, immature sporophyte generations are present in seeds located on the female cones.*

Pine Life Cycle The sporophyte generation is dominant (fig. 26.17b) and the sporangia are located on the scales of the *cones.* There are two types of cones—male and female.

Typically, each scale of the male (pollen) cone has two or more microsporangia on the underside. Within these sporangia, meiosis produces **microspores,** each of which develops into a pollen grain, the *male gametophyte generation.* A pollen grain has a thickened, protective coat and a pair of balloonlike bladders to the sides. These structures give the pollen grains some buoyancy in the wind. Pine trees release so many pollen grains during the pollen season that everything in the area around them may be covered with a dusting of yellow, powdery pine pollen.

Typically, each scale of the female (seed) cone has two ovules that lie on the upper surface. Within each ovule, there is a megasporangium where meiosis produces four **megaspores.** Only one of these spores develops into a *female gametophyte* that has two to six archegonia, each containing a single large egg lying near the ovule opening.

During pollination, pollen grains are transferred from the male to the female cones by the wind. Once enclosed within the female cone, the pollen grain develops a pollen tube that slowly grows into the ovule. Inside the tube are two sperm, one of which fertilizes an egg inside the ovule. The pollen tube discharges its sperm, and fertilization takes place 15 months after pollination. Notice, then, that fertilization is an entirely separate event from pollination, which is simply the transfer of pollen.

After fertilization, the ovule matures and becomes the seed that is composed of the embryo, stored food, and a seed coat. The winged seeds are exposed (or *naked seeds*).

Figure 26.18
Bristlecone pines, perhaps the oldest plants in the world.

Adaptation The life cycle of gymnosperms has advances not seen among the ferns. Swimming sperm are replaced by windblown pollen grains, and fertilization is achieved by formation of the pollen tube. Therefore, there is no need for an outside source of moisture to either transport the sperm to the egg nor for fertilization itself. The sporophyte generation not only protects the female gametophyte but also the developing zygote. The seed protects the embryo until germination and provides it with a source of nutrients to sustain the first period of development. All of these aspects of the life cycle are adaptations to the terrestrial environment that ensure successful reproduction.

Importance The conifers supply much of our lumber for building and wood for the production of paper, turpentine, and other products. The wood of a conifer is usually a softwood while that of flowering trees is usually a hardwood. The wood in a pine tree contains only tracheids and has no vessel elements and accompanying support cells.

As mentioned on page 141, it is possible to examine wood to determine the age of a tree. The oldest living thing on earth is believed to be a bristlecone pine (fig. 26.18) that has been dated at 4,600 years old by the Laboratory of Tree Ring Research, University of Arizona.

The gymnosperms, for example, conifers, are the naked seed plants. In seed plants the formation of pollen and pollination replaces swimming sperm. Following fertilization the seed develops from the ovule, a structure that has been protected within the body of the sporophyte plant. In conifers, the seeds are uncovered and windblown.

Figure 26.19
Pear tree, a representative angiosperm tree.

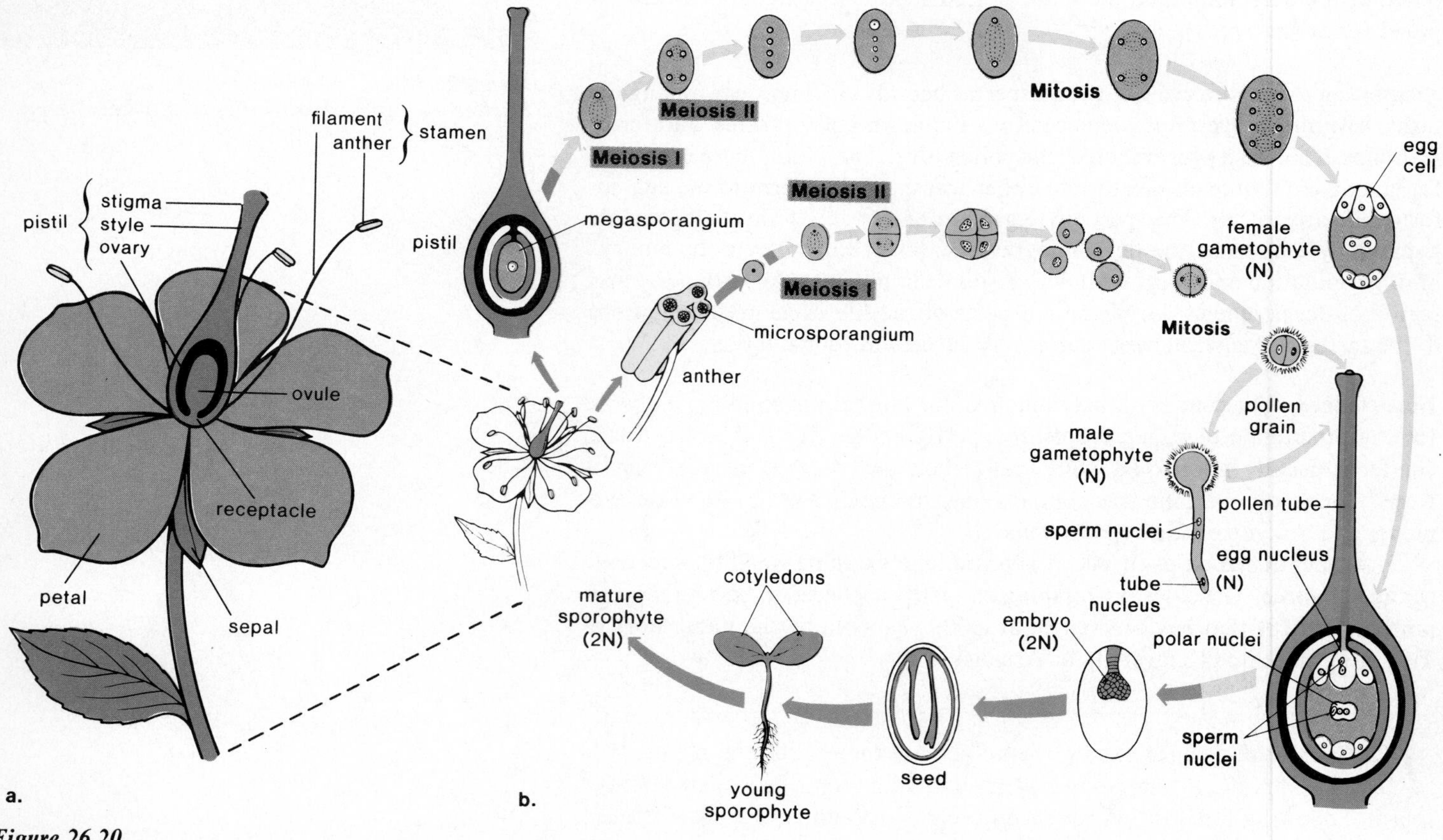

Figure 26.20
Life cycle of flowering plants. Microspores are produced in the anther and megaspores contained within ovules are produced in the ovary. Each microspore develops into a male gametophyte generation (pollen grain) and each megaspore develops into a female gametophyte generation. After fertilization, the embryo is enclosed within a seed that disperses the species.

Table 26.3 Types of Fruits	
Type of Fruit	**Example**
Fleshy fruits	
Simple fruits	
Drupe	Peach, plum, olive
Berry	Grapes, tomato
Pome	Apple, pear
Aggregate fruits	Strawberry, raspberries
Multiple fruits	Pineapple
Dry fruits	
Follicle	Milkweed, peony
Legume	Peas, beans, lentils
Capsule	Poppy
Achene	Sunflower "seeds"
Nuts	Acorns, hickory nuts, chestnuts
Grain	Rice, oats, barley

Angiosperms

Angiosperms (subdivision Anthophyta) are the flowering plants. All hardwood trees (fig. 26.19) including all the deciduous trees of the temperate zone and the broad-leaved evergreen trees of the tropical zone, are angiosperms although sometimes the flowers are inconspicuous. All herbaceous (nonwoody) plants common to our everyday experience, such as grasses and most garden plants (fig. 26.20), are flowering plants. Angiosperms are adapted to every type of habitat, including water (e.g., water lilies and duckweed).

The angiosperms are divided into two groups (fig. 7.8); the *monocots* (e.g., lily) and the *dicots* (e.g., buttercup). The monocots are always herbaceous, with flower parts in threes, parallel leaf veins, scattered vascular bundles in the stem, and one cotyledon or seed leaf in the embryo. The dicots may be woody or herbaceous, with flower parts usually in fours and fives, net veins, vascular bundles forming a circle in the stems, and two cotyledons or seed leaves in the embryo.

Flower

In angiosperms, the microsporangia and megasporangia are located in the **flower** (fig. 26.20). The flower is advantageous in three ways: it attracts insects and birds that aid in pollination (pollination by wind is also possible); it protects the developing female gametophyte; and it produces seeds enclosed by **fruit.** There are many different types of fruits (table 26.3), some of which are fleshy (e.g., apples) and some of which are dry (e.g., peas enclosed by pods). The fleshy fruits are sometimes eaten by animals, which may transport the seeds to a new location and then deposit them during defecation. Fleshy fruits may also provide additional nourishment for the developing embryo. Both fleshy and dry fruits provide protection for the seeds. Many so-called vegetables are actually fruits; for example, tomatoes, string beans, and squash. Nuts and berries and grains of wheat, rice, and oats are also fruits.

Flowering Plant Life Cycle

Stamens, the pollen-producing portion of the flower, consist of a slender *filament* with an *anther* at the tip. In the anther, meiosis within microsporangia produces microspores (fig. 26.20). Each microspore divides and becomes a binucleated pollen grain, the male gametophyte generation. The pollen grains are either blown by the wind or carried by pollinators (fig. 7.27) to the *pistil,* consisting of the *stigma, style,* and *ovary.* The ovary contains from one to many ovules, depending on the species of plant. Each ovule contains a megasporangium where meiosis produces one functional megaspore. The latter develops into a multicelled female gametophyte generation. One of these cells is an egg cell.

When pollen grains are transferred to the stigma, each develops a pollen tube that carries two sperm to the ovule. One sperm fertilizes the egg and the other unites with two other nuclei (polar nuclei) of the female gametophyte generation to form endosperm food for the embryo. This so-called double fertilization is unique to angiosperms. The mature ovule, or seed, contains the embryo and food enclosed with a protective seed coat. The wall of the ovary, and sometimes adjacent parts, develops into a **fruit** that surrounds the seeds. Thus angiosperms are said to have *covered seeds.*

Like the gymnosperms, angiosperms are well adapted to a land environment. Their vascular tissue is more complex and the seeds are enclosed by fruits. Both of these are selective advantages for the angiosperms.

Importance of Angiosperms

The angiosperms are the major producers in most terrestrial ecosystems. They provide the food (fig. 26.21) that sustains most of the animals on land, including humans (table 7.B, p. 135). A study of all the plants that have been cultivated has led two authorities, as discussed in the following reading to suggest that only 12 types of plants stand between humans and starvation. Also, grasses, alfalfa, and clover are used as forage, plant food for livestock.

Figure 26.21
Flowering plants are sources of food for the biosphere.

Humans use angiosperms for many other functions. Their wood becomes the timber used for construction and the making of furniture. It is also used for fuel (firewood), particularly in poorer countries. Flax and cotton are a source of natural fibers for making cloth, but cellulose from any plant can be treated to yield rayon. Plant oils are not only used in cooking, they are also, for example, used in perfumes and as medicines. Spices are from various parts of plants: peppercorns are small berrylike structures from a vine; cinnamon comes from the bark of a tree; cloves are dried-up flower buds. Various drugs are taken from angiosperm plants, including morphine and heroin from the juice of the poppy and marijuana from the leaves of the plant *Cannabis*.

Twelve Plants Standing between Man and Starvation

Since we earlier stressed the point that all our food is ultimately derived from plants, it may come as somewhat of a surprise to learn that relatively few species of plants are involved. Of the 800,000 kinds of plants estimated to be in existence only about 3,000 species have provided food, even in the form of nuts, berries, and other fleshy fruits. Virtually all of these food plants are angiosperms, or flowering plants. This is not surprising when you recall that only the angiosperms have seeds enclosed in a carpel [pistil] and hence only they produce true fruits, many of which are used by man for food.

Of the 3,000 plants noted above, only 150 species have been extensively cultivated and have entered the commerce of the world. And of the 150, only 12 species are really important—indeed it can be said that these 12 plants stand between man and starvation. If all 12 or even if a few of these cultivated plants were eliminated from the earth, millions of people would starve.

Three of these all-important species are cereals—**wheat, corn, and rice;** the last alone supplies the energy required by 50 percent of the people of the world. It is a remarkable fact that each of these cereals, or grains, is associated with a different major culture or civilization—wheat with Europe and the Middle East, corn or maize with the Americas, and rice with the Far East. Three of the 12 food plants are so-called root crops—**white, or Irish, potato** (not a root but a **tuber,** an enlarged tip of a rhizome, or horizontal underground stem); **sweet potato;** and **cassava,** or **manioc** or **tapioca,** from which millions of people in the tropics of both hemispheres derive their basic food. Two of the 12 are sugar-producing plants—**sugar cane** and **sugar beet.** Another pair of species are legumes—the **common bean** and **soybean,** both important sources of vegetable protein and hence sometimes referred to as the "poor man's meat." The final two plants of this august company are tropical tree crops—**coconut and banana. . . .**

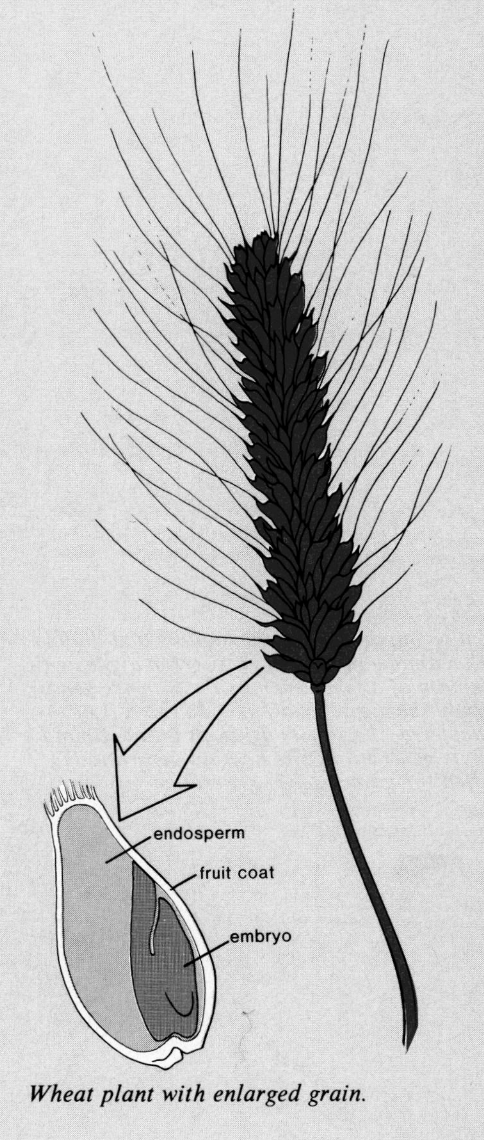

Wheat plant with enlarged grain.

Comparisons

Comparison of Plants

We have seen how plants are adapted to living on land. Not all land plants, however, are *completely* adapted to life on land. Table 26.4 indicates the relative degree of adaptation among bryophyte and tracheophyte plants.

The role of the haploid and diploid stages in the life cycle of various plants may be correlated with their relative adaptation to land (fig. 26.22). Algae, whose life cycle contains only haploid structures except for the diploid zygote, are adapted to a water environment. Mosses with a dominant gametophyte and small, dependent sporophyte have a limited distribution on land. Ferns have a well-developed sporophyte body that has vascular tissue, but they still require very wet conditions for growth of a small, independent nonvascular gametophyte and fertilization by swimming sperm.

The gymnosperms and angiosperms are widely distributed on land because the large, dominant sporophyte is well adapted to terrestrial life. Furthermore, the delicate spores, gametes, zygotes, and embryos are enclosed within protective coverings produced by the sporophyte plant.

Table 26.4 *Adaptation Summary of Terrestrial Plants*

	Vascular Tissue	Sperm
Nonseed plants: Spores disperse the species.		
Bryophytes	Both generations lack vascular tissue.	Swimming sperm require a source of outside moisture.
Psilopsida and ferns	The nonvascular gametophyte generation is separate and independent of the vascular sporophyte.	Same
Seed plants: Seeds disperse the species.		
Gymnosperms (naked seeds)	The nonvascular gametophyte generation is retained by and protected from desiccation by the vascular sporophyte.	Pollen grains replace swimming sperm.
Angiosperms (seeds covered)	Same	Same

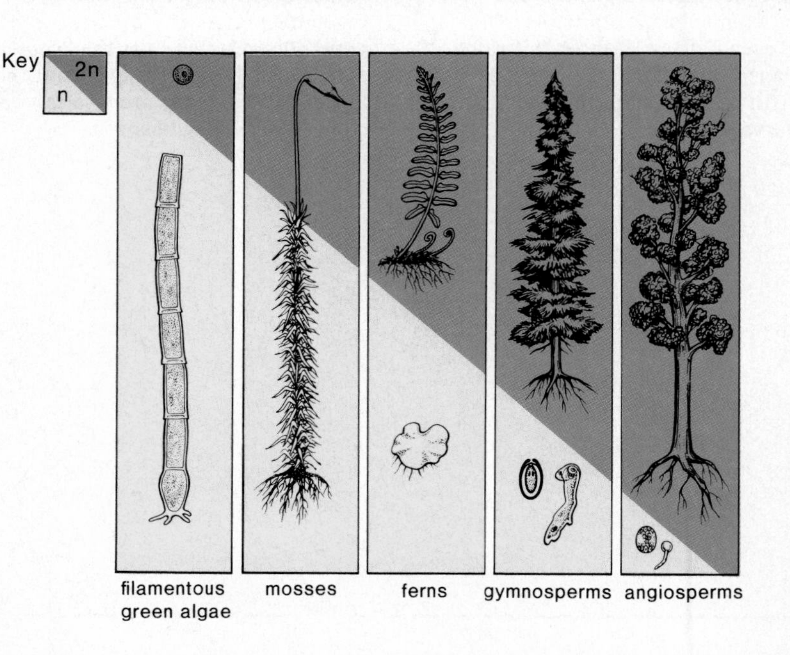

Figure 26.22
The relative importance of the haploid and diploid generation among plants. In most green algae, with the exception of a few—such as those in the genus Ulva*—only the zygote is diploid. In the rest of the plants depicted, the zygote develops into a diploid sporophyte generation. The haploid generation is then called the gametophyte generation.*

A survey of the plant kingdom shows a gradual adaptation to a terrestrial existence, including reproduction on land. It is interesting to compare the manner in which plants have become adapted to the manner in which animals, including humans, have become adapted to life on land.

Comparison of Animal and Plant

As indicated in the first paragraph of this chapter, many animals too, are adapted to life on land. Table 26.5 compares the adaptations of *humans* to that of *trees* and shows that they both have similar adaptations. Two adaptations seem to be of particular interest.

Terrestrial organisms need to protect the gametes and zygote from drying out. In humans, swimming sperm are passed directly from male to female during sexual intercourse. In seed plants, however, the pollen grain replaces swimming sperm. Pollination is the process by which pollen grains are brought to the approximate location of the female gamete. This adaptation seems appropriate for plants in which the structures that produce male and female gametes may remain quite distant from one another.

In regard to maintaining a constant internal environment, higher animals have become somewhat independent of the external environment. Terrestrial plants use an entirely different approach. They often become dormant, during which time their metabolic needs are greatly reduced. We are most familiar with dormancy in deciduous plants because they lose their leaves when dormancy sets in.

***Table 26.5** Comparison of Human to Tree*

	Human	Tree
Protective covering	Skin	Bark or waxy cuticle
Obtain water	Drinking	Absorption by roots
Transport water	Blood vessels	Xylem
Internal support	Skeleton	Woody xylem
Reproduction	Seminal fluid and vaginal secretions provide water for sperm during sexual intercourse.	Pollen grain. Pollen tube allows sperm to reach egg.
Protection of embryo	Internal development in female uterus.	Partial internal development in ovule. Seed coat prevents desiccation.
Constancy of internal environment	Maintains a constant internal environment.	Dormancy during winter, and other unfavorable conditions.

Summary

The plant kingdom includes multicellular photosynthetic organisms and closely related unicellular ones. The green algae are a diverse group in which there are different life cycles ranging from those that only have the zygote diploid to those that show alternation of generations.

Seaweeds include multicellular green, red, and brown algae. Green algae and brown algae are typically found in cool waters along rocky coasts. Red algae are found farther from the shore in warm tropical waters.

There are two main groups of plants adapted to a land existence: (1) the bryophytes (e.g., mosses) are not well adapted; the nonvascular gametophyte (N) is dominant and requires external water for swimming sperm to reach the egg; (2) the tracheophytes (ferns, pines, and flowering plants) vary in their degree of adaptation. The vascular sporophyte (2N) generation is dominant in the fern, but there is a separate water-dependent gametophyte. Swimming sperm need external water to reach the egg. In both mosses and ferns, the species is dispersed by means of spores.

Gymnosperms and angiosperms have microspores and megaspores. The microspores become mature pollen grains (male gametophyte generations). Each megaspore develops into a female gametophyte generation. After pollination, the pollen grain develops a pollen tube within which a sperm travels to an egg. Following fertilization, the ovule becomes a seed. In the pine, the seed is naked and usually windblown. In a flowering plant, the seed is enclosed within fruit and transported in various ways.

Objective Questions

1. In the life cycle of most green algae, the only diploid stage is the ________ .
2. In *Chlamydomonas* the adult is ____________ (choose haploid or diploid).
3. *Volvox* has heterogametes, a definite ____________ and ____________ .
4. In *Spirogyra,* zygotes form following the process of ____________ .
5. *Ulva* has the life cycle ____________ , as do land plants.
6. Among the seaweeds, *Fucus* is a type of ____________ algae.
7. Bryophytes have the ____________ generation dominant, while the tracheophytes have the ____________ generation dominant.
8. The life cycle of the moss is incompletely adapted to life on land because the sperm must ____________ in external water to the egg.
9. In the fern life cycle, the gametophyte generation is independent and ________ from the sporophyte generation.
10. Seed plants have male and female gametophyte generations and therefore produce ____________ .
11. Gymnosperms are fully adapted to life on land; the windblown ____________ replaces the swimming sperm.
12. Angiosperms have ____________ seeds in that they are located inside ____________ .

Answers to Objective Questions

1. zygote 2. haploid 3. sperm, egg 4. conjugation 5. alternation of generations 6. brown 7. gametophyte, sporophyte 8. swim 9. separate 10. heterospores 11. pollen grain 12. covered, fruit.

Study Questions

1. Name the major groups of plants studied in this chapter. (pp. 561, 567)
2. Describe each of the green algae studied, emphasizing their reproductive patterns, both asexual and sexual. (pp. 561–564)
3. Compare the three types of seaweed mentioned in this chapter. (pp. 565–567)
4. Compare and contrast the bryophytes to the tracheophytes in regard to dominant generation and presence of vascular tissue. (p. 567)
5. Draw the diagram of alternation of generations for those terrestrial plants in which spores disperse the species. (p. 561)
6. Draw the diagram of alternation of generations for those terrestrial plants in which the seeds disperse the species. (p. 573)
7. Compare the life cycles of the moss, fern, pine, and flowering plant, emphasizing adaptation to life on land. (pp. 568, 571, 574, 576)
8. Name the reproductive parts of the flower and state a function for each part. (p. 576)
9. Give examples to support the statement: The moss and fern are not fully adapted to life on land, but the pine and flowering plant are fully adapted. (p. 580)
10. With reference to table 26.5, contrast the adaptations of a tree and human to the land environment. (p. 581)

Key Terms

alternation of generations (awl″ter-na′shun uv jen″ĕ-ra′shunz) two phased life cycle displayed by many plants in which there are sporophyte and gametophyte generations. *561*

antheridia (an″ther-id′e-ah) male organs in certain nonseed plants where swimming sperm are produced. *568*

archegonia (ar″kĕ-go′ne-ah) female organs in certain nonseed plants where an egg is produced. *568*

colony (kol′o-ne) an organism that is a loose collection of cells that are specialized and cooperate to a degree. *563*

conjugation (kon″ju-ga′shun) a sexual union in which the nuclear material of one cell enters another. *564*

frond (frond) the large leaf of a fern plant containing many leaflets. *571*

fruit (fro͞ot) a mature ovary enclosing seed(s). *578*

gametophyte generation (gam′ĕ-to-fīt jen″ĕ-ra′shun) the haploid generation that produces gametes in the life cycle of a plant. *561*

heterospores (het′er-o-sporz) nonidentical spores such as microspores and megaspores produced by the same plant. 572

isogametes (i″so-gam′ēts) gametes whose union produces a zygote, but which have a similar appearance. *562*

megaspores (meg′ah-spōrz) in seed plants, spores that develop into the female gametophyte generation. *575*

microspores (mi′kro-spōrz) in seed plants, spores that develop into a pollen grain. *575*

pollen grain (pol′en grān) male gametophyte generation of seed plants. *572*

prothallus (pro-thal′-us) a small, heart-shaped structure; the gametophyte generation of the fern. *571*

sorus (so′rus) a cluster of sporangia found on the underside of fern leaves (plural: sori). *571*

sporophyte generation (spo′ro-fīt jen″ĕ-ra′shun) spore-producing diploid generation of a plant. *561*

27

Animal Kingdom

Chapter Concepts

1 Animals are classified according to certain criteria such as body plan, symmetry, number of germ layers, and level of organization.

2 There is an increase in complexity of organization when the animal groups are surveyed from those first evolved to those latest evolved.

3 Animals are adapted to their way of life; for example, those adapted to an inactive life may be contrasted to those adapted to an active life; those adapted to an aquatic existence may be contrasted to those adapted to a terrestrial existence.

Chapter Outline

Figure 27.1
The animal kingdom is diverse. A hydra (a.) and a wolverine (b.) are both multicellular heterotrophic organisms that must take in preformed food. Note the radial symmetry of the aquatic hydra that has tentacles for capturing small prey. A wolverine lives in arctic and subarctic regions. It captures a reindeer by jumping on its back and holding on with its sharp claws until the animal collapses. Its jaws are powerful enough to kill with a neck bite.

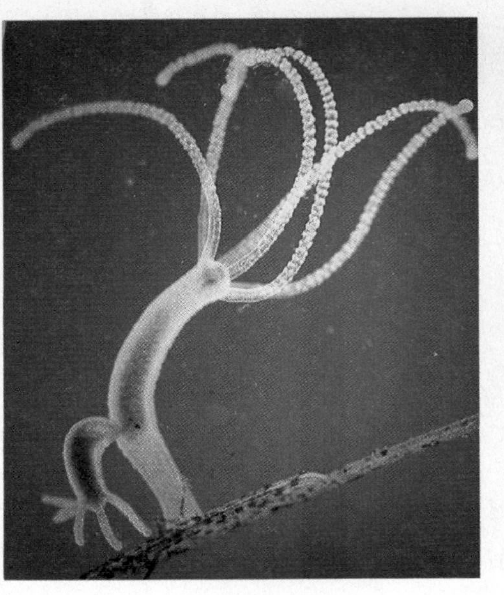

a.

b.

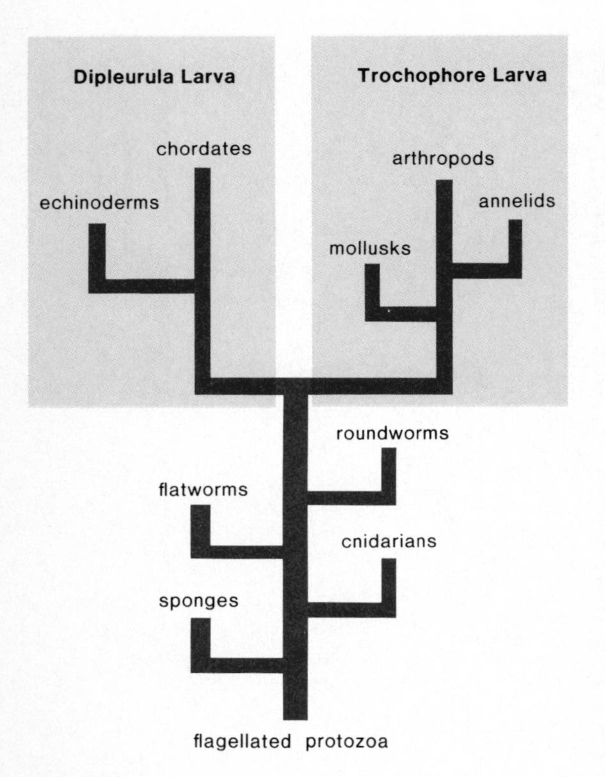

Figure 27.2
Evolutionary tree of animals. The "lower invertebrates" form the trunk of the tree; the "higher invertebrates" are divided into two branches. The marine echinoderms and lower chordates share the dipleurula larva; the marine mollusks, arthropods and annelids share the trochophore larva.

Animals (fig. 27.1) are heterotrophic and must take in food. In contrast to stationary green plants that absorb energy from the sun and make their own organic food, animals are nongreen and possess some means of locomotion that enables them to acquire food.

A predator that actively seeks out and captures food best exemplifies the animal way of life. Predators have bilateral symmetry, good musculature, and a well-developed nervous system including sense organs. All of these help the animal seek prey and escape enemies. Good predators also have a means of seizing and digesting their food.

All animals must digest their food, carry on gas exchange, excrete waste, circulate nutrient and waste products to and from cells, coordinate their movements, protect themselves, and reproduce and disperse the species. The more complex animals have organ systems to carry out these functions; in simple animals, these functions are sometimes carried out by specialized tissues.

Evolution and Classification

Evolution of Animals

The History of Life table (fig. 23.2) shows that all modern phyla of animals[1] had evolved by the beginning of the Paleozoic era some 600 million years ago. The evolutionary tree of animals (fig. 27.2) indicates that animals are believed to have arisen from flagellated protozoans—perhaps a colonial form whose cells have become differentiated into various types of cells.

After animals became moderately complex, there was a split into two main lines. Indeed, the evolutionary tree of animals resembles a tree with two main branches. The animal phyla located on the main trunk of the tree are

[1]The classification system utilized in this text is given on page 764.

Table 27.1 Primitive versus Advanced				
	Most Primitive	**Primitive**	**Advanced**	**Most Advanced**
Body plan	None	Sac plan	Tube within tube	Tube within tube with specialization of parts
Symmetry	None	Radial	Bilateral	Bilateral with cephalization
Germ layers	None	Two	Three	Three
Level of organization	None	Tissues	Organs	Organ systems
Body cavity	Acoelomate	Acoelomate	Pseudocoelom	True coelom
Segmentation	Nonsegmented	Nonsegmented	Segmented	Segmented with specialization of parts

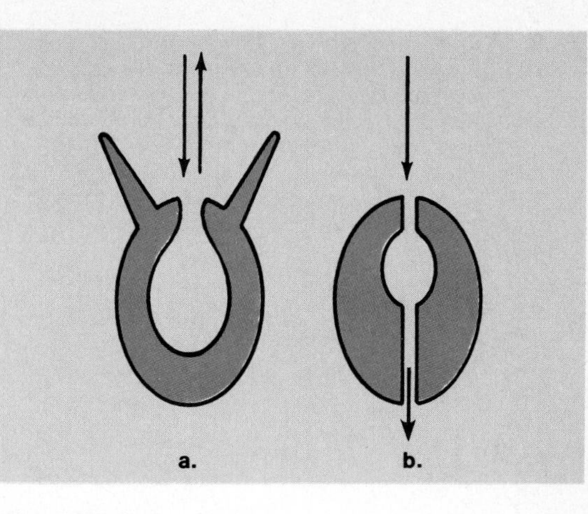

Figure 27.3
Animals have two basic body plans. a. *Sac plan with only one opening.* b. *Tube-within-a-tube plan with two openings.*

referred to as the *lower invertebrates* in this text, and the animals of the upper branches include the *higher invertebrates* and the vertebrates. **Invertebrate** animals lack a dorsal backbone, while **vertebrates** have a backbone made up of vertebrae.

Animals are divided into three groups. The lower invertebrate phyla form the trunk of an evolutionary tree. The higher invertebrates and vertebrates are distributed on two branches of the tree.

Classification of Animals

A study of the evolution of animals indicates that increased complexity of organization can be related to certain anatomical features of structure (table 27.1). Classification is based on the presence or absence of these features which are termed advanced because they evolved later in time.

Two body plans (fig. 27.3) are observed in the animal kingdom: the **sac plan** and the **tube-within-a-tube plan.** Animals with the sac plan have only one opening, which is used both as an entrance for food and an exit for waste. Animals with the tube-within-a-tube plan have an entrance for food and an exit for waste. Two openings allow specialization of parts to occur along the length of the tube.

Asymmetry means that the animal has no particular symmetry. **Radial symmetry** means that the animal is circularly organized and, just as with a wheel, it is possible to obtain two identical halves no matter how the animal is sliced longitudinally. **Bilateral symmetry** means that the animal has a definite left and right half so that only one longitudinal cut down the center of the animal will produce two equal halves (fig. 27.4). Radially symmetrical animals tend to be attached to a substrate, or *sessile.* This type of symmetry is useful to these animals since it allows them to reach out in all directions from one center. Bilaterally symmetrical animals tend to be active and move forward with one anterior end. This end develops a head region (called *cephalization*) that is acutely aware of the environment and aids the animal in its forward progress.

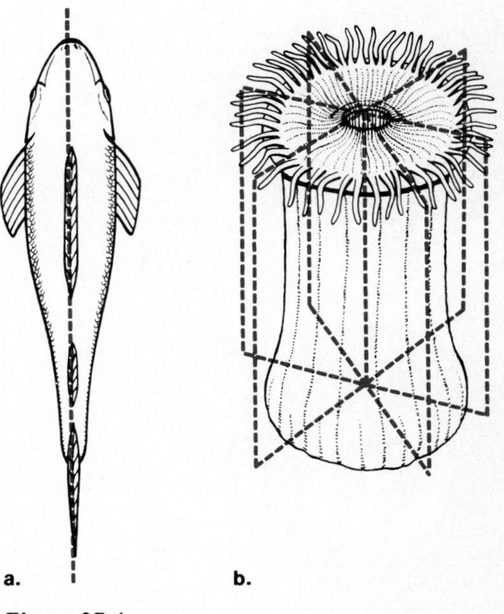
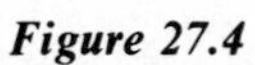

Figure 27.4
Animals have two types of symmetry. a. *Bilateral symmetry. Notice that only the one longitudinal cut shown will give two identical halves of the animal.* b. *Radial symmetry. Notice that any longitudinal cut, such as those shown, will give two identical halves of the animal.*

Figure 27.5
Animals have either no germ layers, two germ layers, or three germ layers. a. *Cross section of animal with two germ layers. In these animals, a packing material called mesoglea is found between the two layers.* b. *Cross section of animal with three germ layers. Endoderm surrounds the gut lumen; mesoderm is the middle layer; and ecotoderm is the outer germ layer.*

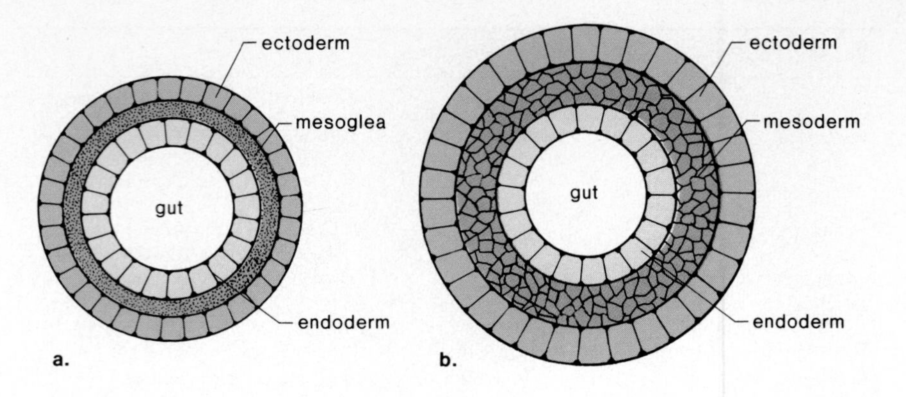

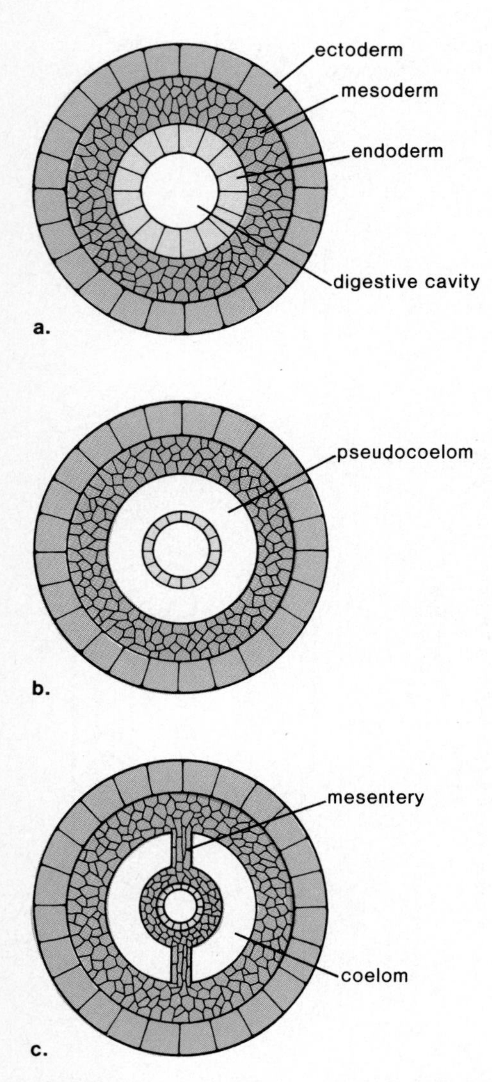

Figure 27.6
Comparison of mesoderm organization. a. *In acoelomate animals, the mesoderm is packed solidly.* b. *Pseudocoelomate animals have mesodermal tissue inside the ectoderm, but not adjacent to the gut endoderm.* c. *In coelomate animals, there is mesodermal tissue both inside the ectoderm and adjacent to the gut endoderm. True coeloms are body cavities completely lined by mesodermal tissue. Mesenteries hold organs in place within the body cavity.*

Although a total of three germ layers is seen in most animals during embryonic development (p. 423), in fact, some animals have two germ layers (fig. 27.5a). Such animals have the tissue level of organization. Animals with three germ layers (fig. 27.5b) have the organ level of organization.

The body cavity we wish to consider is the space surrounding the digestive system. Some animals do not have such a space (fig. 27.6a). Animals that do have this cavity are either pseudocoelomates, having a **pseudocoelom** (fig. 27.6b), or coelomates, having a true **coelom** (fig. 27.6c). A pseudocoelom is lined with mesoderm only beneath the body wall, but a true coelom is lined with mesoderm both beneath the body wall and around the gut.

Some animals are nonsegmented and some have repeating units called segments. It is easy to tell, for example, that an earthworm (fig. 27.22) is segmented because its body appears to be a series of rings. Segmentation leads to specialization of parts in that the various divisions of the body can become differentiated for specific purposes.

Classification of animals considers type of body plan, symmetry, number of germ layers, presence or absence of coelom, and segmentation. Among the invertebrates it is possible to observe an increase in complexity regarding these features.

Lower Invertebrates

The lower invertebrates include the sponges, cnidarians, flatworms, and roundworms. By studying the animals in this order, we will observe an increase in complexity that may reflect the order in which these animals evolved. Nevertheless, it is difficult to determine their exact evolutionary relationship.

Sponges

Most **sponges** (phylum Porifera) are marine and are more abundant in warm ocean water, near the coast. Some sponges grow on rocks and are brightly colored, appearing almost lichen-like when seen at a distance. Sponges are often shaped like vases, with either simple flat walls or convoluted walls containing canals. Regardless, the wall of a sponge is perforated by numerous *pores* surrounded by contractile cells capable of regulating their size.

The wall of a sponge contains three types of cells (fig. 27.7). The outer cells are flattened *epidermal cells.* The inner cells are **collar cells** with flagella whose constant movement produces water currents that flow through the pores into the central cavity and out through the upper opening of the body (called the **osculum**).

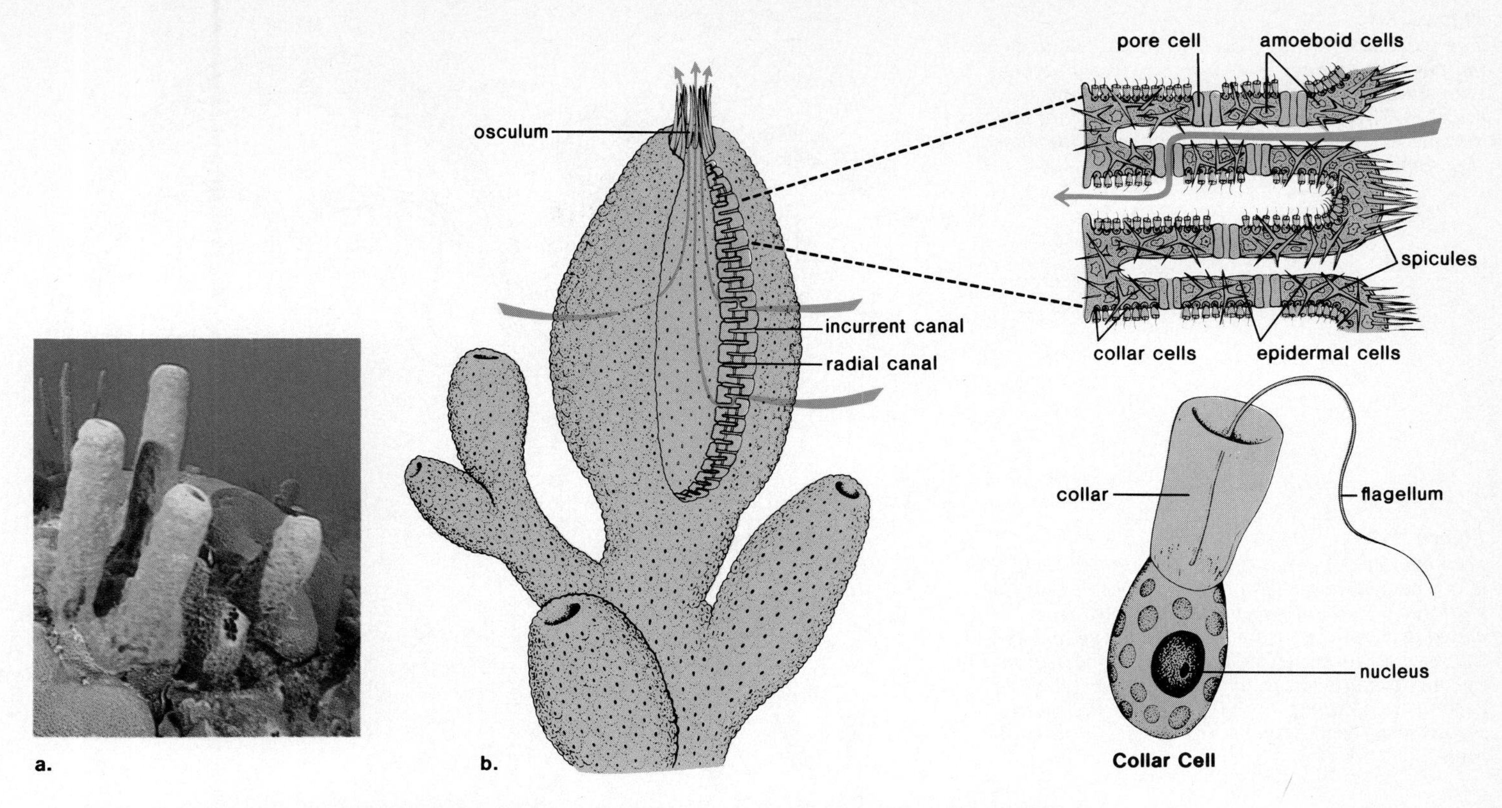

Figure 27.7
a. *Yellow tube sponge.* b. *Generalized anatomy of a sponge. The arrows indicate the flow of water that is kept moving by the beating of the flagella of the collar cells, one of which is drawn enlarged. Sponges are classified according to the chemical makeup of their spicules, which serve as an internal skeleton.*

Sponges are **sessile filter feeders.** This means that they remain in one place as an adult and that the food they acquire filters through the pores. Microscopic food particles brought by the water are engulfed by the collar cells and digested by them in food vacuoles or are passed to the amoeboid cells for digestion.

The *amoeboid cells* within the wall of a sponge not only act as a circulatory device to transport nutrients from cell to cell; they also produce **spicules,** the needle-shaped structures that serve as the internal skeleton of a sponge, and the sex cells, the egg and sperm. Fertilization results in a zygote that develops into a ciliated larva that may swim away to a new location. Sponges also reproduce by budding, and this process produces whole colonies that may become quite large. Like all less-specialized organisms, sponges are capable of **regeneration,** or growth of a whole from a small part.

The cellular organization of sponges is different from that of other animals and the main opening of a sponge is used only as an exit, not an entrance. Further, movement is limited to the beating of the flagella, constriction of the osculum, and larval-stage swimming. Sponges are classified according to type of *spicule.*

Cnidarians

The body of a **cnidarian** (coelenterate) (phylum Cnidaria) is a hollow two layered sac which accounts for the former name of these organisms—coelenterate means hollow sac. The outer layer of the sac, ectoderm, is separated from the inner layer, endoderm, by a jellylike material called **mesoglea** (fig. 27.5). All cnidarians have specialized stinging cells now termed cnidocytes, from a Greek word meaning sea nettles. Within these cells are the **nematocysts,** long, spirally coiled hollow threads. When the trigger of a stinging cell is touched, the discharged thread, which sometimes contains poison, serves to stun either prey or enemy.

Figure 27.8
The two body forms of coelenterates. The drawings indicate the manner in which these germ layers surround the central gastrovascular cavity.
a. *Polyps, with the oral side uppermost, are usually attached to surfaces.* b. *Medusae, with the aboral side uppermost, are free-swimming.*

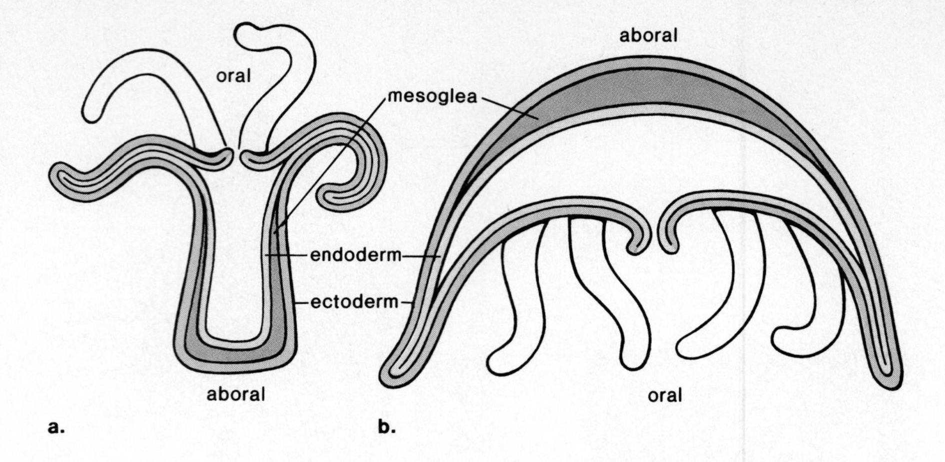

Figure 27.9
Members of the genus Obelia *have a life cycle that is a type of alternation of generations in that the stationary colonial polyp produces medusae asexually and the medusae produce egg and sperm for sexual reproduction. Following fertilization, the zygote develops into a larva that settles down to become a colonial polyp. Notice that the motile medusae serve to disperse the species in this life cycle.*

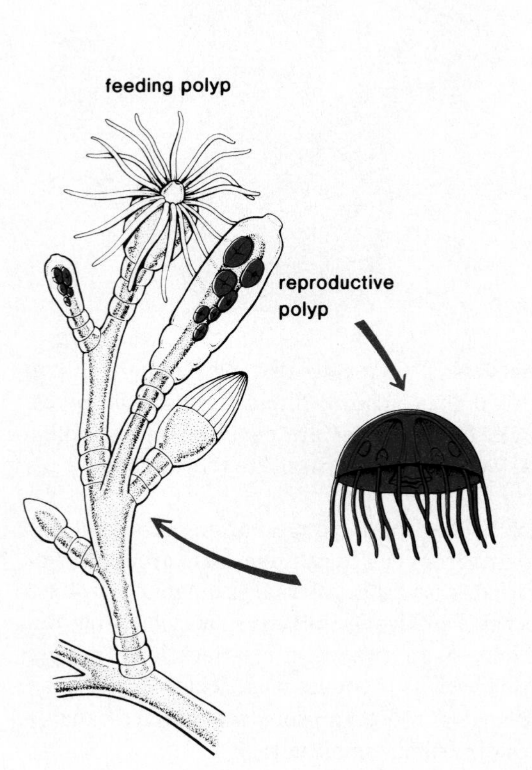

Cnidarians have radial symmetry (fig. 27.4) and there is typically a ring of tentacles surrounding the mouth region. Some cnidarians referred to as **hydroids** or **polyps** (fig. 27.8) have a tubular shape, with the mouth region directed upward. Others, which have a bell shape with the mouth region directed downward, are called jellyfishes or **medusae.** The polyp is adapted to a sessile life, while the medusa is adapted to a floating or free-swimming existence. At one time, both body forms may have been a part of the life cycle of all cnidarians, since today we see an alternation of generations[2] of these two forms in certain cnidarians such as members of the genus *Obelia* (fig. 27.9). When alternation of generations does exist, the polyp stage produces medusae, and the medusae, which produce egg and sperm, disperse the species.

[2]This is not the same as alternation of generations in plants because here both generations are diploid.

a.

b.

c.

Figure 27.10
Cidarian diversity. a. *Medusae of* Aurelia. *A thick layer of jelly (mesoglea) gives the medusae buoyancy and accounts for their common name, jellyfish.* b. *Living coral polyps. Each polyp is individual but has a chalky skeleton that is joined to its neighbor's. The polyps feed on zooplankton but may contain symbiotic algae that contribute to their nutrition.* c. *Sea anemone. These animals are called the "flowers of the sea," but despite their appearance they are carnivorous animals.*
(a.) Carolina Biological Supply Company.
(c.) Carolina Biological Supply Company.

Cnidarians are quite diversified (fig. 27.10). The *Portuguese man-of-war,* whose nematocysts may cause serious or even fatal poisoning in humans, is a colony of polyps suspended from a large medusoid form that serves as a gas-filled float. Many species of jellyfish, such as *Aurelia,* show alternation of generations, but the two generations are not equal—the medusa is the primary stage and the polyp remains quite small and insignificant. *Sea anemones* are solitary polyps with thick walls. They may be brightly colored and look like beautiful flowers. **Corals** are similar to sea anemones, but they have calcium carbonate skeletons. Some corals are solitary, but most are colonial, with either flat, rounded, or upright and branching colonies. The slow accumulation of coral skeletons has formed reefs in the South Pacific, including the Great Barrier Reef along the eastern coast of Australia. An ancient coral reef that now lies beneath Texas is the source of petroleum for that state.

Figure 27.11
Hydra *anatomy. Compare the longitudinal section* (far left) *with the cross section* (lower right). *Before discharge, the nematocyst is tightly coiled within a stinging cell* (upper right) *and after discharge, it is extended.*

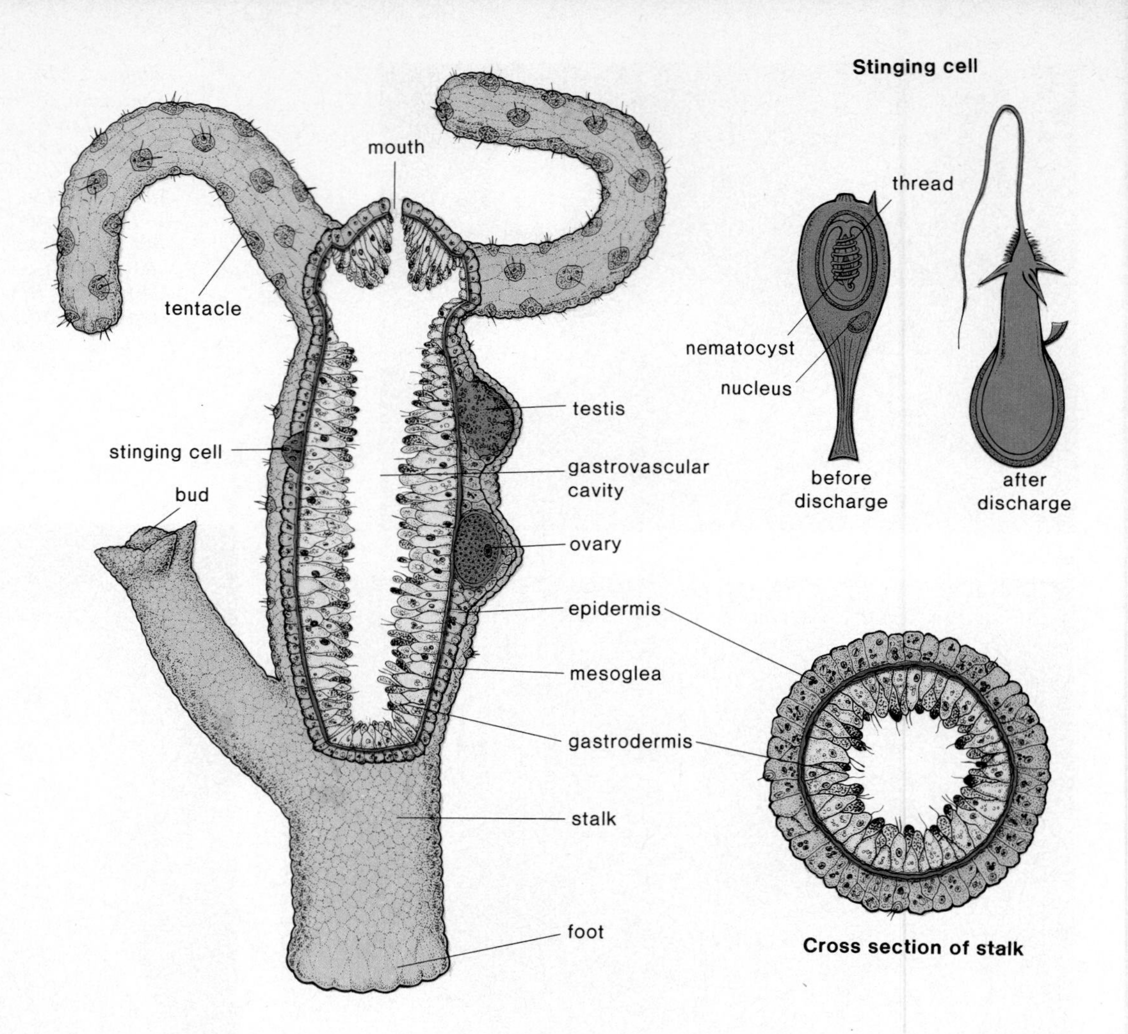

Hydra

Hydras (fig. 27.11) are likely to be found attached to underwater plants or rocks in most lakes and ponds. The body is a small, tubular polyp about one-quarter inch in length. Although hydras usually remain in one place, they may glide along on their base, or even move rapidly by means of somersaulting. Hydras, like other animals capable of locomotion, possess both muscular and nerve cells. The nerve cells form a connecting network throughout the mesoglea known as the **nerve net.** The nerve net makes contact with the outer layer of cells, called the **epidermis,** and the inner layer, called the **gastrodermis.** These cells contain contractile fibers.

The cells of the gastrodermis secrete digestive juices that pour into the central cavity. The enzymes begin the digestive process, which is completed within food vacuoles when small pieces of the prey are engulfed by the cells of the gastrodermis. Nutrient molecules are passed by diffusion to the rest of the cells of the body. The presence of the large inner cavity makes it possible for all cells to exchange gases directly with the surrounding medium. Since this function is carried out by a vascular system in more complex animals, the cavity is known as a **gastrovascular cavity.**

In *Hydra,* the mesoglea contains interstitial, or embryonic cells, capable of becoming other types of cells. For example, they can produce the ovary and testes and probably also account for the animal's great regenerative powers. Like the sponges, a cnidarian can grow whole from a small piece.

The cnidarians are *radially symmetrical* and have a *sac body plan.* There are *two tissue layers,* the epidermis and gastrodermis, derived from the embryonic germ layers, the ectoderm and endoderm. The presence of *nematocysts* is a unique feature.

Flatworms

Flatworms (phylum Platyhelminthes) have three germ layers. The presence of mesoderm not only gives bulk to the animal, it also allows for greater complexity of internal structure. Free-living forms have muscles and excretory, reproductive, and digestive organs. The worms lack respiratory and circulatory organs but since the body is flattened, diffusion alone is adequate for the passage of oxygen and other substances from cell to cell. The presence of bilateral symmetry and good cephalization is very important, along with a well-developed nervous system, including sense organs, in free-living forms. This combination makes them efficient predators.

Flatworms are nonsegmented, lack a coelom, and have the sac plan with only one opening. Therefore, if we evaluate them according to table 27.1, we see that they have a combination of primitive to advanced features.

There are three classes of flatworms: one is free living and two are parasitic. Parasites are degenerate forms of the free-living specimen that, of course, best exemplifies the characteristics of the phylum. Therefore, we will begin with the free-living specimen, the planarian.

Planarians

Freshwater planarians (e.g., *Dugesia;* fig. 27.12) are small (several millimeters to several centimeters), literally flat worms. Some tend to be colorless; others have brown or black pigmentation. Planarians live in lakes, ponds, streams, and springs where they feed on small living or dead organisms, such as worms or crustacea.

Since planarians live in fresh water, water tends to enter the body by osmosis. They have a water-regulating organ that especially rids the body of excess water. The organ consists of a series of interconnecting canals that run the length of the body on each side. The beating of cilia in specialized structures keeps the water moving toward the excretory pores. The beating of the cilia reminded some early investigator of the flickering of a flame and so the excretory organ is called a **flame-cell system** (fig. 27.12b).

Planarians are **hermaphroditic,** which means that they possess both male and female sex organs. The worms practice cross-fertilization; the penis of one is inserted into the genital pore of the other. The fertilized eggs hatch in two to three weeks as tiny worms.

Planarians are often used in biology laboratories to illustrate regeneration. If a worm is cut crosswise, it usually grows a new head or tail as is appropriate. Planarians have also been used in so-called memory experiments.

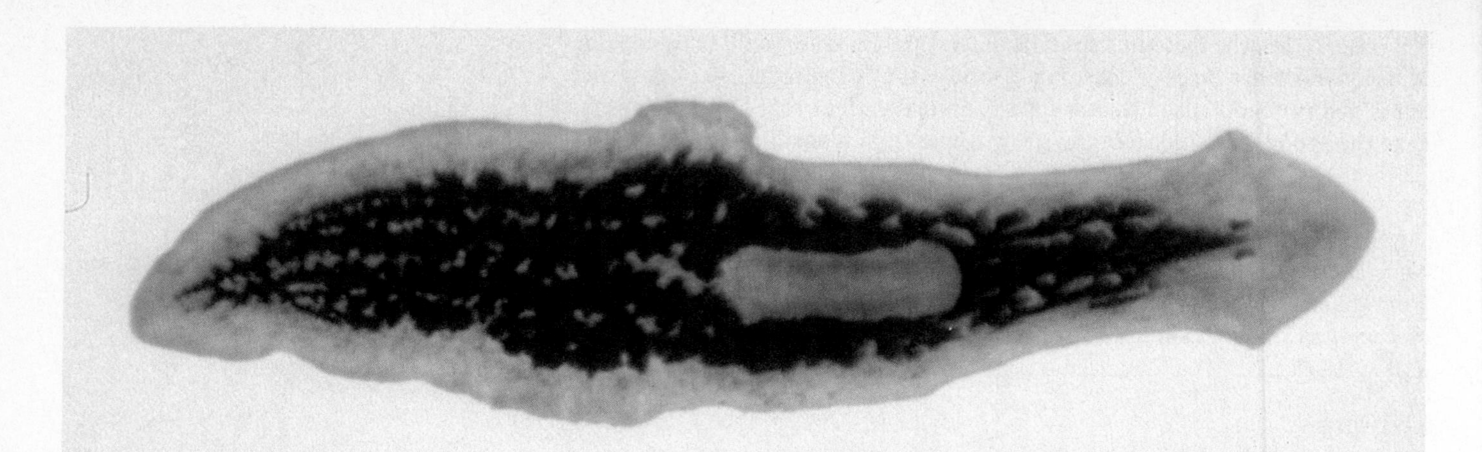

a.

Figure 27.12
Flatworm anatomy as exemplified by a planarian worm.
a. *Photomicrograph of worm. This photo shows the oral side of the animal, the pharynx (pink) leads to the digestive organ, which has been darkly stained to illustrate the manner in which it ramifies throughout the body.*
b. *Details of organ anatomy. (1) Excretory organ is composed of canals that have flame cells* (enlarged drawing) *whose beating cilia draw in fluid that is excreted by way of nephridiopores. (2) The nervous system has a ladder appearance because cross fibers stretch between longitudinal fibers that extend the length of the animal. (3) The digestive organ is highly branched, and since the worms are hermaphroditic there are both male and female reproductive organs.*

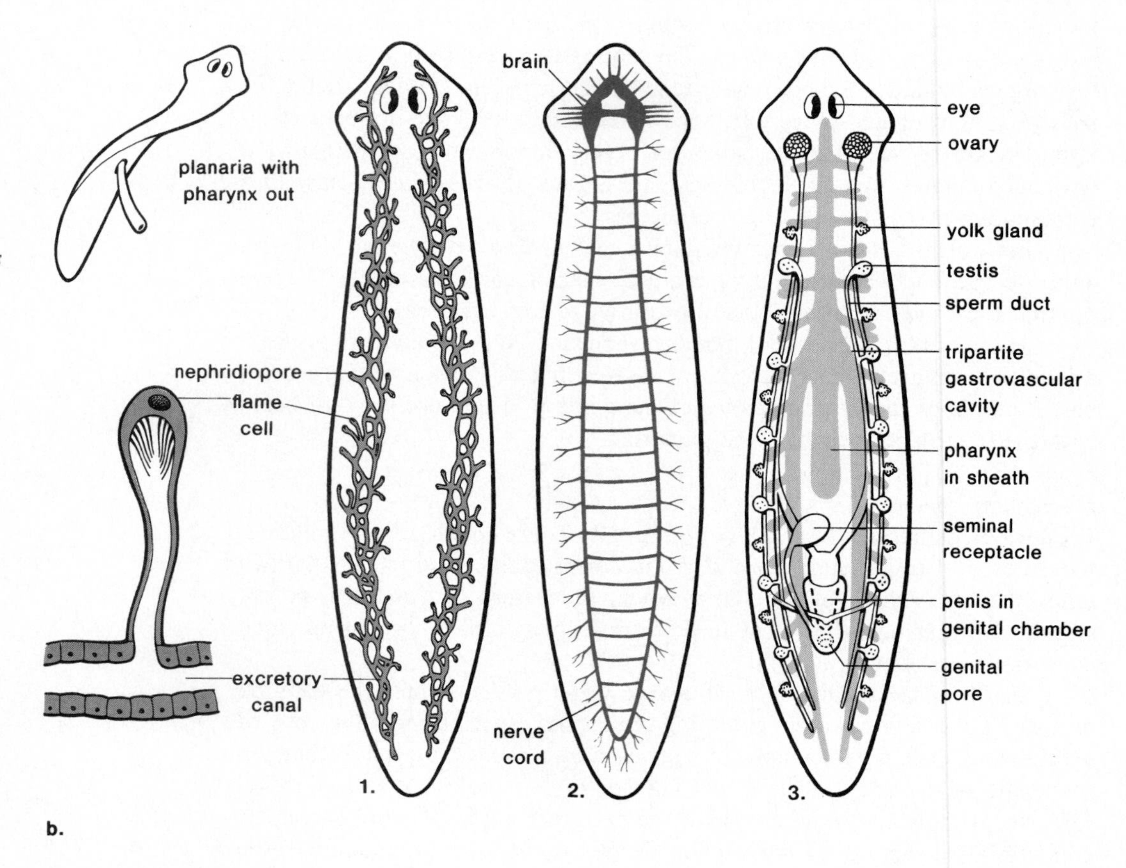

Table 27.2 Free-living versus Parasitic Worms			
	Planarians	**Flukes**	**Tapeworms**
Body wall	Ciliated epidermis	Glycocalyx covers integument	Glycocalyx covers integument
Cephalization	Yes. Eyespots and auricles	No. Oral sucker	No. Scolex with hooks and suckers
Nervous connections	Nerves and brain	Reduced	Reduced
Digestive organ	Ramifies	Reduced	Absent
Reproductive organs	Hermaphroditic	Increased in volume	Greatly increased in volume
Larva	Absent	Present	Present

In these experiments, planarians were trained to swim mazes and then were cut up and fed to untrained planaria. When the cannibals were subsequently taught the same task, they learned faster than the first set. The exact significance of these experiments is debatable, but they have led to some interesting student speculations as to how best to acquire the knowledge of teachers.

Parasitic Flatworms

There are two types of parasitic flatworms: tapeworms (cestodes) and flukes (trematodes). The structure of both these worms illustrates the modifications that occur in parasitic animals (table 27.2). Concomitant with the loss of predation, there is an absence of cephalization; the anterior end notably carries hooks and/or suckers for attachment to the host. The parasite acquires nutrient molecules from the host, and the digestive system is reduced. The presence of a mucopolysaccharide outer coating, called the *glycocalyx,* protects the animal against host attack. The extensive development of the reproductive system, with the production of millions of eggs, may be associated with difficulties in dispersing the species. Both parasites utilize a *secondary host,* or intermediate host, to transport the species from main host to main host. The *primary host* contains the sexually mature adult; the secondary host(s) contain(s) the larval stage or stages.

A **tapeworm** has a head region (fig. 27.13), containing hooks and suckers for attachment to the intestinal wall of the host. Behind the head region, called a **scolex,** there is a short neck and then a long series of proglottids. **Proglottids** are segments, each of which contains a full set of both male and female sex organs. Therefore, the tapeworm is little more than a reproductive factory. There are excretory canals but no digestive system and only the rudiments of nerves.

After fertilization, the proglottids become nothing but a bag filled with developing embryos (larvae). Mature proglottids such as these break off and, as they pass out with the host's feces, the larvae enclosed by a protective covering are released.

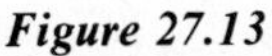

Figure 27.13
Tapeworm of a Taenia *species.* a. *Life cycle, showing mature proglottid* (right) *and gravid proglottid* (left) *in detail. The gravid proglottid is little more than a sac of eggs.* b. *Scanning electron micrograph of tapeworm* (Taenia) *scolex. The scolex contains hooks and suckers that permit the animal to cling to the wall of the digestive tract.*

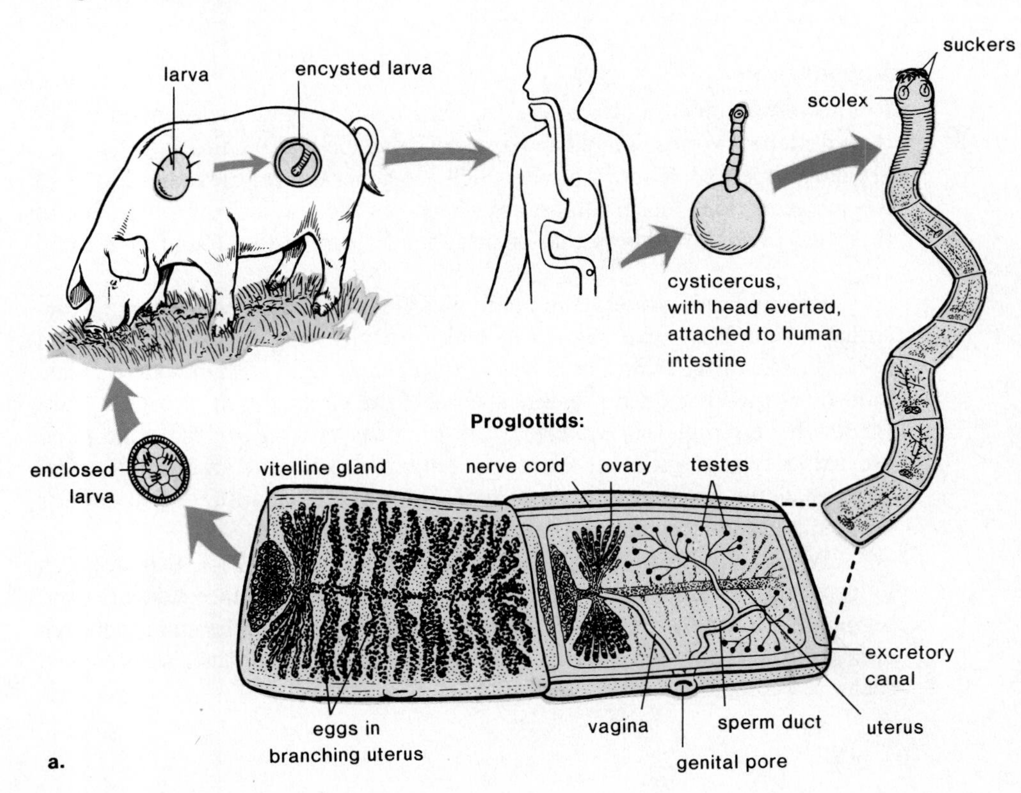

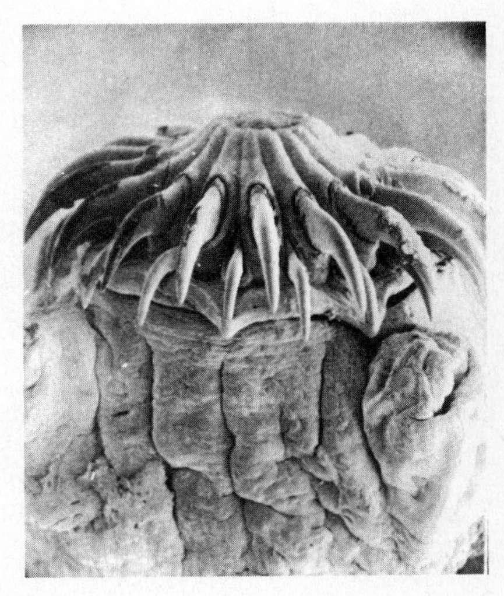

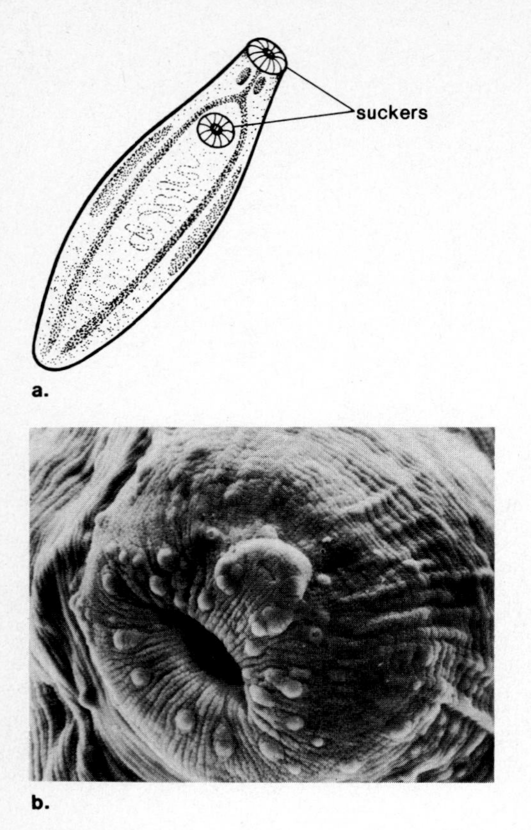

Figure 27.14
Fluke anatomy. a. *Drawing showing general anatomy. Flukes use suckers for attachment to the host.* b. *Scanning electron micrograph of fluke* (Gorgoderina attenuata) *oral sucker. Note upraised structures, which are believed to be sensory in nature.*

If feces-contaminated food is fed to pigs or cattle, the larvae escape when the covering is digested away. They burrow through the intestinal wall and travel in the bloodstream to finally lodge and encyst in muscle. Here a *cyst* means a small, hard-walled structure that contains a small immature worm. When humans eat raw, or rare, infected meat, the worms break out of the cyst, attach themselves to the intestinal wall, and grow to adulthood. Then the cycle begins again (fig. 27.13).

There are many different types of **flukes** (fig. 27.14a), usually designated by the type of vertebrate organ they inhabit; for example, there are blood, liver, and lung flukes. While the structure may vary slightly, in general the fluke body tends to be oval, to elongate with no definite head except that the oral sucker, surrounded by sensory papillae, is at the anterior end (fig. 27.14b). Usually there is at least one other sucker for attachment to the host. Inside, there is a reduced digestive, nervous, and excretory system. There is a well-developed reproductive system, and the adult fluke is usually hermaphroditic although there are exceptions.

A blood fluke causes **schistosomiasis** in Africa and South America. This disease is especially prevalent in areas with irrigation ditches because the secondary host is a freshwater snail. The disease is spread when egg-laden human feces gets into the water and newly hatched larvae enter the snails. Asexual reproduction occurs within the snail and the resulting larvae penetrate human skin to enter blood vessels where they mature.

The Chinese liver fluke requires two hosts: the snail and the fish. Humans contract the disease when they eat uncooked fish. The adults reside in the liver and deposit their eggs in the bile duct, which carries the eggs to the intestine.

Although flatworms have a *sac body plan,* they are more complex than cnidarians because they have *three germ layers* and possess true organs. The free-living forms exhibit *cephalization* and *bilateral symmetry.*

Roundworms

Roundworms (phylum Nematoda), as their name implies, are rounded rather than flattened worms. They have a smooth outside wall, indicating that they are nonsegmented. These worms, which are generally colorless and less than 5 centimeters long, occur almost anywhere—in the sea, in fresh water, and in the soil—in such numbers that thousands of them can be found in a small area.

Roundworms possess two anatomical features not seen before: a tube-within-a-tube body plan (fig. 27.3) and a body cavity. The body cavity is a *pseudocoelom* (fig. 27.6b), or a cavity incompletely lined with mesoderm. This fluid-filled pseudocoelom provides space for the development of organs, substitutes for a circulatory system by allowing easy passage of molecules, and provides a type of skeleton. Worms in general do not have an internal or external skeleton but they do have a *hydrostatic skeleton,* a fluid-filled interior that supports muscle contraction and enhances flexibility.

When roundworms are evaluated according to table 27.1, they are seen to have features associated with advanced animals except that they are nonsegmented. Roundworms are thought to be a side branch to the main evolution of animals and may have arisen from a common ancestor that also produced coelomate animals.

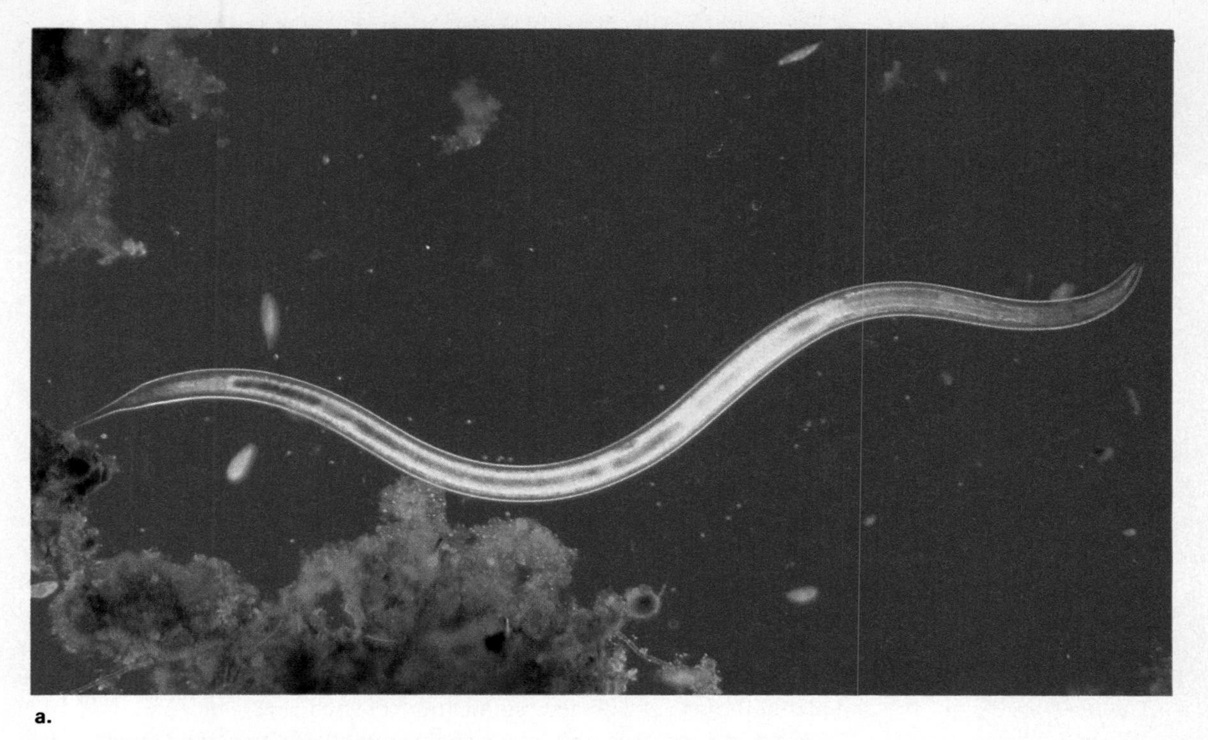

a.

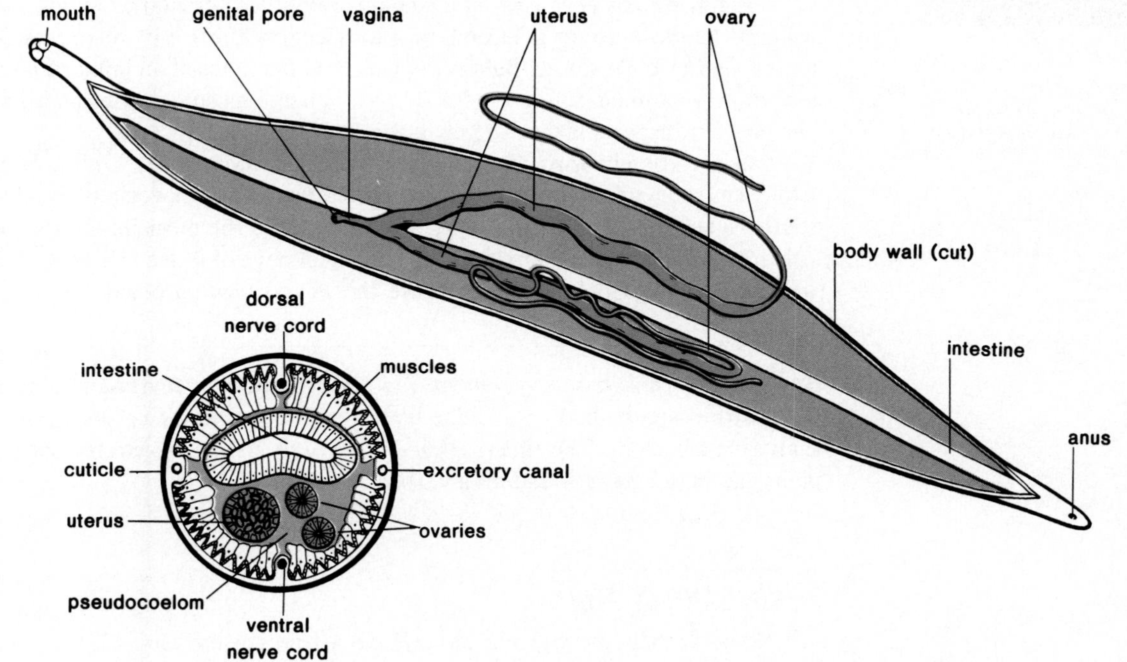

b.

Figure 27.15
Roundworm anatomy. a. *Photomicrograph of a roundworm.* b. *Anatomy of female* Ascaris. *The cross section shows the pseudocoelomate arrangement. There is mesodermally derived muscle tissue inside the epidermis, but no mesodermal tissue adjacent to the intestinal wall.*

Ascaris

Most roundworms are free living, but a few are parasitic. *Ascaris,* a large parasitic roundworm, is often studied as an example of this phylum.

Ascaris (fig. 27.15a) females (20 to 35 centimeters) tend to be larger than males, which have an incurved tail. Both sexes move by means of a characteristic whiplike motion because only longitudinal muscles lie next to the body wall.

Figure 27.16
Elephantiasis. An infection with a filarial worm has caused this individual to experience extreme swelling in regions where the worms have blocked the lymph vessels.

The internal organs (fig. 27.15b), including the tubular reproductive organs, lie within the pseudocoelom. Because mating produces embryos that mature in the soil, the parasite is limited to warmer environments. When larvae within their protective covering are swallowed, they escape and burrow through the intestinal wall. Making their way through the organs of the host, they move from the intestine to the liver, heart, and then the lungs. Within the lungs, molting takes place and after about 10 days, the larvae migrate up the windpipe to the throat, where they are swallowed, allowing them to once again reach the intestine. Then the mature worms mate and the female produces embryo-containing eggs that pass out with the feces. To complete this life cycle, as with other roundworms, feces must reach the mouth of the next host; therefore proper sanitation is the best means to prevent infection with *Ascaris* and other parasitic roundworms.

Other Parasites

Trichinosis is a serious infection of humans caused by *Trichinella.* Humans contract the disease when they eat rare pork containing encysted larvae. After maturation, the female adult burrows into the wall of the small intestine and produces living offspring that are carried by the bloodstream to the skeletal muscles, where they encyst. Since humans are not normally eaten by any other animals, these larvae never reach another host. However, the cycle can be completed if pigs eat infected pig meat or infected rats.

Elephantiasis is caused by a roundworm called the filarial worm, which utilizes the mosquito as a secondary host. Because the adult worms reside in lymph vessels, collection of fluid is impeded and the limbs of an infected human may swell to a monstrous size (fig. 27.16). When a mosquito bites an infected person, it transports larvae to new hosts.

Other roundworm infections are more common in the United States. Children frequently acquire a pinworm infection, and hookworm is seen in the southern states. *Hookworm,* judged by some to be the most important parasitic intestinal worm of humans, is discussed on page 680. An infection by this worm can be very debilitating because the worms feed on blood.

Roundworms possess *two advanced features:* a body cavity and a tube-within-a-tube body plan. The body cavity is a *pseudocoelom* rather than a true coelom. The *tube-within-a-tube plan,* in contrast to the sac plan, has both a mouth and an anus.

Higher Invertebrates

All of the higher invertebrate phyla have a true coelom (fig. 27.6c). Nevertheless, they can be divided into two groups (fig. 27.1) on the basis of embryological evidence. For example, marine mollusks and annelids share larvae of the **trochophore** type; while echinoderms and certain invertebrate chordates share larvae of the **dipleurula** type (fig. 27.17). A larva is an immature stage that is independent and can feed itself. Also, in the mollusks, annelids, and arthropods, the embryonic blastopore, the site of invagination of the endoderm germ layer in the embryo (fig. 27.18), becomes the mouth. In the echinoderms and chordates, the blastopore becomes the anus and the second opening becomes the mouth. Some authorities refer to the former as the **protostomes** and the latter as the **deuterostomes.**

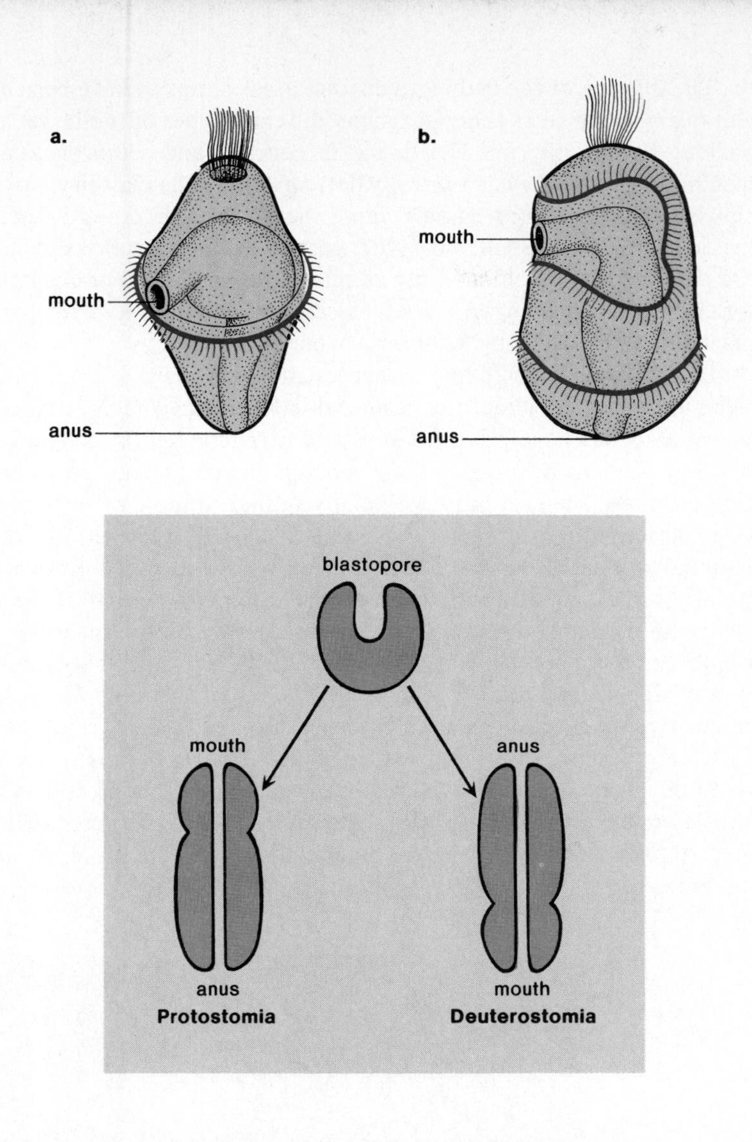

Figure 27.17
The a. *trochophore-type larva is characteristic of marine mollusks, annelids, and arthropods, while the* b. *dipleurula-type larva is characteristic of marine echinoderms and chordates.*

Figure 27.18
Fate of the embryonic blastopore. In the protostomes, the blastopore becomes the mouth. In the deuterostomes, the blastopore becomes the anus.

All higher invertebrates have a true coelom. They are divided into two groups on the basis of embryological evidence; mollusks, annelids, and arthropods are in one group while echinoderms and chordates are in another group.

Mollusks

Mollusks (phylum Mollusca) are a very large and diversified group containing many thousands of living and extinct forms. However, all forms of mollusks have a body composed of at least three distinct parts:

1. **Visceral mass:** the soft-bodied portion that contains internal organs.
2. **Foot:** a strong, muscular portion used for locomotion.
3. **Mantle:** a membranous or sometimes muscular covering that envelops but does not completely enclose the visceral mass. The mantle may secrete a shell.

In addition to these three regions, many mollusks show cephalization and have a head region with eyes and other sense organs.

The division of the body into distinct areas seems to have been a useful evolutionary advance as there are many different types of mollusks, adapted to various ways of life (fig. 27.19). Snails, conches, and nudibranchs are gastropod mollusks that have a ventrally flattened foot. The majority move about by muscular contractions passing along their foot. While nudibranchs, also called sea slugs, lack a shell, most other gastropods have a coiled shell in which the visceral mass spirals. Snails are adapted to life on land. For example, their mantle is richly supplied with blood vessels and functions as a lung when air is moved in and out through respiratory pores.

In cephalopods, including octopuses and squids (fig. 15.5), the foot has evolved into tentacles about the head. Aside from the tentacles that seize the prey, cephalopods have powerful jaws and a radula (toothy tongue) to tear prey apart. Cephalization aids these animals in recognizing prey and in escaping enemies. The eyes are superficially similar to those of vertebrates and have a lens and retina with photoreceptors. The brain is formed from a fusion of ganglia, and nerves leaving the brain supply various parts of the body. An especially large pair of nerves controls the rapid contraction of the mantle allowing these animals to move quickly by a jet propulsion of water. Rapid movement and the secretion of a brown or black pigment from an ink gland help cephalopods escape their enemies. Octopuses have no shell and squids have but a remnant of one concealed beneath the skin.

Clams, oysters, and scallops are all called bivalves because there are two parts to the shell. Because of the ready accessibility of clams, they are often studied as an example of this phylum. However, it should be noted that a clam is adapted to an inactive life whereas other mollusks, such as squid, are adapted to an active life (table 27.3).

Figure 27.19
Molluskan diversity. a. *A scallop with sensory tentacles extended between the valves. Humans eat only the single large muscle that holds the two halves of the shell together.* b. *An octopus moving over the surface of coral in the Pacific Ocean shows that it lacks a protective shell completely.* c. *While most mollusks are marine, snails are adapted to living on land and their mantle tissue is capable of gas exchange with air. During copulation each inserts a penis into the mantle cavity of its partner.*

a.

b.

c.

Clam

In a clam, such as *Anodonta,* the shell is secreted by the mantle and is composed of calcium carbonate with an inner layer of *mother-of-pearl.* If a foreign body is placed between the mantle and the shell, pearls form when concentric layers of shell are deposited about the particle.

Within the mantle cavity, the gills (fig. 27.20) hang down on either side of the visceral mass, which lies about the foot. **Gills** are vascularized, highly convoluted, thin-walled tissue specialized for gas exchange.

The heart of a clam lies just below the hump of the shell within the pericardial sac, the only remains of the coelom. Therefore, the coelom of the clam is said to be *reduced.* The heart pumps blue blood containing blue hemocyanin, instead of red hemoglobin, into vessels that lead to the various organs of the body. Within the organs, however, the blood flows through spaces, or *sinuses,* rather than vessels. Such a circulatory system is called an **open circulatory system** because the blood is not contained within blood vessels all the time. This type of circulatory system may be associated with an inactive animal because it is an inefficient means of transporting blood throughout the body. An active animal needs to have oxygen and nutrients transported quickly to rapidly working muscles, while an inactive animal is able to survive with a sluggish system for transporting these necessities.

Table 27.3 Comparison of Clam and Squid

	Clam	Squid
Food getting	Filter feeder	Active predator
Skeleton	Heavy shell for protection	No external skeleton
Circulation	Open	Closed
Cephalization	None	Marked
Locomotion	Hatchet foot	Jet propulsion
Nervous system	Three separate ganglia	Brain and nerves

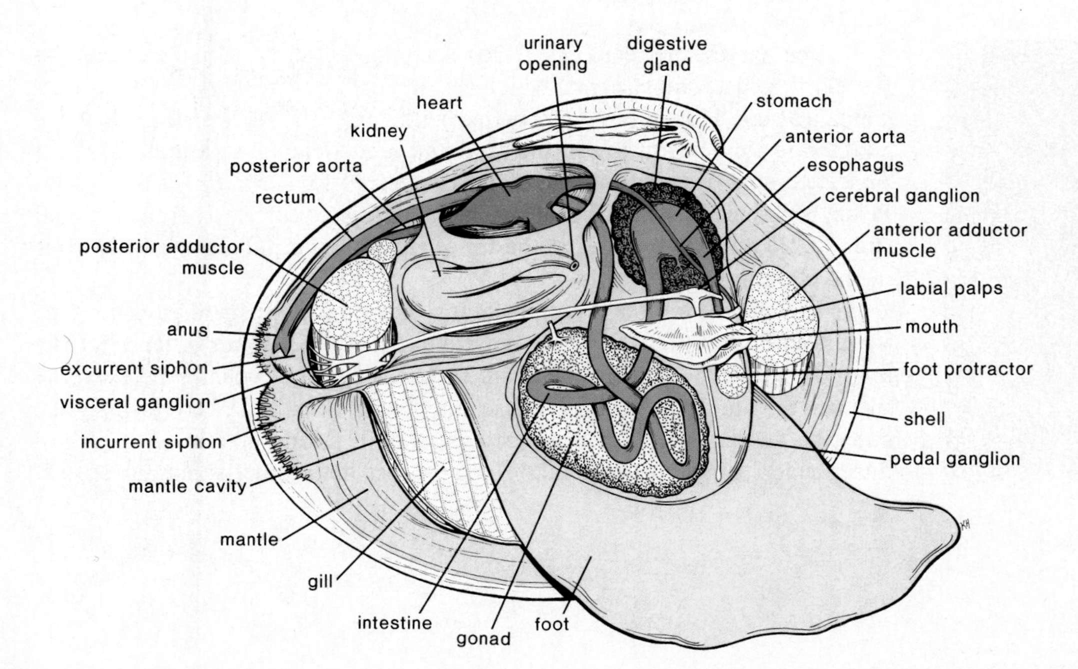

Figure 27.20
Clam anatomy. Trace the path of food from the incurrent siphon, past the gills, to the mouth, esophagus, stomach, intestine, anus, and excurrent siphon. Locate the three ganglia: pedal, cerebral, and visceral. The heart lies in the reduced coelom. Do clams have an open or closed circulatory system?

Importance of Higher Invertebrates and Vertebrates

These are the most familiar animals, chiefly because of their importance to our everyday lives. So-called shellfish, clams, oysters (mollusks) and shrimp (an arthropod), and bony fishes, particularly herring, cod, flounder and tuna, are important sources of protein in the diet. In many western countries, however, domesticated animals, particularly cattle and pigs, which are mammals, and chickens and turkeys, which are birds, supply most of the dietary protein.

Mammals, too,—cattle, horses, and water buffalo—have been exploited as draft animals and for transport. The dung is used as fertilizer and in some cultures as fuel and building plaster.

Before the advent of synthetic materials, wool from sheep, a mammal, and silk produced by silkworms, were more important. Silkworms are insects which have a life cycle that includes complete metamorphosis. The larva spins a cocoon having 1,000 feet of thread. In former days, too, humans made more use of feathers from birds, cowhide for leather goods, and reptilian skin to produce various clothing accessories. Synthetic fur is in use today but mammalian minks and rabbits are raised to supply natural fur.

Some animals transmit diseases to humans and we can only mention a few. For example, among insects mosquitoes transmit malaria, elephantiasis, and yellow fever; fleas carry plague from rats to humans, and the tsetse fly conveys African sleeping sickness. Snails, which are mollusks, are secondary hosts for flukes, and bats, which are mammals, may carry rabies.

In contrast many higher invertebrates and vertebrates have been useful in biological and medical research. Our knowledge of development has been advanced by the study of echinoderms, particularly sea urchin and amphibian embryos, notably frog embryos. Today mammalian rats are especially bred for many physiological experiments. Rats, molluskan octopi, and rhesus monkeys (primates) have contributed much to behavioral studies. It should never be said, "What use is this animal?" because one never knows how a particular animal might someday be useful to humans.* Adult sea urchin skeletons are now used as molds for the production of small artificial blood vessels and armadillos are used in leprosy research.

*These animals are, of course, important in ecosystems as discussed in chapters 31 and 32. Here we mention only their direct relationship to humans.

Figure 27.21
Giant clam with foot extended into a muddy substratum.

The nervous system (fig. 27.20) is composed of *three pairs of ganglia* (cerebral, pedal, and visceral) which are all connected by nerves. Clams lack cephalization. This may be associated with their way of life since they are adapted to slow burrowing in the sand and mud. They have a muscular foot that is compressed and bladelike and is referred to as a *hatchet* foot (class Pelecypoda means "hatchet foot"). The foot projects anteriorly from the shell (fig. 27.21), and by expanding the tip of the foot and pulling the body after it, the clam moves forward.

The clam is a filter feeder, meaning that it feeds on small particles that have been filtered from the water environment. Particles and water enter the mantle cavity by way of the *incurrent siphon,* a posterior opening between the two valves. Mucous secretions cause smaller particles to adhere to the gills, and cilia action sweeps them toward the mouth. Many inactive animals are filter feeders, since this method of feeding does not require rapid movement.

a.

b.

Figure 27.22
The sandworm a. *and the earthworm* b. *are annelids. The former is adapted to living in the sea and the latter is adapted to living on land.*

The digestive system (fig. 27.20) of the clam consists of a mouth, esophagus, stomach, and an intestine, which coils about in the visceral mass and then goes right through the heart before ending in a rectum and anus. The anus empties at an *excurrent siphon,* which lies just above the incurrent siphon. There is also an accessory organ of digestion called a digestive gland. It is readily seen in the clam that the tube-within-a-tube plan does lead to specialization of parts.

There are two excretory kidneys (fig. 27.20), which lie just below the heart and remove waste from the pericardial sac for excretion into the mantle cavity. The clam excretes ammonia (NH_3), a poisonous substance that requires the concomitant excretion of water. Land-dwelling animals tend to excrete a less toxic substance in a more concentrated form.

The male or female gonad (fig. 27.20) of a clam may be found about the coils of the intestine. While all clams have some type of larval stage, only marine clams have a trochophore larva. The presence of the trochophore larva (fig. 27.17a) among some mollusks indicates a relationship to the annelids, some of whose members also have this type of larval stage.

Mollusks are a very diverse group with a body plan always composed of three parts: the foot, mantle, and visceral mass. They possess all advanced features except segmentation.

Annelids

The primary characteristic of annelids (phylum Annelida) compared to the other groups studied is the presence of segmentation (fig. 27.22); obvious rings encircle the body, and the well-developed coelom is even partitioned by membranous septa. Both segmentation and an ample coelom prove to be important advances, facilitating the development of specialization of parts as seen in later phyla.

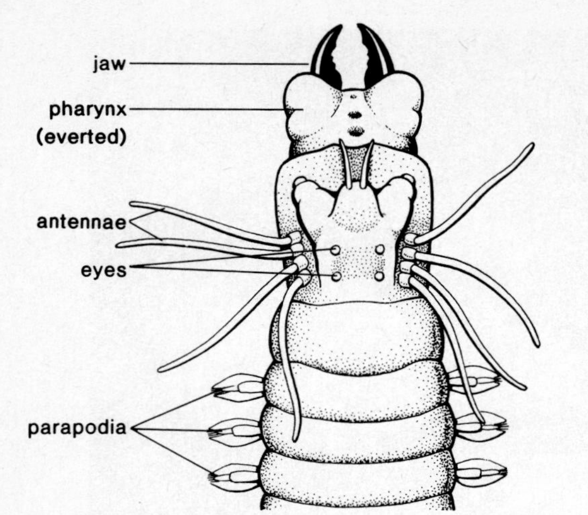

Figure 27.23
Head of Neanthes. *The worm shows cephalization in that there are antennae and eyes on a definite head region. The presence of well-developed jaws indicates that the worm is a predator.*

While we will use the earthworm as an example of this group of animals, marine worms such as sandworms *(Neanthes)* may be more representative. Sandworms are distinguished by the presence of a pair of fleshy lobes, the **parapodia,** on each body segment. These are used not only in swimming but also as respiratory organs where the expanded surface area allows for exchange of gases. Numerous chitinous bristles grow out from the parapodia and hence the name polychetes or "many bristled." These worms are predators. They prey on crustaceans and other small animals, which are captured by a pair of strong chitinous *jaws* that evert with a part of the pharynx when *Neanthes* is feeding. Associated with its way of life, *Neanthes* shows cephalization and has a head region with sense organs including eyes and antennae (fig. 27.23).

Earthworm

The earthworm (*Lumbricus*) (fig. 27.24) is terrestrial, but it is not well adapted to life on land because it is always in danger of drying out. Since the body wall and surrounding cuticle must be kept moist for gas exchange, the worm is protected by burrowing in moist soil and cannot venture forth on a dry, hot day without dire consequences.

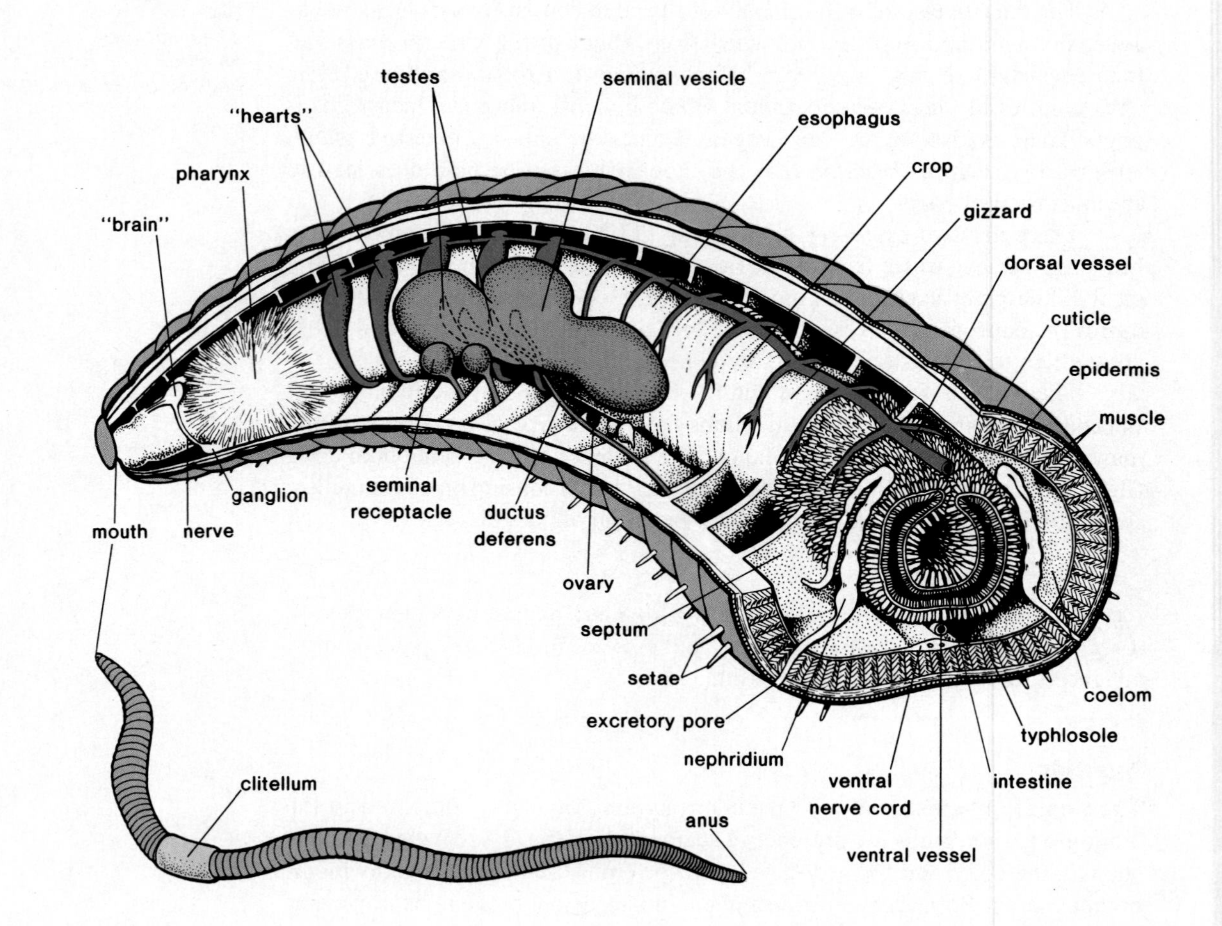

Figure 27.24
Earthworm anatomy. Drawing shows the internal anatomy of the anterior part of an earthworm's body. Segmentation is apparent in that there are pairs of setae for each segment; septa divide up the internal coelom; and there are paired nephridia and branch blood vessels in each segment.

The earthworm lacks obvious cephalization and feeds on leaves or any other organic matter, living or dead, that can conveniently be taken into its mouth along with dirt. Food drawn into the mouth by the action of the muscular pharynx is stored in a crop and ground up in a thick, muscular gizzard. Digestion and absorption occur in a long intestine whose dorsal surface is expanded by a **typhlosole** that allows additional surface for absorption. Notice that the tube-within-a-tube plan has allowed specialization of the digestive system to occur.

Locomotion in the earthworm is suitable to its way of life, and each segment of the body has four pairs of **setae,** or slender bristles. The setae are inserted into the dirt and then the body is pulled forward. Both a circular and longitudinal layer of muscle in the body wall make it possible for the worm to move and change its shape. Muscular contraction is aided by the fluid-filled coelomic compartments that act as a hydrostatic skeleton.

The nervous system (fig. 27.24) consists of an anterior, dorsal, ganglionic mass, or brain, and a long *ventral solid nerve cord* with ganglionic swellings and lateral nerves in each segment. When invertebrates are compared to vertebrates, it is often said that the former have a ventral solid nerve cord, while the latter have a dorsal hollow nerve cord.

The excretory system consists of paired **nephridia** (fig. 27.24), or coiled tubules, in each segment. Nephridia have two openings: one is a ciliated funnel that collects coelomic fluid and the other is an exit in the body wall. Between the two openings is a convoluted region where waste material is removed from the blood vessels about the tubule.

The earthworm has an extensive *closed circulatory* system. Red blood moves anteriorly in a dorsal blood vessel and then is pumped by five pairs of hearts into a ventral vessel. As the ventral vessel takes the blood toward the posterior regions of the worm's body, it gives off branches in every segment.

The worms are *hermaphroditic,* with a complete set of organs for both sexes. The male organs of an earthworm are the testes, seminal vesicles, and sperm ducts; the female organs are the ovaries, oviducts, and seminal receptacles. Copulation occurs when two worms come to lie ventral surface to ventral surface, with the heads pointing in opposite directions. The **clitellum,** a smooth girdle about the worm's body, secretes mucus, which holds the worms together and provides moisture in which the sperm swim from one body to the other. Then the mucus becomes a cocoon from which each worm backs out, releasing both eggs and sperm as they leave. Fertilization results in zygotes (fertilized eggs), which develop directly into miniature earthworms. There is no larval stage.

The annelids show the most obvious segmentation of any phylum of animals. Table 27.4 lists structures that have repeating units illustrating segmentation in earthworms.

Table 27.4 Segmentation in the Earthworm

1. Body rings
2. Coelom divided by septa
3. Setae on each segment
4. Ganglia and lateral nerves in each segment
5. Nephridia in each segment
6. Branch blood vessels in each segment

Annelids are segmented worms and most organ systems show evidence of segmentation. These worms have a well-developed coelom divided by septa, a closed circulatory system, and a ventral solid nerve cord.

Arthropods

The arthropods (phylum Arthropoda) have more species (900,000) than any other group of animals and are often said to be the most successful of all the animals. The phylum includes animals adapted to living in water, such as crayfish, lobsters, and shrimp; and animals adapted to living on land, such as spiders, insects, centipedes, and millipedes (fig. 27.25).

Arthropods have an external skeleton containing **chitin,** a strong, flexible polysaccharide. The skeleton serves many functions such as protection, attachment for muscles, and prevention of desiccation on land. The appendages are also covered by the skeletal material, but they are jointed. The presence of **jointed appendages** is a great advance in the animal kingdom and aids locomotion on land.

An external skeleton is not without difficulties and, since this particular skeleton does not grow larger, arthropods **molt,** or shed the skeleton periodically.

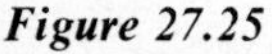

Figure 27.25
Arthropod diversity. a. *Cleaner shrimp. As their name implies these shrimp make a living by removing debris and parasites from other sea animals, particularly fish that line up at cleaning stations.* b. *American lobster. Lobsters typically lie in wait and then spring forward to capture prey by means of well-developed claws.* c. *Giant centipede. Segmentation is obvious in this organism that has a set of appendages on every segment.* d. *Dragonflies mating. Dragonflies begin life at the bottom of a pond and after a series of nymph stages they metamorphose into adults that live only a short time.* e. *Walking stick. Looking like sticks helps these animals survive as they walk about on limbs of trees and bushes.* f. *Garden spider. Spiders spin beautiful webs that vibrate when touched by prey, alerting the spider to its prospective meal.*

b.

c.

a.

Specialization of parts is readily seen in that the arthropod body is not composed of a series of like segments but rather, due to a fusion of segments, is composed of three parts—head, thorax, and abdomen. The head shows good cephalization with sense organs. The sense organs include **antennae** (or feelers) and eyes. The eyes are of two types: **compound** and **simple.** The compound eye is not seen in any other phylum. It is composed of many complete visual units grouped together in a composite structure: each visual unit contains a separate lens and a light-sensitive cell. In the simple eyes, a single lens covers many light-sensitive cells.

The coelom, so well developed in the annelids, is *reduced* in the arthropods and composed chiefly of the space about the reproductive system. Instead of a coelomic cavity, there is a **hemocoel,** or blood cavity, consisting of vessels and open spaces where the blood flows about the organs. The dorsal heart keeps the blood moving around in the sinuses. Arthropods, like most mollusks, have an open circulatory system.

d.

e.

f.

Figure 27.26
Anatomy of the crayfish.
a. *Externally, it is possible to observe the jointed appendages, including the swimmerets, walking legs, and claws. These appendages, plus a portion of the carapace, have been removed from the right side so that the gills are visible.*
b. *Internally, the parts of the digestive system are particularly visible. The circulatory system can be clearly seen. Note also the ventral solid nerve cord.*

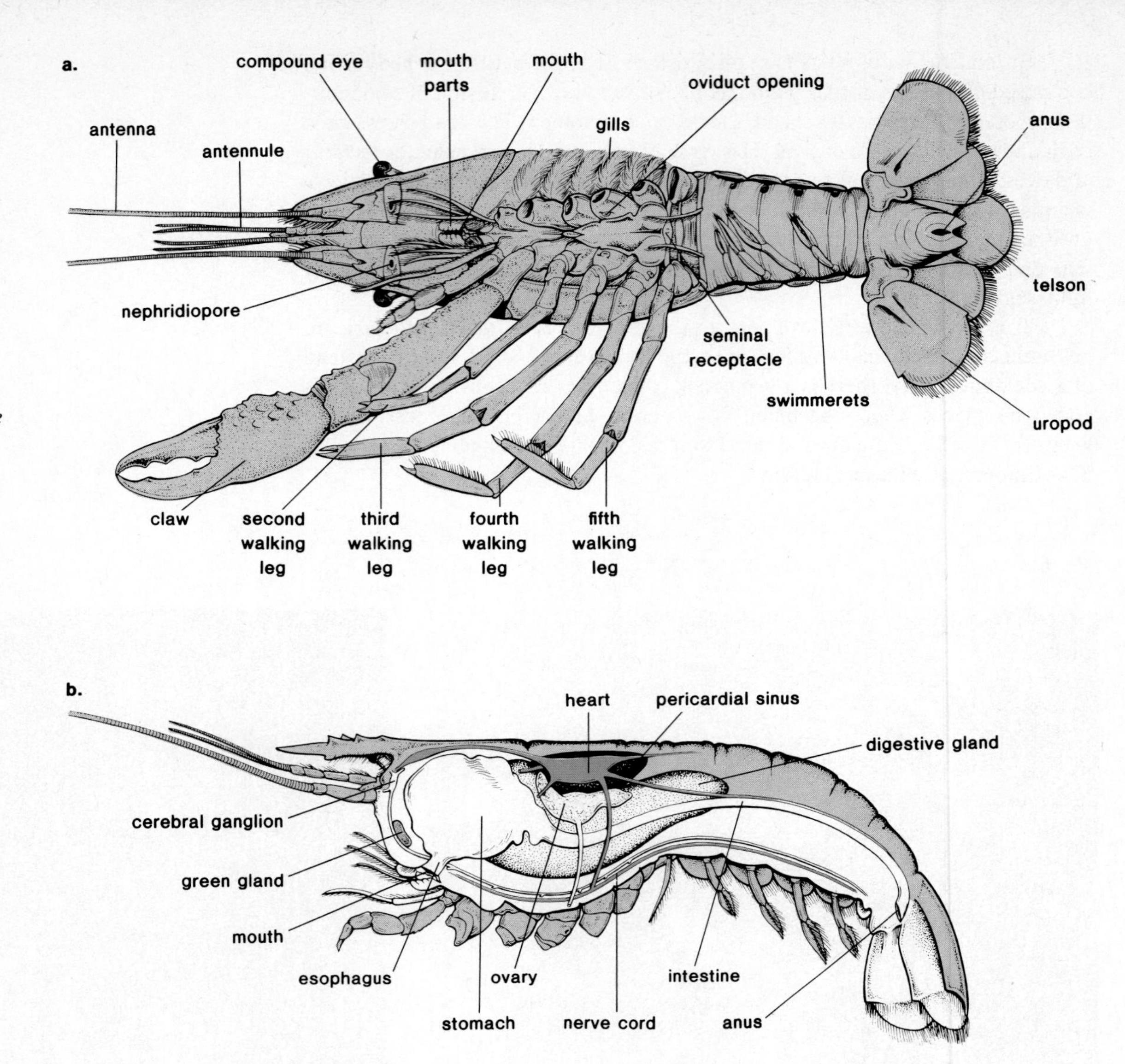

Crayfish

Crayfish are in the class Crustacea along with lobsters, shrimps, copepods, and crabs. Figure 27.26a gives a view of the external anatomy of the crayfish, and it can be seen that the head and thorax are fused into a **cephalothorax,** which is covered on the top and sides by a nonsegmented **carapace.** The abdominal segments, however, are clearly marked off.

On the head are a pair of stalked compound eyes and two pairs of antennae. Chitinous jaws and mouthparts are also present. The appendages in the thorax include accessory mouthparts, *pinching claws,* and four pairs of *walking legs;* the abdominal segments are equipped with *swimmerets,* small paddlelike structures. The first pair of swimmerets in the male are quite strong and are used to pass sperm to the female, reminiscent of the squid, which passes sperm by means of specialized tentacles. The last two segments bear the *uropods* and the *telson,* which make up a fan-shaped tail used for swimming backwards.

Ordinarily, the crayfish lies in wait for prey. It faces out from an enclosed spot with the claws extended and the antennae moving about. If a small animal, dead or alive, happens by, it is quickly seized and carried to the mouth. When a crayfish does move about, it generally crawls slowly but may swim rapidly backwards by using heavy abdominal muscles.

Table 27.5 *Comparison of Crayfish and Grasshopper*

	Crayfish	Grasshopper
Locomotion	Legs and swimmerets	Hopping legs and wings
Respiration	Gills	Tracheae
Excretion	Liquid waste by way of green gland	Solid waste by way of Malpighian tubules
Circulation	Blue blood	Colorless blood
Nervous system	Cephalization	Cephalization with tympanum
Reproduction	Modified swimmerets in male	Penis in male, ovipositor in female

Respiration is by means of *gills* (fig. 27.26a), which lie above the walking legs protected by the carapace. Gills, as we have seen, are typical organs of respiration in water-dwelling animals. The crayfish has blue blood containing the pigment hemocyanin, which aids in the transport of oxygen.

Internally, the digestive system (fig. 27.26b) includes a stomach, which is divided into two main regions: an anterior portion called the *gastric mill,* equipped with chitinous teeth to grind coarse food, and a posterior region, which acts as a filter to sort out food according to consistency.

The nervous system (fig. 27.26b) is quite similar to that of the earthworm. There are anterior ganglia from which a solid *ventral nerve cord* passes posteriorly. Along the length of the nerve cord, periodic ganglia give off lateral nerves.

The excretory system (fig. 27.26b) consists of a pair of *green glands* lying in the head region anterior to the esophagus. Each organ possesses a glandular region for waste removal: a bladder and a duct that opens ventrally at the base of the antennae.

The sexes are separate in the crayfish. The white testes of the male are located just ventral to the pericardial sinus. From each side a coiled ductus deferens passes ventrally and opens to the outside at the base of the fifth walking leg. Sperm transfer is accomplished by the modified first two swimmerets of the abdomen. In the female, the ovaries are located in a position similar to that occupied by the testes and the oviducts pass ventrally, opening near the bases of the third pair of walking legs. There is a cuticular fold between the bases of the fourth and fifth pair that serves as a seminal receptacle.

Table 27.5 compares the crayfish to the grasshopper to illustrate how one is adapted to the water and the other to the land.

Grasshopper

Insects (class Insecta) comprise one of the largest animal groups both in number of species and in number of individuals, perhaps because of the presence of *wings.* Wings enhance the insects' ability to survive by providing a new way of escaping enemies, finding food, facilitating mating, and dispersing the species. Figure 27.25d and e gives representative examples of insects, of which we will study the grasshopper in detail.

Every system of the grasshopper *(Romalea)* (fig. 27.27a) is adapted to life on land. There are *three pairs* of legs and one of these pairs is suited to jumping. There are two pairs of wings; the forewings are tough and leathery and when folded back at rest they protect the broad, thin hindwings. The first abdominal segment bears on its lateral surface a large **tympanum** for the reception of sound waves. The posterior region of the exoskeleton in the female has two pairs of projections that form an ovipositor.

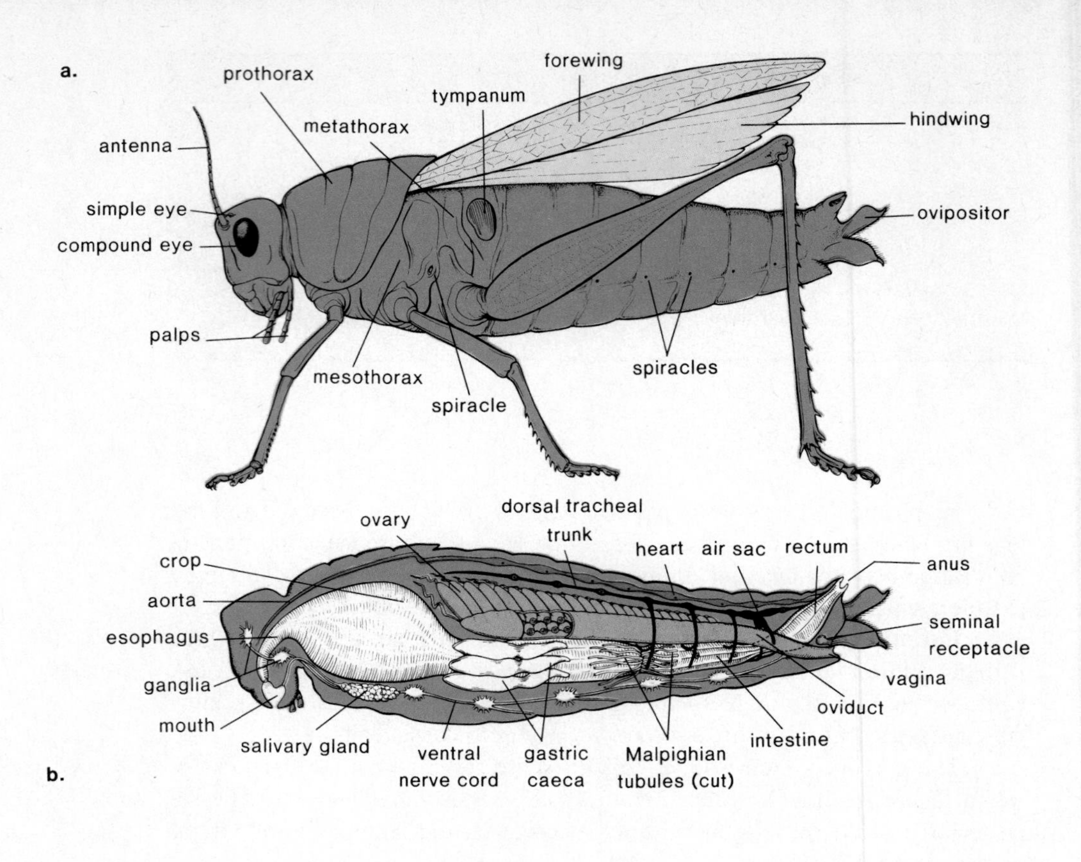

Figure 27.27
Anatomy of the grasshopper.
a. *Externally, it is possible to observe that each side has three legs and a pair of wings, as is typical of most insects. This specimen is a female, as witnessed by the ovipositor for depositing eggs in the soil.* b. *The ovary contains rows of eggs.*

The digestive system (fig. 27.27b) is suitable for a grass diet. The mouth has mouthparts to grind the food and salivary secretions contain enzymes. Food is temporarily stored in the crop before passing into the gizzard, where it is finely ground. Digestion is completed in the stomach with the aid of enzymes secreted by the gastric caeca.

Excretion is carried out by means of **Malpighian tubules** (fig. 27.27b), which extend out into the hemocoel and empty into the digestive tract. A solid nitrogenous waste is excreted, conserving water.

Respiration occurs when air enters small tubules called **tracheae** (fig. 27.27b) by way of openings in the exoskeleton called **spiracles.** The tracheae branch and rebranch, finally ending in moist areas where the actual exchange of gases takes place. The movement of air through this complex of tubules is not a passive process; air is pumped through by a series of several bladderlike structures (air sacs), which are attached to the tracheae near the spiracles. Air enters the anterior four spiracles and exits by the posterior six spiracles. This mechanism of breathing found in insects and arachnids (e.g., spiders and scorpions) is an adaptation to land that requires a drastic modification of the body. It may even account for the small size of insects since the tracheae are so tiny and fragile that they would be crushed by any amount of weight.

The heart is a slender, tubular organ that lies against the dorsal wall of the abdominal exoskeleton. Blood passes into a hemocoel where it circulates before finally returning to the heart again. The blood is colorless and lacks a respiratory pigment since the tracheal system transports gases.

Reproduction is adapted to life on land. The male has two testes and associated ducts that end in the penis. The female has ovaries that occupy the whole dorsal part of the animal, and oviducts that end in the vagina. The sperm received during copulation are stored in the seminal receptacles for future use. Fertilization is internal, usually occurring during late summer or early fall. The female deposits the fertilized eggs in the ground with the aid of her ovipositor.

Figure 27.28
Representative echinoderms. a. *Starfish about to feed on a clam, which it can open by utilizing the suction provided by tube feet.* b. *A sea cucumber whose general overall shape resembles the vegetable of this name.* c. *Sea urchins demonstrating their many external spines.*

Insects are land animals that often have worm-like larval stages and undergo metamorphosis. **Metamorphosis** means a change, usually a drastic one, in form and shape. Some insects undergo what is called complete metamorphosis, in which case they have three stages of development: *larval stage,* the *pupa stage,* and finally the *adult stage.* Metamorphosis occurs during the pupa stage when the animal is enclosed within a hard covering. The animal that is best known for metamorphosis is the butterfly, whose larval stage is called a caterpillar and whose pupa stage is the cocoon; the adult is the butterfly. Grasshoppers undergo incomplete metamorphosis, which is a gradual change in form rather than a drastic change. The immature stages of the grasshopper are called nymphs rather than larvae and they are recognizable as grasshoppers even though they differ somewhat in shape and form. Metamorphosis is controlled by hormones.

Arthropods are the most numerous and varied of all the animal phyla. They have an external skeleton and jointed appendages. Segmentation has led to specialization of parts within the various organ systems.

Echinoderms

The echinoderms (fig. 27.28) include only marine animals—starfish, sea urchin, sea cucumber, feather star, sea lily, and sand dollar. The most familiar of these is the starfish, and we will study this representative. Echinoderms are *radially symmetrical* as adults, with a body plan based on *five parts.* Their other unique feature is the **water vascular system,** which is used as a means of locomotion. They also have a calcareous **endoskeleton,** whose projecting spines give the phylum its name, Echinodermata, meaning "spiny skin."

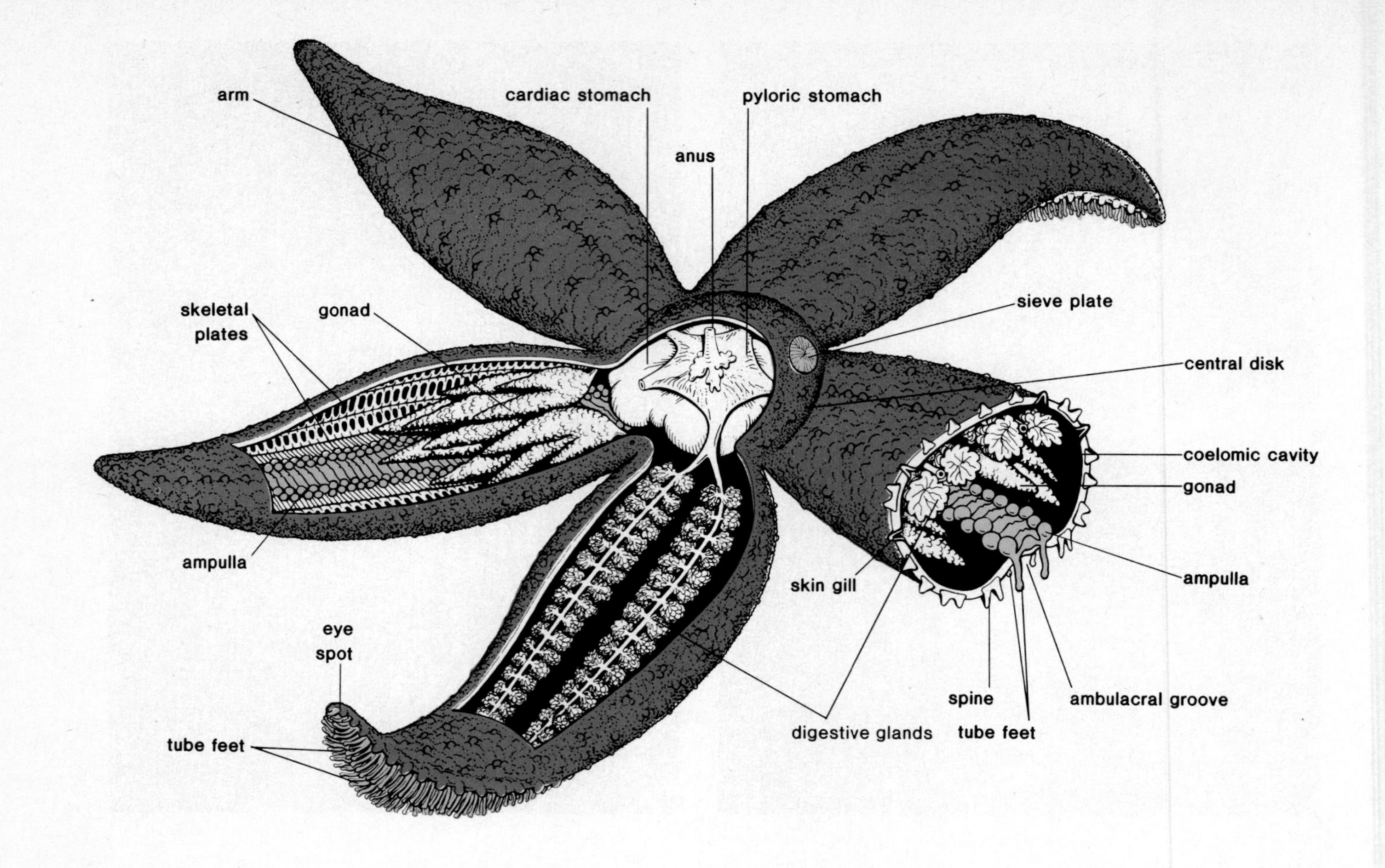

Figure 27.29
Starfish anatomy. Like other echinoderms, starfishes have a water vascular system shown here in color. Water enters the sieve plate and is eventually sent into tube feet by the action of the ampullae. Each arm of a starfish contains digestive glands, gonads, and water vascular system.

Starfish

The starfish *(Asterias)* sometimes called the sea star, is commonly found along rocky coasts. It has a five-rayed body plan with an *oral* (mouth) and *aboral* (anus) side (fig. 27.29). The oral side is actually the underside and the aboral side is the upper side. On the aboral side there are various structures that project through the epidermis: (1) spinelike projections of the endoskeletal plates; (2) pincerlike structures called *pedicellarie,* which keep the surface free of small particles; and (3) skin gills, which serve for respiratory exchange. The mouth is located on the oral surface, where each of the five arms has a groove lined by little **tube feet.**

Starfish feed on mollusks. When a starfish attacks a clam, it arches its body over the shell and by the concerted action of the tube feet forces the clam to open. Then it everts a portion of its stomach to digest the contents of the clam.

The mouth of a starfish opens into a narrow esophagus, which in turn leads to an expanded stomach. The stomach has two portions: the saclike cardiac, which can be everted as described, and the narrower pyloric, which is connected to a short intestine. The anus opens on the aboral or upper side of the animal.

Each of the five arms contains a well-developed coelom, a pair of large **hepatic caeca** that secrete powerful enzymes into the pyloric portion of the stomach, and gonads, which open on the aboral surface by very small pores; the nervous system consists of a central nerve ring that supplies radial nerves to each arm. At the tip of each arm is a light-sensitive eyespot.

Coelomic fluid, circulated by ciliary action, performs many of the normal functions of a circulatory system; the water vascular system is purely for locomotion. Water enters this system through a structure on the aboral side called the **sieve plate,** or madreporite. From there it passes through a short canal, called the *stone canal,* to a *ring canal,* which surrounds the mouth. From the ring canal, five *radial canals* extend into the arms along the ambulacral grooves. From the radial canals many lateral canals extend into the tube feet. One canal goes to each tube foot, where it ends in the *ampulla.* When the ampulla contracts, the water is forced into the tube foot, expanding it and giving it suction. By alternating the expansion and contraction of the tube feet, the starfish moves along slowly.

The adult starfish is anatomically unique, but it is believed that echinoderms and chordates share a common ancestor because of similar embryonic development. The dipleurula-type larva is seen in both the echinoderms and the hemichordates (see following). Therefore, the echinoderms are bilaterally symmetrical as embryos; radial symmetry is found only in the adult.

Echinoderms are radially symmetrical with a spiny skin. They move by tube feet which are part of their water vascular system. Their other body systems are rather primitive.

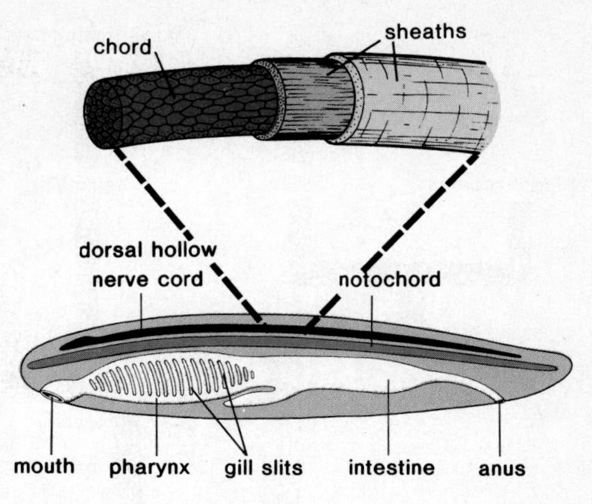

Figure 27.30
An enlargement of the notochord (above) *and a diagram of an idealized chordate* (below). *All chordates at some time in their life history have a dorsal hollow nerve cord, pharyngeal gill pouches, and a notochord. The notochord is a supporting rod covered by an inner fibrous sheath and an outer elastic sheath.*

Chordates

Among the chordates (phylum Chordata) are those animals with which we are most familiar, including human beings themselves. All members of this phylum are observed to have three basic characteristics (fig. 27.30) at some time in their life history:

1. A dorsal supporting rod called a **notochord,** which is replaced by the vertebral column in the adult vertebrates.
2. A *dorsal hollow nerve cord* in contrast to invertebrates, which have a ventral solid nerve cord. By hollow, it is meant that the cord contains a canal that is filled with fluid.
3. *Gill pouches* or *slits,* which may be seen only during embryological development in most vertebrate groups but which persist in adult fish. Water passing into the mouth and pharynx goes through the gill slits, which are supported by gill bars.

Protochordates

Two groups of animals are sometimes called the **protochordates** because they possess all three typical structures in either the larval and/or adult forms as did the first chordates to evolve. They are also called *invertebrate chordates* because they are not vertebrates. These two groups of animals link the vertebrates to the rest of the invertebrates and show how modestly the chordates most likely began.

A **tunicate,** (subphylum Urochordata) or sea squirt (fig. 27.31a), appears to be a thick-walled, squat sac with two openings, an incurrent siphon and an excurrent siphon. Inside the central cavity of the animal and opening into a chamber that opens to the outside are numerous gill slits, the only chordate feature retained by the adult. The larva of the tunicate, however, has a tadpole shape and possesses the three chordate characteristics. It has been suggested that such a larva may have become sexually mature without developing the other adult tunicate characteristics. If so, it may have evolved into a fishlike vertebrate.

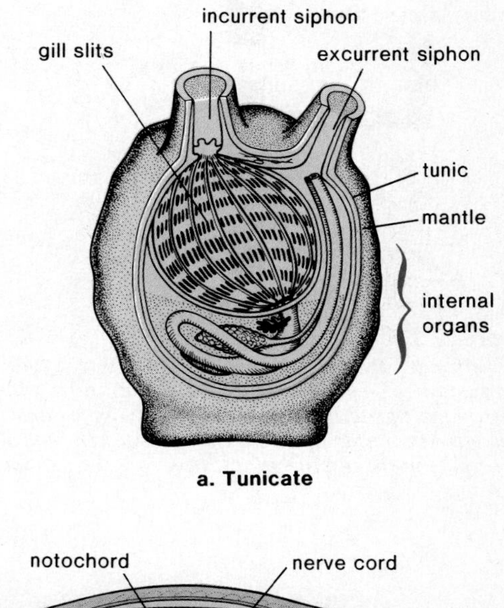

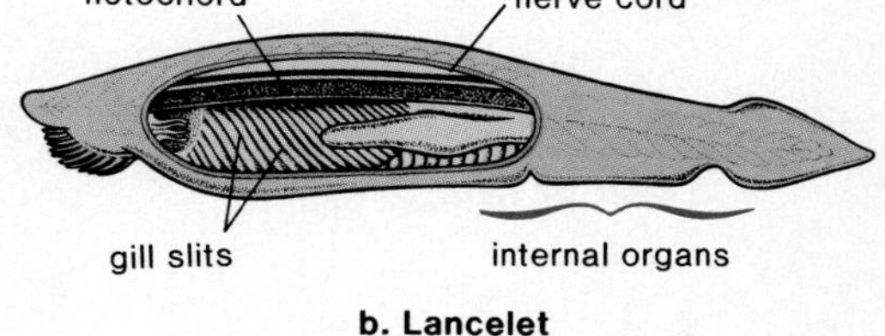

Figure 27.31
Protochordates. a. *Tunicate anatomy. Gill slits are the only chordate feature retained by the adult.* b. *Lancelet anatomy. This animal retains all three chordate characteristics as an adult.*

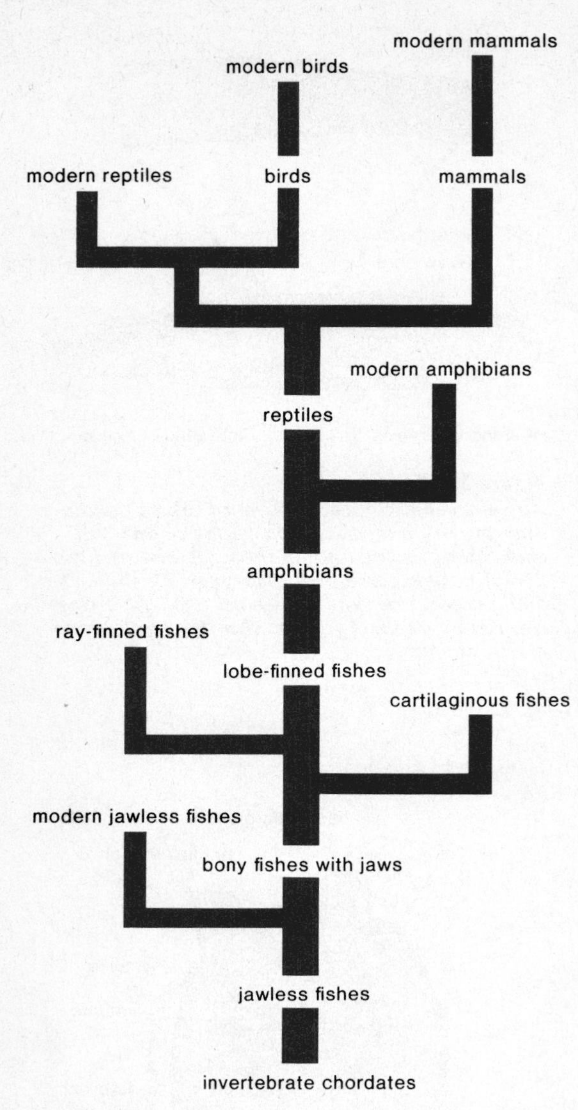

Figure 27.32
Evolutionary tree of vertebrates. Vertebrates evolved in an aquatic environment, but adaptation to a land environment gradually improved as later groups evolved. Mammals and birds possess all the features required for successful adaptation to a terrestrial existence.

A **lancelet** (subphylum Cephalochordata) (fig. 27.31b) is a chordate that shows the *three chordate characteristics as an adult.* In addition, segmentation is present as witnessed by the fact that the muscles are segmentally arranged and the nerve cord gives off periodic branches.

The invertebrate chordates include the tunicates and lancelets. Lancelets are the best example of a chordate that possesses the three chordate characteristics as an adult.

Vertebrate Chordates

Vertebrates (subphylum Vertebrata) have all the advanced characteristics listed in table 27.1. They are segmented chordates in which the notochord is replaced in the adult by a *vertebral column* composed of individual vertebrae. The skeleton is internal, and in all the vertebrates there is not only a backbone but also a skull, or cranium, to enclose and protect the brain. In higher vertebrates, other parts of the skeleton serve as attachment for muscles and for protection of internal organs of the chest and abdomen. All vertebrates have a *closed circulatory system* in which red blood is contained entirely within blood vessels. They show good cephalization with sense organs; the eyes develop as outgrowths of the brain; and the ears serve as equilibrium devices in aquatic vertebrates plus sound wave receivers in land vertebrates. The kidneys are important excretory and water-regulating organs that conserve or rid the body of water as appropriate.

Two comparisons are often made between invertebrates and vertebrates: (1) invertebrates have a *ventral solid* nerve cord, while vertebrates have a dorsal hollow nerve cord called the spinal cord; and (2) invertebrates have an *external* skeleton, while vertebrates have an *internal* skeleton. There are, however, many exceptions among the invertebrates.

As a group, the vertebrates are the dominant animals in the world today. They are found in every habitat from the ocean floor to the mountaintop and in the forest and desert. Included in the group are the three classes of fishes and one class each of amphibian, reptile, bird, and mammal. Figure 27.32 represents an evolutionary tree of the vertebrates and you can see that it is possible to trace the evolution of the vertebrates from fishes to amphibians, to reptiles, to both birds and mammals. All but the fishes are **tetrapods,** meaning that they have four limbs.

Vertebrates are animals in which the vertebral column has replaced the notochord. Vertebrates are often compared to the invertebrates, which lack a vertebral column.

Fishes

There are three classes of fishes: the jawless fishes, the cartilaginous fishes, and the bony fishes. Living representatives of the *jawless fishes* are cylindrical, up to a meter long, with smooth, scaleless skin and no jaws or paired fins. There are two families of jawless fishes: *hagfishes* and *lampreys.* The hagfishes are scavengers, feeding mainly on dead fish, while some lampreys are parasitic. When parasitic, the round mouth of the lamprey serves as a sucker by which it attaches itself to another fish and taps into its circulatory system.

a.

b.

Figure 27.33
Representative cartilaginous and bony fish. a. *The bull shark is a cartilaginous fish.* b. *The freckeled grouper is a bony fish.*

Cartilaginous fishes (fig. 27.33a) are the sharks, rays, and skates, which have skeletons of cartilage instead of bone. The dogfish shark is a small shark often dissected in biology laboratories to show the main features of the vertebrate body. Other sharks are well known to us as vicious predators that attack human swimmers. One of the most dangerous sharks inhabiting both tropical and temperate waters is the hammerhead. The largest of the sharks, the whale sharks, feed on small fishes and marine invertebrates and do not attack humans. Skates and rays are rather flat fishes that live partly buried in the sand and feed on mussels and clams.

Bony fishes (fig. 27.33b) are by far the most numerous and varied of the fishes. Most of the fishes we eat, such as perch, trout, flounder, and haddock, are a type of bony fish called *ray-finned fishes.* These fishes have a *swim bladder* that aids them in changing their depth in the water. By secreting gases into the bladder or by absorbing gases from it, these fishes can change their density and thus go up or down in the water. "Ray-finned" refers to the fact that the fins are thin and supported by bony rays. Another type of bony fish called the *lobe-finned* fish evolved into the amphibians. These fishes not only have fleshy appendages that could be adapted to land locomotion, they also have a lung[1] that is used for respiration. A type of lobe-finned fish called a coelacanth, which exists today, is the only "living fossil" among the fishes.

Fishes are adapted to life in the water. Their streamlined shape, fins, and muscle action are all quite suitable to locomotion in the water. Their bodies are covered by *scales,* which protect the body but do not prevent water loss. Fishes breathe by means of *gills,* respiratory organs that are kept continuously moist by the passage of water through the mouth and out the gill slits. As the water passes over the gills, oxygen is absorbed by the blood and carbon dioxide is given off. The heart of a fish is a simple pump, and the blood flows through the chambers, including a nondivided atrium and ventricle, to the gills only (fig. 27.39a). Oxygenated blood leaves the gills and goes to the body proper.

[1]Actually, the swim bladder of modern-day bony fishes is believed to be derived from an ancient lung.

a.

b.

Figure 27.34
Representative amphibian and reptile. a. *This male bullfrog is an amphibian. Note the thin, moist skin, an indication that adult amphibians are not well adapted to a land existence.* b. *This rough green snake is a reptile. Note the scaly, dry skin, an indication that a snake is well adapted to a land existence.*

Generally speaking, reproduction in the fishes requires external water; sperm and eggs are usually shed into the water where fertilization occurs, and the zygote develops into a swimming larva that can fend for itself until it develops into the adult form.[2]

The most primitive fishes are jawless. Sharks, skates, and rays are cartilaginous, while the bony fishes include all other well-known fishes.

Amphibians

The living amphibians include *frogs, toads, newts,* and *salamanders* (fig. 27.34a). These animals have distinct walking legs, each with five (or fewer) toes. This represents an adaptation to land locomotion. Respiration is accomplished by the use of small, relatively *inefficient lungs,* supplemented by gaseous exchange through the skin. Therefore, the skin is smooth, moist, and glandular. This is a distinct disadvantage on land because of the danger of drying out; therefore, frogs spend most of their time in or near the water. All amphibians possess two nostrils that, unlike those of most fish, are connected directly with the mouth cavity. Air enters the mouth by way of the nostrils and when the floor of the mouth is raised, air is forced into the lungs. Associated with the development of lungs there is a change in the circulatory system. The amphibian heart has a divided atrium but a single ventricle (fig. 27.39b). The right atrium receives impure blood with little oxygen from the body proper and the left atrium receives purified blood from the lungs that has just been oxygenated, but these two types of blood are partially mixed in the single ventricle. Mixed blood is then sent, in part, to the skin where further oxygenation may occur.

Nearly all the members of this class lead an amphibious life—that is, the larval stage lives in the water and the adult lives on the land. The adults must return to the water, however, for the purpose of reproduction. Just as

[2]Some fish, such as sharks, practice internal fertilization and retain their eggs during development. Their young are born alive.

with the fish, the sperm and eggs are discharged into the water and fertilization results in a zygote that develops into the familiar tadpole. The tadpole undergoes metamorphosis into the adult before taking up life on the land.

Amphibians are not fully adapted to life on land. The appendages are not sturdy, the moist skin is a constant threat, and external water is required for reproduction.

Metamorphosis allows amphibians to switch from an aquatic to a land existence. There they are restricted to a moist environment because they use their skin for respiration. Also, they are not well adapted to life on land because they must return to the water for reproduction.

Reptiles

The reptiles living today are the *turtles, alligators,* and *snakes* and *lizards* (fig. 27.34b). Reptiles with limbs, such as lizards, are able to lift their bodies off the ground and the body is covered with hard, *horny scales* that protect the animal from desiccation and from predators. Both of these features are adaptations to life on land.

Reptiles have well-developed lungs with a *rib cage* to protect them. When the rib cage expands, the lungs expand and air rushes in. The creation of a partial vacuum establishes a negative pressure that causes air to rush into the lungs. The atrium of the heart is always separated into right and left chambers, but division of the ventricle varies. There is always at least one interventricular septum but it is incomplete in all but the crocodiles, therefore permitting exchange of oxygenated and deoxygenated blood between the ventricles in all but the latter.

Perhaps the most outstanding adaptation of the reptiles is the fact that they have a means of reproduction suitable to the land. There is usually no need for external water to accomplish fertilization because the penis of the male passes sperm directly to the female. After *internal fertilization* has occurred, the egg is covered by a protective leathery shell and laid in an appropriate location.

The *shelled egg* made development on land possible and eliminated the need for a swimming larva stage during development. It provides the developing embryo with oxygen, food, and water; removes nitrogen wastes; and protects it from drying out and from mechanical injury. This is accomplished by the presence of *extraembryonic membranes* (fig. 20.12).

The reptiles are fully adapted to life on land except for one limitation: they cannot regulate their body temperature. Sometimes animals that cannot maintain a constant temperature, that is, fish, amphibians, and reptiles, are called *cold-blooded.* Actually, however, they take on the temperature of the external environment. If it is cold externally, they are cold internally; and if it is hot externally, they are hot internally. Reptiles try to regulate body temperatures by exposing themselves to the sun if they need warmth or by hiding in the shadows if they need cooling off. This works reasonably well in most areas of the world.

Reptiles are well adapted to a land environment. They have a scaly skin that prevents loss of water and they can reproduce on land because they lay a shelled egg.

a.

b.

c.

Figure 27.35
Bird diversity. a. *A European Bee Eater prepares to swallow a bumblebee.* b. *A Chinstrap Penguin feeding chicks. Penguins are adapted to arctic conditions.* c. *Flamingos are exotic birds of the tropics.*

Birds

Birds (fig. 27.35) are characterized by the presence of feathers, which are actually modified reptilian scales. There are many orders of birds including birds that are flightless (ostrich), web-footed (penguin), divers (loons), fish eaters (pelicans), waders (flamingos), broad-billed (ducks), birds of prey (hawks), vegetarians (fowl), shore birds (sandpipers), nocturnal (owl), small (hummingbirds), and songbirds, the most familiar of the birds.

Nearly every anatomical feature of a bird can be related to its *ability to fly.* The anterior pair of appendages (wings) has become adapted for flight; the posterior is variously modified, depending on the type of bird. Some are adapted to swimming, some to running, and some to perching on limbs. The breastbone is enormous and has a ridge to which the flight muscles are attached. Respiration is efficient since the lobular lungs form *air sacs* throughout the body, including the bones. The presence of these sacs means that the air circulates one way through the lungs during both inspiration and expiration so that "used" air is not trapped in the lungs. Another benefit of air sacs is that the air-filled, hollow bones lighten the body and aid flying. Birds have a four-chambered heart (fig. 27.39c) that completely separates oxygenated from deoxygenated blood.

Birds have well-developed brains, but the enlarged portion seems to be the area responsible for instinctive behavior. Therefore, birds follow very definite patterns of migration and nesting.

Birds are fully adapted to life on land and are *warm-blooded* and, like mammals, are able to maintain a constant internal temperature. This may be associated with their efficient nervous, respiratory, and circulatory systems. Also, the feathers provide insulation.

Birds can maintain a constant internal temperature. All organ systems are adapted to allow birds to fly.

Mammals

The chief characteristics of mammals are the presence of *hair* and *mammary glands* that produce milk to nourish the young. Human mammary glands are called breasts.

Mammals are completely adapted to life on land and have limbs that allow them to move rapidly. In fact, an evaluation of mammalian features leads us to the obvious conclusion that they lead active lives. The brain is well developed; the lungs are expanded not only by the action of the rib cage but also by the contraction of the *diaphragm,* a horizontal muscle that divides the chest cavity from the abdominal cavity; and the heart is *four-chambered* (fig. 27.39c). The internal temperature is constant and hair, when abundant, helps insulate the body.

The mammalian brain is enlarged due to the expansion of the foremost part—the cerebral hemispheres. These have become convoluted and expanded to such a degree that they hide many other parts of the brain from view.

Mammals are classified according to their means of reproduction: there are *egg-laying* mammals, mammals with *pouches* for immature embryos, and **placental mammals.**

Monotremes These are the egg-laying mammals represented by the duck-billed platypus and spiny anteaters (fig. 27.36a). In the same manner as birds, the female incubates the eggs, but after hatching, the young are dependent upon the milk that seeps from glands on the abdomen of the female. Therefore, monotremes have retained the reptilian mode of reproduction while evolving hair and mammary glands. The young are blind, helpless, and completely dependent on the parent for some months. The mouth is variously modified among the monotremes. The *platypus* has a horny, bill-like structure somewhat resembling that of a duck, while the *anteater* has an elongated, cylindrical snout.

Marsupials Another primitive group of mammals are the *marsupials* found in large numbers in Australia, such as the *kangaroo* and *koala* bear (fig. 27.36b). In the Americas, marsupials are represented by the *opossum.* The young in all members of this group are born in a very premature state. Once born, they leave the uterus and crawl to the pouch where each attaches itself to a nipple and continues development for a time.

a.

b.

Figure 27.36
Monotreme and marsupial, two rare types of mammals. a. *Spiny anteater is a monotreme that lays a shelled egg.* b. *Koala bear is a marsupial whose young are born immature and complete their development within a pouch.*

Figure 27.37
a. *White-tail deer feeding fawn. The white of the tail is exposed when the tail is raised as a sign of imminent danger.* b. *Torpedo-shaped harbor seals are excellent swimmers and divers, having short paddlelike flippers and a thick layer of subcutaneous fat (blubber).*

a.

b.

Placental Mammals The vast majority of living mammals are the placental type (fig. 27.37). In these mammals, the extraembryonic membranes have been modified for internal development within the uterus of the female. The chorion contributes to the fetal portion of the placenta, while a portion of the uterine wall contributes to the maternal portion. Here nutrients, oxygen, and waste are exchanged between fetal and maternal blood. These mammals not only

Table 27.6 Some Major Orders of Placental Mammals

Order	Characteristics
Insectivora (moles, shrews)	Primitive; small, sharp-pointed teeth
Chiroptera (bats)	Digits support membranous wings
Carnivora (dogs, bears, cats, sea lions)	Canine teeth long; teeth pointed
Rodentia (mice, rats, squirrels, beavers, porcupines)	Incisor teeth grow continuously
Perissodactyla (horses, zebras, tapirs, rhinoceroses)	Large, long-legged, one or three toes, each with hoof; grinding teeth
Artiodactyla (pigs, cattle, camels, buffalos, giraffes)	Medium to large; two or four toes, each with hoof; many with antlers or horns
Cetacea (whales, porpoises)	Medium to very large; forelimbs paddle-like; hind limbs absent
Primates (lemurs, monkeys, gorillas chimpanzees, humans)	Mostly tree-dwelling; head freely movable on neck; five digits, usually with nails; thumbs and/or large toes usually opposable

have a long embryonic period, they are also dependent on their parents until the nervous system is fully developed and they have learned to take care of themselves.

Placental mammals may be classified into twelve orders, eight of which may be considered major (table 27.6). A study of these reveals that mammals have largely differentiated, or become specialized, according to mode of locomotion and how they get their food.

The anatomy and physiology of human beings exemplifies vertebrate, and especially mammalian, anatomy and physiology. Chapters 8 through 19 may therefore be used for detailed information regarding vertebrate and mammalian anatomy and physiology.

Placental mammals are far more numerous than the monotremes (egg-laying) and marsupial (pouched) mammals. They are adapted to life in the air, water, and on land. Humans are classified as primates, mammals that are adapted to living in trees.

Comparisons between Vertebrates

Vertebrates, like other animals, are adapted to their way of life. Figure 27.38 shows that fish breathe by means of gills, respiratory organs appropriate to life in the water. Amphibians have small, ineffectual lungs that must be supplemented by using the skin as a respiratory organ. Reptiles have more efficient lungs with a rib cage, which not only protects the lungs but helps fill them with air. Birds have lungs expanded by air sacs that allow one-way flow of air, and mammals have highly subdivided lungs surrounded by a rib cage and separated from the abdominal cavity by a diaphragm. These anatomical features make breathing by negative pressure possible in reptiles, birds, and mammals.

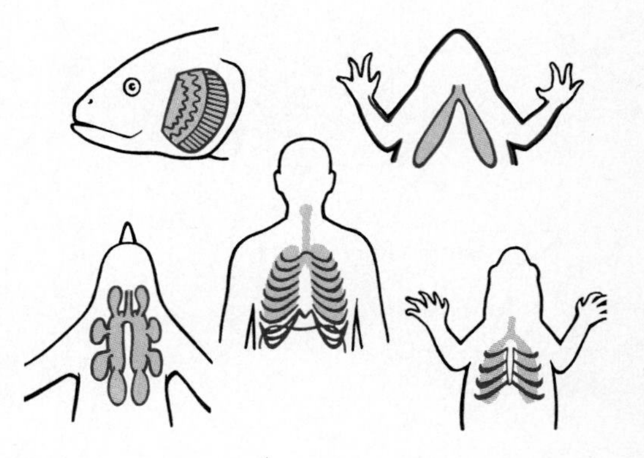

Figure 27.38
Breathing mechanisms of vertebrates. Fish breathe by means of gills; amphibians have poorly developed lungs; reptiles have a rib cage and lungs; birds have lungs with air sacs; and humans have well-developed lungs plus rib cage and diaphragm.

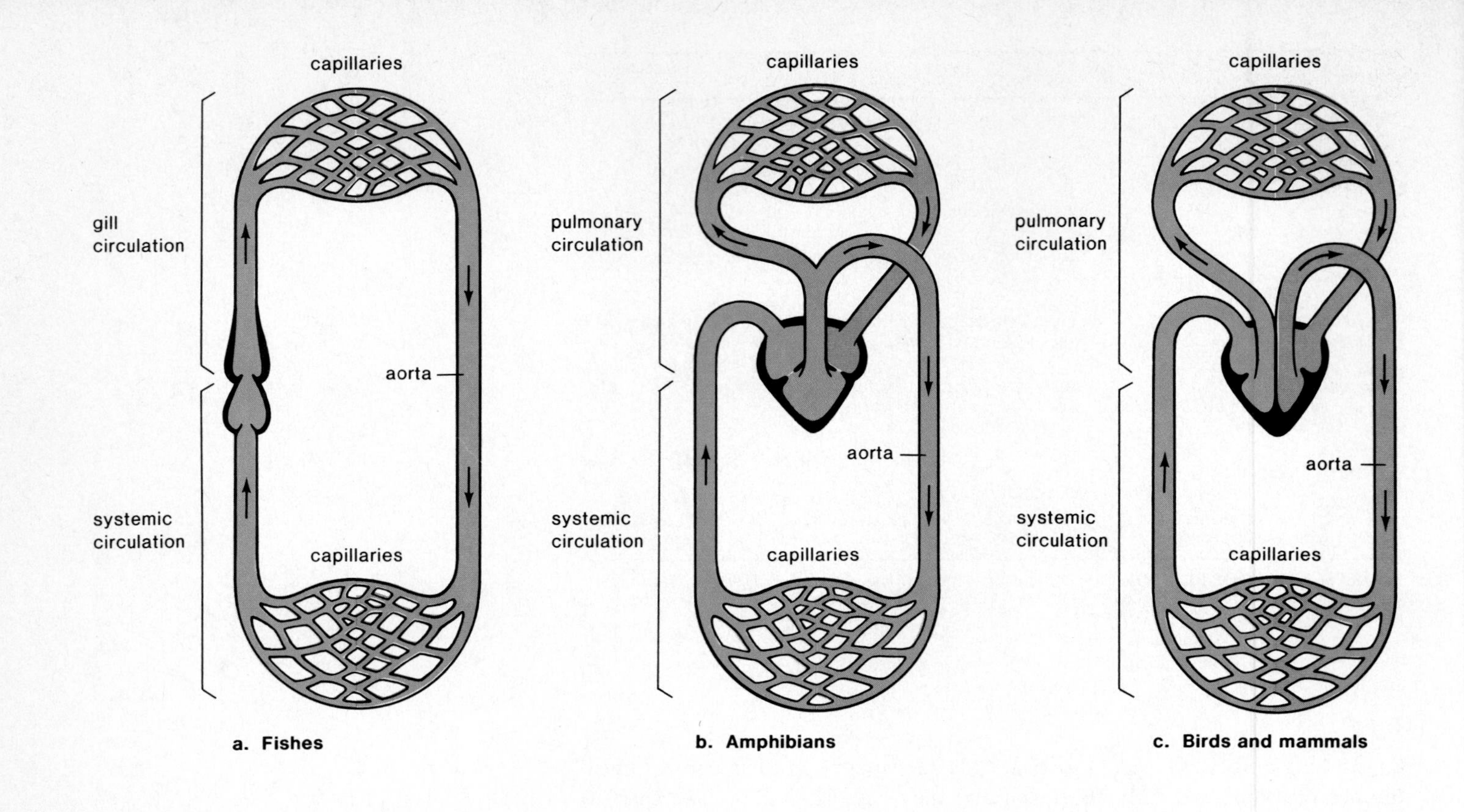

***Figure** 27.39*
A comparison of circulatory paths in vertebrates. a. *A fish heart with only one atrium and one ventricle pumps blood to the gills.* b. *In an amphibian there is a pulmonary and systemic circuit but the heart has two atria and only one ventricle.* c. *The heart of birds and mammals is a double pump; one half of the heart pumps blood to lungs and the other half pumps blood to the body.*

Figure 27.39 compares the circulatory systems of the vertebrates. Fishes have a nondivided atrium and ventricle and the heart pumps blood only to the gills. Amphibians have a heart in which there is a right and left atrium but only a single ventricle where oxygenated and deoxygenated blood are partially mixed before being sent, in part, to the skin for further oxygenation. The reptiles have a right and left atrium and a ventricle that has two partial septa. Some mixing of oxygenated and deoxygenated blood is not a serious disadvantage for the cold-blooded amphibians and reptiles since their oxygen demands are relatively low. Birds and mammals have a four-chambered heart in which division of the atria and ventricles is complete and there is no opportunity for mixing to occur. The right side of the heart pumps blood to the lungs and the left side pumps blood to the rest of the body.

A comparison of the eggs of vertebrates shows that fish and amphibian eggs are generally small with little yolk; these eggs are deposited into the water where they may develop into swimming larva. Both reptiles and birds lay a shelled egg with extraembryonic membranes to take over the functions previously performed by external water. The placental mammals have modified membranes that permit internal development during which time the mother provides for the needs of the developing fetus (chapter 20).

A comparison of the vertebrates shows that the respiratory system becomes increasingly adapted to breathing air; the circulatory system becomes increasingly adapted to maintaining a separation between the pulmonary circuit and the systemic circuit; and there is a gradual development of the ability to reproduce on land (table 27.7).

Table 27.7 Comparison of Vertebrates (in General)

	Fishes	Amphibians	Reptiles	Birds	Mammals
Habitat	Water	Water/land	Land	Land	Land
Heart	Nondivided atrium and ventricle	Two atria and nondivided ventricle	Two atria and partially divided ventricle	Two atria and two ventricles	Two atria and two ventricles
Respiration	Gills	Gills/lungs/skin	Lungs and rib cage	Lungs, rib cage, and air sacs	Lungs, rib cage, and diaphragm
Fertilization	External	External	Internal	Internal	Internal
Egg	Small, no shell, develops externally	Small, no shell, develops externally	Large, with shell, develops externally	Large, with shell, develops externally	Small, no shell, develops internally

Summary

Classification of animals considers type of body plan, symmetry, number of germ layers, presence or absence of coelom, and segmentation. Among the lower invertebrates, it is possible to observe an increase in complexity regarding these features if the various groups are arranged as in table 27.8. While the flatworms and roundworms have specialized organs, all of the lower invertebrates carry out respiration and circulation by diffusion.

Table 27.8 Classification Features

	Sponges	Cnidarians	Flatworms	Roundworms
Body plan	———	Sac	Sac	Tube within a tube
Symmetry	Radial or none	Radial	Bilateral	Bilateral
Germ layers	———	2	3	3
Level of organization	———	Tissues	Organs	Organs
Body cavity	———	———	———	Pseudocoelom
Segmentation	———	———	———	———

All of the higher invertebrates have the tube-within-a-tube body plan, three germ layers, and a true coelom. Embryological evidence suggests that the echinoderms and chordates are related. The former are unique among the higher invertebrates because of their radial symmetry and locomotion by tube feet. At some time in their life history all chordates have a dorsal notochord, dorsal hollow nerve cord, and gill pouches. Lancelets retain all three features as an adult. The vertebrates are chordates in which the notochord is replaced by a vertebral column in the adult. The evolutionary tree indicates that amphibians evolved from the lobe-finned fishes; reptiles evolved from amphibians; and birds and mammals evolved from reptiles.

Adaptation to land life begins poorly in the amphibians, but is almost complete with the reptiles, which are able to reproduce on land due to a shelled egg with extraembryonic membranes. Only birds and mammals are warm-blooded, however, and able to maintain a constant internal temperature; therefore, only they are completely adapted to life on land.

Objective Questions

1. The function of collar cells in a sponge is ______________ .
2. Cnidarians have the ______________ body plan and are ______________ symmetrical.
3. Planarians have the ______________ type of nervous system and a __________ excretory system.
4. The intermediate host for a tapeworm is either ______________ or __________ .
5. Pinworm, trichinosis, hookworm, and elephantiasis are all ______________ worm infections.
6. In protostomes, the first embryonic opening becomes the ______________ .
7. In today's mollusks, the coelom is much ______________ and limited to the region around the ______________ .
8. Earthworms have external rings signifying that they are ______________ animals.
9. The water vascular system of echinoderms consists of canals and ______________ feet.
10. The three chordate characteristics are a ______________ , ______________ , and ______________ .
11. The ______________ and ______________ are primitive chordates.
12. The three classes of fishes, __________ , ______________ , and ______________ , indicate in general their order of evolution.
13. Amphibians evolved from ______________ fishes that had primitive lungs.
14. Whereas amphibians must return to the ______________ to reproduce, reptiles lay ______________ that contain ______ membranes.
15. Both ______________ and mammals maintain a constant internal __________ .
16. There are three types of mammals: ______________ , ______________ , and ______________ .
17. Dogs, cats, horses, mice, rabbits, bats, whales, and humans are all ______________ mammals.
18. Primates are adapted to an arboreal life in the ______________________ .

Answers to Objective Questions

1. to keep water moving through central cavity 2. sac, radially 3. ladder, flame-cell 4. cattle or pigs 5. round 6. mouth 7. reduced, heart 8. segmented 9. tube 10. notochord, dorsal hollow nerve cord, gill pouches 11. tunicates, lancelets 12. jawless, cartilaginous, bony 13. lobe-finned 14. water, eggs, extraembryonic 15. birds, temperature 16. monotremes, marsupials, placental 17. placental 18. trees

Study Questions

1. Which groups of animals comprise the lower invertebrates? (p. 584) Compare the representatives of these animals in regard to body plan, symmetry, germ layers, level of organization, coelom, and segmentation. (pp. 586–596, 621)
2. Compare the representatives of the four lower invertebrate phyla in regard to nervous conduction, musculature, digestion, excretion, and reproduction. (pp. 586–596)
3. Describe the life cycle and structure of a tapeworm. Compare the anatomy of free-living flatworms with that of the fluke and tapeworm. (pp. 592–593)
4. What biological data are used to divide higher animals into two groups? (p. 596)
5. Compare the clam, earthworm, crayfish, grasshopper, and starfish with respect to nervous, digestive, skeletal, excretory, circulatory, and respiratory systems and means of reproduction and locomotion. (pp. 596–620)
6. Compare the adaptations of the clam to those of the squid to show that the clam is adapted to an inactive life and the squid is adapted to an active life. (p. 599)
7. Compare the adaptations of the crayfish to those of the grasshopper to show that the former is adapted to an aquatic existence while the latter is adapted to a terrestrial existence. (p. 607)
8. Name and describe unique features of echinoderm anatomy and physiology. (pp. 609–611)

Key Terms

bilateral symmetry (bi-lat'er-al sim'ĕ-tre) having a right and left half so that only one vertical cut gives two equal halves. *585*

chitin (ki'tin) flexible, strong polysaccharide forming the exoskeleton of arthropods. *604*

compound eye (kom'pownd i) arthropod eyes composed of multiple lenses. *605*

deuterostome (du'ter-o-stōm'') member of a group of animal phyla in which the anus develops from the blastopore and a second opening becomes the mouth. *596*

dipleurula (di-ploor'u-lah) a larval form unique to the deuterostomes that indicates that they are related. *596*

filter feeder (fil'ter fēd'er) an animal that obtains its food, usually in small particles, by filtering it from water. *587*

hermaphroditism (her'maf'ro-di-tizm'') the state of having both male and female sex organs. *591*

invertebrate (in-ver'tĕ-brāt) an animal that lacks a vertebral column. *585*

jointed appendages (joint'ed ah-pen'dij-ez) the flexible exoskeleton extensions found in arthropods that are used as sense organs, mouth parts, and locomotion. *604*

Malpighian tubules (mal-pig'ĭ-an tu'būlz) organs of excretion, notably in insects. *608*

mesoglea (mes''o-gle'ah) a jellylike packing material between the ectoderm and endoderm of cnidarians. *587*

nematocyst (nem'ah-to-sist) a threadlike structure in stinging cells of cnidarians that can be expelled to numb and capture prey. *587*

notochord (no'to-kord) dorsal supporting rod that exists in all chordates sometime in their life history; replaced by the vertebral column in vertebrates. *611*

protostome (pro'to-stōm) member of a group of animal phyla in which the mouth develops from the blastopore. *596*

radial symmetry (ra'de-al sim'ĕ-tre) regardless of the angle of a cut made at the midline of an organism, two equal halves result. *585*

sessile (ses'il) organisms that lack locomotion and remain stationary in one place, such as plants or sponges. *587*

setae (se'te) bristles, especially those of the segmented worms. *603*

tracheae (tra'ke-ah) the air tubes of insects. *608*

trochophore (tro'ko-fōr) a larval form unique to the protostomes that indicates they are related. *596*

vertebrate (ver'tĕ-brāt) animal possessing a backbone composed of vertebrae. *585*

visceral mass (vis'er-al mas) soft-bodied portion of a mollusk which includes internal organs. *597*

28

Human Evolution

Chapter Concepts

1 Humans are primates and many of their physical traits are the result of their ancestors' adaptations to living in trees.

2 Humans share a common ancestor with the apes.

3 All fossils placed in the same family as modern humans were terrestrial tool users.

4 Tool use and intelligence may have evolved together.

5 All human races are classified as *Homo sapiens.*

Chapter Outline

Table 28.1 *Classification of Primates*

Order Primates Subphylum Vertebrata	
Order primates	Prosimians Lemurs Tarsiers Anthropoids Monkeys Apes Humans
Superfamily hominoidea	*Dryopithecus* Modern apes Humans
Family hominidae	*Australopithecus* *Homo habilis* *Homo erectus* *Homo sapiens*
Genus homo *(Humans)*	*Homo habilis?* *Homo erectus* *Homo sapiens*

Humans are mammals in the order Primates (table 27.6). **Primates** were originally adapted to an arboreal life in trees. Long and freely movable arms, legs, fingers, and toes allowed them to reach out and grasp an adjoining tree limb. The opposable thumb and toe, meaning that the thumb and toe could touch each of the other digits, were also helpful. Nails replaced claws; this meant that primates could also easily let go of tree limbs.

The brain became well developed, especially the cerebral cortex and frontal lobes, the highest portions of the brain. Also, the centers for vision and muscle coordination were enlarged. The face became flat so that the eyes were directed forward, allowing the two fields of vision to overlap. The resulting stereoscopic (three-dimensional) vision enabled the brain to determine depth. Color vision aided the ability to find fruit or prey.

One birth at a time became the norm; it would have been difficult to care for several offspring, as large as primates, in trees. The period of postnatal maturation was prolonged, giving the immature young an adequate length of time to learn behavior patterns.

Evolution of Humans

Prosimians

Prosimians (premonkeys) are the first primates (table 28.1) believed to have evolved from primitive shrewlike *insectivores,* arboreal rat-sized animals with sharp canine teeth that most likely resembled today's tree shrew (fig. 28.1a). The prosimians are represented today by several types of animals, among them the *lemurs* (fig. 28.1b), which have a squirrel-like appearance, and the *tarsiers,* curious monkeylike creatures with enormous eyes.

Figure 28.1
Primate origins. a. *The tree shrew is believed to resemble the ancestor of the primates.*
b. *Prosimians (premonkeys) were the first primates, animals adapted to living in trees. These animals today are represented by the lemurs and tarsiers. This is a photo of ring-tailed lemur.*

a.

b.

Anthropoids

Monkeys, along with apes and humans, are **anthropoids.** Monkeys evolved from the prosimians (fig. 28.2) about 38 million years ago. There are two types of monkeys: the New World monkeys, which have long prehensile (grasping) tails and flat noses and Old World monkeys, which lack such tails and have protruding noses. Two of the well-known New World monkeys are the spider monkey and the capuchin, the "organ grinder's" monkey. Some of the better-known Old World monkeys are now ground dwellers such as the baboon, and the rhesus monkey, which has been used in medical research.

Primates evolved from shrewlike insectivores and became adapted to living in trees, as exemplified by skeletal features, good vision, and even reproduction. Primates are represented by prosimians, monkeys, apes, and humans. The latter three are anthropoids.

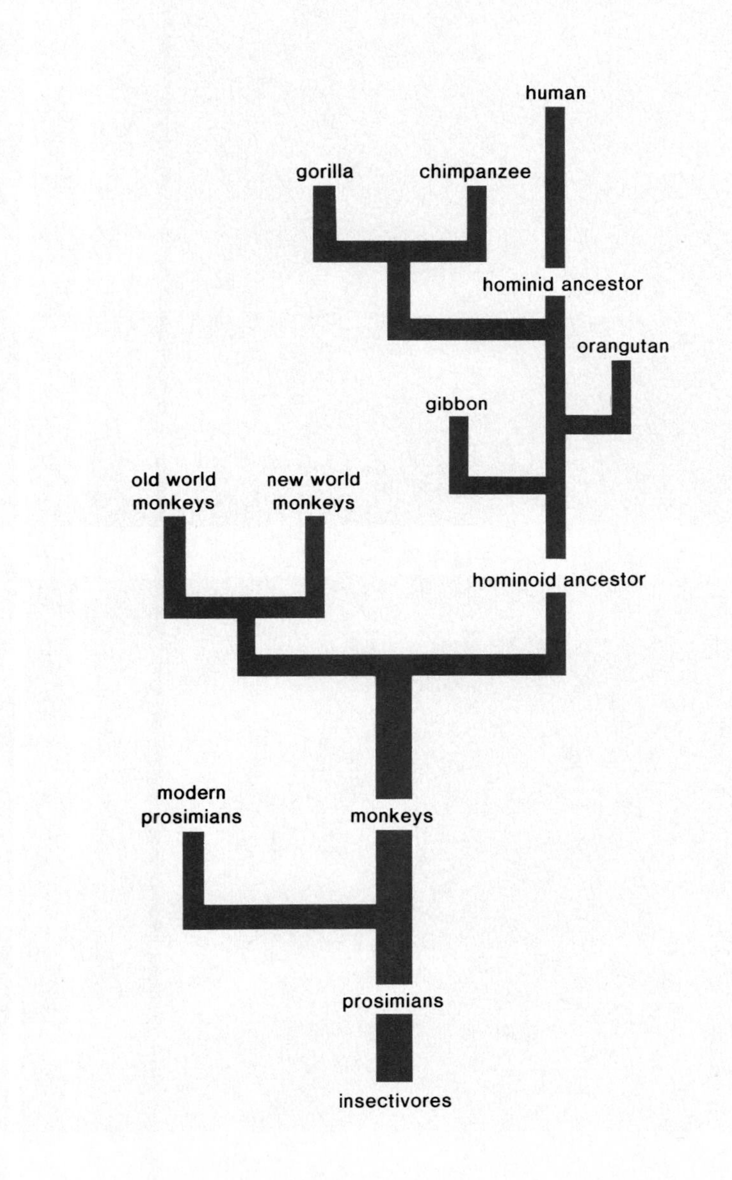

Figure 28.2
Primate evolutionary tree. Primates are believed to have evolved from an insectivore mammal; monkeys are distantly related to humans, but apes are closely related. Just exactly when the ape-man split occurred is not known. Perhaps humans share a recent common ancestor with only certain of the apes. If this is the case, then the split could have occurred within the past several millions of years. This tree suggests that this is the case.

Figure 28.3

Ape diversity. a. *Of the apes, gibbons are the most distantly related to humans. They dislike coming down from trees, even at watering holes. They will extend a long arm into the water and then drink collected moisture from the back of the hand.* b. *Orangutans are solitary except when they come together to reproduce. Their name means "forest man"; early Malayans believed that they were intelligent and could speak but did not because they were afraid of being put to work.* c. *Gorillas are terrestrial and live in groups in which a silver-backed male such as this one is always dominant.* d. *Of the apes, chimpanzees are most closely related to humans.*

a.

b.

c.

d.

Hominoids

Humans are more closely related to apes (fig. 28.3) than to monkeys. **Hominoids** include apes and humans. There are four types of apes: gibbon, orangutan, gorilla, and chimpanzee. The **gibbon** is the smallest of the apes with a body weight ranging from 12 to 25 pounds. Gibbons have extremely long arms that are quite specialized for swinging between tree limbs. The **orangutan** is large (165 lbs.) but nevertheless spends a great deal of time in trees, while the **gorilla,** the largest of the apes (400 lbs.) spends most of its time on the ground. The **chimpanzee,** at home both in trees and on the ground, is the most humanlike of the apes in appearance and is frequently used in psychological experiments.

Hominoid Ancestor

About 25 million years ago, apes became abundant and widely distributed in Africa, Europe, and Asia. Among these, members of the genus ***Dryopithecus*** are of particular interest because they are thought to be a possible hominoid ancestor.

Dryopithecines were forest dwellers that probably spent most of their time in the trees. The bones of their feet, however, indicate that they also may have spent some time on the ground. When they did walk on the ground, however, they probably walked on all fours, perhaps using the knuckles of their hands to support part of their weight. "Knuckle-walking" (p. 629) would have allowed retention of the opposable thumb, a primate characteristic (fig. 28.4).

The skull of *Dryopithecus* had a sloping brow, heavy eyebrow ridges, and jaws that projected forward. These are features that could have led to those of apes and humans (fig. 28.5). In apes the face projects forward forming a muzzle because the canine teeth are much larger than the adjacent teeth.

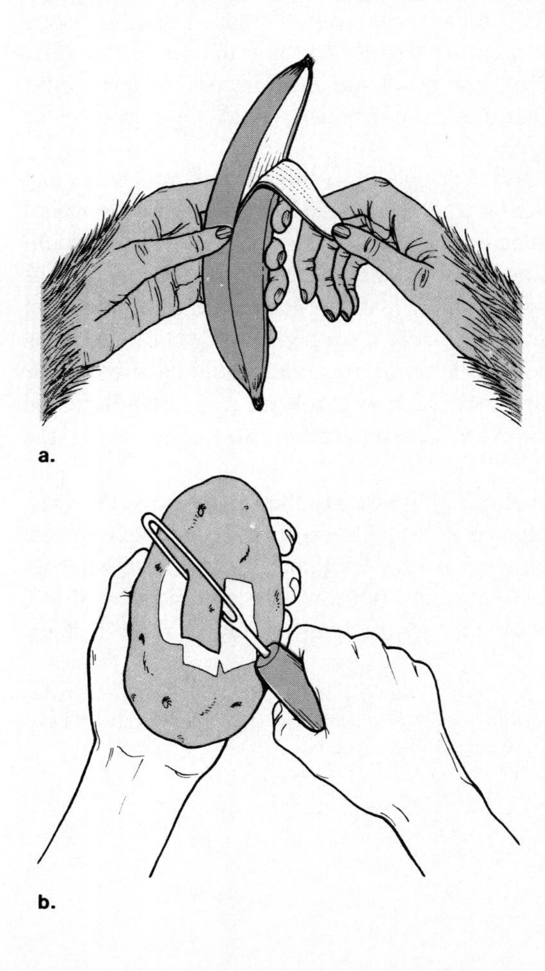

Figure 28.4
The primate hand is capable of grasping objects because of the opposable thumb. In humans, this ability is used to grasp tools.

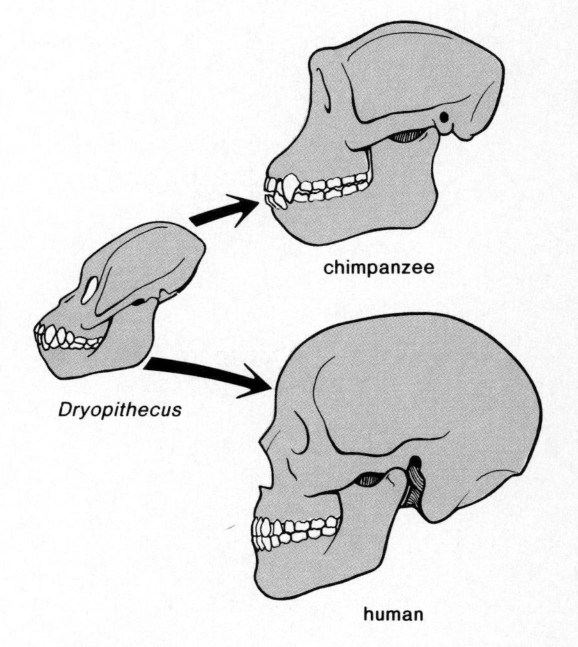

Figure 28.5
Comparison of the skull of Dryopithecus *with that of modern apes and humans.* Dryopithecus *has primitive features; the common chimpanzee has apelike features, including a low brow, heavy eyebrow ridges, and a face that protrudes; while humans have a high brow, reduced eyebrow ridges, and a flat face.*

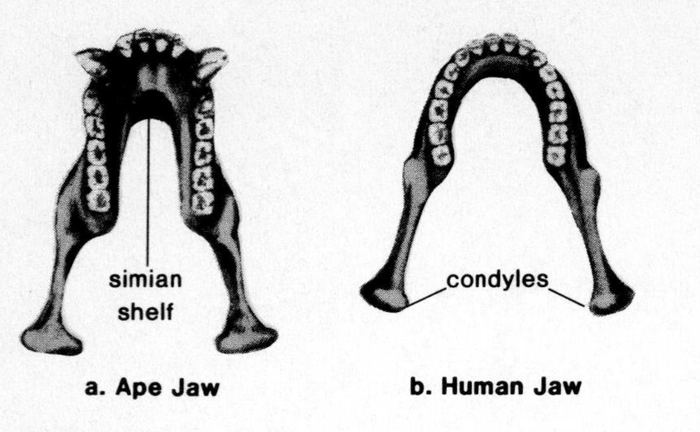

Figure 28.6
Dental arcades of an ape and a human. a. *The two sides of the jaw of an ape are roughly parallel and the canine teeth are much larger than the adjacent teeth; therefore, the dental arcade has a rectangular shape.* b. *In humans, the canine teeth are much reduced and the jaw curves to give a dental arcade that is U-shaped.*

Table 28.2 Comparison of Apes and Humans

Feature	Primitive Ancestor	Apes	Humans
Brain size	Small brain and skull	Slightly enlarged	Very much enlarged
Face	Sloping brow, heavy eyebrow ridges, and projection of face	Same as primitive	High brow, reduced eyebrow ridges, face flat
	Rectangular-shaped jaw with large molars and long canine teeth	Same as primitive	U-shaped jaw with small molars and shortened canine teeth
Locomotion	Quadrupedal locomotion	Same as primitive	Bipedal locomotion
	Limbs of equal length	Forelimbs elongated	Same as primitive
	Opposable thumb and toe	Same as primitive	Opposable thumb retained

In contrast to apes, humans have a high brow and lack the eyebrow ridges of apes. The face is flat and they have no muzzle because the canine teeth are comparable in size to the other human teeth (fig. 28.6). Table 28.2 lists some of the possible characteristics of a common ancestor for apes and humans.

Hominid Ancestor

It is now believed that the ancestors to the modern apes diverged from the human lineage over a period of time that encompasses many millions of years. Biochemists have compared the structure of proteins and DNA in apes and humans. This evidence suggests that the gibbons began to evolve separately about 18 million years ago, and the orangutans[1] split off about 12 million years ago, leaving perhaps an apelike creature that led to the evolution of the other apes and humans (fig. 28.2). The close biochemical similarity between chimpanzees and humans suggests that they may have had a final common ancestor as late as 4 to 5 million years ago.

During the mid-Tertiary period (table 28.3) the weather was becoming cooler and even in Africa, the tropical forests were being replaced by woodlands and grasslands. This change in habitat may have caused the first **hominids** (forms that are ancestral only to humans) to begin walking on two legs as they moved between trees or foraged for food on the ground. Standing erect would have helped assure their survival because then they would have been able to see over tall grass to spot predators or prey, and would have had their hands free to perform other functions, such as throwing rocks. As discussed in the reading on page 630, however, such suppositions are being put to the test of modern investigations.

The transition to erect posture required a number of adaptive changes. Figure 28.7 contrasts the skeleton of a "knuckle-walking" ape with a human skeleton. In the ape, the pelvic region is very long and tilts forward, whereas in the human the pelvic region is short and upright. The ape shoulder girdle is more massive than that of a human, and ape arms are longer than the legs.

[1]Formerly, *Ramapethicus* was considered to be a likely candidate as the hominid ancestor but it is now known that this fossil is an ancestral orangutan.

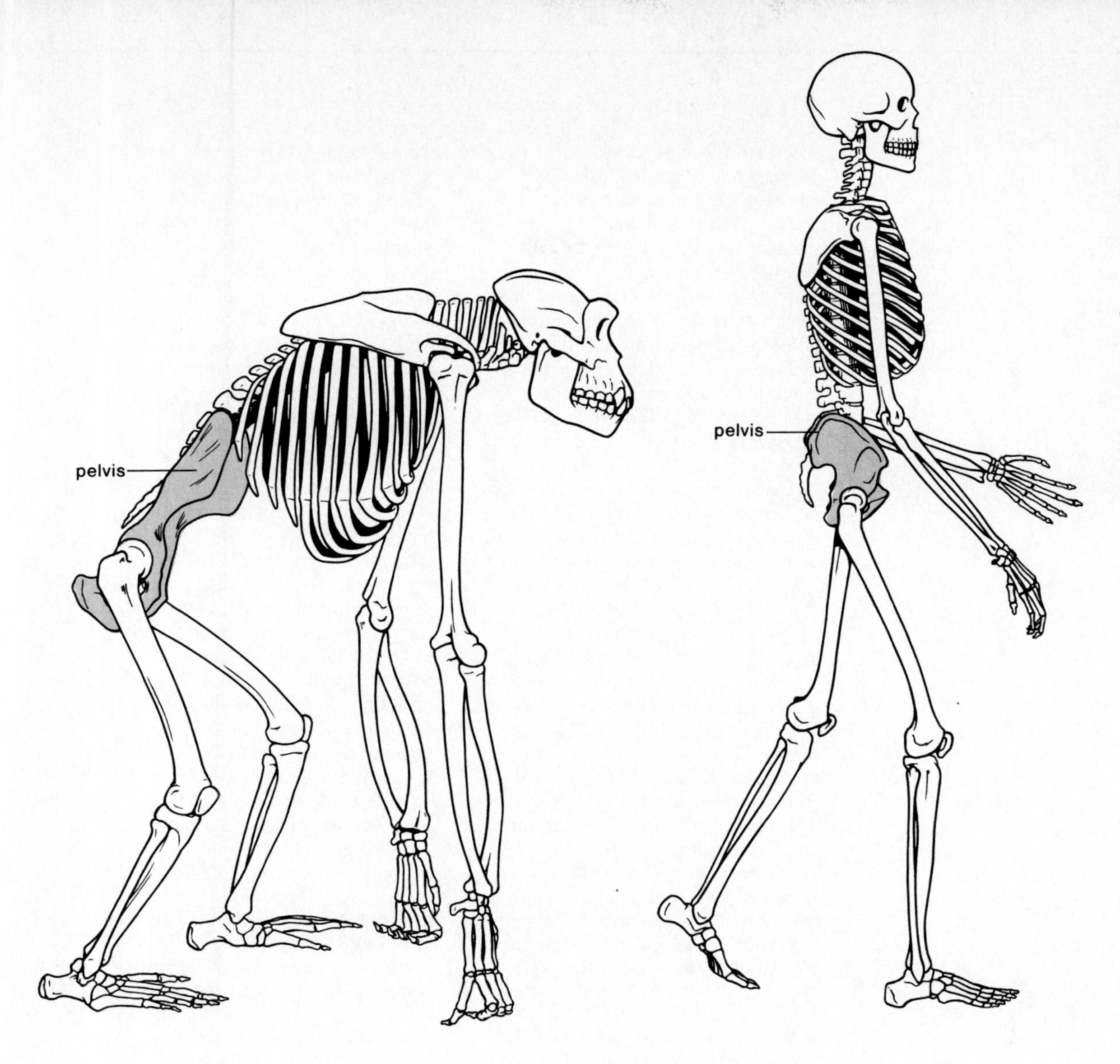

Figure 28.7
Human skeleton compared to ape skeleton. Humans are bipedal while modern-day apes are knuckle-walkers. The human skull sits atop the curved vertebral column. The pelvis is shorter and wider and the legs are vertically attached. The forelimb is shorter and the hindlimb is longer. The hands and feet are not curved and only the thumb (and not the great toe) is opposable.

From "The Antiquity of Human Walking" by John Napier. *Scientific American,* April 1967. Reprinted by permission of Scientific American, Inc.

The head hangs forward because the foramen magnum is well to the rear of the skull. In humans the head is erect because this opening is almost directly in the bottom center of the skull. Also in humans the spine is flexible because of its four curvatures.

Skeletal changes have also occurred in the fingers and toes. In the primitive condition, the thumb and large toe are both opposable, as seen in apes today. Humans have retained only the opposable thumb.

Humans and apes share a common ancestor which may possibly have been *Dryopithecus,* a fossil commonly regarded as the first hominoid. The first hominid arose at a time when a change in weather reduced the size of the African forests making it advantageous to locomote on the ground.

Modern Focus

One of the most striking features of paleoanthropology these days is the multidisciplinary nature of the pursuit. No longer is it a simple alliance between fossil hunters and archaeologists. Instead, a battery of sciences is being focused on the questions of human origins: in addition to geology, which provides the basic backdrop for the recovery of fossils, there is paleoecology, taphonomy (the study of the way bones become buried), primatology, molecular biology, neurophysiology, energetics, and many more.

Instead of pondering the imponderable, such as the causal link between bipedalism, tool-making and use, and expansion of brain size, there is now a growing tendency to ask specific, answerable questions. What are the energetic considerations of upright-walking as opposed to, say, knuckle-walking? Under what ecological circumstances might it be energetically possible to evolve a large brain? What can the surface of fossil teeth tell us about ancient diets? What does one need to know about an assemblage of fossil bones and stone artifacts to be sure that the association is not mere coincidence? What does the history of animals contemporary with our ancestors tell us about, say, important migrations of the past? And so on.

The question of 'why we stood upright' is of course an old one and has engendered responses such as, 'so as to gain more visibility over tall grasses' and, 'so as to free the hands for defensive purposes, because we lost our long, sharp canines.' Neither scenario is very testable. But to ask, 'What are the energetics of bipedalism as compared to quadrupedalism?' as Richard Taylor of Harvard University has is a more promising approach. Other researchers have used Taylor's data and have added observations of their own.

For example, Peter Rodman and Henry McHenry, of the University of California at Davis, and Richard Wrangham, at the University of Michigan, examined the feeding habits of orangutans, which forage mainly for fruit in trees, and chimpanzees, which do the same but usually from sources that are more meagre and more widely dispersed. McHenry and his colleagues wondered whether an ape, faced with food sources dispersed even more thinly than those exploited by chimpanzees, would forage more efficiently in energetic terms by knuckle-walking (see figure) or by bipedalism. It turns out, although bipedalism is not particularly efficient compared with fully developed quadrupedalism in animals moving at high speeds, under the circumstances in which our ancestors evolved, that is, in the forest fringe with an ape-like ancestor, the evolution of upright-walking was a very suitable adaptation. At slow walking speeds, bipedalism compares very favourably with quadrupedalism.

It may be, therefore, that bipedalism arose as an energetic strategy for foraging for widespread food sources. The point here is that the approach adopted by Taylor and others allowed specific questions, predictions and answers, a truly scientific way of tackling the problem.

Another exploration in energetics comes from the work of Robert Martin, at University College, London. Addressing primates in general and humans in particular, he asked the questions, 'What dietary and ecological factors affect a species' energy budget and reproductive strategy?' and, 'How is this related to brain size?' Some strands of the argument are as follows. Animals that adopt the reproductive strategy of having just a few carefully nurtured offspring, as against many that are left to fend for themselves, generally live in environments blessed with a stable food supply. And species that are endowed with large brains are those that feed on high energy diets. How does the requirement for a stable environment and a high energy diet match up to other conclusions about the environment of early hominids? Such questions are, thus, brought into scientific focus. . . .

There has always been much speculation on the putative appearance of the last common ancestor between human and apes, but only infrequently has it been inspired by rigorous scientific questioning. By applying the principles of allometry in the study of body and limb proportions in monkeys, apes and humans, Leslie Aiello, again, of University College, London, has been able to dismiss some of the long-favoured 'models' for the human-ape ancestor because, it turned out, they were biologically incongruent. No, the last common ancestor was not like a chimpanzee; nor was it especially long in the forelimb, an adaptation to an arm-swinging life style; and neither was it like a modern Old World monkey. The comparative study tells us, says Aiello, that the last common ancestor probably moved about in a slow suspensory climbing motion, reminiscent of that of the modern howler monkey, ironically a New World monkey. This is not to suggest an involvement of New World primates in human origins; however it gives some guidance in interpreting the fossils of any putative last common ancestor.

This new approach of asking specific questions often brings unexpected answers, as happened to Alan Walker of Johns Hopkins University, Baltimore. When he imaged under the electron microscope the tooth surface of fossil hominids and compared them with marks on modern teeth he found that the early hominids and *Homo habilis* could be classified with chimpanzees, that is they were fruit-eaters. With *Homo erectus* came an apparently marked change in diet, possibly with the inclusion of underground tubers, possibly with increased meat-eating. This approach, still being developed, brings modern researchers as close as they will ever come to the food actually eaten by the individual whose fossilized remains are in their study. Walker had asked, 'What do the surfaces of chimpanzees' teeth look like?' 'What do baboons' teeth look like?' 'What do pigs' teeth look like?' 'What do hyaenas' teeth look like?' And, 'How does this help us understand better the diet of our ancestors?' He got an answer.

Modern researchers study the energetics of knuckle-walking in order to determine if the hominid ancestor would have walked this way or uprightly.

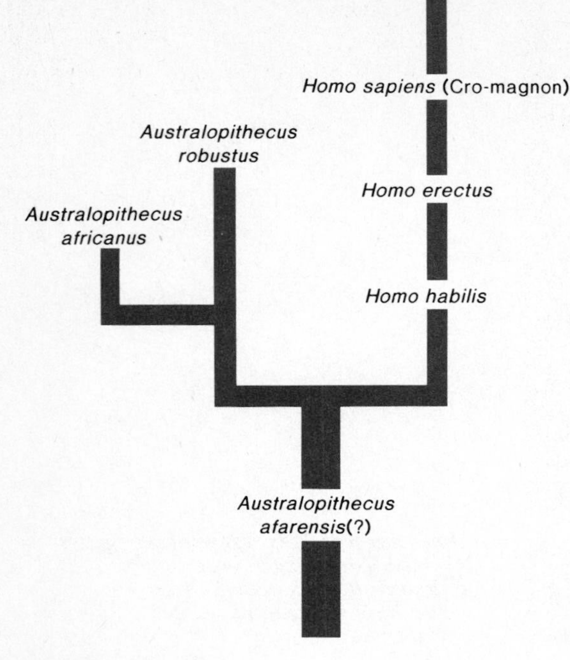

Figure 28.8
Hominid evolutionary tree. Adaptive radiation is evident because this tree indicates that australopithecines and humans coexisted. This is only one possible tree and others have been suggested.

Hominids

New ideas regarding speciation, particularly the concept of adaptive radiation (p. 522), have been used to interpret the hominid fossil record. Whereas scientists formerly attempted to place each hominid fossil in a straight line from the most primitive to the most advanced, it is now reasoned that several hominid species could have existed at the same time. Figure 28.8 indicates a possible evolutionary tree for the known hominids.

Australopithecines

The fossil remains of those classified in the genus *Australopithecus* (Southern Apeman) were found in southern Africa and have been dated at about 4 million years ago. The **australopithecines** (fig. 28.9) were 4 to 5 feet in height, with a brain that ranged in size from 400 to 800 cubic centimeters. The pelvis definitely indicates that they were capable of bipedal locomotion; supporting this contention is the nonopposability of the big toe.

Three species of *Australopithecus* have been identified—*afarensis, africanus, and robustus.* The exact relationship between the three species is in dispute. Because of its small brain size, large canine teeth, and protruding face, it has been suggested that *afarensis* (affectionately known as Lucy from the Beatles' song, "Lucy in the Sky with Diamonds") is the more primitive of the three. Moreover, because of certain skeletal limb features indicating that these peoples walked erect, some believe that *afarensis* is also ancestral to humans, as indicated in figure 28.8. It may also be noted that the combined characteristics of this fossil indicate that an enlarged brain was *not* needed for bipedal locomotion to evolve.

Australopithecus robustus, as its name implies, is larger than *africanus.* At one time it was believed that skull differences might signify a difference in their diet but an analysis of toothwear patterns of the enamel (fig. 28.10) indicates that they both probably fed mostly on fruit, much like that of the modern chimpanzees. Moreover, a study of animal bones found at their campsites shows that marks made by tools cut across those made earlier by a carnivorous predator. It is concluded that *australopithecines* were not hunters; rather, they may have scavenged bones from the kills of other animals.

Figure 28.9
Australopithecus africanus. *Some authorities believe this hominid is directly ancestral to humans, while others believe that they share a common ancestor. This picture gives the impression that the australopithecines were hunters, but evidence based on cut marks left by hominids on animal bones indicates that these hominids were most likely scavengers that sought and confiscated the kill of coexisting predators.*

One possible hominid evolutionary tree shows the bipedal Australopithecus afarensis *as a common ancestor to two other members of this genus and humans.* Australopithecus africanus *was slight of build compared to* Australopithecus robustus.

Humans

The rest of the fossils to be mentioned are in the genus *Homo,* which is the genus for all humans, including ourselves.

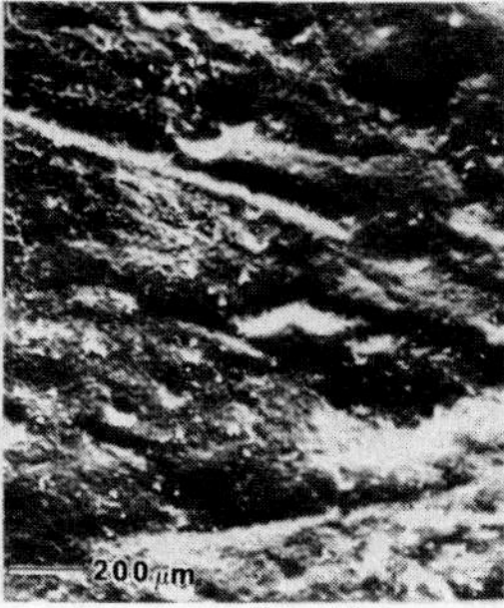

Figure 28.10
Authorities have microscopically studied the toothwear enamel markings of hominids. a. Australopithecines *are believed to have been herbivores that fed mostly on fruit.* b. Homo erectus *is believed to have been carnivorous.*

Homo Habilis

This newly discovered fossil man is dated about 2 million years ago, which may have made him a contemporary of *Australopithecus africanus.* ***Homo habilis*** is significant because formerly, only hominid fossils with a brain capacity of 1,000 cubic centimeters were designated humans. *Homo habilis* had a maximum brain capacity of 800 cubic centimeters and yet he has been placed in the genus *Homo* because he not only used tools, he also made them. *Homo habilis* means handyman; he was given this name because of the quality of the tools (fig. 28.11) found with his bones.

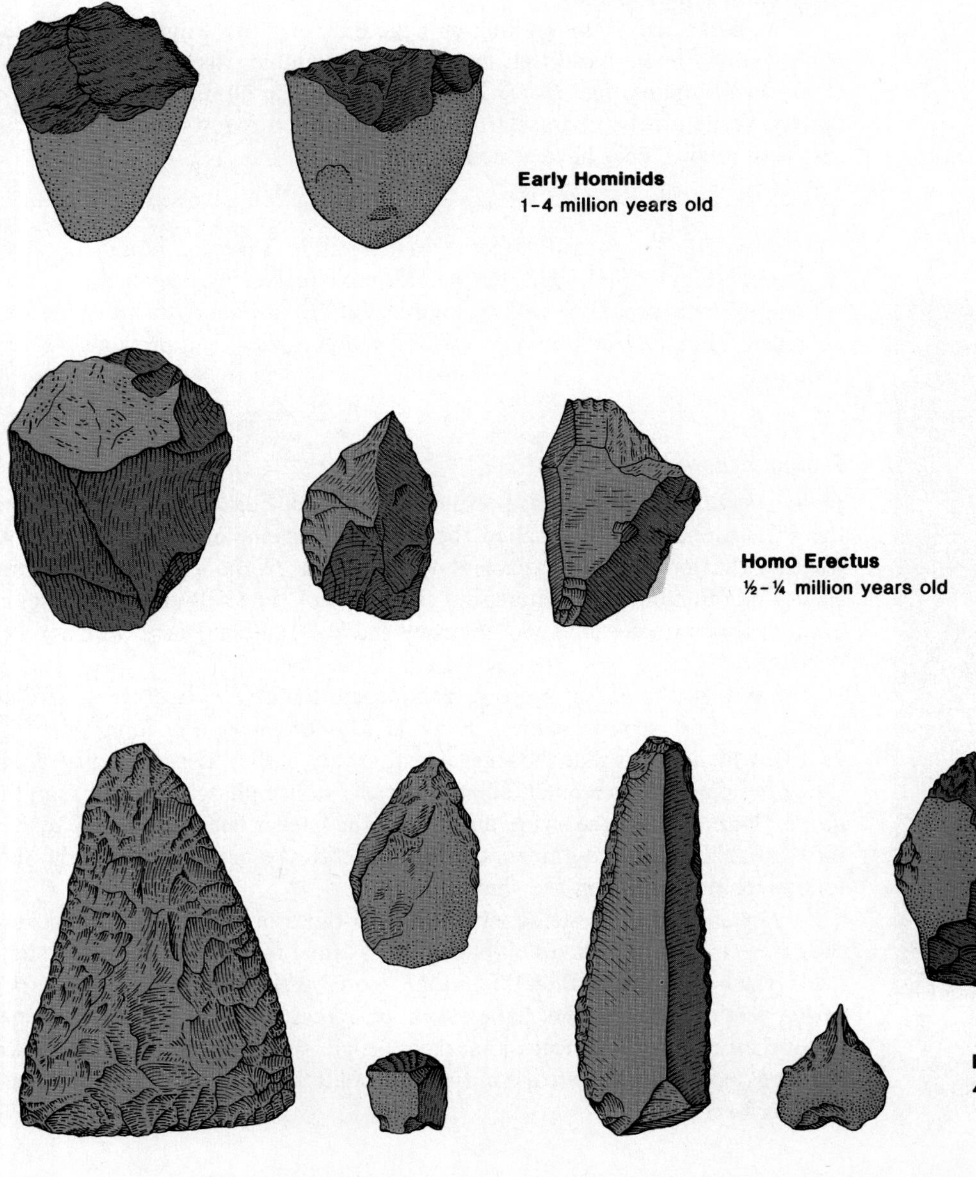

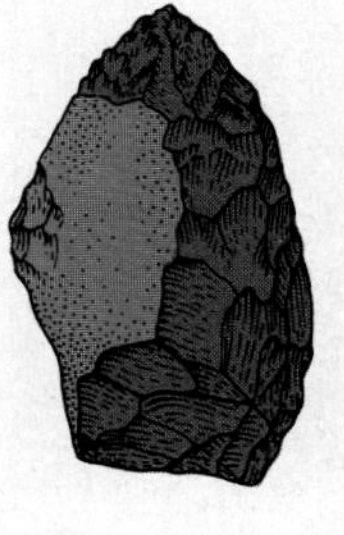

Figure 28.11
Stone tools used by early hominids (e.g., Homo habilis), Homo erectus, *and* Homo sapiens (Cro-Magnon). *Note the increasing amount of refinement in the tools.*

Table 28.3 *Periods and Epochs of the Cenozoic Era**

Era	Period	Epoch	Millions of Years before Present	Biological Events
	Quaternary	Recent	0.01–present	Modern humans
		Pleistocene	2.5	Early humans
Cenozoic	Tertiary	Pliocene	7–2.5	Hominids
		Miocene	25–7	Hominoids
		Oligocene	38	Anthropoids
		Eocene	54	Prosimians
		Paleocene	65	First placental Mammals
Mesozoic			230	

*Compare to fig. 23.2

At one time it was thought that only primitive people with a brain capacity of more than 1,000 cubic centimeters could have made tools. Since *Homo habilis* had a smaller brain and yet made tools, are we to think that the making of tools preceded the evolution of the enlarged brain? This seems to be an unnecessary question. Increased brain capacity, no matter how slight, would have permitted better tool making; this combination would have been selected because tool making would have fostered survival in a grassland habitat. Therefore, as the brain became increasingly larger, tool use would have become more sophisticated.

As described in the reading on page 635, new fossil finds indicate that *Homo habilis* had a build that more closely resembles that of apes than that of modern humans (fig. 28.7). This means that the human form must have evolved very quickly, because *Homo erectus,* which follows *Homo habilis* in the fossil record, does have a modern build.

Homo habilis, a fossil dated at least 2 million years ago, may not have been highly intelligent, but he did make tools. Intelligence and making of tools probably evolved together. If *Homo habilis* is indeed considered human, then humans evolved much earlier than previously thought.

Figure 28.12
Homo erectus. *Evidence indicates that* Homo erectus *was a successful species found throughout Africa and Europe.*

Homo Erectus

Homo erectus (fig. 28.12) was prevalent throughout Eurasia and Africa during the Pleistocene Age, also called the Ice Age because of the recurrent cold weather that produced the glaciers of this epoch. *Homo erectus* had a brain size of 1,000 cubic centimeters, but the shape of the skull indicates that the areas of the brain necessary for memory, intellect, and language were not well developed.

Even so, the grasp, posture, and locomotion of *Homo erectus* were all similar to those seen in modern humans. Humans have a *striding gait* (fig. 28.13), which means that the legs have alternate phases, the stance phase and the swing phase. When one hindlimb is in the stance phase, the other is in the swing phase. During the swing phase, first the knee is bent and then extended forward. The knee straightens, the heel touches the ground, and as the foot follows, that limb enters the stance phase.

Homo erectus made superior tools to those of *Homo habilis* because they were made from flakes chipped from a stone rather than from the stone itself. Each flake was struck off a stone "core" which was previously shaped to give just the right kind of flakes. Analysis of cut marks on the animal bones about their campsites indicates that these people were hunters. Also it is known that they possessed knowledge of fire and would have been able to cook meat in order to tenderize it.

Lucy Gets a Younger Sister

"Whoa. This is a hominid," crowed Anthropologist Tim White when he spotted the first bone fragment, a portion of an elbow, lying on a layer of sand. Looking down, Expedition Leader Donald Johanson shouted, "There's part of a humerus right next to it!" That July 1986 find in Tanzania's Olduvai Gorge marked the beginning of a startling discovery that was formally unveiled last week by White and Johanson. The team of ten U.S. and Tanzanian scientists unearthed 302 fossil bones and teeth that have yielded a more complete picture of modern humans' earliest direct ancestor, *Homo habilis*. The new material could alter the way scientists interpret human evolution.

Until now, most anthropologists have believed that *Homo habilis*, a species that lived in eastern and southern Africa between 2 million and 1.5 million years ago, stood about the same height and had the same body build as *Homo erectus*, its successor. *Homo habilis* (literally, handyman) was the first human ancestor to make stone tools. The new Olduvai Gorge skeleton, however, suggests that *Homo habilis* was much smaller and more apelike than previously thought. If that is the case, says Johanson, the modern body type probably did not evolve until *Homo erectus* emerged some 1.6 million years ago. Moreover, the evolutionary changes leading to *Homo erectus*, which preceded modern man, must have occurred faster than has been supposed.

Earlier discoveries of *Homo habilis* fossils consisted only of skulls, teeth and questionable limb bones, forcing scientists to guess at the creature's size and proportions. But the dramatic new find, which includes skull, arm bones, thigh and shin fragments from a single adult female, permits a more accurate assessment. The length of the thigh bone is a gauge of height, and the relative length of the upper arm bone to the upper leg bone is a vital clue to body build. The remains, described in the British journal *Nature* last week, belong to a creature that lived about 1.8 million years ago and stood no more than 3½ feet tall. Says Johanson, director of the Institute of Human Origins in Berkeley: "This may be the smallest hominid ever found."

The proportions of the skeleton were also a surprise to the scientists. The upper arm bone is about 95% as long as the thigh bone, indicating that the arms dangled to the knees, much as they do in apes. Thus *Homo habilis* closely resembled *Australopithecus afarensis*, of which the best-known example is the famed "Lucy" skeleton, which was discovered by Johanson in 1974. Lucy's ratio is 85%; in modern humans, the figure is about 70% to 75%.

Observes Johanson: "The new specimen suggests that the body pattern we call modern did not appear until *Homo erectus* and that it happened fairly rapidly." Says White, a professor at the University of California, Berkeley: "The question is, Why did they lose those features, and what made them change in just 200,000 years?"

The only thing that seems sure at this point, White adds, "is that we're looking at a major transition in human evolution involving behavior and anatomy. Something major and dramatic happened here."

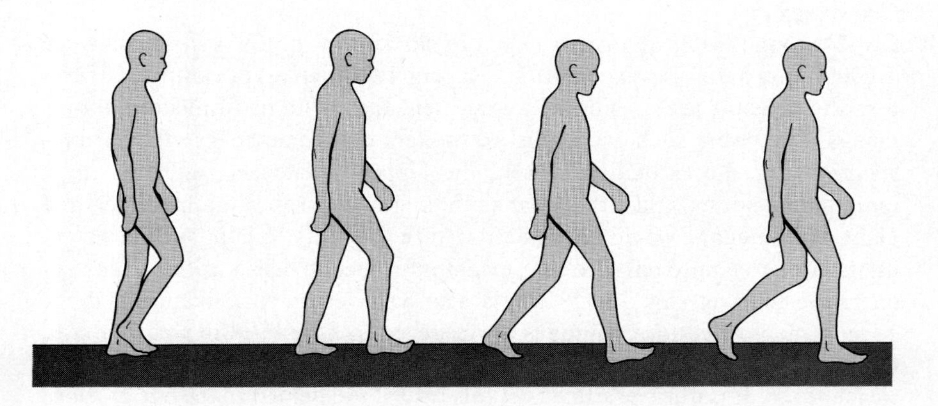

Figure 28.13
The striding gait of modern humans. Each limb alternately goes through a stance and swing phase. In the first drawing, the left limb is in the stationary stance phase and the right is beginning the swing phase. First, the knee is bent and extended forward. Then as the knee straightens, the heel and ball of the foot touch the ground. In the last two drawings, the left limb has entered the swing phase and the right limb is entering the stance phase.

Figure 28.14
Homo sapiens neanderthal. *It is now believed that Neanderthals generally had a more modern appearance than this drawing depicts them.*

The brain of those classified as *Homo erectus* was larger by 50 percent than that of the *australopithecines*. This suggests that superior tool making and increased intelligence may have evolved together during the evolution of humans.

Homo Sapiens

Neanderthal

The **Neanderthals** (*Homo sapiens neanderthalis*) lived over 130,000 years ago during the last ice age and survived 70,000 years. They are called Neanderthal because they were first found in the Neanderthal Valley in Germany. Later they were also found in most of Europe and in the Middle East.

Early investigators believed that Neanderthals were apelike in appearance (fig. 28.14). This has been revised to suggest that they had a more modern-like appearance except that they, like *Homo erectus,* had a low brow, wide nose, and receding chin. Like *Homo erectus,* too, the Neanderthal people were stone tool makers, lived together in a kind of society, and hunted large game together.

The Neanderthals seem to have been more culturally advanced than *Homo erectus,* however. They buried their dead with flowers, as an expression of grief perhaps, and with tools, as if they thought tools would be needed in a future life. Some have even suggested that the Neanderthals had a religion. Great piles of bear skulls have been found in their caves, and the bear may have played a role in this religion.

Cro-Magnon

Cro-Magnon (*Homo sapiens sapiens*) fossils appear in the fossil record some 40,000 years ago. They had a cranial capacity that equals our own and therefore these people are considered to represent the evolution of modern human beings. They were such accomplished hunters that some believe they are responsible for the extinction, during the Upper Pleistocene, of many large mammalian animals, like the giant sloth, mammoth, saber-toothed tiger, and giant ox. Language would have facilitated their ability to hunt such large animals, and it is quite possible that meaningful speech began at this time. Cooperative hunting (fig. 28.15) would also have led to socialization and the advancement of culture. Humans are believed to have lived in small groups, the men going out to hunt by day while the women remained at home with the children. It is quite possible that this way of life helped shape our previous expectations of the traditional role of men and women in modern society.

Figure 28.15
Socialization may have begun when men organized for the hunt. Since they hunted animals larger than themselves, cooperation was required.

Figure 28.16
Cro-Magnon peoples are depicted painting on cave walls and some of these paintings are still available for observation today. Of all the animals, only humans have developed a culture that includes technology and the arts.

Cro-Magnon people had an advanced form of stone technology that included the making of compound tools; the stone was fitted to a wooden handle. They were the first to have a spear thrower (fig. 28.15) enabling them to kill animals from a distance. And they were the first to make blades, knifelike objects that gave no evidence that they had been chipped from a stone (fig. 28.11).

Cro-Magnon peoples lived during the Reindeer Age, a time when great reindeer herds spread across Europe. They used every part of the reindeer including the bones and antlers from which they sculpted many small figurines. They also painted beautiful drawings of animals on cave walls in Spain and France (fig. 28.16). Perhaps they too had a religion and these artistic endeavors were an important part of their form of worship.

Homo erectus had a large brain and walked with a striding gait. He also used fire. *Homo sapiens neanderthal* was not as primitive as formerly thought. The Neanderthals probably evolved directly into Cro-Magnon, the first *Homo sapiens sapiens*. Cro-Magnon was an expert hunter. Hunting promoted language and socialization. All human races belong to the same species.

Figure 28.17
Compared to the total history of the earth, humans have been present a very short period of time. When the history of the earth is compared to a 24-hour day, humans evolved a few seconds before midnight.

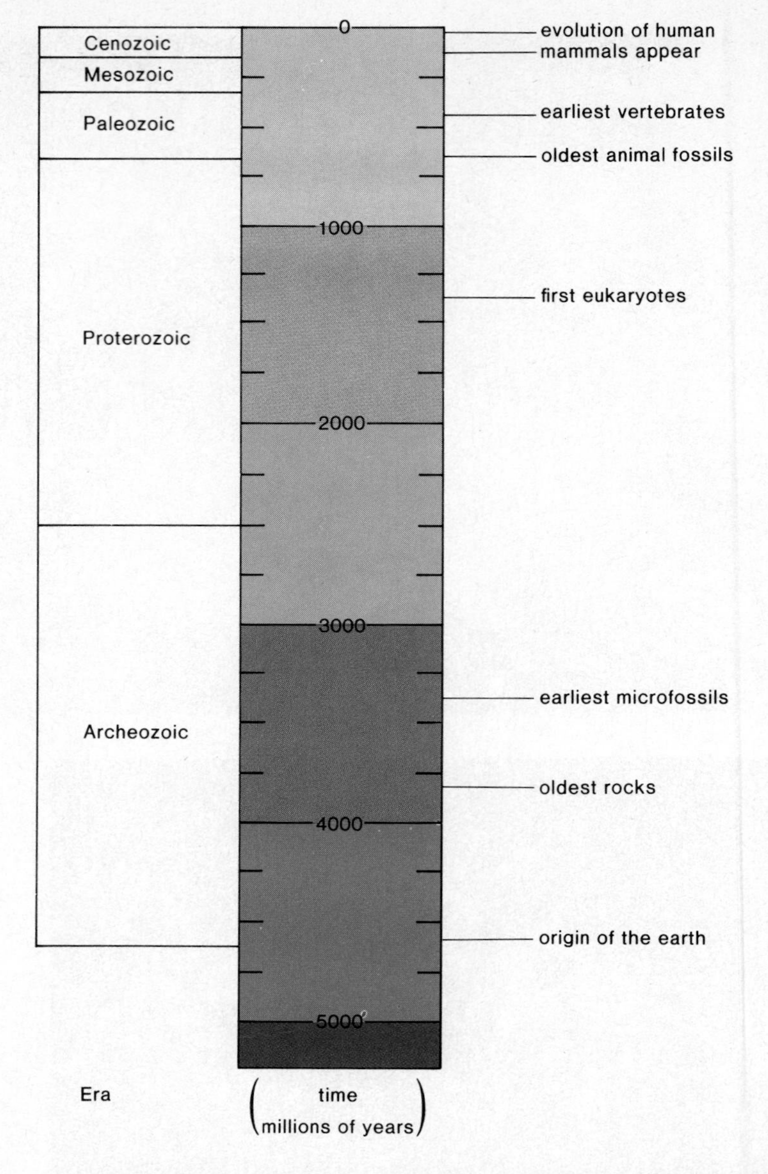

The extinction of many types of animals may account for the transition from a hunting economy to an agricultural economy about 12,000 to 15,000 years ago. Technological progress requiring the use of metals and energy sources led, in an amazingly short time (fig. 28.17), to the Industrial Revolution, which began about 200 years ago. After this time, many people typically lived in cities, in large part divorced from nature and endowed with the philosophy of exploitation and control of nature. Only recently have we begun to realize that the human population, like all other organisms with which we share an evolutionary history, should work with, rather than against, nature. The ecology chapters that follow will stress this theme.

Human Races

All human races of today are also classified as *Homo sapiens sapiens*. This is consistent with the biological definition of species because it is possible for all types of humans to interbreed and bear fertile offspring. The close relationship between the races is supported by biochemical data showing that differences in amino acid sequence between two individuals of the same race are as great as those between two individuals of different races.

It is generally accepted that racial differences developed as adaptations to climate. Although it might seem as if dark skin is a protection against the hot rays of the sun, it has been suggested that it is actually a protection against ultraviolet ray absorption. Dark-skinned persons living in southern regions and

Figure 28.18
All human beings belong to one species, but there are several races such as a. *Negroid,* b. *Mongoloid,* c. *Australoid,* d. *Caucasian, and* e. *American Indian.*

white-skinned persons in northern regions absorb the same amount of radiation. (Some absorption is required for vitamin D production.) Other features that correlate with skin color, such as hair type and eye color, may simply be side effects of pleiotropic genes.

Differences in body shape represent adaptations to temperature. A squat body with shortened limbs and nose retains more heat than an elongated body with longer limbs and nose. Also, the "almond" eyes, flattened nose and forehead, and broad cheeks of the Oriental are believed to be adaptations to the extremely cold weather of the last ice age.

While it has always seemed to some that physical differences might warrant assigning human races to different species, this contention is not borne out by the biochemical data mentioned previously.

Summary

Primates evolved from shrewlike insectivores and became adapted to living in trees. The first primates were prosimians, followed by the monkeys, apes, and hominids. The latter three are all anthropoids, the latter two are hominoids, and the last contains the australopithecines and humans.

Humans and apes share a common ancestor which may possibly have been *Dryopithecus,* a fossil commonly regarded as the first hominoid. The first hominid lived at a time when a change in weather made it advantageous to dwell on the ground.

One possible hominid evolutionary tree shows *Australopithecus afarensis* as a common ancestor to two other species of this genus and *Homo habilis,* the first fossil to be placed in the genus *Homo. Homo habilis* people may not have been highly intelligent but they did make tools. Intelligence and making of tools probably evolved together. If *Homo habilis* is indeed considered human, then humans evolved much earlier than previously thought.

Homo erectus had a large brain and walked with a striding gait. These people used fire and probably were the first true hunters. *Homo sapiens neanderthal* was not as primitive as formerly thought. The Neanderthals probably evolved directly into Cro-Magnon, the first *Homo sapiens sapiens.* Cro-Magnon was an expert hunter. Hunting promoted language and socialization. In a relatively short time, humans developed an advanced culture that has tended to separate them from other organisms in the biosphere. All human races belong to the same species.

Objective Questions

Place each animal below in the highest category possible.

1. Neanderthals	a. prosimian
2. lemurs	b. anthropoid
3. australopithecines	c. hominoid
4. chimpanzees	d. hominid
	e. *Homo*

5. Primates are adapted to an arboreal life in the __________.
6. Among living animals, a __________ best resembles the common ancestor to all primates.
7. The fossil __________ may have been the hominoid ancestor.
8. Based on biochemical evidence, the hominid ancestor is believed to have evolved as late as __________–__________ million years ago.
9. The fossil known as Lucy could probably walk __________, but had a __________ brain.
10. __________ is the first fossil to be classified as *Homo sapiens sapiens.*

Answers to Objective Questions

1. e 2. a 3. d 4. b 5. trees 6. tree shrew 7. *Dryopithecus* 8. 4–5 9. upright, small 10. Cro-Magnon

Study Questions

1. Name several primate characteristics still retained by humans. (p. 624)
2. What animals mentioned in this chapter, whether living or extinct, are anthropoids? Hominoids? Hominids? Humans? (p. 624)
3. Draw an evolutionary tree that includes all primates. Discuss each member of the tree. (p. 627)
4. Which fossil may have been the hominoid ancestor? Contrast the characteristics of this ancestor to those of apes and modern humans. (p. 627)
5. What type of locomotion may have been characteristic of the hominid ancestor? Describe the habitat of this ancestor and discuss how it may have influenced hominid evolution. (p. 628)
6. More than one type of hominid existed at the same time. Explain in terms of adaptive radiation. (p. 632)
7. Contrast the three species of australopithecines. Describe the most recent ideas regarding their diet. (p. 632)
8. Which humans were tool users? Walked erect? Had a striding gait? Used fire? Drew pictures? (pp. 636–637)
9. What evidence do we have that all races of humans belong to the same species? (p. 638) Name several races of humans. (p. 639)

Key Terms

anthropoids (an'thro-poidz) higher primates, including only monkeys, apes, and humans. *625*

australopithecines (aw''strah-lo-pith'e-sīnz) referring to three species of *Australopithecus,* the first generally recognized hominids. *632*

chimpanzee (chim-pan'ze) a small great ape that is closely related to humans and is frequently used in psychological studies. *627*

Cro-Magnon (kro-mag'non) the common name for the first fossils to be accepted as representative of modern humans. *636*

Dryopithecus (dri''o-pith'e-cus) a genus of extinct apes that may have included or resembled a common ancestor to both apes and humans. *627*

gibbon (gib'on) the smallest ape; well known for its arm-swinging form of locomotion. *627*

gorilla (go-ril'ah) the largest of the great apes which is as closely related to humans as to other apes. *627*

hominid (hom'ĭ-nid) member of a family of upright, bipedal primates that includes australopithecines and modern humans. *628*

hominoid (hom'ĭ-noid) member of a superfamily that contains humans and the great apes and humans. *627*

Homo erectus (ho'mo ē-rek'tus) the earliest nondisputed species of humans, named for their erect posture that allowed them to have a striding gait. *634*

Homo habilis (ho'mo hah'bĭ-lis) an extinct species that may include the earliest humans, having a small brain but quality tools. *633*

Neanderthal (ne-an'der-thawl) the common name for an extinct subspecies of humans whose remains are found in Europe, Asia, and Africa. *636*

orangutan (o-rang'oo-tan'') one of the great apes; large with long red hair. *627*

primates (pri'māts) animals that belong to the order Primates, the order of mammals that includes prosimians, monkeys, apes, and humans. *624*

prosimians (pro-sim'e-anz) primitive primates such as lemurs, tarsiers, and tree shrews. *624*

Further Readings for Part Five

Ayala, R. J. 1978. The mechanisms of evolution. *Scientific American* 239(3):56.

Bonatti, Enrico. 1987. The rifting of continents. *Scientific American* 256(3):96.

Cairns-Smith, A. G. 1985. The first organisms. *Scientific American* 252(6):90.

Day, W. 1984. *Genesis on planet earth.* New Haven, Conn.: Yale University Press.

Dickerson, R. E. 1981. Chemical evolution and the origin of life. *Scientific American* 239(3):70.

Eckert, R., and D. Randall. 1983. *Animal physiology,* 2nd ed. San Francisco: W. H. Freeman.

Feder, M. E., and W. W. Burggren. 1985. Skin breathing in vertebrates. *Scientific American* 253(5):126.

Gallo, R. C. 1986. The first human retrovirus. *Scientific American* 255(6):88.

Goreau, T. F., et al. 1979. Corals and coral reefs. *Scientific American* 241(2):124.

Gosline, J. M., and M. E. De Mont. 1985. Jet-propelled swimming in squids. *Scientific American* 252(1):96.

Gould, S. J. The five kingdoms. *Natural History* 85(6):30.

Grant, P. R. 1981. Speciation and the adaptive radiation of Darwin's finches. *American Scientist* 69:653.

Guyton, A. C. 1984. *Physiology of the human body.* 6h ed. Philadelphia: W. B. Saunders.

Hadley, N. F. 1986. The arthropod cuticle. *Scientific American.* 255(1):104.

Hay, R. L., and M. D. Leaky. 1982. The fossil footprints of Laetoli. *Scientific American* 246(2):50.

Hickman, C. P., et al. 1985. *Biology of animals.* 4th ed. St. Louis: C. V. Mosby.

Himmelfarb, G. 1962. *Darwin and the Darwinian Revolution.* New York: W. W. Norton and Co.

Hirsch, M. S. and J. C. Kaplan. 1987. Antiviral therapy. *Scientific American* 256(4):76.

Hogle, J. M. 1987. The structure of poliovirus. *Scientific American* 256(3):42.

Horridge, G. A. 1977. The compound eye of insects. *Scientific American* 237(1):108.

Koehl, M. A. R. 1982. The interaction of moving water and sessile organisms. *Scientific American* 247(6):124.

Levine, J. S., and E. F. MacNichol. 1982. Color vision in fishes. *Scientific American* 246(2):140.

Lewontin, R. C. 1978. Adaptation. *Scientific American* 239(3):212.

Mayr, E. 1977. Darwin and natural selection. *American Scientist.* 65:321–27.

Mayr, E. 1978. Evolution. *Scientific American* 239(3):46.

Moorehead, A. 1979. *Darwin and the beagle.* New York: Harper & Row.

Mossman, D. J., and W. A. S. Sarjeant. 1983. The footprints of extinct animals. *Scientific American* 248(1):74.

Newman, E. A., and P. H. Hartline. 1982. The infrared "vision" of snakes. *Scientific American* 246(3):116.

Racle, F. A. 1979. *Introduction to evolution.* Englewood Cliffs, NJ: Prentice-Hall.

Raven, P. H., et al. 1986. *Biology of plants.* 4th ed. New York: Worth.

Rensberger, B. 1982. Evolution since Darwin. *Science* 82 3(3):40.

Roper, C. F. E., and K. J. Boss. 1982. The giant squid. *Scientific American* 246(4):96.

Russell, Hunter. 1979. *A life of invertebrates.* New York: MacMillan.

Schmidt-Nielson, K. 1981. Countercurrent systems in animals. *Scientific American* 244(5):118.

Schopf, J. W. 1978. The evolution of the earliest cells. *Scientific American* 239(3):110.

Scientific American. 1978. *Evolution.* San Francisco: W. H. Freeman.

Simons, K., et al. 1982. How an animal virus gets into and out of its host cell. *Scientific American* 246(2):58.

Stanier, R. Y., et al. 1976. *The microbial world.* 4th ed. Englewood Cliffs, NJ: Prentice-Hall.

Stebbins, G. L. 1977. *Processes of organic evolution.* 3d ed. Englewood Cliffs, NJ: Prentice-Hall.

Stebbins, G. L., and F. J. Ayala. 1985. The evolution of Darwinism. *Scientific American* 253(1):72.

Stern, K. R. 1985. *Introductory plant biology.* 3d ed. Dubuque, Ia.: Wm. C. Brown Publishers.

Vertebrate structure and function: Readings from Scientific American. 1974. San Francisco: W. H. Freeman.

Vidal, G. 1984. The oldest eukaryotic cells. *Scientific American* 250(2):48.

Volpe, E. P. 1981. *Understanding evolution.* 4th ed. Dubuque, Ia.: Wm. C. Brown Publishers.

Part Six

Behavior and Ecology

The behavior of organisms allows them to interact with their own kind and with other species. These interactions are the framework for ecosystems, units of the biosphere in which energy flows and chemicals cycle. Mature natural ecosystems contain populations that remain constant in size and require the same amount of energy and chemicals each year.

Humans have created their own ecosystem, which differs in that the population constantly increases in size and ever greater amounts of energy and raw materials are needed each year. Since energy is used inefficiently and raw materials are not properly cycled, the human ecosystem is dependent on natural ecosystems to absorb pollutants. Because the natural ecosystems are no longer able to support the human ecosystem in this manner, we must find ways to use energy more efficiently and to recycle materials. Furthermore, the preservation of the natural communities, called biomes, is beneficial to all ecosystems. Preserving the biomes helps to assure the continuance of the biosphere.

Since 1850 the human population has expanded at such a rapid rate that it is doubted by some that there will be sufficient energy and food to permit the same degree of growth in the future. Concomitant with indications that humans desire to preserve the biosphere, the growth rate of the human population has begun to decline.

Flow of energy from one species to another is the norm in the biosphere. A crab spider is feeding on a honeybee caught while gathering nectar from the flower.

29

Behavior within Species

Chapter Concepts

1. Behavioral patterns are inherited and are subject to natural selection.

2. Behavioral patterns may be associated with a continuum that ranges from completely innate at one end to completely learned at the other.

3. The internal state of the animal affects the degree to which the animal performs a certain behavior.

4. Members of a society are able to communicate by chemical, visual, and/or auditory means. They also have ritualistic ways to express aggression.

5. Altruistic behavior by members of a society can be explained on the basis of evolutionary theory.

Chapter Outline

- Evolution of Behavior
  - Adaptation of Behavior
  - Inheritance of Behavior
- Execution of Behavior
  - Innate Behavior
  - Learned Behavior
  - Motivation
- Societies
  - Communication
  - Competition
  - Sociobiology

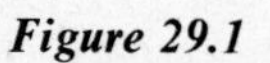

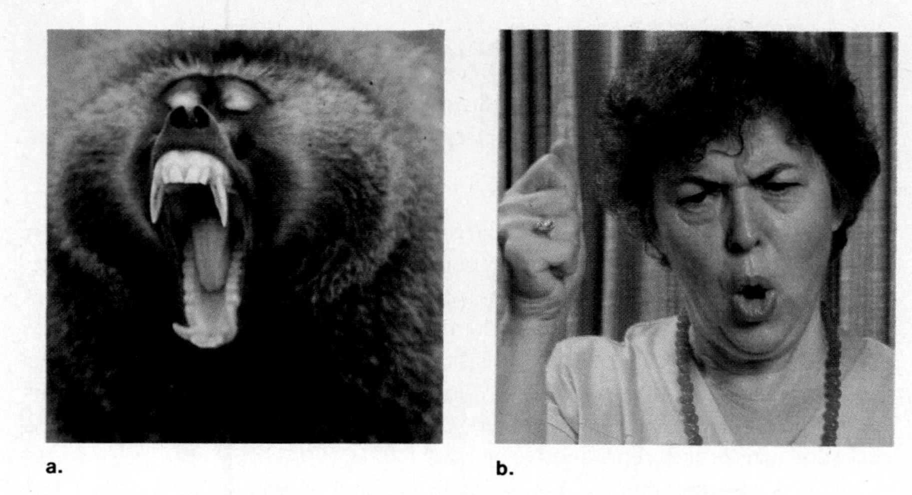

Figure 29.1
Ritualization of aggression among baboons and humans. a. *This male baboon is displaying full threat, a mechanism that is used to establish dominance.* b. *The human being is also displaying full threat, a behavioral pattern that is more likely to occur in a dominant rather than a submissive individual.*

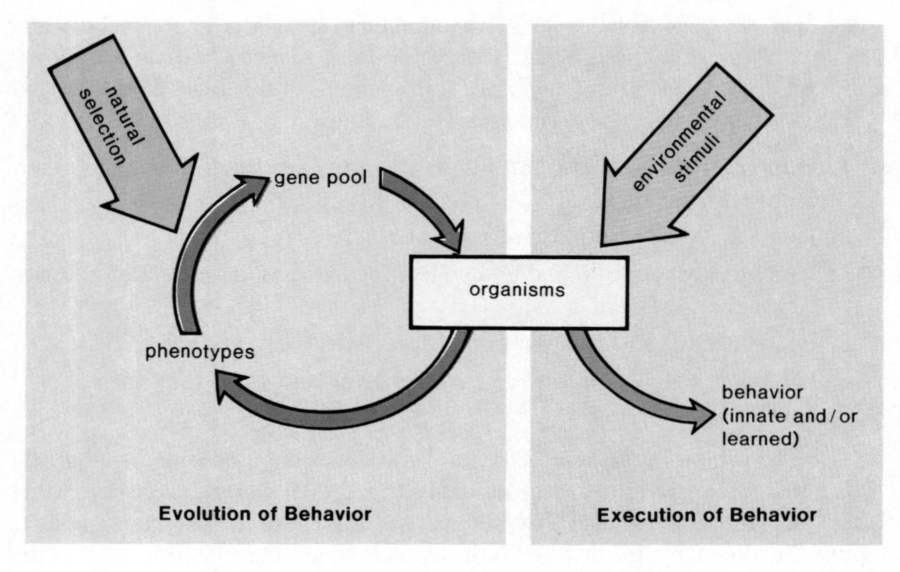

Figure 29.2
A diagram that emphasizes current interests in regard to behavior. On the left-hand side of the diagram we note that selected phenotypes contribute genes to the gene pool of the population and in this way behavior can evolve. On the right-hand side of the diagram we note that environmental stimuli cause organisms to display behavioral patterns which differ as to the degree they are innate or learned. In the latter case, the capacity to learn is a part of the phenotype.

Behavior encompasses the daily activities of an organism. These activities can be analyzed to determine not only their subcomponents but also their possible origination and adaptiveness. For example, it is interesting to compare the aggressive facial appearance of a human and baboon (fig. 29.1) in order to determine if there are any common elements. Ethologists (scientists who study behavior) might ask several types of questions about this behavioral pattern. For one thing, it would be of interest to know how it evolved. And for another, to what degree is the pattern instinctive (innate) or learned. Figure 29.2 diagrammatically presents these two basic aspects of behavioral patterns. We assume that an animal's behavior is adaptive, that is it increases the possibility that the individual's genes will be passed on to the next generation. In other words, then, behavioral patterns are subject to the process of natural selection just as other aspects of the phenotype are also subject to this process. Those individuals with the most adaptive behavioral patterns will be those that will contribute most to the gene pool of the next population of organisms. This is the manner in which the behavior of the species will evolve. The left-hand side of figure 29.2 concerns the evolution of behavior. The right-hand side of the figure concerns the execution of the behavioral pattern. The phenotype of the organism (a typical one for the species) includes a repertoire of behavioral patterns and the particular stimulus determines which pattern will be executed. The behavioral pattern can range from being primarily innate to being primarily learned. Even if the pattern can be modified by learning it should be noted that the capacity to learn is still determined by the phenotype and therefore by the genotype.

Evolution of Behavior

There are two ways to approach the study of the evolution of behavior—to show that the behavioral pattern is adaptive and/or to show that it is inherited.

Adaptation of Behavior

Ethologists sometimes study an isolated behavioral incidence. For example, it has been observed that gulls usually remove broken eggshells from their nests after the young have hatched. Nests with broken shells remaining are subject to more predator attack by crows than those from which shells have been removed. It can be reasoned that gulls that remove broken eggshells from their nests will successfully raise more young than gulls that do not do so. This behavioral pattern has been selected, then, through the evolutionary process because it improves reproductive success.

It is also possible to look at behavior from a broader perspective. For example, the general behavior of an animal can be related to its evolutionary history. A current thesis about the evolution of modern human behavior is termed the "food-sharing" hypothesis. Proponents of this theory list these five behavioral differences between apes and humans:

1. Humans walk erect and this allows them to carry tools, food, and other items with them.
2. Humans have a complicated language.
3. Humans cooperate to find and collect food and sometimes hunt animals larger than they are.
4. Humans postpone consumption of food until they get "home."
5. Humans use tools both while they are away and while they are at home.

The evolution of these behavioral traits assume a division of labor; for example among primitive peoples today (fig. 29.3), women typically gather nearby plant food because they take care of the children. The men go out to hunt. Further they are believed to have encouraged the evolution of a tightly knit social and economic group in which individuals have a sense of duty toward one another.

Inheritance of Behavior

Another example of behavioral evolution is the nest-building habits of the parrot *Agapornis* (fig. 29.4). Three species of this genus have different methods of building nests. Females of the most primitive species use their sharp bills for cutting bits of bark, which they carry in their feathers to a site where they make a padlike nest. Females of a more advanced species cut long regular strips of material, which are placed for transport only in the rump feathers. These strips are used to make elaborate nests with a special section for the eggs. Females of the most advanced species carry stronger materials such as sticks, in their beaks. They construct roofed nests with two chambers and a passageway. The steady progression here from primitive to advanced nest-building suggests that an evolution of behavioral patterns has taken place.

The fact that genes are controlling these traits can be supported by the following experiment. The species that carries material in its rump feathers has been mated to the species that carries material in the beak. The resulting hybrid birds cut strips and try to tuck them in their rump feathers but are unsuccessful. It is as if the genes of one parent require them to tuck the strips, but the genes of the other species prevents them from succeeding. With time, however, these birds learn to carry the cut strips in their bills and only briefly turn the head toward the rump before flying off.

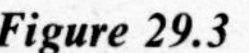

Figure 29.3
The !Kung San (bushmen) of Botswana are believed to behave as humans did when they first evolved. The men hunt cooperatively together and their kill is taken home to be shared by women and children.

Figure 29.4
An Agapornis roseicollis *parrot cuts strips of nesting paper with his beak and then he tucks them in his rump feathers. Birds of this species have been bred with another species that simply carry the strips in their beaks. The resulting hybrids try to tuck the strips but were unable to perform the behavior. Later, they learned to carry the strips in their beaks.*

Recently, investigators have reported on even more in-depth work with the snail, *Aplysia* (fig. 29.5) which lays long strings of more than a million eggs. The snail winds the string into an irregular mass and attaches it to a solid object like a rock. Several years ago, a hormone that caused the snail to lay eggs even though it had not mated was isolated and analyzed. This egg-laying hormone (ELH) was found to be a small protein of 36 amino acids. It acts locally as an excitatory transmitter, augmenting the firing of a particular neuron in the snail's body. It also diffuses into the circulatory system and excites the smooth-muscle cells of the reproductive duct, causing them to contract and expel the egg string. Using recombinant DNA techniques (p. 492), the investigators isolated the ELH gene. Surprisingly, the gene's product turned out to be a protein with 271 amino acids. The protein could possibly be cleaved into as many as 11 possible products and ELH is one of these known to be active. The investigators speculate therefore that all the components of the egg-laying behavior could be mediated by this one gene. This work is quite remarkable because it definitely shows that genes control behavior and further shows the manner in which they do so in a simple organism.

Behavioral patterns are inherited and evolve in the same manner as other animal traits. Breeding experiments with the parrot *Agapornis* and recombinant DNA experimentation with the snail *Aplysia* support this contention.

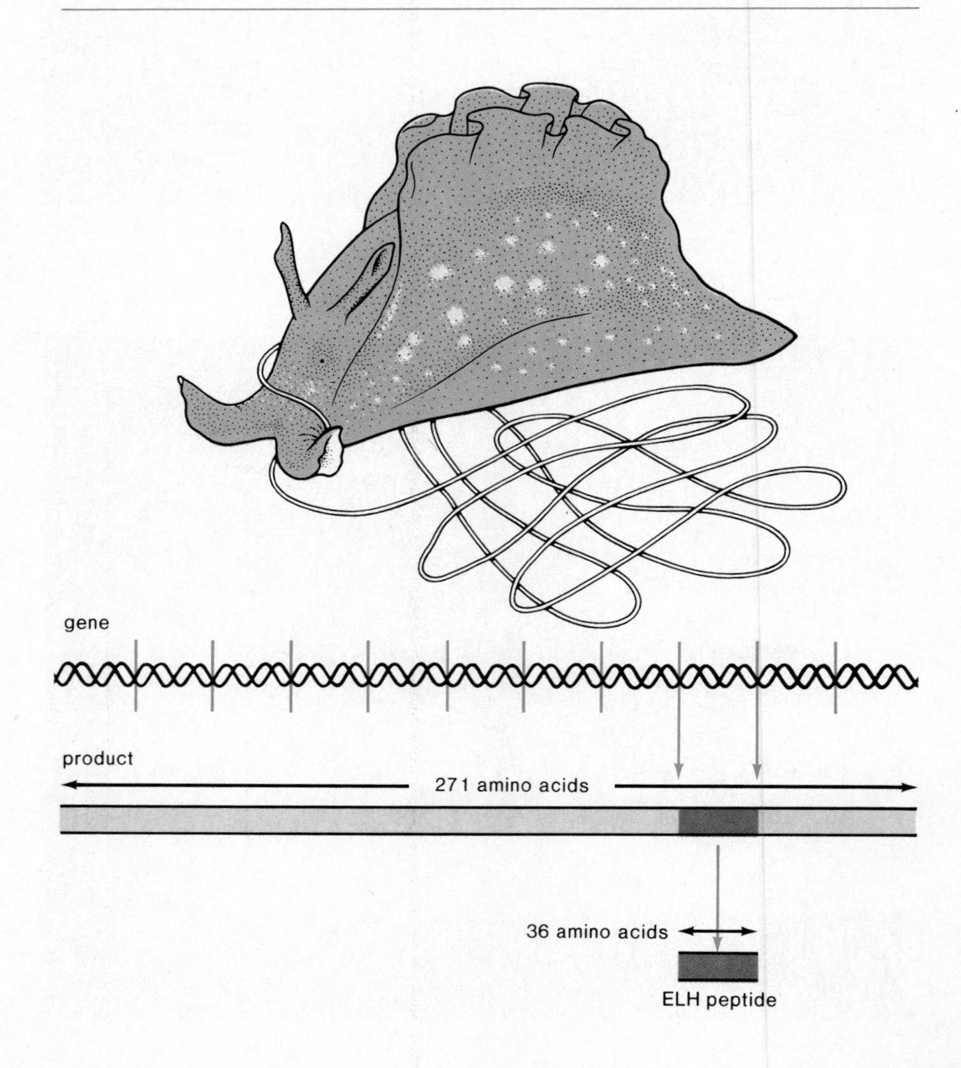

Figure 29.5
The snail Aplysia *lays long strings of eggs which are wound into an irregular mass for deposition on a nearby object. Using modern recombinant DNA techniques, investigators have isolated a gene whose product contains 271 amino acids of which 36 make up the ELH (egg-laying hormone) peptide. This peptide is partially responsible for the egg-laying behavior. Analysis of the gene shows that there are enough signals for cleavage (dots) to permit the product to include ten other peptides, each of which may have some role in the animal's behavior.*

Execution of Behavior

Figure 29.2 tells us that behavior occurs as a response to a stimulus. Therefore, it is dependent on the sense organs, nervous system, and the musculoskeletal system of the organism. Often, however, the animal must be motivated to respond, and motivation is most likely dependent on the physiological or internal state of the organism.

The complexity of behavior increases with the complexity of the nervous system. Animals with simple nervous systems tend to respond automatically to stimuli in a programmed way, whereas animals with complex nervous systems are apt to choose behavior that suits the particular circumstance. The first type of behavior, which is inherited, is called **innate,** or instinctive, while the second type, which requires modification of behavior is said to involve **learning.** All animals have some instinctive behavior, even humans, but as higher animals with complex nervous systems evolved, the capacity to learn behavioral patterns developed (fig. 29.6).

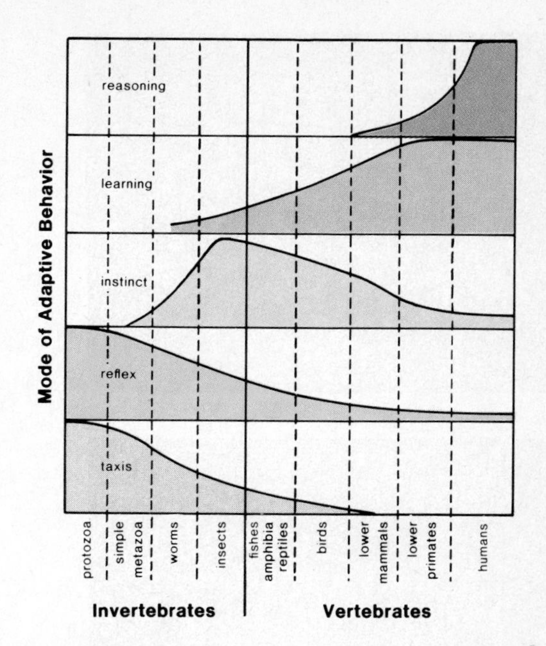

Figure 29.6
Comparative frequency of modes of adaptive behavior among invertebrates and vertebrates. While innate types of behavior (taxes and reflexes) are more frequently found among invertebrates, learned types are more frequently found among vertebrates.

Innate Behavior

In innate behavior, the stimulus appears to trigger a fixed response that does not vary according to the circumstances.

Taxes

Orientation of the body toward or away from a stimulus is termed **taxis** in animals. Animals exhibit a number of different types of taxes; *phototaxis* is movement in relation to light and *chemotaxis* is movement in relation to a chemical. Some insects, for example moths and flies, will fly directly toward a light. Often they orient themselves by shifting the body until the light falls equally on both eyes. If one eye is blind, the animal will move in a spiral, forever trying to find the direction in which the light will be balanced between the two eyes.

Chemotaxes are quite common. Insects are attracted to minute quantities of chemicals, called **pheromones,** given off by members of their species. Army ants are so set on following a pheromone trail that if, by chance, it leads them into a circle, they will continue to circle until they die of exhaustion (fig. 29.7). Vertebrates, too, are sometimes highly responsive to chemicals. A few whiffs of a piece of clothing, and bloodhounds are capable of tracking down a single individual.

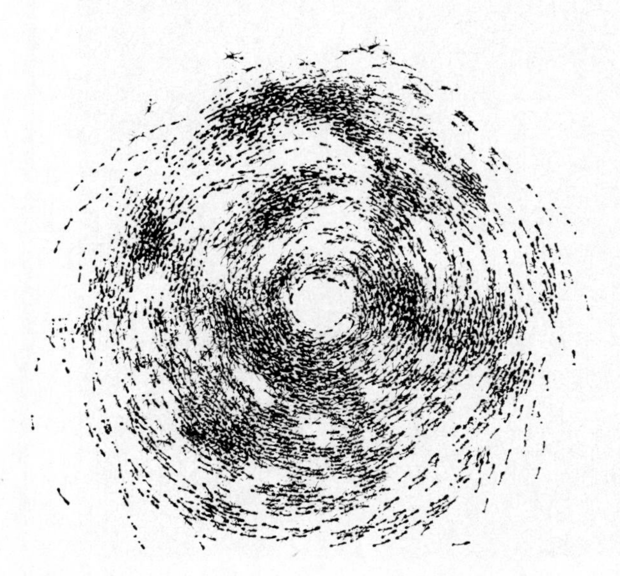

Figure 29.7
These ants are following a pheromone trail in a circle. This is an example of chemotaxis, orientation of the body in relation to a chemical in the environment.

a.

b.

Figure 29.8
Bats use echolocation (sonar soundings) to locate their prey in the dark. Many moths are able to detect the soundings and begin evasive action that often prevents the bat from finding them. This is an example of coevolution.

Only a few animals are able to orientate themselves by means of *sonar* or *echolocation.* Bats send out series of sound pulses and listen for the echoes that come back. The time it takes for an echo to return indicates the location of both inanimate and animate objects and enables a bat to find its way through dark caves and to locate food at night (fig. 29.8). Some moths have evolved the ability to hear the sounds of a bat and they begin evasive tactics when they sense that a bat is near.

Migration and Homing

Migration, which often occurs seasonally, and **homing,** the ability to return home after being transported some distance away, have been studied in birds. Evidence indicates that they use the sun in the day (fig. 29.9) and the stars at night as compasses to determine direction (north, south, east, or west). Like bees, they even allow for the east-to-west movement of the sun during the course of a day. Bees offered a food source can return to their hive no matter where the sun is located in the sky. Therefore, it is believed that they have an internal or *biological clock* that tells them the location of the sun according to the time of day.

Although birds use the sun and stars as compasses, they cannot rely on these to tell them that home is in a particular direction. In other words, suppose you were blindfolded and then transported away from home. Upon being set free, you are given a compass to tell direction; how would you know which direction to select? Some investigators now believe that birds and other animals are sensitive to magnetic lines of force, which are dependent on the earth's magnetic field, and this allows them to determine the direction of home.

Salmon use chemotaxis to pinpoint their exact home. Salmon are hatched in a tributary of a river but grow to maturity in the open sea. At spawning time, mature salmon travel back up the river to the same spot at which they were born. Experiments have shown that the fish appear to find the spawning ground by following the chemical scent of their first home.

Figure 29.9
Migration of Canada geese. Many different types of organisms migrate long distances to breeding grounds where they mate and reproduce. Scientists have long wondered how they find their way. Recently it has been shown that they are capable of using the earth's magnetic field to navigate. A strongly magnetic iron-oxide compound called magnetite has been found in their bodies and it is possible this compound is involved in the process.

Rhythmic or Cyclic Behavior

Certain adaptive behaviors of animals reoccur at regular intervals. Behavior influenced by a **circadian rhythm** occurs on a daily basis. For example, some animals, like humans, are usually active during the day but sleep at night. Others, such as bats, sleep during the day and hunt at night. Behavior controlled by a **circannual rhythm** occurs on a yearly basis, such as when some birds migrate south every fall. Other rhythms are also known; lunar behavior occurs monthly and tidal behavior occurs every 12.4 or 12.8 hours.

Originally, it was assumed that environmental changes, such as day and night, controlled cyclical behavior in animals. But it is now known that such behavior will occur even when the associated stimulus (daylight or darkness) is lacking. For example, fiddler crabs are dark in color during the day but light in color at night, even when kept in a constant environment. But if the crabs are kept in the constant environment indefinitely, the timing of the daily change tends to drift and becomes out of synchronization with the natural cycle. For this reason, it has been suggested that rhythmic behavior is under the control of an innate, internal biological clock that runs on its own but is reset by external stimuli. Aside from keeping time, a biological clock must also be able to bring about the change in behavior. In the fiddler crab, for example, it must be able to stimulate the processes that cause the shell to change color. Therefore, a biological clock system needs (fig. 29.10):

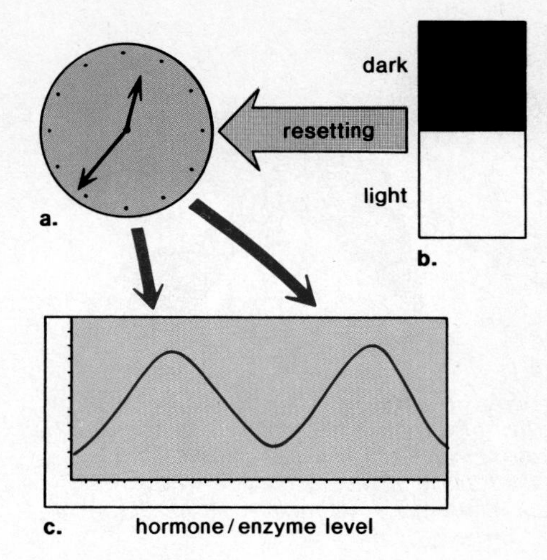

Figure 29.10
Biological clock system. Biological clock systems have three components: a. *an internal timekeeper,* b. *a means of detecting light/dark periods,* c. *a method of communication that eventually brings about a behavioral response.*

1. a time-keeping mechanism that *keeps time* independently of external stimuli (i.e., a minute is always a minute);
2. a receptor that is sensitive to light/dark periods and *can reset* the clock for circadian rhythms or indicate a change in the length of the day/night for circannual rhythms;
3. a communication mechanism by which the clock *induces* the appropriate behavior.

A review of the discussion concerning flowering (p. 150) shows that the last two components of a biological clock system have been tentatively identified: phytochrome is believed to be the receptor sensitive to light and dark periods, and plant hormones are believed to be the means by which flowering is induced. The time-keeping mechanism has not been identified, however.

In animals, experiments with birds suggest that the biological clock resides within the pineal gland. The pineal gland produces *melatonin,* a hormone that lowers body temperature and causes birds to roost, both of which are characteristic of nighttime. Also, there is evidence of neural communication between the eyes and the pineal gland, which is closely associated with the "third eye" present in some lizards, amphibians, and fishes.

Reflexes

Reflexes are simple automatic responses to a stimulus over which the individual appears to have little or no control. When a human knee is hit by a mallet, the lower leg jerks in a characteristic manner. Since reflexes are clearly innate, some investigators believe that it might be possible to explain complex behavior of some lower animals, such as army ants, by a series of reflexes, each one of which acts as a stimulus for the one following.

Fixed Action Pattern The term **fixed action pattern** (stereotyped behavior) has been given to complex behavior that occurs automatically as if it were a composite of reflex actions. A stimulus that initiates a fixed action behavior is called a **sign stimulus.** An animal must possess neural mechanisms (called releasing mechanisms) that are sensitive to the sign stimulus in order to respond in the stereotyped manner. As an example, consider the fact that male

Figure 29.11
A male robin is pecking at a red tuft rather than an exact replica of a robin without a red breast. This shows that the color red is a sign stimulus that initiates the release of the hostile behavioral pattern, an innate fixed action pattern consisting of a series of reflexes.

robins attack a red tuft of feathers in preference to an exact replica of a male robin without the red breast (fig. 29.11). The color red is a sign stimulus that provokes the releasing mechanism, which controls the attacking behavior. This behavior is therefore a fixed action pattern. Animals performing fixed action patterns may seem to be acting in a purposeful manner but, just as with the robin in the previous example, experimentation proves that this is not the case. For example, certain solitary digger wasps dig a hole and then seek out a caterpillar, paralyzing it with a series of stings along the undersurface. The wasp carries the prey to the nest (fig. 29.12) and pulls it in. After laying her egg on the side of the caterpillar, she begins to close the burrow. If, at this point the experimenter removes the caterpillar and puts it on the ground nearby, the wasp will continue to cover the hole even though the caterpillar is in full view.

Gulls, too, have fixed action patterns. When a gull is incubating eggs, it will retrieve any egg that rolls out of the nest. The more speckled the egg, the more strongly is the gull stimulated to roll it into the nest and incubate it. One curious result of experiments like these has been the discovery of **supernormal stimuli.** For example, parent gulls, if given a choice in size, will prefer to retrieve eggs much larger than normal even, if they are then unable to brood them (fig. 29.13).

Fixed Action Patterns and Learning In order to tell if a specific behavior is innate (instinctive), it is customary to determine if it is *(a)* performed by all members of the species in the same manner, and *(b)* performed by animals that have been raised in isolation and/or have been prevented from practicing it.

It is not surprising that some forms of fixed action behavior improve in efficiency after an opportunity to practice has been provided. For example, isolation experiments have been performed to see if songbirds will sing their species' song without having had an opportunity to learn it from another bird. Often they sing a song that is less complicated than the natural song, but they will learn the complete song when they are given the opportunity to hear it, even if many other bird songs are played for them at the same time.

Figure 29.12
A digger wasp is about to pull this paralyzed caterpillar into a hole to serve as food for an offspring that will hatch from an egg she lays on the caterpillar. This is an example of a fixed action pattern because, if by chance an experimenter removes the caterpiller, the wasp will still continue with the behavior and close the hole.

Baby chicks peck at the parent's beak in order to induce the parent to feed them. Experimentation with various models has shown that the chicks are not very discriminatory at first and choose a model that does not resemble the parent to any extent. With time, however, pecking accuracy and efficiency improves and the chicks become progressively more selective and choose a model that more nearly resembles the parent (fig. 29.14).

---

*I*nnate behavior includes taxes and reflexes. Taxes, together with the possession of an internal clock, can, in some cases, offer an explanation for the ability of animals to return to certain locations. Fixed action patterns, which are a series of reflexes, occur as a response to a sign stimulus. These are largely inherited, although they often increase in efficiency with practice.

Figure 29.13
A gull is trying to brood an artificial oversized egg provided by an experimenter, even though the proper size egg is in view. Experimenters have found that fixed action behavioral patterns are more likely to be prompted by supernormal stimuli than by normal stimuli.

Learned Behavior

Learning is a change in behavior as a result of experience. The capacity to learn is inherited and allows an organism to change its behavior to suit the environment. The organism alters its behavior to respond to a specific stimuli in ways that promote its own survival or that of its offspring and/or near relatives.

Kin Recognition

Currently there is a great deal of interest in detecting how organisms learn to recognize close relatives. For instance, tadpoles of American toads can apparently discriminate between siblings and nonsiblings, and Belding's ground squirrels even know the difference between full-siblings and half-siblings.

Kin recognition is beneficial in at least four ways. Parents who can recognize their offspring do not waste time rearing unrelated young; young that recognize their parents avoid the risk of being harmed by a nonparent; giving aid to a sibling rather than a nonsibling increases an animal's fitness (the likelihood of passing on one's genes); and mating with an individual that is to a degree unrelated increases the likelihood of fertile offspring.

Thus far, research has been most extensive in the area of imprinting.

a.

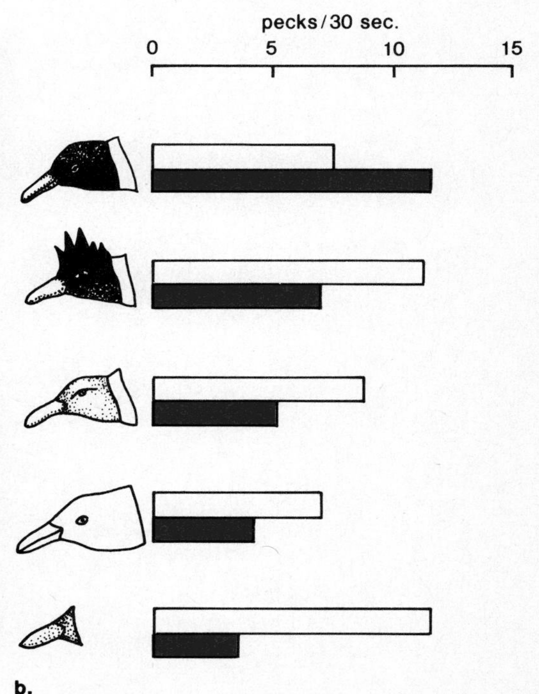

b.

Figure 29.14
Feeding behavior among gulls. a. *Chicks peck at the red spot on a parent's bill when they want to be fed.* b. *In an experiment with laughing gull baby chicks it was shown that with time, chicks peck at a more exact replica of parents' bill (first model). In each example, the top bar represents the pecking frequency of newly hatched chicks; the bottom bar represents the pecking frequency of chicks three-to-five days old.*

Figure 29.15
Baby ducks are following Konrad Lorenz, who, like Niko Tinbergen, did experiments in the field. Lorenz found that ducks learn to follow the first moving object they see which, of course, is normally their mother. Here they have become "imprinted" on an improper object. Normally, the process of imprinting leads to evolutionary success and, therefore, has been selected for.

Imprinting Konrad Lorenz, a famous ethologist, observed that birds become attached to and follow the first moving object they are exposed to. He termed this behavior **imprinting** and suggested that it was a means by which organisms learn to recognize their own species. Ordinarily, the object followed is the mother; however, Lorenz caused chicks and goslings to be imprinted to him (fig. 29.15), and then showed that they would choose him over their own mother when given the opportunity.

A variety of animals, in addition to birds, are susceptible to imprinting during a certain critical period of time following birth. Therefore this term may be used in a larger context to refer to any period of time during which a type of learning is apt to be more successful. For example, puppies take to dog training better than older dogs; and adult humans are better at sports they learned as a child.

Conditioned Learning

In a type of conditioned learning called **associative learning,** an animal learns to give a response to an irrelevant stimulus. Pavlov's dogs learned to expect food and began to salivate when a bell rang because they had learned to associate the ringing of the bell with food (fig. 29.16). Associative learning can explain behavior that seems out of place. For example, the ringing of a bell in and of itself does not have anything to do with food. Even so, associative learning can be useful; for example, it has been suggested that mothers can instill love of learning by holding their children in their laps while reading to them.

Operant Conditioning **Operant conditioning** is trial-and-error learning. An animal faced with several alternatives is *rewarded* (positively reinforced) for making the proper choice and thereafter learns to make this response without hesitation. This type of learning, employed by animal trainers, can be used successfully to make animals learn all sorts of tricks. Using *positive reinforcement,* B. F. Skinner, a well-known Harvard University behaviorist, has even taught pigeons to play a form of Ping-Pong. Skinner believes that humans, too, are primarily controlled by positive reinforcement mechanisms although many disagree with this supposition.

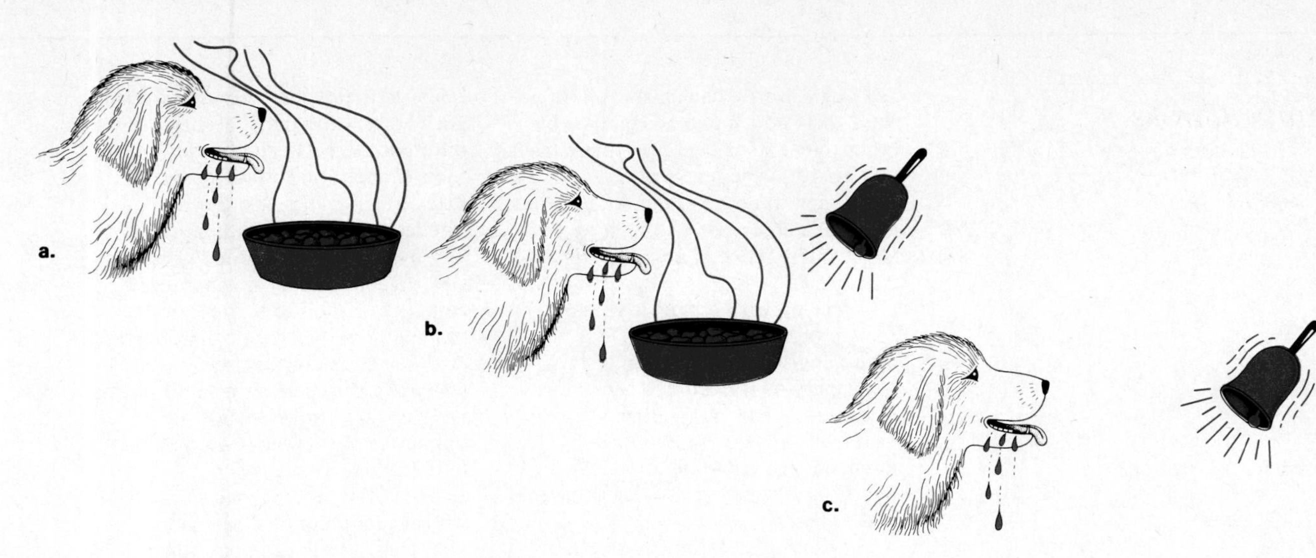

Figure 29.16
Conditioned reflex. The presence of a conditioned reflex shows that the subject has learned to associate two previously unrelated events. For example, (a.) *a dog salivates when presented with food. If* (b.) *a bell is rung when the food is presented, the dog* (c.) *now salivates when the bell is rung, but there is no food. Conditioned reflexes can explain seemingly irrelevant behavioral patterns.*

Insight Learning

Insight, or reasoning, is the ability to solve a problem using previous experiences to think through to a solution. To the observer, it seems as if the animal employing insight needs no practice to successfully reach a goal. For example, apes can devise the means to get to bananas placed out of their arms' reach. They will pile up boxes or use a pole in order to reach the food (fig. 29.17).

Perhaps the ultimate in reasoning ability is learning and using a language. When adult humans use language they are able to put words together in various ways to convey new and creative thoughts. As discussed in the reading on page 656, for the past several years investigators have been trying to determine the degree to which other animals can learn to use language in this way. Most of this work has been done with chimpanzees. Investigators have overcome the biological inability of chimpanzees to articulate many sounds by having them learn a sign language. Chimpanzees have learned as many as four hundred signs of an artificial visual language; however, thus far it has not been possible to demonstrate unequivocally that these animals are capable of putting the signs together to create new sentences and meanings. Some have suggested that the animals are merely copying their trainers and have no true understanding of language use.

Of course, it can be questioned whether cognition, or the ability to think, requires the use of language at all. The reading gives several examples that illustrate that even lower animals, like bees, at times seem capable of analyzing a situation. Perhaps animals only differ in the degree to which they are aware or in the manner in which they are aware of themselves and their environments.

Figure 29.17
Apes are capable of insight learning, reasoning out a solution to a problem without the need for trial and error. On the far left a chimpanzee is unable to reach a banana. In the middle sketch, the chimp seems to suddenly realize a solution to the problem. On the far right, it stands on boxes to reach the banana. If an animal can reason, does it have a cognitive awareness of itself and its surroundings? Investigators are now wrestling with this question.

Learned behavior includes kin recognition, conditioned learning, and insight learning. Much investigative work is being done to determine how aware animals—other than humans—are of themselves and their surroundings, and if these animals can be taught to use a language.

Do Animals Think?

Cognitive psychologists believe their new found acceptance by other scientists is based on their strict adherence to the tenets of experimental observation. And what they have observed has much to say about intelligence in animals. For example:

Working with a pair of young female bottlenose dolphins, psychologist Louis Herman of the University of Hawaii has been able to achieve what many other scientists have not: proof of an animal's understanding not only of vocabulary, but also of communicative grammar and syntax.

Beginning in 1979, the dolphins—which were kept in two 50-foot pools—were taught two artificial languages, each made up of about 35 words. The languages have their own syntactical rules for word order, from which sentences up to five words long can be constructed. Depending on the order of the words, about 1,000 different sentences can be expressed. In one language, the words are made up of electronically generated whistles, created by a computer. In the second language, the words are formed by hand-and-arm gestures performed by the trainer. ("Ball," for instance, is signaled by a human at poolside, raising his hands quickly above his head and then lowering them.) One dolphin specialized in each language.

The evidence that the dolphins appreciate syntax was found in their ability to differentiate correctly between phrases such as "hoop fetch Frisbee" (in other words, "get the hoop and take it to the Frisbee") and "Frisbee fetch hoop" ("get the Frisbee and take it to the hoop") or, "net in basket" ("put the fishing net in the plastic laundry basket") and "basket in net" (the reverse). For research purposes, Herman named the two study dolphins Phoenix (the animal that responds to the trainer's hand gestures) and *Akea Kamai,* which means "Lover of Wisdom" in Hawaiian. "It is a name," remarks Herman, "chosen with hope."

Psychologists R. J. Hermstein of Harvard, Mark Rilling of Michigan State University, and Anthony Wright of the University of Texas have tested pigeons for their ability to make mental discriminations. Some of their most interesting work goes beyond the birds' ability to differentiate between objects; it tests their capacity to form concepts or categories—a higher order of intellectual accomplishment.

To date, the experiments indicate that pigeons can form generalized concepts of such test categories as "human being," "tree," "fish," and "oak leaf." In a typical test, a bird was trained to peck at a small plastic paddle when shown a slide of a human being. It was rewarded with food for a correct answer.

After the bird learned to peck when the picture of a person appeared, and not to peck when photos lacking people were shown, the experiment moved to the test phase. In this stage, the bird was shown a large number of photos—1,200 in all—which it had never seen. Some of the new test transparencies were pictures of human beings. Without food reinforcement, the pigeon correctly pecked when photos of people were shown. It even did so when the new pictures portrayed groups of people. The scientists' conclusion: the pigeons must be able to understand the concept of "people." Similar experiments have also been successfully conducted using trees as a photographic subject. And Wright has achieved the same kind of results using different tests with rhesus monkeys as his subjects instead of pigeons.

Robert Seyfarth and Dorothy Cheney, ethologists from the University of California at Los Angeles, spent more than a year in Kenya studying the vocalizations of vervet monkeys. Their research shows that the animals have at least three distinct alarm cries: a raspy bark for the sighting of a leopard, after which the troop retreats to nearby trees; a short grunt for eagles and hawks, which signals a quick glance skyward and a rush for thick vegetation; and a high chutter for a snake, causing the troop members to rise up on their haunches and survey any tall grass in the vicinity. This sort of *specificity* of communication, "culturally transmitted" from adults to young, is clearly within the province once thought to be reserved only for humans.

The *Schwanzeltanz,* or "waggle dance," with which honey bees tell their sisters back at the hive the precise location of a good nectar source, was first decoded by Karl von Frisch in the 1940s (he was finally awarded the Nobel Prize for the discovery in 1973). More recently, researchers such as James Gould of Princeton University have further investigated the complexity of bee communication. "It is," says Gould, "the second most complex language we know of." (The first is the one you are employing to read this article.) For example, it appears that bee language even has dialects—and that a waggle which in effect means five yards to an Egyptian bee means 50 yards to an Austrian bee.

During the course of his work, however, Gould has chanced upon other evidence of seemingly intelligent behavior. For example, the time-honored method of training bees to fly from a hive to a desired location is to begin by placing a small container of sugar water very close to the hive and then progressively move it farther away. That allows the bees to collect the sugar and return to the hive before flying out again. The standard practice is to relocate the sugar source 25 percent farther away with each move. If it was 20 yards from the hive, it was moved to 25 yards. If it was 200 yards from the hive, it was moved to 250 yards. Using this method, Gould and his associates have successfully trained bees to fly to sources six miles from the hive.

However, during the course of the training, a curious thing began to happen. "The bees seemed to have figured out that we were moving the source in a regular way," recalls Gould. "When we moved to a new, more distant area, they were already there waiting for us—as if they had been able to figure out the rule we were using to move the sugar water. And it's hard to explain this behavior as innate, because there is nothing I know of in the natural history of flowers that has allowed them to move."

Lastly, consider the prodigious feats of memory accomplished by the Clark's nutcracker, a food-hoarding relative of the crow that lives in the southwestern United States. Biologist Russell Balda of Northern Arizona University found that the small bird spends the late summer collecting seeds from the pinon pine and burying as many as 33,000 of them in caches of four or five seeds each. According to Balda's research, the nutcracker finds its caches months after their creation by an elaborate system in which the creature memorizes nearby landmarks. . . .

Herbert Terrace, a psychologist, points to a puppet's mouth and a chimp responds by pointing to its own.

. . . "Now that the legitimacy of animal cognition as a field of inquiry has been demonstrated," says Columbia's Herbert Terrace, a leading cognitive psychologist, "I see little gained by approaching the subject as one would approach human cognition." Maryland's Hodos is even more emphatic: "Attempts to equate animal and human intelligence may be doomed to failure," he argues. "Intelligence is a human concept. Our ideas about it are still evolving. The same behavior may be viewed as intelligence in one situation and not in another. Therefore, it represents a value judgment as much as a biological property. And what we as humans value may not be what an animal values."

The recent move to study animal intelligence on its own terms, in the words of University of Massachusetts psychologist Alan Kamil, "to refocus attention on animals as animals, with less concern for immediate human relevance," is certainly a healthy sign. "It's high time" observes Kamil, "that we stopped looking at animals as proto-humans."

As the search to understand the nature of animal intelligence continues, the quarry is now being viewed in a new light. "I believe," observes Anthony Wright, "that the role of science should be to discover the underlying processes, the foundations, the basic laws."

Most of the researchers are well aware of the scientific dangers. "If you look through the history of this field," says Mark Rilling, "it's littered with corpses rotting in blind alleys." Rutgers' Beer sounds a note of caution: "Most of the work today is quite sober, but you still have to separate the romance from the hard thinking." To which psychologist Roger Thomas of the University of Georgia adds: "We never really know who is more clever, the animal or the experimenter."

Figure 29.18
Ringdove mating behavior. Successful reproductive behavior requires these steps. a. *The male performs courtship behavior,* b. *copulation takes place,* c. *creating of a nest* d. *incubation of eggs, and* e. *baby chicks are fed "crop milk." It can be shown that internal hormonal levels motivate the birds to perform this behavioral pattern. Similarly, in all animals, behavioral patterns are most likely to occur if the animal is ready to perform them.*

Motivation

As mentioned earlier, certain types of behavior recur periodically in animals. These types of behavior seem to require an internal readiness before the animal shows the behavior. Therefore, it is said that the animal is **motivated** to perform this behavior. Motivated behavior seems to include three stages:

1. An appetitive stage during which the animal searches for the goal
2. A consummatory stage or a series of responses directed at the goal
3. A quiescent stage when the animal no longer seeks the goal

A good example of motivated behavior is the need for food. A hungry animal goes out to look for food; once food is found, it eats the food; and then, being satisfied, it no longer seeks food.

Motivated behavior requires the supposition that the animal is readied to perform the behavior because of some internal state. A study of reproduction (fig. 29.18) in ringdoves has indeed shown this to be the case. Ringdoves reproduce in the spring but when male and female ringdoves are separated one from the other, neither shows any tendency toward reproductive behavior. In contrast, when a pair are put together in a cage, the male begins courting by repeatedly bowing and cooing. Since castrated males do not do this, it can be reasoned that the hormone testosterone readies the male for this behavior. The sight of the male courting causes the pituitary gland in the female to release FSH and LH; these in turn cause her ovaries to produce eggs and release estrogen into the bloodstream. Now both male and female are ready to construct a nest, during which time copulation takes place. The hormone progesterone is believed to cause the birds to incubate the eggs and, while they are incubating the eggs, the hormone prolactin causes crop growth so that both parents are capable of feeding their young crop milk.

Reproductive behavior in the ringdove can be explained on the basis of both *external and internal stimuli.* The external stimuli are processed by the central nervous system, which then directs the secretion of hormones. In animals the nervous and endocrine systems work together to produce physiological changes that lead to appropriate behavior patterns. But, animals with complex nervous systems are more likely to be able to control their behavior due to previous learning regardless of their internal state. For example, human beings can decide whether to engage in sexual behavior even if their testosterone or estrogen blood level is high. This decision is probably based on their previous learning experiences in regard to sexual behavior. Also, human beings do not need a hormonal state to be motivated, as when they voluntarily decide to continue the process of learning.

An animal's motivation often determines whether or not it performs a behavior.

Societies

A **society** is a group of individuals belonging to the same species that are organized in a cooperative manner. In order to accomplish cooperation, evidence suggests that members of a society need a means of reciprocal communication and a means of overcoming aggression.

Communication

Communication by chemical, visual, auditory, or tactile (touch) stimuli often includes a *social releaser,* a sign stimulus used between members of the same species that causes the receiver to respond in a certain way.

Chemical Communication

The term *pheromone* is used to designate chemical signals that are passed between members of the same species. Pheromones can have either releaser effects or primer effects. A pheromone with a **releaser effect** evokes an immediate behavioral response, while a pheromone with a **primer effect** alters the physiology of the recipient, leading to a change in behavior.

Sex attractants are good examples of pheromones with releaser effects. For example, female moths secrete chemicals from special abdominal glands. These chemicals are detected downwind by receptors on male antennae (fig. 29.19). This signaling method is extremely efficient since it has been estimated that only 40 out of 40,000 receptors on the male antennae need to be activated in order for the male to respond.

Ants and honeybees provide an example of a pheromone with a *primer effect.* The queen produces a substance that is passed from worker to worker by regurgitation. This substance prevents the workers from raising other queens and it also prevents the ovaries of the workers from maturing. Primer effects have also been seen in mice. Male mice produce a substance that can alter the reproductive cycle of females. When a new male and female are placed together, this substance can cause the female to abort her present pregnancy so that she can then be impregnated by her new mate. Similarly, crowding of female mice causes disturbances or even blockage of their estrous cycles. In both instances, removal of the olfactory lobes prevents these occurrences.

It is well known that male dogs and cats are attracted to the opposite sex by means of scent. Experimentation is suggesting that apes, and even humans, as discussed in the reading on page 662, may also be influenced by pheromones.

Figure 29.19
Example of chemical communication between members of the same species. Mole male antennae are capable of detecting female pheromone from miles away.

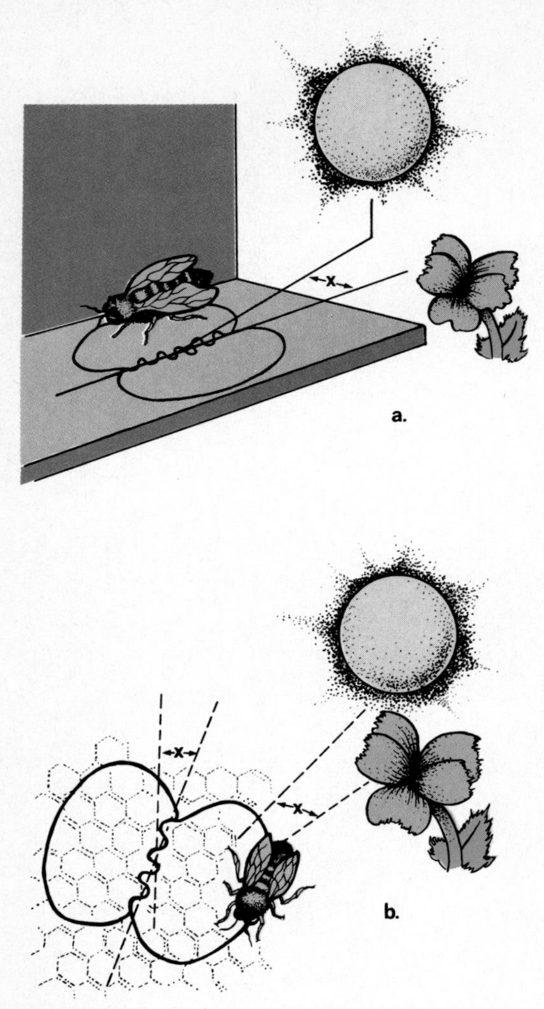

Figure 29.20
Visual communication among bees relates information not about the bee but about the environment. Honeybees do a waggle dance to indicate the direction of food. a. *If the dance is done outside the hive on a horizontal surface, the straight run of the dance will point to the food source.* b. *If the dance is done inside the hive on a vertical surface, the angle of the straight run to that of the direction of gravity is the same as the angle of the food source to the sun.*

Visual Communication

The communication of honeybees is believed to be remarkable because the so-called language of the bees uses not only visual stimuli but other stimuli as well to impart information about the environment and not about the bee itself. When a foraging bee returns to the hive, it performs a routine known as the **waggle dance** (fig. 29.20). The dance, which indicates the distance and direction of a food source, has a figure-eight pattern. As the bee moves between the two loops of the figure eight, it buzzes noisily and shakes its entire body in so-called waggles. *Distance* to the food source is believed to be indicated by the number of waggles and/or the amount of time taken to complete the straight run. The straight run also indicates the *location* of the food; when the dance is performed outside the hive, the straightaway indicates the exact direction of the food, but when it is done inside the hive, the angle of the straightaway to that of the direction of gravity is the same as the angle of the food to the sun. In other words, a 40° angle to the left of vertical means that food is 40° to the left of the sun. The bees can use the sun as a compass to locate food because their biological clocks (p. 651) allow them to compensate for the movement of the sun in the sky.

Visual communication includes many *social releasers*. Male birds and fish sometimes undergo a color change that indicates they are ready to mate. Female baboons show that they are in estrus by a reddening of the sex flesh on the buttocks. On the other hand, visual communication also includes defense and courtship patterns (fig. 29.25) that comprise a type of body language. These patterns are **ritualized,** which means that the behavior, which is stereotyped, exaggerated, and rigid, is always performed in the same way so that its social significance is clear. Ritualized behavior is believed to be derived from body movements associated with such activities as locomotion, feeding, and caring for the young. Facial expressions of human beings (fig. 29.21) seem to be universal and it has been suggested that they, too, are ritualizations of movements that were first used for biological processes.

Auditory Communication

Because auditory (sound) signals are able to reach a larger audience and can be sent even in the dark, they are sometimes favored even by animals with good vision.

Male crickets have calls and male birds have songs for a number of different occasions. For example, birds may have one song for distress, another for courting, and still another for marking territories. Sound stimuli have been shown to be more important than visual stimuli for birds that live in dense

Figure 29.21
Ritualized behavior among humans. Eyebrow flashing is a common human trait displayed by a. *a French woman,* b. *a Balinese man, and* c. *a member of the Woitapmin tribe of New Guinea.*

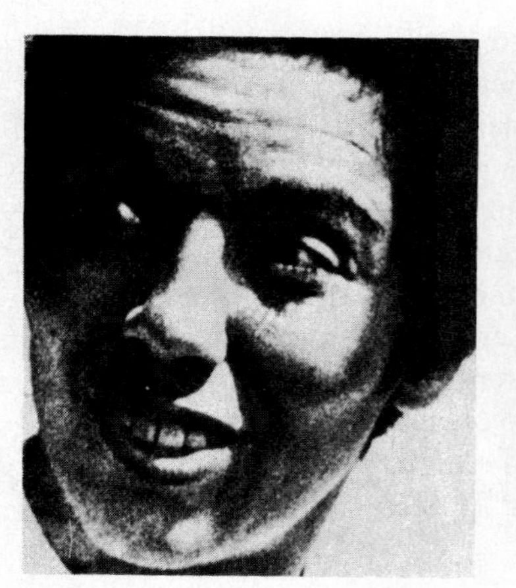
a.

b.

c.

woods where vision is obstructed. In experiments, a male wood thrush attacked models of an unrelated species as long as they were silent. But if the unrelated species' song was played via a loudspeaker, the wood thrush paid the model no heed. Then, too, the wood thrush attacked a model of its own species with vehemency directly proportional to the loudness with which his own species' song was played.

One advantage of auditory communication is that the message can be modified by the sound's intensity, duration, and repetition. In an experiment with rats, an experimenter discovered that an intruder could avoid attack by increasing the frequency with which it made an appeasement sound. Rats isolated from birth could make the sound but had not learned to increase the frequency.

The fact that organisms with good vision often rely on sound communication has been demonstrated in chickens. Hens will react vigorously when they hear a chick peep even if they cannot see the chick. However, they will ignore a chick that is peeping within a soundproof glass container.

Language is the ultimate auditory communication, but only humans have the biological ability to produce a large number of different sounds and to put them together in many different ways. Nonhuman primates have at most only forty different vocalizations, each one having a definite meaning, such as the one meaning "baby on the ground" that is uttered by a baboon when a baby baboon falls out of a tree. As discussed in page 656, although chimpanzees can be taught to use an artificial visual language (fig. 29.22) they never progress beyond the capability level of a two-year-old child. It has also been difficult to prove that chimps understand the concept of grammar or can use their language to reason. It still seems as if humans possess a communication ability unparalleled by other animals.

Figure 29.22
A chimpanzee with an experimenter. Chimpanzees are unable to speak but can learn to use a visual language consisting of symbols. Some researchers believe that chimps are capable of creating their own sentences, but others believe that they only mimic their teachers and never understand the cognitive use of a language. Here the experimenter shows Nim the sign for "drink." Nim copies.

Members of a society communicate and compete with one another. This communication can be chemical, visual, and/or auditory.

Competition

Members of the same population compete with one another for resources, including food and mates. **Aggression** is belligerent behavior that helps an animal compete. Therefore, aggression helps animals establish territories, obtain mates, train and/or defend their young, and maintain status in a group.

Territoriality and Dominance

Territoriality means that a male defends a certain area, preventing other males of the same species from utilizing it (fig. 29.23). Territoriality spaces animals and thereby reduces aggression. It also avoids overcrowding, ensuring that the young will have enough to eat. Since animals without a territory do not mate, it also has the effect of regulating population density to some extent.

A **dominance hierarchy** exists when animals within a society form a sequence in which a higher ranking animal receives food and a chance to mate before a lower ranking animal. Dominant males lead the group and maintain order; therefore a dominance hierarchy assures that the stronger males are in this position.

When the male defends his territory or engages another in contest to determine dominance, rarely is any blood shed. Rather, the animals have a repertoire of sign signals that comprise a threat *display* or *ritual*. As in figure 29.1, the display often includes postures that make the body appear larger and color changes that make the animal more conspicuous. Many of the sign signals in the display are derived from the normal activity of the animal, but now they are used to convey a social message.

Figure 29.23
Two male elephant seals are battling over a piece of beachfront property. This will be the winner's territory where he will collect a harem, females that will willingly mate with him. Territoriality partitions resources according to the supposed fitness of the animal.

The Chemistry between People

Do pheromones affect our behavior toward the opposite sex?

While the idea of human pheromones is intriguing, the dozen or so studies that addressed the possibility in the past 10 years were disappointing: no one established beyond a doubt that human pheromones exist. Now two new studies are stirring up the pheromone debate with the boldest claims yet. Researchers at the University of Pennsylvania and the Monell Chemical Senses Center, a nonprofit research institute in Philadelphia, say that people produce underarm pheromones that can influence menstrual cycles.

The studies, done by chemist George Preti and biologist Winnifred B. Cutler, are not the first of their kind, but they are the first ones rigorous enough to be published in a respected scientific journal, Hormones and Behavior. In one study the researchers collected underarm secretions from men who wore a pad in each armpit. This "male essence" was then swabbed, three times a week, on the upper lips of seven women whose cycles typically lasted less than 26 days or more than 33. By the third month of such treatment, the average length of the women's cycles began to approach the optimum 29.5 days—the cycle length associated with highest fertility. Cutler's conclusion: "Male essence" contains at least one pheromone that "helps promote reproductive health."

The experiment was more rigorous than earlier ones for two reasons. It employed a control group—eight women who were swabbed with alcohol showed no effect—and it was performed in "double-blind" fashion: neither the subjects nor the researchers knew whether alcohol or male essence dissolved in alcohol was being applied until after the study. In spite of such safeguards, skeptics charge that the study was flawed. Richard Doty, a psychologist at the Clinical Smell and Taste Research Center of the University of Pennsylvania, says Cutler and Preti tested too few women for too brief a time to be certain that the observed effects are real. "If they're right, it's fascinating," says Doty, "but I'm not convinced yet."

In Cutler and Preti's second experiment, they studied menstrual synchrony—the phenomenon that

When the animals are facing one another, two opposing responses—*approach response* and *avoidance response*—are simultaneously present in each individual. They are often in *conflict* as to whether to fight or to escape and this conflict causes them to threaten one another rather than to fight outright. Natural selection favors this situation because a contest decided by threat rather than by fighting is more apt to preserve each animal for the purpose of reproducing. The contest result is decided when one animal backs down and flees or submits.

Appeasement, or submission, occurs when an animal actually exposes itself to the attack of another and this gesture prevents further attack. For example, a subordinate baboon of either sex turns away from an aggressor and crouches in the sexual presentation posture. Investigators studying gull behavior have found that the birds have a whole range of postures from actual fighting to appeasement. In gulls, food-begging behavior in adults is appeasement. Appeasement behavior is believed to cause even greater conflict in the aggressor so that inactivity results.

When an aggressor is in conflict, *redirection* of aggression or *displacement* of aggression may occur. As an example of the first of these, a bird might peck at the ground and a human might bang on a table (fig. 29.24). Displacement of aggression is recognized by the fact that the animal performs an irrelevant activity. A bird might preen its feathers, while a human might pull on his chin.

women who live in close quarters tend to have cycles that coincide. And because this experiment's results were more dramatic, several scientists found it more persuasive. This time Cutler and Preti exposed 10 women with normal cycles to *female* underarm sweat. After three months of the same sweat-on-the-lip treatment, the women's cycles were starting roughly in synchrony with those of the women who had donated the sweat. (Again, women exposed to alcohol showed no change.) Menstrual synchrony was first documented in 1970 when psychologist Martha McClintock studied women living in a college dormitory. But this new study is the first to offer solid evidence that pheromones are what mediate the effect. "Pheromone effects are real in human beings," concludes Preti, "and the anecdotal evidence suggests they even occur here in the United States, where we're all deodorized and perfumized."

The researchers don't know how the pheromone exerts its effect on those brain centers that control reproductive hormones, or whether the women inhale it or absorb it through the skin. Cutler and Preti also aren't sure which of the many compounds in human sweat is the active ingredient—but they're focusing now on substances that are already well known for their role in pig sex.

One of the targeted compounds, a steroid known as androstenol, is also found in boar saliva; when a sow in heat detects the musky smell, it promptly assumes the mating stance. A decade ago researchers found that androstenol was also produced in human underarms by bacteria residing there. One firm specializing in mail-order fragrances decided that if the compound was a sex attractant in pigs, it must be one in people, too: SCIENTISTS DISCOVER MYSTERY CHEMICAL THAT SEEMS TO DRIVE WOMEN WILD! it trumpeted in an ad for an androstenol-containing cologne.

Contrary to that claim, repeated studies have shown that androstenol doesn't work as an aphrodisiac in people. Still, Preti and Cutler say it may have other physiological effects. In a third study they found that the amount of androstenol in female underarms changed predictably during the menstrual cycle, peaking just before ovulation. This may help explain how menstrual synchrony occurs: the changes in a woman's androstenol may be what communicate her place in the cycle to other women, the researchers say, and perhaps set off changes in their cycles as well.

If scientists ever do get their hands on an actual human pheromone, it will probably work gradually like the one that Cutler and Preti describe, says McClintock, who is now at the University of Chicago. "It is very unlikely that you'll find a simple push-button effect in people, where a pheromone *compels* somebody to do something," she says. "At the most it will affect how you feel a little bit." No doubt that is why it is so difficult to identify human pheromones. Their messages are not urgent, like a moth's, but subliminal. In people it appears that pheromones are a most subtle body language.

a.

b.

c.

Figure 29.24
Aggressive behavior among humans. Human beings often show conflict in regard to aggressive behavior, including (a.) *vacillation,* (b.) *redirection of aggressiveness, and* (c.) *irrelevant activity.*

Courtship is a time during which aggression must be at least temporarily suspended in order for mating to take place. At this time, conflict within the male may cause him to vacillate between aggression and nonaggression. For example, in the spring the male stickleback stakes out a territory and builds a nest; at this time, his body becomes highly colored, including a red belly.

Figure 29.25
Reproductive behavior of stickleback fish. a. *A red belly on a crude model stimulates females and males more than an exact replica of a fish without a red belly.* b. *The male entices the female to the nest, where he lies flat on his side; the female swims into the nest and lays her eggs when prodded by the male, and the male will later enter the nest to fertilize the eggs. Finally, the male beats his pectoral fins and creates a current of water that flows through the nest. This action supplies oxygen to the developing eggs.*

Any male attempting to enter the territory is attacked as the owner repeatedly darts toward and nips the intruder. (Experiments have shown that the red belly of the male acts as a sign signal.) On the other hand, the owner entices a female to enter the territory by first darting toward her and then away in a so-called zigzag dance (fig. 29.25). Finally, he leads her to the nest, where she deposits her eggs. Investigators have pointed out that the zigzag dance of the male actually contains the same aggressive movements as when the male darts toward and attacks a trespassing male.

The members of a society must overcome aggression so that cooperation results. Territoriality and dominance are two mechanisms by which aggression is controlled. During courtship, aggression is minimized, although aggressive actions are still detectable.

Sociobiology

It can be shown that the degree to which members of a species cooperate with one another is dependent on their own phenotype and by the nature of the environment. For example, an investigator reports that in the Kibale Forest in eastern Africa, chimpanzees usually forage in groups of three or four. The group spreads out and when one spots a tree in fruit, it calls to the others. Since they are able to travel on the ground, it is advantageous for them to forage in a group. On the other hand, orangutans are solitary feeders. They literally have four hands that are adapted to gripping tree limbs. Therefore, orangutans do not travel well on the ground. Under these circumstances, it is best for each one to hoard for itself any tree that is in fruit. Or we might take another example from reproductive behavior. Songbirds are often monogamous while mammals tend to be polygamous. In the case of birds, incubating the eggs and/or gathering enough food requires the effort of both parents. If they did not cooperate, neither would have any offspring. In the case of mammals, the female, not the male, is anatomically and physiologically adapted to care for the offspring. Under these circumstances, it is more adaptive for the male to impregnate as many females as possible.

The basic tenet of **sociobiology,** the biological study of social behavior, is that cooperative behavior among members of a society only exists because such behavior increases the fitness of the individual. *Fitness,* of course, is judged by the frequency that an animal's genes are passed on to the next generation.

Altruism

At times, it may seem as if an animal is acting **altruistically;** that is, for the good of the group rather than for self-interest. For example, in ants and bees, only the queen lays eggs and most of the hive consists of female workers who do not have any offspring at all. How can such a society increase the fitness of the workers? The answer lies in the fact that the male parent is haploid; therefore, siblings have three-fourths of their genes in common (fig. 29.26). Since offspring have only one-half of their genes in common with their parents, it actually increases the fitness of the worker ants to raise siblings rather than offspring.

Altruism, exemplified by the willingness of daughters to help the queen rather than to have their own offspring, can be explained by noting that **kin selection** increases the animal's **inclusive fitness.** In other words, survival of close relatives (*kin*) increases the frequency of an animal's genes in the next generation. Therefore the animal's overall (*inclusive*) fitness is also increased.

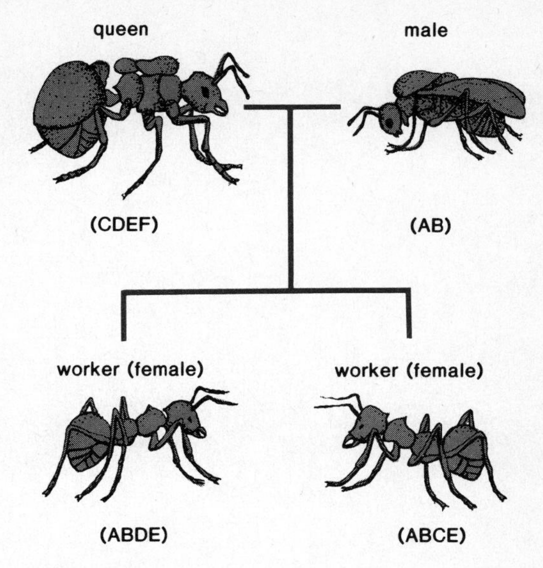

Figure 29.26
Sociobiology and the behavior of ants. It is more genetically advantageous for female ants and bees to assist in caring for siblings rather than their own offspring. Offspring possess one-half of the queen's genes, but since the male parent is haploid, siblings share, on the average, three-fourths of their genes.

Communal Groups The concept of kin selection can possibly apply to other instances, as, for example, in the case of acorn woodpeckers (fig. 29.27). These birds received this name because they store acorns in holes drilled in trees. In California and New Mexico, acorn woodpeckers usually live in a social group consisting of about fifteen birds. Each group defends its storage tree from birds that are not a member of the group.

Investigators have noted that only certain members of a group are *breeders* who share mates. Mate sharing, which is rarely seen in birds, does, in this instance, increase the inclusive fitness of the individual bird. First of all, groups have more offspring per male than do single pairs. For a group of three, investigators found there were 1.16 young per adult male on the average; for pairs there were only .92 young per male. Secondly, among the sharers, the males are usually brothers and the females are usually sisters.

In a large group, there are several *nonbreeders,* birds who only help the breeders raise their young. Since nonbreeders are offspring of the breeders, it is possible that they are also increasing their inclusive fitness by their behavior. However, the investigators also note that the adjoining territories are already occupied and the birds have a very difficult time establishing themselves as breeders. Under these circumstances, it may be that it is simply more advantageous for the nonbreeders to remain in a group where they can feed on the stored acorns and wait until there is a vacancy among the breeding birds.

Figure 29.27
Acorn woodpeckers defend their store of acorns against birds of another group. These birds live and mate in groups but, even so, it can be shown that in doing so they increase their individual fitness.

Alarm Callers Sociobiologists suggest that all animal behavioral patterns should be interpreted as selfish acts. For example, some animals that move in groups give an alarm call when a predator approaches. Careful analysis is expected to reveal that the alarm callers are not putting themselves in danger and may, instead, be protecting *themselves* by this action. Sociobiologists maintain that there is always a biological explanation for social behavior that is consistent with and supports the concept of organic evolution by natural selection of the fittest.

Sociobiology is the study of social behavior according to the tenets of evolutionary theory. While it may seem as if members of a group are altruistic, it can be shown that their motives are most likely selfish and their behavior serves to increase their own fitness. Sometimes it is necessary to utilize the concept of inclusive fitness when interpreting an animal's behavior on this basis.

Humans Sociobiologists interpret human behavior according to these same principles. For example, parental love is clearly selfish in that it promotes the likelihood that an individual's genes will be present in the next generation's gene pool. People tend to keep their aggression under control with those they recognize as blood relatives. Also, we would expect small towns to have less crime than large cities because small towns have a greater percentage of related residents.

Human reproductive behavior can be similarly examined. Human infants are born helpless and have a much better chance of developing properly if both parents contribute to the effort. Perhaps this explains why the human female has evolved to be continuously amenable to sexual intercourse. (Other mammalian females are receptive only during estrus, a physiological state that recurs periodically; sometimes only once a year.) The fact that sex is continuously available may help assure that the human male will remain and help the female raise the young.

Human beings are usually monogamous (one mate at a time); however, there are exceptions that appear to be adaptations to the environment. Among African tribes, for example, one man may have several wives. This is reproductively advantageous to the male, but is also advantageous to the woman. By this arrangement she has fewer children who are thereby assured a more nutritious diet. In Africa, sources of protein are scarce and early weaning poses a threat to the health of the child. In contrast, among the Bihari hillpeople of India, brothers have the same wife. Here the environment is hostile and it takes two men to provide the everyday necessities for one family. Since the men are brothers, they are actually helping each other look after common genes.

It should be stated that not all scientists and laypeople are in favor of applying sociobiological principles to human behavior. Many are concerned that it may lead to the perception that human reproductive behavior is fixed and cannot be changed. Particularly, today, when people are exploring new avenues of behavior, some do not wish to unduly stress any potential obstacles to bringing about a new social order.

Summary

Behavioral patterns are inherited and evolve in the same manner as other animal traits. Behavior occurs as a response to a stimulus and is often divided into two categories: innate (instinctive) and learned. Innate behavior includes taxes and reflexes. Taxes, together with the possession of an internal clock, can in some cases offer an explanation for the ability of animals to return to a previous location. Fixed action patterns are largely inherited, although they often increase in efficiency with practice.

Learned behavior includes kin recognition, conditioned learning, and insight learning. An animal's motivation often determines whether or not it performs a behavior.

Members of a society communicate and compete with one another. If the group is to survive, the members must overcome a certain amount of aggression; enough that cooperation results. Territoriality and dominance are two mechanisms by which aggression is controlled. During courtship aggression is minimized, although aggressive actions are still detectable.

Sociobiology is the study of social behavior according to the tenets of evolutionary theory. While it may seem as if members of a group are altruistic, it can be shown that their motives are most likely selfish and their behavior serves to increase their own fitness or the fitness of the group as a whole.

Objective Questions

1. Behavioral patterns are ____________ and suited to the environment.
2. Breeding experiments with *Agapornis* parrots and biochemical studies with *Aplysia* snails show that behavior is controlled by ____________ .
3. A taxis is the ____________ of the body in relation to an environmental stimulus.
4. A stimulus that initiates a fixed action behavior is called a ____________ stimulus.
5. Experiments have shown that the ability of chaffinches to sing the normal adult song is both ____________ and ____________ .
6. Birds and other animals become ____________ to the first moving object they see.
7. Most animal trainers use ____________ to teach animals to do tricks.
8. Pheromones are an example of ____________ communication.
9. ____________ communication includes various sign stimuli.
10. Both territoriality and dominance hierarchies help reduce outright ____________ between animals.
11. Sociobiologists believe that altruistic acts are actually ____________ acts.
12. Acorn woodpecker offspring help their parents raise young because it actually increases their own ____________ to do so.

Answers to Objective Questions

1. adapted 2. genes 3. orientation 4. sign
5. innate, learned 6. imprinted 7. operant conditioning or positive reinforcement
8. chemical 9. Visual 10. fighting or aggression
11. selfish 12. fitness

Study Questions

1. Draw a diagram showing that behavior evolves and occurs as a result of an environmental stimulus. (p. 645)
2. Describe the breeding experiments with the parrot *Agapornis*, and the results of recombinant DNA work with the snail *Aplysia*. (p. 647)
3. Give several examples of taxes in animals. Include an explanation for migration and homing. (pp. 649–650)
4. What are the three components of a biological clock system? What might these components be in birds? (p. 651)
5. Define a fixed action pattern. Give several examples, and discuss one that is improved by practice. (pp. 651–652)
6. Name three types of learning and give an example of each type. (pp. 653–655)
7. Describe the ringdove reproductive experiment and its significance in regard to motivation. (pp. 658–659)
8. Define a society and name two aspects of behavior that are usually found within a society. (p. 659)
9. Give examples of three common means of communication between members of a society. (pp. 659–661)
10. Name and discuss mechanisms that help reduce overt aggression among members of a society. (pp. 661–663)
11. What is the basic tenet of sociobology? How do sociobiologists explain altruistic behavior? How would they explain human reproductive behavior? (pp. 664–666)

Key Terms

altruism (al'troo-izm) behavior performed for the benefit of others without regard to its possible detrimental effect on the performer. *665*

circadian rhythm (ser''kah-de'an rith'm) a regular physiological or behavioral event that occurs on an approximately 24-hour cycle. *651*

dominance hierarchy (dom'ĭ-nans hi'er-ar''ke) a system in which animals arrange themselves in a pecking order; the animal above takes precedence over the one below. *661*

fixed action pattern (fikst ak'shun pat'ern) a sequence of reflexes that always occurs in the same order and manner. *651*

imprinting (im'print-ing) the tendency of a newborn animal to follow the first moving object it sees. *654*

inclusive fitness (in-kloo͞'siv fit'nes) the fitness of closely related group members is a part of the fitness of the individual. *665*

innate (in'nāt) instinctive, inborn, and not having to be learned. *649*

kin selection (kin sĕ-lek'shun) behavior which functions to protect a group of related organisms. *665*

learning (lern'ing) a change in behavior as a result of experience. *649*

motivated (mo'tĭ-vāt-ed) physiologically orientated to perform a certain behavior. *658*

pheromone (fer'o-mōn) a chemical substance secreted by one organism that influences the behavior of another. *649*

sign stimulus (sīn stim'u-lus) stimulus which releases a fixed action pattern. *651*

sociobiology (so''se-o-bi-ol'ŏ-je) the biology of behavior, particularly social behavior. *664*

taxis (tak'sis) a movement in relation to a stimulus, such as a phototaxis (a movement oriented to a light source). *649*

30

Behavior between Species

Chapter Concepts

1. Competition, predation, and symbiotic relationships are common modes of behavior between species.

2. Competition leads to diversity of species because no two species occupy the same niche.

3. Predator and prey coevolve and therefore a predator rarely overkills its prey population.

4. Symbiotic relationships usually require a close relationship between two species that also coevolve.

Chapter Outline

In the preceding chapter we discussed behavior typical of members of the same species; in this chapter we will be discussing interactions between species. These interactions, such as competition and predation, are important in controlling the sizes of populations. Because of this, these interactions are important in achieving a stable community of populations in which all populations are able to maintain a size appropriate to ensuring their continued existence.

Competition

Similar types of species with the same needs are most apt to compete with one another for resources, such as water, food, sunlight, and space. The outcome of **competition** is often the predominance of one species and the virtual elimination of the other.

Examples

A classic example of competition concerns two species of barnacles. Barnacles are attached to rocks in the intertidal zone and an investigator noticed that one species (genus *Chthamalus*) occupied the upper part of the intertidal zone along a Scottish coast, while another species (genus *Balanus*) occupied the lower intertidal zone. It was found that if one species was removed, each species could live in at least a portion of the other's zone (fig. 30.1). Since *Balanus* could occupy almost the entire zone occupied by *Chthamalus,* it must be competition that prevents it from doing so. In some manner, *Chthamalus* is better adapted to the upper intertidal zone than is *Balanus.*

A study of desert ground squirrels also illustrates how many organisms deal with competition. Both the antelope ground squirrel and the Mohave ground squirrel (fig. 30.2) possess similar anatomical and physiological adaptations that allow them to cope with high temperatures and lack of water. Even so, their behavior is extremely different. The antelope ground squirrel is active the entire year, but the Mohave ground squirrel emerges from its burrow only during the months of March through August when desert vegetation is at its annual peak. It reproduces and fattens and then returns to its burrow, where it is dormant for the rest of the year. Both squirrels eat seeds as a source of water, but the antelope ground squirrel also eats insects and other animals. This adaptation has not been observed in the Mohave ground squirrel. Perhaps this explains why the Mohave ground squirrel is dormant when the desert is driest. It appears that the antelope ground squirrel is the better competitor because it has a wider range; the Mohave ground squirrel is found only in one corner of the Mohave Desert. It is unclear at this time whether the Mohave ground squirrel will become extinct or not. Perhaps its extended dormancy is an adaptation that permits it to coexist with the antelope ground squirrel.

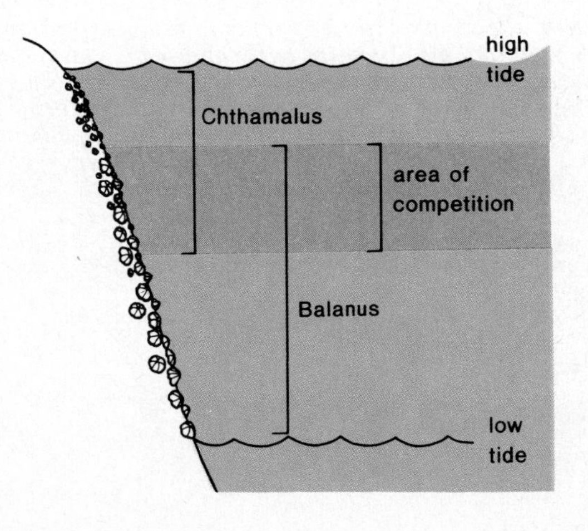

Figure 30.1
Effects of competition. Competition prevents two species of barnacles from occupying as much of the intertidal zone as possible. The area on the farthest right indicates the area of competition between Chthamalus *and* Balanus.

Figure 30.2
Competition between two squirrels. a. *Activity of the antelope ground squirrel on a typical summer day. The squirrel emerges from its burrow. Then it suns and grooms itself. If the squirrel's body temperature rises too high, it retreats to a special burrow to cool off.* b. *In the early afternoon, it stays in the shade and feeds on insects, seeds and dead animals. The squirrel returns to its burrow at night.* c. *Activity of the Mohave ground squirrel during six months of the year. The squirrel emerges from its burrow in March, and the young are born in April.* d. *From May through July it fattens on desert vegetation. In August it again burrows underground where it remains until the following March.*

From "Activity of Antelope Ground Squirrel" by George A. Bartholomew, and "Activity of Mohave Ground Squirrel" by Jack W. Hudson. *Scientific American*, 1961. Reprinted by permission of Scientific American, Inc.

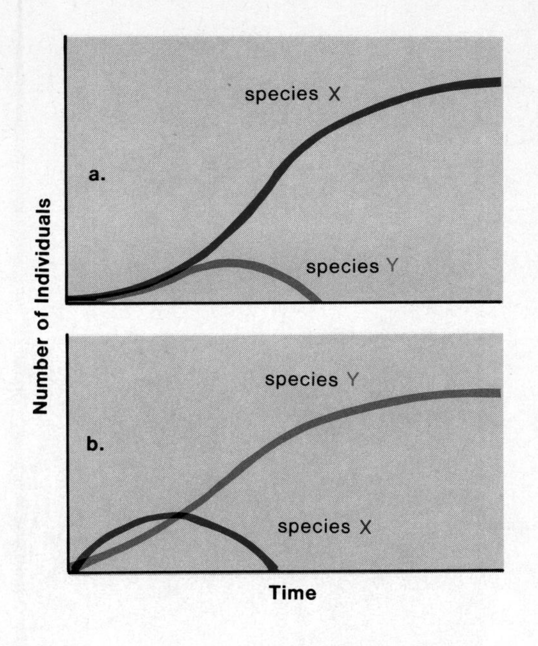

Figure 30.3
Extinction due to competition with a species better adapted to the environment. a. *In this experiment, species X survives and Y becomes extinct because the environmental conditions were favorable to X.* b. *In this experiment, the environmental conditions were changed, with the result that Y survives and X becomes extinct.*

Human Intervention

The fact that successful competition can cause one species to increase in size at the expense of another has been inadvertently demonstrated by human intervention. The carp, a fish imported from the Orient, is able to tolerate polluted water. Therefore, this fish is now often more prevalent than our own native fishes. Melaleuca, an ornamental tree which thrives in wet habitats, became a pest in the Everglades National Park after being introduced into Florida. The burro, which originated in Ethiopia and Somalia and is adapted to a dry environment, is now threatening the existence of deer, pronghorn antelope, and desert bighorn sheep in the Grand Canyon.

Exclusion Principle

The fact that one species can cause another to become extinct has been demonstrated in the laboratory. For example, species X and Y both vie for the same resource when grown together in the same container, and eventually one will replace the other entirely. Which population is successful depends on the environmental conditions. In figure 30.3a, the environmental conditions are favorable to X, while in figure 30.3b, the environmental conditions are favorable to Y.

Such experiments have led biologists to formulate and support a **competitive exclusion principle,** which states that no two species can occupy the same niche at the same time. **Niche** is the term used to refer to the role a species plays in a community of organisms (this is discussed in chapter 31). To describe a species' niche it is necessary to state all the requirements and activities of the species. Table 31.1 lists factors to be included when describing the niche of a particular plant or animal species. Some investigators have suggested that it is possible to represent a niche as a many-sided figure occupying three-dimensional space (fig. 30.4).

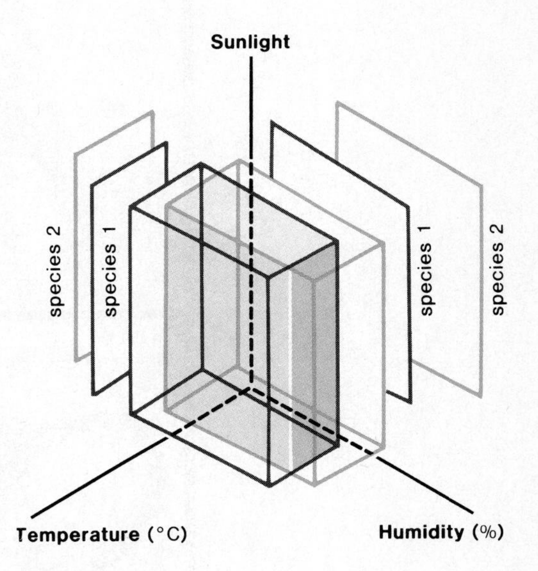

Figure 30.4
Representation of niche by use of a diagram. There is competition if the requirements of two species overlap. For example, the range of sunlight, temperature, and humidity tolerated by species 1 and species 2 overlap in the shaded area.

Figure 30.5
Diversity of monkey species in a tropical rain forest. All of these monkeys can coexist in a tropical rain forest because they fill different niches. Each prefers to live at a different height above ground and each feeds on slightly different foods.

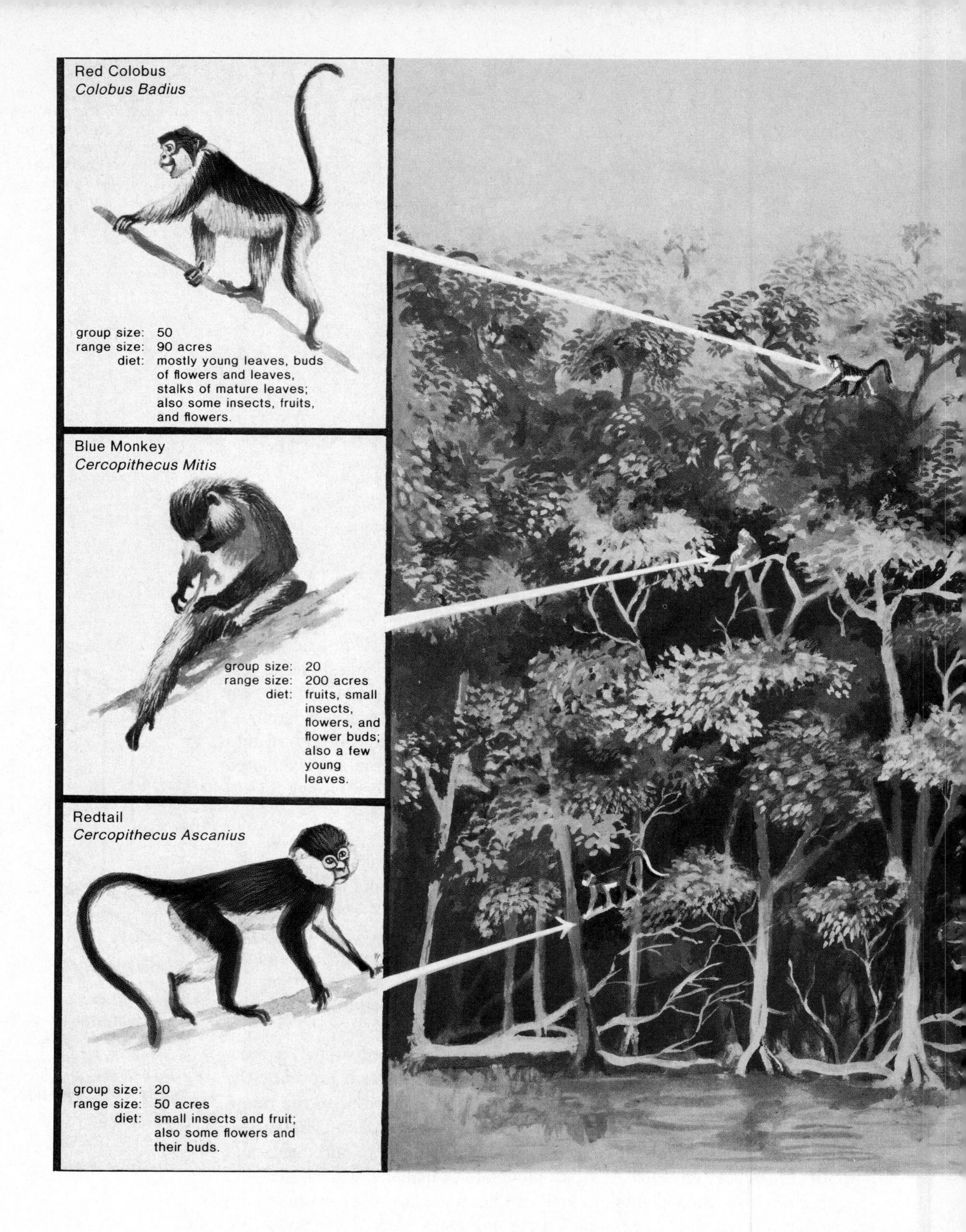

Diversity

Competition leads to diversity of species because similar species will evolve to fill different niches. While it may seem as if several species living in the same area are occupying the same niche, it is usually possible to find slight differences. For example, the six species of monkeys in figure 30.5 have no difficulty living in close proximity because they have different, although sometimes overlapping, habitats and food requirements.

When two species compete for the same resources, one usually elminates or restricts the range of the other. This is in keeping with the exclusion principle which states that no two species can occupy the same niche.

Figure 30.6
The African elephant is a herbivorous browser—it reaches up with trunk to feed on the leaves of a tree.

Predation

Predation, simply defined, is one organism feeding on another. Examples include monkeys feeding on flower bulbs, a bear feeding on salmon, or a lion feeding on a gazelle. Through the evolutionary process, predators become specialized to capture their prey and prey become adapted in escaping their predators. In this manner, predators and prey **coevolve.**

Predation Adaptations

The grassland of all continents support populations of **grazers** that feed on grasses and browsers that feed on shrubs and trees. The anatomy of **browsers** enables them to reach high into trees—giraffes have long necks and elephants have long snouts (fig. 30.6). The most prevalent terrestrial herbivores are the insects, which have evolved efficient and diverse means of eating plants.

Some carnivores, such as birds of **prey,** go out alone and seek their prey. While most birds of prey are specialized for hunting, seizing, and killing small terrestrial animals, some, like the osprey and kingfish are specialized for fishing (fig. 30.7). Instead of seeking prey, some solitary predators lie in wait. The octopus hides within a protective shelter until an unsuspecting prey, such as a crab, should happen by. The octopus then quickly catches the prey with its arms and carries it home to be poisoned and eaten. Sperm whales probably lie suspended and hidden within the darkness of the ocean's depth but are ever ready to dart and snap at luminescent shoals of shrimp.

Predators, such as lions and wolves, prey on animals larger than themselves and therefore they often hunt their prey as a group. Lions typically prey on animals that have been separated from their herd. Although only a small number of lions participate in the kill, the food is shared with the rest of the pride (fig. 30.8).

Figure 30.7
The Eurasian kingfisher is a solitary predator.
a. *Through the rising air bubbles, it catches prey.*
b. *With strong strokes, it rises from water with fish in beak.*

a.

b.

Control of Prey Population Size

A carnivore helps control the size of its prey population. Biologists reasoned that there would be an oscillation in sizes of the predator and prey populations, as illustrated in figure 30.9. The oscillation was expected because it was reasoned that as the number of prey increased, predation would also increase until finally the prey population would suffer a decline. This would be followed by a reduction in the number of predators until the prey population would eventually begin to recover. Then the cycle would begin again. Rarely are such oscillations observed, however. Therefore, it is believed that factors other than predation may have caused the oscillation in the lynx and hare populations observed between 1845 and 1935 (fig. 30.9). For example, the prey population may have been reduced because of some other reason, such as lack of food due to the weather, and this, in turn, may have led to a decline in the predator population.

Figure 30.8
Lions are social animals and share their kill with others. Lions do help keep the zebra population in check, but even so predator-prey cycling such as that shown in figure 30.9 is not expected.

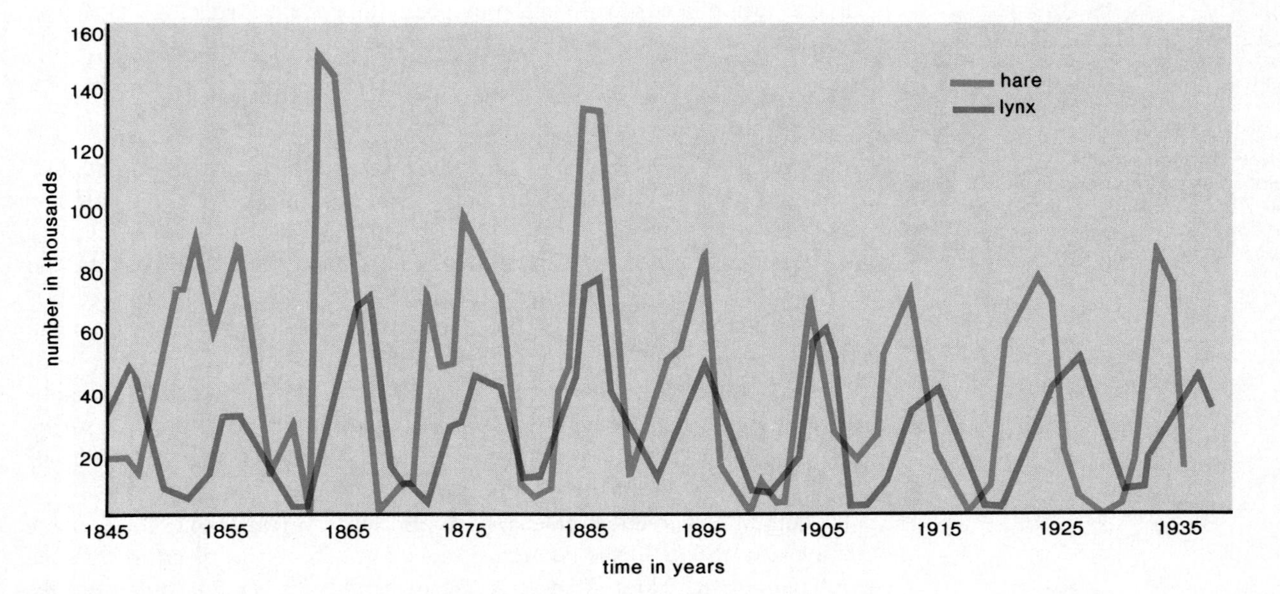

Figure 30.9
Example of predator-prey cycling. As the hare increases in number, so does the lynx, but then the hare, followed by the lynx, suffer a dramatic decrease in number. While it may seem as if this cycling is due to "overkill" by the lynx, other factors could be involved. (These data are based on the number of lynx and snowshoe hare pelts received by the Hudson Bay Company in the years indicated.)

a.

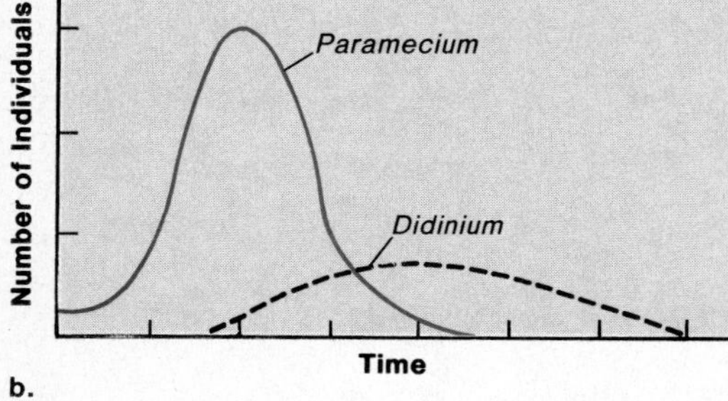

b.

Figure 30.10
Paramecium *and* Didinium *interactions.*
a. *Didinium in the act of engulfing the paramecium.*
b. *If, in the experimental situation, the paramecia have no place to hide, the didinia will sometimes kill off all the paramecia and then die off themselves.*

Usually predators do not overkill the prey population. For example, on Isle Royale, an island in Lake Superior, a population of about 1,000 moose has coexisted with a population of about 24 wolves year after year since 1948. Biologists who have studied this phenomenon tell us that the wolves are able to capture only the old, sick, or very young animals. The inability of the wolves to capture more moose keeps the wolf population in check.

Often predators even have a beneficial effect on the prey population. In 1930, before there were wolves in Isle Royale, the moose overpopulated, overate their food sources, and subsequently suffered a sharp decline in population. Once the vegetation recovered, the moose population again grew to the point of overexploitation and suffered another decline in the 1940s. The chance introduction of wolves has kept the moose population stable ever since.

Whenever predator and prey are coadapted, each population is maintained at proper levels, as can be shown by a laboratory experiment involving two protozoans, *Didinium* and *Paramecium*. Didinia prey on paramecia, and if the paramecia are denied a place to hide, the didinia capture all the paramecia and then they both die out (fig. 30.10). But if debris is provided so that the paramecia can hide, each population remains at a fairly constant level.

Human Intervention

Sometimes humans have neglected to take into consideration that predators help keep prey populations in check. For example, in the past, coyotes have been indiscriminately killed off in the West without regard to the fact that they help keep the prairie dog population under control. Similarly, when the dingo, a wild dog in Australia, was killed off because it attacked sheep, the rabbit and wallaby populations greatly increased. In contrast, humans formerly kept the burro population in the Grand Canyon within reasonable limits by killing them for meat. However, since a federal law was passed in 1971 forbidding the killing of burros, their numbers have increased until they are now considered destructive pests.

Although predators control the prey population size, there is usually no oscillation in carnivore and prey population sizes. Instead, both populations are maintained at constant levels.

Prey Defense Adaptations

While predators have evolved strategies to secure the maximum amount of food with minimal expenditure of energy, prey organisms have evolved strategies to escape predation.

Plants also try to avoid predation. The sharp spines of the cactus; the pointed leaves of the holly; and the tough, leathery leaves of oak trees all discourage predation by insects. Plants even produce poisonous chemicals, some of which interfere with the normal metabolism of the adult insect (table 30.1) and others which act as hormone analogues that interfere with the development of insect larvae. Some insects can still inhabit trees that produce toxins because the insects have either evolved detoxifying enzymes or a method of holding the toxin within their bodies so it is not harmful to them.

Animal **antipredator defenses** are also quite varied (table 30.2). Some animals take the path of passive resistance and attempt to blend in with the background. Stick caterpillars look like twigs (fig. 30.11); katydids look like sprouting green leaves; and moths resemble the barks of trees. Often there is

Table 30.1 *Some Poisonous Green Plants*

Plant	Toxin(s)
	Alkaloids
Poison hemlock *(Conium maculatum)*	Atropine*
Jimson weed *(Datura stramonium)*	Scopolamine*
Deadly nightshade *(Atropa belladonna)*	Hyoscyamine*
	Glycosides
Milkweed *(Asclepias Curassavica)*	Calactin
Foxglove *(Digitalis purpurea)*	Digitoxin
Oleander *(Nerium oleander)*	Oleandrin
Wild cherry (leaves and seeds) *(Prunus)* Manioc or Cassava *(Manihot esculenta)*	Glycoside that produces hydrocyanic acid (HCN)
Rhubarb (leaves) *(Rheum rhaponticum)*	Oxalic acid

*These toxins, in controlled dosages, have important medical uses.

Adapted from *Contemporary Biology*, 2nd edition, by Mary E. Clark. Copyright © 1979 by W. B. Saunders. Reprinted by permission of Holt, Rinehart and Winston.

Table 30.2 *Animal Antipredator Adaptations*

Name	General Behavior	Example
Sensory methods	Cryptic coloration and behavior	Blending in with the background
	Startle display	Sudden unexpected noise or visual effect
	Distraction display	Pretending injury
	Death display	Playing dead
Vigilant method	Detection of predator	Alerting others by alarm calling
Escape behavior	Evasion	Running away
Repellent behavior	Having a chemical or mechanical defense	Injuring predator
	Resembling animals that have a chemical or mechanical defense	Mimicry

Figure 30.11
Some prey use disguises to avoid predators such as the stick caterpillar that resembles a twig.

a backup defense in case of discovery. Many moths have eyelike spots on their underwings that they can flash to startle a bird long enough to escape. Larger animals, too, try to conceal themselves by looking like their environment. Decorator crabs cover themselves with debris, and green heron birds attempt to look like straight tree branches.

Other types of prey immediately try to frighten or attack a would-be predator. Some have a means of making themselves appear much larger than they are, such as the frilled lizard of Australia that can open up folds of skin around its neck. The porcupine sends out arrowlike quills with barbs that dig into the enemy's flesh and penetrate even deeper as the enemy struggles after impact.

Poisonous animals tend to be brightly colored, as a warning to those that may want to prey upon them. The spotted arrow-poison frog (fig. 30.12) found in tropical regions is so poisonous that South American Indians have long used it, as its name implies, to poison their arrow tips. When the arrow strikes, it paralyzes the victim instantly.

Figure 30.12
Arrow-poison frogs are so poisonous that natives use them to make their arrows instant lethal weapons. The coloration of these frogs warns others to beware.

Flocks of birds, schools of fish, and herds of mammals stick together as protection against predators. Grazing herbivores are constantly on the alert; if one begins to dart away, they all run. Baboons who detect predators visually and antelopes who detect predators by smell sometimes forage together, giving double protection against stealthy predators. The gazellelike springboks of southern Africa jump stiff-legged 8 to 12 feet into the air as a warning. Such a jumble of shapes and motions might confuse an attacking lion, and the herd can escape.

One unusual possible defense is discussed in the reading on cicadas (p. 679). It is believed that the life cycle of these animals protects them from predators. They emerge only every 13 to 17 years and thus far a predator adapted to their life cycle has not evolved.

Mimicry

Mimicry occurs when one species resembles another that possesses an overt antipredator defense. For example, the brightly colored monarch butterfly (fig. 30.13a) causes predaceous birds to become sick and sometimes to vomit. Monarch butterflies have this capability because, as larvae, they feed on milkweed plants. These plants contain digitalislike compounds that activate nerve centers controlling vomiting. (Larvae incorporate a modified form of the poison into their own tissues, retaining it even as they develop into adults.) Birds that have eaten a poisonous monarch butterfly avoid all monarch butterflies in the future.

The viceroy butterfly (fig. 30.13b) mimics the monarch butterfly's coloration, but is not toxic. Birds eagerly eat viceroy butterflies unless they have had previous experience with a poisonous monarch. Then, because the butterflies closely resemble each other, birds avoid both types of butterflies. The queen butterfly also mimics the monarch, but the queen butterfly is poisonous.

A mimic that lacks the defense of the organism it resembles is called a **Batesian mimic;** the viceroy butterfly is a Batesian mimic of the monarch. A mimic that also possesses the same defense is called a **Müllerian mimic;** the queen butterfly is a Müllerian mimic of the monarch butterfly.

a.

b.

Figure 30.13
The brightly-colored monarch butterfly (a.) *is poisonous while its mimic, the viceroy butterfly is not.* (b.)

Prey defenses are varied, as listed in table 30.2. Sometimes an animal with a successful antipredator defense is mimicked by other animals that have the same defense (Müllerian mimic) or that lack the defense (Batesian mimic).

Wedding Whirs

Already from the Carolinas to New York, little holes are appearing in lawns and backyards, hillsides and woodlands. Any evening now, out will pop millions of dark little bugs. They will scamper up almost any upright object—trees, poles, buildings—and soon strike up a joyous racket, marking nuptial rites after being buried alive for 17 years.

They are periodical cicadas (pronounced sih-*kay*-duhs) the world's longest-lived insects. Despite a locust-like appearance, they neither bite nor sting nor devastate vegetation. Entomologists currently count 19 separate "broods," which appear at various times in different parts of the country, some once every 13 years. But all follow roughly the same miraculous life cycle. Growing through five skin-shedding molts and sucking nourishing juices from roots, they emerge with uncanny precision, triggered by some still mysterious internal clock.

In the open, they shed their dry, yellowish skins for the last time. Soon the males strike up their cacophony of ticking, buzzing, and shrill whirring sounds. It is all music to the females, who slit open tree bark after they have been impregnated and store their fertilized eggs there. A few weeks later, both parents die. But cicada life goes on as the eggs hatch. The newborn nymphs drop to the ground, burrow, and the age-old cycle starts anew.

Baffled scientists are still unsure why the cicadas behave as they do, but suspect that it may all be a defense against predators like birds. As Entomologist Chris Simon of the State University of New York at Stony Brook writes in *Natural History,* when the cicadas finally emerge, it is in the shadows of dusk. They also gain protection from their monstrous numbers—as many as 1.5 million per acre. Finally, since they appear only once every 13 or 17 years, nature may have endowed them with an unlikely mathematical defense. These are prime numbers, divisible only by themselves, and so parasites would have to live at least as long—a half or a quarter would be improbable—to partake in a 17-year feast.

Cicada sheds its skin for the last time.

Table 30.3 Symbiosis		
	Species 1	**Species 2**
Parasitism	+	−
Commensalism	+	0
Mutualism	+	+

Key: + = benefits
− = harmed
0 = no effect

Symbiosis

Symbiotic relationships (table 30.3) are close relationships between two different species. Here, too, coevolution occurs and the species are closely adapted to one another. The first relationship to be discussed, parasitism, benefits the parasite but harms the host. The next two relationships, commensalism (one species benefits; the other is unaffected) and mutualism (both species benefit), are cooperative relationships.

Parasitism

Parasitism is similar to predation in that the **parasite** derives nourishment from the **host.** Usually, however, the host is larger than the parasite and the parasite does not immediately kill the host. While viruses are the only obligate parasites (p. 538), there are also parasites among bacteria, protista, plants, and animals. Parasites are closely adapted to their host and infect only certain closely related species. Some of the viral infections of humans are listed in table 12.4 and some of the bacterial infections are listed in table 12.5. Malaria is a well-known protozoan infection (fig. 25.15) and athlete's foot is a well-known fungal infection of humans (fig. 25.24). Tapeworms and flukes illustrate typical life cycles of parasitic worms (p. 593).

The hookworm, judged to be the most troublesome parasitic worm, does not require a secondary host. In the New World hookworm, the males are from 5 to 9 millimeters in length and the females are usually about 1 centimeter long. The head is sharply bent in relation to the rest of the body, accounting for its characteristic hooklike appearance. Adult hookworms attach themselves to the intestinal wall (fig. 30.14) of their host and the eggs pass out with the feces. When deposited on moist, sandy soil, the larvae develop and hatch within 24 to 48 hours. After a period of growth and development, the worms then extend their bodies into the air and remain waving about in this position until they come into contact with the skin of a suitable host, such as a human. Penetration usually occurs through the feet. Once in the blood vessels, the worms are passively carried to the lungs where they invade the alveoli. From the lungs, the larvae migrate up the trachea to be swallowed and passed along to the small intestine where they mature. They attach to the intestinal wall by means of their stout mouth parts and suck blood and tissue fluid from the host. Symptoms of hookworm infection include abdominal pains, nausea, diarrhea, and finally, iron deficiency anemia.

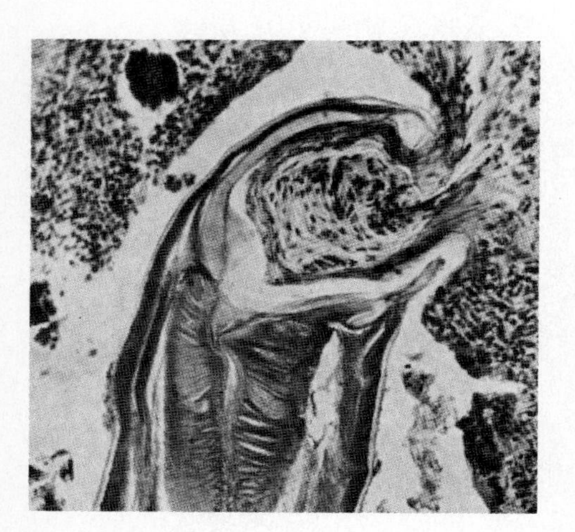

Figure 30.14
Hookworm parasite. Longitudinal section through hookworm attached to intestinal wall. In this position the worms suck blood and tissue fluid from the host.

Many animal parasites also use a secondary host not only for dispersal but also to complete their life cycles. For example, tapeworms utilize cattle or pigs and flukes require snails and sometimes fish as a host for their larvae. When one studies the anatomy and life-style of these animals it seems as if they have traded an active predatory life for an inactive secure life. Table 30.4 contrasts certain features of the animal predator with the animal parasite.

Table 30.4 Predator versus Parasite	
Predator	**Parasite**
Well-developed nervous system	Reduced nervous system
Sense organs, such as eyes	Sense organs, such as touch
Fast moving, with protective devices	Locomotion limited
Well-developed muscles	Minimal muscle fibers
Efficient circulatory system	Reduced circulatory system
Protection of offspring	Complicated life cycle

Just as predators can dramatically reduce the size of a prey population that lacks a suitable defense, so parasites can reduce the size of the host population that has no defense. Thousands of elm trees have died in this country from the inadvertent introduction of Dutch elm disease, which is caused by a parasitic fungus. A new method of treating Dutch elm disease utilizes bacteria that produce a fungus-killing antibiotic when injected into the tree. Another means of curbing this fungal infection of trees is to control the bark beetle that spreads the disease.

Social Parasitism

Social parasitism occurs when one species exploits another species. For example, the cuckoo lays eggs in nests of songbirds and the newly hatched cuckoo ejects its nestmates so that the songbird parents attend only to it. Slave-making Amazon ants of the species *Polyergus rufescens* raid the ant colonies of a slave species *Formica fusca* (fig. 30.15). They destroy any resisting defenders with their mandibles, which are shaped like miniature sabers. *Polyergus* ants are so specialized that they can only groom themselves. In order to eat, they must beg slave workers for food. The slave workers not only provide food for the slave-making ants but also care for the eggs, larvae, and pupae of their captors.

In parasitism, the host species is harmed. Parasites, which help control the size of their host population, may require a secondary host for dispersal. The anatomy and life-style of an animal parasite may be contrasted to that of an active predator. In social parasitism, one society exploits another.

Figure 30.15
Example of social parasitism. Slave-making Amazon ants invade the colony of a slave species in order to carry off cocoons. When ants emerge from the cocoons, they serve as slaves.

Figure 30.16
Example of commensalism. Shark suckers (remora) do not feed on the shark; rather they feed on leftovers from the shark's meal.

Figure 30.17
Example of commensalism. Clownfish live among a sea anemone's tentacles and yet are not seized and eaten as prey. The reason why this relationship is maintained is not known.

Commensalism

In a **commensal** relationship, only one species benefits while the other is neither benefited nor harmed. Often the host species provides a home and/or transportation for the benefited species. Barnacles, which attach themselves to the backs of whales and the shells of horseshoe crabs, are provided with both a home and transportation. Remoras are fish that attach themselves to sharks (fig. 30.16) by means of a modified dorsal fin that acts as a suction cup. The remoras obtain a free ride and also feed on the remains of the shark's prey. *Epiphytes,* such as orchids, grow in the branches of trees where they can receive light, but they take no nourishment from the trees, instead, their roots obtain nutrients and water from the air. Clownfish (fig. 30.17) live within the tentacles and gut of a sea anemone and thereby are protected from predators. Perhaps this relationship is one that borders on mutualism since the clownfish may attract other fish on which the anemone can live.The sea anemone's tentacles quickly paralyze and seize other fish as prey.

Mutualism

Mutualism is a symbiotic relationship in which both members of the association benefit. Mutualistic relationships often help organisms obtain food or avoid predation.

Bacteria that reside in the human intestinal tract are provided with food, but they also provide us with vitamins, molecules we are unable to synthesize for ourselves. Termites would not even be able to digest wood if it were not for the protozoans that inhabit their intestinal tract. These organisms digest cellulose, which termites cannot. Fungi and algae live together as lichens (fig. 25.11). *Lichens* can grow on rocks most likely because the fungi provide moisture and dissolve minerals from the rock. The algae carry on photosynthesis and return carbohydrates to the fungi. *Mycorrhizae* (p. 556) are symbiotic

associations between the roots of plants and fungal hyphae. It is now believed that the roots of most plants form these relationships. Mycorrhizal hyphae, which often penetrate the root but may lie just outside the root, increase the solubility of minerals in the soil, improve the uptake of nutrients for the plant, protect the plant's roots against pathogens, and produce plant growth hormones. In return, the fungus obtains carbohydrates from the plant.

Flowers and their pollinators have coevolved since flowering plants first appeared. For this reason, flowers that attract bees or birds are brightly colored (p. 155) and flowers that attract beetles or bats have strong scents. Sometimes the nectaries, which produce a sugar solution, are located at the base of a tubular corolla so that they are accessible only to the moth or bird that has evolved a long sucking tongue, such as the hummingbird (fig. 30.18). Flowers that attract bees have a landing platform and a structure that requires the bee to brush up against the anther and stigma as it moves toward the nectaries. The coevolution of flowers and their pollinators is of mutual benefit because the pollinator receives food while the flower achieves cross-pollination.

Figure 30.18
Adaptation to a way of life. Organisms are adapted to the manner in which they gather food such as this hummingbird whose long bill allows it to collect the nectar of flowers.

Mutualistic relationships are also common between ants and other organisms. *Leaf cutter ants* keep fungal gardens. They are called leaf cutter ants because they gather flowers and leaves, which they cut into pieces and transport to underground nests. After preparing the leaves, they implant them with fungal mycelia, which then grow in profusion. This relationship helps the fungi compete with other fungal species but also provides the ants with a source of food.

Some ants protect aphids from their predators and in return receive food from the aphids (fig. 30.19). As mentioned previously, aphids remove phloem sap from plants by means of a styletlike mouthpiece. The sap passes through an aphid's body relatively unchanged so that an ant can feed from the rear of an aphid.

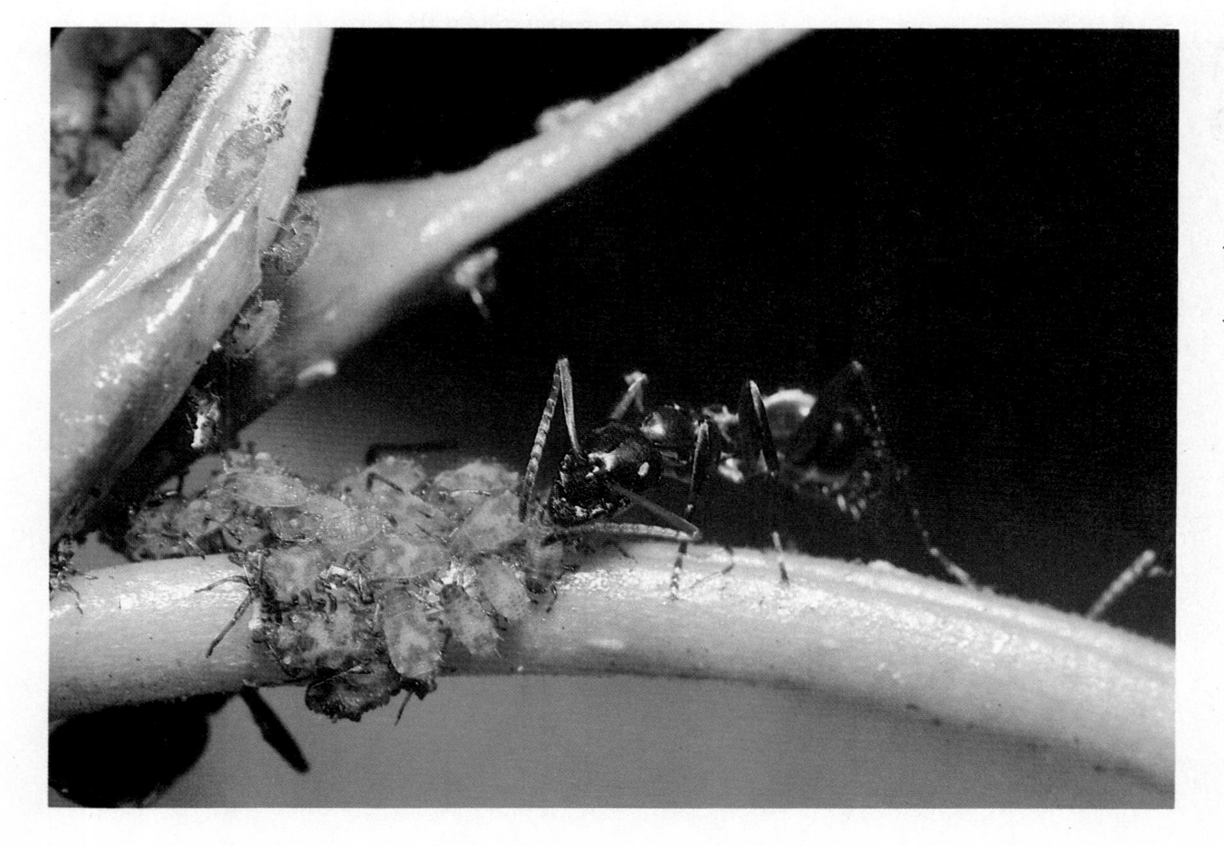

Figure 30.19
Mutualistic relationship between species. Some ants protect and also receive nourishments from aphids. Ants sometimes serve to protect aphids. The sap that aphids suck from plants passes through their digestive system so that an ant can feed from the rear of an aphid.

a. b. c.

Figure 30.20
Mutualistic relationship. The bullhorn acacia is adapted to provide a home for Pseudomyrmex ferruginea, *a species of ant that protects the acacia from other insects.* a. *Thorns are hollow and the ants live inside.* b. *Bases of leaves has nectaries (openings) where ants can feed.* c. *Leaves have bodies at the tips that ants harvest for larvae food.*

In tropical America, the bullhorn acacia is adapted (fig. 30.20) to provide a home for ants of the species *Pseudomyrmex ferruginea.* Unlike other acacias, it has swollen thorns with a soft, pithy interior where ant larvae can grow and develop. In addition to housing the ants, the acacias provide them with food. The ants feed from nectaries at the base of the leaves and eat nodules, called beltian bodies, at the tips of some of the leaves. The ants constantly protect the plant from herbivorous insects because, unlike other ants, they are active 24 hours a day. The plants, on the other hand, have leaves throughout the year, while related acacia species lose their leaves during the dry season.

Cleaning symbiosis (fig. 30.21) is a phenomenon that is believed to be quite common among marine organisms. There are species of small fish and shrimp that specialize in removing parasites from larger fish. The large fish line up at the "cleaning station" and wait their turn, while the small fish feel so secure that they even clean the insides of the mouths of the larger fish. Not everyone plays fair, however, since there are small fish that mimic the cleaners in order to take a bite out of the larger fish, and cleaner fish are sometimes found in the stomachs of the fish they clean.

In commensalism, one species is benefited but the other is unaffected. Often the host simply provides a home and/or transportation. In mutualism, both species benefit and the two species are often closely adapted to one another.

Figure 30.21
Cleaner wrasse within the mouth of a spotted sweetlip is removing debris and parasites.

Summary

Certain interactions are typical between species. When two species compete for the same resources, one usually eliminates or restricts the range of the other. This is in keeping with the competitive exclusion principle.

Predators and prey coevolve. Herbivores prey on plants, and carnivores prey on other animals. But, even so, both predatory and prey population sizes are maintained at a constant level. Antipredator defense adaptations are varied, as listed in table 30.2. Sometimes an animal with a successful antipredator defense is mimicked by other animals that have the same defense or lack the defense.

Symbiotic relationships are of three types. In parasitism, the host species is harmed. Parasites, which help control the size of their host population, may require a secondary host for dispersal. The anatomy and life-style of an animal parasite may be contrasted to that of an active predator. In commensalism, one species is benefited but the other is unaffected. Often the host simply provides a home and/or transportation. In mutualism, both species benefit. The two species are often closely adapted to one another, such as flowers and their pollinators. There are also examples of mutualistic species that live together in the same locale, such as ants that keep fungal gardens. But in cleaning symbiosis, the relationship is transitory.

Objective Questions

1. The competitive exclusion principle states that no two species can occupy the same ____________ .
2. Predators are ____________ to capturing their prey and prey are ____________ to escaping predators.
3. Predators that eat vegetation are ____________ and those that eat other animals are ____________ .
4. Both competition and predation are factors that ____________ population sizes.
5. ____________ is present when one species resembles another that possesses an overt antipredator defense.
6. ____________ mechanisms prevent predators from capturing all the prey and prevent parasites from killing off their hosts.
7. The symbiotic relationship ____________ is exemplified by the habit of barnacles to attach themselves to whales.
8. In the symbiotic relationship known as mutualism ____________ .
9. The adaptations of flowers to attract specific pollinators and the pollinators adaptations to exploit the flower as a source of nutrients is termed ________ .

Answers to Objective Questions
1. niche 2. adapted, adapted 3. herbivores, carnivores 4. control 5. Mimicry 6. Defense 7. commensalism 8. both species benefit 9. coevolution

Study Questions

1. Give two examples to show that competition between species results in the elimination or restriction of the range of the other. (p. 669)
2. Give examples to show that it is sometimes unwise to bring a new competitor into an area. (p. 671)
3. Define the competition exclusion principle and discuss the concept of an organism's niche. (p. 671)
4. Would it be correct to say that the more species variety there is in an area, the more niches there must be? (p. 672)
5. Give examples to show that predators have a beneficial effect, helping to stabilize population sizes in an area. (pp. 675–676)
6. Give examples to show that plants have defenses against herbivores. (p. 676)
7. Give examples of animal antipredator defenses that prevent predator populations from overkilling prey populations. (p. 677)
8. Define a mimic and give two examples in relation to the monarch butterfly. (p. 678)
9. What are the three types of symbiotic relationships? Give several examples of each. (pp. 680–684)
10. Describe the life cycle of the hookworm. (p. 680)
11. Give two examples of social parasitism. (p. 681)

Key Terms

antipredator defense (an″tī-pred′ah-tor de-fens′) a physiological or behavioral activity or a structural modification that protects an organism from its predators. *676*

browsers (browz′erz) animals that feed on higher-growing vegetation such as shrubs and trees. *674*

cleaning symbiosis (klēn′ing sim″bi-o′sis) a mutualistic relationship in which one type organism gains benefit by cleaning another that is benefited by being cleaned of debris and parasites. *684*

coevolve (ko-e-volv′) the interaction of two species such that each influences the evolution of the other species. *674*

commensalism (kō-men′sal-izm) the relationship of two species in which one lives on or with the other without conferring either benefit or harm. *682*

competition (kom″pĕ-tish′un) interaction between members of the same or different species for a mutually required resource. *669*

competitive exclusion principle (kom-pet′ĭ-tiv eks-kloo′zhun prin′sĭ-p′l) an observation that no two species can continue to compete for the same exact resources since one species will eventually become extinct. *671*

grazers (gra′zerz) animals that feed on low-lying vegetation such as grasses. *674*

mimicry (mim′ik-re) the resemblance of an organism to another that has a defense against a common predator. *678*

mutualism (mu′tu-al-izm″) a relationship between two organisms of different species that benefits both organisms. *682*

niche (nich) the functional role and position of an organism in the ecosystem. *671*

parasite (par′ah-sīt) an organism that resides externally on or internally within another organism and does harm to this organism. *680*

parasitism (par′ah-si″tizm) symbiotic relationship in which an organism derives nourishment from and does harm to a host. *680*

predation (pre-da′shun) the eating of one organism by another. *674*

prey (pra) organisms that serve as food for a particular predator. *674*

symbiosis (sim″bi-o′sis) an intimate association of two dissimilar species including commensalism, mutualism, and parasitism. *680*

31

Ecosystems

Chapter Concepts

1 Natural ecosystems, which use solar energy efficiently and chemicals that cycle, produce little pollution and waste.

2 The human ecosystem, which comprises both country and city, utilizes fossil fuel energy inefficently and material resources that do not cycle. Therefore there is much pollution and waste.

3 Humans have begun to address the problem of pollution, which affects the quality of air, water, and land.

Chapter Outline

Ecosystem Composition
- Flow of Energy
- Chemical Cycling

Human Ecosystem
- Country
- City

Pollution
- Air Pollution
- Land Pollution
- Water Pollution

When life first arose it was confined to the oceans, but about 450 million years ago organisms began to colonize the bare land. Eventually the land supported many complex communities of living things. Similarly, today we can also observe a series of sequential events by which bare rock becomes capable of sustaining many organisms. We call this succession (fig. 31.1). During **succession** a sequence of communities replaces one another in an orderly and predictable way until finally there is a **climax community,** a mix of plants and animals typical of that area that remains stable year after year. For example, in the United States not too long ago, most of the Northeast was a deciduous forest, a prairie was common to the Midwest, and a semidesert covered the Southwest. We will learn more about these and other communities in the following chapter.

All of the various climax communities of the world make up the **biosphere,** a narrow sphere or shell that encircles the earth. Although most organisms reside near the surface, others are also found a short distance in the air above and in the waters beneath the surface. Each belongs to a **population,** the members of a species found within the community. The various populations interact with each other and the physical environment. Therefore, they form an **ecosystem** that has both a **biotic** (living) and **abiotic** (nonliving) component. The study of **ecology** is defined as the study of the interactions of organisms among themselves and with the physical environment.

The process of succession leads to a climax community which contains the plants and animals characteristic of the area. The populations within a community interact within themselves and the physical environment, forming an ecosystem.

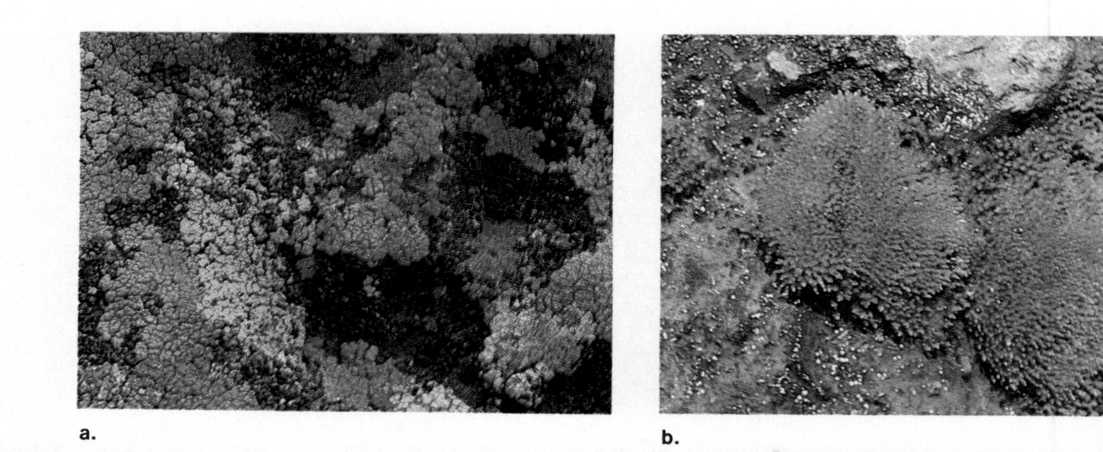

a. b.

c. d.

Figure 31.1
Primary succession includes a series of stages by which bare rock becomes a climax community. These photographs show a possible sequence of events: a. *lichens growing on bare rock,* b. *mosses appear once there is enough soil,* c. *grasses and perennial herbs spread out over the area.* d. *If the climate is suitable, a forest may be the climax community.*

Ecosystem Composition

Each population in an ecosystem has a habitat and a niche. The **habitat** of an organism is its place of residence, that is, the location where it may be found, such as "under a fallen log" or "at the bottom of the pond." The *niche* of an organism is its profession or total role in the community. A description of an organism's niche (table 31.1) includes its interactions with the physical environment and with the other organisms in the community. One important aspect of niche is the manner in which the organism acquires energy and chemicals. In fact, the entire ecosystem has two important aspects: *energy flow and chemical cycling.*

Within an ecosystem, energy flows and chemicals cycle.

Flow of Energy

Energy flow begins when photosynthesizing organisms use the energy of the sun to make their own food. Thereafter, chemicals and energy are passed from one population to another as the populations form food chains (fig. 31.2).

Food Chains

Essentially, food chains are described by telling "who eats whom." Photosynthesizing organisms, which are at the start of a food chain, are called the **producers** because they have the ability to change what was formerly inorganic chemicals to organic food. Producers in a food chain therefore produce food.

***Table 31.1** Aspects of Niche*

Plants	Animals
Season of year for growth and reproduction	Time of day for feeding and season of year for reproduction
Sunlight, water, soil requirements	Habitat requirements
Contribution to ecosystem	Food requirements
Competition and cooperation with other organisms	Competition and cooperation with other organisms
Effect on abiotic environment	Effect on abiotic environment

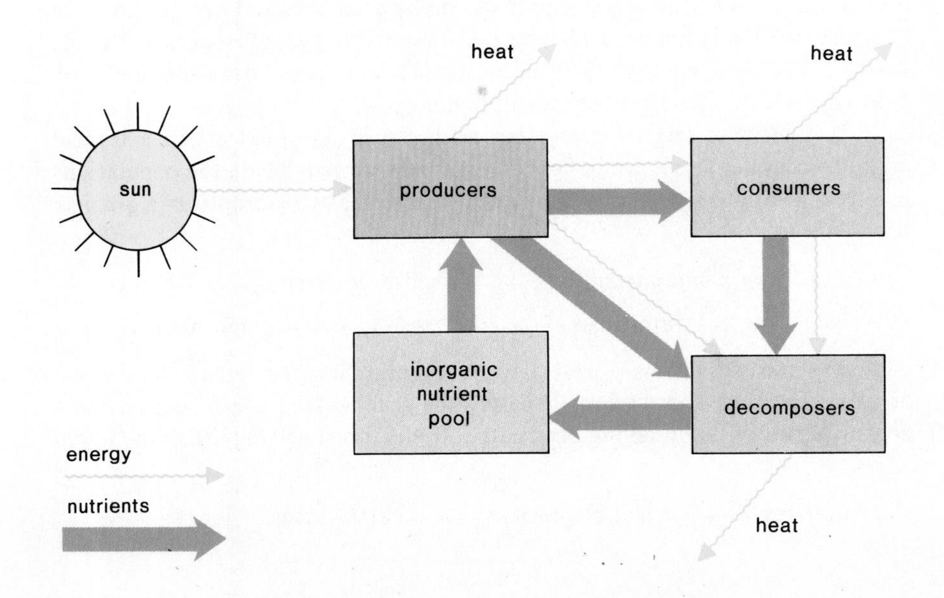

Figure 31.2
A diagram illustrating energy flow and chemical cycling through an ecosystem. Energy does not cycle because all the energy that is derived from the sun eventually dissipates as heat.

Figure 31.3
This drawing depicts populations typical of a deciduous forest ecosystem in the eastern United States. The arrows indicate food chains.

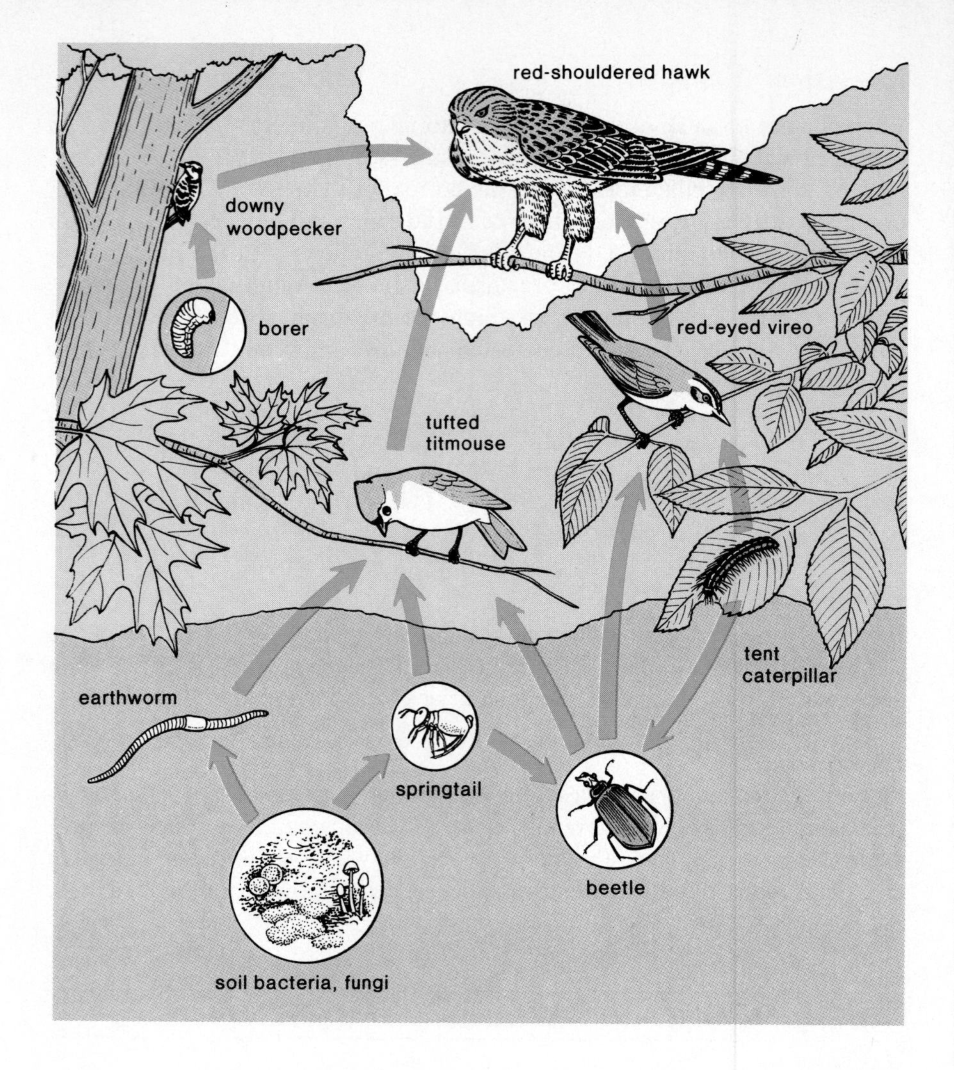

The other two types of populations in the biotic community are the consumers and the decomposers. **Consumers** are organisms that must take in preformed food. Herbivores are primary consumers that feed directly on producers; carnivores are secondary or tertiary consumers that feed only on consumers; omnivores are consumers that feed on both producers and consumers. **Decomposers** are organisms of decay, such as bacteria and fungi, that break down nonliving organic matter (**detritus**) to inorganic matter, which can be used again by producers. In this way the same chemicals can be used over and over again in an ecosystem. This is not true for energy. As detritus decomposes, all that remains of the solar energy taken up by the producer populations dissipates as heat (fig. 31.2). Therefore, energy does not cycle.

A typical terrestrial ecosystem is shown in figure 31.3 and a typical aquatic ecosystem is shown in figure 31.4. The arrows indicate the populations involved in various food chains; for example, here are two examples of a *grazing food chain:*

1. trees ⟶ tent caterpillars ⟶ red-eyed vireos ⟶ hawks
2. algae ⟶ water fleas ⟶ catfish ⟶ green herons

In some ecosystems (forests, rivers, and marshes) the detritus food chain accounts for more energy flow than does the grazing food chain because most organisms die without having been eaten. In the forest one *detritus food chain* is:

detritus ⟶ soil bacteria ⟶ earthworms

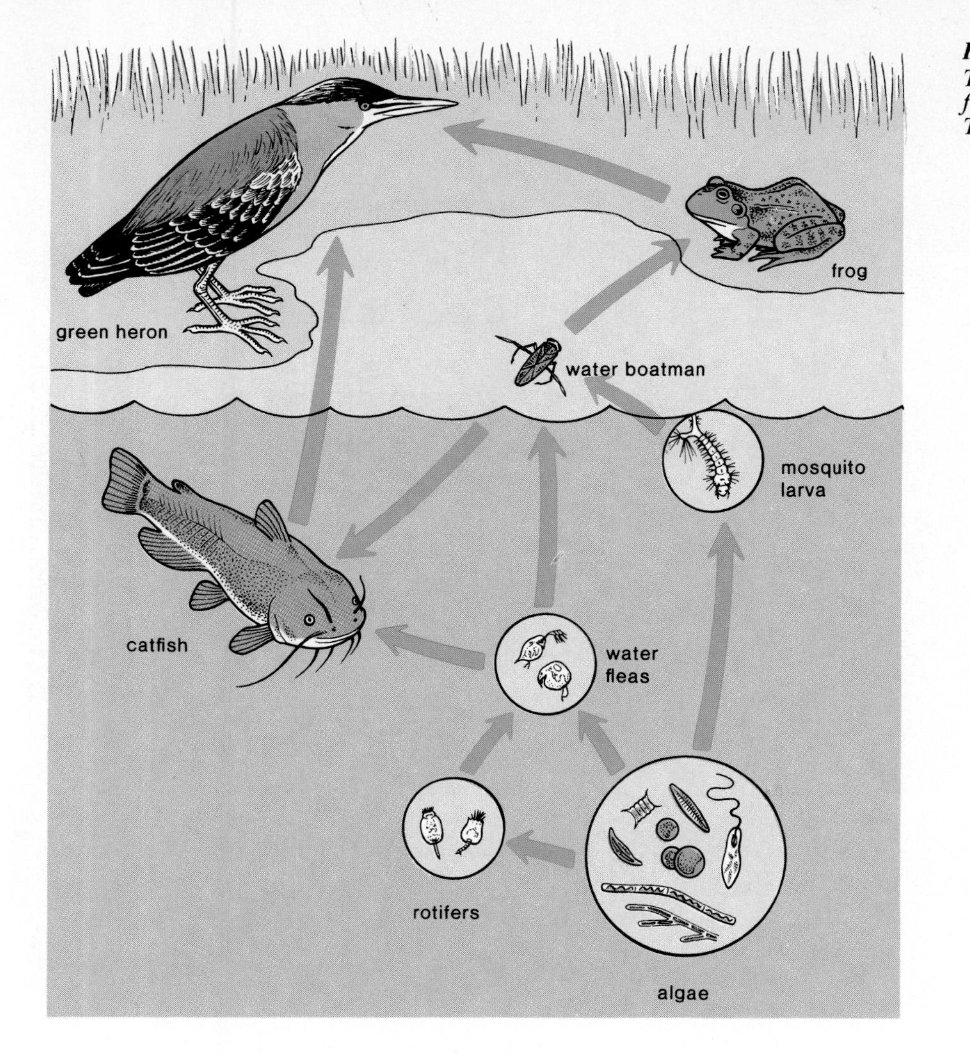

Figure 31.4
This drawing depicts populations typical of a freshwater pond ecosystem in eastern United States. The arrows indicate food chains.

A detritus food chain often ties in with a grazing food chain, as when earthworms are eaten by a tufted titmouse. Although the grazing food chain is most important in aquatic food chains, the detritus food chain is most important in terrestrial food chains. For example, it is estimated that only about 10 percent of the annual leaf production is consumed by herbivores; the rest is either degraded by decomposers or eaten by scavengers, such as soil mites, earthworms, and millipedes.

Food Webs

Each food chain represents one pathway by which chemicals and energy are passed along in an ecosystem. Natural ecosystems have numerous food chains, each linked to form a complex **food web** (figs. 31.3 and 31.4). Since organisms may belong to more than one food chain, energy flow is best described in terms of trophic (feeding) levels. The first level is the producer population, and each successive level is further removed from this population. All animals acting as primary consumers are part of the second level; all animals acting as secondary consumers are part of the third level; and so on. Each succeeding trophic level passes on less energy than was received due to a number of reasons. For example, usable energy is lost at each level because

1. Not all members of the previous trophic level become food; of those that do become food, some portions, such as hair and bones, may be uneaten, or if eaten, may be undigestible;
2. Some food is used for maintenance and never contributes to growth;
3. Energy transformations always result in a loss of usable energy.

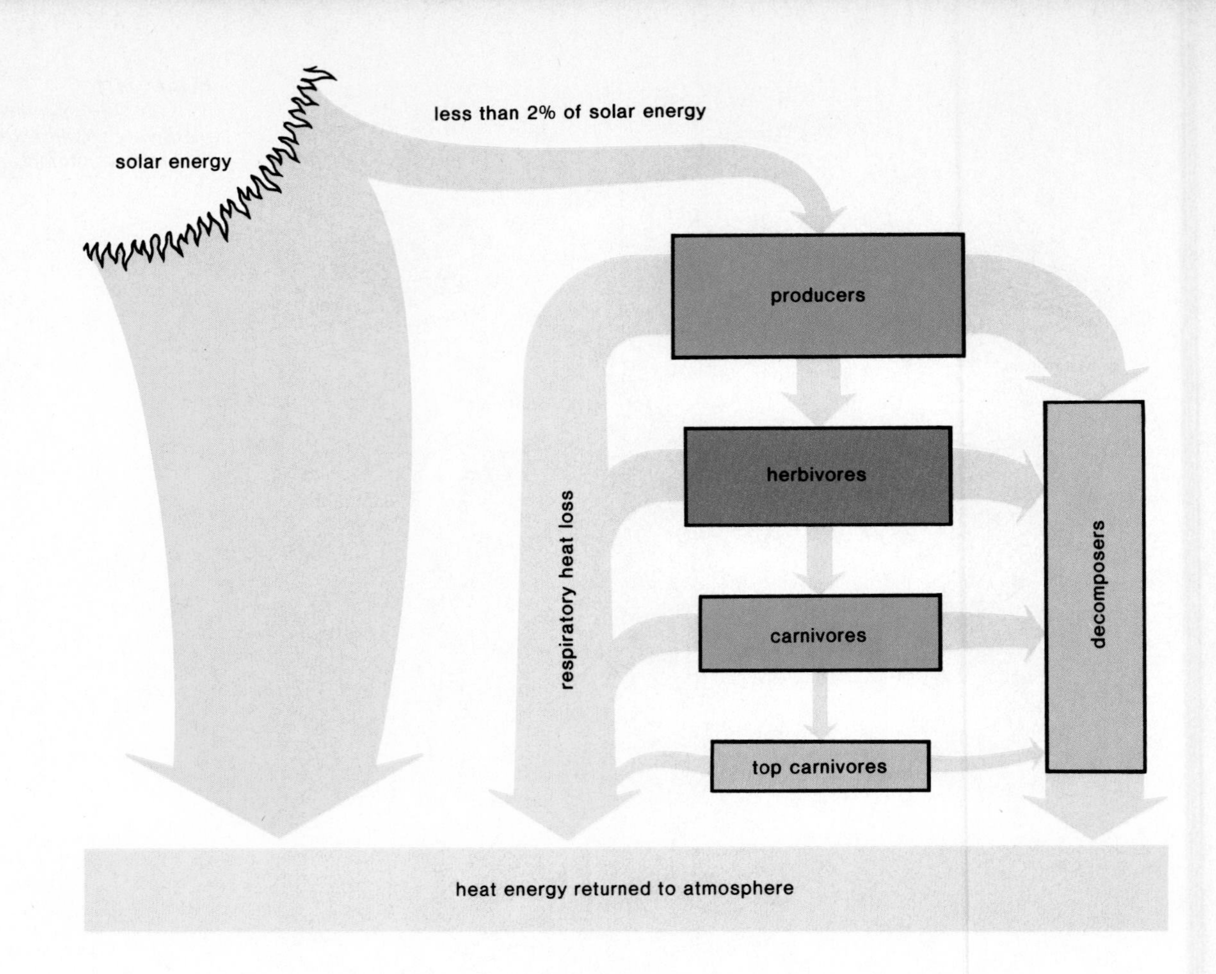

Figure 31.5
Dissipation of energy in an ecosystem. Producers convert inorganic matter to organic food after capturing a small amount of solar energy. When nutrient molecules are used for maintenance, their energy content eventually becomes heat. Energy, therefore, does not cycle, and ecosystems must have a continual supply of solar energy.

To understand that energy is lost due to energy transformations, consider that the conversion of energy in one molecule of glucose to 38 ATP molecules represents only 50 percent of the available energy in a glucose molecule. The rest is lost as heat.

Eventually, as one population feeds on another and as decomposers work on detritus, all of the captured solar energy that was converted to chemical bond energy by algae and plants is returned to the atmosphere as heat (fig. 31.5). Therefore, *energy flows through an ecosystem and does not cycle.*

The populations in an ecosystem form food chains in which the producers produce food for the other populations by being able to capture the energy of the sun. While it is convenient to study food chains, the populations in an ecosystem actually form a food web in which food chains join and overlap with one another.

Food Pyramid

Whether one considers the number of individuals, the biomass (weight of living material), or the amount of energy, each succeeding trophic level is generally smaller than the one preceding. To illustrate this, the various trophic levels are often depicted in the form of a pyramid (fig. 31.6). The producer population is at the base of the pyramid. Obviously it must be the largest because it indirectly produces food for all the other populations.

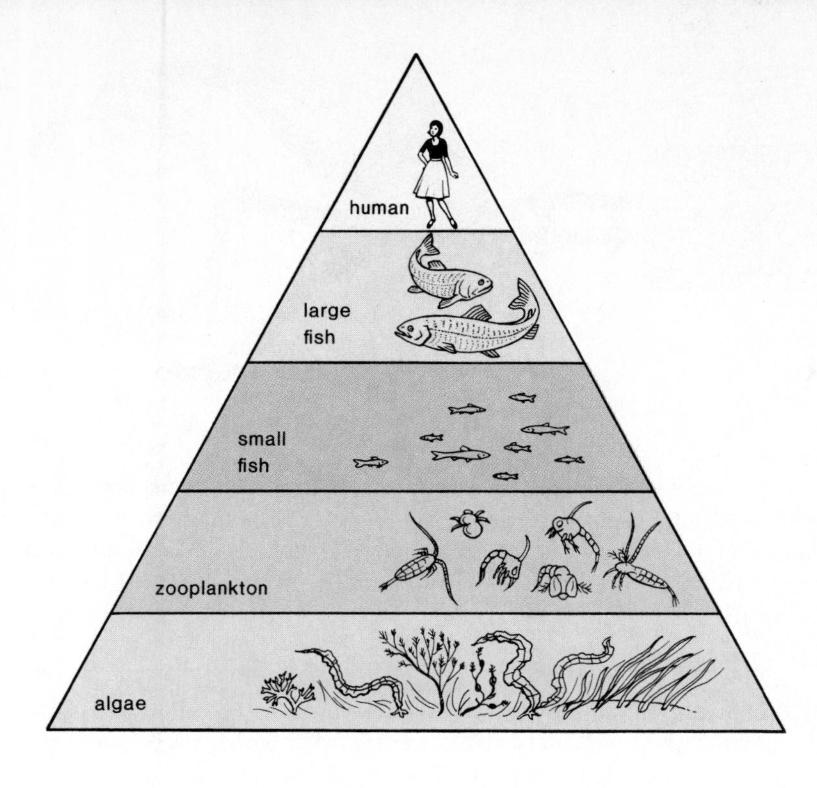

Figure 31.6
Food pyramid. Whether one considers numbers, biomass, or energy content, each trophic level is smaller than the one on which it depends. Therefore the various trophic levels form a pyramid in which the producer population is at the base and the last consumer population is at the peak.

In regard to the pyramid of energy, it is generally stated that only about 10 percent of the energy absorbed by one trophic level can be stored by the next level. About 90 percent is lost for the reasons previously stated. In practical terms, this means that about 10 times the number of people can be sustained on a diet of grain than on a diet of meat.

Table 31.2 *Population Control*
Density independent effects
Climate and weather
Natural disasters
Density dependent effects
Competition
Predation
Parasitism
Emigration

Stability

Mature natural ecosystems tend to be diverse and stable. *First,* the various populations stay at a constant size for the reasons listed in table 31.2. *Density independent* effects are those forces of nature whose magnitude of influence does not depend on the size of the population affected. For example, the severity of a drought or flood has nothing to do with how many plants or animals there are in a particular area. The severity of *density dependent* effects, several of which were studied in the preceding chapter, does depend on the size of the population affected. For example, two large populations compete to a greater degree for the same resource than do two small populations.

Second, diversity assists maintenance of population size. For example, imagine an ecosystem in which hawks eat both rabbits and mice. If the rabbit population suffered an epidemic and declined in size, the mice population would increase in size due to decreased competition for food. The increase would mean that the producer population would still be held in check, and the hawks that eat rabbits or mice would still eventually have the same amount of food. It may take a little time for the new balance to come about; a few hawks may have to migrate or starve, but essentially there would be little change in the size of the hawk population. Further, as the rabbits recover, the mice population would decline and eventually there would be exactly the same balance as before the epidemic. In this manner, variation of species at each level of a food pyramid gives stability to an ecosystem. Actually, we see the same principle at work in business; a company diversifies its products so that as demand fluctuates profits will remain the same.

Figure 31.7
Components of a chemical cycle. The reservoir stores the chemical, the exchange pool makes it available to producers, after which it cycles through food chains. Decomposition returns the chemical to the exchange pool once again if it has not already returned by another process.

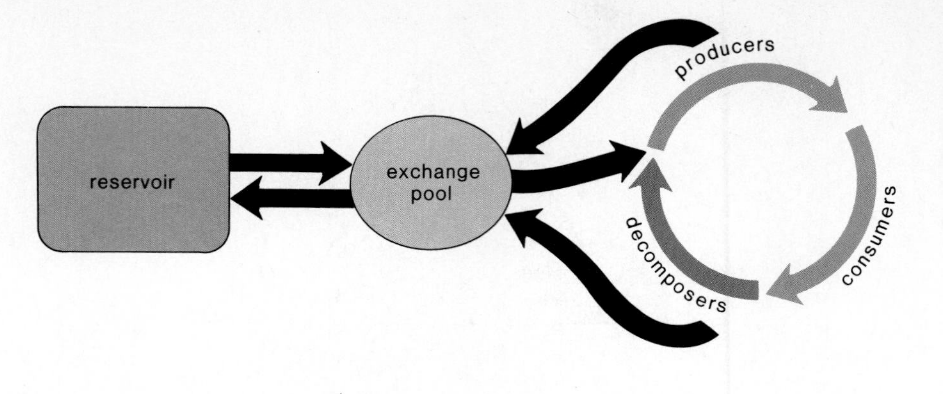

Constancy of population sizes means that most of the solar energy utilized by a biotic community supports stability rather than growth. The same amount of solar energy is required each year in order to maintain a highly diversified community that has little material waste due to the cycling of chemicals.

The food pyramid illustrates that each successive population has a smaller size or biomass because energy does not cycle. Mature natural ecosystems are stable because population sizes are held in check by density independent and density dependent factors. Also, population size is maintained by the diversity of the food web.

Chemical Cycling

In contrast to energy, matter does cycle through ecosystems. Because there is normally no input from outside an ecosystem, the same 30 or 40 elements essential to life are used over and over. For each element the cycling process (fig. 31.7) involves (1) a *reservoir*—that portion of the biosphere that acts as a storehouse for the element; (2) an *exchange pool*—that portion of the biosphere from which the producers take and consumers return inorganic nutrients; and (3) the *biotic community*—through which chemicals move along food chains from and to the reservoir.

Carbon Cycle

The relationship between photosynthesis and respiration should be kept in mind in discussing the carbon cycle. Recall that for simplicity's sake this equation in the forward direction represents respiration and in the other direction may be used to represent photosynthesis:

$$C_6H_{12}O_6 + 6\ O_2 \rightleftharpoons 6\ CO_2 + 6\ H_2O$$

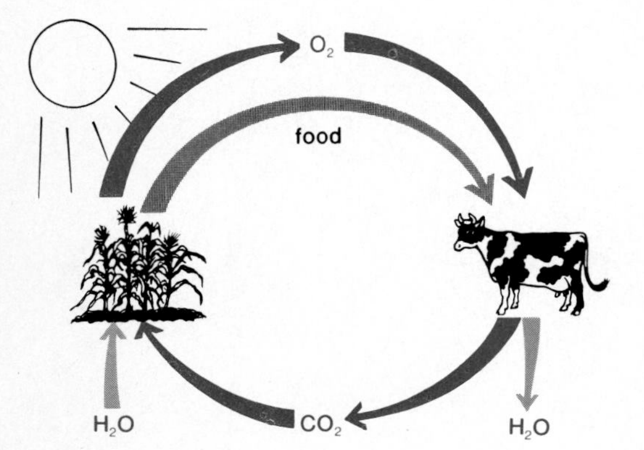

Figure 31.8
Relationship between photosynthesis and respiration. Animals are dependent on plants for a supply of oxygen and plants are dependent on animals for a supply of carbon dioxide.

The equation tells us that respiration releases cabon dioxide, the molecule needed for photosynthesis. However, photosynthesis releases oxygen, the molecule needed for respiration. From figure 31.8, it is obvious that animals are dependent on green organisms, not only to produce organic food and energy but also to supply the biosphere with oxygen.

In the carbon cycle on land (fig. 31.9) animals and animal-like organisms continuously release carbon dioxide into the air, but plants release carbon dioxide only when, as at night, respiration is occurring at a faster rate than is photosynthesis. After organisms die, decomposition also releases carbon dioxide. On the other hand, plants take carbon dioxide up from the air when photosynthesizing.

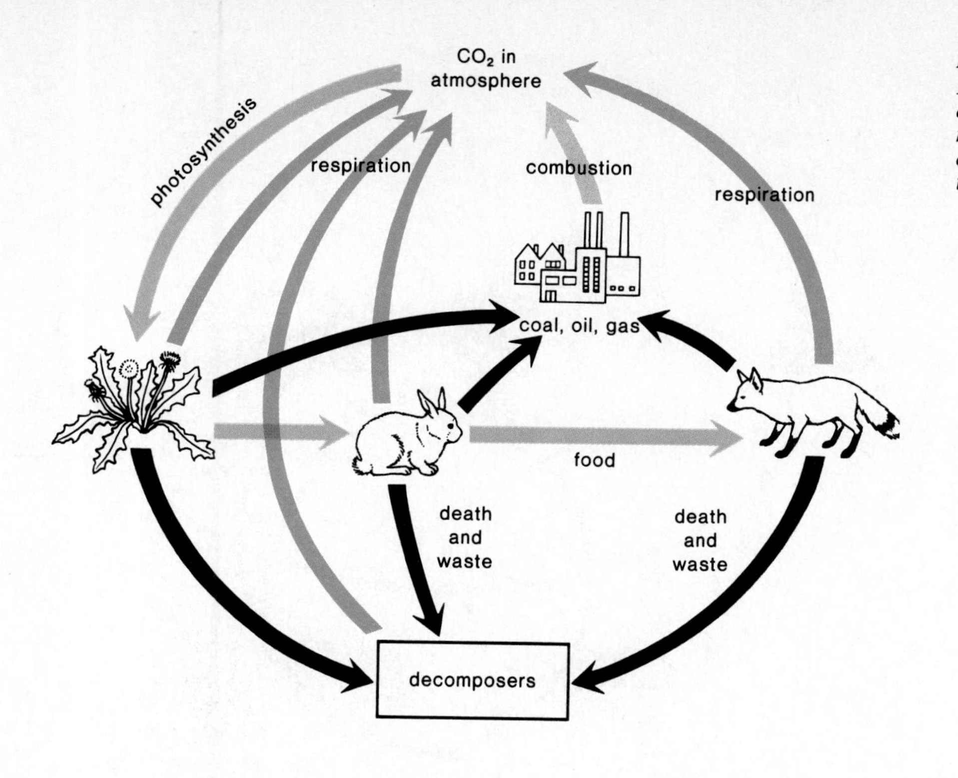

Figure 31.9
Major components of the terrestrial portion of the carbon cycle. Photosynthesizers utilize CO_2 while respiration returns it to the atmosphere. The combustion of fossil fuels (coal, oil, gas) represents the human contribution to the cycle.

The carbon cycle also occurs in aquatic communities, but in this case, carbon dioxide is taken up from and returned to water. Carbon dioxide from the air combines with water to give bicarbonate ions (HCO_3^-) that serve as a source of carbon for algae, which produce food for themselves and then become food for others. Similarly, when aquatic organisms respire, the carbon dioxide they give off becomes bicarbonate ions. The amount of bicarbonate in the water is in equilibrium with the amount in the air.

Reservoirs Living and dead organisms contain organic carbon and serve as one of the reservoirs for the carbon cycle. When we destroy forests and burn wood, we are reducing this reservoir and increasing the amount of CO_2 in the atmosphere.

In prehistoric times, certain plants and animals did not decompose, by chance, and instead were preserved in **fossil fuels** such as coal, oil, and gas. When humans burn fossil fuels, they not only add CO_2 but other combustion products to the air. This contributes greatly to air pollution, which is discussed on page 703.

Another reservoir for the carbon cycle is the formation of inorganic carbonate that accumulates in limestone and in carbonaceous shells. Limestone, particularly, tends to form in the oceans. In this way, the oceans act as a sink for excess carbon dioxide. The burning of fossil fuels in the last 22 years has probably released 78 billion tons of carbon, and yet the atmosphere registers only an increase of 42 billion tons.

---

*I*n the carbon cycle, carbon dioxide is removed from the atmosphere by photosynthesis but is returned by respiration. Living things and dead matter are reservoirs as is the ocean, particularly because it accumulates limestone. When humans cut down trees and burn wood and fossil fuels, they are changing the current balance of the carbon cycle.

Figure 31.10
Major components of the terrestrial portion of the nitrogen cycle. Three types of bacteria are at work: nitrogen-fixing bacteria convert aerial nitrogen to a form usable by plants; nitrifying bacteria, which include both nitrite and nitrate bacteria, convert ammonia to nitrate; and the denitrifying bacteria convert nitrate back to aerial nitrogen again. Humans contribute to the nitrogen cycle by converting aerial nitrogen to fertilizer.

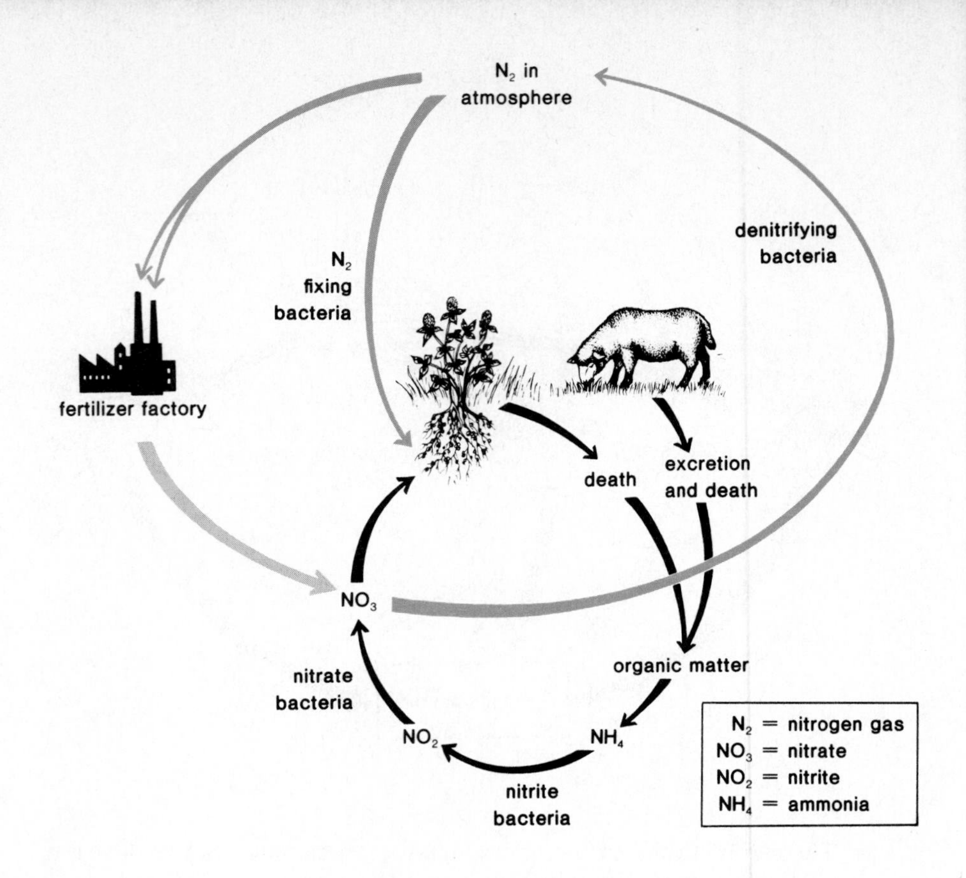

Figure 31.11
Nodules on the roots of a legume (in this case, soybean). The bacteria that live in these nodules are capable of converting aerial nitrogen to a source that the plant can use.

Nitrogen Cycle

That portion of the nitrogen cycle that involves terrestrial organisms is represented by the diagram in figure 31.10. Aerial nitrogen, the reservoir for the nitrogen cycle, is not usable by most organisms, but there are two types of nitrogen-fixing bacteria that make use of it. One type is free-living in the soil, but the other type infects and lives in nodules on the roots of legumes (fig. 31.11). Here the nitrogen-fixing bacteria convert aerial nitrogen to a form that can be used by plants.

Nitrates, in particular, are taken up by the roots of plants and converted to amino acids. Thereafter, animals acquire this organic nitrogen when they eat plants. Plants and animals die, and decomposition produces ammonia that can be converted to nitrates by **nitrifying** (nitrite and nitrate) **bacteria.** To make the cycle complete, there are some bacteria, the **denitrifying bacteria,** that can convert ammonia and nitrates back to aerial nitrogen again.

Other Contributions to the Nitrogen Cycle

There are at least two other ways by which **nitrogen fixation** occurs. Nitrogen is fixed in the atmosphere when cosmic radiation, meteor trails, and lightning provide the high energy needed for nitrogen to react with oxygen. Also, humans make a most significant contribution to the nitrogen cycle when they convert aerial nitrogen to nitrates for use in fertilizers. This industrial process requires an energy input that equals that of the eventual increase in crop yield. The application of fertilizers also contributes to water pollution, as discussed on page 709. Since nitrogen-fixing bacteria do not require fossil fuel energy and do not cause pollution, research is now directed toward finding a way to make all plants capable of forming nodules (fig. 31.11) or, even better, through recombinant DNA research, to possess the biochemical ability to fix nitrogen themselves.

*I*n the nitrogen cycle, nitrogen-fixing bacteria convert aerial nitrogen to nitrates and denitrifying bacteria convert nitrates back to aerial nitrogen. Nitrifying bacteria convert ammonia to nitrate, the form of nitrogen most often used by plants. Humans again tap into a reservoir when they convert aerial nitrogen to nitrate for use in fertilizer.

Human Ecosystem

Mature natural ecosystems tend to be stable and to exhibit the characteristics listed in table 31.3. Each population is of a proper size in relation to other populations; the energy that enters and the amount of matter that cycles is appropriate to support these populations. **Pollution,** defined as any undesirable change in the environment that may be harmful to humans and other life, and excessive waste do not normally occur. The human ecosystem that replaces natural ecosystems is quite different, however.

Human beings have replaced natural ecosystems with one of their own making as is depicted in figure 31.12. This ecosystem essentially has two parts: the *country,* where agriculture and animal husbandry are found, and the *city,* where most people live and industry is carried on. This representation of the human ecosystem, although simplified, allows us to see that the system requires two major inputs: *fuel energy and raw materials* (e.g., metals, wood, synthetic materials). The use of these two necessarily results in *pollution and waste* as outputs.

***Table 31.3** Ecosystems*

Natural	Human
Independent	Dependent
Cyclical (except energy)	Noncyclical
Nonpolluting	Polluting
Renewable solar energy	Nonrenewable fossil fuel energy
Conserves resources	Uses up resources

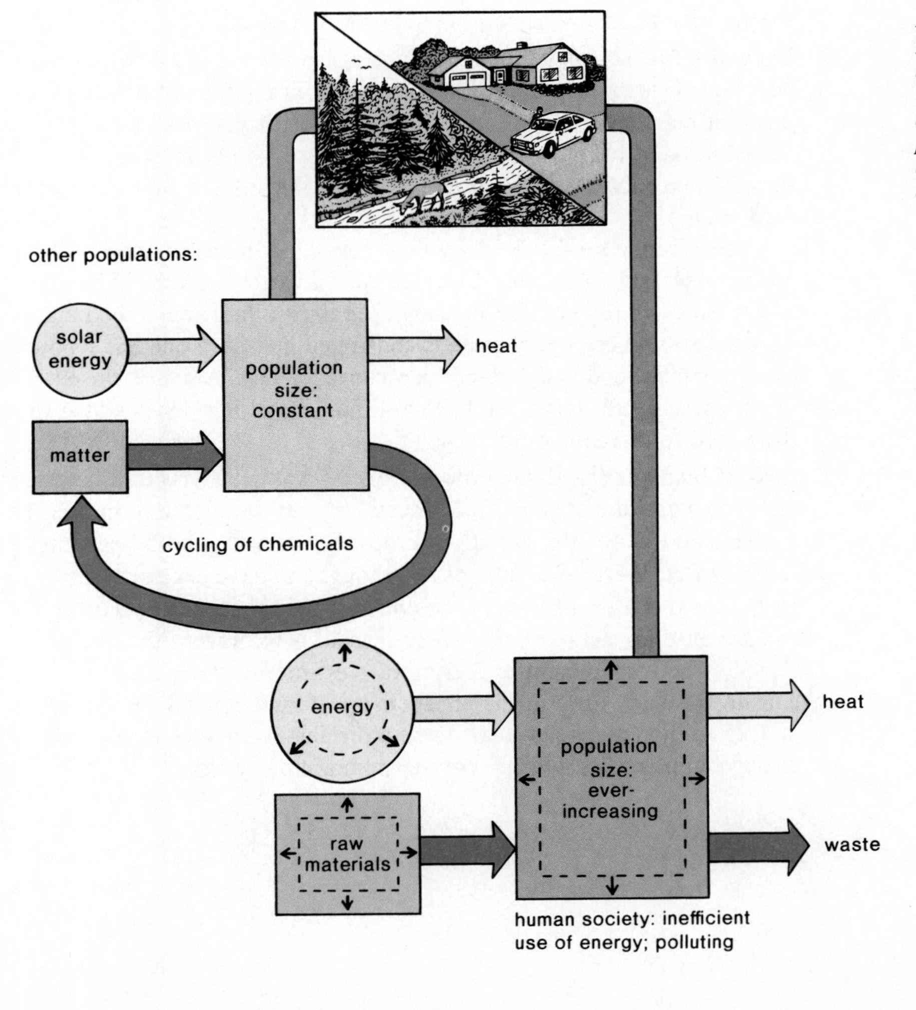

Figure 31.12
Human ecosystem versus a natural ecosystem. In natural ecosystems, population sizes remain about the same year after year; materials cycle and energy is used efficiently. In the human ecosystem, the population size consistently increases, resulting in much pollution because of inadequate cycling of materials and inefficient use of supplemental

Country

Modern U.S. agriculture produces exceptionally high yields per acre, but this bounty is dependent on a combination of five variables given here.

1. **Planting of a few genetic varieties.** The majority of farmers specialize in growing one of these. Wheat farmers plant the same type of wheat, and corn farmers plant the same type of corn. This so-called **monoculture agriculture** is subject to attack by a single type of parasite. For example, a single parasitic mold reduced the 1970 corn crop by 15 percent, and the results could have been much worse because 80 percent of the nation's corn acreage was susceptible.
2. **Heavy use of fertilizers, pesticides, and herbicides.** *Fertilizer* production requires a large energy input, and fertilizer runoff contributes to water pollution (p. 709). *Pesticides* reduce soil fertility because they kill off beneficial soil organisms as well as pests, and some pesticides concentrate in food chains (p. 709), eventually producing toxic effects in predators, possibly even humans. *Herbicides,* especially those containing the contaminant dioxin, have been charged with causing reproductive effects and cancer.
3. **Generous irrigation.** River waters are sometimes redirected for the purpose of irrigation, in which case "used water" returns to the river carrying a heavy concentration of salt. The salt content of the Rio Grande River in the Southwest is so high that the government has built a treatment plant to remove the salt. Water is also sometimes taken from aquifers (underground rivers) whose water content can be so reduced that it becomes too expensive to pump out more water. Farmers in Texas are already facing this situation.
4. **Excessive fuel consumption.** Energy is consumed on the farm for many purposes. Irrigation pumps have already been mentioned but large farming machines also are used to spread fertilizers, pesticides, and herbicides and to sow and harvest the crops. It is not incorrect to suggest that modern farming methods transform fossil fuel energy into food energy.

   Supplemental fossil fuel energy also contributes to animal husbandry yields. At least 50 percent of all cattle are kept in *feedlots* where they are fed grain. Chickens are raised in a completely artificial environment where the climate is controlled and each one has its own cage to which food is delivered on a conveyor belt. Animals raised under these conditions often have antibiotics and hormones added to their feed to increase yield.
5. **Loss of land quality.** Evaporation of excess water on irrigated lands can result in a residue of salt. This process, termed salinization, makes the land unsuitable for the growth of crops. Between 25 and 35 percent of the irrigated Western croplands are thought to have excessive salinity. Soil erosion is also a serious problem. It is said that we are *mining the soil* because farmers are not taking measures to prevent the loss of topsoil. The Department of Agriculture estimates that erosion is causing a steady drop in the productivity of land equivalent to the loss of 1.25 million acres per year. Even more fertilizers, pesticides, and energy supplements will be required to maintain yield.

Energy to Grow Food

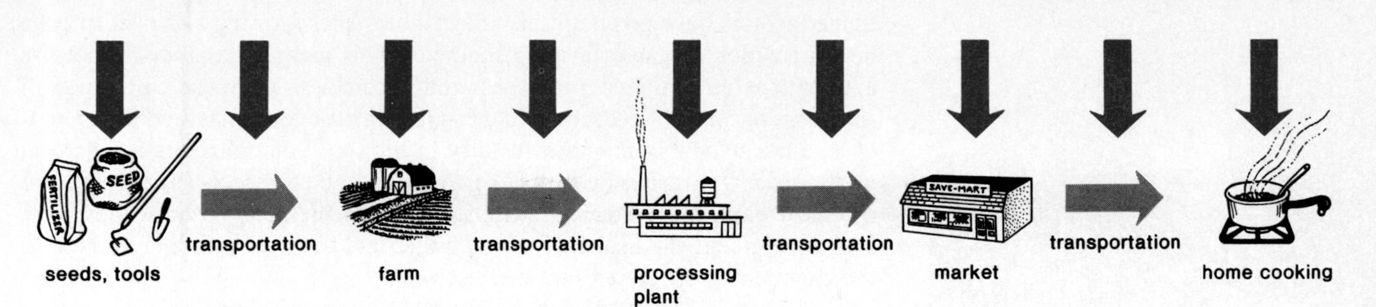

In the United States sunlight provides part of the energy for food production, but it is greatly supplemented by fossil fuel energy. Even before planting time, there is an input of fossil fuel energy for the production of seeds, tools, fertilizers, and pesticides, and for their transportation to the farm *a*. At the farm, fuel is needed to plant the seeds, to apply fertilizers and pesticides, and to irrigate, harvest, and dry the crops. After harvesting, still more fuel is used to process the crops to make the neatly packaged products we buy in the supermarket. Most of the food we eat today has been processed in some way. Even farm families now buy at least some of their food from supermarkets in nearby towns.

Since 1940 the amount of supplemental fuel used in the American food system has increased far more sharply than the caloric content of the food we consume, as shown in *b*. This is partially due to the trend toward producing more food on less land. High-yielding hybrid wheat and corn plants require more care and thus about twice as much supplemental energy as the traditional varieties of wheat and corn. Cattle kept in feedlots and fed grain that has gone through the whole production process need about twenty times the amount of supplemental energy as do range-fed cattle. Our food system has been labeled energy intensive because it requires such a large input of supplemental energy.

The intensive use of fossil fuel energy to grow and provide food in the United States is a matter for concern because the supply of fossil fuel is limited and its cost has risen tremendously. This, in turn, affects the cost of farming and of the food produced. What can be done? First of all, devote as much land as possible to farming and animal husbandry. Plant breeders could sacrifice some yield to develop plants that would need less supplemental energy. And we could depend more on range-fed cattle. If cattle are kept close to farmland, manure can substitute, in part, for chemical fertilizer. Biological control, the use of natural enemies to control pests, would cut down on pesticide use. Solar and wind energy could be used instead of fossil fuel energy, particularly on the farm. For example, wind-driven irrigation pumps are feasible.

Finally, of course, consumers could help matters. We could overcome our prejudice against fruits and vegetables that have slight blemishes. We could consume less processed food. We could eat less meat and buy cheaper cuts of beef, which are more likely to have come from range-fed cattle. And we could avoid using electrically powered gadgets when preparing food at home.

Since the United States food system is so energy-intensive, it is doubtful that needy countries abroad could ever duplicate this system. Indeed, if we are concerned about feeding the hungry of the world we should cut back on our own use of supplemental energy to make more available for use by underdeveloped countries.

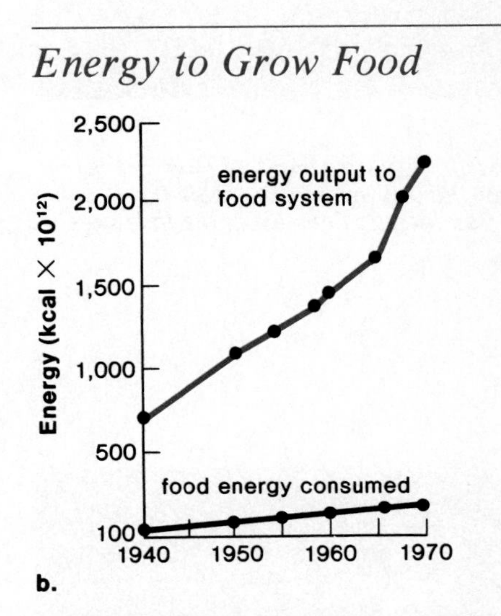

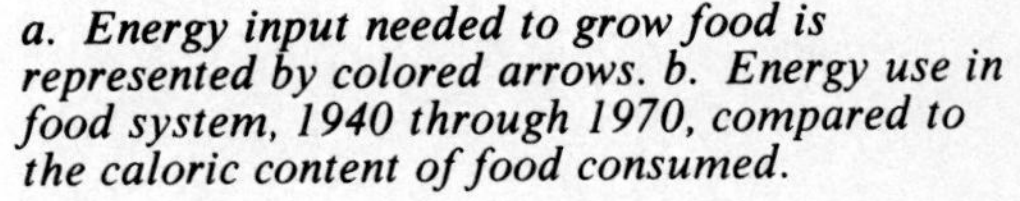

a. Energy input needed to grow food is represented by colored arrows. b. Energy use in food system, 1940 through 1970, compared to the caloric content of food consumed.

Figure 31.13
Biological control of pests. Here ladybugs are being used to control the cottony-cushion scale insect on citrus trees. If an insecticide is used, the ladybugs will be killed.

Figure 31.14
Loss of farmland due to suburban development. Several southern states have enacted "right-to-farm" laws to prevent the loss of farmland in this way.

Organic Farming

Some farmers have given up this modern means of farming and instead have begun to adopt organic farming methods. This means that they do not use applications of fertilizer, pesticides, and herbicides. They use cultivation of row crops to control weeds; crop rotation to combat major pests; and the growth of legumes to supply nitrogen fertility to the soil. Some farmers use natural predators and parasites instead of pesticides to control insects (fig. 31.13). For the most part, these farmers switched farming methods because they were concerned about the health of their familes and livestock and had found that the chemicals were sometimes ineffective.

A study of about 40 farms showed that organic farming for the most part was just as profitable as conventional farming. Crop yields were lower, but so were operating costs. Organic farms required about two-fifths as much fossil energy to produce one dollar's worth of crop. The method of plowing and utilization of crop rotation resulted in one-third less soil erosion. The researchers concluded it would be well to determine how far farmers can move in the direction of reduced agricultural chemical use and still maintain the quality of the product. They noted that a modest application of fertilizer would have improved the protein content of the crop.

City

The city is dependent on the country to meet its needs. For example, each person in the city requires several acres of land for food production. Overcrowding in cities does not mean that less land is needed; each person still requires a certain amount of land to ensure survival. Unfortunately, however, as the population increases, the suburbs and cities tend to encroach on agricultural and range land (fig. 31.14).

The city houses workers for both commercial businesses and industrial plants. Solar and other renewable types of energy are rarely used; cities currently rely mainly on fossil fuel in the form of oil, gas, electricity, and gasoline. The city does not conserve resources. An office building, with constantly burning lights and windows that cannot be opened, is an example of energy waste. Another example is people who drive cars long distances instead of taking public transportation and who drive short distances instead of walking or bicycling. Materials are not recycled and products are designed for rapid replacement.

The burning of fossil fuels for transportation, commercial needs, and industrial processes causes air and water pollution (p. 703). This pollution is compounded by the chemical and solid waste pollution that results from the manufacture of many products. Consider that any product used by the average consumer (house, car, washing machine) causes pollution and waste, both during its production and when it is disposed of. Humans themselves produce much sewage that is discharged into bodies of water, often after only minimal treatment.

Table 31.3 lists the charcteristics of the human ecosystem as it now exists. Just as the city is not self-sufficient and requires the country to supply it with food, so the whole human ecosystem is dependent on the natural ecosystems to provide resources and absorb waste. Fuel combustion by-products, sewage, fertilizers, pesticides, and solid wastes are all added to natural ecosystems in the hope that these systems will cleanse the biosphere of these pollutants. But we have replaced natural ecosystems with our human ecosystem and have exploited natural ecosystems for resources, adding ever more pollutants, to the extent that the remaining natural ecosystems have become overloaded.

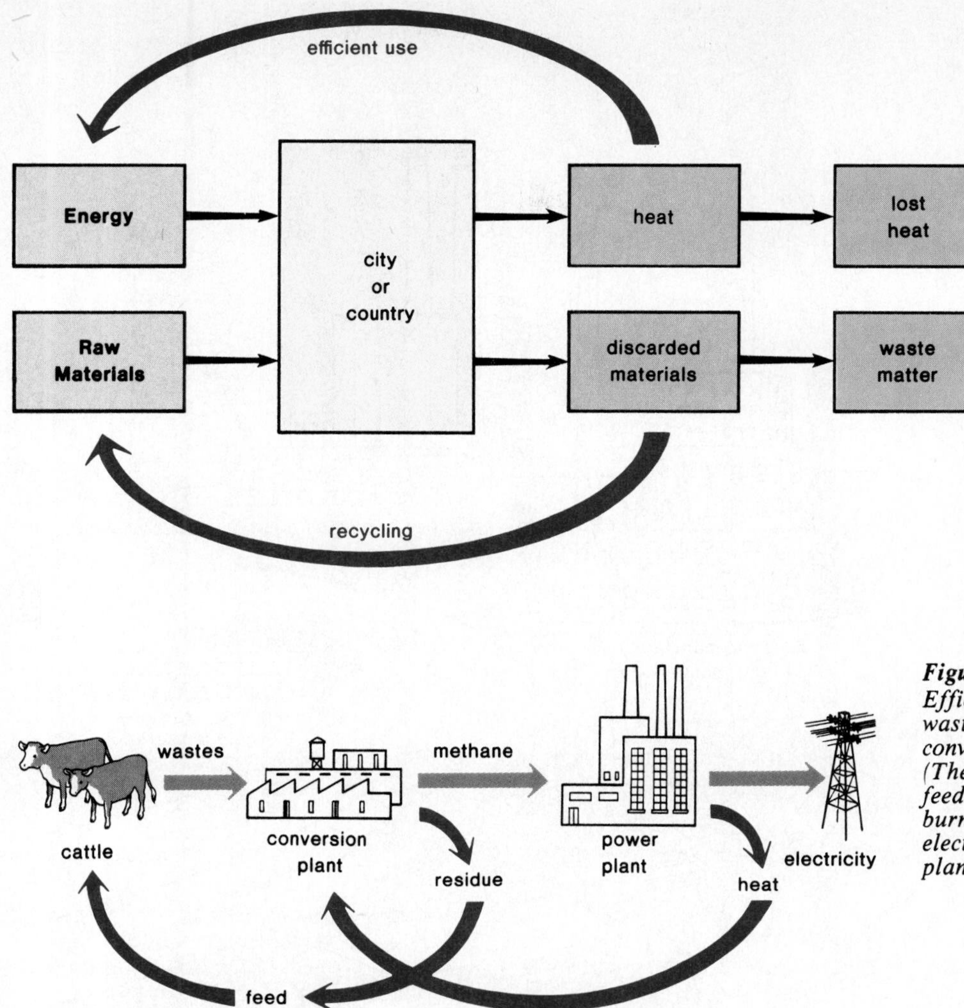

Figure 31.15
Modified human ecosystem. In order to cut down on the amount of lost heat and waste matter, heat could be used more efficiently and discarded materials could be recycled.

Figure 31.16
Efficient use of resources. Instead of allowing cattle waste to enter a water supply, it could be sent to a conversion plant that would produce methane gas. (The remaining residue could be converted back into feed for cattle.) Excess heat, which arises from the burning of methane gas in order to produce electricity, can be cycled back to the conversion plant.

Natural ecosystems have been destroyed and overtaxed because the human ecosystem is noncyclical and because an ever-increasing number of people wish to maintain a standard of living that requires many goods and services. But we can call a halt to this spiraling process if we achieve zero population growth and if we conserve energy and raw materials. Conservation can be achieved in three ways: (1) use wisely only what is actually needed; (2) recycle nonfuel minerals such as iron, copper, lead, and aluminum; and (3) use renewable energy resources (p. 754) and find more efficient ways to utilize all forms of energy. Figure 31.15 presents a diagrammatic representation of what is needed to maintain the delicate balances of the human and natural ecosystems. As a practical example, consider a plant that was built in Lamar, Colorado, which produces methane from feedlot animals' wastes (fig. 31.16). The methane is burned in the city's electrical plant and the heat given off is used to incubate the anaerobic digestion process that produces the methane. In addition, a protein feed supplement is produced from the residue of the digestion process. This system represents a cyclical use of material and efficient use of energy similar to that found in nature. Many other such processes for achieving this end have been and will be devised. However, as long as the human ecosystem on the whole remains inefficient and noncyclical, it will continue to cause pollution.

Figure 31.17
Air pollution. Pollutants enter the atmosphere from various sources, but in most instances air pollution is caused by the burning of fossil fuels (coal, gas, and oil).

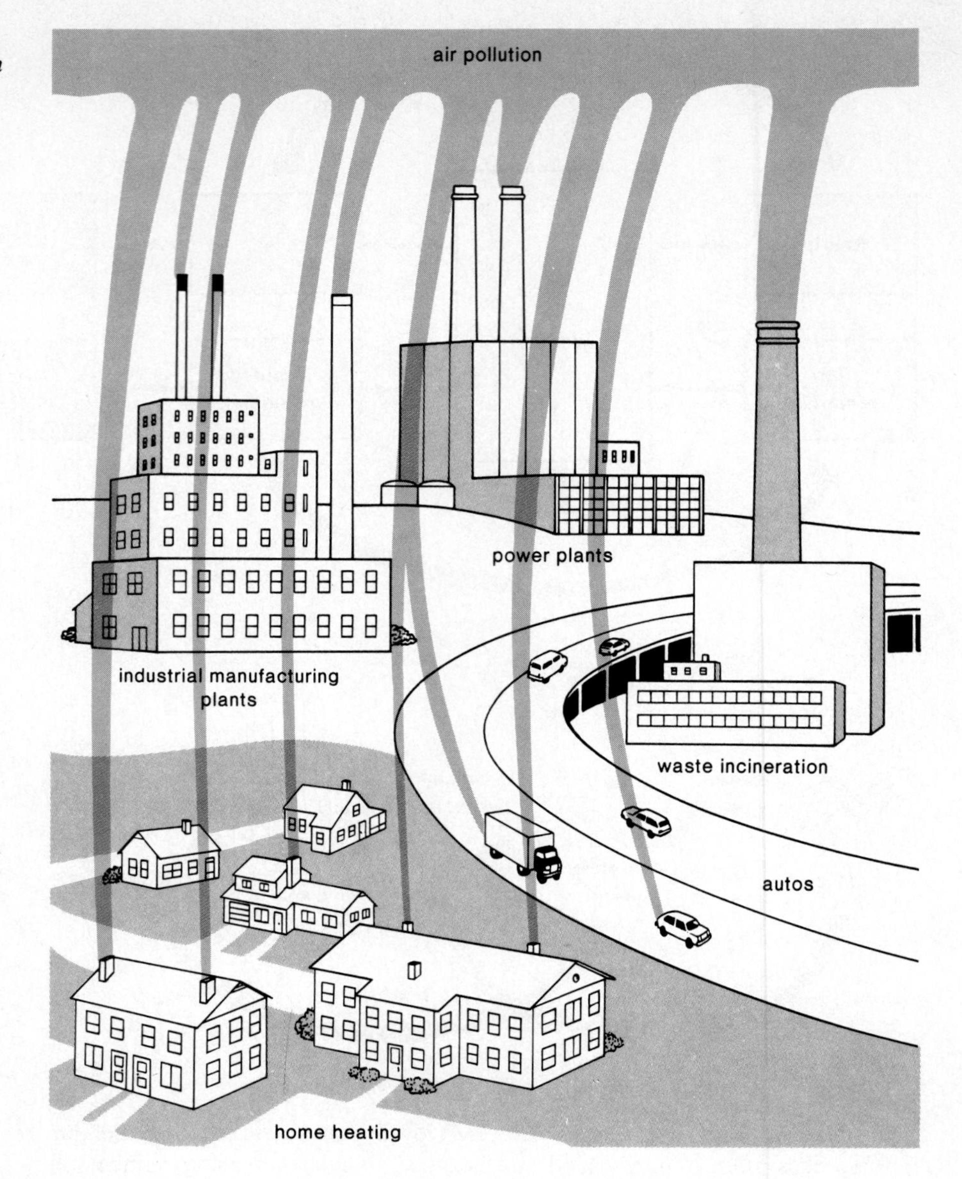

Figure 31.18
Components of air pollution. CO (carbon monoxide), HC (hydrocarbons), NO_x (nitrogen oxides), particulates (solid matter), SO_x (sulfur oxides). Transportation contributes most to air pollution because when gasoline burns it gives off CO, a gas that interferes with the capacity of hemoglobin to carry oxygen.

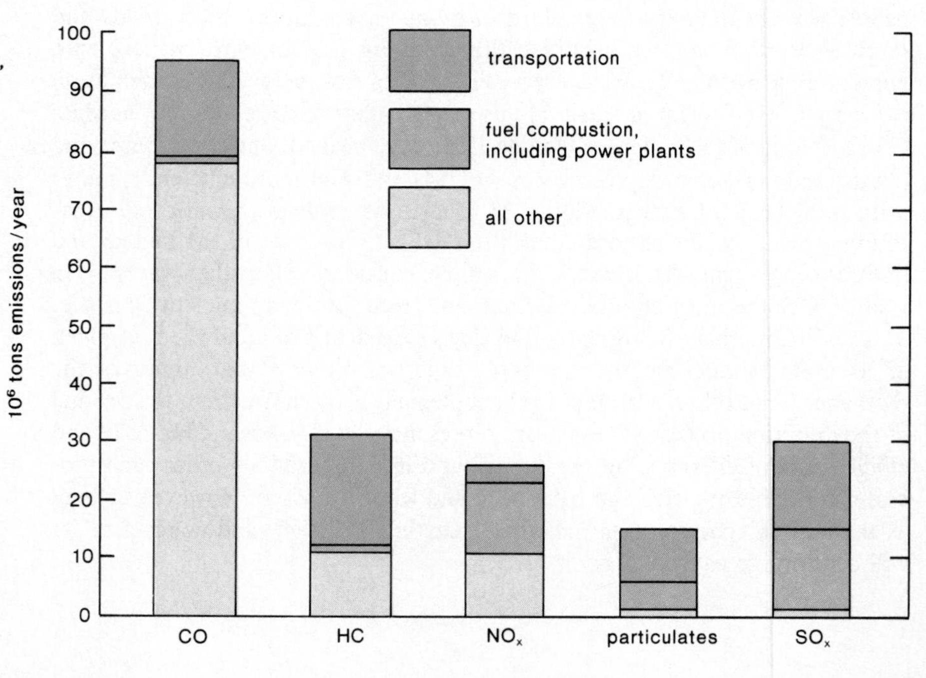

In contrast to mature natural ecosystems, the human ecosystem, consisting of country and city, is not stable. Nonrenewable fossil fuel energy is used inefficiently, and material resources enter the system and do not cycle. Because of these excessive inputs, the outputs to the system are much pollution and waste. While in the past we could rely on natural ecosystems to process our wastes, such as sewage, this is no longer feasible because the size of the natural systems has been reduced and the amount of pollution has been steadily increasing. However, it is possible to change the human ecosystem so that it more nearly corresponds to a natural system by using renewable energy sources and by recycling material resources.

Pollution

Pollution affects all portions of the biosphere: air, land, and water.

Air Pollution

Pollutants enter the atmosphere from various sources (fig. 31.17), but the burning of fossil fuels contributes greatest to the five categories of primary pollutants: carbon monoxide, CO, hydrocarbons, HC, nitrogen oxides, NO_x (NO, NO_2), particulates, and sulfur oxides, SO_x (SO_2, SO_3). The graph in figure 31.18 compares the sources of these pollutants. It is obvious that modes of transportation, especially the automobile, are the main source of carbon monoxide air pollution. This chemical combines preferentially with hemoglobin to prevent circulation of oxygen within the body, causing unconsciousness. In New York City traffic, the blood concentration of carbon monoxide has been shown to reach 5.8 percent, a dangerous level when compared to the 1.5 percent that physicians consider safe. The particulates, dust and soot, can collect in the lungs, and nitrogen oxides and sulfur oxides irritate the respiratory tract. The hydrocarbons (various organic compounds) may be carcinogenic.

Unfortunately, these primary pollutants interact with one another, producing pollutants that are even more dangerous.

Photochemical Smog

Photochemical smog results when two pollutants from automobile exhaust—nitrogen oxide and hydrocarbons—react with one another in the presence of sunlight to produce nitrogen dioxide (NO_2), ozone (O_3), and peroxylacetyl nitrate (PAN). *Ozone* and *PAN* are commonly referred to as oxidants. Breathing ozone affects the respiratory and nervous systems, resulting in respiratory distress, headache, and exhaustion. These symptoms are particularly apt to appear in youngsters; therefore Los Angeles schoolchildren must remain at rest inside the school building whenever the ozone level reaches 0.35 ppm (parts per million by weight). PAN and ozone are especially damaging to plants, resulting in leaf mottling and reduced growth (fig. 31.19).

Normally, warm air near the ground rises, allowing pollutants to be dispersed and carried away by air currents. Sometimes, however, air pollutants, including smog and particulates, are trapped near the earth due to **thermal inversions.** During a temperature inversion, the cold air is at ground level and the warm air is above. (This may occur when a cold front brings in cold air and settles beneath a warm layer.) Cities surrounded by hills, such as Los Angeles and Mexico City, are particularly susceptible to the effects of a temperature inversion.

a.

b.

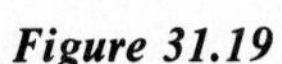

Figure 31.19
The milkweed in (a.) *was exposed to ozone; the milkweed in* (b.) *was grown in an enclosure with filtered air. The plant in* b. *is much healthier than the plant in* a.

Acid Rain

A side effect of air pollution is acid rain. When sulfur oxides (SO_x) and nitrogen oxides (NO_x) are injected into the atmosphere they tend to react with water to produce sulfuric and nitric acid. These acids are eventually deposited on the earth's surface. Here they corrode marble, metal, and stonework, and leach metals from the soil. It is the latter action that leads to the destruction of forests (p. 706) and causes lakes to become devoid of most forms of life.

Coal burning supplies most of the sulfur oxides that lead to acid rain. There are several ways to minimize the problem. Coal could be washed prior to burning. Low-sulfur coal could be substituted for high-sulfur coal. Devices called scrubbers could be installed in the tall stacks to prevent sulfur oxides from entering the air. Under development is a new method of burning coal that uses a mixture of coal and limestone. This technique has reduced both sulfur and nitrogen oxide emissions.

The problem of NO_x emissions is also solvable. Use of the catalytic converter in automobiles reduces the amount of NO_x given off by about three-fourths. A low NO_x burner is also being developed for industrial boilers.

The Weather

It is predicted that the earth's average temperature could rise as much as 8°F over the next 100 to 200 years because of CO_2 buildup due to fossil fuel combustion. CO_2 allows the sun's rays to pass through but absorbs and reradiates heat back toward earth. This may be compared to a greenhouse in that the glass of a greenhouse also allows sunlight to pass through, but traps the heat. This phenomenon also occurs when a car sits in the hot summer sun.

This rise in temperature from the **greenhouse effect** (fig. 31.20a) could have a serious impact on agriculture and eventually on sea levels because of melting polar ice. Favorable climates for the growth of crops would move north, where the soil is not so favorable for agriculture. Even though it is not certain that polar ice would melt, the sea level would rise simply because water expands when it absorbs heat. If the arctic ice should melt substantially, it is predicted that most of the world's cities would be flooded and so would some of our richest farmlands.

Some authorities predict the world's temperature will lower due to suspended particles in polluted air. These prevent the sun's rays from reaching the earth in the first place (fig. 31.20b). Most authorities believe the greenhouse effect will overcome this *refrigerator effect* in the long run.

Indoor Air Pollution

Although much attention has been given to the quality of outdoor air, it has been pointed out that indoor air is often more polluted. NO_2, traced to gas combustion in stoves, has been found indoors at twice the outdoor level; CO, especially in homes with gas, coal, or wood-heating or cooking stoves, often exceeds the current health standards; and radioactive radon gas, emitted naturally from a variety of building materials, has been detected indoors at levels that exceed outdoor levels by factors of 2 to 20. A surprising number of household items continuously give off formaldehyde, including insulation, particleboard, and plywood; asbestos fibers are often dislodged from building materials.

Because of these findings, some have questioned the advisability of tightening up residential buildings to save energy. Reduced ventilation unquestionably increases concentrations of pollutants. Others are calling for research studies so that proper regulatory legislation may be prepared. For example, it would be possible to modify building codes so that safer building materials are used.

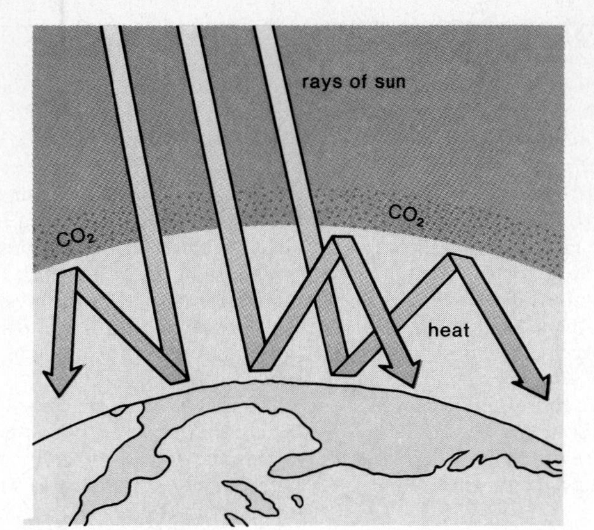

a. greenhouse effect

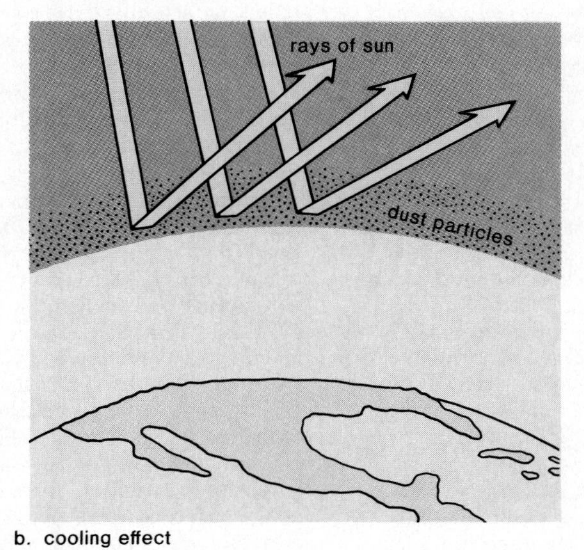

b. cooling effect

Figure 31.20
Effect of pollution on the weather. a. *If the greenhouse effect holds, the earth will get warmer because heat from the sun's rays will be trapped beneath a blanket of carbon dioxide, just as heat is trapped beneath the glass of a greenhouse.* b. *If the refrigerator effect holds, the earth will get colder because the sun's rays will be unable to penetrate the particles produced by air pollution.*

Air Quality Control

Air quality in the United States is controlled by the *Clean Air Act*. The U.S. Environmental Protection Agency (EPA), charged with the task of keeping the environment safe, has established standards for the pollutants listed in figure 31.18. In addition, asbestos, beryllium, mercury, vinyl chloride, and lead are more strictly controlled because they are extremely hazardous to health. Industry is required to use special equipment, such as collectors and scrubbers, to cut down on stack emissions. Automobiles have been equipped with special devices, such as the *catalytic converter,* that chemically changes HC and CO to carbon dioxide and water. The catalytic converter, which necessitates the use of nonleaded gasoline, also contributes to fuel economy. The result has been generally improved air quality throughout most of the nation.

Pollution affects all portions of the biosphere. The greatest single contribution to air pollution is caused by the burning of fossil fuels. Acid rain and smog are caused by air pollutants. It is forecast that carbon dioxide buildup will gradually cause the weather to become warmer.

Fast Cars and Sick Trees

The magnificent spruce rise like giant pillars, their clustered tops dimming even a cloudless sky. Underfoot, a carpet of needles cushions footpaths that vanish mysteriously from sight in the forest darkness. The air is cool and moist, carrying a scent of enchantment, as if a fairy tale might suddenly come to life. This is the legendary Black Forest, West Germany's most beloved woodland, and a source of inspiration for such great cultural figures as Goethe, Beethoven and Hesse. For more than three centuries, it has been admired, protected and lovingly tended.

But now, the 2,300-square-mile forest in the south of Germany is sick, perhaps even dying, afflicted by a disease that is still not completely understood. One major culprit is believed to be acidic fallout from the sulfur dioxide gas emitted by Europe's coal-fired industries. Recently, however, scientists have begun to suspect still another villain, one that many Germans ironically treasure as much as the Black Forest itself—the automobile.

In a country without emission controls on its cars or speed limits on its *autobahns,* motor vehicles spew out more than a million tons of nitrogen oxides (NO_x) a year, creating a deluge of nitric acid and other pollutants that may also be killing trees. As a result, a bitter battle is now being waged between those who are fighting to save the forests and those who want to preserve the freedom of the roads.

. . . In the late 1960s and early 1970s, Scandinavian researchers reported that their lakes were dying because of acid falling from the sky. Soon after, biologists discovered lifeless acidified lakes in the United States' Adirondack Mountains, and reports of forest damage began to trickle in from Poland, East Germany, and Czechoslovakia, where coalburning industry is especially dirty. Scientists examined the evidence and hypothesized that air pollution, in particular acid rain, was responsible for the plight of the lakes and forests.

The notion helped explain the observation that trees at high altitudes suffer the most damage. Fog often shrouds the upper slopes of mountains in the Black Forest and other woodlands, says Schröter. "The fog carries acid. The damage is very heavy in the fog zone."

Many German scientists now believe that there is enough evidence linking the decline of the Black Forest and acid rain to warrant immediate action. "The only way to solve the problem of the forests is to reduce industrial and car emissions," says senior forester Wolfgang Tzschupke of the Baden-Württemberg forest administration. "It must be done in a short time. We don't know how much longer the forests can last.". . .

Economic hardships and cultural tradition pose . . . barriers. The Bonn Parliament originally planned to ban leaded gasoline and impose American-style emission controls in 1986. But the nation's auto manufacturers protested, claiming that they would be at a serious disadvantage in European markets compared to competitors in France and Italy, where a gradual six-year reduction in emissions, agreed to by all the countries of the European community, will not begin until 1988. Germany did, however, institute tax breaks for "environment friendly" cars—tax savings of up to $680 that will go into effect on July 1.

With unemployment still high, the Germans have also been less willing to crack down on sulfurous power-plant emissions. Last summer, in a rare emergency session of Parliament, the government allowed the start-up of a new coal-burning plant that lacked pollution control equipment. Claiming that 3,500 jobs depended on the facility, the government did grant

Land Pollution

Every year, the United States population discards billions of tons of solid wastes, much of it on land. **Solid wastes** include not only household trash but also sewage sludge, agricultural residues, mining refuse, and industrial wastes. Those solid wastes, containing substances that cause human illness and even sometimes death, are called **hazardous wastes.**

environmentalists one concession: it limited the plant to two-percent sulfur coal, instead of the originally planned three-percent.

The government is also resisting strong pressure to impose speed limits, a simple measure that might significantly help the Black Forest. According to scientists at the Heidelberg Institute for Energy and Environment Research, a maximum speed of 62 miles per hour would lower auto emissions all the way down to the levels expected when emission controls finally go into effect. In the land of the *autobahn,* however, voting for speed limits is tantamount to committing political suicide for many members of Parliament. "If there is one freedom that people here are prepared to defend tooth and claw, it is the freedom of being able to hit the gas pedal," explained a recent editorial in the *Rundschau,* a Frankfurt newspaper.

While the politicians balk at even a limited war on airborne pollutants, the grim warnings of German foresters have managed to loosen a generous supply of money for research. Several million dollars from federal and state governments will finance the first comprehensive scientific investigation of the Black Forest and the far more damaged woodlands in other sections of West Germany. In one experiment, researchers will spread lime and potassium over 10,000 acres in the Black Forest to see if trees and soil can be protected from acid rain. If the experiment works, it will be the first good news Germans have heard about their beloved forest in years. Meantime, the situation continues to look grim. Warns Hansjochen Schröter: "Some areas are so damaged they may be gone in five years."

View of distressed tree in Black Forest with auto in the foreground.

Household Trash

In 1920 the per capita population of waste was about 2.75 pounds per day; in 1970 about 5 pounds per day; and in 1980 about 8 pounds per day. The exponential growth of consumer products with "planned obsolescence" and the use of packaging materials to gain a competitive edge account for much of this increase.

Figure 31.21
Hazardous waste dump sites not only cause land and air pollution, they also allow chemicals to enter groundwater, which may later become part of human drinking water.

Open dumping (fig. 31.21), sanitary landfills, or incineration have been the most common practices of disposing trash. These disposal methods have become increasingly expensive and also cause pollution problems. It would be far more satisfactory to recycle materials as much as possible and/or to use organic substances as a fuel to generate electricity. One study showed that it was possible to achieve 70 to 90 percent public participation in recycling by spending only thirty cents per household. A city the size of Washington, D.C., where 500,000 tons of waste are generated each year, could have an increase of 1,300 jobs if the community utilized solid wastes as a resource instead of throwing them away.

Hazardous Wastes

The EPA estimates that in 1980 at least 57 million metric tons of the nation's total wasteload could be classified as hazardous. Hazardous wastes fall into three general categories.

1. *Heavy metals,* such as lead, mercury, cadmium, nickel, and beryllium, can accumulate in various organs, interfering with normal enzymatic actions and causing illness, including cancer.
2. *Chlorinated hydrocarbons,* also called organochlorides, include pesticides and numerous organic compounds, such as PCBs (polychlorinated biphenyls), in which chlorine atoms have replaced hydrogen atoms. Research has often found organochlorides to be cancer producing in laboratory animals.
3. *Nuclear wastes* include radioactive elements that will be dangerous for thousands of years. For example, plutonium must be isolated from the biosphere for 200,000 years before it has lost its radioactivity.

Hazardous wastes are often subject to *bioaccumulation* (fig. 31.22). Bacteria in the soil do not ordinarily break down these wastes, and when other organisms take them up, they remain in the body and are not excreted. Once they enter a food chain, they become more concentrated at each trophic level. Notice in figure 31.22 that the number of dots representing DDT becomes more concentrated as the chemical is passed along from producer to tertiary consumer. Bioaccumulation is most apt to occur in aquatic food chains; there are more trophic levels in aquatic food chains than there are in terrestrial food chains. Humans are the final consumers in both types of food chains, and in some areas, mothers' milk contains a detectable amount of DDT and PCB.

The public has become aware of hazardous dump sites that have polluted nearby water supplies. Chemical wastes buried over a quarter century ago in Love Canal, near Niagara Falls, have seriously damaged the health of some residents there. Similarly, the town of Times Beach, Missouri, is abandoned because a workman spread an organochloride (dioxin)-laced oil on the city streets leading to a myriad of illnesses among its citizens. In other places, such as in Holbrook, Massachusetts, manufacturers have left thousands of drums in abandoned or uncontrolled sites where toxic chemicals are oozing out into the ground and contaminating the water supply. Illnesses, especially forms of cancer, are quite common not only in Holbrook but also in adjoining towns.

Cleanup Although the government established a superfund to clean up toxic dump sites, the enormity of the problem has not as yet led to much cleanup. Specific manufacturers, however, have devised new means to clean up any accidental spills. Very often bacteria are employed. For example, bacteria have now been engineered to eat up PCBs, Agent Orange, and cyanide.

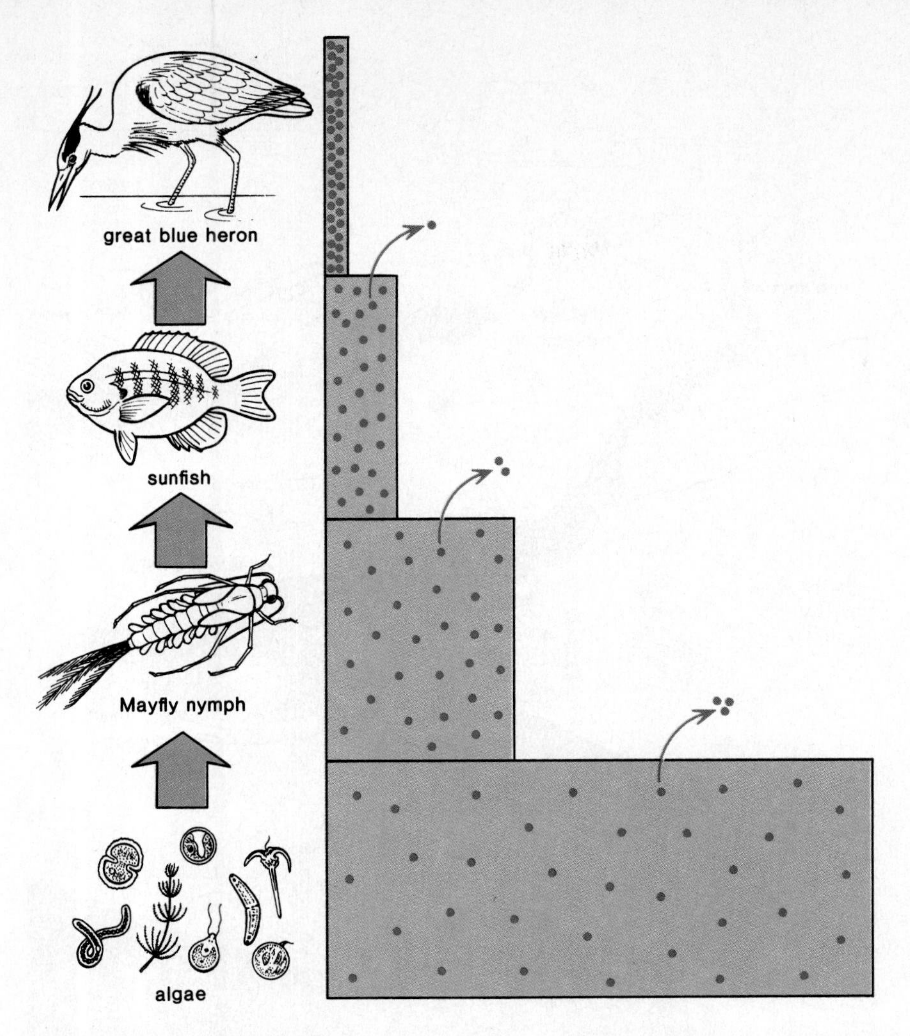

Figure 31.22
Bioaccumulation. A poison (dots) *that is minimally excreted* (arrows) *becomes maximally concentrated as it passes along a food chain due to the reduced size of the trophic levels. Because of this problem, fishermen are warned against consuming fish from the Great Lakes.*

Legislation In an effort to control the level of toxic wastes, Congress passed the *Resource Conservation and Recovery Act,* which empowers the EPA to track all significant quantities of hazardous waste from wherever they are generated to their final disposal. Government regulations are designed to encourage industry to adopt new waste management strategies, including *reduction* (changing a manufacturing process so it does not produce hazardous by-products); *recycling* (reusing waste material); and *resource recovery* (extracting valuable material from waste).

A number of companies have been able to curb their generation of wastes by employing these processes. For example, Minnesota Mining and Manufacturing Company found that it could cut the volume of its toxic wastes associated with adhesive tape production in half by using water-based glues instead of solvent-based glues. And metal processing companies in California have found that an acid they usually discard can be reused to recover zinc-iron compounds.

Water Pollution

Surface Waters

Figure 31.23 shows the many ways that humans cause surface water pollution. Some pollutants, such as fertilizers, sewage, and certain detergents, add nitrates and phosphates to freshwater lakes and ponds. This overabundance of nutrients, called **cultural eutrophication,** speeds up the tendency of bodies of water to fill in and disappear. First, the nutrients cause overgrowth of algae.

Figure 31.23
Water pollution is caused in all the ways shown here. There is much concern of late about Chesapeake Bay because it is dying due to sediments and nutrients that enter the bay, particularly from the Susquehanna River that begins in New York and flows through Pennsylvania.

Source: Adapted from U.S. Environmental Protection Agency, Office of Water Supply and Solid Waste Management Programs, "Waste Disposal Practices and Their Effects on Ground Water" *Executive Summary*. (Washington, D.C., U.S. Government Printing Office, 1977).

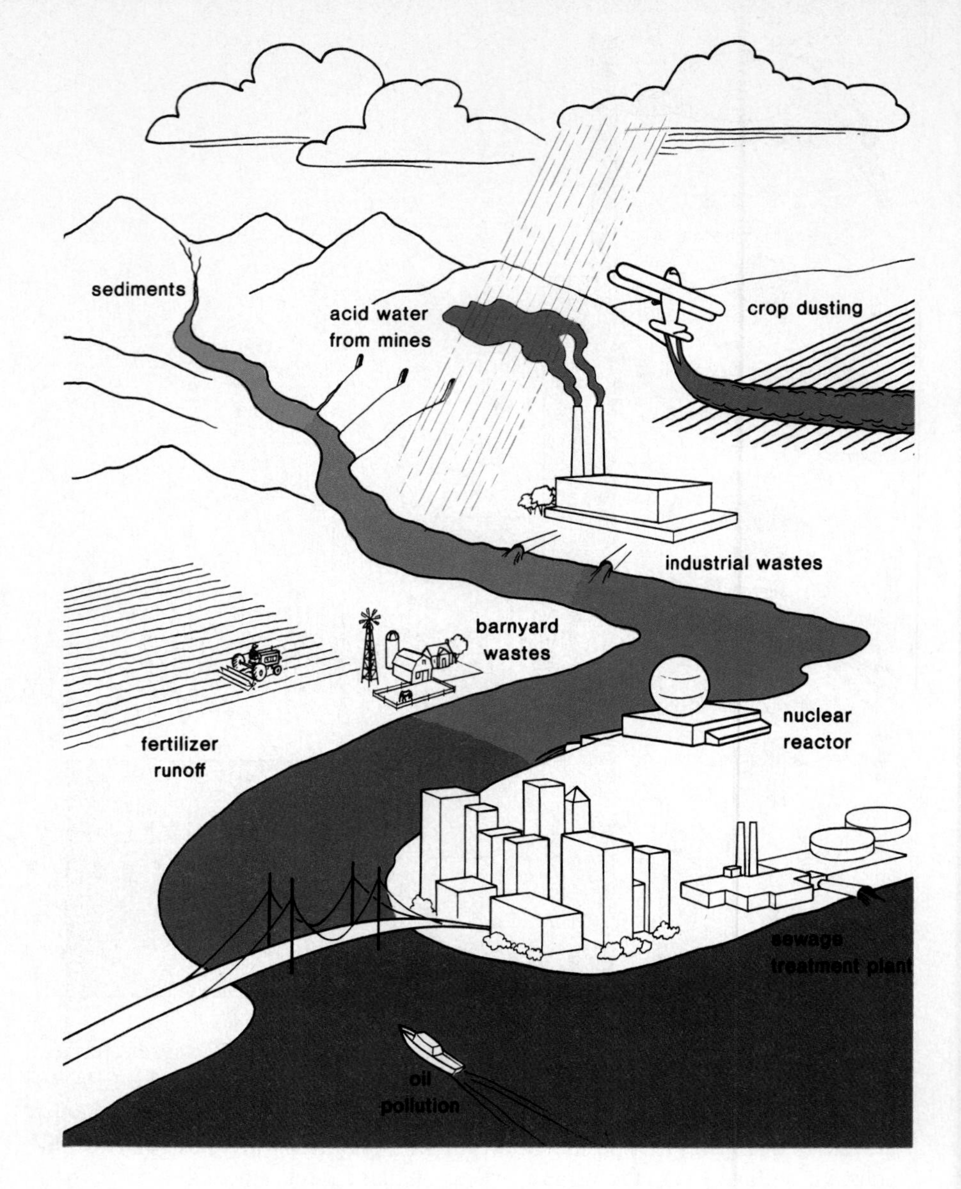

The death of these algae promotes the growth of a very large decomposer population. The decomposers break down the algae, but in so doing they use up oxygen. In addition, algae also consume oxygen during the night when photosynthesis is impossible. Both of these cause a decreased amount of oxygen available to fish, ultimately causing the fish to die. The increased amount of life and dead remains cause the lake to be more eutrophic leading to a reduction in its size.

Water Quality Control The *Water Pollution Control Act* empowers the federal government to set minimum water quality standards for rivers and streams. The building of sewage treatment plants has helped clean up U.S. waters. In addition, large industries are no longer permitted to dump chemicals indiscriminately into water or to send their wastes to local sewage treatment plants. This two-pronged attack has helped clean up the Great Lakes, which previously were said to be dying because of pollution.

Treatment plants capable of not only digesting sewage but also of removing nutrient molecules are very expensive to build. Some communities have devised ingenious ways to accomplish this end, however. Effluent from a sewage treatment plant can be passed through a swampy area or forest before it drains

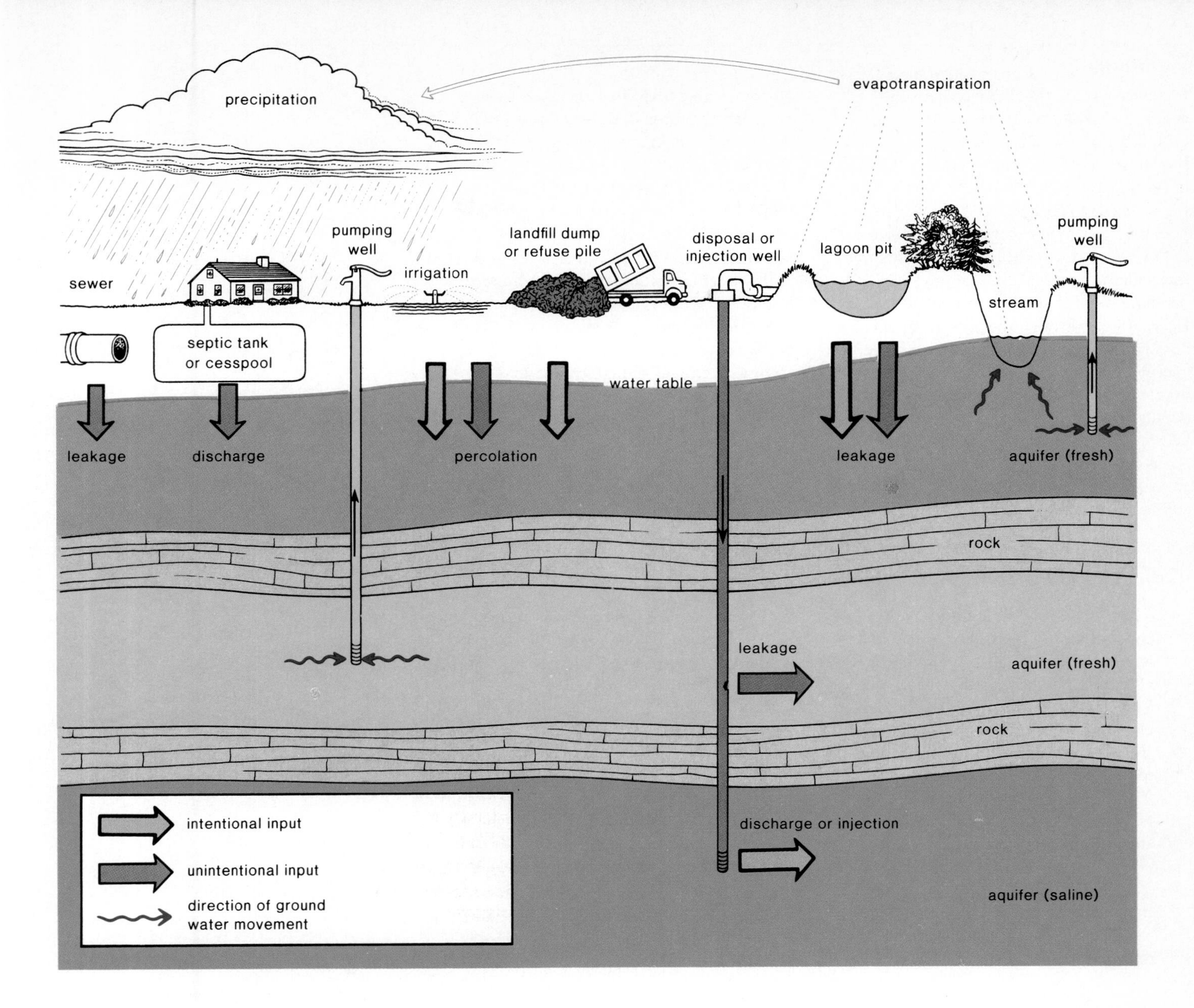

Figure 31.24
Groundwater pollution is caused in all the ways shown here. Discontinuance of these means of disposal for industrial wastes has been difficult to achieve because citizens do not wish to have waste disposal plants located near them.

Source: Adapted from U.S. Environmental Protection Agency, Office of Water Supply and Solid Waste Management Programs, "Waste Disposal Practices and Their Effects on Ground Water" *Executive Summary*, (Washington, D.C., U.S. Government Printing Office, 1977).

into a nearby waterway. Or the effluent can be used to irrigate crops and/or grow algae and aquatic plants in a shallow pond. Since these products can be used as food for animals, this represents a cyclical use of chemicals.

Groundwater

Groundwater is subject to pollution in the ways illustrated in figure 31.24. Chemicals drain from hazardous waste dumping sites, and bacteria and viruses drain from septic tanks and cesspools into the ground and may eventually reach aquifers (underground rivers). Also, previously industry was accustomed to running wastewater into a pit. The pollutants could then seep into the ground. Or pollutants were injected into deep wells from which the pollutants constantly discharged. Both of these customs have been or are in the process of being phased out. However, it is very difficult for industry to find other ways to dispose of wastes. More adequately managed and controlled waste treatment plants are needed, but because citizens do not wish to live near waste treatment plants, towns are often successful in preventing their construction. In the meantime, industries are still employing less approved methods of dealing with industrial wastes.

Summary

In an ecosystem, the biotic component consists of populations that interact with each other and with the abiotic component, the physical environment. The populations in an ecosystem form food chains in which the producers produce food for the other populations. Chemicals cycle through the food chain and, in time, the same inorganic molecules are returned to the producer population for conversion to organic food. Energy flows through the ecosystem but does not cycle, since there is a loss of useful energy at each trophic level. The food pyramid illustrates that each successive population has a smaller size or biomass because energy does not cycle.

In contrast to mature natural ecosystems, the human ecosystem, consisting of country and city, is not stable. Nonrenewable fossil fuel energy is used inefficiently, and material resources enter the system and do not cycle. Because of these excessive inputs, the outputs to the system are much pollution and waste. Pollution affects all portions of the biosphere. The burning of fossil fuels especially causes air pollution, which leads to the development of smog and acid rain. Hazardous wastes, byproducts of various industries, have been deposited on the land but have seeped into surface and groundwater. Unfortunately, due to bioaccumulation, these substances return to humans in concentrated form. While progress has been made in reducing cultural eutrophication, much has to be done in keeping our water supplies free of hazardous wastes.

Objective Questions

1. Chemicals cycle through the populations of an ecosystem, but energy is said to ______________ because it is all eventually dissipated as heat.
2. The first population in a food chain is always a ______________ population.
3. An ecological pyramid illustrates that each succeeding trophic level has less ______________ than the previous level.
4. Natural ecosystems utilize the same amount of energy per year, but the human ecosystem utilizes an ______________ ______________.
5. In the carbon cycle, when organisms ______________ carbon dioxide is returned to the exchange pool.
6. Humans make a significant contribution to the nitrogen cycle when they convert aerial nitrogen to ______________ for use in fertilizers.
7. During the process of denitrification, nitrates are converted to ______________.
8. What are the three categories of hazardous wastes? ______________ ______________.
9. Hazardous wastes are often subject to ______________, a process by which they become concentrated in top consumers, including humans.
10. The burning of ______________ supplies most of the sulfur oxides that lead to acid rain in the United States.

Answers to Objective Questions

1. flow 2. producer 3. energy, biomass 4. ever increasing amount 5. respire 6. nitrates 7. aerial nitrogen 8. heavy metals, chlorinated hydrocarbons, nuclear wastes 9. bioaccumulation 10. coal

Study Questions

1. What is succession and how does it result in a climax community? (p. 688)
2. Give an example of an aquatic and a terrestrial food chain. Name the producer, the consumers, and the decomposers in each chain. Explain the manner in which chemicals cycle. (pp. 689–691)
3. A pyramid describes the size of the various trophic levels within a food web. Discuss the relationship of size and the fact that energy does not cycle. (pp. 691–692)
4. Describe the carbon cycle. How do humans contribute to this cycle? (pp. 694–695)
5. Describe the nitrogen cycle. How do humans contribute to this cycle? (p. 696)
6. Contrast the characteristics of the human ecosystem to those of natural ecosystems. (pp. 697–702)
7. What are the primary components of air pollution? (p. 703) How do they contribute to smog and acid rain? (pp. 703–704)
8. How is air pollution related to the weather? (p. 704)
9. What type of chemicals are termed hazardous wastes? Why? (p. 708)
10. What is cultural eutrophication and how might it be prevented? (pp. 709–711)
11. What causes nonbiodegradable poisons to concentrate as they go from trophic level to trophic level? (p. 708)

Key Terms

abiotic (ab″e-ot′ik) not including living organisms, especially the nonliving portions of the environment. *688*

biosphere (bi′o-sfer) that part of the earth's surface and atmosphere where living organisms exist. *688*

biotic (bi-ot′ik) pertaining to any aspect of life, especially the living portions of the environment. *688*

climax community (kli′maks kŏ-mu′nĭ-te) in succession, the final, stable stage. *688*

consumers (kon-su′merz) a population that feeds on members of other populations in an ecosystem. *690*

cultural eutrophication (kul′tu-ral u″tro-fĭ-ka′shun) enrichment of a body of water causing excessive growth of producers and then death of these and other inhabitants. *709*

decomposers (de-kom-po′zerz) organisms of decay (fungi and bacteria) in an ecosystem. *690*

detritus (de-tri′tus) nonliving organic matter. *690*

ecology (e-kol′o-je) the study of the relationship of organisms between themselves and the physical environment. *688*

ecosystem (ek″o-sis′tem) a biological community together with the associated abiotic environment. *688*

food web (fo͞od web) the complete set of food links between populations in a community. *691*

fossil fuel (fos″l fu′el) the remains of once living organisms that are burned to release energy, such as coal, oil, and natural gas. *695*

greenhouse effect (grēn′hows ĕ-fekt) carbon dioxide buildup in the atmosphere as a result of fossil fuel combustion; retains and reradiates heat, effecting an abnormal rise in the earth's average temperature. *704*

habitat (hab′ĭ-tat) the natural abode of an animal or plant species. *689*

hazardous wastes (haz′er-dus wāsts) wastes containing chemicals hazardous to life. *706*

nitrogen fixation (ni′tro-jen fik-sa′shun) a process whereby free atmospheric nitrogen is converted into compounds, such as ammonia and nitrates, usually by soil bacteria. *696*

producers (pro-du′serz) organisms that produce food and are capable of synthesizing organic compounds from inorganic constituents of the environment; usually the green plants and algae in an ecosystem. *689*

succession (suk-sĕ′shun) a series of ecological stages by which the community in a particular area gradually changes until there is a climax community that can maintain itself. *688*

thermal inversion (ther′mal in-ver′zhun) temperature inversion such that warm air traps cold air and its pollutants near the earth. *703*

32

The Biosphere

Chapter Concepts

1 All life forms exist in major communities called biomes, which are adapted to climate.

2 Humans have exploited and altered both terrestrial and aquatic biomes to the point that representative examples should be preserved as wilderness areas.

3 Complex biomes, such as forests, come into existence by the process of succession, a series of stages leading to a climax or mature stage.

4 While productivity is highest in the early stages of succession, the climax stage is the most stable. Human need for productivity should not cause the abandonment of stability because only a stable biosphere ensures our continued existence.

Chapter Outline

a.

b.

c.

Figure 32.1
Three major biomes in the United States, each containing its own mix of plants and animals. Temperature and amount of rainfall largely determine what type of biome will be found where. a. *A deciduous forest is typical of the eastern United States.* b. *Prairie biome is found in the Midwest and* c. *a semidesert is located in the Southwest.*

Terrestrial Biomes

Each major type of terrestrial ecosystem contains a characteristic community of populations called a **biome** (fig. 32.1). The location of the terrestrial biomes to be studied in this text are indicated in figure 32.2b. The number of biomes could easily be increased by subdividing these. We should also recognize that the various biomes are somewhat artificially imposed upon the biosphere. Actually, one type of biome gradually becomes another type, and it is not surprising to find within any particular biome a region that does not fit the general description.

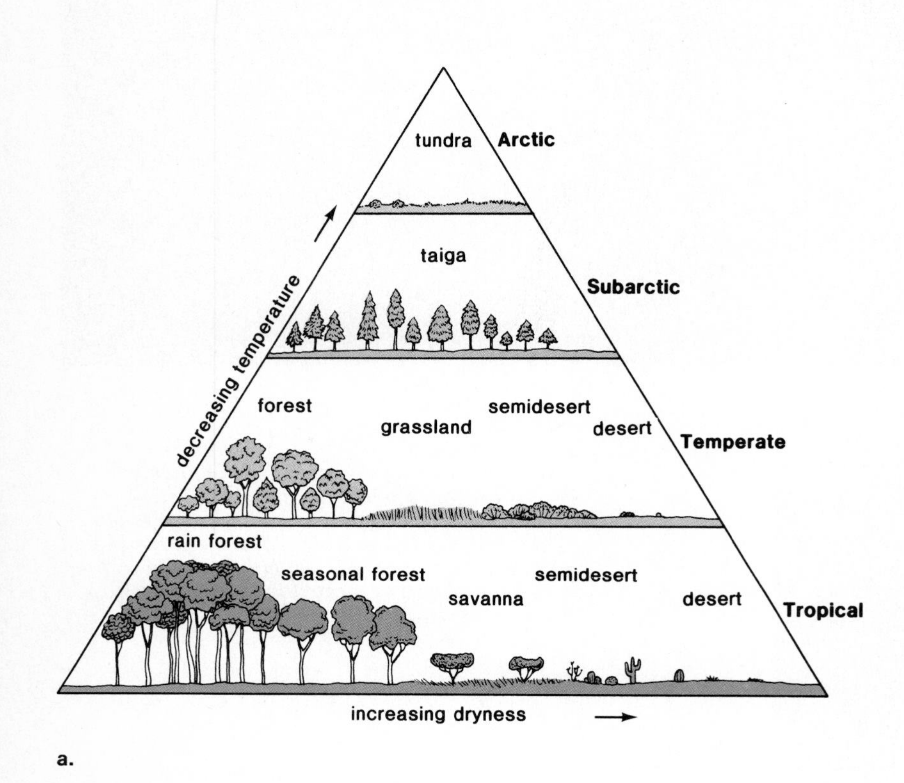

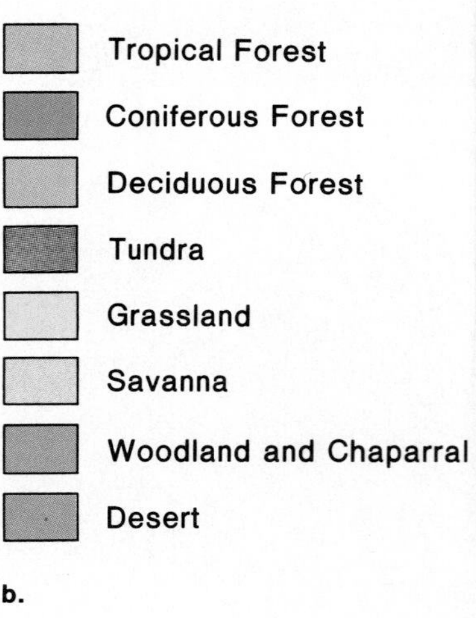

Figure 32.2
a. *Influence of temperature and rainfall on type of biome. Temperature and rainfall determine the biome to a large extent. Deserts occur in regions with minimal rainfall but as the diagram indicates, there are hot deserts and cold deserts. Grasslands occur in regions where there is insufficient water to support trees. The tundra is classified as a grassland.* b. *Distribution of the major biomes of the world. Only remnants of natural, undisturbed biomes remain because humans have superimposed their own ecosystem on all the biomes.*

Types of Biomes

Physical conditions, particularly *climate,* determine the biome of an area. Figure 32.2a shows that the various biomes can be related to temperature and rainfall. Deserts are biomes with the least amount of rainfall.

Deserts

Deserts (fig. 32.1c) are regions of aridity with less than 20 centimeters of rainfall a year. True deserts, with less than 2 centimeters annually, are infrequent, the Sahara in Africa being the largest. **Semideserts,** however, include about one-third of all land areas. In deserts, the days are hot because lack of cloud cover allows the sun's rays to penetrate easily, but the nights are often cold because heat escapes into the atmosphere.

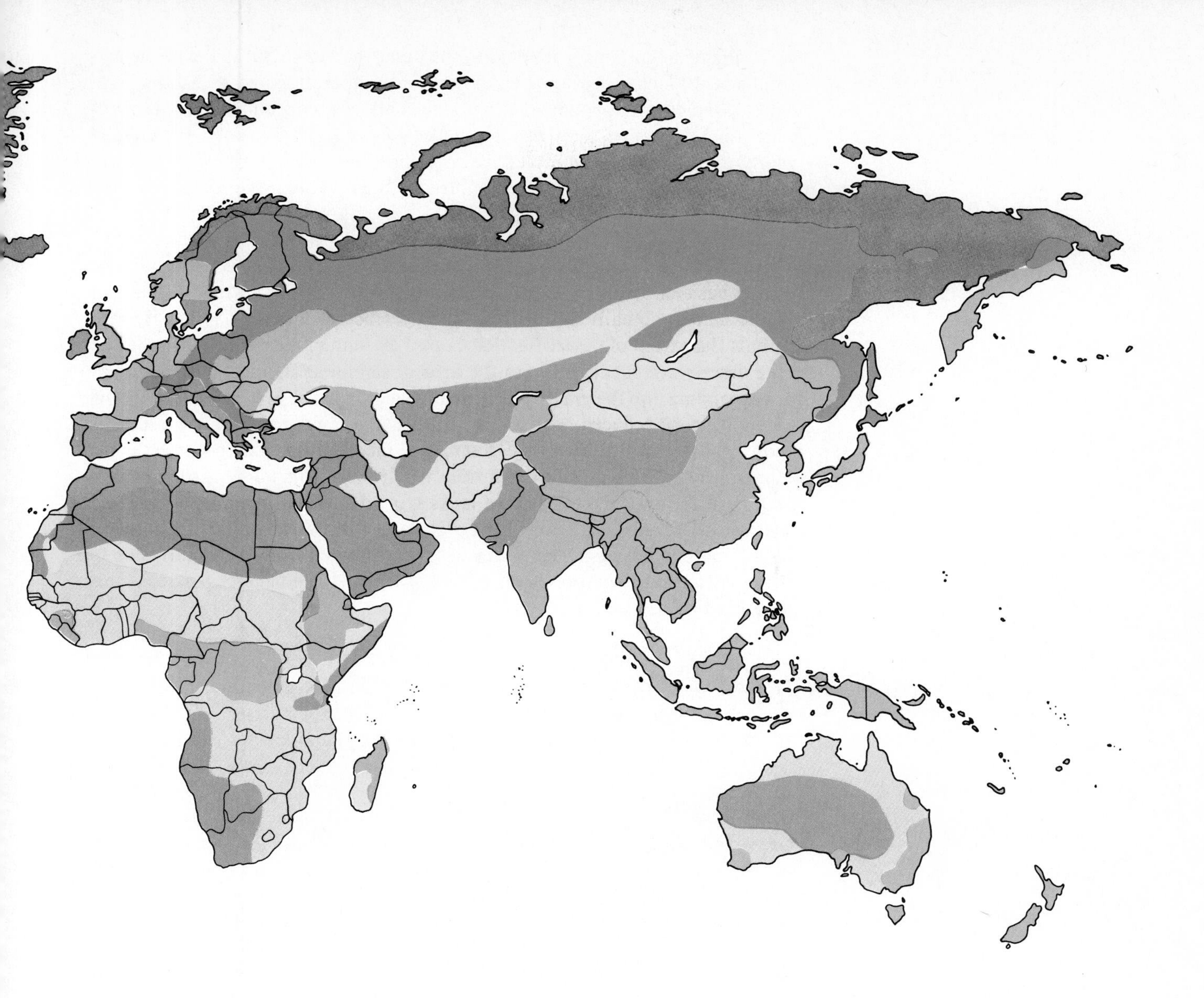

Perennial flowers, including the succulent cacti, non-succulent shrubs, such as sagebrush, and stunted trees are common North American desert vegetation. Reptiles, exemplified by lizards and turtles; rodents, such as the kangaroo and pack rat; and birds, like woodpeckers, hawks, prairie falcons, and the roadrunner, are common small desert animals. The camel is a large herbivorous animal of African deserts, with a reservoir of fat in its hump that allows it to drink and eat infrequently. Large carnivorous animals of the United States desert are the badger, kit fox, and bobcat.

Plants and animals adapted to the desert have structural and/or behavioral adaptations that allow them to prevent evaporation and withstand heat. Plants, and even some animals, reproduce only when adequate water is available. Like the camel, animals often make use of metabolic water. Both desert plants and animals have protective coverings, but animals also hide beneath rocks or burrows in the earth or venture forth only at night to escape the heat of the sun.

Desertification Thirty-six percent of the biosphere is expected to be a desert on the basis of annual rainfall. Yet a world survey indicates that 43 percent of the land is actually desertlike. The difference of 7 percent probably represents the extent of **desertification** caused by human misuse of the land. The true deserts are expanding into the area of semideserts, and the semideserts are expanding into the grasslands. Altogether, a collective area the size of Brazil has undergone reduced productivity. In heavily populated countries, such as those in Africa and India, true deserts are increasing in size because people living in semiarid areas cut down trees for firewood and allow cattle to graze on the shrubs until the entire area lacks ground cover. In the midwestern and southwestern regions of this country, overgrazing of grasslands has caused them to become semideserts. Also, water has been diverted from the countryside to the cities so that irrigation of once fertile farmland is no longer possible. This land has become a desert.

---

*D*eserts are often high-temperature areas with less than 20 centimeters of rainfall a year. Plants and animals living in semideserts need adaptations that protect them from the sun's rays and lack of water.

Grasslands

Grasslands occur where rainfall is greater than 20 centimeters but is generally insufficient to support trees. The extensive root system of grasses allows them to recover quickly from drought, cold, fire, and grazing. Furthermore, these matted roots, which absorb surface water efficiently, prevent invasion by trees.

Grasslands, which occur on all continents, are known by various names. We will be discussing the tundra, an arctic grassland; the savanna, a tropical grassland; and the tallgrass and shortgrass prairies, temperate grasslands in this country (fig. 32.1b).

Arctic Tundra The northernmost biome (fig. 32.3), the **tundra,** is dark most of the year but has 24-hour days in the summer. Since precipitation is only about 20 centimeters a year, it could possibly be considered a desert, but water frozen in the winter is plentiful in the summer because so little of it has evaporated. Only the topmost layer of the earth thaws and beneath this the **permafrost** is forever frozen. Trees are not found in the tundra because their roots cannot penetrate the permafrost and cannot become anchored in the constantly shifting soil.

While the ground is covered with sedges, grassline plants and flowering forbs during the summer months, dwarf woody shrubs and frequent patches of lichens and mosses are always present. A few types of small animals—for example, the arctic fox, the snowshoe hare, and the lemming, which resembles a mouse—live in the tundra the year round. Many migratory birds arrive for the summer, the arctic tern being the most famous of these. At one time, before

a.

b.

Figure 32.3
Tundra biome. a. *Overall view illustrates the lack of trees.* b. *Caribou graze on the low-lying vegetation.*

its virtual extermination by human hunters, the musk-ox was a plentiful year-round resident. Now only the large caribou and reindeer migrate to the tundra in the summer and the wolves follow to prey upon them. Polar bears are common near the coast.

Plants and animals living in the tundra have adaptations that allow them to survive the extreme cold. Low-lying plants are typical because it is warmer near the ground. They have extremely short life cycles, being able to grow and flower within the short summer. In the winter, smaller animals, like the lemmings, burrow, while the large musk-ox has a thick coat and a short, squat body that conserves heat.

The tundra has been the least altered of all the biomes because of its unfavorable climate and soil conditions. However, this biome contains reservoirs of minerals and fossil fuels that only now have become profitable and therefore technologically possible to remove. There is concern that the ecology of the tundra will be disturbed by such factors as the Alaskan pipeline, and that the area may now suffer damaging pollution as a result.

Figure 32.4
Savanna biome. a. *The grasses provide food for many types of herbivores, such as these zebra.* b. *Savanna rose.* c. *The cheetah and her cubs are carnivores that prey on the herbivores.*

Savanna The African **savanna** (fig. 32.4) is a tropical grassland that contains both trees and grasses and therefore supports populations of browsers and grazers. The temperature is warm and there is adequate yearly rainfall, but a severe dry season limits the number of different types of plants. Perhaps the best known of the trees are the flat-topped acacia trees, which shed their leaves during a drought and remain small due to the limited water supply during the dry season.

The large, always warm African savanna supports the greatest number of different types of large herbivores of all the biomes. Elephants and giraffes are browsers. Antelopes, zebras, wildebeest, water buffalo, and rhinoceros are grazers. These are preyed upon by cheetas and lions, whose kill is at times scavenged by hyenas and vultures.

The savanna has been reduced in size due to human encroachment, but those parts that remain, although often misused, have largely retained their usual characteristics. "Conservation by utilization," which means that the native animals are domesticated for dairy and meat products, is favored by far-sighted promoters, especially since imported European cattle are susceptible to tsetse fly infection, a constant threat in many parts of Africa.

Table 32.1 Forest Biomes

Biome	Plants
Coniferous forest	Cone-bearing evergreen trees, such as pine and spruce No understory
Temperate deciduous forest	Broad-leaved trees, such as oak and maple Understories
Tropical rain forest	Broad-leaved evergreen trees Multilevel canopy No understory
Tropical seasonal forest	Mixed broad-leaved evergreen and deciduous Rich understories produce jungle

Figure 32.5
Taiga biome. This biome stretches round the globe in the northern temperate zone and contains evergreen trees that are adapted to harsh conditions.

Prairies As one travels from east to west across the United States, a tallgrass **prairie** (fig. 32.1b) gradually gives way to a shortgrass prairie. Although grasses dominate, they are interspersed by flowering forbs.

The lack of trees places a restriction on the variety of animal life. Insects abound, especially grasshoppers, crickets, leafhoppers, and spiders. Songbirds and prairie chickens sometimes feed off these, but usually they prefer seeds, berries, and fruits. Small mammals, such as mice, prairie dogs, and rabbits, typically burrow in the ground but usually feed aboveground. Hawks, snakes, badgers, coyotes, and kit foxes capture and feed off these. The largest of the herbivores, the buffalo and pronghorn antelope, had few enemies until humans killed them off. Before then large herds of buffalo, in the hundreds of thousands, roamed the prairies and plains, never overgrazing the bountiful vegetation.

Only remnants of the original grasslands remain, and much is now used for farming, especially for growing corn and wheat, or for rangeland and pasture. At times, drier areas have undergone desertification, as discussed on page 718.

Grasslands occur where rainfall is greater than 20 centimeters but is insufficient to support trees. The tundra supports only a limited variety of living things. The savanna, a tropical grassland, supports the greatest number of different types of herbivores that all serve as food for carnivores. U.S. grasslands became the corn and wheat belt of the country.

Forests

The types of forests discussed here are contrasted in table 32.1. Generally speaking, the evergreen coniferous trees are well adapted to the cold because both the leaves and bark have thick coverings. Also, the needlelike leaves can withstand the weight of heavy snow. The broad leaves of deciduous trees carry on a maximum amount of photosynthesis during the short growing season of the temperate zone. Loss of these leaves and dormancy during the winter protect the trees from the danger of cold weather and heavy snow. Trees living in the moist, warm environment of the tropics are both broad-leaved and evergreen because continuous growth is possible and there is no need for dormancy.

Coniferous Forest **Coniferous forests** are found in three locations: in the **taiga,** which extends around the world in the northern part of North America and Eurasia; near mountain tops; and surprisingly enough, along the Pacific coast of North America, as far south as northern California.

The taiga (fig. 32.5) typifies the coniferous forest with its cone-bearing trees, such as pine, fir, and spruce. There is no understory of plants, but the floor is covered by low-lying fungi, mosses, and lichens beneath the layer of needles. Birds harvest the seeds of the conifers, and bears, deer, moose, beaver,

Figure 32.6
Tropical rain forest along a stream. Notice that here the presence of light does allow layers of growth from the ground up.

and muskrat live around the ponds and along the streams. Wolves prey on these larger mammals. In the mountains, the taigalike forests also harbor the wolverine and mountain lion.

The coniferous forest that runs along the west coast of Canada and the United States contains some of the tallest conifer trees ever in existence, including the coastal redwoods. The constant humidity and relatively warm conditions are believed responsible for the unusual growth and development of these trees.

Temperate Deciduous Forests Temperate forests (fig. 32.1a) are found around the world just south of the taiga. They are also found in other areas, such as parts of Japan, Australia, and South America, where there is a moderate climate and well-defined winter and summer seasons and relatively high precipitation (75 to 150 cm per year).

In North America the **deciduous** trees are bare in winter, but awake from dormancy and start to grow again in the spring. They continue growing in the summer, only to lose their leaves and become dormant again in the fall. In the summer the tops of the trees (oak, birch, beech, and maple) form a canopy open enough to allow sunlight to penetrate to the forest floor, thereby allowing several other layers of growth. Beneath the trees are shrubs, grasses, wild flowers, and finally, mosses and liverworts.

Animal life is abundant. Myriads of insects are food for insectivorous birds, such as the red-eyed vireo and woodpeckers. Mammals, such as squirrels, rabbits, deer mice, and white-tailed deer, make their home in the woods and are preyed upon by foxes and wolves.

Tropical Forests The largest tropical **rain forest** (fig. 32.6) is found in the Amazon basin of South America, but such a forest also occurs at the equator in Africa and Australia, where it is always warm and rain is plentiful. The trees are broad-leaved evergreens that form a tall canopy composed of several layers, with characteristic plants and animals in each layer. Woody vines, called **lianas,** reach from the forest floor to the top of the canopy, and **epiphytes,** such as orchids, bromeliads, and ferns, cling to the trees but are not parasitic. Although epiphytes grow on the surface of other plants, they take their nutrients and water from the air. The dense layers of the tropical canopy typically do not allow light to reach the forest floor, and therefore there is no growth here and little litter because decomposition takes place so quickly.

While we usually think of tropical forests as being nonseasonal rain forests, there are tropical forests with wet and dry seasons in India, southeast Asia, West Africa, South and Central America, the West Indies, and northern Australia. Here there are deciduous trees and the canopy does allow light to pass through, helping produce layers of growth beneath the trees. In fact, the tangled mass of growth in tropical seasonal forests is known as jungle.

In both types of tropical forests most animal life is arboreal, living in the trees. There are insects, snakes, lizards, and frogs that spend their whole lives in the treetops and, of interest to us, here we find monkeys in both the Old and New World while apes are found only in the Old World.

Exploitation Forest, particularly tropical rain forests, are being exploited. They are being cleared at an ever-increasing rate and nowhere is this more evident than in Central and South America. First, logging companies enter the forest to extract valuable hardwoods such as mahogany and tropical cedar. Because these trees are tied to others by vines and lianas, many noncommercial trees are damaged. Then landless peasants arrive to cut and burn the vegetation so that enough inorganic nutrients are provided for a few years of farming. Without further fertilization, the soil soon becomes incapable of sustaining crops because tropical rain forests have nutrient-poor soil—all the nutrients are located in the luxurious living matter. Often the land is now converted to pasture by cattlemen who can acquire the capital to plant pasture grasses since it is known that there will be a market for the beef in the United States and Europe. After about seven to ten years, the effects of overgrazing and torrential rains turn the soil into a wasteland.

Many scientists are extremely concerned about exploitation of the tropical rain forests. They point out, as discussed in the reading on page 724, that many species will become extinct because there are more kinds of plants and animals here than in any other biome.

Forests require adequate rainfall. The taiga has the least amount of rainfall. The temperate deciduous forest has trees that gain and lose their leaves because of the alternating seasons of summer and winter. The tropical forests are the least studied and the most complex of all biomes. Humans are beginning to exploit tropical forests so that many fear that thousands of species could become extinct before we even know of their existence.

Latitude versus Altitude

Latitude

If one travels from the Southern to the Northern Hemisphere, it is possible to observe first a tropical rain forest, followed by a temperate deciduous forest, and then the taiga and tundra, in that order. This shows that location of the biomes is influenced by temperature.

Altitude

It is also possible to observe a similar sequence of biomes by traveling from the bottom to the top of mountains (fig. 32.7). These transitions are largely due to decreasing temperature as the altitude increases, but soil conditions and rainfall are also important.

In general the same series of biomes can be observed according to both latitude and altitude.

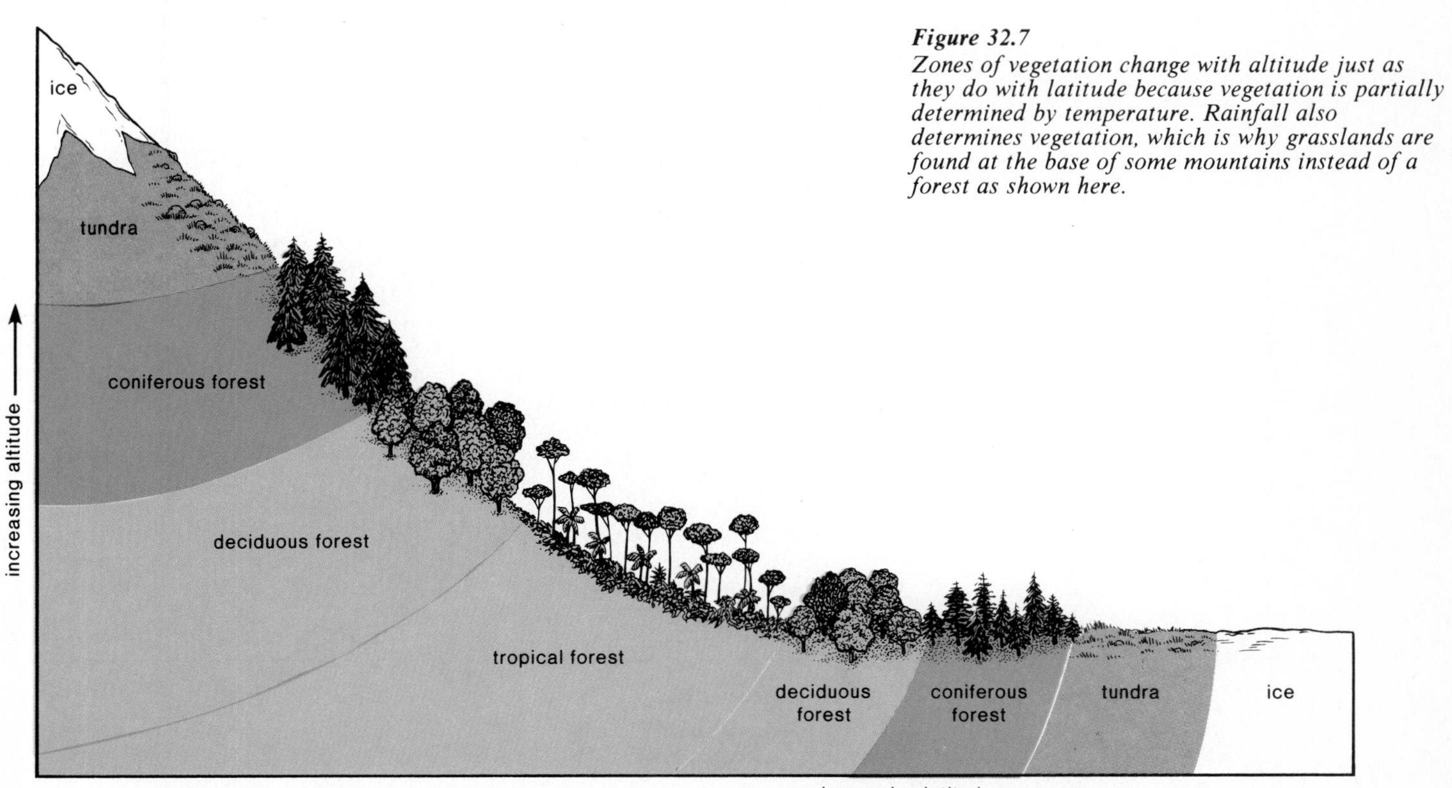

Figure 32.7
Zones of vegetation change with altitude just as they do with latitude because vegetation is partially determined by temperature. Rainfall also determines vegetation, which is why grasslands are found at the base of some mountains instead of a forest as shown here.

The Urgency of Tropical Conservation

Sharing the planet Earth with us are at least four to five million species of animals, plants, and microorganisms, most of which are poorly known. About two-thirds of them have not even been given scientific names.

Of all the world's species, roughly ten percent occur in the United States and Canada together. We can state with confidence that most of these plants, vertebrates, and larger invertebrates and fungi have been studied and classified. Furthermore, we now know that some 5 to 10 percent of North America's species are threatened or endangered, and we are making advances in protecting them.

But to think that by "taking care of our own" we are saving a large proportion of the world's organisms is to delude ourselves. It is to the tropics that we must turn. For instance, about a third of the total number of Earth's plants, animals, and microorganisms occur in Latin America. Of these, only about a sixth have been catalogued. Of the remainder, amounting to nearly 30 percent of all the planet's organisms, we know absolutely nothing. Some scientists, including Terry Erwin of the U.S. National Museum of Natural History, have estimated that the total number of species in Latin America might be much higher. His calculations indicate a regional total of perhaps 15 million species. But whatever the final figure, the number of unknown species is staggering.

For example, South America's fresh waters are inhabited by an estimated 5,000 fish species, only about 3,000 of which have been named. (This amounts to about an eighth of all the world's fish species.) On the eastern slopes of the Andes, 80 or more species of frogs and toads often exist within a single square mile—almost as many as in all of temperate North America. Approximately the size of the state of Colorado, Ecuador harbors more than 1,300 species of birds—roughly twice as many as those inhabiting the United States and Canada.

Moreover, hundreds of *new* species of plants—including many dozen tree species—are being discovered in Latin America every year, as are dozens of new species of terrestrial vertebrates and fishes. Over a third of the nearly 800 species of reptiles and amphibians known to occur in Ecuador have been discovered since 1970, and many more are still being found there. In fact, the three northern Andean countries of Colombia, Ecuador, and Peru combined are home to about a sixth of all the Earth's biota—some 750,000 species that are also the most poorly known organisms on our planet. . . .

The actual and potential importance of species for human welfare cannot be overstated. About 85 percent of our food is derived directly or indirectly from 20 species of plants, and some 60 percent comes from only three: corn, wheat, and rice. It stands to reason that among the world's remaining estimated 235,000 known flowering plant species, there must be many more that could provide important sources of food—not to mention medicines, oils, chemicals, and sources of renewable fuels. Our continued survival depends on our ability to use plants, animals, and microorganisms extensively and wisely. Yet we know absolutely nothing about most tropical organisms, and we certainly haven't examined them for their potential value to humans. By saving the tropical forests and their myriad plant and animal species, we also will be investing wisely in the future survival of our own species.

But despite its riches, tropical vegetation worldwide is being altered or eradicated at an alarming rate. In 1981, the Tropical Forest Resources Assessment Project of the United Nation's Food and Agriculture Organization (FAO) estimated that 44 percent of the tropical rain forests had already been degraded or destroyed by 1980. Currently—according to an international task force sponsored by the World Resources Institute, The World Bank, and the United Nations Development Programme—27 million acres (an area larger than Austria) of tropical forest are being cleared each year. This is some 50 acres per minute. (The task force has launched an $8-billion campaign to reverse tropical deforestation by 1991.) If present

a.

c.

b.

Destroying the tropical rain forests will greatly reduce the variety of life on earth. Already vulnerable are the a. *green iguana,* b. *South American ocelot, and* c. *the blue/yellow macaw.*

trends continue, the rich tropical moist forests of many developing countries will disappear in the next two to three decades. . . .

The greatest consequence of all, however, is that of biological extinction—the extermination of a major fraction of Earth's plants, animals, and microorganisms during the lifetimes of most people living today. Many tropical organisms are very narrow in their geographical ranges and highly specific in their ecological requirements. At least 40 percent of the world's species occur in the very tropical forests that may not survive the next few decades. The loss of only half these organisms would amount to the permanent disappearance from our planet of at least 750,000 species—a far greater number than the *total* number of species found in the United States and Canada. If we wish to preserve the world's biota, can it be doubted that our major concern in the closing years of the 20th century and the early years of the 21st century must be with the tropics? . . .

To find an extinction event of this magnitude in the geological record, one needs to go back some 65 million years to the end of the Cretaceous Period when—possibly spurred by a cloud thrown up by the collision of a gigantic meteorite with the Earth—a large proportion of the Earth's organisms, including the dinosaurs, became extinct. The extinction that is taking place now will occur during our lives and those of our children. But we can ameliorate its effects by learning about tropical organisms and using our knowledge to save them. Our contribution to preserving natural diversity will be minute if we do not look beyond our own borders to the tropics. This task is one of extreme urgency. If we succeed, we shall play an important role in shaping the quality of life for all who come after us.

By Peter H. Raven. Nature Conservancy News, p. 7, 1800 North Kent St., Arlington, Virginia 22209.

Figure 32.8
Simplified overview of succession as it may occur on abandoned farmland along the East Coast in the United States. For the first few years, annual weeds and grass colonize the area. Then shrubs and trees, such as pines that can tolerate direct sunlight, invade the area. These pave the way for the germination and growth of hardwood saplings that grow to finally dominate the area.

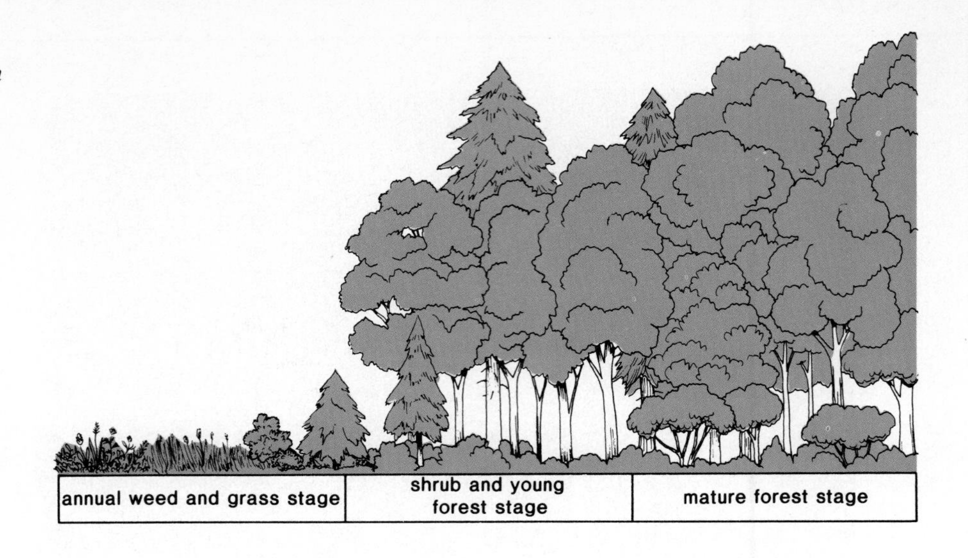

Succession

During the process of *succession,* a sequence of communities replaces one another in an orderly and predictable way. The complete process is called a sere and each stage is a seral stage. Figure 31.1 shows a sere for primary succession by which bare rock becomes soil capable of supporting a climax community.

The stages of succession vary from site to site, but figure 32.8 shows a possible secondary succession beginning with abandoned farmland in the eastern United States. The first community, called the *pioneer community,* includes plants such as weeds that are able to colonize disturbed areas because of their ability to survive under harsh conditions, such as limited soil moisture and direct sunlight. This colonization prepares the way for native plants of the area that most likely have a longer growing period. Notice in figure 32.8, how one plant community replaces another until a climax community has been achieved. A different mix of animals would be associated with each of these communities. Whereas previous communities are replaced, the final stage, a *climax community,* is able to sustain itself indefinitely. We have been studying various climax communities in this chapter.

Stability versus Productivity

There is an ecological theory, based on laboratory and field data, that suggests that stability of a biome is related to its complexity. Certainly the climax, or mature, stage of succession is the most stable since it maintains itself with little change. One reason for this might very well be that the last stage is the most complex in terms of species diversity. Thus, if one species or population of a climax community is reduced in size or eliminated, the other populations can compensate for this loss. For example, the deciduous forests in this country did not disappear when both the chestnut and elm trees succumbed to parasitic disease; the other trees simply filled in.

The early stages of succession are obviously unstable, as witnessed by the fact that they are replaced by later stages. However, investigators find that these stages show the most growth and therefore are the most productive.

Figure 32.9
Selective cutting of trees. In a deciduous forest, selective cutting can help maintain the forest because mature trees always remain.

Knowledge of this relationship between productivity and stability can be utilized by humans when they alter biomes for their own purposes. For example, in forestry, the removal of trees places the biome in an earlier successional stage and causes greater productivity (new trees will grow) but also increases instability (fig. 32.9). Stability can be safeguarded in one of two ways: (*a*) remove only trees of a certain mature age and leave younger trees standing and (*b*) remove trees from certain areas, but leave sections in between standing. Unfortunately, stability is not always safeguarded and many parts of the world that once had forests have them no more.

The term *succession* is applied to the series of stages by which rocks or abandoned land become a climax community. A climax community is the most complex and stable but the early stages of succession are the most productive in terms of new growth. When humans alter biomes, they should balance productivity against stability, never neglecting the latter because only a stable biosphere ensures human existence.

Aquatic Biomes

Aquatic biomes can be divided into two types: (*a*) inland or fresh water and (*b*) ocean or salt water. An **estuary,** however, where a river flows into the ocean, has mixed fresh and salt water, called brackish water. In these biomes, organisms vary according to whether they are adapted to fresh or salt water, warm or cold water, quiet or turbulent water, and the presence or absence of light. In both salt and fresh water, free-drifting microscopic organisms, called **plankton,** are important components of the biome. **Phytoplankton** are photosynthesizing algae that only become noticeable when they reproduce to the extent that a green scum or red tide appears on the water. **Zooplankton** are animals that feed on the phytoplankton.

Figure 32.10
Stratification of a lake. a. *In the summer, a large lake has three distinct layers and the nutrients tend to collect in the hypolimnion, which becomes depleted of oxygen.* b. *In the autumn, a turnover occurs that allows mixing.* c. *In the winter, nutrients again collect in the hypolimnion.* d. *In the spring, another turnover occurs.*

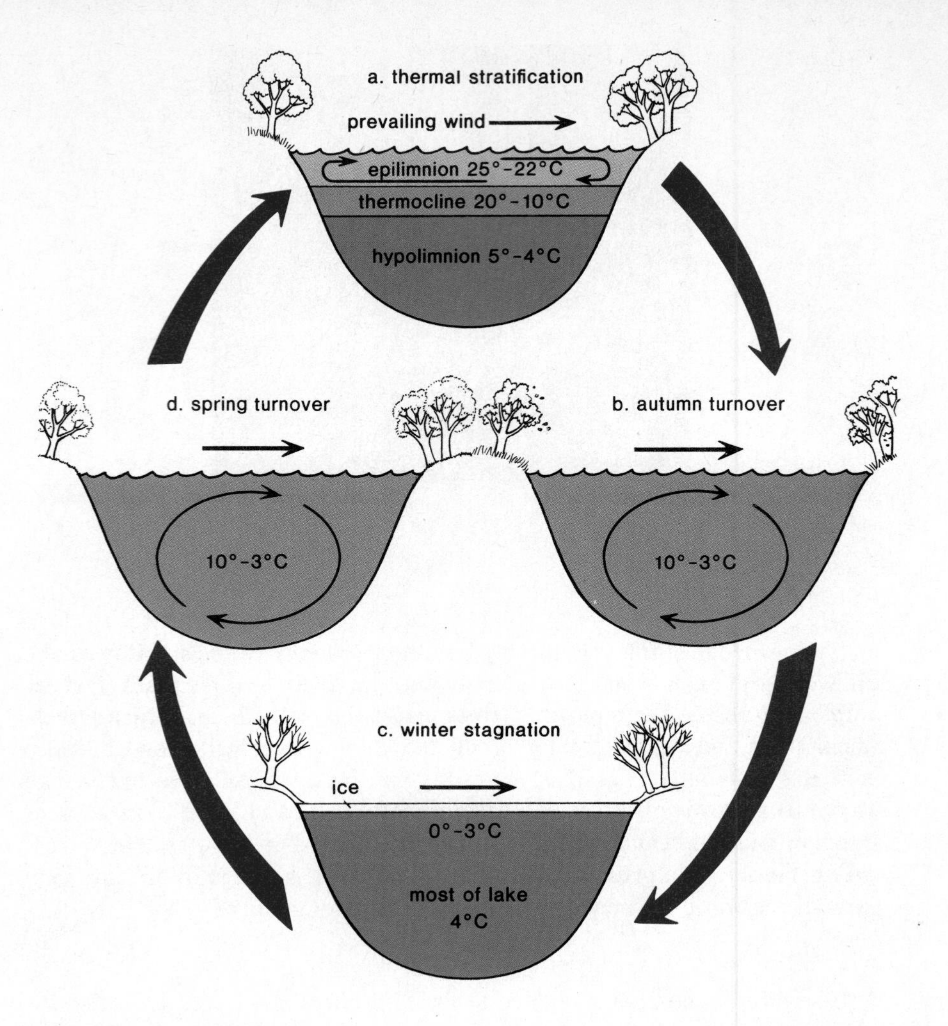

Freshwater Biomes

Lakes and Ponds

Lakes, being larger than ponds, have three layers of water that differ as to temperature (fig. 32.10). In summer, the upper surface layer, the *epilimnion,* is warm; the middle *thermocline* experiences an abrupt drop in temperature; and the *hypolimnion* is cold. This difference in temperature prevents mixing; and the epilimnion lacks nutrients found in the hypolimnion, while the hypolimnion lacks oxygen found in the epilimnion. In the fall, as the epilimnion cools and in the spring as it warms, mixing does occur, causing phytoplankton growth to be most abundant at these times.

Lakes and ponds can be divided into three life zones: the **littoral zone** is closest to the shore, the **limnetic zone** forms the sunlit body of the lake, and the **profundal zone** is below the level of light penetration. Aquatic plants are rooted in the shallow littoral zone of a lake. The rest of the organisms are divided into five groups according to their habitats. The *periphyton* (fig. 32.11) are microscopic or near-microscopic organisms, such as algae or protozoans, that cling to plants, wood, and rocks in the littoral zone. The *plankton* of the limnetic zone includes both phytoplankton and zooplankton, such as rotifers, copepods, and water fleas. *Neuston,* insects that live at the water-air interface, include the water strider and water scorpion, animals that can literally walk on water, and the whirligig beetle and mosquito larvae that prefer a location just beneath the surface of the water. Other insects, such as diving beetles, water boatmen, and backswimmers, are a part of the *nekton,* a group of free-swimming organisms. Most nekton are fish. Minnows and killifish are fish that prefer the littoral zone; trout, whitefish, and cisco prefer the profundal zone, particularly the hypolimnion; pike, bass, and gar can tolerate warmer water.

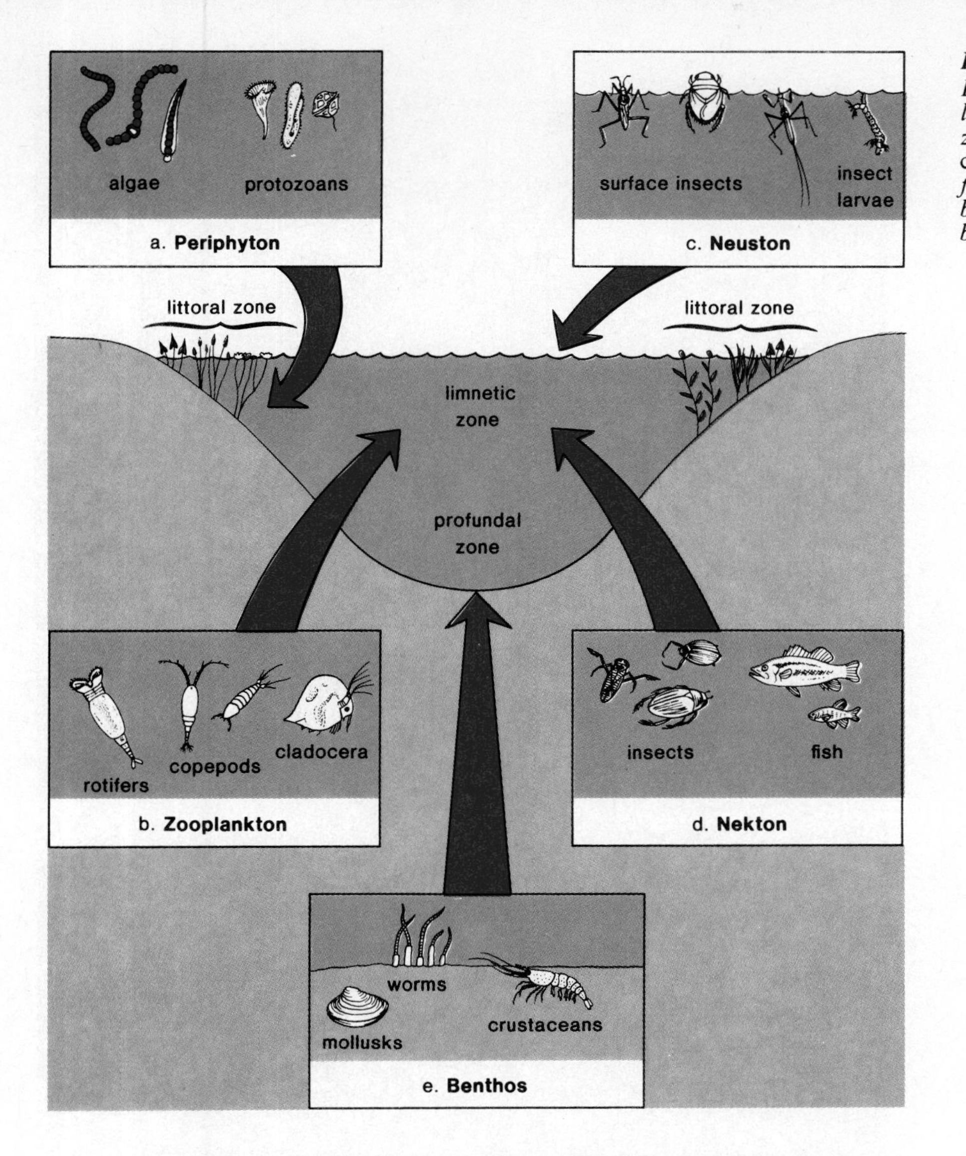

Figure 32.11
Life zones of a lake. a. *Periphyton occur in the littoral zone.* b. *Zooplankton are in the limnetic zone where they provide food for carnivores.* c. *Neuston are at the water's surface.* d. *Nekton are free swimmers in the limnetic zone where food can be found.* e. *Benthic animals are in the benthos, the bottom of the lake.*

The *benthos* are animals that live on the bottom of the benthic zone. In a lake the benthos include crayfish, snails, clams, various types of worms, and insect larvae. Among the insects that spend a large portion of their life cycle as larvae in the benthic zone are the dragonfly, damselfly, and mayfly. The benthic zone may, however, become so depleted of oxygen at times that only such organisms as sludge worms and midge fly larvae, known as bloodworms, can survive.

Rivers and Streams

At first, rivers and streams have rapidly flowing water as they move down out of the mountains. Here insect larvae and water plants are adapted to clinging to rocks as the water passes by. In intermittent pools, various species of fish, including trout, which prefer cold oxygenated water, may be found. As the river nears the ocean, water flow becomes much slower, plankton can now accumulate, and the community begins to resemble that of a lake or pond.

Water is removed from the rivers to grow crops, for use as an industrial coolant, and for various household purposes. At the same time, sewage and pollutants are added to the rivers. With an ever-increasing population, more water is removed and more waste is added to rivers. Since both of these tend to decrease river flow, there is some question if dependable flow can be assured by the year 2000 unless projects are immediately carried out to minimize the use of rivers for these purposes.

Figure 32.12
Brackish water communities. a. *A salt marsh is typical of the temperate zone.* b. *Mangrove swamps are found in the tropical zone.*

Lakes, especially in the summer, are stratified into layers according to temperature, and only a turnover twice a year restores their productivity. Lakes have various life zones, each with typical organisms. Rivers and streams are characterized by rapidly flowing water that gradually moves more slowly as a river approaches the ocean.

Biomes along the Coast

Rivers flow down to the sea to form estuaries, semi-enclosed baylike regions. The silt carried by a river forms mudflats, and within the shallow waters, a **salt marsh** (fig. 32.12a) in the temperate zone and a **mangrove swamp** (fig. 32.12b) in the subtropical and tropical zones are likely to develop. Along either side of the estuary community, the seashores reach out along the coast. It is proper to think of seacoasts and estuaries, including mudflats, salt marshes, and mangrove swamps, as belonging to one ecological system.

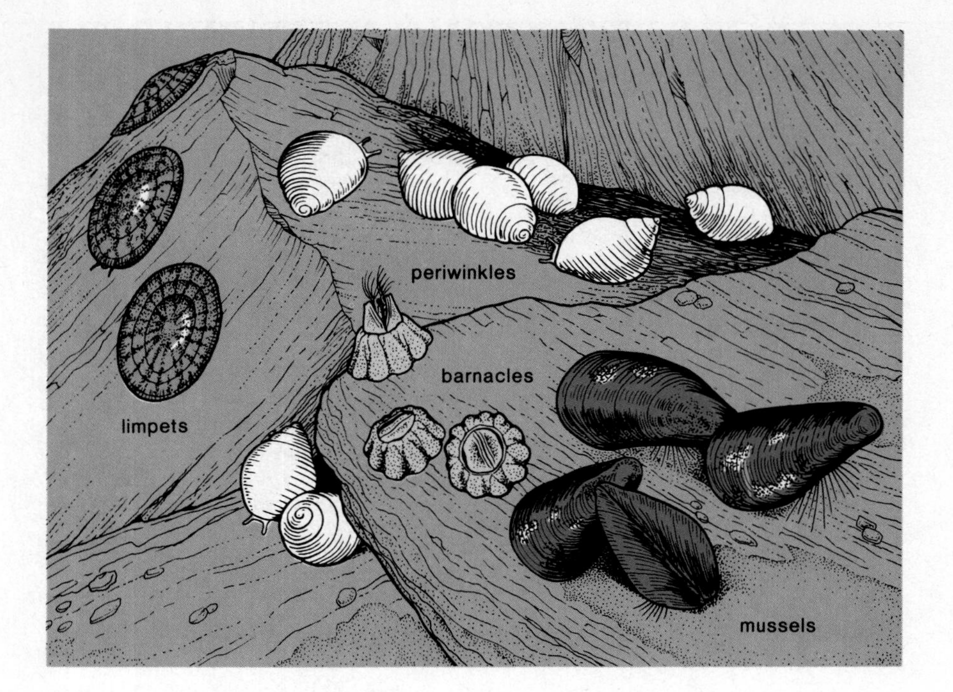

Figure 32.13
Animals typical of a rocky coast in the temperate zone. Limpets, periwinkles, barnacles, and mussels are all adapted to clinging to rocks.

Estuary

A river brings fresh water into the estuary, and the sea, because of the tides, brings salt water. There is a gradation of salinity, or a gradual increase of salt water, from the river to the sea. Organisms living in an estuary must be able to withstand constant mixing of waters and rapid changes in salinity. Not many organisms are suited to this environment, but for those that are suited there is an abundance of nutrients. An estuary acts as a nutrient trap because the tides bring nutrients from the sea and at the same time prevent the seaward escape of nutrients brought by the river. Because of this, estuaries produce much organic food.

Although only a few small fish permanently reside in an estuary, many develop there so that there is always an abundance of larval and immature fish. It has been estimated that well over half of all marine fishes develop in the protective environment of an estuary, which explains why estuaries are called the *nurseries of the sea.* Even so, many estuaries are becoming victims of pollution because of construction and development along the seacoast.

Seashores

Both rocky and sandy shores are constantly bombarded by the sea as the tides roll in and out; therefore, they collect debris that has been dumped into the ocean, as discussed in reading on page 732. The *rocky shore* (fig. 32.13) offers a firm substratum to which organisms can attach themselves and not be swept out to sea. Macroscopic *seaweeds,* which are the main photosynthesizers, anchor themselves to the rocks by holdfasts. Barnacles are glued to the stone by their own secretions so tightly that their calcerous outer plates remain in place even after the enclosed shrimplike animal dies. Oysters and mussels attach themselves to the rocks by filaments called *byssus threads.* Limpets are snails, just as are periwinkles. But whereas periwinkles have a coiled shell and secure themselves by hiding in crevices or under seaweeds, limpets press their single flattened cone tightly to a rock.

The Perils of Plastic Pollution

Cups, Sandwich Bags and Other Debris Threaten Marine Life

Biologist Stewart Fefer and three colleagues from the U.S. Fish and Wildlife Service had eagerly anticipated their assignment: a trip to Laysan, a 960-acre island about 1,000 miles northwest of Honolulu. They were to study some of the 14 million seabirds that nest there, and they looked forward to their stay on what they assumed would be an island paradise with pristine beaches. What they discovered came as a shock. The sands of Laysan were strewed with an unbelievable variety of plastic trash. While doing his bird-watching chores, Fefer cataloged thousands of pellets as well as toy soldiers, disposable lighters and one toy Godzilla—all made of plastic. "This is one of the most remote islands in the world," he says. "I expected it to be just idyllic."

The debris that despoiled Laysan's beaches had been washed ashore by the waters of the Pacific, which like other oceans is becoming increasingly fouled by plastic flotsam. But while the floating and beached plastic is unquestionably an eyesore, the problem goes far beyond aesthetics. At the Sixth International Ocean Disposal Symposium in Pacific Grove, Calif., last month, scientists reported that plastic trash is causing injury and death to countless marine animals that feast on it or become ensnared in it. Says Ecologist David Laist, of the Marine Mammal Commission: "Plastics may be as great a source of mortality among marine mammals as oil spills, heavy metals or other toxic materials."

Prime contributors to the growing tide of plastic pollution are the world's merchant ships, which, according to a study by the National Academy of Sciences, dump at least 6.6 million tons of trash overboard every year. Some 639,000 plastic containers and bags are tossed into the oceans every day. Commercial fishermen are also major offenders. Estimates of the plastic fishing gear lost or discarded at sea every year range as high as 150,000 tons. Boaters and beachgoers add to the marine litter with six-pack yokes, picnic utensils, sandwich bags and Styrofoam cups. Cities and industries discharging waste directly into the water or dumping it at sea are also to blame. On some East Coast beaches near sewage outlets, so many plastic tampon inserters have washed ashore that residents refer to them as "beach whistles."

Perhaps the most ubiquitous form of plastic trash is the tiny polyethylene pellets used in the manufacture of plastic items. In one survey, researchers calculated that, on average, a square mile of the Sargasso Sea, southeast of Florida, contained between 8,000 and 10,000 bobbing pellets. Says Al Pruter, a fishery biologist and partner in a Seattle-based natural-resources consulting firm: "Almost without exception, surveys show plastic to account for over one-half the man-made products on the ocean surface."

The plastic is taking a heavy toll on marine life, particularly on seals, sea lions, turtles and seabirds. By one estimate, as many as 50,000 northern fur seals in the Pribilof Islands die each year after becoming enshrouded in netting. "Young seals get their heads or flippers caught in it," says Laist. "Then they either become exhausted from toting it or their ability to catch food is restricted."

Smaller plastic items are frequently mistaken for prey by turtles and birds, often with fatal results. Leatherback turtles, which feast on jellyfish, are particularly attracted to plastic bags. Says University of Florida Zoologist Archie Carr, an authority on sea turtles: "Any kind of film or semitranslucent material appears to look like jellyfish to them." Trouble is, the bags—or other plastic items like golf tees—can form a lethal plug in the turtle's digestive tract.

At least 42 species of seabirds are known to snack on plastic. Of 50 albatrosses found ill or dead on the Midway Islands, 45 had eaten some form of the substance. In several, the plastic had either obstructed the digestive tract or caused ulcers. Says James Coe, program manager for the Marine Entanglement Research Program at the National Marine Fisheries Service in Seattle: "We have found everything from toy soldiers to pens, fishing bobbers and poker chips in the birds' stomachs." A study of wedge-tailed shearwaters, which breed on central Pacific islands, showed that 60% of the adults surveyed had ingested plastic. Even sea gulls, which are able to

a.

b.

The sea contains plastic trash discarded by us all. a. *California sea lion caught in the tangles of a net either lost or discarded by commercial fishermen.* b. *Herring gull with a six-pack strap over its head.*

disgorge disagreeable food, are not immune to the plastic threat. They have been strangled by six-pack yokes.

Efforts to reduce the amount of plastic jettisoned into the oceans have been largely unsuccessful. Although the U.S. and 59 other nations agreed in 1972 to outlaw the dumping of durable plastics, among other substances, into the oceans, the treaty failed to address the discharge of ordinary garbage, which contains large quantities of plastic items. Ten states are trying to do their part: they have passed legislation requiring that six-pack yokes be made of treated plastic that degrades rapidly in sunlight. Nonetheless, concludes Zoologist Carr, "This junk is growing in abundance year by year. It is just getting outrageous."

Figure 32.14
Animals typical of a sandy beach in the temperate zone. Sandhoppers, ghost crabs, sandworms, and ghost shrimp are all adapted to burrowing in the sand.

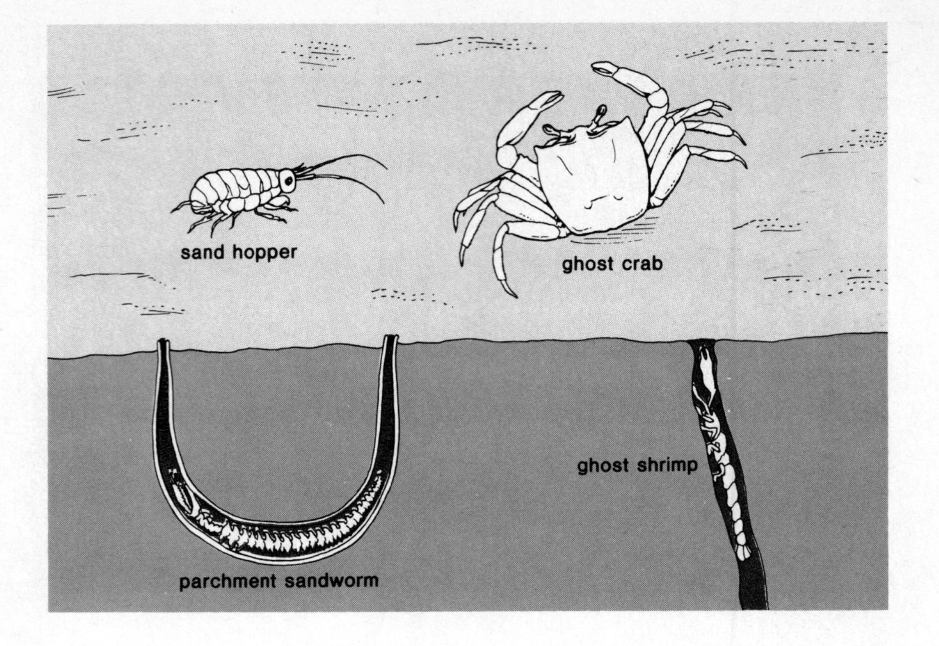

The shifting, unstable sands on a *sandy beach* (fig. 32.14) do not provide a suitable substratum for the attachment of organisms; therefore, nearly all the permanent residents dwell underground. They either burrow during the day and surface to feed at night, or they remain permanently within their burrows and tubes. Ghost crabs and sandhoppers (amphipods) burrow above high tide and feed at night when the tide is out. Sandworms and sand (ghost) shrimp remain within their burrows in the intertidal zone and feed on detritus whenever possible. Still lower in the beach, clams, cockles, and sand dollars are found.

Coral Reefs

Coral reefs (fig. 32.15) are areas of biological abundance found in shallow, warm tropical waters that have a minimum temperature of 70°F. Coral reefs begin on a rocky substratum beneath the surface of the water. Their chief constituents are *stony corals,* which have a calcium carbonate (limestone) exoskeleton, and calcerous red and green algae. Corals do not usually occur individually; rather, they form colonies derived from an individual coral that has reproduced by means of budding. Corals provide a home for a microscopic dinoflagellate called *zooxanthellae.* The corals, which feed at night, and the dinoflagellate, which photosynthesize during the day, are mutualistic and share materials and nutrients.

A reef is densely populated with animal life. There are many types of small fishes (butterfly, damsel, clown, and sturgeon), all beautifully colored. In addition, the large number of crevices and caves provide shelter for filter feeders (sponges, sea squirts, and fan worms) and for scavengers (crabs and sea urchins). The barracuda and moray eel prey on these animals. Some fishes feed on the coral, but the most deadly coral predator of Pacific reefs is the crown-of-thorns starfish, which grows as large as two feet across and has from

Figure 32.15
Coral reef community. Brightly colored fish are seen swimming among the various corals that dominate this biome. Corals are animals that have symbiotic algae living in their walls.

nine to twenty-one arms. Along the northeastern coast of Australia, the very existence of the Great Barrier Reef is threatened by these animals. The giant triton is their natural predator, and some believe that the crown-of-thorns is proliferating because humans have killed off the tritons for its handsome spiral shell.

The coastline includes marshes, swamps, estuaries, rocky and sandy beaches. Although the coast is extremely important to productivity of the ocean, it is the region that has suffered the most pollution. Coral reefs, areas of biological abundance, occur in tropical seas.

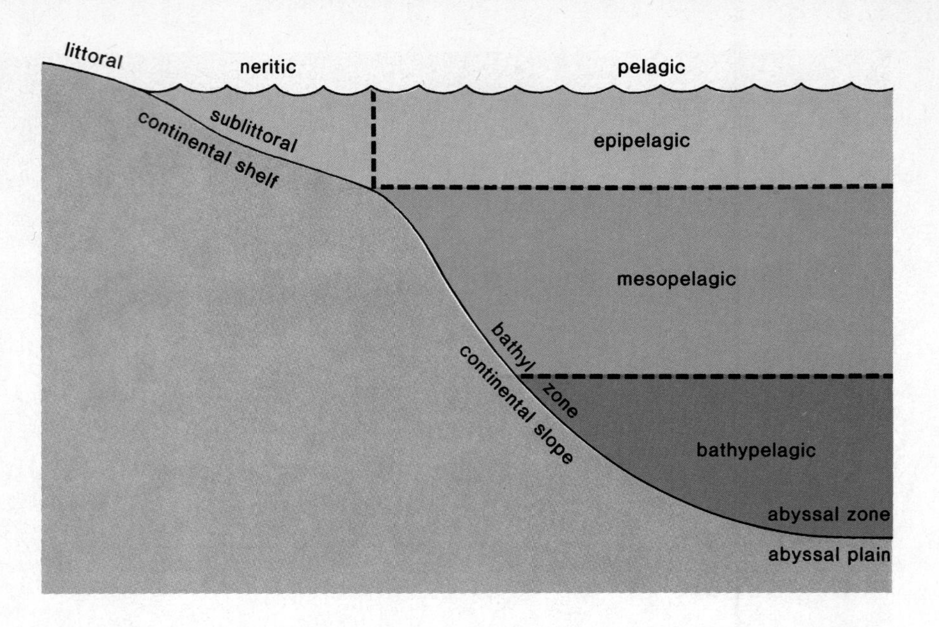

Figure 32.16
Life zones in the ocean. The benthos, comprising the sublittoral, bathyl, and abyssal zones, is found along the ocean floor that is divided into the regions shown. Organisms in the neritic zone live in the shallow waters above the continental shelf; those in the pelagic (open ocean) zone may reside in either the epipelagic, mesopelagic, or bathypelagic zones.

Marine Biomes

Approximately three-quarters of our planet is covered by the oceans. Figure 32.16 shows that the depth of the sea is at first shallow and then abruptly deepens as the *continental shelf* gives way to the *continental slope,* which leads to the *abyssal plain.* Water above the continental shelf has an average depth of about 200 meters, while that above the abyssal plain averages about 4,000 to 6,000 meters, or about two to three miles.

Organisms in the ocean occupy three life zones. *Benthos* are animals that reside in the **benthic zone,** which includes the littoral and sublittoral zones of the continental shelf, the bathyl zone of the continental slope, and the abyssal zone of the abyssal plain. Organisms that are found in the sea above the continental shelf are in the **neritic zone** and those in the sea above the abyssal plain are in the oceanic, or **pelagic zone.**

Coastal Zone

Marine life is most concentrated above and on the continental shelf. Here, seaweed, after it is partially decomposed by bacteria, is a source of food for benthic clams, worms, and sea urchins, which are preyed on by starfish, crabs, and brittle stars, all of which are, in turn, eaten by bottom-dwelling fish. More important, the shallow, sunlit neritic waters, which receive nutrients from the sea and estuaries, produce abundant phytoplankton that grow larger than they would in the open sea. Especially in regions of upwelling where surface waters are blown offshore and replaced by cold, nutrient-laden waters from the deep, ample phytoplankton grow and provide food for zooplankton and small fish. These, in turn, are food for the commercial fishes—herring, cod, and flounder.

Open Sea

The open sea, or *pelagic zone,* is divided into the epipelagic, mesopelagic, and bathypelagic zones (fig. 32.17). Only the *epipelagic zone* is brightly lit, or euphotic; the *mesopelagic zone* is in semidarkness; and the *bathypelagic zone* is in complete darkness.

The open sea, which includes 90 percent of the ocean waters, is not very productive because the surface waters are nutrient poor. Lack of nutrients in the *epipelagic zone* curtails phytoplankton growth to the extent that productivity of the open sea is equivalent to that of a terrestrial desert. Then, too, food chains tend to be longer in the ocean than on land. The small size of the *phytoplankton* and *herbivorous zooplankton* can account for this difference in food chain length, which also contributes to a low fish production. Among

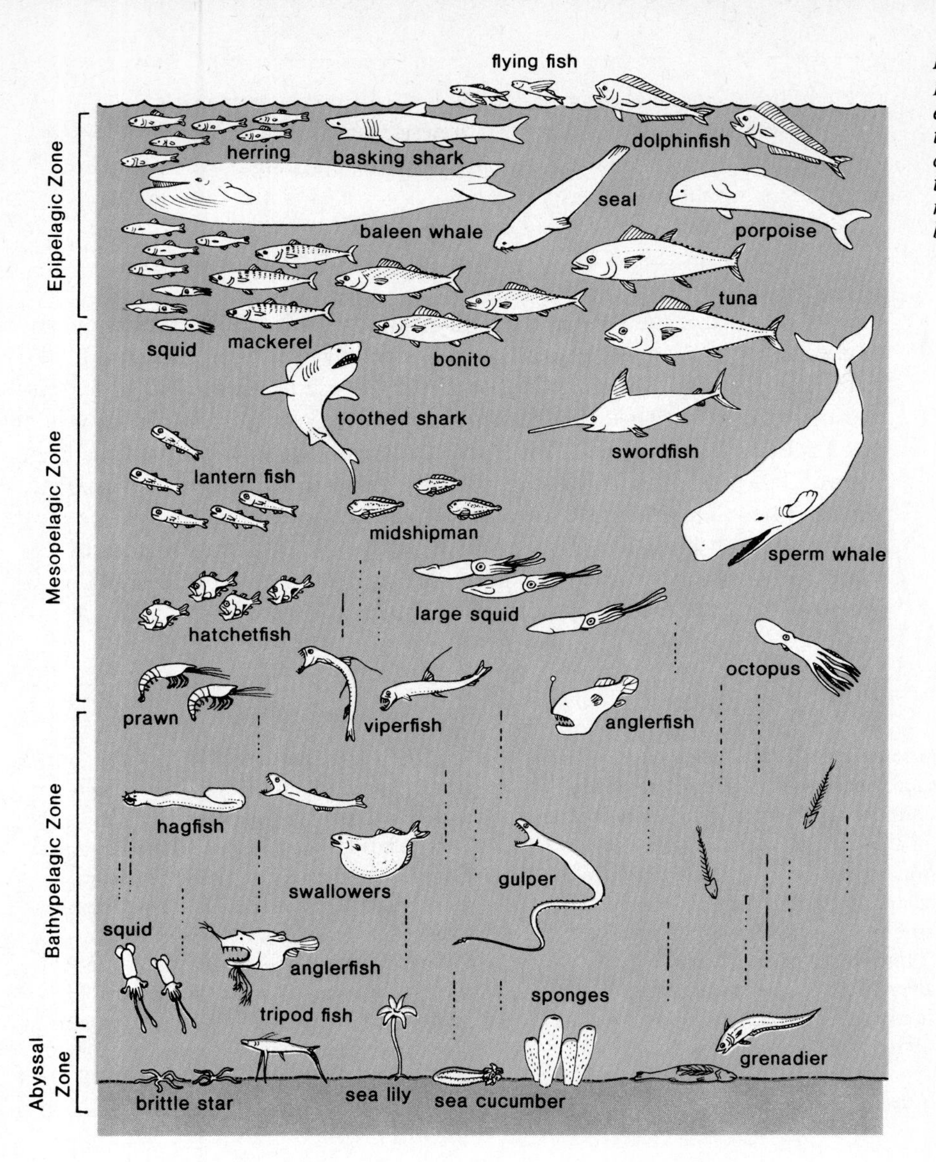

Figure 32.17
Free-swimming nekton of the pelagic zone. The epipelagic zone contains the most animals because there are producers here that support various food chains. Only carnivores and scavengers are found in the other zones. The dotted lines represent organic matter that falls from the upper zones into the bottom zone.

the phytoplankton are diatoms, dinoflagellates, and the smaller coccolithophores, which are less than a thousandth of an inch in size. They have a distinctive chalky shell and swim by moving tiny, whiplike flagella. Copepods, krill, foraminifera, and radiolarians feed on these and in turn are food for the carnivorous zooplankton, such as jellyfishes, comb jellies, wing-footed snails, sea squirts, and arrow worms.

The *nekton* (fig. 32.17) of the epipelagic zone includes herring and bluefish, which are food for the larger mackerel, tuna, and sharks. Flying fishes, which glide above the surface, are preyed upon by dolphin fishes, not to be confused with mammalian porpoises, which are also present. Whales are other mammals found in this zone. Baleen whales strain krill from the water, and the toothed sperm whales feed especially on the common squid.

Animals in the *mesopelagic zone* are adapted to the absence of light and tend to be translucent, red-colored, or even luminescent. Aside from luminescent jellyfishes, sea squirts, copepods, shrimp, and squid, there are luminescent carnivorous fishes, such as lantern and hatchetfishes. The *bathypelagic zone* is in complete darkness except for an occasional flash of bioluminescent light. Strange-looking fishes with distensible mouths and abdomens and small, tubular eyes feed on infrequent prey.

The open seas of the ocean can be divided into the epipelagic, mesopelagic, and bathypelagic zones, each having organisms adapted to different environmental conditions. Only the epipelagic zone receives adequate sunlight to support photosynthesis, which limits the oceans to productivity about equal to that of deserts.

The *abyssal life zone* in the bathypelagic zone is inhabited by those animals that live in or just above the cold, dark abyssal plain. Because of the cold temperature (averaging 2°C) and the intense pressure (300 to 500 atmospheres), it was once thought that only a few specialized animals would live in the abyssal life zone. Yet a diverse assemblage of organisms has been found. Debris from the mesopelagic zone is taken in by filter feeders, such as the sea lilies that rise above the seafloor, and the clams and tubeworms that lie burrowed in the mud. Other animals, such as sea cucumbers and sea urchins, crawl around on the sea bottom, eating detritus and bacteria of decay. They, in turn, are food for predaceous brittle stars and crabs.

Summary

The biosphere can be divided into biomes adapted to climate, especially temperature and rainfall. Deserts are high-temperature areas with less than 20 cm of rainfall a year. Grasslands occur where rainfall is greater than 20 cm but is insufficient to support trees. Forests require adequate rainfall. Tropical rain forests are the most complex of the terrestrial biomes. There is a series of biomes on mountain slopes that mirror the sequence according to latitude because temperature also decreases with altitude. Succession is a series of stages by which rocks or abandoned farmland become one of these climax communities.

Aquatic biomes are divided into fresh water and salt water. Lakes, especially in the summer, are stratified into layers according to temperature, and only a turnover twice a year restores their productivity. Lakes have various life zones, each with typical organisms. Rivers and streams are characterized by rapidly flowing water that gradually moves more slowly as a river approaches the ocean.

The coastline includes marshes, swamps, estuaries, and rocky and sandy beaches. Coral reefs occur in tropical seas. The open seas of the ocean can be divided into different zones. Only the epipelagic zone receives adequate sunlight to support photosynthesis, which limits the open seas to productivity about equal to that of deserts.

Objective Questions

1. Each major type of terrestrial ecosystem contains a characteristic community of populations called a ____________ .
2. A climax community arises due to the process of ____________ , which requires a number of stages.
3. Trees are not plentiful in a desert or a grassland because of the reduced amount of ____________ .
4. The tropical grassland of Africa is called the ____________ .
5. Broad-leaved evergreen trees would be found in a ____________ forest.
6. At the highest altitudes and highest latitudes, a ____________ biome is found.
7. In the fall, as the epilimnion of a lake cools, and in the spring as it warms, a ____________ occurs.
8. An estuary is very productive because it acquires ____________ brought both by river flow and by tidal action.
9. The chief builders of coral reefs are ____________ and calcerous ____________ .
10. The neritic zone occurs along the coast, the pelagic zone is the open ocean, and the benthic zone is the ____________ .

Answers to Objective Questions

1. biome 2. succession 3. rainfall 4. savanna 5. tropical 6. tundra 7. mixing 8. nutrients 9. stony corals, algae 10. seafloor

Study Questions

1. Arrange the terrestrial biomes discussed in this text in a diagram according to temperature and rainfall. (p. 716)
2. Describe the location, climate, and populations of (1) deserts (p. 716), (2) grasslands (tundra, savanna, prairie) (p. 718), and (3) forests (coniferous, deciduous, tropical) (p. 721).
3. Name the terrestrial biomes you would expect to find when going from the base of a mountain to the top. (p. 723)
4. Discuss succession with reference to abandoned farmland. (p. 726)
5. Discuss productivity and stability as they relate to succession. (p. 726)
6. Describe the temperature zones and the life zones of a lake. (pp. 728–729)
7. Describe the coastline biome (including coral reefs) and discuss their importance to the productivity of the ocean. (pp. 730–735)
8. Describe the life zones of the ocean and the organisms you would expect to find in each zone. (pp. 736–737)

Key Terms

biome (bi'ōm) one of the major climax communities present in the biosphere characterized by a particular mix of plants and animals. *715*

coral reef (kor'al rēf) a structure found in tropical waters formed by the buildup of coral skeletons where many and various types of organisms reside. *734*

deciduous (de-sid'u-us) plants that shed their leaves at certain seasons. *722*

desert (dez'ert) an arid biome characterized especially by plants such as cacti that are adapted to receiving less than 20 cm of rain per year. *716*

desertification (dez-ert''ĭ-fĭ-ka'shun) desert conditions caused by human misuse of land. *718*

epiphyte (ep'ĭ-fīt) nonparasitic plant that grows on the surface of other plants, usually above the ground, such as arboreal orchids and Spanish moss. *722*

estuary (es'tu-a-re) an area where fresh water meets the sea; thus, an area with salinity intermediate between fresh water and seawater. *727*

mangrove swamp (man'grōv swahmp) tropical and subtropical coastal community characterized by presence of mangrove trees. *730*

permafrost (perm'ah-frost) earth beneath surface in tundra which remains permanently frozen. *718*

plankton (plank'ton) free-floating microscopic organisms found in most bodies of water. *727*

prairie (prar'e) a biome characterized by the presence of either short stem or long stem grasses. *721*

rain forest (rān for'est) a biome of equatorial forests which remain warm year round and receive abundant rain. *722*

salt marsh (sawlt marsh) coastal grassland exposed to seasonal flooding. *730*

savanna (sah-van'ah) a grassland biome that has occasional trees and is particularly associated with Africa. *720*

taiga (ti'gah) a biome that forms a worldwide northern belt of coniferous trees. *721*

tundra (tun'drah) a biome characterized by lack of trees, due to cold temperatures and the presence of permafrost year round. *718*

33

Human Population Concerns

Chapter Concepts

1 A population undergoing exponential growth has an ever higher growth rate, a shorter doubling time, and may outstrip the carrying capacity of the environment.

2 The world is divided into the developed and developing countries; only the latter are presently undergoing exponential growth.

3 Both types of countries demand an ever greater energy and food supply each year.

4 It would be possible to convert the human ecosystem to a steady state in which population size and resource consumption stay steady and do not increase.

Chapter Outline

Exponential Population Growth
- Growth Rate
- Doubling Time
- Carrying Capacity

World Population
- Developed Countries
- Developing Countries

Resource Consumption
- Present Day
- Steady State

a.

b.

Figure 33.1
Developed versus the developing countries. a. *In the developed countries most people enjoy a high standard of living.* b. *In the developing countries the majority of people are poor and have few amenities.*

The countries of the world today are divided into two groups (fig. 33.1). The developed countries, typified by countries in North America and Europe, are those whose population growth is controlled and in which most people enjoy a good standard of living. The developing countries, typified by countries in South America and Africa, are those whose population growth is out of control and in which the majority of people are poverty stricken. In this chapter we are going to examine the history of the world population growth in order to understand how this dichotomy has arisen. Before we do, however, it is necessary to study exponential population growth in general.

Exponential Population Growth

The human growth curve is an *exponential curve* (fig. 33.2). In the beginning, growth of the population was relatively slow, but as a greater number of reproducing individuals were added, growth increased until the curve began to slope steeply upward. It is apparent from the position of 1987 on the growth curve that growth is now quite rapid. The world population increases at least the equivalent of a medium-sized city every day (200,000) and the combined populations of the United Kingdom, Norway, Ireland, Iceland, Finland, and Denmark every year. These startling figures are a reflection of the fact that a very large world population is undergoing exponential growth.

Mathematically speaking, **exponential growth,** or geometric increase, occurs in the same manner as compound interest; that is, the percentage increase is added to the principal before the next increase is calculated. Referring specifically to populations, we wish to consider the hypothetical population sizes in table 33.1. This table illustrates the circumstances of world population

Table 33.1 *Exponential Growth of Hypothetical Populations*

Population Size	Percentage Increase	Actual Increase in Numbers	Population Size	Percentage Increase	Actual Increase in Numbers	Population Size
500,000,000	2.00	10,000,000	510,000,000	1.99	10,149,000	520,149,000
3,000,000,000	2.00	60,000,000	3,060,000,000	1.99	60,894,000	3,120,894,000
5,000,000,000	2.00	100,000,000	5,100,000,000	1.99	101,490,000	5,201,490,000

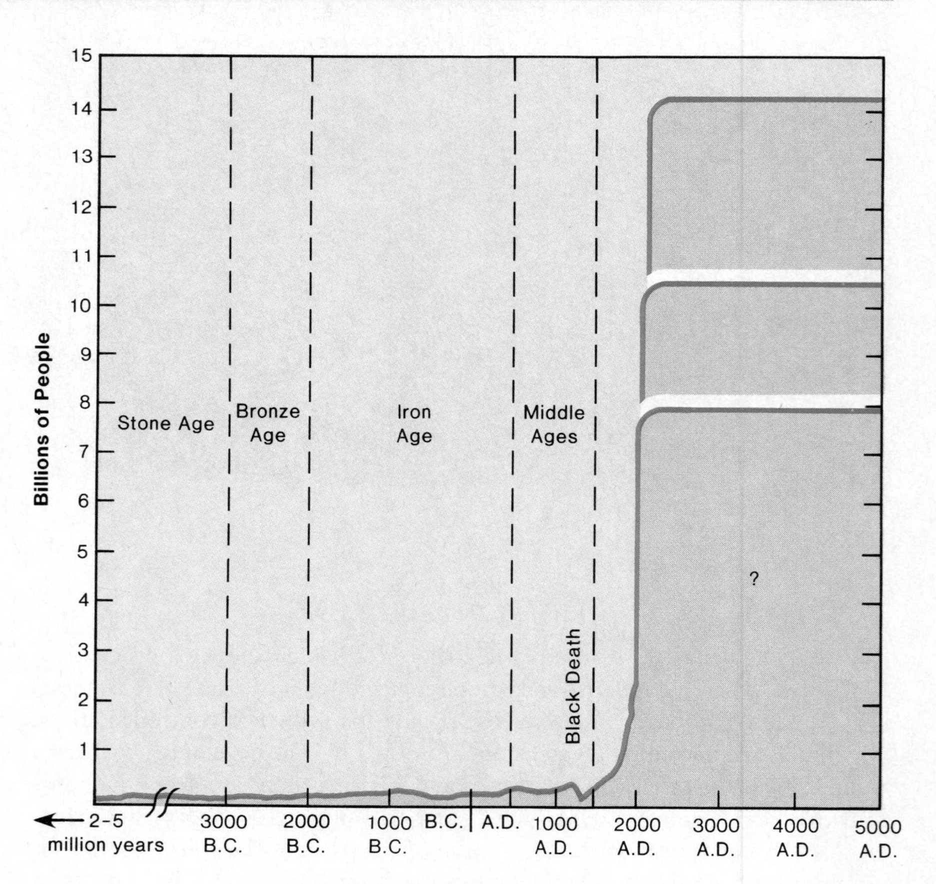

Figure 33.2
Growth curve for human population. The human population is now undergoing rapid exponential growth. Since the growth rate is declining, it is predicted that the population size will level off at 8, 10.5 or 14.2 billion, depending upon the speed with which the growth rate declines.

growth at the moment: the percentage increase has decreased and yet the size of the population grows by a greater amount each year. The increase in size is dramatically large because the world population is very large.

In our hypothetical examples (table 33.1), an initial increase of 2 percent added to the original population size, followed by a 1.99 percent increase, results in the third generation size listed in the last column. Notice that

1. in each instance the second generation has a larger increase than the first generation because the second generation's population was larger than the first;
2. because of exponential growth, the lower percentage increase (i.e., 1.99% compared to 2%) still brings about larger population growth;
3. the larger the population, the larger the increase for each generation.

The percentage increase is termed the **growth rate,** which is calculated per year.

Growth Rate

The growth rate of a population is determined by considering the difference between the number of persons born (birthrate, or natality) and the number of persons who die per year (death rate, or mortality). It is customary to record these rates per 1,000 persons. For example, Russia (USSR) at the present time has a birthrate of 20 per 1,000 per year, but it has a death rate of 10 per 1,000 per year. This means that Russia's population growth, or simply its growth rate, would be:

$$\frac{20 - 10}{1{,}000} = \frac{10}{1{,}000} = \frac{1.0}{100} = 1.0\%$$

Notice that while birth and death rates are expressed in terms of 1,000 persons, the growth rate is expressed per 100 persons, or as a percentage.

After 1750 the world population growth rate steadily increased until it peaked at 2 percent in 1965, but it has fallen slightly since then to 1.7 percent. Yet there is an ever larger increase in the world population each year because of exponential growth. The explosive potential of the present world population can be appreciated by considering the doubling time.

Doubling Time

Table 33.2 shows that the **doubling time** for a population may be calculated by dividing 70 by the growth rate:

$$d = \frac{70}{gr}$$

d = doubling time
gr = growth rate
70 = demographic constant

If the present world growth rate of 1.7 percent should continue, the world population will double in 40 years.

$$d = \frac{70}{1.7} = 40 \text{ years}$$

This means that in 39 years the world would need double the amount of food, jobs, water, energy, and so on if the standard of living is to remain the same.

Table 33.2 *Relationship between Growth Rate and the Doubling Time of a Population*

Growth Rate %	Doubling Time (Years)
0.25	280
0.5	140
1.0	70
2.0	35
3.0	23

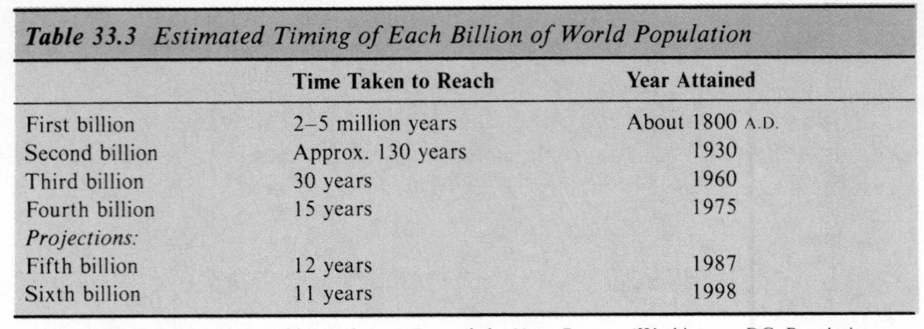

Table 33.3 Estimated Timing of Each Billion of World Population		
	Time Taken to Reach	**Year Attained**
First billion	2–5 million years	About 1800 A.D.
Second billion	Approx. 130 years	1930
Third billion	30 years	1960
Fourth billion	15 years	1975
Projections:		
Fifth billion	12 years	1987
Sixth billion	11 years	1998

Source: Elaine M. Murphy. *World Population: Toward the Next Century* (Washington, DC: Population Reference Bureau, November, 1981), page 3.

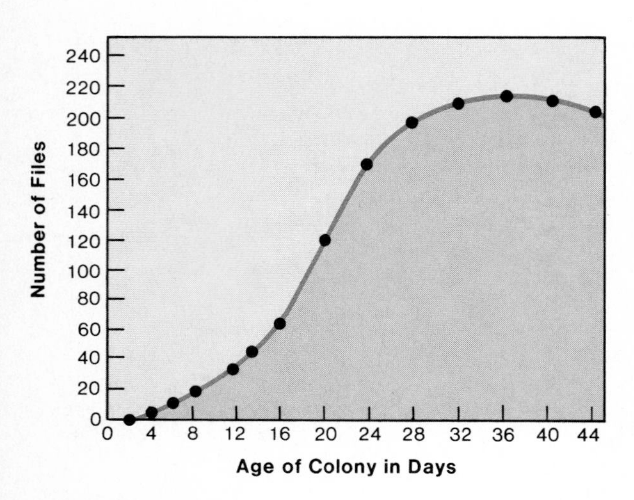

Figure 33.3
Growth curve for a fruit fly colony. The number of fruit flies in a laboratory colony were counted every other day, and when these numbers were plotted, a sigmoidal growth curve resulted. This type of curve is expected for a population that is adjusting to the carrying capacity of the environment.

It is of grave concern to many individuals that the amount of time needed to add each additional billion persons to the world population has taken less and less time (table 33.3). The world reached its first billion around 1800—some 2 million years after the evolution of humans. Adding the second billion took about 130 years, the third billion about 30 years, and the fourth took only about 15 years. However, if the growth rate should continue to decline, this trend would reverse itself and eventually there would be zero population growth. Then population size would remain steady. Therefore, figure 33.2 shows three possible logistic curves: the population may level off at 8, 10.5, or 14.2 billion, depending on the speed with which the growth rate declines.

Carrying Capacity

Examining the growth curves for nonhuman populations reveals that the populations tend to level off at a certain size. For example, figure 33.3 gives the actual data for the growth of a fruit fly population reared in a culture bottle. At the beginning, the fruit flies were becoming adjusted to their new environment and growth was slow. But then, since food and space were plentiful, they began to multiply rapidly. Notice that the curve begins to rise dramatically just as the human population curve does now. At this time, it may be said that the population is demonstrating its **biotic potential.** Biotic potential is the maximum growth rate under ideal conditions. Biotic potential is not usually demonstrated for long because of an opposing force called **environmental resistance.** Environmental resistance includes all the factors that cause early death of organisms and therefore prevents the population from producing as many offspring as it might otherwise have done. As far as the fruit flies are concerned, we can speculate that environmental resistance included the limiting factors of food and space. Also, the waste given off by the fruit flies may have begun to contribute to keeping the population size down.

The eventual size of any population represents a compromise between the biotic potential and the environmental resistance. This compromise occurs at the **carrying capacity** of the environment. The carrying capacity is the maximum population that the environment can support—for an indefinite period.

The carrying capacity of the earth for humans is not certain. Some authorities think the earth is potentially capable of supporting 50 to 100 billion people. Others think we already have more humans than the earth can adequately support.

The human population is expanding exponentially and the doubling time has decreased from 40 to 35 years.

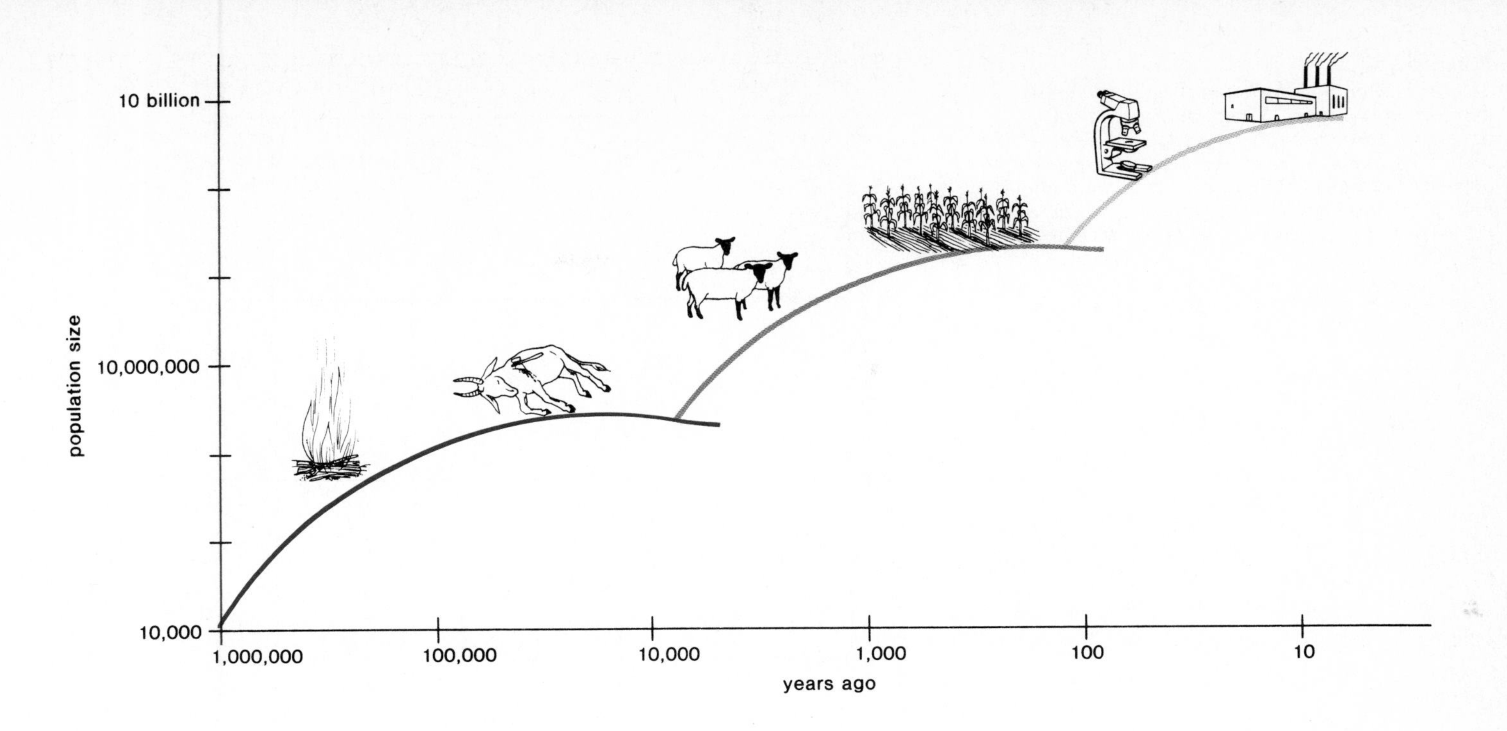

Figure 33.4
History of human population. When the size of the human population is plotted in depth, it can be seen that exponential growth occurred on three occasions: at the time of the cultural, agricultural, and industrial revolutions.

World Population

Figure 33.4 suggests that the human population has undergone three phases of exponential growth. *Toolmaking* may have been the first technological advance that allowed the human population to enter a phase of exponential growth. *Cultivation of plants* and *animal husbandry* may have allowed a second phase of growth; and the *industrial revolution,* which occurred about 1850, promoted the third phase. At first only certain countries of the world became industrialized and these countries are now called the developed countries.

Developed Countries

The industrial revolution, which was also accompanied by a medical revolution, took place in the Western world. In addition to European and North American countries, Russia and Japan also became industrialized. Collectively, these countries are often referred to as the **developed countries.** The developed countries doubled their size between 1850 and 1950 (fig. 33.5), largely due to a decline in the death rate. This decline is attributed to the influence of modern medicine and improved socioeconomic conditions. Industrialization raised personal incomes, and better housing permitted improved hygiene and sanitation. Numerous infectious diseases, such as cholera, typhus, and diphtheria, were brought under control.

The decline in the death rate in the developed countries was followed shortly by a decline in the birthrate. Between 1950 and 1975, populations in the developed countries showed only modest growth (fig. 33.5) because the growth rate fell from an average of 1.1 percent to 0.8 percent.

Figure 33.5
Size of human population in developed versus developing countries. The population of the developed countries (blue) increased between 1850 and 1950, but the size is expected to increase little between 1975 and 2000. In contrast, the population size of the developing countries (yellow) increased in the past and is expected to increase dramatically in the future also.

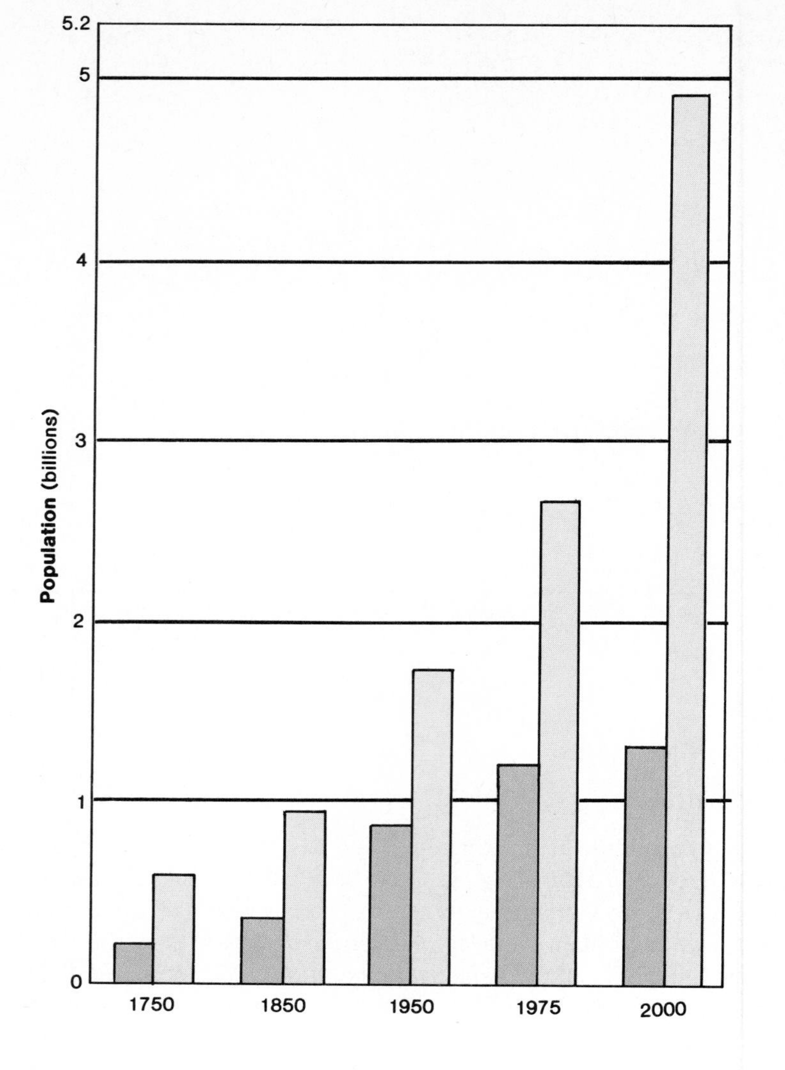

Table 33.4 Analysis of Annual Growth Rates in Developed Countries

Phase	Birthrate	Death Rate	Annual Rate
I	High	High	Low
II	High	Low	High
III	Low	Low	Low

Demographic Transition Overall, the growth rate in developed countries has gone through three phases (table 33.4 and fig. 33.6). In Phase I, prior to 1850, the growth rate was low because a high death rate canceled out the effects of a high birthrate; in Phase II, the growth rate was high because of a lowered death rate; and in Phase III, the growth rate was again low because the birthrate had declined. These phases are now known as the **demographic transition.** In seeking a reason for the transition, it has been suggested that as industrialization occurred, the population became concentrated in the cities. Urbanization may have contributed to the decline in the growth rate because, in the city children were no longer the boon they were in the country. Instead of contributing to the yearly income of the family, they represented a severe drain on its resources. It could also be that urban living made people acutely aware of the problems of crowding and for this reason the birthrate declined. Also, some investigators believe that there was a direct relationship between improvement in socioeconomic conditions and the birthrate. They point out that as the developed nations became wealthier, as infant mortality was reduced, and as educational levels increased, the birthrate declined.

Regardless of the reasons for the demographic transition, it caused the rate of growth to decline in the developed countries. The growth rate for the developed countries is now about 0.6 percent and their overall population size is about one-third that of the developing countries. A few developed countries—Austria, Denmark, East Germany, Hungary, Sweden, West Germany—are not growing or are actually losing population. As discussed in the reading on page 749, the United States has a comparatively high growth rate for a developed country.

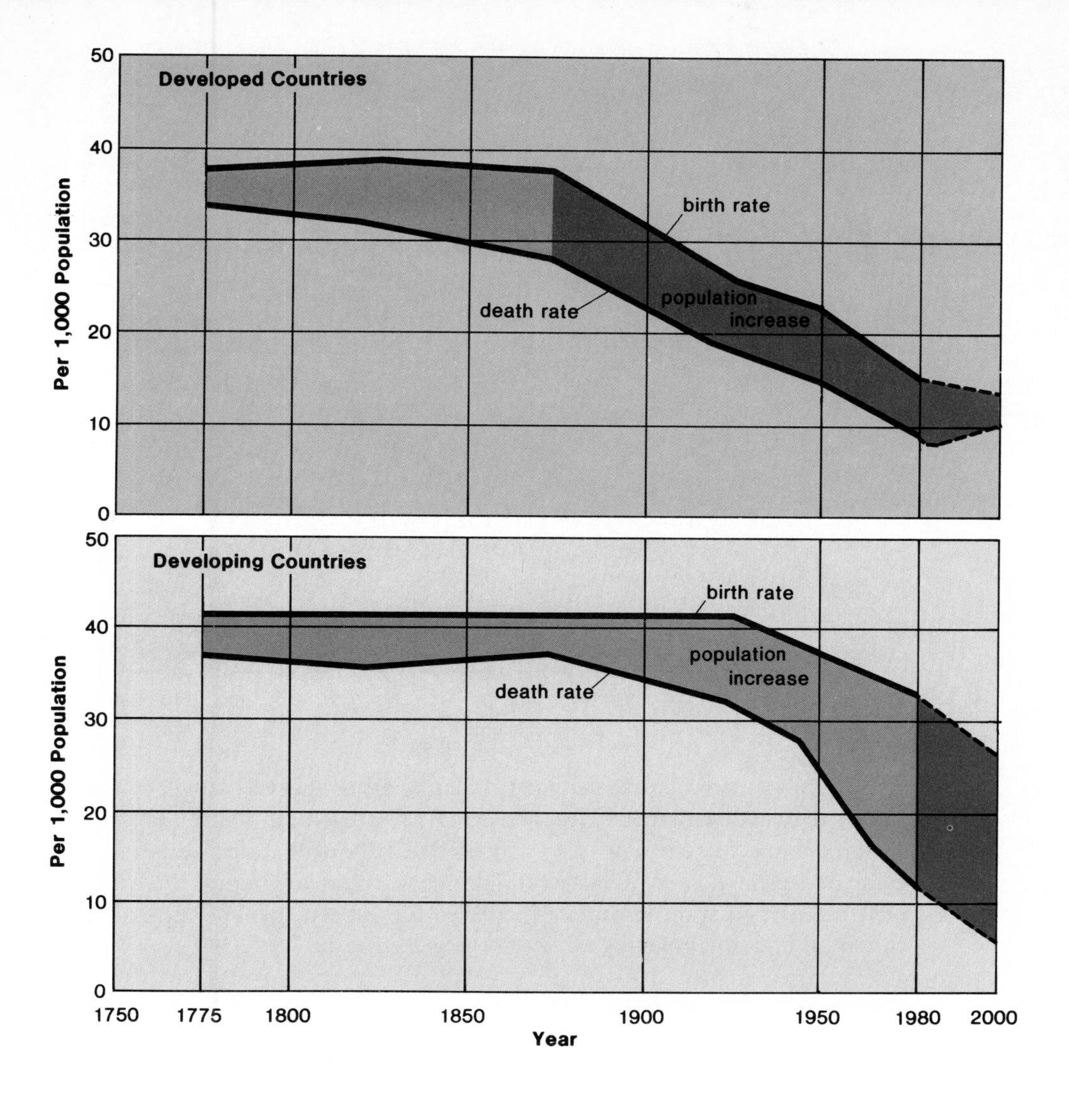

Figure 33.6
Time of demographic transition. In the upper graph it is seen that in the developed countries the demographic transition occurred in the nineteenth century. In the lower graph it is seen that the demographic transition was delayed until the twentieth century.

Developing Countries

Countries, such as those in Africa, Asia, and Latin America, are collectively known as **developing countries** because they have not yet become industrialized. Figure 33.6 indicates that mortality began to decline steeply in these countries following World War II. This decline was prompted not by socioeconomic development but by the importation of modern medicine from the developed countries. Various illnesses were brought under control due to the use of immunization, antibiotics, sanitation, and insecticides. Although the death rate declined, the birthrate did not decline to the same extent (fig. 33.6) and therefore the populations of the developing countries began and today are still increasing dramatically (fig. 33.5). The developing countries were unable to cope adequately with such rapid population expansion so that today many people in these countries are underfed, ill housed, unschooled, and living in abject poverty. Many of these poor have fled to the cities where they live in makeshift shanties on the outskirts.

The growth rate of the developing countries did finally peak at 2.4 percent during 1960–1965. Since that time, the mortality decline has slowed and the birthrate has fallen. The growth rate is expected to decline to 1.8 percent by the end of the century. At that time, about two-thirds of the world population will be in the developing countries.

Investigators are divided as to the cause of the observed growth rate decreases in the developing countries. Previously, it was argued that this would happen only when these countries enjoyed the benefits of an industrialized society. It has now been shown, however, that countries with the greatest decline were those with the best family-planning programs. From this it may be

Figure 33.7
Age structure diagram for developed countries (above) *versus developing countries* (below). *Because the population sizes of the developed countries are approaching stabilization, population growth will be modest, but the developing countries will expand rapidly due to their youthful profile.*

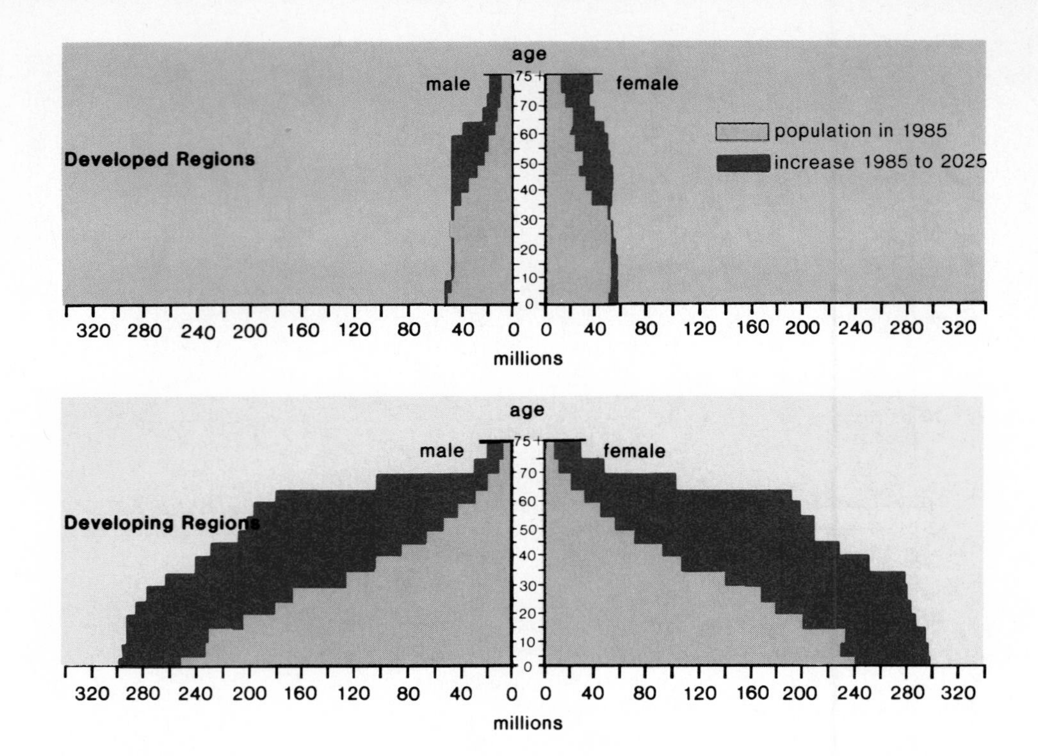

argued that such programs can indeed help to bring about a stable population size in the developing countries. Nevertheless, it has been found that certain socioeconomic factors have also contributed to a decline in the developing countries' growth rate. Relatively high Gross National Product (GNP), urbanization, low infant mortality, increased life expectancy, literacy, and education all had a dampening effect on the growth rate.

Age Structure Comparison

Laypeople are sometimes under the impression that if each couple had two children, zero population growth would immediately take place. However, **replacement reproduction,** as it is called, would still cause most countries today to continue growth due to the age structure of the population. If more young women are entering the reproduction years than there are older women leaving them behind, then replacement reproduction will give a positive growth rate.

Reproduction is at or below replacement level in some 20 developed countries, including the United States. Even so, some of these countries will continue to grow modestly, in part because there was a baby boom after World War II. Young women born in these baby boom years are now in their reproductive years and even if each one has less than two children, the population will still grow. It should also be kept in mind that even the smallest of growth rates can be a considerable number of individuals to a large country. For example, a growth rate of 0.7 percent added over 1.6 million people to the United States population in 1984.

Many developed countries have a stabilized age structure diagram (fig. 33.7), but most developing countries have a youthful profile—a large proportion of the population is below the age of 15. Since there are so many young women entering the reproductive years, the population will still greatly expand even after replacement reproduction is attained. The more quickly replacement reproduction is achieved, however, the sooner zero population growth will result.

The history of the world population shows that the developed countries underwent a demographic transition between 1950 and 1975; the developing countries are just now undergoing demographic transition.

The United States now has a population of over 240 million. The geographic center (the point where there are just as many persons in each direction) has moved steadily westward and recently has also moved southward. Another interesting trend is the shifting emphasis from metropolitan areas to nonmetropolitan areas. In the 1970s cities increased by 9.8 percent, but rural small towns increased by 15.8 percent.

The population size of the United States is not expected to level off any time soon for two primary reasons.

1. A baby boom between 1947 and 1964 has resulted in an unusually large number of reproductive women at this time (see the figure). Thus, although each of these women is having on the average only 1.9 children, the total number of births increased from 3.1 million a year in the mid 1970s to 3.6 million in 1980–81.
2. Many people immigrate to the United States each year. In 1981 immigration accounted for 43 percent of the annual population growth. The number of legal immigrants was about 700,000. Even though ordinarily only 20,000 legal immigrants can come from any one country, we give special permission for large numbers of political refugees to enter the United States.

In recent years, the majority of refugees have come from Latin America (e.g., Cuba) and Asia (e.g., Indonesia).

There is also substantial illegal immigration into the United States, although the exact number is not known. Estimates range from 100,000 to 500,000 or more. About 50 to 60 percent of these illegal immigrants come from Mexico, according to most estimates. There has been an effort to stem the tide of illegal immigration to the United States.

Whether the United States can ever achieve a stable population size depends on the fertility rate (the average number of children each woman bears) and the net annual immigration. If the fertility rate were kept at 1.8 and immigration limited to 500,000, the population would peak at 274 million in 2050, after which it would decline. Those who favor a curtailment of our population size point out that a fertility rate of 1.8 allows couples a great deal of freedom in deciding the number of children they will have. For example, it means that 50 percent of all couples can have two children; 30 percent can have three children; 10 percent can have one child; 5 percent can have no children; and 5 percent can have four children.

United States Population

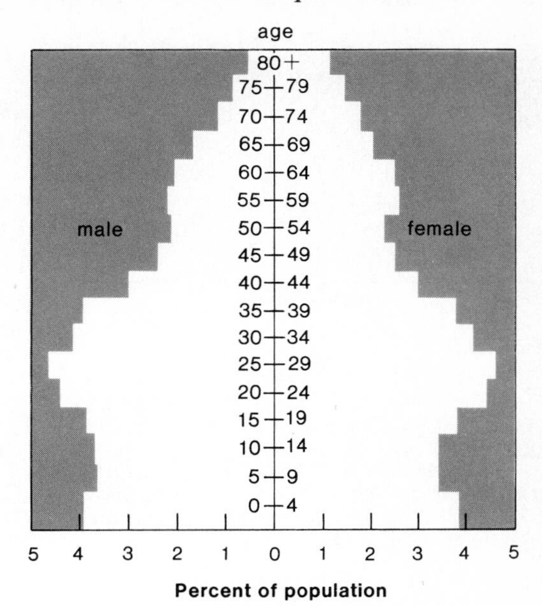

Source: United States Bureau of the Census.

Resource Consumption

Present Day

Today's rapid world population growth puts extreme pressure on the earth's resources, physical environment, and social organization. While it might seem as if population increases in the developing countries are of the gravest concern, this is not necessarily the case since each person in a developed country consumes more resources and is therefore responsible for a greater amount of pollution. Environmental impact (EI) is measured not only in terms of the population size but also in terms of the resource used and the pollution caused by each person in the population.

$$\text{EI} = \text{population size} \times \frac{\text{resource use}}{\text{per person}} \times \frac{\text{pollution per unit of}}{\text{resource used}}$$

Therefore, there are two types of overpopulation. The first type is due to rapidly increased population and occurs mainly in the developing countries. The second type of overpopulation is due to increased resource consumption with its accompanying pollution; this type is most obvious in the developed countries.

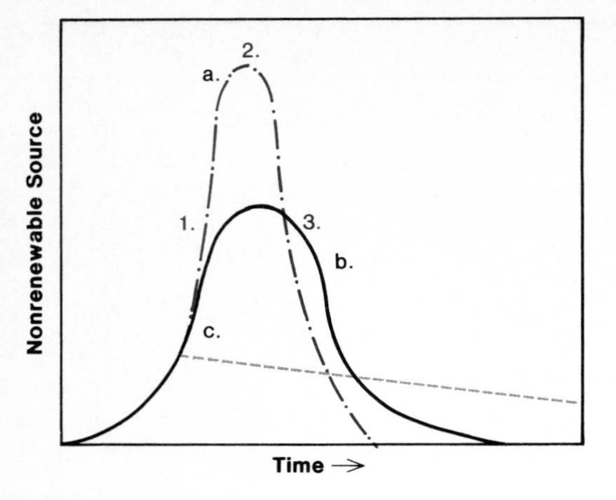

Figure 33.8
Depletion patterns for a nonrenewable resource. a. *Rapid depletion as the resource is used up quickly: (1) exponential consumption of a resource followed by a peak (2) and decline (3) as the resource become difficult to acquire.* b. *Depletion time can be extended with some recycling and less wasteful use.* c. *Efficient recycling extends the depletion curve indefinitely.*

Resources are either renewable or nonrenewable. The supply of **renewable resources** reoccurs. For example, rain periodically brings a renewed supply of fresh water and harvests bring a renewed supply of food. **Nonrenewable resources** are those resources whose supply can be used up, or exhausted. For example, it would be possible to eventually use up all the fossil fuels which have been already deposited in the earth and then there would be no more. There are two points of view, called the Malthusian[1] and Cornucopian[1] views about nonrenewable resource availability. According to the Malthusian view, each nonrenewable resource is subject to depletion, and we are rapidly approaching the time when many, if not most, supplies will be exhausted. Figure 33.8 shows a **depletion curve** for any nonrenewable resource that is consumed at an increasingly rapid rate. The consumption peak is followed by a consumption decline as the resource becomes more expensive to find and process. Those who uphold the Malthusian view believe that, because of exponential consumption, finding new reserves cannot sufficiently extend the depletion curve. They do not believe that technology will ever overcome inevitable shortages. They admonish us, therefore, to cut back on growth, conserve resources, and recycle whenever possible to extend the depletion curve (fig. 33.8b and c).

According to the Cornucopian view, technology will constantly be able to put off the day when no further exploitation is possible. Improved technology will enable us to: (1) find new reserves, (2) exploit new reserves, and (3) substitute one mineral or energy resource for another. Sometimes we are aware of the availability of a resource but must await the development of a technology to exploit it. We now utilize offshore drilling to acquire oil; we might be able to develop the means to mine the ocean floor for minerals, and we can utilize ever poorer grades of mineral ores. For example, previously only 3 percent copper ores were mined, whereas now ores with .3 percent copper are utilized.

To exploit previously inaccessible and/or less concentrated resources, a plentiful supply of energy is required. The Malthusians do not believe that increasing amounts of energy will be available, since fossil fuel reserves are being rapidly depleted. Many Cornucopians are still hoping that nuclear energy will supply the necessary energy. They point out that it is possible to design safe nuclear reactors despite nuclear accidents such as Three-Mile Island and Chernobyl, discussed following. The MHTGR (modular high-temperature gas reactor) is smaller in size than the reactors presently being used; it uses helium gas as coolant instead of water. And even if the gas should stop flowing, it will not get hot enough to melt down because it only generates 80 megawatts of power compared to the 1,000 megawatts of conventional reactors today. Also, because its components are modular, it can be mass produced. A prototype has already been successfully tested in West Germany.

Malthusians and Cornucopians also view the problem of pollution differently. The Malthusians suggest that as resource consumption increases, the amount of pollution control needed to protect the environment becomes prohibitive. Previously, no heed was paid to the cost of pollution. Industries were allowed to pollute the environment, and the general public bore the cost of cleaning up after industry. Now, since controls are required by the government, the cost of pollution is added to the total cost of a product (fig. 33.9).

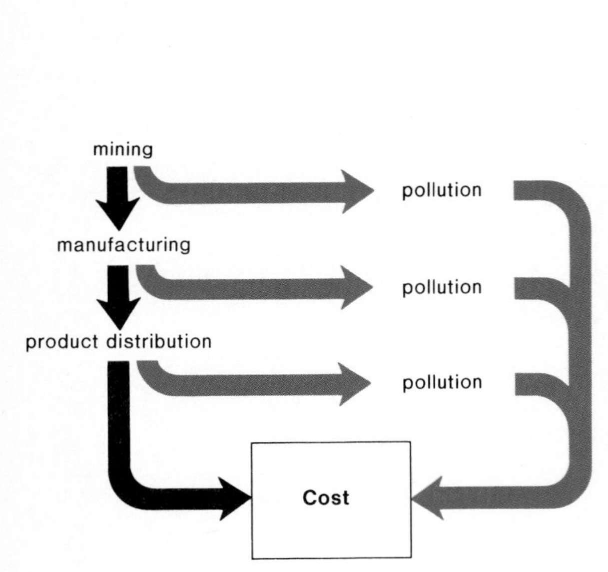

Figure 33.9
The cost of a product includes the cost of controlling pollution due to, for example, mining, manufacturing, and distributing the product.

[1]Malthus was an eighteenth century economist who pointed out that since the size of a population increases geometrically and renewable resources increase only arithmetically, shortages must eventually occur. *Cornucopia* is a Latin word meaning the horn of plenty, a symbol of everlasting abundance.

The profit margin of the producer is thereby cut even as the consumer must pay more for the product. Unfortunately, also, greater and greater efficiency of control is required as output is increased. For example, as table 33.5 shows, if output doubles every fifteen years, pollution control efficiency must increase to 96 percent within 120 years to stay even. Since the cost of pollution control increases manyfold per level of efficiency, it may become unprofitable to continue production.

On the other hand, Cornucopians suggest that whenever strict pollution control regulations are instigated, pollution can be reduced to acceptable levels at relatively modest costs compared to the value of total output. Given reasonable periods of time, all industries and most firms will be able to accommodate themselves to the impact of the new standards. In some instances it may be possible to change industrial processes so there is no waste or to make use of a by-product that formerly was considered a waste.

Table 33.5 Pollution Control Efficiency

Time (Years)	Pollution Level	Pollution Control Efficiency
—	One unit	—
15	Two units	50%
30	Four units	75%
60	Eight units	88%
120	Sixteen units	99%

Resource consumption is dependent on population size and on consumption per individual. The developing countries are responsible for the first type of environmental impact and the developed countries for the second type.

Energy

A larger world population will require more energy in the years ahead. Developing nations are expected to be responsible for most of the increased demand for energy (fig. 33.10). The opposite trend seems to be occurring in the United States where already high demand is not increasing and might even diminish by almost 25 percent. Surprisingly, it has been suggested that decreased energy consumption need not mean a change in U.S. life-style. Instead, it is possible that the GNP could rise even as less energy is used because when less money is spent on energy, more funds can be devoted to growth.

The energy sources currently available or being investigated are listed in table 33.6.

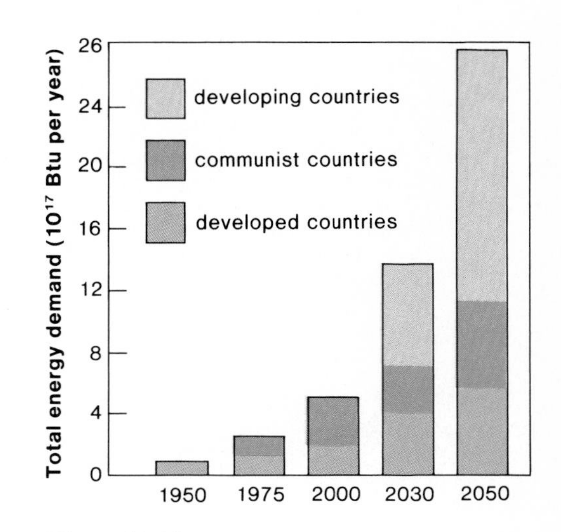

Figure 33.10
History of world energy utilization. The developing countries are expected to account for much of the increase in the future.
Source: Oak Ridge National Laboratory, operated by Martin Marietta Energy Systems, Inc., for the U.S. Department of Energy.

Table 33.6 Energy Sources

	Advantages	Disadvantages
Nonrenewable	*Technology well established*	*Finite fuel supply*
Fossil fuels		
Coal, oil shale, and tar sands	Plentiful supply	Surface mining Air and water pollution
Petroleum	Cleaner burning	Limited supply
Natural gas	Cleanest burning	Limited supply
Nuclear		
Light water	Fuel availability	Thermal pollution Radiation pollution
Renewable	*Infinite fuel supply*	*Technology under development*
Nuclear		
Breeder	Fuel availability	Radiation pollution Thermal pollution Nuclear weapons proliferation
Fusion	Fuel availability	Radiation pollution
Geothermal	Less pollution	Availability limited
Solar and wind	Nonpolluting Large and small scale possible	Noncompetitive cost
Ocean	Nonpolluting	Applicable only in certain areas
Biomass	Utilizes wastes	Air and water pollution

Figure 33.11
Aerial view of extensive surface mine spoil. Most coal today is surface-mined, a process that reduces the quality of the land unless proper careful measures are taken to restore it as it was before.

Nonrenewable Energy Sources The depletion curve (fig. 33.8) for *petroleum* is now in the decline phase, and the depletion curve for *natural gas* has reached its peak. Supplies of these favored fossil fuels are not expected to last more than thirty years. Petroleum became the favored fuel during the present generation because it is easily transportable; is versatile, serving as the raw material for gasoline and many organic compounds; and is cleaner-burning than coal.

Coal is in plentiful supply, and the United States probably will depend primarily on its use to make up for decreased use of petroleum. There are environmental drawbacks to using coal in place of petroleum, however. Thousands of acres of land are strip-mined (fig. 33.11) producing huge piles of residue having the potential to leach hazardous chemicals into underground aquifers. Coal burning adds many pollutants to the air that damage human lungs, poison lakes, and erode buildings. It is even possible that the CO_2 given off will eventually cause a global warming trend. All these concerns were discussed in chapter 31.

Three possible types of *nuclear energy* (table 33.6) can be used to generate electricity. Thus far, only light water **fission reactors,** which split ^{235}uranium, have been utilized in the United States. Approximately one hundred operating nuclear plants exist and twenty-five are under construction. Many proposed plants have been canceled, and there have been no new orders for the past several years. Some predict that the nuclear power industry is dead in this country. Possible reasons for this are numerous; for example, there is little need for new plants because demand for electricity is not increasing; it is cheaper to build and operate coal-firing plants; and the public is very much concerned about nuclear power safety.

In May, 1986, the Chernobyl nuclear power plant in Russia suffered an accident that released so much radioactivity it triggered an alarm system in Sweden. Due to operator error, the nuclear reaction that powers the plant increased beyond control, causing a fuel meltdown.The molten fuel reacted with water, and generated an immense burst of steam that shattered the roof of the plant. The damaged reactor core and the surrounding graphite (a material that normally functions to control a nuclear reaction) began to burn, sending a plume of radioactivity into the atmosphere. Western nations were quick to point out that the West has no graphite-utilizing plants and that the Chernobyl plant lacked a containment tower that keeps radioactivity from entering the atmosphere should an accident occur. Even so, the drive to control the nuclear power industry was given new impetus by the Chernobyl accident, which caused several immediate deaths and may lead to hundreds more.

The public is also aware of the need to store radioactive wastes that have accumulated since the start of the nuclear age. Spent fuel rods contain low-, medium-, and high-level wastes, depending on the amount and character of the radiation they emit. High-level wastes must be stored hundreds of thousands of years before they lose their radioactivity. As yet, the government has not sanctioned any particular permanent method of disposal. Most probably, the wastes will finally be incorporated into glass or ceramic beads, packaged in metal canisters, and buried in stable salt beds, red clay deposits in the center of the ocean, or in stable rock formations. Meanwhile, some wastes have been temporarily stored above ground in tanks that on occasion have developed leaks.

Another source of radioactive waste, not often mentioned, occurs when uranium is mined. Uranium **mill tailings** are the sand that remains after uranium has been removed from mined material. The tailings contain ^{226}radium, which has a half-life[2] of 1,620 years. As ^{226}radium decays, it gives off radon gas, a substance that has been associated with the development of lung cancer. At first the government was slow to regulate proper disposal of the tailings, and in the meantime, they were used as fill at many construction sites. This procedure has now been stopped, and mill operators are required to bury tailings beneath a thick layer of soil that is replanted with vegetation to prevent erosion of radioactive dust into the water and air.

Breeder fission reactors do not have the same environmental problems as light water nuclear power plants. They need less raw uranium, and they have little waste because they use ^{239}plutonium, a fuel that is actually generated from what would be reactor waste.

Plutonium, however, is a very toxic element that readily causes lung cancer; it is also the element used to make nuclear weapons. (The chances of a nuclear explosion are much greater with breeder than with light water reactors.) The fear of nuclear weapons proliferation, in particular, has thus far prevented the start-up of a breeder reactor in this country.

Nuclear fusion requires that two atoms, usually deuterium and tritium, be fused, and the heat needed is so great that there is no conventional container for the reaction. Scientists are now perfecting laser-beam ignition and magnetic containment for the reaction. Since the fusion reaction gives off neutrons that can change uranium to plutonium, the best use of fusion plants would be to provide fuel for breeder reactors. The latest idea is to have hybrid fusion plants that would combine both fusion and breeder fission plants. But, in any case, the fusion process is still experimental and not expected to be ready for production any time soon.

---

*E*xploitation of the environment is dependent on a plentiful energy source. Among the nonrenewable sources, coal is in ready supply and can be utilized without developing new technologies. Each of the three types of nuclear power has a drawback that has prevented full-scale operation in this country.

---

[2]Length of time required for one-half of the radiation to dissipate.

Figure 33.12
Solar energy can be utilized by various community institutions, such as Betatakin National Monument in Arizona.

Renewable Energy Sources **Solar energy** is diffuse energy that must be collected and concentrated before utilization is possible. The public, rather than the government, has provided most of the impetus for practical solutions, including solar heating of homes and offices (fig. 33.12). **Solar collectors** placed on rooftops absorb radiant energy but do not release the resulting heat. A fluid within the solar collector heats up and is pumped to other parts of the building for space heating, cooling, or the generation of electricity. Passive systems are also possible; specially constructed glass can be used for the south wall of a building, and building materials can be designed to collect the sun's energy during the day and release it at night.

The U.S. government subsidized the building of Solar One, the world's largest solar-powered electrical generating facility (fig. 33.13), which began operation in 1982 at Daggett, California, in the Mojave Desert. A single boiler is located atop a large tower twenty stories high, and a large field of mirrors (called **heliostats,** which are capable of tracking the sun) reflects the sun's rays onto the tower. The water is heated to 500° C, and the steam is used to produce electricity in a conventional generator. Systems such as this require much land and cannot be placed just anywhere. For this reason, **photovoltaic (solar) cells** that produce electricity directly may be a more promising energy source. It has even been suggested that cells might be placed in orbit around the earth where they would collect intense solar energy, generate electricity, and send it back to earth via microwaves.

Solar energy is clean energy. It does not produce air or water pollution. Nor does it add additional heat to the atmosphere, since radiant energy is eventually converted to heat anyway. The problem of storage can be overcome in a number of ways, including the use of solar energy to produce hydrogen by means of hydrolysis of water. Hydrogen as either a gas or a liquid can be piped into existing pipelines and used as fuel for automobiles and airplanes. When it is burned, it forms fog, not smog (fig. 33.14).

Some types of renewable energy sources have been utilized for quite some time. *Falling water* is used to produce electricity in hydroelectric plants. *Geothermal energy* is trapped heat produced by radioactive material deep beneath the surface of the earth. Water converted to steam by this heat, may be pumped up and used to heat buildings or to generate electricity. *Wind power* provides enough force to turn vanes, blades, or propellers attached to a shaft, which, in turn, spins the motor of a generator that produces electricity. The government has allocated a small amount of money for wind research, particularly the promotion of large windmills, such as those being constructed in the windy

Figure 33.13
Solar electric power plant. a. *Overview of Solar One, which has a field of 1,818 heliostats.*
b. *Closeup of one heliostat, consisting of six sun-tracking mirrors on each side of a central support.*

a.

b.

Columbia River valley in the Pacific Northwest. Many others believe it would be best to build numerous small windmills, such as those found in the first wind farm now operating in New Hampshire.

All of the renewable resources (falling water, geothermal, wind, etc.) still hold some promise. Solar One is now in operation and solar cells are being perfected. Solar energy can be used to hydrolyze water to produce hydrogen gas or liquid, which can be substituted for fossil fuels.

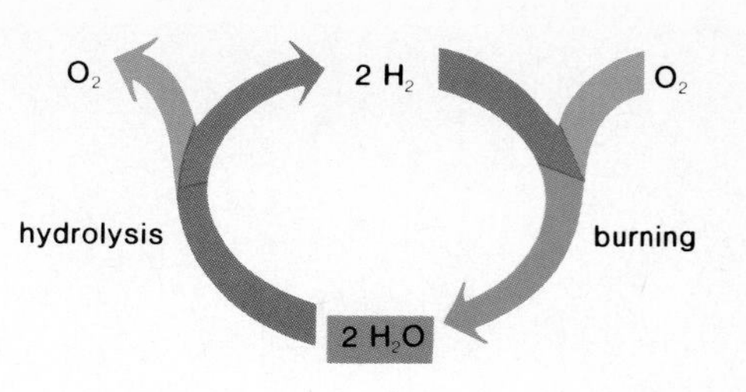

Figure 33.14
Hydrogen as a fuel. When hydrogen is burned, the product is water and water can be hydrolyzed, using solar energy, to give hydrogen again. This cycle makes hydrogen an attractive fuel.

Food

There is great concern that it will not be possible to provide enough food to feed the population increases that are expected well into the twenty-first century.

Figures 33.15a and 33.15c show that the *production of food* has increased in the developed countries and the per capita share has also increased. This means that food production has kept pace with population increase in the developing countries where the majority of people can afford the price of food and are eating better than formerly. Figure 33.15b and figure 33.15d show that the production of food has increased in the developing countries, but the per capita share has not substantially increased because increased food production must be divided by a large population increase. Further, in the developing countries only a segment of the population can afford the price of food and are eating well; a larger portion cannot afford the price of food and are not eating adequately. (Since the per capita share is calculated by dividing the total production by the total population size, it does not necessarily mean that all persons are eating an equal amount.) While some undernourished individuals live in the developed countries, the majority live in the developing countries. In particular, Africa, with the fastest population growth ever recorded, has had difficulty providing enough food for its populace. Of the many factors involved in producing enough food, the following are most often discussed.

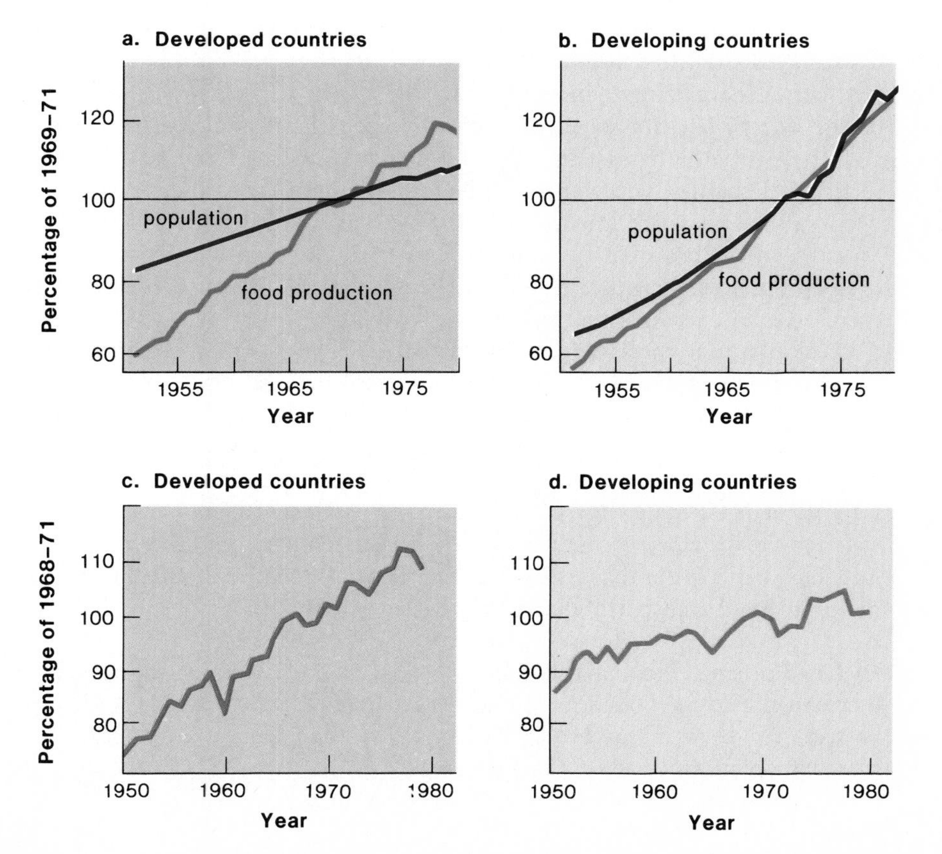

Figure 33.15
Food production in developed versus developing countries. a. *and* b. *Food production has increased at similar rates in the developed and developing countries over the last three decades, but population growth has continued at a faster rate in developing than in developed countries; therefore* (c *and* d), *there has been a per capita gain in the developed countries but not in the developing countries.*

Figure 33.16
Experimental rice plots. Hybridization studies are continuing in order to produce high-yield rice plants that are also resistant to pests and environmental stresses.

Factors in Food Production Today, most land is already being used for one purpose or another, and it would be difficult to expand the amount of farmland. Only in the tropics (sub-Saharan Africa and the Amazon Basin of Brazil) are there still sizable tracts of land not presently utilized that have enough water to grow crops. Slash-and-burn agriculture has been traditional in this area because the soil contains so few nutrients. The jungle is cut and the debris burned, providing the nutrients to allow a few years' crops. When the soil can no longer support continued cultivation, a new area is selected for cultivation. Attempts to cultivate the land in a traditional manner have proven difficult because the soil tends to become hard and compact when exposed to the sun. If these problems could be solved, the food supply would be multiplied manyfold. Even so, many believe that tropical forests should remain as they are because they help control global pollution.

Since virtually all readily available, relatively fertile cropland is already in use, more effort should be directed toward safekeeping the land already being cultivated. Millions of acres of fertile lands are lost each year from encroachment by cities and suburbs, industrial development, soil erosion, and desertification. The United Nations estimates that by the end of this century twice as much land will be lost due to these factors than will be developed as farmland.

The developed countries, especially the United States, have had spectacular success in increasing yields by utilizing monoculture agriculture and this method of agriculture, along with its many environmental problems, is now being exported to the developing countries. The term "Green Revolution" was coined to describe the introduction and rapid spread of high-yielding wheat and rice plants (fig. 33.16) that are especially developed to grow in warmer countries even when they are not supplied with generous amounts of fertilizer and water. Research efforts are currently striving toward providing plants with improved internal efficiency. The new focus is on greater photosynthetic efficiency, more efficient nutrient and water uptake, improved biological nitrogen fixation, and genetic resistance to pests and environmental stress.

The application of fertilizer has contributed greatly to increased yield. When fertilizer is first applied to soil, there is a dramatic increase in yield, but later, as more and more fertilizer is used, the increase in yield drops off. This suggests that the developed countries could very well do with less fertilizer, but the developing countries need to apply more. Unfortunately, the rising cost of energy has caused fertilizer cost to more than double and developing countries will most likely find it difficult to make or buy adequate amounts of fertilizer. Legumes, you will recall, increase the nitrogen content of the soil because their roots are infected with nitrogen-fixing bacteria. Research goes forward on the possibility of infecting more plants with these bacteria and even on the possibility of transferring the nitrogen-fixing genes to plant cells themselves by means of recombinant DNA methods.

High agricultural yields are extremely dependent on both indirect and direct uses of energy. In the United States, fossil fuels have supplied the supplemental energy, as is apparent when one considers that nearly all of the large farm machinery utilizes fossil fuel energy. In Japan, however, much of the supplemental energy required for high yields is provided by human labor. Since we are now entering a time when fossil fuels are in short supply, it would be wise for the developing countries to rely on human energy to increase yield. Keeping people in rural areas close to the source of food would also help solve the problem of transportation and packaging.

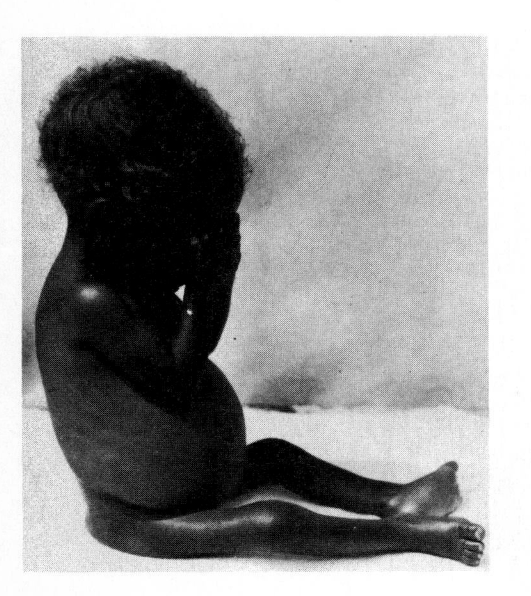

Figure 33.17
Kwashiorkor is an illness caused by lack of protein. Characteristically, there is edema due to lack of plasma proteins. The nervous system of these children may also fail to develop, which frequently results in mental retardation.

Dietary Protein Providing sufficient calories will not assure an adequate diet; there must also be enough protein in the food eaten. Due to a lack of dietary protein, children in the developing countries often have the symptoms of kwashiorkor in which the entire body is bloated, the skin is discolored, a rash is present, and the hair has an orange-reddish tinge (fig. 33.17).

Plants are low-quality, nutritionally incomplete protein sources because, although they contain some amino acids, they lack others. It is possible, however, to eat a combination of plants so that together they contain an acceptable level of all the essential amino acids. For example, wheat and beans complement each other to give a balance that is comparable to a high-quality protein such as meat and cheese.

In the developed countries most of the grain produced is fed to cattle, pigs, or chickens, which then serve as a source of high-quality protein. In many developing countries most grain must be consumed directly in order to provide sufficient calories. Therefore, beef is not expected to supply the necessary protein. Another possibility is fish consumption. Between 1950 and 1970, fish supplied an increasing portion of the human diet until the catch averaged some 18.5 kilograms per person annually. Since that time, although the catch has increased, the per capita share has decreased due to population growth. Also, overexploitation of the ocean's fisheries and pollution of coastal waters may eventually cause a decline in the yearly catch. Therefore, the expansion of aquaculture—the cultivation of fish—is highly desirable.

The production of food has become increasingly dependent on supplements, such as fossil fuel energy and fertilizer, because the amount of new agricultural land that can be cultivated is limited. Development of new high-yield plants may help the developing countries feed expected increases in population, but there may still be a problem regarding dietary protein.

Steady State

Presently, the world's population increases in size each year and an ever greater supply of energy and food is needed each year. Does this pattern have to continue in order to assure economic growth and our well-being? On the contrary, it seems unlikely that this exponential increase can long continue. The pollution caused by energy consumption and the production of goods has increased so much that it can no longer be ignored, and the once-hidden cost of pollution has now become apparent. The human population, like other populations of the biosphere, could exist in a steady state, with no increase in number of people or in resource consumption.

A stable population size would be a new experience for humans. Figure 33.18 shows the age structure for a hypothetical stable population. As you can see,

1. Over 40 percent of the people would be fairly youthful, with only about 15 percent being in the senior citizen category. Further, a combination of good health habits and advances in medical science could well mean that people would remain more youthful and more productive for a longer time than is common today.
2. There would be proportionately fewer children and teenagers than in a rapidly expanding population. This might mean a reduction in automobile accidents and certain types of crime statistically related to the teenage years.
3. There might be increased employment opportunities for women and a generally less competitive workplace, since newly qualified workers would enter the job market at a more moderate rate.
4. Creativity need not be impaired. A study of Nobel Prize winners showed that the average age at which the prize-winning work was done was over 30.
5. The quality of life for children might increase substantially, since fewer unwanted babies might be born, and the opportunity would exist for each child to receive the loving attention it needs.

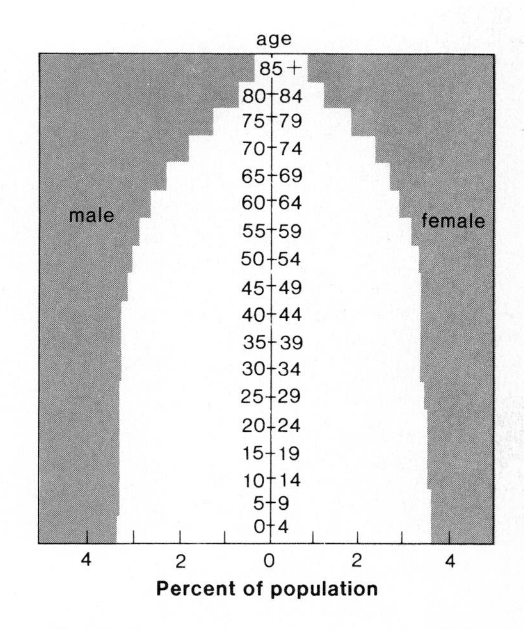

Figure 33.18
This diagram shows the age and sex structure of a hypothetical human population that remains the same size each year.
Source: United States Bureau of the Census.

Figure 33.19
In the steady state, environmental preservation will be an important consideration.

The economy, like the population, should show no growth. For example, if the population size was to increase, the resource consumption per person should decline. In this way the amount of resource consumption would remain constant. In order to ensure that people have the goods and services they need for a comfortable but not necessarily luxurious life, goods should have a long lifetime expectancy. Frugality is envisioned as absolutely necessary to the steady-state economy. Also, materials must be recycled so that they can be used over and over again. In order to provide jobs, technology could sometimes be labor intensive instead of energy intensive. The steady state would have need of very sophisticated technology, but it is hoped that the technology will not exploit the environment; instead, it is hoped that the technology will work with the environment.

Environmental preservation would be the most important consideration in a steady-state world. Renewable energy sources, such as solar energy, would play a greater role in providing energy needs. Pollution would be minimized. Ecological diversity would be maintained and overexploitation would cease (fig. 33.19). Ecological principles would serve as guidelines for specialists in all fields, creating a unified approach to the environment. In a steady-state world all people would strive to be aware of the environmental consequences of their actions, consciously working toward achieving balance in the ecosystems of their planet.

In a steady-state society, there would be no yearly increase in population nor resource consumption. It is forecast that under these circumstances the quality of life would improve.

What would our culture be like if we had steady-state manufacturing and a steady-state population? Perhaps it would be greatly improved. Certainly there are no limits to growth in knowledge, education, art, music, scientific research, human rights, justice, and cooperative human interactions. In a steady-state world the general sense of fearful competition among peoples might diminish, allowing human compassion and creativity to prosper as never before.

Summary

The human population is expanding exponentially and even though the growth rate has declined, there is a large increase in the population each year—the doubling time is now about 40 years. The developed countries underwent a demographic transition between 1950 and 1975, but the developing countries are just now undergoing demographic transition. In these countries where the average age is less than 15, it will be many years before reproduction replacement will mean zero population growth.

Resource consumption is dependent on population size and on consumption per individual. The developing countries are responsible for the first type of environmental impact and the developed countries for the second type. According to the Malthusian view, we are running out of nonrenewable resources and only conservation can help extend their depletion curves. According to the Cornucopian view, we will continue to find new ways to exploit the environment.

Exploitation of the environment is dependent on a plentiful energy source. Of the nonrenewable sources, coal is in ready supply and can be utilized without developing new technologies. Each of the three types of nuclear power has a drawback that has prevented full-scale operation in this country. All of the renewable resources hold some promise but haven't been fully developed yet. Solar One is now in operation and solar cells are being perfected.

The production of food has become increasingly dependent on supplements, such as fossil fuel energy and fertilizer. Development of new high-yield plants have helped developing countries feed their increasing population, but an increased supply of protein remains a problem.

Objective Questions

1. After a country has undergone the demographic transition, both the death rate and the birthrate are ____________ (choose high or low).
2. In contrast to the developed countries, the developing countries are not as yet ____________.
3. The people of the developed countries have a higher standard of living and consume more ____________ than do those of developing countries.
4. Those who hold a Cornucopian view concerning nonrenewable resources believe that ____________ ____________.
5. For the past several years, the nonrenewable energy source of choice has been ____________.
6. The steady accumulation of ____________ wastes is a deterrent to the use of nuclear power.
7. Solar power can be used to hydrolyze water and the resulting ____________ can be a gaseous or liquid fuel.
8. The continent least able to feed its ever-growing populace is ____________.
9. One of the chief problems in regard to a proper diet in developing countries is supplying adequate ____________.
10. Overexploitation and pollution may very well cause a ____________ in the size of the yearly fish catch.

Answers to Objective Questions
1. low 2. industrialized 3. resources 4. there is no limit to the amount of resources that can be consumed 5. oil 6. radioactive 7. hydrogen 8. Africa 9. protein 10. decline

Study Questions

1. Define exponential growth. (p. 742) Draw a growth curve to represent exponential growth and explain why a curve representing population growth usually levels off. (p. 744)
2. Calculate the growth rate and doubling time for a population in which the birthrate is 20 and the death rate is 2 per 1000. (p. 743)
3. Define demographic transition. When did the developed countries undergo demographic transition? When did the developing countries undergo demographic transition? (pp. 746–747)
4. Give at least three differences between the developed countries and the developing countries. (pp. 745–749)
5. Contrast the Malthusian view and the Cornucopian view toward nonrenewable supplies. (p. 750)
6. Draw a typical depletion curve and relate it to the consumption of fossil fuels. (pp. 750, 752)
7. Name the three types of nuclear power and give at least one related drawback to each. (pp. 752–753)
8. Name at least four types of renewable energy resources. What types of fuels might be produced from these sources? (p. 754)
9. Give at least three reasons why intensive monoculture does not seem to be a feasible solution to the food crises of developing countries. (pp. 755–756)
10. Discuss the various characteristics of a steady state and list several advantages to having a human ecosystem with these characteristics. (pp. 757–758)

Key Terms

biotic potential (bi-ot'ik po-ten'shal) the maximum population growth rate under ideal conditions. *744*

breeder fission reactor (brēd'er fish'un re-ak'tor) a nuclear reactor that produces more nuclear fuel than it consumes by converting nuclear wastes into ^{239} plutonium. *753*

carrying capacity (kar'e-ing kah-pas'ĭ-te) the largest number of organisms of a particular species that can be maintained indefinitely in an ecosystem. *744*

demographic transition (dem-o-graf'ik tran-zĭ'shun) the change from a high birthrate to a low birthrate so that the growth rate is lowered. *746*

depletion curve (de-ple'shun kerv) graphical depiction of a resource's dwindling amount over time. *750*

developed countries (de-vel'opt kun'trēz) industrialized nations that typically have a strong economy and a low rate of population growth. *745*

developing countries (de-vel'op-ing kun'trēz) nations that are not yet industrialized and have a weak economy and a high rate of population growth. *747*

doubling time (dŭ'b'l-ing tīm) the number of years it takes for a population to double in size. *743*

environmental resistance (en-vi''ron-men'tal re-zis'tans) sum total of factors in the environment that limit the numerical increase of a population in a particular region. *744*

exponential growth (eks''po-nen'shal grōth) growth, particularly of a population, in which the increase occurs in the same manner as compound interest. *742*

fission reactor (fish'un re-ak'tor) a nuclear reactor that splits ^{235}uranium in order to release energy. *752*

growth rate (grōth rāt) percentage of increase or decrease in the size of a yearly population. *743*

heliostat (he'le-o-stat'') large-field mirrors that track the sun and reflect its energy onto a mounted boiler for solar heating. *754*

mill tailings (mil tāl'ingz) radioactive material in the waste produced by mining and processing uranium ores. *753*

nonrenewable resource (non-re-nu'ah-b'l re'sors) a resource that can be used up or at least depleted to such an extent that further recovery is too expensive. *750*

nuclear fusion (nu'kle-ar fu'zhun) a process in which the nuclei of two atoms are forced together to form the nucleus of a heavier atom with the release of substantial amounts of energy. *753*

photovoltaic (solar) cells (fo''to-vol-táik selz) a manufactured mechanism that uses sunlight to produce an electromotive force. *754*

renewable resources (re-nu'ah-b'l re'sors) a resource that is not used up because it is continually produced in the environment. *750*

replacement reproduction (re-plās'ment re''pro-duk'shun) a population whose average number of children is one per person. *748*

solar collector (so'lar kō-lek'ter) any manufactured device that absorbs radiant energy. *754*

Further Readings for Part Six

Batie, S. S., and R. G. Healy. 1983. The future of American agriculture. *Scientific American* 248(2):45.

Beddington, J. R., and R. M. May. 1982. The harvesting of interacting species in a natural ecosystem. *Scientific American* 247(5):62.

Bergerud, A. T. 1983. Prey switching in a simple ecosystem. *Scientific American* 249(6):130.

Borgese, E. M. 1983. The law of the sea. *Scientific American* 248(3):42.

Brown, L. R., et al. 1986. *State of the world; 1986.* New York: W. W. Norton and Co.

Cloud, P. 1983. The biosphere. *Scientific American* 249(3):176.

Donaldson, L. R., and T. Joyner. 1983. The Salmonid fishes as a natural livestock. *Scientific American* 249(1):50.

Environment. All issues of this journal contain articles covering modern ecological problems.

Gose, J. R., R. T. Holmes, G. E. Likens, and F. H. Bormann. 1978. The flow of energy in a forest ecosystem. *Scientific American* 238(3):92.

Gwatkin, D. R. 1982. Life expectancy and population growth in the third world. *Scientific American* 246(5):57.

Hamakawa, Y. 1987. Photovoltaic power. *Scientific American* 256(4):86.

Hinman, C. W. 1986. Potential new crops. *Scientific American.* 255(1):32.

Keely, C. B. 1982. Illegal migration. *Scientific American* 246(3):41.

Kormondy, E. J. 1984. *Concepts of ecology.* 3d ed. Englewood Cliffs, NJ: Prentice-Hall.

Lester, R. K. 1986. Rethinking nuclear power. *Scientific American* 254(3):31.

Lewontin, R. C. 1978. Adaptation. *Scientific American* 239(3):212.

Martin, P. I. 1983. Labor-intensive agriculture. *Scientific American* 249(4):54.

Miller, G. T. 1981. *Living in the environment.* 3d ed. Belmont, CA: Wadsworth.

Moran, J. M., et al. 1980. *Introduction to environmental science.* San Francisco: W. H. Freeman.

Moretti, P. M., and V. D. Louis. 1986. Modern windmills. *Scientific American.* 254(6):110.

Nebel, B. J. 1986. *Environmental science: The way the world works.* 2d ed. Englewood Cliffs, NJ: Prentice-Hall.

Odum, E. P. 1983. *Basic ecology.* New York: CBS College Publishing.

Power, J. F., and R. F. Follett. 1987. Monoculture. *Scientific American.* 256(3):78.

Putman, R. J., and S. D. Wratten. 1984. *Principles of ecology.* Berkeley and Los Angeles: University of California Press.

Rasmussen, E. D. 1982. The mechanization of agriculture. *Scientific American* 247(3):76.

Revelle, R. 1982. Carbon dioxide and world climate. *Scientific American* 247(2):35.

Ricklefs, Robert E. 1983. *The economy of nature: A textbook in basic ecology.* New York: Chiron Press.

Shaw, R. W. 1987. Pollution by particles. *Scientific American.* 257(2):96.

Smith, R. L. 1985. *Ecology and field biology.* 3d ed. New York: Harper & Row.

Swaminathan, M. S. 1984. Rice. *Scientific American* 250(1):80.

Turk, J., and A. Turk. 1984. *Environmental science.* New York: CBS College Publishing.

Whittaker, R. H. 1975. *Communities and ecosystems.* 2d ed. New York: Macmillan.

Appendix A

Table of Chemical Elements

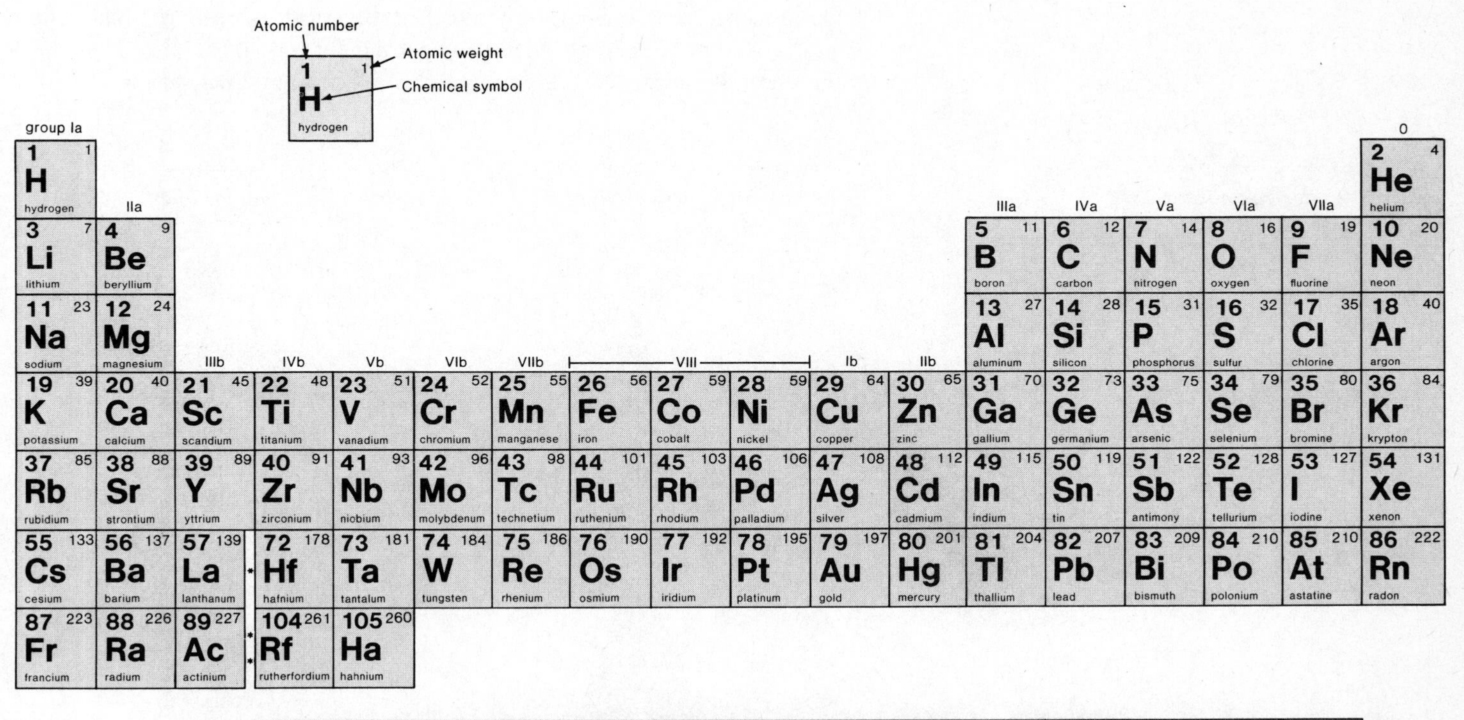

Appendix B

Microscopic Observations

Light versus Electron Microscope

In a light microscope, a concentrated beam of light passes through the object on the stage in such a way that the emerging rays of light indicate the areas of light and dark in the object. Very often the prepared specimen has been treated with a stain to enhance these contrasts. A magnified image of the object is achieved when these rays of light are focused by the objective lens; the image is further magnified by means of the ocular lens. Since eyes are sensitive to light rays, the image may be viewed directly by the experimenter.

In an electron microscope, a concentrated beam of electrons passes through a vacuum to bombard the object. The vacuum is created when air is pumped out of the tube transmitting the electrons. The electrons are scattered by the object in such a way as to indicate areas of light and dark. The electrons are focused, and a magnified image is formed as the electrons pass the lenses of the electron microscope. The lenses of an electron microscope are not glass; they are coils of wire about which a magnetic field exists because of an electric current that travels within the wire. This field can focus electrons because they carry a negative charge that makes them sensitive to magnetic fields.

Since eyes are not sensitive to electrons, the image cannot be viewed directly; instead, it is projected onto a screen at the foot of the microscope where the electrons excite the chemical coating of the screen, producing light rays that can be seen by the viewer. A permanent record can also be made on a photographic plate, and this record is an **electron micrograph.**

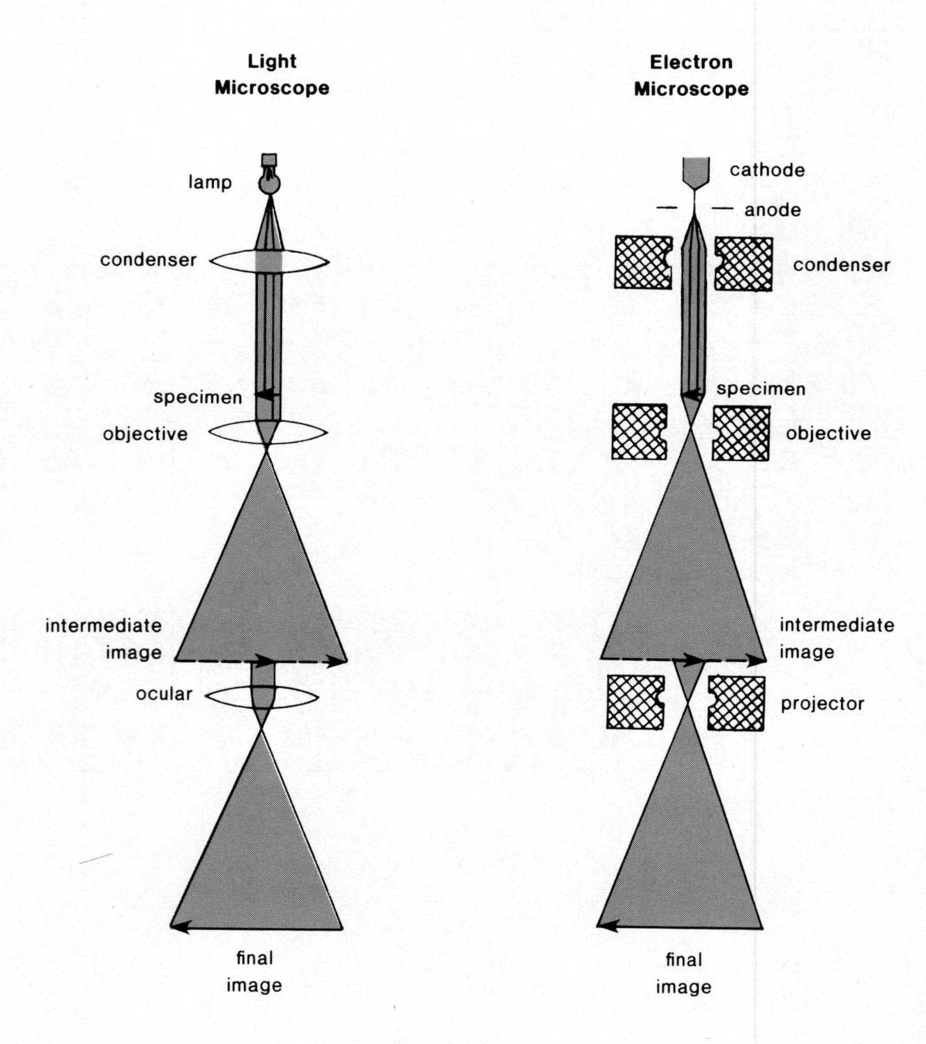

Figure B.1
Light versus electron microscope.

A comparison of these two microscopes shows that they both illuminate the object, magnify the object, and produce an image that eventually can be viewed by the observer. The most important difference between these two instruments is not the degree to which they magnify but lies instead in their resolving power. This term is used to indicate the amount of detail that can be distinguished by a microscope. The physical laws of optics tell us that resolving power is dependent on the wavelength of the light or electron beam. The theoretical limit of the resolving power of the light microscope is 200 nm(nanometers), while that of the electron microscope is 0.5 nm. This means that any two structures separated by less than 200 nm in the light microscope and 0.5 nm in the electron microscope will appear as one object. Thus, the electron microscope allows us to see much more.

One drawback to the electron microscope has been that the specimen could be viewed only after drying, because a vacuum is needed to produce the electron beam. We do not know the full extent to which this drying process distorts the true appearance of the specimen. Methods are now being investigated to allow the observation of living materials in their natural state.

The Metric System

As mentioned in the text, biologists describe the sizes of objects such as cell organelles in terms of the metric system. The linear units of measurement most often used are given in table B.1. The ruler (fig. B.2) allows you to visualize the relationship of these units.

Table B.1 *Units of Measurement*

Unit	Symbol	Seen by
Centimeter	cm = .01 m	Naked eye
Millimeter	mm = 0.1 cm	Naked eye
Micrometer	μm = 0.001 mm	Light microscope
Nanometer	nm = 0.001 μm	Electron microscope

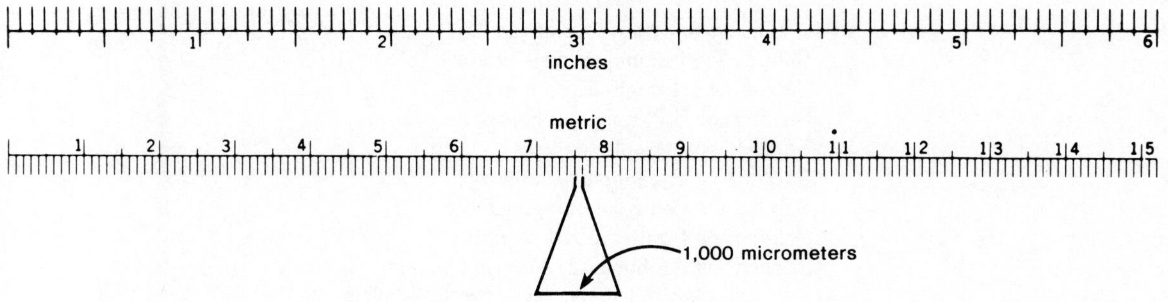

Figure B.2
Six-inch ruler with equivalent centimeter units.

Appendix C

Classification of Organisms

This classification system given here has been simplified but contains the major kingdoms and divisions (called phyla in the kingdom Protista and kingdom Animalia). The text does not discuss all the divisions and phyla listed here.

Kingdom Monera

Prokaryotic unicellular organisms. Nutrition principally by absorption, but some are photosynthetic or chemosynthetic.

Division Archaebacteria: methanogens, halophiles, and thermoacidophiles
Division Eubacteria: true bacteria including cyanobacteria (formerly called blue-green algae)

Kingdom Protista

Eukaryotic unicellular organisms (and the most closely related multicellular forms). Nutrition by photosynthesis, absorption, or ingestion.

Phylum Chrysophyta: diatoms
Phylum Euglenophyta: *Euglena* and relatives
Phylum Pyrrophyta: dinoflagellates
Phylum Zoomastigina: flagellated protozoans
Phylum Sarcodina: amoeboid protozoans
Phylum Ciliophora: ciliated protozoans
Phylum Sporozoa: parasitic protozoans

Kindgom Fungi

Eukaryotic, usually having haploid or multinucleated hyphal filaments: spore formation during both asexual and sexual reproduction. Nutrition principally by absorption.

Division Oomycota: water molds
Division Zygomycota: black bread molds
Division Ascomycota: sac fungi
Division Basidiomycota: club fungi
Division Deuteromycota: imperfect fungi—i.e., not known to reproduce sexually.
Division Gymnomycota: slime molds

Kingdom Plantae

Eukaryotic multicellular organisms (and the most closely related unicellular forms) with rigid cellulose cell walls and chlorophyll a and b. Nutrition principally by photosynthesis. Starch serves as the food reserve.

Division Rhodophyta: red algae
Division Phaeophyta: brown algae
Division Charophyta: stoneworts
Division Chlorophyta: green algae
Division Bryophyta: mosses and liverworts
Division Tracheophyta: vascular plants
- Subdivision Psilophyta: whisk ferns
- Subdivision Lycophyta: club mosses
- Subdivision Sphenophyta: horsetails
- Subdivision Pterophyta: ferns
- Subdivision Cycadophyta: cycads
- Subdivision Ginkgophyta: gingkoes
- Subdivision Coniferophyta: conifers
- Subdivision Anthophyta: flowering plants
  - Class Dicotyledonae: most flowering plants
  - Class Monocotyledonae: grasses, lilies, orchids

Kingdom Animalia

Eukaryotic, usually motile, multicellular organisms without cell walls or chlorophyll. Nutrition principally ingestive, with digestion in an internal cavity.

Phylum Porifera: sponges
Phylum Cnidaria: radially symmetrical marine animals
- Class Hydrozoa: *Hydra,* Portuguese man-of-war
- Class Scyphozoa: jellyfish
- Class Anthozoa: sea anemones and corals

Phylum Platyhelminthes: flatworms
- Class Turbellaria: free-living flatworms
- Class Trematoda: parasitic flukes
- Class Cestoda: parasitic tapeworms

Phylum Nematoda: roundworms
Phylum Rotifera: rotifers
Phylum Mollusca: soft-bodied, unsegmented animals
- Class Polyplacophora: chitons
- Class Monoplacophora: *Neopilina*
- Class Gastropoda: snails and slugs
- Class Bivalvia: clams and mussels
- Class Cephalopoda: squids and octopuses

Phylum Annelida: segmented worms
- Class Polychaeta: sandworms
- Class Oligochaeta: earthworms
- Class Hirudinea: leeches

Phylum Arthropoda: joint-legged animals; exoskeleton
- Class Crustacea: lobsters, crabs, barnacles
- Class Arachnida: spiders, scorpions, ticks
- Class Chilopoda: centipedes
- Class Diplopoda: millipedes
- Class Insecta: grasshoppers, termites, beetles

Phylum Echinodermata: marine; spiny, radially symmetrical animals
- Class Crinoidea: sea lilies and feather stars
- Class Asteroidea: sea stars
- Class Ophiuroidea: brittle stars
- Class Echinoidea: sea urchins and sand dollars
- Class Holothuroidea: sea cucumbers

Phylum Hemichordata: acorn worms
Phylum Chordata: dorsal supporting rod (notochord) at some stage; dorsal hollow nerve cord; pharyngeal pouches or slits
- Subphylum Urochordata: tunicates
- Subphylum Cephalochordata: lancelets
- Subphylum Vertebrata: vertebrates
  - Class Agnatha: jawless fishes (lampreys, hagfishes)
  - Class Chondrichthyes: cartilaginous fishes (sharks, rays)
  - Class Osteichthyes: bony fishes
    - Subclass Crossopterygii: lobe-finned fishes
    - Subclass Dipnoi: lungfishes
    - Subclass Actinopterygii: ray-finned fishes
  - Class Amphibia: frogs, toads, salamanders
  - Class Reptilia: snakes, lizards, turtles
  - Class Aves: birds

Class Mammalia: mammals
- Subclass Prototheria: egg-laying mammals
  - Order Monotremata: duckbilled platypus, spiny anteater
- Subclass Metatheria: marsupial mammals
  - Order Marsupialia: opossums, kangaroos
- Subclass Eutheria: placental mammals
  - Order Insectivora: shrews, moles
  - Order Chiroptera: bats
  - Order Edentata: anteaters, armadillos
  - Order Rodentia: rats, mice, squirrels
  - Order Lagomorpha: rabbits and hares
  - Order Cetacea: whales, dolphins, porpoises
  - Order Carnivora: dogs, bears, weasels, cats, skunks
  - Order Proboscidea: elephants
  - Order Sirenia: manatees
  - Order Perissodactyla: horse, hippopotamus, zebra
  - Order Artiodactyla: pigs, deer, cattle
  - Order Primates: lemurs, monkeys, apes, humans
    - Suborder Prosimii: lemurs, tree shrews, tarsiers, lorises, pottos
    - Suborder Anthropoidea: monkeys, apes, humans
      - Superfamily Ceboidea: New world monkeys
      - Superfamily Cercopithecoidea: Old world monkeys
      - Superfamily Hominoidea: apes and humans
        - Family Hylobatidae: gibbons
        - Family Pongidae: chimpanzee, gorilla, orangutan
        - Family Hominidae: *Australopithecus,** *Homo erectus,** *Homo sapiens* neanderthalis,* *Homo sapiens sapiens*

*extinct

Glossary

A

abiotic (ab″e-ot′ik) not including living organisms; especially the nonliving portions of the environment. *688*

abortion (ah-bor′shun) artificial removal of an embryo or fetus from the womb. *408*

accommodation (ah-kom″o-da′shun) lens adjustment in order to see close objects. *358*

acetylcholine (as″ĕ-til-ko′lēn) ACh; a neurotransmitter substance secreted at the ends of many neurons; responsible for the transmission of a nerve impulse across a synaptic cleft. *314*

acetylcholinesterase (as″ĕ-til-ko″lin-es′ter-ās) AChE; an enzyme that breaks down acetylcholine. *314*

acid (as′id) a solution in which pH is less than 7; a substance that contributes or liberates hydrogen ions (protons) in a solution. *25*

acromegaly (ak″ro-meg′ah-le) condition resulting from an increase in growth hormone production after adult height has been achieved. *376*

acrosome (ak′ro-sōm) covering on the tip of a sperm that contains enzymes necessary for fertilization. *394*

ACTH (adrenocorticotropic hormone); hormone secreted by the anterior lobe of the pituitary gland that stimulates activity in the adrenal cortex. *377*

actin (ak′tin) one of the two major proteins of muscle; makes up thin filaments in myofibrils of muscle fibers. *See* myosin. *346*

action potential (ak′shun po-ten′shal) the change in potential propagated along the membrane of a neuron; the nerve impulse. *313*

active acetate (ak′tiv as′ĕ-tāt) an acetyl group attached to coenzyme A; a product of the transition reaction that links glycolysis to the Krebs cycle. *104*

active site (ak′tiv sīt) the region on the surface of an enzyme where the substrate binds and where the reaction occurs. *95*

active transport (ak′tiv trans′port) transfer of a substance into or out of a cell from a lesser to a greater concentration by a process that requires a carrier and expenditure of energy. *69*

adaptation (ad″ap-ta′shun) the fitness of an organism for its environment, including the process by which it becomes fit, in order that it may survive and reproduce; also the adjustment of sense receptors to a stimulus so that the stimulus no longer excites them. *353, 516*

adenosine triphosphate (ah-den′o-sēn tri-fos′fāt) *See* ATP. *38*

adrenalin (ah-dren′ah-lin) a hormone produced by the adrenal medulla that stimulates "fight or flight" reactions. Also called epinephrine. *380*

adrenocorticotropic hormone (ah-dre″no-kor″te-ko-trop′ik hor′mōn) *See* ACTH. *377*

adventitious root (ad″ven-tish′us ro͞ot) roots that develop from, for example, the stem which are not part of the embryonic root system. *136*

aerobic (a″er-ōb′ik) growing or metabolizing only in the presence of oxygen, as in aerobic respiration. *100, 534*

agglutination (ah-gloo″ti-na′shun) clumping of cells, particularly in reference to red cells involved in an antigen-antibody reaction. *246*

aggression (ah-gresh′un) belligerent behavior. *661*

agranulocytes (ah-gran′u-lo-sīts″) white blood cells that do not contain distinctive granules. *240*

aldosterone (al″do-ster′ōn) a hormone, secreted by the adrenal cortex that functions in regulating sodium and potassium concentrations of the blood. *381*

allantois (ah-lan′to-is) one of the extraembryonic membranes; in reptiles and birds, a pouch serving as a repository for nitrogenous waste; in mammals a source of blood vessels to and from the placenta. *427*

allele (ah-lēl′) an alternative form of a gene that occurs at a given chromosome site (locus). *445*

all-or-none response (al′or-nun′ re-spons′) phenomenon in which a muscle fiber contracts completely when it is exposed to a stimulus of threshold strength. *342*

alpha helix (al′fah he′liks) common right-handed spiral found in proteins and DNA. *28*

alternation of generations (awl″ter-na′shun uv jen″ĕ-ra′shunz) two-phased life cycle displayed by many plants in which there are sporophyte and gametophyte generations. *153, 561*

altruism (al′troo-izm) behavior performed for the benefit of others without regard to its possible detrimental effect on the performer. *665*

alveoli (al-ve′o-lī) saclike structures that are the air sacs of a lung. *274*

amino acid (ah-me′no as′id) a unit of protein that takes its name from the fact that it contains an amino group (NH_2) and an acid group (COOH). *28*

ammonia (ah-mo′ne-ah) NH_3, a nitrogenous waste product resulting from deamination of amino acids. *291*

amnion (am′ne-on) an extraembryonic membrane, a sac around the embryo that contains fluid. *427*

amoeba (ah-me′bah) protozoans that move by means of pseudopodia. *548*

ampulla (am-pul′lah) base of a semicircular canal in the inner ear. *364*

amylase (am′i-lās) an enzyme that catalyzes chemical breakdown of starch to maltose; salivary amylase works in the mouth, and pancreatic amylase works in the small intestine. *000*

anaerobic respiration (an-a″er-ōb′ik res″pī-ra′shun) growing or metabolizing in the absence of molecular oxygen. *531*

analogous (ah-nal′o-gus) similar in function but not in structure; particularly in reference to similar adaptations. *508*

anaphase (an′ah-fāz) stage in mitosis during which chromatids separate, forming chromosomes. *82*

annual rings (an′u-al ringz) yearly rings formed by buildup of secondary xylem in trees. *141*

antennae (an-ten′e) sensory organs located on arthropod head. *605*

anterior pituitary (an-te′re-or pī-tu′ĭ-tār″e) a portion of the pituitary gland which produces six hormones and is controlled by a hypothalamic releasing hormone. *373*

anther (an′ther) that portion of a stamen in which pollen is formed. *152*

antheridium (an″ther-id′e-um) male organ in certain nonseed plants where swimming sperm are produced. *568*

anthropoids (an′thro-poidz) higher primates, including only monkeys, apes, and humans. *625*

antibody (an′tī-bod″e) a protein produced in response to the presence of some foreign substance in the blood or tissues. *241*

anticodon (an″tī-ko′don) a "triplet" of three nucleotides in transfer RNA that pairs with a complementary triplet (codon) in messenger RNA. *486*

antidiuretic hormone (an″tī-di″u-ret′ik hōr′mōn) (ADH); sometimes called vasopressin, a hormone secreted by the posterior pituitary that controls the rate at which water is reabsorbed by the kidneys. *303, 374*

antigen (an′tī-jen) a foreign substance, usually a protein, that stimulates the immune system to produce antibodies. *241*

antipredator defense (an″tī-pred′ah-tor de-fens′) a physiological or behavioral activity or a structural modification that protects an organism from its predators. *676*

anus (a′nus) inferior outlet of the digestive tube. *193*

aorta (a-or′tah) major systemic artery that receives blood from the left ventricle. *212*

apoenzyme (ap″o-en′zīm) protein portion of an enzyme. *97*

appeasement (ah-pēz′ment) submissive behavior usually causing a cessation of aggression. *662*

appendicular skeleton (ap″en-dik′u-lar skel′ĕ-ton) portion of the skeleton forming the upper extremities, pectoral girdle, lower extremities, and pelvic girdle. *335*

appendix (ah-pen′diks) a small, tubular appendage that extends outward from the cecum of the large intestine. *194*

aqueous humor (a′kwe-us hu′mor) watery fluid that fills the anterior and posterior chambers of the eye. *356*

archaebacteria (ar′ke-bak-te′re-ah) monerans that are able to live under adverse circumstances and represent an early branch of living organisms. *547*

archegonium (ar″kē-go′ne-um) female organ in certain nonseed plants where an egg is produced. *568*

archenteron (ar-ken′ter-on) central cavity or primitive gut in the animal embryo. *422*

arterial duct (ar-te′re-al dukt) ductus arteriosus; fetal connection between the pulmonary artery and the aorta. *438*

arteriole (ar-te′re-ōl) a vessel that takes blood from an artery to a capillary. *208*

artery (ar′ter-e) a vessel that takes blood away from the heart; characteristically possessing thick elastic walls. *208*

ascending reticular activating system (ah-send′ing rē-tik′u-lar ak′tī-vāt″ing sis′tem) ARAS; system by which the thalamus is connected to various parts of the brain and composed of the diffuse thalamic projection system and the reticular formation. *323*

asci (as′i) saclike cells that produce ascospores during sexual reproduction of sac fungi. *554*

associative learning (ah-so′se-ah-tiv″ lern′ing) the learning of a set response to irrelevant stimuli. *654*

aster (as′ter) short rays of microtubules that appear at the ends of the spindle apparatus in animal cells during cell division. *90*

asymmetry (a-sim′ĕ-tre) lacking symmetry. *585*

atom (at′om) smallest unit of matter nondivisible by chemical means. *17*

atomic number (ah-tom′ik num′ber) the number of protons within the nucleus of an atom. *18*

atomic weight (ah-tom′ik wāt) the number of protons plus the number of neutrons within the nucleus of an atom. *18*

ATP (adenosine triphosphate); a compound containing adenine, ribose, and three phosphates, two of which are high-energy phosphates. It is the "common currency" of energy for most cellular processes. *38*

atria (a′tre-ah) chambers; particularly the upper chambers of the heart that lie above the ventricles. (*sing.* atrium) *210*

atrioventricular (a″tre-o-ven-trik′u-lar) a structure in the heart that pertains to both the atria and ventricles; for example, an atrioventricular valve is located between an atrium and a ventricle. *210*

atrioventricular node (a″tre-o-ven-trik′u-lar nōd) *See* AV node. *213*

auditory canal (aw′dī-to″re kah-nal′) a tube in the external ear that lies between the pinna and the tympanic membrane. *362*

australopithecines (aw″strah-lo-pith′e-sīnz) referring to three species of *Australopithecus,* the first generally recognized hominids. *632*

autosome (aw′to-sōm) a chromosome other than a sex chromosome. *76*

autotroph (aw′to-trōf) an organism that is capable of making its food (organic molecules) from inorganic molecules. *115, 531*

AV node (atrioventricular node); a small region of neuromuscular tissue that transmits impulses received from the SA node to the ventricular walls. *213*

axial skeleton (ak′se-al skel′ĕ-ton) portion of the skeleton that supports and protects the organs of the head, neck, and trunk. *335*

axillary bud (ak′sī-lar″e bud) group of meristem cells located at junction of a leaf and stem which become branches or flowers; lateral bud. *137*

axon (ak′son) process of a neuron that conducts nerve impulses away from the cell body. *310*

B

bacteria (bak-te′re ah) prokaryotes that lack the organelles of eukaryotic cells; archaebacteria and eubacteria. *544*

bacteriophage (bak-te′re-o-fāj″) a virus that infects a bacterial cell. *539*

bark (bark) outer tissues of a tree containing cork, cork cambium, cortex, and phloem. *142*

basal bodies (ba′sal bod′ēz) short cylinders having a circular arrangement of 9 microtubule triplets (9 + 0 pattern) located within the cytoplasm at the bases of cilia and flagella. *55*

base (bās) a solution in which pH is more than 7; a substance that contributes or liberates hydroxide ions in a solution; alkaline; opposite of acidic. Also, in genetics the chemicals adenine, guanine, cytosine, thymine, and uracil that are found in DNA and RNA. *25, 38*

basement membrane (bās′ment mem′brān) thin interior non-cellular surface of epithelium which attaches to connective tissue. *168*

basidia (bah-sid′e-ah) club-shaped structures that produce basidiospores during sexual reproduction of club fungi. *556*

Batesian mimic (bāt′se-an mim′ik) mimic lacking the defense of its model. *678*

benthic zone (ben′thik zōn) a region containing those organisms that reside at the bottom of bodies of water. *736*

bilateral symmetry (bi-lat′er-al sim′ĕ-tre) the condition of having a right and left half so that only one vertical cut gives two equal halves. *585*

bile (bīl) a secretion of the liver that is temporarily stored in the gallbladder before being released into the small intestine where it emulsifies fat. *188*

binary fission (bi′na-re fish′un) reproduction by division into two equal parts by a process that does not involve a mitotic spindle. *545*

binomial system (bi-no′me-al sis′tem) the assignment of two names to each organism, the first of which designates the genus and the second of which designates the species. *508*

biological evolution (bi-o-loj′ĕ-kal ev″o-lu′shun) changes that have occurred in life forms from the origination of the first cell to the many diverse forms in existence today. *530*

biome (bi′ōm) one of the major climax communities present in the biosphere characterized by a particular mix of plants and animals. *715*

biosphere (bi′o-sfer) that part of the earth's surface and atmosphere where living organisms exist. *688*

biotic (bi-ot′ik) pertaining to any aspect of life, especially to characteristics of entire populations or ecosystems. *688*

biotic potential (bi-ot′ik po-ten′shal) the maximum population growth rate under ideal conditions. *744*

blade (blād) the main portion of a leaf. *142*

blastocyst (blas′to-sist) An early stage of embryonic development that consists of a hollow ball of cells. *421*

blastula (blas′tu-lah) an early stage in animal development; usually a hollow sphere of cells about a central cavity called the blastocoele. *421*

blind spot (blīnd spot) area containing no rods and cones where the optic nerve passes through the retina. *357*

blood (blud) connective tissue composed of cells separated by plasma. *170*

B lymphocyte (lim′fo-sīt) lymphocyte that reacts against foreign substances in the body by producing and secreting antibodies. *252*

bone (bōn) connective tissue having a hard matrix of calcium salts deposited around protein fibers. *170*

Bowman's capsule (bo'manz kap'sūl) a double-walled cup that surrounds the glomerulus at the beginning of the kidney tubule. *297*

bradykinin (brad''e-ki'nin) a substance found in damaged tissue which initiates nerve impulses resulting in the sensation of pain. *243*

breathing (brēth'ing) entrance and exit of air into and out of the lungs. *270*

breeder fission reactor (brēd'er fish'un re-ak'tor) a nuclear reactor that produces more nuclear fuel than it consumes by converting nuclear wastes into ^{239}plutonium. *753*

bronchi (brong'ki) the two major divisions of the trachea leading to the lungs. *274*

bronchiole (brong'ke-ōl) the smaller air passages in the lungs of mammals. *274*

browsers (browz'erz) animals that feed on higher-growing vegetation such as shrubs and trees. *674*

buffer (buf'er) a substance or compound that prevents large changes in the pH of a solution. *26*

bundle sheath (bun'd'l shēth) tightly packed parenchyma cells enclosing larger leaf veins. *000*

C

C_3 photosynthesis (se thre fo''to-sin'thē-sis) photosynthesis that utilizes the Calvin cycle to take up and then fix carbon dioxide; so named because the first molecule detected after CO_2 uptake is a C_3 molecule. *123*

C_4 photosynthesis (se for fo''to-sin'thē-sis) photosynthesis in which the first detected molecule following CO_2 uptake is a C_4 molecule. Later, this same CO_2 is made available to the Calvin cycle. *125*

calcitonin (kal''si-to'nin) hormone secreted by the thyroid gland that helps to regulate the level of blood calcium. *378*

calorie (kal'o-re) the amount of heat required to raise one kilogram of water one degree centigrade. *196*

Calvin-Benson cycle (kal'vin ben'sun si'kl) a circular series of reactions by which CO_2 fixation occurs within chloroplasts. *122*

capillaries (kap'ĭ-lar''ēz) microscopic vessels connecting arterioles to venules through whose thin walls molecules either exit or enter the blood. *208*

capsid (kap'sid) the outer coat of a virus composed of protein subunits. *538*

carapace (kar'ah-pās) upper covering (shell) of some animals. *606*

cardiac muscle (kar'de-ak mus'l) specialized type of muscle tissue found only in the heart. *173*

carpals (kar'palz) bones that are located in the human wrist. *336*

carrier (kar'e-er) a molecule that combines with a substance and actively transports it through the cell membrane; an individual that transmits an infectious or genetic disease. *69*

carrying capacity (kar'e-ing kah-pas'ĭ-te) the largest number of organisms of a particular species that can be maintained indefinitely in an ecosystem. *744*

cartilage (kar'tĭ-lij) a connective tissue, in which the cells lie within lacunae embedded in a flexible matrix. *170*

Casparian strip (kas-par'e-an strip) band of waxy material found on endodermal cells of plants; prevents passage of molecules outside of cells. *136*

CCK (chole cyst okinin); hormone produced by duodenum that stimulates release of bile from gallbladder. *190*

cell (sel) the structural and functional unit of an organism; the smallest structure capable of performing all the functions necessary for life. *43*

cell body (sel bod'e) portion of a nerve cell that includes a cytoplasmic mass and a nucleus and from which the nerve fibers extend. *310*

cell cycle (sel si'kl) a cyclical series of phases that includes cellular events before, during, and after mitosis. *78*

cell-first hypothesis (sel ferst hi-poth'ĕ-sis) a suggestion that only the first cell or cells contained true proteins and nucleic acids. *530*

cell membrane (sel mem'brān) a membrane that surrounds the cytoplasm of cells and regulates the passage of molecules into and out of the cell. *43, 62*

cell plate (sel plāt) a structure that forms between two plant cells during telophase and marks the location of the cell membrane and cell wall. *82*

cell theory (sel the'o-re) theory that all living things are composed of cells and that existing cells are derived from pre-existing cells. *43*

cellular respiration (sel'u-lar res''pĭ-ra'shun) the reactions of glycolysis, Krebs cycle, and electron transport system that provide energy and accompanying reactions that use this energy to produce ATP. *51, 270*

cellulose (sel'u-lōs) a polysaccharide composed of glucose molecules; the chief constituent of a plant's cell wall. *32*

central canal (sen'tral kah-nal') tube within the spinal cord that is continuous with the ventricles of the brain and contains cerebrospinal fluid. *322*

central nervous system (CNS) (sen'tral ner'vus sis'tem) the brain and spinal cord in vertebrate animals. *309*

centriole (sen'tre-ōl) a short, cylindrical organelle in animal cells that contains microtubules in a 9 + 0 pattern and is associated with the formation of the spindle during cell division. *54*

centromere (sen'tro-mēr) a region of attachment for a chromosome to a spindle fiber that is generally seen as a constricted area. *76*

cephalothorax (sef''ah-lo-tho'raks) fusion of head and thoracic regions displayed by some arthropods. *606*

cerebellum (ser''ĕ-bel'um) the part of the vertebrate brain that controls muscular coordination. *323*

cerebral hemisphere (ser'ĕ-bral hem'ĭ-sfēr) one of the large, paired structures that together constitute the cerebrum of the brain. *324*

cerebrospinal fluid (ser''ĕ-bro-spi'nal floo'id) fluid present in ventricles of brain and in central canal of spinal cord. *321*

cerebrum (ser'ĕ-brum) the enlarged portion of the vertebrate brain responsible for consciousness. *324*

chemical evolution (kem'ĭ-kal ev''o-lu'shun) a gradual increase in the complexity of chemical compounds that is believed to have brought about the origination of the first cell or cells. *528*

chemosynthesis (ke''mo-sin'the-sis) the process of making food by using energy derived from the oxidation of reduced molecules in the environment. *127*

chimpanzee (chim-pan'ze) a small great ape that is closely related to humans and is frequently used in psychological studies. *627*

chitin (ki'tin) flexible, strong polysaccharide forming the exoskeleton of arthropods. *604*

chlorophyll (klo'ro-fil) the green pigment found in photosynthesizing organisms that is capable of absorbing energy from the sun's rays. *52, 117*

chloroplast (klo'ro-plast) a membrane-bounded organelle which contains chlorophyll and where photosynthesis takes place. *52*

cholecystokinin (ko''le-sis''to-ki'nin) *See* CCK. *190*

cholesterol (ko-les'ter-ol) a lipid that is used in the synthesis of steroid hormones. *37*

chorion (ko're-on) an extraembryonic membrane that forms an outer covering around the embryo; in reptiles and birds, it functions in gas exchange; in mammals it contributes to the formation of the placenta. *427*

choroid (ko'roid) the vascular, pigmented middle layer of the wall of the eye. *356*

chromatids (kro'mah-tidz) the two identical parts of a chromosome following replication of DNA. *76*

chromatin (kro'mah-tin) threadlike network in the nucleus that is made up of DNA and proteins. *47*

chromosomes (kro'mo-sōmz) rod-shaped bodies in the nucleus, particularly during cell division, that contain the hereditary units or genes. *47*

cilia (sil'e-ah) hairlike projections that are used for locomotion by many unicellular organisms and have various purposes in higher organisms. *54*

ciliary muscle (sil'e-er''e mus'el) a muscle that controls the curvature of the lens of the eye. *356*

ciliates (sil'e-āts) protozoans that move by means of cilia. *549*

circadian rhythm (ser''kah-de'an rith'm) a regular physiological or behavioral event that occurs on an approximately 24-hour cycle. *651*

circannual rhythm (serk-an'u-al rith''m) a behavior that recurs; having a yearly cycle. *651*

circumcision (ser''kum-sizh'un) removal of the foreskin of the penis. *396*

citric acid cycle (sit'rik as'id si'kl) *See* Krebs cycle. *105*

class (klas) in taxonomy, the category below phylum and above order. *508*

clavicle (klav'ĭ-k'l) a slender, rodlike bone located at the base of the neck that runs between the sternum and the shoulders. *335*

cleaning symbiosis (klēn'ing sim″bi-ō'sis) a mutualistic relationship in which one type of organism gains benefit by cleaning another that is benefited by having been cleaned of debris and parasites. *684*

cleavage (klēv'ij) cell division of the fertilized egg that is unaccompanied by growth so that numerous small cells result. *420*

climax community (kli'maks kō-mu'ni-te) in succession, the final, stable stage. *688*

clone (klōn) asexually produced organisms having the same genetic makeup; also DNA fragments from an external source that have been reproduced by *E. coli*. *84, 492*

clotting (klot'ing) process of blood coagulation, usually when injury occurs. *238*

cnidarian (ni-dah're-an) small aquatic animals having radial symmetry and bearing stinging cells with nematocysts. *587*

coacervate (ko-as'er-vāt) a mixture of polymers that may have preceded the origination of the first cell or cells. *531*

cochlea (kok'le-ah) that portion of the inner ear that resembles a snail's shell and contains the organ of Corti, the sense organ for hearing. *364*

cochlear canal (kok'le-ar kah-nal') canal within the cochlea bearing small hair cells which function as hearing receptors. *364*

codon (ko'don) a "triplet" of three nucleotides in messenger RNA that directs the placement of a particular amino acid into a polypeptide chain. *484*

coelom (se'lom) body cavity of higher animals that contains internal organs such as those of the digestive system. *177, 586*

coenzyme (ko-en'zīm) a nonprotein molecule that aids the action of an enzyme, to which it is loosely bound. *97*

coenzyme A (ko-en'zīm a) coenzyme which participates in the transition reaction and carries the organic product to the Kreb's cycle. *105*

coevolve (ko-e-volv') the interaction of two species such that each determines the evolution of the other species. *674*

cohesion-tension theory (ko-he'zhun ten'shun the'o-re) explanation for upward transportation of water in xylem based upon transpiration created tension and the cohesive properties of water molecules. *147*

collar cells (kol'ler selz) flagella-bearing cells found in inner layer of the wall of sponges. *586*

collecting duct (kō-lekt'ing dukt) a tube that receives urine from several distal convoluted tubules. *298*

colon (ko'lon) the large intestine of vertebrates. *193*

colony (kol'o-ne) an organism that is a loose collection of cells that are specialized and cooperate to a degree. *563*

colostrum (kō-los'trum) watery, yellowish-white fluid produced by the breasts. *407*

columnar epithelium (ko-lum'nar ep″ī-the le-um) pillar-shaped cells usually having the nuclei near the bottom of each cell and found lining the digestive tract, for example. *163*

commensalism (kō-men'sal-izm) the relationship of two species in which one lives on or with the other without conferring either benefit or harm. *682*

common ancestor (kō'mun an'ses-tor) an ancestor to two or more branches of evolution. *508*

compact bone (kom-pakt'bōn) hard bone consisting of Haversian systems cemented together. *170, 336*

companion cell (kom-pan'yun sel) a specialized cell that lies adjacent to a sieve-tube cell. *147*

competition (kom″pē-tish'un) interaction between members of the same or different species for a mutually required resource. *669*

competitive exclusion principle (kom pet'ī-tiv eks-kloo'zhun prin'sī-p'l) an observation that no two species can continue to compete for the same exact resources since one species will eventually become extinct. *671*

complement (kom'plē-ment) a group of proteins in plasma that produce a variety of effects once an antigen-antibody reaction has occurred. *255*

complementary base pairing (kom″plē-men'tā-re bās pār'ing) pairing of bases found in DNA and RNA; adenine is always paired with either thymine (DNA) or uracil (RNA) and cytosine is always paired with guanine. *477*

compound eye (kom'pownd i) arthropod eyes composed of multiple lenses. *605*

cones (kōnz) bright-light receptors in the retina of the eye that detect color and provide visual acuity; specialized structures composed of scale-shaped leaves in conifers. *356, 573*

conidia (ko-nid'e-ah) spores produced by sac and club fungi during asexual reproduction. *554*

conifer (kon'ī-fer) a cone-bearing seed plant, mostly trees, such as pines. *721*

conjugation (kon″ju-ga'shun) a sexual union in which the nuclear material of one cell enters another. *564*

connective tissue (kō-nek'tiv tish'u) a type of tissue, characterized by cells separated by a matrix, that often contains fibers. *168*

consumers (kon-su'merz) organisms of one population that feed on members of other populations in an ecosystem. *690*

control (kon-trōl') in experimentation a sample that undergoes all the steps in the experiment except the one being tested. *8*

coral (kor'al) a cnidarian that has a calcium carbonate skeleton whose remains accumulate to form reefs. *589*

coral reef (kor'al rēf) a structure found in tropical waters formed by the buildup of coral skeletons where many and various types of organisms reside. *734*

cork (kork) a tissue made up of dead hollow cells forming a protective covering in woody stems. *142*

cork cambium (kork kam'be-um) meristem tissue which produces cork. *142*

coronary artery (kor'ō-na-re ar'ter-e) an artery that supplies blood to the wall of the heart. *216*

corpus callosum (kor'pus kah-lo'sum) a mass of white matter within the brain, composed of nerve fibers connecting the right and left cerebral hemispheres. *325*

corpus luteum (kor'pus lut'e-um) a body, yellow in color, that forms in the ovary from a follicle that has discharged its egg. *400*

cortex (kor'teks) in animals, the outer layer of an organ; in plants, the tissue beneath the epidermis in certain stems. *135, 296*

cortisol (kor'tī-sol) a glucocorticoid secreted by the adrenal cortex. *381*

cotyledon (kot″ī-le'don) the seed leaf of the embryo of a plant. *156*

courtship (kort'ship) behavior performed for the purpose of acquiring a mate. *663*

Cowper's glands (kow'perz glandz) two small structures located below the prostate gland in males. *395*

cranial nerve (kra'ne-al nerv) nerve that arises from the brain. *316*

creatine phosphate (kre'ah-tin fos'fāt) a compound unique to muscles that contains a high-energy phosphate bond. *346*

creatinine (kre-at'ī-nin) excretion product from creatine phosphate breakdown. *292*

cretinism (kre'tin-izm) a condition resulting from a lack of thyroid hormone in an infant. *378*

Cro-Magnon (kro-mag'non) the common name for the first fossils to be accepted as representative of modern humans. *636*

crossing over (kros'ing o'ver) the exchange of corresponding segments of genetic material between chromatids of homologous chromosomes during synapsis of meiosis I. *85*

cuboidal epithelium (ku-boid'al ep″ī-the'le-um) cube-shaped cells found lining, for example, the kidney tubules. *163*

cultural eutrophication (kul'tu-ral u″tro-fī-ka'shun) enrichment of a body of water causing excessive growth of producers and then death of these and other inhabitants. *709*

cuticle (ku'te-kl) waxlike epidermal covering on nonwoody stems which prevents water loss. *138*

cyanobacteria (si″ah-no-bak-te're-ah) photosynthetic prokaryotes that contain chlorophyll and release O_2; formerly called blue-green algae. *547*

cyclic photophosphorylation (sik'lik fo″to-fos″for-ī-la'shun) the synthesis of ATP by a cytochrome system within thylakoid membranes utilizing electrons received from water. *122*

cytochrome system (si'to-krōm sis'tem) *See* respiratory chain. *120*

cytokinesis (si″to-ki-ne'sis) division of the cytoplasm of a cell. *82*

cytoplasm (si'to-plazm) the ground substance of cells located between the nucleus and the cell membrane. *43*

cytoskeleton (si″to-skel'ē-ton) filamentous protein structures found throughout the cytoplasm that help maintain the shape of the cell. *53*

cytotoxic T cells (si″to-tok'sik te selz) T lymphocytes which attack cells bearing foreign bodies. *257*

D

data (da'tah) experimentally derived facts. *8*

daughter cells (daw'ter selz) cells formed by division of a mother cell. *77*

deamination (de-am''ĭ-na'shun) removal of an amino group ($-NH_2$) from an amino acid or other organic compound. *103, 192*

deciduous (de-sid'u-us) plants that shed their leaves at certain seasons. *722*

decomposers (de-kom-po'zerz) organisms of decay (fungi and bacteria) in an ecosystem. *690*

dehydration synthesis (de''hi-dra'shun sin'thĕ-sis) the joining together of molecules to form macromolecules by removing components that form water. *27*

dehydrogenase (de-hi'dro-jen-ās) an enzyme that accepts hydrogen atoms, speeding up the process of dehydrogenation. *98*

demographic transition (dem-o-graf'ik tran-zĭ'shun) the change from a high birthrate to a low birthrate so that the growth rate is lowered. *746*

dendrite (den'drīt) process of a neuron, typically branched, that conducts nerve impulses toward the cell body. *310*

denitrifying bacteria (de-ni'trī-fi-ing bak-te're ah) bacteria that convert nitrate to atmospheric nitrogen. *696*

deoxyribonucleic acid (de-ok''se-ri''bo-nu-kle'ik as'id) *See* DNA. *37*

depletion curve (de-ple'shun kerv) graphical depiction of a resource's dwindling amount over time. *750*

depolarization (de-po''lar-ĭ-za'shun) a loss in polarization of a membrane as when the nerve impulse or action potential occurs. *313*

dermis (der'mis) the thick skin layer that lies beneath the epidermis. *175*

desert (dez'ert) an arid biome characterized especially by plants such as cacti that are adapted to receiving less than 20 cm rain per year. *716*

desertification (dez-ert''ĭ-fĭ-ka'shun) desert conditions caused by human misuse of land. *718*

detritus (de-tri'tus) non-living organic matter. *690*

deuterostome (du'ter-o-stōm'') member of a large group of animal phyla in which the anus develops from the blastopore and a second opening becomes the mouth. *596*

developed countries (de-vel'opt kun'trēz) industrialized nations that typically have a strong economy and a low rate of population growth. *745*

developing countries (de-vel'op-ing kun'trēz) nations that are not yet industrialized and have a weak economy and a high rate of population growth. *747*

diabetes insipidus (di''ah-be'tēz in-sip'ĭ-dus) condition characterized by an abnormally large production of urine due to a deficiency of antidiuretic hormone. *374*

diabetes mellitus (di''ah-be'tez mē-li'tus) condition characterized by a high blood glucose level and the appearance of glucose in the urine due to a deficiency of insulin production or uptake by cells. *382*

diaphragm (di'ah-fram) a sheet of muscle that separates the chest cavity from the abdominal cavity in higher animals. Also, a birth control device inserted in front of the cervix in females. *276*

diastole (di-as'to-le) relaxation of heart chambers. *212*

diastolic blood pressure (di''ah-stol'ik blud presh'ur) arterial blood pressure during the diastolic phase of the cardiac cycle. *219*

diatoms (di'ah-tomz) a large group of fresh and marine unicellular algae having a cell wall consisting of two silica impregnated valves that fit together as in a pill box. *552*

dicot (di'kot) (dicotyledon); a type of angiosperm distinguished particularly by the presence of two cotyledons in the seed. *138*

differentiation (dif''er-en''she-a'shun) the process and developmental stages by which a cell becomes specialized for a particular function. *421*

diffusion (dĭ-fu'zhun) the movement of molecules from an area of greater concentration to an area of lesser concentration. *64*

digit (dij'it) a finger or toe. *336*

dihybrid (di-hi'brid) the offspring of parents who differ in two ways; shows the phenotype governed by the dominant alleles but carries the recessive alleles. *452*

dinoflagellates (di''no-flaj'ĕ-lāts) a large group of marine unicellular algae that have two flagella; one circles the body while the other projects posteriorly. *552*

dipeptide (di-pep'tīd) a molecule consisting of only two amino acids joined by a peptide bond. *28*

dipleurula larva (di-ploor'u-lah lar'vah) a larval form unique to the deuterostomes that indicates that they are related. *596*

diploid (dip'loid) the 2N number of chromosomes; twice the number of chromosomes found in gametes. *77*

directional selection (dĭ-rek'shun-al sĕ-lek'shun) natural selection which favors an atypical phenotype. *517*

disaccharide (di-sak'ah-rid) a sugar such as maltose that contains two units of a monosaccharide. *31*

dissociate (dis-so'she-āt) the breakdown of a compound into its ionic or elemental components. *25*

distal convoluted tubule (dis'tal kon'vo-lūt-ed tu'būl) highly coiled region of a nephron that is distant from Bowman's capsule. *297*

division (dĭ-vizh'un) in taxonomy, the category applied to plants and fungi that follows kingdom and lies above class. *508*

DNA (deoxyribonucleic acid); a nucleic acid, the genetic material that directs protein synthesis in cells. *37*

dominance hierarchy (dom'ĭ-nans hi'er-ar''ke) a system in which animals arrange themselves in a pecking order; the animal above takes precedence over the one below. *661*

dominant (dom'ĭ-nant) hereditary factor that expresses itself even when the genotype is heterozygous. *445*

dorsal root ganglion (dor'sal rōōt gang'gle-on) mass of sensory neuron cell bodies located in the dorsal root of a spinal nerve. *316*

double helix (dŭ'b'l he'liks) a double spiral often used to describe the three-dimensional shape of DNA. *477*

doubling time (dŭ-b'ling tim) the number of years it takes for a population to double in size. *743*

Down's syndrome (downz sin'drōm) human congenital disorder associated with an extra 23rd chromosome. *464*

Dryopithecus (dri''o-pith'e-cus) a genus of extinct apes that may have included or resembled a common ancestor to both apes and humans. *627*

ductus deferens (duk'tus def'er-enz) tube connecting epididymis to ejaculatory duct; sperm duct, also called vas deferens. *395*

duodenum (du''o-de'num) the first portion of the small intestine in vertebrates into which ducts from the gallbladder and pancreas enter. *188*

dyad (di'ad) a chromosome having two chromatids held together at a centromere. *78*

E

ecology (e-kol'o-je) the study of the relationship of species between themselves and the physical environment. *688*

ecosystem (ek''o-sis'tem) a biological community together with the associated abiotic environment. *688*

ectoderm (ek'to-derm) the outer germ layer of the embryonic gastrula; it gives rise to the skin and nervous system. *422*

ectopic pregnancy (ek-top'ik preg'nan-se) a pregnancy that begins at a location other than the uterus. *414*

edema (ĕ-de'mah) swelling due to tissue fluid accumulation in the intercellular spaces. *227*

effector (ĕ-fek'tor) a structure that allows an organism to respond to environmental stimuli such as the muscles and glands. *318*

elastic cartilage (e-las'tik kar'tĭ-lij) cartilage composed of elastic fibers allowing greater flexibility. *170*

electrocardiogram (e-lek''tro-kar'de-o-gram'') ECG or EKG; a recording of the electrical activity associated with the heartbeat. *217*

electroencephalogram (e-lek''tro-en-sef'ah-lo-gram'') EEG; graphic recording of the brain's electrical activity. *325*

electron (e-lek'tron) a subatomic particle that has almost no weight and carries a negative charge; travels in an orbital, called a shell, about the nucleus of an atom. *17*

electron transport system (e-lek'tron trans'port sis'tem) series of metabolic reactions in which electrons are passed along a chain of carrier molecules with the concurrent production of ATP; also called the respiratory chain. *107*

element (el'ĕ-ment) the simplest of substances consisting of only one type of atom; i.e., carbon, hydrogen, oxygen. *18*

elephantiasis (el″ĕ-fan-ti′ah-sis) disease caused by a parasitic nematode that blocks a lymphatic vessel and characterized by extreme swelling of a limb. *596*

emulsification (e-mul″si-fi′ka′shun) the act of dispersing one liquid in another. *36*

endocrine gland (en′do-krin gland) a gland that secretes hormones directly into the blood or body fluids. *163, 372*

endocytosis (en″do-si-to′sis) a process in which a vesicle is formed at the cell membrane to bring a substance into the cell. *70*

endoderm (en′do-derm) an inner layer of cells that line the primitive gut of the gastrula. It becomes the lining of the digestive tract and associated organs. *422*

endodermis (en″do-der′mis) a plant tissue consisting of a single layer of cells that surrounds and regulates the entrance of materials into particularly the vascular cylinder of roots. *133*

endometrium (en″do-me′tre-um) the lining of the uterus that becomes thickened and vascular during the uterine cycle. *401*

endoplasmic reticulum (en-do-plaz′mic rĕ-tik′u-lum) a complex system of tubules, vesicles, and sacs in cells; sometimes having attached ribosomes. *48*

endoskeleton (en″do-skel′ĕ-ton) calcareous supportive internal tissue of echinoderms and vertebrates. *609*

energy of activation (en′er-je uv ak″ti-va′shun) the amount of energy that a molecule must gain to become sufficiently "excited" to enter into a chemical reaction. *96*

energy sources (en′er-je sorsz) ways by which energy can be made available to organisms from the environment. *529*

environmental resistance (en-vi″ron-men′tal re-zis′tans) sum total of factors in the environment that limit the numerical increase of a population in a particular region. *744*

enzyme (en′zīm) a protein catalyst that speeds up a specific reaction or a specific type of reaction. *27*

epicotyl (ep″i-kot′il) plant embryo portion above the cotyledon(s) that contributes to stem development. *156*

epidermis (ep″i-der′mis) the outer layer of cells of organisms including plants. Also the outer layer of skin composed of stratified squamous epithelium. *132, 174, 590*

epididymis (ep″i-did′i-mis) coiled tubules next to the testes where sperm mature and may be stored for a short time. *395*

epiglottis (ep″i-glot′is) a structure that covers the glottis during the process of swallowing. *185*

epiphyte (ep′i-fīt) nonparasitic plant that grows on the surface of other plants, usually above the ground, such as arboreal orchids and Spanish moss. *722*

epithelial tissue (ep″i-the′le-al tish′u) a type of tissue that lines cavities and covers the external surface of the body. *163*

equilibrium (e″kwi-lib′re-um) a state of balance; a steady state where forces are equalized. *365*

erection (ĕ-rek′shun) referring to a structure such as the penis when it is turgid and erect as opposed to flaccid and lacking turgidity. *396*

erythrocyte (ĕ-rith′ro-sīt) red blood cell that contains hemoglobin and carries oxygen from the lungs to the tissues in vertebrates. *170*

esophagus (ĕ-sof′ah-gus) a tube that transports food from the mouth to the stomach. *186*

essential fatty acids (ĕ-sen′shal fat′e as′idz) linolenic and linoleic fats essential to diet because the body is unable to manufacture. *199*

estuary (es′tu-a-re) an area where fresh water meets the sea; thus, an area with salinity intermediate between fresh water and seawater. *727*

euglenoids (u-gle′noidz) a small group of unicellular algae that are bounded by a flexible pellicle and move by flagella. *551*

eukaryotic (u″kar-e-ot′ik) possessing the membranous organelles characteristic of complex cells. *43*

evolutionary tree (ev″o-lu′shun-ar-e tre) a diagram describing the evolutionary relationship of groups of organisms. *508*

excretion (eks-kre′shun) removal of metabolic wastes. *291*

exocrine gland (ek′so-krin gland) secreting externally; particular glands with ducts whose secretions are deposited into cavities, such as salivary glands. *163*

exocytosis (eks″o-si-to′sis) a process in which an intracellular vesicle fuses with the cell membrane so that the vesicle's contents are released outside the cell. *70*

exophthalmic goiter (ek″sof-thal′mik goi′ter) an enlargement of the thyroid gland accompanied by an abnormal protrusion of the eyes. *378*

expiration (eks″pi-ra′shun) process of expelling air from the lungs; exhalation. *270*

exponential growth (eks″po-nen′shal grōth) growth, particularly of a population, in which the total number increases in the same manner as compound interest. *742*

external respiration (eks-ter′nal res″pi-ra′shun) exchange between air and blood of oxygen and carbon dioxide. *270*

extraembryonic membranes (eks″trah-em″bre-on′ik mem′brānz) in embryology, membranes that are not a part of the embryo but are necessary to the continued existence and health of the embryo. *427*

F

facilitated transport (fah-sil′i-tāt-ed trans′port) transfer of a substance into or out of a cell along a concentration gradient by a process that requires a carrier. *69*

FAD a coenzyme of oxidation; a dehydrogenase that participates in hydrogen (electron) transport within the mitochondria. *105*

family (fam′i-le) in taxonomy the category above genus and below order. *508*

fatigue (fah-tēg′) muscle relaxation in the presence of stimulation due to energy reserve depletion. *342*

fatty acid (fat′e as′id) an organic molecule having a long chain of carbon atoms and ending in an acidic group. *34*

feces (fe′sēz) indigestible wastes expelled from the digestive tract; excrement. *193*

femur (fe′mur) the thighbone found in the upper leg. *336*

fermentation (fer″men-ta′shun) anaerobic breakdown of carbohydrates that results in organic end products such as alcohol and lactic acid. *108*

fetal erythroblastosis (fe′tal ĕ-rih″ro-blas-to′sis) destruction of Rh+ fetal red cells due to antibodies produced by a mother who is Rh−. *247*

fibrin (fi′brin) insoluble, fibrous protein formed from fibrinogen during blood clotting. *238*

fibrinogen (fi-brin′o-jen) plasma protein that is converted into fibrin threads during blood clotting. *238*

fibroblasts (fi′bro-blasts) cells that form fibers in connective tissues. *168*

fibrocartilage (fi″-bro-kar′ti-lij) cartilage with a matrix of strong collagenous fibers. *170*

fibrous connective tissue (fi′brus kō-nek′tiv tish′u) tissue composed mainly of closely packed collagenous fibers and found in tendons and ligaments. *169*

fibrous root (fi′brus rōōt) root system in which each root is equal in size and which grows in an intertwining manner. *136*

fibula (fib′u-lah) a long slender bone located on the lateral side of the tibia. *336*

filament (fil′ah-ment) a threadlike structure such as the thick (myosin) and thin (actin) filaments found in myofibrils of muscle fibers; in flowering plants, the stalk that supports the anther within a stamen. *153, 344*

filter feeder (fil′ter fēd′er) an animal that obtains its food, usually in small particles, by filtering it from water. *600*

filtrate (fil′trāt) the filtered portion of blood that is contained within Bowman's capsule. *298*

fimbriae (fim′bre-ā) fingerlike extensions from the oviduct near the ovary. *401*

fission reactor (fish′un re-ak′tor) a nuclear reactor that splits ^{235}uranium in order to release energy. *752*

fixed action pattern (fikst ak′shun pat′ern) a sequence of reflexes that always occurs in the same order and manner. *651*

flagella (flah-jel′ah) slender, long processes used for locomotion by the flagellate protozoans, bacteria, and sperm. *54*

flame-cell system (flām sel sis′tem) excretory organ of flatworms. *591*

flower (flow′er) the blossom of a plant that contains the reproductive organs of higher plants. *577*

fluid mosaic model (floo′id mo-za′ik mod′el) proteins form a mosaic pattern within a bilayer of lipid molecules having a fluid consistency. *46*

fluke (flōōk) a parasitic flatworm; member of Class Trematoda. *594*

focusing (fo-kus-ing) manner by which light rays are bent by the cornea and lens, creating an image on the retina. *357*

follicle (fol′i-kl) a structure in the ovary that produces the egg and particularly the female sex hormone, estrogen. *399*

follicle-stimulating hormone (fol'ĭ-k'l stim'u-lāt''ing hor'mōn) *See* FSH. *398*

fontanel (fon''tah-nel') soft spot in skull of newborn where membrane rather than bone is present. *335*

food web (fōōd web) the complete set of food links between populations in a community. *691*

foreskin (fōr'skin) skin covering the glans penis in uncircumcised males. *396*

formed element (form'd el'ĕ-ment) those elements that make up the solid portion of blood; they are either cells (erythrocytes and leukocytes) or derived from a cell (platelets). *231*

fossil fuel (fos''l fu'el) the remains of once living organisms that are burned to release energy, such as coal, oil, and natural gas. *695*

fossils (fos''lz) any remains of an organism that have been preserved in the earth's crust. *505*

fovea centralis (fo've-ah sen-tral'is) region of the retina, consisting of densely packed cones, which is responsible for the greatest visual acuity. *357*

frond (frond) the large leaf of a fern plant containing many leaflets. *571*

frontal lobe (fron'tal lōb) area of the cerebrum responsible for voluntary movements and higher intellectual processes. *324*

fruit (frōōt) a mature ovary enclosing seed(s). *577*

fruiting bodies (frōōt'ing bod'es) a spore-bearing structure found in certain types of fungi, such as mushrooms. *554*

FSH (follicle-stimulating hormone); a hormone secreted by the anterior pituitary gland that stimulates the development of an ovarian follicle in a female or the production of sperm in a male. *398*

fungus (fung'gus) an organism, usually composed of strands called hyphae, that lives chiefly on decaying matter; e.g., mushroom and mold. *553*

furrowing (fur'o-ing) a constriction of the cell membrane that accompanies cytokinesis in animal cells. *82*

G

gallbladder (gawl'blad-er) saclike organ associated with the liver that stores and concentrates bile. *188*

gallstones (gawl'stōnz) precipitated crystals of cholesterol or calcium carbonate formed from bile within the gallbladder or bile duct. *192*

gamete (gam'ēt) a reproductive cell that joins with another in fertilization to form a zygote; most often an egg or sperm. *76*

gametophyte generation (gam'ĕ-to-fīt jen''ĕ-ra'shun) the haploid generation that produces gametes in the life cycle of a plant. *561*

ganglion (gang'gle-on) a collection of neuron cell bodies outside the central nervous system. *316*

gap junction (gap junk'shun) cell junction in which protein molecules form cell-to-cell channels allowing molecule exchange. *168*

gastric gland (gas'trik gland) gland within the stomach wall that secretes gastric juice. *186*

gastric inhibitory peptide (gas'trik in-hib'i-tor''e pep'tīd) *See* GIP. *190*

gastrin (gas'trin) a hormone secreted by stomach cells to regulate the release of pepsin by the stomach wall. *190*

gastrodermis (gas''tro-der'mis) layer of cells found lining the body cavity in cnidarians. *590*

gastrovascular cavity (gas''tro-vas'ku-lar kav'ĭ-te) a central cavity, with only one opening, of a lower animal in which digestion takes place and where nutrients are distributed to the cells lining the cavity. *590*

gene flow (jēn flo) the movement of genes from one population to another via reproduction between members of the populations. *515*

gene pool (jēn pōōl) the total of all the genes of all the individuals in a population. *512*

genetic drift (jĕ-net'ik drift) evolution by chance processes alone. *516*

genotype (je'no-tīp) the genetic makeup of any individual. *449*

genus (je'nus) in taxonomy, the category above species and below family. *508*

gibbon (gib'on) the smallest ape; well known for its arm-swinging form of locomotion. *627*

gills (gilz) gas exchange organs found in fishes and other types of marine and freshwater animals. *599*

GIP (gastic inhibitory peptide); a hormone produced by the small intestine that inhibits the release of gastric secretions. *190*

glomerulus (glo-mer'u-lus) a cluster; for example, the cluster of capillaries surrounded by Bowman's capsule in a kidney tubule. *298*

glottis (glot'is) slitlike opening between the vocal cords. *185*

glucagon (gloo'kah-gon) hormone secreted by the pancreatic islets of Langerhans that causes the release of glucose from glycogen. *382*

glucose (gloo'kōs) the most common six-carbon sugar. *31*

glycerol (glis'er-ol) an organic compound that serves as a building block for fat molecules. *34*

glycogen (gli'ko-jen) the storage polysaccharide found in animals that is composed of glucose molecules joined in a linear-type fashion but having numerous branches. *32*

glycolysis (gli-kol'i-sis) the metabolic pathway that converts glucose to pyruvic acid. *101*

Golgi apparatus (gol'ge ap''ah-ra'tus) an organelle that consists of concentrically folded membranes and functions in the packaging and secretion of cellular products. *49*

gonad (go'nad) an organ that produces sex cells; the ovary, which produces eggs, and the testis, which produces sperm. *77*

gonadotropic hormone (go-nad''o-trŏp'ik hor'mon) a type of hormone that regulates the activity of the ovaries and testes; principally FSH and LH (ICSH). *377*

gorilla (go-ril'ah) the largest of the great apes which are as closely related to humans as to other apes. *627*

Graafian follicle (graf'e-an fol'lĭ-k'l) mature follicle within the ovaries which houses a developing egg. *400*

grana (gra'nah) stacks of flattened membranous vesicles in a chloroplast where chlorophyll is located and photosynthesis begins. *117*

granulocytes (gran'u-lo-sīts) white blood cells that contain distinctive granules. *240*

grazers (gra'zerz) animals that feed on low-lying vegetation such as grasses. *674*

greenhouse effect (grēn'hows ĕ-fekt) carbon dioxide buildup in the atmosphere as a result of fossil fuel combustion; retains and re-radiates heat, creating an abnormal rise in the earth's average temperature. *704*

growth (grōth) increase in the number of cells and/or the size of these cells. *420*

growth hormone (grōth hor'mōn) GH or somatotropin; hormone released by the anterior lobe of the pituitary gland that promotes the growth of the organism. *376*

growth rate (grōth rāt) percentage of increase or decrease in the size of a population. *743*

guard cell (gahrd sel) a bean-shaped, epidermal cell; one found on each side of a leaf stoma; their activity controls stoma size. *144*

H

habitat (hab'ĭ-tat) the natural abode of an animal or plant species. *689*

haploid (hap'loid) the N number of chromosomes; half the diploid number; the number characteristic of gametes that contain only one set of chromosomes. *77*

hard palate (hard pal'at) anterior portion of the roof of the mouth which contains several bones. *184*

hazardous wastes (haz'er-dus wāsts) wastes containing chemicals hazardous to life. *706*

HCG (human chorionic gonadotropic); a gonadotrophic hormone produced by the chorion that functions to maintain the uterine lining. *405*

heart (hart) muscular organ located in thoracic cavity responsible for maintenance of blood circulation. *210*

heliostat (he'le-o-stat'') large-field mirrors that track the sun and reflect its energy onto a mounted boiler for solar heating. *754*

helper T cells (hel'per te selz) T lymphocytes that stimulate certain other T and B lymphocytes to perform their respective functions. *256*

heme (hēm) the iron-containing portion of a hemoglobin molecule. *235*

hemocoel (he'mo-sēl) residual coelom found in anthropods that is filled with blood. *605*

hemoglobin (he''mo-glo'bin) a red, iron-containing pigment in blood that combines with and transports oxygen. *281*

hepatic caeca (hĕ-pat'ik se'kah) enzyme secreting organs of echinoderms. *610*

hepatic portal vein (he-pat'ik por'tal vān) vein leading to the liver formed by the merging of blood vessels from the villi of the small intestine. *191*

herbaceous stem (her-ba'shus stem) non-woody stem. *140*

hermaphroditism (her-maf'ro-di-tizm'') the state of having both male and female sex organs. *591*

heterospores (het'er-o-sporz) nonidentical spores such as microspores and megaspores produced by the same plant. *572*

heterotroph (het'er-o-trōf) an organism that takes in preformed food. *115, 531*

heterotroph hypothesis (het'er-o-trof'' hi-poth'e-sis) the suggestion that the protocell and first cell(s) were heterotrophs. *531*

heterozygous (het''er-o-zi'gus) having two different alleles (as *Aa*) for a given trait. *449*

hexose (hek'sōs) a six-carbon monosaccharide. *31*

histamine (his'tah-min) substance produced by basophil-derived mast cells in connective tissue which causes capillaries to dilate and release immune and other substances. *243*

homeostasis (ho''me-o-sta'sis) the constancy of conditions, particularly the internal environment of birds and mammals: constant temperature, blood pressure, pH, and other body conditions. *178*

homing (hōm'ing) the ability of an animal to return to a homesite. *650*

hominid (hom'ĭ-nid) member of the family of upright, bipedal primates (family Hominidae) that includes modern humans. *628*

hominoid (hom'ĭ-noid) member of a superfamily containing humans and the great apes. *627*

Homo erectus (ho'mo ē-rek'tus) the earliest nondisputed species of humans, named for their erect posture that allowed them to have a striding gait. *634*

Homo habilis (ho'mo hah'bi-lis) an extinct species that may include the earliest humans, having a small brain but making quality tools. *633*

homologous (ho-mol'o-gus) similarly constructed; homologous chromosomes have the same shape and contain genes for the same traits; homologous structures in animals share a common ancestry. *508*

homozygous (ho''mo-zi'gus) having identical alleles (as *AA* or *aa*) for a given trait; pure breeding. *449*

host (hōst) an organism on or in which another organism lives. *680*

humerus (hu'mer-us) a heavy bone that extends from the scapula to the elbow. *335*

hyaline cartilage (hi'ah-līn kar'ti-lij) cartilage composed of very fine collagenous fibers and a matrix of a clear milk-glass appearance. *170*

hybrid (hi'brid) an offspring resulting from the crossing of genetically different strains, populations, or species. *449*

hybridomas (hi''brid-o'mahz) fused lymphocytes and cancer cells used in the manufacture of monoclonal antibodies. *265*

hydrogen bond (hi'dro-jen bond) a weak attraction between a hydrogen atom carrying a partial positive charge and an atom of another molecule carrying a partial negative charge. *24*

hydroid (hi'droid) tubular-shaped polyp displayed by some cnidarians. *588*

hydrolysis (hi-drol'i-sis) the splitting of a bond within a larger molecule by the addition of the components of water. *27*

hydrolytic enzyme (hi-dro-lit'ik en'zīm) an enzyme that catalyzes a reaction in which the substrate is broken down with the addition of water. *185*

hypertonic solution (hi''per-ton'ik so-lu'shun) one that has a greater concentration of solute, a lesser concentration of water than the cell. *67*

hypha (hi'fah) one filament of a mycelium that constitutes the body of a fungus. *553*

hypocotyl (hi''po-kot'il) plant embryo portion below the cotyledon(s) that contributes to stem development. *156*

hypothalamus (hi''po-thal'ah-mus) a region of the brain, the floor of the third ventricle, that helps maintain homeostasis. *322*

hypothesis (hi-poth'ĕ-sis) a scientific theory that is capable of explaining present data and that may be used to predict the outcome of future experimentation. *7*

hypotonic solution (hi''po-ton'ik so-lu'shun) one that has a greater concentration of water, a lesser concentration of solute than the cell. *67*

I

immune complex (ĭ-mūn' kom'pleks) the product of an antigen-antibody reaction. *255*

implantation (im''plan-ta'shun) the attachment of the embryo in the lining (endometrium) of the uterus. *405*

impotency (im'po-ten''se) failure of the penis to achieve erection. *396*

imprinting (im'print-ing) the tendency of a newborn animal to follow the first moving object it sees. *654*

inclusive fitness (in-kloo͞'siv fit'nes) the fitness of closely related group members is a part of the fitness of the individual. *665*

inducible operon model (in-dūs'ĭ-b'l op'er-on mod'el) a model in which the operon is normally inactive due to presence of a repressor. The inducer inactivates the repressor and transcription takes place. *490*

induction (in-duk'shun) a process by which one tissue controls the development of another, as when the embryonic notochord induces the formation of the neural tube. *425*

inflammatory reaction (in-flam'ah-to''re re-ak'shun) a tissue response to injury that is characterized by dilation of blood vessels and an accumulation of fluid in the affected region. *242*

innate (in'nāt) instinctive, inborn, and not having to be learned. *649*

inner cell mass (in'er sel mas) mass of cells within a blastocyst which will develop into the individual. *422*

inner ear (in'er ēr) portion of the ear consisting of the vestibule, semicircular canals and the cochlea where balance is maintained and sound is transmitted. *364*

innervate (in'er-vāt) to activate an organ, muscle or gland by motor neuron stimulation. *310*

innominate (i-nom'ĭ-nāt) one of two hipbones that form the pelvis. *336*

inorganic (in''or-gan'ik) refers to chemical reactions and compounds which do not contain both carbon and hydrogen. Not organic. *24*

insertion (in-ser'shun) the end of a muscle that is attached to a movable part. *339*

insight (in'sīt) the use of higher mental abilities and previous experiences to arrive at a new idea or conclusion. *655*

inspiration (in''spi-ra'shun) the act of breathing in. *270*

insulin (in'su-lin) a hormone produced by the pancreas that regulates carbohydrate storage. *382*

integration (in''tĕ-gra'shun) the summing up of negative and positive stimuli within a dendrite or cell body. *314*

interferon (in''ter-fēr'on) a protein formed by a cell infected with a virus that can increase the resistance of other cells to the virus. *265*

internal respiration (in-ter'nal res''pi-ra'shun) exchange between blood and tissue fluid of oxygen and carbon dioxide. *270*

interneuron (in''ter-nu'ron) a neuron that is found within the central nervous system and takes nerve impulses from one portion of the system to another. *310*

internodes (in'ter-nōdz) regions between stem nodes. *137*

interphase (in'ter-fāz) the interval between successive cell divisions; during this time the chromosomes are in an extended state and are active in directing protein synthesis. *79*

interstitial cells (in''ter-stish'al selz) hormone-secreting cells located between the seminiferous tubules of the testes. *395*

invertebrate (in-ver'tĕ-brāt) an animal that lacks a vertebral column. *585*

ion (i'on) an atom or group of atoms carrying a positive or negative charge. *19*

ionic reaction (i-on'ik re-ak'shun) chemical reaction in which atoms acquire or lose electrons. *19*

islets of Langerhans (i'lets uv lahng'er-hanz) distinctive groups of cells within the pancreas that secrete insulin and glucagon. *382*

isogametes (i''so-gam'ēts) gametes whose union produces a zygote, but which have a similar appearance. *562*

isotonic solution (i''so-ton'ik so-lu'shun) one that contains the same concentration of solute and water as does the cell. *66*

isotopes (i'so-tōps); atoms with the same number of protons and electrons but differing in the number of neutrons and therefore in weight. *18*

J

jointed appendages (joint'ed ah-pen'dij-ez) the flexible exoskeleton extensions found in arthropods that are used as sense organs, mouth parts, and locomotion. *604*

K

karyotype (kar'e-o-tīp) the arrangement of all the chromosomes within a nucleus by pairs in a fixed order. *75*

keratin (ker'ah-tin) an insoluble protein present in the epidermis and in epidermal derivatives such as hair and nails. *175*

kidneys (kid'nēz) organs in the urinary system which concentrate and excrete urine. *294*

kingdom (king'dum) the largest taxonomic category into which organisms are placed: Monera, Protista, Fungi, Plants, and Animals. *508*

kin selection (kin sē-lek'shun) behavior which functions to protect a group of related organisms. *665*

Klinefelter's syndrome (klīn'fel-terz sin'drōm) a condition caused by the inheritance of a chromosome abnormality in number; an XXY individual. *464*

Krebs cycle (krebz si'kl) a series of reactions found within the matrix of mitochondria that give off carbon dioxide. Also called the citric acid cycle because the reactions begin and end with citric acid. *101*

L

labium (la'be-um) a fleshy border or liplike fold of skin, as in the labia majora and labia minora of the female genitalia. *402*

lacteal (lak'te-al) a lymph vessel in a villus of the intestinal wall of mammals. *191*

lactogenic hormone (lak"to-jen'ik hor'mōn) LTH; a hormone secreted by the anterior pituitary that stimulates the production of milk from the mammary glands. 376

lacuna (lah-ku'nah) a small pit or hollow cavity, as in bone or cartilage where a cell or cells are located. *170*

lancelet (lans'let) type of protochordate that has the three chordate characteristics as an adult; formerly called amphioxus. *612*

lanugo (lah-nu'go) downy hair with which a fetus is born; fetal hair. *438*

larynx (lar'ingks) structure that contains the vocal cords; voice box. *185*

lateral bud *See* axillary bud. *137*

leaf veins (lēf vānz) structures that contain vascular tissue in a leaf. *142*

learning (lern'ing) a change in behavior as a result of experience. *649*

lens (lenz) a clear membranelike structure found in the eye behind the iris. The lens brings objects into focus. *356*

lenticel (len'ti-sel) pocket of loosely arranged cells in cork that permit gas exchange. *142*

leukemia (lu-ke'me-ah) form of cancer characterized by uncontrolled production of leukocytes in red bone marrow or lymphoid tissue. *242*

leukocyte (lu'ko-sīt) white blood cell of which there are several types each having a specific function in protecting the body from invasion by foreign substances and organisms. *170*

lianas (le-ah'nahz) woody vines found in rain forests. *722*

lichen (li'ken) fungi and algae coexisting in a mutualistic relationship. *548*

ligament (lig'ah-ment) dense connective tissue that joins bone to bone. *169, 338*

limb buds (lim budz) protrusions in the vertebrate embryo that will develop into the limbs, such as the arms and legs in humans. *433*

limbic system (lim'bik sis'tem) an area of the forebrain implicated in visceral functioning and emotional responses; involves many different centers of the brain. *326*

limnetic zone (lim-net'ik zōn) the sunlit body of a lake. *728*

linkage (lingk'ij) alleles on the same chromosome are linked in the sense that they tend to move together to the same gamete; crossing over interferes with linkage. *458*

lipase (li'pās) a fat-digesting enzyme secreted by the pancreas. *188*

lipids (lip'idz) a group of organic compounds that are insoluble in water; notably fats, oils, and steroids. *34*

littoral zone (lit'or-al zōn) in a lake the portion closest to the shore; at the seashore, the intertidal zone. *728*

local excitation (lo'kal ek"si-ta'shun) postsynaptic membrane stimulation that is insufficient to produce a nerve impulse. *314*

loose connective tissue (lo͞os kō-nek'tiv tish'u) tissue composed mainly of fibroblasts which are separated by collagen and elastin fibers and found beneath epithelium. *168*

lumen (lu'men) the cavity inside any tubular structure, such as the lumen of the gut. *186*

luteinizing hormone (lu'te-in-īz"ing hor'mōn) hormone produced by the anterior pituitary gland that stimulates the development of the corpus luteum in females and the production of testosterone in males. *398*

lymph (limf) fluid having the same composition as tissue fluid and carried in lymph vessels. *227*

lymphatic system (lim-fat'ik sis'tem) vascular system which takes up excess tissue fluid and transports it to the bloodstream. *227*

lymph node (limf nōd) a mass of lymphoid tissue located along the course of a lymphatic vessel. *228*

lymphokines (lim'fo-kīnz) chemicals secreted by T cells that have the ability to affect the characteristics of monocytes. *256*

lysosome (li'so-sōm) an organelle in which digestion takes place due to the action of hydrolytic enzymes. *50*

M

macrophage (mak'ro-fāj) an enlarged monocyte that ingests foreign material and cellular debris. *243*

Malpighian tubules (mal-pig'i-an tu'būlz) organs of excretion, notably in insects. *608*

maltose (mawl'tōs) a disaccharide composed of two glucose units. *31*

mangrove swamp (man'grōv swahmp) tropical and subtropical coastal communities characterized by presence of mangrove trees. *730*

mantle (man't'l) fleshy fold that envelops the visceral mass of mollusks. *597*

matrix (ma'triks) the secreted basic material or medium of biological structures, such as the matrix of cartilage or bone. *168*

medulla (mē-dul'ah) the inner portion of an organ; for example, the adrenal medulla. *296*

medulla oblongata (mē-dul'ah ob"long-gah'tah) the lowest portion of the brain that is concerned with the control of internal organs. *322*

medusa (mē-du'sah) a bell-shaped, free-swimming stage resembling a jellyfish that is capable of sexual reproduction in the life cycle of some sessile cnidarians. *588*

megaspore (meg'ah-spōr) in seed plants, a spore that develops into the female gametophyte generation. *575*

meiosis (mi-o'sis) type of cell division that occurs during the production of gametes or spores by means of which the daughter cells receive the haploid number of chromosomes. *76*

meiosis I (mi-o'sis wun) that portion of meiosis during which homologous chromosomes come together and then later separate. Includes prophase I, metaphase I, anaphase I, and telophase I. *84*

meiosis II (mi-o'sis too) that portion of meiosis during which sister chromatids separate resulting in four haploid daughter cells. Includes prophase II, metaphase II, anaphase II and telophase II. *84*

melanin (mel'ah-nin) a pigment found in the skin and hair of humans that is responsible for their coloration. *175*

memory B cells (mem'o-re be selz) cells derived from B cells that are long present within the body that produce a specific antibody and account for the development of active immunity. *253, 257*

meninges (mē-nin'jēz) protective membranous coverings about the central nervous system. *321*

meniscus (mē-nis'kus) a piece of fibrocartilage that separates the surfaces of bones in the knee. *338*

menopause (men'o-pawz) termination of the ovarian and uterine cycle in older women. *407*

menstruation (men"stroo-a'shun) loss of blood and tissue from the uterus at the end of a female uterine cycle. *403*

meristem tissue (mer'i-stem tish'u) plant tissue that always remains undifferentiated and capable of dividing to produce new cells. *132*

mesoderm (mes'o-derm) the middle germ layer of an animal embryo that gives rise to the muscles, connective tissue, and circulatory system. *423*

mesoglea (mes"o-gle'ah) a jellylike packing material between the ectoderm and endoderm of cnidarians. *587*

mesophyll (mes'o-fil) the middle tissue of a leaf made up of parenchyma cells. *142*

messenger RNA (mes'en jer) mRNA; a nucleic acid (ribonucleic acid) complementary to genetic DNA and bearing a message to direct cell protein synthesis at the ribosome. *483*

metabolism (mĕ-tab'o-lizm) all of the chemical changes that occur within cells. *93*

metacarpal (met"ah-kar'pal) bones found in the palm of the hand. *336*

metal (met'al) class of elements which in reactions characteristically lose electrons and become positively charged ions. *21*

metamorphosis (met"ah-mor'fo-sis) change in form as when a tadpole becomes an adult frog or as when an insect larva develops into the adult. *609*

metaphase (met'ah-fāz) stage in mitosis during which chromosomes assemble at mitotic spindle equator. *82*

metatarsal (met"ah-tar'sal) bones found in a foot between the ankle and the toes. *336*

MHC protein (major histocompatibility protein); a surface molecule that serves as a genetic marker. *256*

microfilament (mi"kro-fil'ah-ment) an extremely thin fiber found within the cytoplasm that is involved in the maintenance of cell shape and movement of cell contents. *53*

microspore (mi'kro-spōr) in seed plants, a spore that develops into a pollen grain. *575*

microtubule (mi"kro-tu'būl) an organelle composed of 13 rows of globular proteins; found in multiple units in several other organelles such as the centriole, cilia, and flagella. *54*

middle ear (mid"l ēr) portion of the ear consisting of the tympanic membrane, the oval and round windows, and the ossicles where sound is amplified. *362*

migration (mi-gra'shun) the movement of organisms to and from a geographical site. *650*

mill tailings (mil tāl'ingz) radioactive material in the waste produced by mining and processing uranium ores. *753*

mimicry (mim'ik-re) the resemblance of an organism to another that has a defense against a common predator. *678*

mineral (min'er-al) an inorganic, homogeneous substance. *204*

mitochondrion (mi"to-kon'dre-on) an organelle in which aerobic respiration produces the energy molecule, ATP. *51*

mitosis (mi-to'sis) type of cell division in which the daughter cells receive the exact chromosome and genetic makeup of the mother cell; occurs during growth and repair. *76*

mixed nerve (mikst nerv) nerve that contains both the long dendrites of sensory neurons and the long axons of motor neurons. *316*

mold (mōld) a type of fungus that produces woolly or cottony growth. *556*

molecule (mol'ĕ-kūl) a chemical unit in which two or more atoms share electrons. *21*

molting (mōlt'ing) shedding all or part of an outer covering; in arthropods, periodic shedding of parts of the exoskeleton to allow increase in size. *604*

monoclonal antibodies (mon"o-klon'al an'ti-bod"ēz) antibodies of one type that are produced by cells that are derived from a lymphocyte that has fused with a cancer cell. *265*

monocot (mon'o-kot) (monocotyledon); a type of angiosperm in which the seed has only one cotyledon, such as corn and lily. *138*

monohybrid (mon"o-hi'brid) the offspring of parents who differ in one way only; shows the phenotype of the dominant allele but carries the recessive allele. *450*

mononucleosis (mon"o-nu"kle-o'sis) viral disease characterized by the presence of an increase in atypical lymphocytes in the blood. *242*

monosaccharide (mon"o-sak'ah-rīd) a simple sugar; a carbohydrate that cannot be decomposed by hydrolysis. *31*

morphogenesis (mor"fo-jen'ĭ-sis) the movement of cells and tissues to establish the shape and structure of an organism. *420*

mother cell (muth'er sel) cell which divides producing daughter cells. *77*

motivated (mo'tĭ-vāt-ed) physiologically orientated to perform a certain behavior. *658*

motor nerve (mo'tor nerv) nerve containing only the long axons of motor neurons. *316*

motor neuron (mo'tor nu'ron) a neuron that takes nerve impulses from the central nervous system to the effectors. *310*

Müllerian mimic (mil-e're-an mim'ik) mimic that possesses the same defense as the model. *678*

multicellular (mul"tĭ-sel'u-lar) composed of many cells. *43*

muscle action potential (mus'el ak'shun po-ten'shal) an electrochemical change due to increased sarcolemma permeability that is propagated down the T system and results in muscle contraction. *347*

muscle spindle (mus'el spin'dul) modified skeletal muscle fiber that can respond to changes in muscle length. *343*

muscle tissue (mus'el tish'u) a type of tissue that contains cells capable of contracting; skeletal muscles are attached to the skeleton, smooth muscle is found within walls of internal organs, and cardiac muscle comprises the heart. *172*

mutagen (mu'tah-jen) an agent, such as a chemical, that increases the rate of mutations. *499*

mutualism (mu'tu-al-izm") a relationship between two organisms of different species that benefits both organisms. *682*

mycelium (mi-se'le-um) a mass of hyphae that make up the body of a fungus. *553*

myelin (mi'ĕ-lin) the fatty cell membranes that cover long neuron fibers and give them a white, glistening appearance. *310*

myocardium (mi"o-kar'de-um) heart muscle. *210*

myofibrils (mi"o-fi'brilz) the contractile portions of muscle fibers. *344*

myogram (mi'o-gram) a recording of a muscular contraction. *342*

myosin (mi'o-sin) one of two major proteins of muscle; makes up thick filaments in myofibrils and is capable of breaking down ATP. *See* actin. *346*

myxedema (mik" sĕ-de'mah) a condition resulting from a deficiency of thyroid hormone in an adult. *378*

N

NAD a coenzyme of oxidation; a dehydrogenase that frequently participates in hydrogen transport. *98*

NADP a coenzyme of reduction; a hydrogenase that frequently donates hydrogen atoms to metabolites. *121*

natural selection (nat'u-ral sĕ-lek'shun) the process by which better adapted organisms are favored to reproduce to a greater degree and pass on their genes to the next generation. *516*

Neanderthal (ne-an'der-thawl) the common name for an extinct subspecies of humans whose remains are found in Europe, Asia, and Africa. *636*

negative feedback (neg'ah-tiv fēd'bak) mechanism that is activated by a surplus imbalance and acts to correct it by stopping the process that brought about the surplus. *179*

nematocyst (nem'ah-to-sist) a threadlike structure in stinging cells of cnidarians that can be expelled to numb and capture prey. *587*

nephridia (nĕ-frid'e-ah) excretory tubules found in invertebrates; notably the segmented worms. *603*

nephron (nef'ron) the anatomical and functional unit of the vertebrate kidney; kidney tubule. *297*

neritic zone (ner-it'ik zōn) the sunlit water containing those organisms that reside above the continental shelves. *736*

nerve (nerv) a bundle of long nerve fibers that run to and/or from the central nervous system. *173*

nerve impulse (nerv im'puls) an electrochemical change due to increased membrane permeability that is propagated along a neuron from the dendrite to the axon following excitation. *311*

nerve net (nerv net) network of nerve cells found in cnidarians. *590*

neuroglia cells (nu-rog'le-ah selz) supporting cells within brain and spinal cord that perform functions other than transmission of nerve impulse. *173*

neuromuscular junction (nu"ro-mus'ku-lar jungk'shun) the point of contact between a nerve cell and a muscle cell. *347*

neuron (nu'ron) nerve cell that characteristically has three parts: dendrite, cell body, axon. *173, 309*

neurotransmitter substance (nu"ro-trans-mit'er sub'stans) a chemical made at the ends of axons that is responsible for transmission across a synapse. *314*

neurula (nu′roo-lah) the early embryonic stage during which the primitive nervous system forms. *423*

neutral fat (nu′tral fat) lipid composed of glycerol and fatty acid units joined in such a way that there are no charged nor polar groups. *34*

neutron (nu′tron) a subatomic particle that has a weight of one atomic mass unit, carries no charge, and is found in the nucleus of an atom. *17*

niche (nich) the functional role and position of an organism in the ecosystem. *671*

nitrifying bacteria (ni′trī-fi″ing bak-te′re-ah) bacteria active in the nitrogen cycle that oxidize ammonia (NH_4^+) to nitrite (NO_2^-) and nitrate (NO_3^-). *696*

nitrogen fixation (ni′tro-jen fik-sa′shun) a process whereby free atmospheric nitrogen is converted into compounds, such as ammonia and nitrates, usually by soil bacteria. *696*

node (nōd) area of a stem where a leaf is attached. *137*

noncyclic photophosphorylation (non-sik′lik fo″to-fos″for-i-la′shun) the passage of electrons within thylakoid membranes from water (by way of Photosystem II, a cytochrome system, Photosystem I) to NADP; results in the generation of ATP and $NADPH_2$. *122*

nondisjunction (non″dis-jungk′shun) the failure of homologous chromosomes or sister chromatids to separate during the formation of gametes. *464*

nonmetal (non-met′al) class of elements which characteristically react to gain electrons forming negatively charged ions. *21*

nonrenewable resource (non-re-nu′ah-b′l re′sors) a resource that can be used up or at least depleted to such an extent that further recovery is too expensive. *750*

noradrenalin (nor″ah-dren′ah-lin) NA; excitatory neurotransmitter active in the peripheral and central nervous systems; norepinephrine. *314, 380*

notochord (no′to-kord) dorsal supporting rod that exists in all chordates sometime in their life history; replaced by the vertebral column in vertebrates. *611*

nuclear envelope (nu′kle-ar en′vĕ-lōp) the double membrane that surrounds the nucleus and is continuous with the endoplasmic reticulum. *46*

nuclear fusion (nu′kle-ar fu′zhun) a process in which the nuclei of two atoms are forced together to form the nucleus of a heavier atom with the release of substantial amounts of energy. *753*

nucleic acid (nu-kle′ik as′id) a large organic molecule made up of nucleotides joined together; for example, DNA and RNA. *37*

nucleolus (nu-kle′o-lus) an organelle found inside the nucleus; composed largely of RNA for ribosome formation. *47*

nucleotide (nu′kle-o-tīd) a molecule consisting of three subunits: phosphoric acid, a five-carbon sugar, and a nitrogenous base; a building block of a nucleic acid. *38*

nucleus (nu′kle-us) a large organelle containing the chromosomes and acting as a control center for the cell; center of an atom. *17, 46*

O

occipital lobe (ok-sip′ĭ-tal lōb) area of the cerebrum responsible for vision, visual images and other sensory experiences. *324*

olfactory cells (ol-fak′to-re selz) sense organs whose stimulation results in the sensation of smelling. *354*

oncogene (ong′ko-jēn) a gene that contributes to the transformation of a cell into a cancerous cell. *499*

oogenesis (o″o-jen′ĕ-sis) production of egg in females by the process of meiosis and maturation. *88*

open circulatory system (o′pen ser′ku-lah-to″re sis′tem) transport of blood that is not always contained in vessels. *599*

operant conditioning (op′ĕ-rant kon-dish′un ing) trial-and-error learning in which desired responses are rewarded. *654*

operon (op′er-on) an operator gene and the adjacent group of genes it controls. *490*

optic nerve (op′tik nerv) nerve that carries nerve impulses from the retina of the eye to the brain. *356*

orangutan (o-rang′oo-tan″) one of the great apes; large with long red hair. *627*

order (or′der) in taxonomy, the category below class and above family. *508*

organelle (or″gah-nel′) specialized structures within cells such as the nucleus, mitochondria, endoplasmic reticulum. *43*

organic (or-gan′ik) pertaining to any aspect of living matter. *24*

organic soup (or-gan′ik sōōp) an expression used to refer to the ocean before the origin of life when it contained newly formed organic compounds. *530*

organ of Corti (or′gan uv kor′ti) a portion of the inner ear that contains the receptors for hearing. *364*

orgasm (or′gazm) physical and emotional climax during sexual intercourse; results in ejaculation in the male. *397*

origin (or′ĭ-jin) end of a muscle that is attached to a relatively immovable bone. *339*

osculum (os′ku-lum) opening to the exterior of a sponge central cavity. *586*

osmosis (oz-mo′sis) the movement of water from an area of greater concentration of water to an area of lesser concentration of water across a semipermeable membrane. *65*

osmotic pressure (oz-mot′ik presh′ur) pressure generated by the osmotic flow of water. *66*

ossicles (os′-ĭ-k′lz) the tiny bones found in the middle ear: hammer, anvil, and stirrup. *364*

osteoblast (os′te-o-blast″) a bone-forming cell. *337*

osteoclast (os′te-o-klast″) a cell that causes the erosion of bone. *337*

osteocyte (os′te-o-sīt) a mature bone cell. *337*

otoliths (o′to-liths) granules associated with ciliated cells in the utricle and saccule. *364*

outer ear (out′er ēr) portion of the ear consisting of the pinna and the auditory canal. *362*

oval opening (o′val o′pen-ing) foramen ovale; an opening between the two atria in the fetal heart. *438*

oval window (o′val win′do) opening between the stapes and the inner ear. *362*

ovarian cycle (o-va′re-an si′k′l) monthly occuring changes in the ovary that effects the level of sex hormones in the blood. *403*

ovaries (o′var-ez) female gonads, the organs that produce eggs, and estrogen and progesterone; the base of the pistil in angiosperms. *152, 399*

ovulation (o″vu-la′shun) the discharge of a mature egg from the follicle within the ovary. *400*

ovule (o′vūl) structure that contains megasporangium in seed plants where meiosis occurs and the female gametophyte is produced. *153*

oxidation (ok″si-da′shun) the loss of electrons (inorganic) or the removal of hydrogen atoms (organic). *23*

oxidative decarboxylation (ok″si-da′tiv de″kar-bok″si-la′shun) a reaction that involves the release of carbon dioxide as oxidation occurs. *105*

oxidizing atmosphere (ok′si-dīz-ing at′mos-fēr) an atmosphere that contains oxidizing molecules such as O_2 rather than reducing molecules such as H_2. *534*

oxygen debt (ok′si-jen det) oxygen that is needed to metabolize lactic acid, a compound that accumulates during vigorous exercise. *346*

oxyhemoglobin (ok″se-he″mo-glo′bin) hemoglobin bound to oxygen in a loose, reversible way. *235*

oxytocin (ok″se-to′sin) hormone released by posterior pituitary that causes contraction of uterus and milk letdown. *374*

ozone shield (o′zōn shēld) a layer of O_3 present in the upper atmosphere which protects the earth from damaging U.V. light. Nearer the earth, ozone is a pollutant. *534*

P

pacemaker (pās′māk-er) *See* SA node. *214*

palisade layer (pal′ĭ-sād la′er) the upper layer of the mesophyll of a leaf that carries on photosynthesis. *142*

parapodia (par″ah-po′de-ah) footlike fleshy lobes found on the segments of marine annelids. *602*

parasite (par′ah-sīt) an organism that resides externally on or internally within another organism and does harm to this organism. *680*

parasitism (par′ah-si″tizm) symbiotic relationship in which an organism derives nourishment from and does harm to a host. *680*

parasympathetic nervous system (par″ah-sim″pah-thet′ik ner′vus sis′tem) that part of the autonomic nervous system that usually promotes those activities associated with a normal state. *321*

parathyroid hormone (par″ah-thi′roid hor′mōn) (PTH); a hormone secreted by the parathyroid glands that affect the level of calcium and phosphate in the blood. *378*

parenchyma (pah-reng'kĭ-mah) relatively unspecialized cells that make up the fundamental tissue of plants. *132*

parietal lobe (pah-ri'ĕ-tal lōb) area of the cerebrum responsible for sensations involving temperature, touch, pressure, and pain, as well as speech. *324*

parturition (par"tu-rish'un) the processes that lead to and include the birth of a mammal and the expulsion of the extraembryonic membranes through the terminal portion of the female reproductive tract. *440*

pectoral girdle (pek'to-ral ger'd'l) portion of the skeleton that provides support and attachment for the arms. *335*

pelagic zone (pe-laj'ik zōn) the main portion of an ocean containing those organisms that reside in either the upper, middle, or lower levels of the open sea. *736*

pelvic girdle (pel'vik ger'd'l) portion of the skeleton to which the legs are attached. *336*

pelvic inflammatory disease (pel'vik in-flam'ah-to"re dĭ-zēz) PID; a disease state of the reproductive organs caused by an organism that is sexually transmitted. *414*

pelvis (pel'vis) a bony ring formed by the innominate bones. Also a hollow chamber in the kidney that lies inside the medulla and receives freshly prepared urine from the collecting ducts. *296, 336*

penis (pe'nis) male copulatory organ. *396*

pentose (pen'tōs) a 5-carbon sugar; deoxyribose is a pentose found in DNA; ribose is a pentose found in RNA. *31*

pepsin (pep'sin) a protein-digesting enzyme secreted by gastric glands. *187*

peptide bond (pep'tīd bond) the bond that joins two amino acids. *28*

pericycle (per"i-si'kl) a single layer of tissue next to the endodermis that produces secondary roots. *136*

periodontitis (per"e-o-don-tī'tis) inflammation of the gums. *184*

peripheral nervous system (pĕ-rif'er-al ner'vus sis'tem) PNS; nerves and ganglia that lie outside the central nervous system. *309*

peristalsis (per"ĭ-stal'sis) a rhythmical contraction that serves to move the contents along in tubular organs such as the digestive tract. *186*

peritubular capillary (per"i-tu'bu-lar kap'ĭ-lar"e) capillary that surrounds a nephron and functions in reabsorption during urine formation. *298*

permafrost (perm'ah-frost) earth beneath surface in tundra which remains permanently frozen. *718*

peroxisome (pĕ-roks'ĭ-sōm) organelle involved in oxidation of molecules and other metabolic reactions. *49*

petals (pet'alz) leaves of a flower that are often colored. *152*

petiole (pet'e-ōl) structure connecting a leaf to a stem. *142*

PG *See* prostaglandins. *385*

pH a measure of the hydrogen ion concentration; any pH below 7 is acid and any pH above 7 is basic. *25*

phagocytosis (fag"o-si-to'sis) the taking in of bacteria and/or debris by engulfing; cell eating. *70, 243*

phalanges (fah-lan'jēz) bones of the finger and thumb. *336*

pharynx (far'ingks) a common passageway (throat) for both food intake and air movement. *185*

phenotype (fe'no-tīp) the outward appearance of an organism caused by the genotype and environmental influences. *449*

pheromone (fer'o-mōn) a chemical substance secreted by one organism that influences the behavior of another. *649*

phloem (flo'em) the vascular tissue in plants that transports organic nutrients. *136*

phospholipid (fos"fo-lip'id) lipids containing phosphorus that are particularly important in the formation of cell membranes. *36*

photoperiodism (fo"to-pe're-od-izm) a response to light and dark; particularly in reference to flowering in plants. *150*

photophosphorylation (fo"to-fos"for-ĭ-la'shun) ATP production utilizing light energy. *122*

photosynthesis (fo"to-sin'thĕ-sis) the process of making carbohydrate from carbon dioxide and water by using the energy of the sun. *52, 115*

Photosystem I and II (fo'to-sis"tem) molecular units located within the membrane of a thylakoid that capture solar energy and making photophosphorylation possible. *119*

photovoltaic cell (fo"to-vol-ta'ik sel) a manufacture mechanism that uses sunlight to produce an electromotive force. *754*

phylum (fi'lum) in taxonomy, the category applied to animals that is below kingdom and above class. *508*

physiograph (fiz'e-o-graf) instrument used to record a myogram. *342*

phytochrome (fi'to-krōm) a plant pigment that is involved in photoperiodism in plants. *151*

phytoplankton (fi"to-plank'ton) photosynthesizing algae which drift freely in aquatic communities. *727*

pinna (pin'nah) outer, funnellike structure of the ear that picks up sound waves. *362*

pinocytosis (pin"o-si-to'sis) the taking in of fluid along with dissolved solutes by engulfing; cell drinking. *70*

pistil (pis't'l) part of the flower that contains a stigma, style, and ovary. *152*

pith (pith) a plant tissue located in the central portion of dicot stems. *136*

placenta (plah-sen'tah) a region formed from the chorion of the fetus and the uterine lining where nutrients pass from the mother's blood to fetal blood and wastes pass in the opposite direction. *405*

placental mammals (plah-sen'tal mam'alz) mammals displaying internal fetal development supported by the presence of a placenta. *617*

plankton (plank'ton) free-floating microscopic organisms found in most bodies of water. *727*

plasma cell (plaz'mah sel) a cell derived from a B cell lymphocyte that is specialized to mass produce antibodies. *252*

plasmid (plaz'mid) a circular DNA segment that is present in bacterial cells but is not part of the bacterial chromosome. *492*

plasmolysis (plaz-mol'ĭ-sis) contraction of the cell contents due to the loss of water. *68*

plastids (plas'tidz) organelles of plants that are specialized for various functions, including photosynthesis. *52*

platelet (plāt'let) a formed element that is necessary to blood clotting. *238*

pleural membranes (ploo'ral mem'brānz) serous membranes that enclose the lungs. *276*

plumule (ploo'mūl) the shoot tip and first two leaves of a plant. *156*

polar (po'lar) condition in a compound when electrons are unequally shared by the atoms involved. *24*

polar bodies (po'lar bod'ēz) nonfunctioning daughter cells that have little cytoplasm and are formed during oogenesis. *88*

pollen grain (pol'en grān) male gametophyte generation of seed plants. *572*

pollination (pol"ĭ-na'shun) the delivery by wind or animals of pollen to the stigma of a pistil in flowering plants and leading to fertilization. *155*

pollution (pŏ-lu'shun) detrimental alteration of the normal constituents of air, land, and water due to human activities. *697*

polymer (pol'ĭ-mer) a large molecule made up of many identical subunits. *27*

polymerization (pol"ĭ-mer"ĭ-za'shun) the joining together of unit molecules to form large organic molecules such as nucleic acid and protein molecules. *530*

polyp (pol'ip) the sedentary stage in the life cycle of cnidarians; a benign growth. *588*

polypeptide (pol"e-pep'tīd) a molecule composed of many amino acids linked together by peptide bonds. *28*

polysaccharide (pol"e-sak'ah-rīd) a macromolecule composed of many units of sugar. *31*

polysome (pol'e-sōm) a cluster of ribosomes all attached to the same mRNA molecule and thus all participating in the synthesis of the same polypeptide. *49*

population (pop"u-la'shun) all the organisms of the same species in one place (area/space). *512, 688*

posterior pituitary (pos-tēr'e-or pĭ-tu'ĭ-tār"e) portion of the pituitary gland connected by a stalk to the hypothalamus. *373*

postganglionic axon (pōst"gang-gle-on'ik ak'son) axon located posterior to a ganglion. *319*

postmating isolating mechanisms (pōst'māt-ing i'so-lāt-ing mek'ah-nizmz) isolating mechanisms which prevent reproductive viability of hybrids. *522*

postsynaptic membrane (pōst"sĭ-nap'tik mem'brān) in a synapse the membrane of the neuron opposite the presynaptic membrane. *314*

prairie (prar'e) a biome characterized by the presence of either short stem or long stem grasses. *721*

predation (pre-da'shun) the eating of one organism on another. *674*

preganglionic axon (pre"gang-gle-on'ik ak'son) axon located anterior to a ganglion. *319*

premating isolating mechanisms (pre'māt-ing i'so-lāt-ing mek'ah-nizmz) isolating mechanisms which prevent reproduction between species. *521*

pressure filtration (presh'ur fil-tra'shun) the movement of small molecules from the glomerulus into Bowman's capsule due to action of blood pressure. *298*

pressure-flow hypothesis (presh'ur flo hi-poth'ĕ-sis) theory explaining phloem transport; solutes actively transported into phloem at a source brings about an osmotic pressure that causes flow of solutes to a sink where they are actively transported out of phloem. *149*

presynaptic membrane (pre"sī-nap'tik mem'brān) the membrane of a synaptic ending. *314*

prey (pra) organisms that serve as food for a particular predator. *674*

primate (pri'māt) an animal that belongs to the order Primates, the order of mammals that includes prosimians, monkeys, apes, and humans. *624*

primer effect (prīm'er ĕ-fekt) change in behavior produced by some pheromones which act to alter physiology. *659*

primitive atmosphere (prim'ī-tiv at'mos-fēr) the gases that were found in the atmosphere when the earth first arose. *528*

prions (pri'onz) nervous system disease-causing agents consisting only of a glycoprotein. *542*

producers (pro-du'serz) organisms that produce food and are capable of synthesizing organic compounds from inorganic constituents of the environment; usually the green plants and algae in an ecosystem. *689*

profundal zone (pro-fun'dal zōn) that portion of a lake below the level of light penetration. *728*

proglottids (pro-glot'idz) the body sections of a tapeworm. *593*

prokaryotic (pro"kar-e-ot'ik) lacking the organelles found in complex cells; bacteria including cyanobacteria are prokaryotes. *43*

prolactin (pro-lak'tin) *See* lactogenic hormone. *376*

prophase (pro'fāz) early stage in mitosis during which chromosomes appear. *79*

proprioceptor (pro"pre-o-sep'tor) sensory receptor that assists the brain in knowing the position of the limbs. *353*

prosimians (pro-sim'e-anz) primitive primates such as lemurs, tarsiers, and tree shrews. *624*

prostaglandins (pros"tah-glan'dinz) hormones that have various and powerful effects often within the cells that produce them. *385*

prostate gland (pros'tāt gland) a gland in males that is located about the urethra at the base of the bladder; produces most of the seminal fluid. *395*

protein (pro'te-in) a macromolecule composed of one or several long polypeptides. *28*

proteinoid microsphere (pro'te-in-oid mi-kro-sfer') a phase, consisting of polypeptides only, during the chemical evolution of the first cell or cells. *530*

prothallus (pro'thal-us) a small, heart-shaped structure; the gametophyte generation of the fern. *571*

prothrombin (pro-throm'bin) plasma protein that functions in the formation of blood clots. *238*

protocell (pro'to-sel) a structure that precedes the evolution of the true cell in the history of life. *530*

protochordates (pro"to-kor'dāts) the invertebrate chordates that possess the three chordate characteristics in either the larval or adult form; tunicates and lancelets. *611*

proton (pro'ton) a subatomic particle found in the nucleus of an atom that has a weight of one atomic mass unit and carries a positive charge; a hydrogen ion. *17*

protonema (pro"to-ne'mah) in the moss life cycle, single row of algalike cells resulting from spore germination. *568*

protostome (pro'to-stōm) member of a large group of animal phyla in which the mouth develops from the blastopore. *596*

protozoans (pro"to-zo'anz) animal-like protists that are classified according to means of locomotion: amoebas, flagellates, ciliates. *548*

proximal convoluted tubule (prok'sī-mal kon'vo-lūt-ed tu'būl) highly coiled region of a nephron near Bowman's capsule. *297*

pseudocoelom (su"do-se'lom) a coelom incompletely lined by mesoderm. *586*

pseudostratified (su"-do strat'e-fīd) the appearance of layering in some epithelial cells when actually each cell touches a base line and true layers do not exist. *163*

pulmonary artery (pul'mo-ner"e ar'ter-e) artery involved in transport of blood from heart to the lungs. *212*

pulmonary circuit (pul'mo-ner"e ser'kit) that part of the circulatory system that takes deoxygenated blood to, and oxygenated blood away from, the lungs. *215*

pulmonary vein (pul'mo-ner"e vān) vein involved in transporting blood from the lungs to heart. *212*

pulse (puls) vibration felt in arterial walls due to expansion of the aorta following ventricle contraction. *218*

Punnett square (pun'et skwār) a gridlike device that enables one to calculate the results of simple genetic crosses by lining gametic genotypes of two parents on the outside margin and their recombination in boxes inside the grid. *450*

pure (pūr) *See* homozygous. *449*

purines (pu'rinz) nitrogenous bases found in DNA and RNA that have two interlocking rings, as in adenine and guanine. *477*

pus (pus) thick yellowish fluid composed of dead phagocytes, dead tissue, and bacteria. *244*

pyrimidines (pi-rim'ī-dinz) nitrogenous bases found in DNA and RNA that have just one ring, as in cytosine and thymine. *477*

pyruvate (pi'roo-vāt) the end product of glycolysis; pyruvic acid. *104*

R

radial symmetry (ra'de-al sim'ĕ-tre) regardless of the angle of a cut made at the midline of an organism, two equal halves result. *585*

radicle (rad'ī-k'l) the embryonic root of a plant. *156*

radioactive (ra"de-o-ak'tiv) a property of certain elements or isotopes by which the nucleus emits particles and/or rays in order to stabilize itself. *19*

radius (ra'de us) an elongated bone located on the thumb side of the lower arm. *336*

rain forest (rān for'est) a biome of equatorial forests which remains warm year round and receives abundant rain. *722*

receptor (re-sep'tor) a sense organ specialized to receive information from the environment. Also a structure found in the membrane of cells that combines with a specific chemical in a lock and key manner. *62, 352*

recessive allele (re-ses'iv ah-lēl') hereditary factor that expresses itself only when the genotype is homozygous. *445*

recombinant DNA (re-kom'bī-nant) DNA having genes from two different organisms, often produced in the laboratory by introducing foreign genes into a bacterial plasmid. *492*

rectum (rek'tum) the terminal portion of the intestine. *193*

red marrow (red mar'o) blood-cell-forming tissue located in spaces within bones. *337*

reduced hemoglobin (re-dūst' he"mo-glo'bin) hemoglobin which has released its oxygen. *235*

reducing atmosphere (re-dūs'ing at'mos-fer) an atmosphere that contains reducing molecules such as H_2 rather than oxidizing ones such as O_2. *528*

reduction (re-duk'shun) the gain of electrons (inorganic); the addition of hydrogen atoms (organic). *23*

reflex (re'fleks) an inborn autonomic response to a stimulus that is dependent on the existence of fixed neural pathways. *318, 651*

regeneration (re-jen"er-a'shun) regrowth of tissue; formation of a complete organism from a small portion. *587*

regulatory gene (reg'u-lah-tor"e jēnz) genes which code for proteins that are involved in regulating the activity of structural genes. *490*

releaser effect (re-lēs'er ĕ-fekt') the immediate behavioral response provoked by some pheromones. *659*

REM (rapid eye movement); a stage in sleep that is characterized by eye movements and dreaming. *325*

renewable resource (re-nu'ah-b'l re'sors) a resource is not used up because it is continually produced in the environment. *750*

replacement reproduction (re-plās'ment re"pro-duk'shun) a population whose average number of children is one per person. *748*

replication (re"plī-ka'shun) the duplication of DNA; occurs when the cell is not dividing. *78, 480*

repolarization (re-po″lar-i-za′shun) the recovery of a neuron's polarity to the resting potential after it ceases transmitting impulses. *313*

repressible operon model (re-pres′i-b'l op′er-on mod′el) a model in which the operon is normally active. When a corepressor combines with a repressor, transcription is halted. *490*

resolving power (re-solv′ing pow′er) in microscopy, the capacity to distinguish between two points. *43*

respiratory chain (re-spi′rah-to″re chan) a series of carriers within the inner mitochondrial membrane that pass electrons one to the other from a higher level to a lower energy level; the energy released is used to build ATP; also called the electron transport system; the cytochrome system. *101*

resting potential (rest′ing po-ten′shal) the voltage recorded from inside a neuron when it is not conducting nerve impulses. *312*

retina (ret′i-nah) the innermost layer of the eyeball that contains the rods and cones. *356*

retinene (ret′i-nēn) substance used in the production of rhodopsin. *359*

retroviruses (ret″ro-vi′rus-ez) viruses that contain only RNA and carry out RNA to DNA transcription. *542*

Rh factor (ar′āch fak′tor) a type of antigen on the red cells. *246*

rhodopsin (ro-dop′sin) visual purple, a pigment found in the rods of one type of receptor in the retina of the eye. *359*

ribonucleic acid (ri″bo-nu-kle′ik as′id) *See* RNA. *37*

ribosomal RNA (ri′bo-sōm″al) rRNA; RNA occurring in ribosome structures involved in protein synthesis. *486*

ribosomes (ri′bō-sōmz) minute particles, found attached to endoplasmic reticulum or loose in the cytoplasm, that are the site of protein synthesis. *49*

ribs (ribz) bones hinged to the vertebral column and sternum which, with muscle, define the top and sides of the chest cavity. *276, 335*

ribulose biphosphate (ri′bu-lōs bi-fos′fāt) RuBP; the molecule that acts as the acceptor for carbon dioxide within the Calvin-Benson cycle. *122*

ritualized behavior (rich′u-al-īzd be-hāv′yor) in the evolution of behavior certain patterns which have become stereotyped. *660*

RNA (ribonucleic acid); a nucleic acid important in the synthesis of proteins that contains the sugar ribose; the bases uracil, adenine, guanine, cytosine; and phosphoric acid. *37*

rods (rodz) dim-light receptors in the retina of the eye that detect motion but no color. *356*

root cap (ro͞ot kap) thimble-shaped mass of parenchyma cells which protects the apical meristem of a root. *134*

root hairs (ro͞ot hārz) hairs on root epidermal cells which function in absorption. *135*

rough endoplasmic reticulum (ruf en″do-plas′mik rē-tik′u-lum) RER; endoplasmic reticulum having attached ribosomes. *48*

round window (rownd win′do) a membrane-covered opening between the inner ear and the middle ear. *362*

RuBP *See* ribulose biphosphate. *122*

S

SA node (sinoatrial node); a small region of neuromuscular tissue that initiates the heartbeat. Also called the pacemaker. *213*

saccule (sak′ūl) a saclike cavity that makes up part of the membranous labyrinth of the inner ear; receptor for static equilibrium. *364*

sac plan (sak plan) body plan possessed by animals having a single opening. *585*

salivary gland (sal′i-ver-e gland) a gland associated with the mouth that secretes saliva. *184*

salt marsh (sawlt marsh) coastal grassland exposed to seasonal flooding. *730*

saprophyte (sap′ro-fit) a heterotrophic organism such as bacteria and fungi that externally breaks down dead organic matter before absorbing the products. *546*

sarcolemma (sar″ko-lem′ah) the membrane that surrounds striated muscle cells. *344*

sarcoplasmic reticulum (sar″ko-plaz′mik rē-tik′u-lum) membranous network of channels and tubules within a muscle fiber, corresponding to the endoplasmic reticulum of other cells. *344*

savanna (sah-van′ah) a grassland biome that has occasional trees and is particularly associated with Africa. *720*

scapula (skap′u-lah) a broad somewhat triangular bone located on either side of the back. *335*

schistosomiasis (skis″to-so-mi′ah-sis) disease caused by a blood fluke. *594*

scientific method (si″en-tif′ik meth′ud) characteristic process by which scientists test their conclusions. Consists of hypothesis generation, observation and experimentation and results in testable theories. *7*

sclera (skle′rah) white fibrous outer layer of the eyeball. *355*

sclerenchyma (skle-reng′ki-mah) a support tissue in plants made of hollow cells with thickened walls. *132*

scolex (sko′leks) head region of a tapeworm. *593*

scotopsin (sko-top′sin) protein portion of rhodopsin. *359*

scrotal sacs (skro′tal saks) the sac that contains the testes. *000*

secretin (se-kre′tin) hormone secreted by the small intestine that stimulates the release of pancreatic juice. *190*

selective reabsorption (sē-lek′tiv re″ab-sorp′shun) movement of nutrient molecules as opposed to waste molecules, from the contents of the kidney tubule into the blood. *299*

semen (se′men) the sperm-containing secretion of males; seminal fluid plus sperm. *396*

semicircular canals (sem″e-ser′ku-lar kah-nal′z) tubular structures within the inner ear that contain the receptors responsible for the sense of dynamic equilibrium. *364*

semideserts (sem″e-dez′ert) arid regions that support plants adapted to conserve water such as cacti. *716*

semilunar (sem″e-lu′nar) a heart valve that consists of three cusps, i.e., the pulmonary semilunar and aortic semilunar valves. *210*

seminal fluid (sem′i-nal floo′id) fluid produced by various glands situated along the male reproductive tract. *395*

seminal vesicle (sem′i-nal ves′i-k′l) a convoluted saclike structure attached to ductus deferens near the base of the bladder in males. *395*

seminiferous tubules (sem″i-nif′er-us tu′būlz) highly coiled ducts within the male testes that produce and transport sperm. *394*

sensory nerve (sen′so-re nerv) nerve containing only sensory neuron dendrites. *316*

sensory neuron (sen′so-re nu′ron) a neuron that takes nerve impulses to the central nervous system; afferent neuron. *310*

sepals (se′palz) leafy divisions of the calyx found in a whorl at the base of petals. *152*

septum (sep′tum) partition or wall such as the septum in the heart, which divides the right half from the left half. *210*

serum (se′rum) light-yellow liquid left after clotting of the blood. *239*

sessile (ses′il) organisms that lack locomotion and remain stationary in one place, such as plants or sponges. *000*

setae (se′te) bristles, especially those of the segmented worms. *603*

sex chromosome (seks kro′mo-sōm) a chromosome responsible for the development of characteristics associated with maleness or femaleness; an *X* or *Y* chromosome. *76*

sex linked (seks lingkt) alleles located on sex chromosomes which determine traits unrelated to sex. *461*

sieve plate (siv plāt) madreporite; a large pore through which water enters echinoderms. *611*

sieve-tube cell (siv tūb sel) specialized cells (elements) that form a linear array running vertically through phloem that functions in transport of organic nutrients. *147*

sign stimulus (sīn stim′u-lus) stimulus which releases a fixed action pattern. *651*

simple goiter (sim′p′l goi′ter) condition in which an enlarged thyroid produces low levels of thyroxin. *378*

sinoatrial node (si″no-a′tre-al nōd) *See* SA node. *213*

sinus (si′nus) a cavity, as the sinuses in the human skull and the blood sinuses of some animals with open circulatory systems. *335, 599*

skeletal muscle (skel′ē-tal mus′el) the contractile tissue that comprises the muscles attached to the skeleton; also called striated muscle. *172*

skull (skul) structure that consists of the cranial and facial bones of the head. *335*

sliding filament theory (slīd′ing fil′ah-ment the′o-re) the movement of actin in relation to myosin in explaining the mechanics of muscle contraction. *346*

smooth endoplasmic reticulum (smo͞oth en″do-plas′mik rē-tik′u-lum) SER; endoplasmic reticulum without attached ribosomes. *48*

smooth muscle (smo͞oth mus′el) the contractile tissue that comprises the muscles found in the walls of internal organs. *172*

social parasitism (so′shal par′ah-si″tizm) the utilization of one population by another such that the first is harmed and the second is benefited. *681*

society (so-si'ĕ-te) members of a population that are specialized to perform specific duties and cooperate with one another for the good of the group. *659*

sociobiology (so''se-o-bi-ol'ŏ-je) the biology of behavior, particularly social behavior. *664*

soft palate (sŏft pal'at) entirely muscular posterior portion of the roof of the mouth. *184*

solar collector (so'lar kŏ-lek'ter) any manufactured structure that absorbs radiant energy. *754*

solar energy (so'lar en'er-je) energy derived from the sun's rays. *754*

solid wastes (sol'id wāsts) household trash, sewage sludge, agricultural residues, mining and industrial wastes. *706*

solute (sol'ūt) a substance dissolved in a solvent to form a solution. *65*

solvent (sol'vent) a fluid such as water that dissolves solutes. *65*

somatic nervous system (so-mat'ik ner'vas sis'tem) that portion of the PNS containing motor neurons that control skeletal muscles. *318*

sorus (so'rus) a cluster of sporangia found on the underside of fern leaves (plural: sori). *571*

species (spe'shēz) a group of similarly constructed organisms that are capable of interbreeding and producing fertile offspring; organisms that share a common gene pool. *5, 508*

spermatogenesis (sper''mah-to-jen'ĕ-sis) production of sperm in males by the process of meiosis and maturation. *88*

sphincter (sfingk'ter) a muscle that surrounds a tube and closes or opens the tube by contracting and relaxing. *186*

spicules (spik'ūlz) needle-shaped structures produced by some sponges which function as supportive inner skeleton. *587*

spinal nerve (spi'nal nerv) nerve that arises from the spinal cord. *316*

spindle fibers (spin'd'l fi'berz) microtubule bundles involved in the movement of chromosomes during mitosis and meiosis. *80*

spiracles (spir'ah-k'lz) respiratory openings in arthropods. *608*

spleen (splēn) a large, glandular organ located in the upper left region of the abdomen that serves as a reservoir for blood and also functions as a lymphatic organ. *228*

sponge (spunj) member of the phylum Porifera. *586*

spongy bone (spun'je bōn) porous bone found at the ends of long bones. *170, 336*

spongy layer (spun'je la'er) the lower layer of the mesophyll of a leaf that carries out gas exchange. *142*

sporangium (spo-ran'je-um) a structure within which spores are produced. *554*

sporophyte generation (spo'ro-fit jen''ĕ-ra'shun) spore-producing diploid generation of a plant. *153, 561*

sporozoa (spo''ro-zo'ah) nonmotile parasitic protozoans. *551*

spot desmosome (spot des'mo-sōm) joining of two cells by intercellular filaments that penetrate cytoplasmic plaques, one in each cell. *168*

squamous epithelium (skwa'mus ep''ĭ-the'le-um) flat cells found lining the lungs and blood vessels. *163*

stabilizing selection (sta'bil-īz''ing sĕ-lek'shun) effect of natural selection eliminating atypical phenotypes. *517*

stamen (sta'men) part of flower, composed of filament and anther, where pollen grains are produced. *152*

starch (starch) the storage polysaccharide found in plants that is composed of glucose molecules joined in a linear-type fashion. *32*

stereoscopic vision (ste''re-o-skop'ik vizh'un) the product of two eyes and both cerebral hemispheres functioning together allowing depth perception. *359*

sternum (ster'num) the breastbone to which the ribs are ventrally attached. *335*

steroid (ste'roid) lipid soluble, biologically active molecules having four interlocking rings; examples are cholesterol, progesterone, testosterone. *37*

stigma (stig'mah) the uppermost part of a pistil. *152*

stoma (sto'mah) opening in the leaves of plants through which gas exchange takes place (pl. stomata). *117, 144*

stratified (strat'ĭ-fīd) layered, as in stratified epithelium, which contains several layers of cells. *163*

stretch receptors (strech re-sep'torz) muscle fibers which upon stimulation cause muscle spindles to increase the rate at which they fire. *353*

striated (stri'āt-ed) having bands; cardiac and skeletal muscle are striated with bands of light and dark. *172*

stroma (stro'mah) the interior portion of a chloroplast. *117*

stroma lamellae (stro'mah lah-mel'e) membranous connections between adjacent thylakoids of the grana. *117*

structural genes (struk'tūr-al jēnz) genes which code for enzymes or proteins otherwise necessary to the structure and function of the cell. *490*

style (stīl) the long slender part of the pistil. *152*

subcutaneous (sub''ku-ta'ne-us) a tissue layer found in vertebrate skin that lies just beneath the dermis and tends to contain adipose tissue. *175*

substrate (sub'strāt) a reactant in a reaction controlled by an enzyme. *94*

succession (suk-sĕ'shun) a series of ecological stages by which the community in a particular area gradually changes until there is a climax community that can maintain itself. *688*

summation (sum-ma'shun) ever greater contraction of a muscle due to constant stimulation that does not allow complete relaxation to occur. *342*

superfemale (su''per-fe'māl) a female that has three X chromosomes. *464*

supernormal stimuli (su''per-nor'mal stim'u-li) greater than normal sign stimuli. *652*

supressor T cell (su-pres'or T sel) T lymphocyte that suppresses certain other T and B lymphocytes from continuing to divide and perform their respective functions. *000*

symbiosis (sim''bi-o'sis) an intimate association of two dissimilar species including commensalism, mutualism, and parasitism. *680*

sympathetic nervous system (sim''pah-thet'ik ner'vus sis'tem) that part of the autonomic nervous system that usually causes effects associated with emergency situations. *319*

synapse (sin'aps) the region between two nerve cells where the nerve impulse is transmitted from one to the other; usually from axon to dendrite. *314*

synapsis (sĭ-nap'sis) the attracting and pairing of homologous chromosomes during prophase I of meiosis. *85*

synaptic cleft (sĭ-nap'tik kleft) small gap between presynaptic and postsynaptic membranes. *314*

synaptic ending (sĭ-nap'tik end'ing) the knob at the end of an axon at a synapse. *314*

synovial joint (si-no've-al joint) a freely movable joint. *338*

systemic circuit (sis-tem'ik ser'kit) that part of the circulatory system that serves body parts other than the gas-exchanging surfaces in the lungs. *215*

systole (sis'to-le) contraction of the heart chambers. *212*

systolic blood pressure (sis-tol'ik blud presh'ur) arterial blood pressure during the systolic phase of the cardiac cycle. *219*

T

taiga (ti'gah) a biome that forms a worldwide northern belt of coniferous trees. *721*

tapeworm (tāp werm) parasitic flatworm; member of class cestoda. *593*

taproot (tap'rōōt) root system which consists of one large root which grows straight down. *136*

tarsal (tahr'sal) a bone of the ankle in humans. *336*

taste bud (tāst bud) organ containing the receptors associated with the sense of taste. *354*

taxis (tak'sis) a movement in relation to a stimulus, such as a phototaxis (a movement oriented to a light source). *649*

taxonomy (tak-son'o-me) the science of naming and classifying organisms. *508*

tectorial membrane (tek-te're-al mem'brān) membrane associated with the organ of Corti which transmits nerve impulses to the brain. *364*

telophase (tel'o-fāz) stage of mitosis during which diploid number of chromosomes are located at each pole. *82*

template (tem'plāt) a pattern that serves as a mold for the production of an oppositely shaped structure; one strand of DNA is a template for the complementary strand. *479*

temporal lobe (tem'po-ral lōb) area of the cerebrum responsible for hearing and smelling, the interpretation of sensory experience and memory. *324*

tendon (ten'don) dense connective tissue that joins muscle to bone. *169, 339*

territoriality (ter″ī-tor″ī-al′ŭ-te) behaviorial display by animals to closely guard a given space needed for reproduction from other intruders. *661*

testcross (test kros) the crossing of a heterozygote with an organism homozygous recessive for the characteristic(s) in question in order to determine the genotype. *451*

testes (tes′tēz) the male gonads, the organs that produce sperm and testosterone. *393*

testosterone (tes-tos′tĕ-rōn) the most potent androgen. *398*

tetanic contraction (tĕ-tan′ik kon-trak′shun) sustained muscle contraction without relaxation. *342*

tetany (tet′ah-ne) severe twitching caused by involuntary contraction of the skeletal muscles due to a lack of calcium. *378*

tetrad (tet′rad) a set of four chromatids resulting from the pairing of homologous chromosomes during prophase I of meiosis. *85*

tetrapod (tet′rah-pod) having four limbs. *612*

thalamus (thal′ah-mus) a mass of gray matter, located at the base of the cerebrum in the wall of the third ventricle, that receives sensory input. *322*

thermal inversion (ther′mal in-ver′zhun) temperature inversion such that warm air traps cold air and its pollutants near the earth. *703*

thrombin (throm′bin) enzyme that converts fibrinogen to fibrin threads during blood clotting. *238*

thylakoid (thi′lah-koid) an individual flattened vesicle found within a granum (pl. grana). *117*

thymus (thi′mus) an organ that lies in the neck and chest area and is absolutely necessary to the development of immunity. *228*

thyroid-stimulating hormone (thi′roid stim′u-lāt″ing hor′mōn) *See* TSH. *377*

thyroxin (thi-rok′sin) the hormone produced by the thyroid that speeds up the metabolic rate. *378*

tibia (tib′e-ah) the shinbone found in the lower leg. *336*

tight junction (tīt junk′shun) cell membranes from two cells interlock in a zipperlike fashion preventing leakage into or out of a tissue. *168*

tissue fluid (tish′u floo′id) fluid found about tissue cells containing molecules that enter from or exit to the capillaries. *227*

T lymphocyte (lim′fo-sīt) lymphocytes that interact directly with antigen-bearing cells and are responsible for cell-mediated immunity. *252*

tone (tōn) the continuous partial contraction of muscle; also the quality of a sound. *343*

trachea (tra′ke-ah) in vertebrates, the windpipe; in insects, the tracheae are the air tubes. *273, 608*

tracheids (tra′ke-idz) a component of xylem made of long, tapered nonliving cells. *146*

tract (trakt) a bundle of neurons forming a transmission pathway through the brain and spinal cord. *322*

trait (trāt) specific term for a distinguishing feature studied in heredity. *445*

transcription (trans-krip′shun) the process that results in the production of a strand of mRNA that is complementary to a segment of DNA. *483, 484*

transduction (trans-duk′shun) the transfer by bacteriophages of DNA from one cell to another. *545*

transfer RNA (trans′fer tRNA; molecule of RNA that carries an amino acid to a ribosome engaged in the process of protein synthesis. *486*

transformation (trans″for-ma′shun) a change in the phenotype of a cell due to an introduction of genes from the environment. *545*

transition reaction (tran-zish′un re-ak′shun) a reaction within aerobic cellular respiration during which hydrogen and carbon dioxide are removed from pyruvic acid; connects glycolysis to Krebs cycle. *101*

translation (trans-la′shun) the process involving mRNA, ribosomes, and tRNA molecules that results in a synthesis of a polypeptide having an amino acid sequence dictated by the sequence of codons in mRNA. *483*

transpiration (tran″spī-ra′shun) the evaporation of water from a leaf; pulls water from the roots through a stem to leaves. *145*

trichinosis (trik″ī-no′sis) disease caused by a roundworm in which the larva are found in cysts within muscle cells. *596*

trichocyst (trik′o-sist) threadlike darts released by some ciliates that may help in defense against predators or in capturing prey. *549*

trochophore (tro′ko-fōr) a larval form unique to the protostomes that indicates they are related. *596*

trophoblast (trof′o-blast) the outer membrane that surrounds the human embryo and, when thickened by a layer of mesoderm, becomes the chorion, an extraembryonic membrane. *428*

trypsin (trip′sin) a protein-digesting enzyme secreted by the pancreas. *188*

TSH (thyroid-stimulating hormone); hormone which causes the thyroid to produce thyroxin. *377*

tube feet (tūb fēt) rows of small tube-shaped appendages in echinoderms which are used in locomotion. *610*

tube-within-a-tube plan (tūb with-in′ ah tūb plan) body plan of animals having two openings. *585*

tubular excretion (tu′bu-lar eks-kre′shun) the movement of certain molecules from the blood into the distal convoluted tubule so that they are added to urine. *302*

tundra (tun′drah) a biome characterized by lack of trees, due to cold temperatures and the presence of permafrost year round. *718*

tunicate (tu′nī-kāt) a type of protochordate in which only the larval stage has the three chordate characteristics; and of these only gills are found in the adult. *611*

turgor pressure (tur′gor presh′ur) osmotic pressure that adds to the strength of the cell. *68*

Turner's syndrome (tur′nerz sin′drōm) a condition caused by the inheritance of an abnormality in chromosome number; an X chromosome lacks a homologous counterpart-XO. *464*

twitch (twich) a brief muscular contraction followed by relaxation. *342*

tympanic membrane (tim-pan′ik mem′brān) membrane located between outer and middle ear that receives sound waves; the eardrum. *362*

tympanum (tim′pah-num) sound receptor found in terrestrial animals, for example, the grasshopper. *607*

typhlosole (tif′lo-sōl) surface of long intestine in annelids which enhances absorption capacity. *603*

U

ulna (ul′nah) an elongated bone found within the lower arm. *336*

umbilical arteries and vein (um-bil′ī-kal ar′ter-ēz and vān) fetal blood vessels which travel to and from the placenta. *438*

umbilical cord (um-bil′ī-kal kord) cord connecting the fetus to the placenta through which blood vessels pass. *432*

urea (u-re′ah) primary nitrogenous waste of mammals derived from amino acid breakdown. *292*

ureters (u-re′terz) tubes that take urine from the kidneys to the bladder. *294*

urethra (u-re′thrah) tube that takes urine from bladder to outside. *294*

uric acid (u′rik as′id) waste product of nucleotide breakdown. *292*

urinalysis (u″rī-nal′ī-sis) a medical procedure in which the composition of a patient's urine is determined. *304*

urinary bladder (u′rī-ner″e blad′der) an organ where urine is stored before being discharged by way of the urethra. *294*

uterine cycle (u′ter-in si′k′l) monthly occurring changes in the characteristics of the uterine lining. *403*

uterus (u′ter-us) the organ in females in which the fetus develops. *401*

utricle (u′tre-k′l) saclike cavity that makes up part of the membranous labyrinth of the inner ear receptor for static equilibrium. *364*

V

vaccine (vak′sēn) antigens prepared in such a way that they can promote active immunity without causing disease. *262*

vacuole (vak′u-ōl) a membrane-bounded cavity, usually fluid filled. *50*

vagina (vah-ji′nah) copulatory organ and birth canal in females. *402*

valve (valv) an opening that opens and closes, insuring one-way flow only; common to the systemic veins, the lymphatic veins, and to the heart. *210, 220, 227*

vascular (vas'ku-lar) containing or concerning vessels that conduct fluid, for example, vascular bundles in plant stems and vascular cylinders in plant roots contain xylem and phloem. *136, 138*

vascular bundle (vas'ku-lar bundle) tissues that include xylem and phloem enclosed by a sheath and typically found in herbaceous plant stems. *138*

vascular cambium (vas'ku-lar kam'be-um) a cylindrical sheath of meristematic tissue that produces secondary xylem and phloem. *140*

vascular cylinder (vas'ku-lar sil'in-der) a central region of dicot roots that contains the vascular tissues, xylem and phloem. *136*

vascular tissue (vas'ku-lar tish'u) a transport tissue; circulatory vessels in animals and xylem and phloem in plants. *145, 208*

vasopressin (vas''o-pres'in) secreted by the posterior pituitary; promotes reabsorption of water by the kidneys; also called antidiuretic hormone (ADH). *374*

vegetative propagation (vej'e-ta''tiv prop''ah-ga'shun) asexual reproduction in plants in which a portion of a plant produces an entire new plant. *149*

vein (vān) a vessel that takes blood to the heart. *208*

venae cavae (ve'nah ka'vah) large systemic veins that return blood to the right atrium of the heart. Inferior vena cava collects blood from lower body regions. Superior vena cava collects blood from upper body regions. *212*

venous duct (ve'nus dukt) ductus venosus; fetal connection between the umbilical vein and the inferior vena cava. *438*

ventilation (ven''tĭ-la'shun) breathing; the process of moving air into and out of the lungs. *275*

ventricle (ven'trĭ-k'l) a cavity in an organ such as the ventricles of the heart or the ventricles of the brain. *210, 322*

venule (ven'ūl) a vessel that takes blood from capillaries to veins. *208*

vernix caseosa (ver'niks ka''se-o'sah) cheese-like substance covering the skin of the fetus. *438*

vertebral column (ver'te-bral kol'um) the backbone of vertebrates composed of individual bones called vertebrae. *335*

vertebrate (ver'tĕ-brāt) animal possessing a backbone composed of vertebrae. *585*

vessel cell (ves'el sel) an individual vessel element in xylem. During development, vessel cells lose their contents and end walls so that they form a continuous pipeline in xylem. *145*

vestibule (vest'tĭ-būl) space or cavity at the entrance of a canal such as within the inner ear at the base of the semicircular canals. *364*

vestigial (ves-tij'e-al) the remains of a structure that was functional in some ancestor but is no longer functional in the organism in question. *510*

villi (vil'i) fingerlike projections that line the small intestine and function in absorption. *191*

viroids (vi'roidz) short chains of RNA that are capable of causing diseases in plants. *542*

visceral mass (vis'er-al mas) soft-bodied portion of a mollusk which includes internal organs. *597*

vitamin (vi'tah-min) essential requirement in the diet. Needed in small amounts, that are often part of coenzymes. *202*

vitreous humor (vit're-us hu'mor) the substance that occupies the space between the lens and retina of the eye. *356*

vocal cords (vo'kal kordz) folds of tissue within the larynx that create vocal sounds when they vibrate. *272*

vulva (vul'vah) the external genitalia of the female that lie near the opening of the vagina. *402*

W

waggle dance (wag'l dans) the movements performed by bees to indicate to other bees the location of food. *660*

water vascular system (wah'ter vas'ku-lar sis'tem) a series of canals that take water to the tube feet of echinoderms allowing them to expand. *609*

wood (wood) interior tissue of a tree consisting of secondary xylem. *142*

woody stem (wood'e stem) stem of a plant, for example a tree, which contains rings of secondary xylem. *140*

X

X-linked (eks lingkt) an allele located in an X chromosome that determines a characteristic unrelated to the sex of the individual. *461*

xylem (zi'lem) the vascular tissue in plants that transports water and minerals. *136*

XYY male (eks wi wi māl) a male that has an extra Y chromosome. *464*

Y

yellow marrow (yel'o mar'o) fat storage tissue found in the cavities within certain bones. *337*

yolk sac (yōk sak) one of the extraembryonic membranes within which yolk is found. *427*

Z

zooflagellates (zo''o-flaj'ĕ-lāts) protozoans that move by means of flagella. *550*

zooplankton (zo''o-plank'ton) animal members of plankton community. *727*

zygote (zi'gōt) diploid cell formed by the union of two gametes, the product of fertilization. *76*

Credits

Line Art/Tables

Part Openers

Part One: © Francis Leroy.
Part Three: © Francis Leroy.
Part Four: © Francis Leroy.

Chapter Openers

Introduction and Chapters 6, 7, 9, 10, 13, 14, 15, 17, 23, and 26: Morgan & Morgan, Inc.
Chapters 16, 20, 25, 27, 28, 29, 30, and 32: Dover Publications, Inc.

Chapter 1

Illustration, p. 30: © Francis Leroy.
Figs. 1.21b, 1.32, 1.33, and 1.34: From Mader, Sylvia S., *Biology, Evolution, Diversity, and the Environment,* 2d ed. © 1985, 1987 Wm. C. Brown Publishers, Dubuque, Iowa. All Rights Reserved. Reprinted by permission.

Chapter 2

Fig. 2.4b: © Francis Leroy.
Figs. 2.12, 2.13 b, c, & d, 2.16, and 2.17: From Mader, Sylvia S., *Biology: Evolution, Diversity, and the Environment,* 2d ed. © 1985, 1987 Wm. C. Brown Publishers, Dubuque, Iowa. All Rights Reserved. Reprinted by permission.

Chapter 3

Fig. 3.17: © Francis Leroy.

Chapter 4

Fig. 4.9: © Francis Leroy.
Fig. 4.21: From Mader, Sylvia S., *Biology: Evolution, Diversity, and the Environment,* 2d ed. © 1985, 1987 Wm. C. Brown Publishers, Dubuque, Iowa. All Rights Reserved. Reprinted by permission.

Chapter 5

Figs. 5.2 and 5.17: From Mader, Sylvia S., *Biology: Evolution, Diversity, and the Environment,* 2d ed. © 1985, 1987 Wm. C. Brown Publishers, Dubuque, Iowa. All Rights Reserved. Reprinted by permission.

Chapter 6

Fig. 6.11 and Table 6.1: From Mader, Sylvia S., *Biology: Evolution, Diversity, and the Environment,* 2d ed. © 1985, 1987 Wm. C. Brown Publishers, Dubuque, Iowa. All Rights Reserved. Reprinted by permission.

Chapter 7

Fig. 7.3b: Carolina Biological Supply.
Figs. 7.4, 7.20, 7.21, and 7.22b: From Mader, Sylvia S., *Biology: Evolution, Diversity, and the Environment,* 2d ed. © 1985, 1987 Wm. C. Brown Publishers, Dubuque, Iowa. All Rights Reserved. Reprinted by permission.
Fig. 7.26: © Francis Leroy.

Chapter 8

Figs. 8.4a, 8.10, and 8.15: From Hole, John W., Jr., *Human Anatomy and Physiology,* 4th ed. © 1978, 1981, 1984, 1987 Wm. C. Brown Publishers, Dubuque, Iowa. All Rights Reserved. Reprinted by permission.
Fig. 8.5: From Mader, Sylvia S., *Human Biology.* © 1988 Wm. C. Brown Publishers, Dubuque, Iowa. All Rights Reserved. Reprinted by permission.

Chapter 9

Fig. 9.3: From Van De Graaff, Kent M., *Human Anatomy.* © 1984 Wm. C. Brown Publishers, Dubuque, Iowa. All Rights Reserved. Reprinted by permission.
Fig. 9.10: From Hole, John W., Jr., *Human Anatomy and Physiology,* 4th ed. © 1978, 1981, 1984, 1987 Wm. C. Brown Publishers, Dubuque, Iowa. All Rights Reserved. Reprinted by permission.
Fig. 9.13 and Box Fig. 9.1: From Mader, Sylvia S., *Human Biology.* © 1988 Wm. C. Brown Publishers Dubuque, Iowa. All Rights Reserved. Reprinted by permission.

Chapter 10

Figs. 10.1 and 10.5: From Van De Graaff, Kent M. and Stuart Ira Fox, *Concepts of Human Anatomy and Physiology.* © 1986 Wm. C. Brown Publishers, Dubuque, Iowa. All Rights Reserved. Reprinted by permission.
Fig. 10.6: From Van De Graaff, Kent M., *Human Anatomy.* © 1984 Wm. C. Brown Publishers, Dubuque, Iowa. All Rights Reserved. Reprinted by permission.
Fig. 10.8: From Crouch, James E., *Functional Human Anatomy,* 2d ed. © 1972 Wm. C. Brown Publishers, Dubuque, Iowa. All Rights Reserved. Reprinted by permission.
Fig. 10.14: From Fox, Stuart I., *Laboratory Guide to Human Physiology: Concepts and Applications,* 4th ed. © 1976, 1980, 1984, 1987 Wm. C. Brown Publishers, Dubuque, Iowa. All Rights Reserved. Reprinted by permission.
Fig. 10.15: From Mader, Sylvia S., *Human Biology.* © 1988 Wm. C. Brown Publishers, Dubuque, Iowa. All Rights Reserved. Reprinted by permission.
Fig. 10.16: From Fox, Stuart Ira, *Human Physiology,* 2d ed. © 1984, 1987 Wm. C. Brown Publishers, Dubuque, Iowa. All Rights Reserved. Reprinted by permission.
Fig. 10.17: From Mader, Sylvia S. *Human Biology.* © 1988 Wm. C. Brown Publishers, Dubuque, Iowa. All Rights Reserved. Reprinted by permission.
Box Fig. 10.1: From Mader, Sylvia S., *Biology: Evolution, Diversity, and the Environment,* 2d ed. © 1985, 1987 Wm. C. Brown Publishers, Dubuque, Iowa. All Rights Reserved. Reprinted by permission.

Chapter 11

Illustrations, pgs. 232 and 243: © Francis Leroy.
Figs. 11.3a and 11.7: From Hole, John W., Jr., *Human Anatomy and Physiology,* 4th ed. © 1978, 1981, 1984, 1987 Wm. C. Brown Publishers, Dubuque, Iowa. All Rights Reserved. Reprinted by permission.
Fig. 11.3b: © Francis Leroy.
Figs. 11.6, 11.11b, d, 11.13, and 11.15: From Mader, Sylvia S., *Human Biology.* © 1988 Wm. C. Brown Publishers, Dubuque, Iowa. All Rights Reserved. Reprinted by permission.
Fig. 11.16: From Mader, Sylvia S., *Biology: Evolution, Diversity, and the Environment,* 2d ed. © 1985, 1987 Wm. C. Brown Publishers, Dubuque, Iowa. All Rights Reserved. Reprinted by permission.

Chapter 12

Figs. 12.2 and 12.8: From Fox, Stuart Ira, *Human Physiology,* 2d ed. © 1984, 1987 Wm. C. Brown Publishers, Dubuque, Iowa. All Rights Reserved. Reprinted by permission.
Figs. 12.3 and 12.6 b, c: From Mader, Sylvia S., *Human Biology.* © 1988 Wm. C. Brown Publishers, Dubuque, Iowa. All Rights Reserved. Reprinted by permission.
Fig. 12.7: © Francis Leroy.

Chapter 13

Figs. 13.4 and 13.12: From Hole, John W., Jr. *Human Anatomy and Physiology,* 4th ed. © 1978, 1981, 1984, 1987 Wm. C. Brown Publishers, Dubuque, Iowa. All Rights Reserved. Reprinted by permission.
Fig. 13.5: From Van De Graaff, Kent M. and Stuart Ira Fox, *Concepts of Human Anatomy and Physiology.* © 1986 Wm. C. Brown Publishers, Dubuque, Iowa. All Rights Reserved. Reprinted by permission.
Figs. 13.8 and 13.10: From Mader, Sylvia S., *Human Biology.* © 1988 Wm. C. Brown Publishers, Dubuque, Iowa. All Rights Reserved. Reprinted by permission.
Fig. 13.13: From Vander, A. J., et al., *Human Physiology.* © 1979 McGraw-Hill Book Company, New York. Reprinted by permission.

Chapter 14

Figs. 14.5 and 14.7: From Van De Graaff, Kent M., *Human Anatomy.* © 1984 Wm. C. Brown Publishers, Dubuque, Iowa. All Rights Reserved. Reprinted by permission.
Fig. 14.14: From Van De Graaff, Kent M. and Stuart Ira Fox, *Concepts of Human Anatomy and Physiology.* © 1986 Wm. C. Brown Publishers, Dubuque, Iowa. All Rights Reserved. Reprinted by permission.
Fig. 14.18: From Mader, Sylvia S., *Human Biology.* © 1988 Wm. C. Brown Publishers, Dubuque, Iowa. All Rights Reserved. Reprinted by permission.

Fig. 14.19: From *Time,* December 18, 1978, p. 82. Copyright © Time, Inc. All Rights Reserved. Reprinted by permission of *Time.*

Chapter 15

Fig. 15.11: Reprinted with permission of the Macmillan Publishing Company from *Anatomy and Physiology,* 14th ed. by D. C. Kimber, C. Gray, C. Stackpole, and L. Leavell. Copyright © 1961 by Macmillan Publishing Company.
Fig. 15.15: From Mader, Sylvia S., *Biology, Evolution, Diversity, and the Environment,* 2d ed. © 1985, 1987 Wm. C. Brown Publishers, Dubuque, Iowa. All Rights Reserved. Reprinted by permission.
Fig. 15.19: From Hole, John W., Jr., *Human Anatomy and Physiology,* 4th ed. © 1978, 1981, 1984, 1987 Wm. C. Brown Publishers, Dubuque, Iowa. All Rights Reserved. Reprinted by permission.
Box Fig. 15.1: From Mader, Sylvia S., *Human Biology.* © 1988 Wm. C. Brown Publishers, Dubuque, Iowa. All Rights Reserved. Reprinted by permission.

Chapter 16

Figs. 16.6 and 16.7: From Hole, John W., Jr., *Human Anatomy and Physiology,* 4th ed. © 1978, 1981, 1984, 1987 Wm. C. Brown Publishers, Dubuque, Iowa. All Rights Reserved. Reprinted by permission.
Fig. 16.16 b, c: From Katz, Bernard, *Nerve, Muscle, and Synapse.* © 1966 McGraw-Hill Book Company, New York. Reprinted by permission.

Chapter 17

Fig. 17.4a: From Mader, Sylvia S., *Human Biology.* © 1988 Wm. C. Brown Publishers, Dubuque, Iowa. All Rights Reserved. Reprinted by permission.
Figs. 17.4b and 17.16a: From Hole, John W., Jr., *Human Anatomy and Physiology,* 4th ed. © 1978, 1981, 1984, 1987 Wm. C. Brown Publishers, Dubuque, Iowa. All Rights Reserved. Reprinted by permission.
Fig. 17.8a: From Crouch, James E., *Functional Human Anatomy,* 2d ed. © 1972 Lea & Febiger, Philadelphia. Reprinted by permission.
Fig. 17.13b: From Hole, John W., Jr., *Human Anatomy and Physiology,* 3d ed. © 1978, 1981, 1984 Wm. C. Brown Publishers, Dubuque, Iowa. All Rights Reserved. Reprinted by permission.
Figs. 17.17, 17.18, and 17.19a: From Hole, John W., Jr., *Human Anatomy and Physiology,* 4th ed. © 1978, 1981, 1984, 1987 Wm. C. Brown Publishers, Dubuque, Iowa. All Rights Reserved. Reprinted by permission.

Chapter 18

Figs. 18.3 and 18.5: From Van De Graaff, Kent M. and Stuart Ira Fox, *Concepts of Human Anatomy and Physiology.* © 1986 Wm. C. Brown Publishers, Dubuque, Iowa. All Rights Reserved. Reprinted by permission.
Fig. 18.14a: From Mader, Sylvia S., *Biology: Evolution, Diversity, and the Environment,* 2d ed. © 1985, 1987 Wm. C. Brown Publishers, Dubuque, Iowa. All Rights Reserved. Reprinted by permission.
Fig. 18.16a: From Hole, John W., Jr., *Human Anatomy and Physiology,* 4th ed. © 1978, 1981, 1984, 1987 Wm. C. Brown Publishers, Dubuque, Iowa. All Rights Reserved. Reprinted by permission.
Fig. 18.20: From Van De Graaff, Kent M., *Human Anatomy.* © 1984 Wm. C. Brown Publishers, Dubuque, Iowa. All Rights Reserved. Reprinted by permission.

Chapter 19

Figs. 19.2 and 19.7: From Hole, John W., Jr., *Human Anatomy and Physiology,* 4th ed. © 1978, 1981, 1984, 1987 Wm. C. Brown Publishers, Dubuque, Iowa. All Rights Reserved. Reprinted by permission.
Fig. 19.3 a, b: From Mader, Sylvia S. *Human Reproductive Biology.* © 1980 Wm. C. Brown Publishers, Dubuque, Iowa. All Rights Reserved. Reprinted by permission.
Fig. 19.4: From Mader, Sylvia S., *Biology: Evolution, Diversity, and the Environment,* 2d ed. © 1985, 1987 Wm. C. Brown Publishers, Dubuque, Iowa. All Rights Reserved. Reprinted by permission.
Fig. 19.5a: © Francis Leroy.
Fig. 19.14: From Mader, Sylvia S., *Human Biology.* © 1988 Wm. C. Brown Publishers, Dubuque, Iowa. All Rights Reserved. Reprinted by permission.
Fig. 19.15: From Phil Simone, "Profile of Infected Population". © *Discover Magazine* 9/86.
Table 19.4: From Mader, Sylvia S., *Biology: Evolution, Diversity, and the Environment,* 2d ed. © 1985, 1987 Wm. C. Brown Publishers, Dubuque, Iowa. All Rights Reserved. Reprinted by permission.

Chapter 20

Fig. 20.16: From Van De Graaff, Kent M. and Stuart Ira Fox, *Concepts of Human Anatomy and Physiology.* © 1986 Wm. C. Brown Publishers, Dubuque, Iowa. All Rights Reserved. Reprinted by permission.
Fig. 20.17: From Mader, Sylvia S., *Biology: Evolution, Diversity, and the Environment,* 2d ed. © 1985, 1987 Wm. C. Brown Publishers, Dubuque, Iowa. All Rights Reserved. Reprinted by permission.
Fig. 20.18: From Van De Graaff, Kent M., *Human Anatomy.* © 1984 Wm. C. Brown Publishers, Dubuque, Iowa. All Rights Reserved. Reprinted by permission.
Fig. 20.20: From Mader, Sylvia S., *Human Reproductive Biology.* © 1980 Wm. C. Brown Publishers, Dubuque, Iowa. All Rights Reserved. Reprinted by permission.

Chapter 21

Figs. 21.10, 21.15, 21.18, and 21.23: From Mader, Sylvia S., *Human Biology.* © 1988 Wm. C. Brown Publishers, Dubuque, Iowa. All Rights Reserved. Reprinted by permission.
Fig. 21.16b: From Mader, Sylvia S., *Biology: Evolution, Diversity, and the Environment,* 2d ed. © 1985, 1987 Wm. C. Brown Publishers, Dubuque, Iowa. All Rights Reserved. Reprinted by permission.

Chapter 22

Figs. 22.2, 22.5, 22.6, 22.11, 22.20, 22.21, and 22.24: From Mader, Sylvia S., *Biology: Evolution, Diversity, and the Environment,* 2d ed. © 1985, 1987 Wm. C. Brown Publishers, Dubuque, Iowa. All Rights Reserved. Reprinted by permission.

Chapter 23

Box Fig. 23.2 and Fig. 23.14: From Mader, Sylvia S., *Biology: Evolution, Diversity, and the Environment,* 2d ed. © 1985, 1987 Wm. C. Brown Publishers, Dubuque, Iowa. All Rights Reserved. Reprinted by permission.
Fig. 23.6: From Storer, et al., *General Zoology,* 6th ed. © McGraw-Hill Book Company, New York. Reprinted by permission.
Fig. 23.7: From Kohne, et al., in *Journal of Human Evolution,* 1:627, 1972. © 1972 Academic Press, Inc. (London) Limited. Reprinted by permission.

Chapter 25

Fig. 25.4: From Mader, Sylvia S., *Human Biology.* © 1988 Wm. C. Brown Publishers, Dubuque, Iowa. All Rights Reserved. Reprinted by permission.
Fig. 25.6a: © Francis Leroy.
Fig. 25.10a: © Francis Leroy.
Fig. 25.15: From Mader, Sylvia S., *Biology: Evolution, Diversity, and the Environment,* 2d ed. © 1985, 1987 Wm. C. Brown Publishers, Dubuque, Iowa. All Rights Reserved. Reprinted by permission.

Chapter 26

Figs. 26.3, 26.5, and 26.20a: From Mader, Sylvia S., *Biology: Evolution, Diversity, and the Environment,* 2d ed. © 1985, 1987 Wm. C. Brown Publishers, Dubuque, Iowa. All Rights Reserved. Reprinted by permission.
Figs. 26.4 and 26.6: © Kendall/Hunt Publishing Company.

Chapter 27

Figs. 27.5, 27.6, and 27.39: From Mader, Sylvia S., *Biology: Evolution, Diversity, and the Environment,* 2d ed. © 1985, 1987 Wm. C. Brown Publishers, Dubuque, Iowa. All Rights Reserved. Reprinted by permission.
Figs. 27.7, 27.11, 27.20, 27.26, 27.27, and 27.29: Courtesy Carolina Biological Supply Company, Burlington, N.C.

Chapter 28

Fig. 28.4: From Mader, Sylvia S., *Biology: Evolution, Diversity, and the Environment,* 2d ed. © 1985, 1987 Wm. C. Brown Publishers, Dubuque, Iowa. All Rights Reserved. Reprinted by permission.
Figs. 28.5 and 28.6: From *Fossil Man* by Michael H. Day. © 1970 by Grosset & Dunlap, Inc. © 1969 by the Hamlyn Group. Used by permission of Grosset & Dunlap, Inc.
Fig. 28.11: From John Alcock, *Animal Behavior: An Evolutionary Approach,* Second edition (1979). Reprinted with permission.

Chapter 29

Figs. 29.2 and 29.5: From Mader, Sylvia S., *Biology: Evolution, Diversity, and the Environment,* 2d ed. © 1985, 1987 Wm. C. Brown Publishers, Dubuque, Iowa. All Rights Reserved. Reprinted by permission.
Fig. 29.6: From Dethier, V. G. and E. Stellar, *Animal Behavior,* 3d ed. © 1970 Prentice-Hall, Inc., Englewood Cliffs, New Jersey. Reprinted by permission.
Fig. 29.14: From Hailman, J. P., in *Behaviour Supplement #15.* © E. J. Brill, Publisher. Leiden, Holland. Reprinted by permission.
Fig. 29.17: From *Biology Today,* Second Edition by David L. Kirk. © 1972, 1975 by Random House, Inc. Reprinted by permission of the publisher.
Fig. 29.25a: From Helena Curtis, *Biology,* 3d ed. © 1979 Worth Publishers, New York. All Rights Reserved. Reprinted by permission.
Fig. 29.25b: From Helena Curtis and N. S. Barnes, *Invitation to Biology,* 3d ed. © 1981 Worth Publishers, New York. All Rights Reserved. Reprinted by permission.

Chapter 30

Fig. 30.4: From Kelly and McGrath: *Biology: Evolution and Adaptation to the Environment.* © 1975 Houghton Mifflin Company. Used with permission.
Fig. 30.5: Adapted from *Web of Life.* Used with permission of Aldus Books, Ltd., London.
Fig. 30.9: From Eugene P. Odum, *Fundamentals of Ecology,* 3d ed. © 1971 Holt, Rinehart and Winston. Reprinted by permission of Holt, Rinehart and Winston.
Fig. 30.15: From Mader, Sylvia S., *Biology: Evolution, Diversity, and the Environment,* 2d ed. © 1985, 1987 Wm. C. Brown Publishers, Dubuque, Iowa. All Rights Reserved. Reprinted by permission.

Chapter 31

Figs. 31.2, 31.3, 31.4, 31.7, and 31.12: From Mader, Sylvia S., *Biology: Evolution, Diversity, and the Environment,* 2d ed. © 1985, 1987 Wm. C. Brown Publishers, Dubuque, Iowa. All Rights Reserved. Reprinted by permission.
Figs. 31.9, 31.17, and 31.30: From Mader, Sylvia S., *Human Biology.* © 1988 Wm. C. Brown Publishers, Dubuque, Iowa. All Rights Reserved. Reprinted by permission.
Box Fig. 31.1: From Mader, Sylvia S., *Biology: Evolution, Diversity, and the Environment,* 2d ed. © 1985, 1987 Wm. C. Brown Publishers, Dubuque, Iowa. All Rights Reserved. Reprinted by permission.

Chapter 32

Figs. 32.2, a, b, 32.7, 32.8, 32.13, and 32.14: From Mader, Sylvia S., *Biology: Evolution, Diversity, and the Environment,* 2d ed. © 1985, 1987 Wm. C. Brown Publishers, Dubuque, Iowa. All Rights Reserved. Reprinted by permission.

Chapter 33

Fig. 33.5: From Mader, Sylvia S., *Biology: Evolution, Diversity, and the Environment,* 2d ed. © 1985, 1987 Wm. C. Brown Publishers, Dubuque, Iowa. All Rights Reserved. Reprinted by permission.
Fig. 33.6: Source: Population Reference Bureau, based on United Nations estimates. Used by permission.
Fig. 33.7: Source: Thomas W. Merrick, with PRB staff, "World Population in Transition" in *Population Bulletin,* Vol. 41, No. 2, April 1986. Copyright © 1986 Population Reference Bureau.
Fig. 33.8: After Cloud, 1969, from Odum, *Fundamentals of Ecology,* 3d ed. © 1971 by the W. B. Saunders Company, Philadelphia, Pa. Reprinted by permission.
Fig. 33.9: From Mader, Sylvia S., *Biology: Evolution, Diversity, and the Environment,* 2d ed. © 1985, 1987 Wm. C. Brown Publishers, Dubuque, Iowa. All Rights Reserved. Reprinted by permission.
Fig. 33.15: From Barr, Terry N., "The World Food Situation and Global Grain Prospects" in *Science, 214,* 1087–1095, 1981, Fig. 5. Copyright © 1981 American Association for the Advancement of Science. Reprinted by permission.

Readings

Introduction

Reading Part I, p. 10: "The Everglades." From Mader, Sylvia S., *Human Biology.* © 1988 Wm. C. Brown Publishers, Dubuque, Iowa. All Rights Reserved. Reprinted by permission.

Chapter 4

Reading and figure, p. 80: "What's in a Chromosome?" Mader, Sylvia S., *Biology: Evolution, Diversity, and the Environment,* 2d ed. © 1985, 1987 Wm. C. Brown Publishers, Dubuque, Iowa. All Rights Reserved. Reprinted by permission.

Chapter 10

Reading, p. 224: "Hypertension." Mader, Sylvia S., *Biology: Evolution, Diversity, and the Environment,* 2d ed. © 1985, 1987 Wm. C. Brown Publishers, Dubuque, Iowa. All Rights Reserved. Reprinted by permission.

Chapter 22

Reading and figure, p. 478: From Mader, Sylvia S., *Biology: Evolution, Diversity, and the Environment,* 2d ed. © 1985, 1987 Wm. C. Brown Publishers, Dubuque, Iowa. All Rights Reserved. Reprinted by permission.

Chapter 26

Reading, p. 579: "Twelve Plants Standing Between Man and Starvation." From Tippo and Stern, *Humanistic Botany.* © 1977 W. W. Norton & Co., New York. Reprinted by permission.

Chapter 31

Reading, p. 699: "Energy to Grow Food." From Mader, Sylvia S., *Biology: Evolution, Diversity, and the Environment,* 2d ed. © 1985, 1987 Wm. C. Brown Publishers, Dubuque, Iowa. All Rights Reserved. Reprinted by permission.

Chapter 32

Reading, p. 724: "The Urgency of Tropical Conservation." Reprinted from *The Nature Conservancy News,* January–March 1986. © 1986 The Nature Conservancy. Used with permission.

Photo Credits

Part Openers

Part Two: © BioMedia Associates.
Part Five: © Alex Kerstitch.
Part Six: © Hans Pfletschinger/Peter Arnold.

Introduction

I.1: © M. D. Kahl/Bruce Coleman
I.2b: © Carolina Biological Supply Company
I.2c: © Edwin A. Reschke
I.3: © Ron Austing/Photo Researchers, Inc.
I.4a: © Merlin D. Tuttle and J. Scott Altenbach, Bat Conversation International
I.4b: © E. R. Degginger
I.4c: © Hugo Van Lawick/Nature Photographers, Ltd.
I.6: © Leonard Lessin/Peter Arnold, Inc.
I.7: Earth Scenes/© Ted Levin
I.9: © Ray Coleman/Photo Researchers, Inc.
page 11 (top, bottom left): © Glenn Van Nimwegen **(bottom right):** © Jack Fields/Photo Researchers, Inc.

Chapter 1

1.1: © John Cann/Australasian Nature Transparencies
1.5b: © Journalism Services
1.7: © Bob Coyle
1.11: © Van Bucher/Photo Researchers, Inc.
1.17: © Gerry Cranhams/Photo Researchers, Inc.
1.25b: © Jeremy Burgess/Photo Researchers, Inc.
1.26b: © Don Fawcett/Photo Researchers, Inc.
1.27b: © Farrell Grehan/Photo Researchers, Inc.
page 35: © E. R. Degginger.

Chapter 2

2.1a: © Biophoto Associates/Photo Researchers, Inc.
2.1b: © Stephen L. Wolfe
2.1c: © CNRI/Science Photo Library/Photo Researchers, Inc.
2.1d: Earth Scenes/© Doug Wechsler
2.4a: © Dr. Stephen Wolfe
2.6a: © Dr. Keith Porter
2.6c: James A. Lake, *Journal of Molecular Biology* 105, 131–159 (1976). © Academic Press, Inc., (London) Ltd.
2.7a: © Dr. Keith Porter
2.8: © Dr. L. Andrew Staehelin/Univ. of Colorado, Boulder
2.10: VU/© KG Murti
2.11a: © Dr. Keith Porter
2.12a: © Dr. L. K. Shumway/Washington State University
2.13a: © Dr. Keith Porter
2.14: © Dr. Jean Paul Revel
2.15: © John Walsh/Photo Researchers, Inc.
2.16c: © Dr. Keith Porter
2.18, page 58: left © Stephen L. Wolfe, **right** © Keith Porter; **figure 2.18, page 59, top to bottom:** © Keith Porter, © Biophoto Associates, © K. G. Murti/Visuals Unlimited, © Keith Porter, © Keith Porter.

Chapter 3

3.1d: © Dr. Daniel Branton
3.2b: © Biophoto Associates/Photo Researchers, Inc.
3.8, 3.9, 3.10: © S. J. Singer
3.11a, b: © Dwight Kuhn.

Chapter 4

4.1: © Donald Yeager/Camera M. D. Studios
4.2b: © Biophoto Associates/Photo Researchers, Inc.
4.8, 4.11, 4.12, 4.13: © Edwin A. Reschke
4.14: © G. G. Maul and Academic Press, Inc. from *J. Ultrastr. Research* 31:375(1970)
4.16: © John D. Cunningham/VU.

Chapter 5

5.1: © L. S. Stepanowicz/Panographics
5.17b: © H. Fernandes-Morán
page 109(all): © Bob Coyle.

Chapter 6

6.3: © Gordon Leedale/Biophoto Associates
6.6c: © Kenneth Miller
6.12: © R. R. Hessler
page 126: © Dr. Raymond Chollet.

Chapter 7

7.1a: © John D. Cunningham/VU
7.3b: © Carolina Biological Supply Company
7.5a: © G. R. Roberts
7.5b: © E. R. Degginger
7.5c: VU/© David Newman
7.6c: © Carolina Biological Supply Company
7.7b: © Carolina Biological Supply Company
7.9b: © Thomas Eisner
7.10: © Carolina Biological Supply Company
7.11b: © J. H. Troughton and F. B. Sampson
7.12b: © J. H. Troughton and L. Donaldson
7.17: © John Shaw/Bruce Coleman, Inc.
7.19: © Michael Wotton/Weyerhaeuser Company
7.22a: © David M. Stone/PHOTO/NATS
7.23: © Frank B. Salisbury
7.27a: © Bob Coyle
7.27b: © Bob Coyle
7.30: © Michael Godfrey
page 139: © Bob Coyle.

Chapter 8

8.2a, 8.2b: © Edwin A. Reschke
8.2c: © Manfred Kage/Peter Arnold, Inc.
8.3: Kessel, R. G. and Kardon, R. H.: *Tissues and Organs: A Test-Atlas of Scanning Electron Microscopy.* © 1979 by W. H. Freeman & Co.
8.4b, 8.6a, 8.7, 8.8, 8.9, 8.11 a-c, 8.12: © Edwin Reschke
page 176: © Kennon Cooke/Valan Photos.

Chapter 9

9.5: Kessel, R. G. and Kardon, R. H.: *Tissues and Organs: A Test-Atlas of Scanning Electron Microscopy.* © 1979 by W. H. Freeman & Co.
9.7: © Edwin Reschke
9.8: © Martin M. Rotker/Taurus Photos
9.9 c, d: Kessel, R. G. and Kardon, R. H.: *Tissues and Organs: A Test-Atlas of Scanning Electron Microscopy.* © 1979 by W. H. Freeman & Co.
9.11a: © Carroll H. Weiss, Camera M. D. Studios
9.11b: © Dr. Sheril D. Burton
9.16 a, b: WHO
9.16 c, d: Centers for Disease Control, Atlanta, GA
page 200: © Mark Sherman/Bruce Coleman, Inc.

Chapter 10

10.9: © Igaku Shoin, Ltd.
page 225: © Lewis Lainey.

Chapter 11

11.1a: © Biophoto Associates/Photo Researchers, Inc.
11.1b: © Lennart Nilsson: *Behold Man,* Little, Brown and Company, Boston
11.5: © Alfred T. Lamme/Camera M. D. Studios
11.9: © Manfred Kage/Peter Arnold, Inc.
11.11a: © Dr. Victor A. Najjar
11.11c: © Dr. Etienne deHarven/University of Toronto and Miss Nina Lampen, Sloan-Kettering Institute, NY
11.12: Almeida, J. D., Cinader, B., and Howatson, A. F.: The Structure of Antigen-Antibody Complexes: A Study by Electron Microscopy. *J. Exptl. Med.* 118: 327–340. © 1963 Rockefeller University Press
11.16a (both): © Stuart I. Fox
page 237a: © Robert Eckert/EKM-Nepenthe; **page 237b:** C. Anthony Hunt: *Science,* Dec. 6, 1985, p. 1166/ © AAAS.

Chapter 12

12.5 a, b: © Dr. Burton D. Goldberg, from *J. Exptl. Med.* 109 (1959): 505 and *Scientific American* 229 (1973): 54
12.6a: © Dr. Kirk Ziegler
12.9: © Guy Gillette/Photo Researchers, Inc.
12.11: Earth Scenes/© G. I. Bernard
page 258: © Lennart Nilsson/Boehringer Ingelheim International, GMBH.

Chapter 13

13.1: © Bob Coyle
13.6: © John Watney Photo Library
13.11: © Bruce Russell/BioMedia Associates
13.14: Richard Feldman/National Institutes of Health
13.16: © Biophoto Associates/Photo Researchers, Inc.
13.17 a, b: © Oscar Auerbach, M.D., Veterans Administration Hospital, East Orange, New Jersey
page 286 (both): © Martin M. Rotker/Taurus Photos, Inc.

Chapter 14

14.1: © Gennaro/Photo Researchers, Inc.
14.3: © Phil Degginger
14.4: © E. R. Degginger
14.10: © CNRI/Science Photo Library/Photo Researchers, Inc.
14.13: Bloom, W., and Fawcett, D. W.: *A Textbook of Histology.* © W. B. Saunders Company, 1966. Photo by Dr. Ruth Bulger.

Chapter 15

15.1: © David Madison/Bruce Coleman, Inc.
15.4b: © J. D. Robertson
15.6: © Linda Bartlett, 1981
15.9c: © Randolph E. Perkins
15.20: © Glenn Short/Bruce Coleman, Inc.
15.24: © Dan McCoy/Rainbow.

Chapter 16

16.1: © Lois Greenfield/Bruce Coleman, Inc.
16.9: © NARCO Bio-Systems
16.12d: © Edwin A. Reschke
16.14a: © H. E. Huxley
16.16a: © John D. Cunningham/VU
page 341: © Mickey Pfleger.

Chapter 17

17.1: © Kaiser Porcelain, Ltd.
17.6: © Courtesy of Palmer House Hotel
17.8b: © Thomas Sims
17.9b: © Frank S. Werblin
17.13a: © Bob Coyle
17.15c: © Ken Touchton
page 367 (both): © Robert S. Preston/courtesy of Professor J. E. Hawkins, Kresge.

Chapter 18

18.6: John Iacono for *Sports Illustrated* © Time, Inc.
18.7: from *Clinical Endocrinology and Its Physiological Basis,* by Arthur Grollman, 1964. Used by permission of J. B. Lippincott Company
18.9: Kessel, R. G., and Kardon, R.: *Tissues and Organs: A Text-Atlas of Scanning Electron Microscopy.* © 1979 W. H. Freeman & Company
18.10: © Lester V. Bergman & Associates, Inc.
18.11, 18.12: from *Clinical Endocrinology & Its Physiological Basis* by Arthur Grollman, 1964. Used by permission of J. B. Lippincott Company
18.13: © Lester V. Bergman & Associates, Inc.
18.15: © Alice J. Garik/Peter Arnold, Inc.
18.18: from *Clinical Endocrinology & Its Physiological Basis* by Arthur Grollman, 1964. Used by permission of J. B. Lippincott Company
18.19: © F. A. Davis Co., Philadelphia, PA and R. H. Kampmeier
page 387: © Agence Vandystatt/Photo Researchers, Inc.

Chapter 19

19.3c: © Biophoto Associates/Photo Researchers, Inc.
19.5b: © Dr. Gerald Schatten
19.13: © Alan Carey/Image Works
19.16a: © Charles Lightdale/Photo Researchers, Inc.
19.16b: © Robert Setinerra/Sierra Productions South
19.17: © Eric Grave.

Chapter 20

20.1 a, b: © Claude Edelmann, Petit Format et Guigorz from the book *First Days of Life*/Black Star
20.11a: © Lennart Nilsson
20.21: © Lennart Nilsson/*A Child Is Born*, Dell Publishing, Co., NY.

Chapter 21

21.1: © John Hicks/Australasian Nature Transparencies
21.3 a, b: © Bob Coyle
21.9: © London Express
21.16a: © Jill Cannefax/EKM-Nepenthe
21.17a: © F. A. Davis Company, Philadelphia, and R. H. Kampmeier
21.17b: © From M. Bartalos & T. A. Baramski: *Medical Cytogenetics.* © 1967 Williams and Wilkins Company
21.19: AP/Wide World Photos
21.21b: © Bill Longcore/Photo Researchers
21.22: © S. Schneider
21.24: © Baylor College of Medicine
page 466: from the collection of The Library of Congress.

Chapter 22

22.8 a, b: © Bill Longcore/Photo Researchers, Inc.
22.13b: © Alexander Rich
22.17 a, b: © Steve Uzzell III, 1982
22.18a: © Biophoto Associates/Photo Researchers, Inc.
22.18b: © Dr. Robert Cocking
22.19: © Robert Noonan
22.22 a & b: © Dr. G. Stephen Martin
page 478a (top): © Biophysics Department, Kings College, London
page 478b (bottom): Original model of the double helix from J. D. Watson: *The Double Helix,* Atheneum, NY, p. 206. © 1968 by J. D. Watson.

Chapter 23

23.1a: © Thomas Taylor/Photo Researchers, Inc.
23.1b: © William E. Ferguson
23.3a: © American Museum of Natural History, Dept. of Library Services
23.3b: © Edward S. Ross
23.5 a, b: © Carolina Biological Supply Co.
23.8: Bettmann Newsphotos
23.9a: © Christina Lake/Photo Researchers, Inc.
23.9b: © Leonard Lee Rue III/Photo Researchers, Inc.
23.11: © D. P. Hershkowitz/Bruce Coleman, Inc.
23.15: © J. A. Bishop and L. M. Cook
23.18b_1: © F. B. Gill/VIREO
23.18b_2: © Miguel Castro/Photo Researchers, Inc.
23.18b_3: © Alan Root/Bruce Coleman, Inc.

Chapter 24

24.1: Earth Scenes/© Zig Leszczynski
24.3a: © Steven Brooke
24.3b: © Sidney W. Fox, Institute for Molecular & Cellular Evolution
24.4: © Dr. Tatiana Evreinova
24.5: © Professor John M. Hayes
page 532: © Miryam Glickson.

Chapter 25

25.1a: © Carl Zeiss, Inc.
25.1b (top): Richard Feldman/National Institutes of Health
25.1b (bottom): © Biophoto Associates/Photo Researchers, Inc.
25.1c: © Lee D. Simon/Photo Researchers, Inc.
25.1d: © Omikron/Photo Researchers, Inc.
25.2: © Dr. E. R. Degginger
25.6 b, c: © David M. Phillips/Visuals Unlimited
25.8: © T. J. Beveridge, University of Guelph/Biological Photo Service
25.9: © John D. Cunningham/Visuals Unlimited
25.10b: © David Hall/Photo Researchers, Inc.
25.12b: © Biology Media/Photo Researchers, Inc.
25.13b: R. G. Kessel and C. Y. Shih: *Scanning Electron Microscopy,* © 1976. Springer Verlag, Berlin
25.14a: © Eric V. Grave/Photo Researchers, Inc.
25.16b: Visuals Unlimited/© T. E. Adams
25.17: © Biophoto Associates/Photo Researchers, Inc.
25.18a: © Biophoto Associates/Photo Researchers, Inc.
25.18 b, c: © Eric Grave
25.20 a, c: © Bob Coyle
25.20b: © Kitty Kohout/Root Resources
25.20d: © Harold V. Green/Valan Photos
25.21a: © Philip Zito
25.21b: © Pat O'Hara
25.22b: © Gordon Leedale/Biophoto Associates
25.23 a, b: © Donald H. Marx, Institute for Mycorrhizal Research and Development
25.24 a, b: from Volk and Wheeler: *Basic Microbiology;* 1973. Used with permission of J. B. Lippincott Company
page 543: © Pfizer, Inc.

Chapter 26

26.1: © Fred and Marian Nickerson
26.6 (left): © Carolina Biological Supply Company
26.5b: © Bruce Russell/BioMedia Associates
26.7a: © J. Robert Waaland, University of Washington/Biological Photo Service
26.8b: © John S. Flannery/Bruce Coleman, Inc.
26.9: © P. Gates, University of Durham/Biological Photo Service
26.10a: © Biophoto Associates
26.11: © Bob Coyle
26.12a: © Jerg Kroener
26.12b: © G. R. Roberts
26.12c: © Dr. E. R. Degginger
26.13: © Roger Conant Wilde/Photo Researchers, Inc.
26.14a: © Foss/Photo Researchers, Inc.
26.15: © Field Museum of Natural History
26.17a_1, a_2: © Dwight Kuhn
26.18: USDA, Forest Service
26.19: © Walter Hodge/Peter Arnold, Inc.
26.21 (top): © E. S. Ross
26.21 (left): © Bob Coyle
26.21 (right): © Leonard Lee Rue III.

Chapter 27

27.1a: © Bruce Russell/BioMedia Associates
27.1b: © Alan D. Carey/Photo Researchers, Inc.
27.7a: © Bud Higdon
27.10 a, c: © Carolina Biological Supply Company
27.10b: © Paul Janosi/Valan Photos
27.12a: © Michael DiSpezio
27.13b: © Dr. Fred H. Whittaker, Biology Dept., University of Louisville
27.14b: © Paul Nollen and Matthew Nadakavukaren
27.15a: © L. S. Stepanowicz/Panographics
27.16: From Markell, E. K. & Voge, M.: *Medical Parasitology,* 3rd edition. © W. B. Saunders Co., 1971. Photo courtesy of Dr. John F. Kessel
27.19 a, b: © Biophoto Associates/NHPA
27.19c: © Rick Poley
27.21: © Richard Mariscal/Bruce Coleman, Inc.
27.22a: © Michael DiSpezio
27.22b: © Biophoto Associates
27.25a: © Bruce Russell/BioMedia Associates
27.25b: © Zig Leszczynski/Animals, Animals
27.25c: © Tom McHugh/Photo Researchers, Inc.
27.25d: © Bob Coyle
27.25e: © Richard Humbert, Stanford University/Biological Photo Service
27.25f: © Philip Zito
27.28a: © Michael DiSpezio
27.28b: © Bud Higdon
27.28c: © Robert A. Ross
27.33 a, b: © Douglas Faulkner/Sally Faulkner Collection
27.34a: © Leonard Lee Rue III
27.34b: © Andrew Odum/Peter Arnold, Inc.
27.35a: © J. C. Carton/Bruce Coleman, Inc.
27.35b: © Paul R. Erlich, Stanford University/Biological Photo Service
27.35c: © Bob Coyle
27.36a: © Oxford Scientific Films/Animals, Animals
27.36b: © Dallas Heaton/Uniphoto
27.37a: © Bob Coyle
27.37b: © E. S. Ross
page 600: © Guy Gillette/Photo Researchers, Inc.

Chapter 28

28.1a: © Tom McHugh/Photo Researchers, Inc.
28.1b: © Russ Kinne/Photo Researchers, Inc.
28.3a: © Martha E. Reeves/Photo Researchers, Inc.
28.3 b, d: © Tom McHugh/Photo Researchers, Inc.
28.3c: © George Holton/Photo Researchers, Inc.
28.9: © American Museum of Natural History
28.10 (both): © Dr. Alan Walker
28.12: © American Museum of Natural History
28.14: © American Museum of Natural History
28.15: © British Museum of Natural History
28.16: © American Museum of Natural History
28.18a: © Linda Bartlett/Photo Researchers, Inc.
28.18b: © Delta Willis/Bruce Coleman, Inc.
28.18c: © Jocelyn Burt/Bruce Coleman, Inc.
28.18d: © Larry Voigt/Photo Researchers, Inc.
28.18e: © Stephen Trimble
page 631: © Russ Kinne/Photo Researchers, Inc.

Chapter 29

29.1a: © Irven DeVore/AnthroPhoto
29.1b: © Bob Coyle
29.3: © Anthony Bannister/Animals, Animals
29.4: © David G. Allen
29.8 (both): © Frederick Webster
29.9: © Bob Coyle
29.12: © Dwight Kuhn
29.13: Thomas McAvoy/*Life* Magazine © 1955 Time, Inc.
29.14a: © Lynn Stone
29.15: Nina Leen/*Life* Magazine © 1964 Time, Inc.
29.18 a, b: © Dr. Rae Silver, Barnard College
29.19: © Herbert Parsons/PHOTO/NATS
29.21 a-c: © Professor I. Eibl-Eibesfeldt
29.22: © Susan Kuklin/Photo Researchers, Inc.
29.23: © R. J. Tomkins/Australasian Nature Transparencies
29.24 a-c: © Bob Coyle
29.27: © John D. Cunningham
page 657: © Susan Kuklin
page 662: © Nancy Anne Dawe.

Chapter 30

30.6: © G. I. Bernard/Animals, Animals
30.7 a, b: © Fritz Polking
30.8: © Jim & Julie Bruton/Photo Researchers, Inc.
30.10a: © H. S. Wessenberg and G. A. Antipa
30.11: © Hans Pfletschinger/Peter Arnold, Inc.
30.12: © Zig Leszczynski/Animals, Animals
30.13 a, b: © J. A. L. Cooke/Oxford Scientific Picture Library
30.14: Hunter, G. W., Swartzwelder, J. C., and Frye, W. W.: *Tropical Medicine*, 5th ed. © 1976 W. B. Saunders Co.
30.16: © Douglas Faulkner/Sally Faulkner Collection
30.17: © C. B. Frith/Bruce Coleman, Inc.
30.18: © Robert Lee/Photo Researchers, Inc.
30.19: © Dwight Kuhn/Bruce Coleman, Inc.
30.20 a-c: © Dr. Daniel Janzen
30.21: © Bill Wood/Bruce Coleman, Inc.
page 679: © L. West/Photo Researchers, Inc.

Chapter 31

31.1a: © Stephen Kraseman/Peter Arnold, Inc.
31.1b: © Richard Ferguson/William Ferguson
31.1c: © Mary Thacher/Photo Researchers, Inc.
31.1d: © Larry West
31.11: © The Nitragin Company, Inc.
31.13: © Jane Windsor, Division of Plant Industry, Florida Department of Agriculture, Gainesville, Florida
31.14: © William E. Ferguson
31.19 a, b: © Dr. John Skelly
31.21: Tim McCabe/USDA Soil Conservation Service
page 707: © Tom McHugh/Photo Researchers, Inc.

Chapter 32

32.1a: © Peter Kaplan/Photo Researchers, Inc.
32.1b: © Frank Miller/Photo Researchers, Inc.
32.1c: © Pat O'Hara
32.3a: © Charlie Ott/Photo Researchers, Inc.
32.3b: © S. J. Krasemann/Peter Arnold, Inc.
32.4a: © Fred and Marian Nickerson
32.4b: © Walter Hodge
32.4c: © Anthony Mercieca/Root Resources
32.5: © Norman Owen Tomalin/Bruce Coleman, Inc.
32.6: © Richard D. Estes/Photo Researchers, Inc.
32.9: © George D. Lepp/BIO-TEC IMAGES
32.12a: © John Bova/Photo Researchers, Inc.
32.12b: © Jack Fields/Photo Researchers, Inc.
32.15: © Douglas Faulkner/Sally Faulkner Collection
page 725(a): © Stephen Dalton/Photo Researchers, Inc.
page 725(b): © Erwin and Peggy Bauer/Bruce Coleman, Inc.
page 725(c): © Bruce Coleman/Bruce Coleman, Inc.
page 733(a): © Frans Lanting/Photo Researchers, Inc.
page 733(b): © T. C. Dickinson/Photo Researchers, Inc.

Chapter 33

33.1a: © Dr. E. R. Degginger
33.1b: © Ingeborg Lippman/Peter Arnold, Inc.
33.11: USDA Soil Conservation Service
33.12: © Gary Milburn/Tom Stack & Associates
33.13 a, b: © George Gerster/Photo Researchers, Inc.
33.16: © William E. Ferguson
33.17: UNICEF
33.19: © Bob Coyle.

Index

A

B

C

D

E

F

G

H

I

N

O

P

Q

R

S

T

U

V

W

X

Y

Z

Year	Name	Country	Contribution
1903	Karl Landsteiner	Austria	Discovers ABO blood types.
1904	Ivan Pavlov	Russia	Shows that conditioned reflexes affect behavior based on experiments with dogs.
1905	Paul Ehrlich	Germany	Discovers first antibiotic, named Salvarsan, to cure syphilis.
1910	Thomas H. Morgan	United States	States that each gene has a locus on a particular chromosome, based on experiments with *Drosophila*.
1914	Robert Feulgen	Germany	Devises a stain for DNA and shows that chromosomes contain DNA.
1922	Sir Fredrick Banting Charles Best	Canada	Isolate insulin from the pancreas.
1924	Hans Spemann Hilde Mangold	Germany	Show that induction occurs during development, based on experiments with frog embryos.
1927	Hermann J. Muller	United States	Proves that X rays cause mutations.
1929	Sir Alexander Fleming	Britain	Discovers the toxic effect of a mold product he called penicillin on certain bacteria.
1931	Fred Griffith	Britain	Discovers that nonvirulent bacteria can be transformed and become virulent if exposed to dead virulent bacteria within mice.
1932	Albert Szent-Györgyi von Nagyrapolt	Hungary	Isolates ascorbic acid from tissues and proves it is the anti-scurvy vitamin C.
1937	Konrad Z. Lorenz	Austria	Founds the study of ethology and shows the importance of imprinting as a form of early learning.
1937	Sir Hans A. Krebs	Britain	Discovers the reactions of a cycle that produces carbon dioxide during cellular respiration.
1940	George Beadle Edward Tatum	United States	Develop the one gene-one enzyme theory, based on red bread mold studies.
1944	O. T. Avery Maclyn McCarty Colin MacLeod	United States	Demonstrate that DNA alone from virulent bacteria can transform nonvirulent bacteria.
1945	Melvin Calvin Andrew A. Benson	United States	Discover the individual reactions of a cycle that reduces carbon dioxide during photosynthesis.
1950	Barbara McClintock	United States	Discovers transposons (jumping genes) while doing experiments with corn.
1950	A. L. Hodgkin A. F. Huxley	Britain	Present a model to explain the movement of the nerve impulse along a neuron.
1952	Alfred D. Hershey Martha Chase	United States	Find that only DNA from viruses enters cells and directs the reproduction of new viruses.